PHOTOMETRIC ATLAS OF NORTHERN BRIGHT GALAXIES

PHOTOMETRIC ATLAS OF NORTHERN BRIGHT GALAXIES

Edited by
Keiichi Kodaira
National Astronomical Observatory
Sadanori Okamura
Institute of Astronomy, University of Tokyo
Shin-ichi Ichikawa
Institute of Astronomy, University of Tokyo

UNIVERSITY OF TOKYO PRESS

Printed in Japan

Published by:
UNIVERSITY OF TOKYO PRESS, Tokyo
ISBN 0-86008-461-2
ISBN 4-13-066107-8

The 105-cm Schmidt telescope in its dome.

Aerial view of the Kiso Observatory.

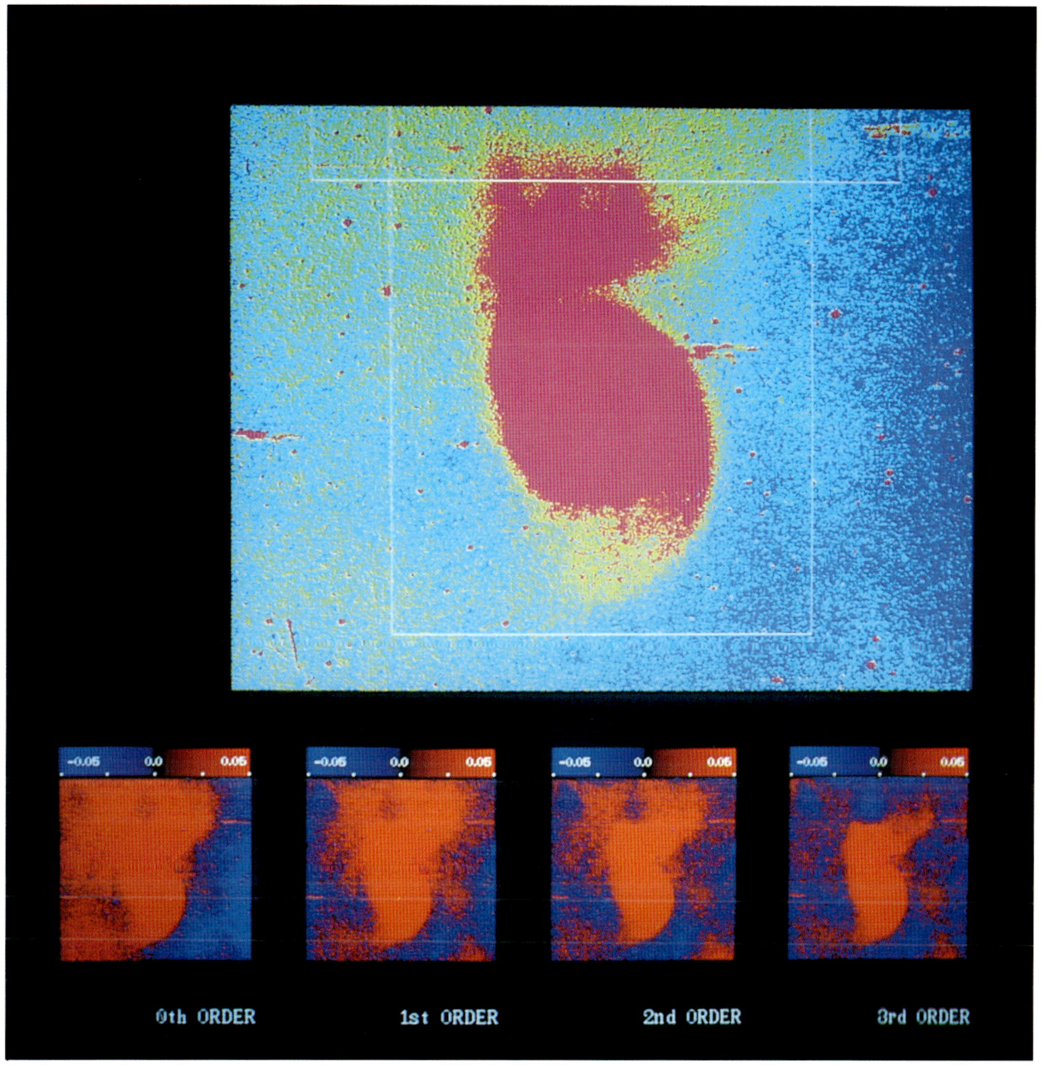

Data processing with SPIRAL: Sky subtraction.
(top) Display of the density frame of M51 showing a highly non-uniform sky background on a hypersensitized IV-N plate. (bottom) Residual frames in relative intensity units when sky background is approximated by zeroth-, first-, second-, and third-order polynomials (from left to right).

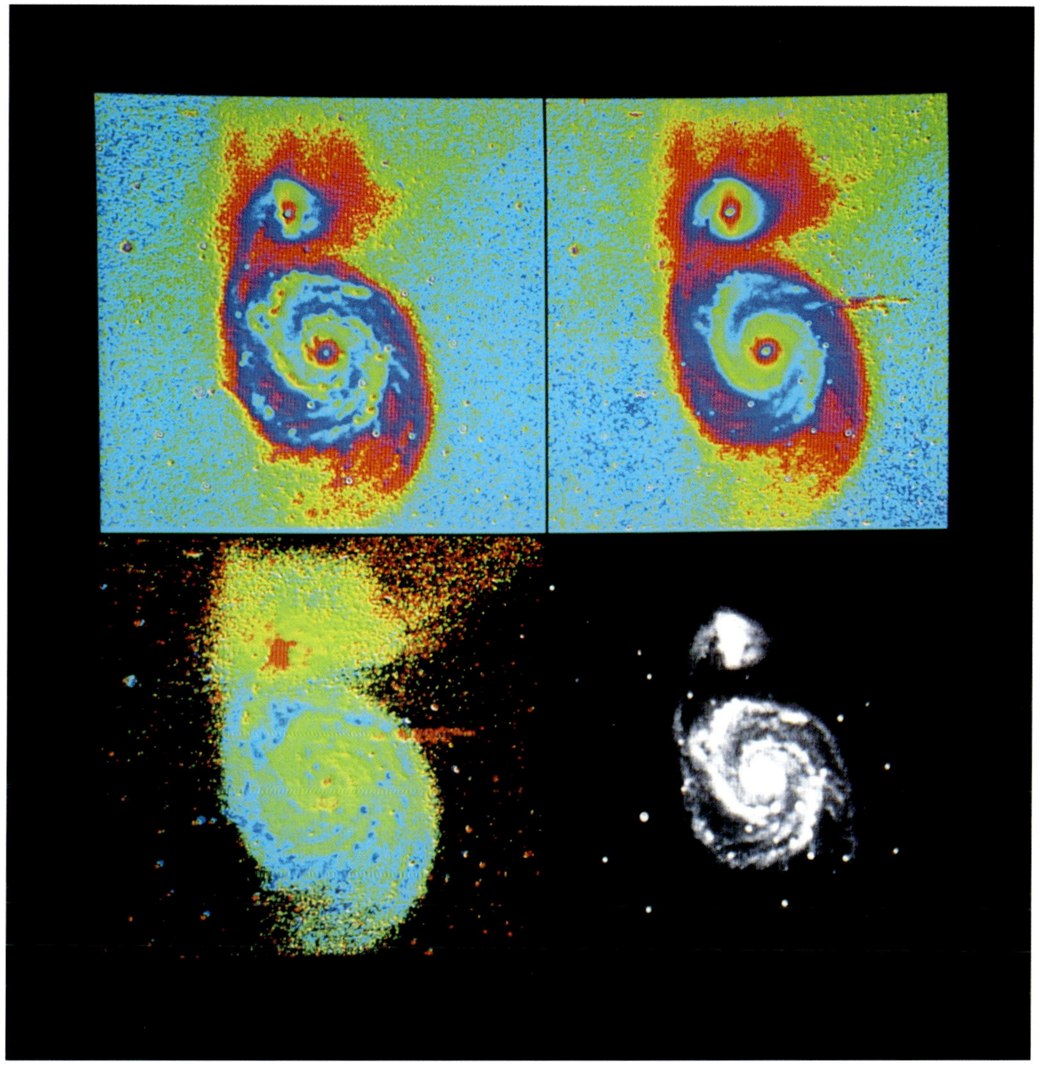

Data processing with SPIRAL: Visual presentation.
(top left) False-color presentation of the surface brightness distribution of M51 in the blue (B) band. (top right) Same as top left, but in the near infrared (I) band. (bottom left) Pseudo-color presentation of (B-I) color index distribution. (bottom right) Same as top left, but in gray scale. Although the examples shown here are images in various color bands, the following Atlas is based on the images in the V band.

Table of Contents

List of Authors

Masaru HAMABE
Kiso Observatory, Institute of Astronomy, The University of Tokyo,
Mitake-mura, Kiso-gun, Nagano-ken, 397-01 Japan

Shin-ichi ICHIKAWA
Kiso Observatory, Institute of Astronomy, The University of Tokyo,
Mitake-mura, Kiso-gun, Nagano-ken, 397-01 Japan
(present address)
National Astronomical Observatory, Mitaka, Tokyo, 181 Japan

Keiichi KODAIRA
National Astronomical Observatory, Mitaka, Tokyo, 181 Japan

Sadanori OKAMURA
Kiso Observatory, Institute of Astronomy, The University of Tokyo,
Mitake-mura, Kiso-gun, Nagano-ken, 397-01 Japan

Masaaki WATANABE
Department of Astronomy, The University of Tokyo, Bunkyo-ku,
Tokyo, 113 Japan
(present address)
AdIn Research Institute, 1-5-8, Jin-nan, Shibuya, Tokyo, 150 Japan

Acknowledgments

This Atlas is the product of a long-term project to compile a database of the two-dimensional surface brightness distribution of bright galaxies using the 105cm Schmidt telescope of the Kiso Observatory, Institute of Astronomy, the University of Tokyo. A large number of people were involved in the project and made significant contributions at various stages.

We express our special thanks to the visiting astronomers at the Kiso Observatory: Dr. Walter Saurer of the Institute for Astronomy, University of Innsbruck, Austria, and Mr. Hu Fu-Xing of the Purple Mountain Observatory, Academia Sinica, Nanjing, China, for their devoted engagement in the measurement of the plates and in the intensive data reduction.

We greatly appreciate the warm continuous support and technical assistance provided by the staff members of the Kiso Observatory, headed by Professor K. Ishida, Director: Drs. H. Maehara, and Y. Taniguchi, Messrs. T. Noguchi, T. Aoki, T. Soyano, K. Tarusawa, and W. Tanaka, and Mrs. Y. Tanaka. A considerable number of the photographic plates analyzed in this project were taken by the staff and the visiting observers at the Kiso Observatory.

We sincerely thank Messrs. T. E. Aoki, M. Doi, K. Shimasaku, Y. Hayano, and K. Mochizuki of the Department of Astronomy, the University of Tokyo, Messrs. M. Nakagiri, A. Miyashita, and Y. Torii, and Mrs. T. Mashiyama of the National Astronomical Observatory, Dr. M. Othman of the University Kebangsaan Malaysia, Dr. Ann H.-B of the Pusan National University, Korea, Mmes. Y. Tarusawa and S. Okamura for their help in the preparation of this Atlas, Professor K. Wakamatsu for useful comments, and Professor Emeritus B. Takase for his continuous encouragement and for paving the way for this project.

This project was supported by the Grant-in-Aid for Scientific Research (Specially Promoted Research: No. 59065002) from the Ministry of Education, Science and Culture during the period 1984–1988. The publication of this Atlas was also supported by the Grant-in-Aid for Publication of Scientific Research Result from the same Ministry. S. Ichikawa and M. Watanabe were supported by JSPS Fellowships for Japanese Junior Scientists from the Japan Society for the Promotion of Science during the preparation of this atlas.

Finally, it is a pleasure to thank Mr. Mitsuo Inoue of the University of Tokyo Press. Without his enthusiasm and patience, the publication of this Atlas would not have become a reality.

December 1989

Keiichi Kodaira
Sadanori Okamura
Shin-ichi Ichikawa
Masaru Hamabe
Masaaki Watanabe

Part I Description of the Atlas

1.1 Purpose of the Atlas

Galaxies are the most massive, luminous astronomical objects bound by self-gravity, and they comprise the major fraction of luminous mass in the universe. Studies of their physical structure, together with their origin and evolution, and of their distribution and kinematics are essentially important for an understanding of the universe.

Recent investigations of the structure of galaxies, cosmic distance scale using various empirical relations, and the kinematics of clusters of galaxies require increasingly accurate photometric data as well as detailed spectroscopic data, in both the optical and the radio wavelength regions. Comprehensive investigations of this kind inevitably involve statistical analyses based on a large body of observational data, for which a high degree of homogeneity is imperative.

In the late 1970s, we started a project to obtain homogeneous surface photometric data in the visual wavelength region for as many nearby bright galaxies as possible using the 105cm Schmidt telescope at the Kiso Observatory. Initial observational targets of the project were, among others, the Virgo and the Ursa Major clusters. The surface photometric data accumulated in the initial phase were published in Watanabe et al. (1982), Watanabe (1983) and Takase et al. (1984). Statistical analysis of the data resulted in a series of papers (Kodaira et al. 1983, 1986; Okamura et al. 1984; Watanabe et al. 1985; Kodaira and Watanabe 1989; Kodaira 1989), which encouraged us to extend the project.

Beginning in 1984, the project was extended to cover most of the northern bright galaxies whose structures can be reasonably resolved on the Kiso Schmidt plates. Rigorous completeness was not pursued. However, efforts were made to secure a homogeneous database for as many galaxies as possible which can be used for studies of galaxy structure, distance indicator relations, cosmic distance scale, and kinematics of the Local Supercluster.

The present Atlas includes 791 galaxies observed with the 105cm Schmidt telescope at the Kiso Observatory. Previous publications of a large body of galaxy data in the optical wavelength region tend to be in the form of either photographic atlases with or without isophotal maps, or catalogs of photometric parameters alone. The increasing accuracy required in recent statistical analyses, however, necessitates a database that includes both the illustrative information and the processed photometric parameters in a consistent system. The present Atlas will serve as the first of such atlases, which enable investigators to refer to the atlas part whenever they need additional information on the photometric parameters listed in the catalog part. This Atlas will also serve as a basic reference in future observations of the galaxies in optical as well as other wavelength regions such as radio, infrared, and X-ray.

1.2 Description of the Sample

A Revised Shapley-Ames Catalog of Bright Galaxies (Sandage and Tammann 1981: RSA) was used as the source of program galaxies. A total of 911 galaxies in the catalog with declinations north of $-25°$ were selected as the primary targets. They are distributed over 354 celestial areas covered by Kiso Schmidt plates. When the observation conditions were good, efforts were made to extend the survey slightly south of $-25°$. By August 1988, when the intensive data reduction started, plates of good quality had been secured for 256 celestial areas. The present sample is a subset of the RSA sample, consisting of 791 galaxies that appear on these Schmidt plates.

It should be noted that the present sample is not complete in any respect. The distributions of morphological types, apparent magnitudes, and radial velocities of the sample galaxies are shown in Figs. 1.1a–c.

1.3 Observation

Observations were made with the 105cm Schmidt telescope at the Kiso Observatory, Institute of Astronomy, the University of Tokyo (Takase et al. 1977). The focal length of the telescope is 3300mm, giving an image scale of 62.5 arcsec mm^{-1}. A 6-degree square field is covered by a 14-inch photographic plate.

A Kodak IIa-D emulsion was combined with a Schott GG495 glass filter to define the photographic *V* band (5000–6500Å) used in the observation. The standard celestial areas for the Kiso Schmidt telescope had been set up to allow for slight overlaps on the plates between adjacent areas. Some galaxies were photographed on more

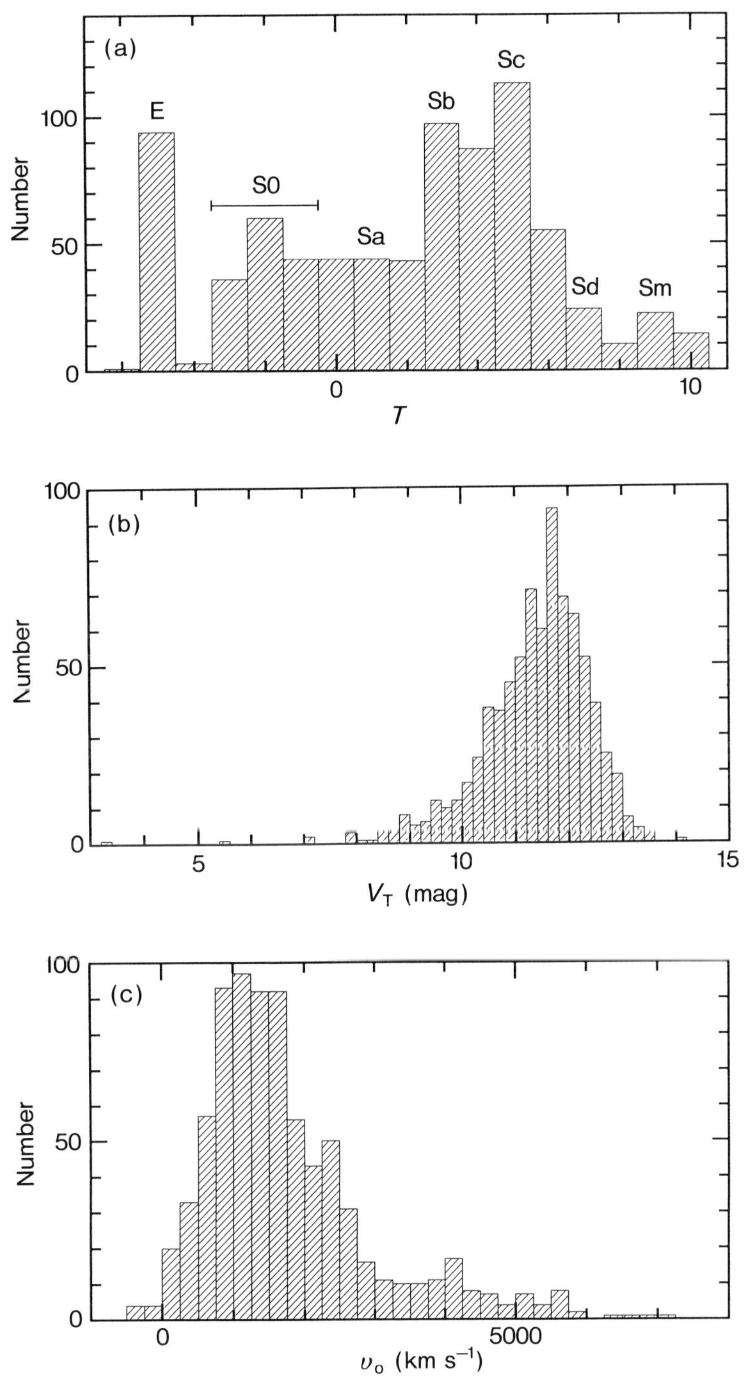

Fig. 1.1 The distributions of (a) morphological types, (b) total magnitudes V_T, and (c) corrected radial velocities υ_0, in the present sample. The morphological type index T is taken from RC 2, V_T is derived in the present study, and υ_0 is taken from RSA.

than two plates. Further, several areas were photographed more than twice in order to provide an estimate of internal accuracy.

An exposure time of 50 minutes was taken as standard. However, the actual exposure time ranged from about 30 minutes to 60 minutes, depending upon sky brightness and plate sensitivity. The seeing size was measured visually using a 20cm guiding telescope during the exposure. After an area was photographed on a plate, four sets of step wedges, each consisting of 15 steps, were exposed on the margin of the plate for the purpose of calibration using the same filter and for the same exposure time as for the observation. The plates were developed for 8 minutes in Fuji Pandol developer diluted 1:1 with water. A small number of plates, those taken in the periods when the sky was darkest, were developed for 5 minutes in Kodak D-19 developer. The temperature of the developer was kept at 20°C.

The record of the plates is given in Table 4.1, together with the sky brightness determined in the process of absolute calibration described in section 1.5.

1.4 Plate Measurement

The sample galaxies were identified on the plates, and the scanning area was defined according to the apparent size of each galaxy. Care was taken to scan a sufficiently wide area to allow for accurate determination of the background sky level. In this process, many non-RSA galaxies were recognized that have the magnitudes and/or sizes comparable to the original sample galaxies in RSA. These galaxies were also included in the measuring list for future study. About 1600 galaxies on 280 plates were measured in total, including duplications of the same galaxies on different plates. Most of the measurement was done during the period April–August 1988.

The galaxy image was scanned with the PDS 2020GMS microdensitometer at the Kiso Observatory (Tarusawa et al. 1986). A logarithmic amplifier was devised which worked five times faster than the Perkin-Elmer amplifier originally built in the PDS (c.f. Anderson et al. 1983). A control program of the PDS was developed that allowed us to scan many galaxies continuously on a Schmidt plate. A square aperture of 17 microns and the sampling pitch of 16 microns (1 arcsec on the plate) were adopted with two exceptions,

M31 and M33. The 170-micron aperture was used for them, and the sampling pitch was 160 microns for M31 and 96 microns for M33.

The output of a galaxy scan consisted of a digital array of photographic densities, i.e., a *density frame*. A digital array of the image and its element will be called hereafter a *frame* and a *pixel*, respectively.

The step wedges were scanned for each plate before and after the galaxy scans for monitoring the possible drift in the response of the PDS. Mean densities of the same steps measured before and after the galaxy scans differed by less than $0.002D$ except for the occasion when the lamp of the PDS was nearly burnt out. Experience showed that the lifetime of the lamp was between 1200 and 1500 hours. We replaced the lamp with a new one after every 1000 hours or so of continuous operation.

1.5 Data Reduction

Data reduction was carried out using a software library SPIRAL *(Surface Photometry Interactive Reduction and Analysis Library)* developed at the Kiso Observatory for a superminicomputer FACOM S-3500 (Ichikawa et al. 1987). The basic procedure of data reduction was the same as that described in our previous papers (e.g., Watanabe 1983; Takase et al. 1984; Ichikawa et al. 1987; Okamura 1988). Only the outline and special treatments applied to the present sample are summarized here.

(1) Relative Intensity Frame

After we confirmed that there was no significant difference larger than $\Delta D = 0.005$ between the wedge data measured before and after the galaxy scans, we merged them to define the characteristic curve for each plate. Goad function (e.g., Hamabe and Okamura 1982) was fitted to the plot of mean densities of the steps versus their wedge constants by the least squares method. The typical value of the standard deviation of the residuals was 0.05 in log I. Examples of the characteristic curve are shown in Fig. 1.2.

Pixel values of the density frame were converted, pixel by pixel, to intensity values using the characteristic curve of the plate to produce the *intensity frame*. The *galaxy frame* was scissored out from the intensity frame. The size of the galaxy frame was defined according to the apparent size of the galaxy. Since we were dealing with a large number of frames, some standardization of the frame size appeared appropriate. The size of the galaxy frame was chosen to be either 512 or 256 pixels on a side, with few exceptions. For large galaxies, the original density frame was binned by pixels to match the standard size. Accordingly, the pixel size of the galaxy frame differed galaxy to galaxy, but it was always a multiple of 16 microns, i.e., 1 arcsec. The frame size and the pixel size are given in Table 4.2 for each galaxy, together with the source plate number.

The sky background was determined by fitting a two-dimensional polynomial to the intensity values of the pixels in the whole intensity frame except for the regions covered by the galaxy itself and by bright stars and/or neighboring galaxies. First, trial sky subtractions

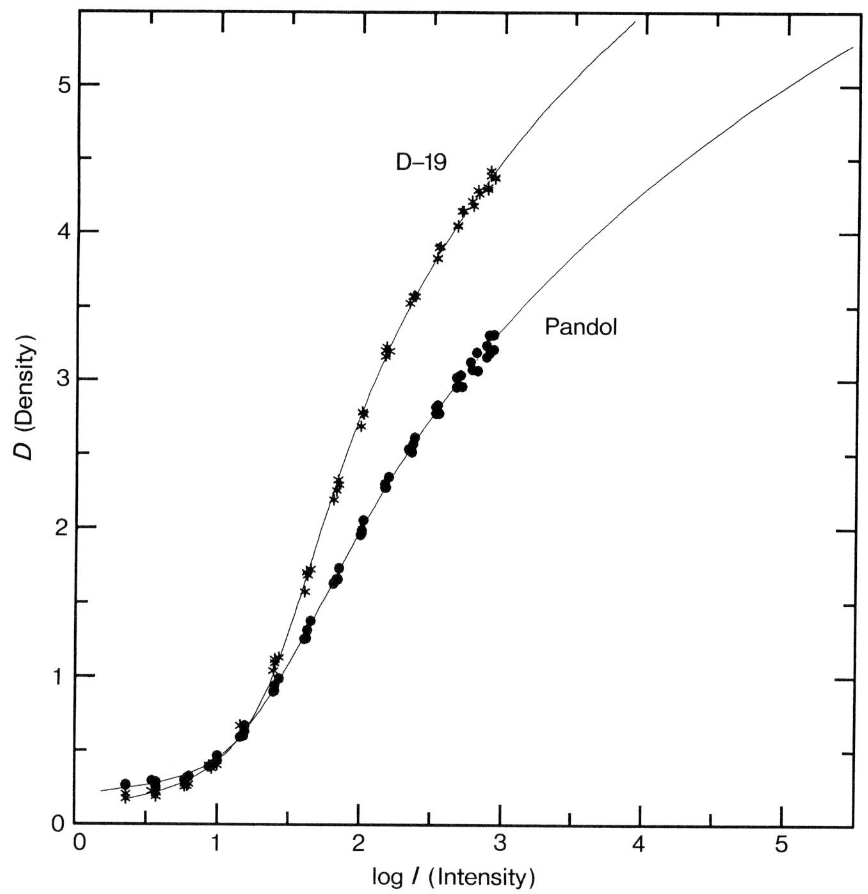

Fig. 1.2 Examples of the characteristic curve of the plates developed with Pandol 1:1 and D-19.

were made using the 0-th, the 1st, the 2nd, and the 3rd order polynomials. The resulting residual frames were displayed on the image display. On the basis of visual inspection of the residual frames, a decision was made on the order of polynomial to be adopted, from 0 to 5. The first order polynomial, i.e., an inclined plane, was usually accepted for small galaxies (size < 10 arcmin). Large galaxies required higher order polynomials. Then, the sky background approximated by the polynomial of the adopted order was subtracted, pixel by pixel, from the galaxy frame. The resultant galaxy frame was divided, again pixel by pixel, by the sky background to yield the *relative intensity frame* of the galaxy.

(2) Absolute Calibration

The unit of the relative intensity, i.e., the sky brightness, was determined in a process called the absolute calibration. The photoelectric magnitudes of a galaxy measured with various apertures were compared with photographic magnitudes integrated

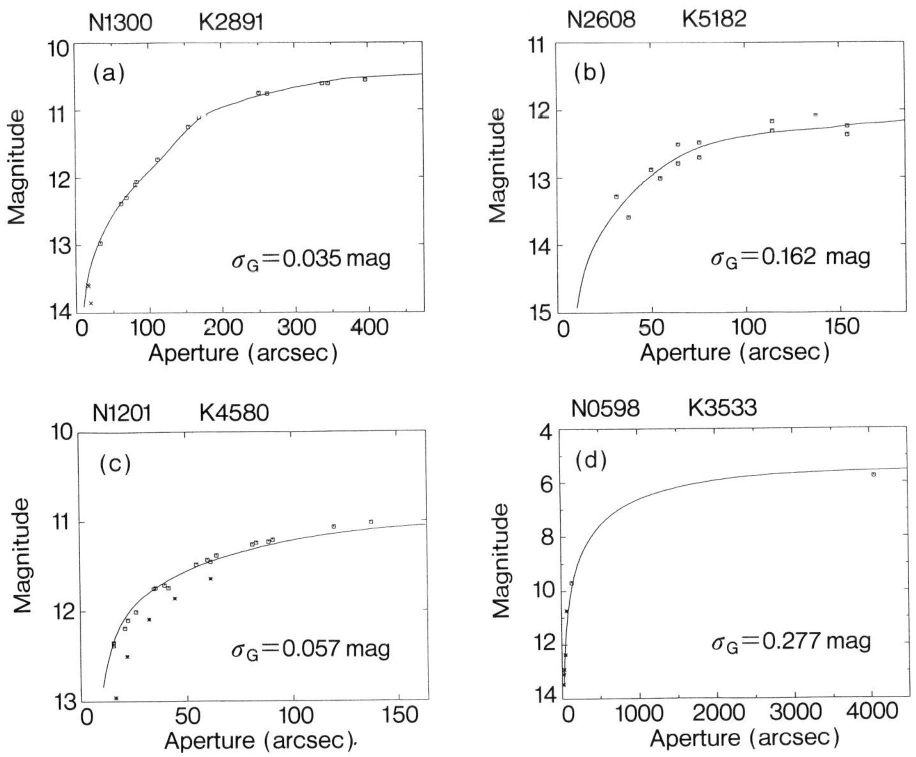

Fig. 1.3 Examples of absolute calibration. The solid curve is the growth curve constructed from the present photographic photometry. The sky brightness was computed by fitting the growth curve to the plots of photoelectric magnitudes (squares) by the least squares method. The crosses and asterisks are photoelectric data rejected in the fitting. Examples are given for (a) an excellent case, (b) a case of large dispersion (0.16 mag) in photoelectric magnitudes, (c) a case of systematic errors found in a series of photoelectric magnitudes, and (d) a case of few photoelectric magnitudes with sufficiently large apertures. Case (d) is common to apparently large nearby galaxies. The magnitudes converted from old photographic photometry tend to have heavy weights in the absolute calibration of such galaxies.

within the same apertures. In practice, the run of photographic magnitudes as a function of the aperture size, i.e., the growth curve, was fitted by the least squares method to the plots of photoelectric magnitudes. The magnetic tape version of the catalog of photoelectric magnitudes compiled by Longo and de Vaucouleurs (1983) was used as the source of photoelectric magnitudes. Since the present photographic photometry is based on the Schmidt plate with a small image scale, data in the very central region of a galaxy may subject to errors due to the lack of resolution and to occasional saturation effects. Photoelectric magnitudes measured with apertures smaller than 10 pixels were automatically rejected for safety in the fitting. Deviant magnitudes were also rejected manually in the interactive processing. Several examples of the absolute calibration are shown in Fig. 1.3.

First, the absolute calibration was performed galaxy by galaxy to yield the sky brightnesses for all the sample galaxies that were included in Longo and de Vaucouleurs (1983). They will be called the *calibrators*. Absolute calibration could not yield a reliable sky brightness for a small fraction of calibrators because of the severe contamination of field stars. When very bright stars were included in the integrating aperture, the photographic magnitude was unreliable due to the saturation effect of the plate. Even for field stars unsaturated on the plate, it was difficult to apply proper corrections for their effects on the growth curve in the least squares fitting. One difficulty was that the details of the correction for field stars in the measuring aperture of photoelectric photometry could be traced back only for a small number of bright stars with reference to the notes given in the catalog. It appeared that magnitudes listed in Longo and de Vaucouleurs (1983) which were transformed from old photographic photometry were corrected for field stars, while many of those which came from photoelectric photometry were not.

Sky brightnesses were obtained for 732 calibrators in total, including the duplications of the same calibrators on different plates. When more than two photoelectric magnitudes were available for a calibrator, the standard deviation of the residuals in the least squares fitting, σ_G, was computed. The sky brightness and σ_G for each calibrator thus obtained are given in Table 4.2. Figure 1.4a shows the histogram of the standard deviations σ_G. The median value is 0.063 mag. For quite a few galaxies, photoelectric magnitudes themselves were found to be poorly determined with the standard deviation larger than 0.1 magnitude (Fig. 1.3b).

Next, the mean sky brightness for a plate was computed as the weighted mean of the sky brightnesses determined for the calibrators included in the plate. When more than two calibrators were included in a plate, the standard deviation, σ_P, was also computed. The implicit assumption for this procedure was that the actual sky brightness was constant over the 6-degree field covered by a Schmidt plate. The weight was assigned to each calibrator in proportion to the number of photoelectric magnitudes used in the determination of the sky brightness for the calibrator.

Calibrators with deviant values of sky brightness were rejected in computing the mean sky brightness. They are noted in Table 4.2. Except for several unexplained cases, causes of the deviation were identified as one or a combination of the following reasons: First, too few (often only one) photoelectric magnitudes were available. Second, the galaxy was contaminated by field stars and/or neighboring galaxies. Third, the galaxy had a small apparent size and/or low surface brightness.

The resulting mean sky brightness is given for each plate in Table 4.1. *This mean sky brightness was finally used to set the unit of the relative intensity of all the galaxies included in the plate, including calibrators as well.* The number of calibrators used in computing the mean sky brightness, the total number of photoelectric magnitudes used, and the standard deviation σ_P are also given in Table 4.1. In the case where the uncertainty of the mean sky brightness was estimated to be larger than the nominal σ_P, the value is followed by a colon in Table 4.1. The sky brightness of the plate lacking calibrators was determined by adjusting the total magnitude of a common galaxy so that it was coincident between the two plates for adjacent areas.

The histogram of σ_P for 245 plates is shown in Fig. 1.4b. The distribution of σ_P peaks at a larger value than that of σ_G shown in Fig. 1.4a, the median value being 0.101 mag. This may be due to the variation of actual sky brightness over the 6-degree field, and in part due to possible errors, yet unidentified, that gave a false variation in the sky brightness. It is found that the standard deviations of the plates that include more than ten calibrators all exceed 0.1 mag. Smaller standard deviations for the plates with smaller number of calibrators may be in part spurious.

We recognize that this size of error in the absolute calibration

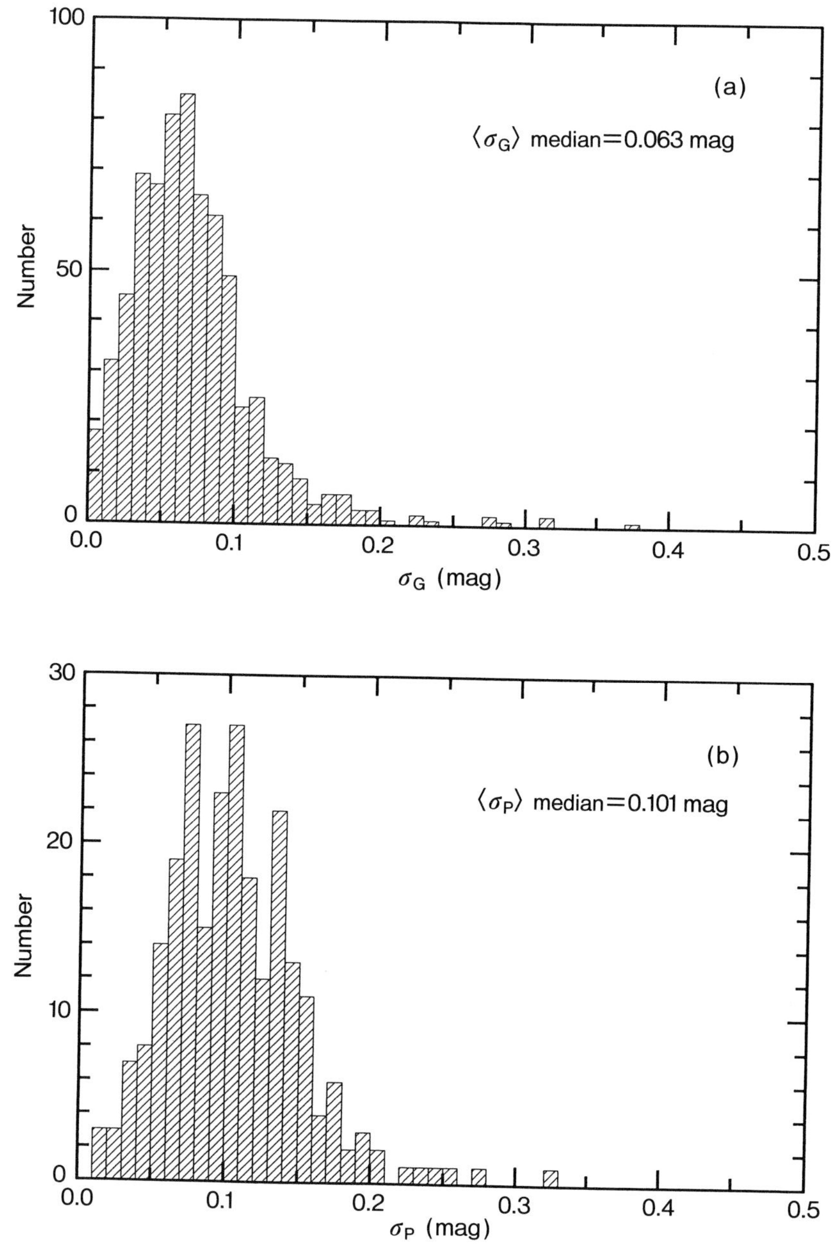

Fig. 1.4 The distribution of the standard deviations in the absolute calibration. Histograms are shown (a) for σ_G of the calibrators with more than two photoelectric magnitudes, and (b) for σ_P of the plates including more than two calibrators.

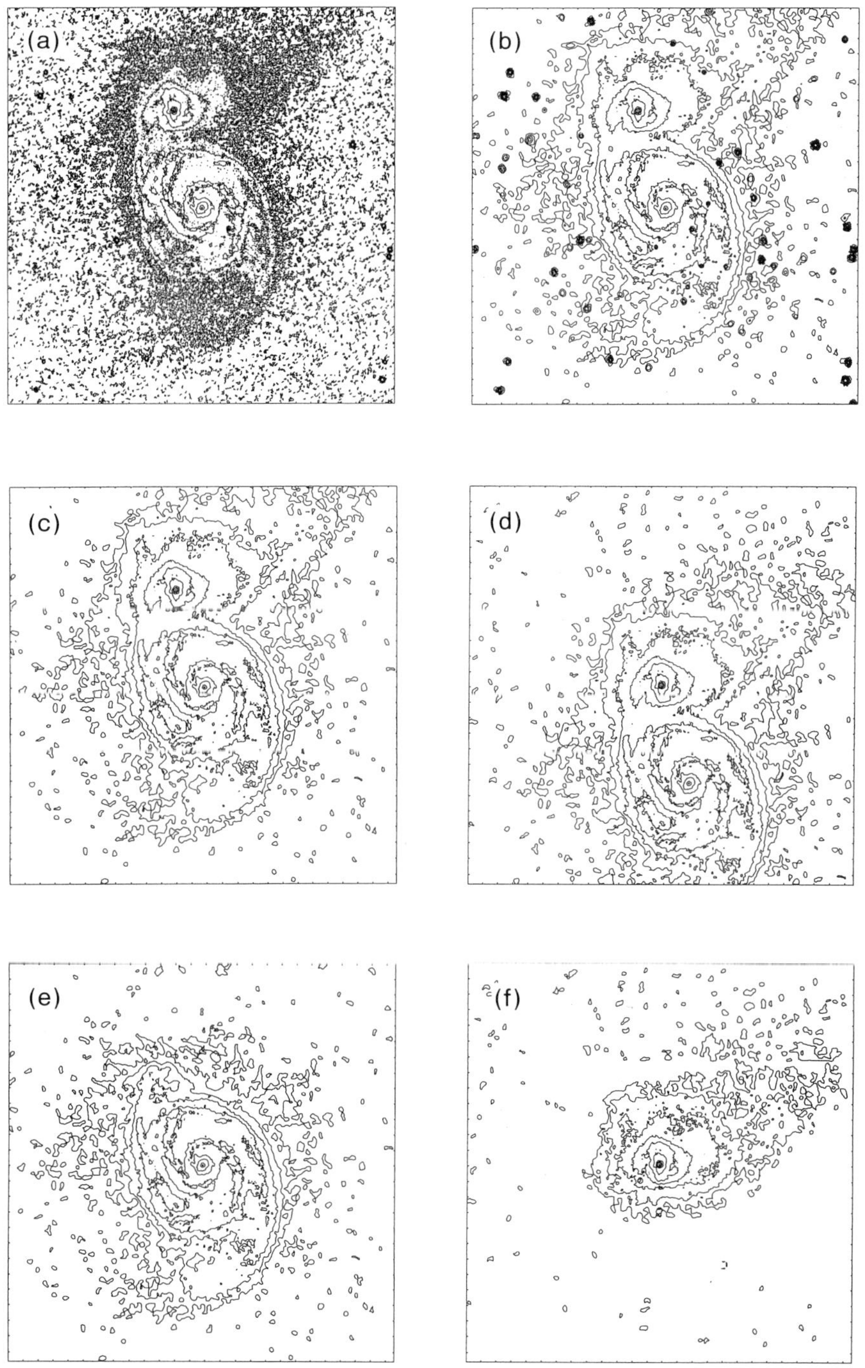

Fig. 1.5 Processes of data reduction (NGC5194 and NGC5195). The relative intensity frame of NGC5194 (a) was smoothed to yield the smoothed frame (b). Using the cleaned frames of NGC 5194 (c) and NGC5195 (d), the two galaxies were decomposed to produce the frames (e, f) to be used in the surface photometric analysis.

($\pm 0.10-0.15$ mag) is the major source of error in the *total magnitude* determined in the present photographic photometry because the growth curves of the same galaxy obtained from different plates usually gave much better agreement, as we will show in section 1.8.

(3) Smoothing

The relative intensity frame was smoothed twice to produce the *smoothed frame*. In the first step, smoothing with a Gaussian beam was applied with the standard deviation of 0.5 pixels. As the next step, the smoothing with the variable-width Gaussian beam with a threshold relative intensity of 10 (Ichikawa et al. 1987) was applied. In this smoothing process, the width of the Gaussian beam for a given pixel was increased until the sum of relative intensities of the pixels covered by the beam reached the threshold value. In short, regions of lower intensities were more heavily smoothed than those of higher intensities in the second smoothing.

(4) Cleaning

The smoothed frame was cleaned of field stars and neighboring galaxies, if any, to produce a *cleaned frame*. The values of the pixels that were lying in a stellar image or in any unwanted image were replaced with the values interpolated from the surrounding pixels by means of two-dimensional polynomials. The 2nd, the 1st, and the 0-th order polynomials were used on a case-by-case basis.

The cleaning program automatically cleaned the field stars that were not superposed on the galaxy image and left those superposed for further interactive cleaning. This process, although automatic, was monitored on the image display, and it was confirmed for each frame whether the automatic cleaning was performed properly. In the interactive mode, decisions were made for each superposed image, whether or not it should be cleaned. Plate faults, scratches, and trails of artificial satellites seen occasionally in the frame were also cleaned out in the interactive mode.

In the case where a galaxy of considerable size overlapped on the objective galaxy, the galaxy was decomposed interactively using the cleaned frame. In the decomposition, the sum of luminosities of the two galaxies was kept unchanged. Some galaxies were deeply embedded in the envelope of a neighboring bright galaxy, or in the photographic halo of a bright star. Photometric parameters of such galaxies were inevitably subject to large uncertainties, although every effort was made to recover their uncontaminated surface brightness distribution. Galaxies to which special treatments were applied in the cleaning process are noted in Part II.

Figure 1.5 illustrates the processes of data reduction described in this section.

1.6 Total Magnitude and Effective Radius

Using the cleaned frame, the growth curve, i.e., the plot of the integrated magnitude versus the radius of the integration circle was constructed for each galaxy. The radius was expressed in logarithmic scale for convenience of the *sliding fit* described below.

Growth curves were divided into two inclination classes according to log R_{25} given in the *Second Reference Catalogue of Bright Galaxies* (de Vaucouleurs et al. 1976: RC 2). The division was made at log $R_{25} = 0.30$. Within each inclination class, growth curves were sorted according to the morphological type index T with a bin size of $\Delta T = 1$. The sorted growth curves were graphically overlapped by sliding along both coordinate axes, and deviant ones were rejected by visual inspection to produce the average curve for each type bin.

The average curves for respective type bins were again graphically overlapped, and the similarity among adjacent type bins was inspected. Finally, the average growth curves were binned into the following six type categories to yield the template growth curves to be used in measuring the total (asymptotic) magnitude and the effective radius of individual galaxies: E ($T = -5, -4$), S0 ($T = -3, -2, -1$), Sa ($T = 0$, 1), Sb ($T = 2$, 3, 4), Sc ($T = 5$, 6, 7), and Sm ($T = 8$, 9, 10), each being divided into the two inclination classes. Table 1.1 gives the number of galaxies used to produce the templates. The templates are shown in Fig. 1.6.

Table 1.1 Number of galaxies used to produce the templates of growth curves for six type categories

	E	S0	Sa	Sb	Sc	Sm
T=	−5, −4	−3, −2, −1	0, 1	2, 3, 4	5, 6, 7	8, 9, 10
log$R_{25} \leq 0.3$	94	105	60	182	165	40
log$R_{25} > 0.3$	3	54	28	73	62	13

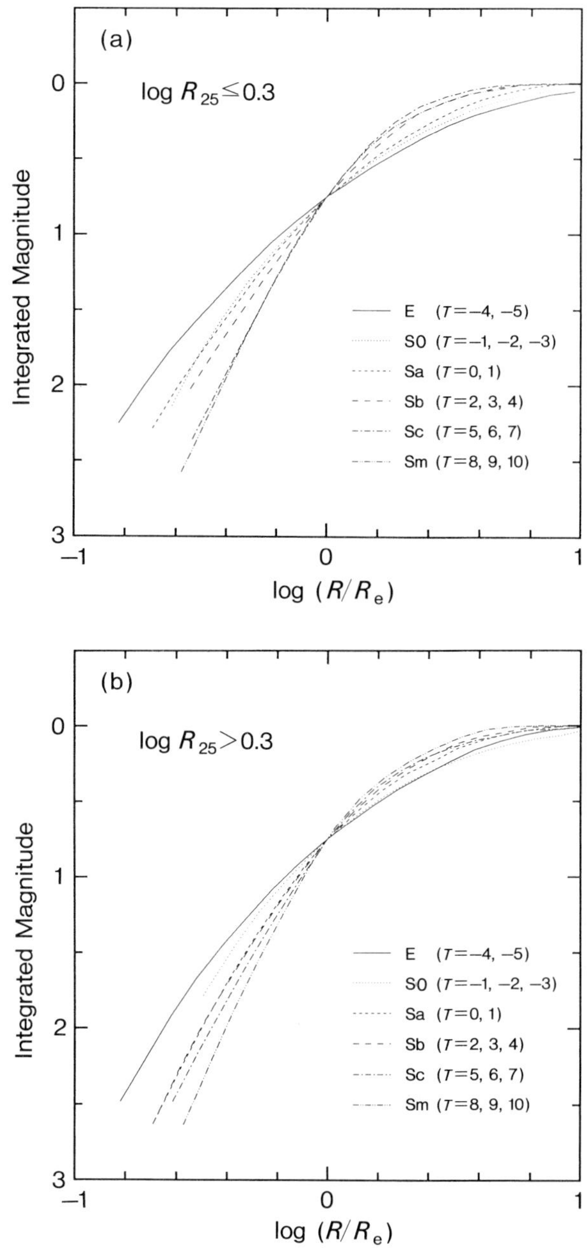

Fig. 1.6 Templates of the growth curve for inclination class log $R_{25} \leq 0.3$ (a) and for inclination class log $R_{25} > 0.3$ (b). The abscissa is the logarithm of the radius of the integration circle normalized by the effective radius and the ordinate is the integrated magnitude.

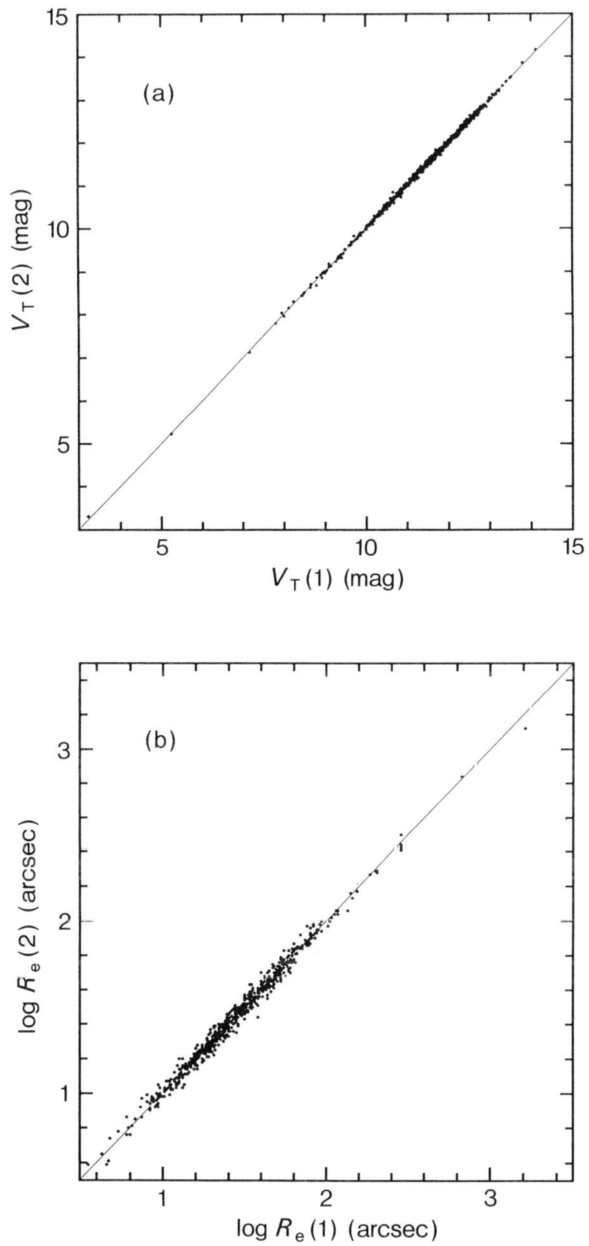

Fig. 1.7 Comparison of the total magnitude V_T (a) and effective radius log R_e (b) between two independent determinations.

The growth curve constructed from the cleaned frame of each galaxy was fitted by one of the templates in the inclination class to which the galaxy belongs. Preference was given to the template of the same type category as the galaxy in question. However, when the template of a different type category yielded a much better fit than that of the same type category, the former was used. Fitting was done graphically by sliding the template along both coordinate axes, and the total magnitude (V_T) and the logarithm of effective radius (log R_e) were read on the graph. The effective radius is defined as the radius where the integrated magnitude is $V_T + 0.75$ mag, i.e., the radius in which a half of the total luminosity is included. Fitting was done twice, by different persons independently. The comparison of the two determinations is shown in Fig. 1.7. The standard deviations of the differences were $\sigma(V_T) = 0.039$ mag and σ (log R_e) = 0.034. These values can be considered as the measure of uncertainties in V_T and log R_e due to the arbitrariness in growth curve fitting.

The mean value of the two determinations is given for V_T and log R_e in Part II. Since the present surface photometry was based on Schmidt plates with a small image scale, central regions ($R < 15$ arcsec) of galaxies with bright nuclei could not be resolved properly. The value of R_e smaller than 15 arcsec is followed by a colon.

1.7 Surface Photometric Parameters

Cleaned frames were subjected to surface photometric analysis. An ellipse fitting program similar to that described in Lauer (1985) was developed to derive surface photometric parameters.

(1) Definition of the Galactic Center

The apparent galactic center was defined in the early stage of data reduction by the visual inspection of the galaxy image on the image display as the position of the nucleus, or the approximate center of luminosity distribution when no appreciable nucleus was recognized. The galactic center was redetermined later as the center of gravity within the window of 21 × 21 pixels centered on the apparent galactic center. The location of the adopted galactic center can be identified in the isophotal map given in Part III as the crossing point of the two lines that connect the two pairs of solid and the broken marks shown in the margin of the frame.

(2) Isophotal Parameters

The *isophotal* parameters were derived by fitting an ellipse to the isophote of 25 mag arcsec^{-2} level. The center of this ellipse was left free in the fitting. However, it was forced to be at the galactic center when the fitting algorithm did not converge. This ellipse will be called hereafter the *envelope ellipse*. The parameters of the envelope ellipse are given in Part II. They are the integrated magnitude within the ellipse (V_{25}), logarithm of the semi-major axis ($\log a_{25}$), the axis ratio ($\epsilon_{25} = b_{25}/a_{25}$), and the position angle of the major axis (θ_{25}), where b_{25} is the semi-minor axis of the ellipse. The envelope ellipse is superposed on the isophotal map of each galaxy in Part III. The isophotal map were made from the *smoothed frame before cleaning* for the sake of the identification of the field. However, it should be noted that the analysis was made using the *cleaned frame*.

(3) Metric Parameters

In addition to the *isophotal* parameters, the *metric* parameters independent of any particular isophote were derived. Average surface brightness of a galaxy was computed as a function of the distance from the galactic center for 180 radial sectors of a vertex angle 2° to produce the smoothed surface brightness distribution. Concentric ellipses were fitted to the isophotes of this smoothed surface brightness distribution. The center of the ellipses was fixed at the galactic center.

The parameters of the ellipses that encircled the 1/4, 1/2, and 3/4 of the total luminosity were computed. They included the semi-major axis (a), the semi-minor axis (b), the axis ratio (ϵ), the square root of the area ($S = (\pi ab)^{1/2}$), the position angle of the major axis (θ), and the surface brightness (μ). The parameters of the three ellipses were discriminated by the subscripts '1', 'e', and '3', respectively. The subscript 'e' denotes *effective*. Using S_1, S_e, and S_3, two luminosity concentration indices were computed as $S_{13} = S_1/S_3$ and $S_{23} = S_e/S_3$. The parameters of the *effective* ellipse and the concentration indices are given in Part II. When $(a_e b_e)^{1/2} < 15$ arcsec and/or $(a_1 b_1)^{1/2} < 15$ arcsec, the values of S_{23} and/or S_{13} are followed by a colon.

(4) Surface Brightness Profiles

An ellipse was fitted to the isophote of 25 mag arcsec^{-2} level of the *smoothed surface brightness distribution with its center fixed at the galactic center*. This ellipse will be called hereafter the *centered ellipse*. Surface brightness profiles were measured along the major and the minor axes of the centered ellipse. The directions of the axes are indicated by the pairs of the solid (major axis) and the broken (minor axis) marks shown in the margin of the isophotal maps in Part III. It is noted that the directions are not necessarily the same as those of the envelope ellipse shown overlapped on the isophotal map, which was fitted to the isophote of 25 mag arcsec^{-2} level of the cleaned frame with its center not fixed at the galactic center as described above. The profiles were obtained by running the measuring apertures of varying diameters from the galactic center outward and by sampling the data at every two pixels. The diameter of the aperture was two pixels at the galactic center and increased outward in such a way that the tangents of the series of sampling apertures form a sector of vertex angle 5°. The varying apertures were adopted to compensate for the low signal-to-noise ratio in the outer region of the galaxy. The profiles are shown in Part III.

(5) Geometric Information

The envelope ellipse, the circle with the effective radius, and the three ellipses that encircled the 1/4, 1/2, and 3/4 of the total luminosity are illustrated as the inset to the figure of the surface brightness profiles in Part III, together with the position of the adopted galactic center. The inset is on the scale of 1/2 of the isophotal map.

1.8 Photometric Accuracy

(1) Plate Noise

Effects of the plate noise were assessed by analyzing two frames of the same galaxy with the same coverage on the same plate but with different pixel sizes. Such pairs of the frames were produced for 134 galaxies on 85 plates by changing the binning factor in creating the relative intensity frame (section 1.5).

Fig. 1.8 Effects of plate noise. The differences in V_T and log R_e measured with a pair of frames of the same galaxy with the same coverage on the same plate but with different pixel sizes are plotted as functions of $<V_T>$ (a) and $<\log R_e>$ (b), respectively.

Fig. 1.9 Internal accuracy of the total magnitude and the effective radius. The differences in V_T and log R_e measured with a pair of frames of the same galaxy with the same frame size and the same pixel size but produced from different plates are plotted as functions of $<V_T>$ (a) and $<\log R_e>$ (b), respectively.

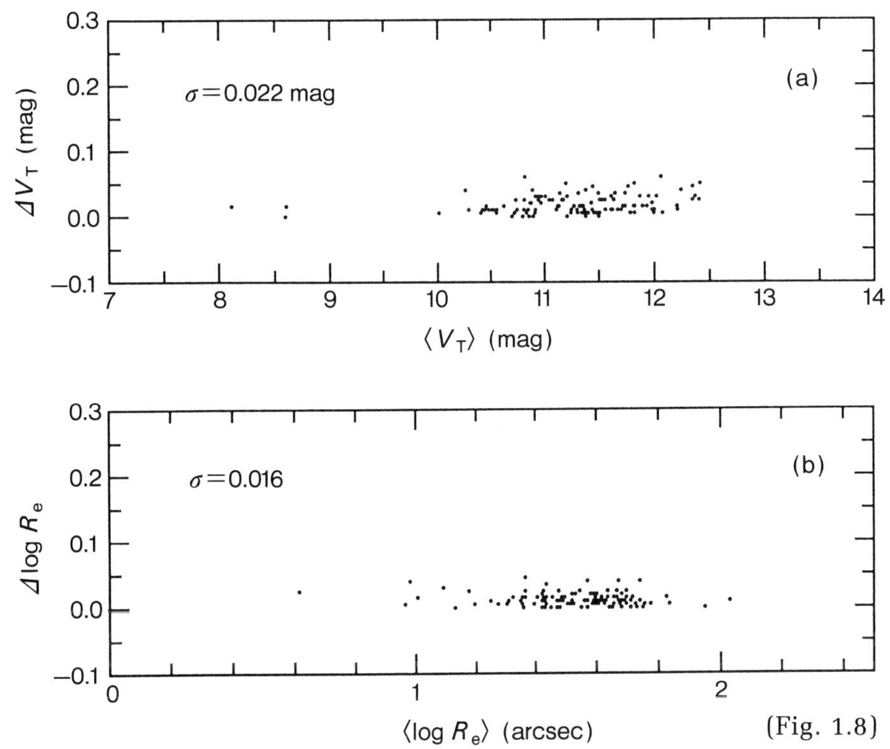

(Fig. 1.8)

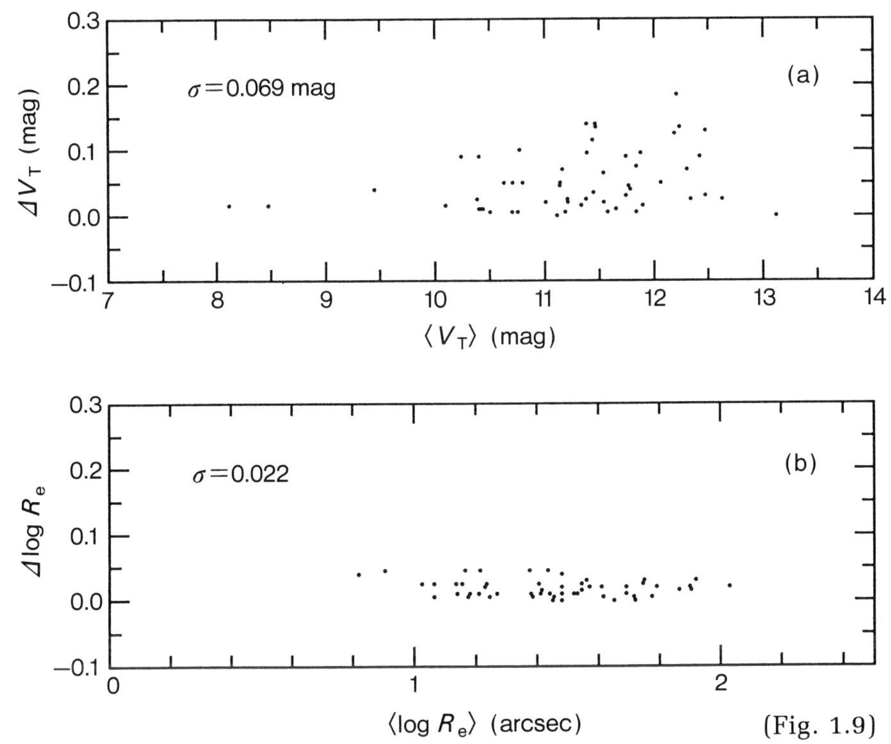

(Fig. 1.9)

The pairs were processed in the same manner to yield two total magnitudes (V_T (1) and V_T (2)) and two effective radii (log R_e (1) and log R_e (2)) for each galaxy. The average values, $<V_T>$ and $<\log R_e>$, and the absolute values of the differences, $\Delta V_T = |<V_T> - V_T(1)|$ and $\Delta \log R_e = |<\log R_e> - \log R_e(1)|$, were computed for each galaxy. The difference computed in this way corresponds to the distance perpendicular to the regression line of slope 1.0, and is smaller by a factor of $2^{1/2} = 1.414$ than the residual usually quoted in the regression analysis.

The distributions of the differences were examined with reference to the uncertainties, $\sigma(V_T) = 0.039/1.414 = 0.028$ and $\sigma(\log R_e) = 0.034/1.414 = 0.024$, due to the arbitrariness of the growth curve fitting (section 1.6). The plates that gave differences greater than twice the uncertainties, i.e., those with $\Delta V_T > 0.06$ mag and $\Delta \log R_e > 0.05$, were judged to have excessively high plate noise, and the frames produced from the plates were discarded. By the visual inspection of all the isophotal maps, the plates showing a similar degree of noise were identified among the remaining about 200 plates. The frames from those plates were also discarded.

The differences, ΔV_T and $\Delta \log R_e$, are plotted as functions of $<V_T>$ and $<\log R_e>$ in Figs. 1.8a and 1.8b, respectively. The standard deviations are $\sigma(V_T) = 0.022$ mag and $\sigma(\log R_e) = 0.016$. These values are smaller than the uncertainties of the respective parameters due to the arbitrariness of the growth curve fitting. It is concluded that no significant errors were introduced by plate noise within the accuracy of the present photometry.

(2) Internal Accuracy

Fifty galaxies were photographed on two plates, and one galaxy was photographed on four plates. These were used to provide the estimate of the internal accuracy. Fifty-six pairs of frames were produced for the 51 galaxies which originated from independent observations and passed through independent data reductions. Each pair of frames for the same galaxy had the same frame size and the same pixel size, covering the same area on the plate. These pairs were analyzed in the same manner as that used for the pairs described in the preceding subsection.

The differences, ΔV_T and $\Delta \log R_e$, are plotted as functions of $<V_T>$ and $<\log R_e>$ in Figs.1.9a and 1.9b, respectively. The standard deviation for V_T, 0.069 mag, is more than twice as large as the uncertainty of 0.028 mag due to the arbitrainess in the growth curve fitting, while that for $\log R_e$, 0.022, is comparable to the corresponding uncertainty of 0.024. This demonstrates that the uncertainty in the absolute calibration is the major source of the error in the total magnitude of galaxies. These values, $\sigma(V_T) = 0.07$ mag and $\sigma(\log R_e) = 0.02$, or $\sigma(V_T) = 0.10$ mag and $\sigma(\log R_e) = 0.03$ when multiplied by 1.414, can be considered to be typical internal accuracies of the two parameters.

Among the fifty-six pairs of frames, twenty-four pairs were subjected to surface photometric analysis. The standard deviations of the differences were computed for all the surface photometric parameters in order to provide estimates of their internal accuracies. The standard deviations multiplied by 1.414 are summarized in Table 1.2.

Table 1.2 Internal accuracy

Parameters	σ		N
V_T	0.10	(mag)	56
$\log R_e$	0.03		56
V_{25}	0.10	(mag)	24
$\log a_{25}$	0.03		24
ϵ_{25}	0.04		24
θ_{25}	4.5[1)]	(degree)	18
μ_e	0.19	(mag arcsec^{-2})	24
μ_e	0.16[2)]	(mag arcsec^{-2})	22
$\log a_e$	0.04		24
ϵ_e	0.03		24
S_{13}	0.01		24
S_{23}	0.03		24

1) only for $\epsilon_{25} < 0.8$
2) only for $\mu_e > 19.5$ mag arcsec^{-2}

It was found that the internal accuracy of μ_e was lower for galaxies with $\mu_e < 19.5$ mag arcsec^{-2} than for those with $\mu_e > 19.5$ mag arcsec^{-2}. Most of the galaxies with $\mu_e < 19.5$ mag arcsec^{-2} had effective radii smaller than 10 arcsec. The low internal accuracies for such galaxies may be attributed to errors due to the lack of resolution and the saturation effect of the plate, which become significant for galaxies with high surface brightness and high degrees of central concentration. The values of μ_e brighter than 19.5 mag arcsec^{-2} given in Part II are followed by a colon.

The average values of V_T and $\log R_e$ computed for the 51 multiply observed galaxies are summarized in Table 4.3. In order to preserve homogeneity, we present the parameters in Part II that were derived from a single observation for the 51 galaxies in the same manner as for other galaxies.

(3) External Consistency

To our knowledge, there are few catalogs based upon homogeneous photometry of a large sample of galaxies in the *V* band which we can use to evaluate the external accuracy of the present photometry. The compilation by Sandage and Visvanathan (1978) lists aperture magnitudes of 405 early-type galaxies in the *V* band. However, their apertures are too small to allow a meaningful comparison of their magnitudes with those of the present photometry.

In order to examine the consistency of the present photometry we relied on RC 2, which contains the largest body of photometric data compiled from the literature and *transformed statistically* to a homogeneous system. We compare four parameters—total magnitude (V_T), logarithm of the effective radius ($\log R_e$), logarithm of the semi-major axis of the envelope ellipse ($\log a_{25}$), and axis ratio (ϵ_{25})—with the corresponding parameters, V_T, $A_e = 2R_e$, $\log D_{25}$, and $R_{25} = (a/b)_{25} = \epsilon_{25}^{-1}$, given in RC 2. It is noted that RC 2 is based primarily on the *B*-band surface photometry. The parameters D_{25} and $(a/b)_{25}$ in RC 2 are the isophotal major diameter and the isophotal axis ratio referred to the level of $\mu_B = 25$ mag arcsec^{-2}. This level corresponds to $\mu_V \sim 24$–24.5 mag arcsec^{-2}, depending on the $B-V$ color of the galaxy. The parameters of the present study, a_{25} and ϵ_{25}, were measured at $\mu_V = 25$ mag arcsec^{-2}, which is fainter than the isophotal level adopted in RC 2. It should be noted that we are not comparing identical parameters except for V_T.

For each pair of parameters, regression analysis was carried out for the total sample and for the sample subjected to one cycle of 2-sigma rejection, i.e., the sample from which the galaxies with residuals larger than twice the standard deviation were rejected. For each sample, two regression lines, one with the slope fixed at 1.0

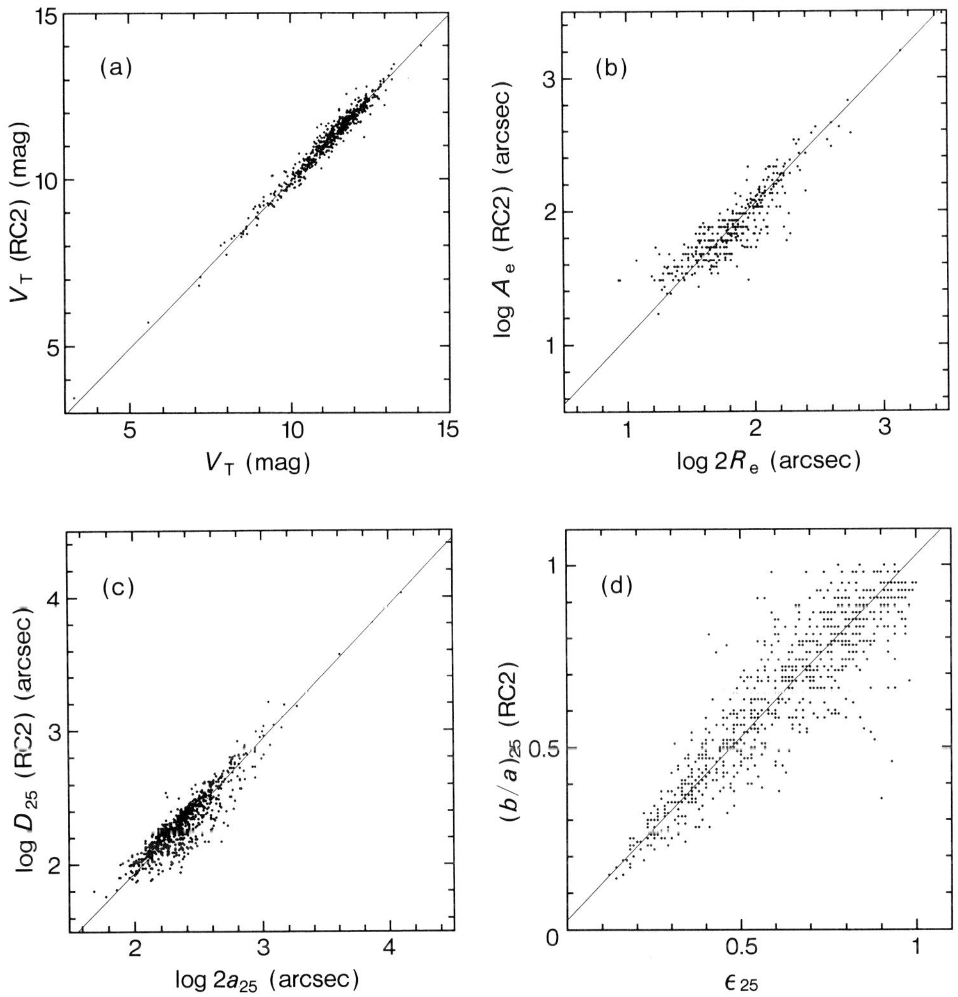

Fig. 1.10 External consistency of the present photometry. The total magnitude (a), the logarithm of the effective diameter (b), the logarithm of the major axis at 25 mag arcsec^{-2} (c), and the axis ratio at 25 mag arcsec^{-2} (d), are compared with the corresponding data given in RC 2.

and the other with a free slope, were derived. The comparisons for the total samples are illustrated in Figs. 1.10a–d, where the regression lines with fixed slopes are shown. The results of the regression analysis are summarized in Table 1.3.

The following trends are observed in the comparison. No parameters derived in the present study show a significant non-linear systematic difference from RC 2 parameters, except for the effective radius in the range of log $2R_e < 1.5$, i.e., $R_e < 15$ arcsec. In this range, the present R_e tends to be smaller than the value in RC 2. This may be due to systematic errors in the present photographic photometry based on Schmidt plates with the small image scale, as discussed above in sections 1.5 and 1.7. The present total magnitude is

Table 1.3 Summary of the regression analysis in the comparison of photometric parameters between the present study and RC 2[1)]

X (1)	Y (2)	a (3)	b (4)	σ_Y (5)	c.c. (6)	N (7)	Remarks
V_T	V_T	1.0	−0.063	0.209	0.982	555	total: Fig.1.10a
V_T	V_T	1.0	−0.071	0.169	0.988	530	2σ rejection
V_T	V_T	0.971	+0.260	0.207	0.982	555	total
V_T	V_T	0.981	+0.139	0.168	0.988	530	2σ rejection
log $2R_e$	log A_e	1.0	+0.060	0.143	0.890	371	total: Fig.1.10b
log $2R_e$	log A_e	1.0	+0.064	0.114	0.926	353	2σ rejection
log $2R_e$	log A_e	0.796	+0.430	0.128	0.890	371	total
log $2R_e$	log A_e	0.867	+0.306	0.106	0.926	353	2σ rejection
log $2a_{25}$	log D_{25}	1.0	−0.062	0.101	0.907	790	total: Fig.1.10c
log $2a_{25}$	log D_{25}	1.0	−0.049	0.079	0.943	747	2σ rejection
log $2a_{25}$	log D_{25}	0.900	+0.176	0.098	0.907	790	total
log $2a_{25}$	log D_{25}	0.934	+0.107	0.077	0.943	747	2σ rejection
ϵ_{25}	$(b/a)_{25}$	1.0	+0.027	0.098	0.893	790	total: Fig.1.10d
ϵ_{25}	$(b/a)_{25}$	1.0	+0.030	0.077	0.934	752	2σ rejection
ϵ_{25}	$(b/a)_{25}$	0.889	+0.096	0.095	0.893	790	total
ϵ_{25}	$(b/a)_{25}$	0.931	+0.073	0.076	0.934	752	2σ rejection

[1)] Regression analysis of the form $Y = aX + b$ was carried out.
Column (1): independent variable X (parameters from the present study)
Column (2): dependent variable Y (parameters from RC 2)
Column (3): slope of the regression line
Column (4): zero point of the regression line
Column (5): standard deviation of the residuals in Y
Column (6): correlation coefficient
Column (7): number of data

systematically brighter by 0.07 mag, and the effective radius is systematically smaller by 0.06 dex (15 percent), than the corresponding values given in RC 2. The present semi-major diameter is systematically larger by 0.05 dex (12 percent) and the axis ratio is systematically smaller by 0.03—i.e., the shape of the isophote is rounder—than the corresponding values given in RC 2. The sense of the differences in the semi-major diameter and the axis ratio is explained by the fact that the present values refer to a fainter isophote level than RC 2, as described above.

The average mean errors quoted in RC 2 for V_T, log A_e, log D_{25}, and log $(a/b)_{25}$ are 0.085 mag (range 0.03 to 0.16), 0.04 (range 0.02 to 0.1), 0.04 (range 0.02 to 0.32), and 0.04 (range 0.013 to 0.18), respectively. The error for V_T was computed as the square root of the sum of the squared errors for B_T and $(B-V)_T$. The mean errors of the parameters of individual galaxies given in RC 2 scatter over considerably wide ranges, although the average errors are comparable to the internal accuracies of the corresponding parameters derived in the present study (Table 1.2). The standard deviations of V_T, log R_e, log a_{25}, and ϵ_{25} for the samples after 2-sigma rejection are 1.5–2 times larger than the internal accuracies of their respective parameters given in Table 1.2.

Finally, the surface brightness profiles were compared with the data in the literature. Few galaxies were available for which the profiles in the V band had been measured in previous studies. We were forced to compare the profiles in different color bands. Elliptical galaxies are suitable to this purpose because they show smooth light distribution and a smaller degree of color gradient than spiral galaxies. We selected two elliptical galaxies, NGC3379 and NGC4472, and extracted the profiles from the cleaned frames. The comparisons are shown in Figs. 1.11a–d. Since the color bands are different, the results of the comparison should be taken as only indicative of general trends. Except for the very central region ($R<10$–15 arcsec) the agreement is reasonable with a difference of less than ± 0.1 mag in the bright region ($\mu_V \sim 19$–24 mag arcsec^{-2}) and about ± 0.3 mag in the faint region ($\mu_V \sim 24$–25 mag arcsec^{-2}). The accuracy of the present photometry is comparable to the typical error of photographic photometry found in the literature (e.g., Okamura 1988).

In conclusion, the present surface photometry has produced a homogeneous database of two-dimensional luminosity distribution of

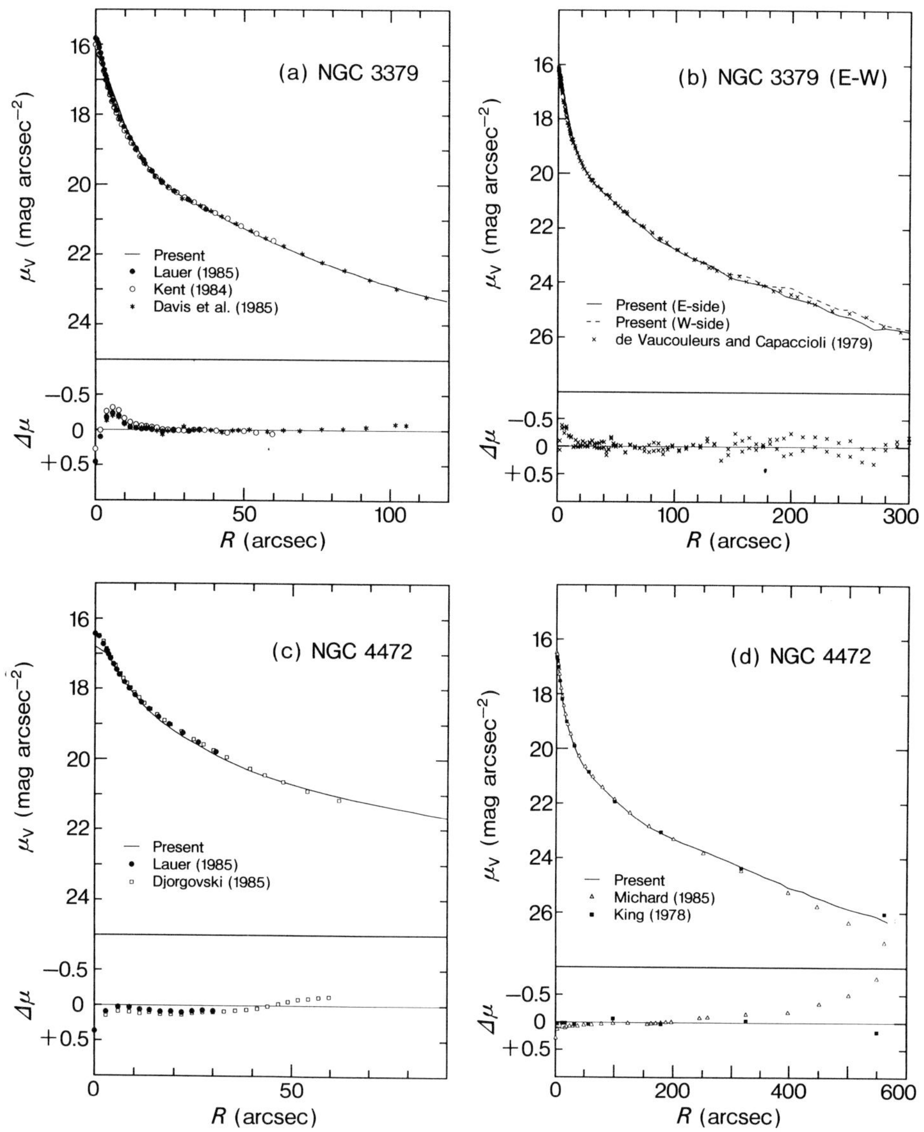

Fig. 1.11 Comparison of surface brightness profiles of two representative galaxies. In each figure, the top panel shows the profiles and the bottom panel shows the difference between the present photometry and the previous data. The previous data are taken from the literature; they come from photographic photometry (pg), photoelectric photometry (pe), and CCD (Charge Coupled Device) photometry (CCD) (Davis et al. 1985, CCD, R-band; de Vaucouleurs and Capaccioli 1979, pg + pe, B-band; Djorgovski 1985, CCD, R-band; King 1978, pg, B-band; Kent 1984, CCD, R-band; Lauer 1985, CCD, R-band; Michard 1985, pg, B-band). The symbols and their data sources are given in the figures. The profiles are shifted along the ordinate to give the best coincidence, since the color bands are different. Comparisons are made for the mean profile (a), i.e., the profile along the 45° direction from the major axis, and the east-west profile (b) of NGC3379, and for the mean profile of NGC4472 ((c) for central region and (d) for whole region).

the largest sample of northern bright galaxies to date, with an accuracy near the limit attainable by photographic photometry based on Schmidt plates. The photometric data presented in this Atlas provide a new basis for statistical analyses in various subjects including the structure of galaxies, distance indicater relations, cosmic distance scale, and kinematics of the Local Supercluster.

1.9 Notes to the Catalog

The catalog presented in Part II of this volume is composed of 21 columns. Columns 1–7 give the data taken from Sandage and Tammann (1981: RSA), de Vaucouleurs et al. (1976: RC 2) and Burstein and Heiles (1984). Columns 8–19 give the parameters derived in the present photometry. No corrections are made for the Galactic extinction and the internal absorption to any parameter. Column 20 is reserved for notes that are given at the end of the catalog. Column 21 gives, for every five galaxies, the page in the atlas in Part III where the preceding five galaxies are presented. The descriptions of the column entries are given below:

Column 1: The galaxy name taken from RSA.
Column 2: The right ascension in hours, minutes, and seconds at the epoch of 1950.0 taken from RSA.
Column 3: The declination in degrees and arcminutes at the epoch of 1950.0 taken from RSA.
Column 4: The morphological type taken from RC 2 and decoded by the authors.
Column 5: The morphological type index taken from RC 2.
Column 6: The corrected radial velocity in kms^{-1} relative to the centroid of the *Local Group* taken from RSA.
Column 7: Galactic absorption in the V band in units of magnitudes calculated as $A_V = 0.75A_B$, using the A_B tabulated in Burstein and Heiles (1984).
Column 8: The apparent total (asymptotic) magnitude in the V band derived in the growth curve fitting.
Column 9: The logarithm of the effective radius derived in the growth curve fitting in units of arcsec. Values smaller than $R_e = 15$ arcsec are followed by a colon.
Column 10: The type category of the template of the growth curve used in the derivation of the apparent total magnitude and the effective radius given in columns 8 and 9. The type category is indicated only when the template of a different category from that to which the galaxy belongs was used. The type categories are given in Table 1.1.
Column 11: The apparent magnitude in the V band integrated within the *envelope ellipse* explained in section 1.7.
Column 12: The logarithm of the semi-major axis of the envelope ellipse in units of arcsec.
Column 13: The axis ratio of the envelope ellipse.
Column 14: The position angle of the major axis of the envelope ellipse in units of degrees. The position angle is measured counter-clockwise from upward direction with respect to the *frame*. The angle between north and the upward direction of the frame is given in Table 4.2.
Column 15: The effective surface brightness in units of mag arcsec^{-2} at the effective ellipse, i.e., the ellipse that encircles half of the total luminosity. Values brighter than $\mu_e = 19.5$ mag arcsec^{-2} are followed by a colon.
Column 16: The logarithm of the semi-major axis of the effective ellipse in units of arcsec. Values smaller than $a_e = 15$ arcsec are followed by a colon.
Column 17: The axis ratio of the effective ellipse.
Column 18: The luminosity concentration index computed as the ratio of $(\pi ab)^{1/2}$ of the ellipses that encircle 1/4 and 3/4 of the total luminosity, as explained in section 1.7. The colon which follows the values indicates that $(a_1 b_1)^{1/2}$ is smaller than 15 arcsec, where a_1 and b_1 are the semi-major and the semi-minor axes of the ellipse that encircles 1/4 of the total luminosity.
Column 19: The luminosity concentration index computed as the ratio of $(\pi ab)^{1/2}$ of the ellipses that encircle 1/2 and 3/4 of the total luminosity. The colon which follows the values indicates that $(a_e b_e)^{1/2}$ is smaller than 15 arcsec, where a_e and b_e are the semi-major and the semi-minor axes of the effective ellipse.
Column 20: The asterisk in this column indicates that a note is given on the galaxy at the end of the catalog.
Column 21: Page of the atlas in Part III where illustrations of the preceding five galaxies are presented.

1.10 Notes to the Atlas

In the atlas presented in Part III of this volume, the following three figures are shown for each of the galaxies listed in Part II.

(1) The photograph of the smoothed frame displayed on the screen of the image display. The uniform background corresponding to the sky brightness of 22 mag arcsec^{-2} was added to the smoothed frame to simulate the appearance of the galaxy on the plate. All the galaxies were displayed on the logarithmic gray scale with the same display range of 19 to 22 mag arcsec^{-2}.

(2) The isophotal map of the smoothed frame. The level of the outermost isophote is $\mu_V = 25$ mag arcsec^{-2}, and the interval of the isophotes is 1 mag arcsec^{-2}. The *envelope ellipse* explained in section 1.7 is superposed. Two pairs of solid and broken marks are shown in the margin of the map. They indicate the directions of the major and the minor axes of the *centered ellipse* explained in section 1.7. The adopted galactic center can be identified as the crossing point of the two lines that connect the pairs of the marks. The tick marks are given on the side at every 50 pixels. The galaxy name, plate number, pixel size, sky brightness, and the frame size are given.

(3) Surface brightness profiles in the logarithmic coordinate along the major and the minor axes of the centered ellipse, i.e., along the directions toward the solid and the broken marks in the margin of the isophotal map, are shown by the solid and broken curves, respectively. Both halves of the profiles with respect to the galactic center are folded. The horizontal bar on the ordinate shows the surface brightness at the adopted galactic center. Illustrated in the inset to the figure are the envelope ellipse (broken ellipse), the circle with the effective radius given in column (9) in Part II (broken circle), the three ellipses that encircle 1/4, 1/2, and 3/4 of the total luminosity (solid ellipses), and the adopted galactic center (dot). The inset is on the scale of 1/2 of the isophotal map.

The photographs and the isophotal map are shown on the same scale, either 0.1 or 0.2 millimeters to a pixel. The lengths of the photographs are slightly shortened due to the distortion of the screen of the image display. Approximately, north is up, and east is to the left. Since wide-field Schmidt plates were used in the present study, exact orientation of the frame with respect to north and east depends on the location of the galaxy on a plate. The angle between north and the upward direction of each frame is given in Table 4.2.

In Fig. 3.1 in Part III (p. 203), surface brightness profiles of two model galaxies, one with a $R^{1/4}$-law surface brightness distribution and the other with an exponential surface brightness distribution, are shown on the same scale as the profiles of the sample galaxies. These model profiles may be used as the template in measuring the qualitative behaviors of the profiles of the sample galaxies.

References

Anderson, C.M., Slovak, M.H., and Michalski, D.E. 1983, *Publ. A. S. Pacific*, **95**, 792.
Burstein, D., and Heiles, C. 1984, *Astrophys. J. Suppl.*, **54**, 33.
Davis, L.E., Cawson, M., Davies, R.L., and Illingworth, G. 1985, *Astron. J.*, **90**, 169.
de Vaucouleurs, G., and Capaccioli, M. 1979, *Astrophys. J. Suppl.*, **40**, 699.
de Vaucouleurs, G., de Vaucouleurs, A., and Corwin, H.G., Jr. 1976, *Second Reference Catalogue of Bright Galaxies* (Austin: Univ. Texas Press) (RC 2)
Djorgovski, S. 1985, Ph D. thesis, University of California, Berkeley.
Hamabe, M., and Okamura, S. 1982, *Annals Tokyo Astron. Obs.*, **18**, 191.
Ichikawa, S., Okamura, S., Watanabe, M., Hamabe, M., Aoki, T., and Kodaira, K. 1987, *Annals Tokyo Astron. Obs.*, **21**, 285.
Kent, S.M. 1984, *Astrophys. J. Suppl.*, **56**, 105.
King, I.R. 1978, *Astrophys. J.*, **222**, 1.
Kodaira, K. 1989, *Astrophys. J.*, **342**, 122.
Kodaira, K., and Watanabe, M. 1989, *Astron. J.*, **96**, 1593.
Kodaira, K., Okamura, S., and Watanabe, M. 1983, *Astrophys. J. (Letters)*, **274**, L49.
Kodaira, K., Watanabe, M., and Okamura, S. 1986, *Astrophys. J. Suppl.*, **62**, 703.
Lauer, T.R. 1985, *Astrophys. J. Suppl.*, **57**, 473.
Longo, G., and de Vaucouleurs, A. 1983, *The University of Texas Monograph in Astronomy*, No. 3 (Austin: Univ. Texas).
Michard, R. 1985, *Astron. Astrophys. Suppl.*, **59**, 205.
Okamura, S. 1988, *Publ. A. S. Pacific*, **100**, 524.
Okamura, S., Kodaira, K., and Watanabe, M. 1984, *Astrophys. J.*, **280**, 7.
Sandage, A., and Visvanathan, N. 1978, *Astrophys. J.*, **223**, 707.
Sandage, A., and Tammann, G.A. 1981, *A Revised Shapley-Ames Catalog of Bright Galaxies* (Washington: Carnegie Institution of Washington). (RSA) (Second edition was published in 1987)
Takase, B., Ishida, K., Shimizu, M., Maehara, H., Hamajima, K., Noguchi, T., and Ohashi, M. 1977, *Annals. Tokyo Astron. Obs.*, **16**, 74.
Takase, B., Kodaira, K., and Okamura, S. (eds.) 1984, *An Atlas of Selected Galaxies* (Tokyo and Utrecht: Univ. Tokyo Press and VNU Science Press).
Tarusawa, K., Soyano, T., Noguchi, T., and Okamura, S. 1986, *Tokyo Astron. Obs. Report*, **20**, 674. *(in Japanese)*
Watanabe, M. 1983, *Annals Tokyo Astron. Obs.*, **19**, 121.
Watanabe, M., Kodaira, K., and Okamura, S. 1982, *Astrophys. J. Suppl.*, **50**, 1.
Watanabe, M., Kodaira, K., and Okamura, S. 1985, *Astrophys. J.*, **292**, 72.

Part II The Catalog

Galaxy (1)	α_c (1950) (2)	δ_c (3)	Type (4)	T (5)	v_0 (6)	A_V (7)	V_T (8)	log R_e (9)	GCT (10)	V_{25} (11)	log a_{25} (12)	ϵ_{25} (13)	θ_{25} (14)	μ_e (15)	log a_e (16)	ϵ_e (17)	S_{13} (18)	S_{23} (19)	Note (20)	Page (21)
NGC7814	00 00 41	+15 52.0	SA(s)ab:/	2	1289	0.08	10.39	1.71	S0	10.57	2.36	0.45	134	21.87	1.79	0.54	0.21	0.45		
NGC 16	00 06 29	+27 27.2	SAB0$^-$/	−3	3382	0.13	12.00	0.75:	E	12.03	1.85	0.67	27	18.89:	0.74:	0.95	0.15:	0.36:		
NGC 23	00 07 19	+25 38.8	SB(s)a	1	4836	0.11	12.19	0.89:	S0	12.22	1.94	0.50	167	19.90	0.88:	0.75	0.15:	0.38:		
NGC 147	00 30 28	+48 13.8	E5p	−5	136	0.53	9.15	2.45	Sc	9.48	2.76	0.51	34	24.07	2.52	0.56	0.30	0.60		
NGC 151	00 31 30	−09 58.9	SB(r)bc	4	3871	0.08	11.57	1.48		11.65	2.05	0.48	69	21.74	1.53	0.68	0.27:	0.60		45
NGC 148	00 31 48	−32 03.7	S0$^+$:/	−1:	1916	−.03*	12.13	0.96:		12.25	1.82	0.45	84	20.05	1.00:	0.69	0.21:	0.47:		
NGC 150	00 31 49	−28 04.7	SB(rs)b:	3:	1620	0.05	11.42	1.47	Sc	11.49	2.07	0.48	113	21.47	1.53	0.72	0.38:	0.74		
NGC 157	00 32 14	−08 40.3	SAB(rs)bc	4	1813	0.09	10.45	1.66	Sm	10.53	2.16	0.60	46	20.98	1.79	0.36	0.42	0.59		
NGC 185	00 36 11	+48 03.7	E3p	−5	67	0.59	9.05	2.27	Sb	9.29	2.64	0.75	48	23.26	2.31	0.72	0.26	0.54		
NGC 205	00 37 38	+41 24.9	E5p	−5	49	0.11	7.81	2.30	Sb	7.85	2.85	0.54	149	21.67	2.37	0.48	0.33	0.59	*	46
NGC 221	00 39 59	+40 35.5	cE2	−6	86	0.23	8.15	1.51	E	8.22	2.50	0.76	175	19.21:	1.55	0.81	0.11:	0.32	*	
NGC 224	00 40 00	+40 59.7	SA(s)b	3	−10	0.24	3.28	3.16	E	3.47	3.80	0.32	37	21.84	3.28	0.34	0.21	0.51	*	
NGC 227	00 40 04	−01 48.3	E3?	−5:	5478	0.05	12.08	1.10:		12.23	1.95	0.71	156	21.36	1.17:	0.69	0.08:	0.29:		
NGC 237	00 40 55	−00 24.0	SAB(rs)cd	6	4308	−.01	13.23	1.12:	Sb	13.35	1.71	0.66	174	21.39	1.20:	0.52	0.27:	0.49:		
NGC 245	00 43 32	−01 59.6	SA(rs)bp?	3?	4275	0.07	12.59	1.26	Sc	12.72	1.68	0.91	11	21.40	1.38	0.42	0.34:	0.55		47
NGC 254	00 45 03	−31 41.7	SAB(r)0/a?	0	1458	−.03*	11.69	1.14:	E	11.78	1.96	0.67	139	20.93	1.26:	0.58	0.13:	0.40:		
NGC 253	00 45 08	−25 33.7	SAB(s)c	5	293	0.00	7.17	2.36		7.23	2.97	0.39	48	20.93	2.45	0.26	0.40	0.64		
NGC 255	00 45 16	−11 44.5	SAB(rs)bc	4	1726	0.08	12.02	1.46	Sc	12.14	1.95	0.79	4	22.01	1.48	0.88	0.28:	0.56		
NGC 268	00 47 37	−05 28.1	SAB(rs)b?	3	5659	0.11	12.97	1.26	Sm	13.10	1.73	0.57	90	21.79	1.33	0.62	0.35:	0.63		
NGC 274	00 48 30	−07 19.7	SAB(r)0$^-$p	−3	1894	0.15	12.07	1.15:		12.22	1.83	0.93	58	21.15	1.15:	0.92	0.15:	0.41:	*	48
NGC 275	00 48 33	−07 20.1	SB(rs)cdp	6	1889	0.15	12.40	1.32	Sm	12.51	1.88	0.79	33	21.65	1.36	0.75	0.26:	0.55	*	
NGC 278	00 49 15	+47 16.8	SAB(rs)b	3	932	0.55	10.87	1.35		10.94	2.01	0.88	113	20.33	1.36	0.85	0.29:	0.55		
NGC 289	00 50 17	−31 28.7	SB(rs)bc	4	1834	−.01*	11.16	1.45	Sc	11.24	2.03	0.79	119	21.30	1.51	0.77	0.27:	0.61		
NGC 309	00 54 13	−10 11.3	SAB(r)c	5	5786	0.13	12.11	1.56		12.25	1.88	0.95	42	22.48	1.59	0.77	0.36	0.63		
NGC 337	00 57 19	−07 50.7	SB(s)d	7	1776	0.24	11.69	1.48		11.79	1.91	0.72	121	21.63	1.53	0.65	0.37:	0.70		49
NGC 357	01 00 50	−06 36.3	SB(r)0/a:	0:	2674	0.18	12.02	1.27		12.13	1.86	0.82	9	21.85	1.36	0.70	0.18:	0.48		
New 1	01 02 33	−06 28.6	SAB(rs)d	7	1116	0.18	11.47	1.95		11.72	2.25	0.67	109	23.97	1.98	0.81	0.35	0.65	*	
NGC 404	01 06 39	+35 27.1	SA(s)0$^-$:	−3	231	0.17	9.44	1.98		9.68	2.47	0.79	70	22.73	2.01	0.89	0.15	0.44	*	
NGC 428	01 10 23	+00 42.9	SAB(s)m	9	1311	0.02	11.62	1.63		11.76	2.05	0.72	100	22.29	1.70	0.56	0.31	0.54		
NGC 473	01 17 15	+16 17.0	SAB(r)0/a:	0:	2345	0.08	12.56	1.07:		12.64	1.78	0.62	159	20.79	1.11:	0.71	0.23:	0.48:		50
NGC 514	01 21 25	+12 39.5	SAB(rs)c	5	2675	0.06	12.15	1.61		12.26	2.02	0.63	109	22.77	1.63	0.74	0.35	0.64		
NGC 521	01 22 00	+01 28.2	SB(r)bc	4	5223	0.06	11.71	1.55		11.84	1.97	0.98	32	22.58	1.65	0.84	0.28	0.62		
NGC 524	01 22 10	+09 16.7	SA(rs)0$^+$	−1	2566	0.10	10.43	1.61		10.63	2.18	0.95	93	21.77	1.61	0.95	0.15	0.39		
NGC 533	01 22 56	+01 30.3	E3:	−5:	5664	0.05	11.20	1.77		11.57	2.22	0.55	43	23.51	1.87	0.68	0.14	0.43		
NGC 598	01 31 03	+30 23.9	SA(s)cd	6	69	0.14	5.56	2.84		5.65	3.31	0.56	20	22.22	2.89	0.61	0.35	0.64		51
NGC 628	01 34 01	+15 31.6	SA(s)c	5	861	0.09	9.48	2.03		9.53	2.46	0.87	106	22.43	2.06	0.86	0.33	0.63		
NGC 670	01 44 37	+27 38.1	SA0	−2	4023	0.15	12.45	1.03:		12.58	1.85	0.35	170	20.40	1.10:	0.52	0.24:	0.50:		
NGC 672	01 45 05	+27 11.1	SB(s)cd	6	647	0.18	11.03	1.78	Sm	11.14	2.28	0.43	69	21.86	1.91	0.20	0.42	0.71		
NGC 718	01 50 37	+03 57.0	SAB(s)a	1	1954	0.06	11.64	1.15:	S0	11.73	1.92	0.85	6	21.02	1.23:	0.64	0.08:	0.34:		
NGC 741	01 53 44	+05 23.1	E0:	−5:	5723	0.11	11.28	1.70		11.64	2.12	0.79	70	23.47	1.79	0.77	0.13	0.42		52

Galaxy (1)	α_c (1950) (2)	δ_c (3)	Type (4)	T (5)	v_0 (6)	A_V (7)	V_T (8)	log R_e (9)	GCT (10)	V_{25} (11)	log a_{25} (12)	ϵ_{25} (13)	θ_{25} (14)	μ_e (15)	log a_e (16)	ϵ_e (17)	S_{13} (18)	S_{23} (19)	Note (20)	Page (21)
NGC 750	01 54 38	+32 58.0	Ep	−5	5373	0.11	12.22	1.41		12.44	2.06	0.41	148	22.70	1.44	0.81	0.13:	0.36	*	
NGC 753	01 54 46	+35 40.3	SAB(rs)bc	4	5145	0.17	12.39	1.34	Sc	12.48	1.91	0.66	120	21.99	1.41	0.65	0.26:	0.52		
NGC 777	01 57 21	+31 11.2	E1	−5	5248	0.11	11.78	1.22		11.92	1.96	0.75	151	21.35	1.25	0.82	0.12:	0.36		
NGC 890	02 19 02	+33 02.3	SAB(r)0^{-}?	−3	4273	0.20	11.29	1.43		11.51	2.06	0.74	58	21.87	1.53	0.56	0.11:	0.36		
NGC 908	02 20 47	−21 27.7	SA(s)c	5	1563	0.00	10.32	1.78		10.38	2.28	0.48	74	21.55	1.85	0.51	0.45	0.71		53
NGC 925	02 24 18	+33 21.1	SAB(s)d	7	792	0.20	10.46	2.01		10.55	2.45	0.53	106	23.01	2.12	0.40	0.32	0.67		
NGC 936	02 25 05	−01 22.7	SB(rs)0^{+}	−1	1512	0.04	10.24	1.53		10.31	2.19	0.78	135	21.38	1.65	0.55	0.13:	0.44		
NGC 941	02 25 55	−01 22.5	SAB(rs)c	5	1717	0.03	12.68	1.45	Sm	12.85	1.89	0.70	167	22.85	1.49	0.85	0.30:	0.57		
NGC 949	02 27 39	+36 54.7	SA(rs)b:?	3?	855	0.13	11.91	1.38		11.98	2.07	0.42	147	21.27	1.39	0.52	0.24:	0.45		
NGC 955	02 28 00	−01 19.8	Sab:/	2:	1641	0.03	11.87	1.09:	S0	12.01	1.97	0.32	20	20.06	1.18:	0.40	0.18:	0.39:		54
NGC 976	02 31 11	+20 45.4	SA(rs)c:	5:	4550	0.29	12.42	1.08:	Sb	12.49	1.80	0.67	157	20.84	1.11:	0.82	0.22:	0.48:		
NGC 972	02 31 17	+29 05.5	I0	0	1761	0.26	11.22	1.47	Sb	11.33	2.10	0.55	150	21.41	1.54	0.50	0.22:	0.54		
NGC 991	02 33 03	−07 22.0	SAB(rs)c	5	1607	0.00	12.22	1.60		12.38	1.92	0.87	130	22.82	1.60	0.91	0.35	0.64		
NGC1022	02 36 04	−06 53.4	(R′) SB(s)a	1	1575	0.03	11.31	1.52		11.49	1.96	0.97	25	21.71	1.52	0.83	0.21:	0.49		
NGC1035	02 37 01	−08 20.8	SA(s)c?	5?	1307	0.02	12.23	1.39	Sm	12.35	1.94	0.33	145	20.92	1.48	0.25	0.49:	0.72		55
NGC1023	02 37 16	+38 50.9	SB(rs)0^{-}	−3	895	0.19	9.44	1.65	E	9.51	2.45	0.44	90	20.74	1.73	0.52	0.21	0.52		
NGC1042	02 37 56	−08 38.8	SAB(rs)cd	6	1436	0.05	11.25	1.85		11.40	2.18	0.76	9	22.98	1.86	0.80	0.33	0.64		
NGC1052	02 38 37	−08 28.1	E4	−5	1539	0.05	10.53	1.36		10.66	2.17	0.75	126	20.68	1.43	0.67	0.13:	0.37		
NGC1055	02 39 11	+00 13.7	SBb:/	3	1098	0.05	10.55	1.89	Sm	10.70	2.38	0.52	103	22.94	1.97	0.62	0.24	0.61		
NGC1068	02 40 07	−00 13.5	(R)SA(rs)b	3	1234	0.04	8.99	1.49	E	9.05	2.43	0.86	85	20.04	1.52	0.96	0.11:	0.36		56
NGC1073	02 41 05	+01 09.9	SB(rs)c	5	1318	0.05	11.15	1.87		11.34	2.17	0.87	2	23.22	1.89	0.94	0.26	0.66		
NGC1084	02 43 32	−07 47.1	SA(s)c	5	1478	0.04	10.73	1.48		10.80	2.09	0.87	71	20.47	1.52	0.58	0.35	0.59		
NGC1087	02 43 52	−00 42.5	SAB(rs)c	5	1628	0.08	11.13	1.63		11.25	2.09	0.62	179	21.69	1.66	0.71	0.39	0.65		
NGC1090	02 44 01	−00 27.4	SB(rs)bc	4	2835	0.08	12.08	1.49	Sc	12.20	2.05	0.41	97	21.99	1.61	0.42	0.34	0.60		
NGC1140	02 52 08	−10 13.9	Im	10	1560	0.08	12.42	0.82:	S0	12.48	1.81	0.49	172	19.15:	0.83:	0.66	0.20:	0.43:		57
NGC1156	02 56 47	+25 02.4	IB(s)m	10	558	0.50	11.77	1.52		11.85	1.99	0.83	37	21.83	1.63	0.43	0.33	0.55		
NGC1172	02 59 16	−15 01.8	E1:	−5:	1566	0.08	11.79	1.56		12.06	2.10	0.57	20	23.01	1.66	0.69	0.15:	0.46		
NGC1169	03 00 10	+46 11.1	SAB(r)b	3	2614	0.74	12.02	1.55		12.13	2.03	0.64	34	22.84	1.65	0.71	0.27	0.63		
NGC1187	03 00 24	−23 03.8	SB(r)c	5	1424	−.02	10.72	1.78		10.86	2.24	0.73	126	22.22	1.85	0.59	0.30	0.55		
NGC1175	03 01 16	+42 08.7	SA(r)0^{+}	−1	5685	0.42	12.42	1.10:		12.57	1.89	0.38	155	20.86	1.21:	0.47	0.21:	0.48:		58
NGC1199	03 01 19	−15 48.6	E3:	−5:	2686	0.08	11.40	1.47		11.66	2.06	0.70	65	22.32	1.54	0.73	0.11:	0.36		
NGC1201	03 01 58	−26 15.9	SA(r)0:	−2	1694	−.05	10.93	1.31		11.01	2.06	0.64	5	20.67	1.35	0.68	0.18:	0.46		
NGC1209	03 03 44	−15 48.2	E6:	−5:	2652	0.08	11.39	1.25		11.51	2.13	0.45	83	20.81	1.31:	0.53	0.14:	0.36:		
NGC1241	03 08 49	−09 06.6	SB(rs)b	3	4072	0.17	12.10	1.38		12.16	1.99	0.49	151	21.76	1.45	0.63	0.28:	0.59		
NGC1255	03 11 23	−25 54.7	SAB(rs)bc	4	1656	−.09	11.21	1.75	Sc	11.41	2.13	0.63	116	22.38	1.79	0.71	0.32	0.64		59
NGC1292	03 16 08	−27 47.8	SA(s)c	5	1390	−.04	12.19	1.45		12.27	1.98	0.45	10	21.68	1.53	0.45	0.40:	0.67		
NGC1275	03 16 30	+41 19.8	P	−2	5433	0.53	11.14	1.71	E	11.48	2.17	0.75	81	23.39	1:81	0.76	0.14	0.42		
NGC1297	03 16 59	−19 17.0	SA0^{-}:	−3	1592	0.06	11.61	1.67		11.95	2.01	0.90	179	23.08	1.66	0.90	0.16	0.41		
NGC1300	03 17 25	−19 35.5	SB(rs)bc	4	1526	0.06	10.48	1.93	Sc	10.59	2.28	0.87	124	22.47	2.05	0.26	0.31	0.51	*	
NGC1325	03 22 13	−21 43.1	SA(s)bc	4	1574	0.02	11.50	1.73		11.74	2.21	0.32	52	22.01	1.79	0.30	0.30	0.51		60

Galaxy (1)	α_c (1950) (2)	δ_c (3)	Type (4)	T (5)	v_0 (6)	A_V (7)	V_T (8)	log R_e (9)	GCT (10)	V_{25} (11)	log a_{25} (12)	ϵ_{25} (13)	θ_{25} (14)	μ_e (15)	log a_e (16)	ϵ_e (17)	S_{13} (18)	S_{23} (19)	Note (20)	Page (21)
NGC1332	03 24 04	−21 30.5	S(s)0$^-$:/	−3	1455	0.03	10.41	1.36	E	10.48	2.23	0.48	110	20.16	1.40	0.55	0.19:	0.50		
NGC1337	03 25 40	−08 33.7	SA(s)cd	6	1270	0.11	11.91	1.67		12.01	2.24	0.25	148	22.35	1.75	0.31	0.41	0.66		
NGC1339	03 26 06	−32 27.3	E4	−5	1216	−.05	11.77	1.02:		11.87	1.85	0.81	0	20.19	1.06:	0.76	0.14:	0.38:		
NGC1344	03 26 18	−31 14.4	E5	−5	1200	−.04	10.45	1.52		10.62	2.23	0.66	156	21.22	1.58	0.63	0.14:	0.38		
NGC1353	03 29 49	−20 59.3	SB(rs)b:	3:	1521	0.02	11.41	1.50		11.52	2.11	0.43	138	21.29	1.59	0.43	0.33:	0.57		61
NGC1357	03 30 56	−13 50.0	SA(s)ab	2	2101	0.08	11.22	1.67	Sa	11.42	2.12	0.80	77	22.67	1.70	0.79	0.21	0.49		
NGC1358	03 31 11	−05 15.4	SAB(r)0/a	0	4114	0.08	12.09	1.31	S0	12.24	1.88	0.81	178	22.66	1.41	0.93	0.12:	0.53		
IC 1953	03 31 29	−21 38.6	SB(rs)d	7	1856	0.02	12.00	1.64		12.14	2.00	0.66	123	22.66	1.66	0.83	0.30	0.71		
NGC1359	03 31 33	−19 39.5	SB(s)m?p	9:	1972	0.05	12.60	1.48		12.79	1.81	0.85	149	22.22	1.64	0.21	0.26:	0.40	*	
NGC1366	03 31 53	−31 21.6	S0$^+$/	−1	821	−.07	11.78	0.78:		11.82	1.83	0.50	8	18.49:	0.76:	0.87	0.20:	0.40:		62
NGC1371	03 32 53	−25 06.0	SAB(rs)a	1	1418	−.02	10.78	1.64		10.90	2.22	0.73	129	22.10	1.70	0.74	0.18	0.49		
NGC1376	03 34 37	−05 12.3	SA(s)cd	6	4198	0.07	12.60	1.44	Sm	12.77	1.77	0.96	83	22.56	1.46	0.90	0.34	0.62		
NGC1385	03 35 20	−24 39.9	SB(s)cd	6	1968	−.03	11.15	1.52	Sb	11.23	2.10	0.65	174	22.02	1.61	0.66	0.28	0.61		
NGC1395	03 36 19	−23 11.4	E2	−5	1664	0.01	9.78	1.72		9.95	2.42	0.80	149	22.13	1.80	0.80	0.10	0.35		
NGC1398	03 36 45	−26 29.9	(R′)SB(r)ab	2	1471	−.07	9.79	1.72	S0	9.88	2.37	0.71	94	21.73	1.72	0.93	0.14	0.45		63
NGC1400	03 37 15	−18 50.9	SA0$^-$	−3	504	0.10	10.93	1.36		11.03	2.08	0.92	108	21.06	1.36	0.88	0.11:	0.38		
NGC1406	03 37 23	−31 29.1	SB(s)bc?/	4	998	−.10	11.84	1.52		11.97	2.10	0.23	15	21.15	1.60	0.24	0.43:	0.66		
NGC1407	03 37 57	−18 44.4	E0	−5	1764	0.12	9.64	1.93		9.95	2.37	0.94	118	22.93	2.01	0.91	0.14	0.42		
NGC1415	03 38 47	−22 43.4	(R)SAB(s)0/a	0	1526	0.01	11.49	1.38		11.58	2.16	0.40	153	21.35	1.47	0.53	0.18:	0.48		
NGC1417	03 39 28	−04 51.9	SAB(rs)b	3	4139	0.11	12.38	1.25		12.41	1.88	0.52	8	21.62	1.35	0.51	0.24:	0.57		64
NGC1421	03 40 09	−13 38.9	SAB(rs)bc:	4	2080	0.08	11.51	1.58	Sm	11.56	2.13	0.32	179	21.15	1.66	0.28	0.51	0.75		
NGC1425	03 40 10	−30 03.3	SA(s)b	3	1440	−.10	11.06	1.55	Sa	11.12	2.28	0.39	131	21.88	1.68	0.40	0.22:	0.53		
NGC1426	03 40 38	−22 16.0	E4	−5	1405	0.02	11.37	1.43		11.58	2.07	0.69	107	21.89	1.48	0.63	0.11:	0.34		
NGC1439	03 42 38	−22 04.6	E1	−5	1632	0.05	11.30	1.55		11.59	2.05	0.91	102	22.74	1.62	0.89	0.11:	0.39		
NGC1440	03 42 48	−18 25.1	SB(rs)0?	−2	1511	0.14	11.35	1.35		11.54	1.97	0.92	31	21.67	1.44	0.68	0.12:	0.40		65
NGC1452	03 43 08	−18 47.3	(R′)SB(r)0/a	0	1880	0.11	11.81	1.26	S0	11.90	1.98	0.59	111	21.44	1.32	0.68	0.15:	0.40		
NGC1453	03 43 57	−04 07.6	E2–3	−5	3979	0.18	11.39	1.47		11.65	2.04	0.92	22	22.49	1.54	0.82	0.09:	0.35		
NGC1461	03 46 11	−16 32.4	SA(r)0	−2	1433	0.06	11.64	1.17:	S0	11.79	2.04	0.29	156	20.76	1.27:	0.59	0.22:	0.51:		
NGC1507	04 01 56	−02 19.5	SB(s)mp?	9:	898	0.29	12.54	1.43		12.55	2.05	0.24	13	21.79	1.51	0.33	0.41:	0.67		
NGC1518	04 04 38	−21 18.7	SB(s)dm	8	914	0.02	11.97	1.48		12.04	2.05	0.37	30	21.64	1.57	0.34	0.34:	0.65		66
NGC1521	04 06 08	−21 11.1	E3:	−5:	4170	0.01	11.59	1.27		11.71	2.11	0.55	6	21.75	1.34	0.75	0.09:	0.34		
NGC1537	04 11 44	−31 46.3	S0$^-$?	−3?	1254	−.04	10.68	1.37		10.76	2.23	0.61	111	20.79	1.43	0.60	0.12:	0.34		
NGC1569	04 26 05	+64 44.4	IBm	10	144	1.52	11.19	1.34		11.24	2.13	0.46	122	20.73	1.38	0.50	0.23:	0.52		
NGC1600	04 29 12	−05 11.5	E3	−5	4734	0.06	10.75	1.80		11.14	2.27	0.61	2	23.50	1.91	0.69	0.11	0.38		
NGC1625	04 34 35	−03 24.2	SB(rs)b:	3:	3032	0.08	12.82	1.31		12.97	1.86	0.37	131	21.55	1.39:	0.35	0.36:	0.61:		67
NGC1637	04 38 58	−02 57.1	SAB(rs)c	5	715	0.09	10.90	1.73		11.02	2.15	0.76	35	22.21	1.75	0.90	0.35	0.68		
NGC1638	04 39 05	−01 54.1	SAB(rs)0?	−2:	3308	0.07	11.81	1.51	E	12.12	1.99	0.73	57	22.65	1.59	0.65	0.13:	0.39		
NGC1640	04 40 04	−20 31.8	SB(r)b	3	1600	0.00	11.71	1.45		11.81	1.94	0.94	91	21.66	1.52	0.50	0.22:	0.47		
NGC1659	04 44 02	−04 52.7	SA(r)bcp	4	4522	0.10	12.76	1.14:		12.80	1.76	0.60	46	21.38	1.25:	0.59	0.29:	0.59:		
NGC1667	04 46 10	−06 24.4	SAB(r)c	5	4562	0.17	12.10	1.18		12.15	1.82	0.88	6	20.79	1.28:	0.58	0.30:	0.58:		68

Galaxy (1)	α_c (1950) (2)	δ_c (3)	Type (4)	T (5)	v_0 (6)	A_V (7)	V_T (8)	log R_e (9)	GCT (10)	V_{25} (11)	log a_{25} (12)	ϵ_{25} (13)	θ_{25} (14)	μ_e (15)	log a_e (16)	ϵ_e (17)	S_{13} (18)	S_{23} (19)	Note (20)	Page (21)
NGC1700	04 54 28	−04 56.5	E4	−5	3929	0.09	11.29	1.18		11.39	2.03	0.66	109	20.70	1.22:	0.72	0.11:	0.35:		
NGC1726	04 57 18	−07 49.8	SA(s)0:	−2:	4034	0.12	11.56	1.48	E	11.85	2.05	0.69	172	22.77	1.59	0.69	0.10:	0.36		
NGC1744	04 57 56	−26 05.8	SB(s)d	7	639	−.01	11.61	1.85		11.83	2.23	0.48	165	23.15	1.93	0.40	0.32	0.56		
NGC1784	05 03 07	−11 56.4	SB(r)c	5	2254	0.30	11.72	1.60	Sb	11.81	2.09	0.71	109	22.63	1.67	0.64	0.23	0.53		
HA85—1	05 09 25	−14 51.0	SAB(rs)c	5	2063	0.23	12.59	1.57		12.69	1.93	0.75	79	23.23	1.61	0.84	0.37	0.68	*	69
NGC1832	05 09 48	−15 44.8	SB(r)bc	4	1855	0.14	11.60	1.33	Sc	11.63	1.91	0.71	177	21.18	1.39	0.69	0.28:	0.61		
NGC1964	05 31 14	−21 58.9	SAB(s)b	3	1579	0.02	11.09	0.92:		11.19	2.14	0.34	29	17.26:	0.62:	0.91	0.07:	0.15:		
NGC1961	05 36 34	+69 21.3	SAB(rs)c	5	4147	0.29	11.41	1.70	Sb	11.54	2.13	0.73	75	22.75	1.75	0.66	0.28	0.57		
NGC2139	05 59 04	−23 40.3	SAB(rs)cd	6	1688	0.00	11.83	1.45		11.96	1.88	0.84	132	21.86	1.52	0.66	0.29:	0.57		
NGC2179	06 05 47	−21 44.3	SA(s)0/a	0	2614	0.20	12.17	1.23	S0	12.29	1.91	0.79	173	21.64	1.28	0.76	0.12:	0.39		70
NGC2196	06 10 04	−21 47.8	(R′)SA(s)a	1	2184	0.32	11.19	1.49		11.34	2.11	0.70	74	21.94	1.58	0.69	0.18:	0.48		
NGC2146	06 10 45	+78 22.5	SB(s)abp	2	1106	0.23	10.66	1.74	Sc	10.74	2.22	0.61	119	21.95	1.85	0.38	0.35	0.59		
NGC2207	06 14 14	−21 21.2	SAB(rs)bcp	4	2590	0.37	11.23	1.51	Sm	11.28	2.05	0.70	126	22.29	1.67	0.59	0.23:	0.62	*	
NGC2217	06 19 41	−27 12.5	(R)SB(rs)0^+	−1	1434	0.11	10.49	1.54		10.56	2.19	0.93	18	22.21	1.70	0.63	0.11:	0.43		
NGC2223	06 22 31	−22 48.7	SAB(r)b	3	2529	0.26	11.82	1.53	Sc	11.92	1.94	0.98	61	22.51	1.60	0.89	0.29	0.64		71
NGC2280	06 42 50	−27 35.2	SA(s)cd	6	1709	0.58	11.18	1.64	Sb	11.29	2.25	0.49	167	22.23	1.71	0.54	0.21	0.50		
NGC2325	07 00 42	−28 37.4	E4	−5	2042	0.84	11.28	1.57		11.38	2.18	0.65	9	22.20	1.65	0.57	0.20:	0.47		
NGC2268	07 00 49	+84 27.8	SAB(r)bc	4	2458	0.16	11.60	1.44		11.69	2.06	0.53	70	21.63	1.50	0.60	0.25:	0.53		
NGC2314	07 03 54	+75 24.4	E3	−5	4079	0.16	12.12	1.01:		12.25	1.88	0.70	5	20.68	1.06:	0.79	0.12:	0.33:		
NGC2339	07 05 25	+18 51.7	SAB(rs)bc	4	2229	0.43	11.82	1.41	Sc	11.89	1.91	0.82	168	21.74	1.44	0.93	0.33:	0.62		72
NGC2276	07 10 31	+85 50.9	SAB(rs)c	5	2648	0.17	11.68	1.61		11.80	1.92	0.95	76	22.27	1.61	0.91	0.39	0.69		
NGC2347	07 11 16	+64 48.1	(R′)SA(r)b:	3:	4692	0.11	12.67	1.09:		12.70	1.80	0.61	7	20.97	1.14:	0.61	0.25:	0.53:		
NGC2300	07 15 47	+85 48.6	SA0	−2	2190	0.17	10.73	1.65	E	10.95	2.30	0.60	81	22.71	1.71	0.82	0.10:	0.38		
NGC2336	07 18 28	+80 16.6	SAB(r)bc	4	2424	0.10	10.83	1.81	Sc	10.88	2.28	0.53	2	22.52	1.88	0.61	0.38	0.70		
NGC2366	07 23 37	+69 19.1	IB(s)m	10	281	0.13	11.23	2.06	Sm	11.61	2.37	0.40	30	22.96	2.20	0.13	0.28	0.40		73
NGC2403	07 32 03	+65 42.7	SAB(s)cd	6	299	0.12	8.64	2.15		8.69	2.71	0.58	121	21.98	2.19	0.71	0.33	0.63		
NGC2441	07 46 20	+73 09.4	SAB(r)b:	3:	3815	0.07	12.43	1.45	Sm	12.59	1.79	0.91	77	22.43	1.52	0.78	0.35:	0.69		
NGC2460	07 52 36	+60 29.0	SA(s)a	1	1593	0.10	11.70	1.26		11.86	1.98	0.67	55	21.08	1.30	0.70	0.16:	0.38		
NGC2500	07 58 08	+50 52.6	SB(rs)d	7	615	0.10	11.45	1.71		11.63	2.03	0.88	64	22.54	1.71	0.92	0.39	0.66		
NGC2525	08 03 15	−11 17.1	SB(s)c	5	1395	0.29	11.50	1.60	Sm	11.61	2.01	0.75	89	21.80	1.66	0.67	0.36	0.68		74
NGC2523	08 09 15	+73 43.8	SB(r)bc	4	3638	0.08	12.10	1.44		12.14	1.95	0.64	59	22.17	1.61	0.32	0.25:	0.52		
NGC2537	08 09 43	+46 08.7	IB(s)mp	10	513	0.11	11.76	1.38		11.88	1.90	0.95	22	23.01	1.51	0.83	0.36	0.65		
NGC2541	08 11 02	+49 13.0	SA(s)cd	6	646	0.13	11.85	1.78		12.05	2.17	0.52	165	23.23	1.88	0.44	0.31	0.56		
NGC2545	08 11 20	+21 30.4	(R)SB(r)ab	2	3312	0.09	12.50	1.07:	S0	12.57	1.82	0.56	172	21.13	1.19:	0.51	0.14:	0.47:		
NGC2549	08 14 57	+57 57.6	SA(r)0/	−2	1207	0.09	11.26	0.99:		11.31	2.11	0.30	3	19.39:	1.04:	0.65	0.19:	0.44:		75
NGC2552	08 15 42	+50 10.1	SA(s)m?	9	607	0.13	12.37	1.72		12.60	2.03	0.65	55	23.54	1.76	0.63	0.33	0.59		
NGC2551	08 19 13	+73 34.6	SA(s)0/a	0	2484	0.05	12.31	0.95:	S0	12.38	1.79	0.60	50	20.25	0.96:	0.75	0.16:	0.39:		
NGC2613	08 31 11	−22 48.0	SA(s)b	3	1446	0.26	10.59	1.64		10.60	2.35	0.23	114	21.02	1.73	0.25	0.31:	0.57		
NGC2608	08 32 15	+28 38.8	SB(s)b:	3:	2112	0.08	12.19	1.46	Sc	12.32	1.92	0.66	41	21.86	1.52	0.52	0.35:	0.59		
NGC2639	08 40 04	+50 23.2	(R)SA(r)a:?	1:	3270	0.08	11.66	1.16:		11.78	1.90	0.72	134	20.18	1.19:	0.57	0.21:	0.45:		76

Galaxy (1)	α_c (1950) (2)	δ_c (3)	Type (4)	T (5)	v_0 (6)	A_V (7)	V_T (8)	log R_e (9)	GCT (10)	V_{25} (11)	log a_{25} (12)	ϵ_{25} (13)	θ_{25} (14)	μ_e (15)	log a_e (16)	ϵ_e (17)	S_{13} (18)	S_{23} (19)	Note (20)	Page (21)
NGC2633	08 42 34	+74 17.0	SB(s)b	3	2416	0.04	11.99	1.17:	S0	12.06	1.88	0.66	7	21.06	1.27:	0.44	0.11:	0.35:		
NGC2646	08 45 00	+73 38.9	SB(r)0:	−2	3731	0.03	12.54	1.03:		12.66	1.71	0.88	11	21.00	1.08:	0.72	0.15:	0.41:		
NGC2654	08 45 11	+60 24.4	SBab:/	2	1468	0.12	12.11	1.18		12.17	2.09	0.18	59	20.63	1.23:	0.46	0.27:	0.57:		
NGC2672	08 46 31	+19 15.6	E1−2	−5	3907	0.04	11.51	1.48		11.80	2.05	0.75	125	22.60	1.54	0.84	0.11:	0.35		
NGC2655	08 49 08	+78 24.8	SAB(s)0/a	0	1592	0.02	9.90	1.60	E	10.06	2.28	0.88	93	21.68	1.71	0.72	0.11:	0.38		77
NGC2683	08 49 35	+33 36.5	SA(rs)b	3	399	0.06	9.70	1.69		9.71	2.45	0.29	42	20.21	1.74	0.32	0.31	0.54		
NGC2681	08 49 58	+51 30.2	(R′)SAB(rs)0/a	0	800	0.07	10.52	1.21	E	10.58	2.04	0.96	123	20.36	1.30	0.71	0.09:	0.34		
NGC2685	08 51 41	+58 55.5	(R)SB0p	−2	1001	0.11	11.32	1.30		11.47	2.04	0.60	40	20.75	1.35	0.53	0.15:	0.35		
NGC2693	08 53 25	+51 32.3	E3:	−5:	5041	0.05	11.79	1.24		11.99	1.95	0.68	156	21.57	1.30	0.74	0.11:	0.35		
NGC2713	08 54 44	+03 06.8	SB(rs)ab	2	3690	0.11	11.69	1.27	Sa	11.72	2.10	0.32	109	21.19	1.32	0.71	0.19:	0.51		78
NGC2701	08 55 28	+53 58.3	SAB(rs)c:	5:	2421	0.04	12.24	1.43	Sm	12.41	1.80	0.69	28	21.91	1.46	0.75	0.37:	0.65		
NGC2712	08 56 09	+45 06.6	SB(r)b:	3:	1892	0.03	12.16	1.37		12.26	1.95	0.48	178	21.68	1.46	0.59	0.34:	0.65		
NGC2715	09 01 50	+78 17.2	SAB(rs)c	5	1540	0.02	11.61	1.60	Sm	11.66	2.15	0.36	22	21.99	1.65	0.58	0.42	0.68		
NGC2749	09 02 33	+18 30.8	E3	−5	4143	0.06	12.00	1.13:	E	12.21	1.86	0.84	85	21.39	1.20:	0.77	0.10:	0.31:		
NGC2742	09 03 38	+60 40.9	SA(s)c:	5:	1422	0.14	11.76	1.54		11.84	1.98	0.56	86	21.50	1.62	0.41	0.42	0.62		79
NGC2764	09 05 27	+21 38.7	S0:	−2:	2636	0.10	13.03	1.06:	Sb	13.13	1.70	0.60	26	20.66	1.12:	0.42	0.27:	0.52:		
NGC2732	09 06 54	+79 23.6	S0/	−2	2325	0.02	11.77	0.91:		11.86	1.81	0.47	66	19.28:	0.94:	0.69	0.23:	0.49:		
NGC2775	09 07 41	+07 14.5	SA(r)ab	2	1195	0.08	10.09	1.64	Sa	10.21	2.27	0.88	168	21.52	1.70	0.79	0.17	0.46		
NGC2768	09 07 45	+60 14.5	E6:	−5:	1531	0.12	10.29	1.69		10.40	2.36	0.45	86	21.27	1.71	0.47	0.21	0.47		
NGC2748	09 08 01	+76 40.9	SAbc	4	1684	0.01	11.93	1.31	Sm	11.99	1.93	0.49	38	20.76	1.42:	0.32	0.29:	0.53:		80
NGC2776	09 08 56	+45 09.6	SAB(rs)c	5	2673	0.03	11.76	1.50	Sb	11.88	1.93	0.97	125	22.08	1.49	0.96	0.29	0.57		
NGC2784	09 10 06	−23 57.9	SA(s)0:	−2	451	0.50:	10.14	1.39		10.16	2.33	0.40	75	20.32	1.47	0.62	0.17:	0.44		
NGC2782	09 10 54	+40 19.3	SAB(rs)ap	1	2574	−.01	11.66	1.41		11.76	2.15	0.43	89	21.80	1.46	0.73	0.17:	0.48		
NGC2793	09 13 43	+34 38.4	SB(s)mp	9	1674	0.00	13.53	1.24	Sc	13.65	1.58	0.89	97	23.02	1.32	0.78	0.23:	0.57		
NGC2811	09 13 50	−16 06.2	SB(rs)a	1	2279	0.09	11.40	1.13:		11.45	1.97	0.40	19	20.12	1.18:	0.67	0.23:	0.54:		81
NGC2815	09 14 05	−23 25.5	SB(r)b:	3:	2333	0.41	11.90	1.35	Sc	11.91	2.05	0.34	12	21.33	1.36	0.67	0.31:	0.61		
NGC2798	09 14 10	+42 12.7	SB(s)ap	1	1773	0.00	12.36	1.06:		12.39	1.93	0.35	158	20.50	1.06:	0.73	0.21:	0.49:		
NGC2787	09 14 50	+69 24.9	SB(r)0^+	−1	828	0.12	10.72	1.34		10.81	2.09	0.56	116	20.52	1.39	0.66	0.19:	0.47		
NGC2835	09 15 37	−22 08.8	SB(rs)c	5	624	0.31	10.56	1.91		10.70	2.29	0.70	11	22.88	1.93	0.83	0.34	0.67		
NGC2832	09 16 45	+33 57.6	E^+2:	−4	6935	−.01	11.76	1.42		11.94	2.12	0.51	156	22.65	1.56	0.66	0.13:	0.45		82
NGC2848	09 17 49	−16 18.8	SAB(s)c:	5:	1795	0.10	12.09	1.56		12.26	1.96	0.78	14	22.67	1.66	0.57	0.30	0.58		
NGC2841	09 18 35	+51 11.3	SA(r)b:	3	714	0.00	9.31	1.76	S0	9.41	2.46	0.51	146	21.18	1.88	0.51	0.22	0.57		
NGC2844	09 18 37	+40 22.0	SA(r)a:	1:	1507	−.02	12.86	1.09:		12.95	1.80	0.57	6	21.15	1.14:	0.59	0.22:	0.47:		
NGC2855	09 19 02	−11 41.8	(R)SA(rs)0/a	0	1689	0.11	10.86	1.60		11.02	2.17	0.85	155	22.17	1.66	0.79	0.18	0.45		
NGC2865	09 21 14	−22 56.8	E3−4	−5	2457	0.20	11.33	1.16:		11.43	2.02	0.72	135	20.57	1.20:	0.76	0.12:	0.34:		83
NGC2859	09 21 15	+34 43.7	(R)SB(r)0^+	−1	1590	0.00	10.94	1.35		11.03	2.15	0.83	86	21.31	1.39	0.74	0.08:	0.30		
NGC2889	09 24 32	−11 25.3	SAB(rs)c	5	3194	0.11	11.89	1.35		11.96	1.82	0.92	16	21.48	1.37	0.91	0.30:	0.64		
NGC2880	09 25 42	+62 42.7	$SB0^-$	−3	1646	0.07	11.69	1.22		11.86	1.95	0.67	159	21.24	1.26	0.75	0.13:	0.37		
NGC2902	09 28 30	−14 31.0	SA(s)0:	−2	1831	0.20	12.14	1.16:		12.32	1.83	0.85	9	21.24	1.17:	0.96	0.14:	0.38:		
NGC2903	09 29 20	+21 43.2	SAB(rs)bc	4	472	0.05	8.91	1.90	Sc	8.94	2.56	0.48	18	20.73	1.97	0.44	0.31	0.56		84

Galaxy (1)	α_c (1950) (2)	δ_c (3)	Type (4)	T (5)	v_0 (6)	A_V (7)	V_T (8)	log R_e (9)	GCT (10)	V_{25} (11)	log a_{25} (12)	ϵ_{25} (13)	θ_{25} (14)	μ_e (15)	log a_e (16)	ϵ_e (17)	S_{13} (18)	S_{23} (19)	Note (20)	Page (21)
NGC2907	09 29 20	−16 30.9	SA(s)a?/	1	1824	0.17	11.47	1.43	S0	11.67	2.04	0.65	113	21.79	1.50	0.56	0.13:	0.38		
NGC2962	09 38 17	+05 23.6	(R)SAB(rs)0$^+$	−1	1958	0.07	12.03	1.24		12.13	1.91	0.74	5	21.40	1.34	0.57	0.17:	0.45		
NGC2950	09 38 59	+59 04.8	(R)SB(r)0	−2	1488	0.02	10.91	0.99:		10.97	2.03	0.66	120	19.10:	0.98:	0.84	0.13:	0.33:		
NGC2967	09 39 29	+00 33.9	SA(s)c	5	2065	0.12	11.67	1.45	Sc	11.77	1.91	0.97	174	21.61	1.44	0.93	0.32:	0.59		
NGC2964	09 39 56	+32 04.6	SAB(r)bc:	4	1292	0.02	11.32	1.37	Sc	11.38	2.01	0.53	97	20.89	1.44	0.78	0.39:	0.69		85
NGC2968	09 40 15	+32 09.6	I0	0	1576	0.02	11.57	1.46		11.71	2.05	0.71	74	21.87	1.50	0.65	0.18:	0.41		
NGC2976	09 43 10	+68 08.9	SAcp	5	168	0.08	10.31	1.74	Sm	10.34	2.29	0.56	141	21.15	1.84	0.40	0.45	0.66		
NGC2997	09 43 27	−30 57.7	SAB(rs)c	5	799	0.38	9.59	1.98	Sm	9.60	2.46	0.68	115	22.30	2.07	0.78	0.42	0.75		
NGC2990	09 43 40	+05 56.6	Sc:	5:	3006	0.04	12.87	1.07:	Sm	12.91	1.59	0.66	83	20.46	1.15:	0.50	0.38:	0.63:		
NGC3001	09 44 07	−30 12.4	SAB(rs)bc:	4	2171	0.36	11.23	1.67	Sa	11.48	2.15	0.63	176	22.40	1.71	0.58	0.20	0.41		86
NGC3003	09 45 37	+33 39.2	Sbc?	4?	1461	0.00	11.92	1.61	Sc	11.99	2.21	0.25	80	21.65	1.68	0.24	0.38:	0.61		
NGC2985	09 45 54	+72 30.7	(R′)SA(rs)ab	2	1451	0.05	10.28	1.63	E	10.40	2.30	0.85	7	22.05	1.70	0.81	0.14:	0.44		
NGC3021	09 47 59	+33 47.1	SA(rs)bc:	4	1520	−.01	12.16	1.15:	Sc	12.23	1.72	0.63	105	20.35	1.23:	0.50	0.40:	0.69:		
NGC3038	09 49 05	−32 31.1	SA(s)a	1	2412	0.40	11.50	1.44		11.62	2.03	0.73	94	21.75	1.50	0.63	0.20:	0.50		
NGC3032	09 49 14	+29 28.3	SAB(r)0	−2	1492	0.02	12.07	1.05:		12.18	1.83	0.79	83	20.77	1.00:	0.97	0.12:	0.33:	*	87
NGC3041	09 50 23	+16 54.8	SAB(rs)c	5	1296	0.06	11.93	1.61		11.98	2.04	0.60	85	22.46	1.68	0.68	0.44	0.72		
NGC3044	09 51 06	+01 49.0	SB(s)c?/	5:	1206	0.04	11.88	1.59		11.95	2.20	0.16	113	20.98	1.62	0.16	0.32:	0.55		
NGC3031	09 51 30	+69 18.3	SA(s)ab	2	124	0.11	7.12	2.18		7.17	2.87	0.54	152	20.98	2.28	0.55	0.25	0.57		
NGC3034	09 51 41	+69 54.9	I0/	0	409	0.10	8.52	1.84	Sm	8.51	2.57	0.42	59	20.06	1.94	0.28	0.40	0.62		
NGC3056	09 52 18	−28 03.8	(R)SA(s)0$^+$:	−1	770	0.13	11.29	1.13:		11.35	2.00	0.53	26	20.44	1.18:	0.72	0.15:	0.43:	*	88
NGC3043	09 52 42	+59 32.7	Sb:/	3:	3048	0.02	13.41	1.14:	Sm	13.47	1.72	0.37	79	21.14	1.22:	0.39	0.42:	0.65:		
NGC3067	09 55 26	+32 36.5	SAB(s)ab?	2?	1429	0.05	12.16	1.28	Sm	12.21	1.88	0.40	102	20.82	1.35	0.45	0.41:	0.67		
NGC3089	09 57 22	−28 05.1	SAB(rs)b	3	2375	0.14	12.37	1.33	Sc	12.48	1.82	0.68	147	21.63	1.34	0.84	0.34:	0.61		
NGC3065	09 57 35	+72 24.6	SA(r)0	−2	2181	0.08	12.03	0.98:		12.09	1.78	0.88	8	20.16	0.95:	0.95	0.13:	0.33:	*	
NGC3095	09 57 53	−31 18.8	SAB(rs)c	5	2564	0.24	11.48	1.69		11.63	2.12	0.55	124	22.34	1.76	0.44	0.30	0.57		89
NGC3077	09 59 22	+68 58.5	I0p	0	165	0.16	9.85	1.87		10.06	2.36	0.76	54	22.17	1.90	0.73	0.21	0.47		
NGC3115	10 02 44	−07 28.5	S0$^-$/	−3	437	0.08	8.84	1.74	E	8.92	2.55	0.61	42	20.40	1.81	0.38	0.14:	0.39		
NGC3125	10 04 18	−29 41.5	E3?	−5?	827	0.19	12.88	0.98:	Sc	12.91	1.71	0.65	83	20.15	1.03:	0.55	0.30:	0.51:		
NGC3145	10 07 43	−12 11.3	SB(rs)bc	4	3416	0.16	12.02	1.35		12.07	1.95	0.54	23	21.65	1.45	0.55	0.25:	0.61	*	
NGC3158	10 10 53	+39 00.7	E3:	−5:	7029	0.03	11.63	1.60		11.97	2.04	0.80	170	23.33	1.66	0.86	0.11:	0.39		90
NGC3166	10 11 10	+03 40.5	SAB(rs)0/a	0	1166	0.01	10.49	1.30	S0	10.50	2.21	0.57	87	20.25	1.29	0.98	0.13:	0.40		
NGC3169	10 11 38	+03 43.2	SA(s)ap	1	1067	0.03	10.14	1.80	S0	10.35	2.33	0.70	64	22.32	1.84	0.64	0.11	0.41		
NGC3175	10 12 24	−28 37.2	SAB(s)a?	1?	843	0.18	10.94	1.69		11.10	2.26	0.33	56	21.29	1.73	0.32	0.32	0.56		
NGC3177	10 13 49	+21 22.4	SA(rs)b	3	1220	0.05	12.45	0.93:	Sc	12.47	1.67	0.84	163	19.96	0.93:	0.92	0.26:	0.51:		
NGC3185	10 14 54	+21 56.3	(R)SB(r)a	1	1151	0.07	12.25	1.39	Sb	12.40	1.92	0.65	140	22.00	1.50	0.52	0.28:	0.58		91
NGC3184	10 15 17	+41 40.0	SAB(rs)cd	6	607	−.03	10.13	1.99		10.21	2.35	0.83	111	22.70	2.04	0.78	0.38	0.69		
NGC3190	10 15 21	+22 05.1	SA(s)ap/	1	1384	0.07	11.02	1.22		10.99	2.17	0.40	122	19.85	1.25:	0.45	0.18:	0.42:		
NGC3193	10 15 40	+22 08.8	E2	−5	1307	0.06	10.97	1.36		11.15	2.06	0.90	12	21.20	1.35	0.90	0.10:	0.32		
NGC3226	10 20 43	+20 09.2	E2:p	−5:	1260	0.02	11.25	1.64		11.59	2.07	0.87	41	22.91	1.70	0.78	0.13	0.38	*	
NGC3227	10 20 47	+20 07.1	SAB(s)ap	1	1102	0.02	11.00	1.53		11.10	2.15	0.79	5	21.60	1.70	0.40	0.22:	0.47	*	92

Galaxy (1)	α_c (1950) (2)	δ_c (3)	Type (4)	T (5)	v_0 (6)	A_V (7)	V_T (8)	log R_e (9)	GCT (10)	V_{25} (11)	log a_{25} (12)	ϵ_{25} (13)	θ_{25} (14)	μ_e (15)	log a_e (16)	ϵ_e (17)	S_{13} (18)	S_{23} (19)	Note (20)	Page (21)
NGC3245	10 24 30	+28 45.8	SA(r)0:?	−2	1210	0.04	10.92	1.31		11.06	2.03	0.64	175	20.58	1.36	0.58	0.17:	0.45		
NGC3254	10 26 32	+29 44.9	SA(s)bc	4	1219	0.03	11.83	1.47	Sa	11.86	2.16	0.27	44	22.04	1.58	0.40	0.26:	0.64		
NGC3274	10 29 30	+27 55.6	SABcd:	6	486	0.04	12.82	1.28	Sm	12.93	1.86	0.47	95	21.82	1.32	0.65	0.33:	0.60		
NGC3277	10 30 08	+28 46.2	SA(r)ab	2	1365	0.02	11.55	1.35	S0	11.70	2.02	0.78	32	21.55	1.33	0.86	0.14:	0.36		
NGC3287	10 32 04	+21 54.5	SB(s)d	7	1219	0.02	12.64	1.42		12.72	1.92	0.45	20	22.00	1.47	0.46	0.35:	0.69		93
NGC3294	10 33 24	+37 35.1	SA(s)c	5	1566	−.01	11.55	1.52		11.58	2.06	0.46	117	21.32	1.62	0.41	0.42	0.66		
NGC3301	10 34 13	+22 08.4	(R′)SB(rs)0/a	0	1246	0.02	11.54	1.16:	S0	11.61	2.07	0.28	51	20.25	1.18:	0.66	0.21:	0.51:		
NGC3309	10 34 15	−27 15.8	E3	−5	3812	0.13	11.58	1.40		11.84	2.00	0.74	56	22.01	1.44	0.87	0.12:	0.34	*	
NGC3312	10 34 42	−27 18.3	SA(s)bp?	3	2494	0.13	11.61	1.54	Sa	11.75	2.06	0.46	174	22.11	1.64	0.47	0.26:	0.67	*	
NGC3310	10 35 39	+53 45.9	SAB(r)bcp	4	1073	−.02	10.99	1.01:	Sc	10.87	2.05	0.97	58	18.71:	1.03:	0.85	0.33:	0.55:		94
NGC3319	10 36 14	+41 56.8	SB(rs)cd	6	776	−.01	11.33	1.89		11.43	2.28	0.55	41	23.15	2.00	0.35	0.29	0.64		
NGC3320	10 36 37	+47 39.4	Scd:	6:	2380	0.01	12.79	1.26		12.88	1.81	0.48	28	21.55	1.33:	0.49	0.33:	0.60:		
NGC3338	10 39 29	+14 00.6	SA(s)c	5	1171	0.05	11.04	1.74	Sb	11.16	2.23	0.55	103	22.17	1.82	0.46	0.31	0.56		
NGC3344	10 40 47	+25 11.1	(R)SAB(r)bc	4	627	0.02	10.26	1.83	Sc	10.40	2.24	0.96	4	22.29	1.88	0.88	0.32	0.62		
NGC3346	10 40 59	+15 08.1	SB(rs)cd	6	1138	0.05	11.89	1.61	Sm	11.95	1.96	0.81	114	22.46	1.62	0.97	0.46	0.74		95
NGC3351	10 41 19	+11 58.1	SB(r)b	3	641	0.04	9.85	1.74		9.87	2.36	0.69	11	21.72	1.84	0.72	0.22	0.54		
NGC3353	10 42 17	+56 13.6	Sb?p	3?	1089	.04	12.79	0.79:	Sc	12.79	1.67	0.68	85	19.56	0.82:	0.85	0.27:	0.50:		
NGC3367	10 43 56	+14 00.8	SB(rs)c	5	2906	0.04	11.56	1.51	Sm	11.68	1.88	0.93	68	21.73	1.59	0.60	0.32	0.63		
NGC3368	10 44 08	+12 05.1	SAB(rs)ab	2	758	0.05	9.41	1.76		9.47	2.45	0.62	174	21.09	1.79	0.68	0.19	0.46		
NCC3370	10 44 23	+17 32.3	SA(s)c	5	1289	0.03	12.38	1.27		12.40	1.91	0.53	140	21.22	1.32:	0.50	0.27:	0.51:		96
NGC3377	10 45 03	+14 15.0	E5–6	−5	591	0.05	10.21	1.52		10.35	2.26	0.82	38	21.22	1.64	0.51	0.11:	0.34		
NGC3379	10 45 11	+12 50.8	E1	−5	759	0.04	9.36	1.65		9.49	2.37	0.92	77	21.08	1.69	0.86	0.12:	0.38	*	
NGC3384	10 45 38	+12 53.7	SB(s)0$^-$:	−3	638	0.05	10.01	1.40		10.03	2.28	0.50	59	20.56	1.42	0.82	0.15:	0.45	*	
NGC3390	10 45 44	−31 16.2	Sb/	3	2561	0.22	12.32	1.30		12.31	2.09	0.18	178	20.89	1.35:	0.42	0.39:	0.61:		
NGC3389	10 45 50	+12 47.9	SA(s)c	5	1127	0.05	12.01	1.47		12.05	1.97	0.45	108	21.50	1.54	0.40	0.39:	0.67		97
NGC3395	10 47 02	+33 14.7	SAB(rs)cdp:	6:	1599	0.01	12.21	1.24		12.30	1.83	0.57	44	21.15	1.31	0.61	0.36:	0.65	*	
NGC3396	10 47 08	+33 15.3	IBmp	10	1619	0.00	12.35	1.24	Sa	12.39	2.07	0.30	105	21.31	1.33:	0.34	0.23:	0.55:	*	
NGC3412	10 48 15	+13 40.5	SB(s)0	−2	731	0.05	10.67	1.26		10.72	2.06	0.59	151	20.54	1.33	0.76	0.15:	0.50		
NGC3414	10 48 32	+28 14.5	S0p	−2	1364	−.03	10.83	1.37	E	11.02	2.09	0.79	18	21.30	1.44	0.76	0.11:	0.36		
NGC3423	10 48 38	+06 06.3	SA(s)cd	6	845	0.06	11.15	1.72	Sm	11.33	2.06	0.78	34	22.50	1.78	0.75	0.37	0.67		98
NGC3415	10 48 52	+43 59.0	SA0$^+$:	−1	3330	−.01	12.38	0.98:		12.49	1.86	0.54	9	20.30	1.02:	0.67	0.16:	0.37:		
NGC3430	10 49 25	+33 12.9	SAB(rs)c	5	1555	0.01	11.79	1.50		11.88	2.02	0.58	31	21.75	1.56	0.53	0.36	0.61		
NGC3432	10 49 43	+36 53.1	SB(s)m/	9	607	−.07:	11.49	1.70	Sc	11.54	2.35	0.19	35	21.52	1.79	0.14	0.46:	0.73		
NGC3437	10 49 54	+23 12.0	SAB(rs)c:	5:	1201	−.02	12.12	1.24		12.23	1.92	0.34	116	20.71	1.33:	0.41	0.33:	0.58:		
NGC3449	10 50 34	−32 39.7	SAB(r)0/a:	0	2975	0.20	11.81	1.36	E	11.92	2.13	0.38	153	21.65	1.44	0.50	0.20:	0.47		99
NGC3445	10 51 34	+57 15.3	SAB(s)m	9	2083	−.04	12.79	1.29		12.93	1.67	0.87	120	21.88	1.39	0.51	0.30:	0.54		
NGC3448	10 51 40	+54 34.5	I0	0	1442	−.04	12.23	1.31	Sm	12.25	1.95	0.37	59	21.37	1.41	0.46	0.37:	0.67		
NGC3455	10 51 52	+17 33.1	(R′)SAB(rs)b	3	991	0.02	13.12	1.28	Sm	13.24	1.83	0.49	64	21.95	1.33	0.61	0.35:	0.59		
NGC3458	10 52 58	+57 23.0	SAB0:	−2	1899	−.04	12.04	0.57:		12.07	1.59	0.81	176	17.44:	0.54:	0.98	0.19:	0.38:		
NGC3478	10 56 35	+46 23.4	SB(rs)bc	4	6730	−.01	12.90	1.39		13.03	1.84	0.49	128	22.31	1.46	0.46	0.36:	0.66		100

Galaxy (1)	α_c (1950) (2)	δ_c (3)	Type (4)	T (5)	v_0 (6)	A_V (7)	V_T (8)	log R_e (9)	GCT (10)	V_{25} (11)	log a_{25} (12)	ϵ_{25} (13)	θ_{25} (14)	μ_e (15)	log a_e (16)	ϵ_e (17)	S_{13} (18)	S_{23} (19)	Note (20)	Page (21)
NGC3485	10 57 24	+15 06.6	SB(r)b:	3:	1395	0.01	12.02	1.49	Sc	12.17	1.89	0.90	40	21.63	1.70	0.21	0.29:	0.49		
NGC3489	10 57 41	+14 10.2	SAB(rs)0$^+$	−1	532	0.02	10.36	1.21		10.41	2.14	0.54	71	19.69	1.24	0.79	0.19:	0.49		
NGC3486	10 57 42	+29 14.6	SAB(r)c	5	636	0.01	10.62	1.64	Sa	10.72	2.27	0.76	87	22.22	1.72	0.77	0.19	0.47		
NGC3504	11 00 29	+28 14.5	(R)SAB(s)ab	2	1480	0.03	11.01	1.24	S0	11.09	1.95	0.94	177	20.55	1.34	0.49	0.12:	0.45		
NGC3510	11 01 01	+29 09.3	SB(s)m/	9	660	0.02	12.97	1.35	Sb	13.00	2.07	0.18	164	21.23	1.43:	0.21	0.31:	0.52:		101
NGC3512	11 01 20	+28 18.3	SAB(rs)c	5	1340	0.02	12.50	1.18		12.53	1.69	0.88	145	21.24	1.23:	0.75	0.30:	0.59:		
NGC3549	11 08 03	+53 39.6	SA(s)c:	5:	2921	−.03	12.27	1.48	Sm	12.35	1.97	0.38	39	21.60	1.55	0.31	0.45:	0.71		
NGC3556	11 08 37	+55 56.7	SB(s)cd/	6	790	−.03	10.41	1.92	Sm	10.50	2.39	0.31	83	22.10	1.98	0.36	0.49	0.77		
NGC3593	11 12 00	+13 05.4	SA(s)0/a:	0:	493	−.03	10.84	1.52		10.92	2.23	0.45	87	21.38	1.64	0.46	0.21:	0.55		
NGC3596	11 12 29	+15 03.5	SAB(rs)c	5	1072	−.01	11.49	1.51		11.57	1.98	0.85	147	21.79	1.62	0.54	0.34	0.63		102
NGC3605	11 14 08	+18 17.6	E4–5	−5	587	−.04	12.70	0.81:	Sb	12.70	1.61	0.78	17	19.37:	0.77:	0.82	0.22:	0.48:	*	
NGC3607	11 14 17	+18 19.7	SA(s)0:	−2	828	−.04	10.04	1.65		10.20	2.29	0.90	140	21.79	1.67	0.86	0.11:	0.37	*	
NGC3608	11 14 21	+18 25.6	E2	−5	1105	−.04	10.91	1.40		11.08	2.10	0.79	91	21.45	1.46	0.75	0.09:	0.37	*	
NGC3611	11 14 55	+04 49.7	SA(s)ap	1	1450	0.11	12.25	1.13:		12.33	1.86	0.61	24	20.96	1.16:	0.64	0.15:	0.43:		
NGC3610	11 15 31	+59 03.6	E5:	−5:	1873	−.01	10.85	0.91:		10.89	1.98	0.84	86	18.50:	0.92:	0.82	0.16:	0.35:		103
NGC3613	11 15 42	+58 16.3	E6	−5	2158	−.03	11.00	1.27		11.15	2.14	0.54	99	20.63	1.35	0.56	0.12:	0.33		
NGC3623	11 16 19	+13 21.9	SAB(rs)a	1	675	0.03	9.37	1.74		9.44	2.45	0.32	171	20.76	1.83	0.46	0.31	0.61		
NGC3619	11 16 28	+58 02.0	(R)SA(s)0$^+$:	−1	1662	−.04	11.60	1.46		11.81	2.04	0.77	66	22.45	1.49	0.86	0.10:	0.39		
NGC3626	11 17 26	+18 37.8	(R)SA(rs)0$^+$	−1	1403	−.04	11.00	0.95:		11.01	2.00	0.69	164	19.14:	0.88:	0.82	0.10:	0.28:		
NGC3627	11 17 38	+13 15.8	SAB(s)b	3	593	0.04	8.96	1.85		9.00	2.52	0.57	166	20.91	1.94	0.44	0.25	0.55		104
NGC3628	11 17 40	+13 52.1	Sbp/	3	719	0.02	9.43	2.14	Sc	9.55	2.68	0.27	103	22.26	2.21	0.28	0.40	0.62		
NGC3630	11 17 42	+03 14.3	S0/	−2	1340	0.08	11.89	0.84:		11.91	1.81	0.45	37	18.85:	0.81:	0.83	0.22:	0.45:		
NGC3631	11 18 13	+53 26.7	SA(s)c	5	1238	−.02	10.80	1.73		10.91	2.15	0.79	104	22.59	1.85	0.60	0.28	0.63		
NGC3640	11 18 33	+03 30.6	E3	−5	1178	0.08	10.45	1.54		10.59	2.19	0.89	94	21.63	1.58	0.82	0.14:	0.40	*	
NGC3646	11 19 05	+20 26.7	Ring	4P	4100	−.01	11.59	1.59	Sc	11.64	2.08	0.52	52	22.55	1.69	0.69	0.34	0.77		105
NGC3642	11 19 25	+59 21.0	SA(r)bc:	4	1733	−.01	11.70	1.59	S0	12.11	1.94	0.84	58	23.04	1.58	0.97	0.13:	0.35		
NGC3655	11 20 17	+16 51.7	SA(s)c:	5	1378	−.02	11.81	1.16:	Sb	11.86	1.81	0.98	63	20.53	1.26:	0.56	0.27:	0.56:		
NGC3659	11 21 08	+18 05.4	SB(s)m?	9?	1109	−.04	12.87	1.27		13.00	1.82	0.51	52	20.93	1.35:	0.27	0.24:	0.44:		
NGC3664	11 21 51	+03 36.3	SB(s)mp	9	1231	0.08	13.01	1.51		13.16	1.75	0.97	33	22.70	1.68	0.18	0.29:	0.41		
NGC3665	11 22 01	+39 02.2	SA(s)0	−2	2006	−.01	10.27	1.79		10.53	2.31	0.79	34	22.69	1.84	0.78	0.13	0.38		106
NGC3672	11 22 31	−09 31.2	SA(s)c	5	1633	0.05	11.33	1.62		11.42	2.11	0.46	9	21.54	1.68	0.40	0.38	0.65		
NGC3681	11 23 54	+17 08.3	SAB(r)bc	4	1135	−.04	11.76	1.29		11.85	1.87	1.00	38	21.24	1.32	0.94	0.21:	0.53		
NGC3684	11 24 35	+17 18.3	SA(rs)bc	4	1065	0.01	12.10	1.37	Sc	12.17	1.98	0.65	122	21.78	1.46	0.62	0.30:	0.57		
NGC3683	11 24 52	+57 09.2	SB(s)c?	5?	1785	−.04	12.37	1.21		12.51	1.80	0.63	127	20.82	1.29:	0.38	0.32:	0.54:		
NGC3686	11 25 07	+17 30.0	SB(s)bc	4	1034	0.02	11.52	1.53	Sc	11.58	1.98	0.81	16	21.98	1.59	0.85	0.36	0.64		107
NGC3691	11 25 35	+17 11.8	SBb?	3	947	0.01	12.88	1.24	Sm	12.98	1.72	0.65	12	21.34	1.35	0.45	0.35:	0.61	*	
NGC3690	11 25 42	+58 50.0	IBmp	9	3096	−.04	11.91	1.20		11.97	1.78	0.87	37	21.11	1.34	0.53	0.23:	0.57	*	
NGC3705	11 27 33	+09 33.2	SAB(r)ab	2	870	0.10	10.97	1.42	Sa	10.99	2.18	0.45	120	20.92	1.47	0.61	0.19:	0.55		
NGC3717	11 29 04	−30 01.9	SAb:/	3:	1443	0.17	11.46	1.48		11.45	2.29	0.23	32	21.63	1.55	0.43	0.27:	0.54		
NGC3720	11 29 48	+01 04.8	Sab?	2?	5831	0.06	13.21	0.94:	Sc	13.23	1.56	0.88	15	20.64	0.96:	0.89	0.31:	0.57:		108

Galaxy (1)	α_c (1950) (2)	δ_c (3)	Type (4)	T (5)	v_0 (6)	A_V (7)	V_T (8)	log R_e (9)	GCT (10)	V_{25} (11)	log a_{25} (12)	ϵ_{25} (13)	θ_{25} (14)	μ_e (15)	log a_e (16)	ϵ_e (17)	S_{13} (18)	S_{23} (19)	Note (20)	Page (21)
NGC3718	11 29 50	+53 20.7	SB(s)ap	1	1082	0.00	10.75	1.77		10.85	2.44	0.31	3	22.40	1.84	0.72	0.27	0.61		
NGC3729	11 31 04	+53 24.0	SB(r)ap	1	1116	−.01	11.56	1.49		11.68	1.97	0.69	170	21.53	1.58	0.59	0.30:	0.61		
NGC3738	11 33 04	+54 47.9	Im	10	312	−.03	11.56	1.56	Sa	11.77	2.02	0.87	153	22.62	1.64	0.64	0.25	0.48		
NGC3756	11 34 05	+54 34.3	SAB(rs)bc	4	1372	−.04	11.60	1.63	Sm	11.67	2.10	0.48	1	21.73	1.69	0.43	0.41	0.63		
NGC3769	11 35 02	+48 10.3	SB(r)b:	3:	791	0.02	12.01	1.31	Sc	12.07	1.99	0.37	152	20.82	1.38	0.43	0.32:	0.62		109
NGC3773	11 35 38	+12 23.4	SA0:	−2:	851	0.05	12.89	1.15:		13.13	1.69	0.80	146	22.08	1.18:	0.94	0.14:	0.42:		
NGC3780	11 36 38	+56 33.0	SA(s)c:	5	2481	−.02	11.76	1.53		11.83	1.96	0.74	83	22.01	1.65	0.51	0.37	0.61		
NGC3810	11 38 24	+11 44.9	SA(rs)c	5	860	0.11	10.84	1.55		10.86	2.11	0.66	24	21.48	1.61	0.67	0.28	0.56		
NGC3813	11 38 40	+36 49.4	SA(rs)b:	3	1459	0.01	11.81	1.34	Sm	11.88	1.86	0.69	86	20.76	1.41	0.47	0.44:	0.70		
NGC3818	11 39 24	−05 52.7	E5	−5	1286	0.08	11.79	1.26		11.95	2.01	0.64	97	21.67	1.36	0.62	0.11:	0.35		110
NGC3877	11 43 29	+47 46.2	SA(s)c:	5	939	0.01	11.15	1.58	Sb	11.17	2.24	0.23	35	21.13	1.64	0.34	0.39:	0.66		
NGC3885	11 44 15	−27 38.7	SA(s)0/a	0	1530	0.26	11.84	1.18		11.88	1.99	0.44	121	20.50	1.22:	0.60	0.23:	0.47:		
NGC3887	11 44 33	−16 34.6	SB(r)bc	4	915	0.04	10.85	1.63	Sc	10.93	2.11	0.71	6	21.49	1.68	0.67	0.34	0.63		
NGC3888	11 44 55	+56 14.9	SAB(rs)c	5	2454	−.03	12.34	1.15:		12.36	1.76	0.70	116	20.67	1.23:	0.56	0.32:	0.58:		
NGC3893	11 46 01	+48 59.4	SAB(rs)c:	5:	1026	0.02	10.76	1.52		10.81	2.13	0.60	162	21.40	1.58	0.73	0.28	0.58		111
NGC3900	11 46 34	+27 18.1	SA(r)0$^+$	−1	1744	0.06	11.62	1.33		11.67	2.06	0.50	4	21.18	1.39	0.63	0.24:	0.55		
NGC3898	11 46 36	+56 21.8	SA(s)ab	2	1258	−.03	10.68	1.27	E	10.78	2.18	0.59	109	20.74	1.37	0.63	0.10:	0.36		
NGC3904	11 46 42	−28 59.9	E2–3:	−5:	1333	0.14	10.82	1.36		10.99	2.13	0.69	12	21.13	1.44	0.69	0.12:	0.37		
NGC3912	11 47 30	+26 45.3	SAB(s)b?p	3	1733	0.04	12.69	1.19	Sm	12.78	1.85	0.45	175	20.92	1.26:	0.45	0.27:	0.53:		
NGC3917	11 48 08	+52 06.2	SAcd:	6:	1041	−.01	11.74	1.65		11.85	2.21	0.23	78	21.54	1.75	0.20	0.40	0.60		112
NGC3923	11 48 30	−28 31.7	E4–5	−5	1509	0.18	9.41	1.95		9.64	2.52	0.76	20	22.89	2.07	0.75	0.13	0.42		
NGC3936	11 49 48	−26 37.6	SB(s)b?	3	1738	0.29	12.51	1.53	Sc	12.60	2.10	0.19	63	21.57	1.55	0.23	0.38:	0.60		
NGC3938	11 50 13	+44 24.0	SA(s)c	5	844	−.03	10.49	1.77		10.58	2.20	0.92	39	22.17	1.80	0.92	0.33	0.61		
NGC3941	11 50 20	+37 15.9	SB(s)0	−2	942	−.02	10.41	1.26		10.49	2.03	0.74	9	19.99	1.34	0.62	0.18:	0.44		
NGC3945	11 50 36	+60 57.3	SB(rs)0$^+$	−1	1340	0.05	10.67	1.37	E	10.73	2.29	0.54	153	21.04	1.41	0.83	0.09:	0.32		113
NGC3949	11 51 05	+48 08.3	SA(s)bc:	4	857	0.02	11.11	1.38	Sc	11.19	1.98	0.74	117	20.50	1.48	0.53	0.37:	0.63		
NGC3952	11 51 07	−03 43.1	I0p?	0	1424	0.04	13.05	1.18		13.15	1.83	0.35	82	20.96	1.23:	0.34	0.21:	0.51:		
NGC3953	11 51 13	+52 36.5	SB(r)bc	4	1036	0.01	10.33	1.72		10.38	2.30	0.48	10	21.71	1.83	0.51	0.31	0.62		
NGC3955	11 51 25	−22 53.2	I0	0	1060	0.11	11.94	1.36		12.01	2.01	0.39	165	21.21	1.43	0.43	0.36:	0.66		
NGC3956	11 51 28	−20 17.3	SA(s)c:	5	1394	0.05	12.38	1.47		12.50	2.02	0.35	59	21.78	1.57	0.30	0.33:	0.54		114
NGC3957	11 51 33	−19 17.3	SA0$^+$:/	−1	1583	0.05	11.95	1.24	Sa	12.03	2.01	0.24	174	20.29	1.30:	0.38	0.32:	0.54:		
NGC3962	11 52 07	−13 41.8	E1	−5	1584	0.03	10.60	1.52		10.77	2.19	0.79	9	21.93	1.60	0.80	0.12:	0.40		
NGC3976	11 53 23	+07 01.7	SAB(s)b	3	2333	−.03	11.75	1.20	E	11.82	2.04	0.36	50	20.97	1.31:	0.43	0.18:	0.50:		
NGC3981	11 53 35	−19 36.9	SA(rs)bc	4	1554	0.06	11.46	1.66		11.70	2.22	0.40	26	22.25	1.76	0.32	0.19	0.38		
NGC3982	11 53 53	+55 24.0	SAB(r)b:	3:	1195	−.01	11.83	1.22	Sc	11.88	1.78	0.85	176	20.65	1.26	0.83	0.35:	0.62		115
NGC3985	11 54 06	+48 36.9	SB(s)m:	9	1004	0.04	12.57	1.19		12.70	1.71	0.66	88	21.60	1.29	0.61	0.30:	0.65		
NGC3992	11 55 01	+53 39.3	SB(rs)bc	4	1134	0.01	10.27	1.81	Sm	10.31	2.33	0.57	64	22.30	1.95	0.58	0.29	0.68		
NGC3995	11 55 10	+32 34.3	SAmp	9	3327	0.03	12.40	1.31		12.49	1.90	0.38	26	21.54	1.41	0.41	0.33:	0.58		
NGC3998	11 55 20	+55 44.1	SA(r)0?	−2	1214	−.03	10.53	1.07:	E	10.61	2.03	0.78	130	19.19:	1.02:	0.85	0.10:	0.28:		
NGC4008	11 55 43	+28 28.2	E5	−5	3524	−.01	11.80	1.22		11.97	1.90	0.77	173	21.21	1.26	0.70	0.14:	0.38		116

Galaxy (1)	α_c (1950) (2)	δ_c (3)	Type (4)	T (5)	v_0 (6)	A_V (7)	V_T (8)	log R_e (9)	GCT (10)	V_{25} (11)	log a_{25} (12)	ϵ_{25} (13)	θ_{25} (14)	μ_e (15)	log a_e (16)	ϵ_e (17)	S_{13} (18)	S_{23} (19)	Note (20)	Page (21)
NGC4013	11 55 58	+44 13.4	Sb/	3	866	−.04	11.39	1.65		11.53	2.21	0.34	66	21.75	1.72	0.30	0.37	0.55		
NGC4024	11 55 59	−18 04.2	SAB0$^-$	−3	1444	0.04	11.70	1.11:		11.81	1.83	0.84	69	20.63	1.18:	0.68	0.15:	0.40:		
IC 749	11 56 00	+43 00.8	SB(rs)c	5	827	−.04	12.32	1.52	Sm	12.47	1.88	0.82	117	22.54	1.59	0.73	0.43	0.67		
IC 750	11 56 17	+43 00.1	Sab:/	2:	742	−.04	11.86	1.22	Sa	11.96	1.94	0.50	43	20.56	1.29:	0.41	0.26:	0.50:		
NGC4026	11 56 50	+51 14.4	S0/	−2	951	0.03	10.60	1.06:	E	10.60	2.17	0.29	174	19.07:	1.11:	0.65	0.18:	0.44:		117
NGC4027	11 56 56	−18 59.4	SB(s)dm	8	1419	0.05	11.15	1.51	Sc	11.23	2.01	0.80	162	21.59	1.61	0.56	0.28	0.56		
NGC4030	11 57 50	−00 49.4	SA(s)bc	4	1322	0.02	10.50	1.51	Sc	10.54	2.11	0.82	29	21.09	1.56	0.83	0.28	0.56		
NGC4033	11 58 01	−17 34.0	E6	−5	1273	0.05	11.56	0.94:		11.64	1.90	0.52	47	19.37:	0.97:	0.68	0.17:	0.39:		
NGC4037	11 58 50	+13 40.7	SB(rs)b:	3?	811	0.01	12.12	1.69	Sc	12.42	2.00	0.80	165	23.47	1.72	0.89	0.27	0.57		
NGC4036	11 58 54	+62 10.5	S0$^-$	−3	1509	0.02	10.73	1.31		10.84	2.12	0.45	79	20.04	1.40	0.46	0.23:	0.49		118
NGC4038	11 59 19	−18 35.1	SB(s)mp	9	1391	0.04	10.76	1.67		10.94	2.14	0.55	128	21.54	1.73	0.48	0.36	0.57	*	
NGC4045	12 00 09	+02 15.5	SAB(r)a	1	1765	0.00	11.71	1.45		11.88	2.01	0.67	84	22.12	1.46	0.89	0.22:	0.50	*	
NGC4047	12 00 18	+48 55.2	(R)SA(rs)b:	3:	3416	0.01	12.17	1.24	Sc	12.26	1.88	0.71	83	21.23	1.27	0.79	0.27:	0.55		
NGC4050	12 00 20	−16 05.4	SB(r)ab	2	1661	0.07	11.81	1.66	Sc	11.92	2.06	0.63	92	22.26	1.76	0.33	0.35	0.56		
NGC4051	12 00 37	+44 48.7	SAB(rs)bc	4	746	−.04	10.41	1.78	Sc	10.52	2.23	0.81	136	21.85	1.95	0.32	0.30	0.53		119
NGC4062	12 01 31	+32 10.4	SA(s)c	5	745	−.01	11.08	1.61	Sm	11.13	2.15	0.42	99	21.18	1.70	0.35	0.42	0.63		
NGC4073	12 01 54	+02 10.6	SA0$^-$:	−3	5670	0.02	11.31	1.73		11.57	2.24	0.46	103	23.10	1.79	0.60	0.17	0.46		
NGC4085	12 02 51	+50 38.0	SAB(s)c:?	5	823	0.02	12.39	1.31	Sm	12.43	1.92	0.28	75	20.97	1.38:	0.33	0.38:	0.67:		
NGC4088	12 03 02	+50 49.1	SAB(rs)bc	4	820	0.02	10.67	1.68	Sm	10.68	2.22	0.36	52	21.57	1.76	0.50	0.33	0.72		
IC 2995	12 03 13	−27 39.7	SB(s)c:	5:	1566	0.26	12.60	1.49	Sm	12.71	2.00	0.29	118	22.26	1.57	0.35	0.46:	0.72		120
NGC4094	12 03 18	−14 15.1	SA(rs)c:	5	1184	0.07	11.99	1.66		12.14	2.13	0.36	64	22.19	1.73	0.30	0.32	0.55		
NGC4096	12 03 29	+47 45.7	SAB(rs)c	5	616	−.03	10.81	1.75		10.88	2.35	0.26	19	21.65	1.84	0.33	0.41	0.69		
NGC4100	12 03 37	+49 51.4	(R′)SA(rs)bc	4	1135	−.01	11.23	1.60	Sm	11.22	2.20	0.31	166	21.17	1.66	0.33	0.48	0.73		
NGC4105	12 04 06	−29 28.9	E3	−5	1623	0.18	10.79	1.46		10.96	2.17	0.83	133	21.53	1.52	0.73	0.10:	0.33	*	
NGC4106	12 04 11	−29 29.4	SB(s)0$^+$	−1	1905	0.18	11.59	1.05:		11.67	1.91	0.78	84	19.64	1.01:	0.87	0.16:	0.36:	*	121
NGC4111	12 04 32	+43 20.7	SA(r)0$^+$:/	−1:	824	−.03	10.59	0.97:		10.62	2.16	0.23	151	18.20:	1.00:	0.58	0.21:	0.45:		
NGC4116	12 05 04	+02 58.2	SB(rs)dm	8	1140	−.01	11.88	1.66		12.06	2.08	0.61	153	22.37	1.78	0.33	0.20:	0.47		
NGC4124	12 05 36	+10 39.5	SA(r)0$^+$	−1	1501	−.01	11.34	1.59	Sc	11.42	2.19	0.36	111	21.37	1.63	0.37	0.32:	0.59		
NGC4123	12 05 38	+03 09.3	SB(r)c	5	1157	−.01	11.24	1.76	Sm	11.42	2.15	0.80	127	22.18	1.89	0.24	0.26	0.42		
NGC4136	12 06 46	+30 12.3	SAB(r)c	5	409	0.02	11.55	1.69	Sm	11.74	2.05	0.97	101	22.88	1.71	0.92	0.31	0.58		122
NGC4138	12 06 59	+43 57.8	SA(r)0$^+$	−1	1075	−.02	11.34	1.27	Sa	11.41	1.99	0.61	146	20.59	1.37	0.57	0.24:	0.53		
NGC4143	12 07 05	+42 48.8	SAB(s)0	−2	814	−.01	11.01	1.07:		11.09	1.94	0.68	150	19.78	1.12:	0.74	0.16:	0.45:		
NGC4144	12 07 28	+46 44.1	SAB(s)cd?/	6:	316	−.05	11.67	1.70		11.76	2.27	0.26	100	22.08	1.77	0.26	0.34	0.61		
NGC4145	12 07 30	+40 09.7	SAB(rs)d	7	1030	−.01	11.31	1.81		11.39	2.24	0.59	97	23.23	1.94	0.57	0.39	0.72		
IC 764	12 07 39	−29 27.5	SA(s)c?	5?	1851	0.19	12.41	1.65		12.64	2.16	0.27	178	22.53	1.74	0.24	0.28	0.46		123
NGC4151	12 08 00	+39 40.9	(R′)SAB(rs)ab:	2:	978	−.04	10.56	1.34	E	10.68	2.18	0.58	145	21.24	1.36	0.94	0.10:	0.41		
NGC4150	12 08 01	+30 40.9	SA(r)0?	−2	211	0.04	11.61	1.14:		11.73	1.85	0.76	146	20.63	1.21:	0.70	0.17:	0.48:		
NGC4152	12 08 05	+16 18.7	SAB(rs)c	5	2055	0.08	12.29	1.21		12.31	1.82	0.73	124	21.06	1.24	0.77	0.32:	0.60		
NGC4157	12 08 35	+50 45.8	SAB(s)b?/	3?	852	0.00	11.35	1.59		11.39	2.28	0.20	69	21.53	1.68	0.26	0.33:	0.60		
NGC4162	12 09 20	+24 24.0	(R)SA(rs)bc	4	2481	0.07	11.65	1.32	Sc	11.69	1.91	0.54	171	20.69	1.37	0.59	0.35:	0.63	*	124

Galaxy (1)	α_c (1950) (2)	δ_c (3)	Type (4)	T (5)	v_0 (6)	A_V (7)	V_T (8)	log R_e (9)	GCT (10)	V_{25} (11)	log a_{25} (12)	ϵ_{25} (13)	θ_{25} (14)	μ_e (15)	log a_e (16)	ϵ_e (17)	S_{13} (18)	S_{23} (19)	Note (20)	Page (21)
NGC4168	12 09 44	+13 29.0	E2	−5	2269	0.04	11.46	1.44		11.59	2.05	0.77	133	21.88	1.45	0.90	0.19:	0.45		
NGC4178	12 10 14	+11 08.8	SB(rs)dm	8	224	0.00	11.67	1.69		11.73	2.20	0.34	31	22.04	1.81	0.18	0.30	0.55		
NGC4179	12 10 19	+01 34.7	S0/	−2	1105	0.03	11.05	1.18		11.13	2.11	0.34	142	19.95	1.26:	0.53	0.20:	0.46:		
NGC4183	12 10 47	+43 58.6	SA(s)cd?/	6	968	−.04	12.39	1.65		12.58	2.15	0.18	164	22.31	1.70	0.23	0.36:	0.61		
NGC4190	12 11 13	+36 54.6	Imp	10	231	−.01	12.85	1.45		13.03	1.83	0.78	68	22.70	1.50	0.66	0.31:	0.57		125
NGC4189	12 11 14	+13 42.2	SAB(rs)cd?	6	1980	0.02	11.99	1.52	Sm	12.08	1.89	0.75	75	21.91	1.57	0.65	0.40	0.66		
NGC4192	12 11 15	+15 10.8	SAB(s)ab	2	−251	0.11	10.28	1.82		10.35	2.44	0.26	156	21.39	1.91	0.27	0.44	0.70		
NGC4203	12 12 34	+33 28.7	SAB0$^-$:	−3:	1072	−.01	10.60	1.38		10.66	2.07	0.94	8	21.20	1.42	0.88	0.13:	0.43		
NGC4212	12 13 07	+14 10.8	SAbc?	4	−165	0.06	11.33	1.47	Sm	11.39	1.98	0.65	77	21.04	1.59	0.44	0.39	0.59		
NGC4214	12 13 08	+36 36.5	IAB(s)m	10	290	−.04	10.11	1.85		10.26	2.32	0.92	160	22.35	1.92	0.68	0.22	0.48		126
NGC4216	12 13 21	+13 25.4	SAB(s)b:	3	−8	0.02	10.13	1.56	S0	10.15	2.42	0.24	22	20.91	1.66	0.41	0.20:	0.51		
NGC4217	12 13 21	+47 22.3	Sb/	3	1084	−.01	11.19	1.77	Sc	11.33	2.29	0.29	47	22.07	1.82	0.32	0.33	0.64		
NGC4215	12 13 22	+06 41.0	SA(r)0$^+$:/	−1	1942	−.09:	12.00	0.94:		12.07	1.95	0.33	174	20.05	1.05:	0.61	0.23:	0.55:		
NGC4220	12 13 43	+48 09.5	SA(r)0$^+$	−1	1039	−.04	11.40	1.38	Sa	11.47	2.07	0.34	137	20.58	1.46	0.37	0.33:	0.57		
NGC4224	12 14 01	+07 44.4	SA(s)a:/	1	2442	−.04	11.92	1.28	Sa	11.93	2.05	0.38	53	21.39	1.33	0.68	0.23:	0.61		127
NGC4233	12 14 35	+07 54.1	S0$^+$	−1	2079	−.03	11.94	0.99:		12.01	1.94	0.44	175	20.21	1.01:	0.91	0.19:	0.42:		
NGC4234	12 14 36	+03 57.7	SB(s)dm	8	1981	0.01	12.97	1.29		13.10	1.68	0.91	57	21.99	1.33	0.76	0.30:	0.61		
NGC4235	12 14 37	+07 28.1	SA(s)a/	1	2380	−.04	11.64	1.38	S0	11.73	2.15	0.25	51	21.19	1.45	0.44	0.27:	0.57		
NGC4237	12 14 39	+15 36.1	SAB(rs)bc	4	814	0.08	11.78	1.35	Sc	11.88	1.93	0.62	113	21.07	1.42	0.61	0.36:	0.62		
NGC4244	12 15 00	+38 05.2	SA(s)cd:/	6	249	−.01	10.23	2.05	Sb	10.30	2.70	0.12	48	22.09	2.08	0.23	0.37	0.63		128
NGC4242	12 15 01	+45 53.8	SAB(s)dm	8	564	−.06	11.20	1.91	Sc	11.43	2.21	0.71	31	23.16	1.96	0.66	0.37	0.63		
NGC4245	12 15 05	+29 52.9	SB(r)0/a:	0	855	0.06	11.16	1.52	S0	11.38	2.06	0.79	168	21.76	1.59	0.59	0.15:	0.40		
NGC4251	12 15 36	+28 27.1	SB0?/	−2	971	0.06	10.76	1.19		10.83	2.11	0.60	93	19.71	1.22:	0.65	0.15:	0.40:		
NGC4254	12 16 17	+14 41.7	SA(s)c	5	2301	0.11	10.05	1.72		10.08	2.22	0.84	81	21.64	1.74	0.95	0.31	0.62		
NGC4258	12 16 29	+47 35.0	SAB(s)bc	4	520	.08	8.46	2.05	Sa	8.40	2.79	0.36	149	21.22	2.11	0.48	0.27	0.52		129
NGC4260	12 16 49	+06 22.6	SB(s)a	1	1806	−.02	11.78	1.34		11.84	2.08	0.36	65	21.16	1.45	0.40	0.25:	0.52		
NGC4261	12 16 50	+06 06.1	E2–3	−5	2075	−.01	10.33	1.61		10.52	2.29	0.70	165	22.04	1.68	0.84	0.12:	0.39		
NGC4262	12 16 59	+15 09.3	SB(s)0$^-$?	−3	1254	0.07	11.41	0.98:		11.48	1.84	0.85	167	19.18:	0.95:	0.91	0.19:	0.38:		
NGC4267	12 17 13	+13 04.6	SB(s)0$^-$?	−3	1140	0.02	10.96	1.45		11.09	2.05	0.92	105	21.86	1.45	0.98	0.12:	0.43		
NGC4270	12 17 17	+05 44.5	S0	−2	2193	0.00	12.11	1.10:		12.24	1.87	0.48	109	20.62	1.19:	0.56	0.23:	0.53:		130
NGC4274	12 17 20	+29 53.3	(R)SB(r)ab	2	878	0.07	10.50	1.62		10.54	2.33	0.34	99	21.50	1.69	0.69	0.32	0.67		
NGC4273	12 17 23	+05 37.3	SB(s)c	5	2232	0.00	12.03	1.28		12.05	1.88	0.58	7	21.07	1.37	0.50	0.30:	0.56		
NGC4278	12 17 36	+29 33.6	E1–2	−5	594	0.08	10.06	1.59		10.28	2.31	0.76	51	21.68	1.63	0.89	0.09:	0.31	*	
NGC4281	12 17 49	+05 39.9	S0$^+$:/	−1:	2448	0.00	11.21	1.30		11.32	2.07	0.61	90	20.68	1.39	0.47	0.16:	0.43		
NGC4283	12 17 50	+29 35.2	E0	−5	1053	0.08	11.99	0.96:		12.06	1.88	0.82	132	19.77	0.96:	0.93	0.16:	0.35:	*	131
NGC4291	12 18 07	+75 38.8	E2–3	−5	2016	0.05	11.57	0.95:		11.66	1.96	0.67	118	19.74	0.95:	0.84	0.11:	0.29:		
NGC4293	12 18 42	+18 39.7	(R)SB(s)0/a	0	841	0.06	10.51	1.75	Sc	10.63	2.29	0.50	65	21.27	1.86	0.26	0.31	0.51		
NGC4294	12 18 45	+11 47.4	SB(s)cd	6	227	0.02	12.26	1.43	Sm	12.33	1.97	0.43	155	21.84	1.51	0.45	0.35:	0.71		
NGC4298	12 19 00	+14 53.1	SA(rs)c	5	1004	0.08	11.48	1.59		11.55	2.12	0.50	132	21.82	1.63	0.63	0.38	0.67	*	
NGC4299	12 19 08	+11 46.8	SAB(s)dm:	8	107	0.02	12.62	1.38		12.74	1.75	0.97	61	22.13	1.42	0.84	0.41:	0.67		132

Galaxy (1)	α_c (1950) (2)	δ_c (3)	Type (4)	T (5)	v_0 (6)	A_V (7)	V_T (8)	log R_e (9)	GCT (10)	V_{25} (11)	log a_{25} (12)	ϵ_{25} (13)	θ_{25} (14)	μ_e (15)	log a_e (16)	ϵ_e (17)	S_{13} (18)	S_{23} (19)	Note (20)	Page (21)
NGC4302	12 19 10	+14 52.5	Sc:/	5:	1004	0.05	11.46	1.84		11.65	2.33	0.25	178	22.36	1.91	0.20	0.44	0.69	*	
NGC4303	12 19 22	+04 45.1	SAB(rs)bc	4	1404	0.01	9.87	1.76	Sc	9.90	2.27	0.89	147	21.15	1.90	0.36	0.29	0.51		
NGC4307	12 19 33	+09 19.1	Sb/	3	1168	−.01	11.92	1.53	Sc	12.06	2.07	0.30	24	21.22	1.57	0.25	0.34:	0.60		
NGC4314	12 20 03	+30 10.1	SB(rs)a	1	850	0.06	10.69	1.53		10.76	2.12	0.92	67	21.46	1.65	0.49	0.24:	0.54		
NGC4321	12 20 23	+16 06.0	SAB(s)bc	4	1464	0.03	9.59	1.94	Sc	9.67	2.39	0.81	23	22.00	2.01	0.67	0.32	0.60		133
NGC4324	12 20 34	+05 31.7	SA(r)0$^+$	−1	1508	−.01	11.71	1.10:	Sa	11.73	1.97	0.38	56	20.29	1.16:	0.71	0.23:	0.60:		
NGC4346	12 21 01	+47 16.2	S0/	−2	906	−.04	10.99	1.00:		11.05	2.07	0.37	96	19.26:	1.03:	0.72	0.17:	0.42:		
NGC4339	12 21 02	+06 21.6	E0	−5	1128	−.05	11.63	1.29	S0	11.72	1.95	0.92	54	21.59	1.32	0.91	0.15:	0.44		
NGC4340	12 21 04	+17 00.0	SB(r)0$^+$	−1	775	0.02	11.31	1.43	Sa	11.42	2.10	0.55	93	21.76	1.51	0.71	0.18:	0.50		
NGC4342	12 21 07	+07 20.0	S0$^-$/	−3	569	−.02	11.79	0.64:	Sm	11.80	1.63	0.64	159	17.05:	0.54:	0.89	0.31:	0.55:	*	134
NGC4348	12 21 20	−03 10.0	Sb:/	3:	1964	0.08	12.33	1.32		12.40	2.00	0.26	38	20.88	1.37:	0.31	0.34:	0.61:		
NGC4350	12 21 26	+16 58.3	SA0/	−2	1085	0.02	10.91	1.16:		10.99	1.99	0.90	36	19.54	1.21:	0.58	0.25:	0.50:		
NGC4365	12 21 56	+07 35.7	E3	−5	1086	−.02	9.51	1.90		9.78	2.53	0.56	39	22.64	2.00	0.76	0.12	0.38		
NGC4369	12 22 09	+39 39.7	(R)SA(rs)a	1	1069	−.04	11.65	1.19		11.71	1.87	0.95	86	21.24	1.22	0.97	0.14:	0.43		
NGC4386	12 22 22	+75 48.3	SAB0$^-$?	−3:	2001	0.05	11.82	1.15:	E	11.98	1.91	0.69	135	20.96	1.25:	0.58	0.13:	0.37:		135
NGC4371	12 22 23	+11 58.8	SB(r)0$^+$	−1	854	0.02	10.81	1.41		10.89	2.21	0.47	89	21.16	1.44	0.81	0.16:	0.46		
NGC4374	12 22 31	+13 09.8	E1	−5	835	0.10	9.15	1.81		9.33	2.48	0.88	78	22.09	1.89	0.93	0.11	0.38		
NGC4377	12 22 40	+15 02.4	SA0$^-$	−3	1227	0.06	11.80	1.01:		11.85	1.77	0.81	180	19.89	1.01:	0.91	0.22:	0.45:		
NGC4379	12 22 43	+15 53.0	S0$^-$p?	−3	926	0.02	11.75	1.14:	Sa	11.81	1.89	0.75	97	20.51	1.17:	0.81	0.22:	0.47:		
NGC4378	12 22 45	+05 12.1	(R)SA(s)a	1	2391	−.01	11.40	1.39	E	11.61	2.02	0.86	6	22.03	1.46	0.78	0.10:	0.33		136
NGC4380	12 22 50	+10 17.6	SA(rs)b:?	3:	832	−.01	11.64	1.61	Sc	11.74	2.10	0.51	157	22.28	1.71	0.52	0.34	0.66		
NGC4382	12 22 53	+18 28.0	SA(s)0$^+$p	−1	648	0.03	9.02	1.84		9.12	2.52	0.73	62	21.41	1.94	0.69	0.19	0.47		
NGC4383	12 22 54	+16 44.8	Sa?p	1?	1509	0.02	12.03	1.18		12.15	1.91	0.65	6	20.90	1.20:	0.70	0.19:	0.41:		
NGC4389	12 23 10	+45 57.7	SB(rs)bcp:	4	767	−.04	11.83	1.47	Sc	11.94	1.97	0.61	90	21.27	1.59	0.21	0.38:	0.66		
NGC4385	12 23 12	+00 50.9	SB(rs)0$^+$:	−1:	1969	0.05	12.45	1.29		12.65	1.86	0.53	78	21.95	1.42	0.43	0.17:	0.48		137
NGC4388	12 23 14	+12 56.3	SA(s)b:/	3:	2369	0.08	11.14	1.64		11.20	2.35	0.25	92	21.48	1.68	0.35	0.32	0.56		
NGC4395	12 23 20	+33 49.5	SA(s)m:	9	304	0.00	10.05	2.30		10.32	2.56	0.74	144	24.34	2.36	0.70	0.32	0.64		
NGC4394	12 23 25	+18 29.4	(R)SB(r)b	3	853	0.03	10.89	1.57	Sa	11.02	2.09	0.92	92	21.92	1.69	0.48	0.16:	0.45		
NGC4406	12 23 40	+13 13.4	E3	−5	−367	0.08	8.89	2.10		9.10	2.64	0.61	128	22.74	2.16	0.73	0.17	0.49	*	
NGC4414	12 23 57	+31 29.9	SA(rs)c?	5	702	0.02	10.23	1.46	Sa	10.27	2.19	0.75	152	20.52	1.57	0.51	0.23:	0.53		138
NGC4412	12 24 02	+04 14.3	SB(r)b?p	3	1577	−.01	12.61	1.31	Sm	12.70	1.73	0.87	77	21.56	1.36	0.68	0.36:	0.64		
NGC4417	12 24 18	+09 51.7	SB0:/	−2	710	−.01	11.11	1.10:		11.19	2.05	0.53	47	19.71	1.17:	0.58	0.18:	0.43:		
NGC4419	12 24 25	+15 19.4	SB(s)a/	1	−335	0.05	11.21	1.27		11.24	2.05	0.36	130	20.34	1.34	0.52	0.30:	0.62		
NGC4420	12 24 25	+02 46.3	SB(r)bc:	4:	1515	0.01	12.47	1.23	Sm	12.51	1.81	0.46	9	20.71	1.30:	0.47	0.47:	0.70:		
NGC4424	12 24 40	+09 41.8	SB(s)a:	1	323	−.02	11.69	1.52	Sb	11.78	2.11	0.51	89	21.79	1.65	0.31	0.27:	0.53		139
NGC4425	12 24 42	+13 00.7	SB0:/	−2	1765	0.08	11.97	1.29	Sa	12.05	1.98	0.42	28	20.90	1.37	0.45	0.32:	0.58		
NGC4428	12 24 53	−07 53.9	SAB(rs)c	5	2828	0.03	12.58	1.28		12.73	1.82	0.50	85	21.45	1.35	0.51	0.35:	0.62		
NGC4429	12 24 54	+11 23.1	SA(r)0$^+$	−1	989	0.02	10.08	1.64	Sa	10.17	2.35	0.51	89	21.08	1.76	0.44	0.24	0.54		
NGC4433	12 25 03	−08 00.3	SAB(s)ab	2	2771	0.05	12.60	1.22	Sm	12.66	1.87	0.39	180	21.03	1.31:	0.40	0.40:	0.65:		
NGC4435	12 25 08	+13 21.4	SB(s)0	−2	753	0.09	10.67	1.35		10.76	2.10	0.94	14	20.44	1.38	0.62	0.15:	0.36	*	140

Galaxy (1)	α_c (1950) (2)	δ_c (3)	Type (4)	T (5)	v_0 (6)	A_V (7)	V_T (8)	log R_e (9)	GCT (10)	V_{25} (11)	log a_{25} (12)	ϵ_{25} (13)	θ_{25} (14)	μ_e (15)	log a_e (16)	ϵ_e (17)	S_{13} (18)	S_{23} (19)	Note (20)	Page (21)
NGC4438	12 25 14	+13 17.1	SA(s)0/ap:	0	−115	0.08	10.12	1.79		10.22	2.48	0.48	32	22.68	1.93	0.47	0.15	0.41	*	
NGC4442	12 25 32	+10 04.7	SB(s)0	−2	449	−.01	10.54	1.18		10.57	2.17	0.39	85	19.58	1.22:	0.70	0.19:	0.45:		
NGC4448	12 25 46	+28 53.8	SB(r)ab	2	655	0.23:	11.21	1.38		11.21	2.13	0.33	93	20.88	1.45	0.56	0.32:	0.67		
NGC4449	12 25 47	+44 22.3	IBm	10	250	−.04	9.69	1.71		9.75	2.35	0.65	66	21.04	1.80	0.50	0.25	0.60	*	
NGC4450	12 25 59	+17 21.7	SA(s)ab	2	1862	0.00	10.10	1.69		10.20	2.28	0.68	169	21.55	1.76	0.63	0.23	0.53		141
NGC4452	12 26 12	+12 02.0	Sa?/	1?	30	0.03	11.95	1.26		12.01	2.04	0.32	38	20.20	1.29:	0.29	0.31:	0.55:		
NGC4454	12 26 16	−01 39.7	(R)SB(r)0/a	0	2231	0.05	12.02	1.45	Sb	12.17	1.90	0.97	64	22.13	1.53	0.64	0.23:	0.49		
NGC4460	12 26 19	+45 08.6	SB(s)0$^+$?/	−1?	605	−.04	11.61	1.50		11.80	2.13	0.28	39	21.19	1.54	0.33	0.25:	0.48		
NGC4457	12 26 26	+03 50.8	(R)SAB(s)0/a	0	709	0.02	10.68	1.29	E	10.74	2.12	0.94	169	20.72	1.34	0.75	0.09:	0.36		
NGC4459	12 26 29	+14 15.3	SA(r)0$^+$	−1	1000	0.06	10.36	1.47		10.46	2.17	0.86	98	20.99	1.50	0.81	0.17:	0.42		142
NGC4461	12 26 31	+13 27.7	SB(s)0$^+$:	−1:	1772	0.06	11.19	1.22	Sa	11.25	2.12	0.36	11	20.41	1.27	0.70	0.23:	0.51		
NGC4469	12 26 56	+09 01.5	SB(s)0/a?/	0	372	−.02	11.22	1.53		11.35	2.21	0.29	90	21.10	1.61	0.38	0.34:	0.60		
NGC4472	12 27 14	+08 16.7	E2	−5	822	−.01	8.43	2.06		8.65	2.64	0.80	150	22.33	2.13	0.84	0.14	0.41		
NGC4473	12 27 17	+13 42.4	E5	−5	2162	0.03	10.22	1.34		10.30	2.23	0.65	91	20.31	1.41	0.64	0.14:	0.38		
NGC4474	12 27 22	+14 20.7	S0p:	−2	1416	0.06	11.54	1.30		11.69	1.94	0.88	78	20.95	1.35	0.57	0.18:	0.40		143
NGC4476	12 27 28	+12 37.5	SA(r)0$^-$:	−3	1860	0.07	12.35	1.15:	Sa	12.48	1.83	0.77	42	21.11	1.18:	0.75	0.20:	0.42:		
NGC4477	12 27 31	+13 54.7	SB(s)0:?	−2:	1151	0.05	10.45	1.49		10.57	2.12	0.94	47	21.45	1.55	0.83	0.16:	0.46		
NGC4478	12 27 46	+12 36.3	E2	−5	1364	0.07	11.44	1.10:	Sb	11.47	1.82	0.92	148	19.58	1.11:	0.79	0.30:	0.54:		
NGC4485	12 28 05	+41 58.5	IB(s)mp	10	817	−.04	12.04	1.38		12.14	1.90	0.74	162	21.57	1.42	0.71	0.30:	0.57		
NGC4483	12 28 08	+09 17.1	SB(s)0/a:	0	741	−.03	12.27	1.05:		12.34	1.83	0.56	62	20.60	1.09:	0.73	0.21:	0.50:		144
NGC4490	12 28 10	+41 54.9	SB(s)dp	7	601	−.04	9.88	1.69		9.89	2.28	0.50	122	20.72	1.79	0.41	0.36	0.61		
NGC4486	12 28 17	+12 40.1	E$^+$0–1p	−4	1136	0.07	8.94	1.88		9.05	2.56	0.73	146	21.61	1.90	0.89	0.16	0.43		
NGC4487	12 28 30	−07 46.5	SAB(rs)cd	6	831	0.05	11.40	1.70		11.52	2.11	0.64	76	22.65	1.75	0.72	0.36	0.68		
NGC4494	12 28 55	+26 03.1	E1–2	−5	1259	0.05	10.00	1.55		10.14	2.26	0.85	168	21.22	1.60	0.84	0.14:	0.40		
NGC4496	12 29 07	+04 12.9	SB(rs)m	9	1568	0.01	11.86	1.68		12.03	2.02	0.86	57	22.92	1.70	0.90	0.30	0.62	*	145
NGC4501	12 29 28	+14 41.7	SA(rs)b	3	2161	0.07	9.62	1.77		9.68	2.38	0.48	140	21.13	1.85	0.50	0.33	0.62		
NGC4503	12 29 34	+11 27.2	SB0$^-$:	−3:	1269	0.00	11.20	1.24	Sa	11.20	2.08	0.44	4	20.43	1.31	0.58	0.20:	0.50		
NGC4504	12 29 42	−07 17.3	SA(s)cd	6	794	0.03	11.75	1.68		11.90	2.11	0.58	135	22.62	1.74	0.54	0.35	0.58		
R 80	12 29 55	+00 39.9	SB(rs)dm:	8:	1349	0.02	12.33	1.81		12.69	2.08	0.57	27	24.06	1.87	0.69	0.34	0.63	*	
NGC4517	12 30 12	+00 23.3	SA(s)cd:/	6	956	0.05	10.68	1.99		10.78	2.52	0.17	81	21.69	2.05	0.12	0.36	0.55		146
NGC4519	12 30 58	+08 55.8	SB(rs)d	7	1094	−.01	12.22	1.49		12.32	1.95	0.65	143	22.62	1.54	0.85	0.34	0.66		
NGC4522	12 31 08	+09 27.0	SB(s)cd:/	6	2186	−.03	12.60	1.51	Sm	12.69	2.06	0.28	35	21.83	1.60	0.22	0.36:	0.55		
NGC4526	12 31 31	+07 58.5	SAB(s)0:	−2	309	−.01	9.86	1.61	E	9.99	2.40	0.46	115	20.87	1.68	0.51	0.19:	0.45		
NGC4527	12 31 35	+02 55.7	SAB(s)bc	4	1577	0.00	10.54	1.69		10.62	2.29	0.42	68	21.68	1.79	0.36	0.26	0.62		
NGC4532	12 31 47	+06 44.7	IBm	10	1858	−.03	11.80	1.39		11.87	2.07	0.32	161	21.05	1.48	0.31	0.31:	0.52		147
NGC4535	12 31 48	+08 28.6	SAB(s)c	5	1818	−.01	10.29	1.93		10.40	2.31	0.80	173	22.53	1.96	0.85	0.33	0.68		
NGC4536	12 31 54	+02 27.7	SAB(rs)bc	4	1646	−.01	10.70	1.77		10.75	2.37	0.39	125	22.39	1.84	0.57	0.30	0.62		
NGC4540	12 32 20	+15 49.6	SAB(rs)cd	6	1184	0.03	11.83	1.48		11.97	1.95	0.80	30	22.05	1.51	0.81	0.26:	0.55		
NGC4546	12 32 55	−03 31.1	SB(s)0$^-$:	−3	849	0.05	10.44	1.31		10.56	2.22	0.53	82	20.18	1.41	0.55	0.17:	0.41		
NGC4548	12 32 55	+14 46.4	SB(rs)b	3	366	0.05	10.16	1.75		10.23	2.28	0.76	154	22.21	1.88	0.62	0.24	0.57		148

Galaxy (1)	α_c (1950) (2)	δ_c (3)	Type (4)	T (5)	v_0 (6)	A_V (7)	V_T (8)	log R_e (9)	GCT (10)	V_{25} (11)	log a_{25} (12)	ϵ_{25} (13)	θ_{25} (14)	μ_e (15)	log a_e (16)	ϵ_e (17)	S_{13} (18)	S_{23} (19)	Note (20)	Page (21)
NGC4550	12 32 59	+12 29.8	SB0$^+$?/	−1	261	0.09	11.67	1.02:	Sa	11.69	2.01	0.26	180	19.28:	1.09:	0.46	0.28:	0.53:		
NGC4552	12 33 08	+12 50.0	E0	−5	203	0.11	9.64	1.70		9.82	2.41	0.84	126	22.03	1.76	0.93	0.09:	0.35		
NGC4559	12 33 29	+28 14.1	SAB(rs)cd	6	771	0.05	10.32	1.88	Sm	10.42	2.44	0.34	144	21.89	1.96	0.35	0.38	0.61		
NGC4561	12 33 38	+19 36.1	SB(rs)dm	8	1325	0.05*	12.98	1.32	Sc	13.12	1.73	0.84	36	22.21	1.35	0.79	0.33:	0.61		
NGC4565	12 33 52	+26 15.6	SA(s)b?/	3	1171	0.03	9.70	1.85	E	9.76	2.65	0.14	137	21.85	1.88	0.50	0.21	0.50		149
NGC4564	12 33 55	+11 42.9	E6	−5	894	0.03	11.07	1.07:		11.13	2.02	0.62	46	19.64	1.13:	0.57	0.14:	0.41:		
NGC4567	12 34 01	+11 32.0	SA(rs)bc	4	2136	0.01	11.27	1.62	Sm	11.41	2.08	0.78	79	21.82	1.69	0.57	0.31	0.55	*	
NGC4568	12 34 02	+11 30.9	SA(rs)bc	4	2199	−.01	11.05	1.66		11.19	2.20	0.44	26	21.67	1.72	0.43	0.31	0.65	*	
NGC4569	12 34 19	+13 26.4	SAB(rs)ab	2	−373	0.06	9.66	1.80		9.70	2.48	0.40	24	21.47	1.95	0.35	0.29	0.65		
NGC4570	12 34 21	+07 31.4	S0/	−2	1590	−.02	10.50	1.02:		10.54	2.11	0.46	157	18.55:	1.01:	0.74	0.15:	0.41:		150
NGC4571	12 34 25	+14 29.6	SA(r)d	7	236	0.05	11.31	1.74		11.43	2.11	0.83	33	22.71	1.78	0.84	0.37	0.67		
NGC4578	12 34 59	+09 49.8	SA(r)0?	−2	2153	−.01	11.52	1.35		11.64	2.00	0.75	33	21.77	1.42	0.70	0.13:	0.43		
NGC4579	12 35 12	+12 05.6	SAB(rs)b	3	1687	0.11	9.72	1.69	S0	9.84	2.31	0.78	99	21.51	1.81	0.63	0.16	0.48		
NGC4580	12 35 16	+05 38.6	SAB(rs)ap	1	1105	−.07	11.77	1.47	Sm	11.91	1.93	0.73	158	21.67	1.53	0.60	0.33	0.57		
NGC4589	12 35 28	+74 28.0	E2	−5	2157	0.03	10.99	1.43		11.19	2.15	0.66	72	21.85	1.52	0.78	0.11:	0.36		151
NGC4586	12 35 55	+04 35.6	SA(s)a:/	1:	660	0.01	11.69	1.53		11.81	2.14	0.33	114	21.74	1.62	0.39	0.34:	0.65		
NGC4592	12 36 45	−00 15.4	SA(s)dm:	8	903	0.02	12.18	1.52		12.22	2.15	0.28	96	21.70	1.58	0.32	0.33:	0.61		
NGC4593	12 37 05	−05 04.2	(R)SB(rs)b	3	2505	0.03	11.12	1.58		11.23	2.14	0.73	103	22.13	1.71	0.45	0.18:	0.46		
NGC4595	12 37 21	+15 34.3	SAB(rs)b?	3	559	0.02	12.44	1.32	Sm	12.54	1.83	0.63	111	21.55	1.37	0.70	0.36:	0.65		
NGC4594	12 37 23	−11 21.0	SA(s)a/	1	873	0.09	8.23	1.88		8.34	2.62	0.93	70	21.01	1.97	0.57	0.11	0.37		152
NGC4596	12 37 24	+10 27.0	SB(r)0$^+$	−1	1834	−.02	10.58	1.52	Sa	10.66	2.17	0.83	125	21.57	1.64	0.56	0.17:	0.43		
NGC4597	12 37 38	−05 31.5	SB(rs)m	9	851	0.02	12.00	1.76		12.26	2.14	0.40	44	23.19	1.80	0.59	0.37	0.61		
NGC4605	12 37 47	+61 53.1	SB(s)cp	5	273	−.01	10.30	1.67	Sm	10.35	2.27	0.55	122	20.77	1.76	0.33	0.34	0.56		
NGC4602	12 38 03	−04 52.0	SAB(rs)bc	4	2347	0.02	11.87	1.45	Sm	11.96	2.04	0.35	99	21.48	1.54	0.46	0.41:	0.71		
NGC4608	12 38 42	+10 25.7	SB(r)0	−2	1738	−.01	11.07	1.43	E	11.24	2.06	0.85	117	21.90	1.52	0.60	0.11:	0.35		153
NGC4612	12 39 00	+07 35.3	(R)SAB0	−2	1745	0.02	11.13	1.30		11.22	2.11	0.66	139	20.90	1.32	0.83	0.11:	0.38		
NGC4618	12 39 08	+41 25.6	SB(rs)m	9	563	−.04	10.85	1.67		10.91	2.09	0.84	30	22.40	1.73	0.90	0.24	0.64		
NGC4621	12 39 31	+11 55.2	E5	−5	320	0.05	9.96	1.50		10.02	2.29	0.92	153	20.73	1.56	0.63	0.12:	0.38	*	
NGC4623	12 39 38	+07 57.0	SB0$^+$?/	−1	1827	0.01	12.45	1.18		12.56	1.92	0.36	175	20.98	1.27:	0.40	0.25:	0.51:		
NGC4631	12 39 41	+32 48.8	SB(s)d/	7	606	0.02	9.15	2.10	Sm	9.24	2.71	0.21	85	20.91	2.13	0.29	0.30	0.65		154
NGC4630	12 39 58	+04 13.9	IB(s)m:	10	533	0.00	12.35	1.38		12.51	1.89	0.69	7	22.19	1.44	0.68	0.08:	0.52		
NGC4632	12 39 59	+00 11.4	SAc	5	1557	0.02	11.80	1.48		11.92	2.01	0.41	61	21.48	1.57	0.38	0.36:	0.62		
NGC4635	12 40 09	+20 13.2	SAB(s)d	7	902	0.10	13.04	1.49		13.21	1.85	0.65	171	22.99	1.58	0.56	0.39	0.63		
NGC4638	12 40 16	+11 42.9	S0$^-$/	−3	962	0.03	11.15	0.97:		11.21	1.93	0.72	112	18.69:	1.01:	0.50	0.16:	0.35:		
NGC4636	12 40 17	+02 57.7	E0−1	−5	821	0.01	9.38	1.94	Sa	9.52	2.52	0.69	147	22.35	2.03	0.74	0.20	0.51		155
NGC4639	12 40 21	+13 31.9	SAB(rs)bc	4	860	0.04	11.61	1.32		11.68	1.97	0.64	124	21.08	1.45	0.53	0.26:	0.53		
NGC4643	12 40 48	+02 15.1	SB(rs)0/a	0	1271	−.01	10.40	1.42	S0	10.51	2.19	0.69	53	20.86	1.48	0.65	0.12:	0.34		
NGC4647	12 41 01	+11 51.2	SAB(rs)c	5	1331	0.03	11.52	1.51		11.58	1.99	0.83	121	21.84	1.57	0.84	0.34	0.66	*	
NGC4649	12 41 09	+11 49.5	E2	−5	1142	0.03	8.95	1.82		9.09	2.52	0.81	105	21.55	1.90	0.80	0.14	0.40	*	
NGC4651	12 41 13	+16 40.1	SA(rs)c	5	723	0.02	10.78	1.49	Sb	10.83	2.23	0.51	75	21.12	1.53	0.67	0.25:	0.49		156

Galaxy (1)	α_c (1950) (2)	δ_c (3)	Type (4)	T (5)	v_0 (6)	A_V (7)	V_T (8)	log R_e (9)	GCT (10)	V_{25} (11)	log a_{25} (12)	ϵ_{25} (13)	θ_{25} (14)	μ_e (15)	log a_e (16)	ϵ_e (17)	S_{13} (18)	S_{23} (19)	Note (20)	Page (21)
NGC4653	12 41 17	−00 17.2	SAB(rs)cd	6	2433	0.02	12.32	1.57		12.50	1.93	0.80	165	22.91	1.61	0.72	0.31	0.58		
NGC4654	12 41 26	+13 24.0	SAB(rs)cd	6	926	0.04	10.65	1.75		10.75	2.22	0.55	122	21.76	1.83	0.43	0.31	0.60		
NGC4656	12 41 32	+32 26.5	SB(s)mp	9	624	0.02	10.99	1.93	Sb	11.11	2.50	0.21	38	23.36	2.06	0.26	0.34	0.69		
NGC4660	12 42 01	+11 27.6	E5–6:	−5	898	−.01	11.06	0.77:		11.09	1.90	0.77	97	17.86:	0.79:	0.75	0.17:	0.38:		
NGC4658	12 42 02	−09 48.7	SB(s)bc	4	2198	0.08	12.69	1.22	Sm	12.78	1.78	0.53	6	21.52	1.30	0.58	0.33:	0.64		157
NGC4665	12 42 33	+03 19.8	SB(s)0/a	0	629	0.04	10.47	1.58		10.56	2.18	0.83	141	21.73	1.70	0.53	0.17:	0.45		
NGC4666	12 42 35	−00 11.2	SABc:	5	1474	0.01	10.64	1.51	Sa	10.72	2.24	0.39	43	20.88	1.61	0.36	0.28:	0.62		
NGC4670	12 42 50	+27 24.0	SB(s)0/ap:	0	1036	0.02	12.70	0.94:	Sb	12.73	1.65	0.81	100	20.10	0.94:	0.75	0.24:	0.47:		
NGC4668	12 42 58	−00 15.7	SB(s)d:	7:	1530	0.00	13.11	1.23	Sm	13.22	1.72	0.68	7	21.54	1.30:	0.52	0.38:	0.62:		
NGC4682	12 44 40	−09 47.4	SAB(rs)c	5	2099	0.08	12.77	1.41	Sm	12.80	1.93	0.46	83	22.15	1.51	0.48	0.43:	0.70		158
NGC4689	12 45 15	+14 02.1	SA(rs)bc	4	1449	0.03	10.99	1.75	Sc	11.11	2.19	0.79	158	22.36	1.80	0.78	0.35	0.60		
NGC4691	12 45 39	−03 03.6	(R)SB(s)0/ap	0	942	0.02	10.90	1.51		11.04	2.09	0.81	35	21.22	1.59	0.38	0.15:	0.36		
NGC4694	12 45 44	+11 15.4	SB0p	−2	1059	−.01	11.62	1.35		11.77	2.03	0.60	143	21.38	1.46	0.44	0.17:	0.41		
NGC4698	12 45 52	+08 45.6	SA(s)ab	2	856	−.03	10.76	1.47	Sa	10.88	2.12	0.80	162	21.35	1.56	0.67	0.17:	0.48		
NGC4697	12 46 01	−05 31.7	E6	−5	1033	0.03	9.27	1.84		9.45	2.46	0.80	51	21.91	1.93	0.65	0.13	0.39		159
NGC4699	12 46 26	−08 23.6	SAB(rs)b	3	1309	0.07	9.47	1.38	E	9.59	2.32	0.85	35	19.79	1.40	0.76	0.08:	0.28		
NGC4700	12 46 32	−11 08.2	SB(s)c?/	5	1193	0.06	12.29	1.44		12.39	2.03	0.24	48	21.23	1.48	0.24	0.32:	0.59		
NGC4701	12 46 39	+03 39.7	SA(s)cd	6	572	0.03	12.20	1.36	Sa	12.39	1.95	0.60	43	21.95	1.36	0.78	0.21:	0.41		
New 3	12 46 47	−09 50.8	SB(s)m	9	1105	0.08	12.43	1.72		12.61	2.07	0.72	96	23.75	1.79	0.71	0.36	0.63	*	
NGC4712	12 47 08	+25 44.5	SA(s)bc	4	4332	0.03	12.95	1.39	Sm	13.05	1.88	0.40	161	22.25	1.49	0.45	0.43:	0.70		160
NGC4710	12 47 09	+15 26.3	SA(r)0$^+$?/	−1?	1019	0.02	10.86	1.50	Sb	10.90	2.23	0.27	27	20.40	1.57	0.31	0.40:	0.64		
NGC4713	12 47 25	+05 35.0	SAB(rs)d	7	512	−.01	11.79	1.46		11.92	1.91	0.68	95	21.65	1.53	0.67	0.38	0.65		
NGC4725	12 48 00	+25 46.5	SAB(r)abp	2	1167	0.02	9.30	1.95		9.40	2.51	0.70	40	21.87	2.12	0.37	0.25	0.51		
NGC4750	12 48 20	+73 08.8	(R)SA(rs)ab	2	1829	0.03	11.39	1.36		11.51	1.95	0.84	158	20.99	1.39	0.75	0.27:	0.48		
NGC4731	12 48 26	−06 07.3	SB(s)cd	6	1303	0.03	11.57	1.85		11.70	2.29	0.40	88	22.98	1.94	0.25	0.23	0.54		161
NGC4736	12 48 32	+41 23.6	(R)SA(r)ab	2	345	−.09	7.93	1.80	E	8.00	2.71	0.83	125	20.56	1.84	0.82	0.12	0.36		
NGC4742	12 49 12	−10 11.0	E4:	−5:	1114	0.07	10.94	0.64:	S0	10.95	1.92	0.71	76	16.47:	0.58:	0.90	0.18:	0.35:		
NGC4747	12 49 19	+26 02.8	SBcd?/p	6:	1154	−.01	12.07	1.72		12.38	2.08	0.47	34	22.71	1.81	0.27	0.27	0.45		
NGC4754	12 49 47	+11 35.1	SB(r)0$^-$:	−3	1346	0.03	10.67	1.27	E	10.73	2.17	0.53	21	20.41	1.30	0.80	0.12:	0.37		
NGC4753	12 49 48	−00 55.7	I0	0	1117	0.02	9.80	1.74		9.90	2.45	0.51	76	21.43	1.80	0.60	0.21	0.48	*	162
NGC4762	12 50 25	+11 30.1	SB(r)0?/	−2	822	0.03	10.33	1.55	E	10.37	2.47	0.22	29	20.89	1.60	0.33	0.19:	0.47		
NGC4760	12 50 31	−10 13.3	E0?	−5?	4451	0.08	11.50	1.54		11.76	2.07	0.74	7	22.71	1.61	0.83	0.13:	0.41		
NGC4763	12 50 35	−16 44.1	Sa:	1:	3961	0.04	12.83	1.09:		12.94	1.70	0.64	114	21.21	1.13:	0.72	0.25:	0.53:		
NGC4771	12 50 48	+01 32.4	SAd?/	7?	1042	0.02	11.90	1.52		12.02	2.15	0.24	132	21.48	1.58	0.31	0.34:	0.58		
NGC4772	12 50 56	+02 26.4	SA(s)a	1	886	0.00	11.11	1.54		11.26	2.20	0.49	146	22.15	1.64	0.62	0.17:	0.49		163
NGC4775	12 51 11	−06 21.2	SA(s)d	7	1375	0.02	11.96	1.46		12.10	1.84	0.92	59	22.02	1.50	0.82	0.33	0.63		
NGC4781	12 51 47	−10 15.9	SB(rs)d	7	689	0.07	11.55	1.51	Sm	11.62	2.04	0.47	113	21.63	1.60	0.54	0.39	0.70		
NGC4786	12 51 57	−06 35.3	E3p	−5	4454	0.03	11.92	1.27		12.16	1.91	0.73	158	21.68	1.32	0.77	0.13:	0.35		
NGC4782	12 51 59	−12 18.1	E0p	−5	3773	0.05	11.52	1.36		11.77	1.97	0.78	157	21.74	1.42	0.76	0.14:	0.35	*	
NGC4783	12 52 00	−12 17.4	E0p	−5	4429	0.05	11.95	1.38		12.21	1.98	0.69	172	22.29	1.36	0.88	0.11:	0.32	*	164

Galaxy (1)	α_c (1950) (2)	δ_c (3)	Type (4)	T (5)	v_0 (6)	A_V (7)	V_T (8)	log R_e (9)	GCT (10)	V_{25} (11)	log a_{25} (12)	ϵ_{25} (13)	θ_{25} (14)	μ_e (15)	log a_e (16)	ϵ_e (17)	S_{13} (18)	S_{23} (19)	Note (20)	Page (21)
NGC4790	12 52 16	−09 58.6	SB(rs)c:?	5	1154	0.09	12.54	1.28	Sm	12.59	1.75	0.65	84	21.38	1.36	0.60	0.44:	0.71		
NGC4793	12 52 16	+29 12.5	SAB(rs)c	5	2460	0.02	11.89	1.28		11.92	1.97	0.54	41	20.74	1.35	0.53	0.35:	0.62		
NGC4800	12 52 20	+46 48.0	SA(rs)b	3	808	−.04	11.80	1.16:		11.84	1.80	0.81	10	20.63	1.23:	0.71	0.26:	0.56:		
New 4	12 52 39	+00 23.2	SB(s)d	7	1160	0.00	12.40	1.68		12.58	2.05	0.48	164	23.10	1.76	0.45	0.40	0.62	*	
IC 3896	12 53 51	−05 04.6	E1	−5	1983	0.76	13.50	1.47	Sc	13.74	1.82	0.59	98	23.53	1.51	0.70	0.31:	0.59		165
NGC4818	12 54 13	−08 15.1	SAB(s)ab	2	848	0.06	11.38	1.52		11.48	2.14	0.37	2	21.66	1.61	0.45	0.29:	0.63		
NGC4826	12 54 17	+21 57.1	(R)SA(rs)ab	2	350	0.11	8.60	1.83		8.66	2.51	0.51	115	20.43	1.92	0.57	0.25	0.58		
NGC4825	12 54 36	−13 23.9	SA0$^-$	−3	4236	0.03	11.56	1.47		11.77	2.08	0.63	128	22.16	1.52	0.69	0.14:	0.40		
NGC4845	12 55 28	+01 50.8	SA(s)ab/	2	904	0.02	11.20	1.60		11.25	2.25	0.26	80	21.13	1.66	0.35	0.42	0.67		
NGC4861	12 56 40	+35 07.9	SB(s)m:	9:	836	−.02*	12.99	1.64		13.17	2.05	0.36	18	23.52	1.72	0.43	0.34	0.64		166
NGC4868	12 56 49	+37 34.7	SAb?	3	4748	−.01	12.34	1.22		12.40	1.69	0.93	111	20.93	1.24	0.75	0.35:	0.61		
NGC4889	12 57 43	+28 14.7	E$^+$4	−4	6482	0.04	11.08	1.75		11.37	2.29	0.52	77	23.52	1.87	0.70	0.14	0.48		
NGC4900	12 58 06	+02 46.1	SB(rs)c	5	823	−.01	11.32	1.48	Sm	11.44	1.90	0.94	67	21.33	1.57	0.74	0.37	0.67		
NGC4891	12 58 15	−13 10.9	SB(r)bc:	4	2418	0.03	12.17	1.44	Sa	12.32	1.93	0.86	152	22.88	1.54	0.65	0.17:	0.49		
NGC4899	12 58 19	−13 40.6	SAB(rs)c:	5	2437	0.05	12.21	1.45		12.35	1.93	0.54	6	22.00	1.50	0.64	0.36:	0.63		167
NGC4902	12 58 22	−14 14.7	SB(r)b	3	2426	0.05	11.15	1.55		11.30	1.97	0.99	10	21.78	1.57	0.88	0.25	0.58		
NGC4914	12 58 23	+37 35.0	SA0$^-$	−3	4796	0.00	11.32	1.25		11.44	2.09	0.56	152	20.85	1.30	0.65	0.13:	0.37		
NGC4904	12 58 25	+00 14.5	SB(s)cd	6	1123	0.00	12.07	1.44		12.20	1.88	0.70	23	22.02	1.46	0.84	0.27:	0.63		
NGC4915	12 58 53	−04 16.6	E0	−5	2971	0.03	11.86	0.95:		11.94	1.86	0.85	43	19.84	0.94:	0.95	0.15:	0.32:		
NGC4928	13 00 26	−07 49.0	SA(s)bcp	4	1434	0.07	12.77	1.14:	Sm	12.88	1.66	0.82	31	21.01	1.20:	0.66	0.38:	0.62:		168
NGC4933	13 01 20	−11 13.8	S0/ap	0	3070	0.08	11.77	1.44		11.98	1.98	0.70	43	21.89	1.51	0.54	0.18:	0.41	*	
NGC4941	13 01 37	−05 17.0	(R)SAB(r)ab:	2:	878	0.01	11.38	1.59		11.49	2.12	0.68	23	22.02	1.67	0.56	0.25	0.57		
NGC4939	13 01 38	−10 04.4	SA(s)bc	4	2903	0.08	11.40	1.64		11.52	2.24	0.40	8	22.82	1.74	0.64	0.25	0.58		
NGC4951	13 02 31	−06 13.7	SAB(rs)cd:	6	993	0.05	12.20	1.40		12.26	2.07	0.38	87	21.78	1.47	0.47	0.28:	0.53		
NGC4958	13 03 12	−07 45.1	SB(r)0?/	−2	1323	0.06	10.57	1.09:		10.60	2.26	0.25	9	18.84:	1.07:	0.65	0.19:	0.43:		169
NGC4961	13 03 24	+28 00.1	SB(s)cd	6	2508	0.04	13.28	1.18		13.37	1.69	0.69	95	22.08	1.23:	0.70	0.29:	0.58:		
NGC4981	13 06 13	−06 30.8	SAB(r)bc	4	1492	0.06	11.66	1.48		11.79	1.96	0.67	152	21.75	1.56	0.58	0.29:	0.58		
NGC4984	13 06 23	−15 15.0	(R)SAB(rs)0$^+$	−1	1042	0.05	10.92	1.36		10.98	2.22	0.57	22	21.08	1.41	0.78	0.12:	0.39		
NGC4999	13 07 00	+01 56.4	SB(r)b	3	2954	0.03	11.99	1.47		12.12	1.89	0.80	127	22.21	1.55	0.76	0.28:	0.64		
NGC4995	13 07 04	−07 34.0	SAB(rs)b	3	1645	0.08	11.47	1.41	Sc	11.54	1.96	0.72	96	21.28	1.45	0.89	0.35:	0.65		170
NGC5005	13 08 37	+37 19.4	SAB(rs)bc	4	1042	−.04	9.94	1.47	Sa	9.98	2.25	0.54	67	20.42	1.58	0.43	0.20:	0.49		
NGC5017	13 10 15	−16 30.1	E2:	−5:	2323	0.08	12.52	0.92:		12.63	1.73	0.83	38	20.46	0.93:	0.91	0.15:	0.36:		
NGC5018	13 10 20	−19 15.2	E3:	−5:	2669	0.17	10.95	1.20		11.06	2.05	0.71	87	20.26	1.25	0.73	0.14:	0.37		
NGC5033	13 11 08	+36 51.8	SA(s)c	5	897	−.01	10.43	1.65	S0	10.49	2.46	0.36	177	21.85	1.78	0.40	0.15:	0.43		
NGC5055	13 13 35	+42 17.8	SA(rs)bc	4	550	−.09	8.62	2.04		8.70	2.64	0.61	100	21.51	2.11	0.52	0.26	0.53		171
NGC5074	13 16 05	+31 44.0	SAbp?	3	5675	−.01	14.14	0.94:	Sa	14.20	1.39	0.95	167	21.29	0.94:	0.74	0.30:	0.55:		
NGC5068	13 16 13	−20 46.6	SAB(rs)cd	6	443	0.23	10.37	1.95		10.47	2.30	0.97	96	22.88	1.99	0.87	0.34	0.65		
NGC5084	13 17 34	−21 33.9	S0/	−2	1507	0.25	10.74	1.57		10.86	2.36	0.38	78	21.95	1.67	0.50	0.13:	0.41		
NGC5087	13 17 43	−20 20.9	SA0:	−3	1604	0.22	10.99	1.34	E	11.17	2.08	0.87	147	20.84	1.36	0.59	0.09:	0.25		
NGC5112	13 19 41	+38 59.8	SB(rs)cd	6	998	−.07	11.61	1.67	Sm	11.73	2.05	0.69	122	22.87	1.77	0.61	0.31	0.69		172

Galaxy (1)	α_c (1950) (2)	δ_c (3)	Type (4)	T (5)	v_0 (6)	A_V (7)	V_T (8)	log R_e (9)	GCT (10)	V_{25} (11)	log a_{25} (12)	ϵ_{25} (13)	θ_{25} (14)	μ_e (15)	log a_e (16)	ϵ_e (17)	S_{13} (18)	S_{23} (19)	Note (20)	Page (21)
NGC5134	13 22 36	−20 52.6	SA(s)b?	3?	1027	0.21	11.51	1.48		11.67	2.04	0.58	147	21.62	1.57	0.51	0.26:	0.55		
NGC5170	13 27 05	−17 42.3	SA(s)c:/	5:	1285	0.16	11.18	1.75	E	11.32	2.43	0.16	127	22.74	1.84	0.32	0.19	0.48		
NGC5204	13 27 44	+58 40.7	SA(s)m	9	329	0.01	11.36	1.65		11.50	2.12	0.62	170	22.26	1.72	0.58	0.31	0.57		
NGC5194	13 27 46	+47 27.3	SA(s)bcp	4	541	−.01	8.51	1.98		8.56	2.55	0.63	10	21.54	2.09	0.79	0.36	0.68	*	
NGC5195	13 27 53	+47 31.8	I0p	0	552	−.01	9.51	1.76		9.64	2.44	0.61	108	21.64	1.80	0.67	0.13	0.39	*	173
NGC5198	13 28 04	+46 55.7	E1–2:	−5:	2605	−.03	11.91	1.17:	Sa	11.99	1.87	0.87	52	20.90	1.19:	0.85	0.18:	0.45:		
NGC5273	13 39 55	+35 54.5	SA(s)0	−2	1070	−.06	11.65	1.38	Sa	11.78	1.90	0.96	30	21.85	1.45	0.88	0.19:	0.54		
NGC5297	13 44 19	+44 07.4	SAB(s)bc:/	4	2654	−.04	11.77	1.50		11.87	2.15	0.25	144	21.44	1.57	0.36	0.40:	0.67		
NGC5301	13 44 22	+46 21.4	SA(s)b:/	3	1709	−.01	11.92	1.61		12.09	2.11	0.28	149	21.96	1.67	0.31	0.35:	0.62		
NGC5313	13 47 37	+40 14.0	Sb?	3?	2588	−.04	12.09	1.20	Sc	12.12	1.81	0.53	45	20.70	1.30:	0.46	0.32:	0.62:		174
NGC5326	13 48 42	+39 49.2	SAa:	1	2603	−.03	11.70	1.11:	E	11.84	2.00	0.52	140	20.47	1.16:	0.55	0.11:	0.32:		
NGC5324	13 49 30	−05 48.4	SA(rs)c:	5	2853	0.05	12.28	1.41	Sm	12.33	1.84	0.89	51	21.94	1.45	0.81	0.38	0.63		
NGC5347	13 51 04	+33 44.2	(R′)SB(s)ab	2	2394	−.02	12.60	1.35		12.78	1.78	0.77	128	22.14	1.46	0.50	0.29:	0.57		
NGC5350	13 51 14	+40 36.7	SB(r)b	3	2305	−.04	11.62	1.47	Sc	11.75	1.92	0.88	46	21.71	1.53	0.77	0.30:	0.62	*	
NGC5351	13 51 19	+38 09.5	SA(r)b:	3:	3663	−.05	12.34	1.43	Sc	12.37	1.93	0.53	97	21.93	1.50	0.59	0.43:	0.74		175
NGC5353	13 51 21	+40 31.5	S0/	−2	2224	−.04	10.83	1.24		10.91	2.09	0.85	91	19.69	1.25:	0.52	0.15:	0.32:	*	
NGC5362	13 52 48	+41 33.7	Sb?p	3?	2321	−.07	12.94	1.26	Sc	13.00	1.86	0.38	84	21.56	1.35:	0.42	0.35:	0.61:		
NGC5371	13 53 33	+40 42.4	SAB(rs)bc	4	2616	−.04	10.82	1.72	Sc	10.88	2.15	0.72	23	21.99	1.78	0.60	0.34	0.67		
NGC5363	13 53 37	+05 30.0	I0?	0?	1027	0.02	9.85	1.64	E	10.01	2.35	0.68	122	21.54	1.70	0.76	0.12:	0.39		
NGC5364	13 53 42	+05 15.6	SA(rs)bcp	4	1140	0.02	10.57	1.88	Sc	10.71	2.31	0.65	38	22.80	1.96	0.69	0.31	0.62		176
NGC5377	13 54 17	+47 28.9	(R)SB(s)a	1	1883	0.01	11.39	1.42	S0	11.50	2.11	0.58	27	21.49	1.53	0.47	0.19:	0.51		
NGC5380	13 54 49	+37 51.1	SA0⁻	−3	3059	−.04	12.45	1.05:		12.56	1.78	0.91	99	21.48	1.09:	0.94	0.11:	0.35:		
NGC5383	13 55 01	+42 05.6	SB(rs)b:p	3	2322	−.05	11.67	1.52		11.82	1.94	0.88	97	22.16	1.65	0.47	0.27:	0.56		
NGC5395	13 56 30	+37 38.2	SA(s)bp	3	3584	−.04	11.88	1.48	Sc	11.95	1.98	0.43	173	21.60	1.59	0.34	0.39:	0.62	*	
NGC5406	13 58 14	+39 09.2	SAB(rs)bc	4	5241	−.01	12.48	1.24		12.56	1.80	0.72	107	21.75	1.34	0.61	0.22:	0.55		177
NGC5422	13 58 57	+55 24.3	S0/	−2	1961	0.00	12.00	1.12:		12.06	2.02	0.29	151	20.66	1.21:	0.51	0.18:	0.47:		
NGC5426	14 00 48	−05 49.8	SA(s)cp	5	2455	0.06	12.37	1.41	Sb	12.47	2.00	0.43	170	22.23	1.51	0.49	0.28:	0.56	*	
NGC5427	14 00 49	−05 47.5	SA(s)cp	5	2565	0.06	11.58	1.55		11.71	1.92	0.92	57	22.05	1.60	0.75	0.36	0.66	*	
NGC5448	14 00 56	+49 24.8	(R)SAB(r)a	1	2121	0.03	11.75	1.47		11.82	2.09	0.44	113	21.43	1.57	0.34	0.33:	0.59		
NGC5444	14 01 13	+35 22.5	SA0⁻:	−3	4008	−.02	11.80	1.25		11.95	1.92	0.82	78	21.45	1.29	0.81	0.14:	0.40		178
NGC5457	14 01 28	+54 35.6	SAB(rs)cd	6	372	−.02	7.99	2.43		8.24	2.75	0.94	94	22.98	2.45	0.87	0.31	0.56		
NGC5473	14 02 59	+55 07.9	SAB(s)0⁻:	−3:	2147	0.00	11.35	1.28		11.52	1.98	0.78	144	21.28	1.32	0.87	0.11:	0.40		
NGC5474	14 03 15	+53 54.0	SA(s)cdp	6	394	−.03	11.17	1.77	Sm	11.32	2.12	0.87	19	23.52	1.87	0.81	0.36	0.68		
NGC5468	14 03 58	−05 12.8	SAB(rs)cd	6	2696	0.06	12.22	1.55	Sm	12.32	1.87	0.94	154	22.60	1.61	0.74	0.39	0.67		
NGC5480	14 04 31	+50 57.7	SA(s)c:	5:	1963	0.04	12.20	1.38		12.35	1.88	0.76	48	21.53	1.44	0.59	0.32:	0.56		179
NGC5485	14 05 27	+55 14.2	SA0p	−2	2110	−.01	11.20	1.60	E	11.45	2.18	0.61	163	22.54	1.63	0.79	0.13:	0.37		
NGC5493	14 08 53	−04 48.5	S0p/	−2	2482	0.06	11.38	0.99:		11.42	1.96	0.72	42	19.07:	0.99:	0.67	0.17:	0.36:		
NGC5496	14 09 04	−00 55.4	Sd:/	7:	1398	0.12	12.76	1.66		12.99	2.14	0.18	173	22.65	1.74	0.21	0.38:	0.65		
NGC5494	14 09 29	−30 24.8	SA(s)c	5	2461	0.21	11.77	1.51		11.87	1.92	0.84	64	21.92	1.55	0.81	0.37	0.68		
NGC5523	14 12 35	+25 33.0	SA(s)cd:	6:	1040	−.03	13.09	1.47	Sm	13.13	2.09	0.22	93	22.73	1.54	0.33	0.36:	0.64		180

Galaxy (1)	α_c (1950) (2)	δ_c (3)	Type (4)	T (5)	v_0 (6)	A_V (7)	V_T (8)	log R_e (9)	GCT (10)	V_{25} (11)	log a_{25} (12)	ϵ_{25} (13)	θ_{25} (14)	μ_e (15)	log a_e (16)	ϵ_e (17)	S_{13} (18)	S_{23} (19)	Note (20)	Page (21)
NGC5533	14 14 00	+35 34.5	SA(rs)ab	2	3903	−.01	11.78	1.43	S0	11.97	2.05	0.60	32	22.09	1.49	0.61	0.14:	0.41		
NGC5534	14 15 01	−07 11.2	(R′)SAB(s)abp:	2	2483	0.09	12.59	1.05:	Sa	12.66	1.70	0.84	142	21.04	1.13:	0.63	0.21:	0.51:		
NGC5548	14 15 43	+25 22.0	(R′)SA(s)0/a	0	4964	0.00	12.73	0.63:	E	12.78	1.68	0.82	101	18.03:	0.53:	0.92	0.13:	0.24:		
NGC5557	14 16 20	+36 43.4	E1	−5	3243	−.05	11.09	1.39		11.28	2.07	0.74	74	21.43	1.42	0.84	0.13:	0.38		
NGC5566	14 17 49	+04 09.7	SB(r)ab	2	1468	0.05	10.58	1.42	E	10.64	2.33	0.35	32	21.17	1.48	0.80	0.13:	0.45		181
NGC5585	14 18 12	+56 57.5	SAB(s)d	7	441	0.00	10.98	1.84		11.17	2.25	0.64	37	23.01	1.93	0.57	0.30	0.57		
NGC5574	14 18 25	+03 28.0	SB:0$^-$?/	−3?	1612	0.03	12.50	1.04:	Sb	12.57	1.73	0.71	39	20.41	1.10:	0.53	0.28:	0.56:		
NGC5576	14 18 33	+03 29.9	E3	−5	1424	0.03	10.71	1.40		10.87	2.17	0.71	75	21.46	1.48	0.68	0.08:	0.33		
NGC5584	14 19 50	−00 09.6	SAB(rs)cd	6	1518	0.11	11.91	1.63	Sm	12.09	2.02	0.75	165	23.06	1.73	0.80	0.37	0.68		
NGC5600	14 21 26	+14 51.8	Scp	5	2297	−.04	12.10	1.18	Sm	12.17	1.71	0.95	93	20.85	1.21:	0.74	0.29:	0.55:		182
NGC5595	14 21 27	−16 30.0	SAB(rs)c	5	2501	0.26	12.21	1.35		12.27	1.85	0.60	55	21.63	1.40	0.65	0.40:	0.71		
NGC5597	14 21 42	−16 32.2	SAB(s)cd	6	2444	0.26	12.35	1.35	Sm	12.40	1.83	0.77	107	22.17	1.56	0.33	0.24:	0.57		
NGC5614	14 22 01	+35 05.1	SA(r)abp	2	3915	−.04	11.65	1.32	S0	11.81	1.94	0.80	109	21.57	1.32	0.92	0.16:	0.42		
NGC5605	14 22 25	−12 56.3	(R′)SAB(rs)c:	5	3196	0.23	12.42	1.39		12.58	1.76	0.85	84	22.10	1.44	0.81	0.38:	0.68		
NGC5641	14 27 05	+29 02.6	(R′)SAB(r)ab	2	4485	−.04	12.50	1.32	S0	12.68	1.89	0.53	154	21.84	1.41	0.39	0.18:	0.43		183
NGC5638	14 27 09	+03 27.3	E1	−5	1592	0.04	11.13	1.50		11.41	2.07	0.78	7	22.12	1.54	0.88	0.12:	0.35		
NGC5653	14 28 01	+31 26.3	(R′)SA(rs)b	3	3587	−.04	12.42	1.04:		12.50	1.73	0.81	128	20.36	1.06:	0.80	0.24:	0.48:		
NGC5660	14 28 04	+49 50.8	SAB(rs)c	5	2433	0.00	12.01	1.47		12.10	1.90	0.91	44	22.24	1.54	0.75	0.35	0.63		
NGC5645	14 28 10	+07 29.8	SB(s)d	7	1375	0.02	12.16	1.43		12.25	1.97	0.55	69	22.00	1.51	0.56	0.31:	0.60		
NGC5665	14 29 58	+08 18.0	SAB(rs)cp?	5	2206	0.03	12.16	1.31	Sm	12.24	1.90	0.57	154	21.40	1.33	0.84	0.36:	0.61		184
NGC5668	14 30 54	+04 40.2	SA(s)d	7	1491	0.04	11.72	1.63		11.87	2.02	0.83	117	22.73	1.64	0.87	0.31	0.58		
NGC5676	14 31 02	+49 40.8	SA(rs)bc	4	2239	0.02	11.35	1.47	Sc	11.39	2.06	0.45	43	21.26	1.55	0.52	0.38:	0.68		
NGC5687	14 33 16	+54 41.6	S0$^-$?	−3?	2253	0.00	11.40	1.26	E	11.58	2.06	0.63	94	21.56	1.40	0.62	0.10:	0.34		
NGC5689	14 33 44	+48 57.6	SB(s)0/a:	0	2274	0.09	11.66	1.12:	S0	11.68	2.11	0.24	82	20.12	1.12:	0.62	0.21:	0.49:		
NGC5690	14 35 09	+02 30.4	Sc?/	5?	1653	0.09	11.73	1.59		11.97	2.05	0.48	145	21.61	1.64	0.27	0.20:	0.39		185
NGC5691	14 35 20	−00 10.9	SAB(s)a:p	1	1768	0.09	12.26	1.35		12.45	1.85	0.77	112	21.79	1.37	0.63	0.21:	0.45	*	
NGC5701	14 36 41	+05 34.8	(R)SB(rs)0/a	0	1424	0.07	10.99	1.62		11.15	2.13	0.99	66	22.46	1.67	0.69	0.14:	0.39		
NGC5713	14 37 38	−00 04.5	SAB(rs)bcp	4	1777	0.08	11.37	1.42	Sc	11.41	1.96	0.90	11	21.07	1.43	0.66	0.25:	0.52		
NGC5728	14 39 37	−17 02.3	SAB(r)a:	1:	2800	0.26	11.39	1.46		11.48	2.08	0.74	5	21.70	1.59	0.42	0.19:	0.51		
NGC5739	14 40 34	+42 03.3	SAB(r)0$^+$:	−1	5692	0.02	12.14	1.25	E	12.31	1.94	0.72	19	22.09	1.34	0.77	0.11:	0.41		186
NGC5740	14 41 52	+01 53.4	SAB(rs)b	3	1490	0.11	12.01	1.30		12.07	1.96	0.49	159	21.28	1.39	0.55	0.30:	0.59		
NGC5746	14 42 23	+02 09.9	SAB(rs)b?/	3	1638	0.09	10.64	1.55	Sa	10.67	2.38	0.18	170	21.47	1.62	0.49	0.25:	0.55		
NGC5750	14 43 37	−00 00.8	SB(r)0/a	0	1922	0.08	11.64	1.39	Sb	11.71	2.07	0.45	61	21.43	1.41	0.85	0.29:	0.64		
NGC5756	14 44 48	−14 38.4	SAB(s)b?/	3	2025	0.26	12.30	1.34	Sa	12.39	1.99	0.39	60	21.28	1.42	0.37	0.31:	0.57		
NGC5757	14 44 57	−18 52.1	(R)SB(r)b	3	2598	0.26	12.12	1.24		12.20	1.83	0.78	35	21.38	1.39	0.39	0.20:	0.48		187
NGC5768	14 49 33	−02 19.7	SA(rs)c:	5:	1960	0.16	12.73	1.30		12.81	1.79	0.84	31	21.71	1.33	0.72	0.32:	0.58	*	
NGC5775	14 51 27	+03 44.8	SBc?/	5:	1502	0.10	11.28	1.63	Sa	11.52	2.11	0.41	146	21.66	1.74	0.26	0.25:	0.51		
NGC5792	14 55 48	−00 53.4	SB(rs)b	3	1889	0.12	11.43	1.55		11.42	2.35	0.24	82	21.99	1.60	0.68	0.33	0.65		
NGC5796	14 56 36	−16 25.4	E0–1	−5	2790	0.36	11.31	1.24		11.49	2.02	0.75	60	21.42	1.30	0.87	0.09:	0.32		
NGC5820	14 57 11	+54 05.1	S0/	−2	3410	0.02	12.09	1.02:		12.18	1.91	0.55	103	20.01	1.08:	0.60	0.22:	0.43:	*	188

Galaxy (1)	α_c (1950) (2)	δ_c (3)	Type (4)	T (5)	v_0 (6)	A_V (7)	V_T (8)	log R_e (9)	GCT (10)	V_{25} (11)	log a_{25} (12)	ϵ_{25} (13)	θ_{25} (14)	μ_e (15)	log a_e (16)	ϵ_e (17)	S_{13} (18)	S_{23} (19)	Note (20)	Page (21)
NGC5806	14 57 28	+02 05.4	SAB(s)b	3	1270	0.11	11.32	1.61	Sa	11.57	2.12	0.57	167	21.87	1.66	0.45	0.18	0.37		
NGC5812	14 58 17	−07 15.5	E0	−5	1946	0.23	11.00	1.32		11.12	2.17	0.72	35	21.48	1.36	0.94	0.09:	0.32		
NGC5813	14 58 39	+01 53.9	E1–2	−5	1800	0.11	10.55	1.65		10.76	2.29	0.59	127	22.25	1.76	0.72	0.16	0.44		
NGC5831	15 01 34	+01 24.9	E3	−5	1602	0.10	11.35	1.43		11.53	2.07	0.85	132	22.06	1.46	0.93	0.12:	0.38		
NGC5838	15 02 54	+02 17.6	SA0$^-$	−3	1349	0.09	10.97	1.15:	E	10.99	2.14	0.38	40	20.16	1.20:	0.71	0.15:	0.42:		189
NGC5846	15 03 56	+01 47.8	E0–1	−5	1674	0.11	9.80	1.97		10.17	2.39	0.83	78	23.30	2.02	0.95	0.13	0.39		
NGC5850	15 04 35	+01 44.2	SB(r)b	3	2430	0.11	11.05	1.73		11.17	2.23	0.70	158	22.67	1.85	0.43	0.19	0.46		
NGC5866	15 05 07	+55 57.3	SA0$^+$/	−1	824	0.00	10.03	1.55	Sa	10.14	2.30	0.54	124	20.46	1.62	0.45	0.22:	0.45		
NGC5854	15 05 16	+02 45.6	SB(s)0$^+$/	−1	1654	0.08	11.90	1.13:		11.98	1.97	0.34	55	20.30	1.15:	0.70	0.25:	0.52:		
NGC5861	15 06 33	−11 07.9	SAB(rs)c	5	1725	0.31	11.63	1.59	Sm	11.80	1.98	0.76	143	22.07	1.68	0.56	0.41	0.68		190
NGC5864	15 07 02	+03 14.6	SB(s)0?/	−2	1548	0.08	11.77	1.31		11.93	2.07	0.31	71	20.75	1.36	0.43	0.27:	0.52		
NGC5879	15 08 29	+57 11.4	SA(rs)bc:?	4:	929	0.02	11.70	1.27		11.71	2.12	0.36	4	20.78	1.36:	0.42	0.22:	0.45:		
NGC5878	15 11 00	−14 05.1	SA(s)b	3	1974	0.37	11.67	1.41	E	11.79	2.11	0.35	5	21.74	1.54	0.37	0.21:	0.59		
F 703	15 11 00	−15 16.7	SA(s)d	7	2128	0.34	12.46	1.66		12.60	1.94	0.97	163	23.43	1.66	0.93	0.38	0.67	*	
NGC5899	15 13 14	+42 14.0	SAB(rs)c	5	2657	0.06	11.78	1.38	Sb	11.86	2.02	0.43	15	21.26	1.47	0.49	0.35:	0.65		191
NGC5905	15 14 02	+55 42.1	SB(r)b	3	3544	0.02	11.93	1.55	Sa	12.15	2.07	0.55	130	22.90	1.59	0.83	0.20:	0.48		
NGC5907	15 14 37	+56 30.4	SA(s)c:/	5:	779	0.01	10.47	1.91	Sb	10.54	2.55	0.14	153	21.94	1.97	0.23	0.33	0.64		
NGC5898	15 15 17	−23 55.0	E0	−5	2098	0.38	11.36	1.36		11.56	2.04	0.79	124	21.88	1.39	0.94	0.10:	0.35		
NGC5908	15 15 24	+55 35.4	SA(s)b:/	3	3467	0.01	11.96	1.35		12.02	1.93	0.56	157	21.37	1.42	0.45	0.24:	0.53		
NGC5903	15 15 40	−23 53.1	E2	−5	2378	0.38	11.19	1.62		11.50	2.11	0.75	165	22.82	1.69	0.77	0.12:	0.38		192
NGC5915	15 18 48	−12 54.9	SB(s)abp	2	2146	0.41	12.40	1.05:		12.48	1.66	0.81	171	20.30	1.11:	0.53	0.25:	0.50:		
NGC5921	15 19 28	+05 14.9	SB(r)bc	4	1428	0.05	11.32	1.59	Sc	11.38	2.13	0.76	147	22.29	1.68	0.65	0.23	0.54		
NGC5949	15 27 19	+64 56.0	SA(r)bc?	4	624	0.05	12.07	1.37	Sm	12.17	1.87	0.53	140	21.11	1.45	0.51	0.41:	0.69		
NGC5936	15 27 39	+13 09.6	SB(rs)b	3	3995	0.07	12.36	1.31	Sc	12.49	1.78	0.75	158	21.24	1.34	0.63	0.34:	0.56		
NGC5970	15 36 08	+12 21.0	SB(r)c	5	2047	0.05	11.61	1.37		11.63	1.94	0.66	91	21.09	1.42	0.57	0.26:	0.55		193
NGC5982	15 37 38	+59 31.1	E3	−5	3046	0.02	10.98	1.41		11.15	2.13	0.81	116	21.71	1.50	0.69	0.09:	0.34		
NGC5985	15 38 36	+59 29.6	SAB(r)b	3	2694	−.06*	11.29	1.67		11.29	2.20	0.47	19	21.94	1.76	0.48	0.39	0.68		
NGC6015	15 50 39	+62 27.5	SA(s)cd	6	1018	0.02	11.18	1.66	Sm	11.28	2.23	0.40	29	21.90	1.73	0.57	0.39	0.67		
NGC6052	16 03 01	+20 40.5	Sc	5	4777	0.08	13.22	0.93:		13.27	1.48	0.75	6	20.59	0.98:	0.75	0.31:	0.63:		
NGC6070	16 07 26	+00 50.4	SA(s)cd	6	1979	0.27	11.79	1.56		11.87	2.06	0.45	59	22.01	1.62	0.59	0.38	0.65		194
NGC6118	16 19 13	−02 10.1	SA(s)cd	6	1535	0.35	11.69	1.66		11.74	2.17	0.40	53	22.17	1.80	0.32	0.43	0.64		
NGC6181	16 30 10	+19 55.9	SAB(rs)c	5	2438	0.17	11.85	1.21		11.89	1.88	0.46	177	20.37	1.29:	0.47	0.31:	0.55:		
NGC6217	16 35 03	+78 18.0	(R)SB(rs)bc	4	1598	0.11	11.43	1.49		11.59	1.94	0.92	119	21.52	1.60	0.43	0.29:	0.54		
NGC6207	16 41 18	+36 55.7	SA(s)c	5	984	0.04	11.61	1.31		11.73	1.96	0.52	17	20.97	1.41	0.43	0.24:	0.54		
NGC6340	17 11 16	+72 21.9	SA(s)0/a	0	1469	0.15	11.07	1.57		11.27	2.11	0.86	44	22.23	1.60	0.98	0.17:	0.46		195
NGC6384	17 29 59	+07 05.8	SAB(r)bc	4	1735	0.31	10.76	1.78		10.84	2.30	0.62	40	22.73	1.87	0.63	0.23	0.58		
NGC6482	17 49 44	+23 05.0	E3:	−5:	4061	0.32	12.34	1.06:	S0	12.44	1.81	0.80	71	20.66	1.10:	0.71	0.19:	0.41:		
NGC6503	17 49 58	+70 09.5	SA(s)cd	6	303	0.11	10.26	1.63	Sm	10.29	2.37	0.30	118	20.45	1.69	0.36	0.38	0.63		
NGC6643	18 21 14	+74 32.7	SA(rs)c	5	1743	0.17	11.12	1.57		11.18	2.08	0.55	34	21.43	1.62	0.64	0.42	0.71		
NGC6814	19 39 55	−10 26.6	SAB(rs)bc	4	1643	0.45	11.35	1.53	Sc	11.39	2.04	0.90	29	21.89	1.56	0.98	0.30	0.60		196

Galaxy (1)	α_c (1950) (2)	δ_c (3)	Type (4)	T (5)	v_0 (6)	A_V (7)	V_T (8)	log R_e (9)	GCT (10)	V_{25} (11)	log a_{25} (12)	ϵ_{25} (13)	θ_{25} (14)	μ_e (15)	log a_e (16)	ϵ_e (17)	S_{13} (18)	S_{23} (19)	Note (20)	Page (21)
NGC6822	19 42 07	−14 55.7	IB(s)m	10	15	0.65	8.81	2.48		9.08	2.69	0.95	102	23.82	2.52	0.72	0.29	0.62		
NGC6907	20 22 07	−24 58.3	SB(s)bc	4	3192	0.23	11.21	1.60	Sc	11.28	2.06	0.74	66	21.99	1.69	0.45	0.30	0.58		
NGC6925	20 31 14	−32 09.2	SA(s)bc	4	2780	0.18	11.34	1.44		11.39	2.14	0.33	4	21.17	1.51	0.50	0.31:	0.65		
NGC6946	20 33 48	+59 59.0	SAB(rs)cd	6	336	1.22	8.68	2.30	Sm	8.97	2.58	0.94	39	22.76	2.36	0.60	0.30	0.52		
NGC6951	20 36 37	+65 55.9	SAB(rs)bc	4	1710	0.66	11.31	1.63		11.42	2.10	0.73	155	22.30	1.71	0.57	0.24	0.54		197
NGC7137	21 45 54	+21 55.6	SAB(rs)c	5	1951	0.23	12.40	1.34	Sm	12.54	1.76	0.96	23	21.42	1.33	0.96	0.41:	0.61		
NGC7177	21 58 18	+17 29.9	SAB(r)b	3	1419	0.17	11.34	1.18		11.39	1.93	0.65	86	20.26	1.22:	0.75	0.23:	0.53:		
NGC7171	21 58 20	−13 30.6	SB(rs)b	3	2758	0.06	12.37	1.44	Sc	12.49	1.88	0.53	120	21.96	1.50	0.55	0.38:	0.70		
NGC7217	22 05 36	+31 07.0	(R)SA(r)ab	2	1234	0.32	9.85	1.79	Sa	10.03	2.33	0.88	90	21.89	1.80	0.89	0.18	0.44		
NGC7218	22 07 29	−16 54.6	SB(rs)cd	6	1781	0.06	12.06	1.38		12.13	1.96	0.42	18	21.32	1.43	0.60	0.40:	0.68		198
NGC7300	22 28 21	−14 15.7	SAB(rs)b:	3	5021	0.12	12.86	1.27	Sc	12.96	1.83	0.41	165	21.64	1.33	0.50	0.33:	0.59		
NGC7309	22 31 42	−10 37.0	SAB(rs)c	5	4082	0.14	12.40	1.40	Sm	12.53	1.81	0.81	7	22.10	1.46	0.66	0.31:	0.58		
NGC7331	22 34 47	+34 09.5	SA(s)bc	4	1114	0.25	9.49	1.71	Sa	9.57	2.51	0.39	169	21.26	1.80	0.46	0.14:	0.42		
NGC7332	22 35 01	+23 32.3	S0p/	−2	1483	0.08	11.14	1.05:		11.22	2.06	0.33	154	19.58	1.15:	0.55	0.20:	0.48:		
NGC7361	22 39 31	−30 19.2	S(r)c:?	5:	1276	−.08	12.39	1.62		12.58	2.10	0.25	5	22.17	1.67	0.25	0.36	0.54		199
NGC7371	22 43 25	−11 16.0	(R)SA(r)0/a:	0:	2826	0.13	11.92	1.47	Sb	12.08	1.90	0.90	109	22.04	1.47	0.98	0.29	0.57		
NGC7377	22 45 05	−22 34.6	SA(s)0^+	−1	3501	0.03	10.99	1.65		11.20	2.14	0.82	128	22.45	1.70	0.78	0.17	0.46		
NGC7448	22 57 35	+15 42.8	SA(rs)bc	4	2485	0.11	11.66	1.31	Sc	11.76	1.87	0.56	175	20.80	1.37	0.54	0.33:	0.61		
NGC7469	23 00 44	+08 36.3	(R′)SAB(rs)a	1	5115	0.09	11.99	0.64:	Sm	11.93	1.76	0.76	125	16.93:	0.50:	0.81	0.21:	0.38:		
NGC7479	23 02 26	+12 03.1	SB(s)c	5	2630	0.12	11.02	1.60	Sm	11.11	2.08	0.77	30	22.28	1.77	0.42	0.26	0.56		200
NGC7541	23 12 11	+04 15.6	SB(rs)bc:p	4	2887	0.16	11.83	1.47	Sm	11.87	2.03	0.33	100	20.80	1.53	0.24	0.44:	0.64		
NGC7619	23 17 43	+07 56.0	E2	−5	4020	0.12	10.78	1.72	Sa	10.96	2.24	0.81	45	22.62	1.76	0.84	0.18	0.46		
NGC7626	23 18 10	+07 56.6	E1p:	−5	3625	0.12	10.86	1.69	Sa	11.09	2.14	0.91	61	22.40	1.71	0.83	0.19	0.44		
NGC7640	23 19 43	+40 34.2	SB(s)c	5	669	0.32	11.14	1.87	Sb	11.21	2.48	0.19	166	22.43	1.94	0.26	0.38	0.69		
NGC7679	23 26 14	+03 14.2	SB0p:	−2	5397	0.18	12.62	0.71:		12.69	1.68	0.76	103	18.92:	0.68:	0.94	0.19:	0.35:		201
NGC7716	23 33 58	+00 01.3	SAB(r)b:	3:	2735	0.08	12.12	1.30	Sa	12.29	1.87	0.81	24	21.95	1.39	0.77	0.18:	0.49		
NGC7741	23 41 23	+25 47.9	SB(s)cd	6	1030	0.11	11.34	1.74	Sm	11.45	2.14	0.65	165	22.55	1.78	0.76	0.28	0.66		
NGC7743	23 41 49	+09 39.3	(R)SB(s)0^+	−1	2027	0.14	11.52	1.33	Sa	11.58	1.98	0.96	97	21.83	1.52	0.57	0.16:	0.51	*	
NGC7769	23 48 31	+19 52.3	(R)SA(rs)b	3	4606	0.08	12.17	1.23		12.25	1.88	0.59	114	21.22	1.25	0.90	0.25:	0.55		
NGC7782	23 51 21	+07 41.5	SA(s)b	3	5584	0.14	12.44	1.35		12.55	1.88	0.55	167	21.79	1.43	0.57	0.35:	0.65		202
NGC7785	23 52 46	+05 38.3	E5–6	−5	4038	0.13	11.76	1.18		11.91	1.86	0.77	134	20.69	1.24:	0.56	0.16:	0.43:	*	203

Note:

NGC 205: envelope of NGC 224 cleaned out
NGC 221: envelope of NGC 224 cleaned out
NGC 224: NGC 205 and NGC 221 cleaned out; non-standard frame size
NGC 274 + 275: decomposed
New 1 = A0102 − 06
NGC 404: halo of a bright star cleaned
NGC 750: NGC 751 (V_T = 12.82, logR_e = 1.32) decomposed
NGC1300: non-standard frame size
NGC1359: low signal-to-noise ratio
HA85-1 = A0509 − 14
NGC2207: overlapping galaxy (IC2163) cleaned out
NGC3038: low signal-to-noise ratio
NGC3056: low signal-to-noise ratio
NGC3089: low signal-to-noise ratio
NGC3145: envelope of a bright star cleaned out
NGC3226 + 3227: decomposed
NGC3309: NGC3311 (V_T = 10.71, log R_e = 1.92) decomposed
NGC3312: overlapping galaxy (NGC3314) cleaned out
NGC3379: overlapping galaxy (NGC3384) cleaned out
NGC3384: overlapping galaxies (NGC3379 and 3389) cleaned out
NGC3395 + 3396: decomposed
NGC3605: envelope of NGC 3607 was cleaned out
NGC3607: overlapping galaxies (NGC3605 and NGC3608) cleaned out
NGC3608: overlapping galaxy (NGC3607) cleaned out
NGC3640: overlapping galaxy (NGC3641) cleaned out
NGC3691: low signal-to-noise ratio
NGC3690: IC694 and an anonymous galaxy at 1′ north-west included
NGC4038: NGC4039 (V_T = 11.35, log R_e = 1.62) decomposed
NGC4045: overlapping galaxy (NGC4045A) cleaned out
NGC4105 + 4106: decomposed
NGC4162: low signal-to-noise ratio
NGC4278 + 4283: decomposed
NGC4298 + 4302: decomposed
NGC4342 = IC3256
NGC4406: non-standard frame size
NGC4435 + 4438: decomposed
NGC4449: non-standard frame size
NGC4496 = NGC4496A: overlapping galaxy (NGC4496B) cleaned out
R 80 = NGC4517A
NGC4567 + 4568: decomposed
NGC4621: non-standard frame size
NGC4647 + 4649: decomposed
New 3 = A1246 − 09
NGC4753: non-standard frame size
NGC4782 + 4783: decomposed
New 4 = A1252 + 00
NGC4933 = NGC4933A: overlapping galaxy (NGC4933B) cleaned out
NGC5194 + 5195: decomposed
NGC5350: diffraction image of a bright star cleaned out
NGC5353: overlapping galaxy (NGC5354) cleaned out
NGC5395: overlapping galaxy (NGC5394) cleaned out
NGC5426 + 5427: decomposed
NGC5691: low signal-to-noise ratio
NGC5768: low signal-to-noise ratio
NGC5820: low signal-to-noise ratio
F 703 = A1511 − 15
NGC7743: low signal-to-noise ratio
NGC7785: low signal-to-noise ratio

Part III The Atlas

NGC7814

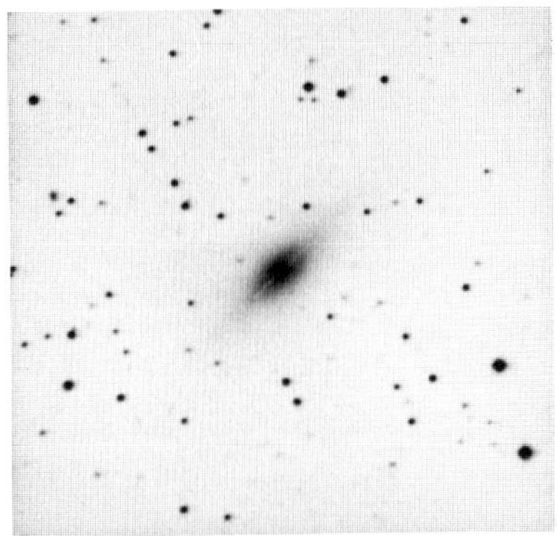

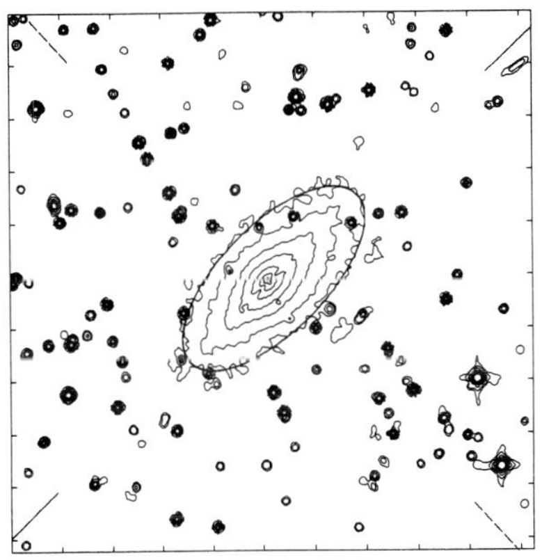

N7814 K2488 2 21.12 512X512

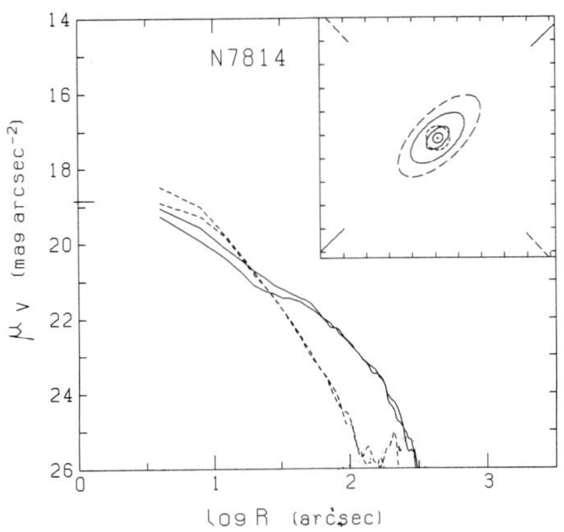

NGC 16

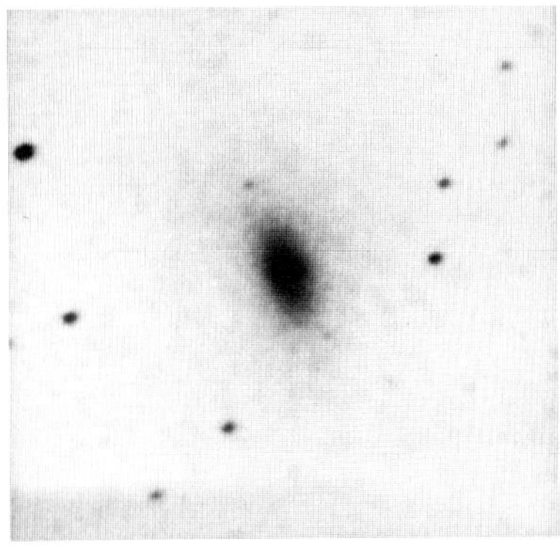

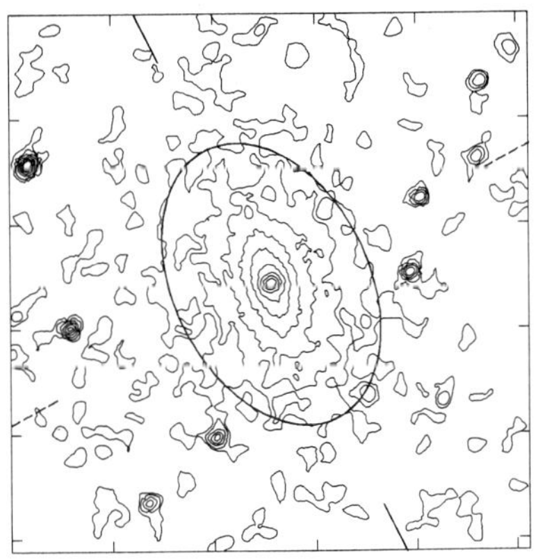

N0016 K4779 1 21.76 256X256

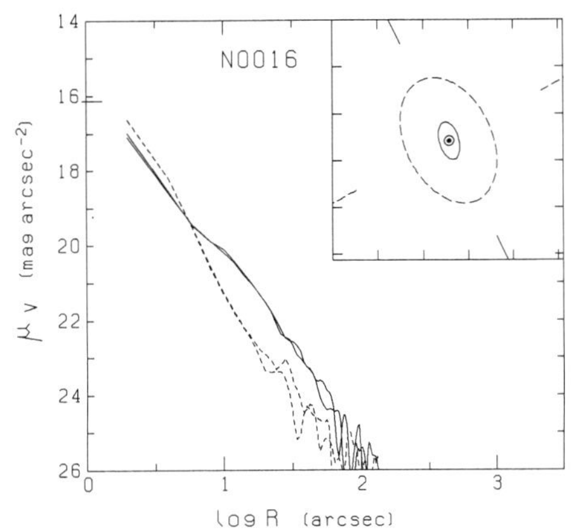

NGC 23

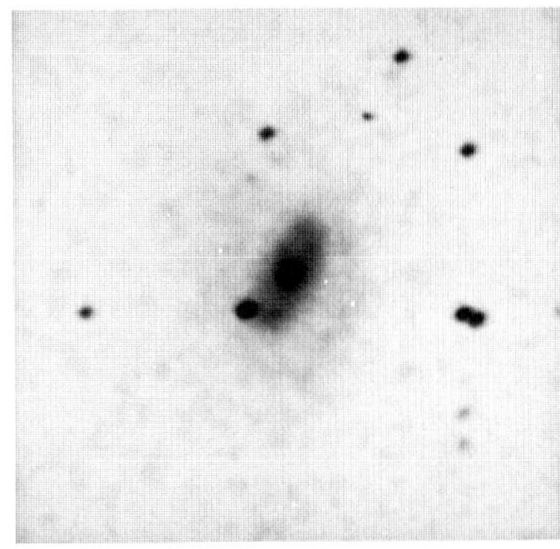

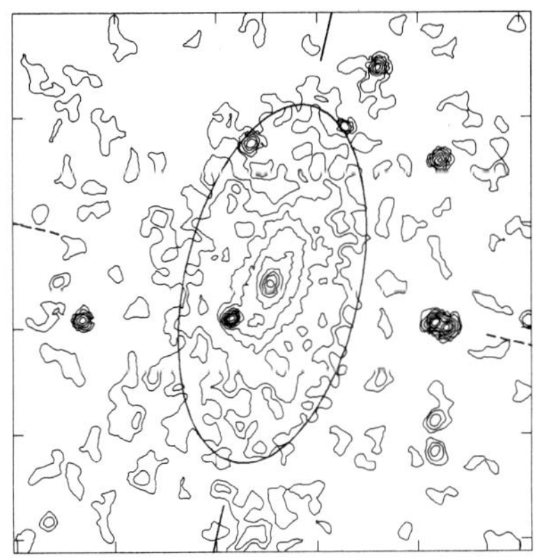

N0023 K4779 1 21.76 256X256

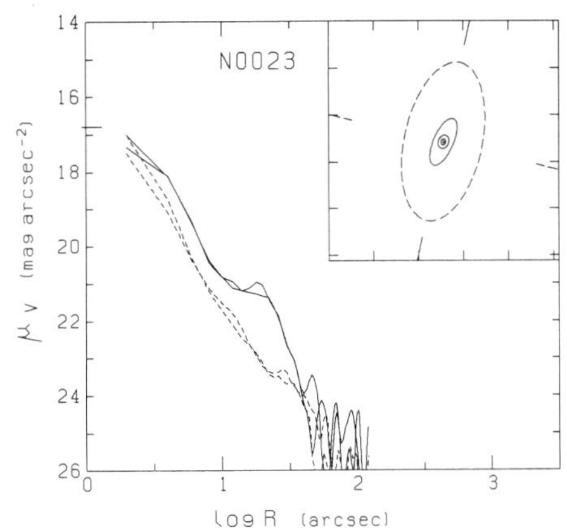

NGC 147

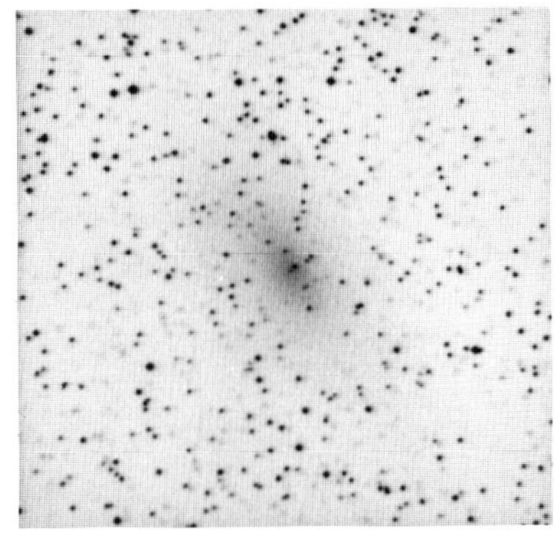

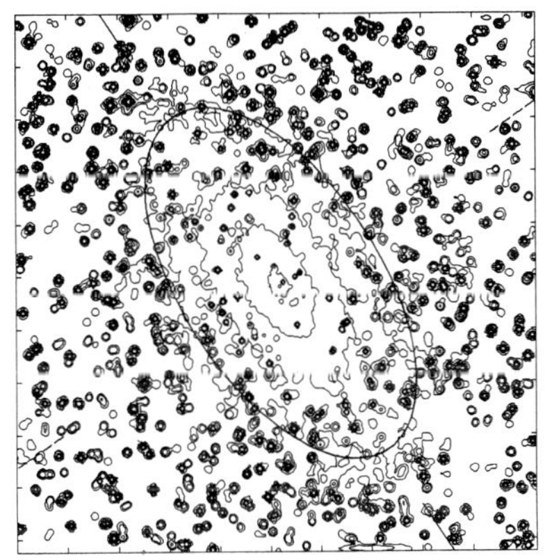

N0147 K2913 3 21.12 512X512

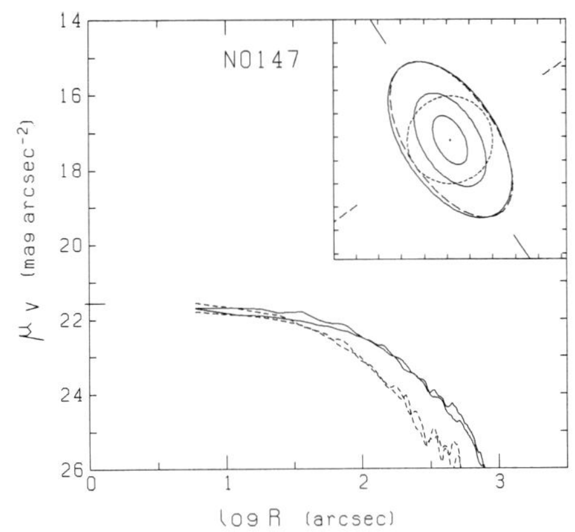

NGC 151

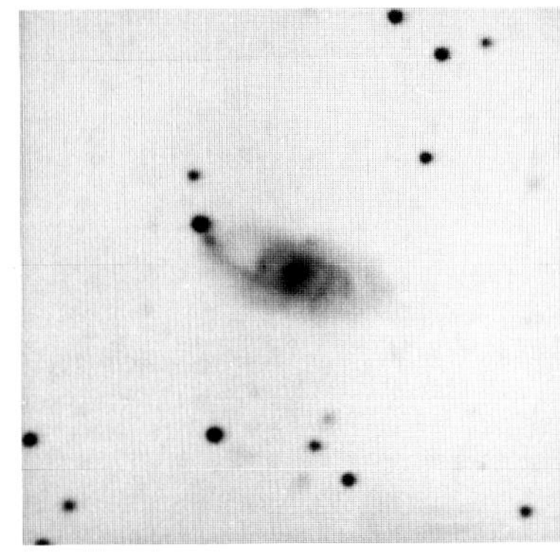

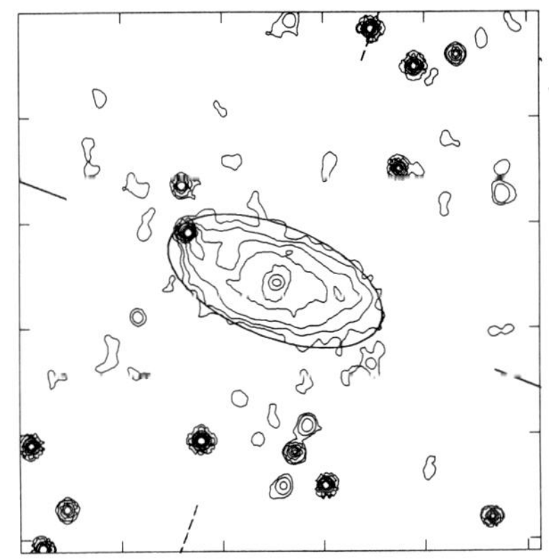

N0151 K4518 2 20.96 256X256

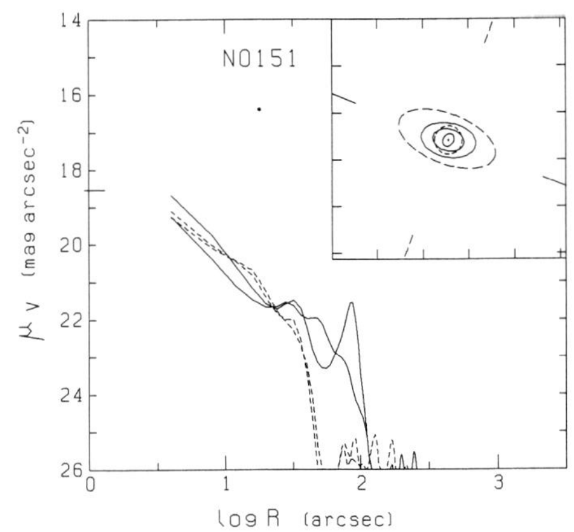

NGC 148

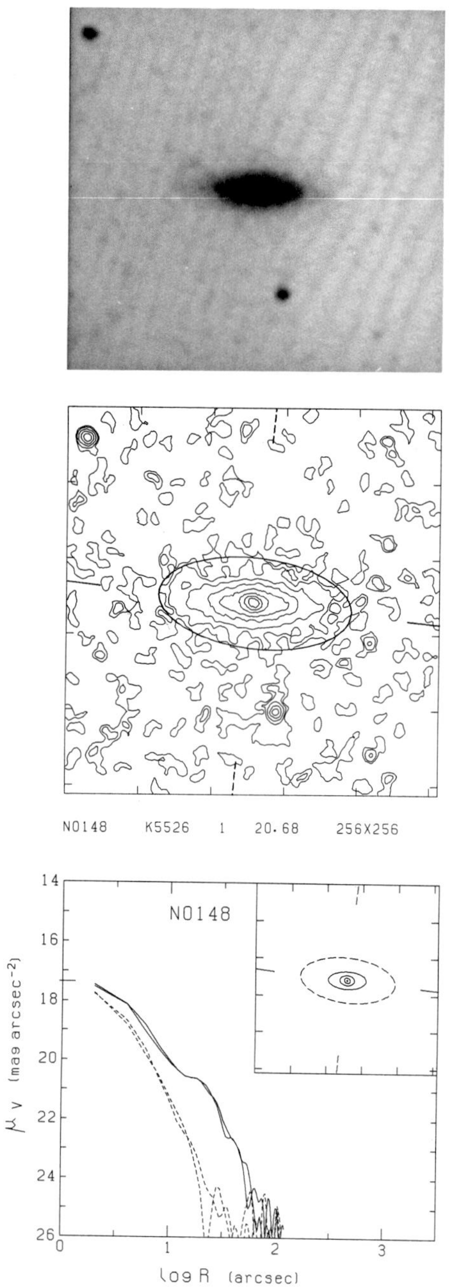

NGC 150

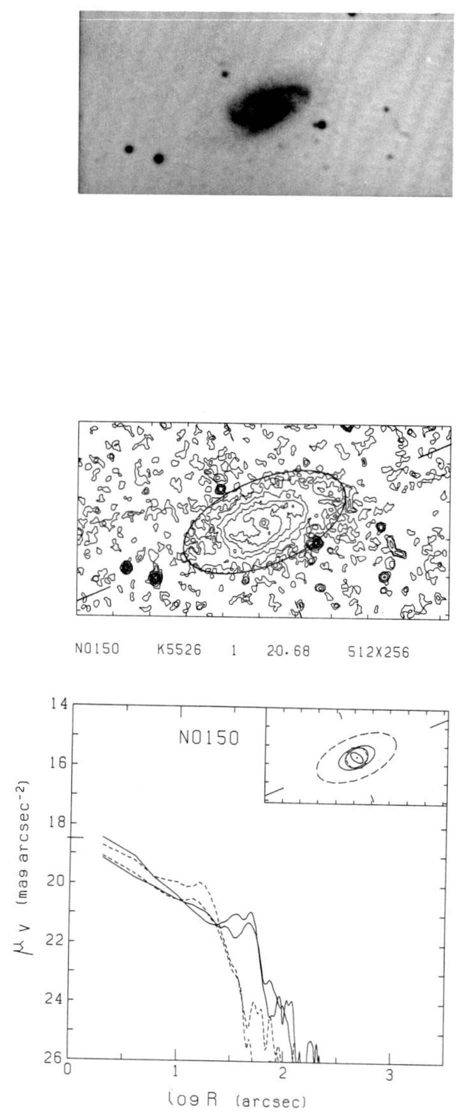

NGC 157

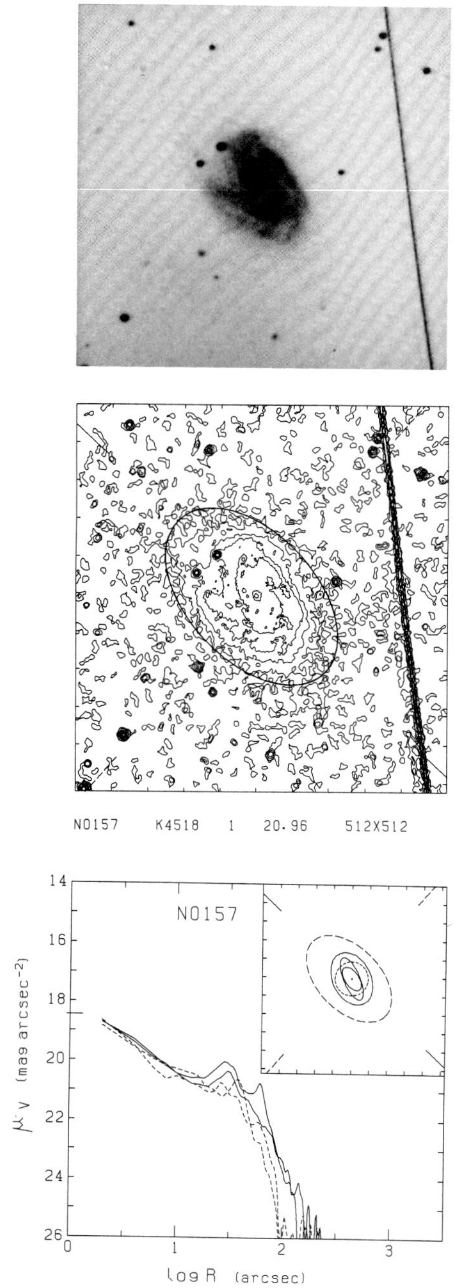

NGC 185

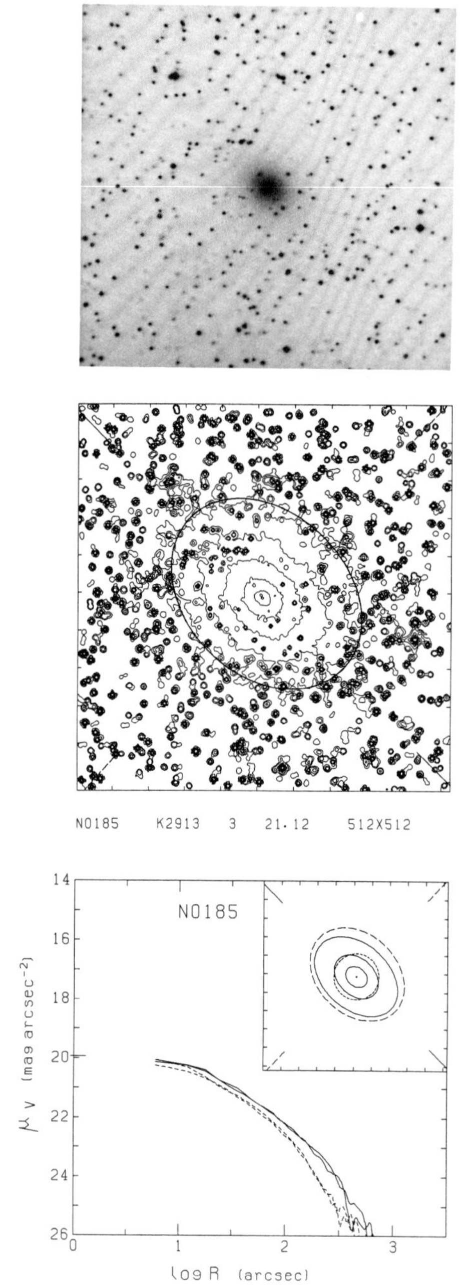

NGC 205

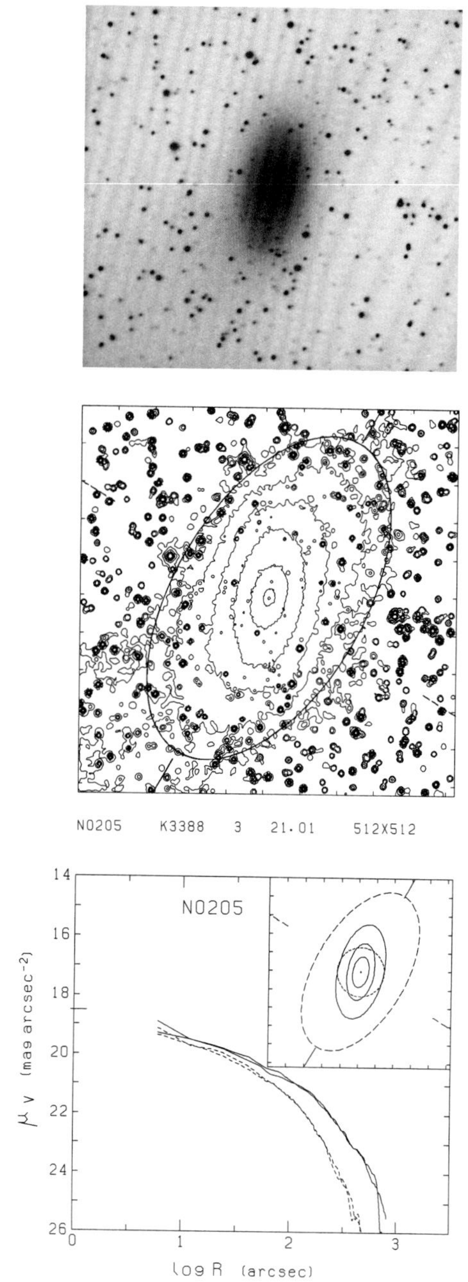

NGC 221

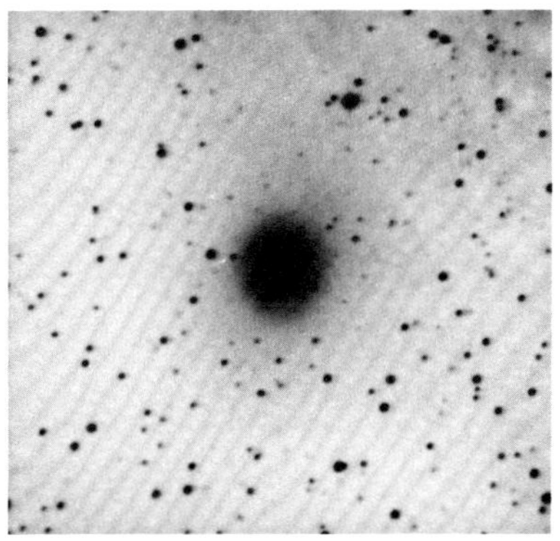

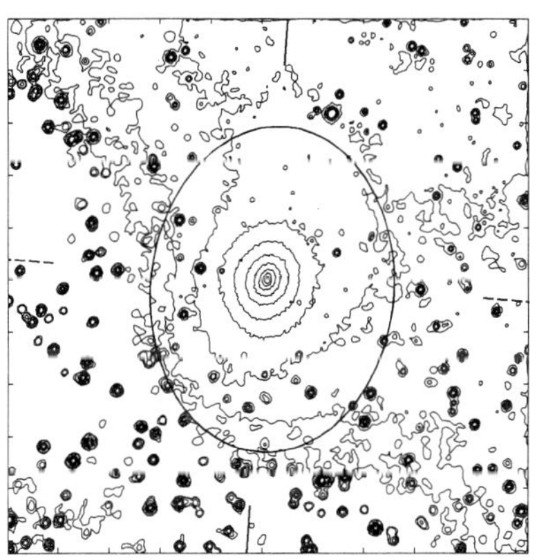

N0221 K3388 2 21.01 512X512

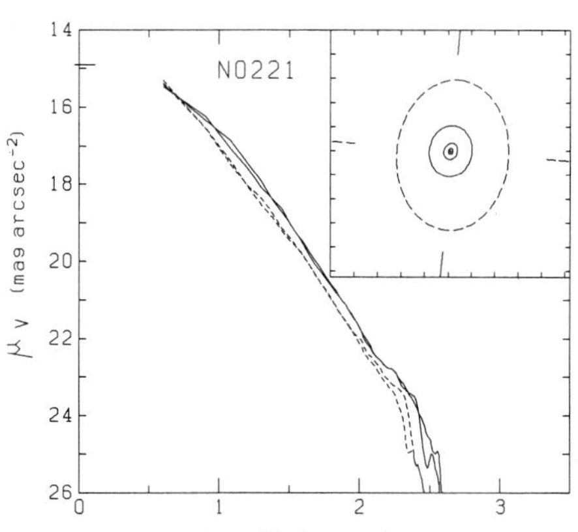

NGC 224

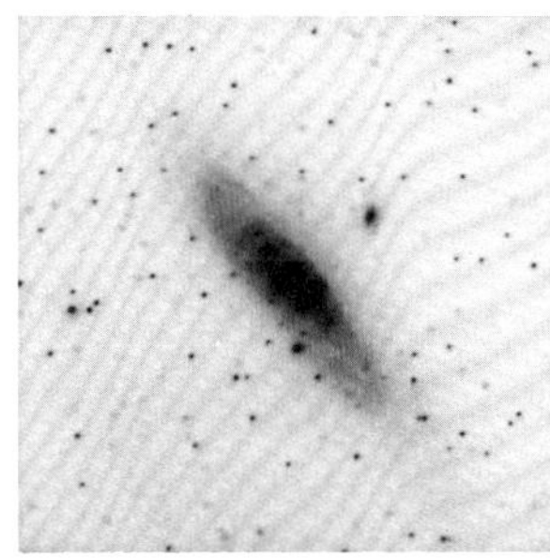

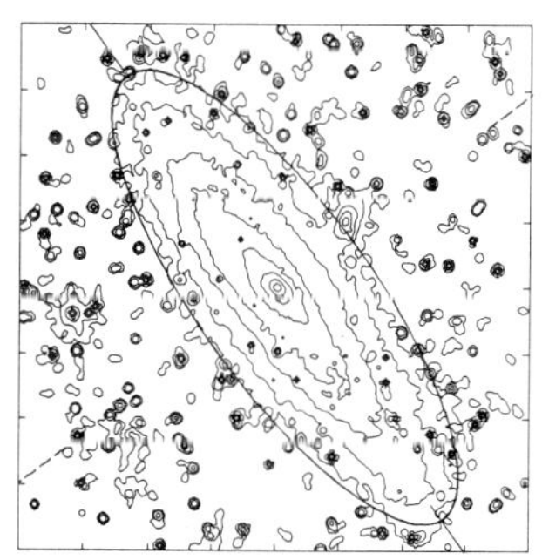

N0224 K3388 30 21.01 400X400

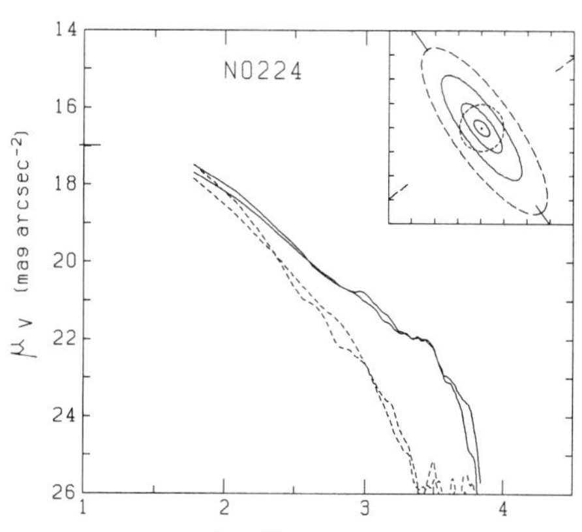

NGC 227

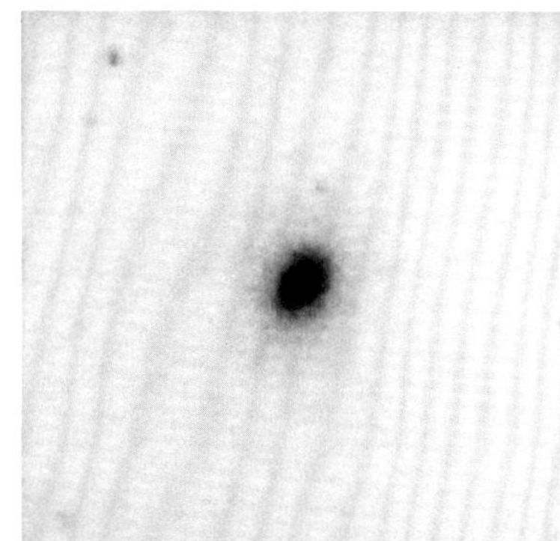

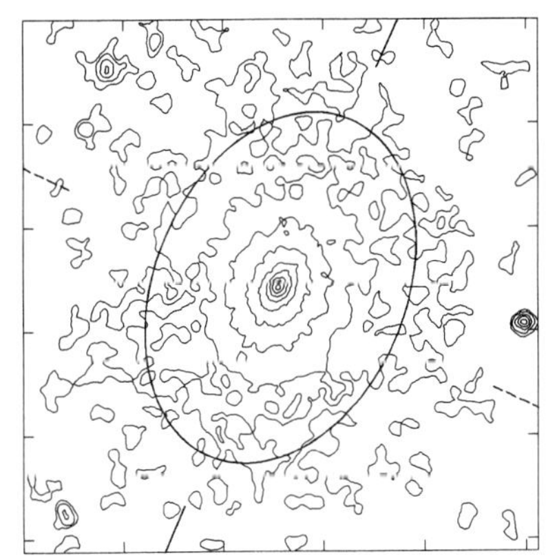

N0227 K4487 1 21.48 256X256

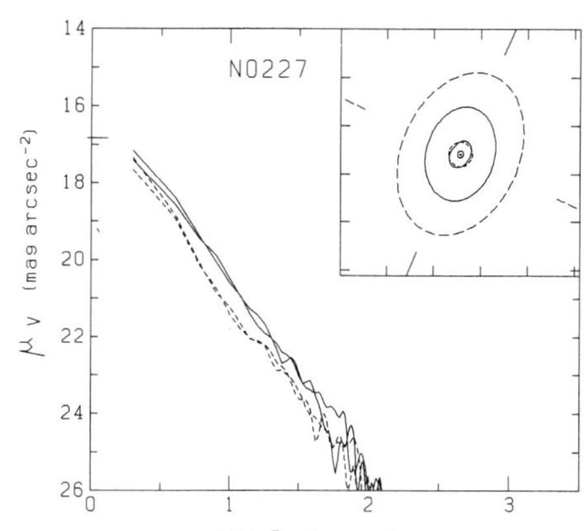

NGC 237

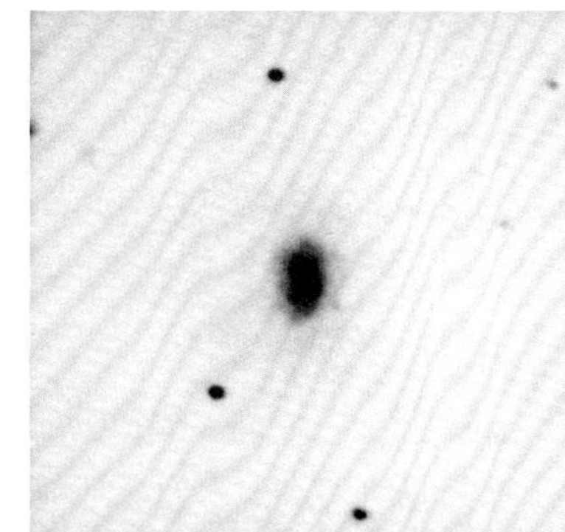

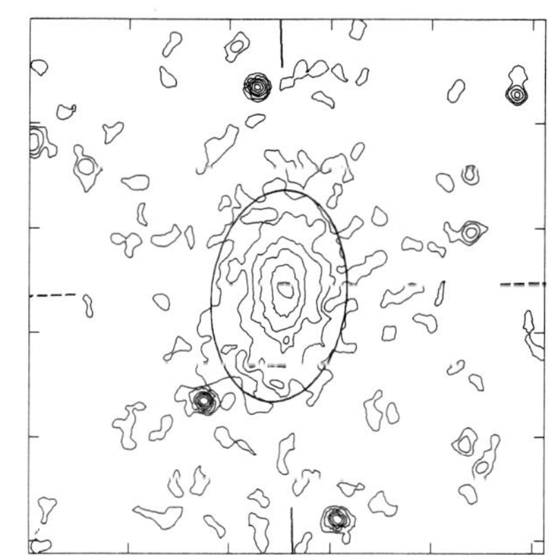

N0237 K4487 1 21.48 256X256

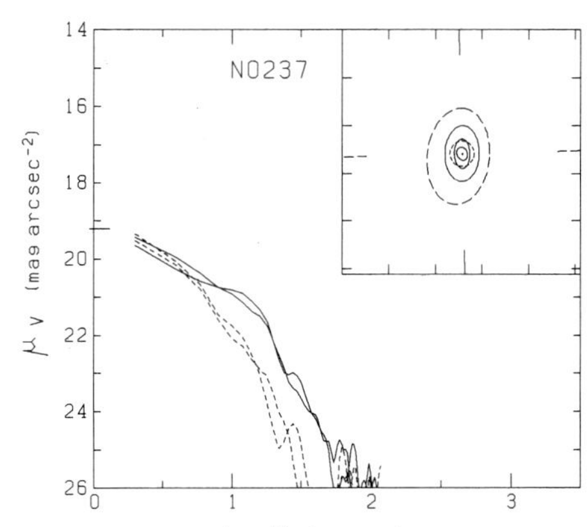

NGC 245

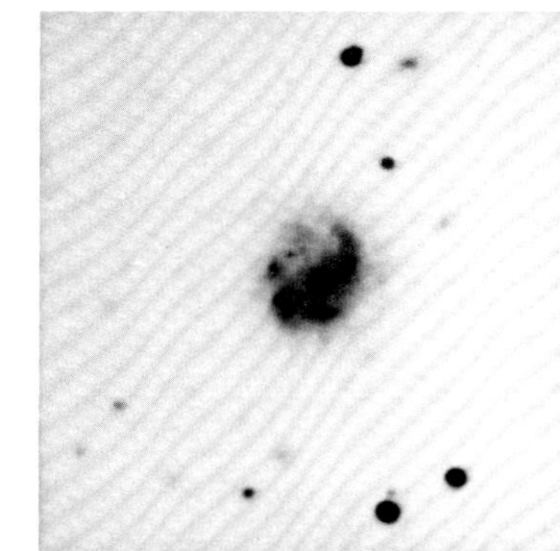

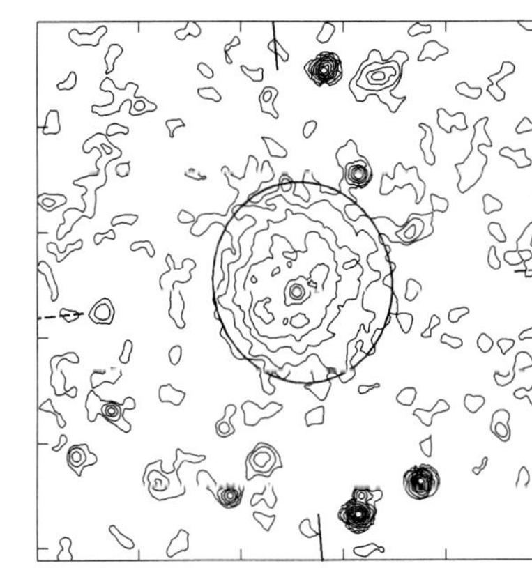

N0245 K4487 1 21.48 256X256

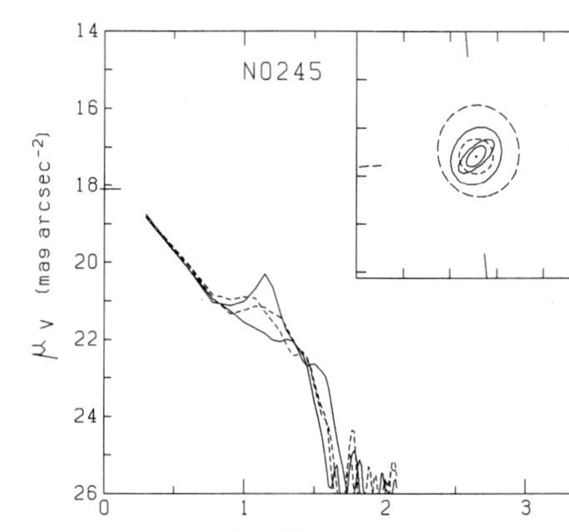

NGC 254

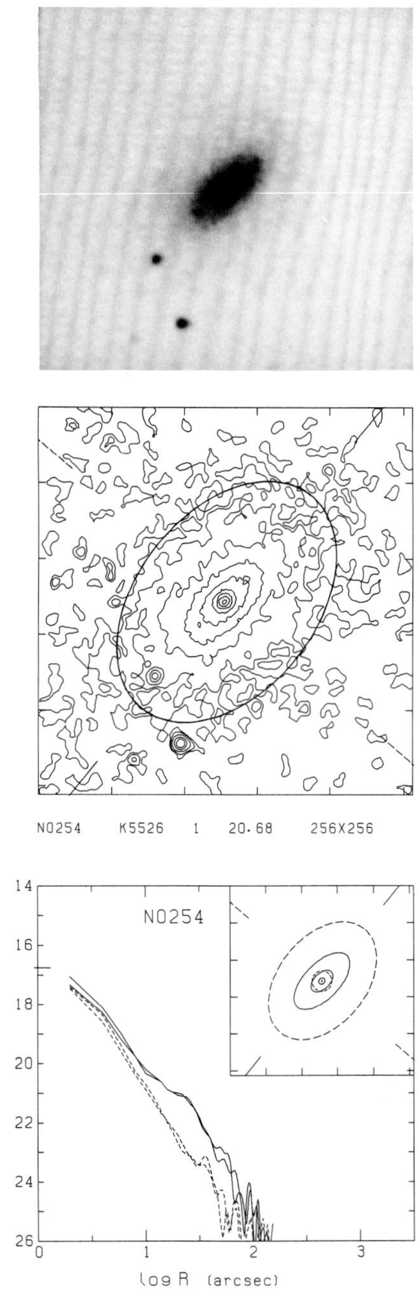

NGC 253

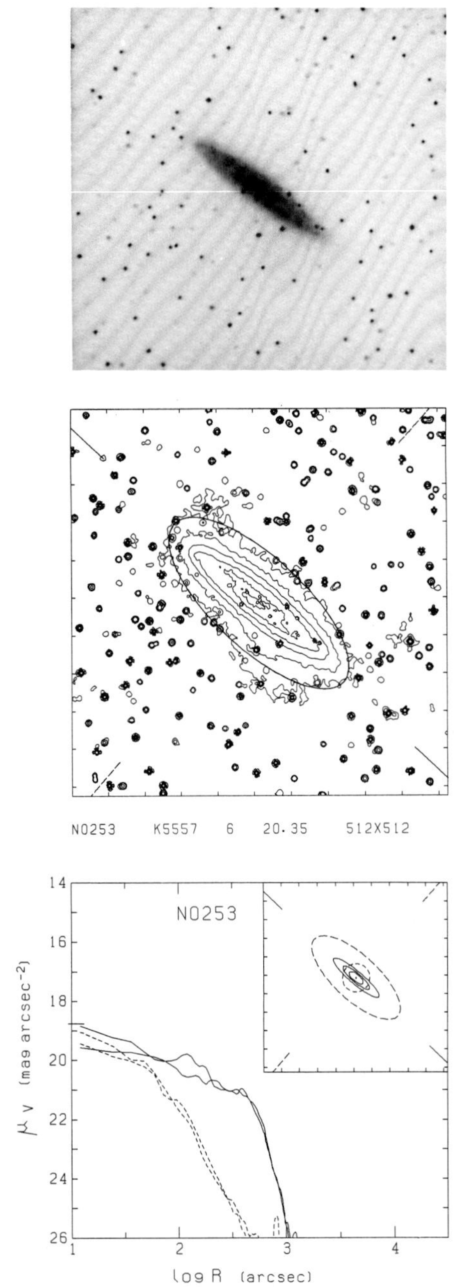

NGC 255

NGC 268

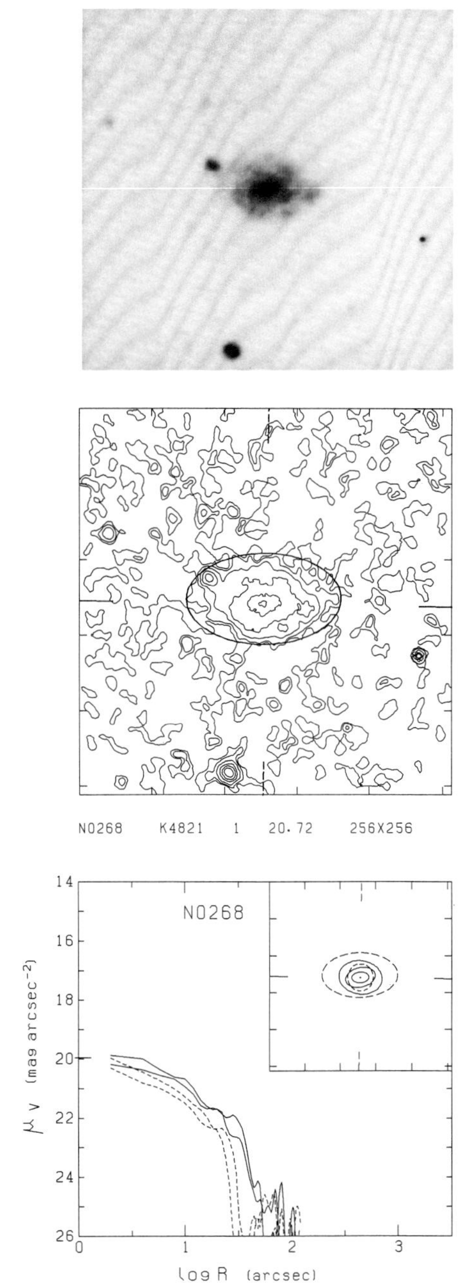

NGC 274

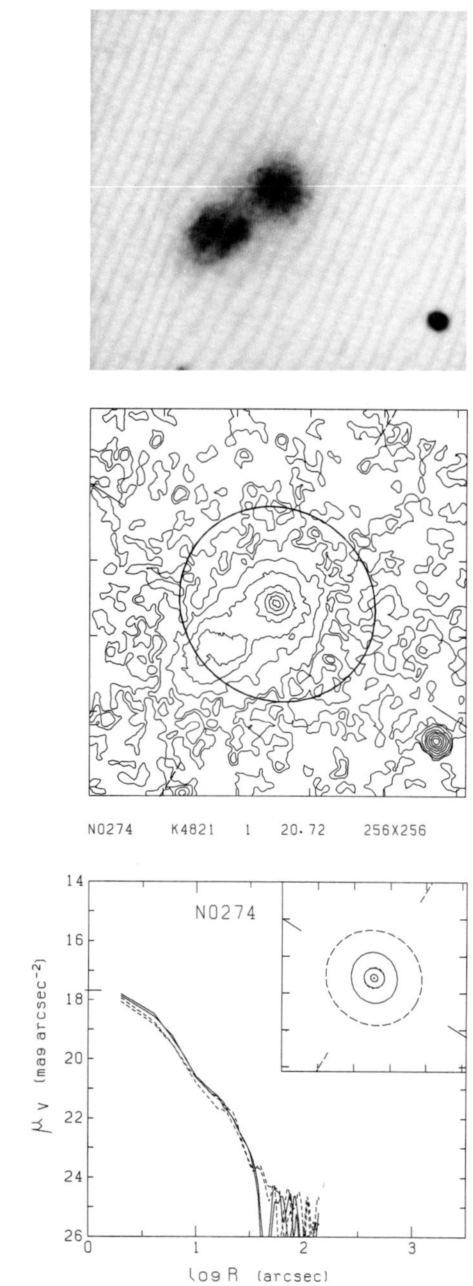

NGC 275

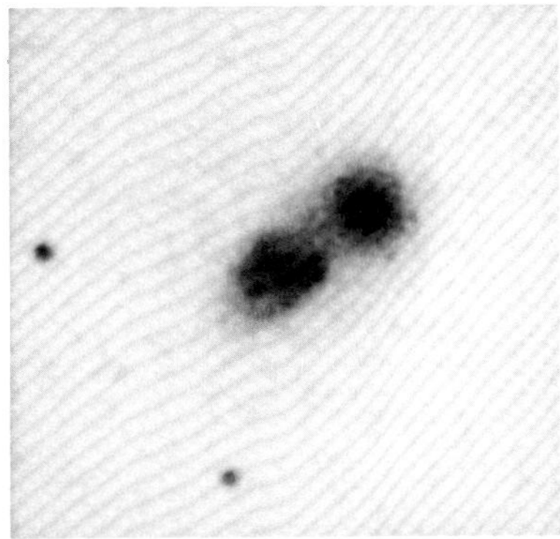

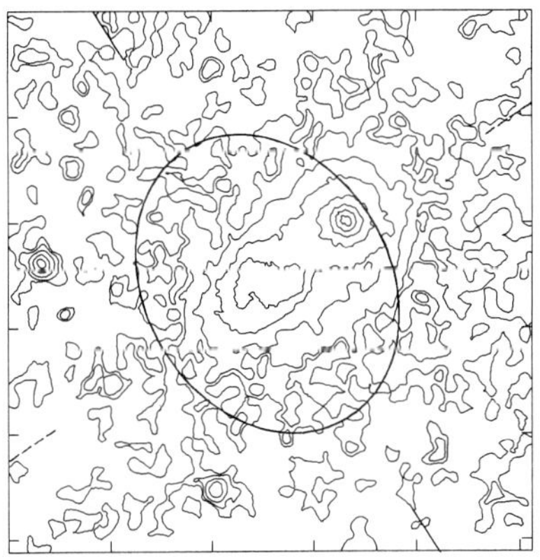

N0275 K4821 1 20.72 256X256

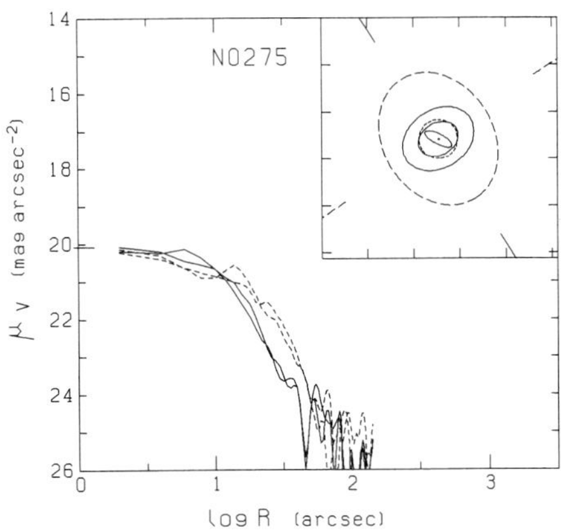

NGC 278

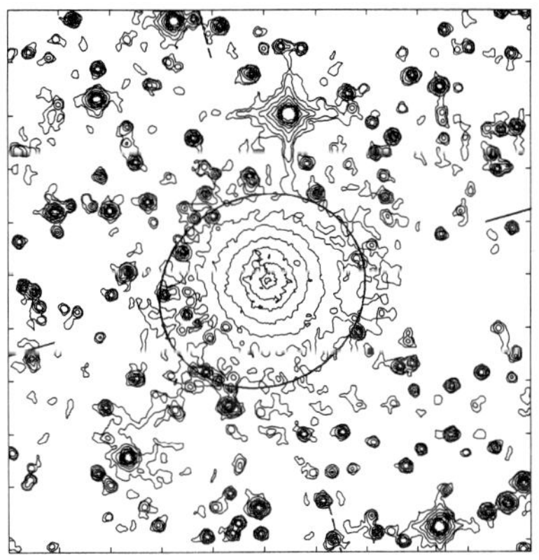

N0278 K2913 1 21.12 512X512

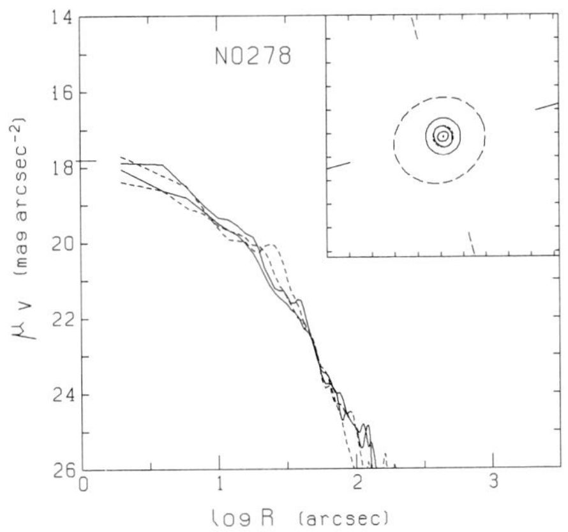

NGC 289

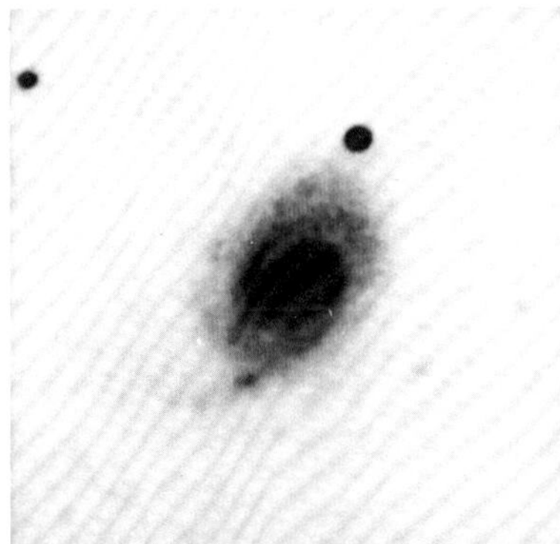

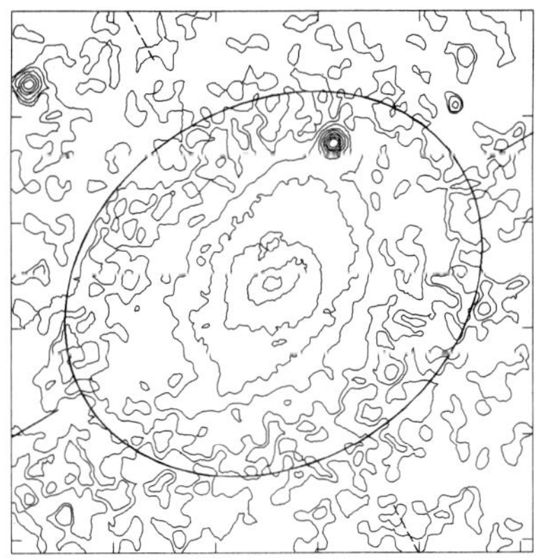

N0289 K5526 1 20.68 256X256

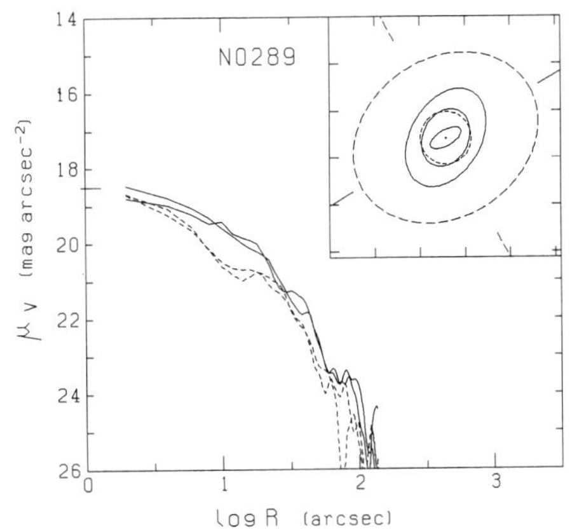

NGC 309

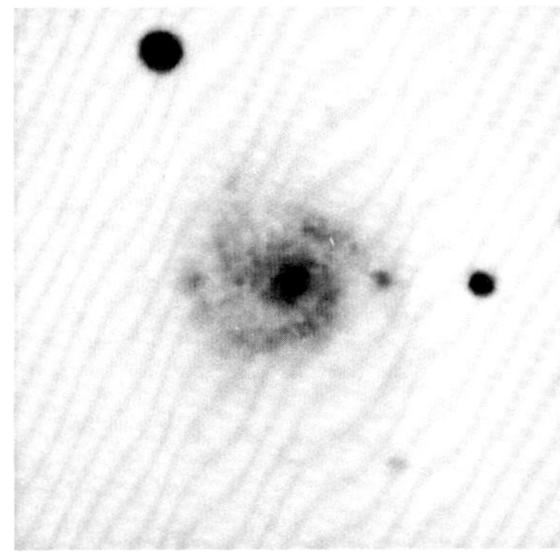

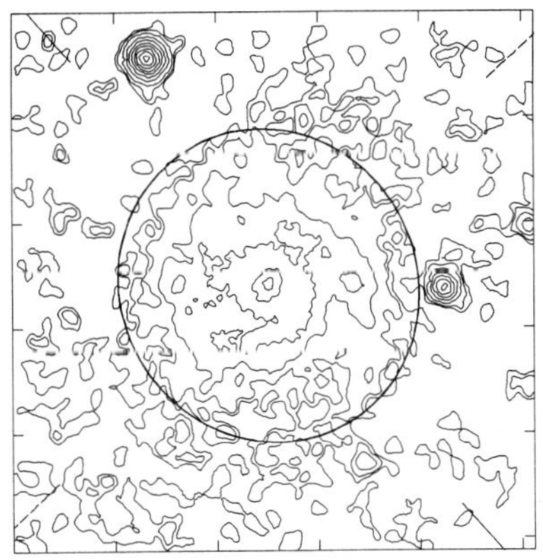

N0309 K4821 1 20.72 256X256

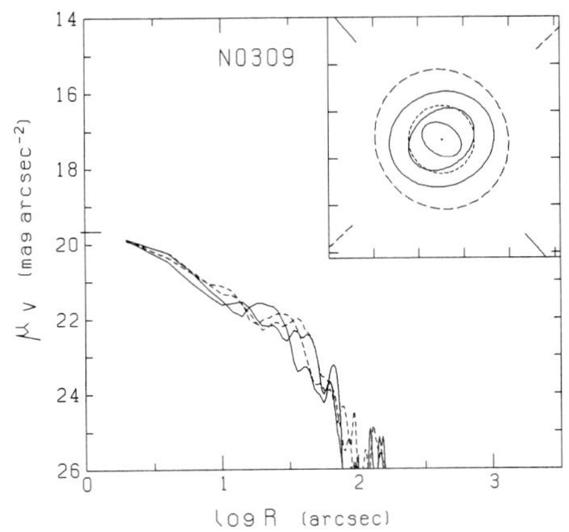

NGC 337

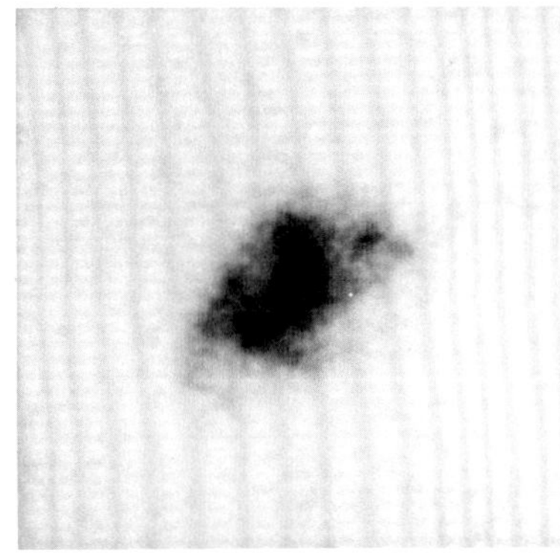

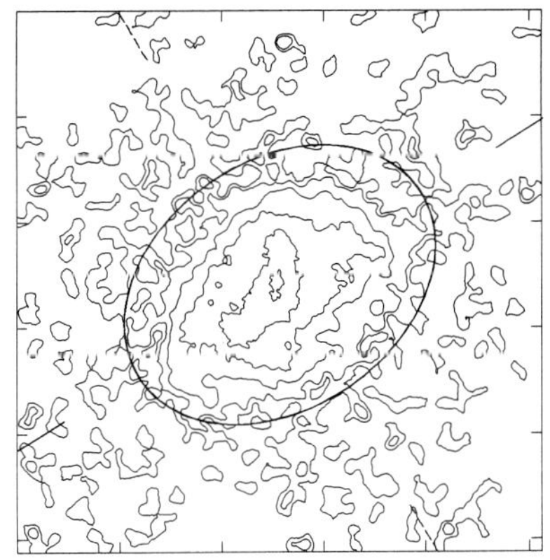

N0337 K4821 1 20.72 256X256

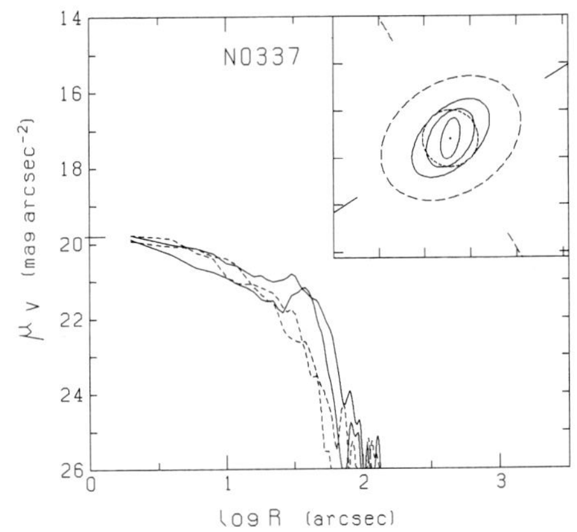

NGC 357

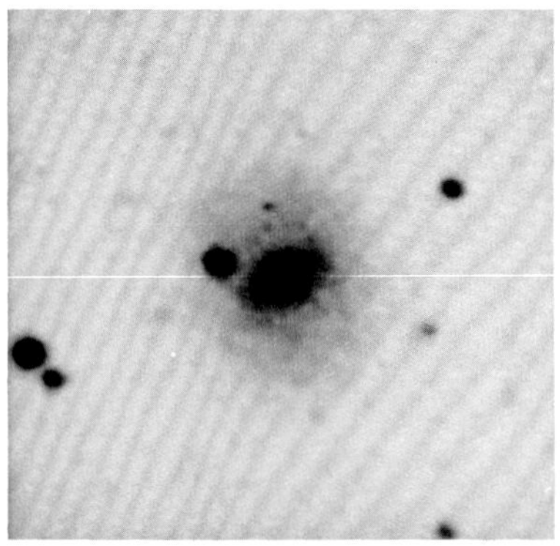

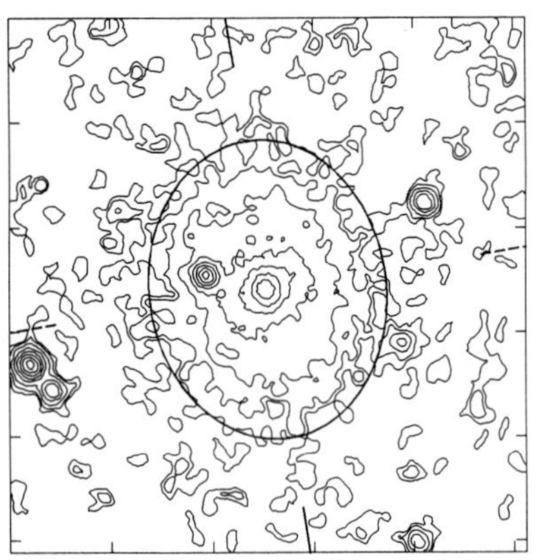

N0357 K4821 1 20.72 256X256

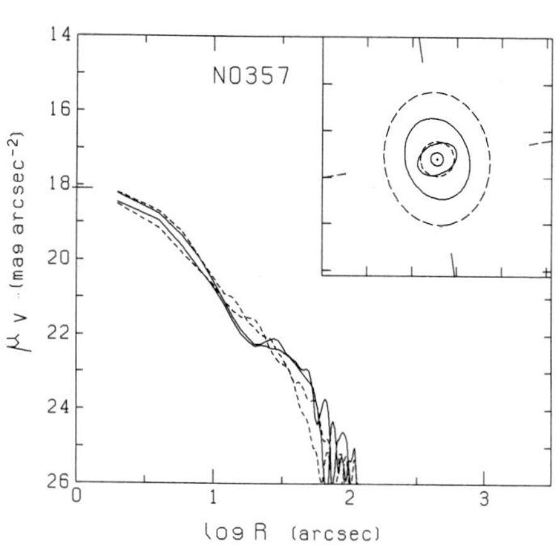

New 1

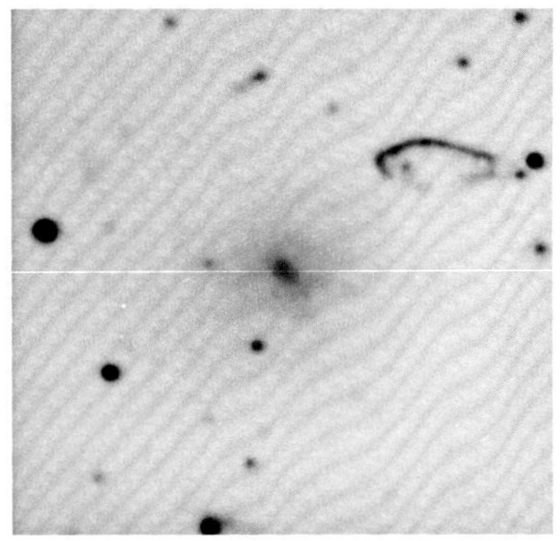

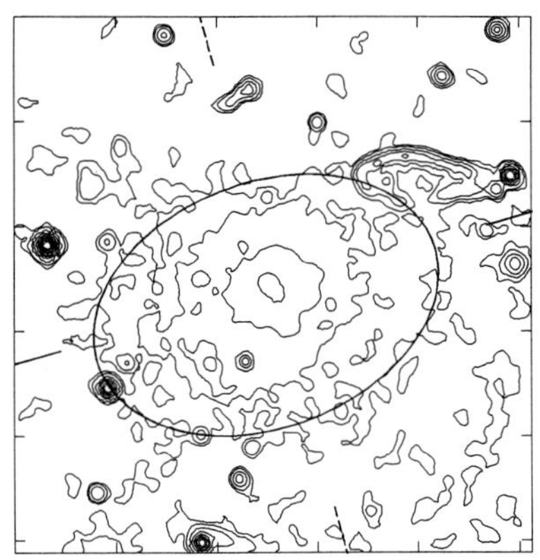

NEW1 K4821 2 20.72 256X256

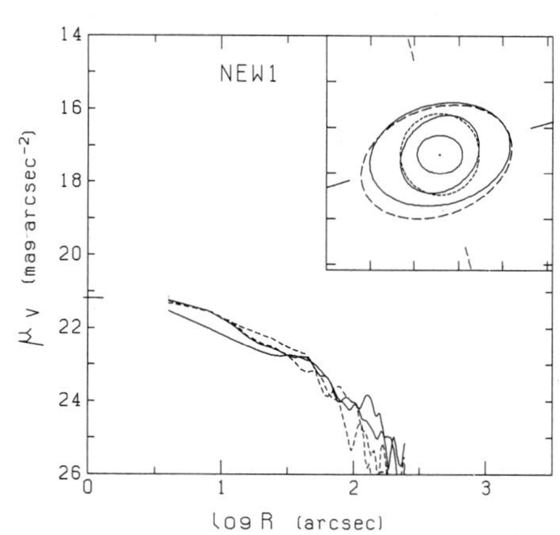

NGC 404

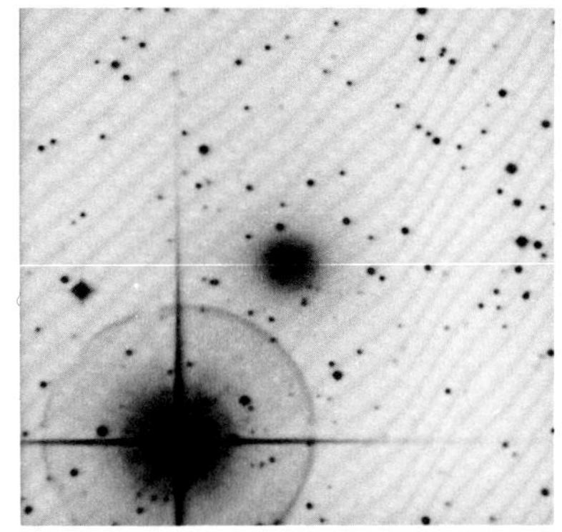

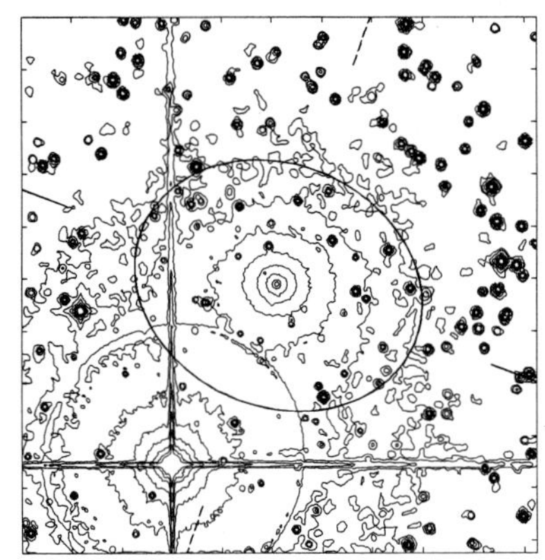

N0404 K5141 2 21.36 512X512

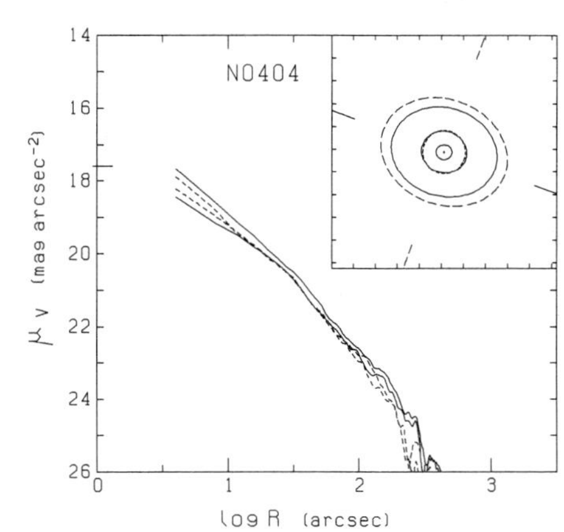

NGC 428

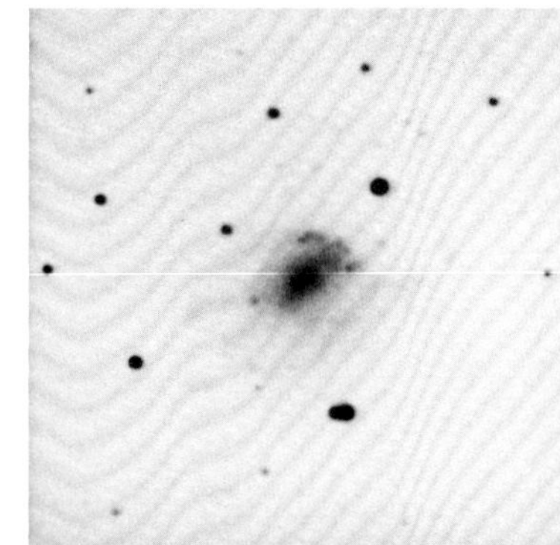

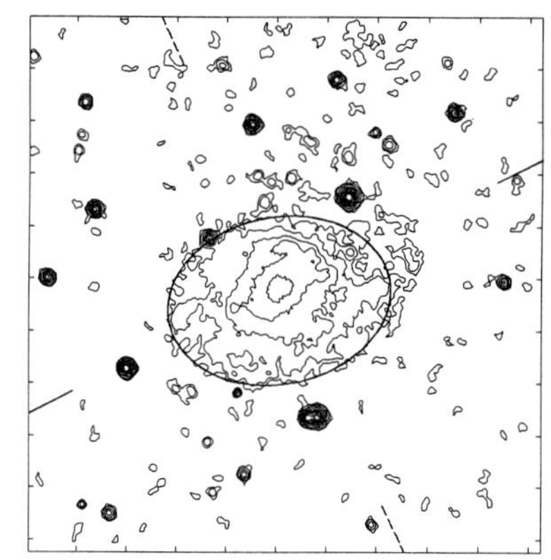

N0428 K2416 1 21.08 512X512

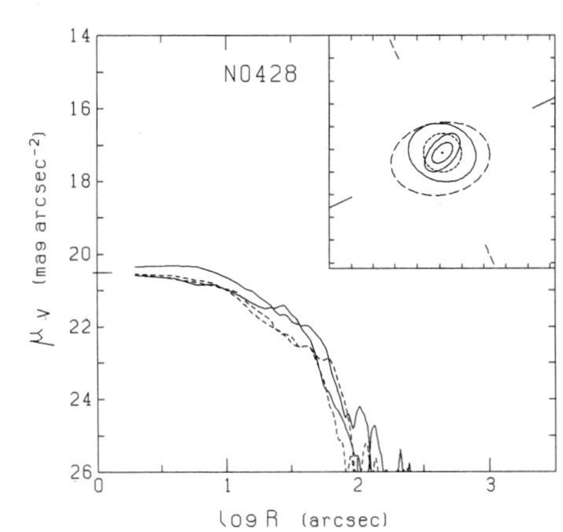

NGC 473

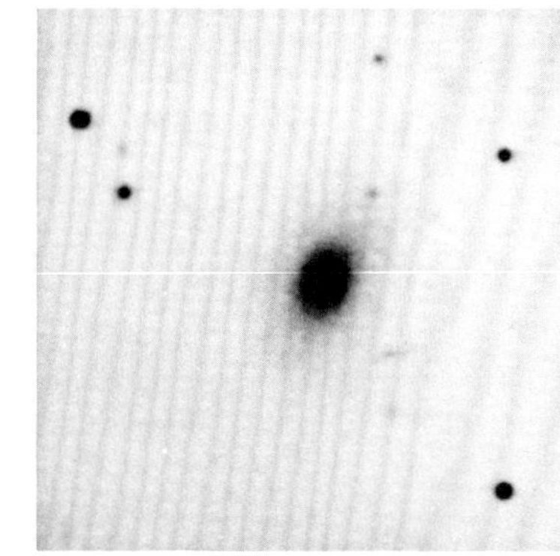

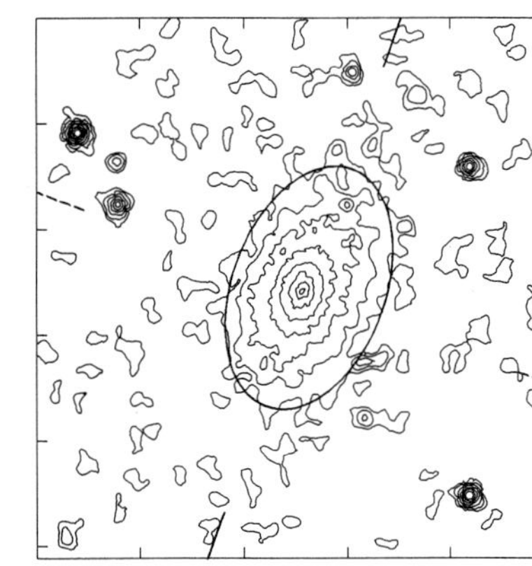

N0473 K5518 1 21.76 256X256

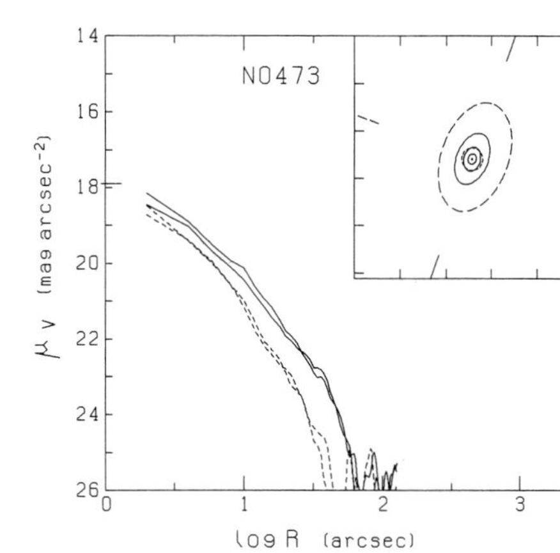

NGC 514

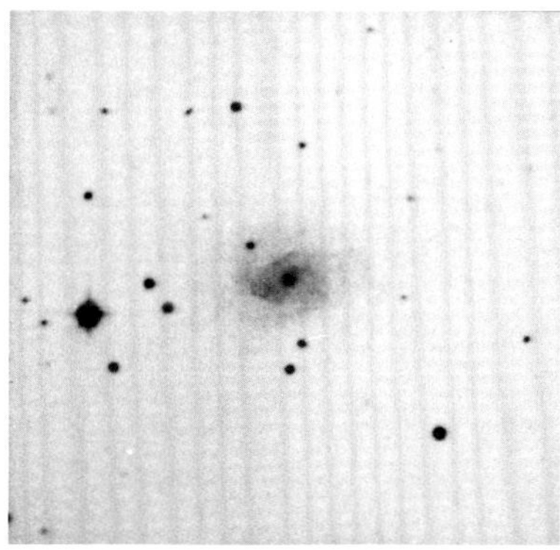

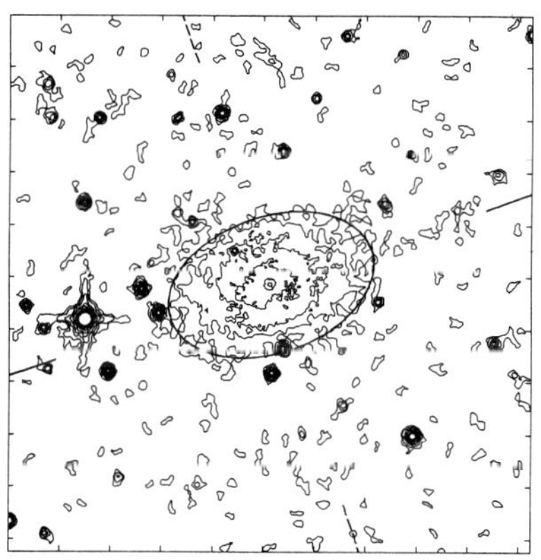

N0514 K5518 1 21.76 512X512

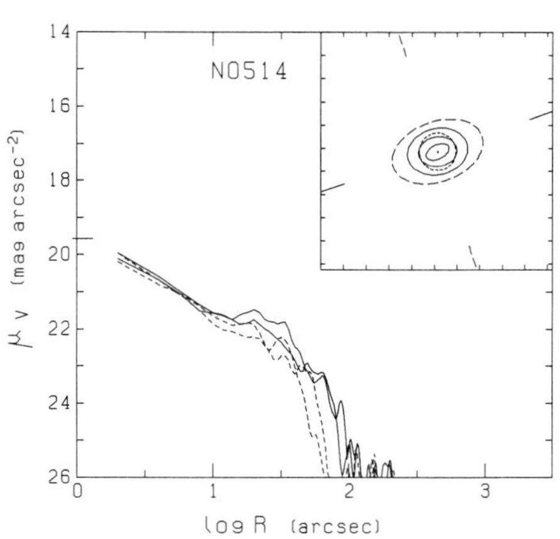

NGC 521

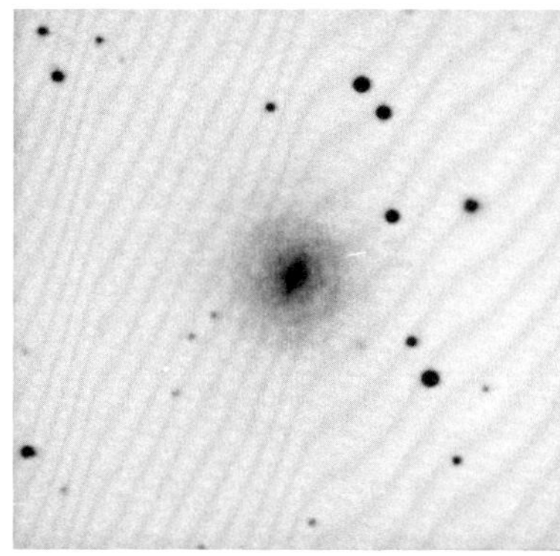

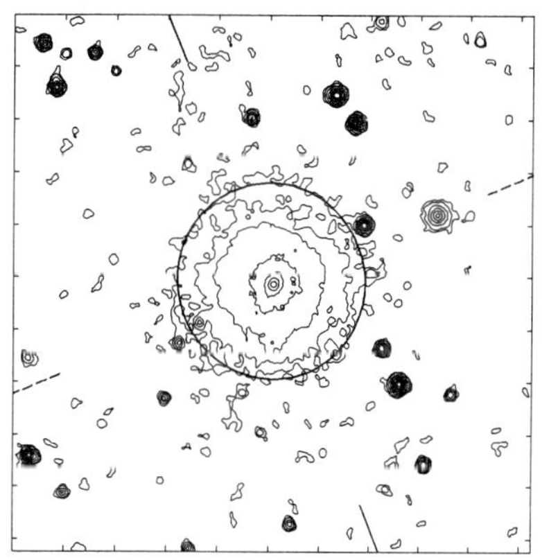

N0521 K2416 1 21.08 512X512

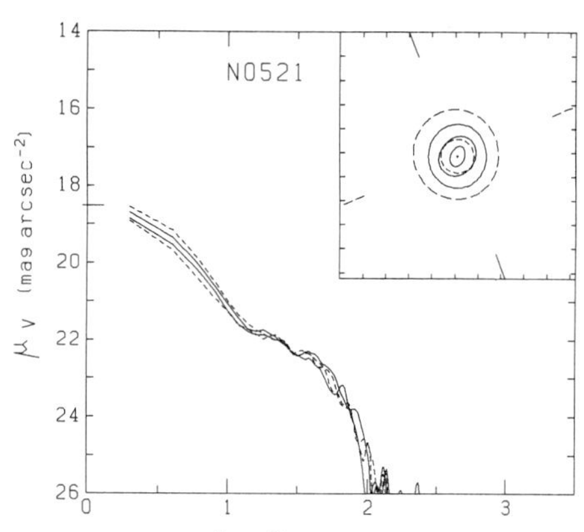

NGC 524

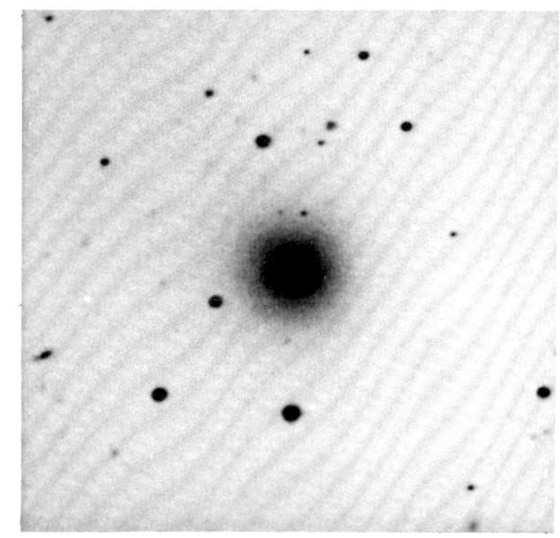

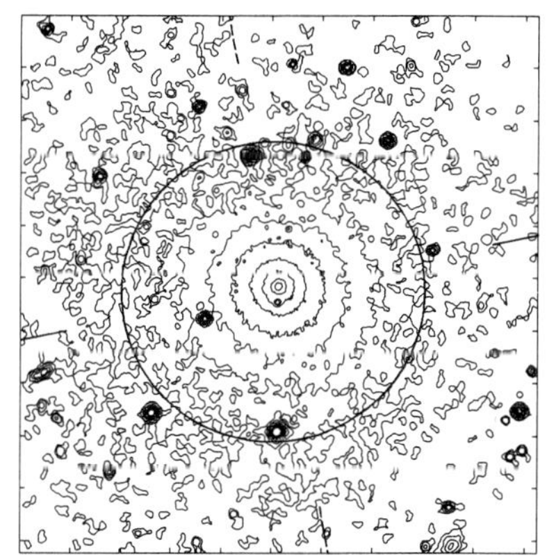

N0524 K3567 1 20.88 512X512

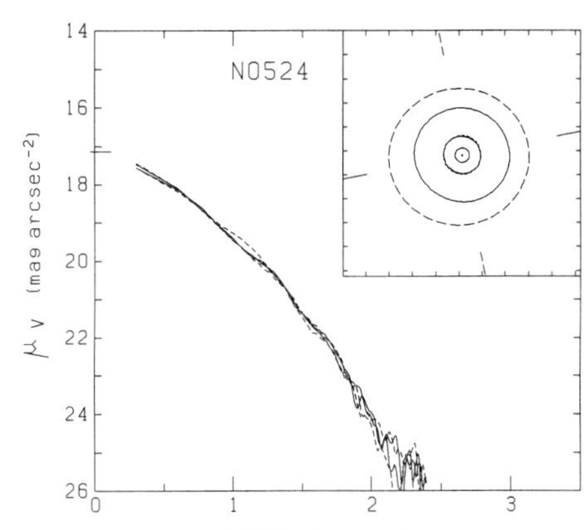

NGC 533

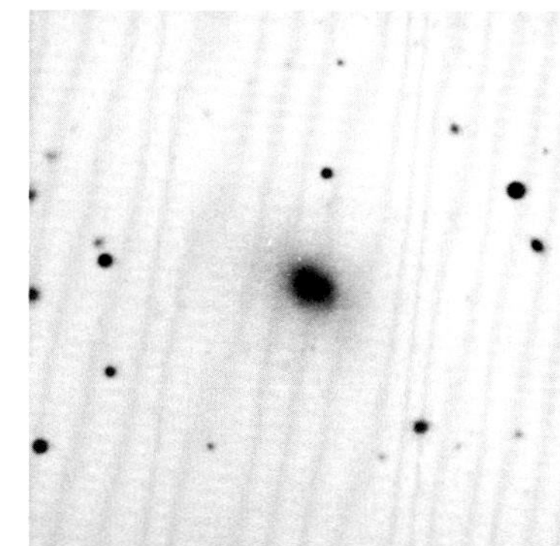

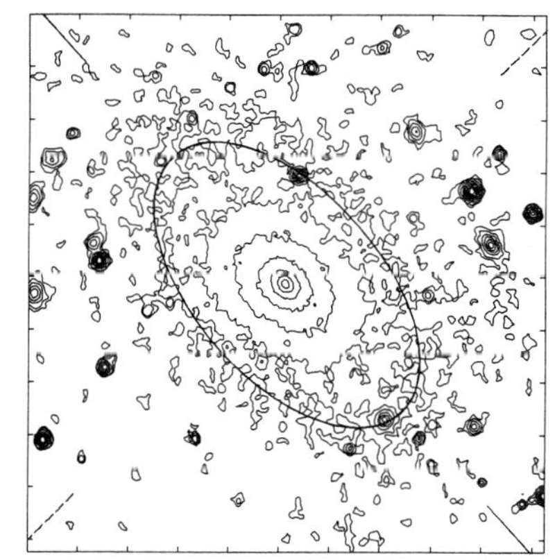

N0533 K2416 1 21.08 512X512

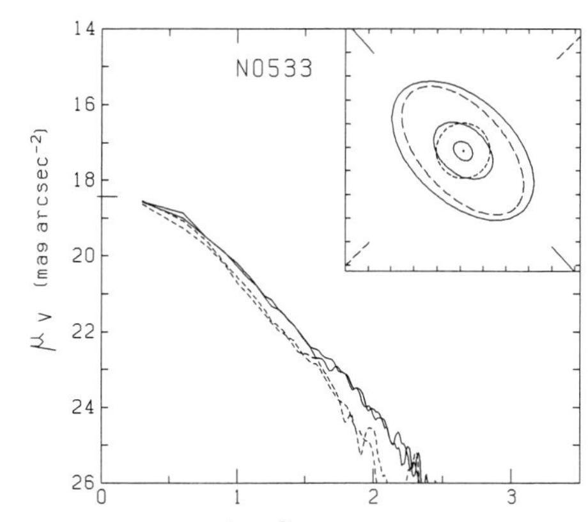

NGC 598

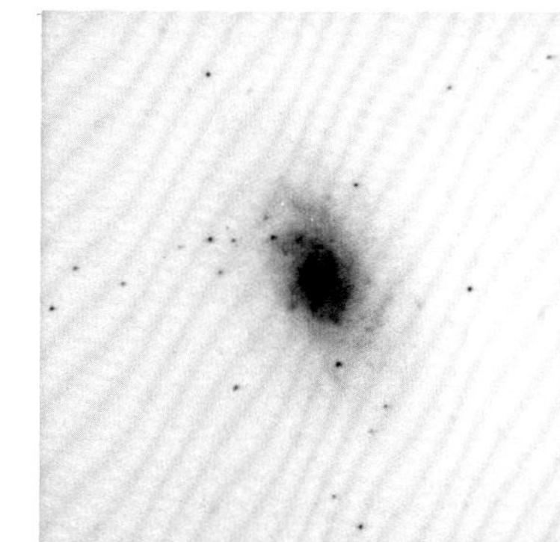

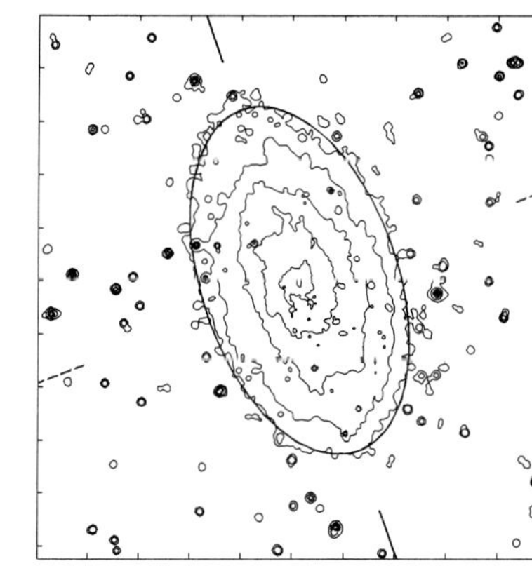

N0598 K3533 12 21.32 512X512

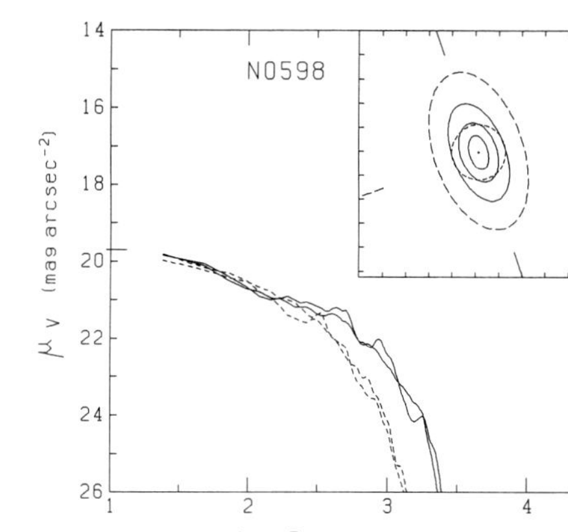

NGC 628

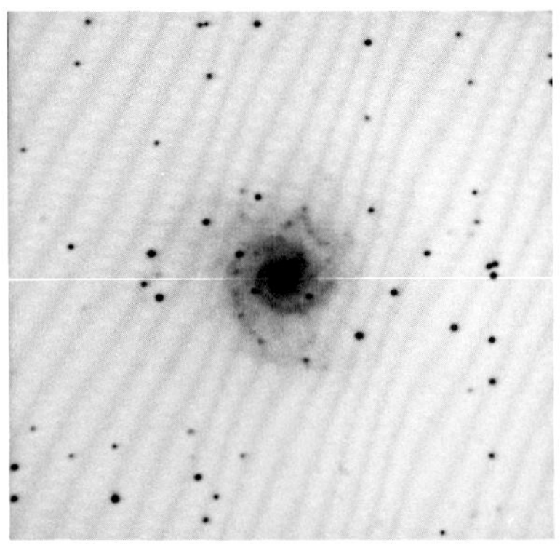

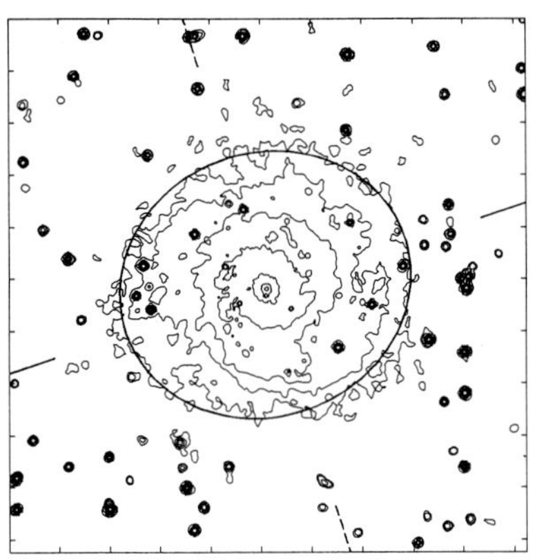

N0628 K3398 2 20.89 512X512

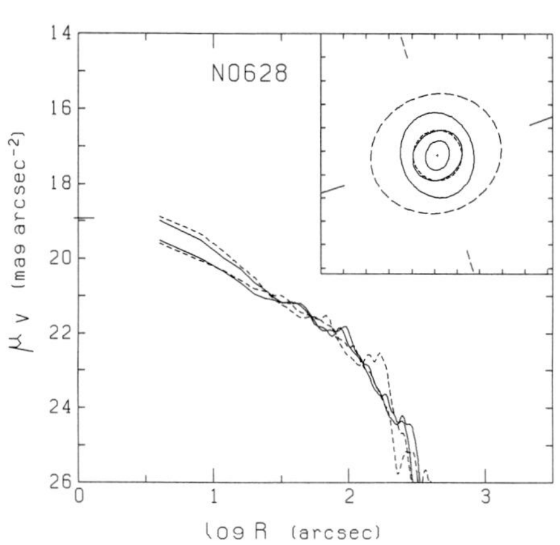

NGC 670

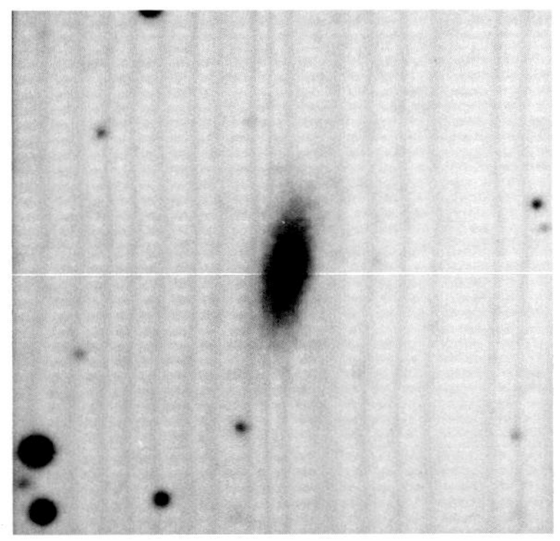

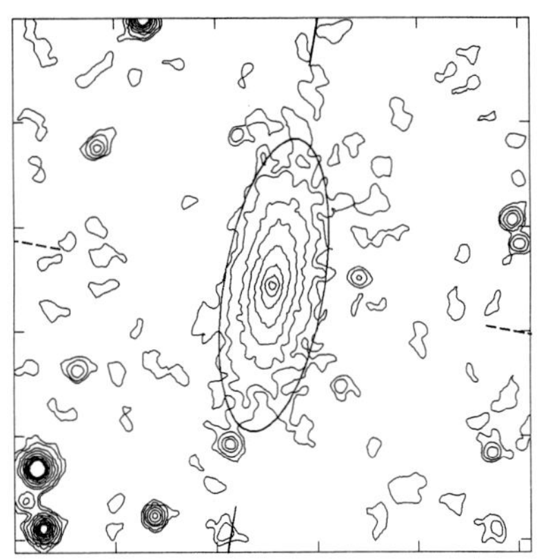

N0670 K5575 1 21.52 256X256

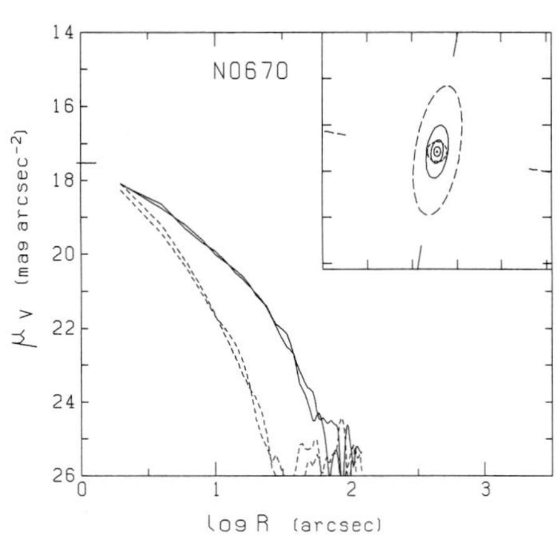

NGC 672

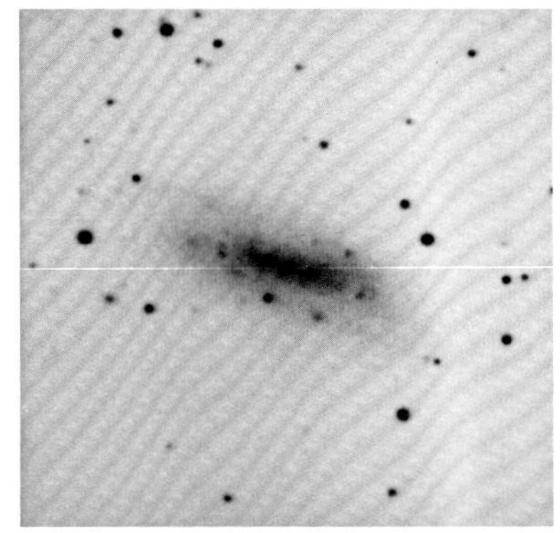

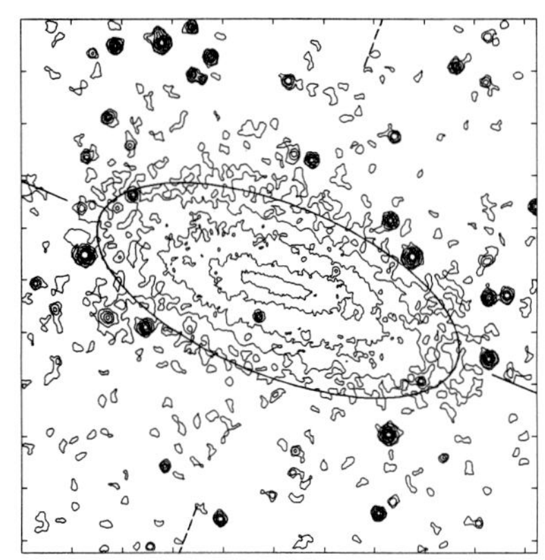

N0672 K5575 1 21.52 512X512

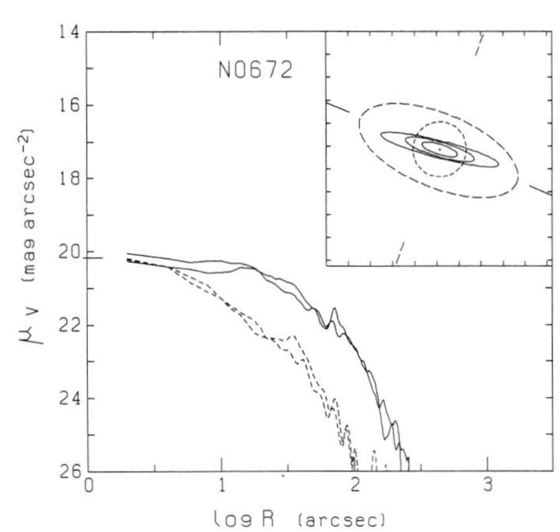

NGC 718

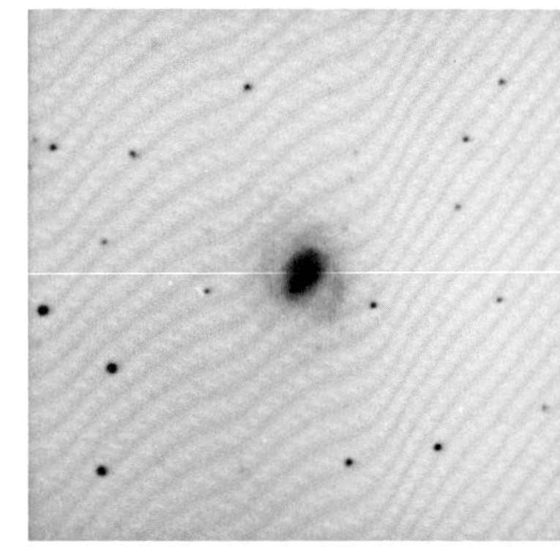

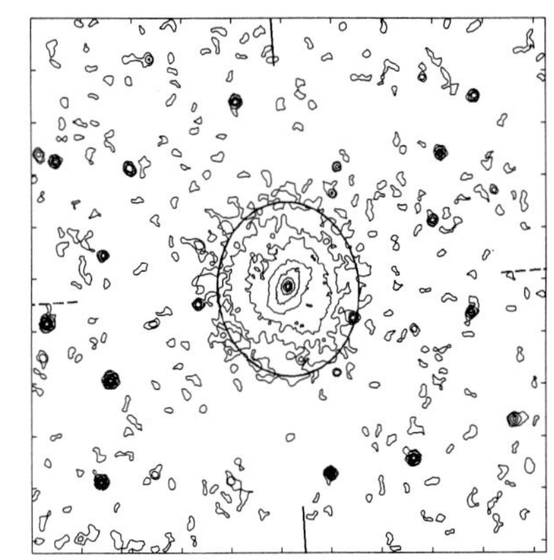

N0718 K5667 1 21.47 512X512

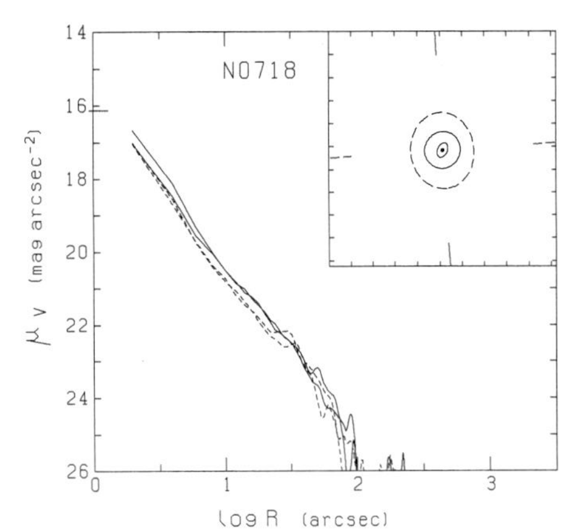

NGC 741

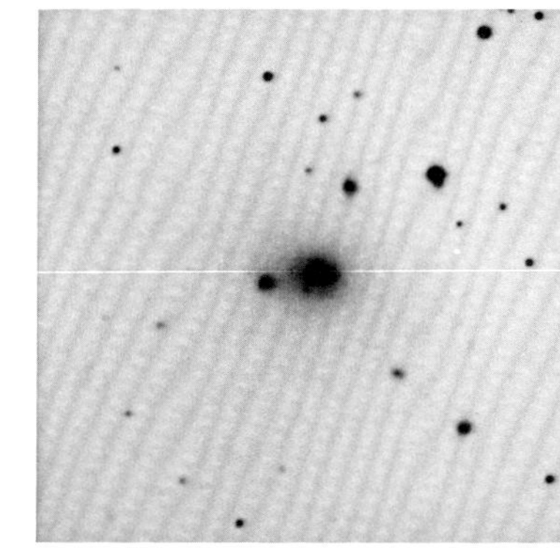

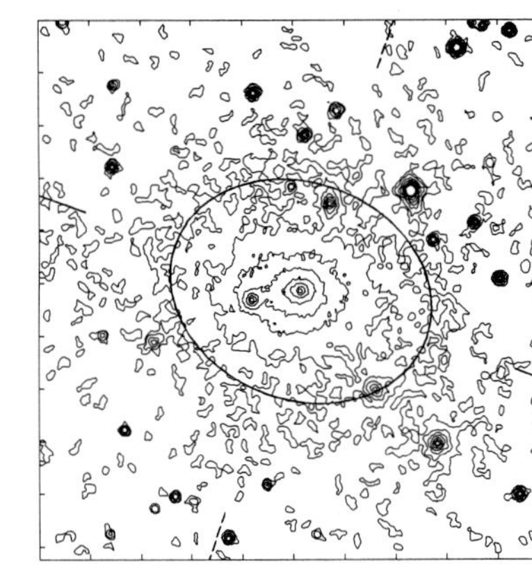

N0741 K5667 1 21.47 512X512

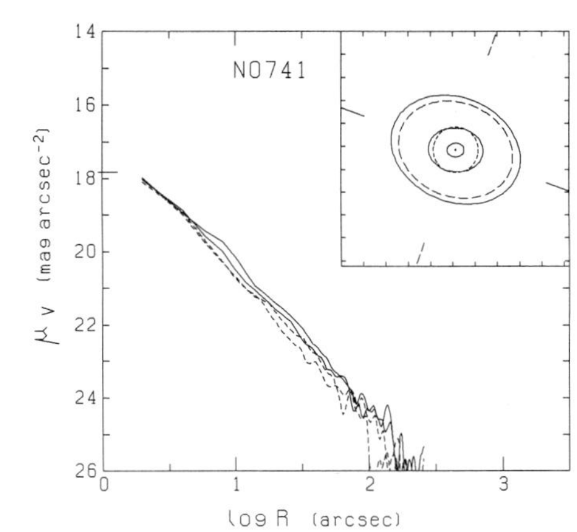

NGC 750

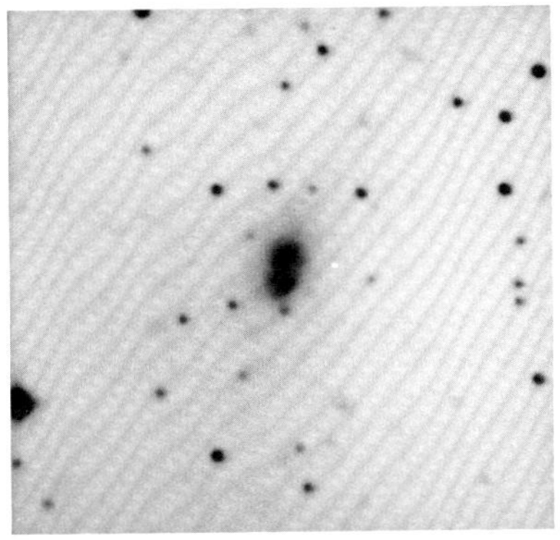

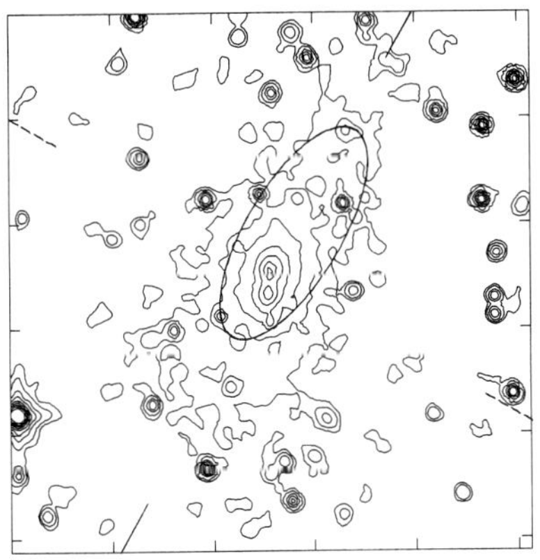

N0750 K5142 2 21.42 256X256

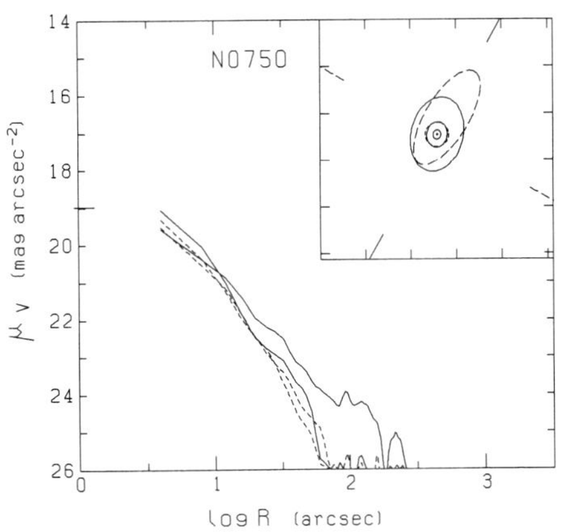

NGC 753

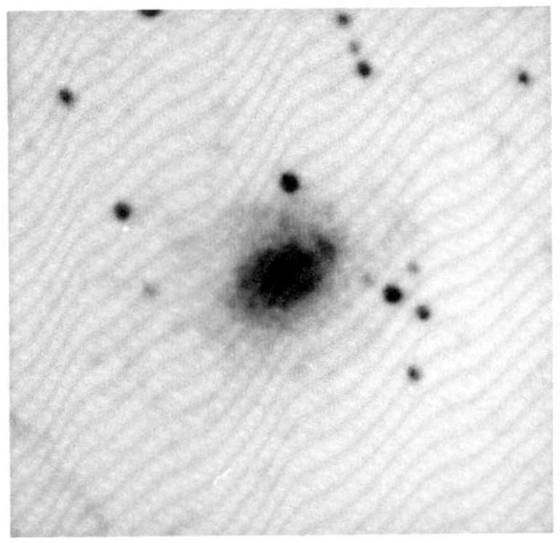

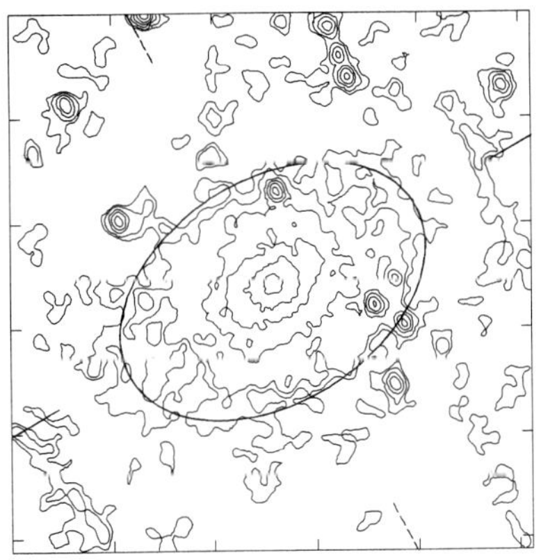

N0753 K5142 1 21.42 256X256

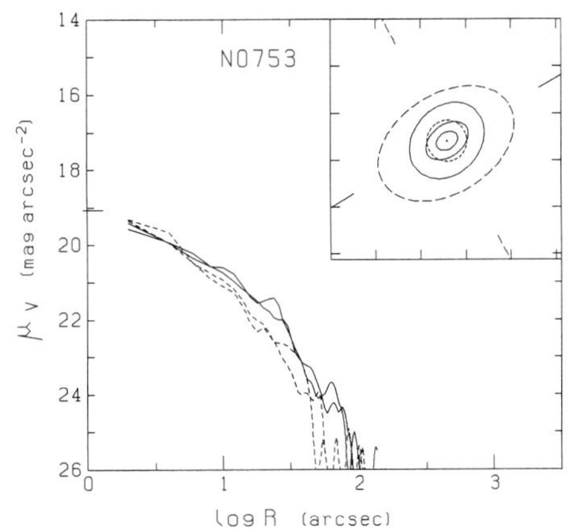

NGC 777

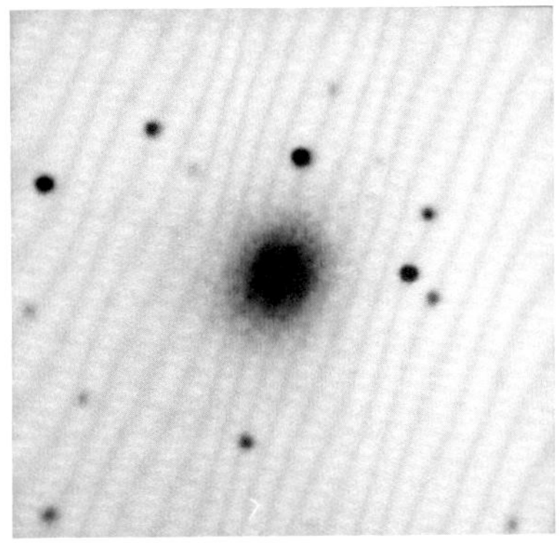

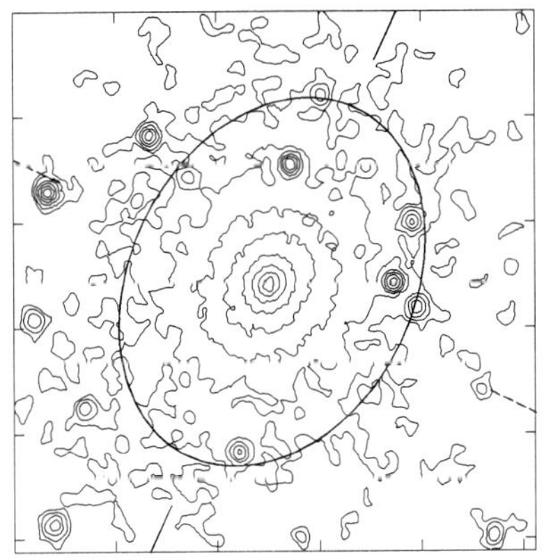

N0777 K5576 1 21.63 256X256

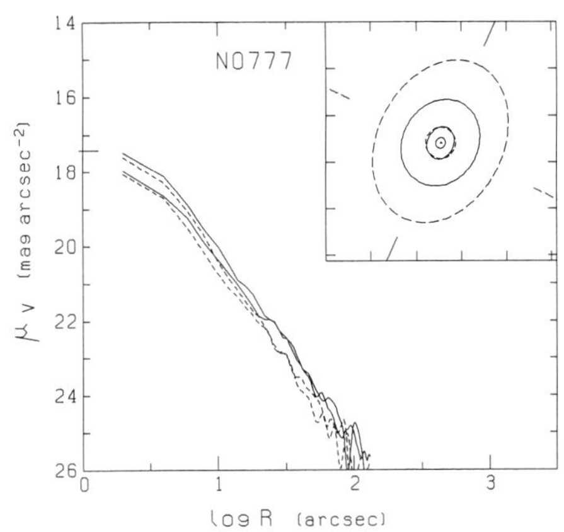

NGC 890

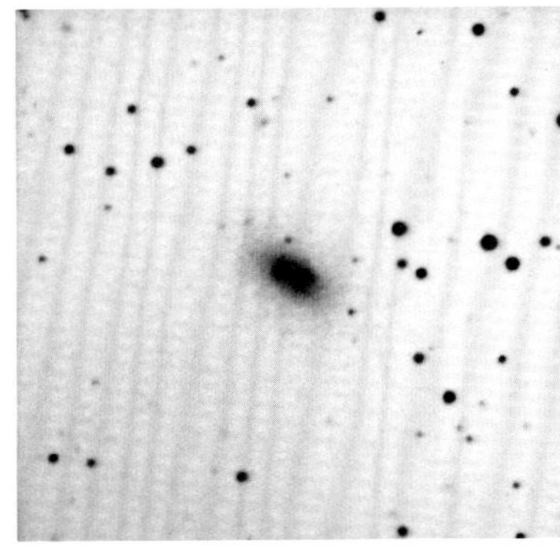

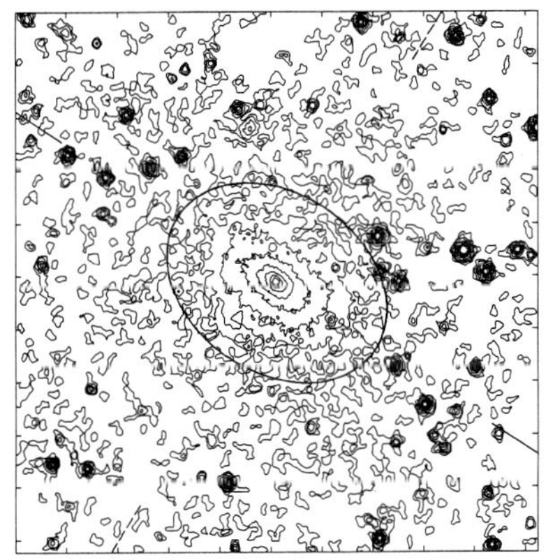

N0890 K4782 1 21.76 512X512

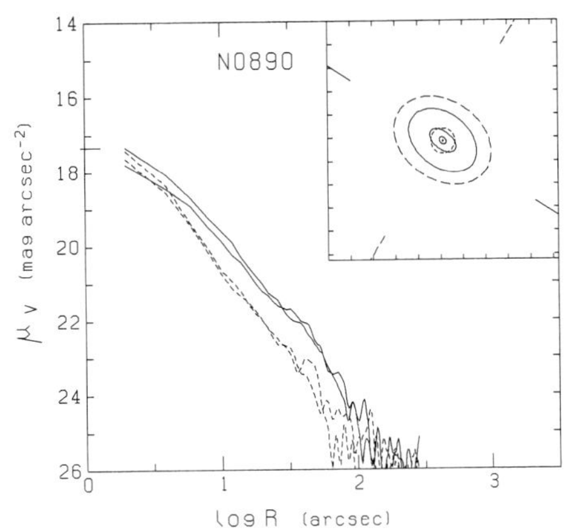

NGC 908

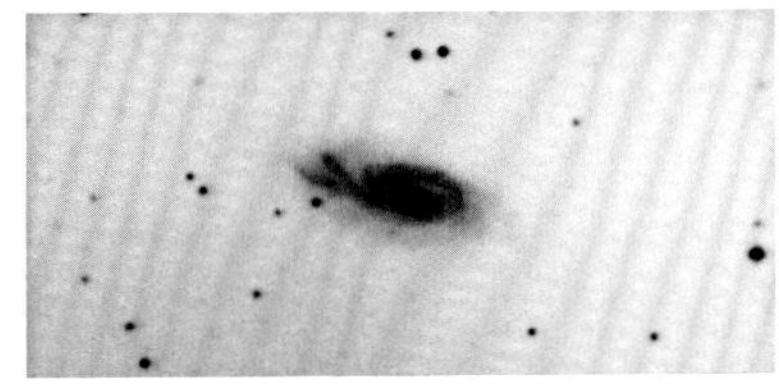

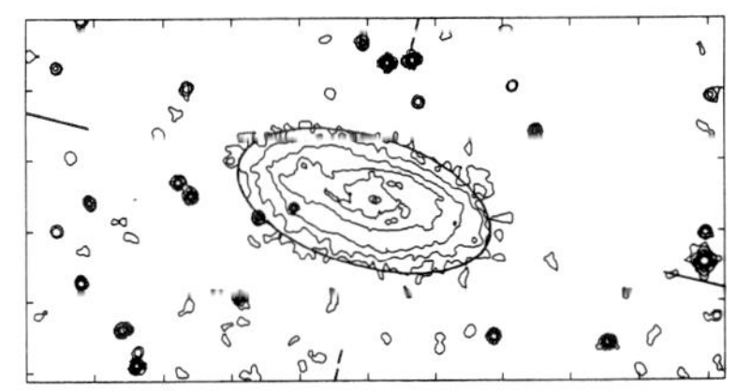

N0908 K5534 2 21.08 512X256

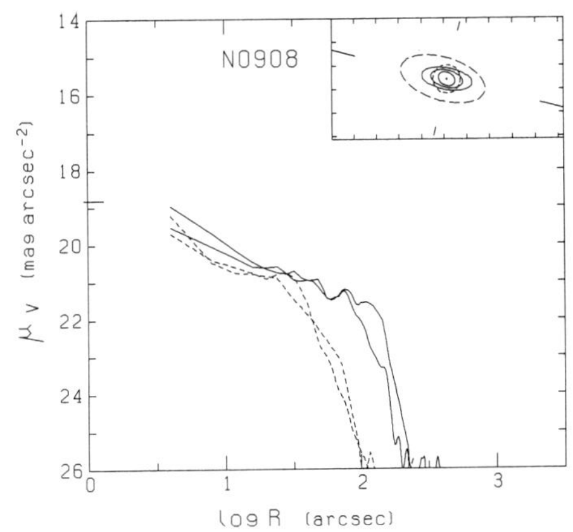

NGC 925

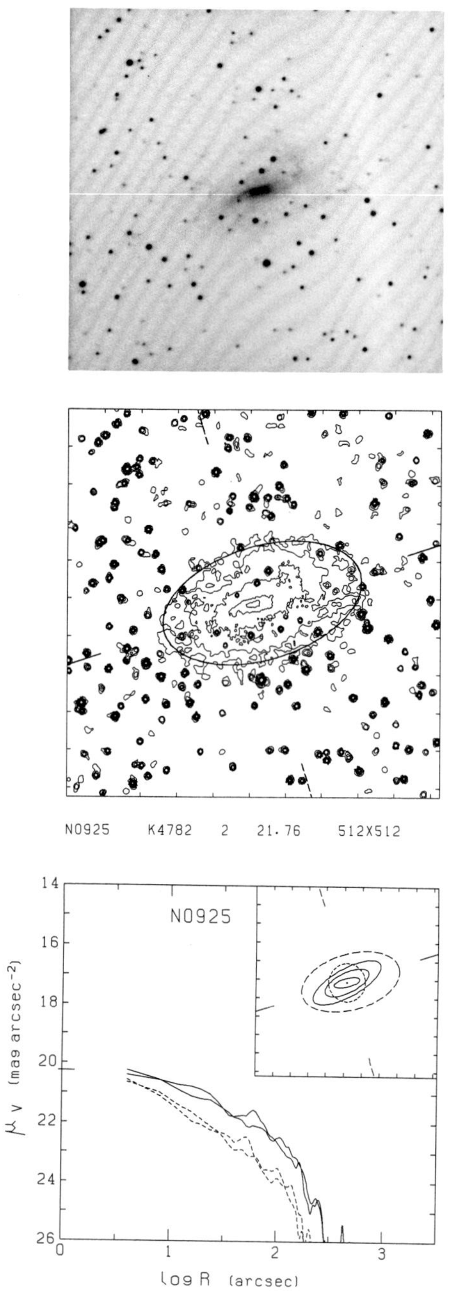

NGC 936

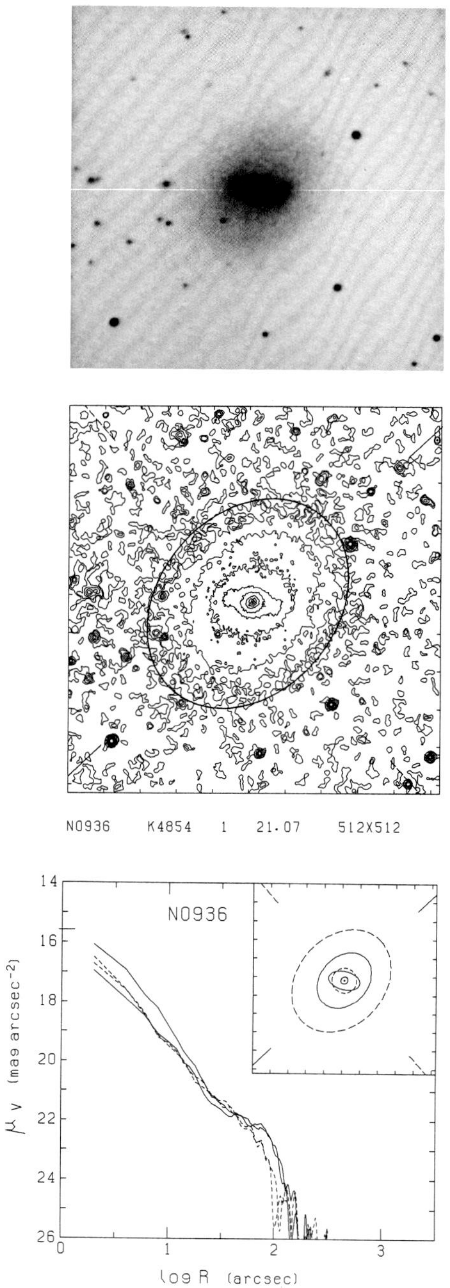

NGC 941

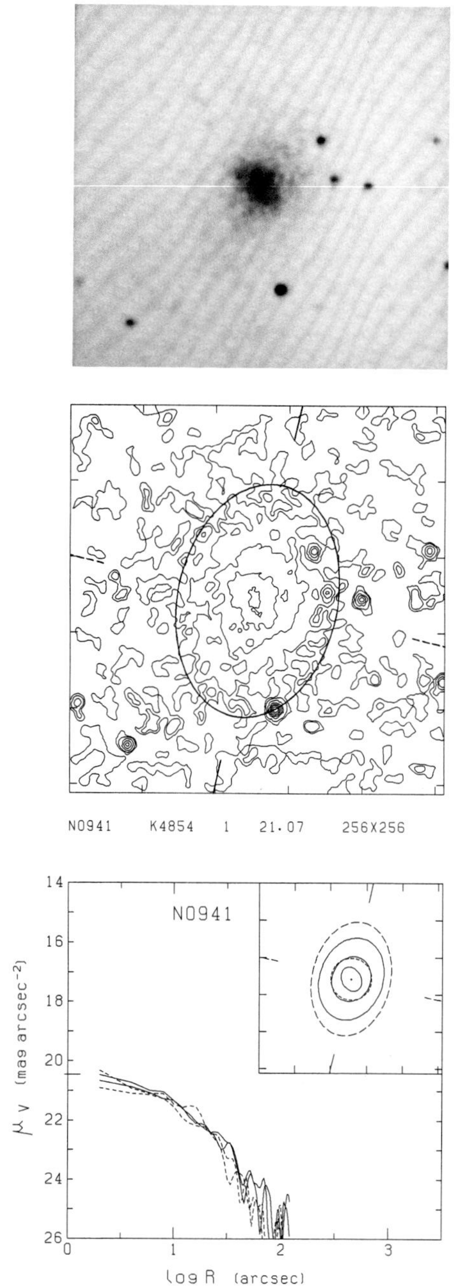

NGC 949

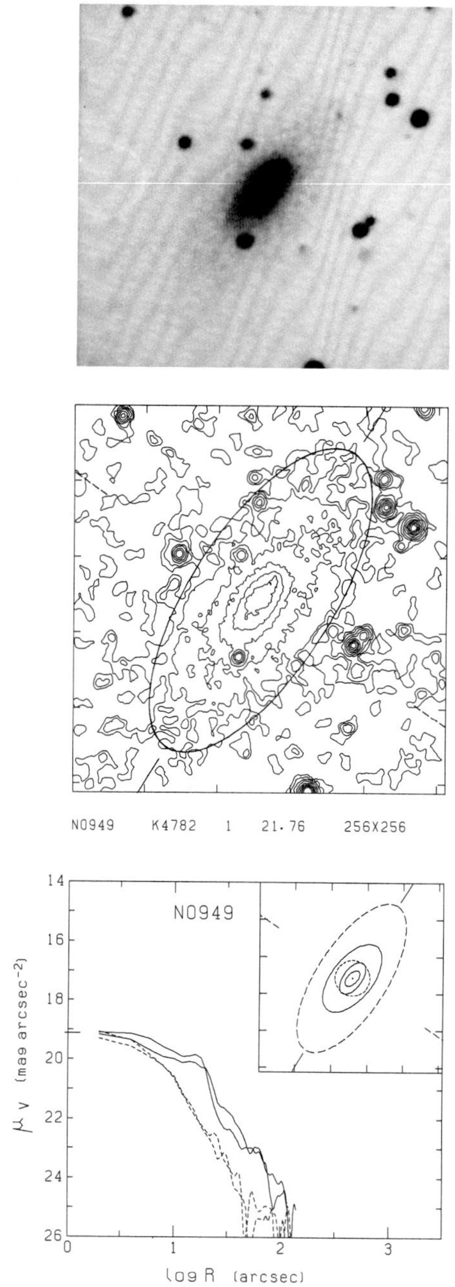

NGC 955

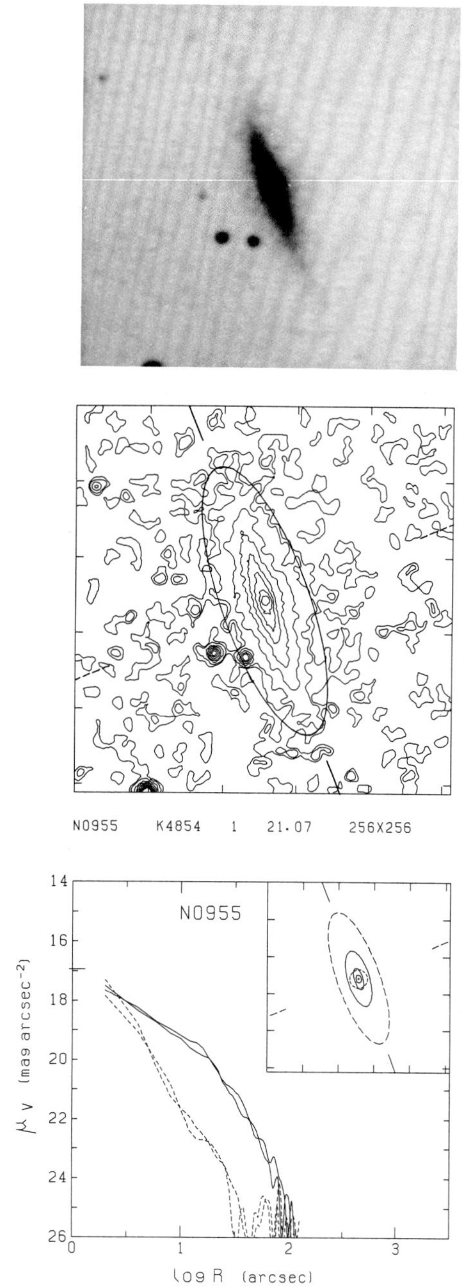

NGC 976

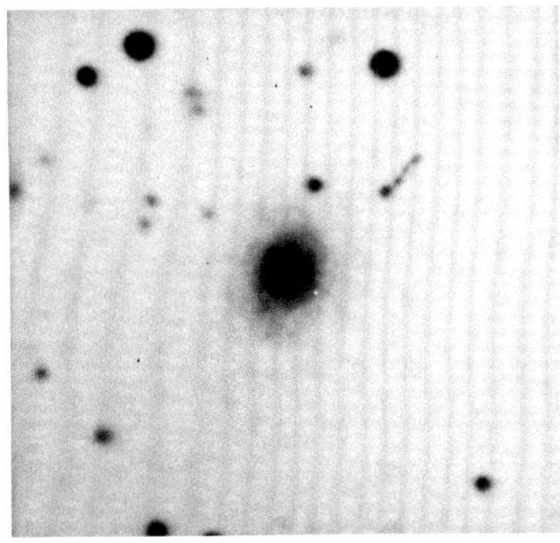

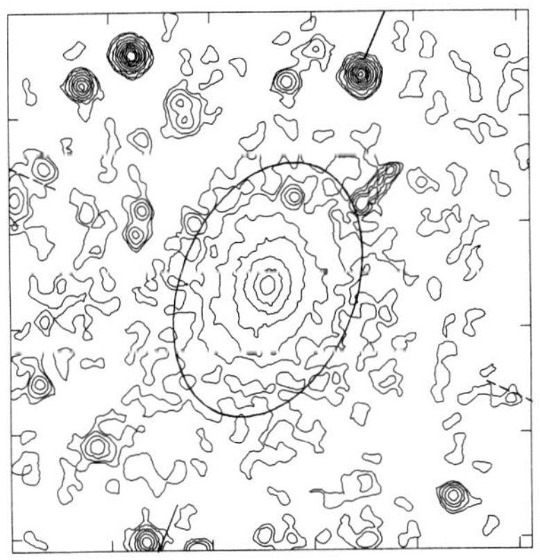

N0976 K5578 1 21.24 256X256

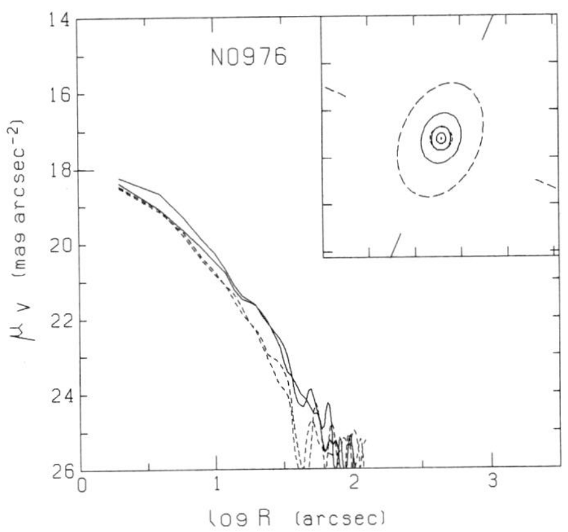

NGC 972

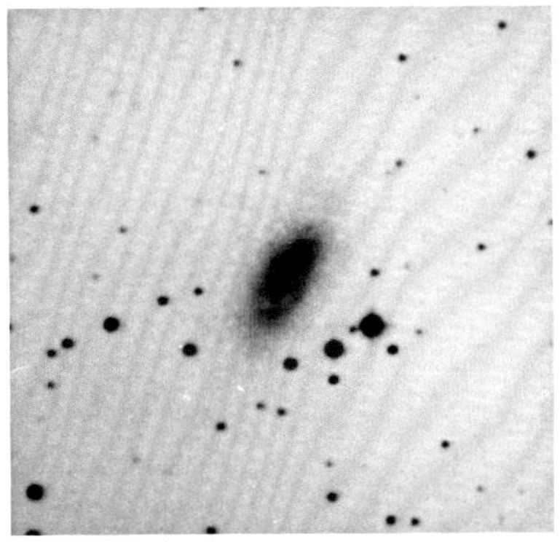

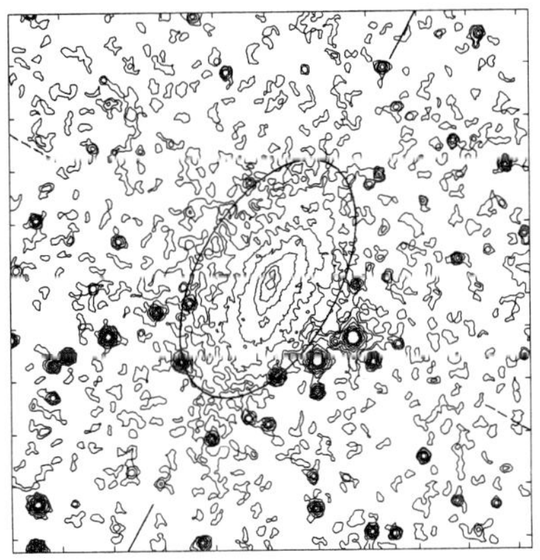

N0972 K5527 1 21.59 512X512

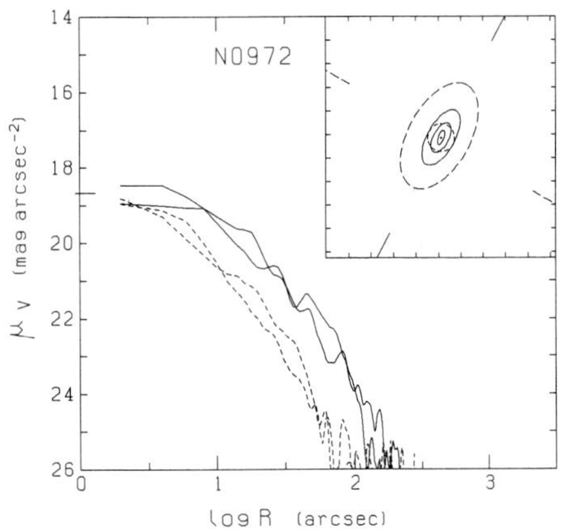

NGC 991

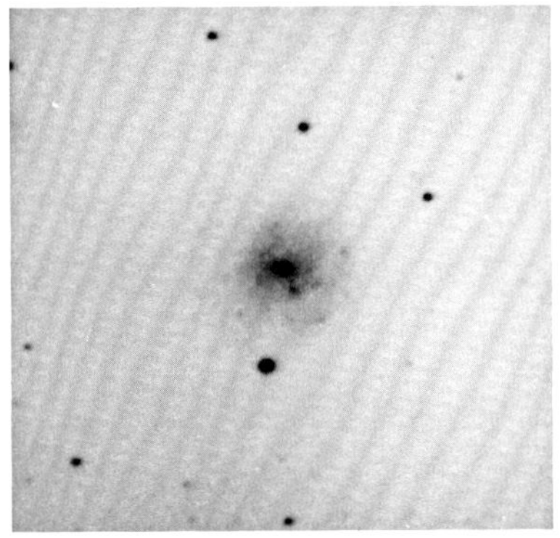

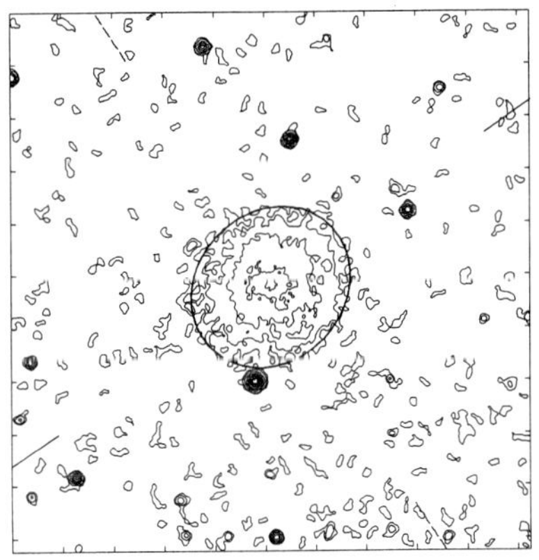

N0991 K4221 1 21.29 512X512

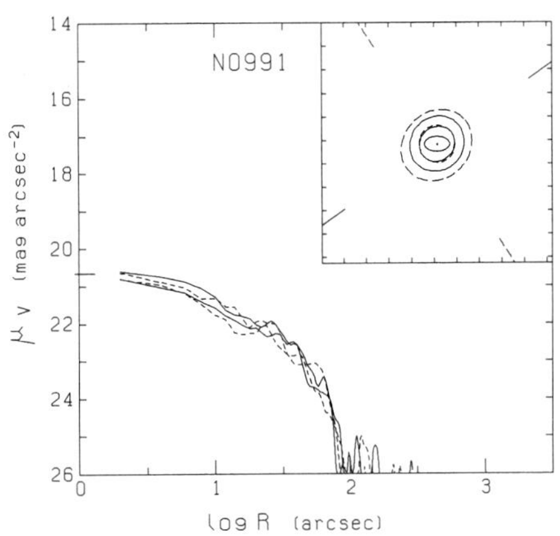

NGC1022

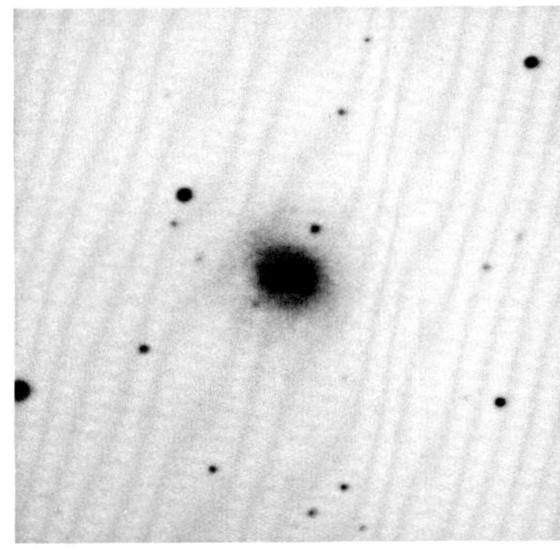

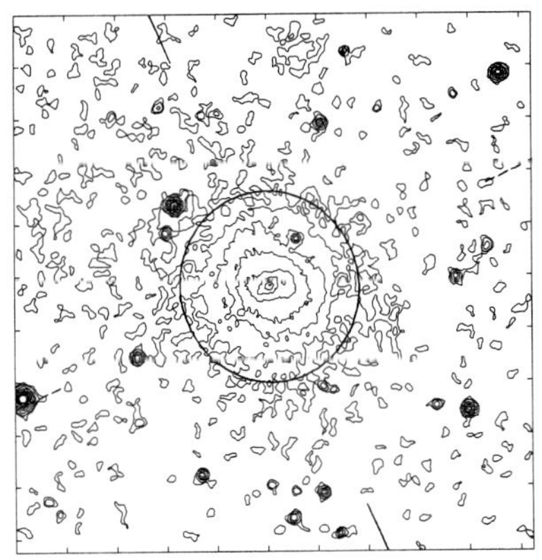

N1022 K4221 1 21.29 512X512

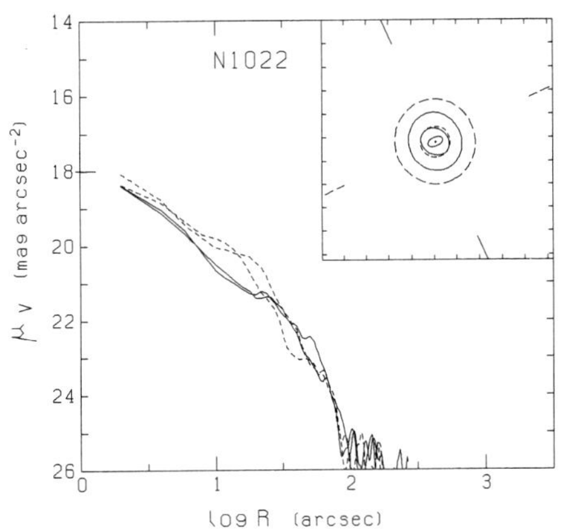

NGC1035

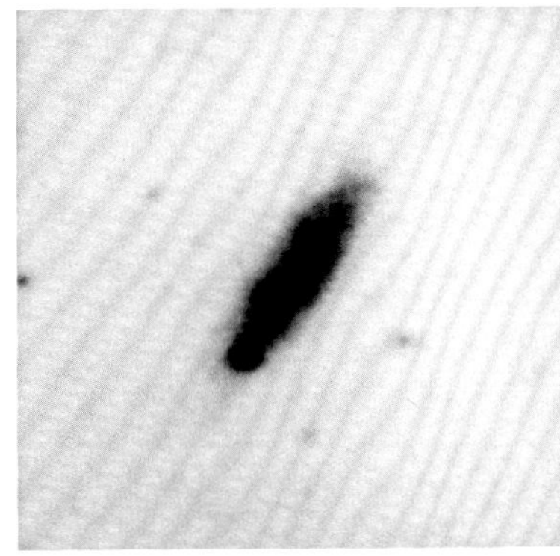

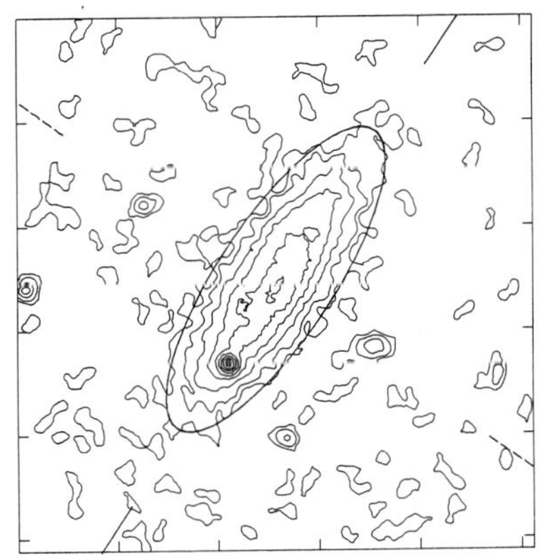

N1035 K4221 1 21.29 256X256

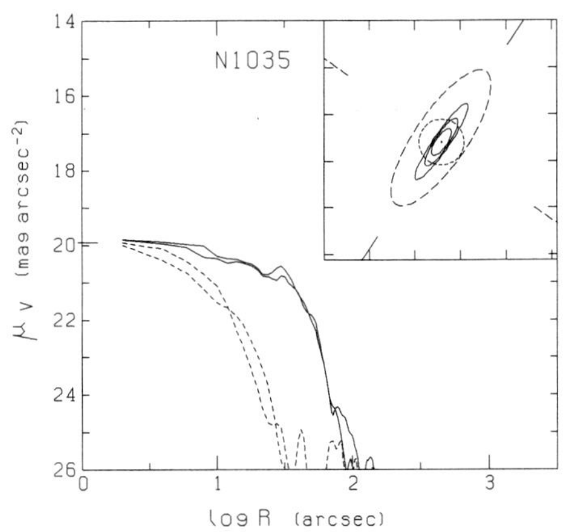

NGC1023

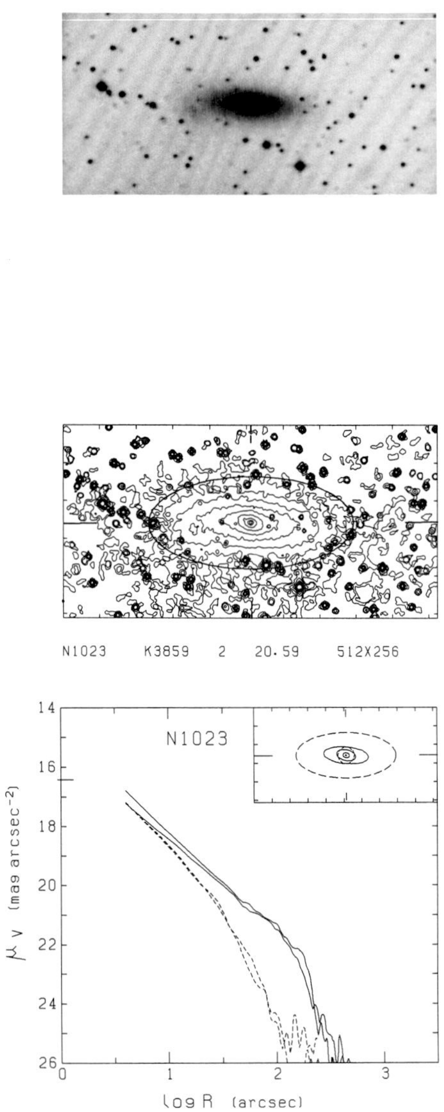

NGC1042

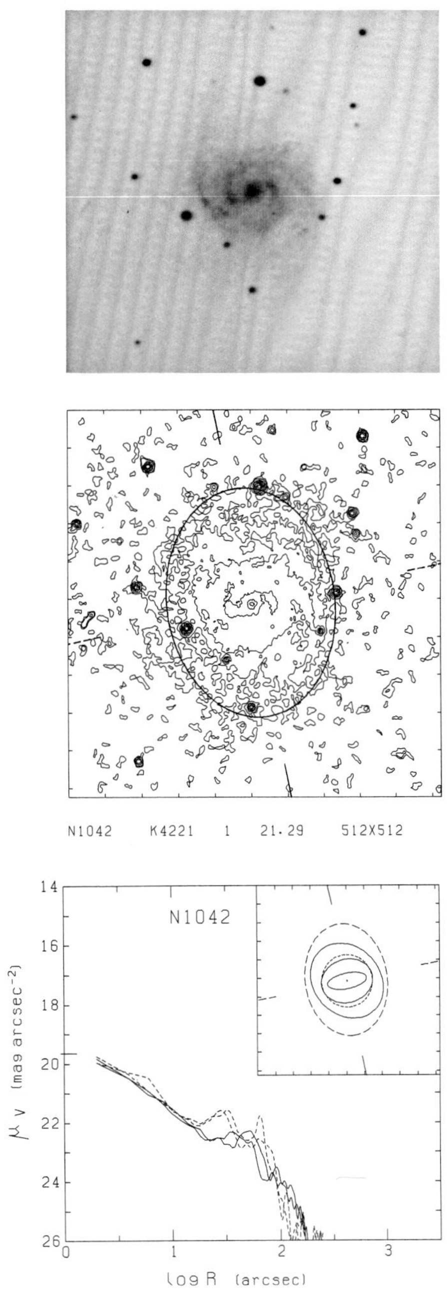

NGC1052

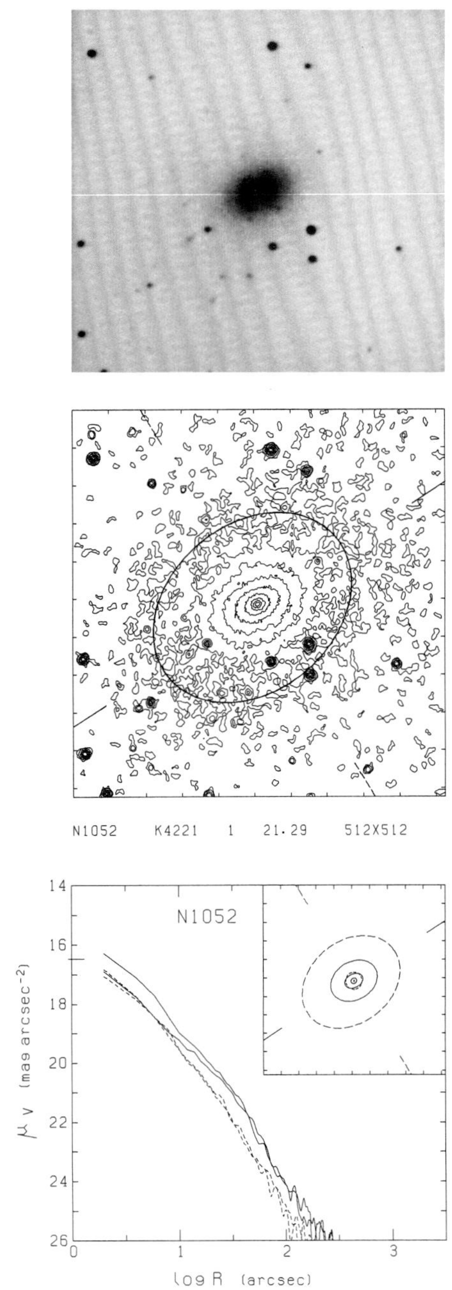

NGC1055

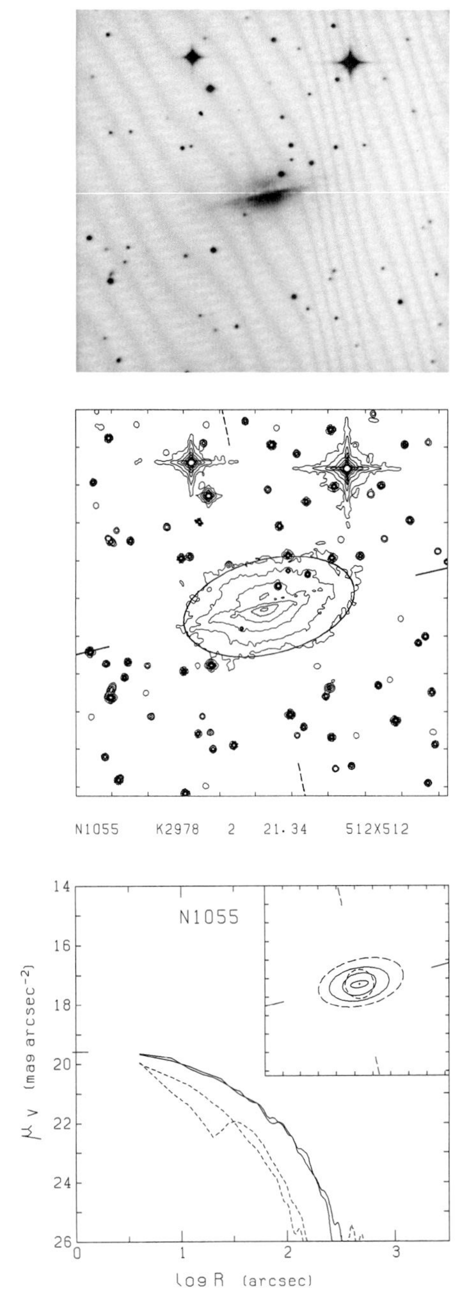

NGC1068

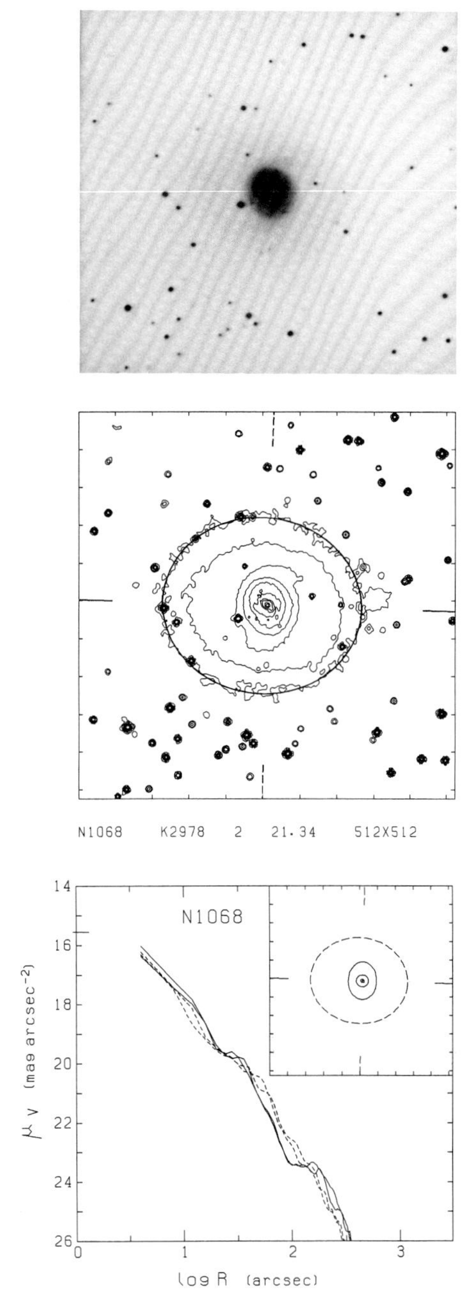

NGC1073

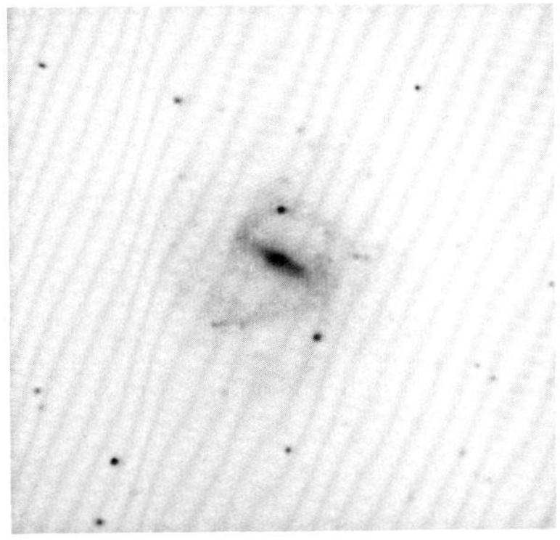

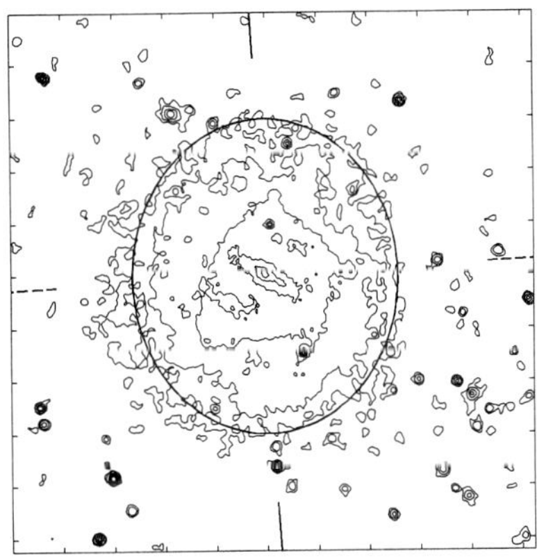

N1073 K2978 1 21.34 512X512

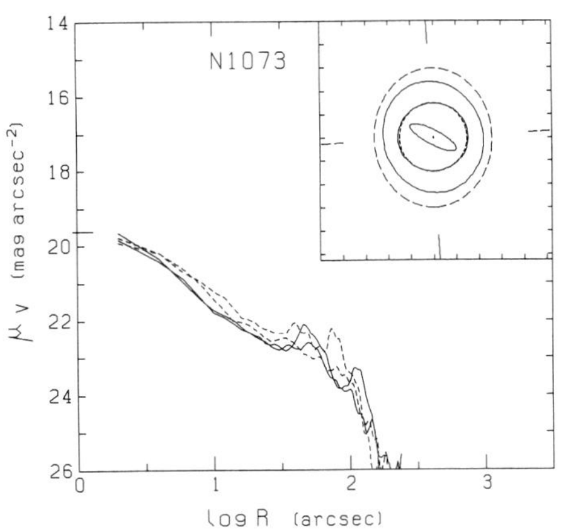

NGC1084

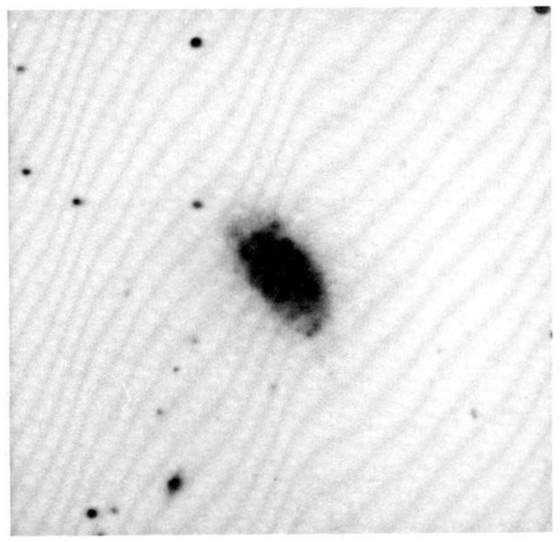

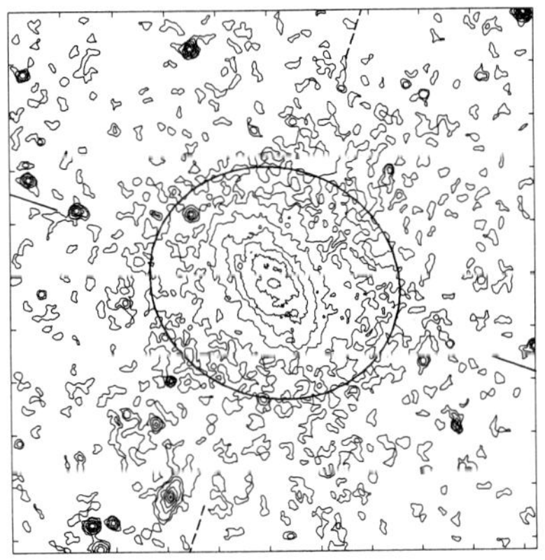

N1084 K4221 1 21.29 512X512

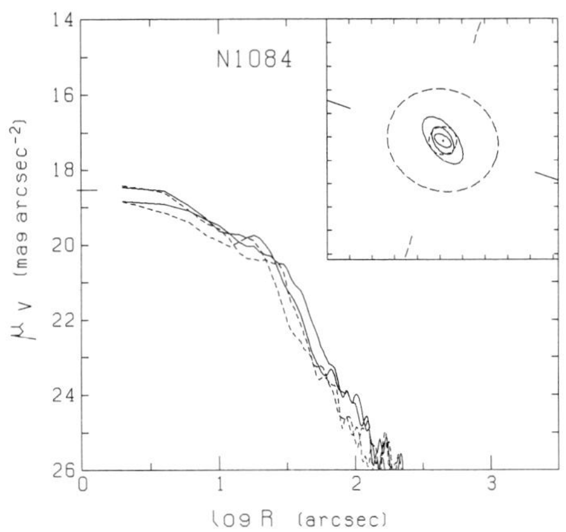

NGC1087

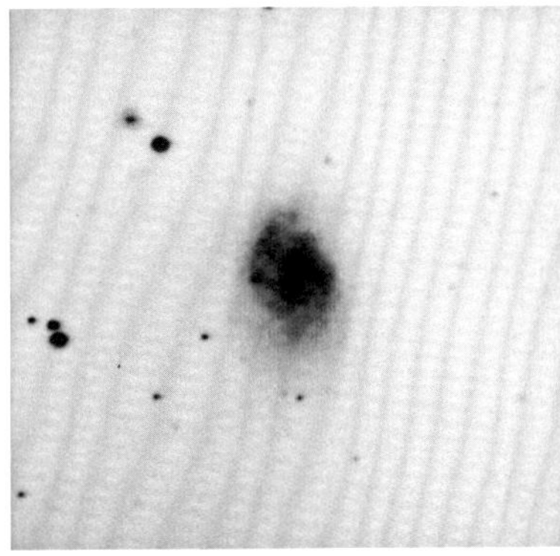

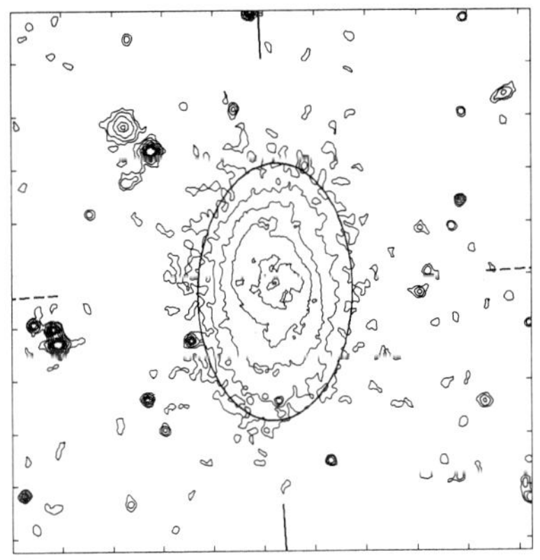

N1087 K2978 1 21.34 512X512

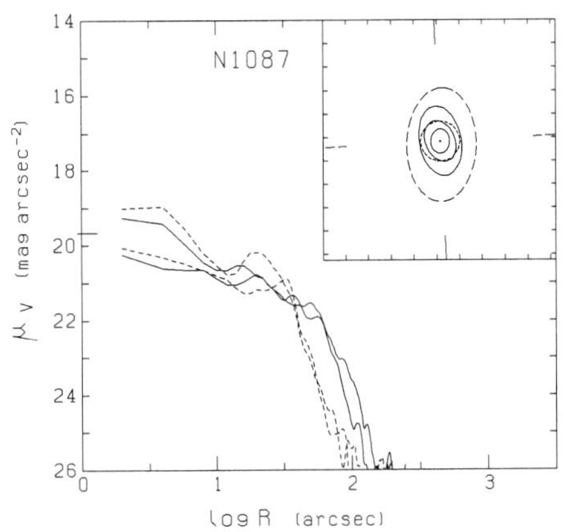

NGC1090

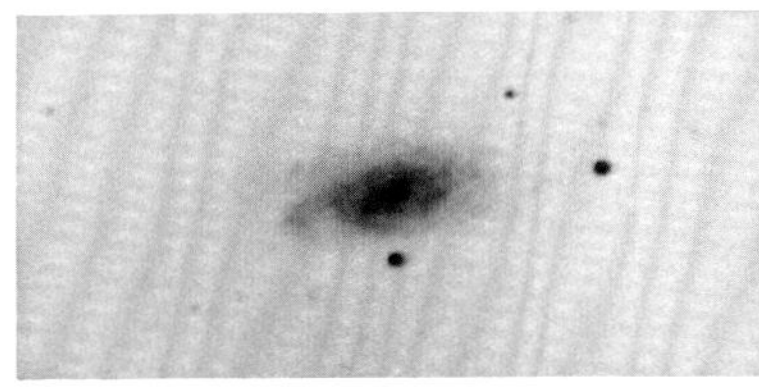

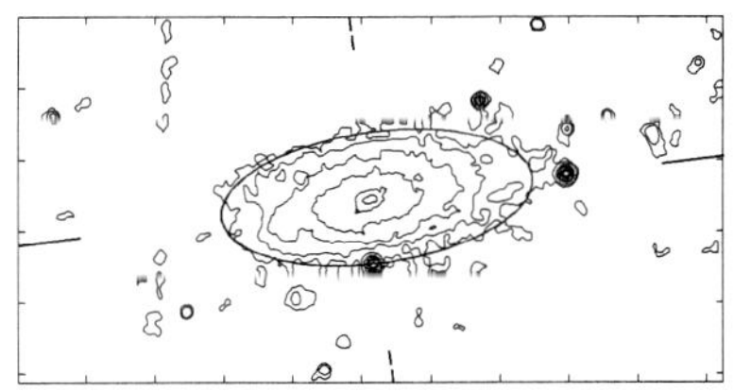

N1090 K2978 1 21.34 512X256

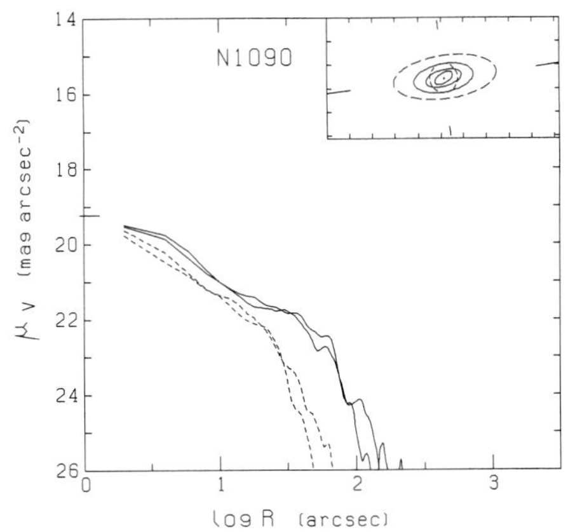

NGC1140

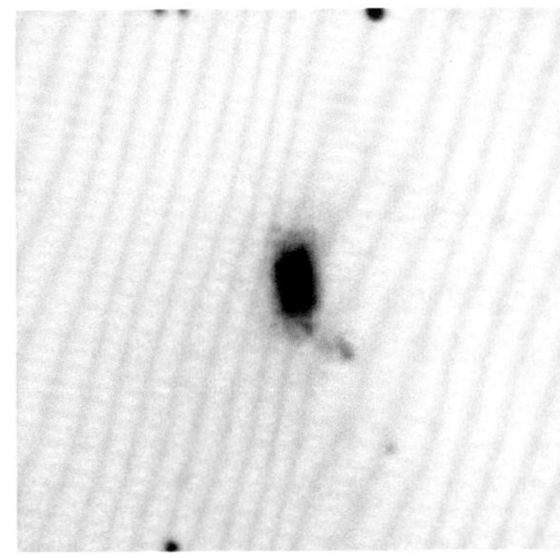

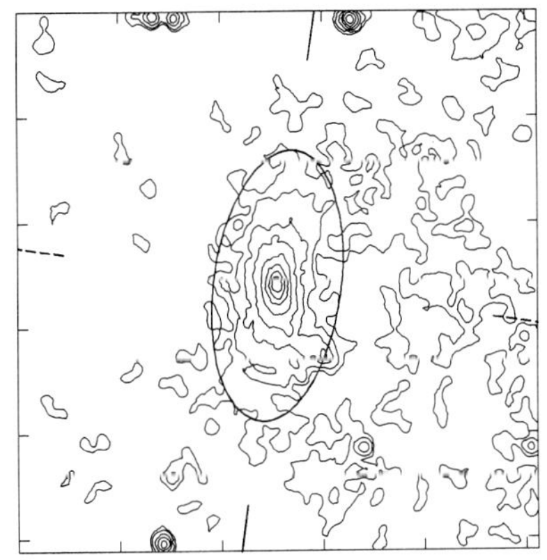

N1140 K5668 1 21.49 256X256

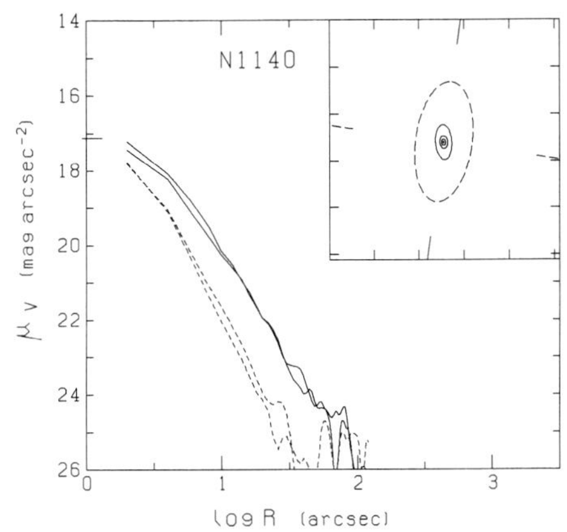

NGC1156

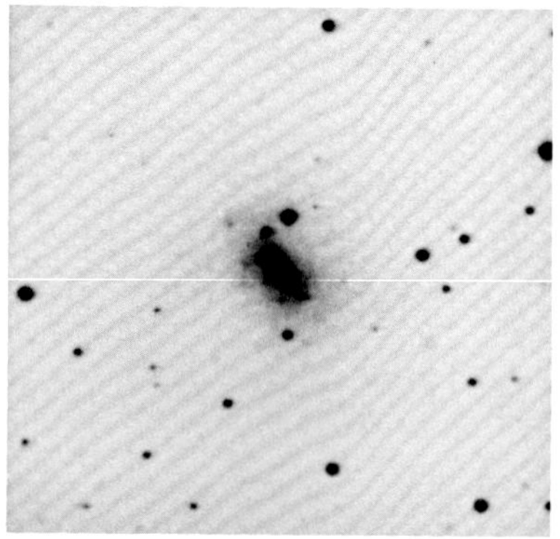

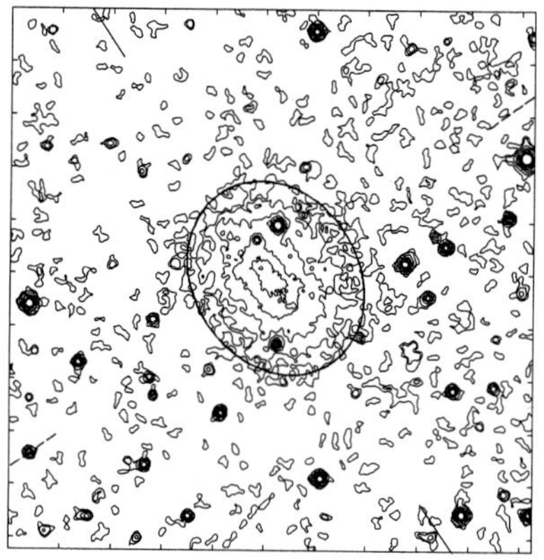

N1156 K5643 1 20.99 512X512

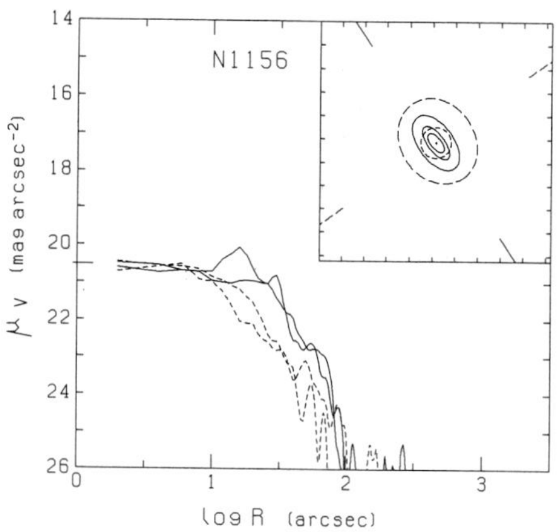

NGC1172

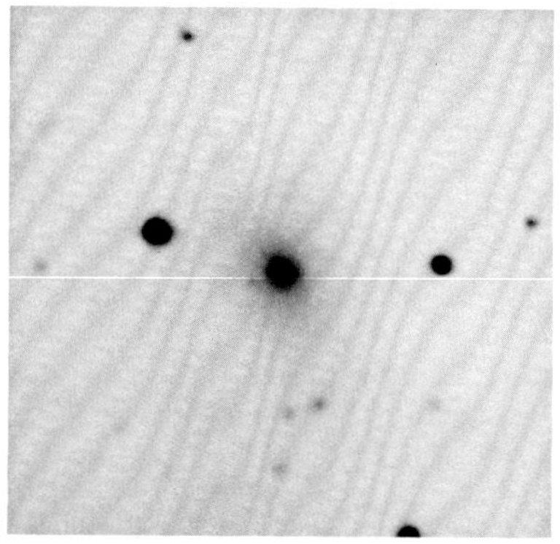

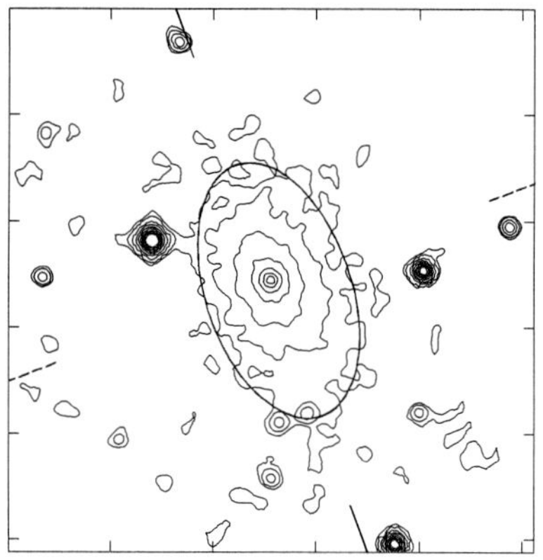

N1172 K4823 2 21.07 256X256

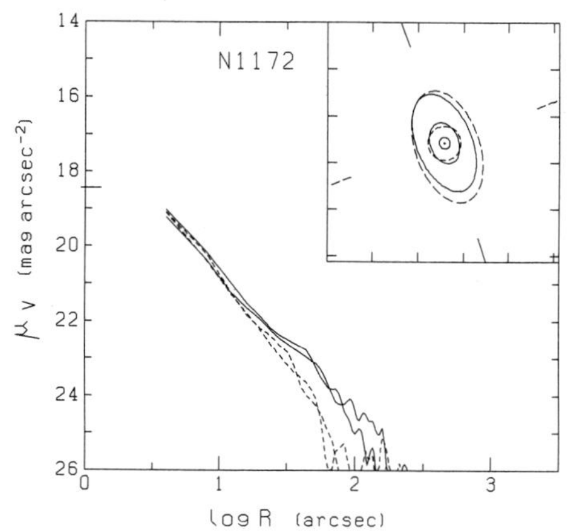

NGC1169

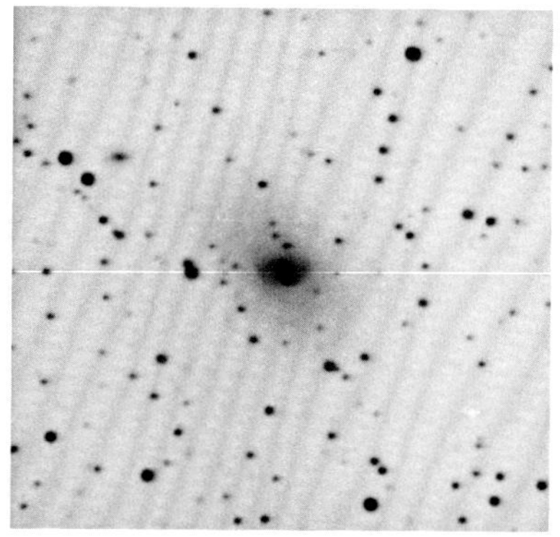

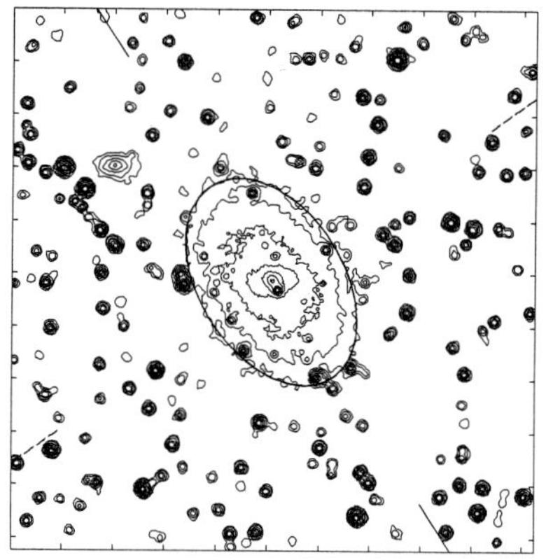

N1169 K2007 1 21.71 512X512

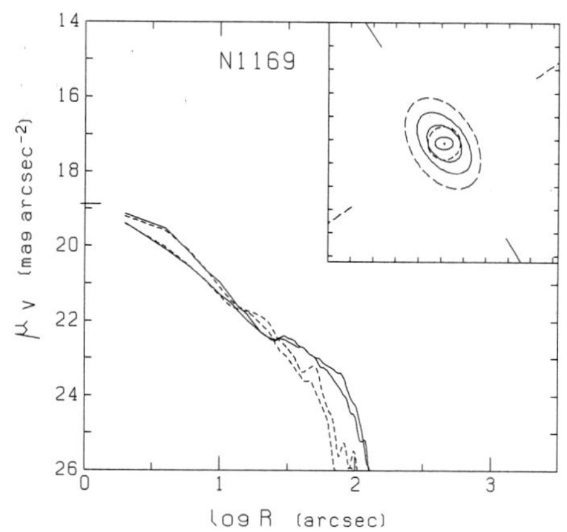

NGC1187

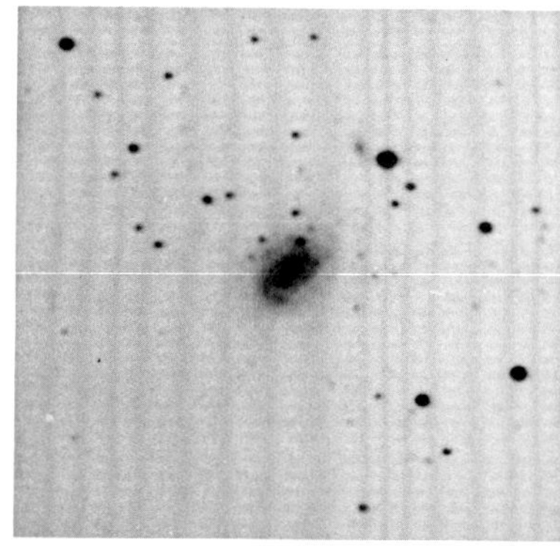

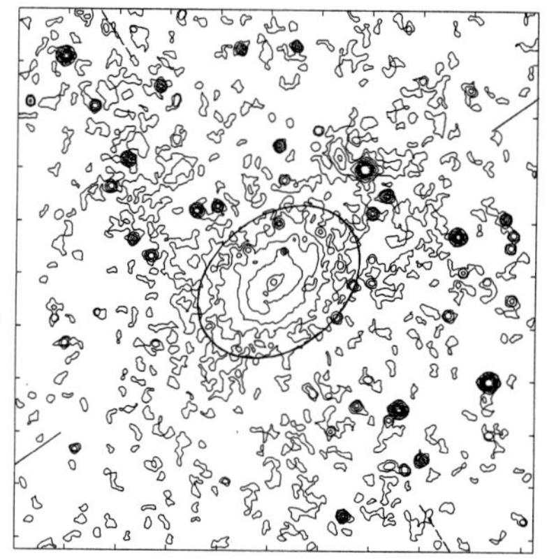

N1187 K4580 2 20.48 512X512

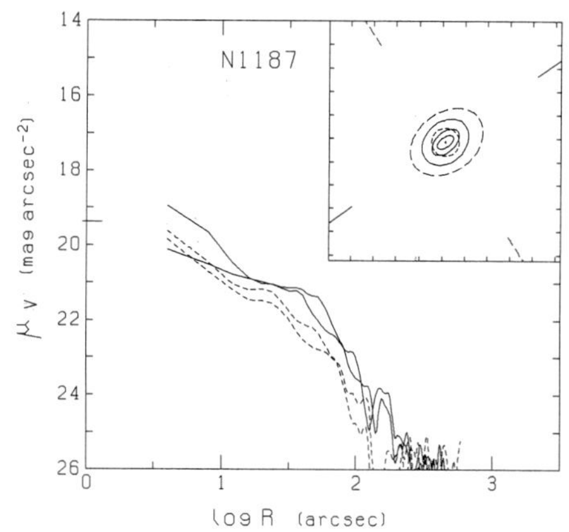

NGC1175

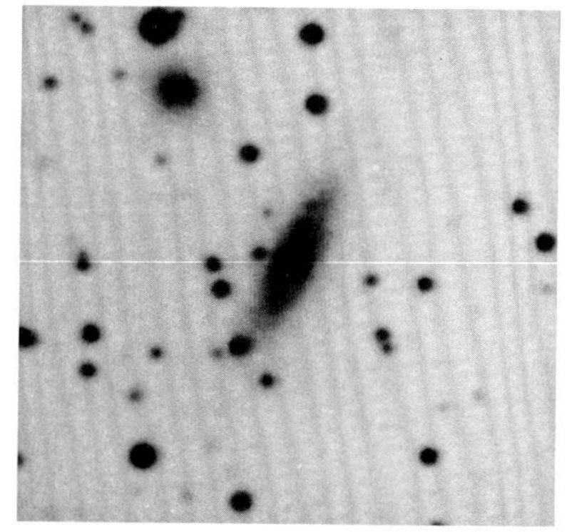

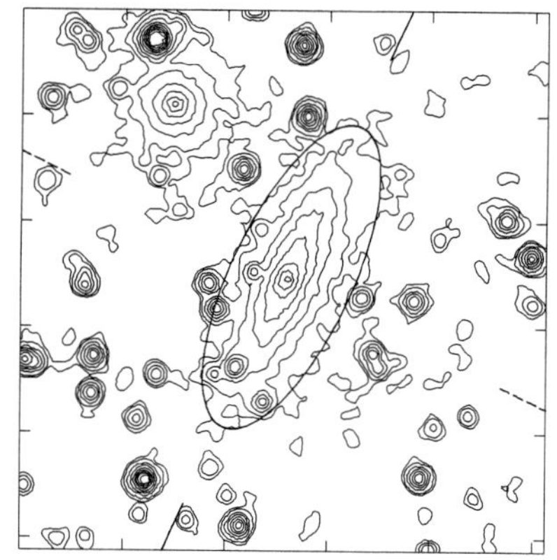

N1175 K2970 1 21.21 256X256

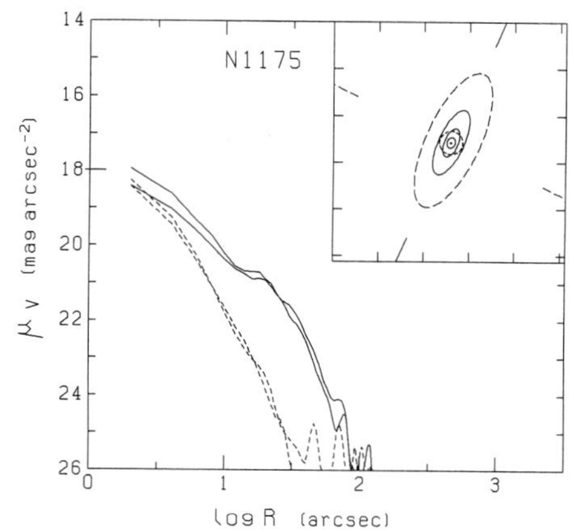

NGC1199

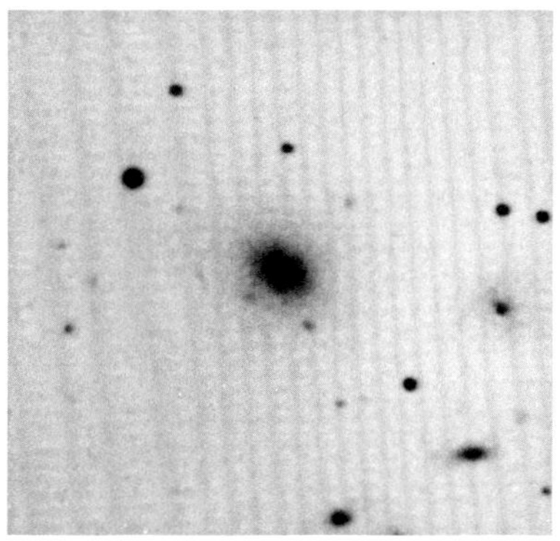

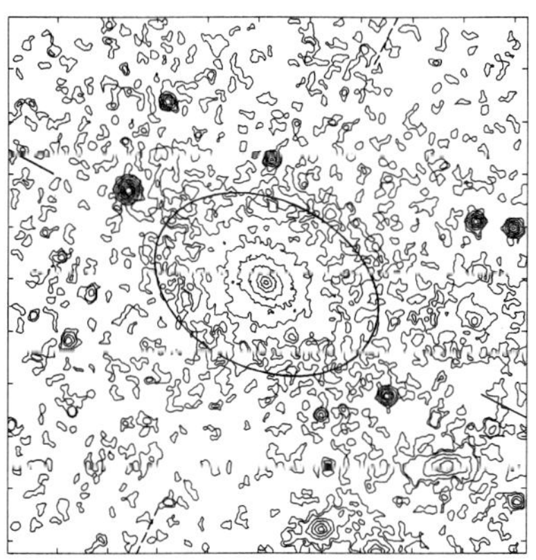

N1199 K4823 1 21.07 512X512

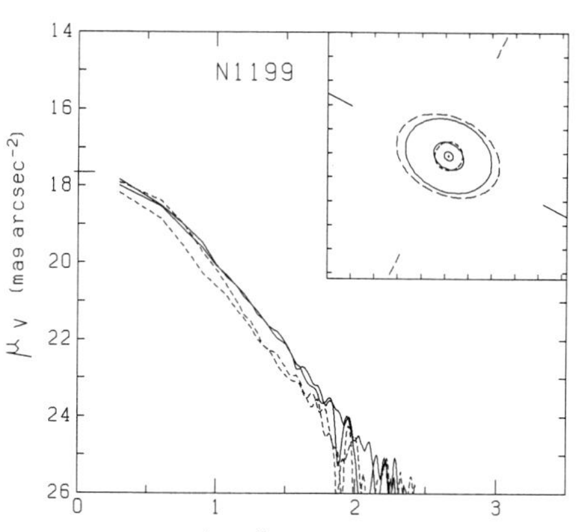

NGC1201

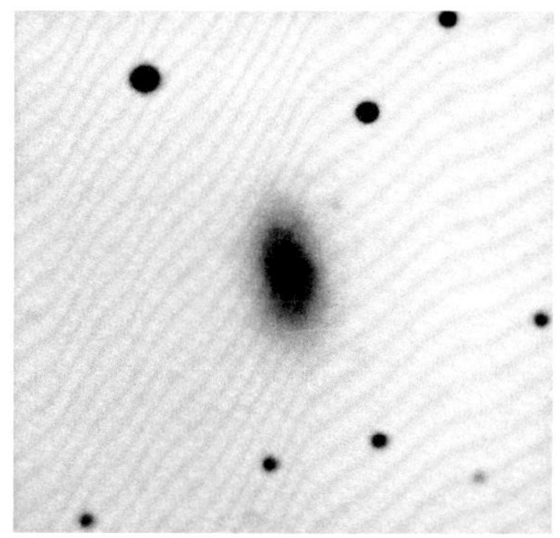

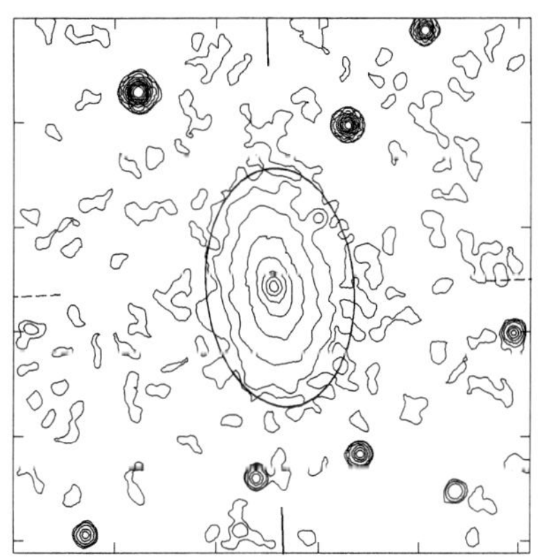

N1201 K4580 2 20.48 256X256

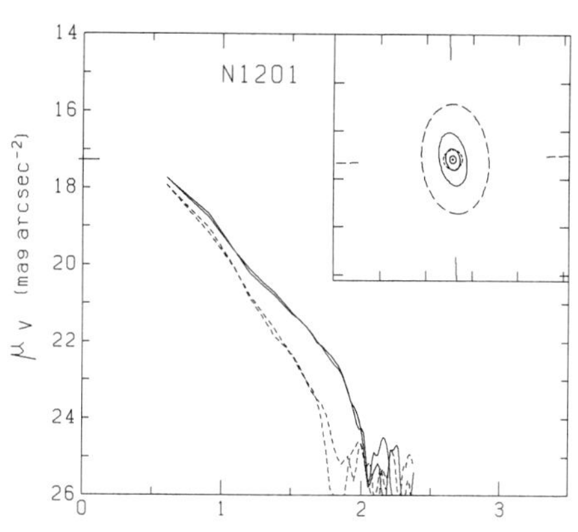

NGC1209

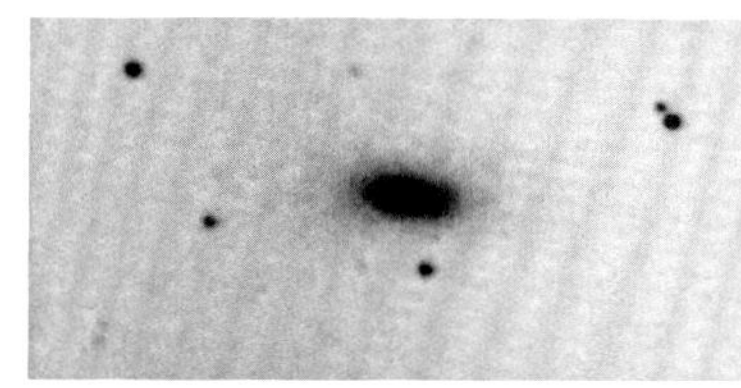

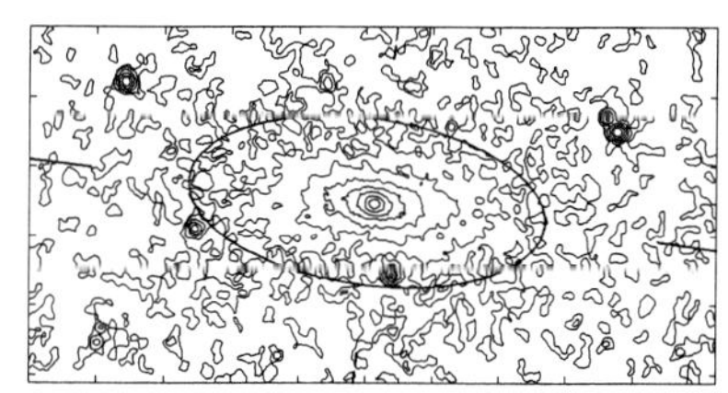

N1209 K4823 1 21.07 512X256

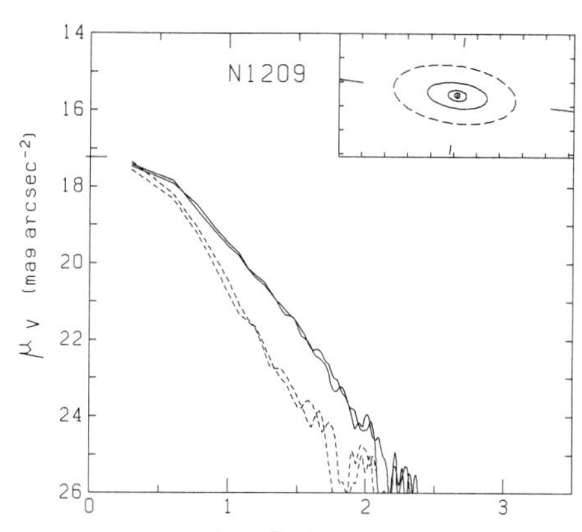

NGC1241

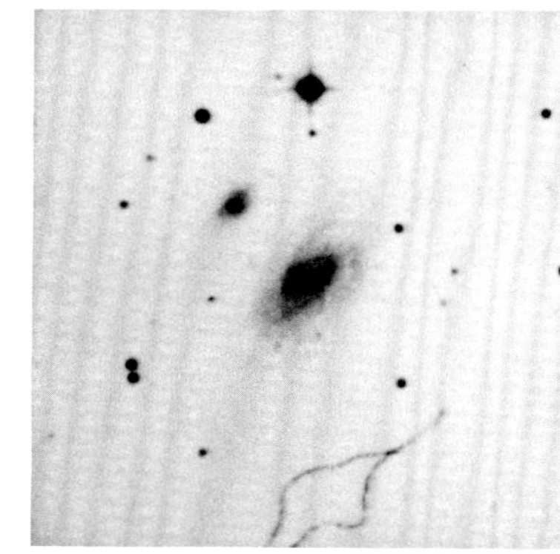

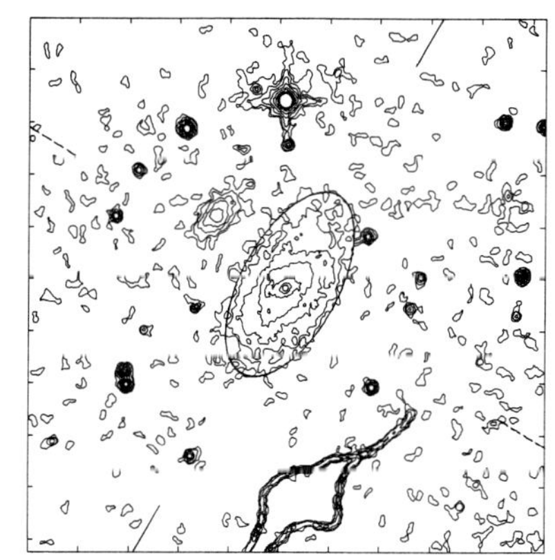

N1241 K5668 1 21.49 512X512

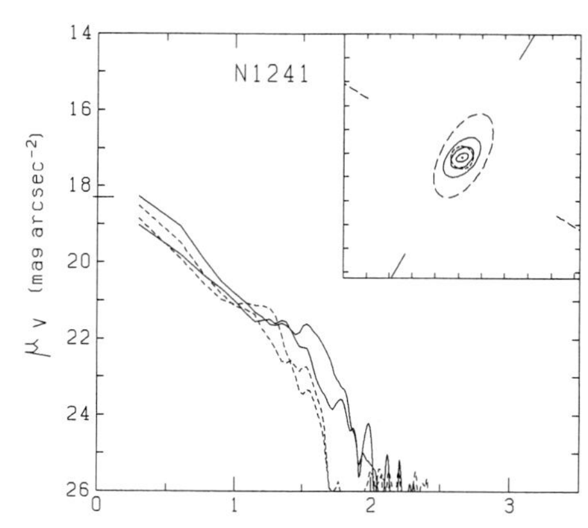

NGC1255

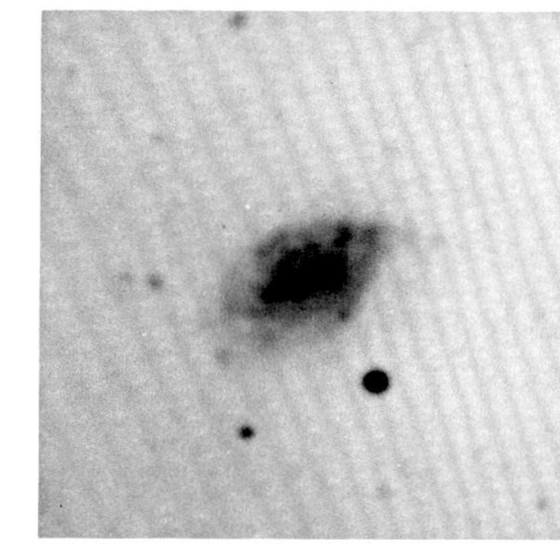

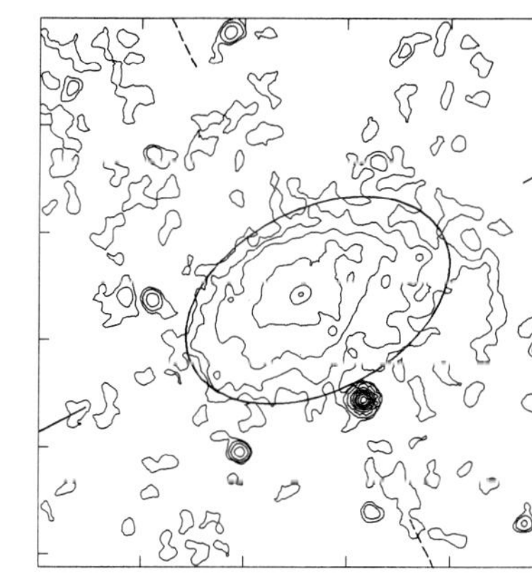

N1255 K4580 2 20.48 256X256

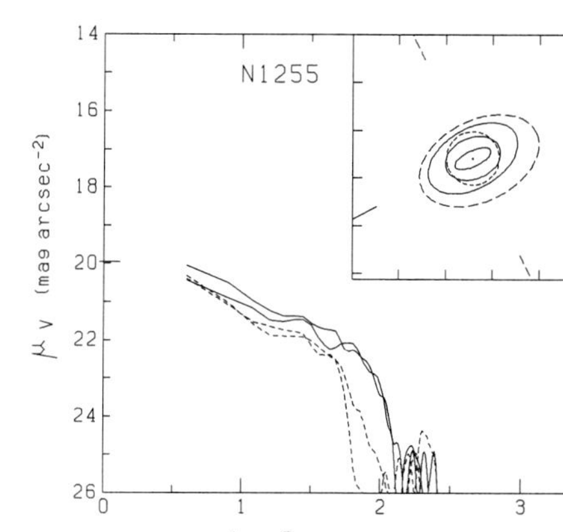

NGC1292

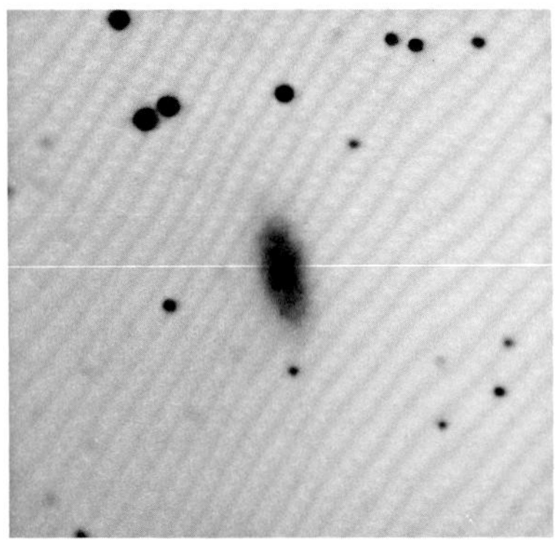

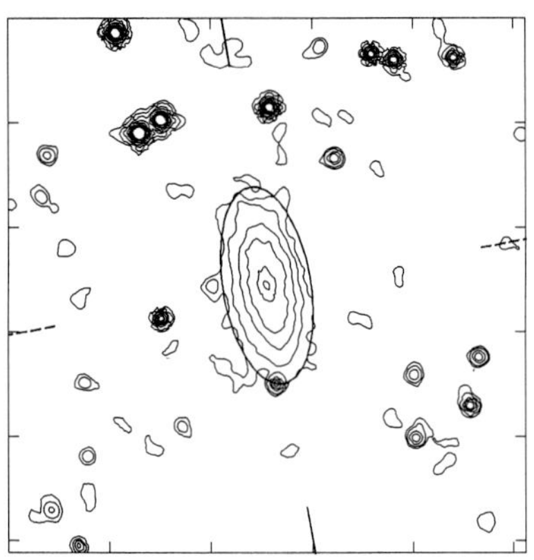

N1292 K4519 2 21.11 256X256

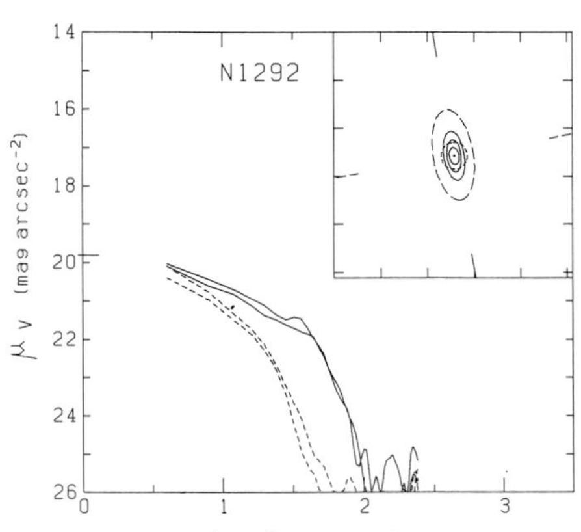

NGC1275

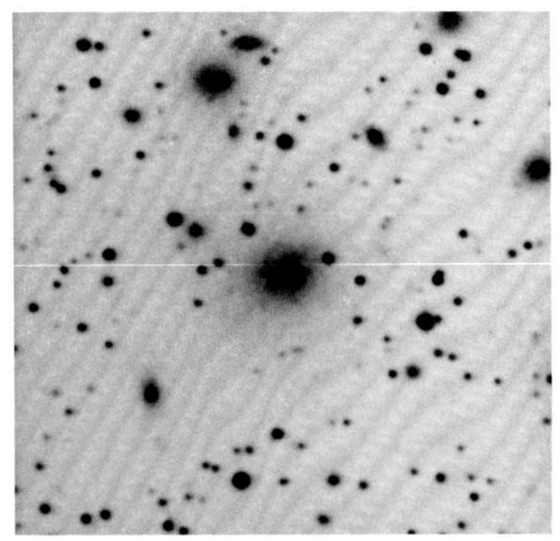

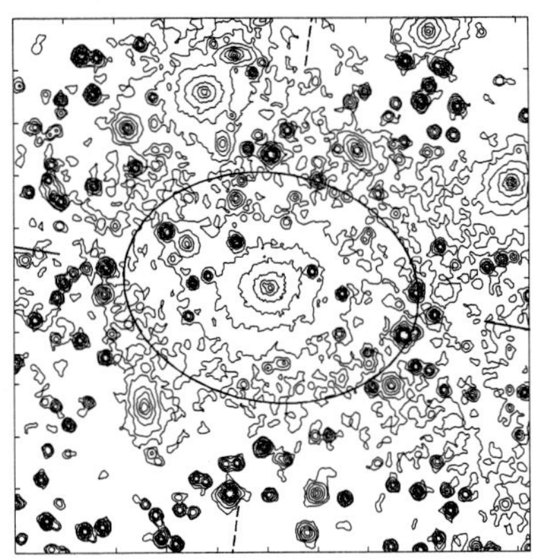

N1275 K2970 1 21.21 512X512

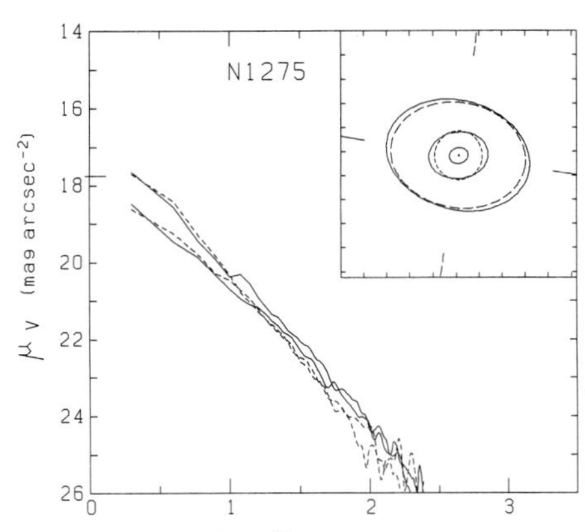

NGC1297

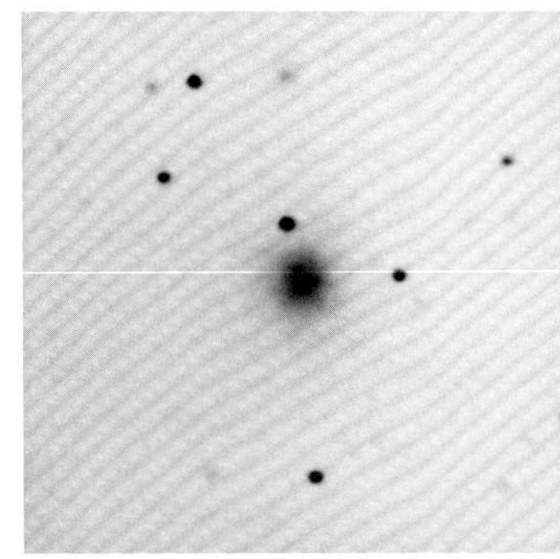

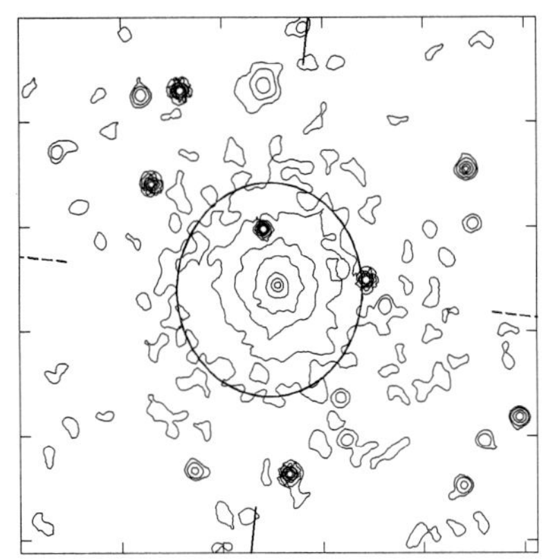

N1297 K2891 2 21.01 256X256

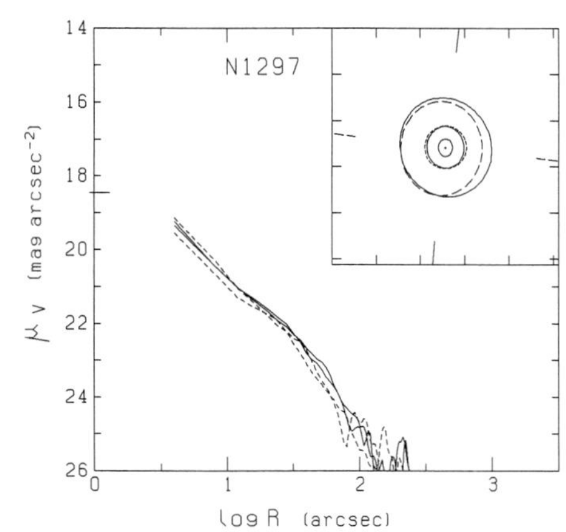

NGC1300

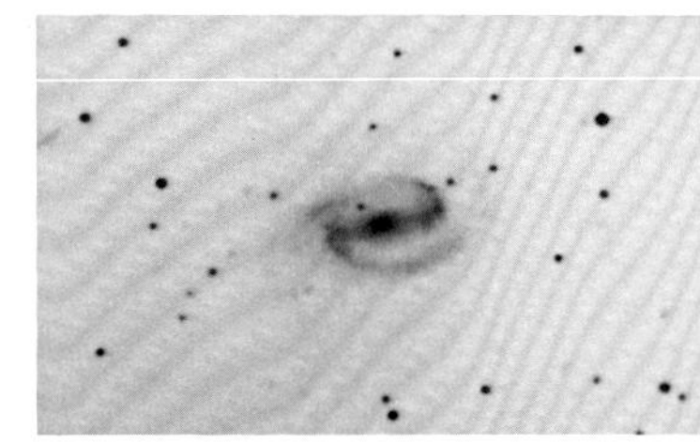

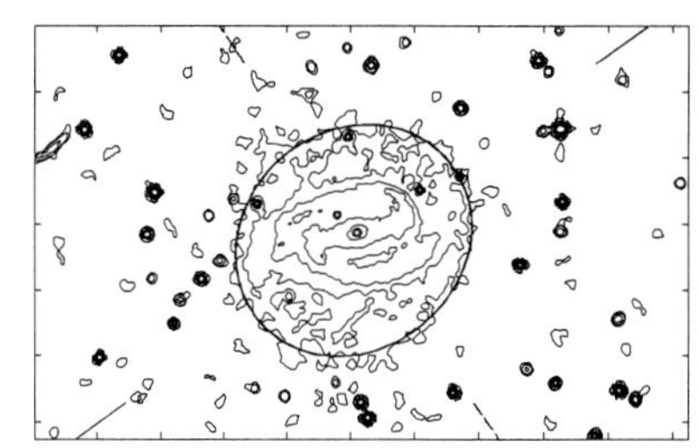

N1300 K2891 2 21.01 512X313

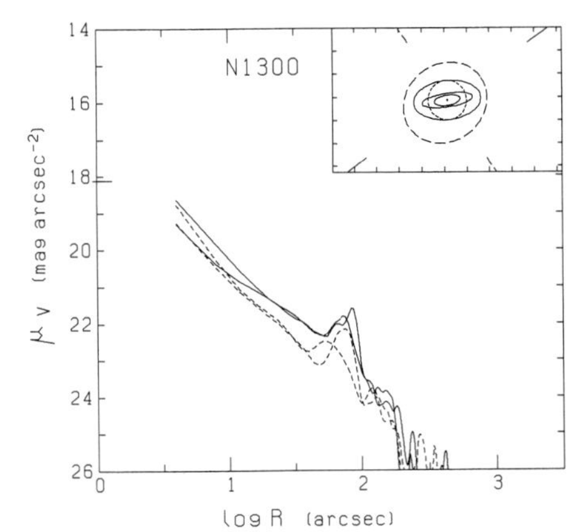

NGC1325

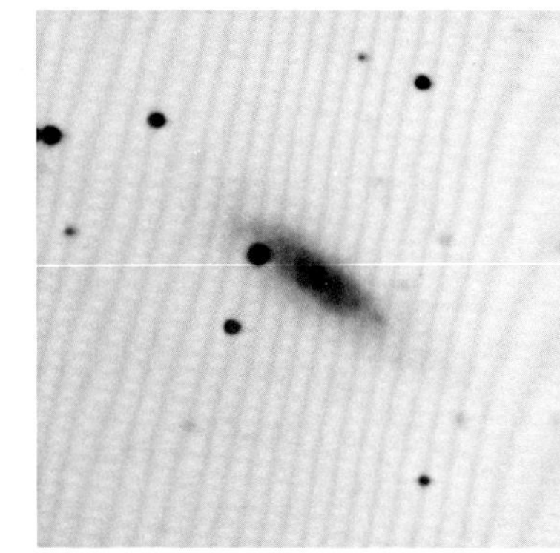

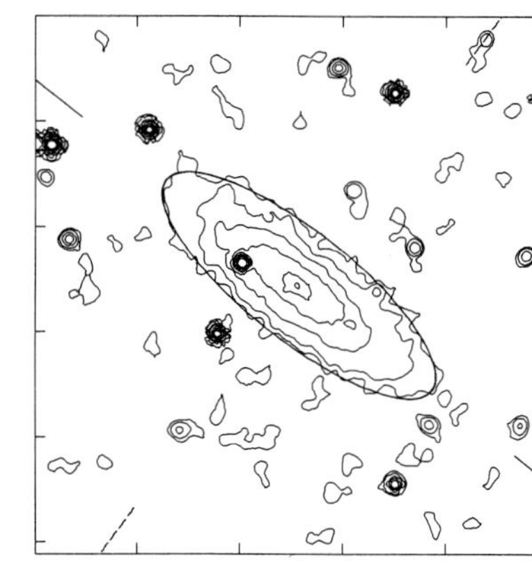

N1325 K2891 2 21.01 256X256

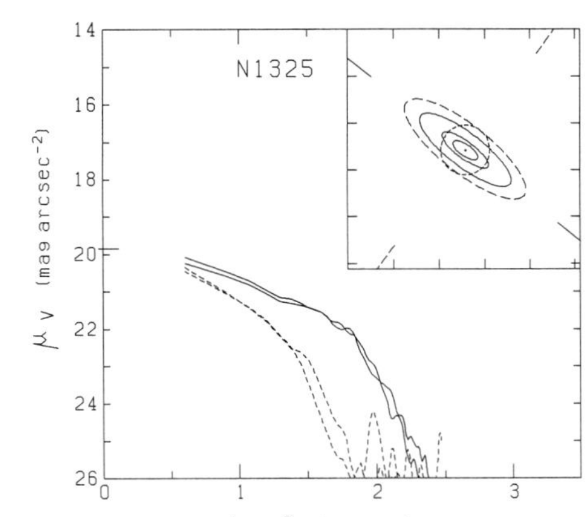

NGC1332

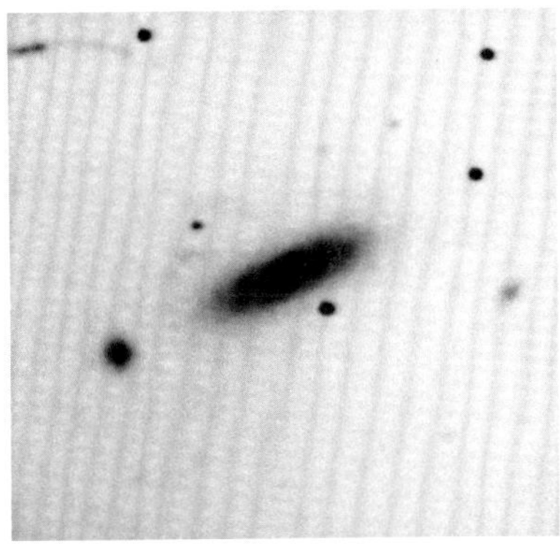

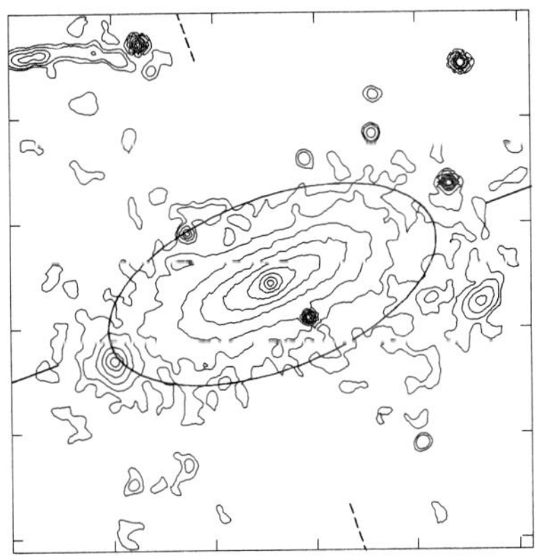

N1332 K2891 2 21.01 256X256

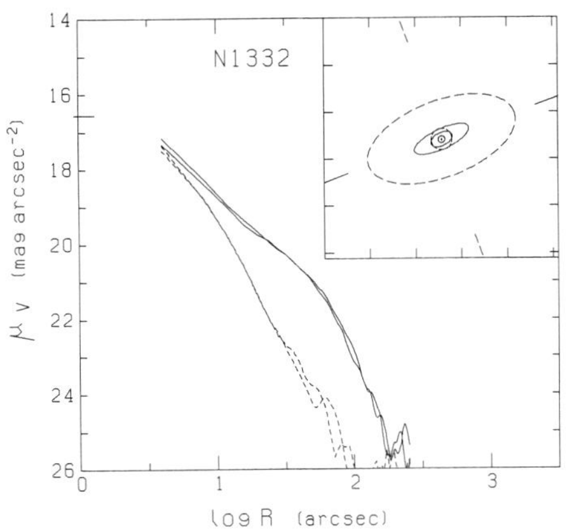

NGC1337

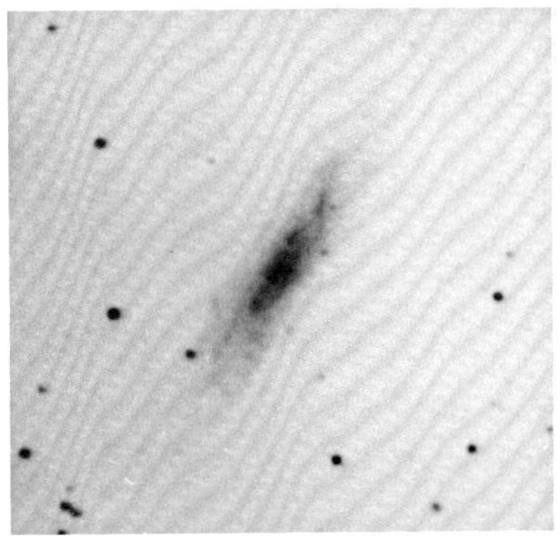

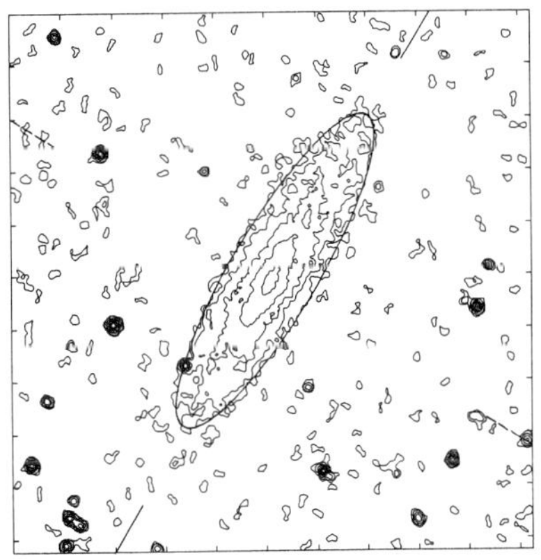

N1337 K5535 1 21.48 512X512

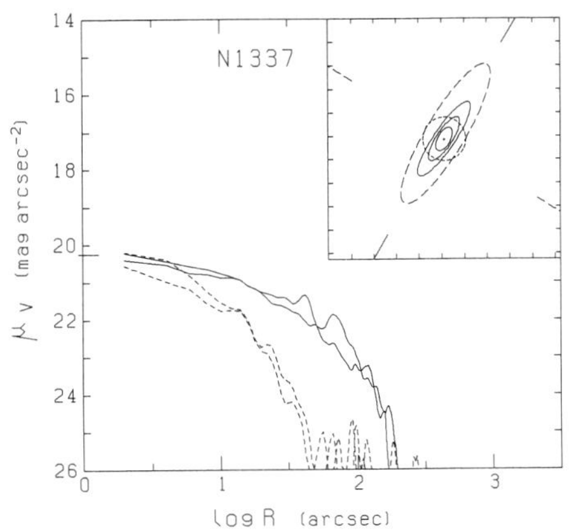

NGC1339

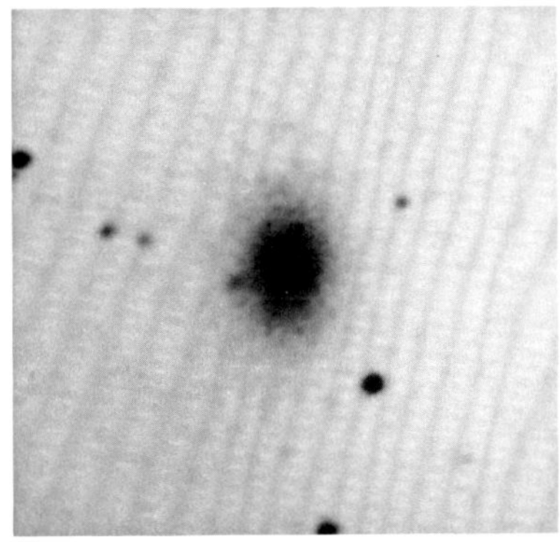

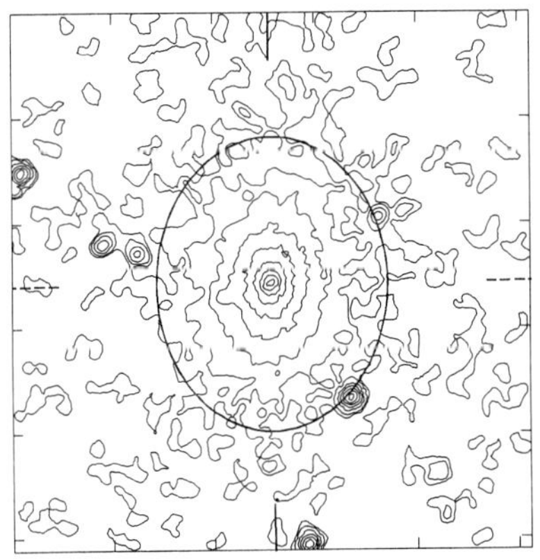

N1339 K4519 1 21.11 256X256

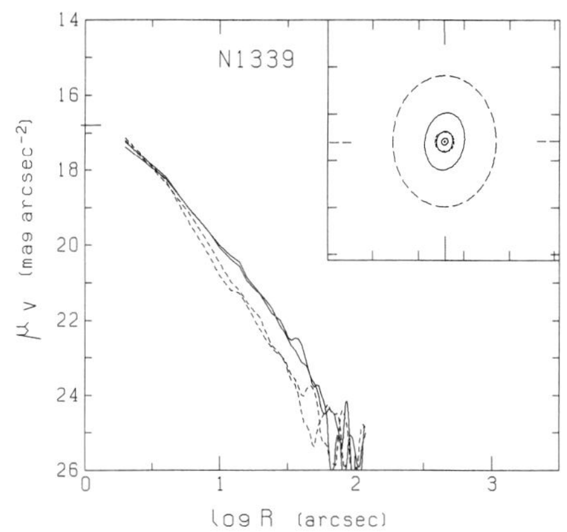

NGC1344

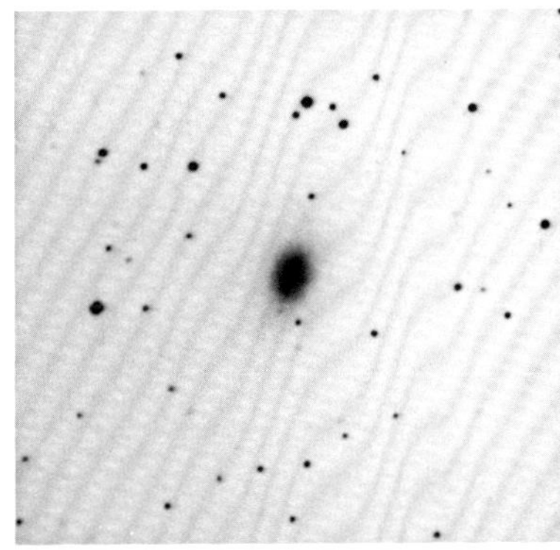

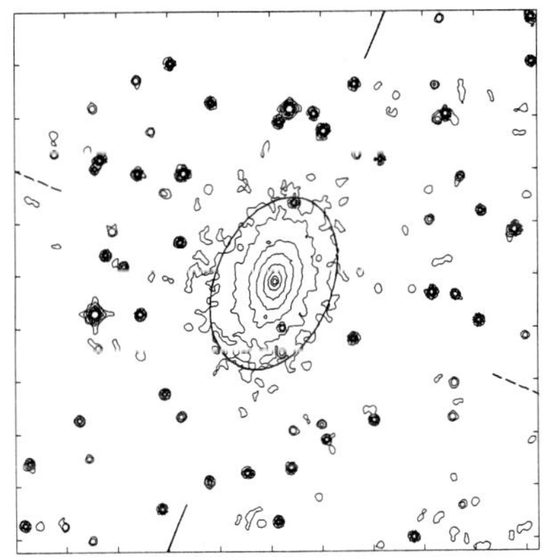

N1344 K4519 2 21.11 512X512

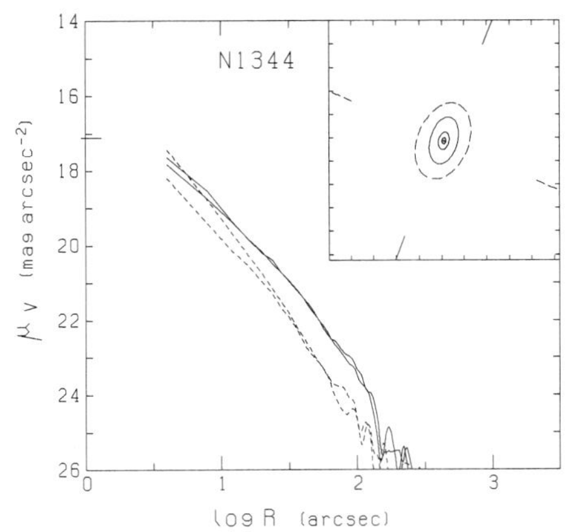

NGC1353

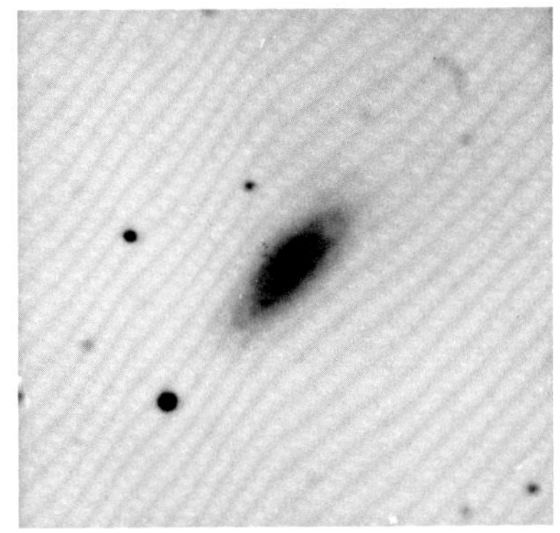

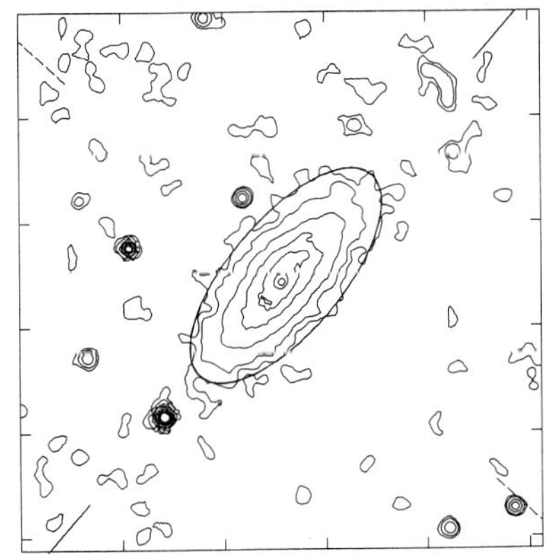

N1353 K2891 2 21.01 256X256

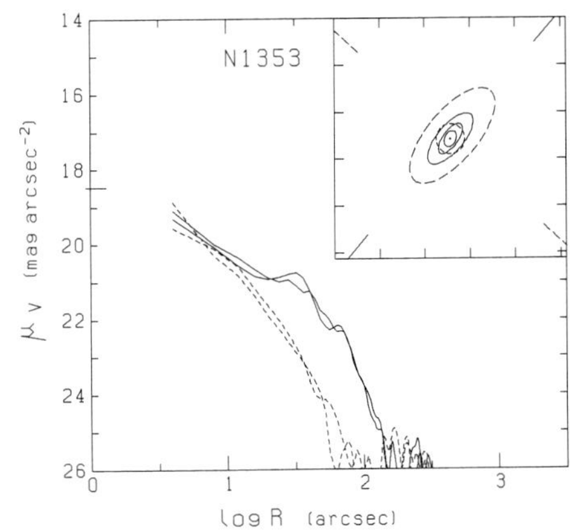

NGC1357

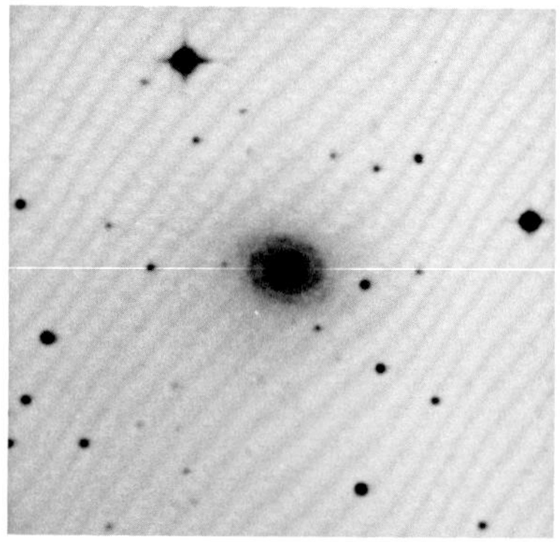

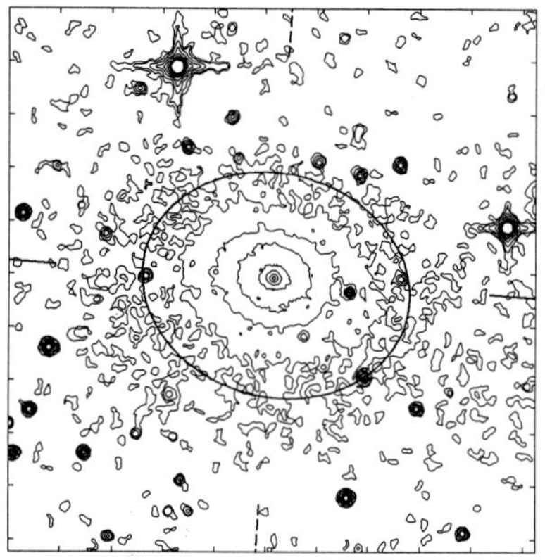

N1357 K2071 1 20.82 512X512

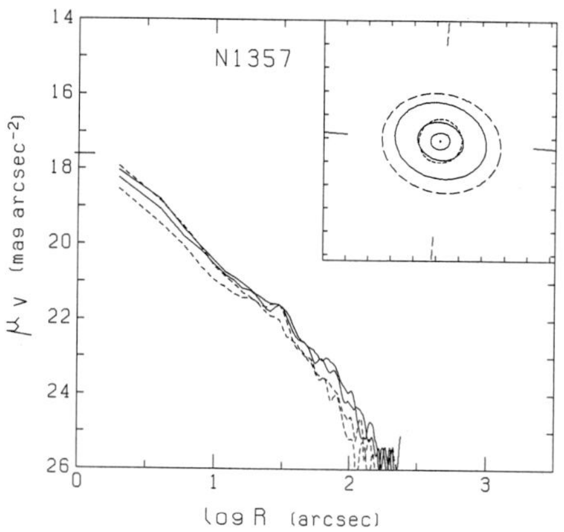

NGC1358

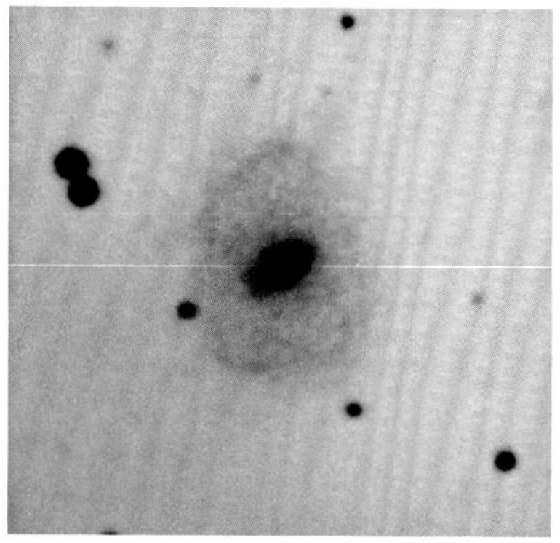

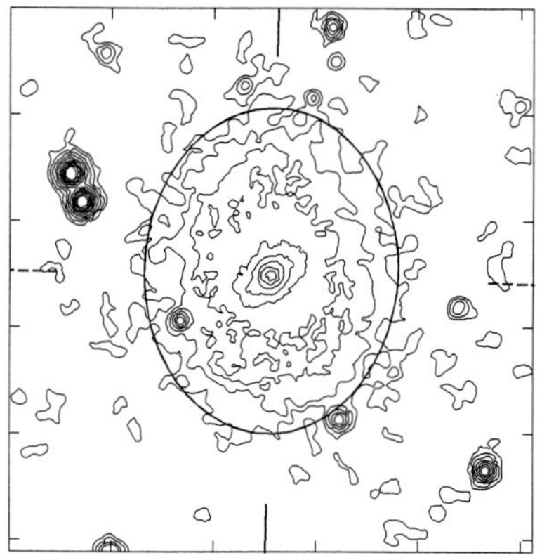

N1358 K4520 1 21.77 256X256

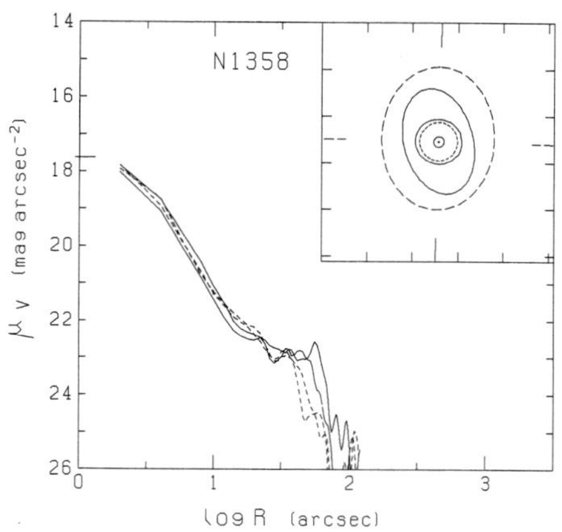

IC1953

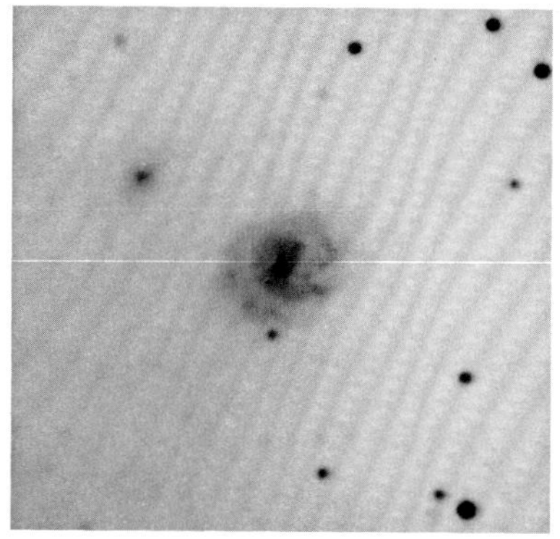

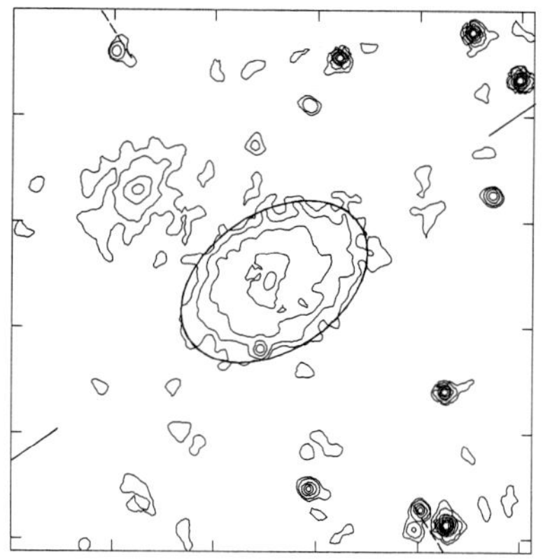

I1953 K4783 2 21.07 256X256

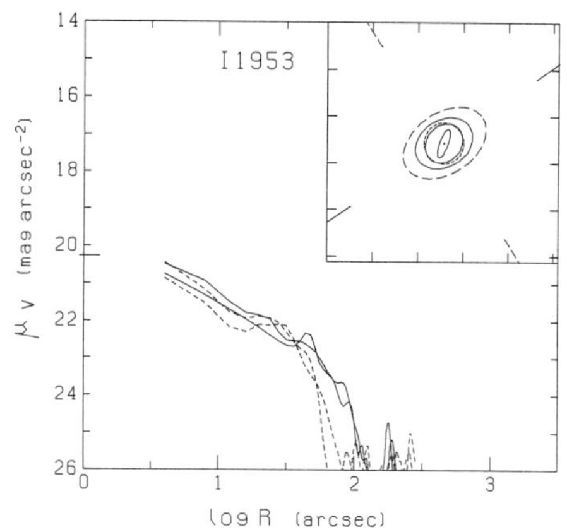

NGC1359

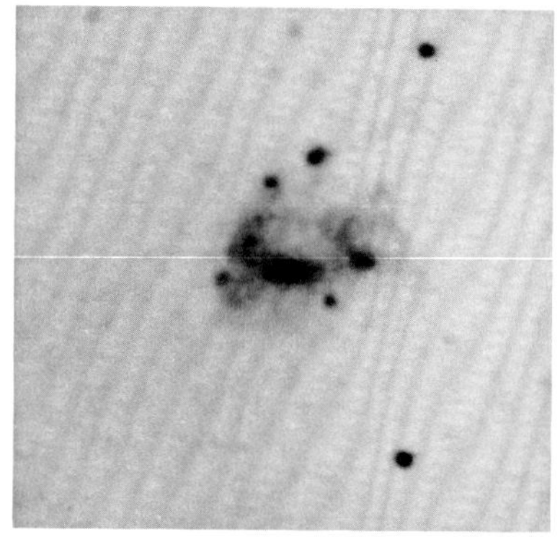

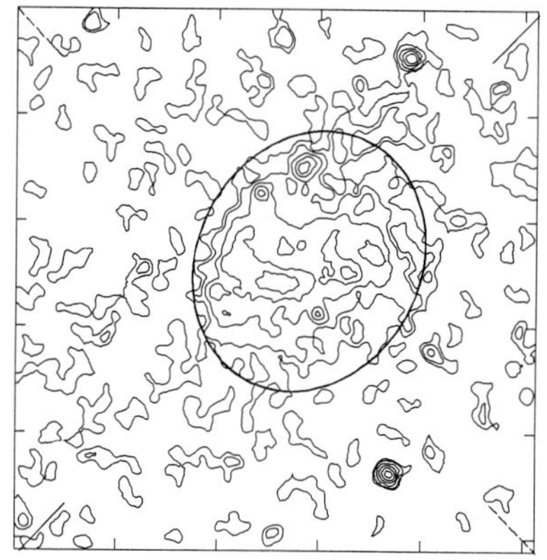

N1359 K2891 1 21.01 256X256

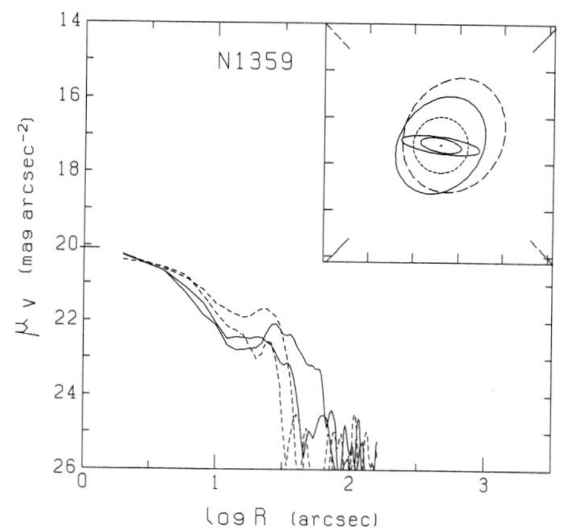

NGC1366

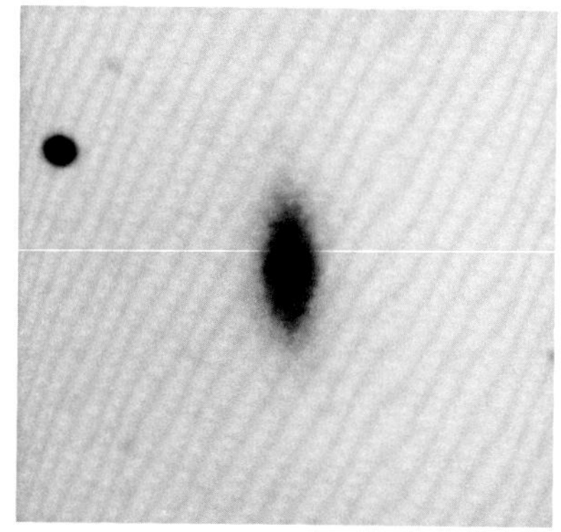

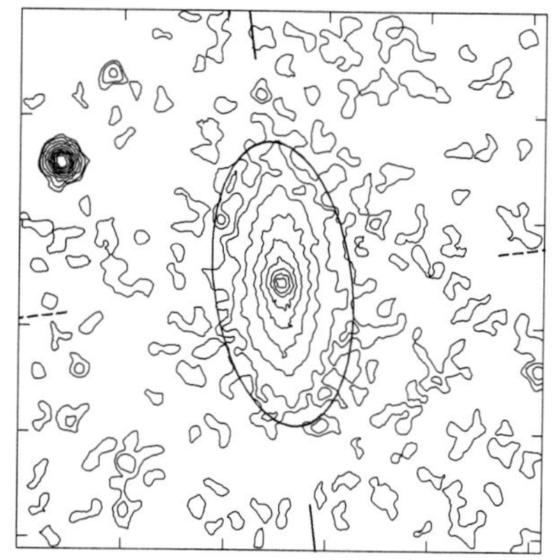

N1366 K4525 1 21.02 256X256

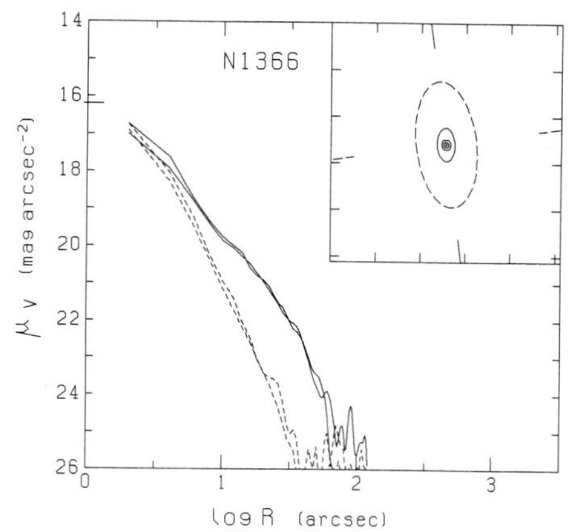

NGC1371

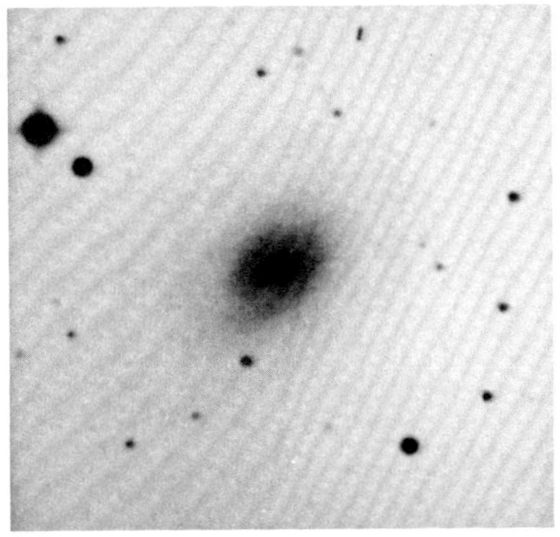

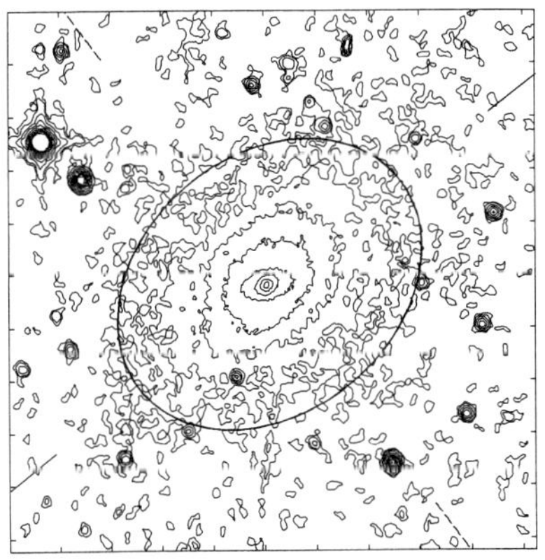

N1371 K4513 1 21.19 512X512

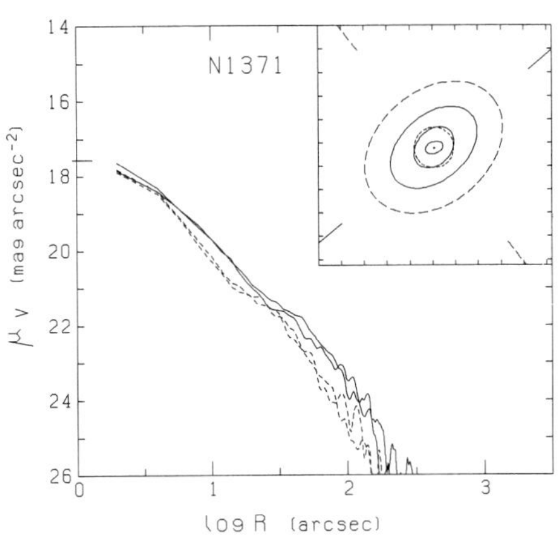

NGC1376

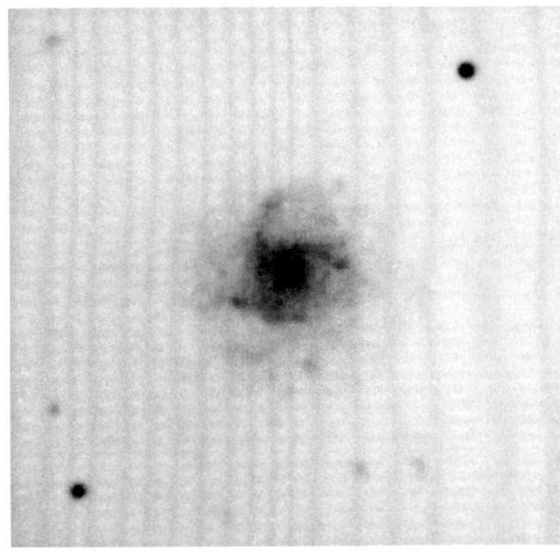

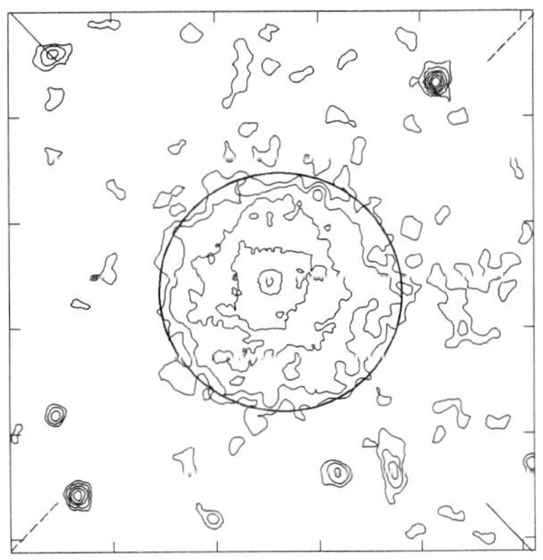

N1376 K4520 1 21.77 256X256

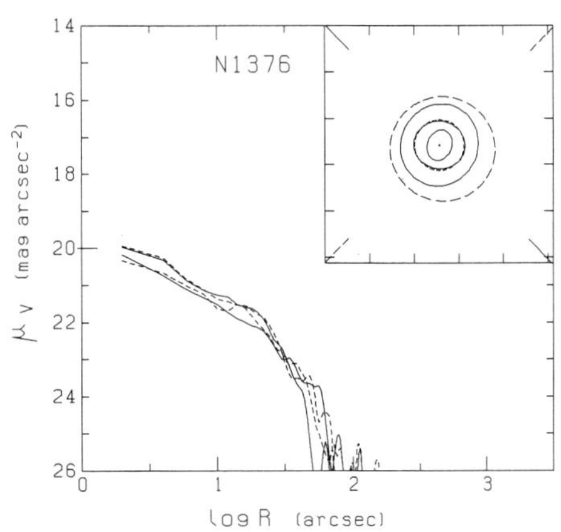

NGC1385

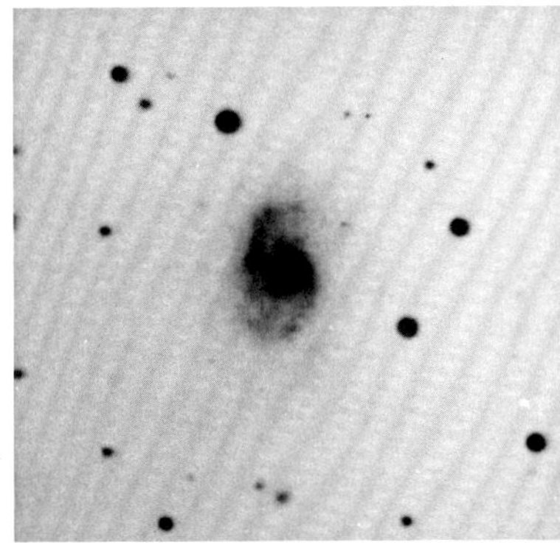

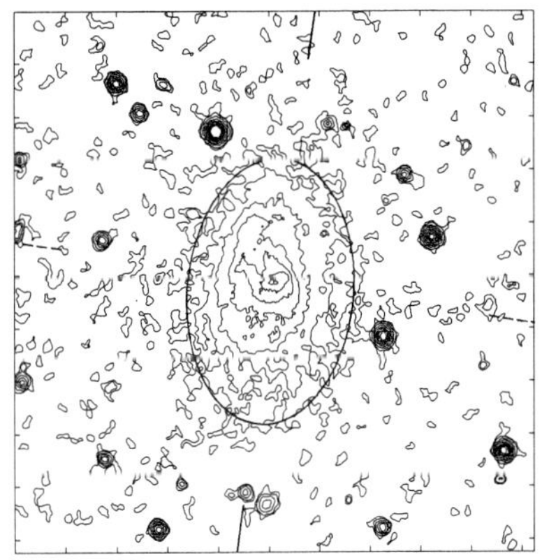

N1385 K4513 1 21.19 512X512

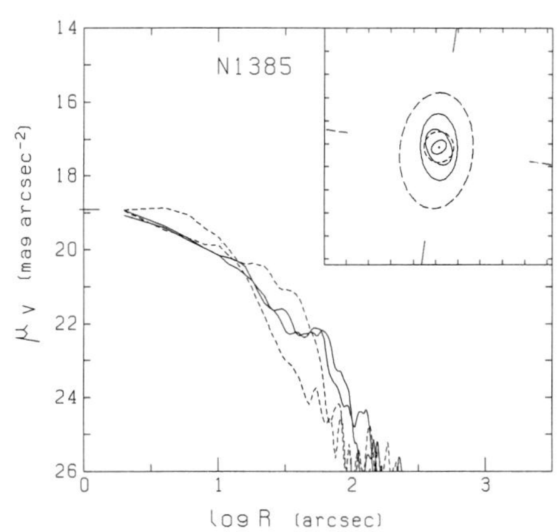

NGC1395

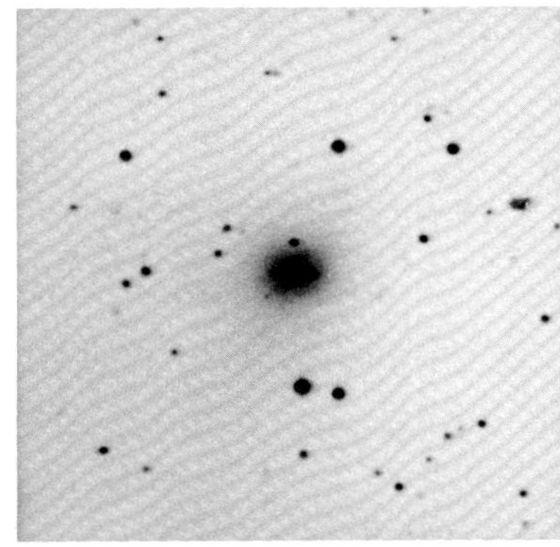

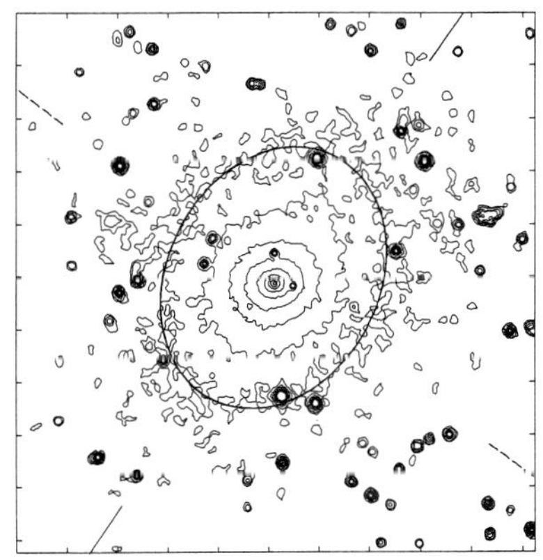

N1395 K4513 2 21.19 512X512

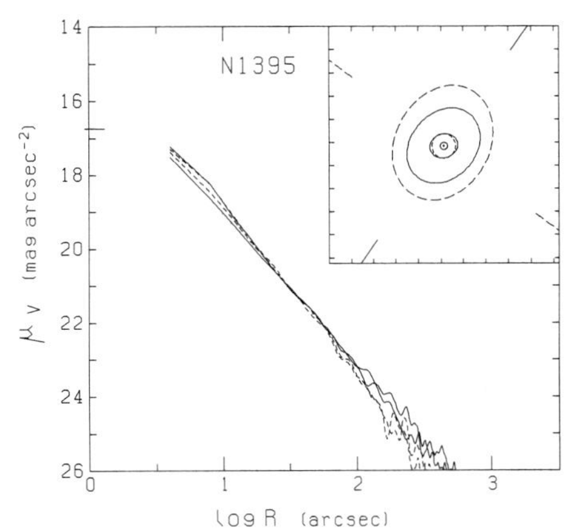

NGC1398

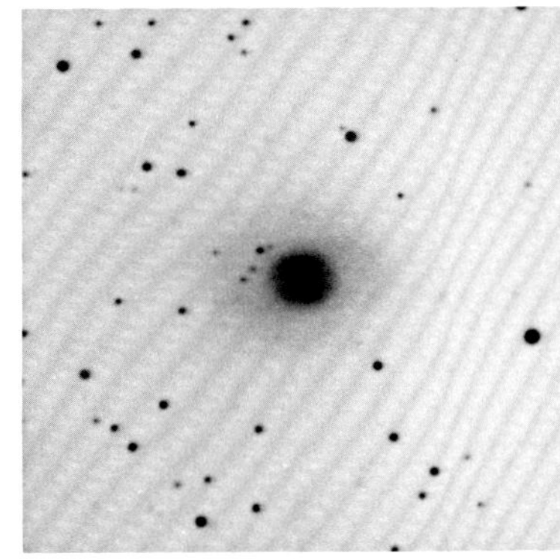

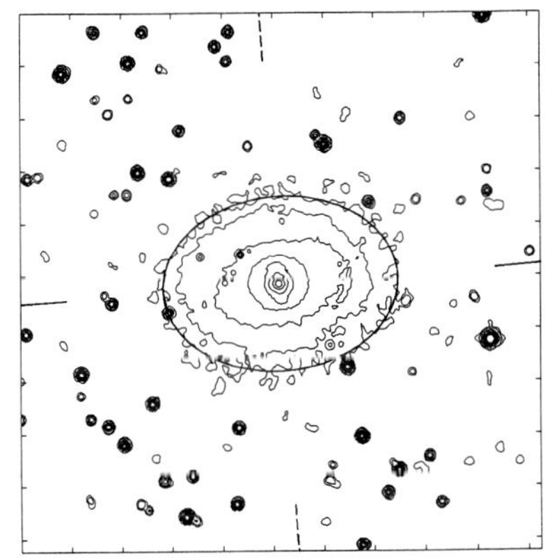

N1398 K4513 2 21.19 512X512

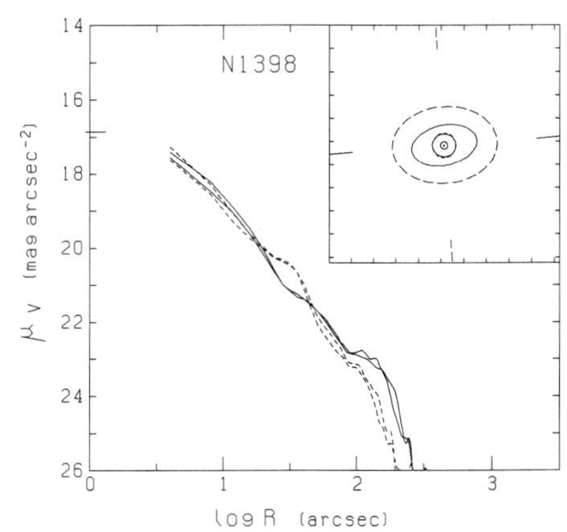

NGC1400

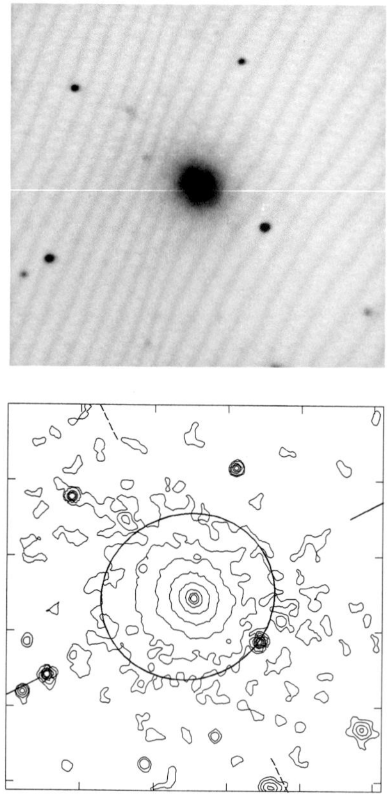

N1400 K4783 2 21.07 256X256

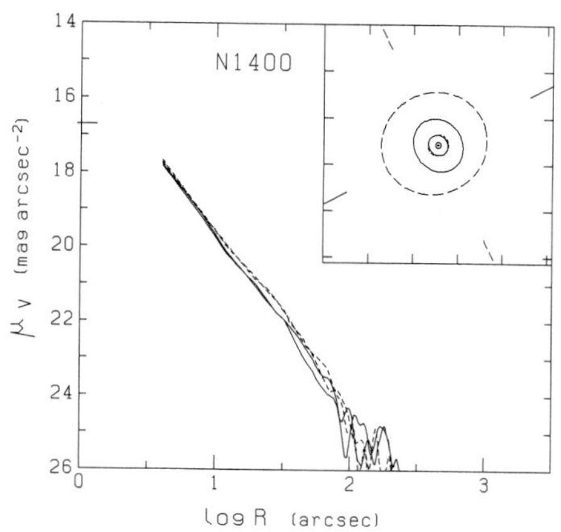

NGC1406

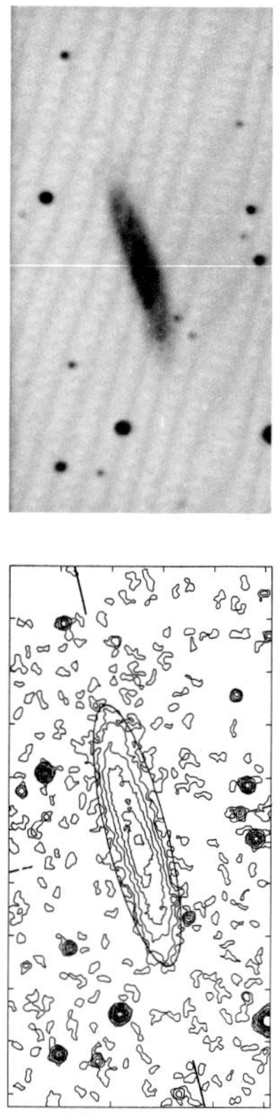

N1406 K4525 1 21.02 256X512

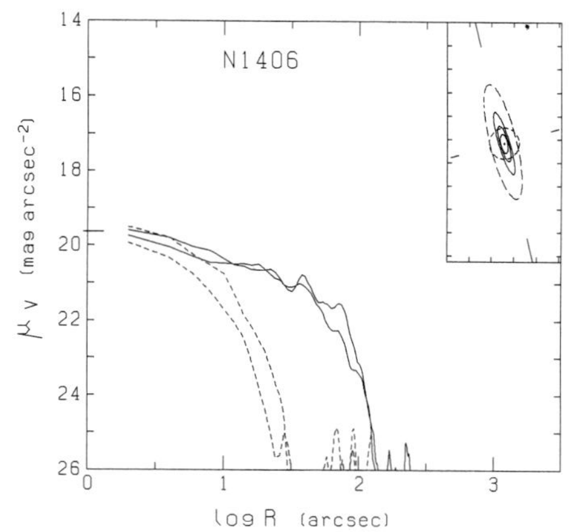

NGC1407

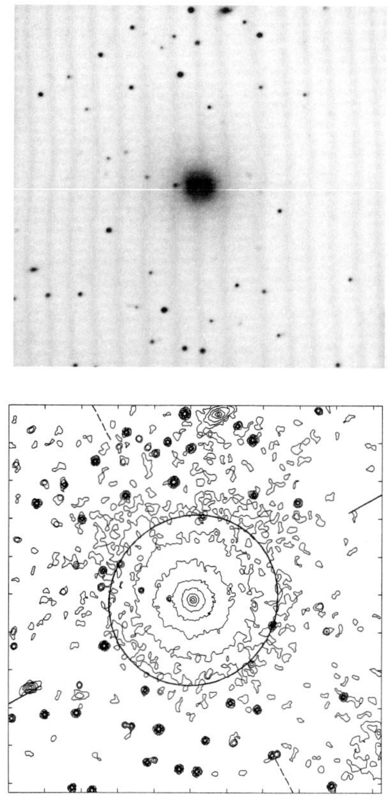

N1407 K4783 2 21.07 512X512

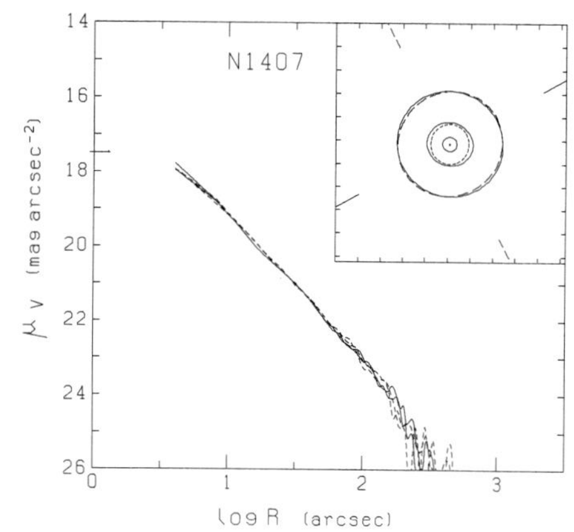

NGC1415

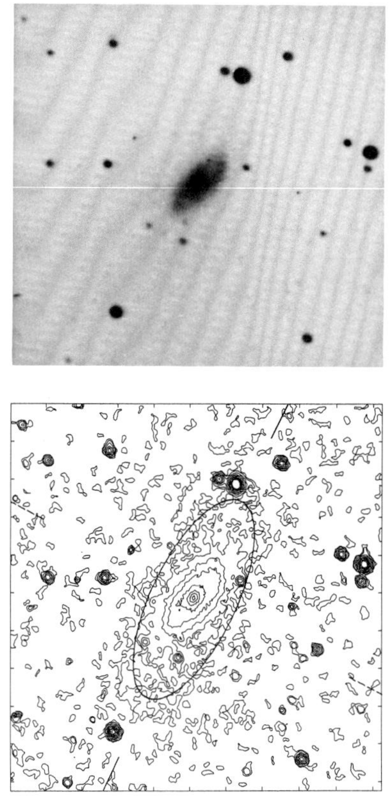

N1415 K4513 1 21.19 512X512

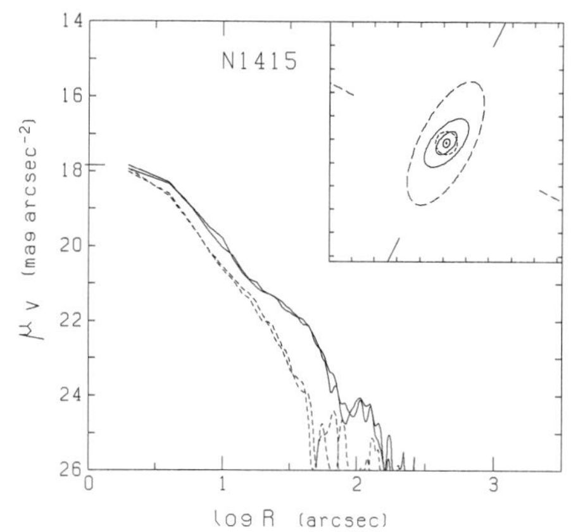

NGC1417

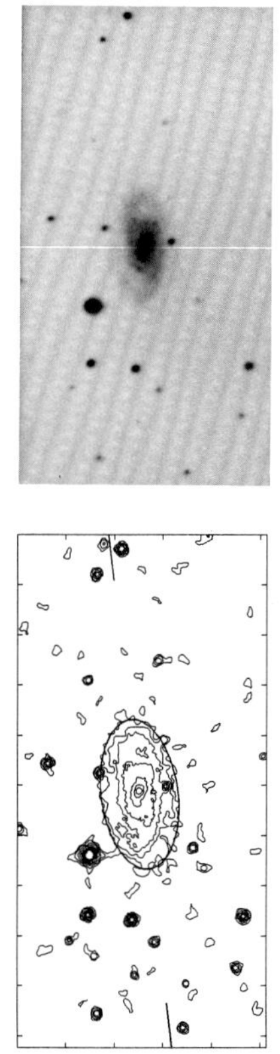

N1417 K4520 1 21.77 256X512

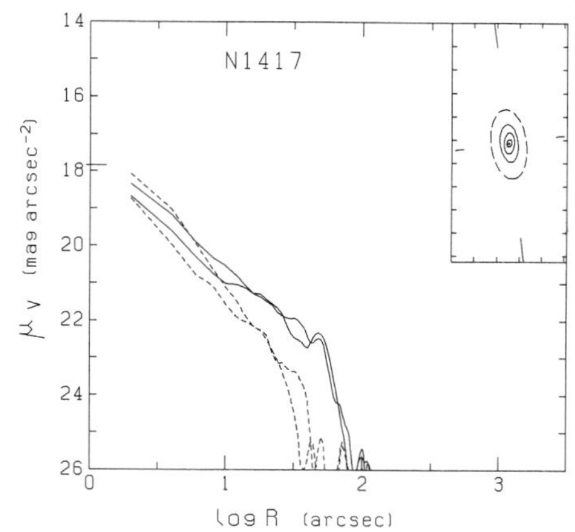

NGC1421

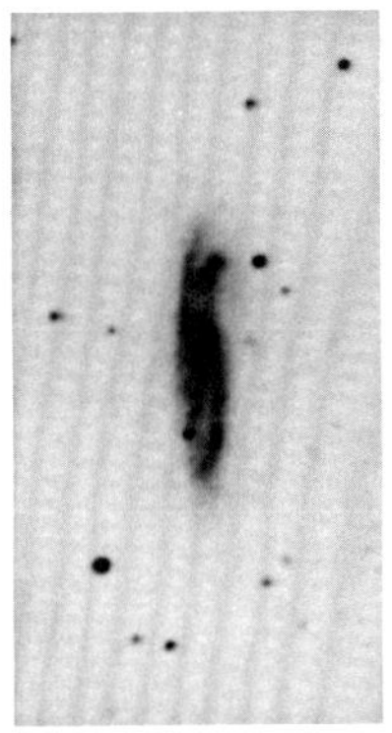

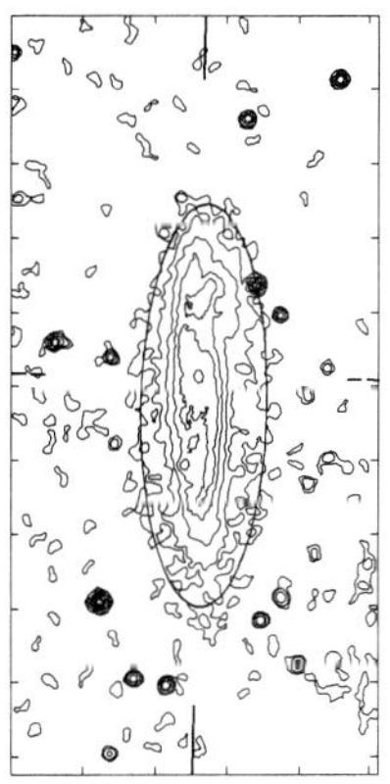

N1421 K2071 1 20.82 256X512

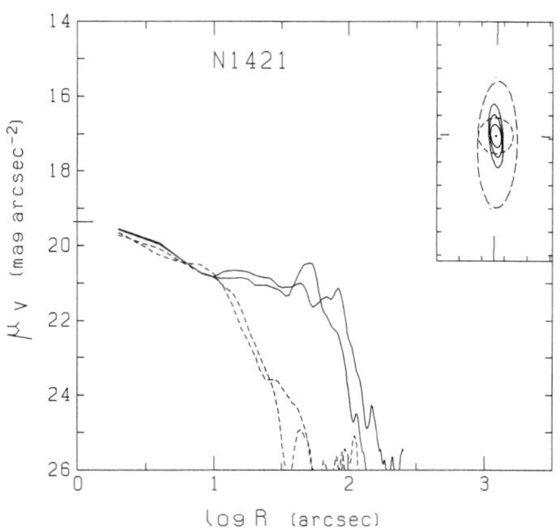

NGC1425

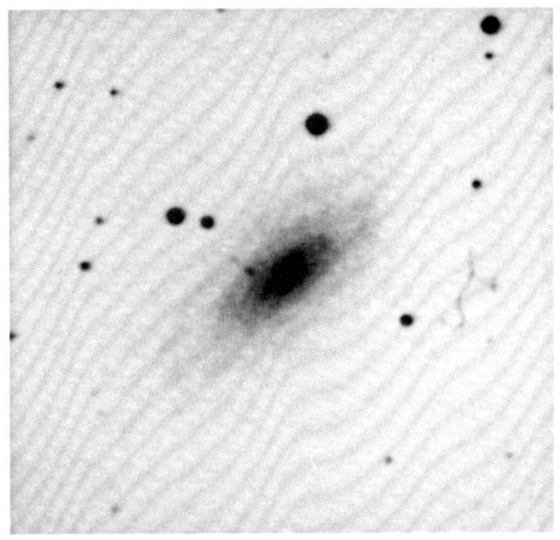

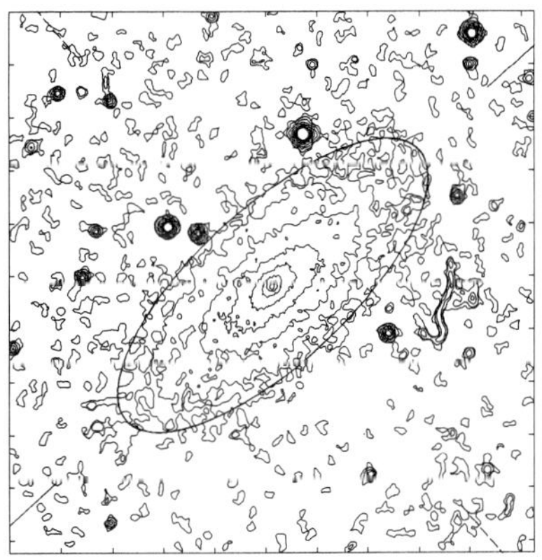

N1425 K4525 1 21.02 512X512

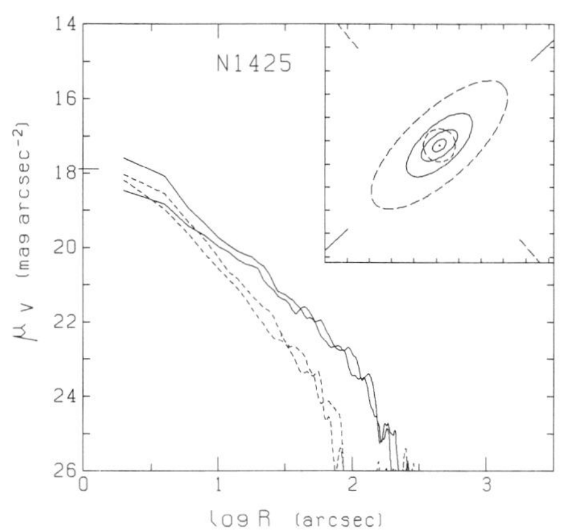

NGC1426

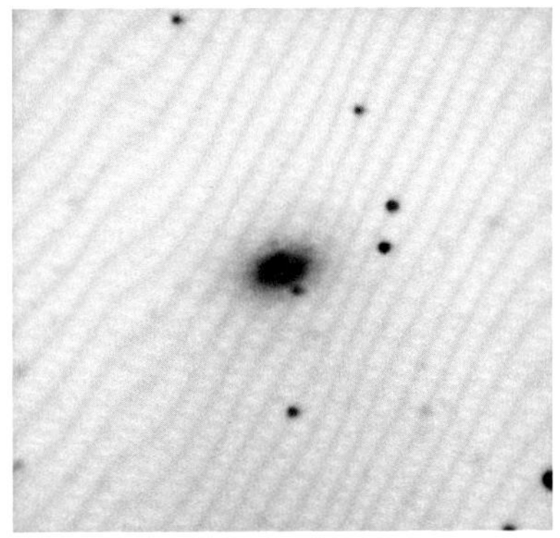

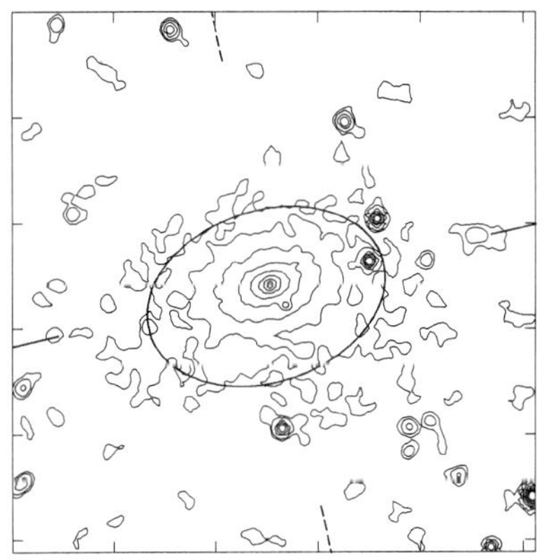

N1426 K4783 2 21.07 256X256

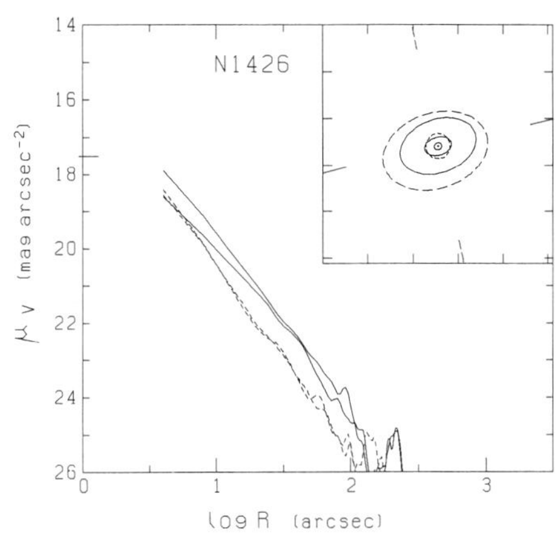

NGC1439

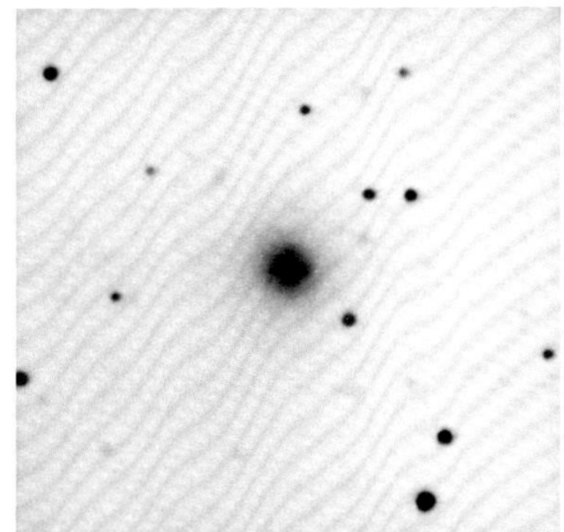

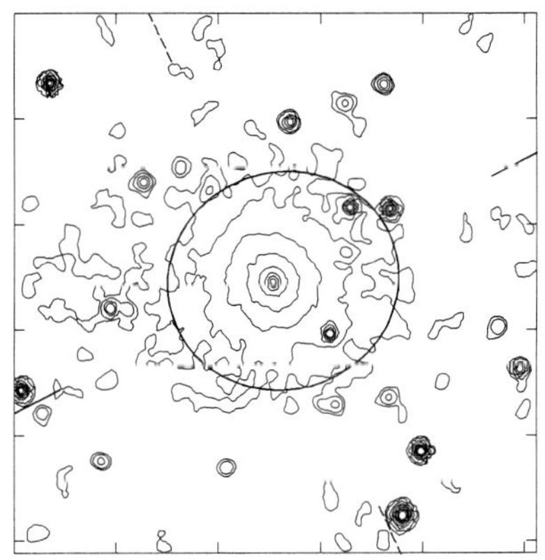

N1439 K4783 2 21.07 256X256

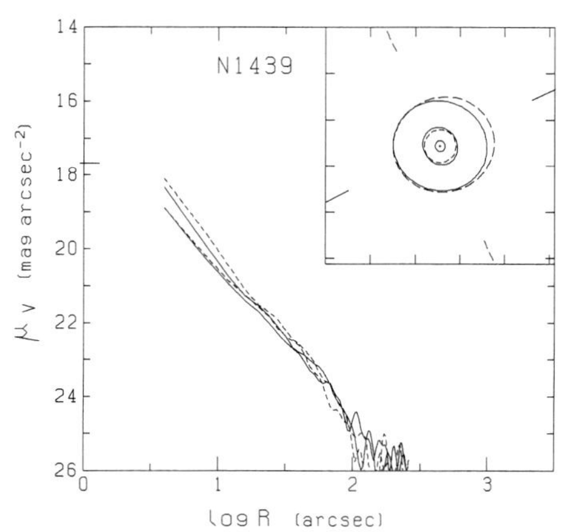

NGC1440

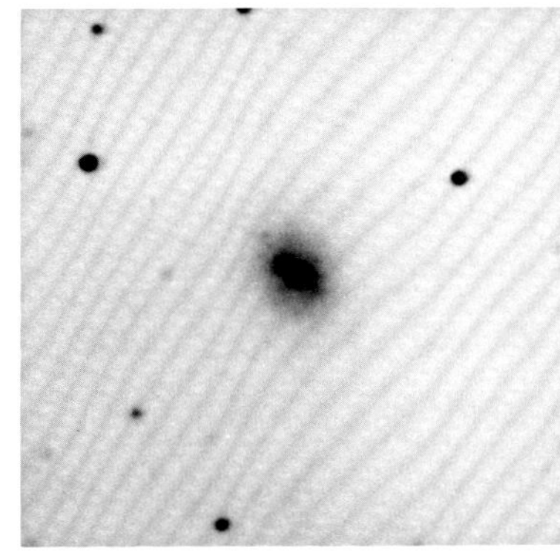

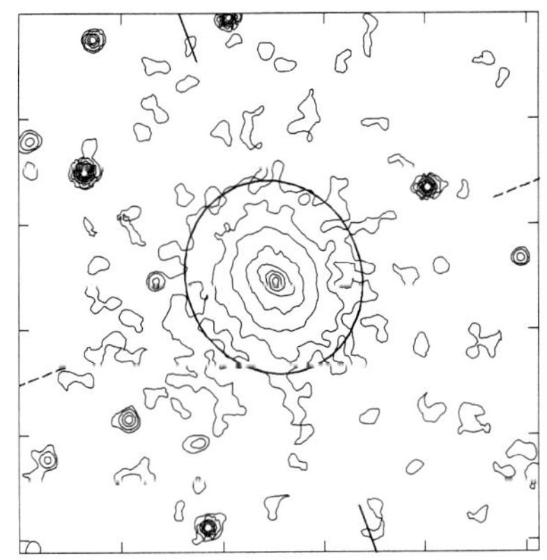

N1440 K4783 2 21.07 256X256

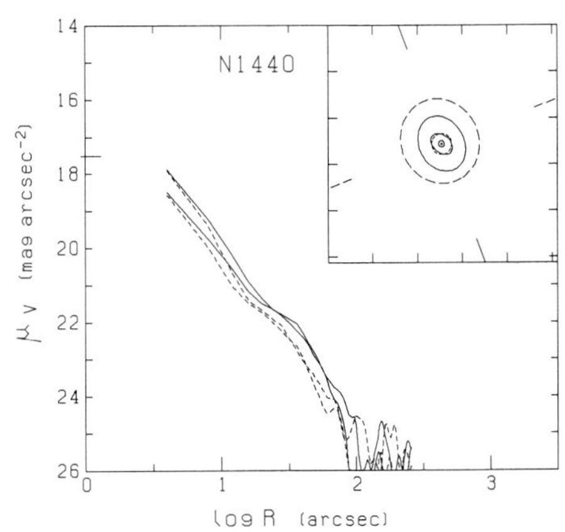

NGC1452

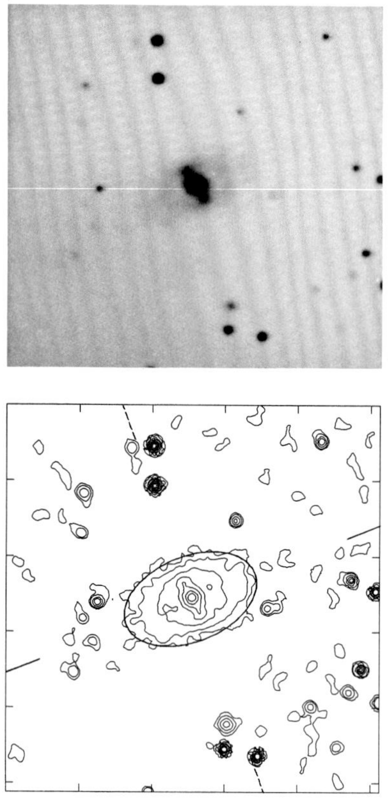

N1452 K4783 2 21.07 256X256

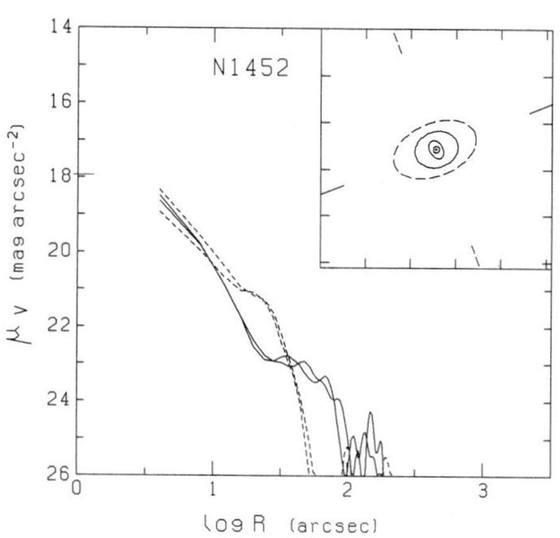

NGC1453

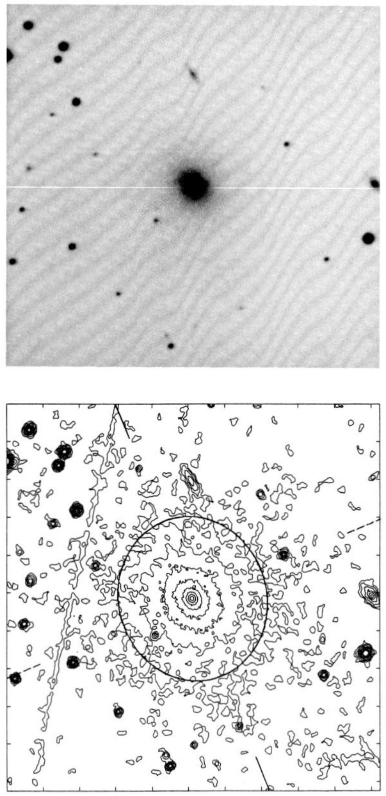

N1453 K4520 1 21.77 512X512

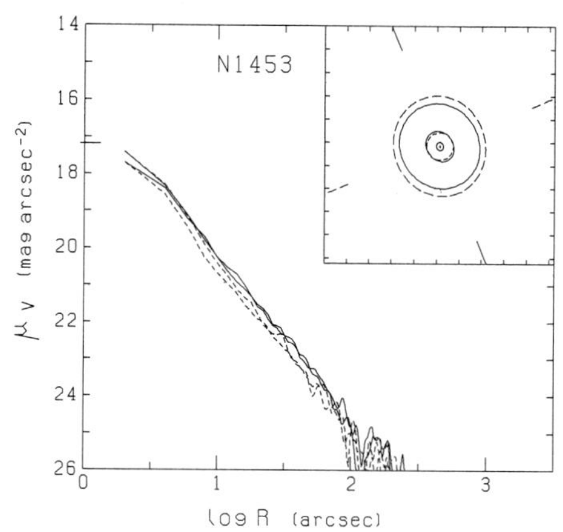

NGC1461

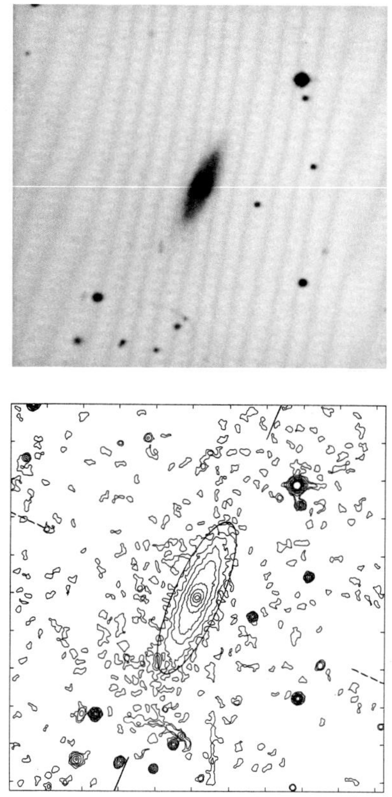

N1461 K2071 1 20.82 512X512

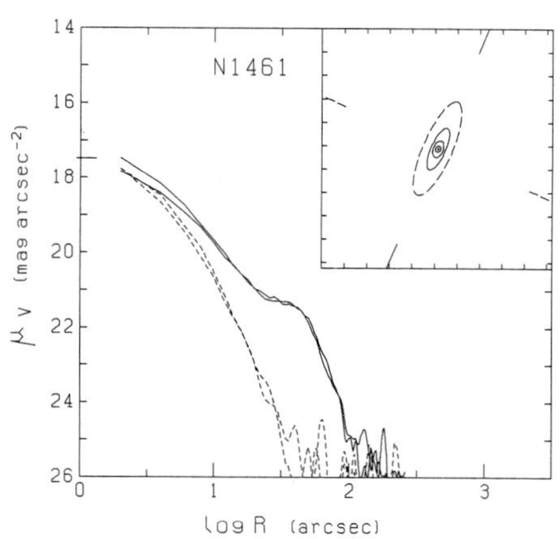

NGC1507

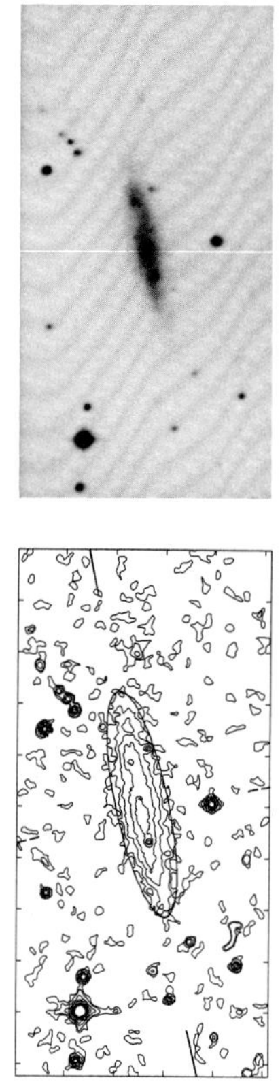

N1507 K4549 1 21.20 256X512

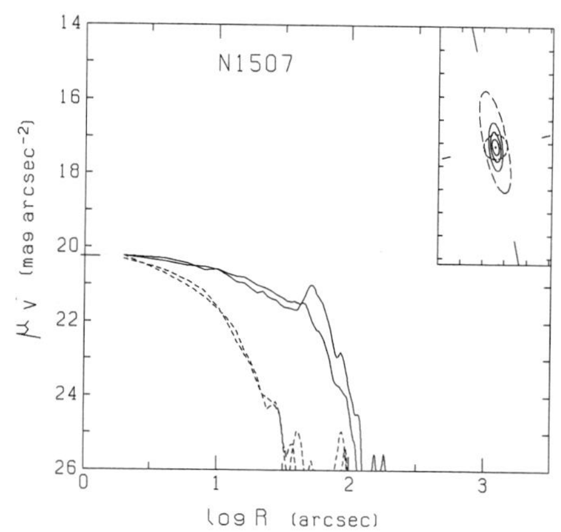

NGC1518

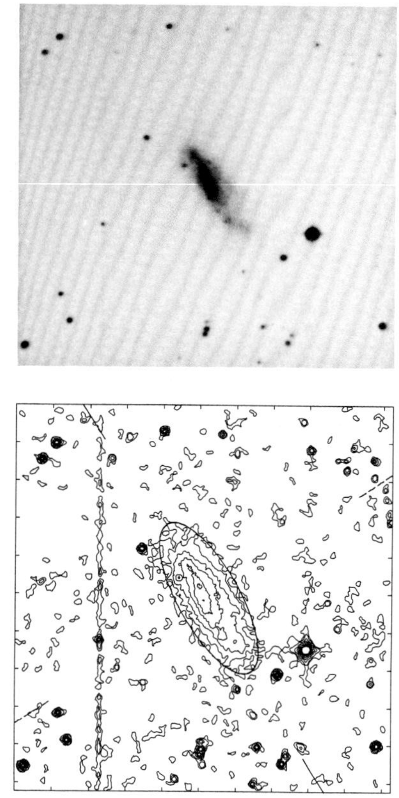

N1518 K4531 1 21.40 512X512

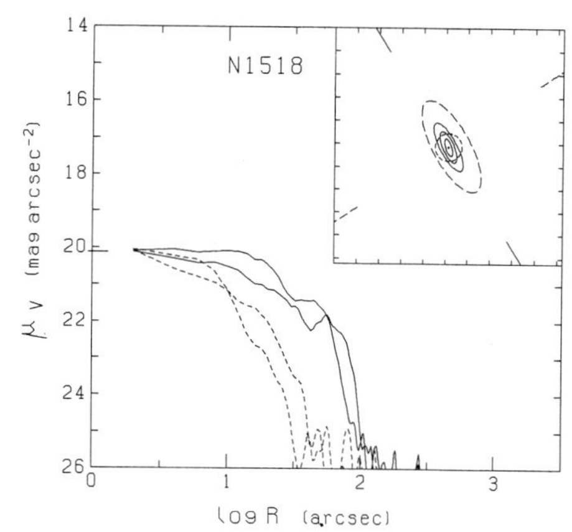

NGC1521

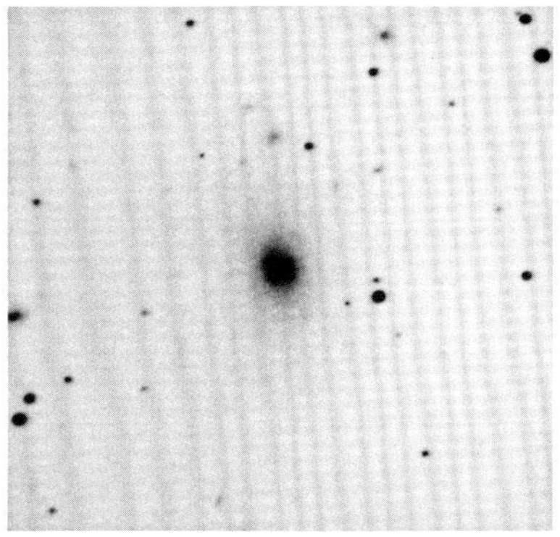

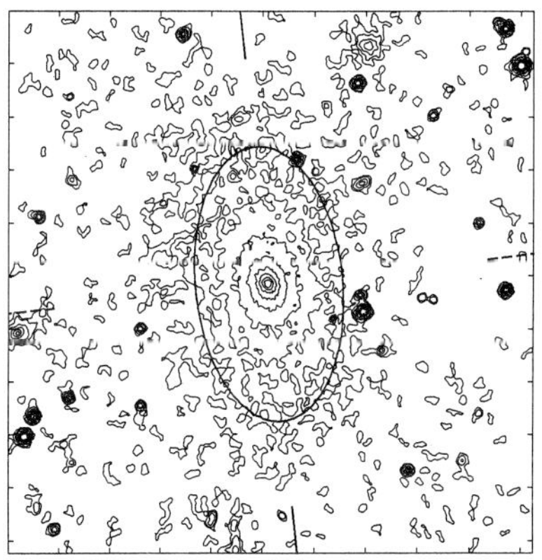

N1521 K4531 1 21.40 512X512

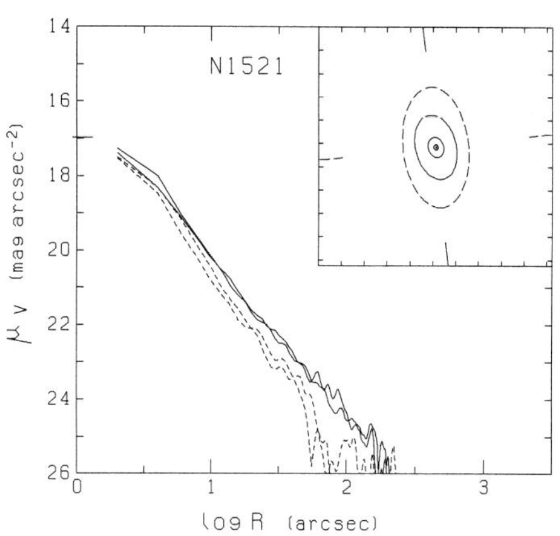

NGC1537

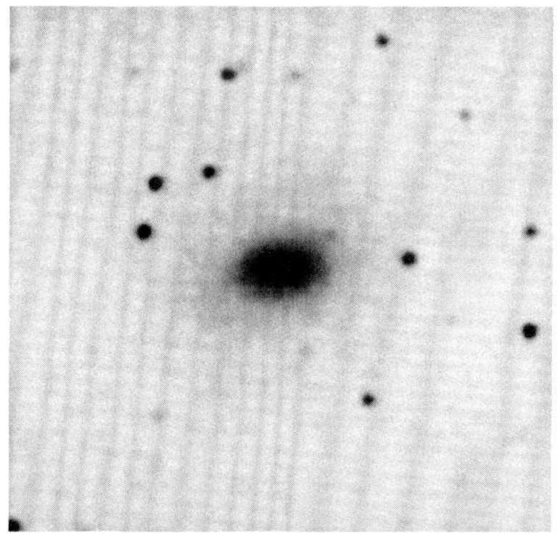

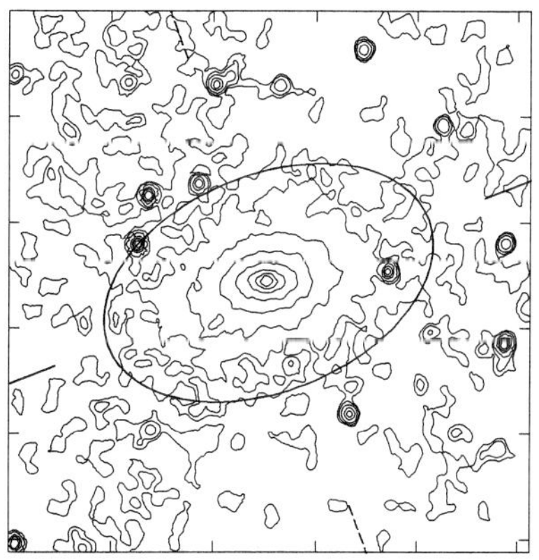

N1537 K4939 2 20.23 256X256

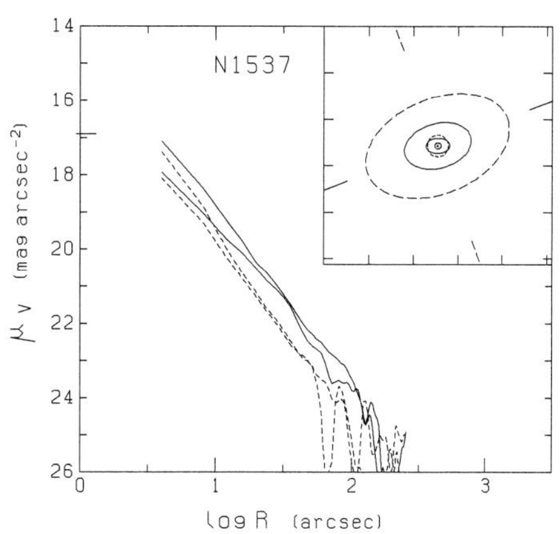

NGC1569

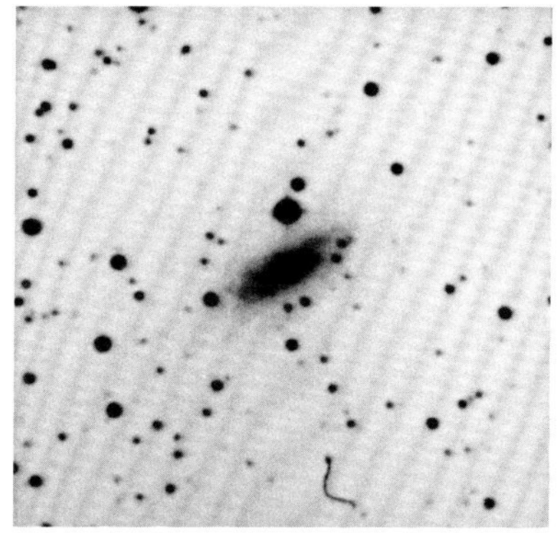

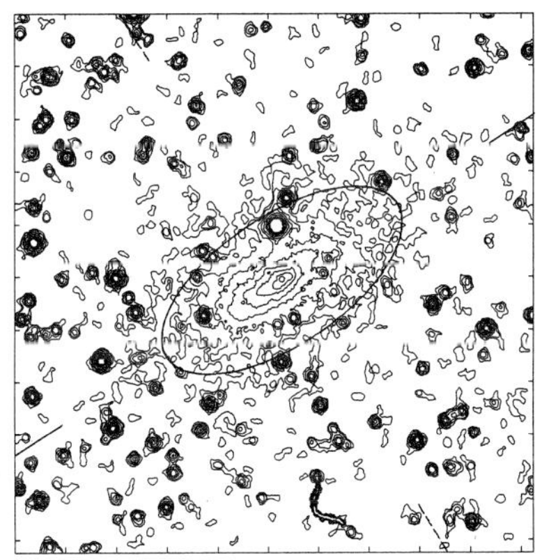

N1569 K5579 1 21.74 512X512

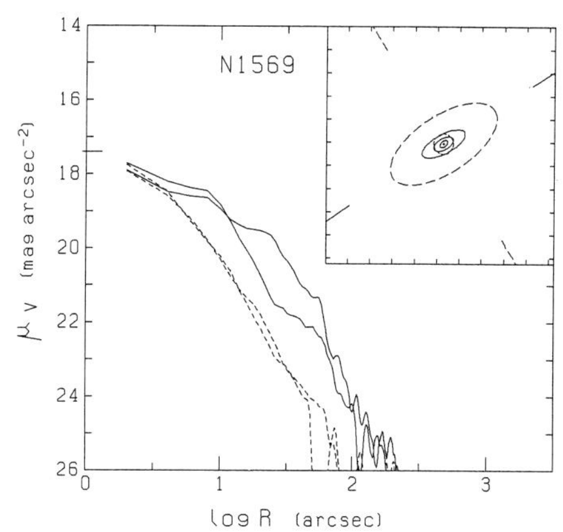

NGC1600

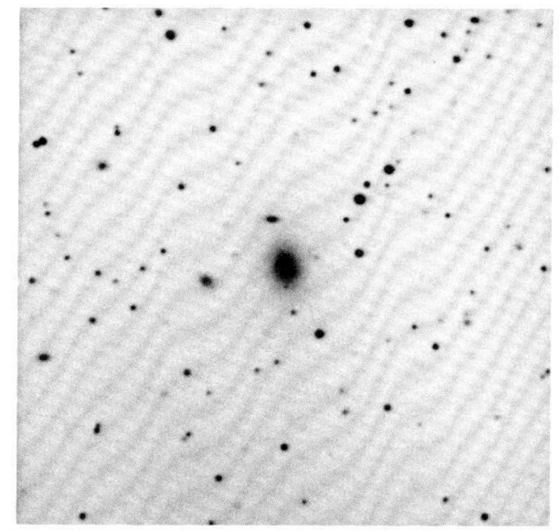

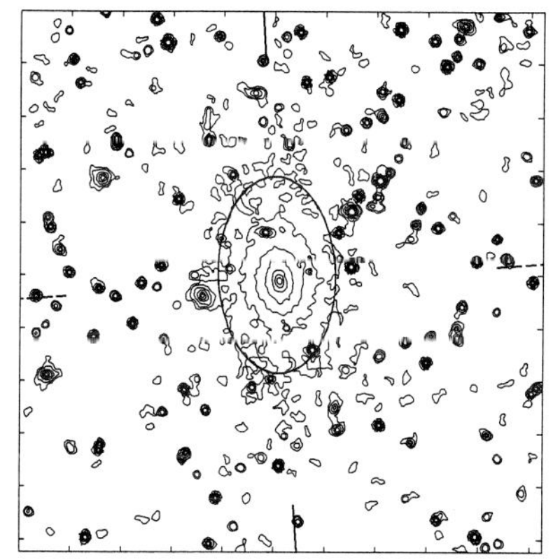

N1600 K4784 2 21.37 512X512

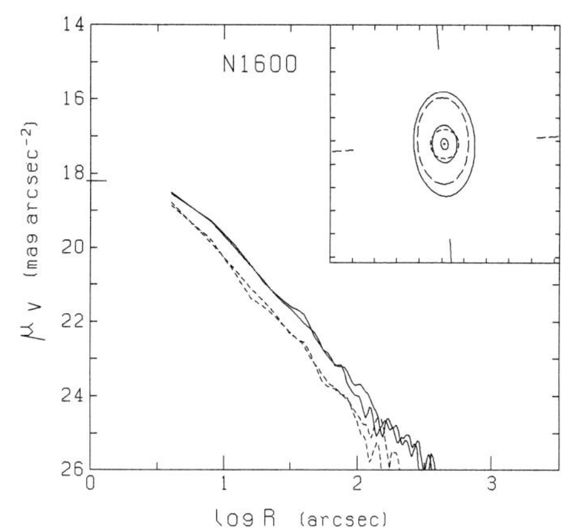

NGC1625

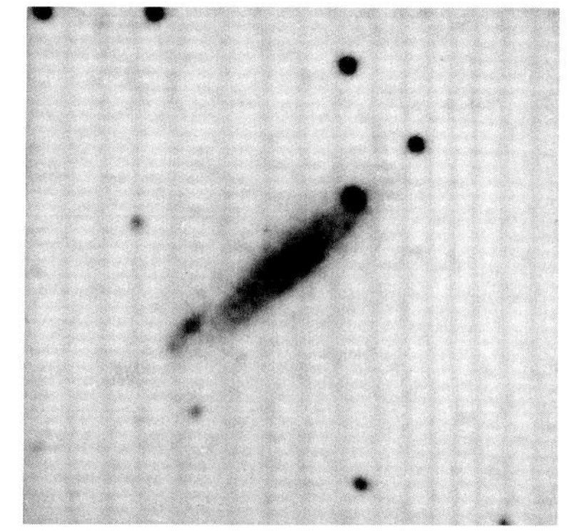

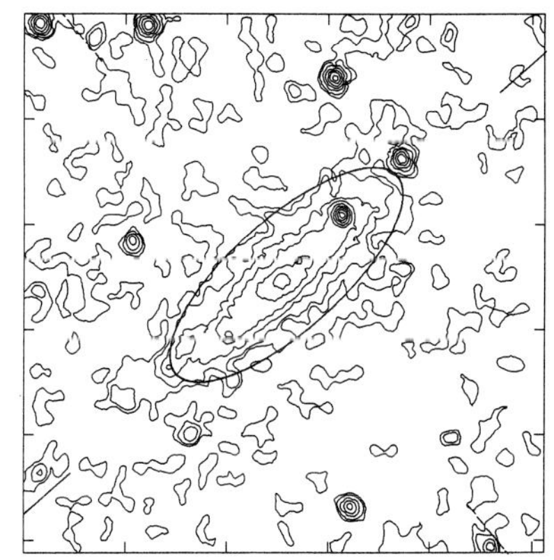

N1625 K4784 1 21.37 256X256

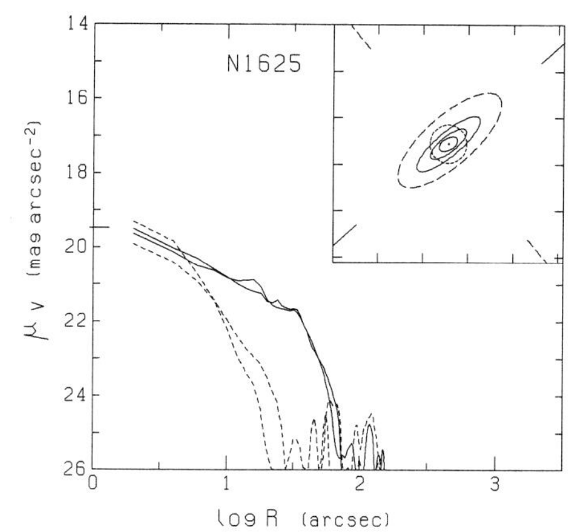

NGC1637

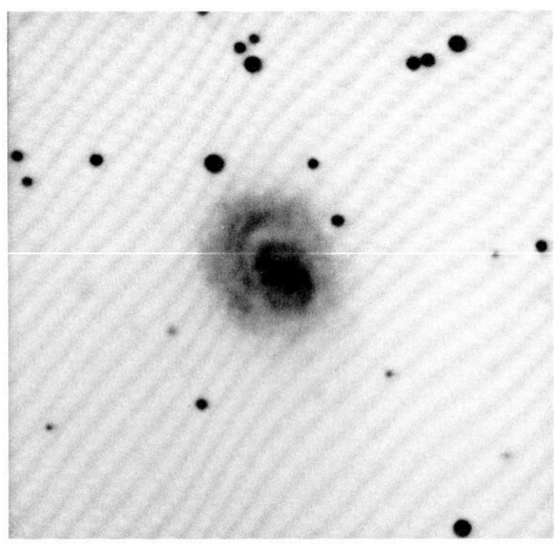

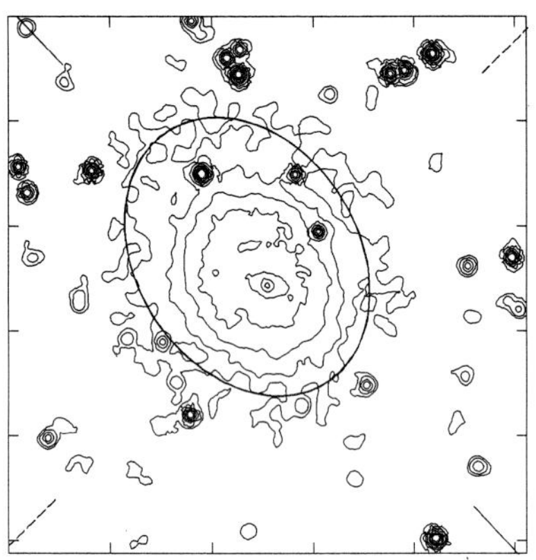

N1637 K4784 2 21.37 256X256

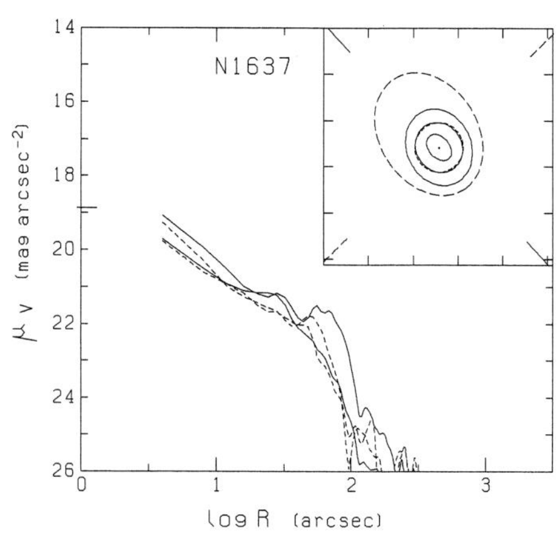

NGC1638

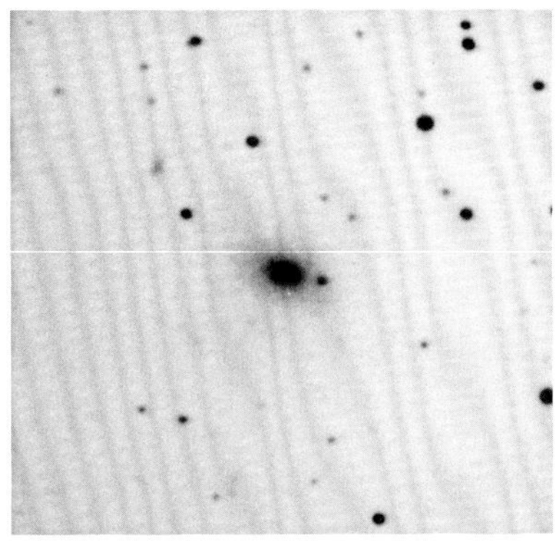

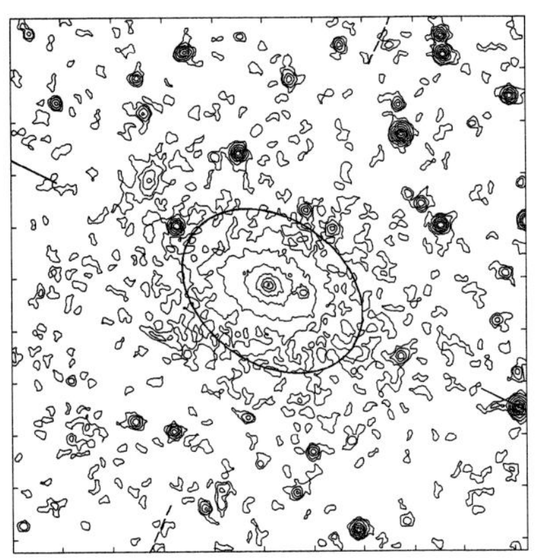

N1638 K5180 1 21.27 512X512

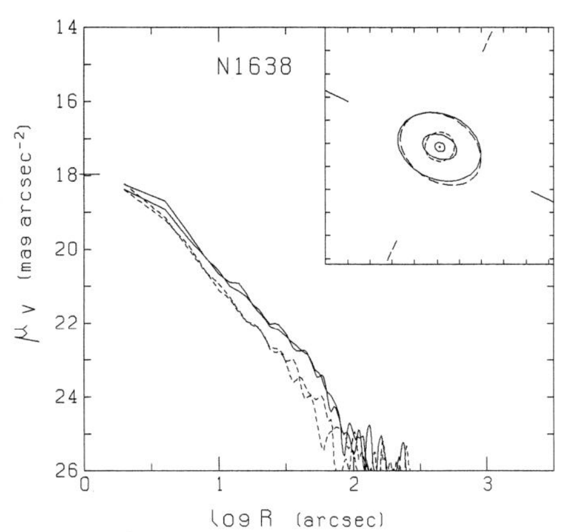

NGC1640

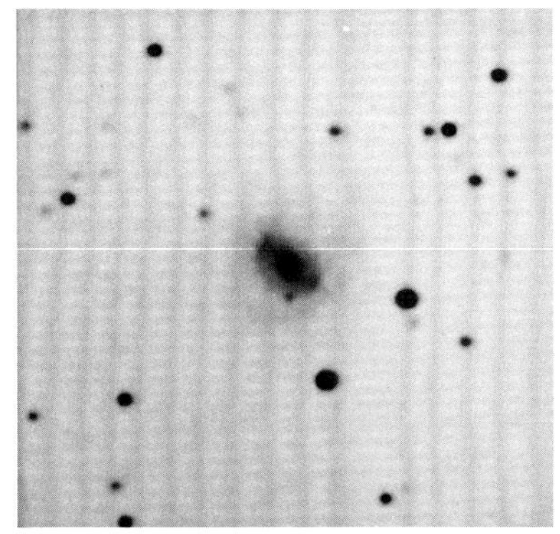

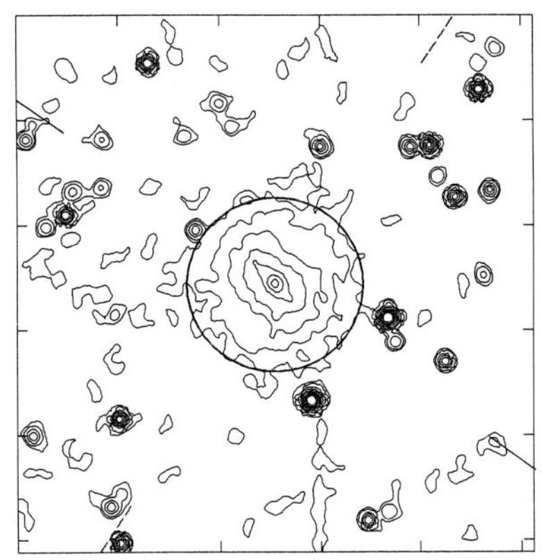

N1640 K4508 2 21.05 256X256

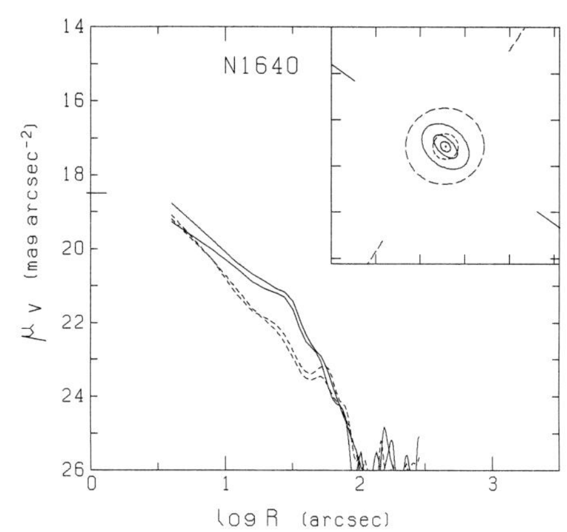

NGC1659

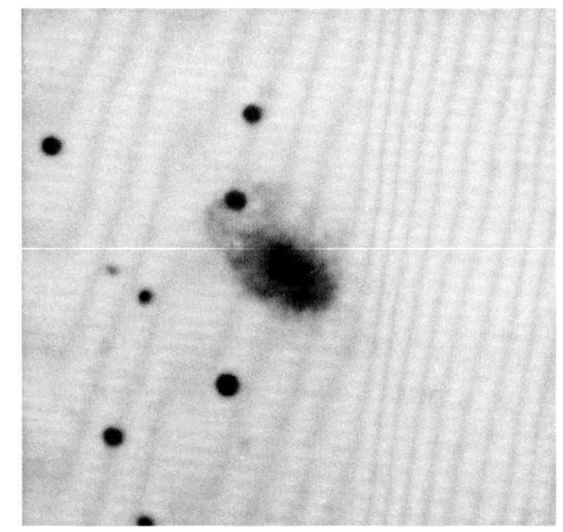

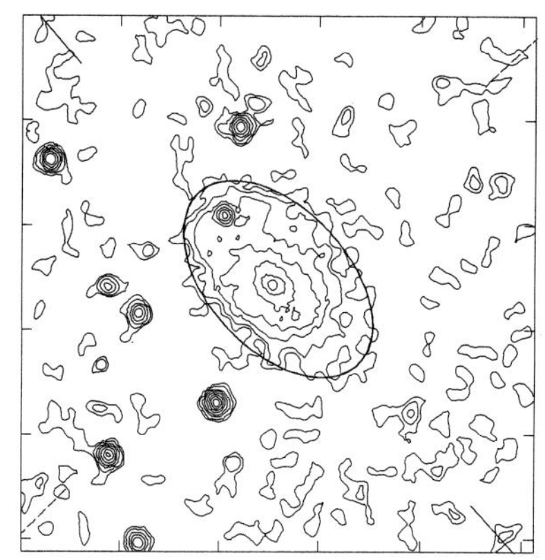

N1659 K4784 1 21.37 256X256

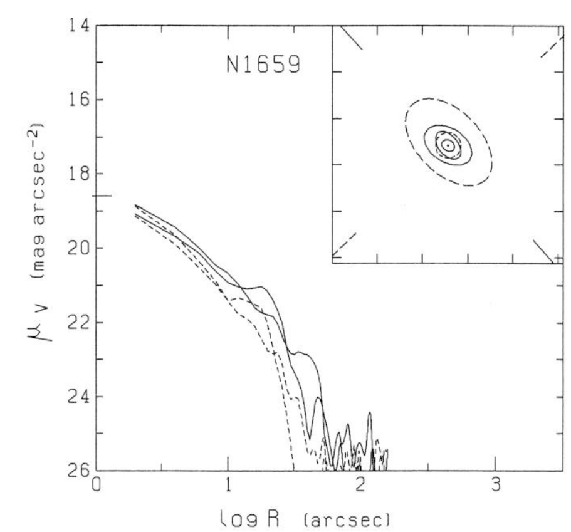

NGC1667

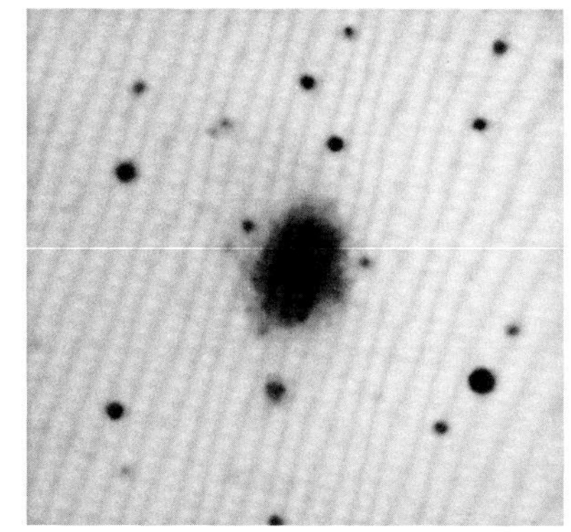

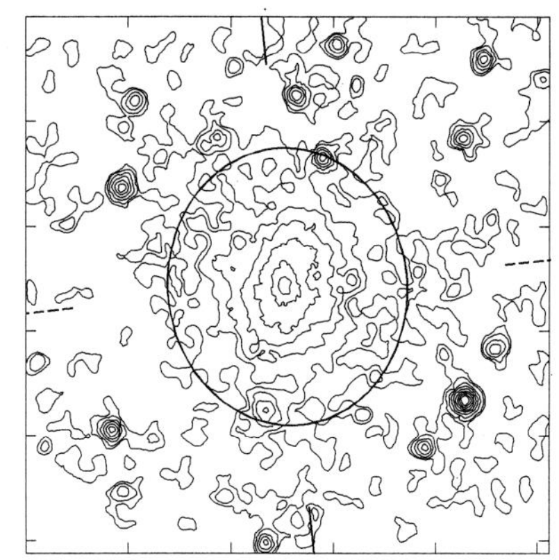

N1667 K4784 1 21.37 256X256

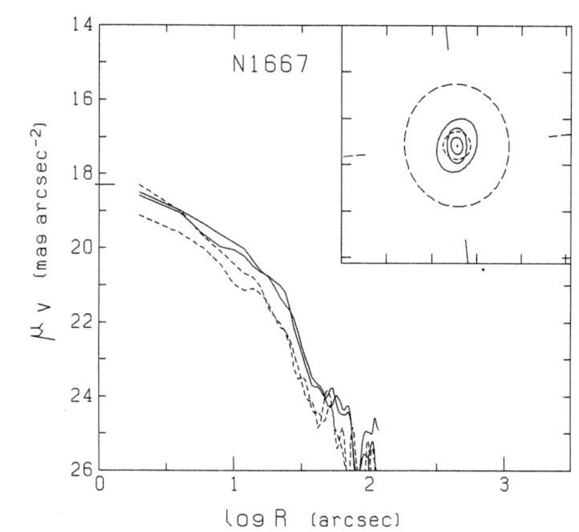

NGC1700

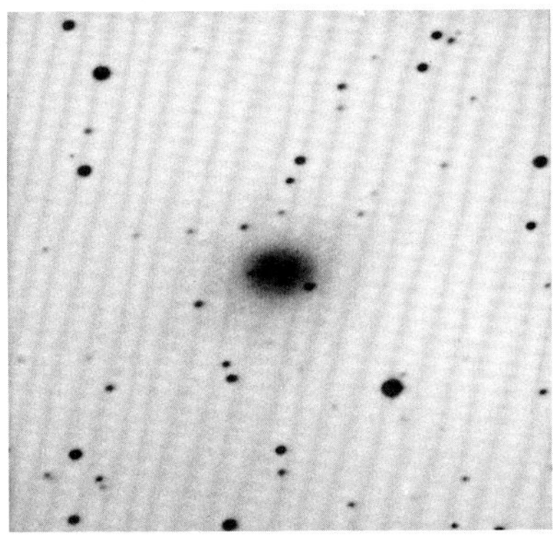

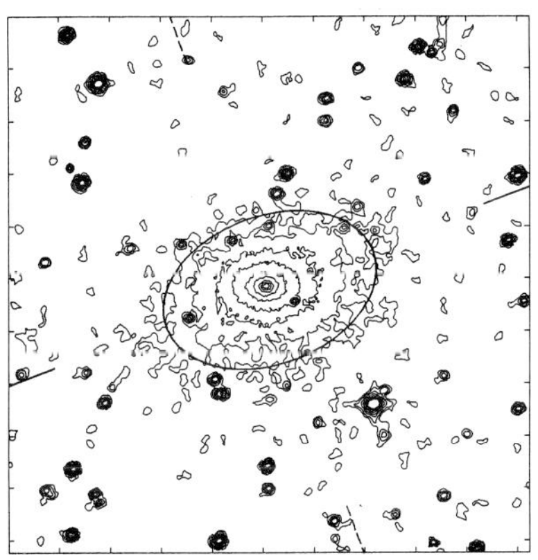

N1700 K5679 1 21.82 512X512

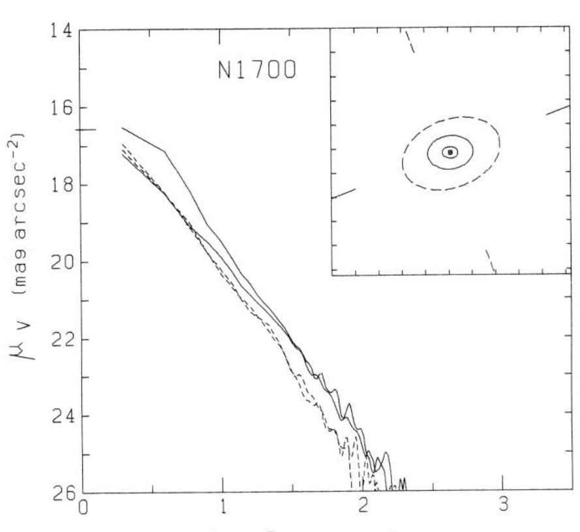

NGC1726

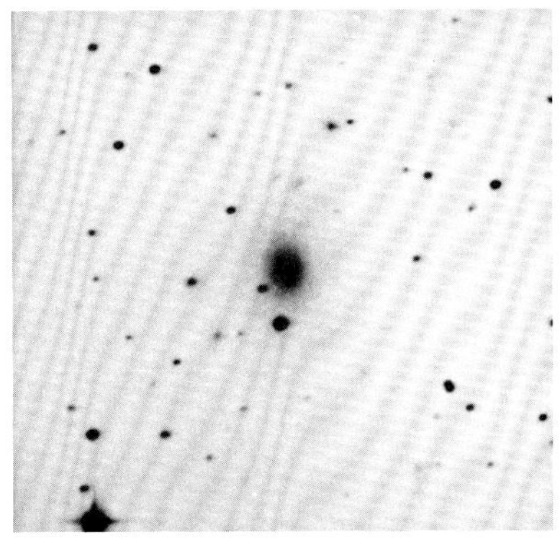

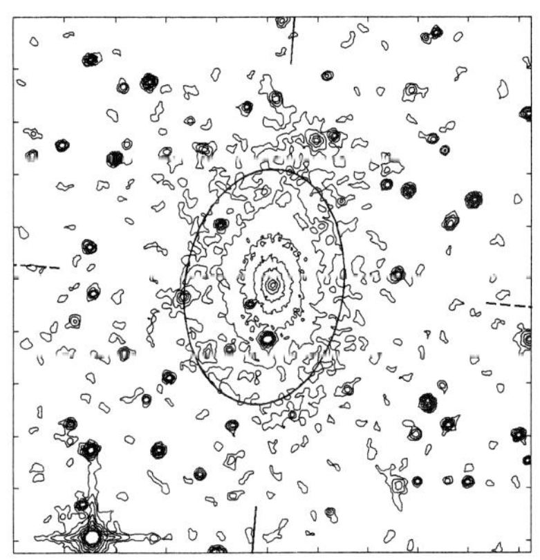

N1726 K5679 1 21.82 512X512

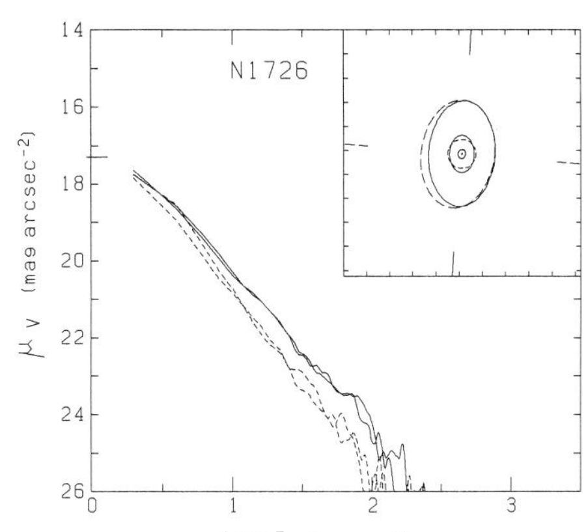

NGC1744

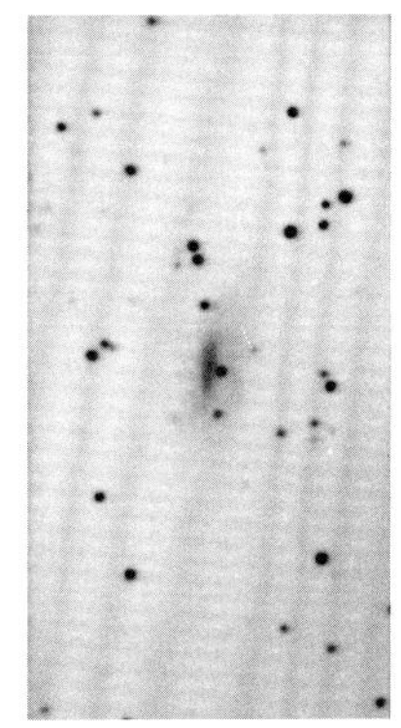

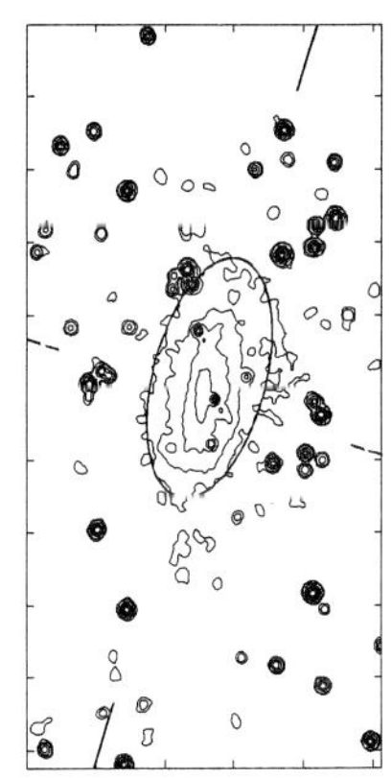

N1744 K5235 2 20.98 256X512

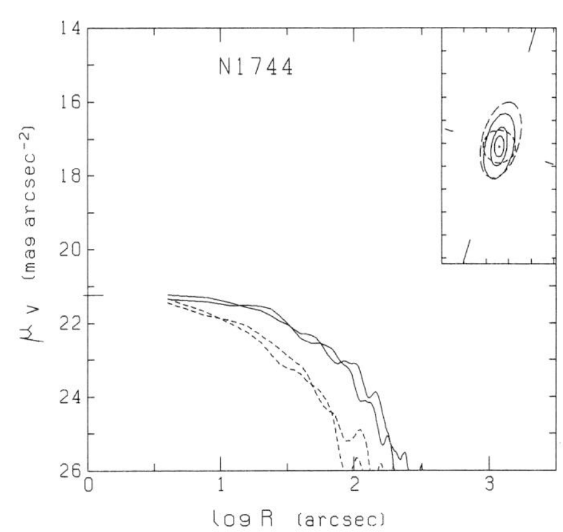

NGC1784

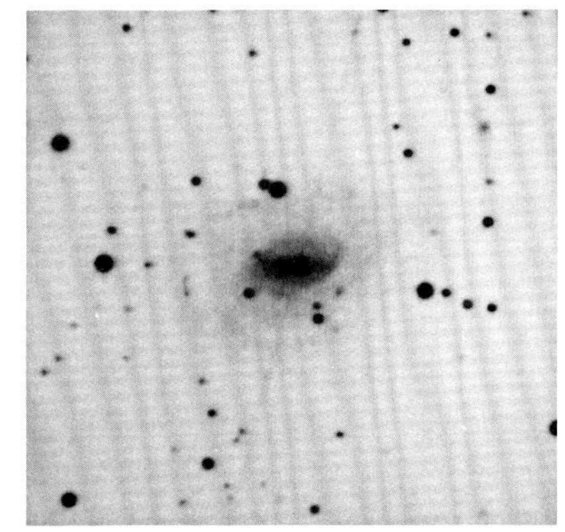

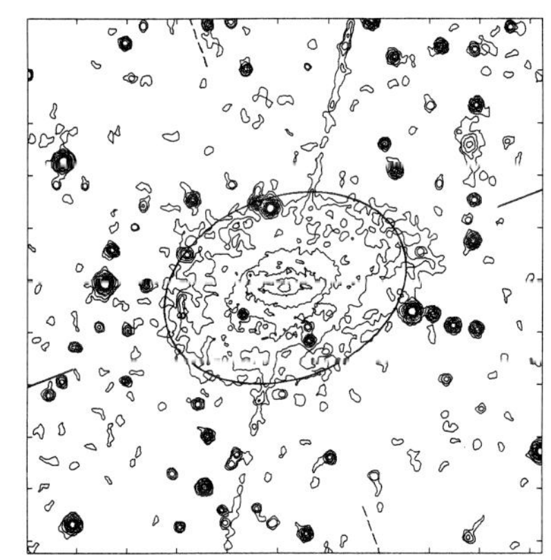

N1784 K4521 1 21.64 512X512

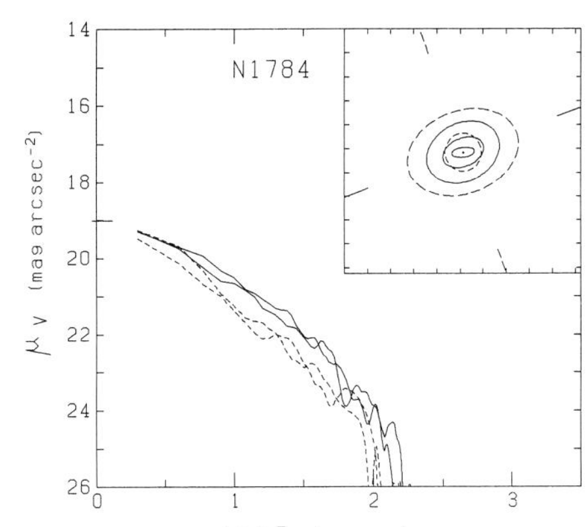

HA85—1

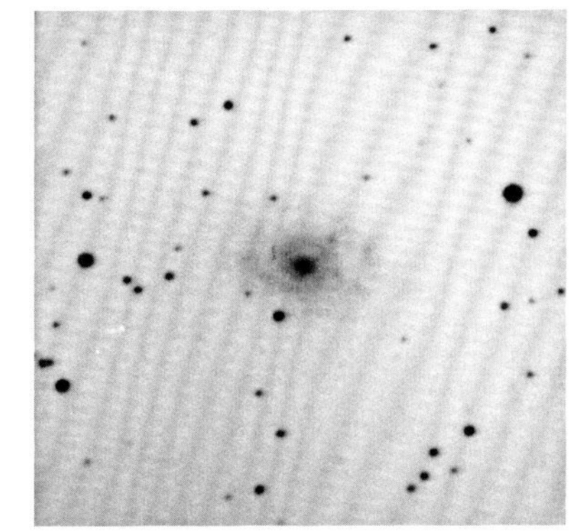

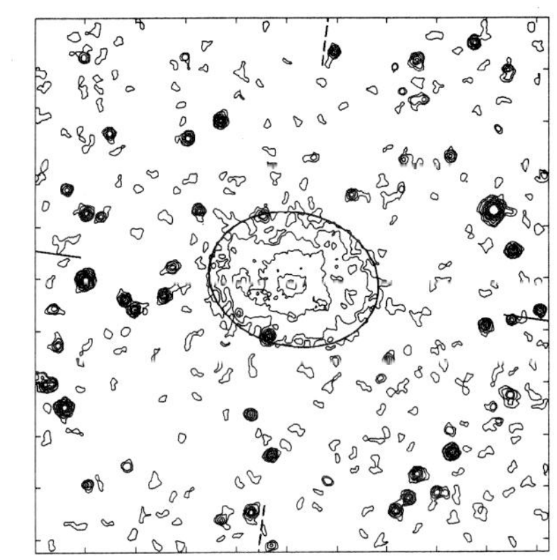

HA085 K4526 1 21.54 512X512

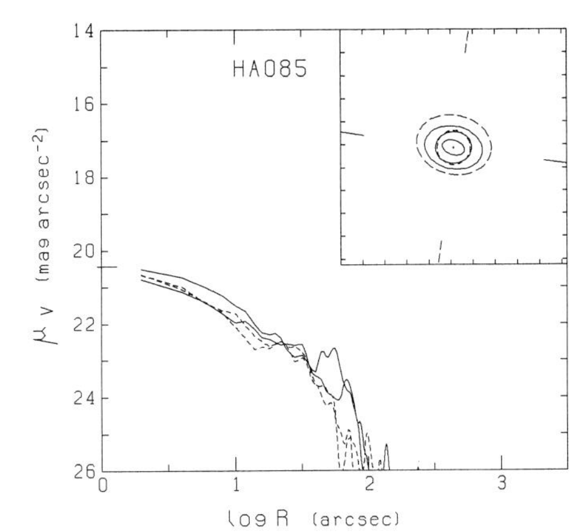

NGC1832

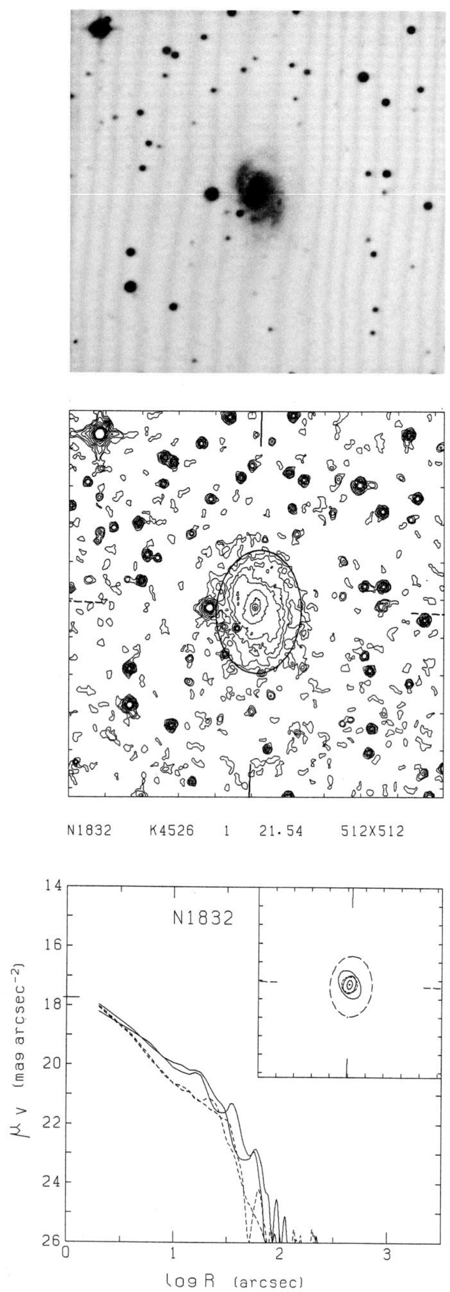

NGC1964

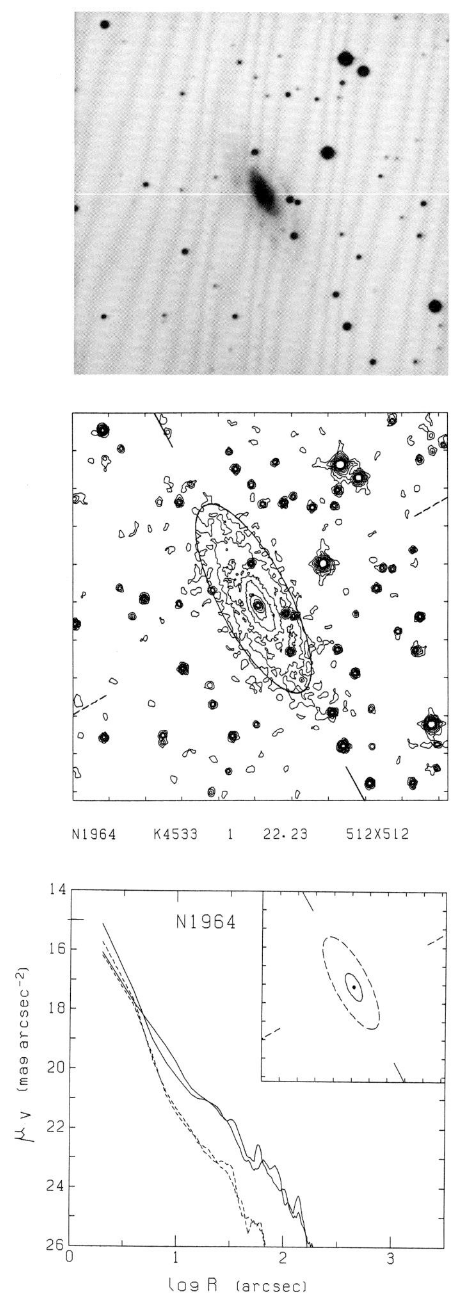

NGC1961

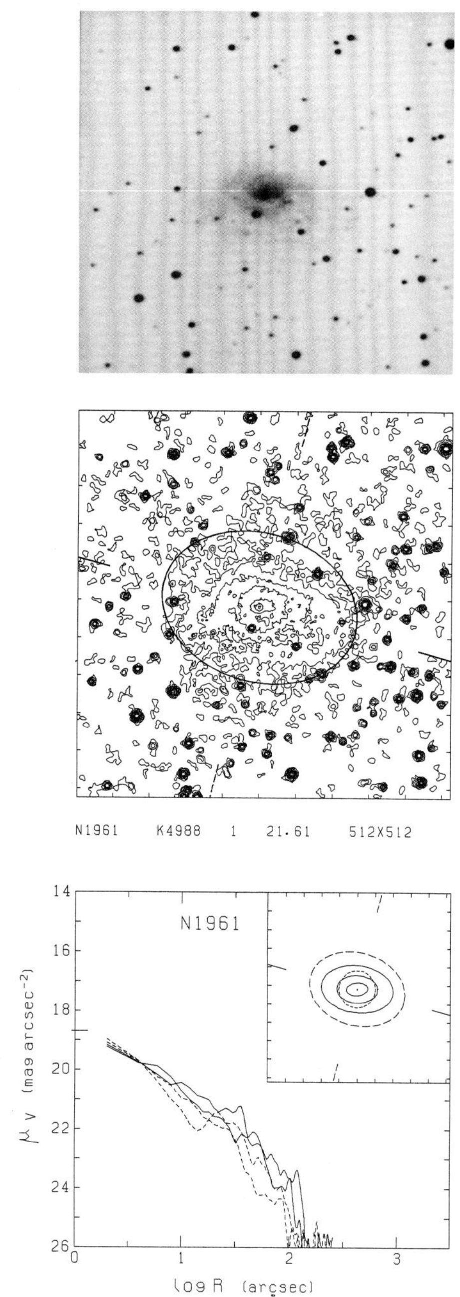

NGC2139

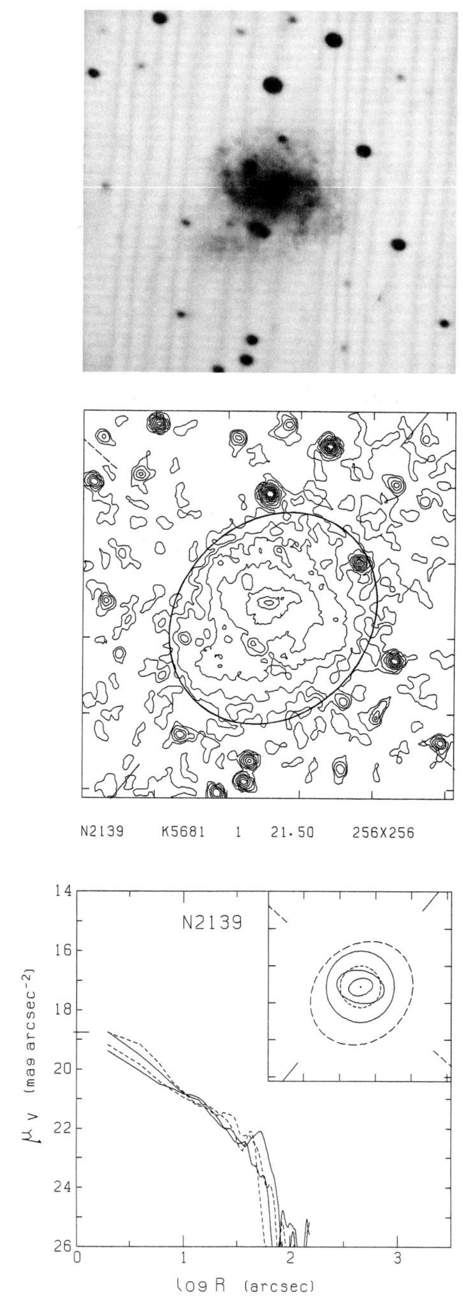

NGC2179

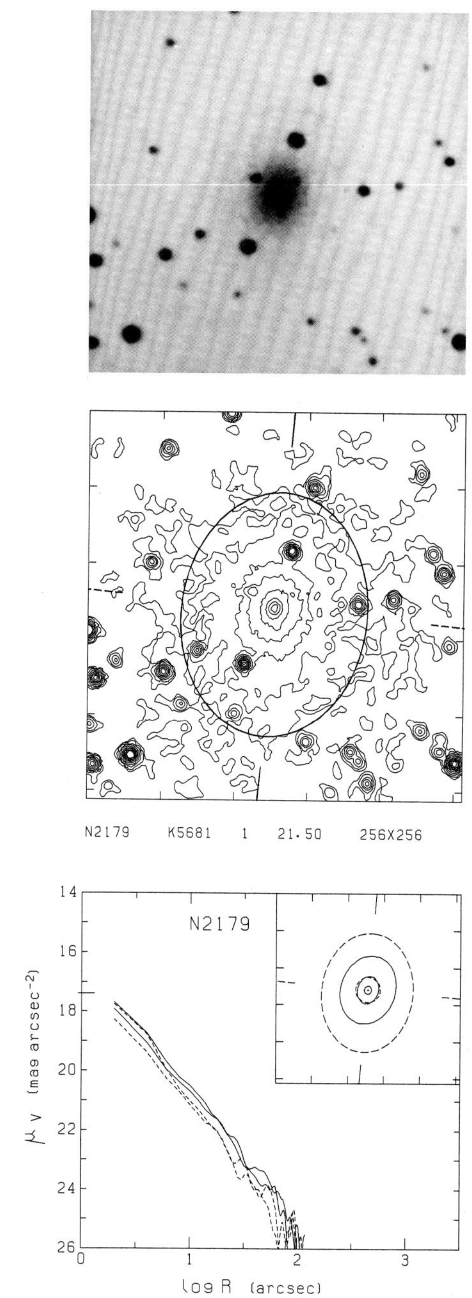

NGC2196

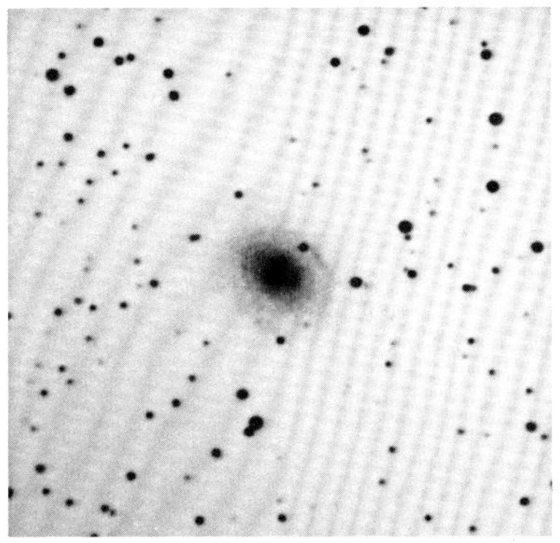

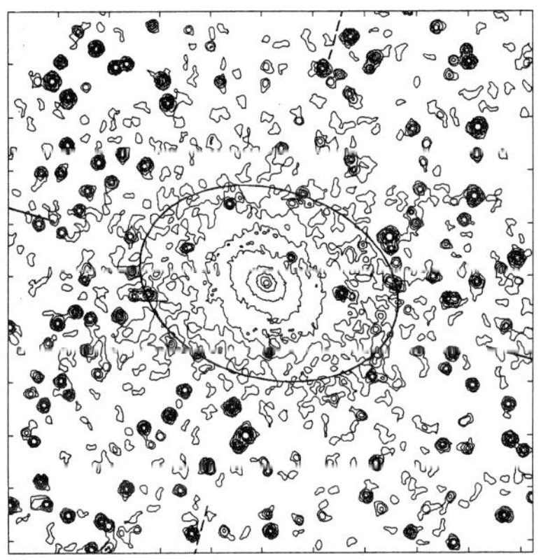

N2196 K5681 1 21.50 512X512

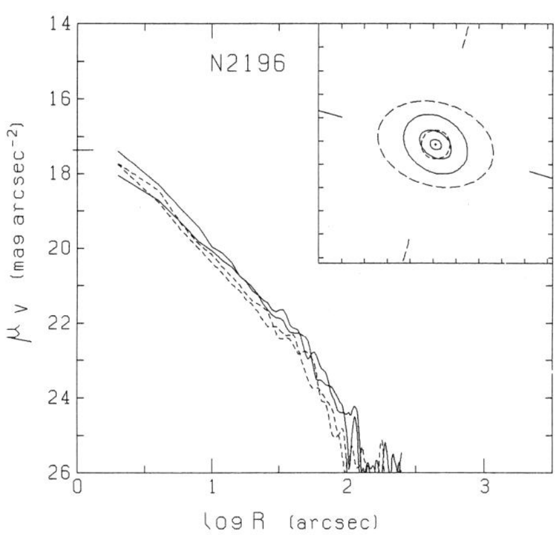

NGC2146

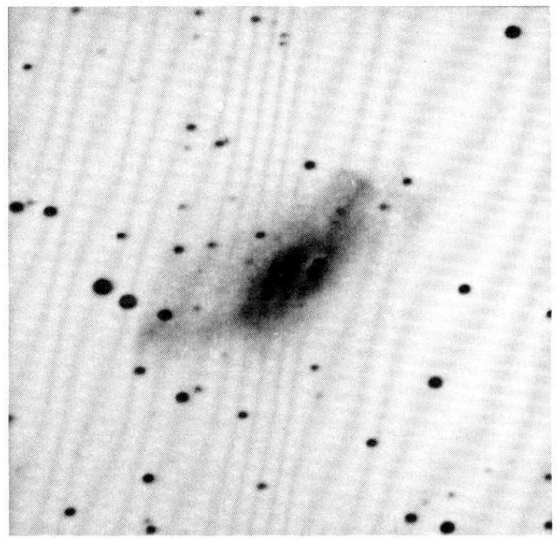

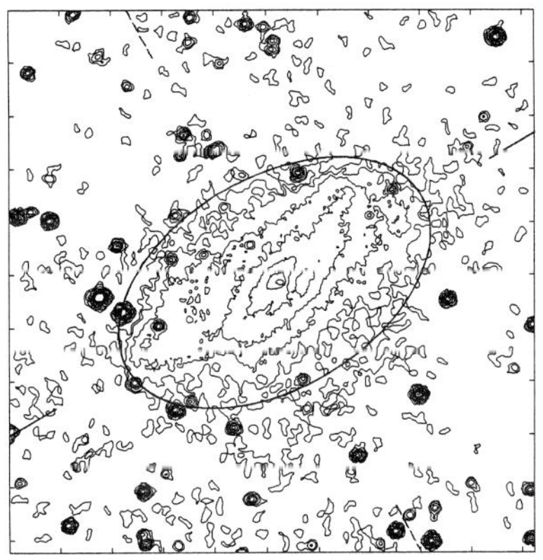

N2146 K4536 1 21.61 512X512

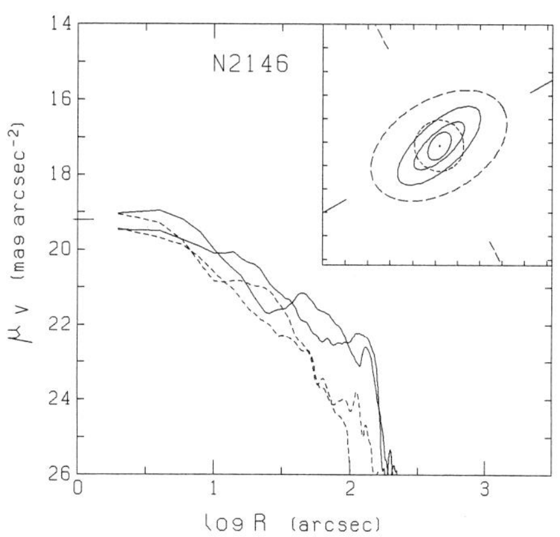

NGC2207

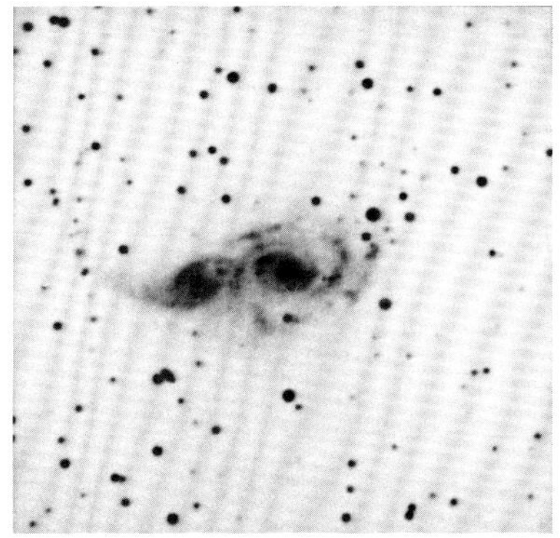

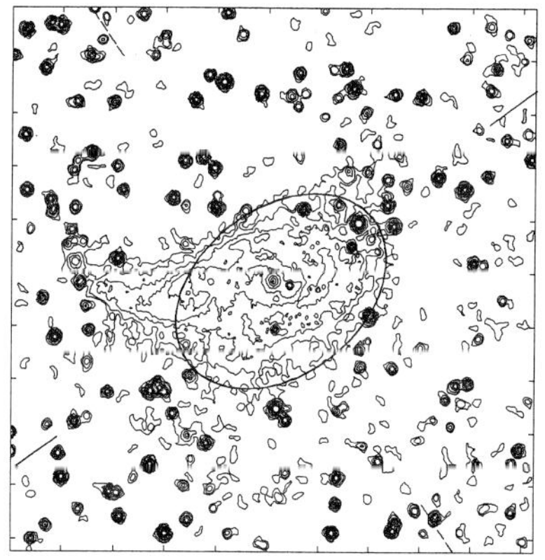

N2207 K5681 1 21.50 512X512

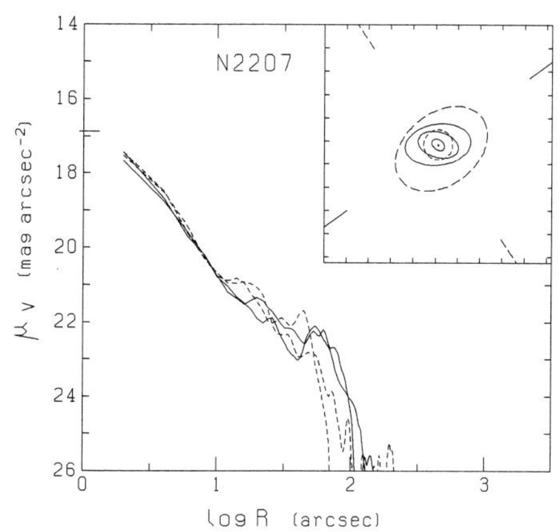

NGC2217

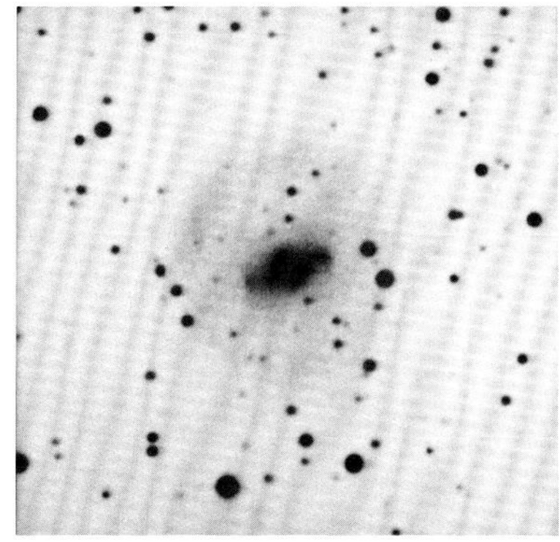

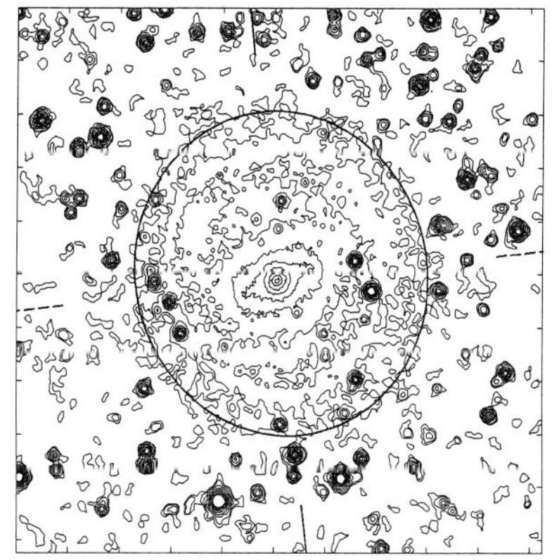

N2217 K4522 1 21.23 512X512

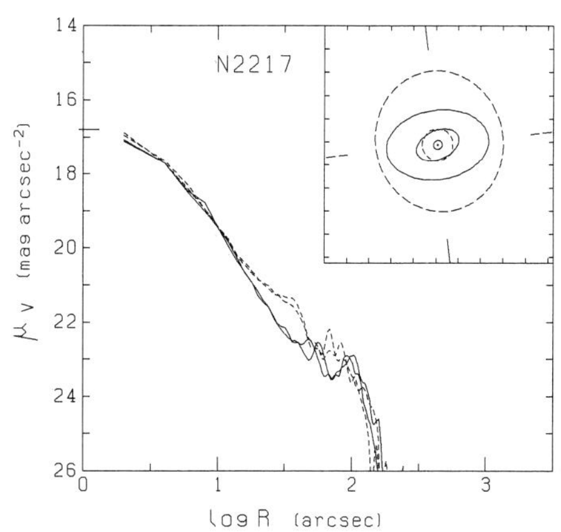

NGC2223

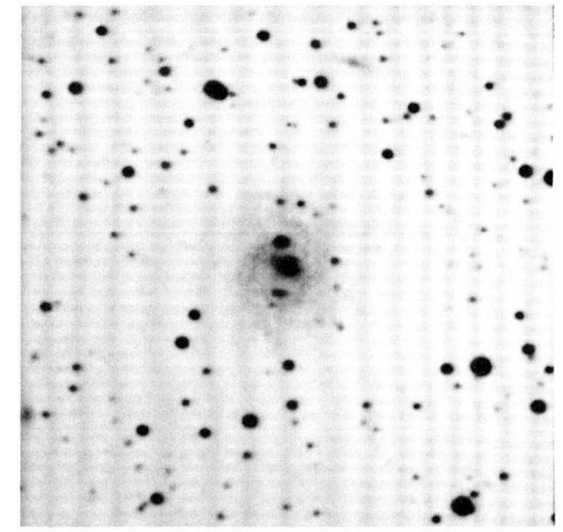

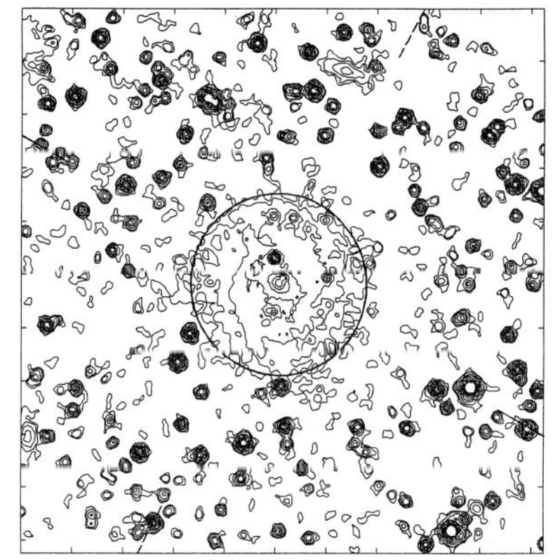

N2223 K4522 1 21.23 512X512

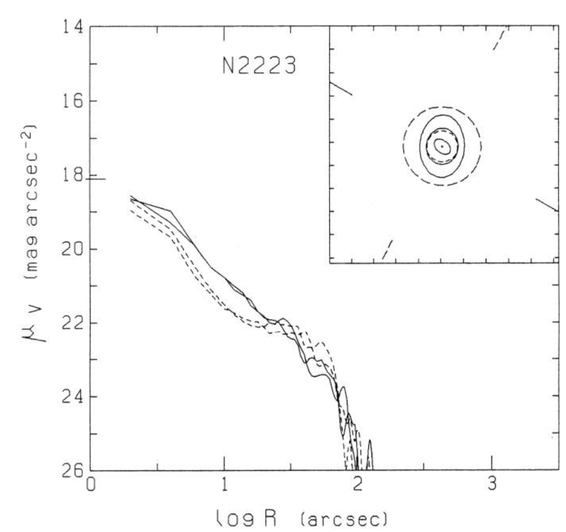

NGC2280

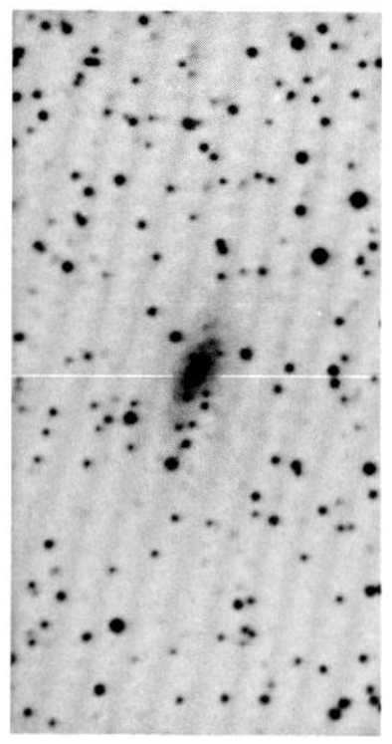

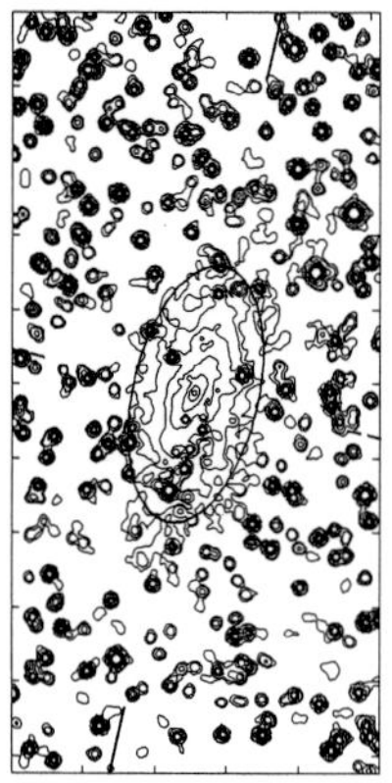

N2280 K5653 2 20.90 256X512

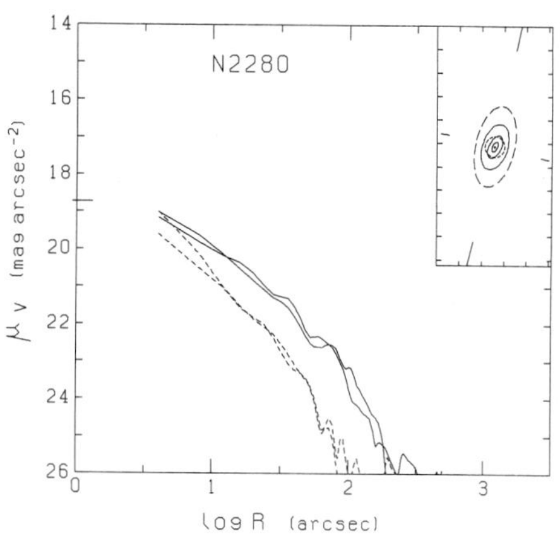

NGC2325

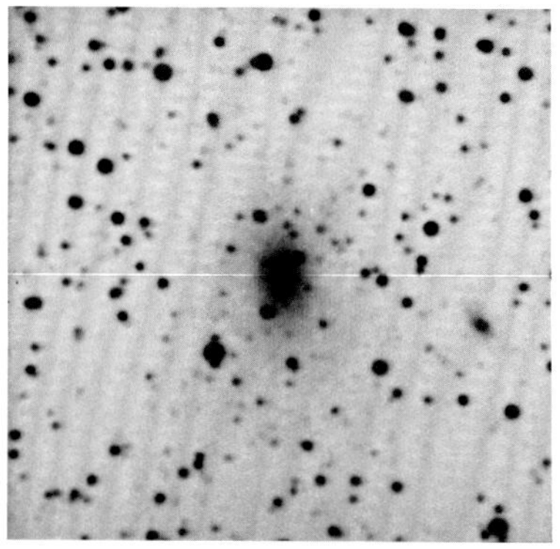

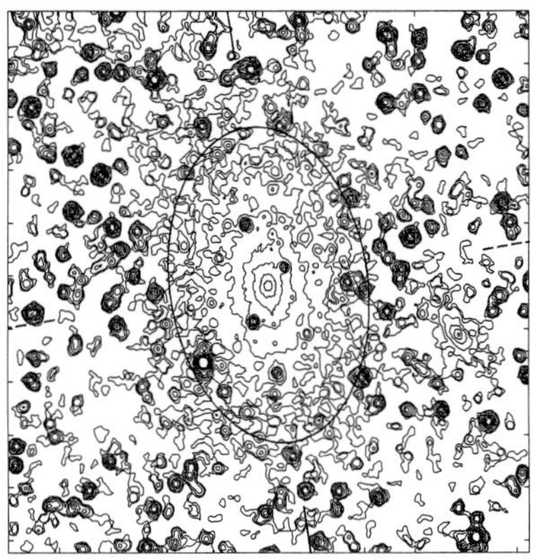

N2325 K5648 1 20.60 512X512

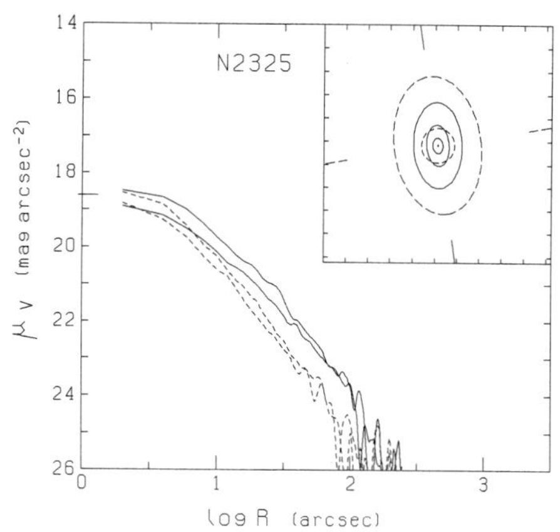

NGC2268

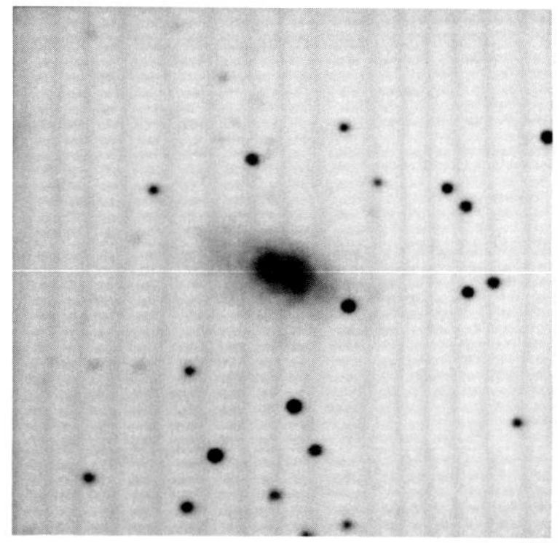

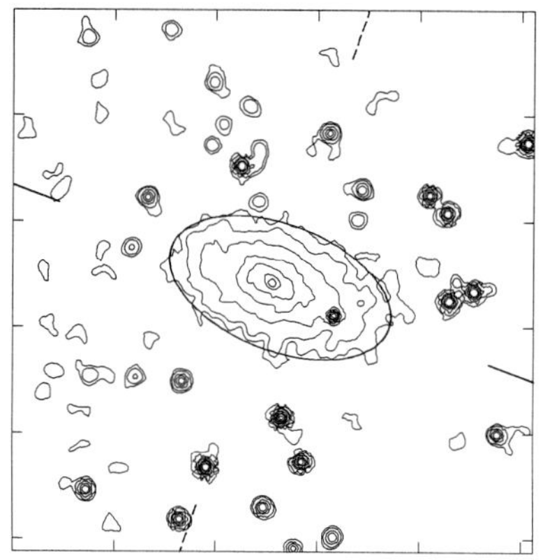

N2268 K4786 2 21.00 256X256

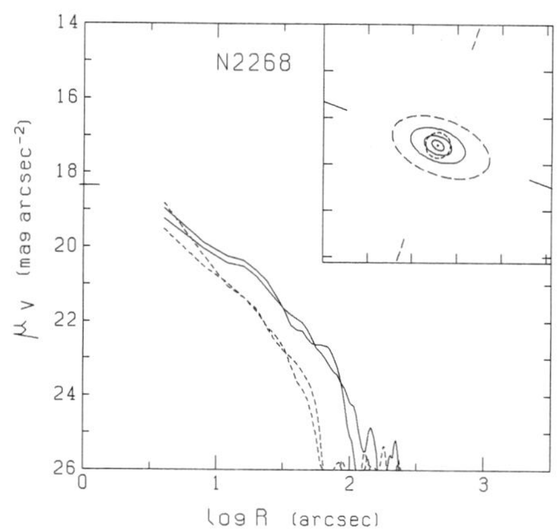

NGC2314

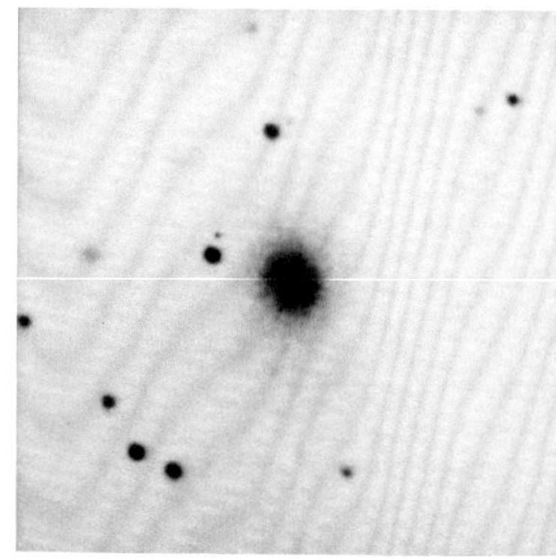

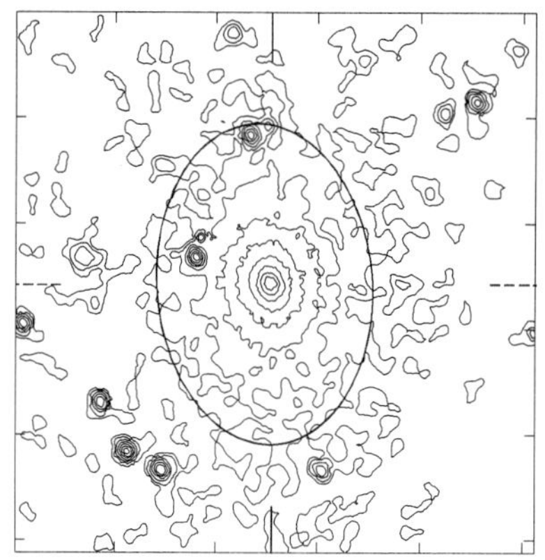

N2314 K4941 1 21.77 256X256

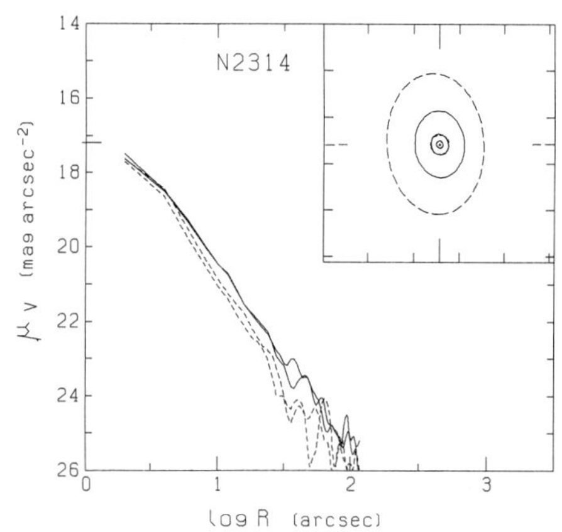

NGC2339

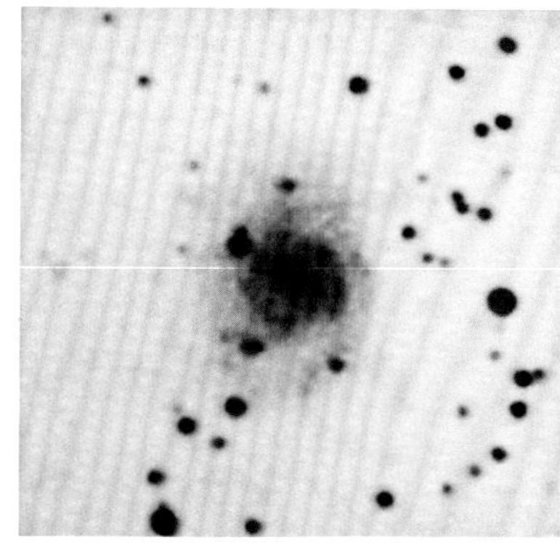

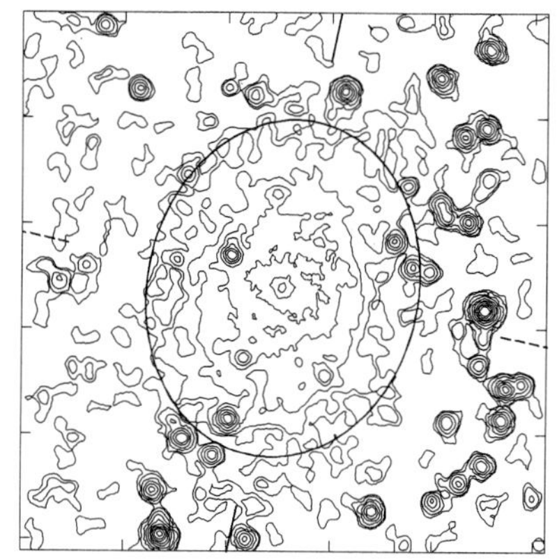

N2339 K4787 1 20.98 256X256

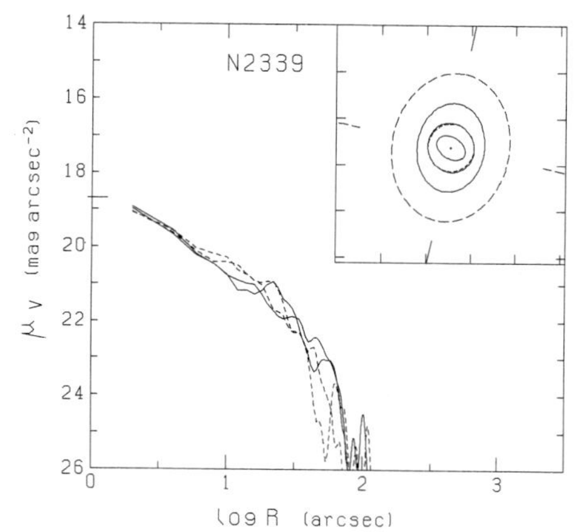

NGC2276

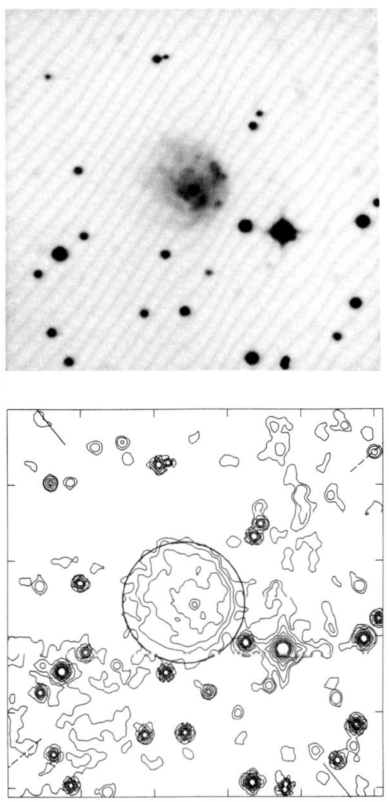

N2276 K4786 2 21.00 256X256

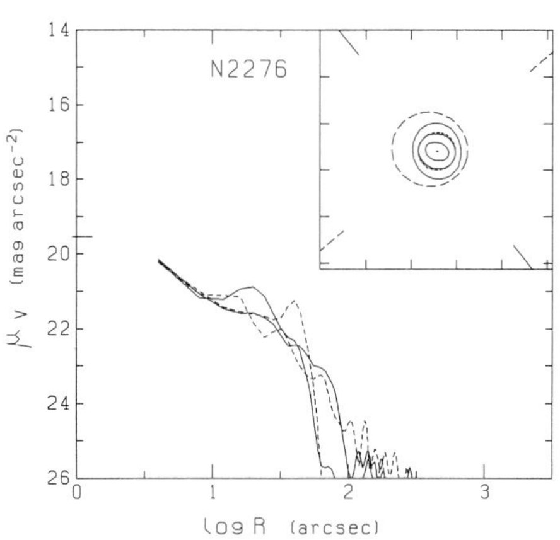

NGC2347

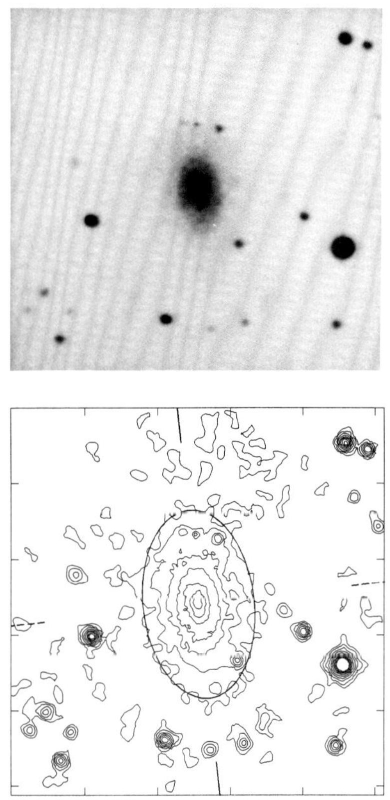

N2347 K5721 1 21.78 256X256

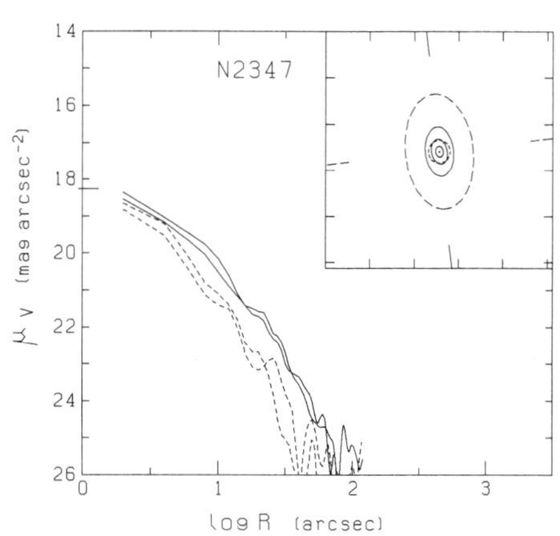

NGC2300

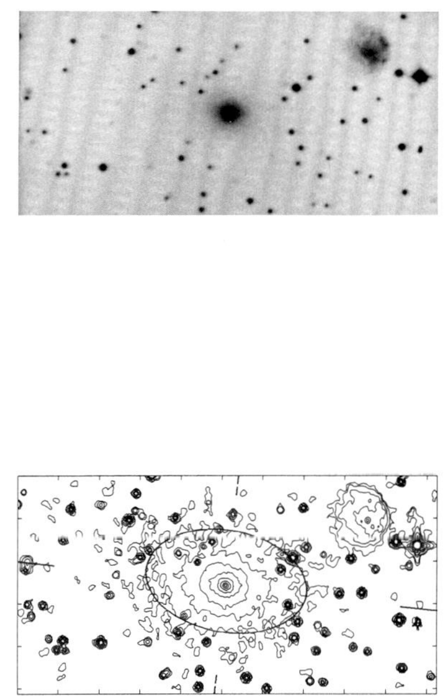

N2300 K4786 2 21.00 512X256

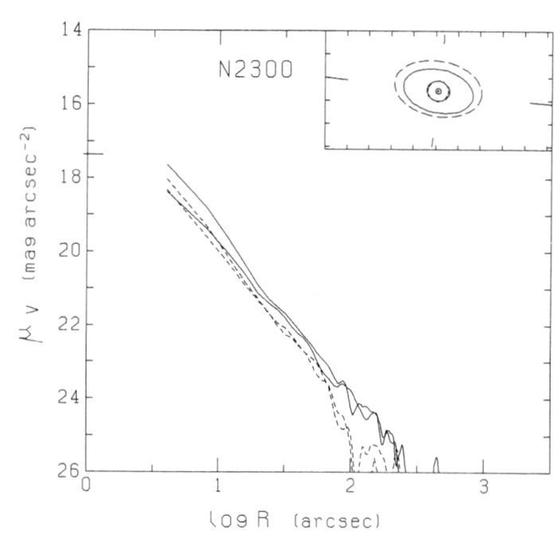

NGC2336

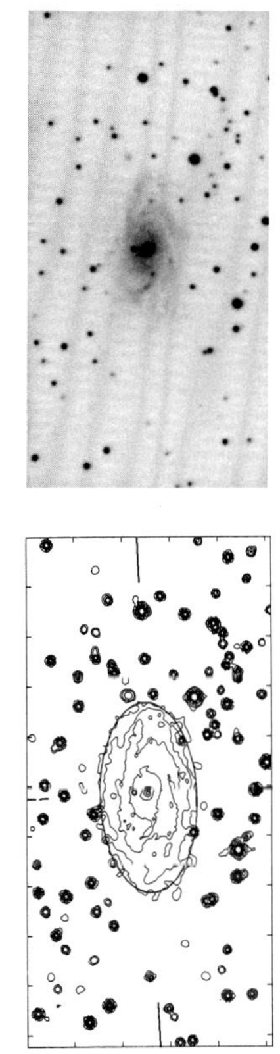

N2336 K5734 2 21.57 256X512

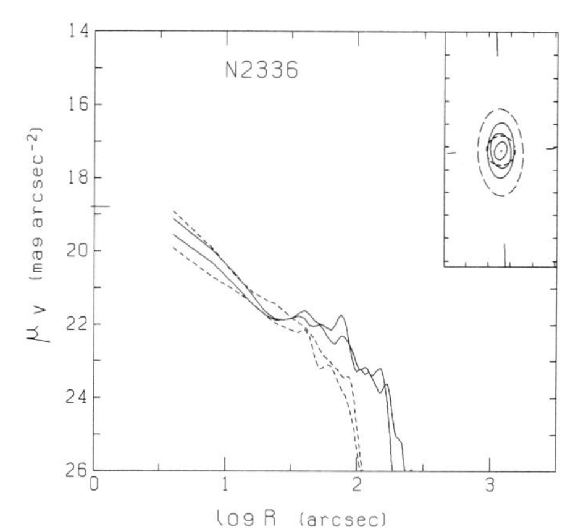

NGC2366

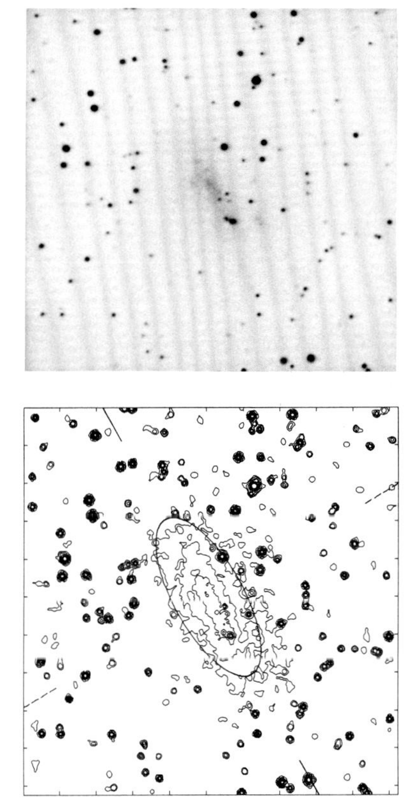

N2366 K4945 2 21.53 512X512

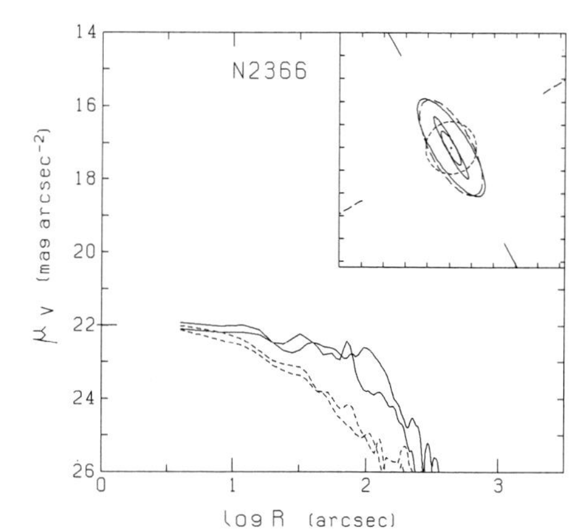

NGC2403

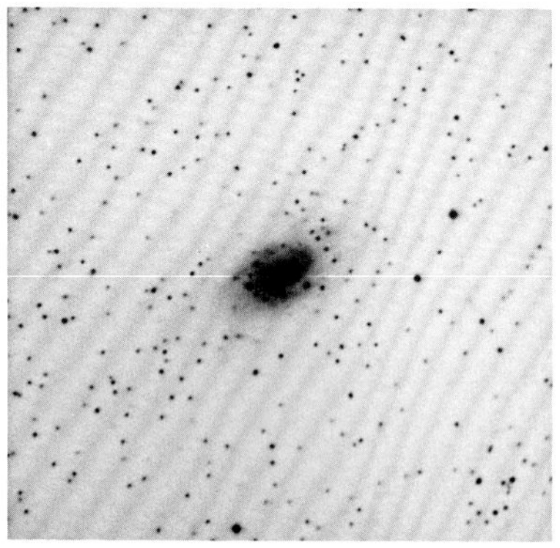

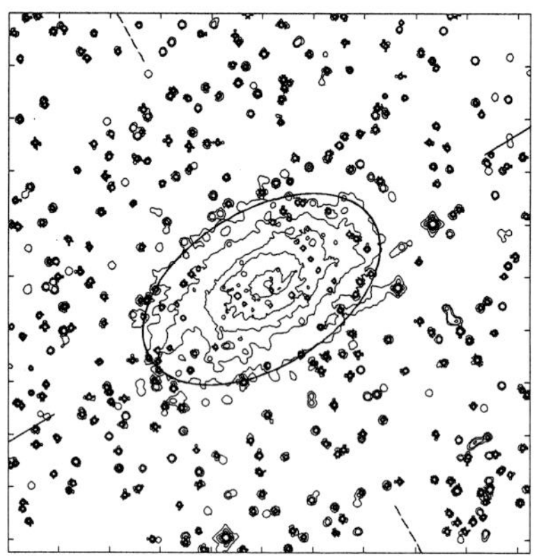

N2403 K5721 4 21.78 512X512

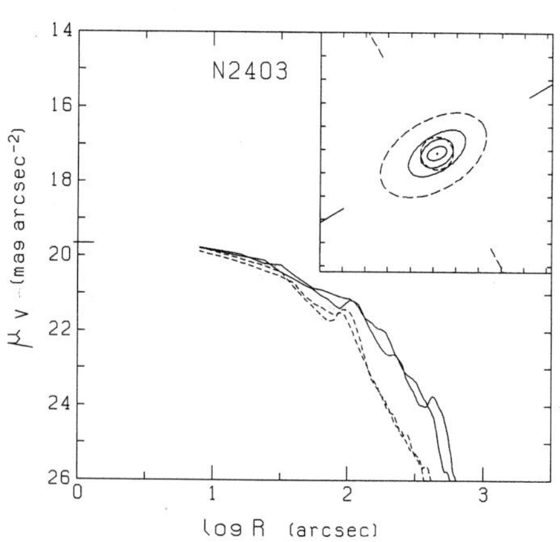

NGC2441

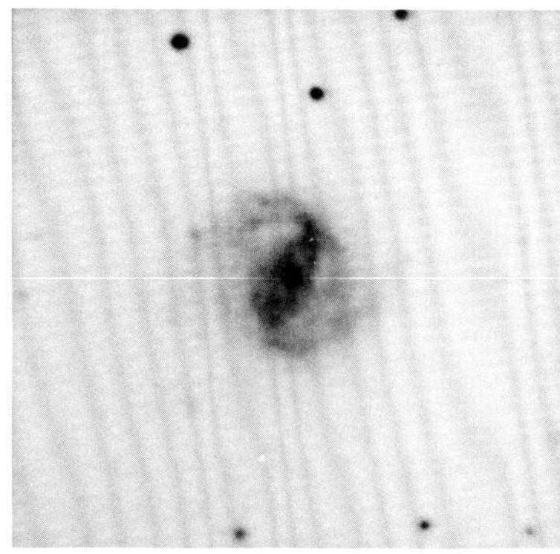

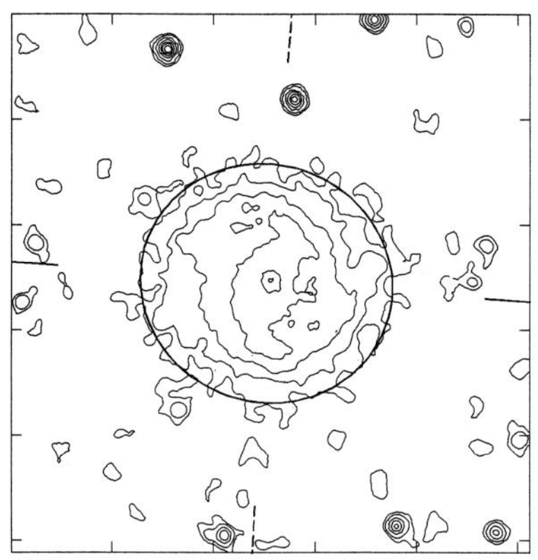

N2441 K3615 1 21.20 256X256

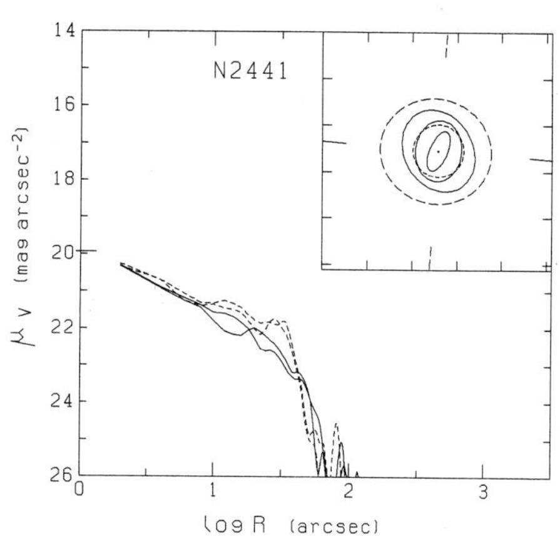

NGC2460

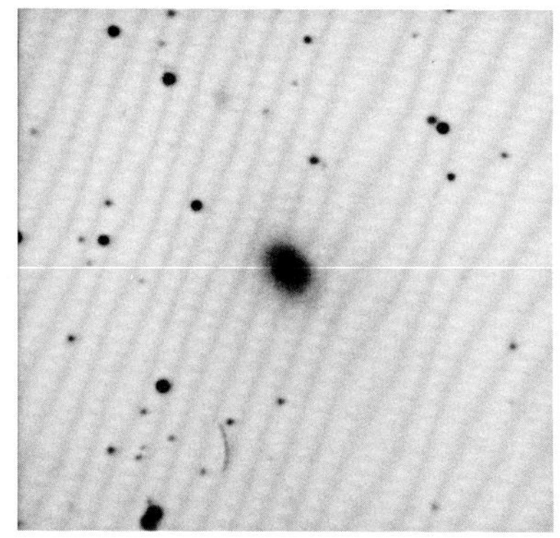

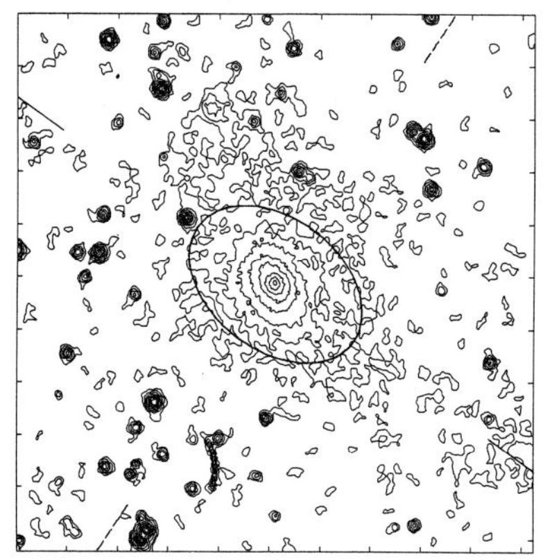

N2460 K4933 1 21.98 512X512

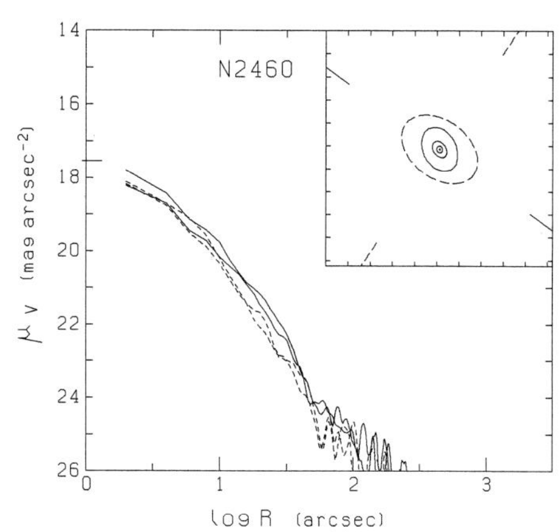

NGC2500

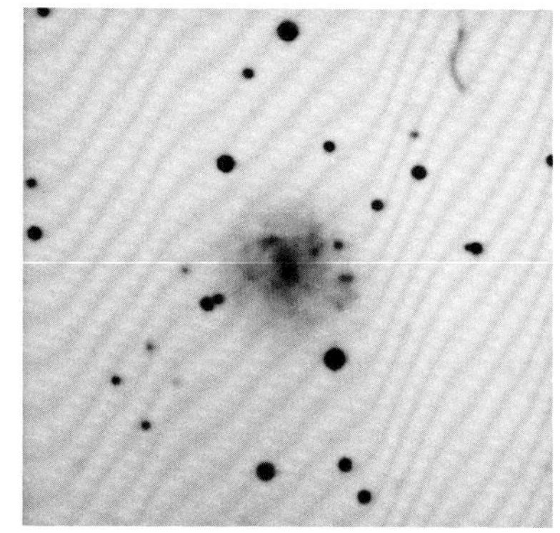

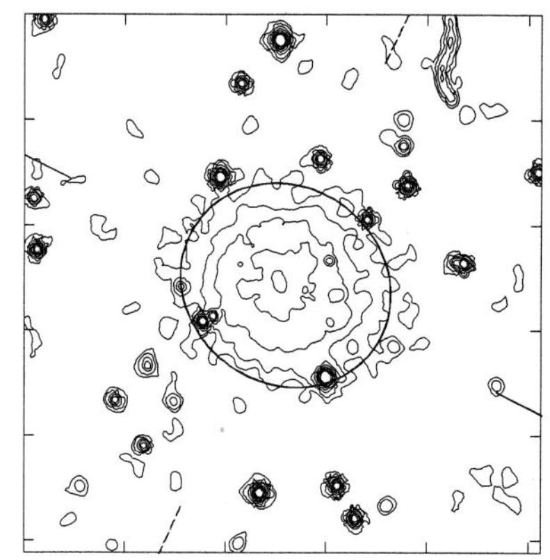

N2500 K4942 2 21.44 256X256

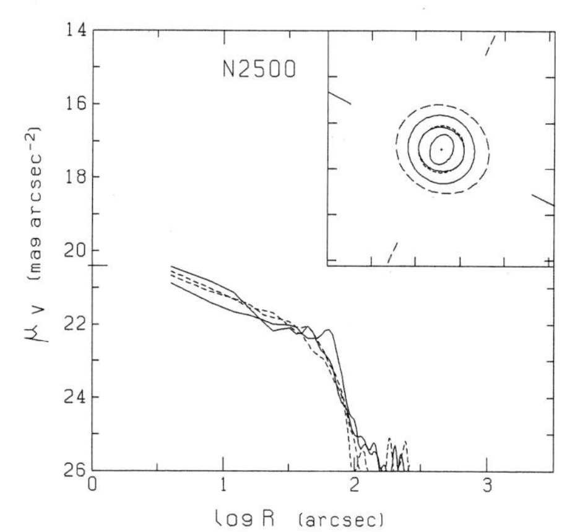

NGC2525

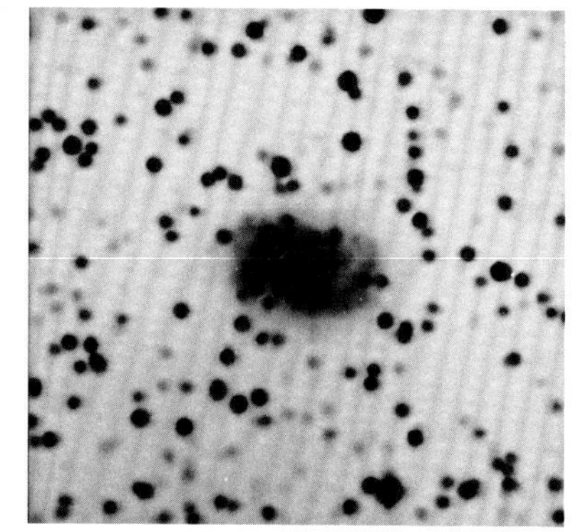

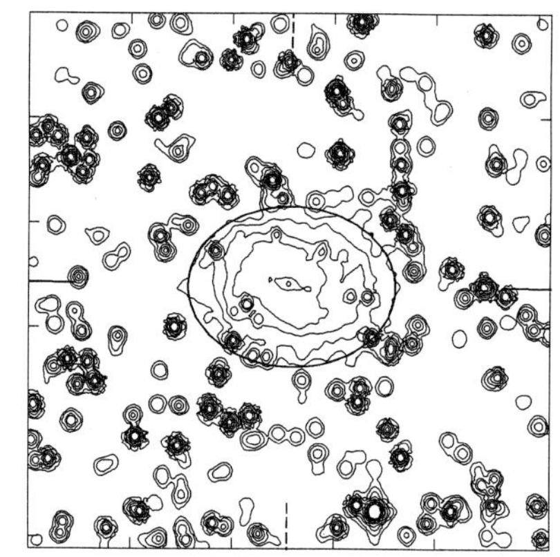

N2525 K5735 2 21.34 256X256

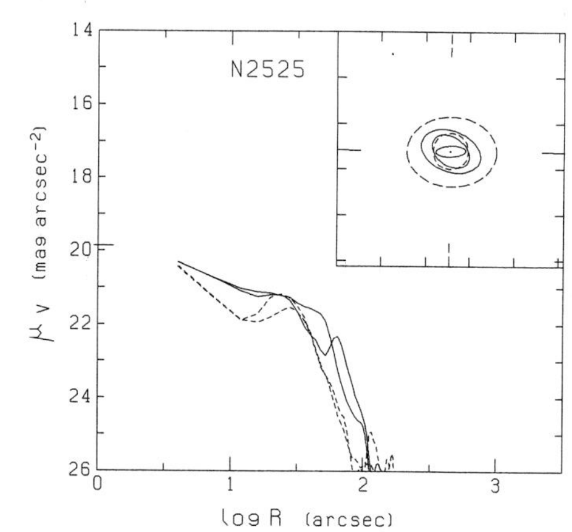

NGC2523

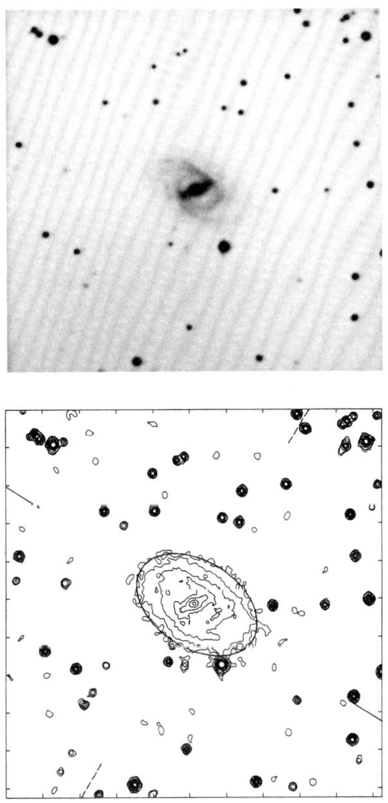

N2523 K3615 1 21.20 512X512

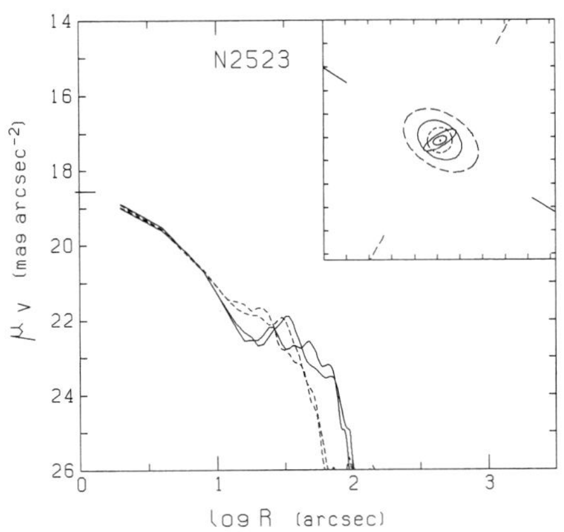

NGC2537

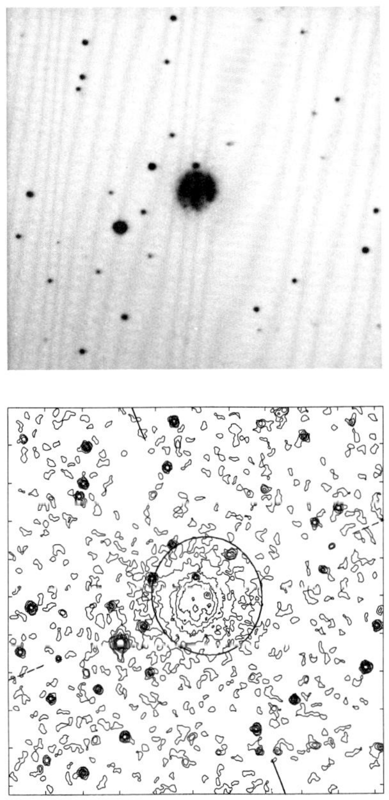

N2537 K4946 1 21.66 512X512

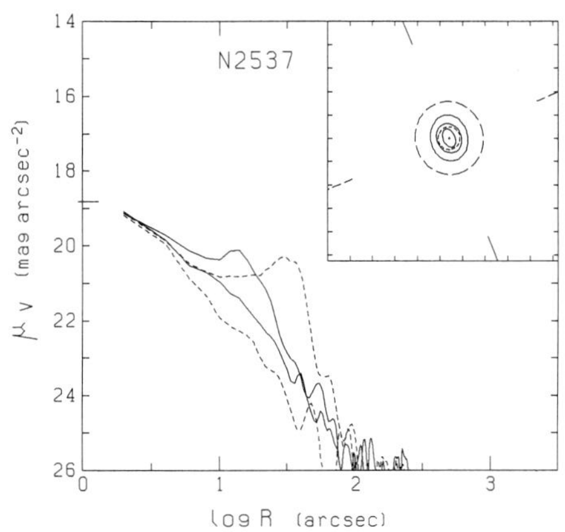

NGC2541

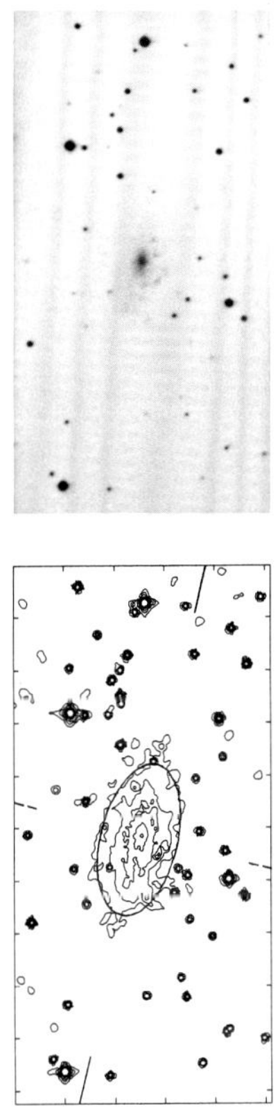

N2541 K4524 2 22.02 256X512

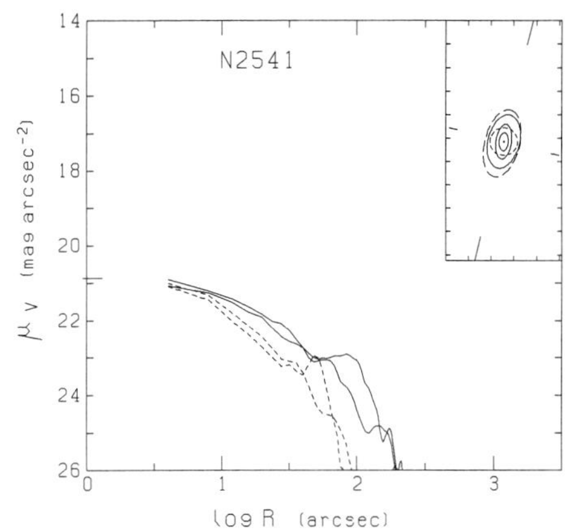

NGC2545

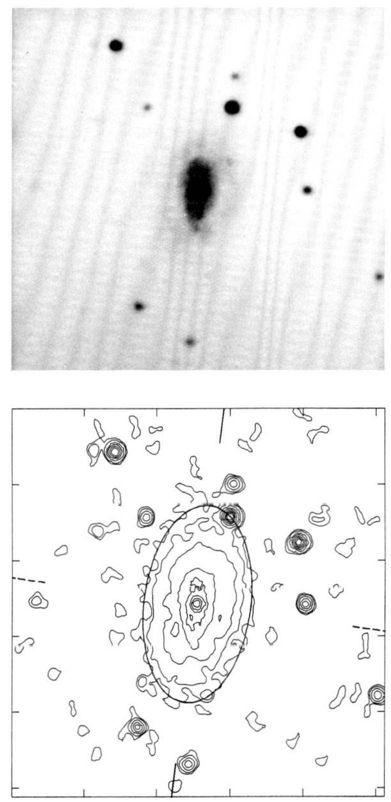

N2545 K2953 1 20.93 256X256

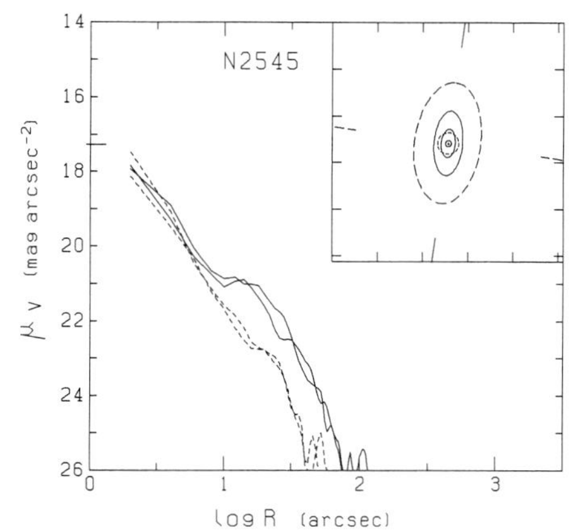

NGC2549

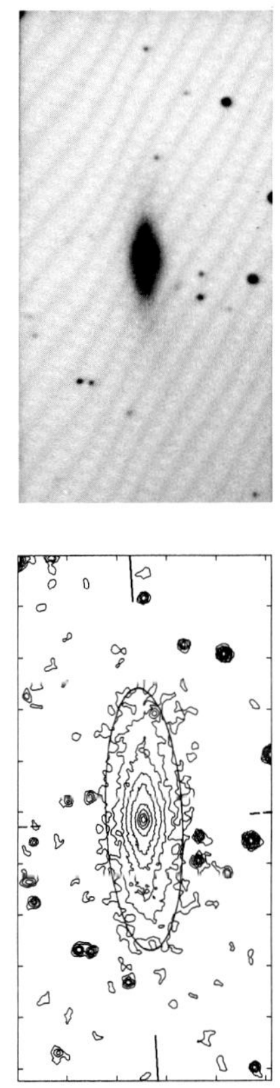

N2549 K4528 1 22.14 256X512

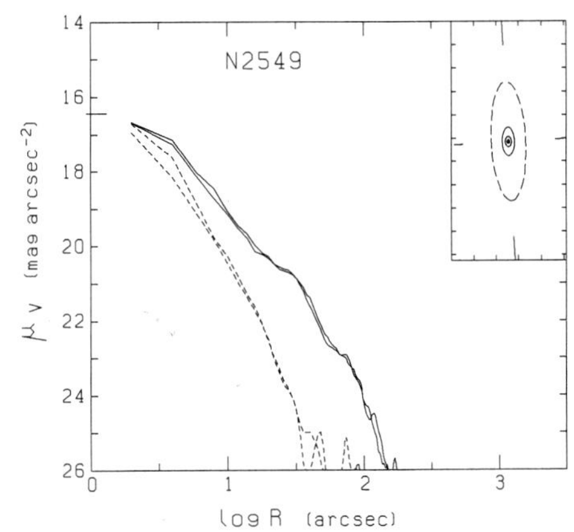

NGC2552

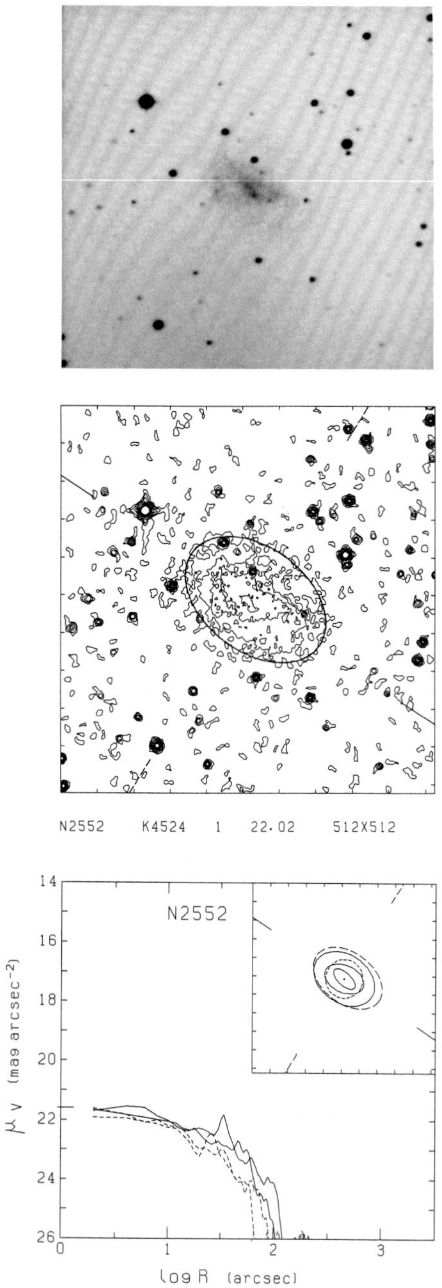

NGC2551

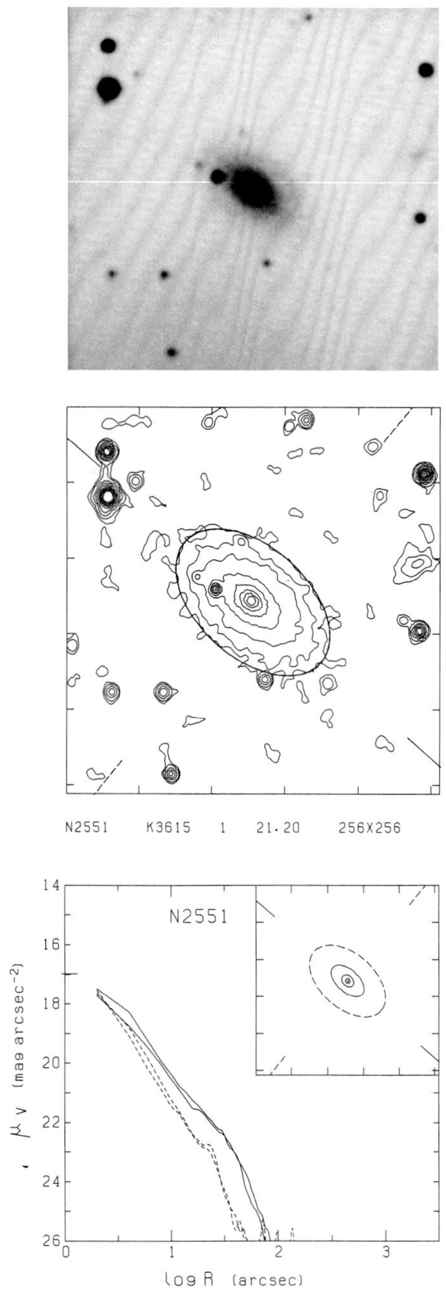

NGC2613

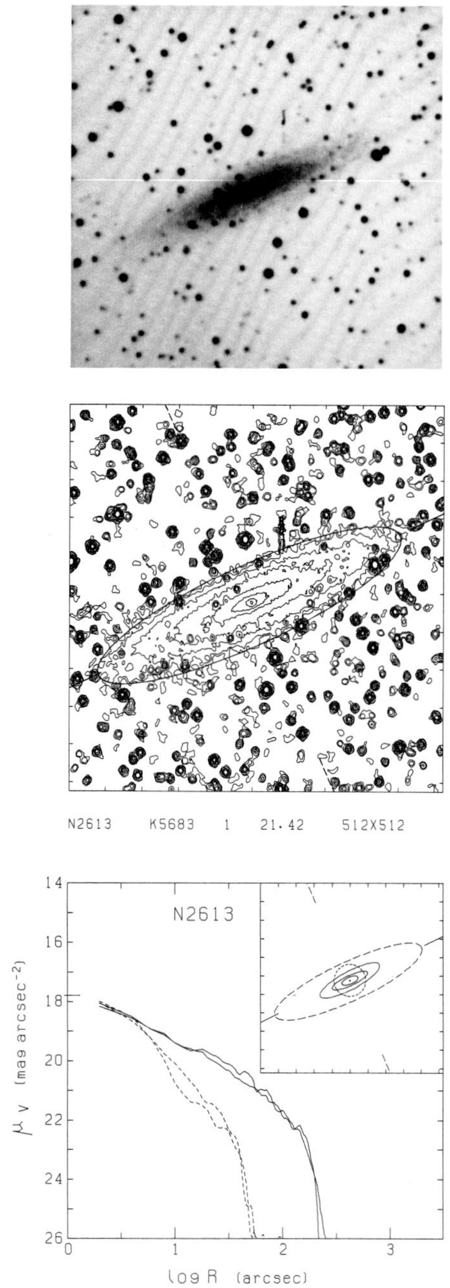

NGC2608

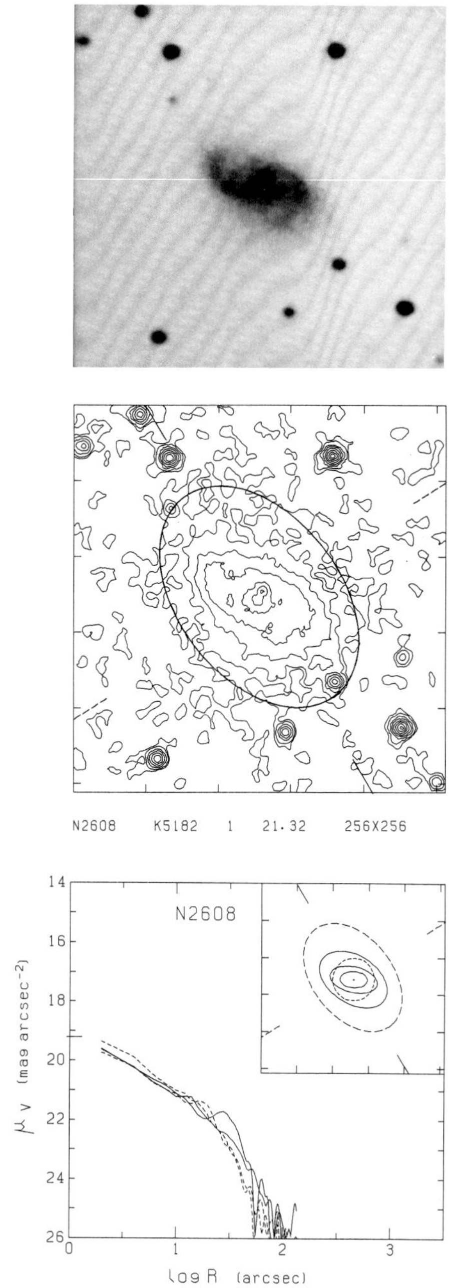

NGC2639

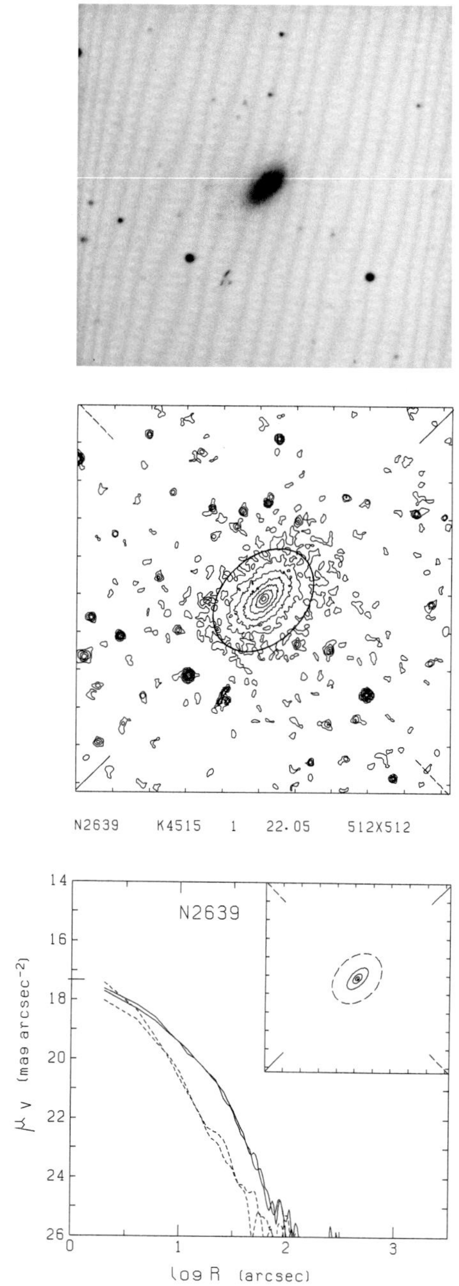

NGC2633

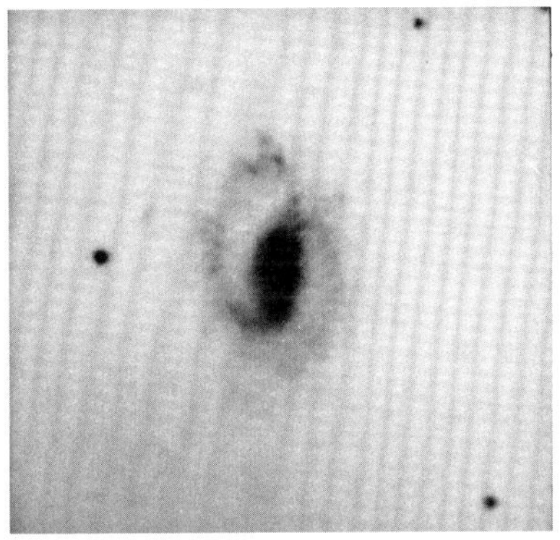

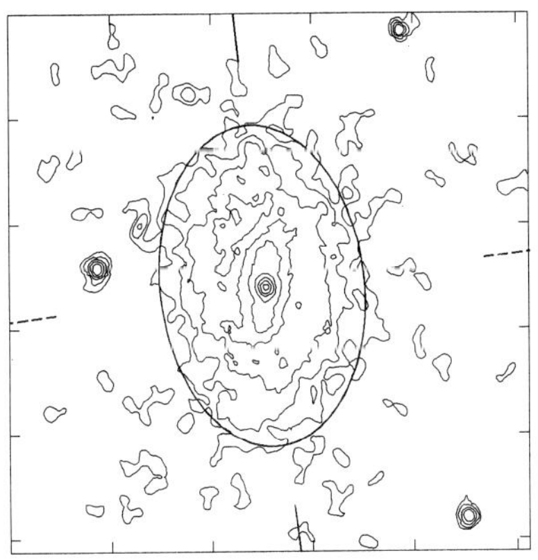

N2633 K4523 1 21.85 256X256

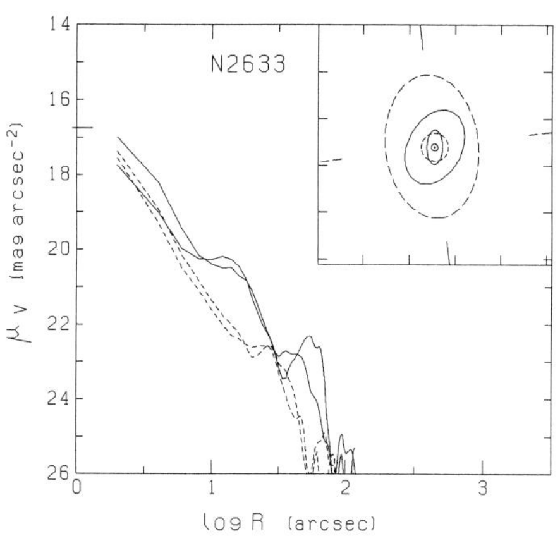

NGC2646

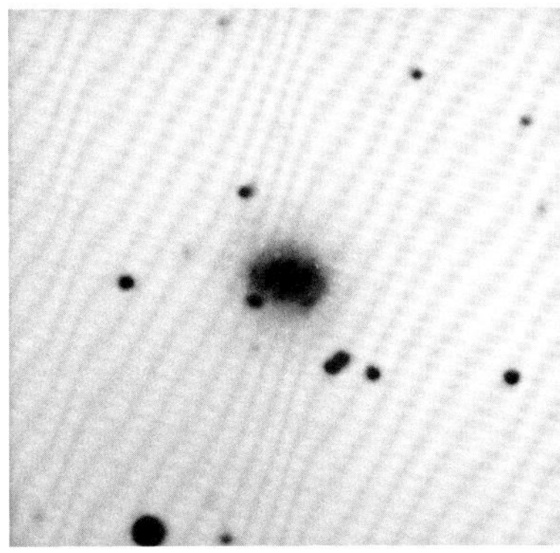

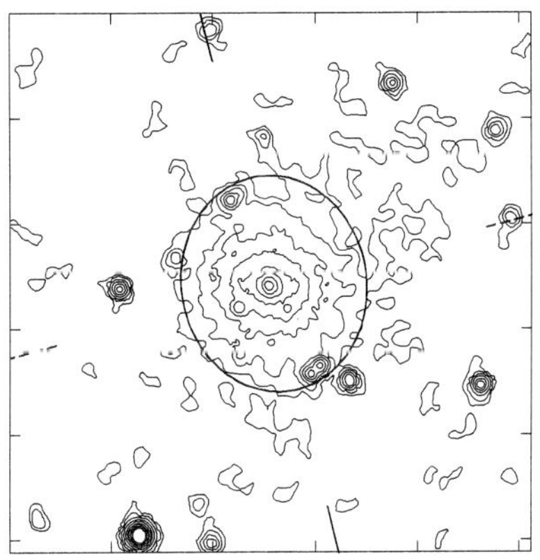

N2646 K4523 1 21.85 256X256

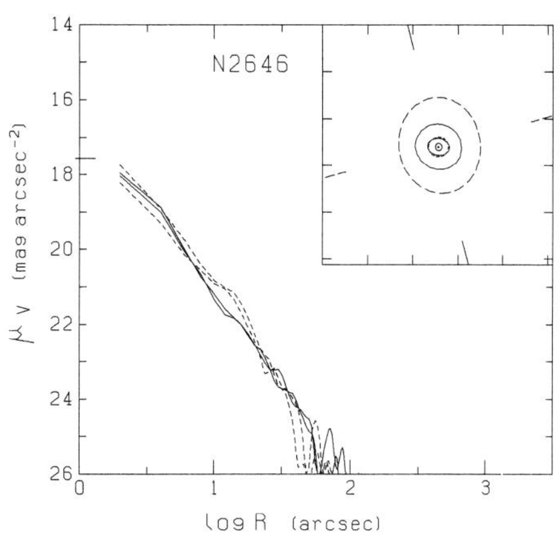

NGC2654

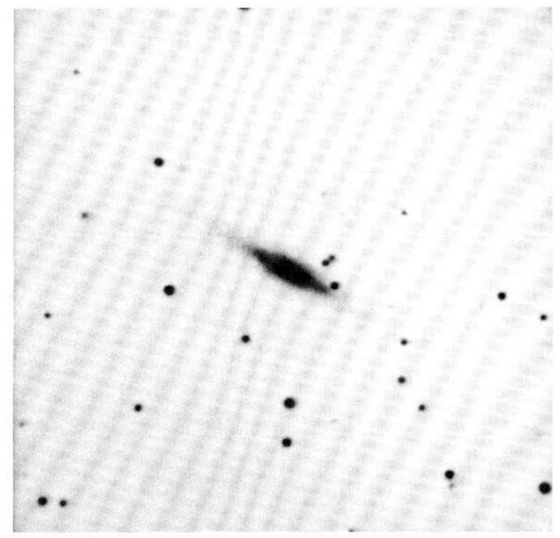

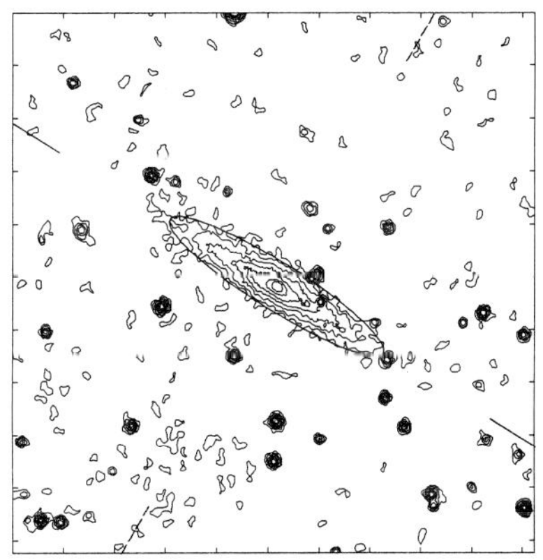

N2654 K4528 1 22.14 512X512

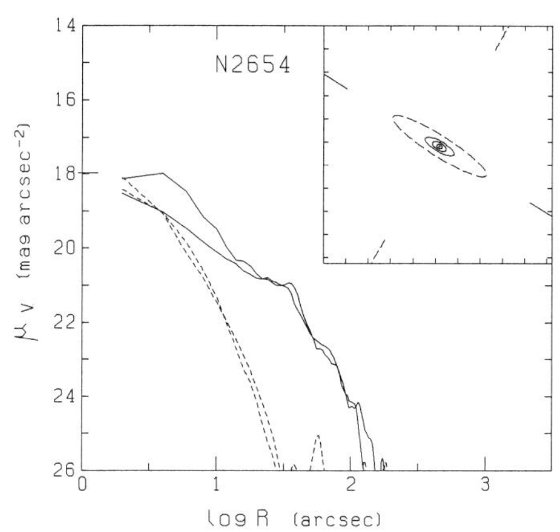

NGC2672

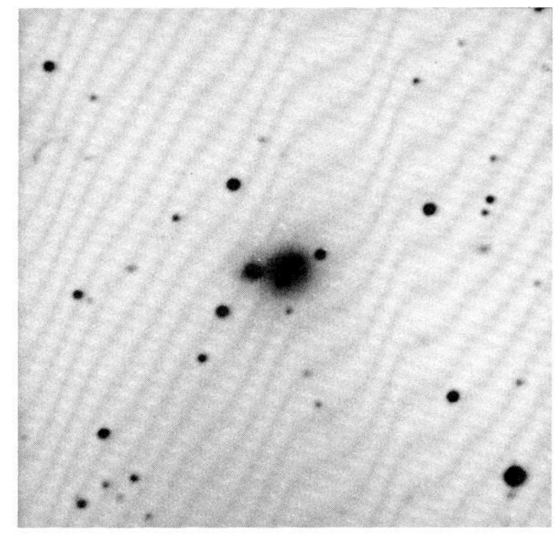

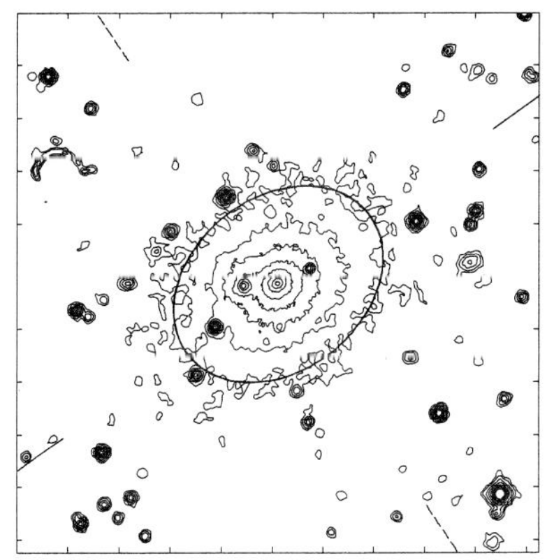

N2672 K2536 1 21.86 512X512

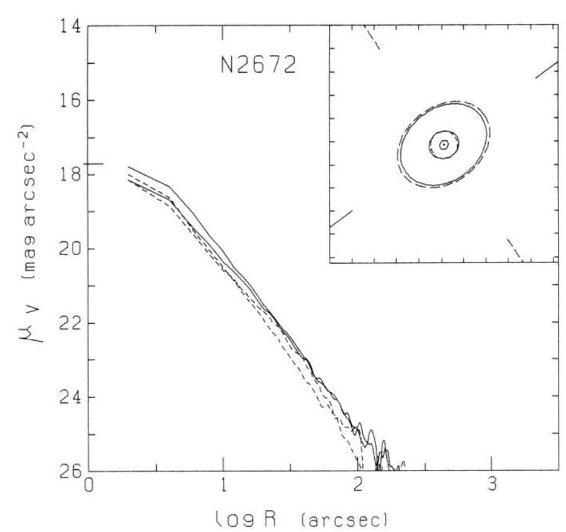

NGC2655

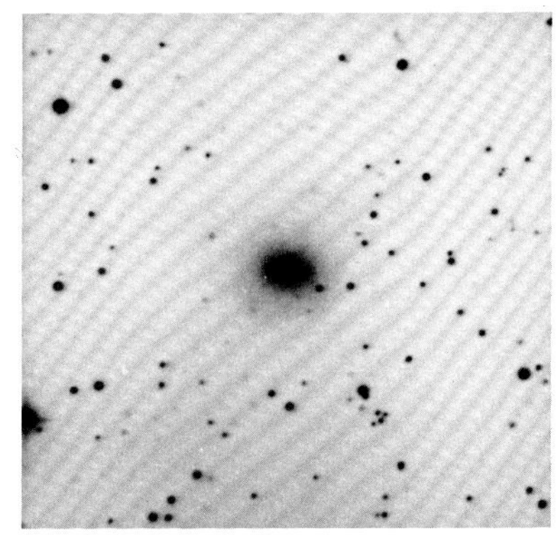

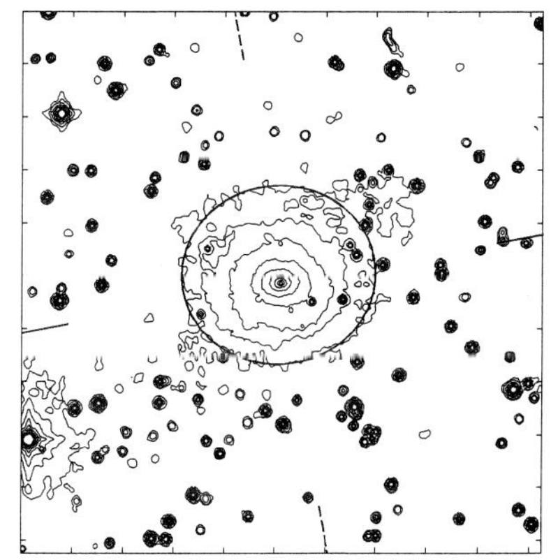

N2655 K5389 2 21.38 512X512

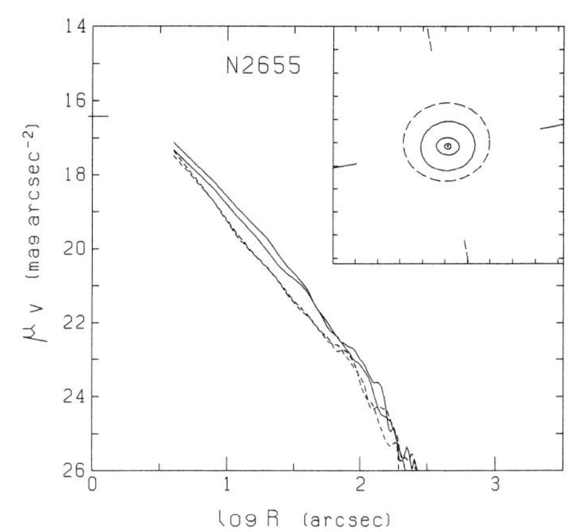

NGC2683

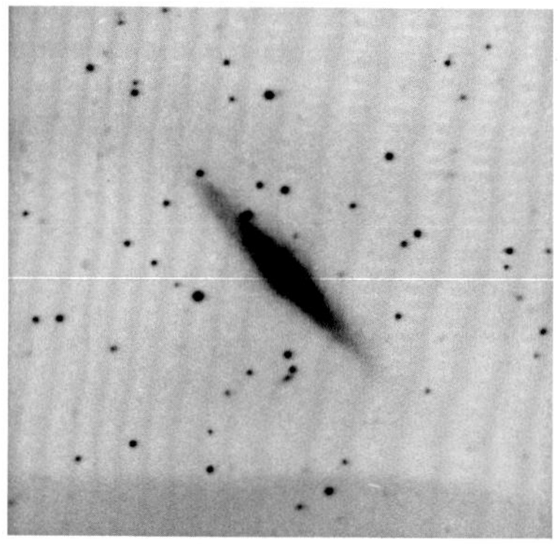

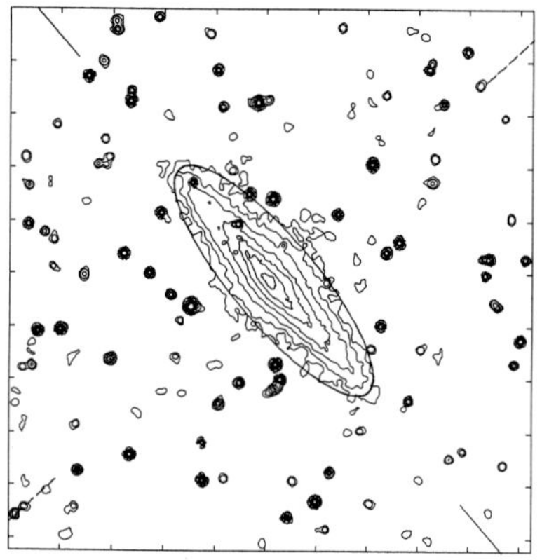

N2683 K5652 2 21.19 512X512

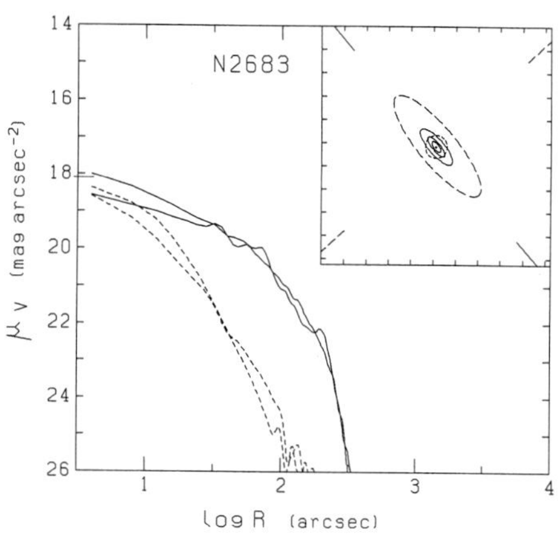

NGC2681

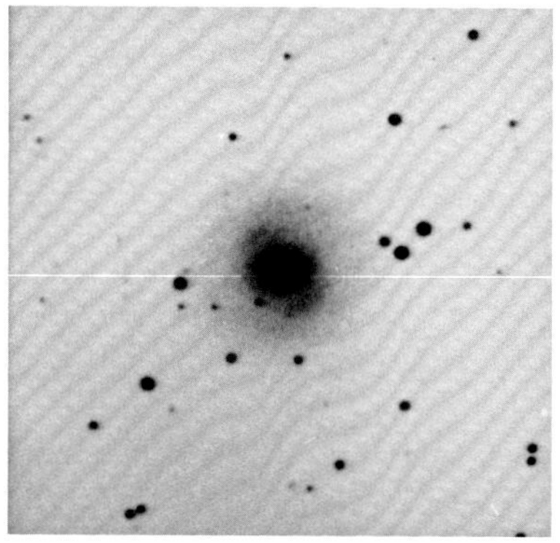

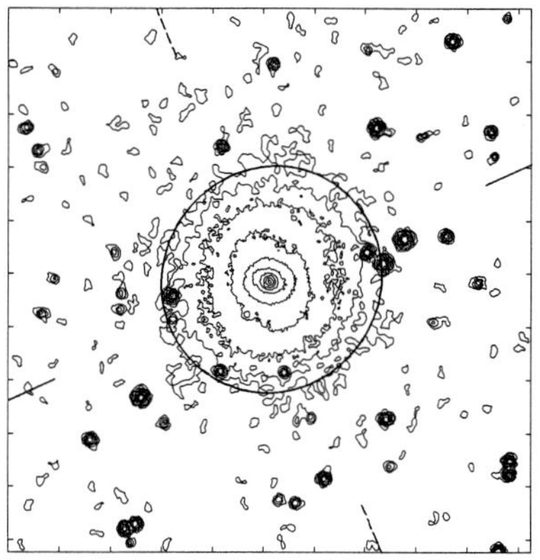

N2681 K4515 1 22.05 512X512

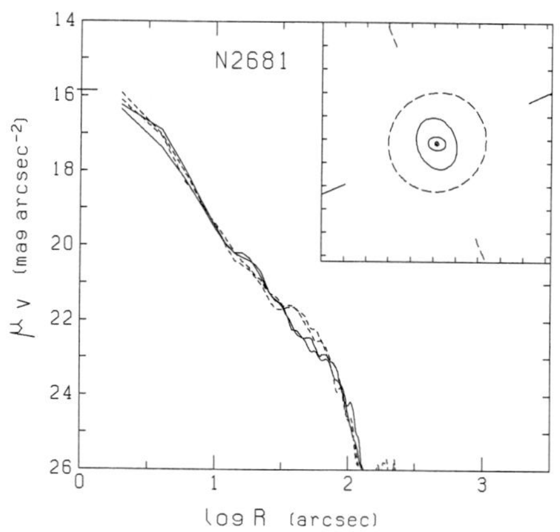

NGC2685

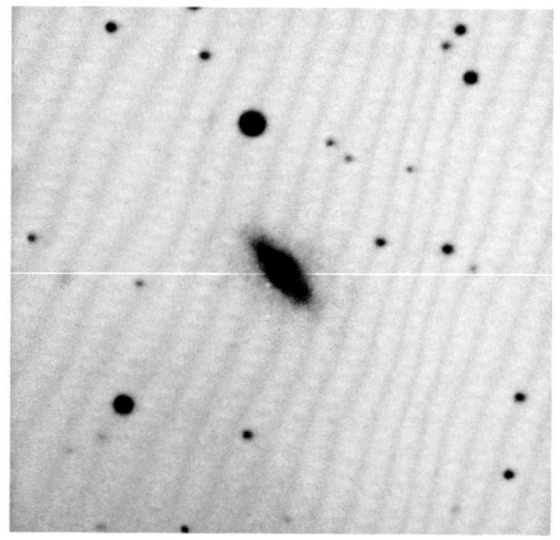

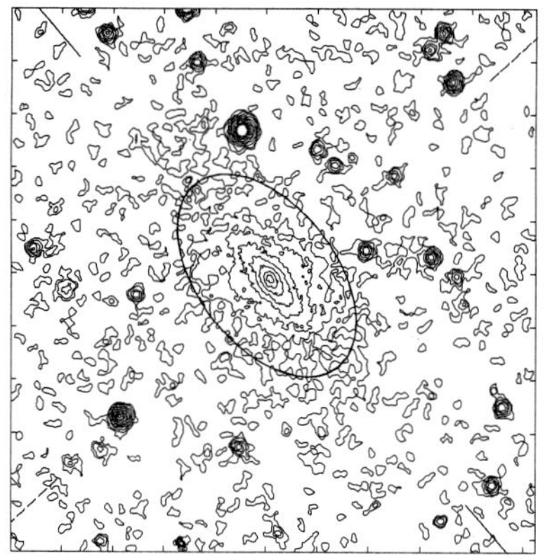

N2685 K4581 1 21.79 512X512

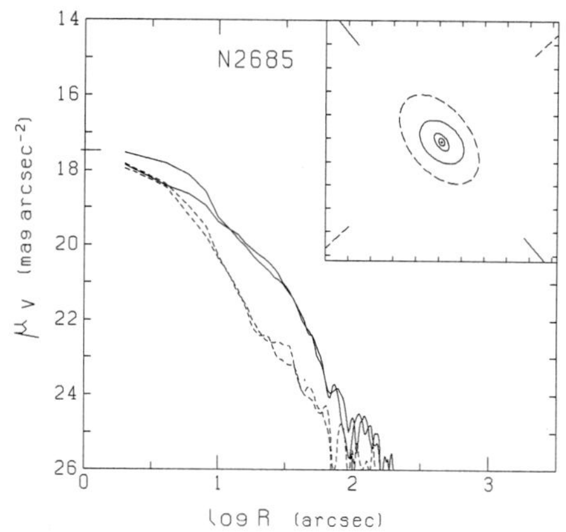

NGC2693

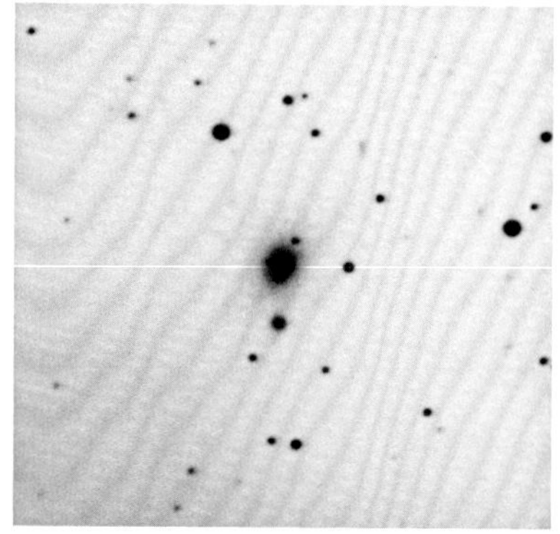

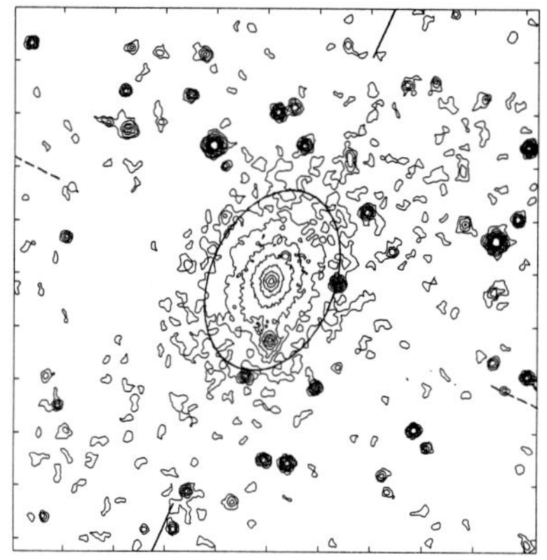

N2693 K4515 1 22.05 512X512

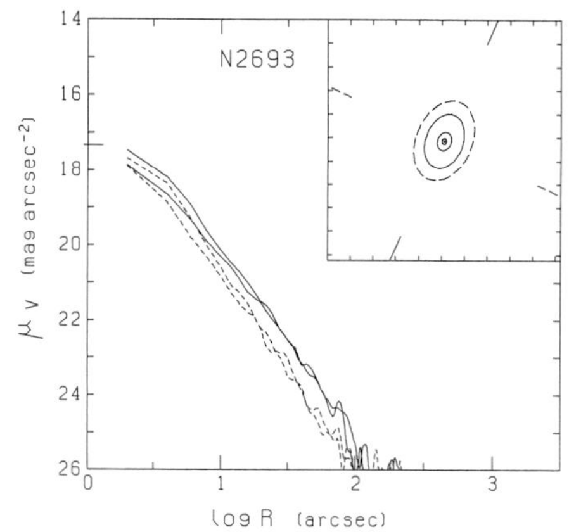

NGC2713

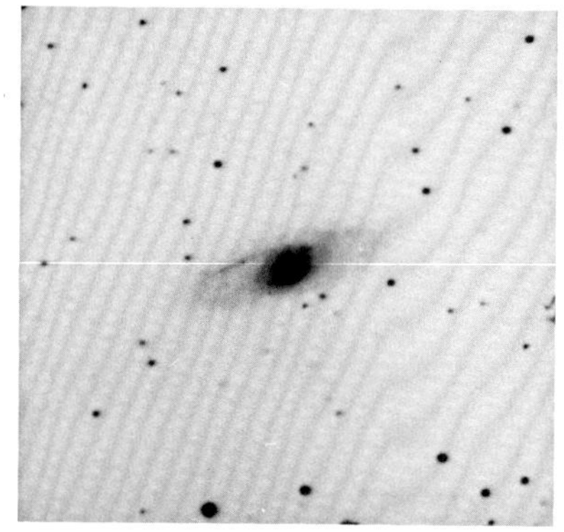

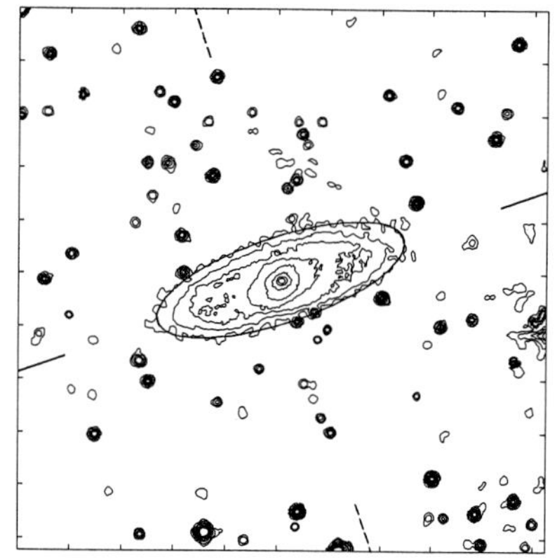

N2713 K2034 1 21.66 512X512

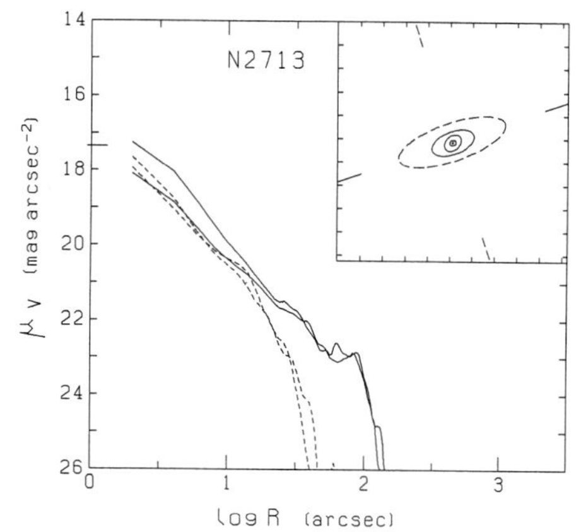

NGC2701

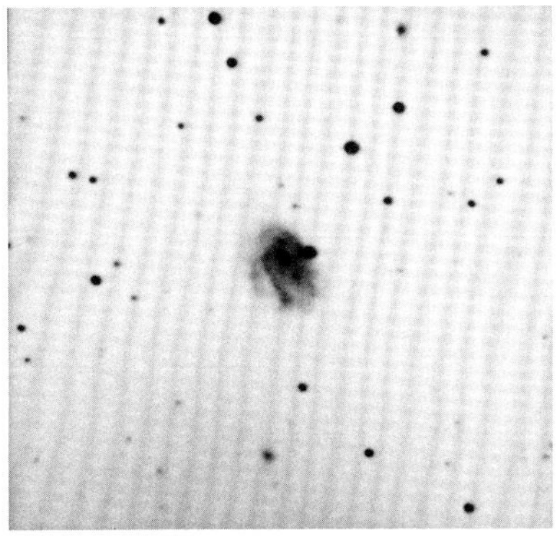

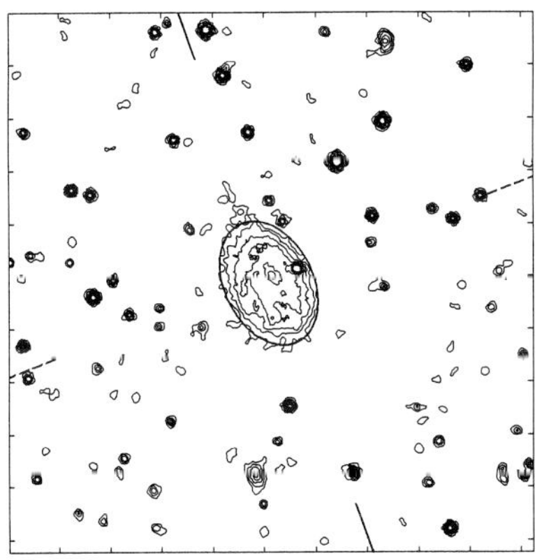

N2701 K5361 1 22.22 512X512

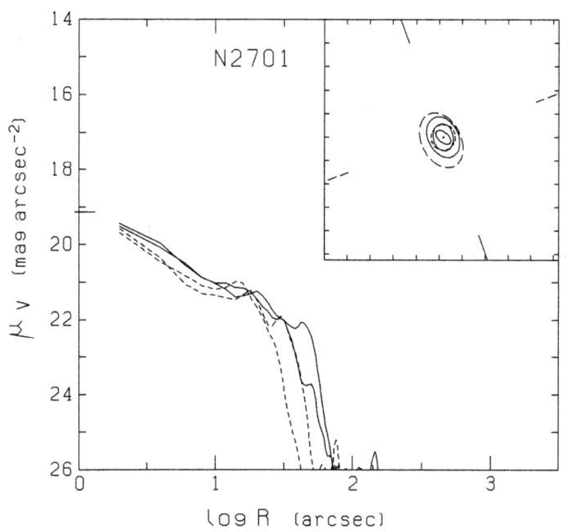

NGC2712

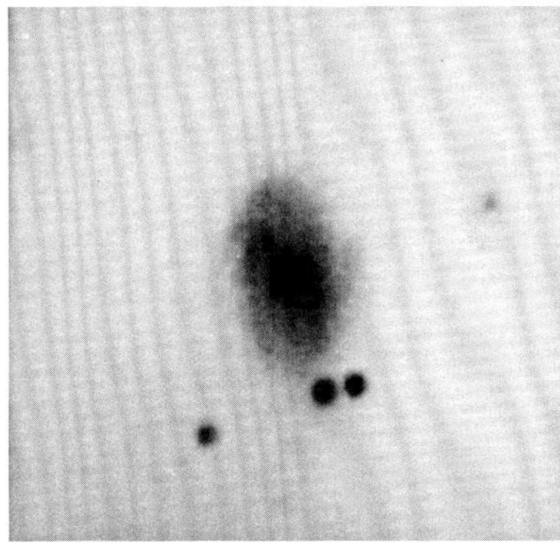

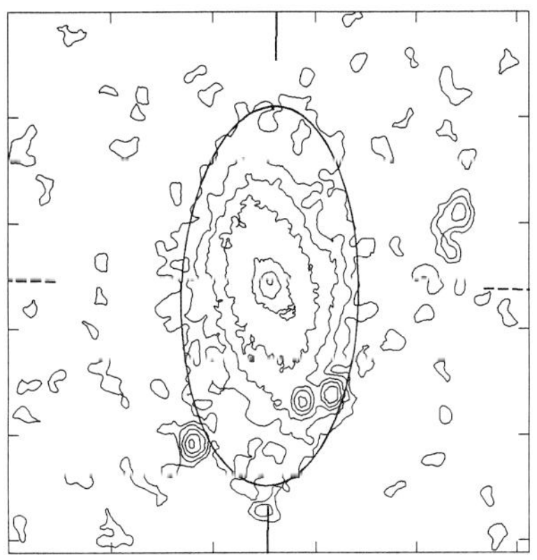

N2712 K4593 1 21.70 256X256

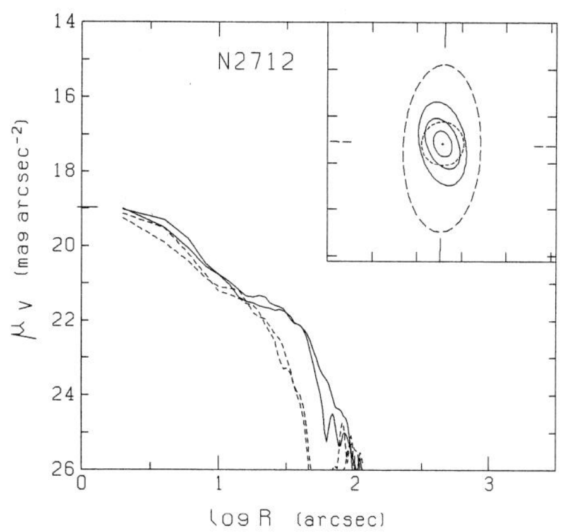

NGC2715

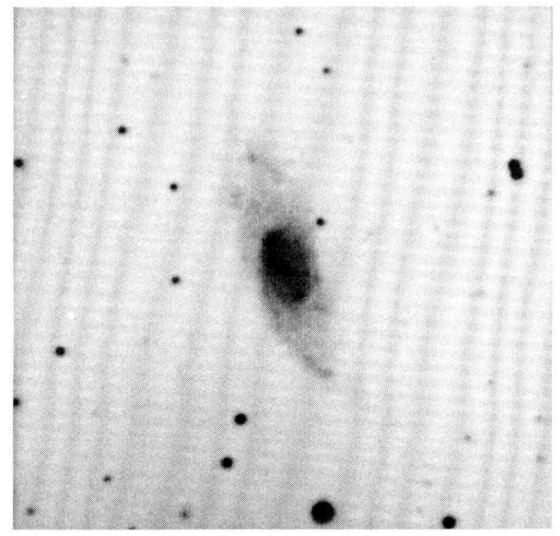

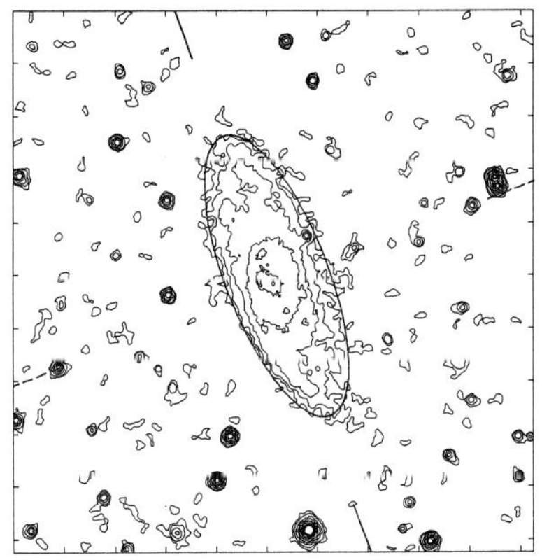

N2715 K5389 1 21.38 512X512

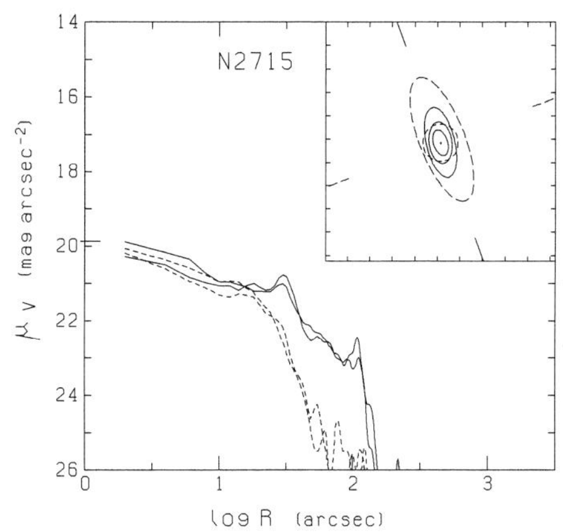

NGC2749

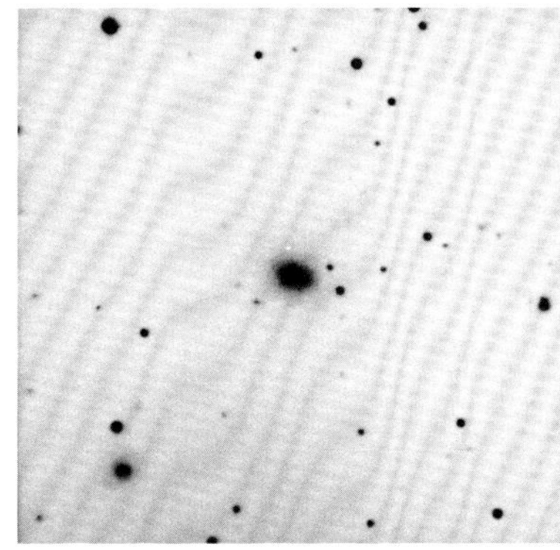

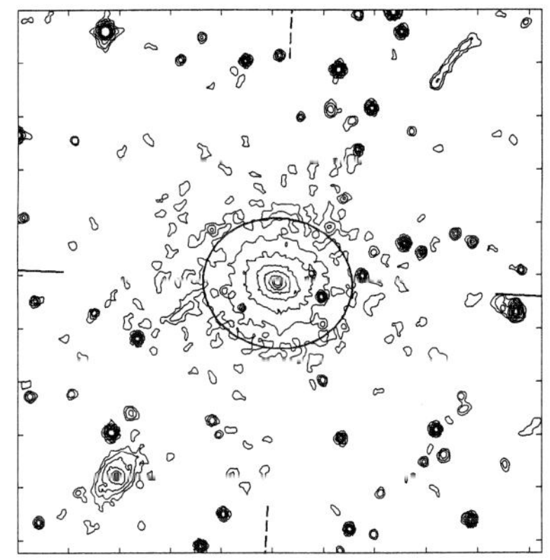

N2749 K5356 1 21.75 512X512

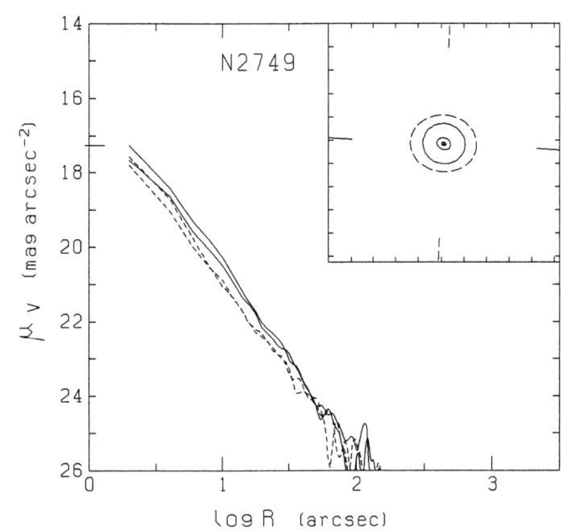

NGC2742

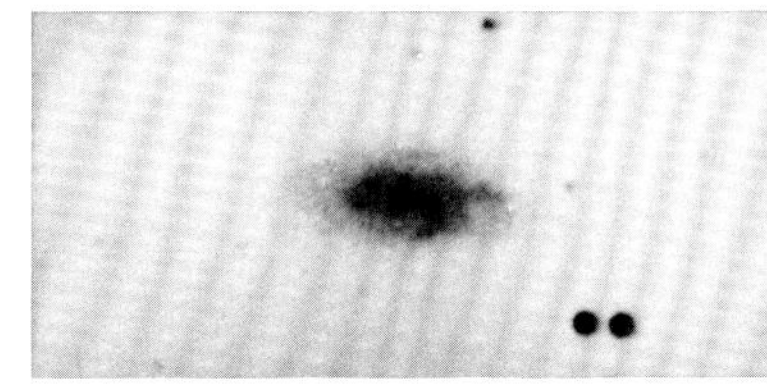

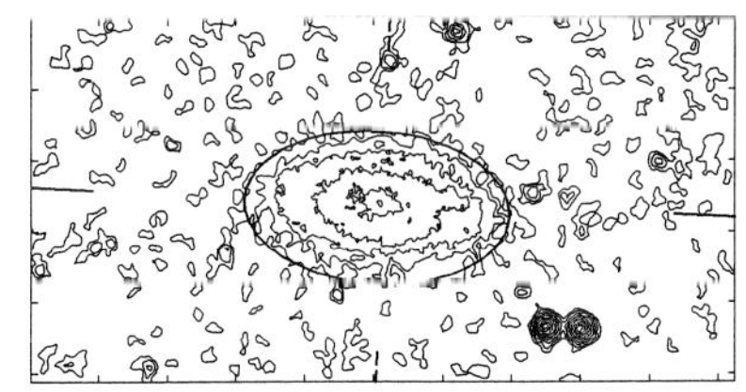

N2742 K4581 1 21.79 512X256

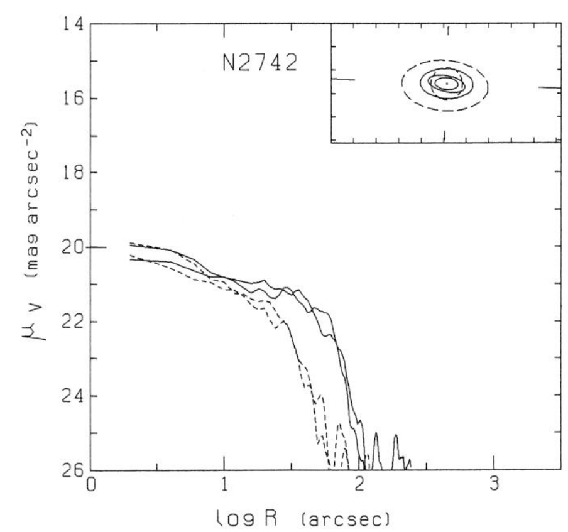

NGC2764

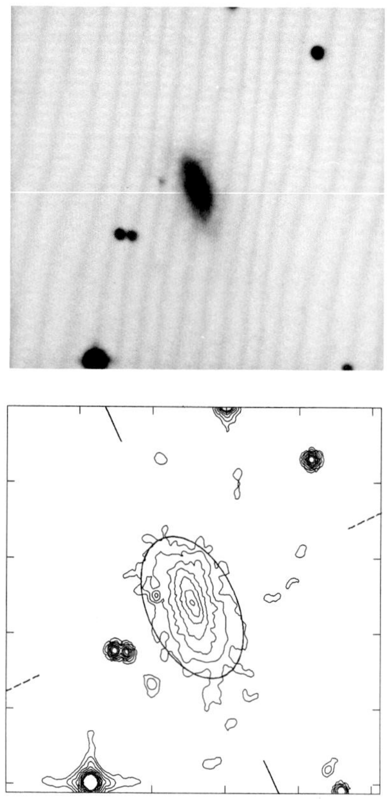

N2764 K5356 1 21.75 256X256

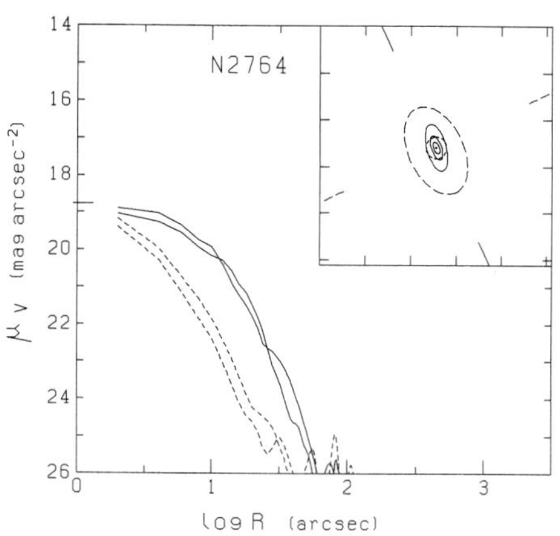

NGC2732

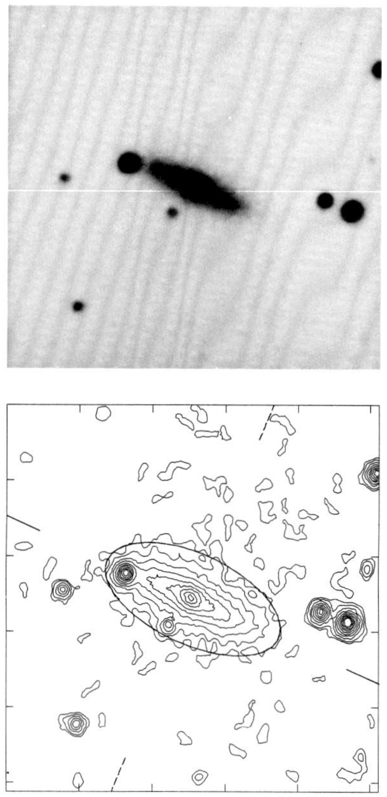

N2732 K5389 1 21.38 256X256

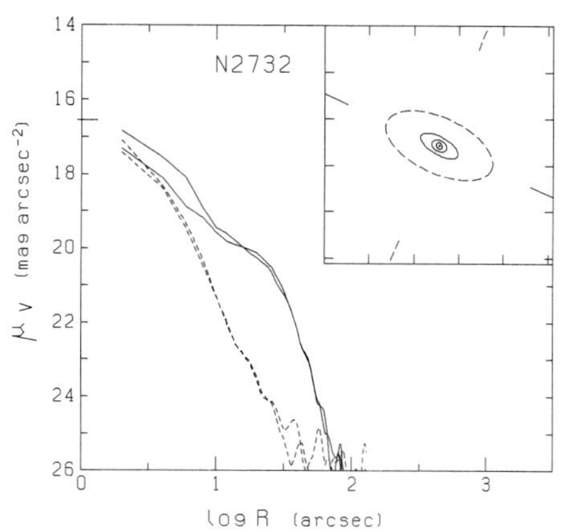

NGC2775

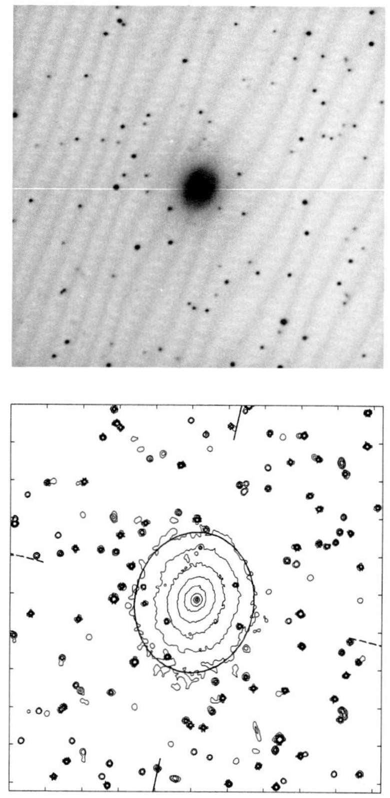

N2775 K2034 2 21.66 512X512

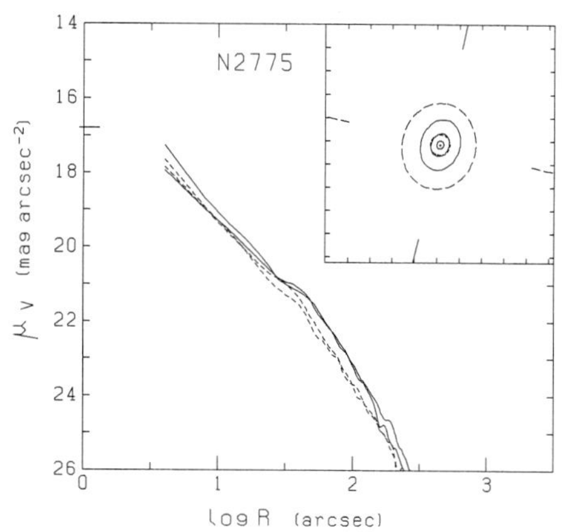

NGC2768

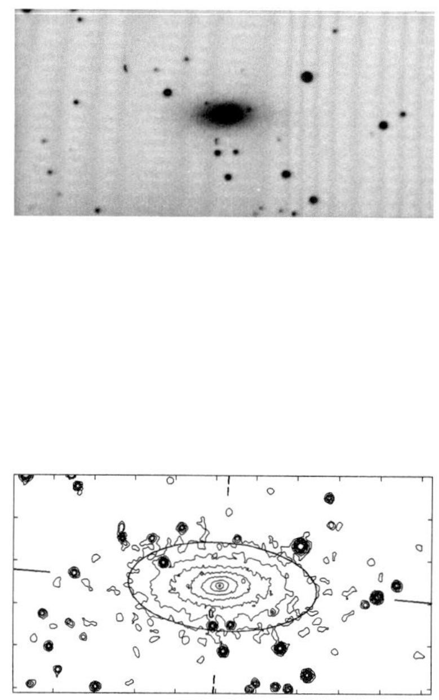

N2768 K4581 2 21.79 512X256

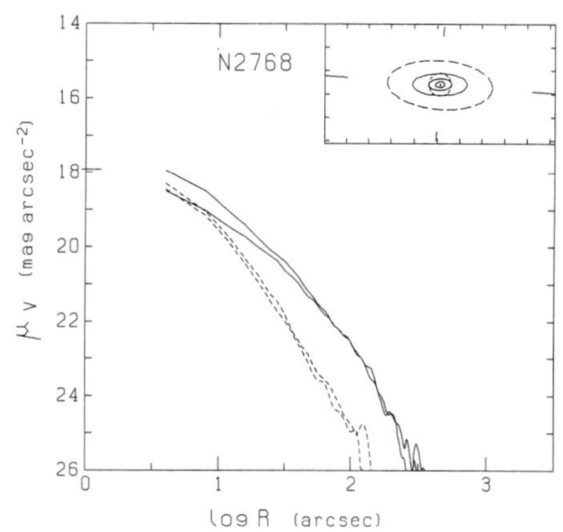

NGC2748

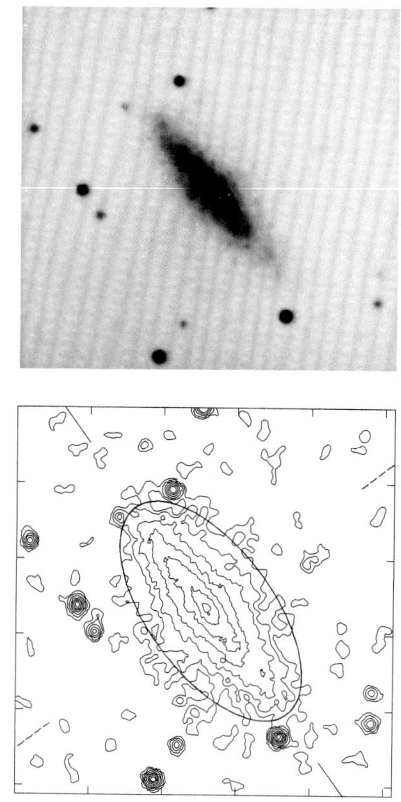

N2748 K4523 1 21.85 256X256

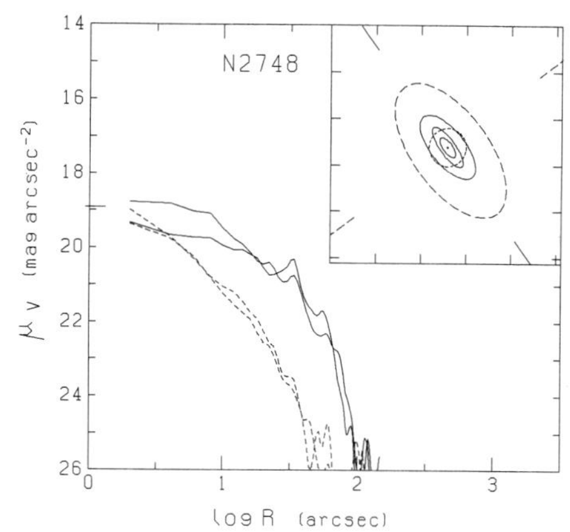

NGC2776

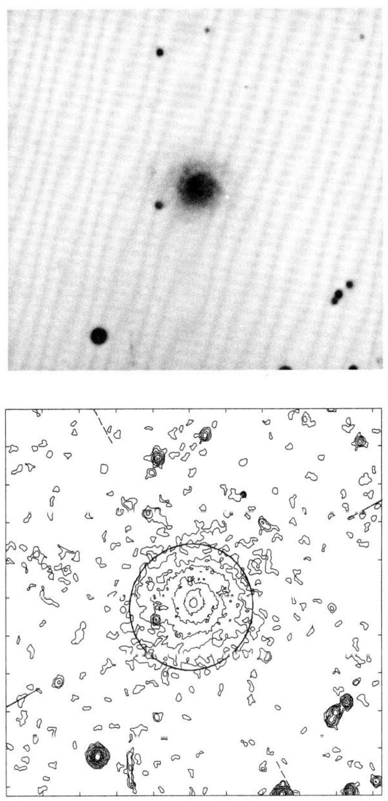

N2776 K4593 1 21.70 512X512

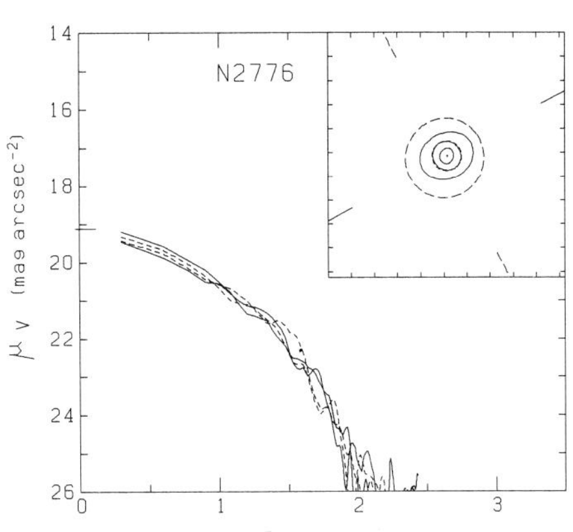

NGC2784

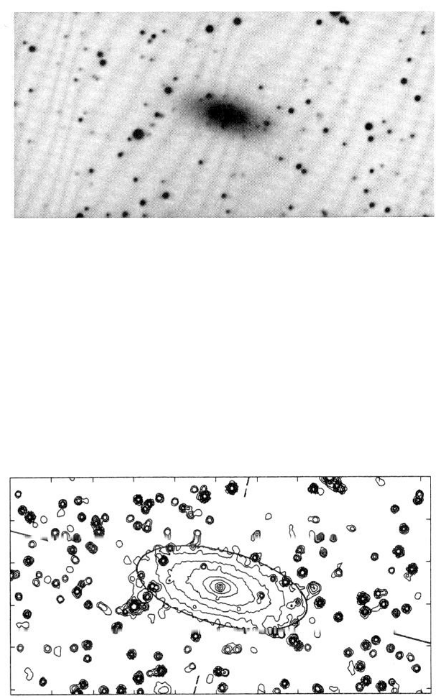

N2784 K5348 2 21.65 512X256

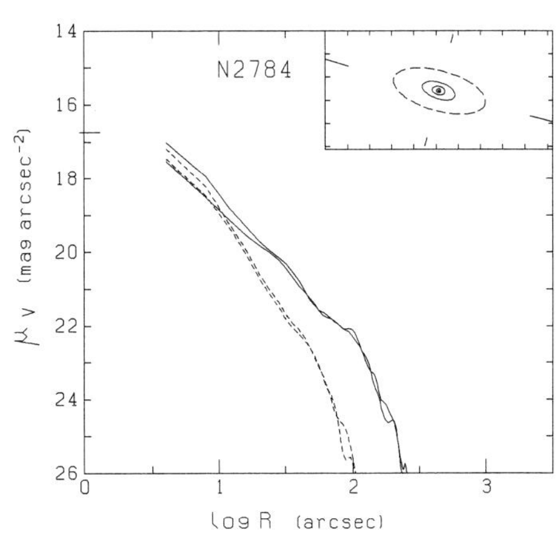

NGC2782

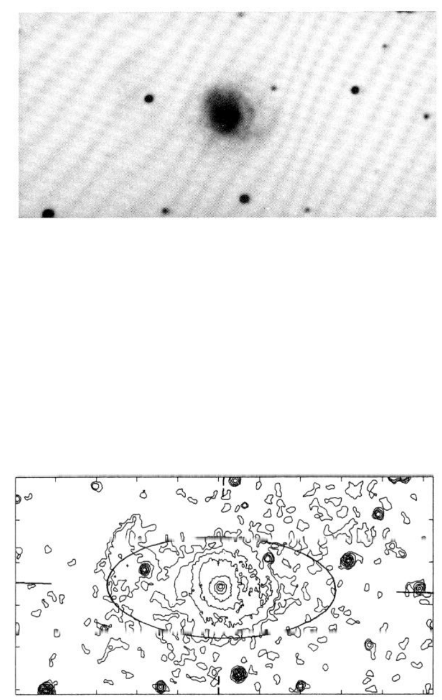

N2782 K5657 1 21.67 512X256

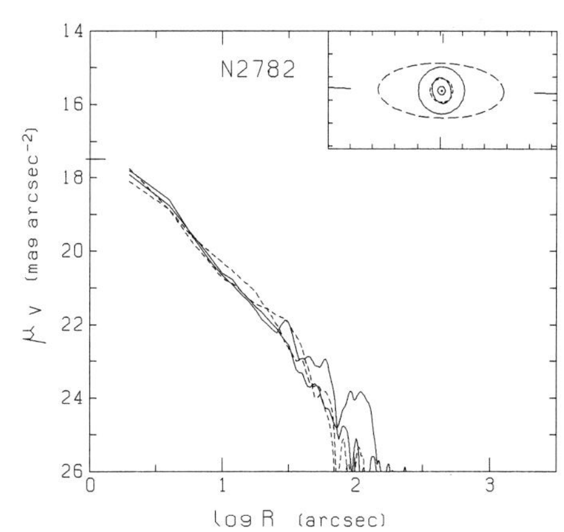

NGC2793

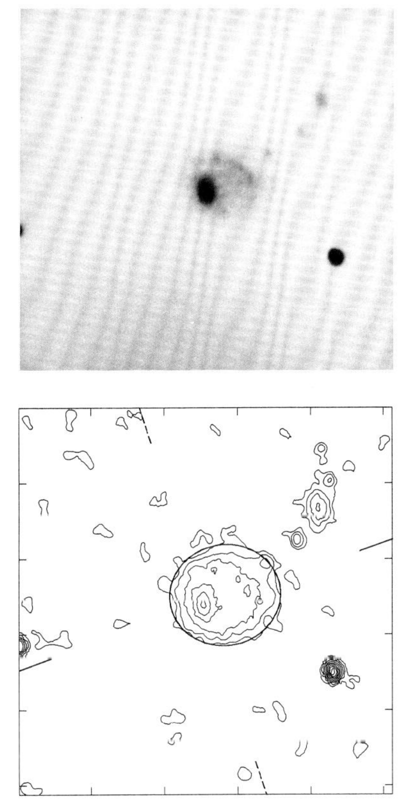

N2793 K5649 1 21.81 256X256

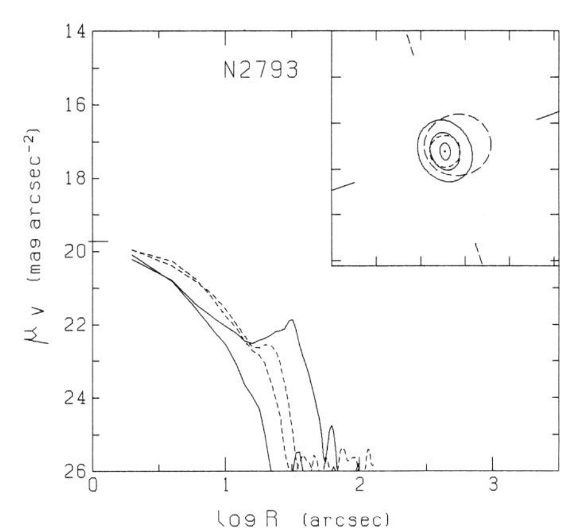

NGC2811

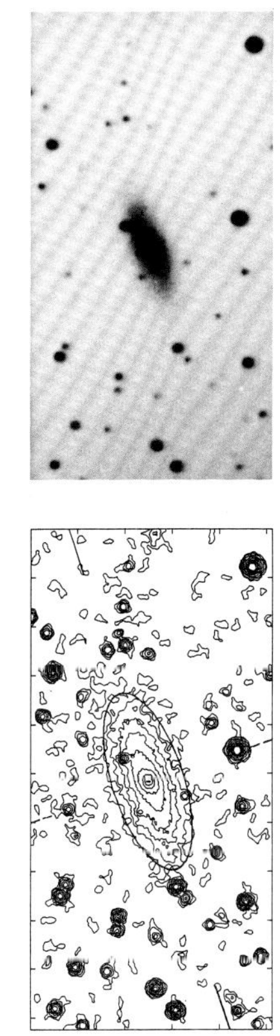

N2811 K4582 1 21.81 256X512

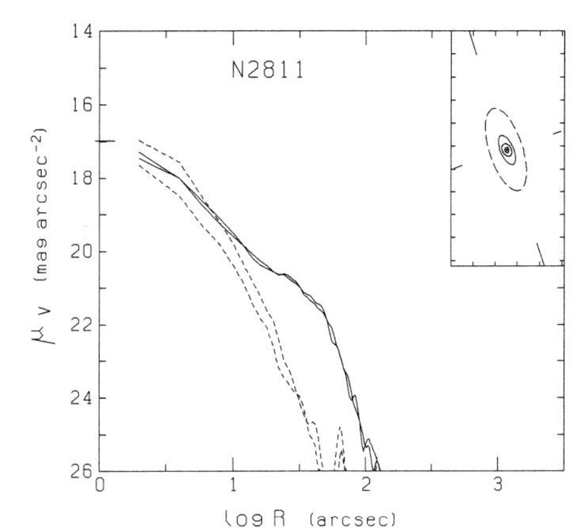

NGC2815

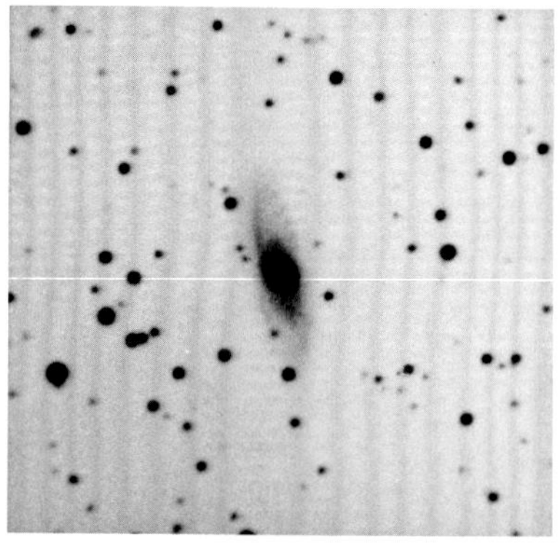

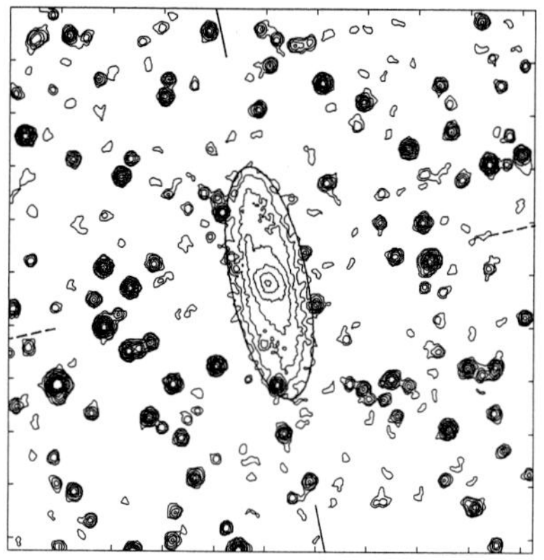

N2815 K5348 1 21.65 512X512

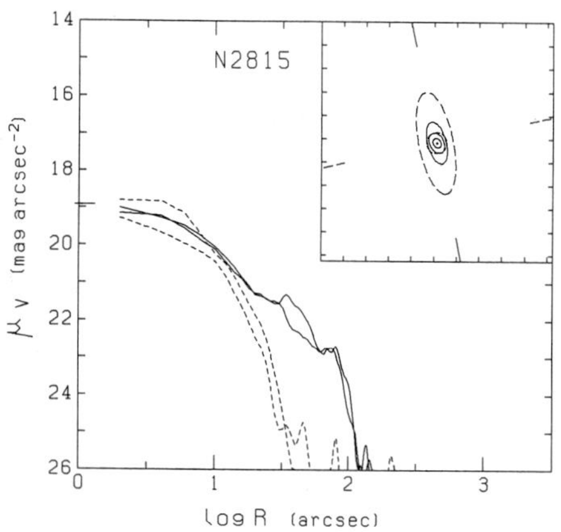

NGC2798

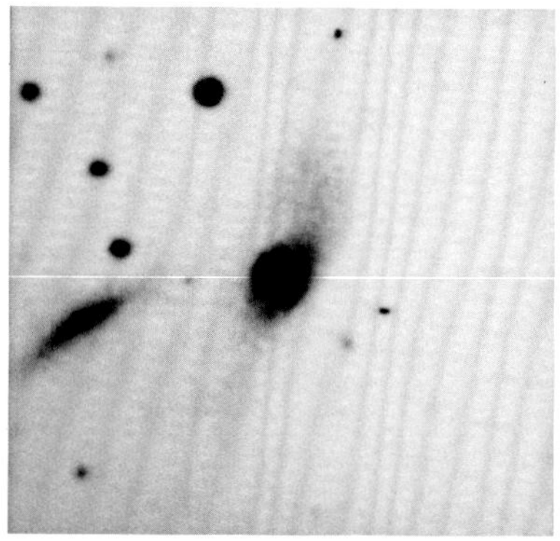

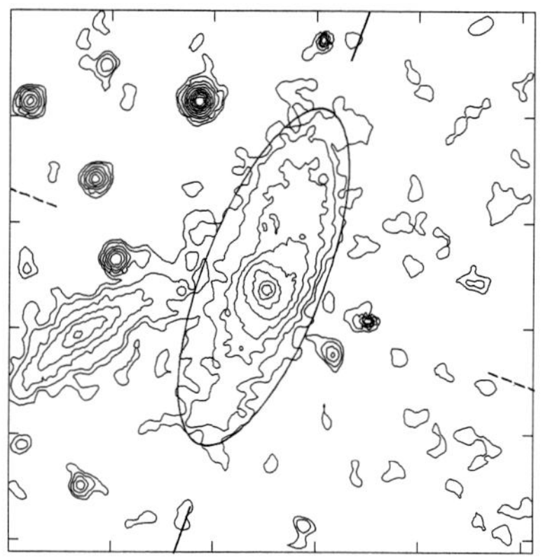

N2798 K5657 1 21.67 256X256

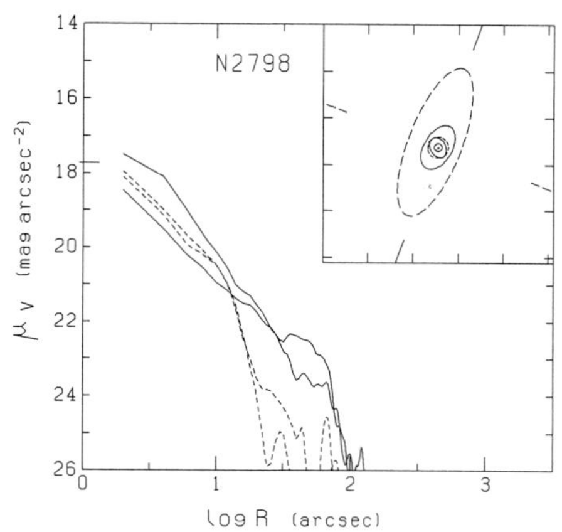

NGC2787

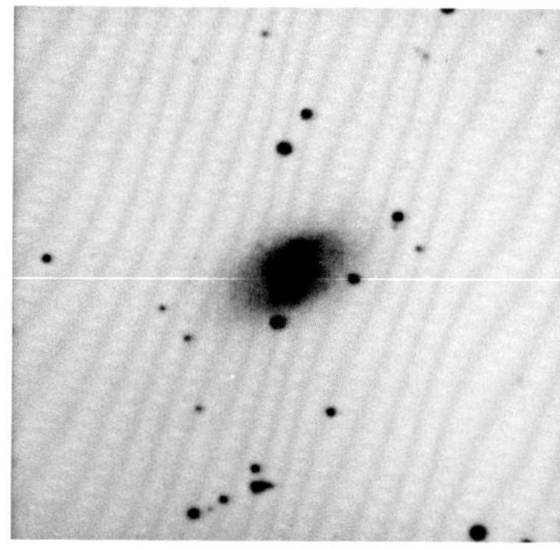

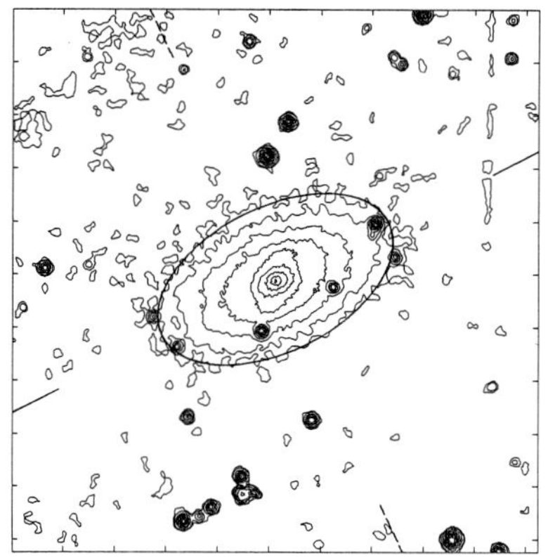

N2787 K3193 1 21.21 512X512

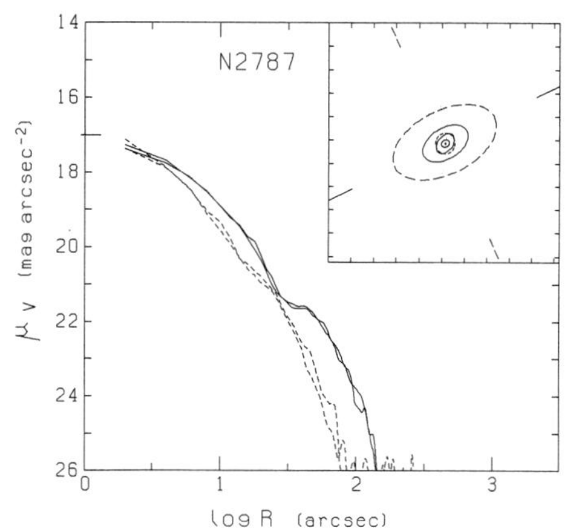

NGC2835

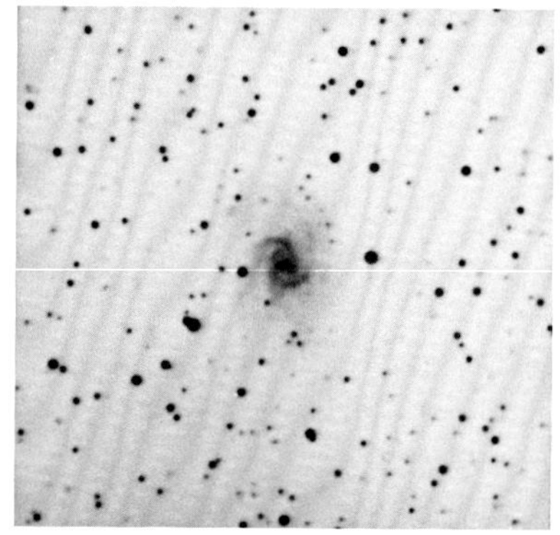

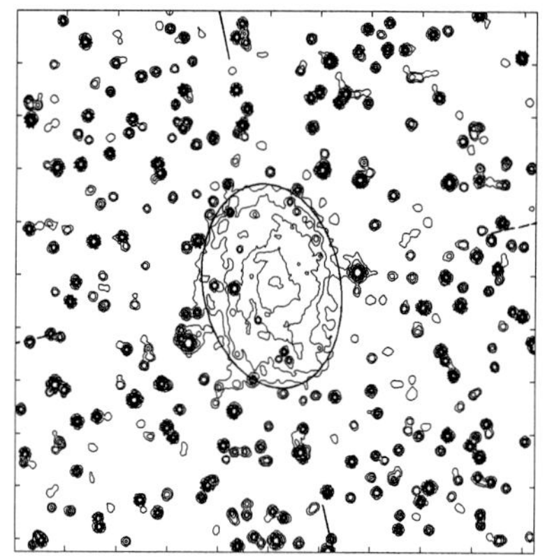

N2835 K5348 2 21.65 512X512

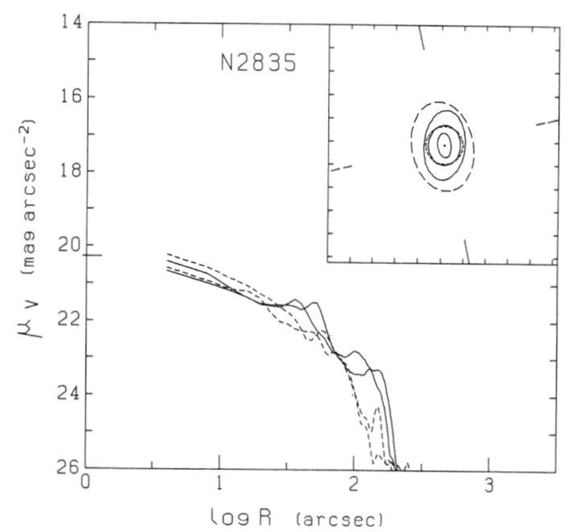

NGC2832

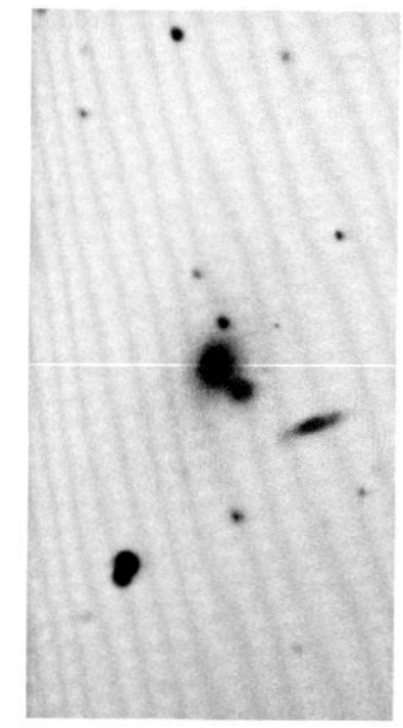

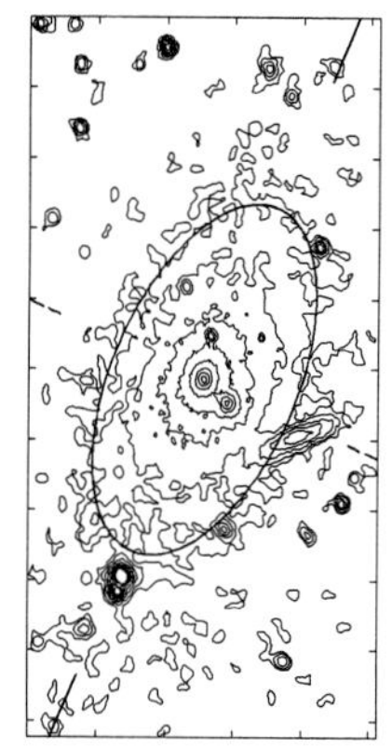

N2832 K5649 1 21.81 256X512

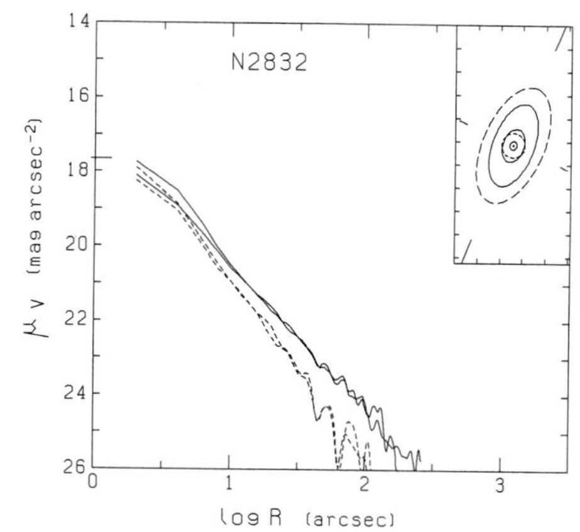

NGC2848

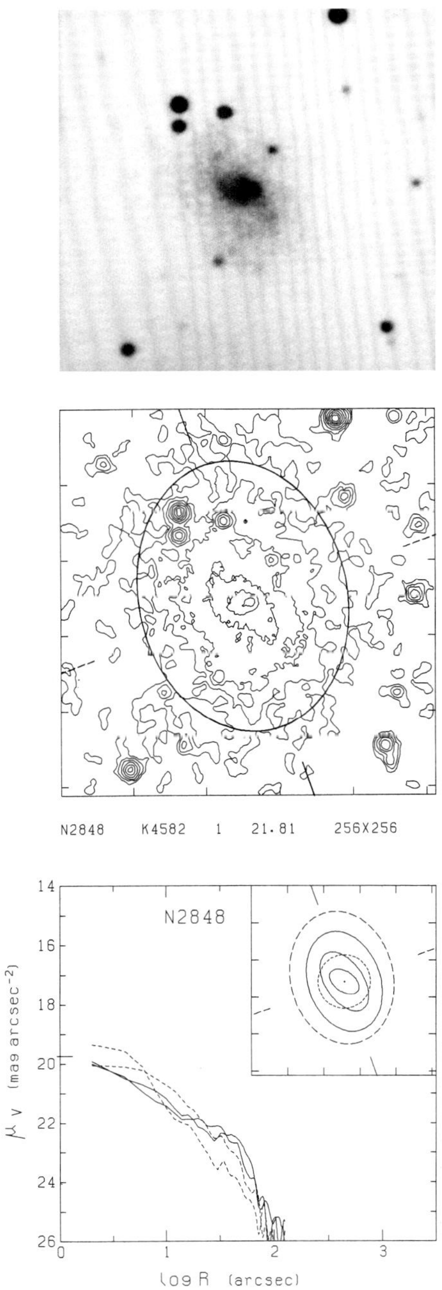

NGC2841

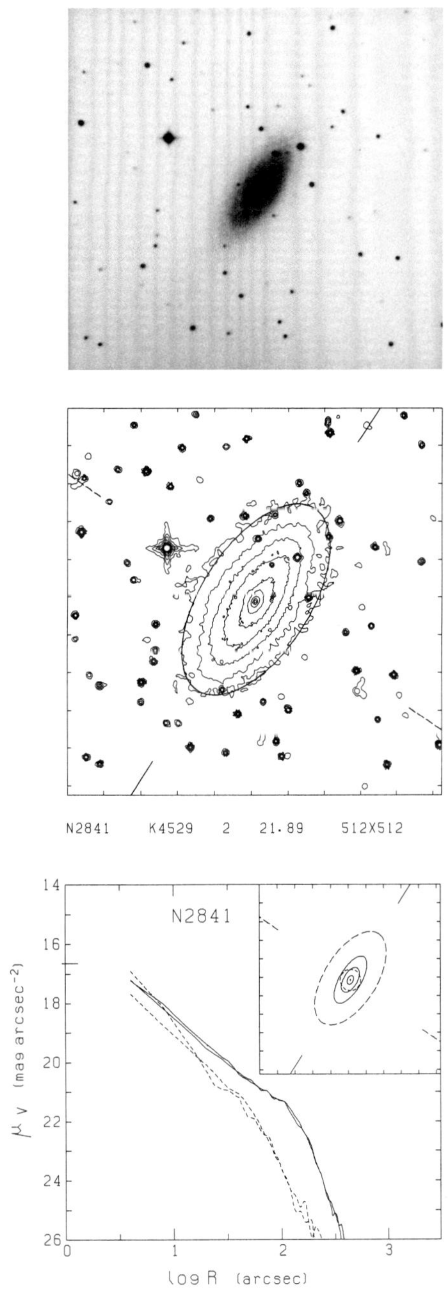

NGC2844

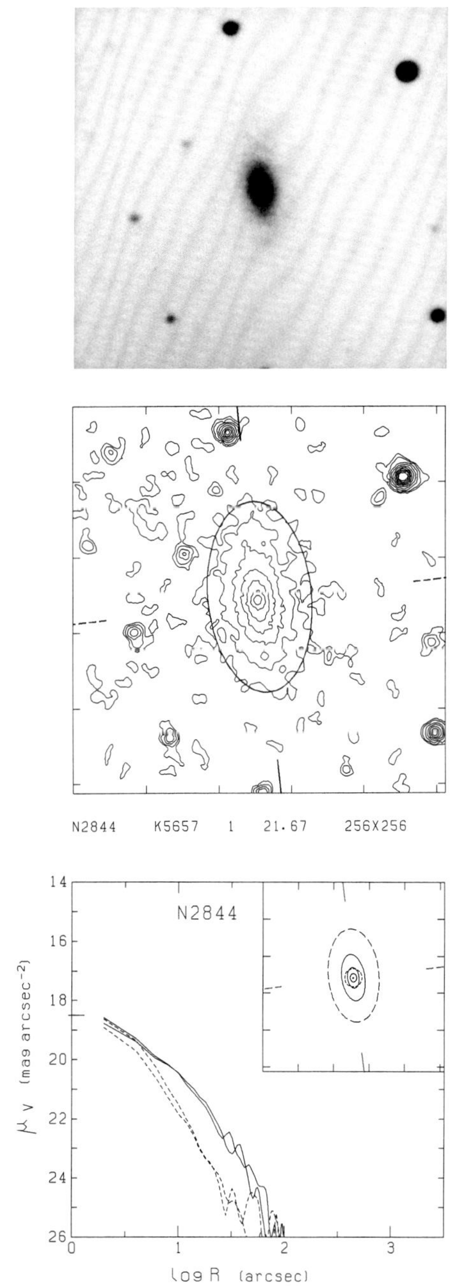

NGC2855

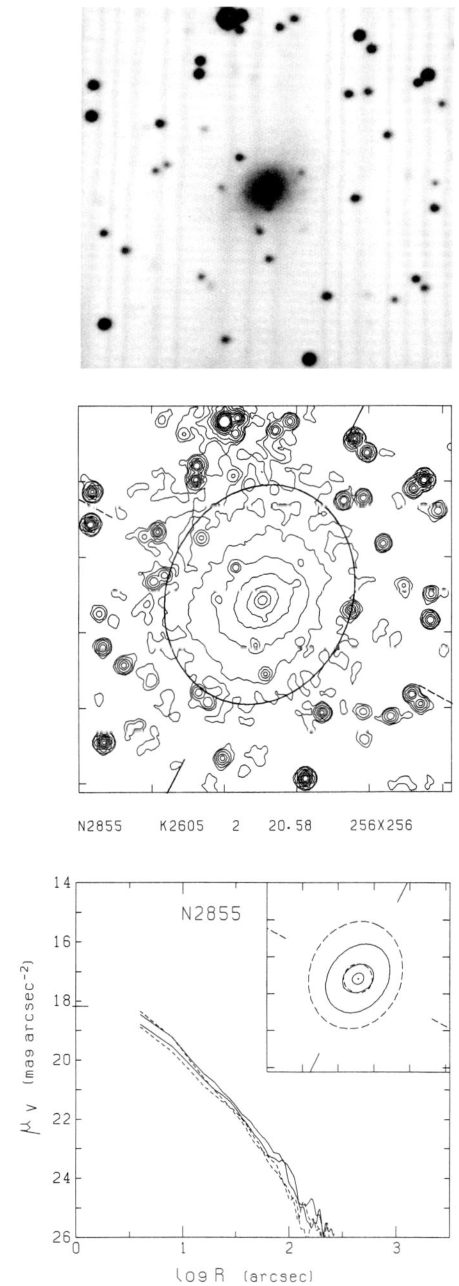

NGC2865

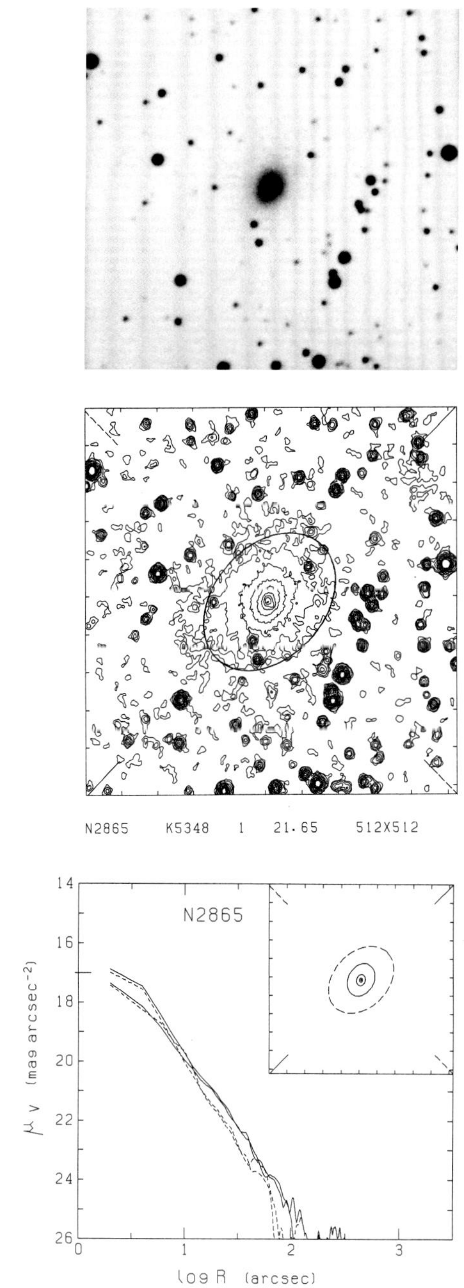

NGC2859

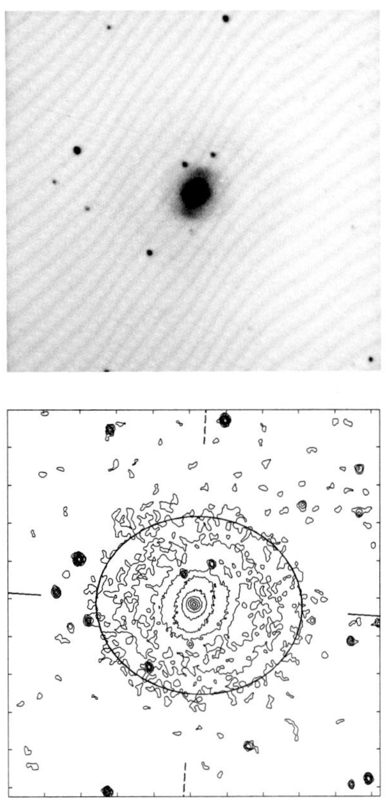

N2859 K5649 1 21.81 512X512

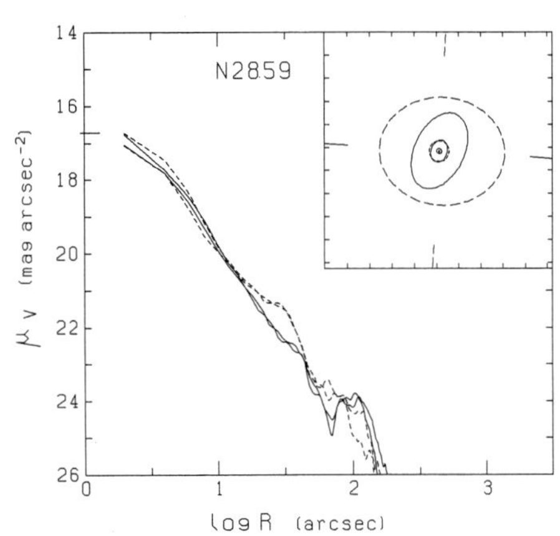

NGC2889

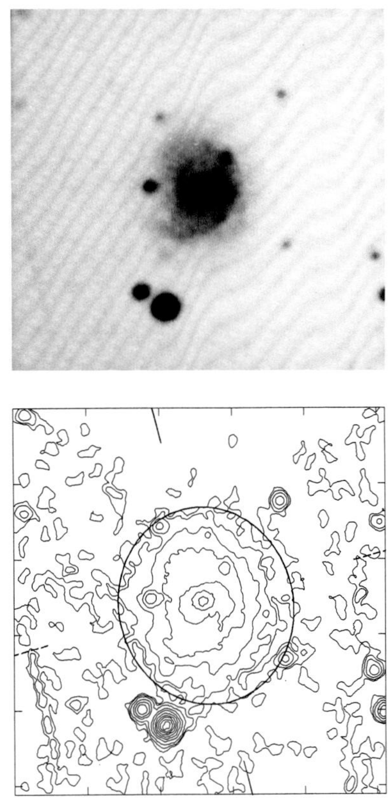

N2889 K2605 1 20.58 256X256

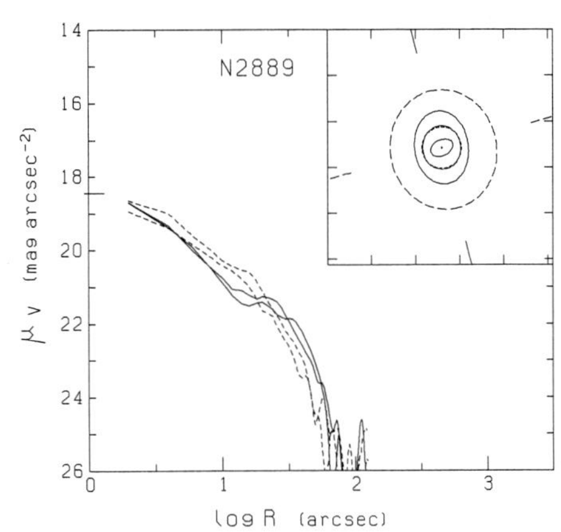

NGC2880

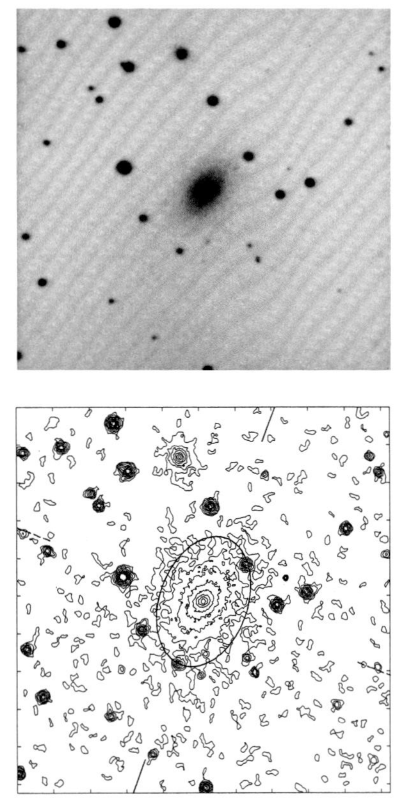

N2880 K4947 1 21.89 512X512

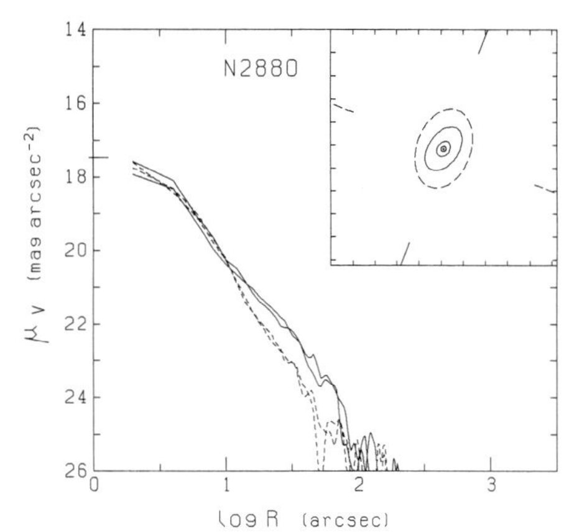

NGC2902

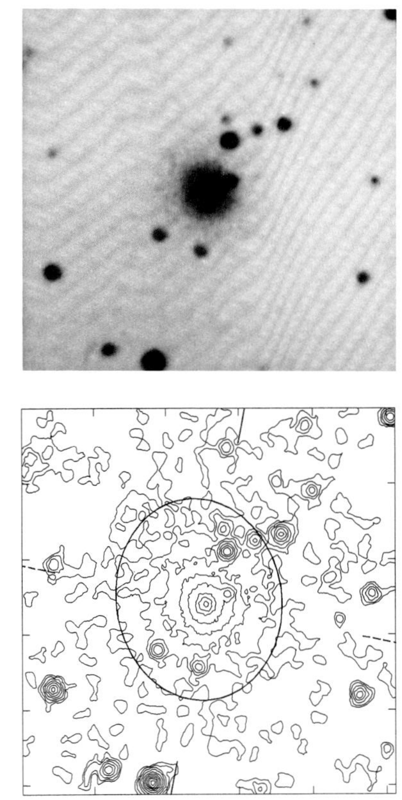

N2902 K4582 1 21.81 256X256

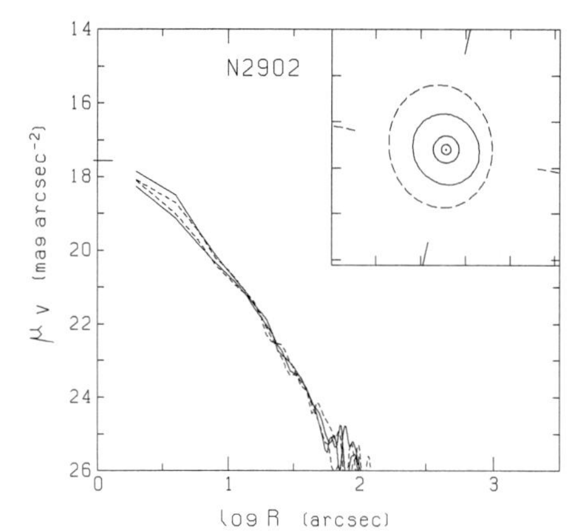

NGC2903

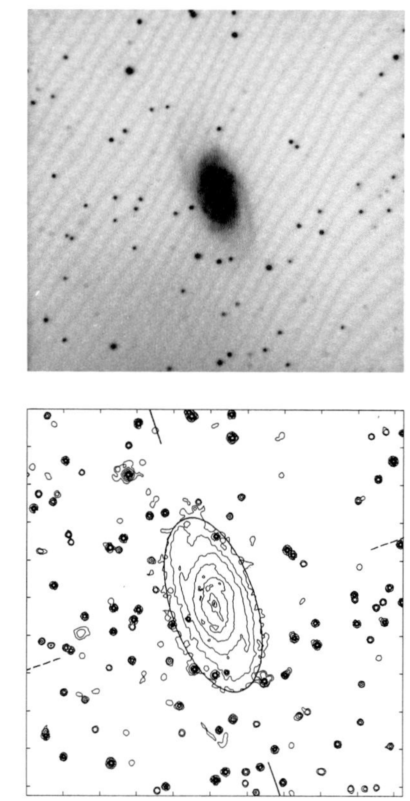

N2903 K3681 3 20.84 512X512

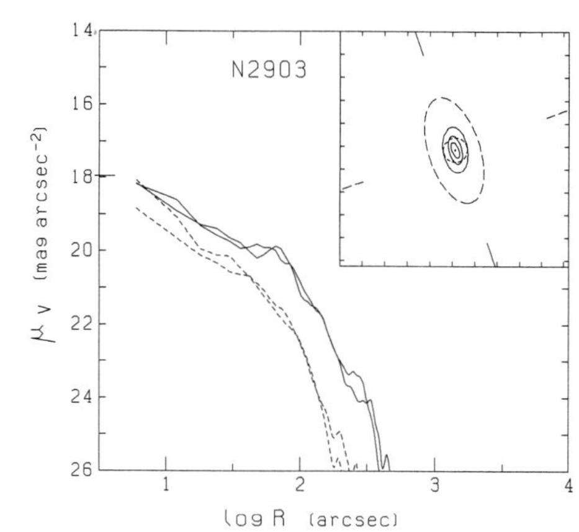

NGC2907

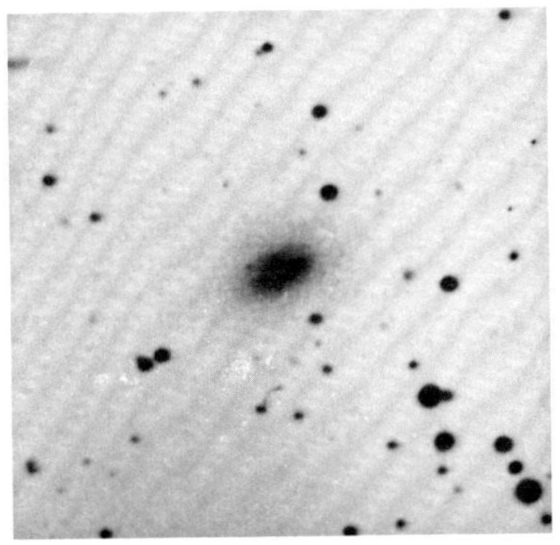

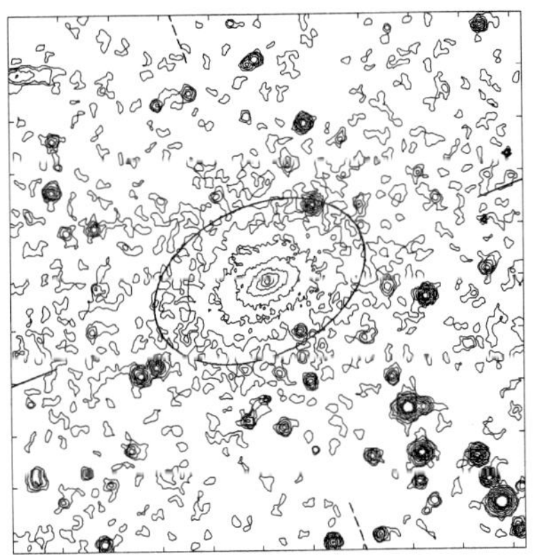

N2907 K4582 1 21.81 512X512

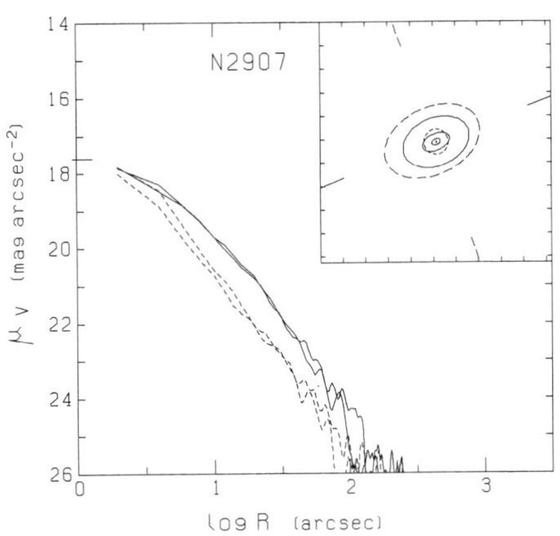

NGC2962

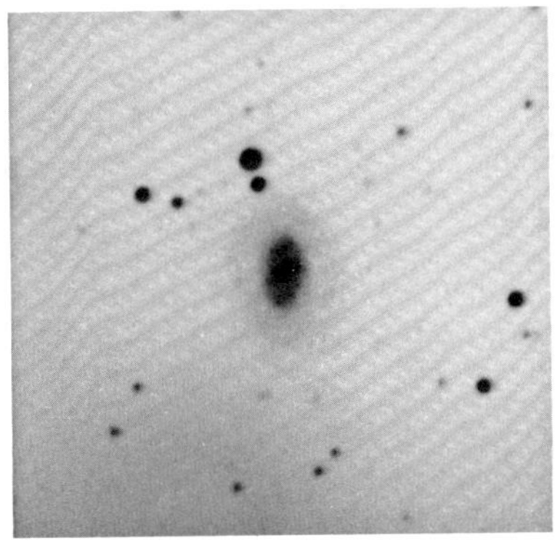

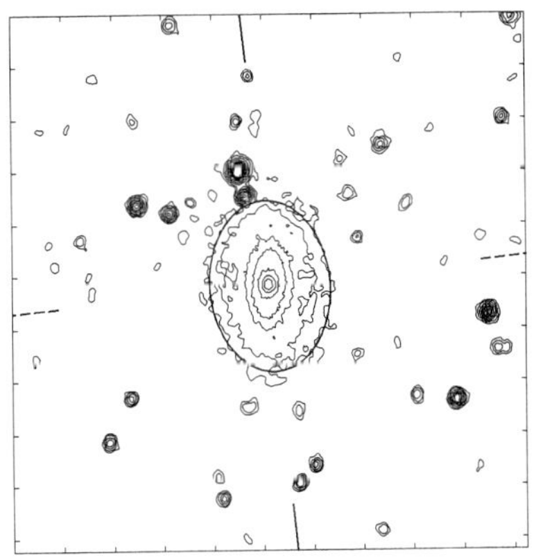

N2962 K5369 1 21.93 512X512

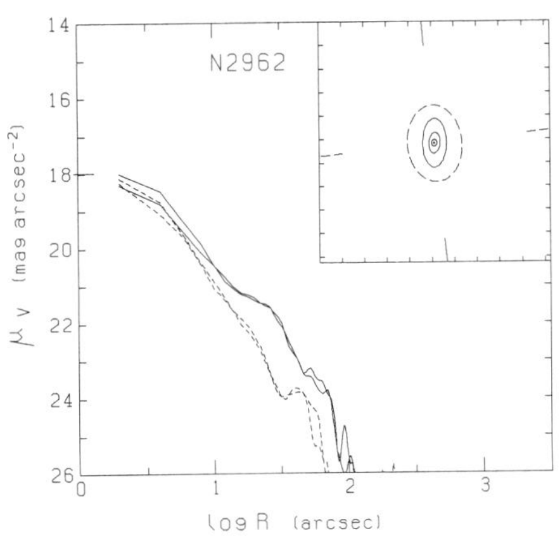

NGC2950

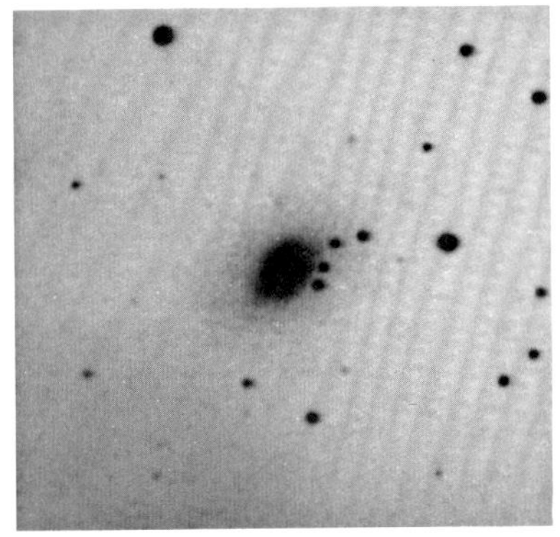

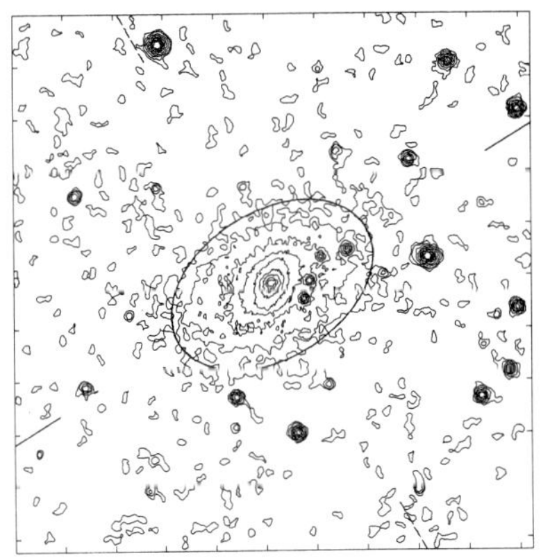

N2950 K4947 1 21.89 512X512

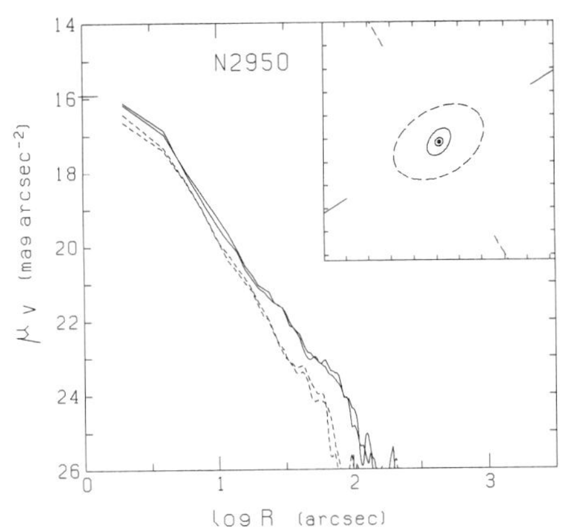

NGC2967

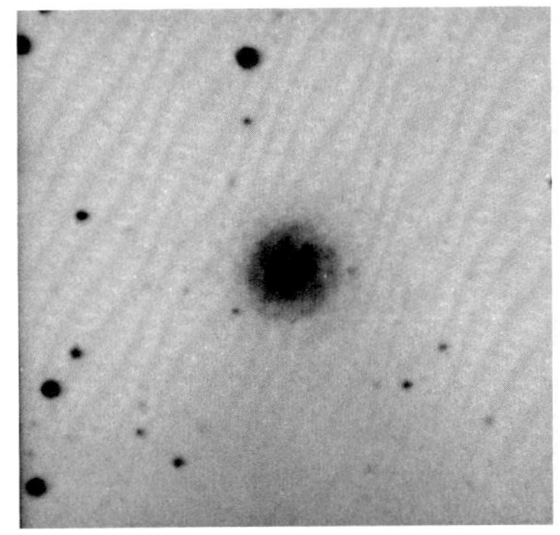

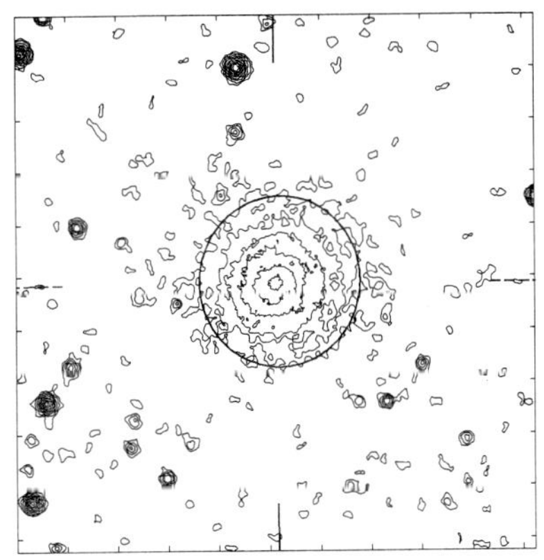

N2967 K5740 1 21.94 512X512

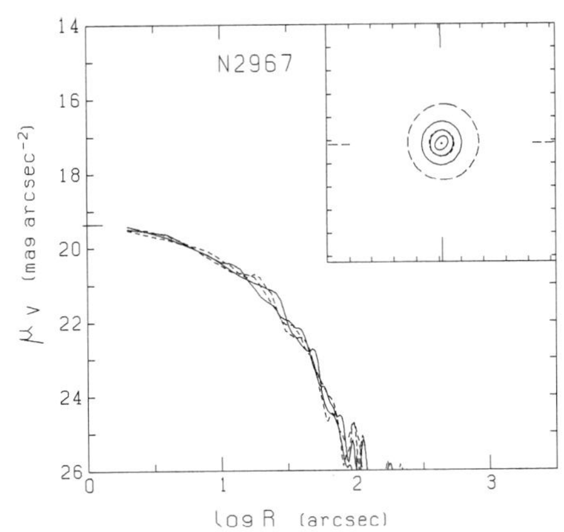

NGC2964

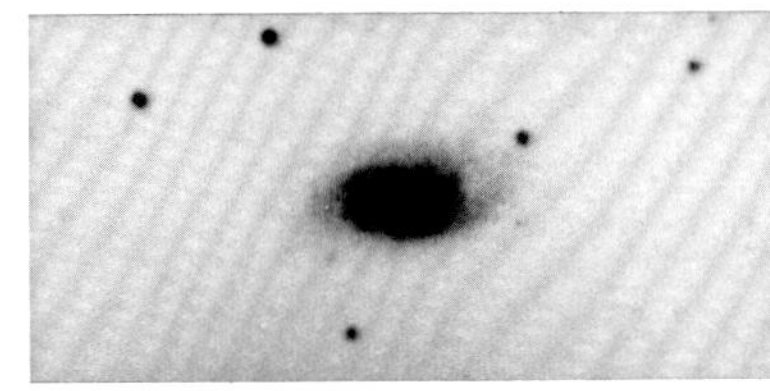

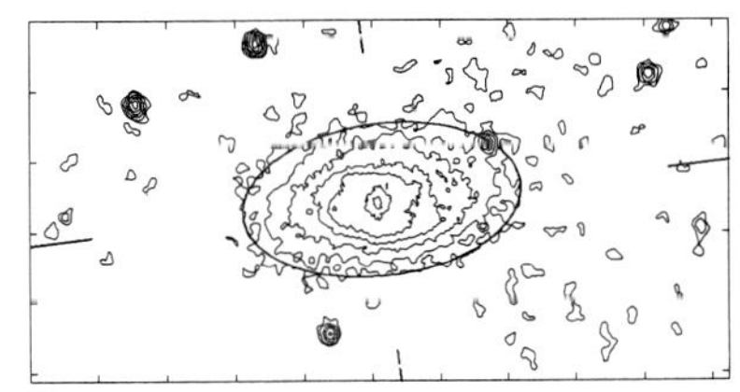

N2964 K5742 1 21.72 512X256

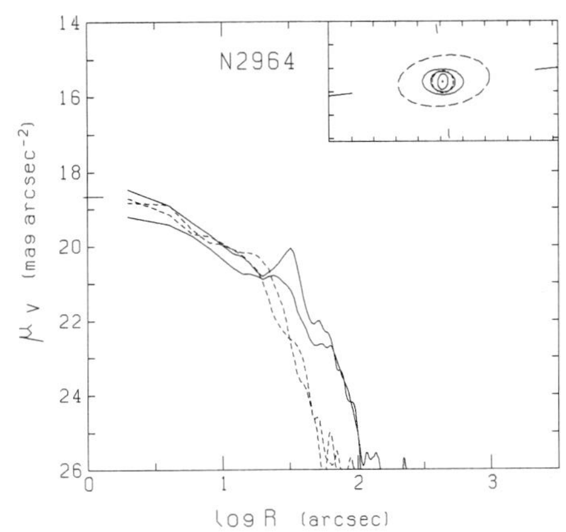

NGC2968

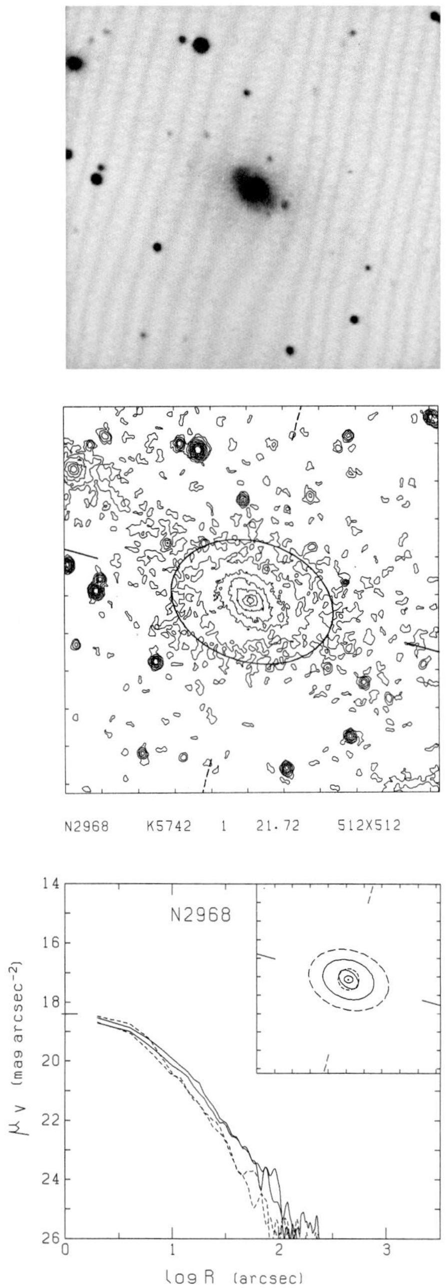

NGC2976

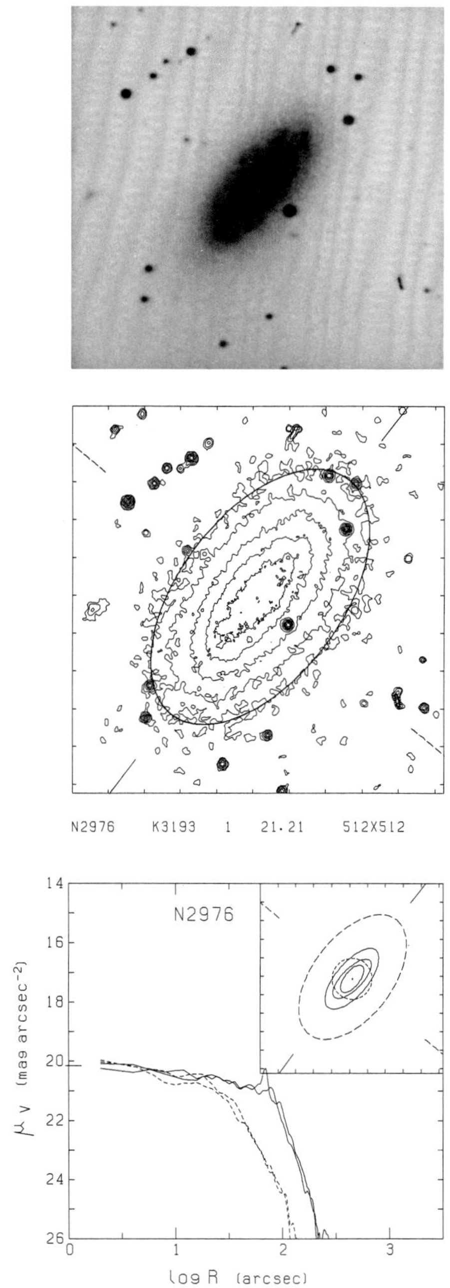

NGC2997

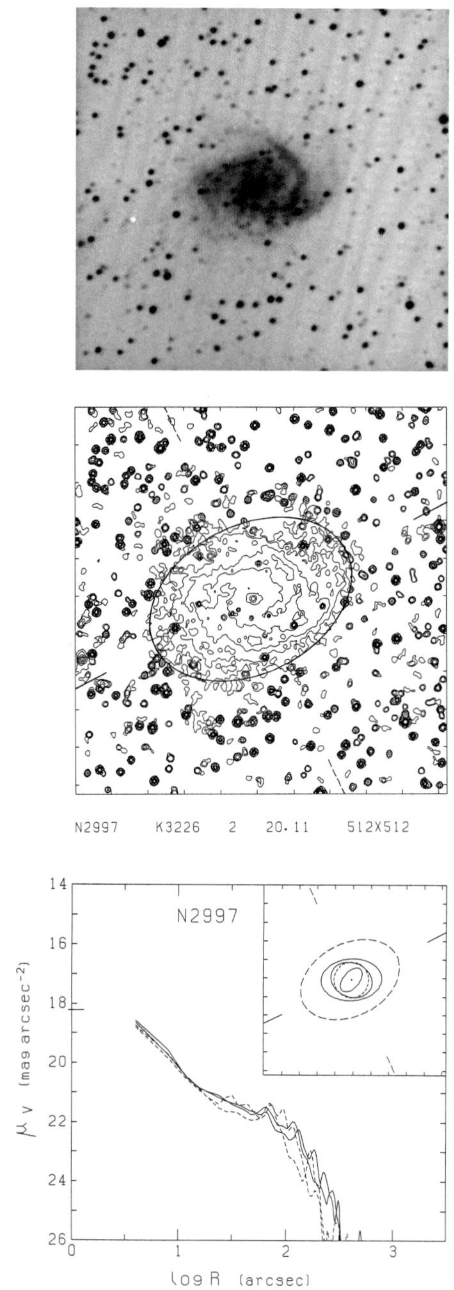

NGC2990

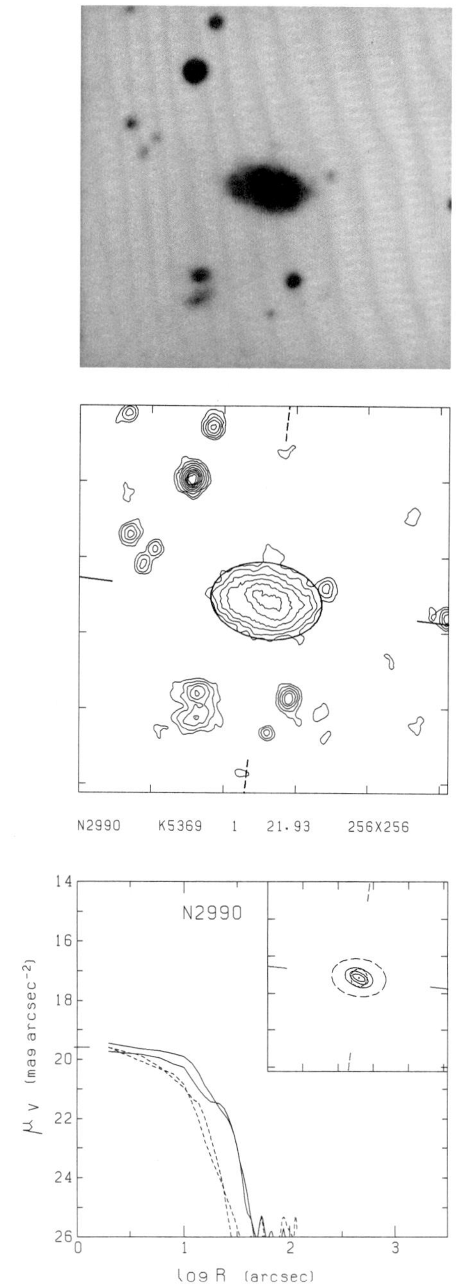

NGC3001

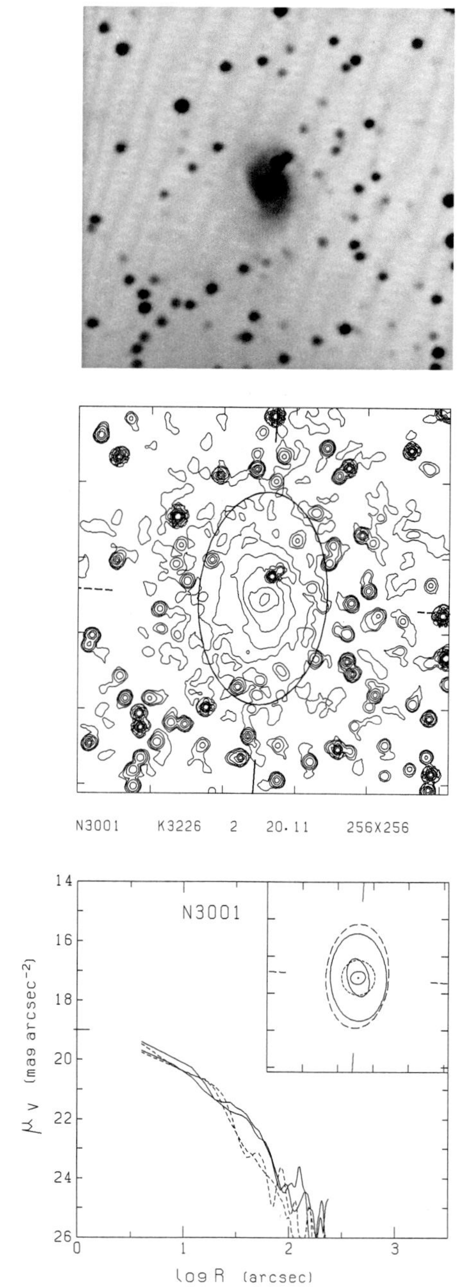

NGC3003

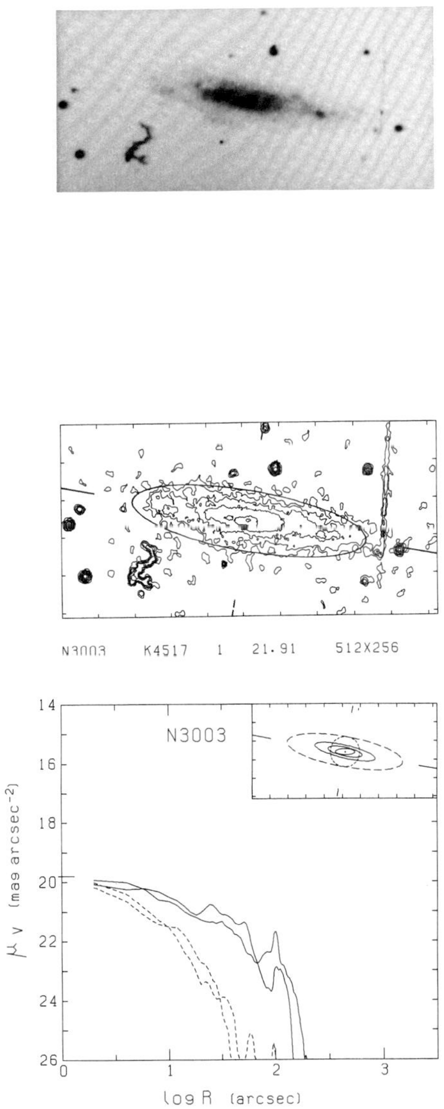

NGC2985

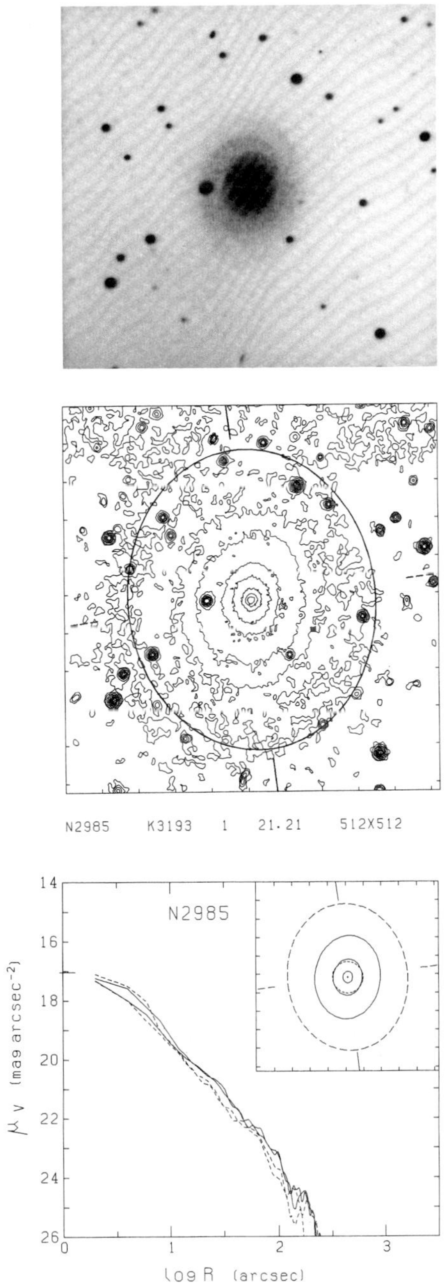

NGC3021

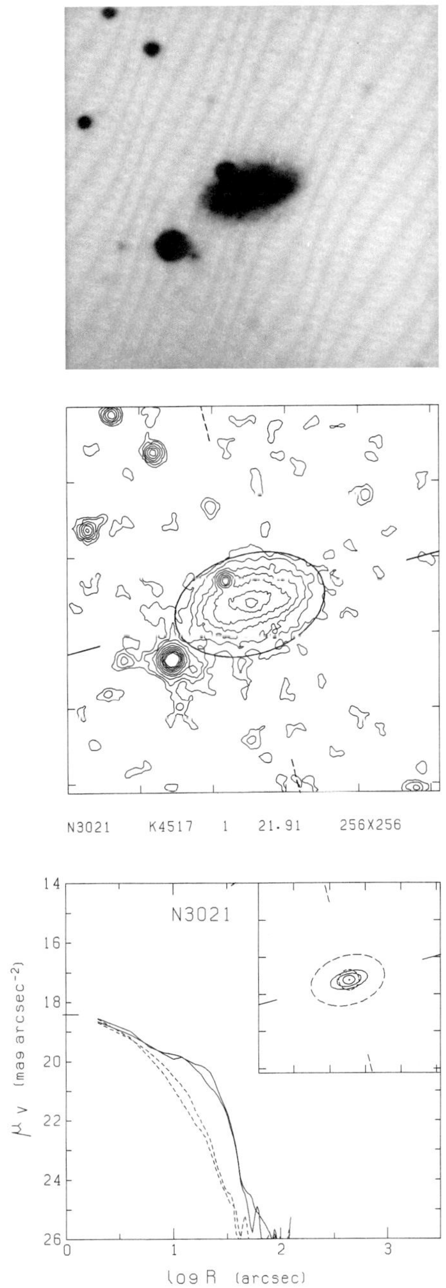

NGC3038

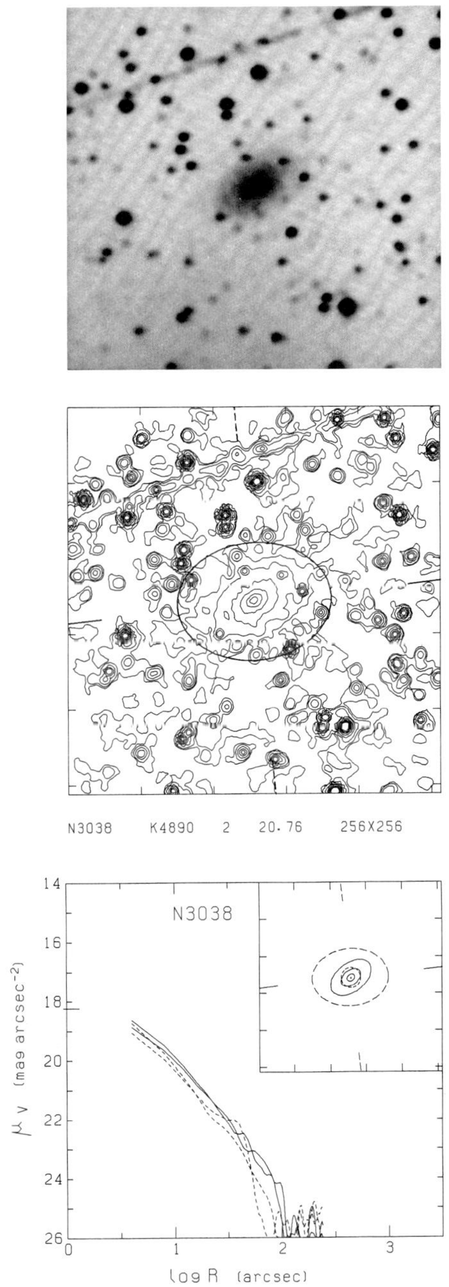

NGC3032

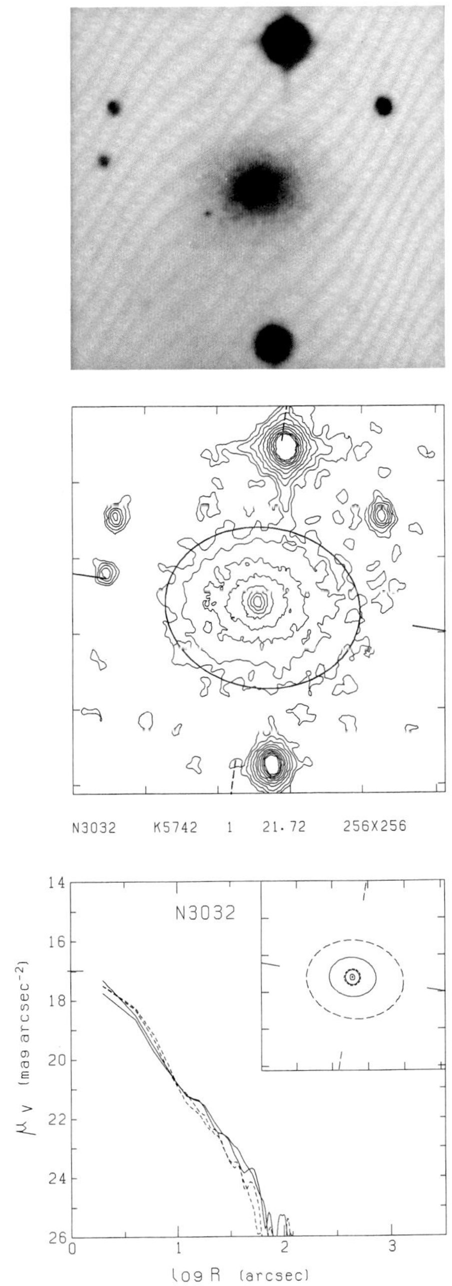

NGC3041

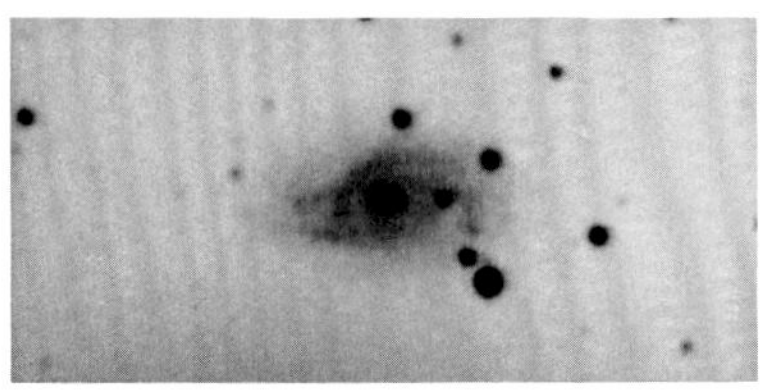

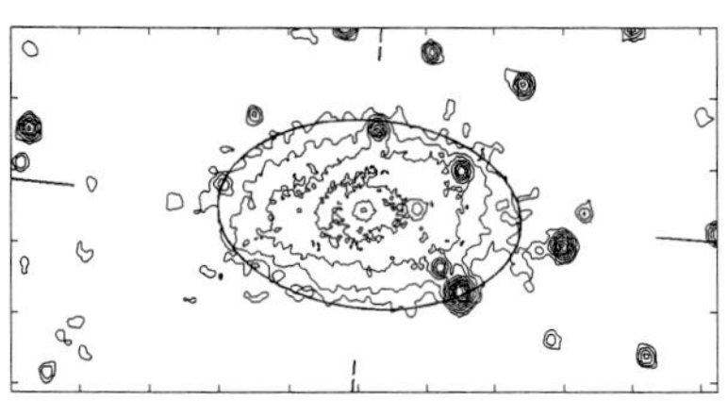

N3041 K5741 1 21.94 512X256

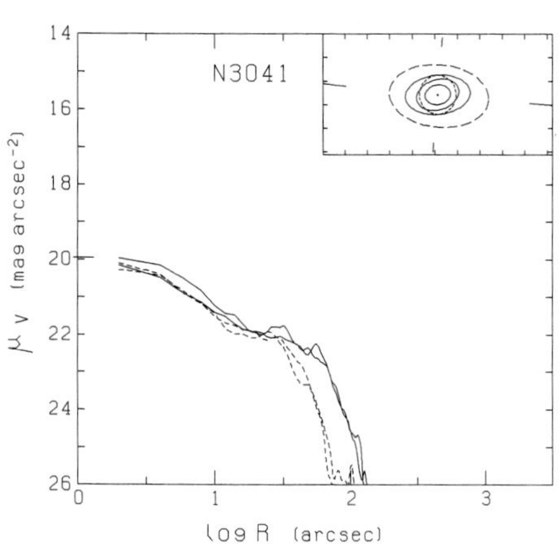

NGC3044

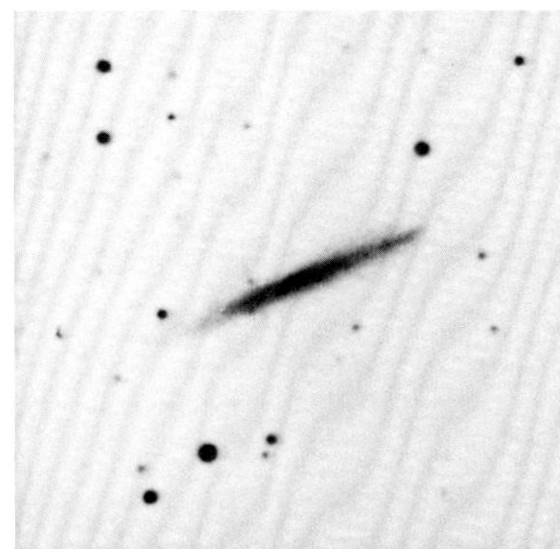

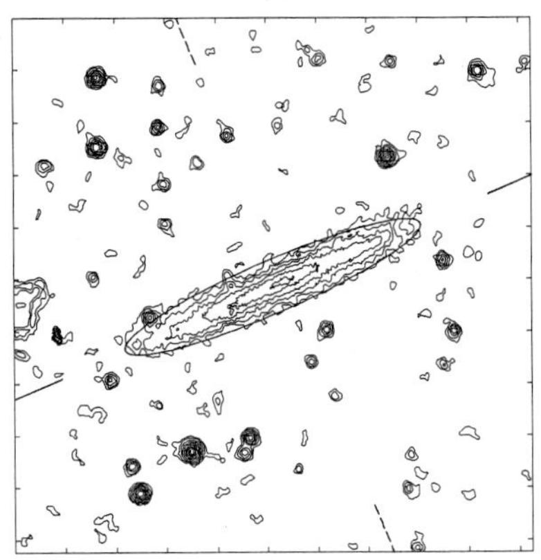

N3044 K5740 1 21.94 512X512

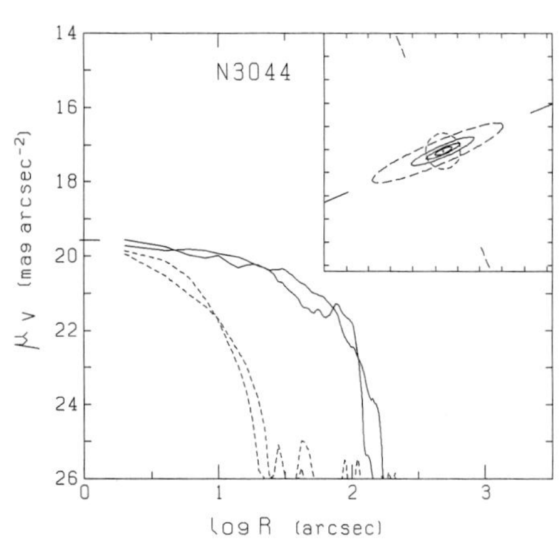

NGC3031

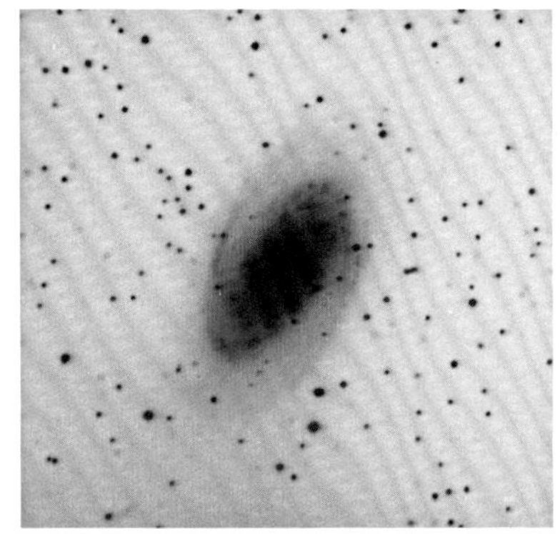

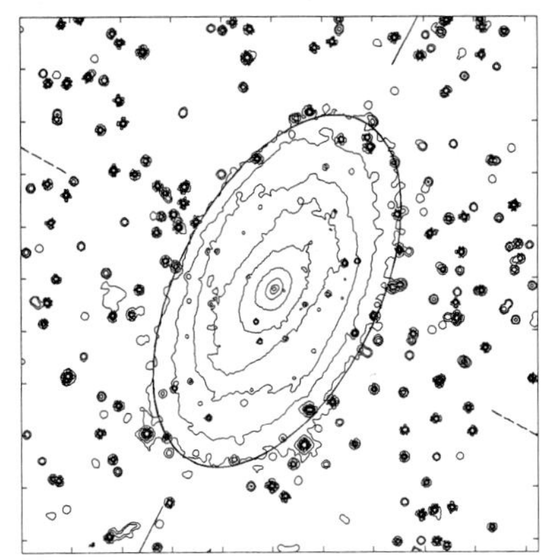

N3031 K3193 4 21.21 512X512

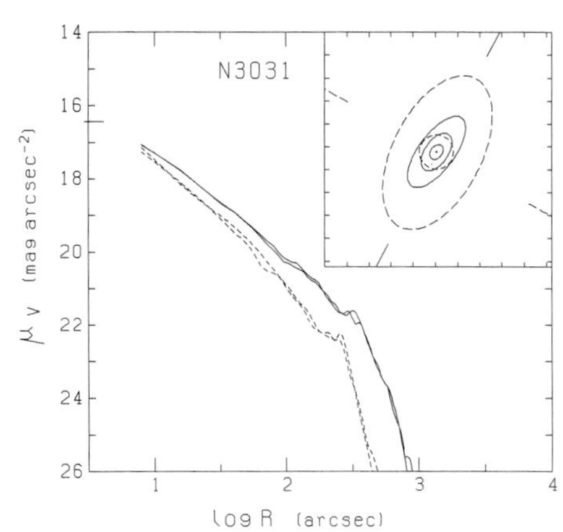

NGC3034

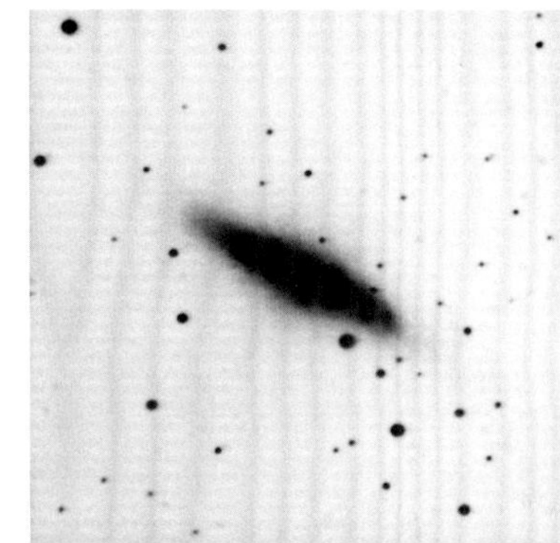

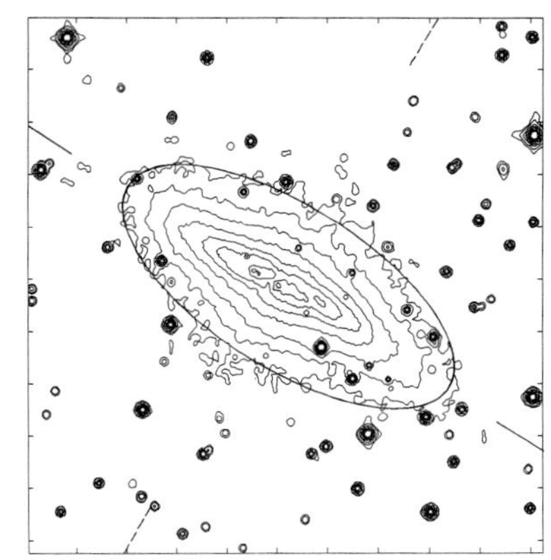

N3034 K3193 2 21.21 512X512

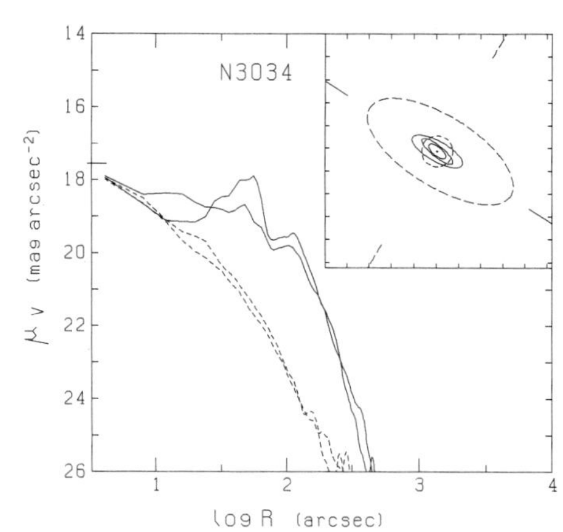

NGC3056

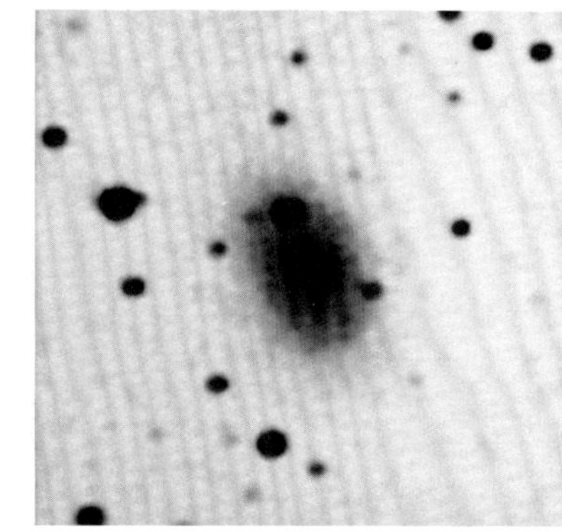

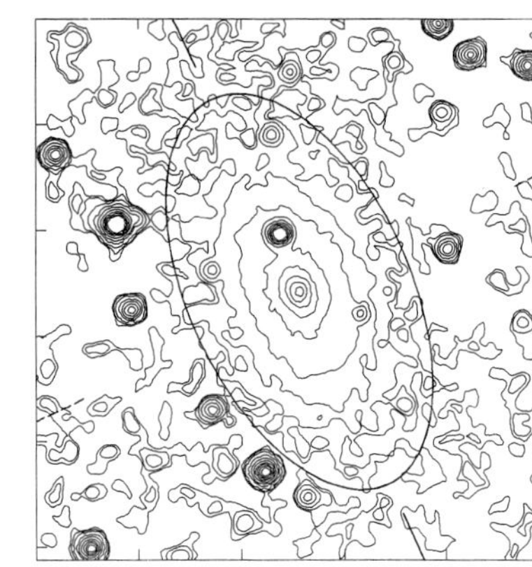

N3056 K3226 1 20.11 256X256

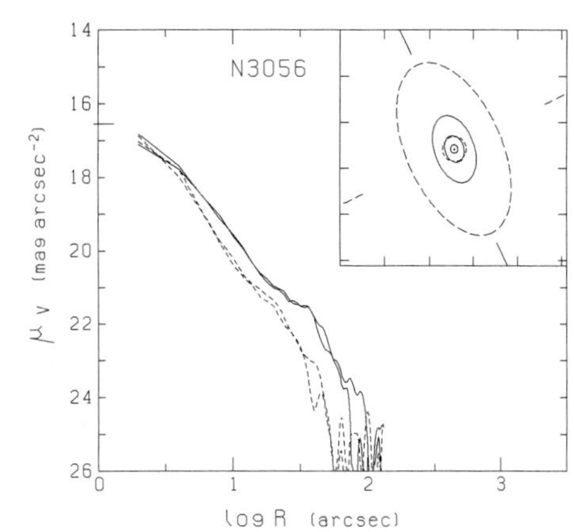

NGC3043

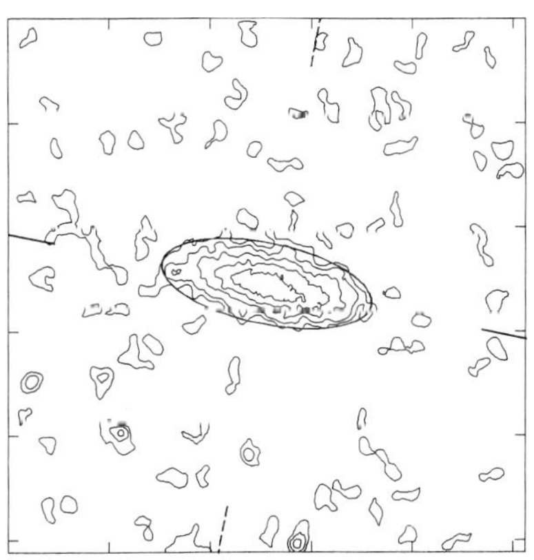

N3043 K4917 1 21.89 256X256

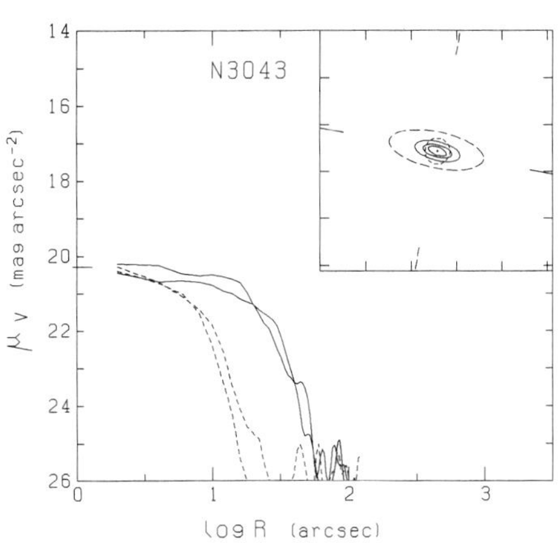

NGC3067

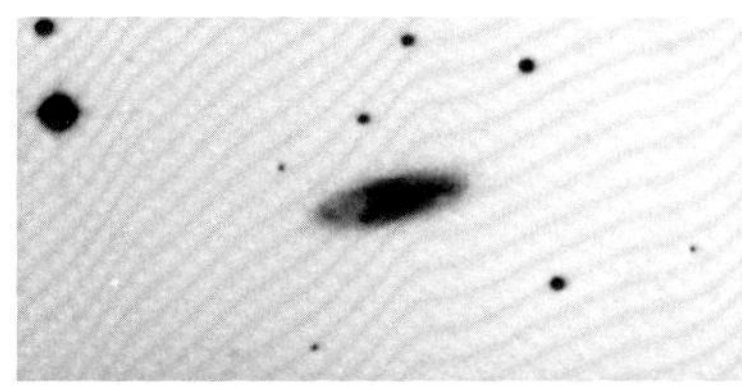

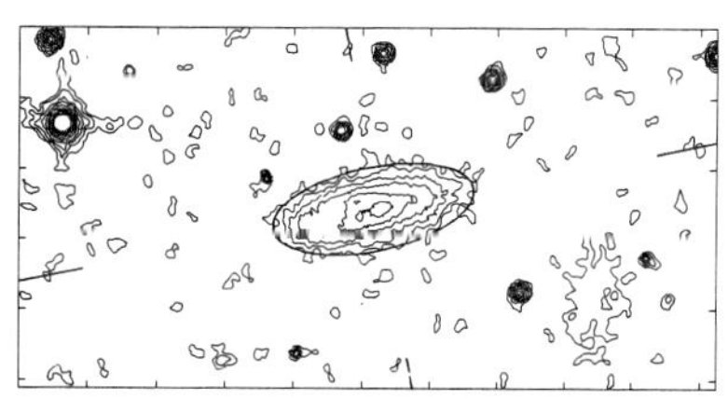

N3067 K4517 1 21.91 512X256

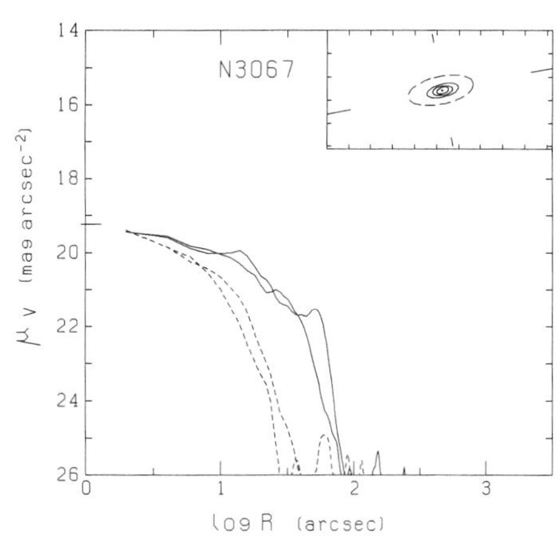

NGC3089

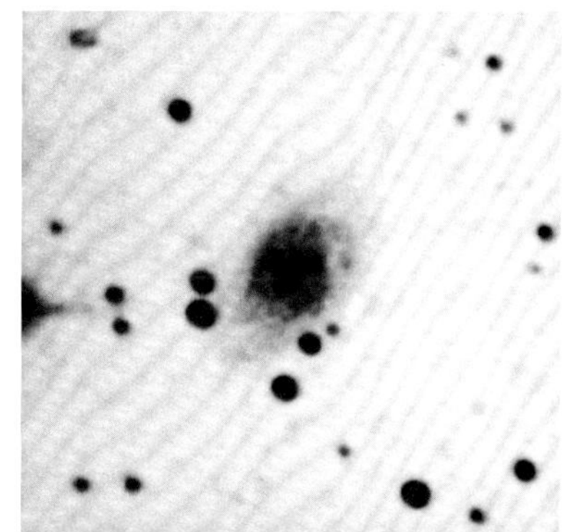

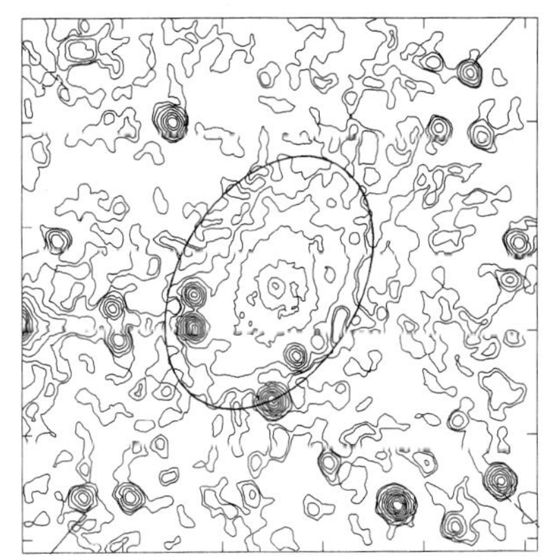

N3089 K4890 1 20.76 256X256

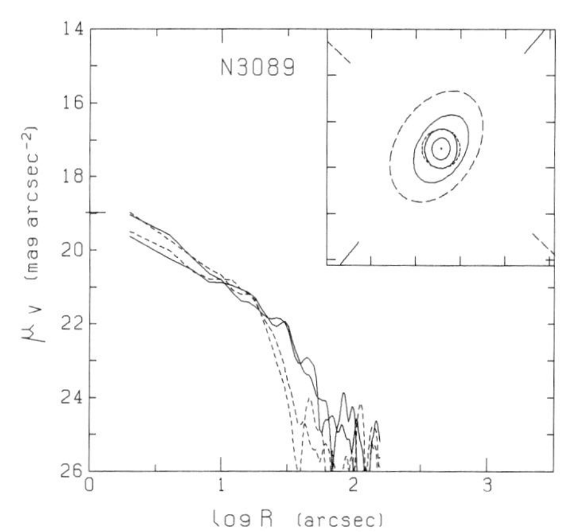

NGC3065

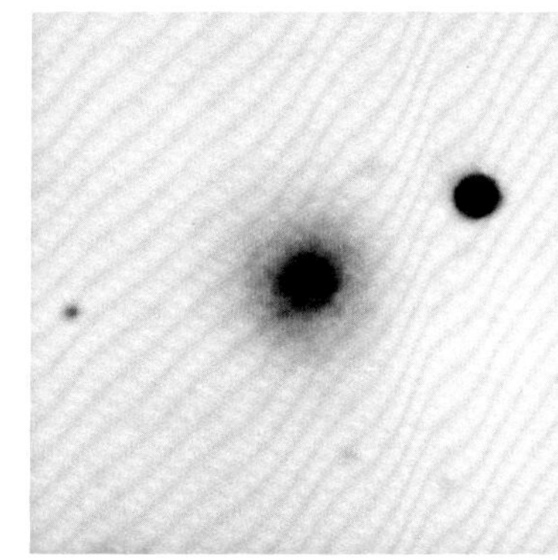

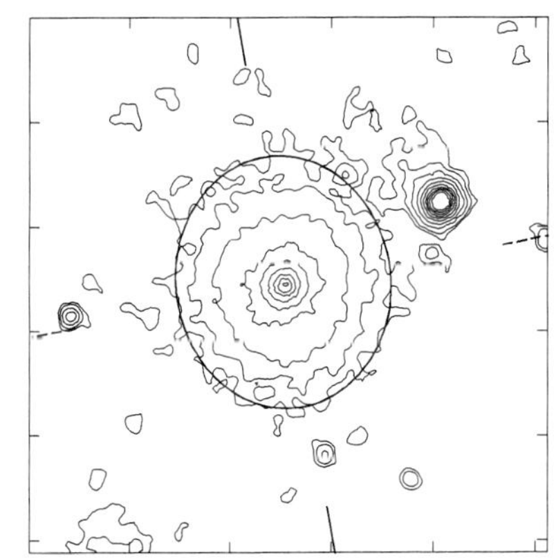

N3065 K3193 1 21.21 256X256

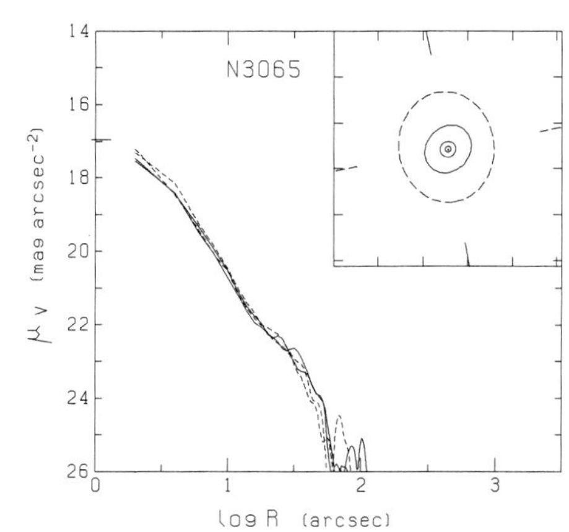

NGC3095

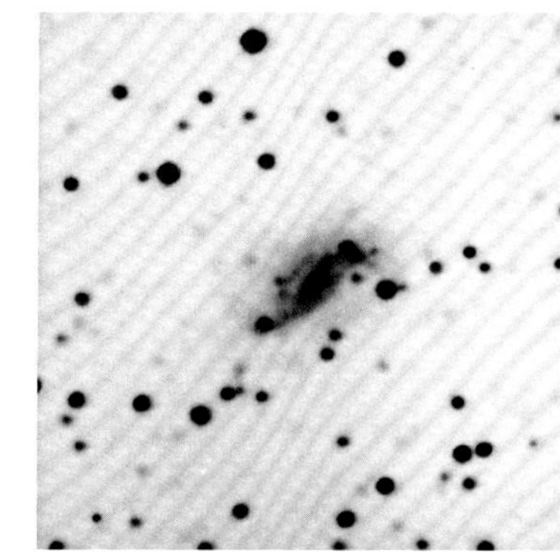

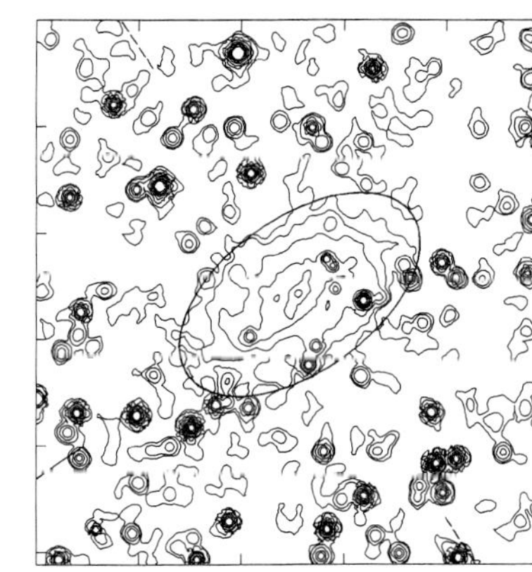

N3095 K4890 2 20.76 256X256

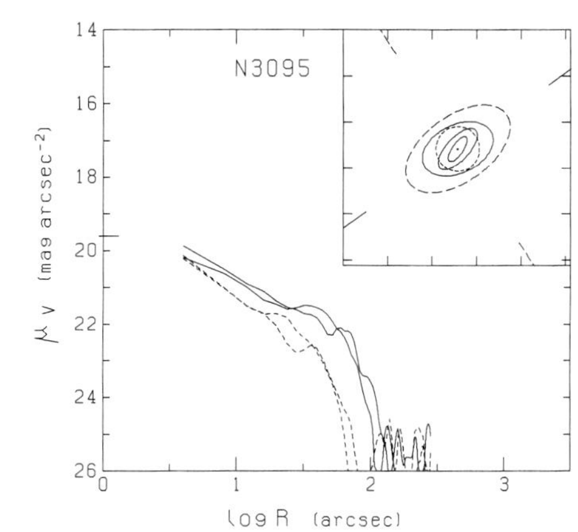

NGC3077

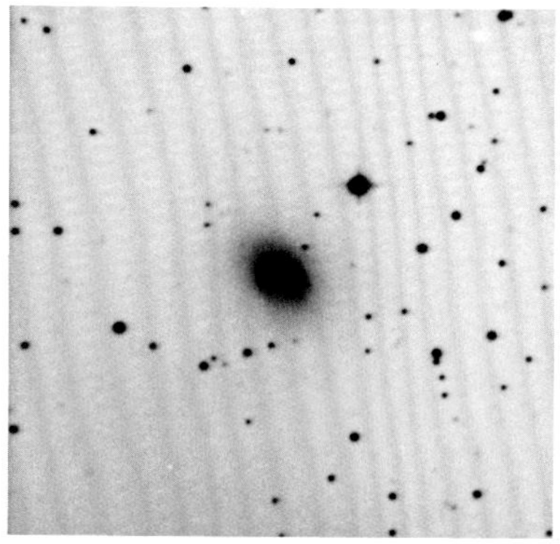

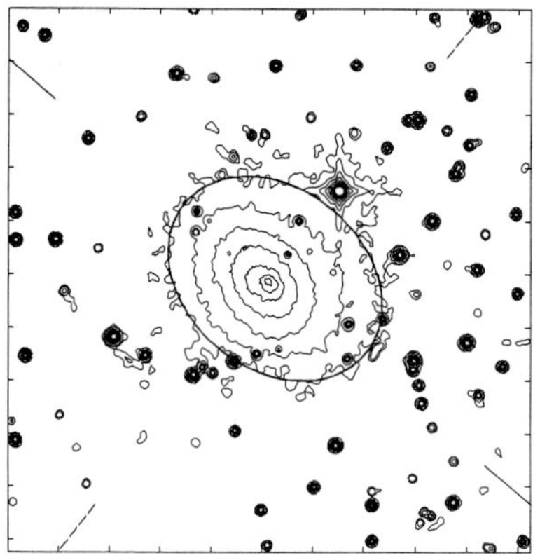

N3077 K3193 2 21.21 512X512

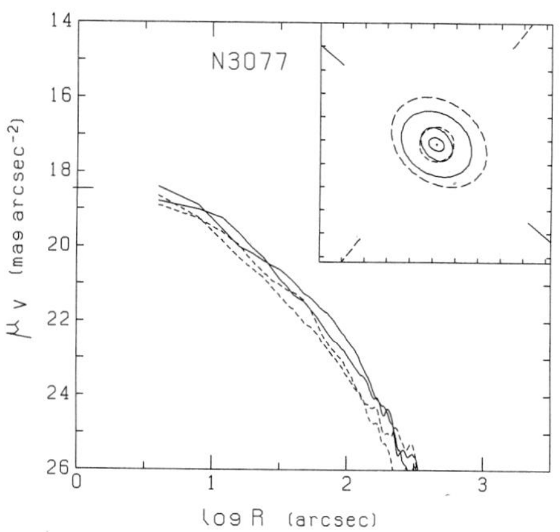

NGC3115

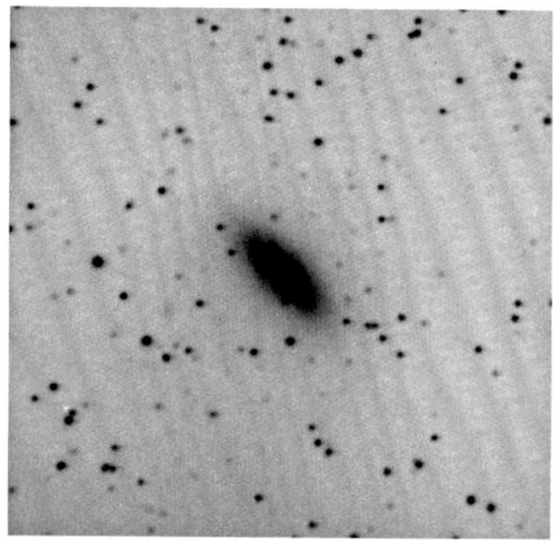

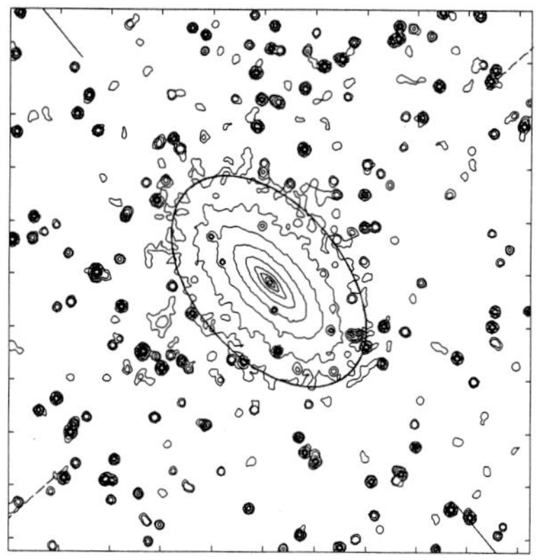

N3115 K3680 3 20.55 512X512

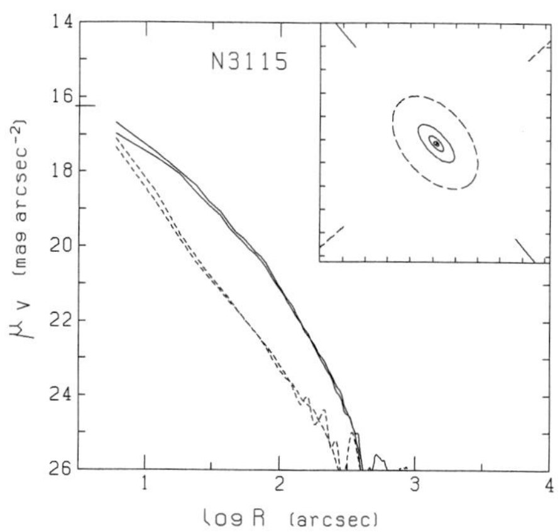

NGC3125

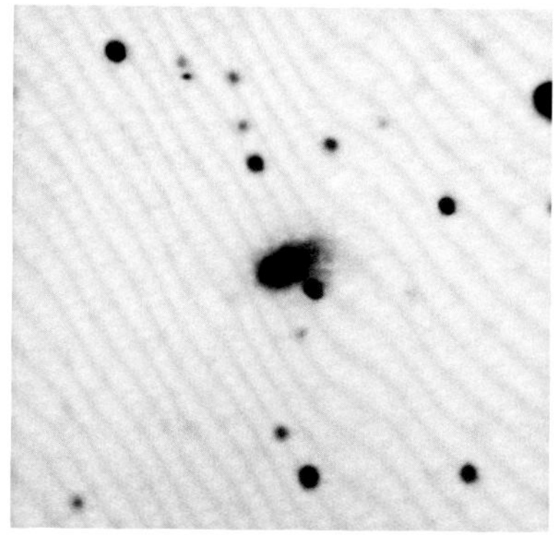

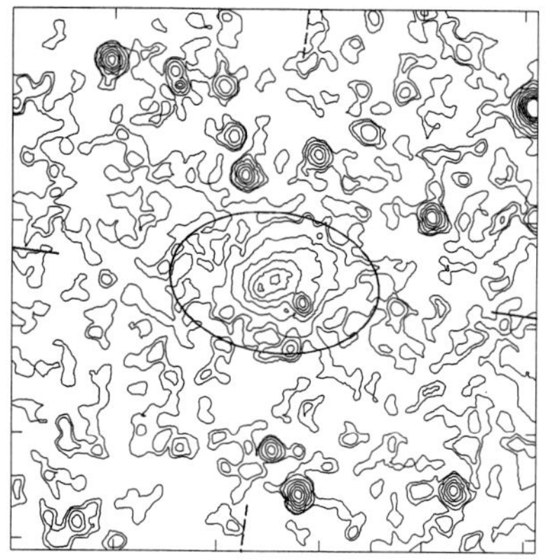

N3125 K4890 1 20.76 256X256

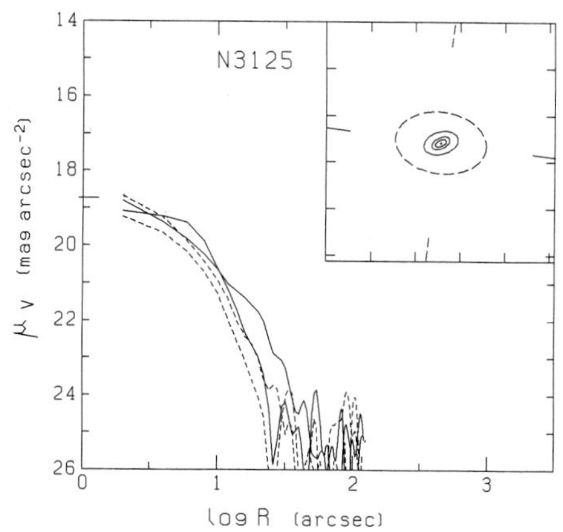

NGC3145

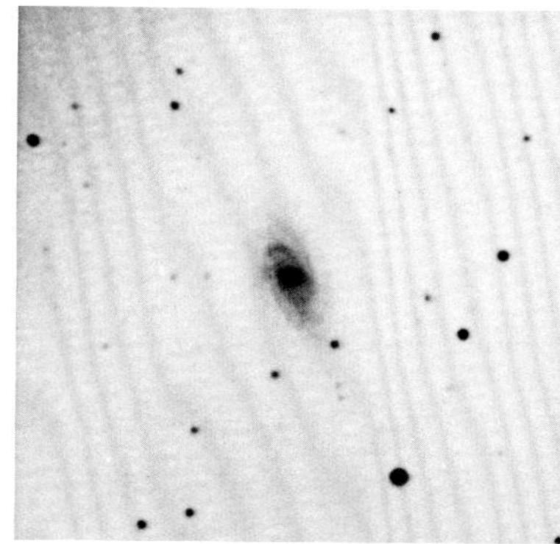

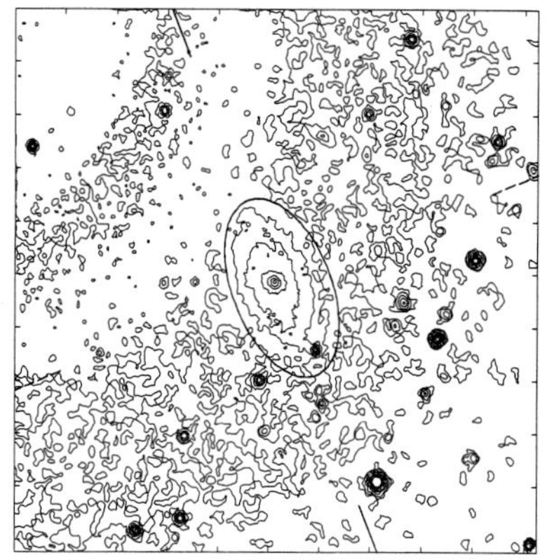

N3145 K5357 1 21.31 512X512

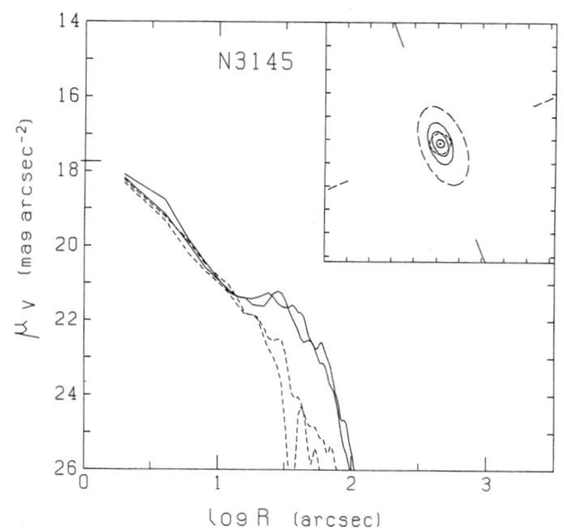

NGC3158

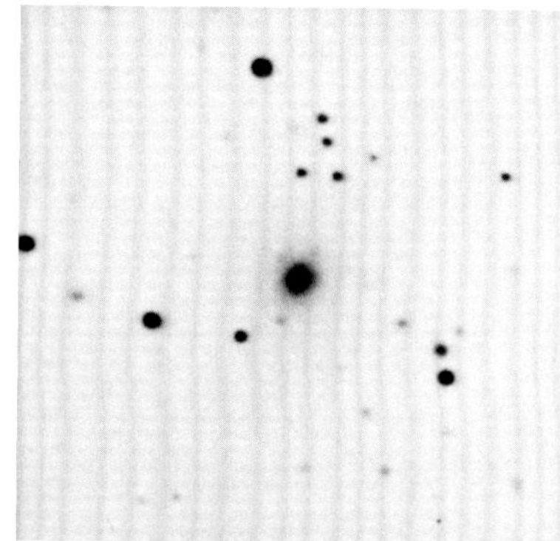

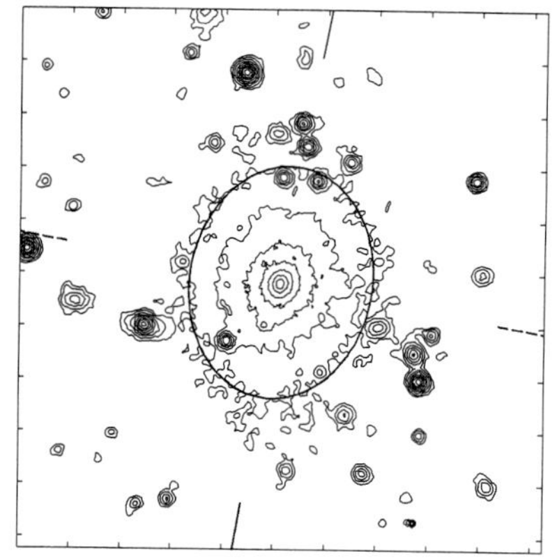

N3158 K1962 1 21.85 512X512

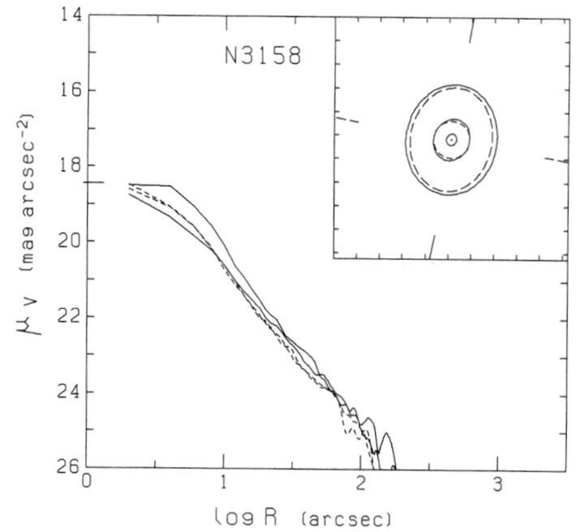

NGC3166

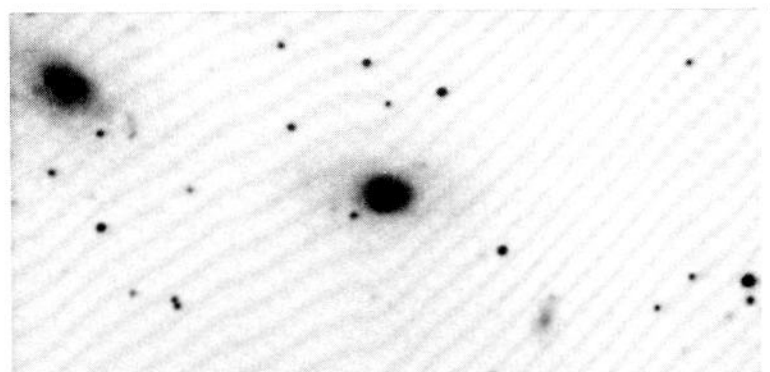

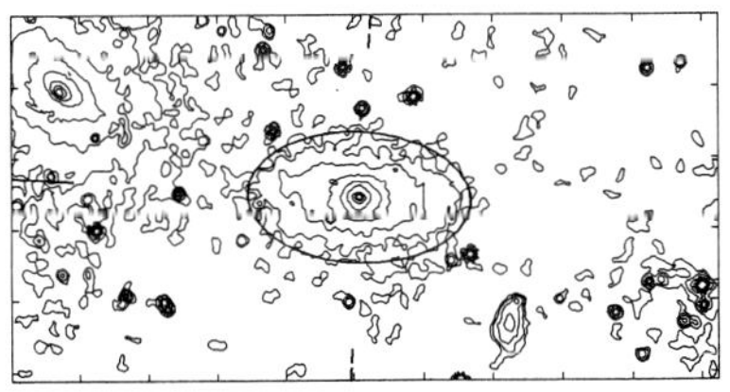

N3166 K5768 2 20.53 512X256

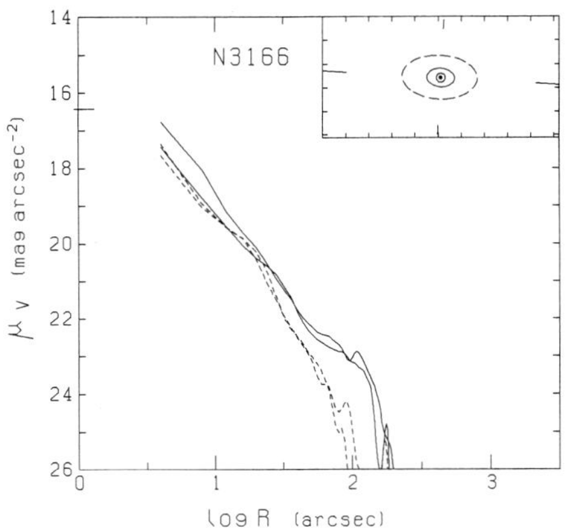

NGC3169

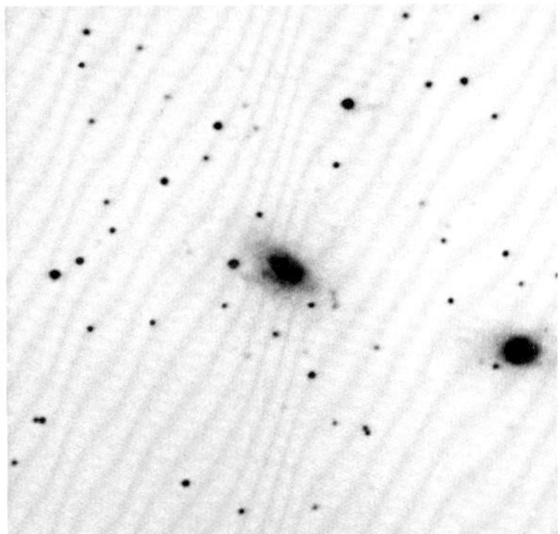

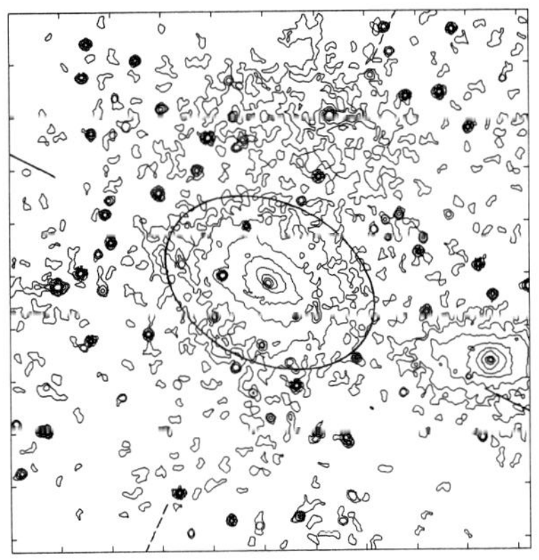

N3169 K5768 2 20.53 512X512

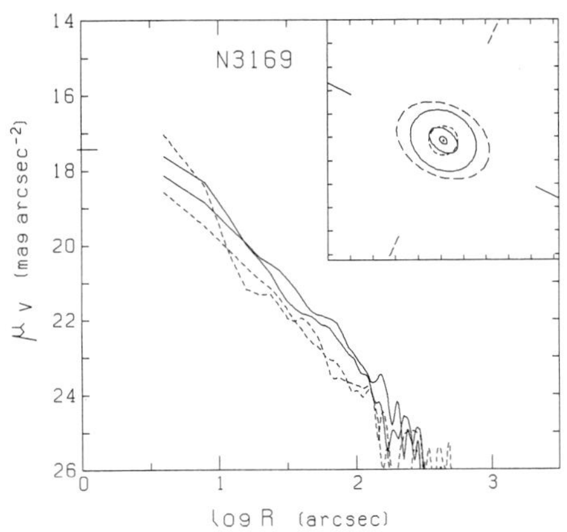

NGC3175

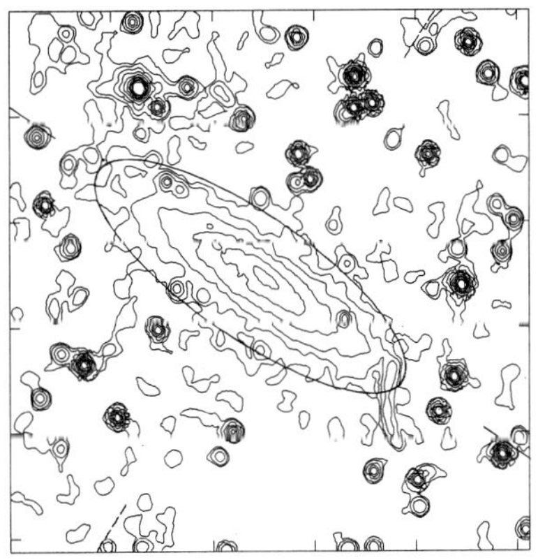

N3175 K4890 2 20.76 256X256

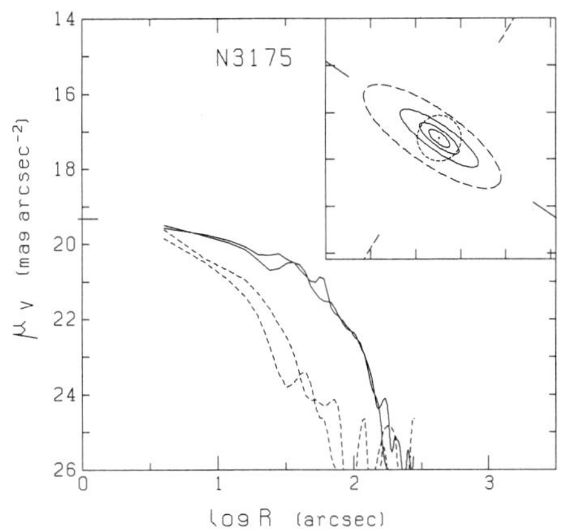

NGC3177

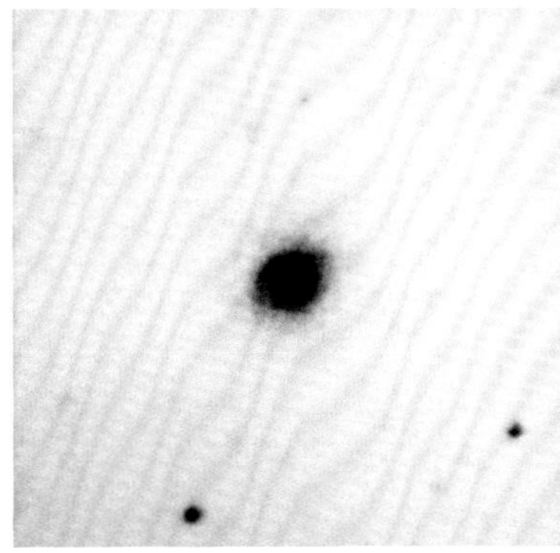

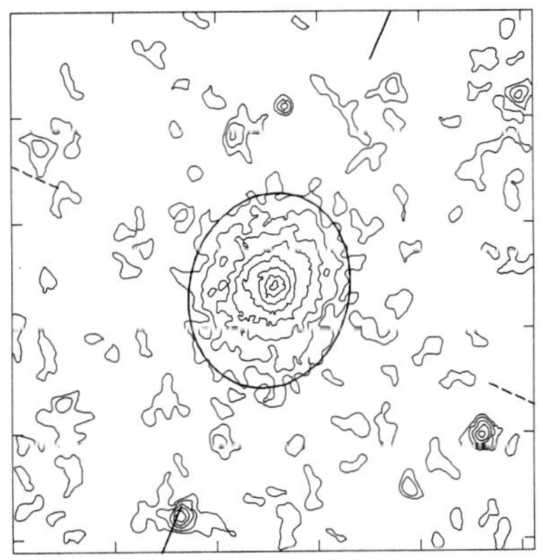

N3177 K4883 1 21.88 256X256

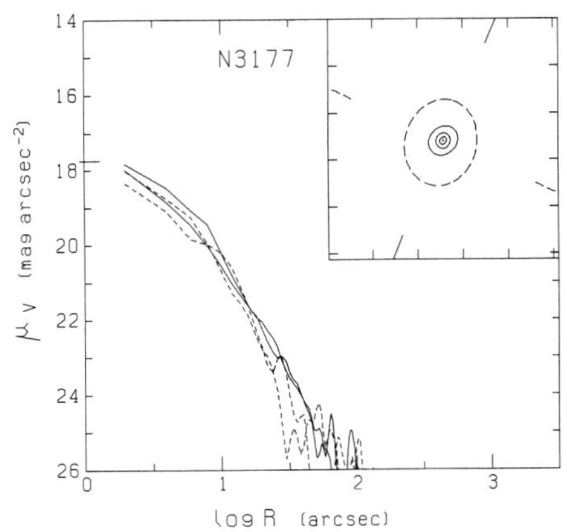

NGC3185

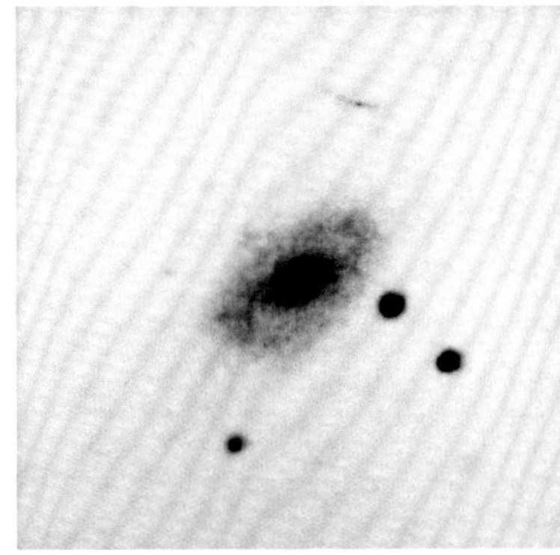

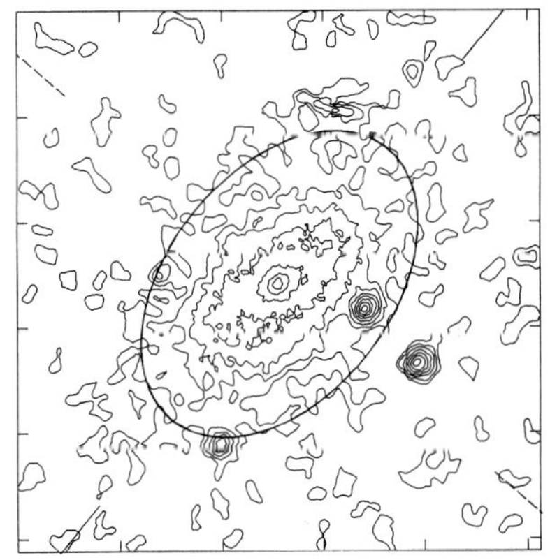

N3185 K4883 1 21.88 256X256

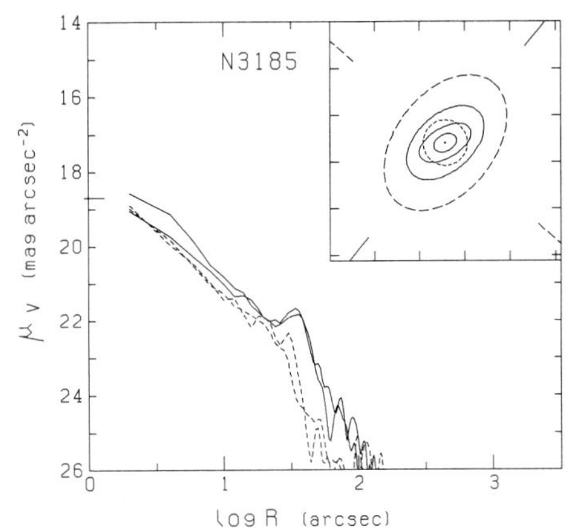

NGC3184

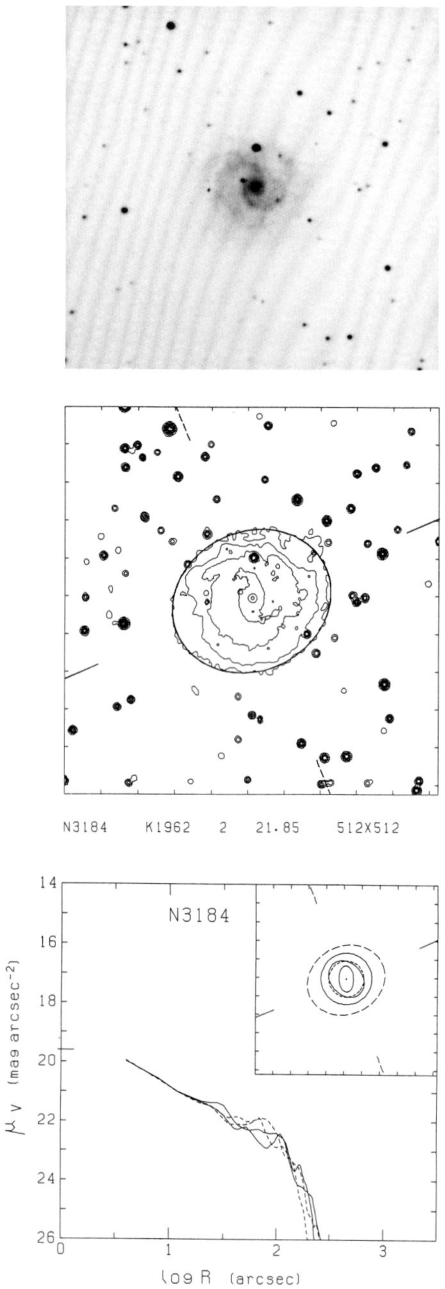

NGC3190

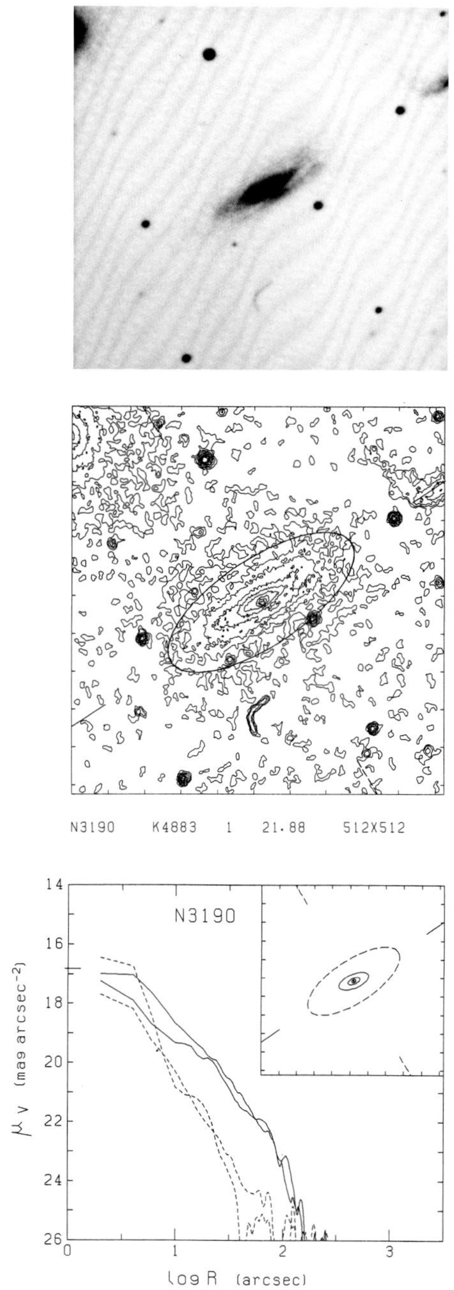

NGC3193

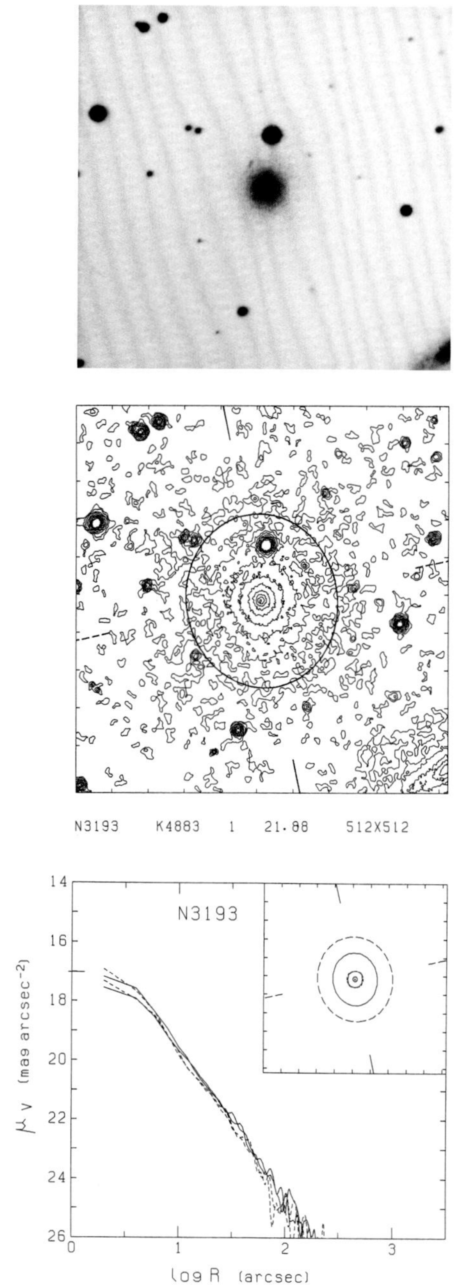

NGC3226

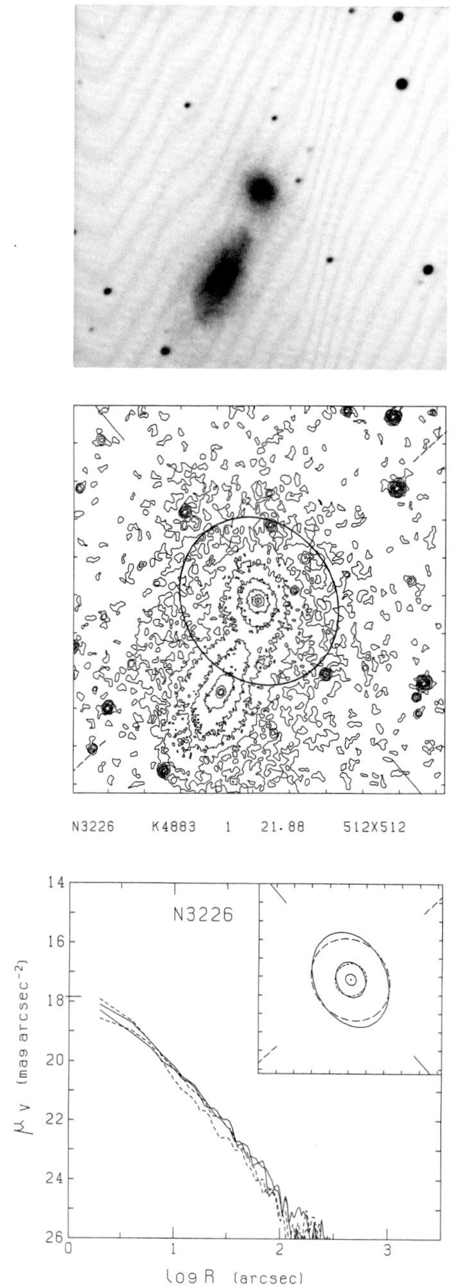

NGC3227

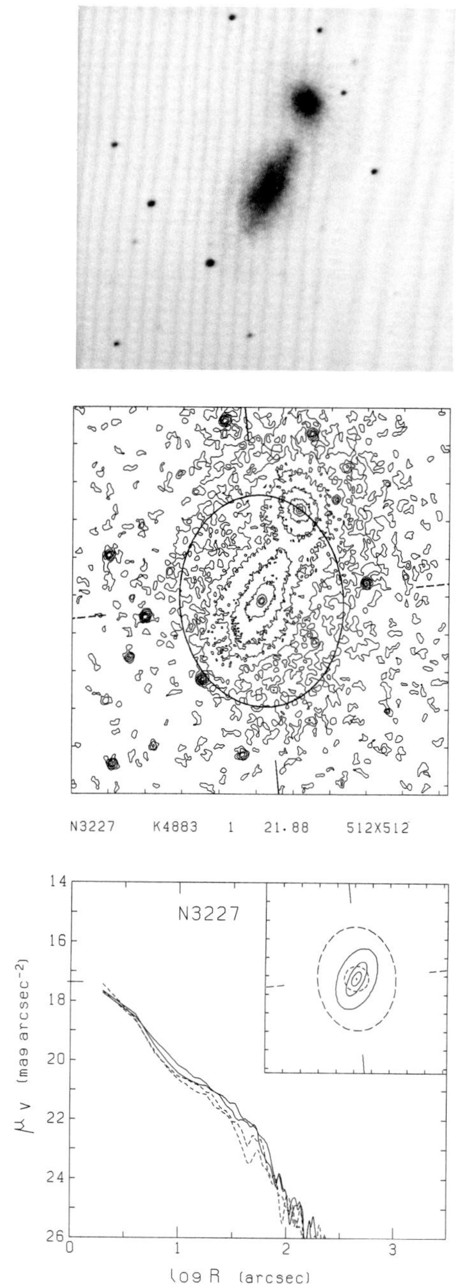

NGC3245

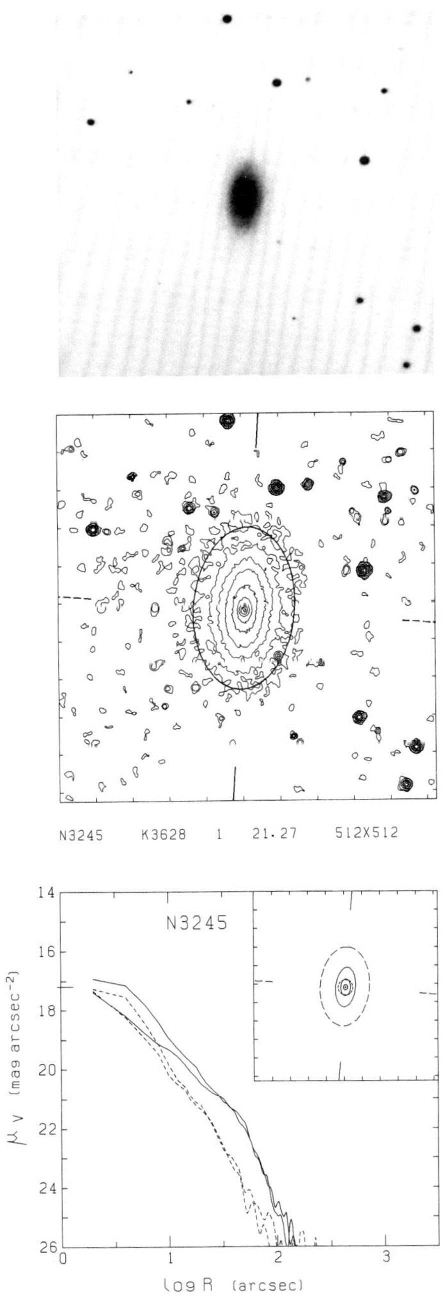

NGC3254

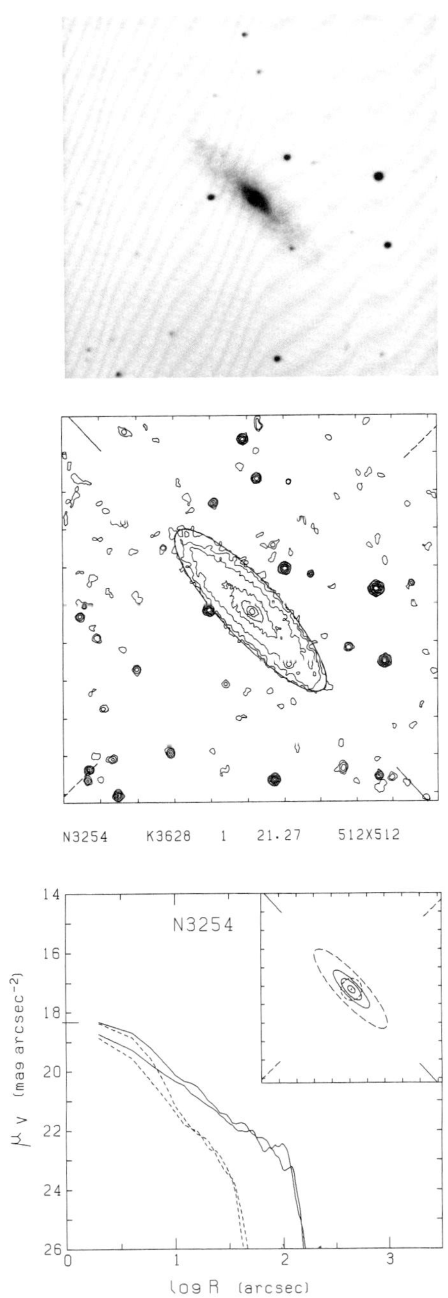

NGC3274

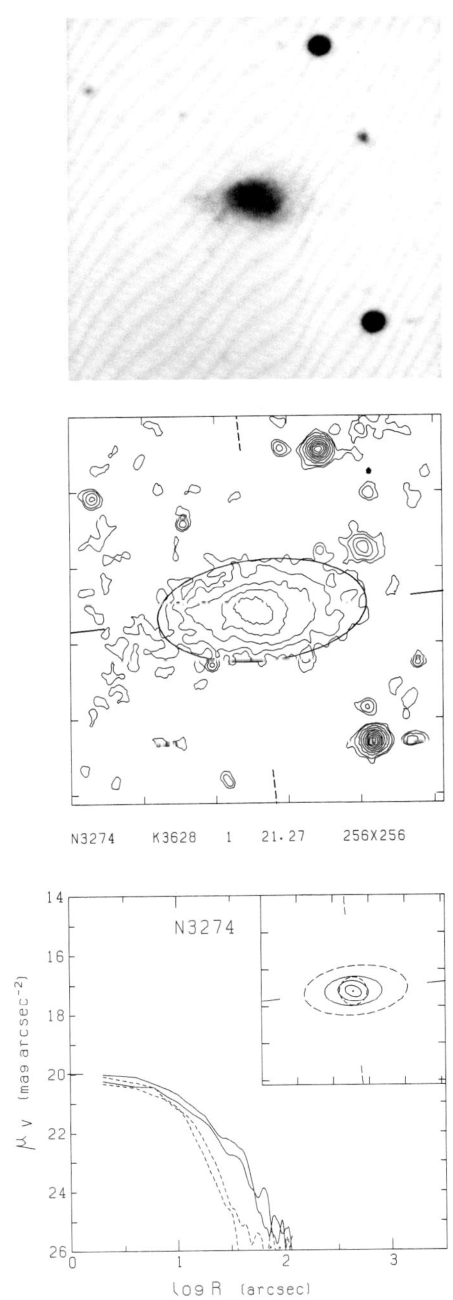

NGC3277

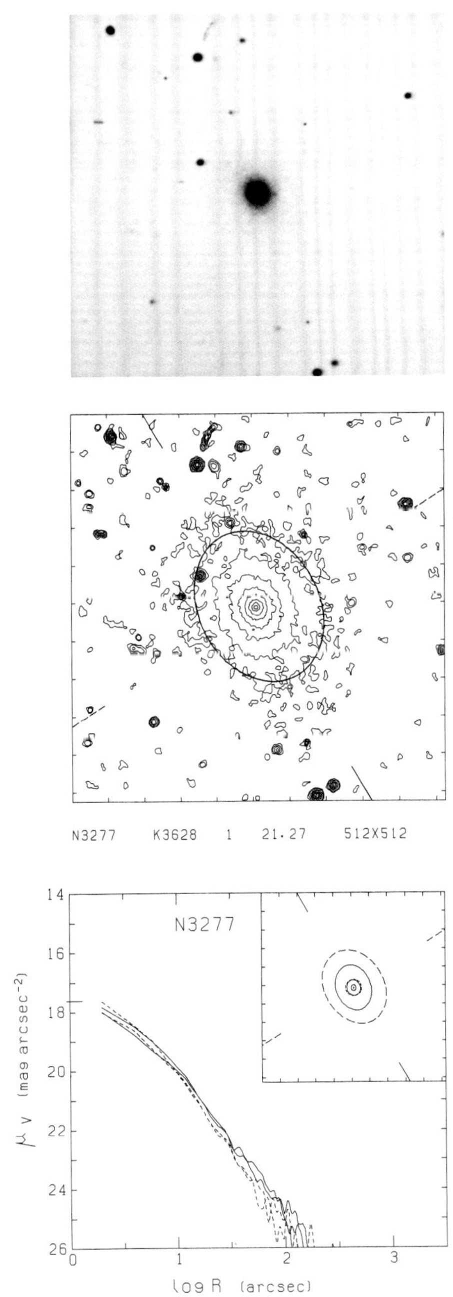

NGC3287

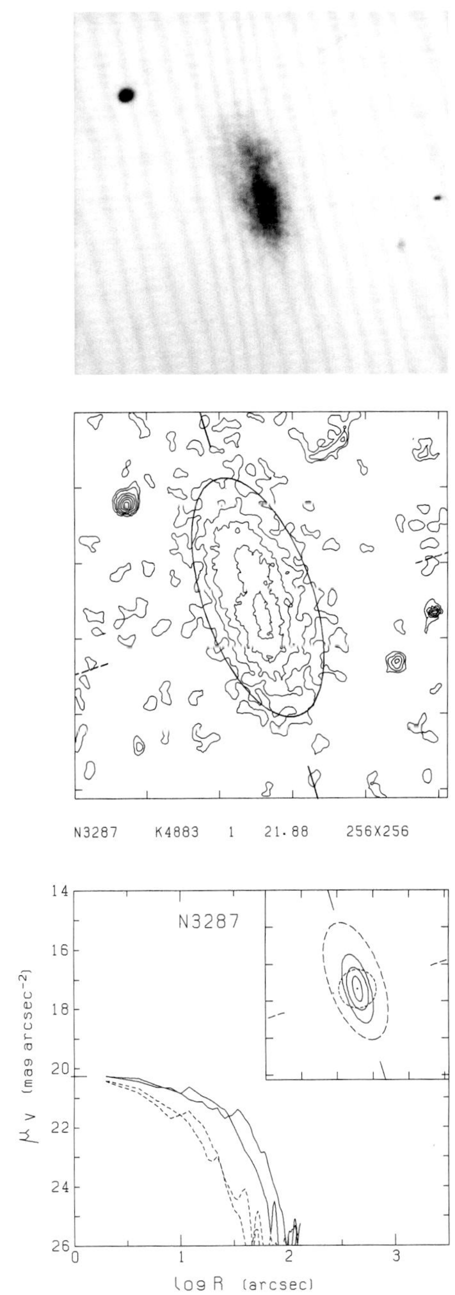

NGC3294

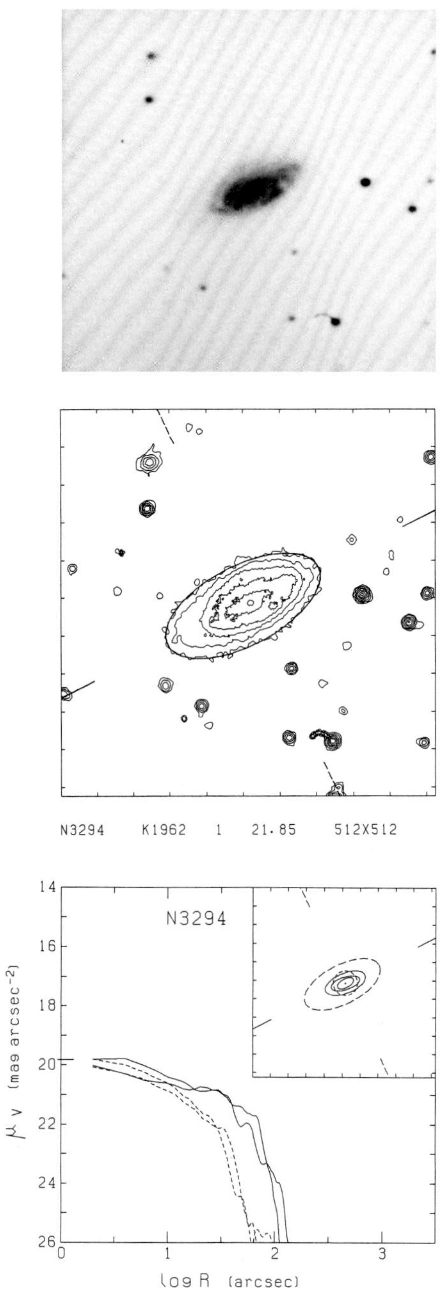

NGC3301

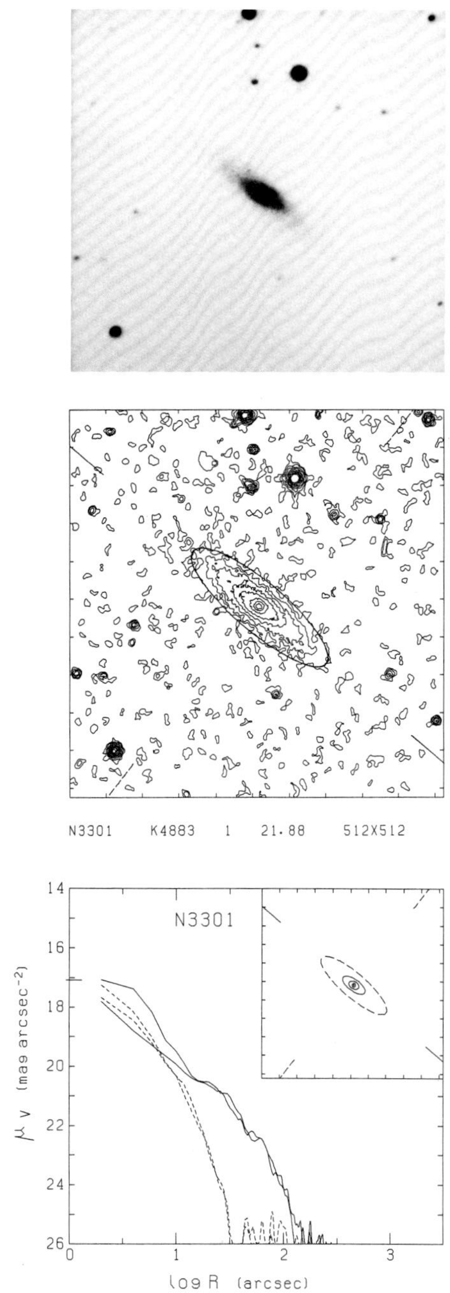

NGC3309

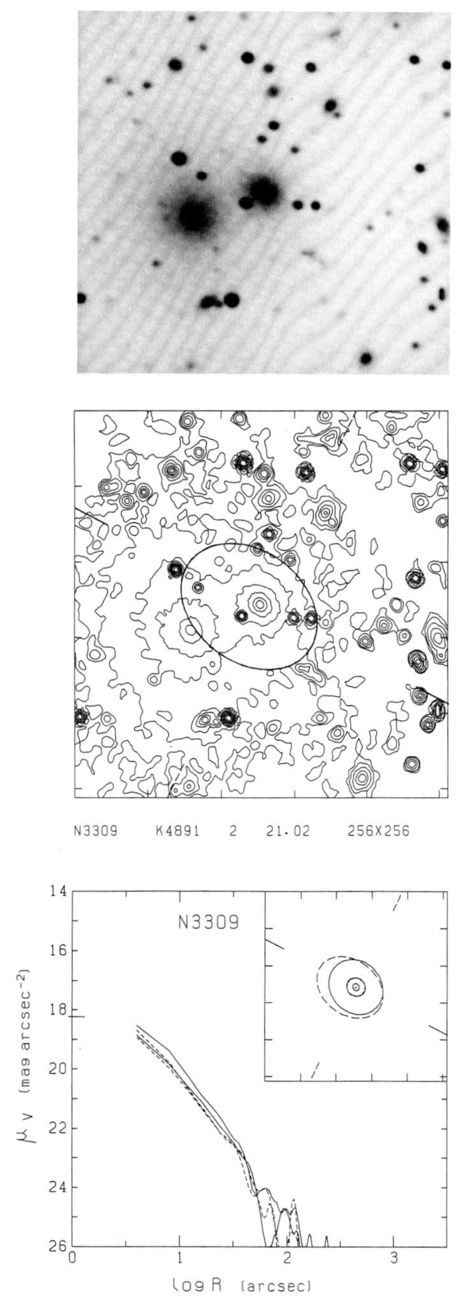

NGC3312

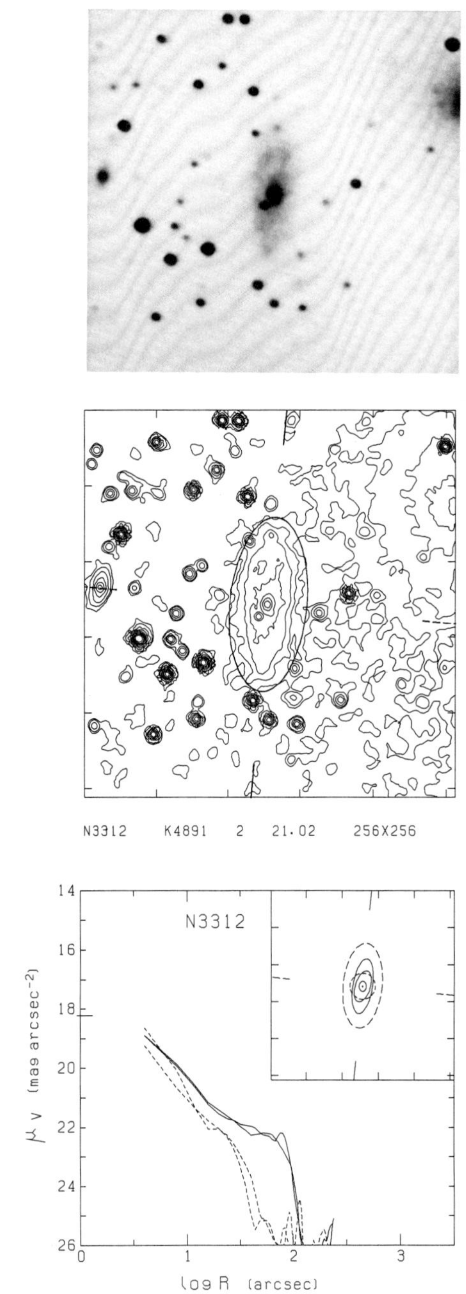

NGC3310

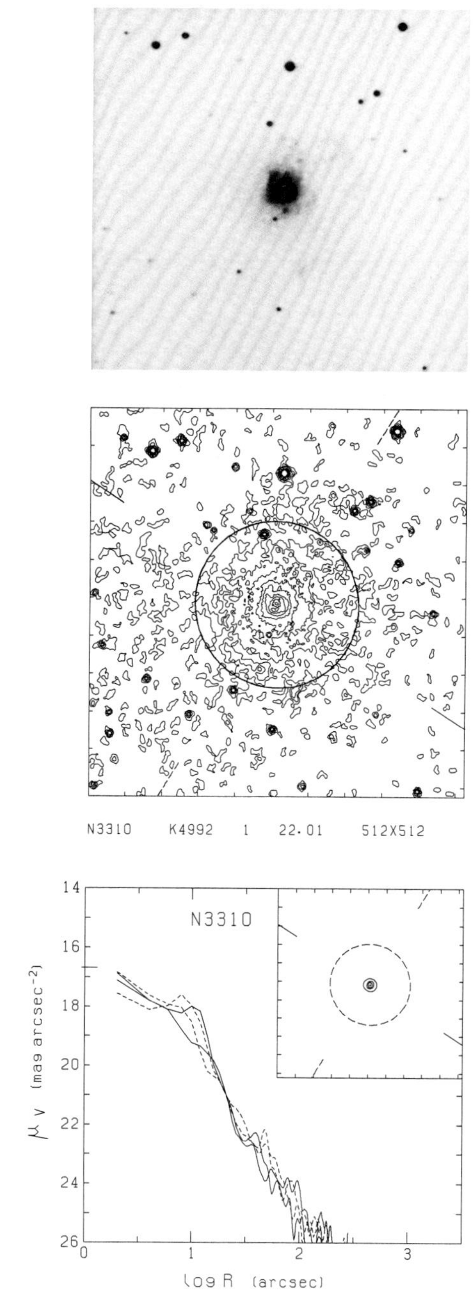

NGC3319

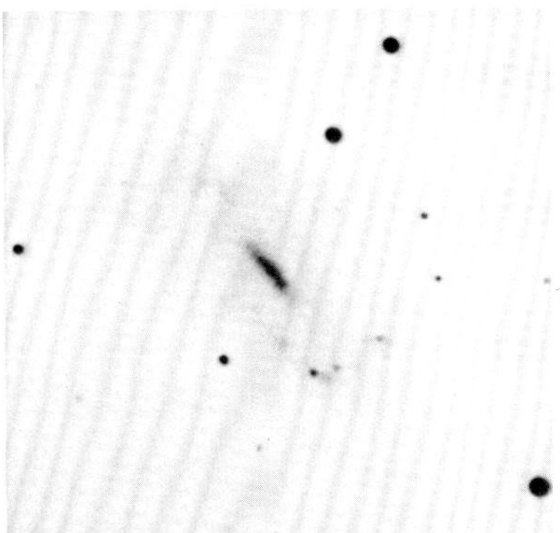

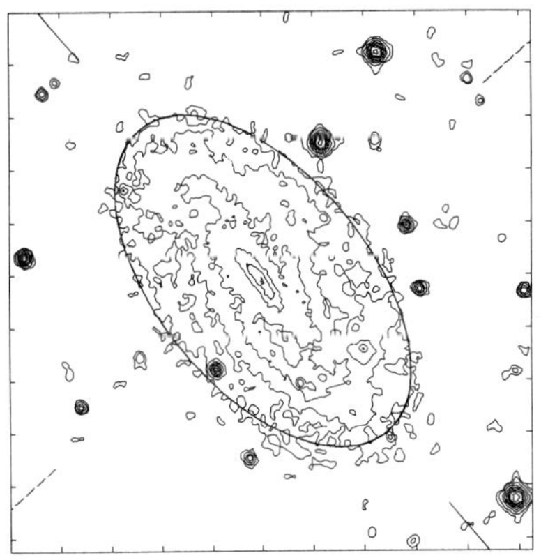

N3319 K3135 1 21.41 512X512

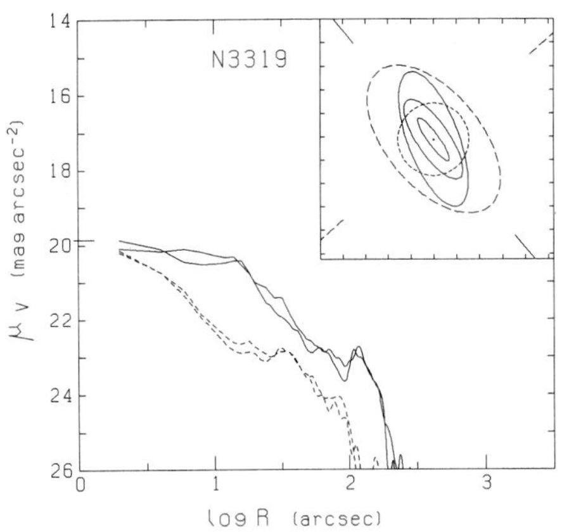

NGC3320

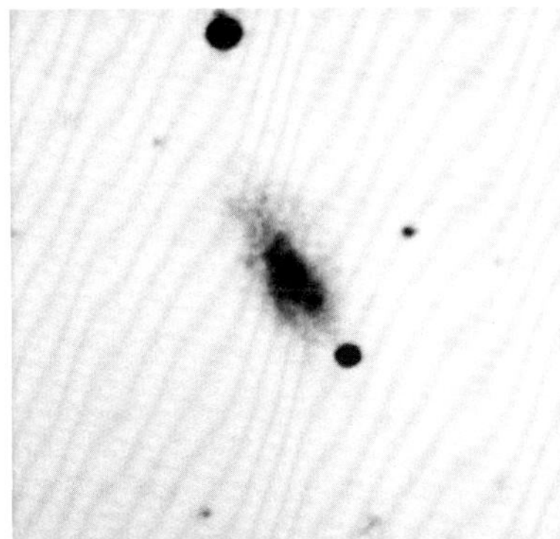

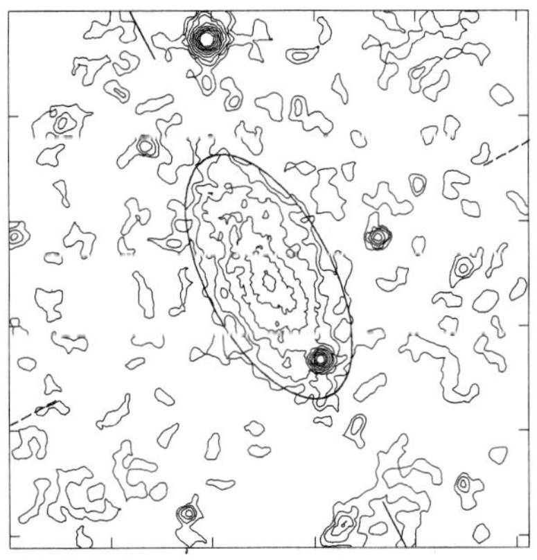

N3320 K4880 1 21.87 256X256

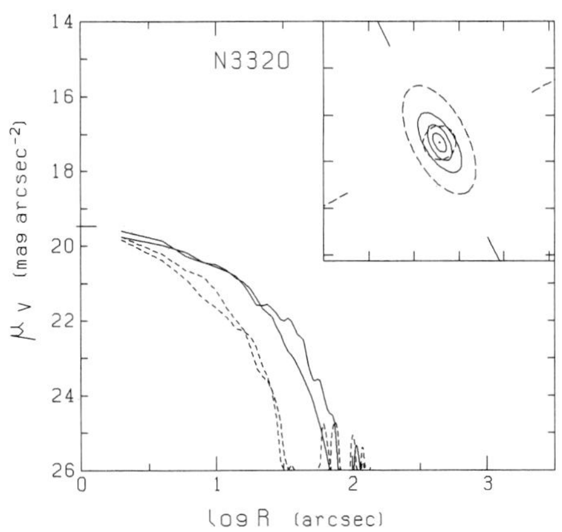

NGC3338

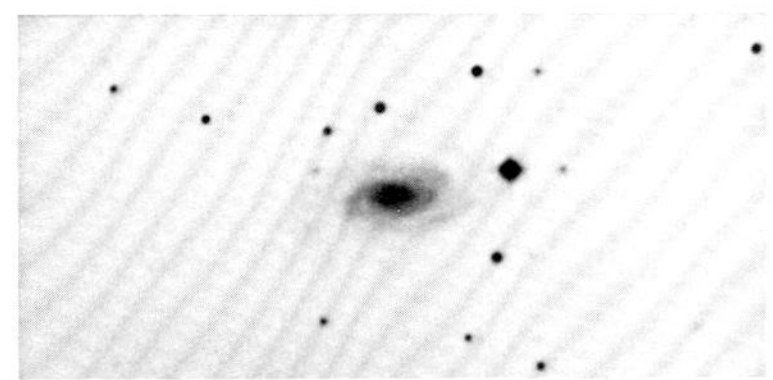

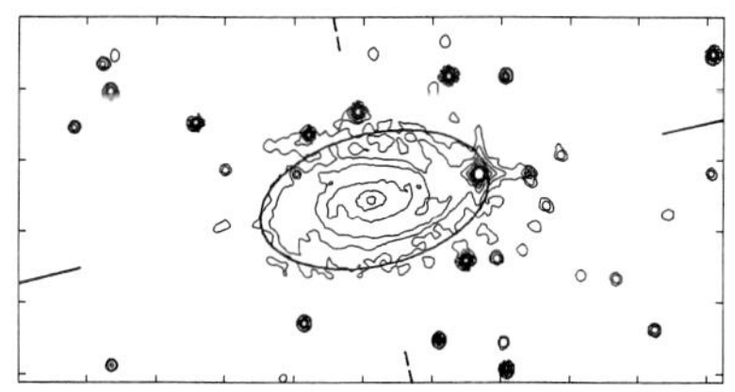

N3338 K3194 2 21.12 512X256

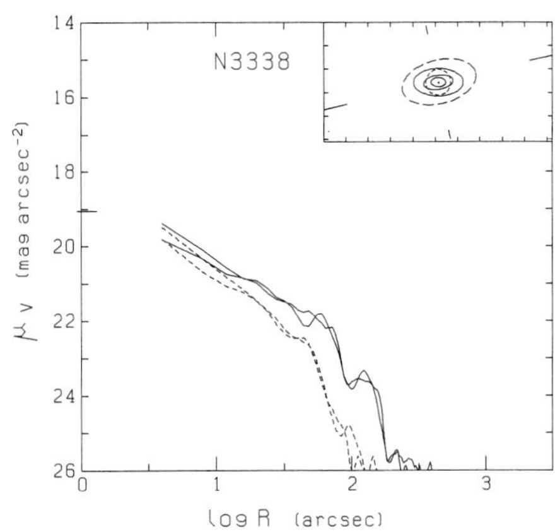

NGC3344

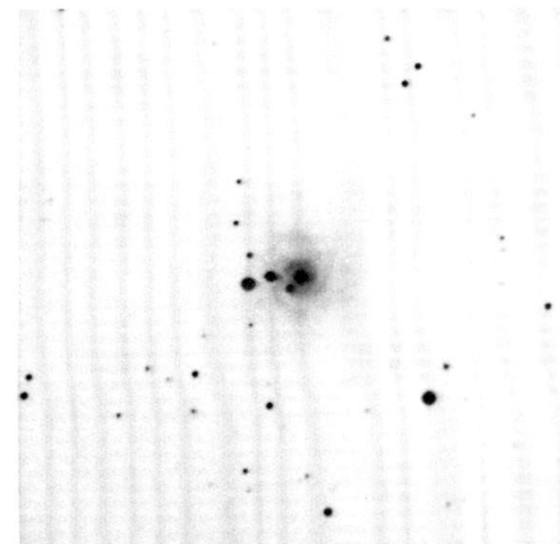

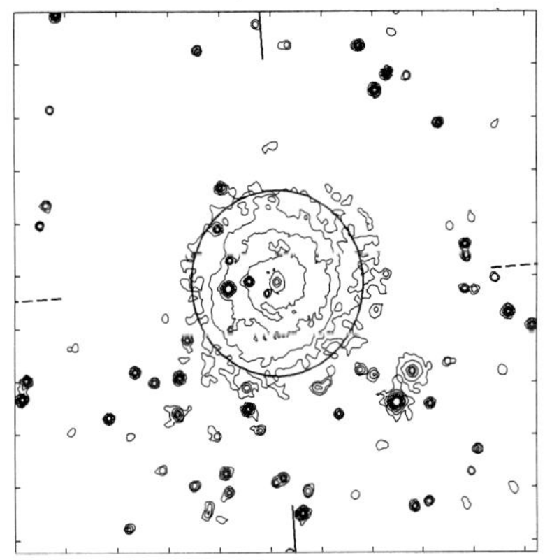

N3344 K4886 2 21.62 512X512

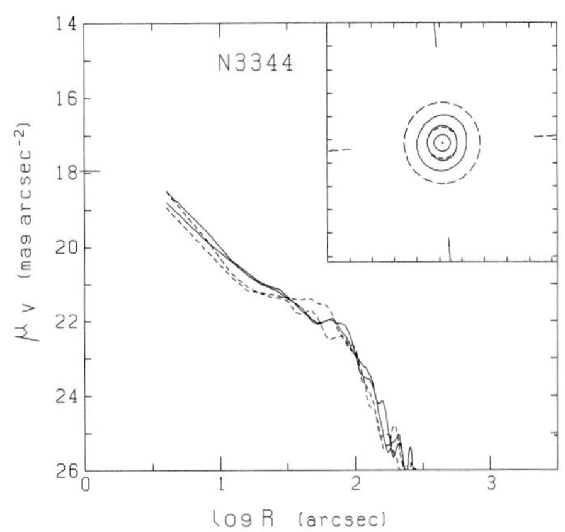

NGC3346

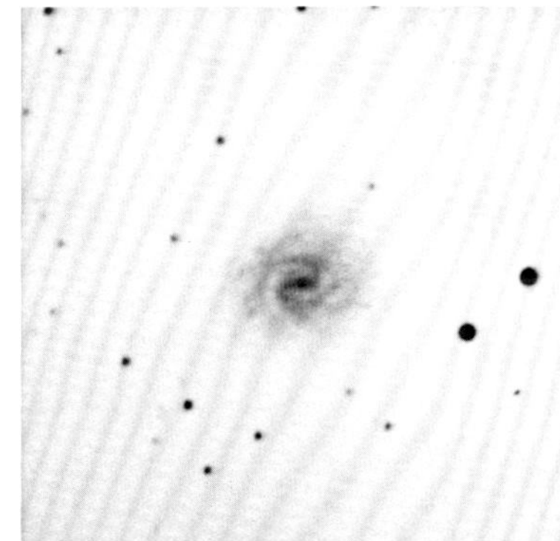

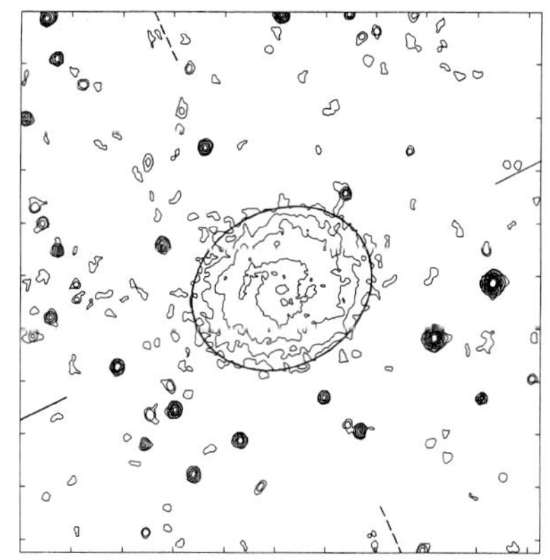

N3346 K3194 1 21.12 512X512

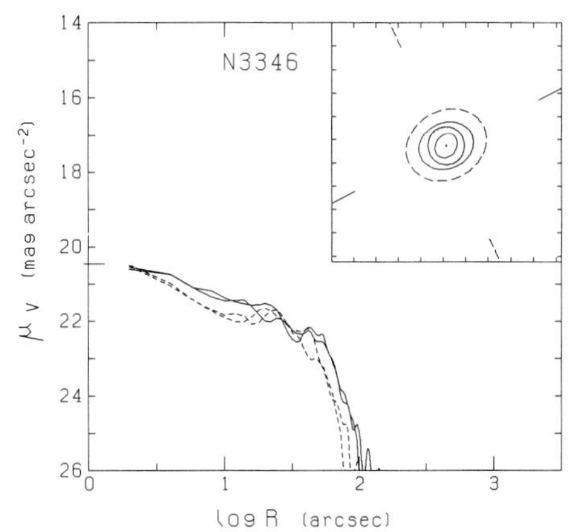

NGC3351

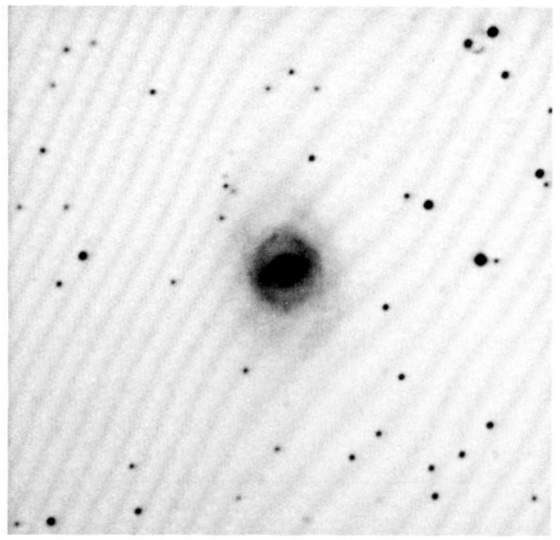

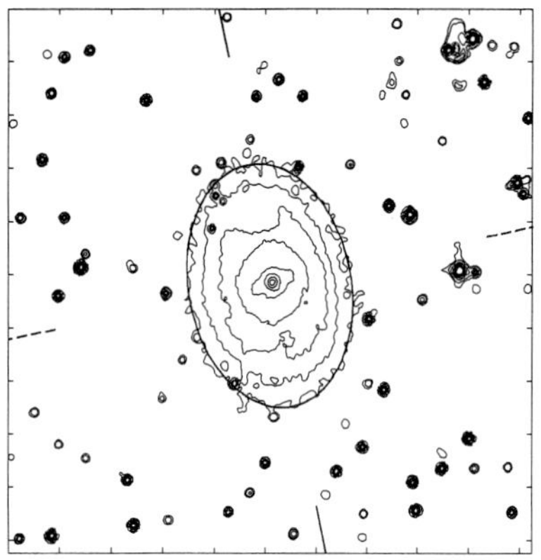

N3351 K3194 2 21.12 512X512

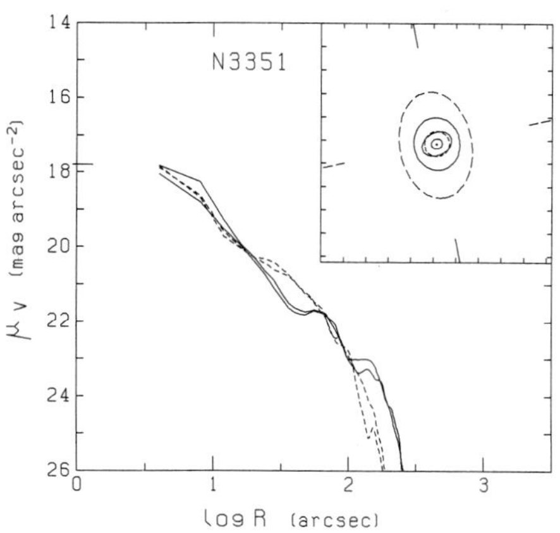

NGC3353

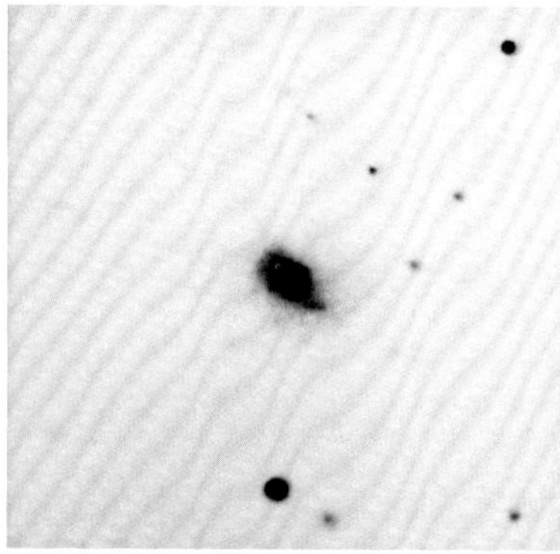

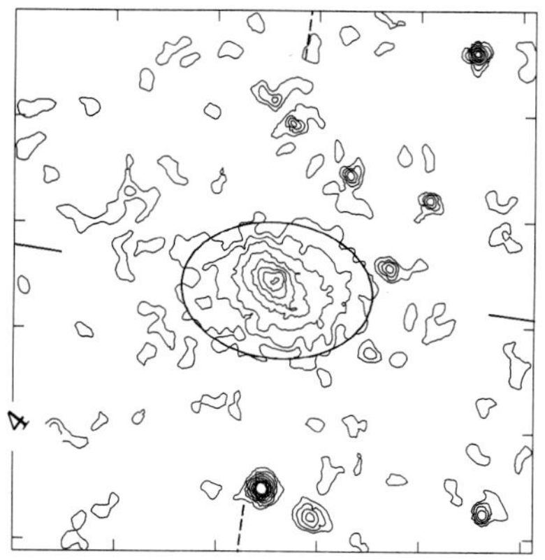

N3353 K4992 1 22.01 256X256

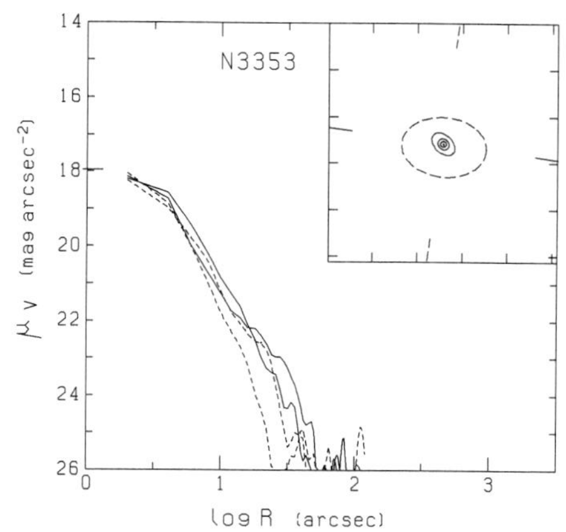

NGC3367

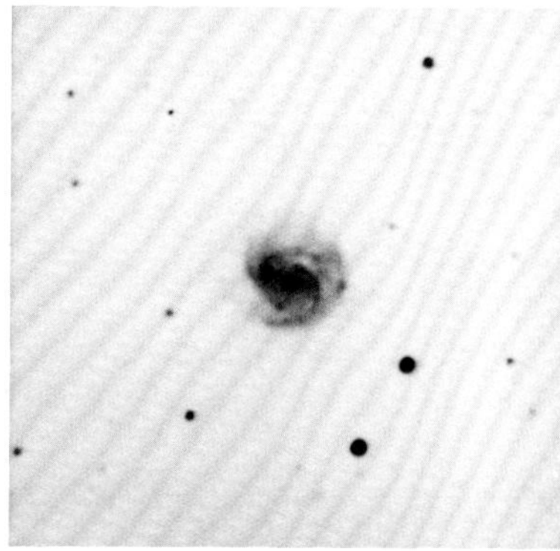

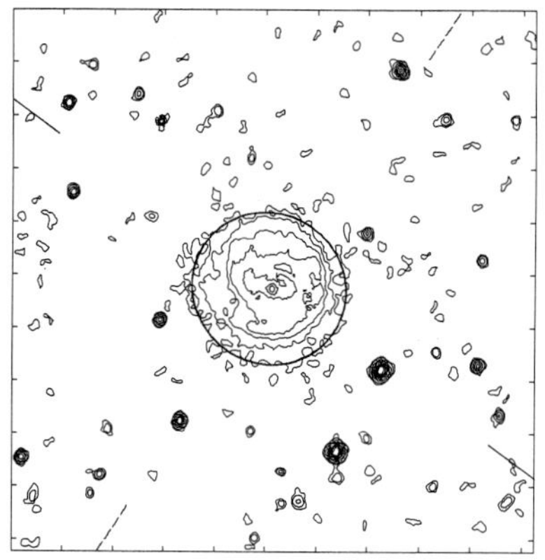

N3367 K3194 1 21.12 512X512

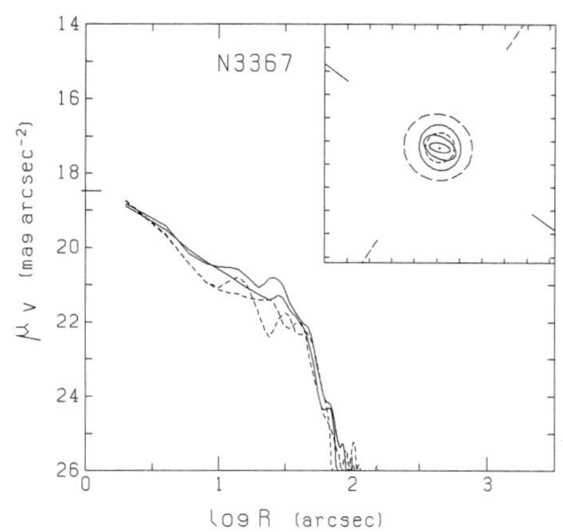

NGC3368

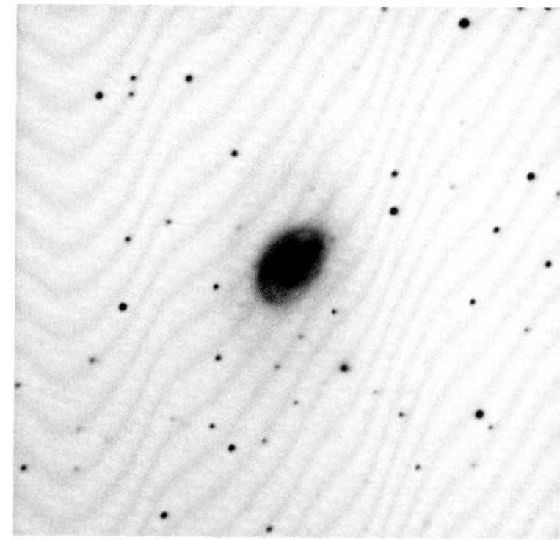

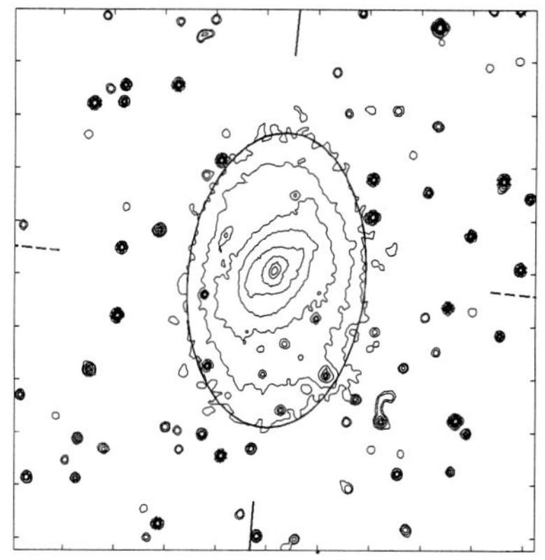

N3368 K3194 2 21.12 512X512

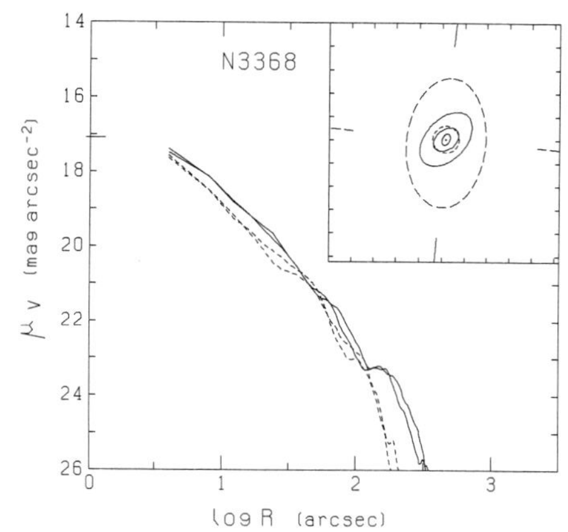

NGC3370

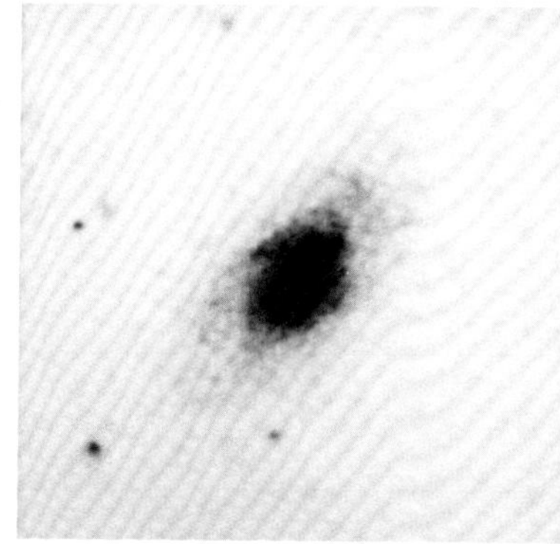

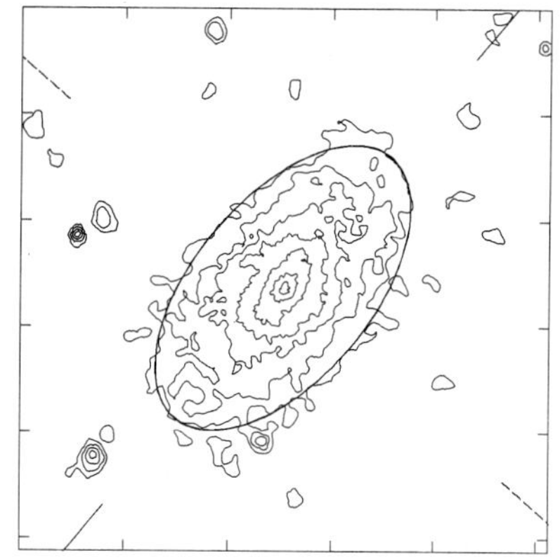

N3370 K5011 1 22.26 256X256

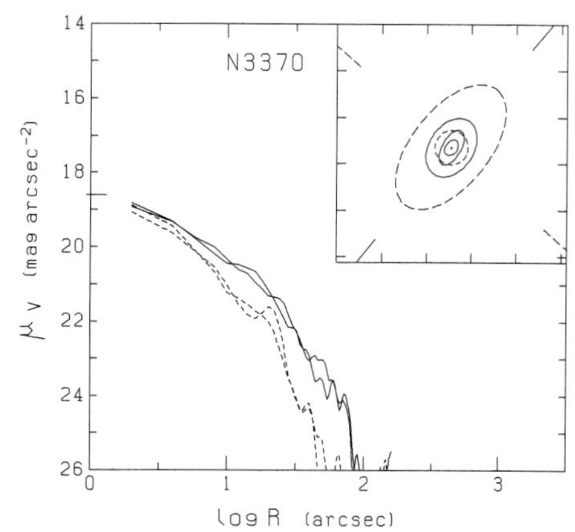

NGC3377

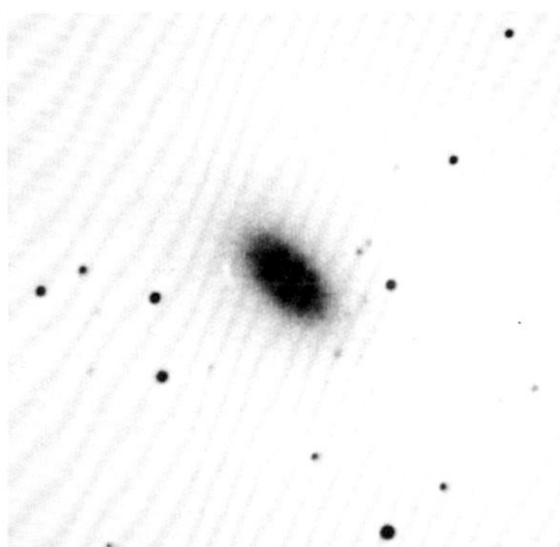

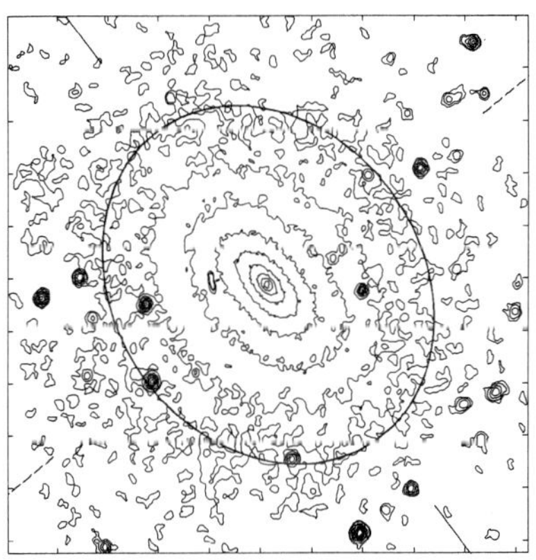

N3377 K3194 1 21.12 512X512

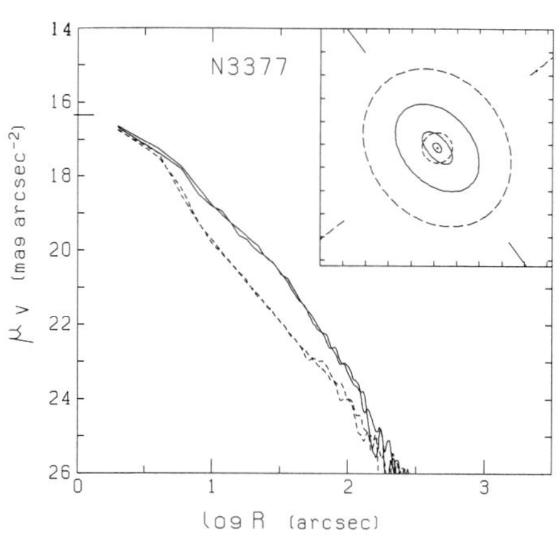

NGC3379

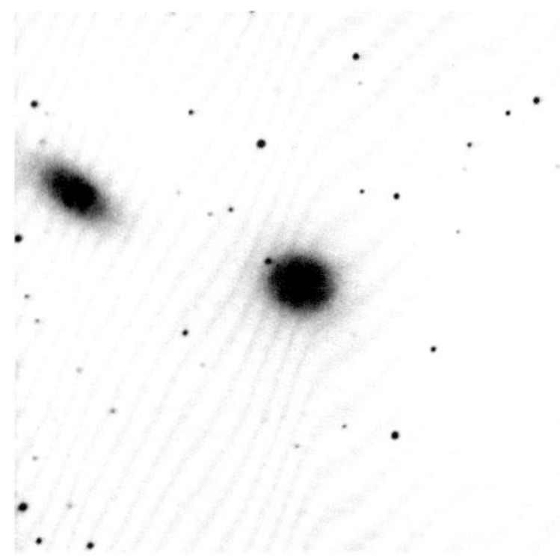

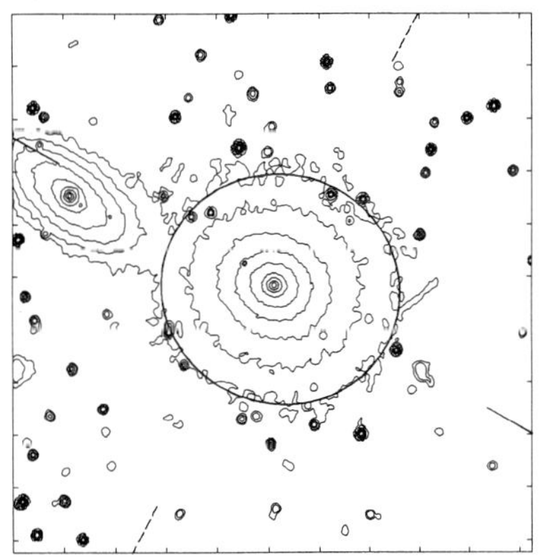

N3379 K3194 2 21.12 512X512

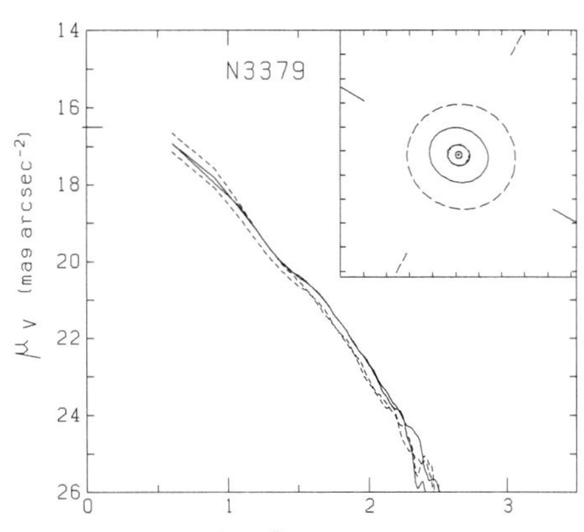

NGC3384

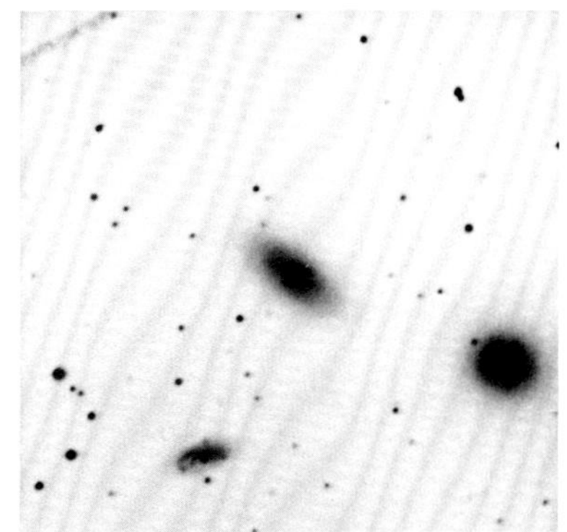

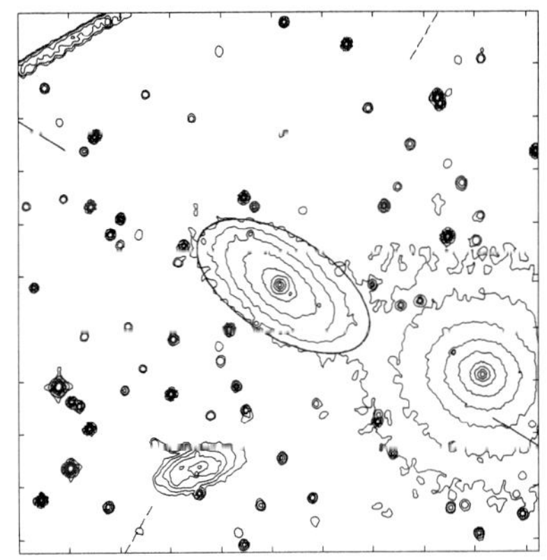

N3384 K3194 2 21.12 512X512

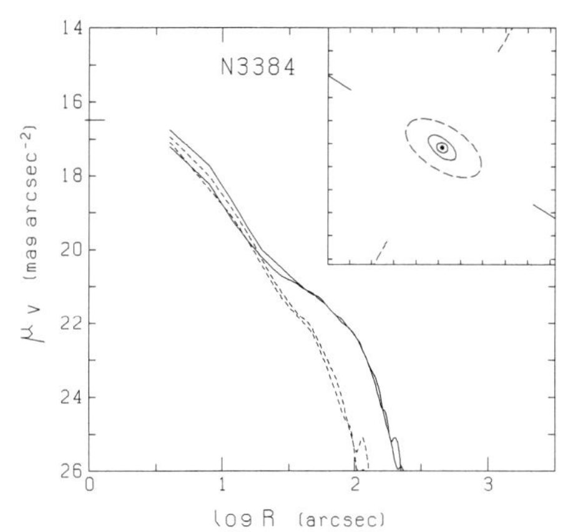

NGC3390

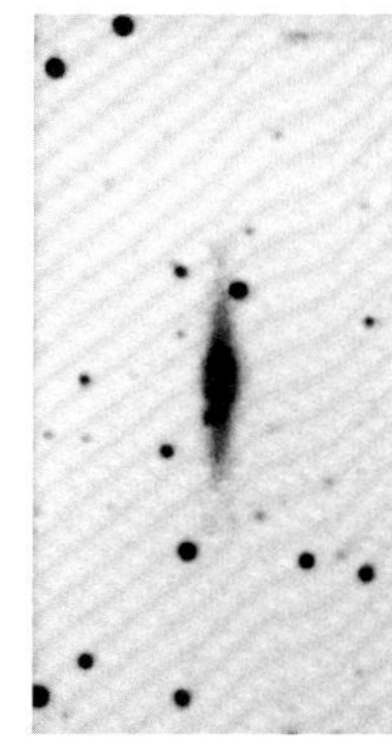

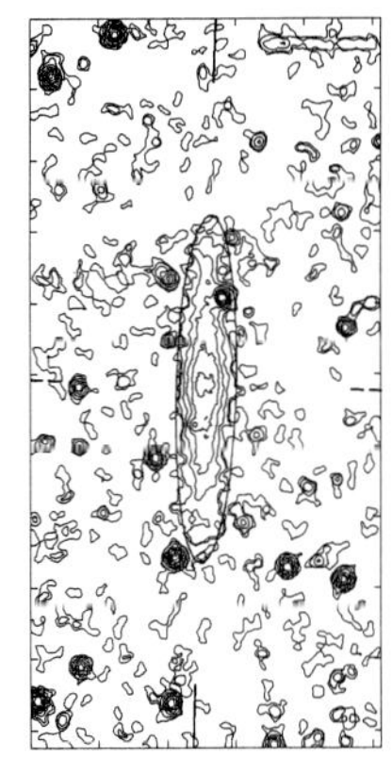

N3390 K4891 1 21.02 256X512

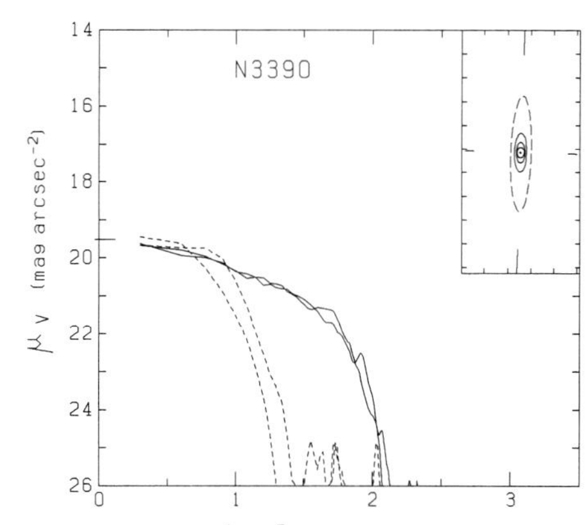

NGC3389

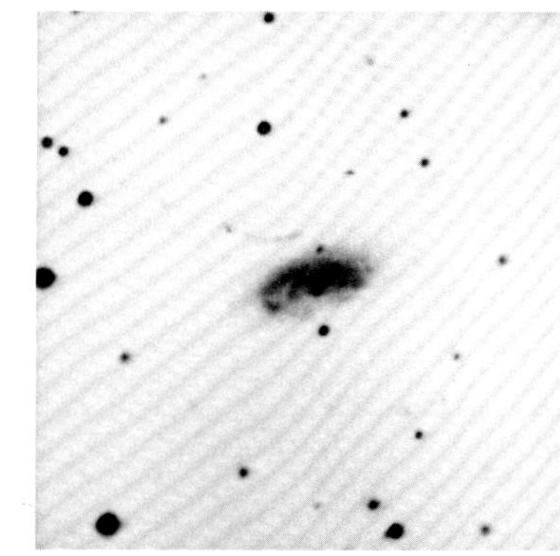

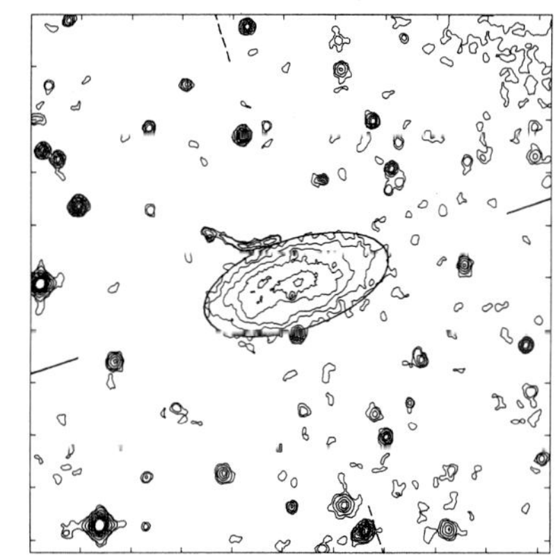

N3389 K3194 1 21.12 512X512

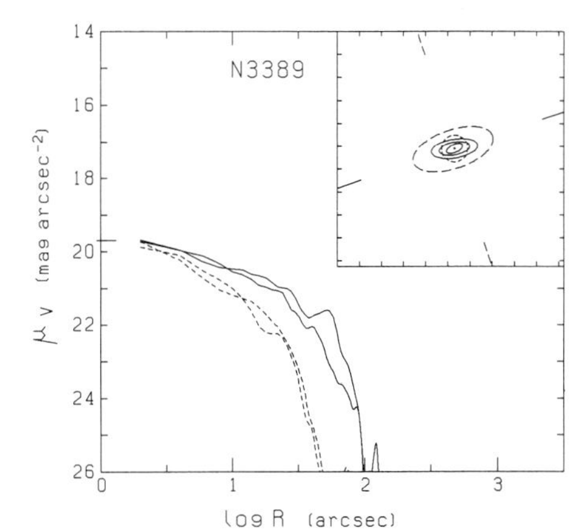

NGC3395

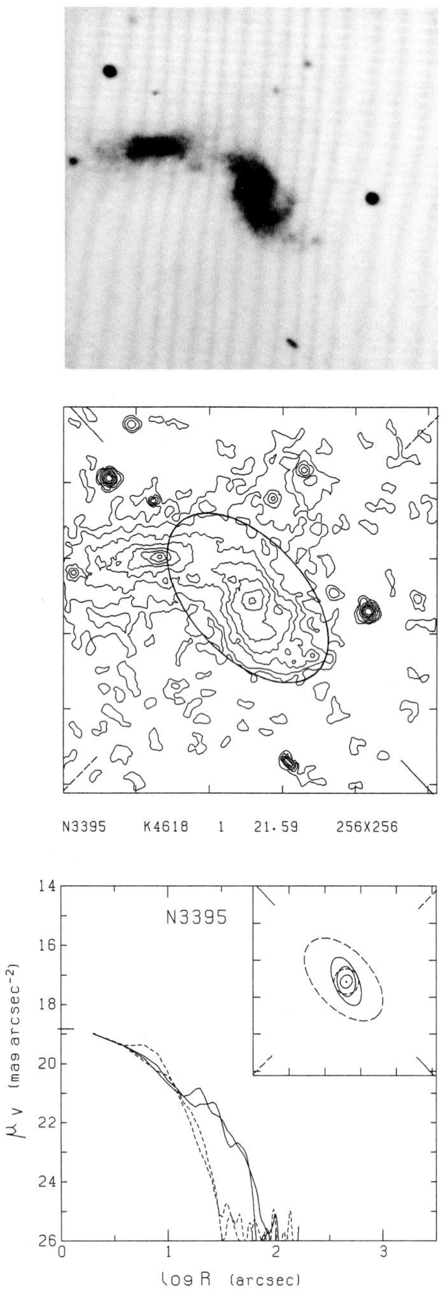

NGC3396

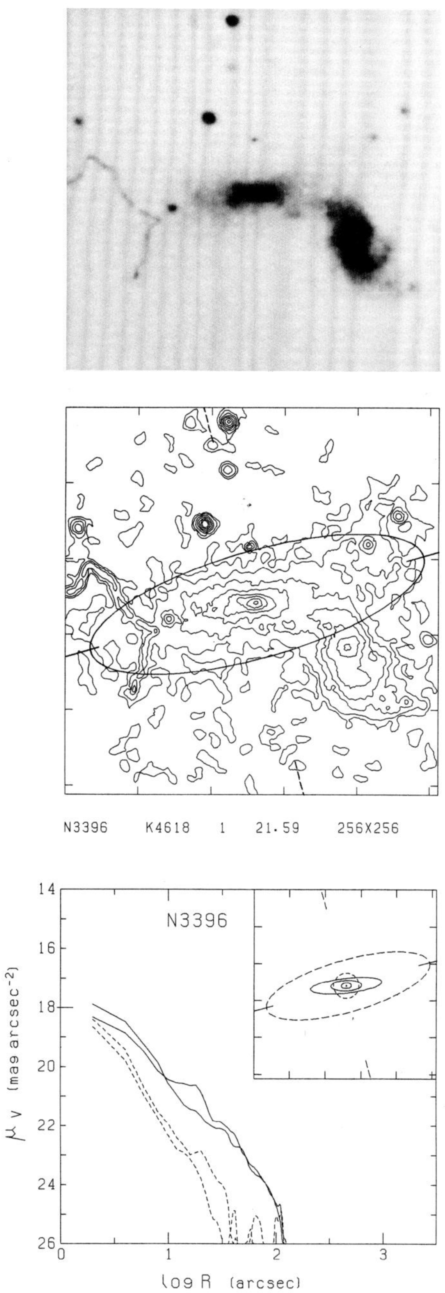

NGC3412

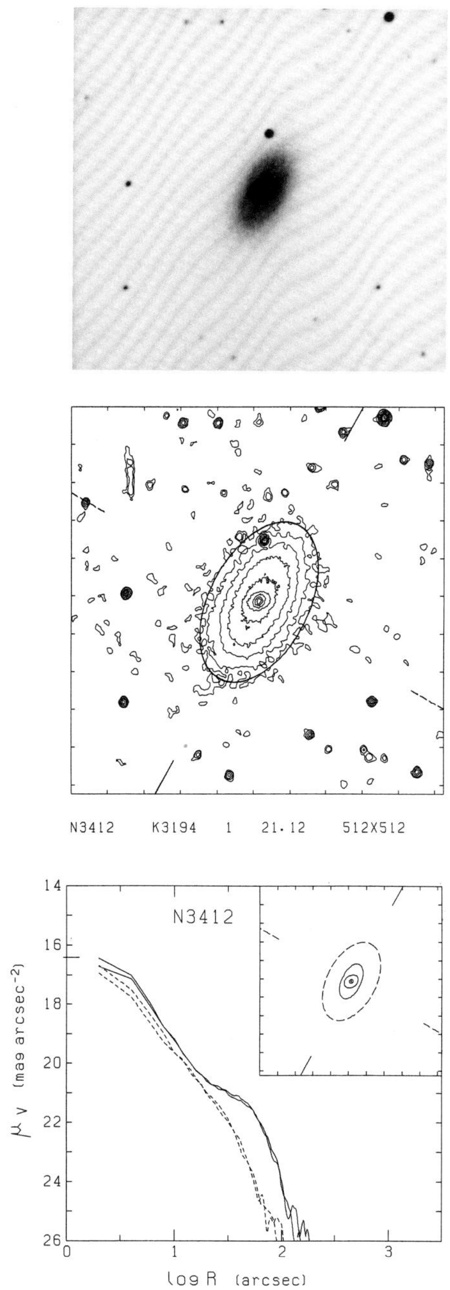

NGC3414

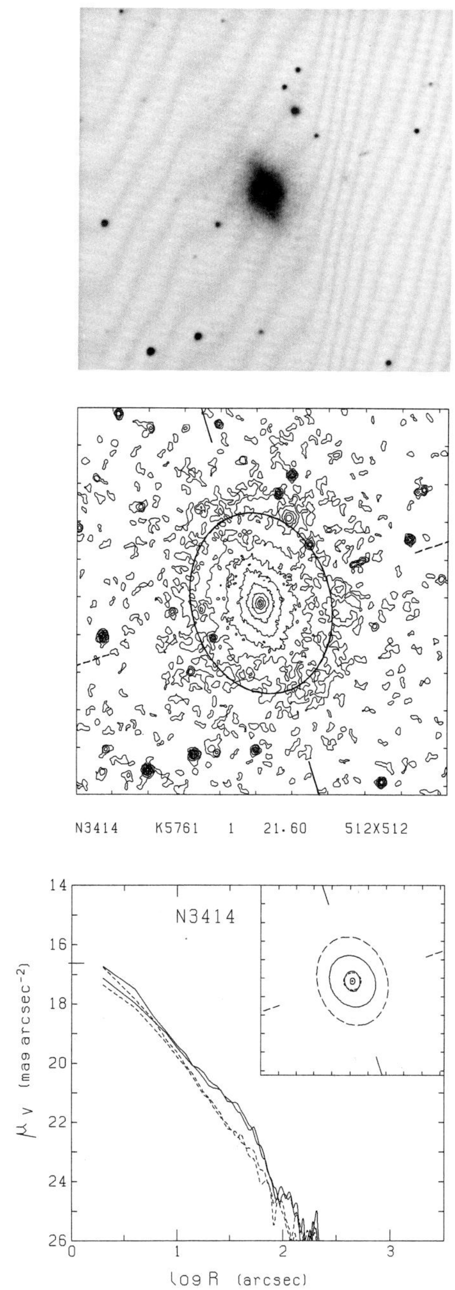

NGC3423

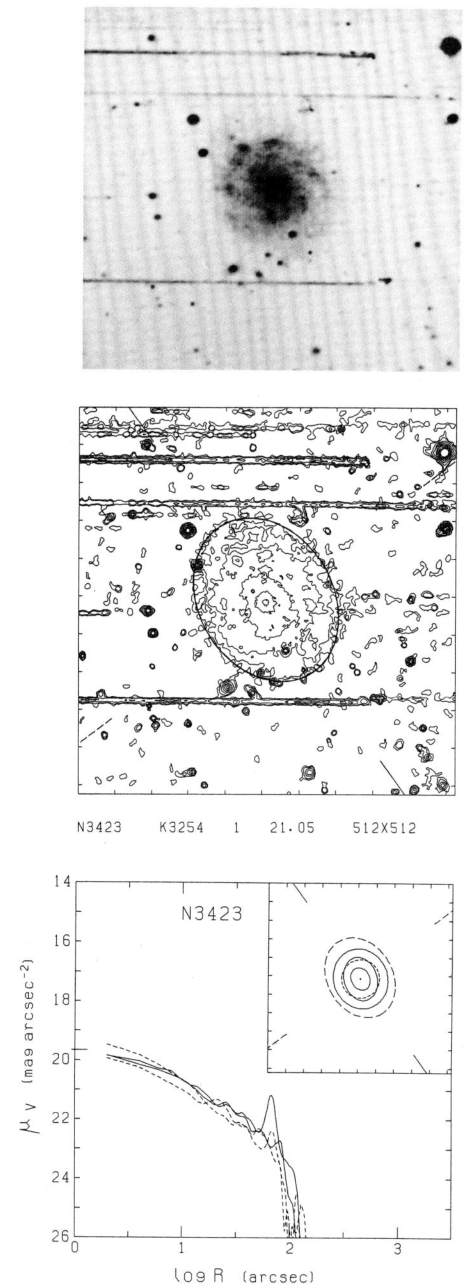

NGC3415

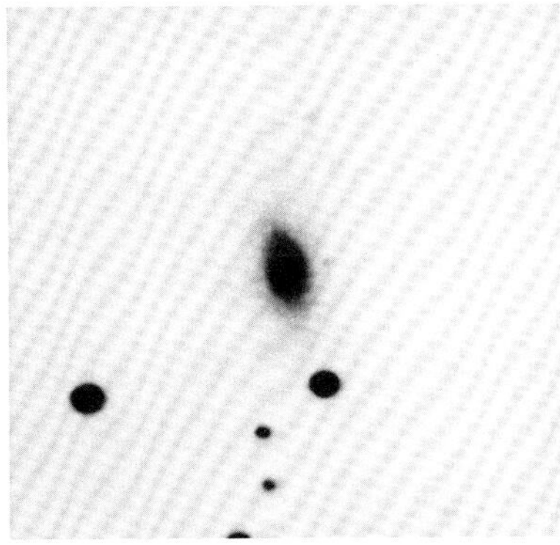

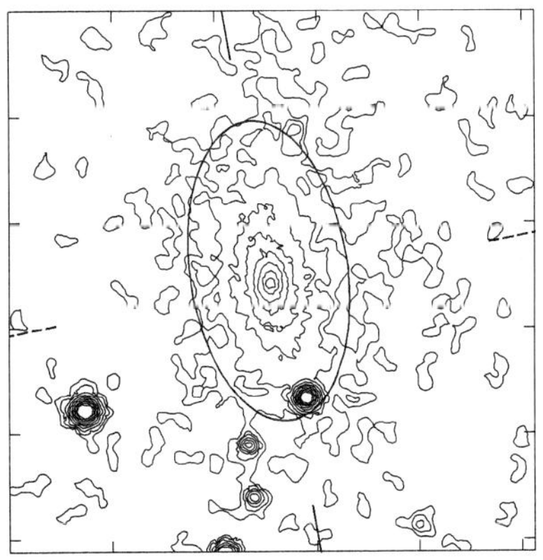

N3415 K4880 1 21.87 256X256

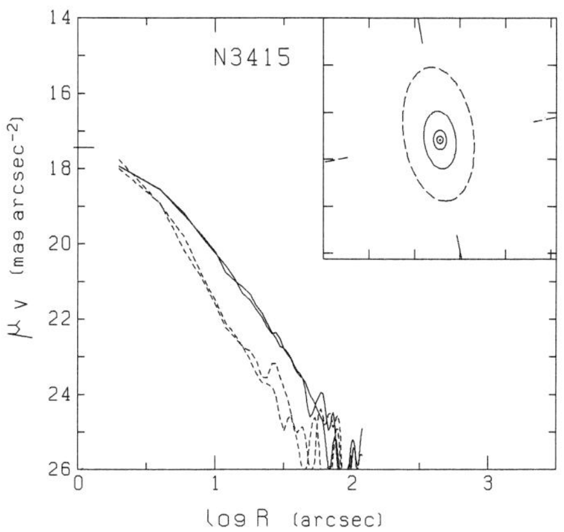

NGC3430

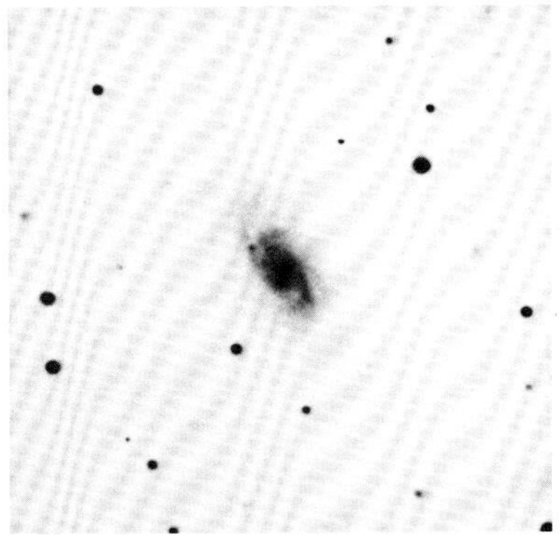

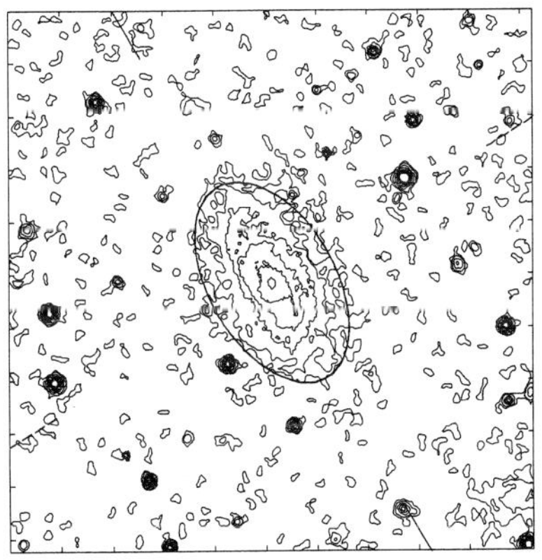

N3430 K4618 1 21.59 512X512

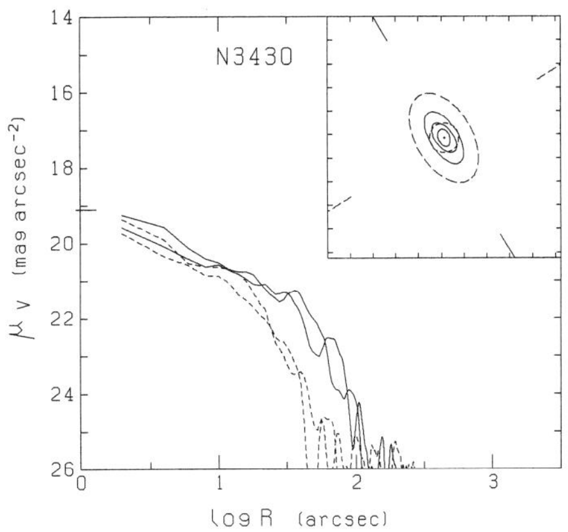

NGC3432

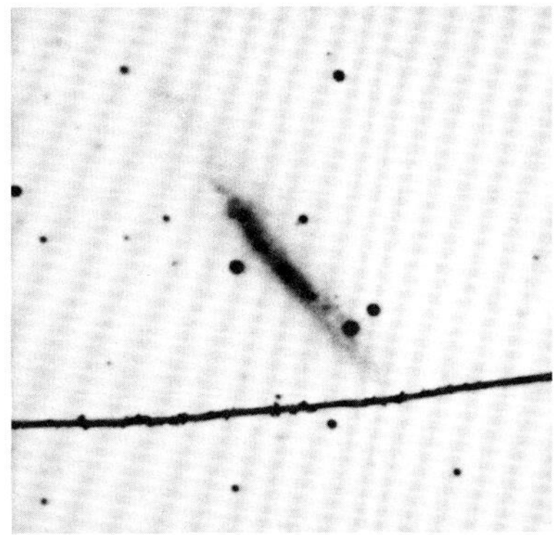

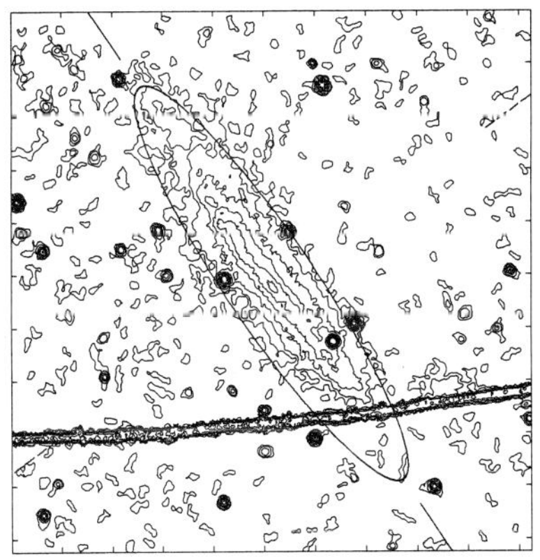

N3432 K4618 1 21.59 512X512

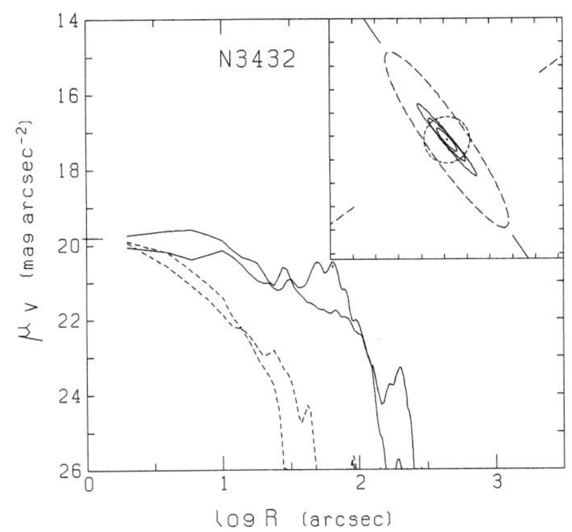

NGC3437

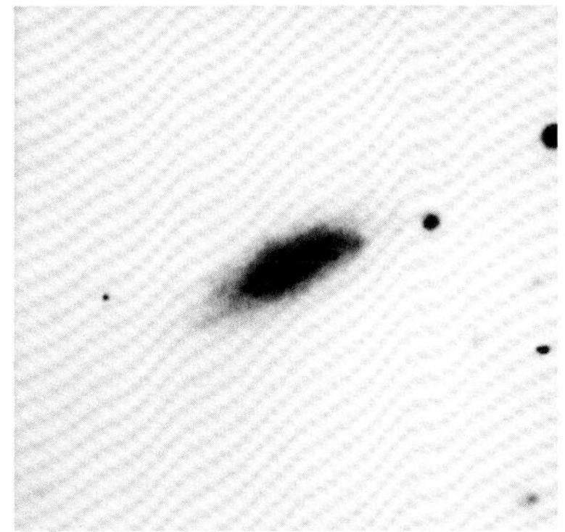

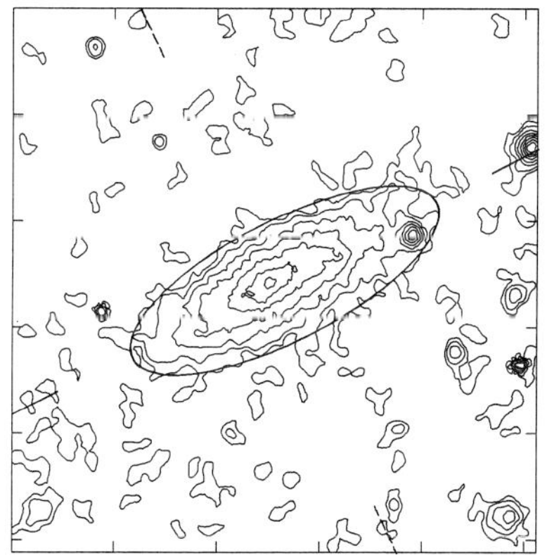

N3437 K4886 1 21.62 256X256

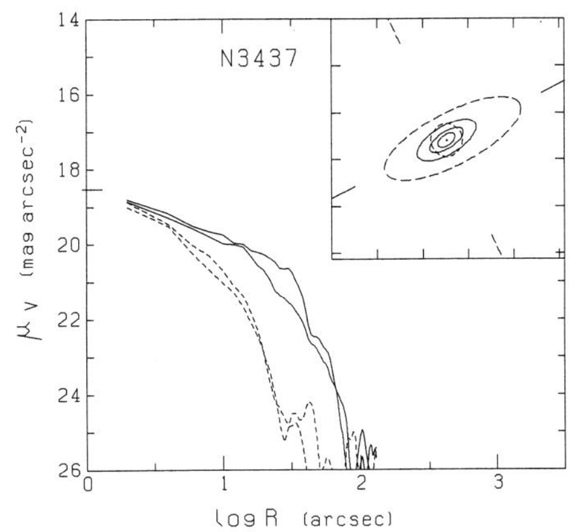

NGC3449

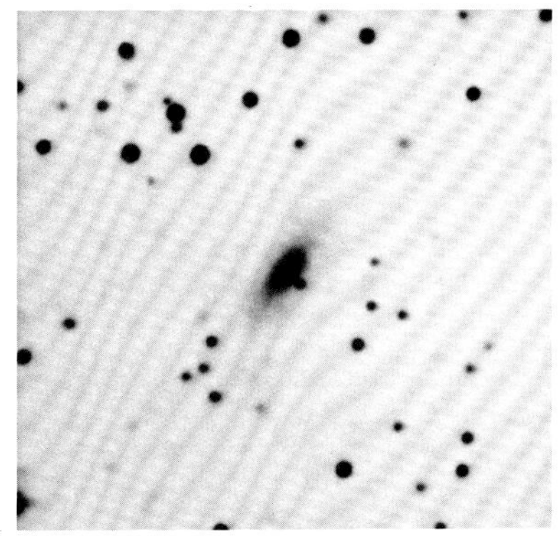

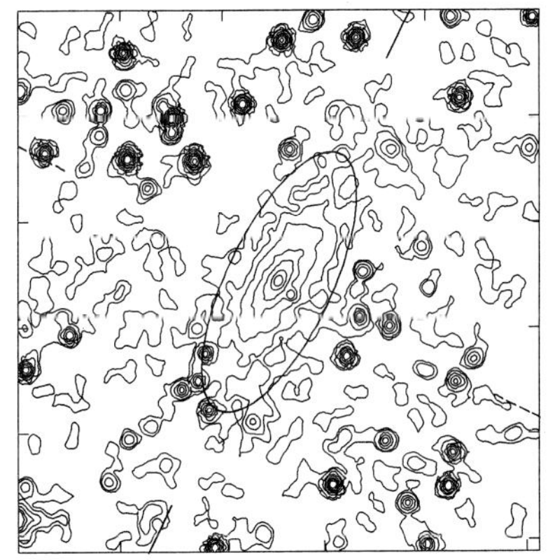

N3449 K4891 2 21.02 256X256

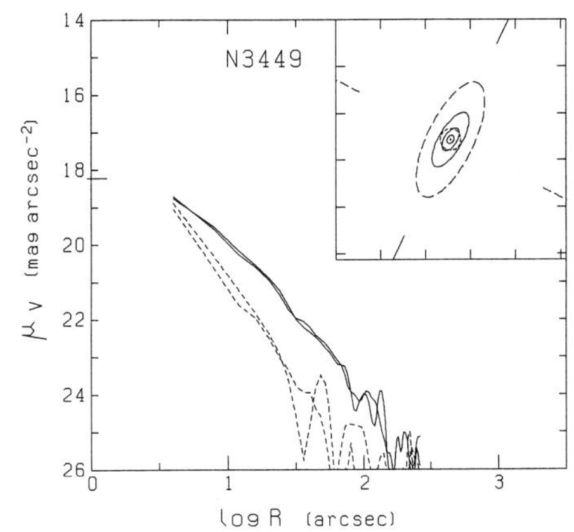

NGC3445

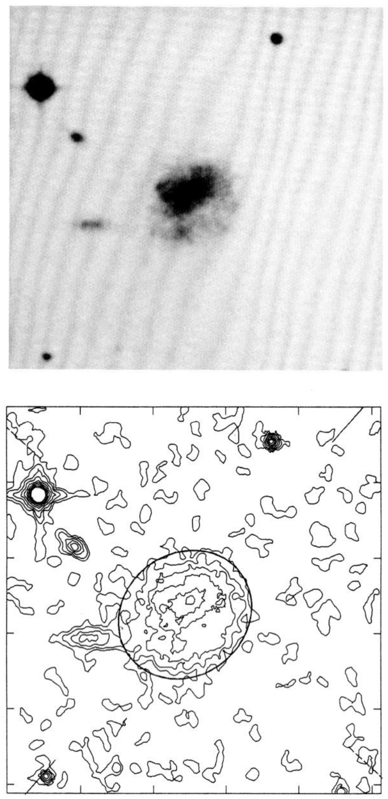

N3445 K4992 1 22.01 256X256

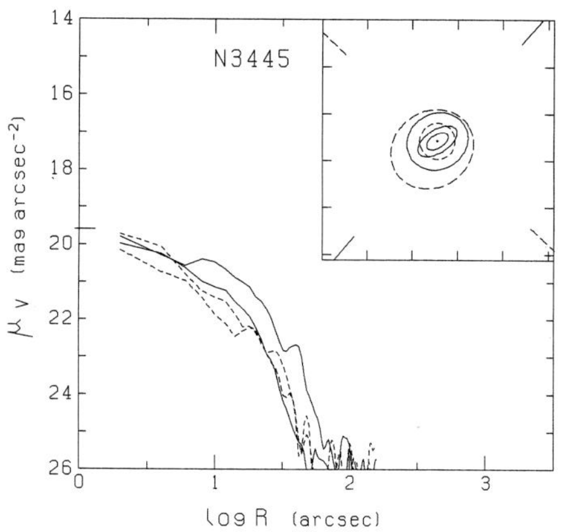

NGC3448

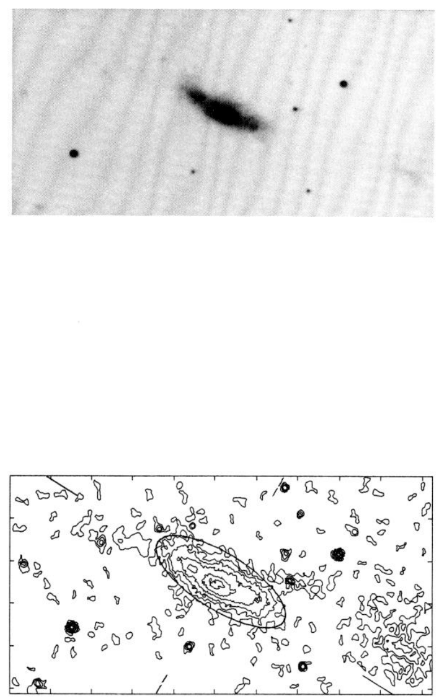

N3448 K4992 1 22.01 512X256

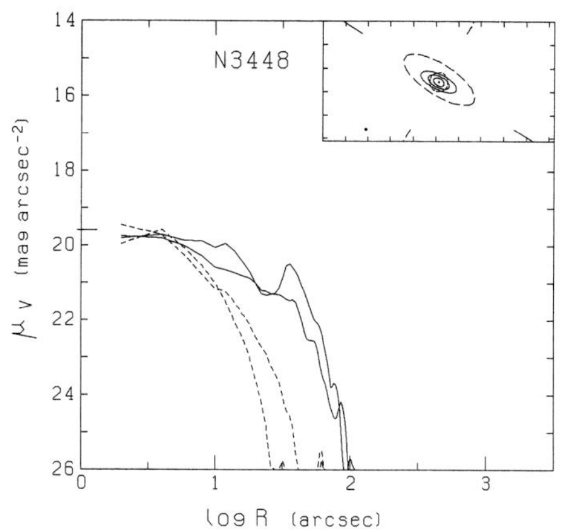

NGC3455

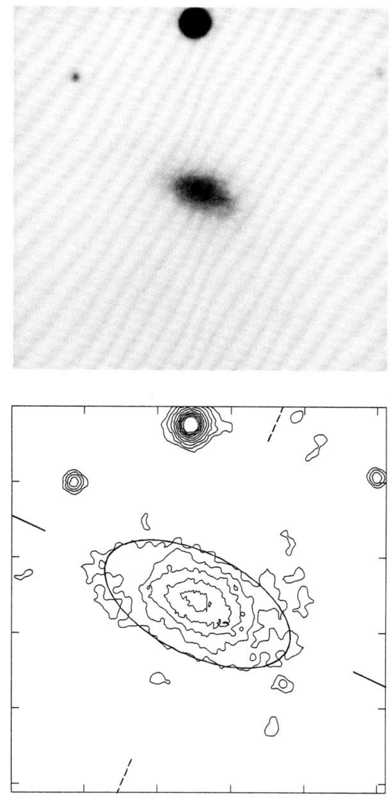

N3455 K5370 1 22.01 256X256

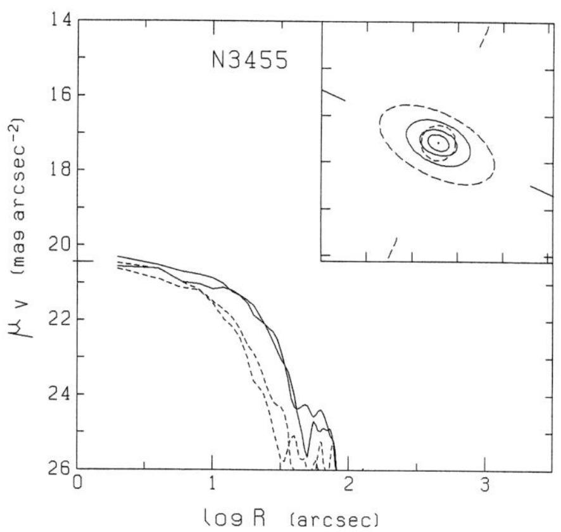

NGC3458

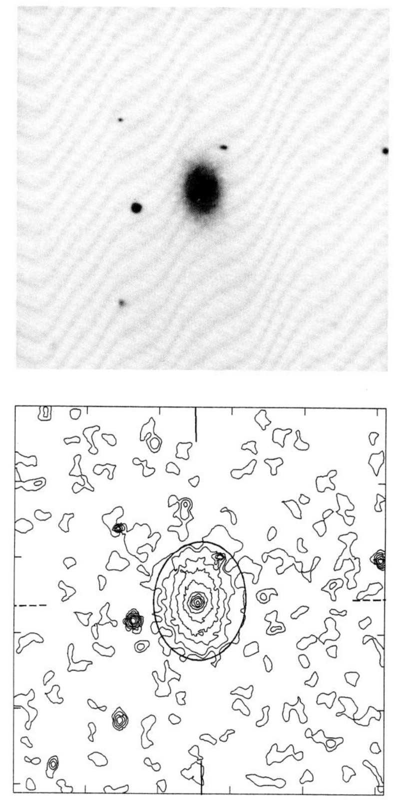

N3458 K4992 1 22.01 256X256

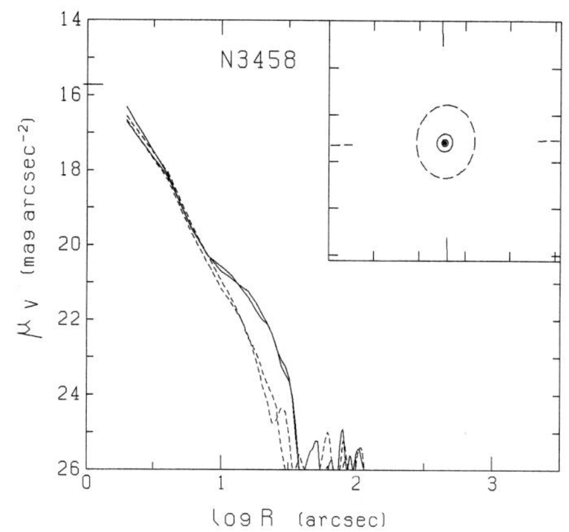

NGC3478

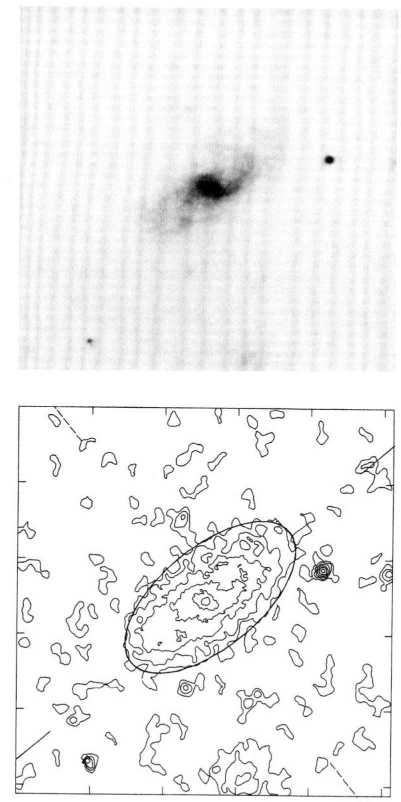

N3478 K4880 1 21.87 256X256

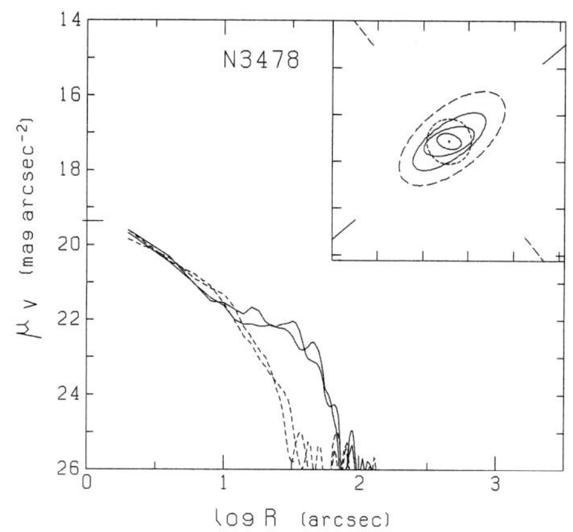

NGC3485

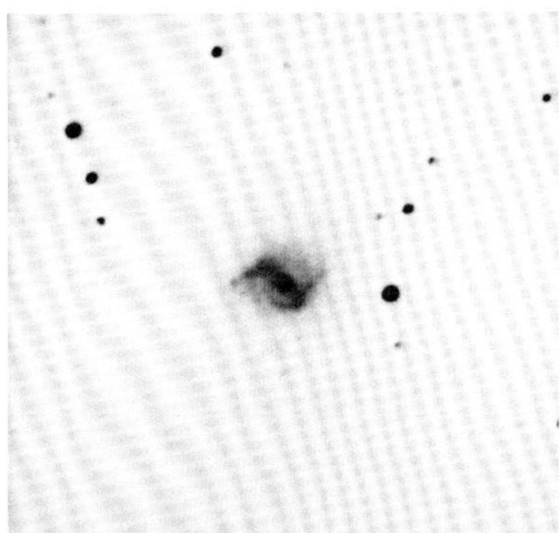

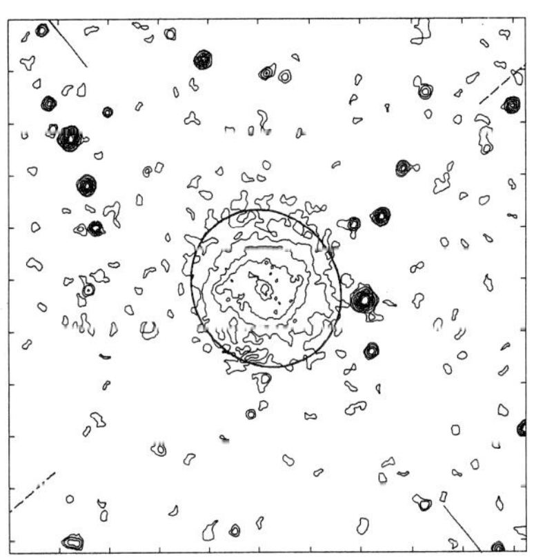

N3485 K3194 1 21.12 512X512

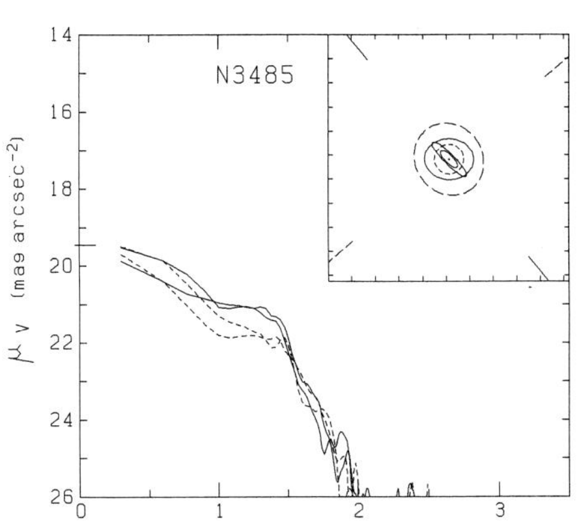

NGC3489

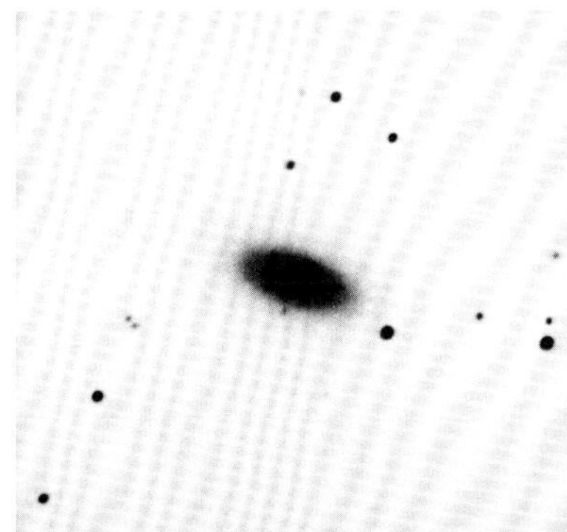

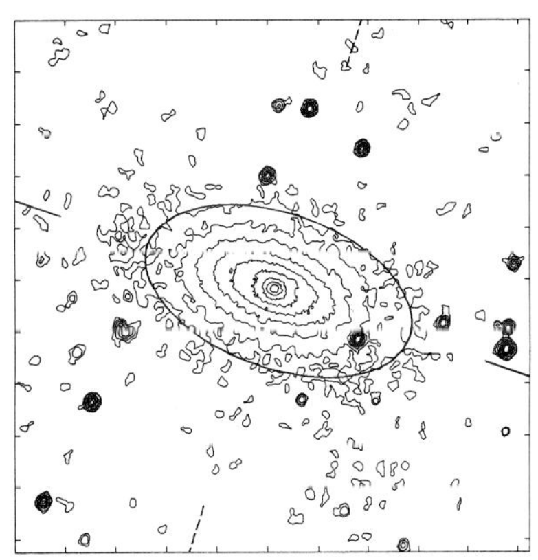

N3489 K3194 1 21.12 512X512

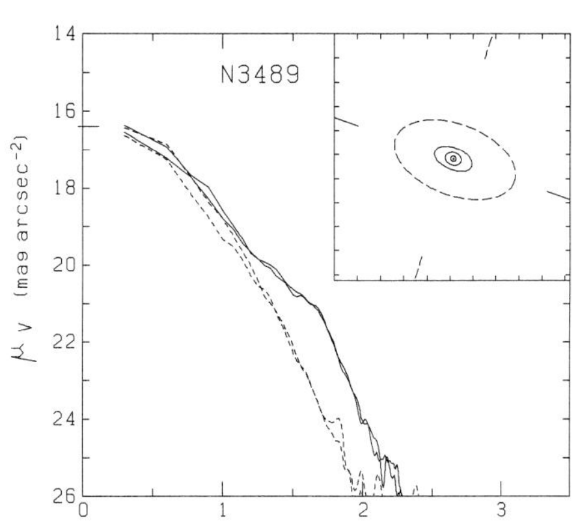

NGC3486

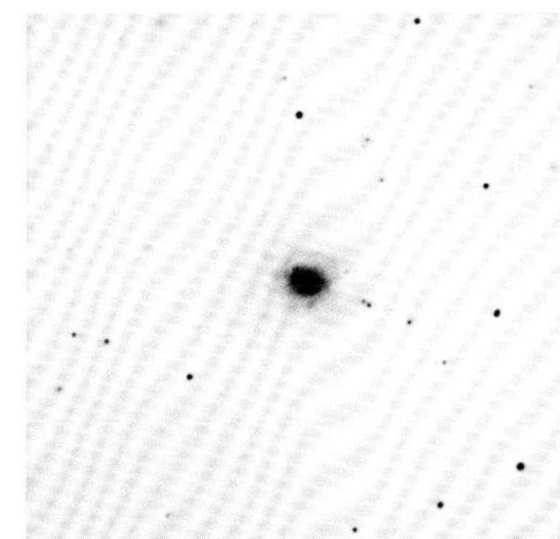

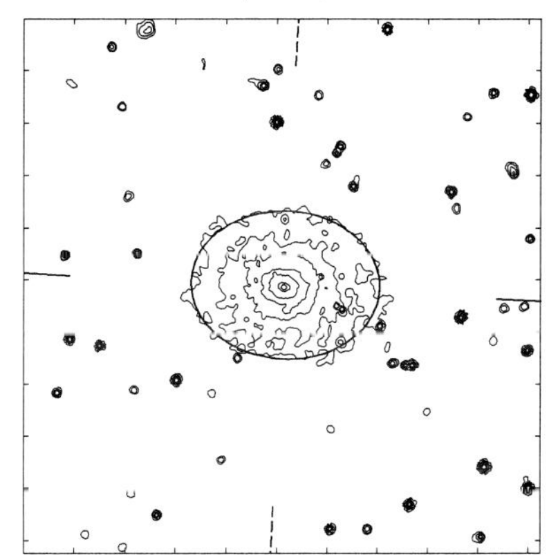

N3486 K5761 2 21.60 512X512

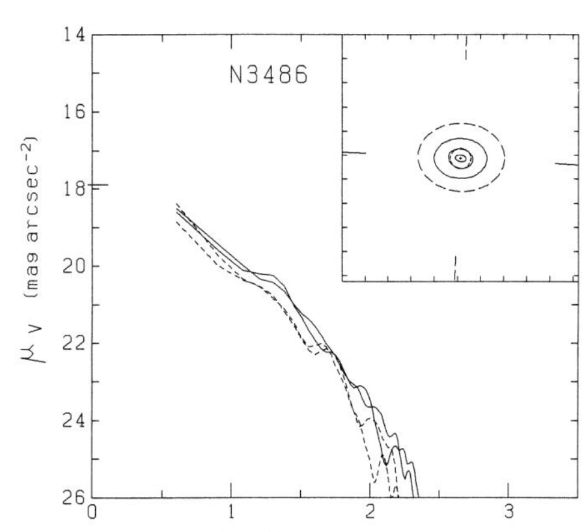

NGC3504

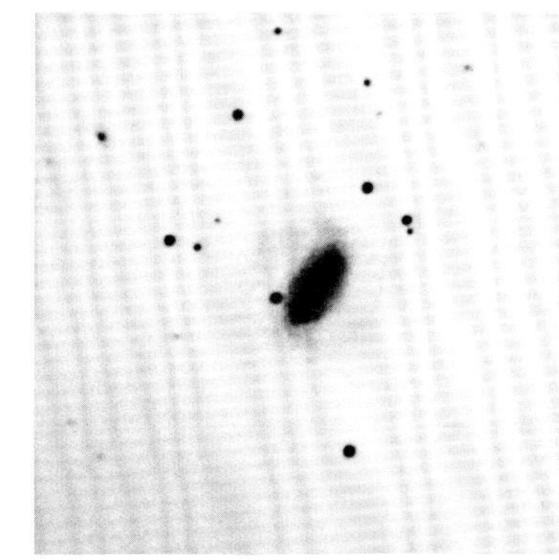

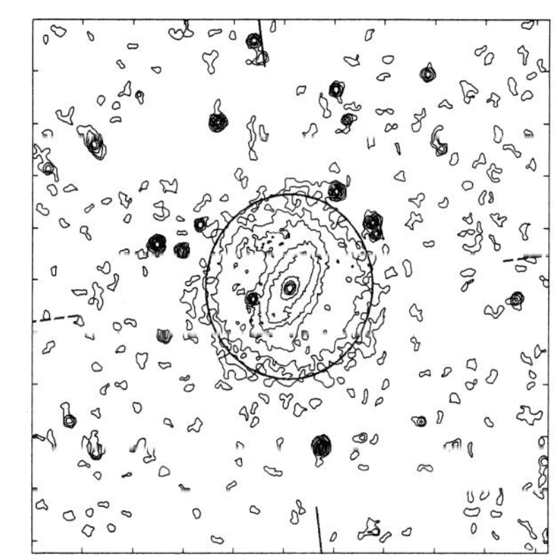

N3504 K5761 1 21.60 512X512

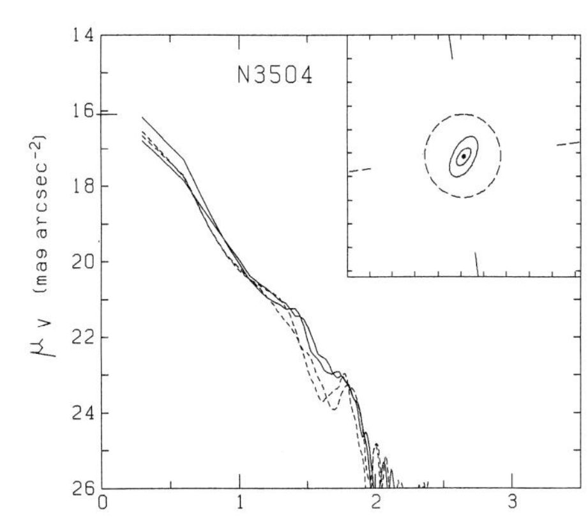

NGC3510

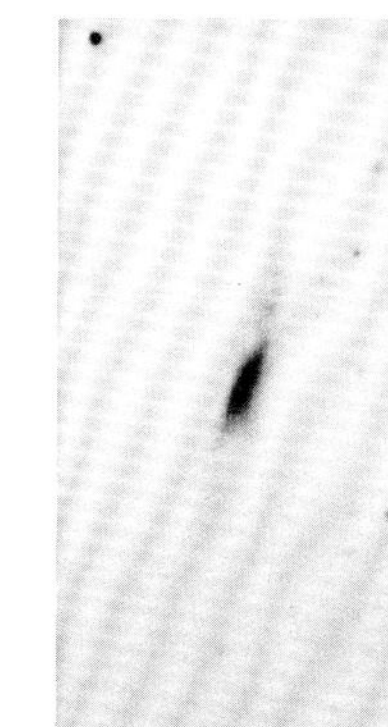

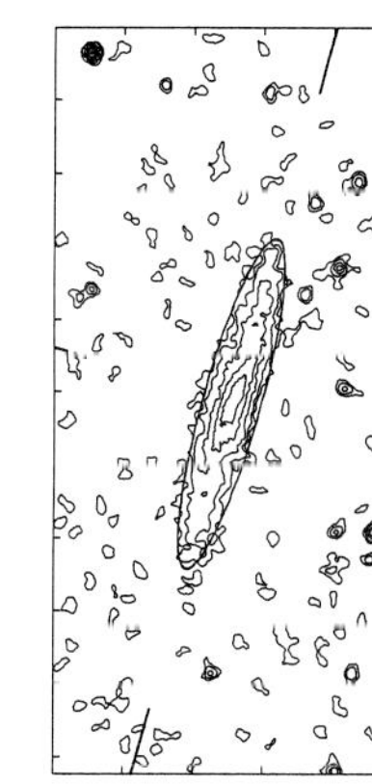

N3510 K5761 1 21.60 256X512

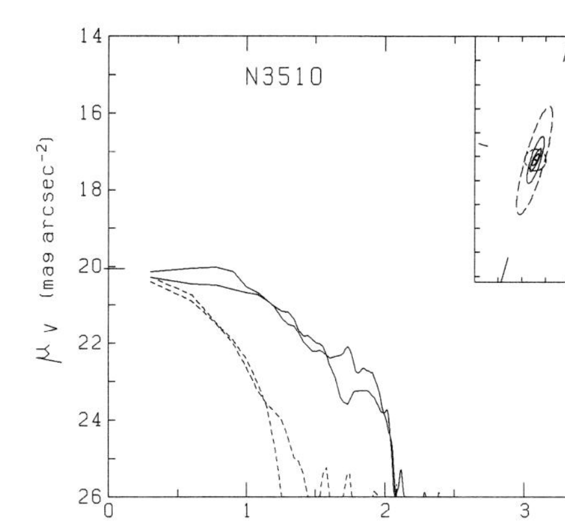

NGC3512

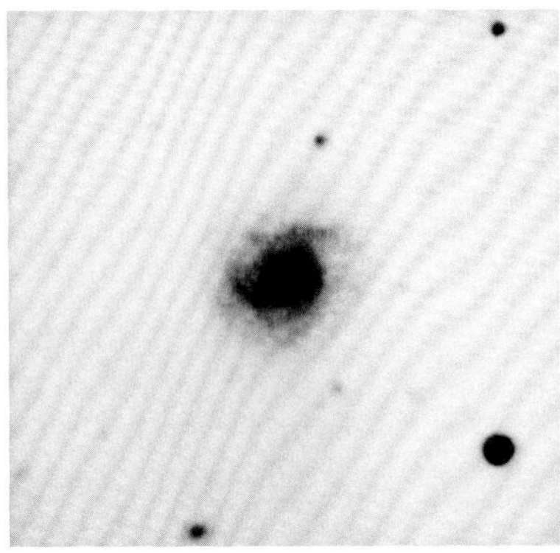

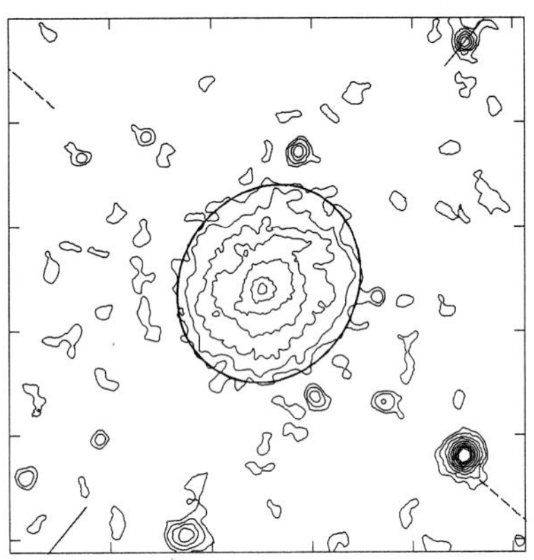

N3512 K5761 1 21.60 256X256

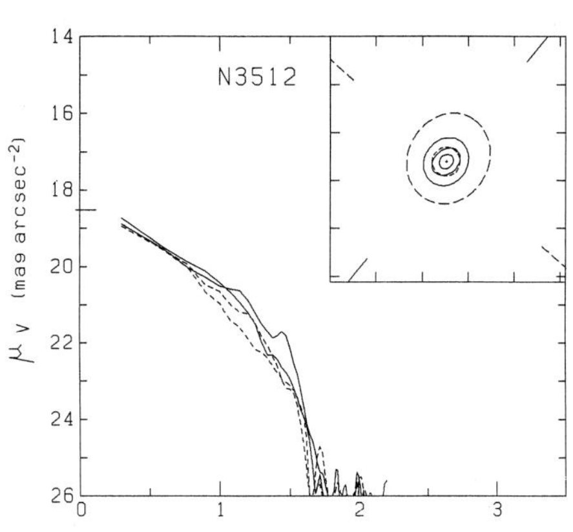

NGC3549

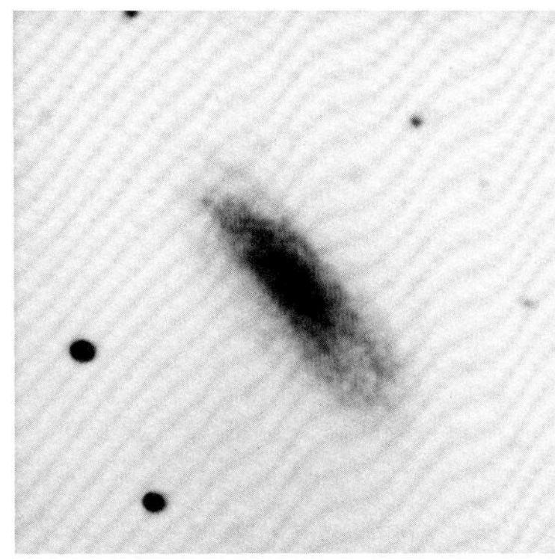

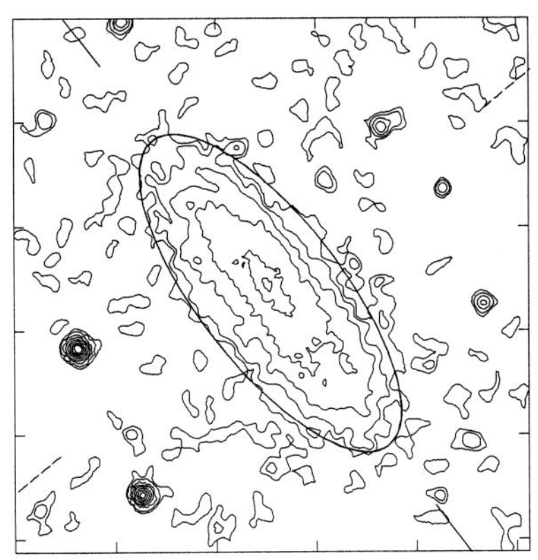

N3549 K4058 1 21.01 256X256

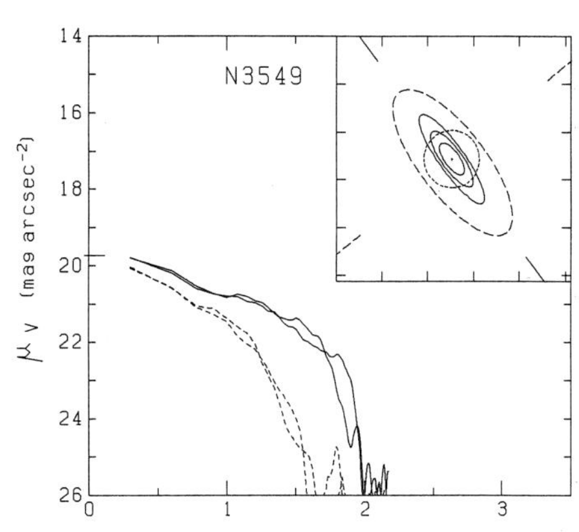

NGC3556

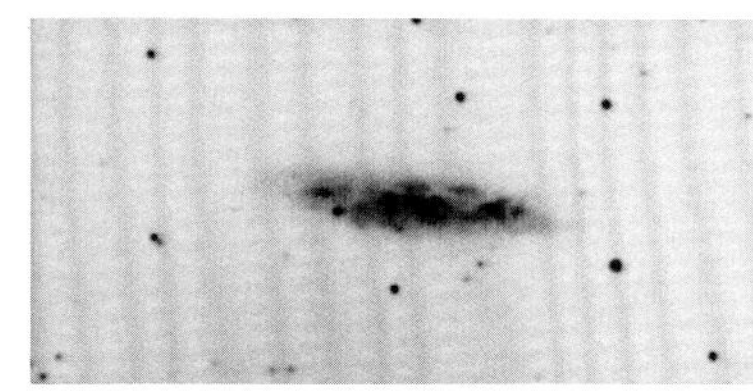

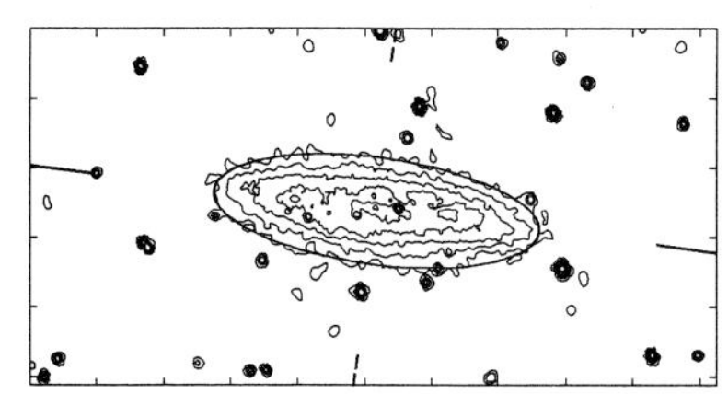

N3556 K4914 2 22.28 512X256

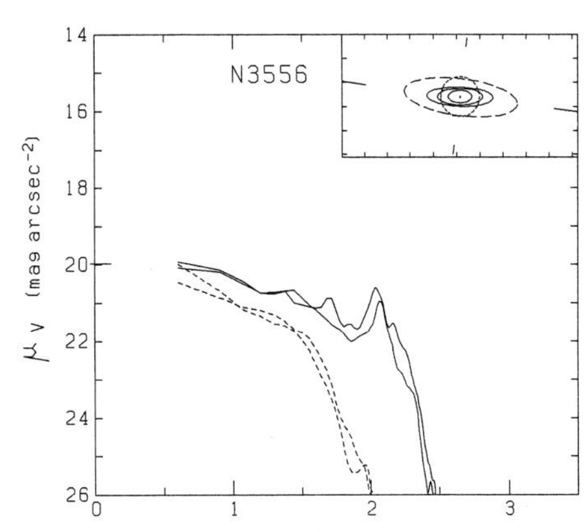

NGC3593

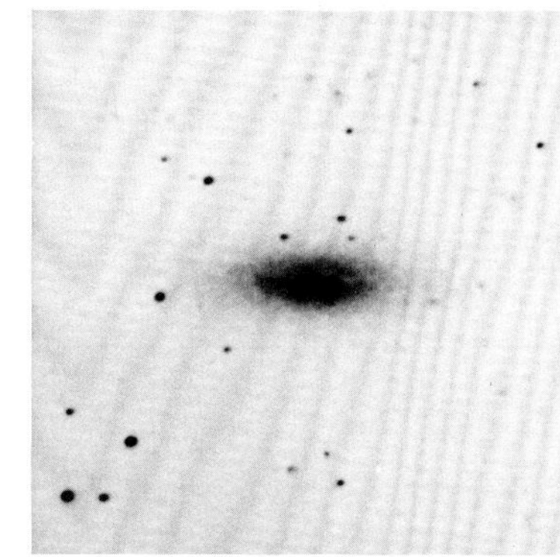

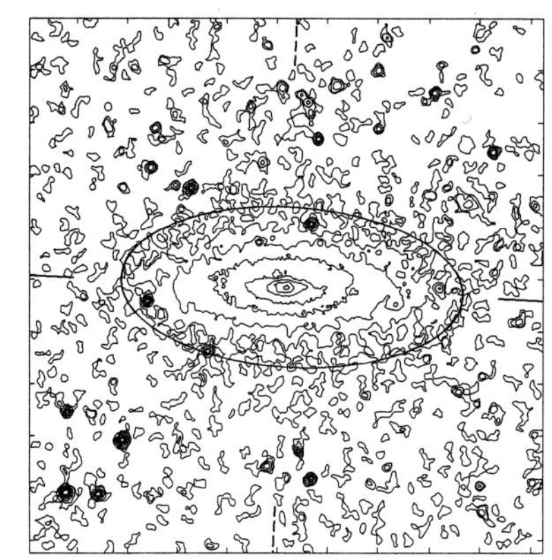

N3593 K3737 1 20.70 512X512

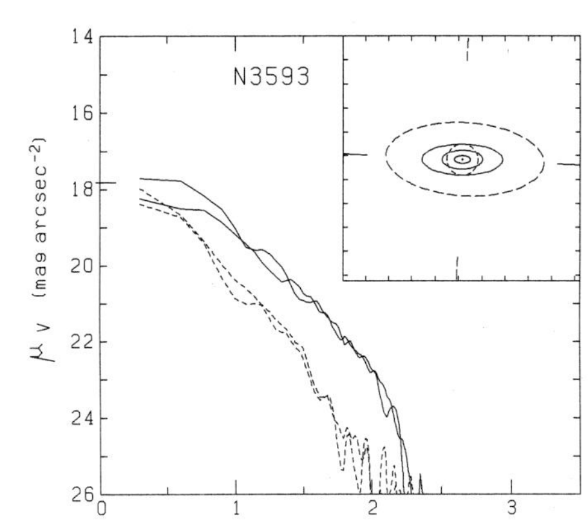

NGC3596

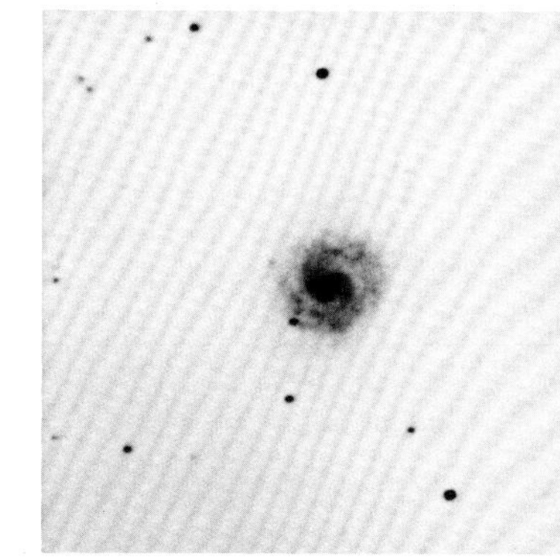

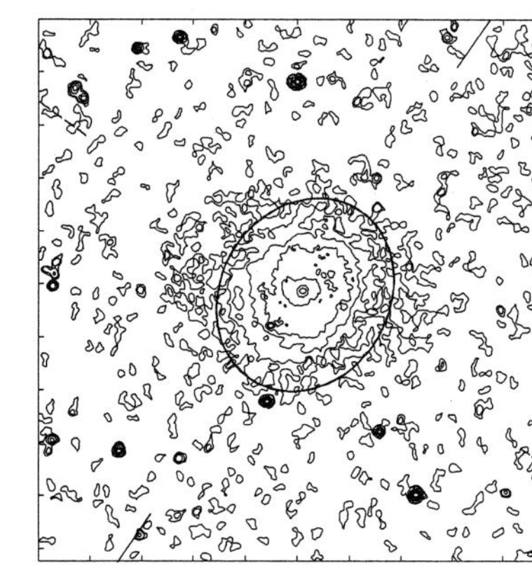

N3596 K3737 1 20.70 512X512

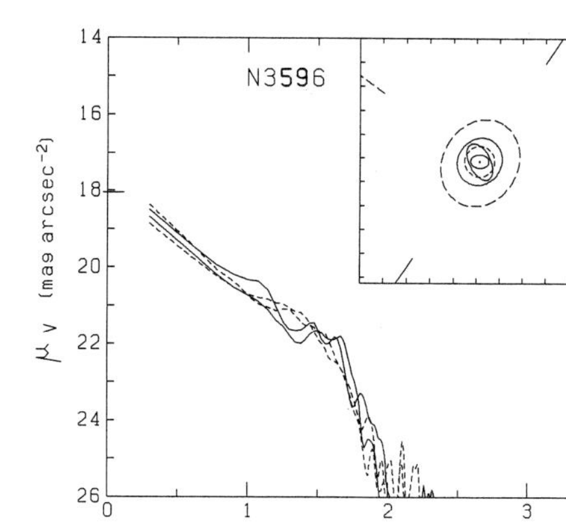

NGC3605

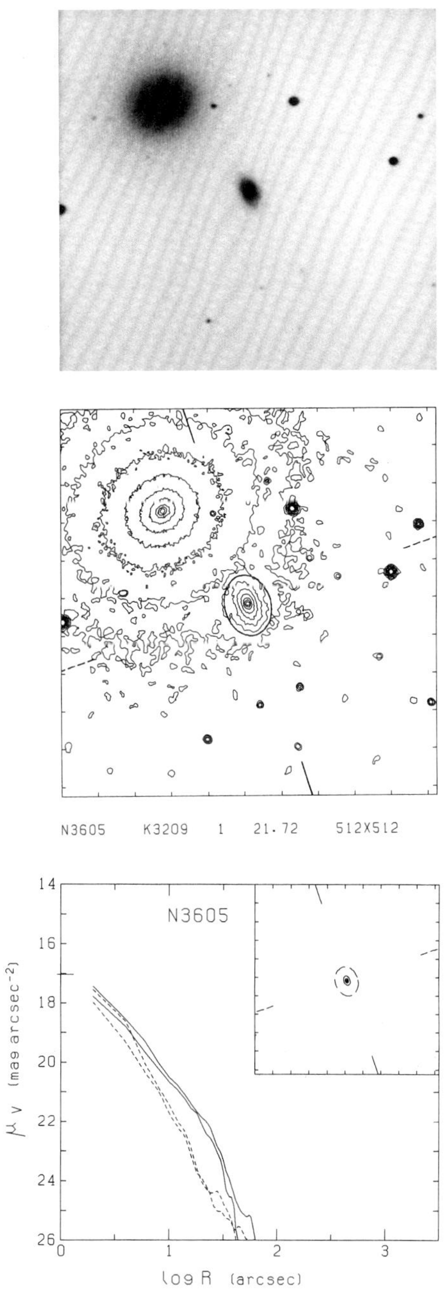

NGC3607

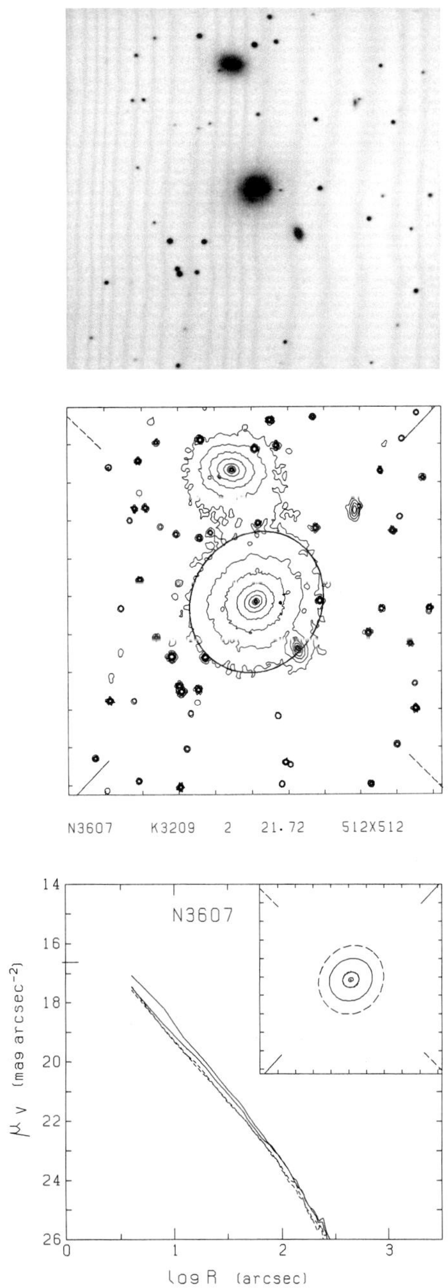

NGC3608

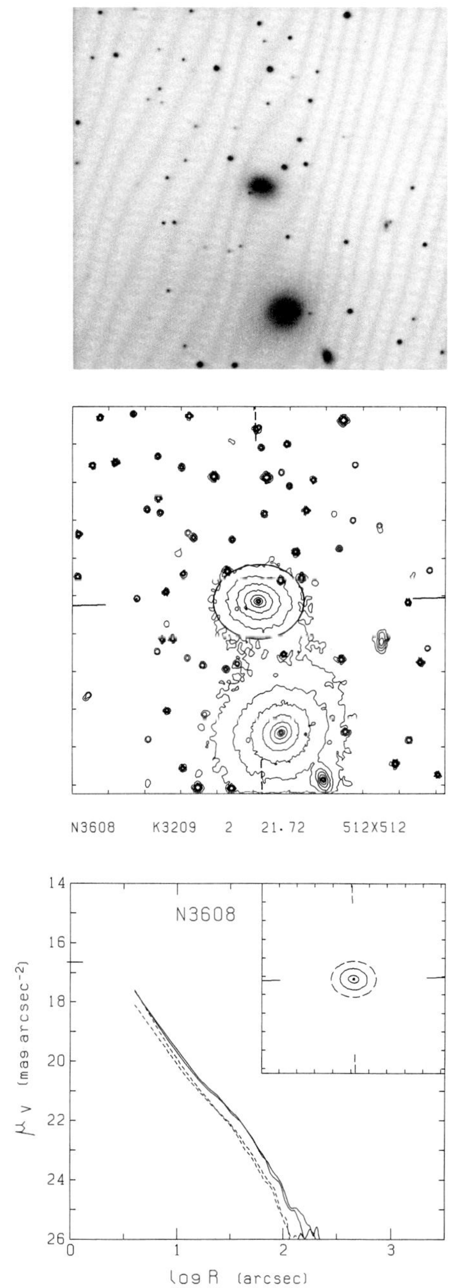

NGC3611

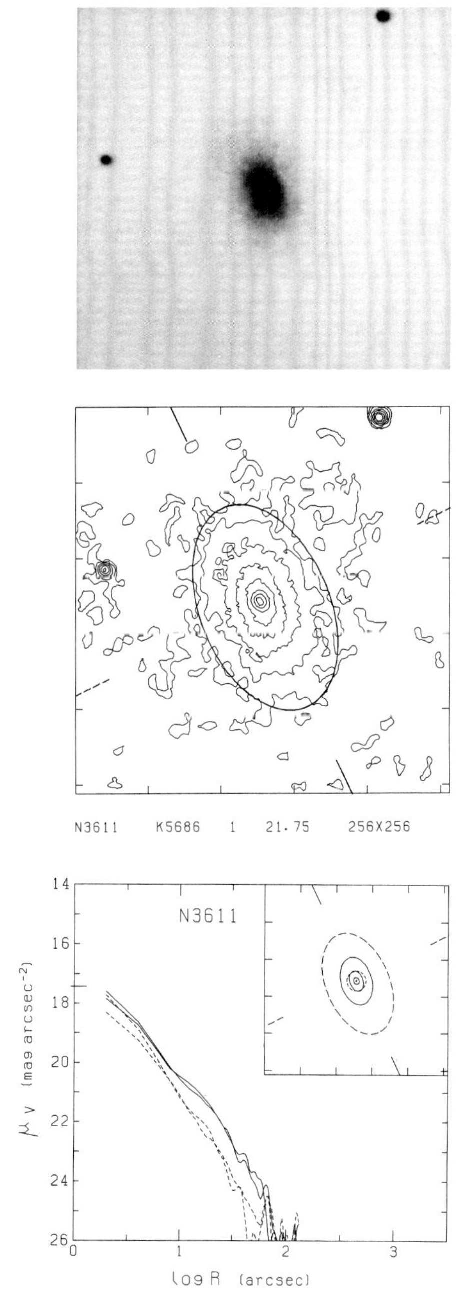

NGC3610

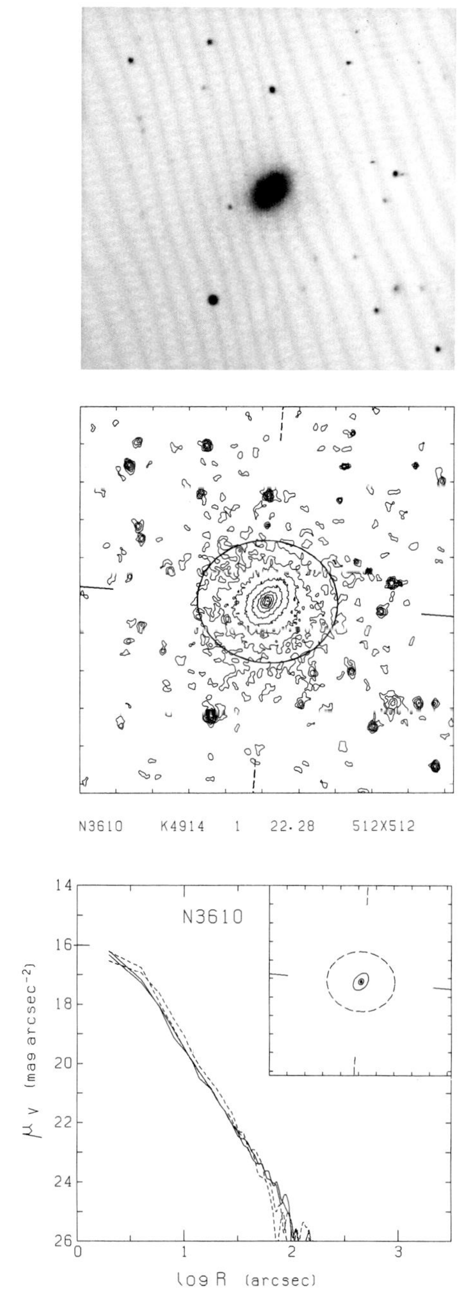

NGC3613

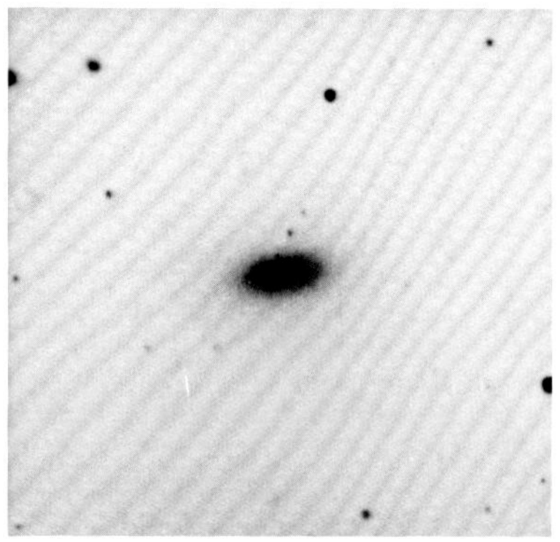

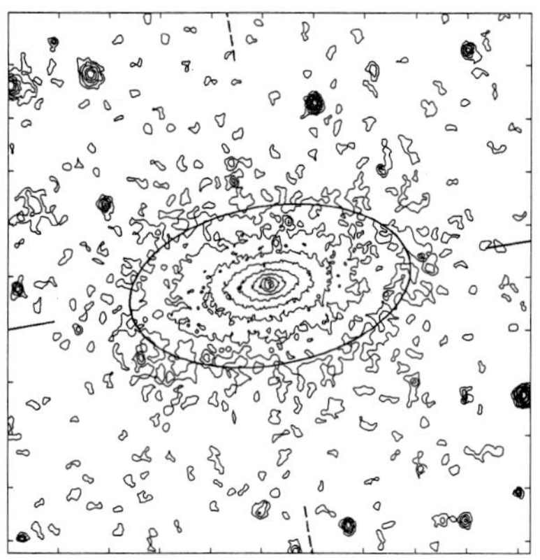

N3613 K4914 1 22.28 512X512

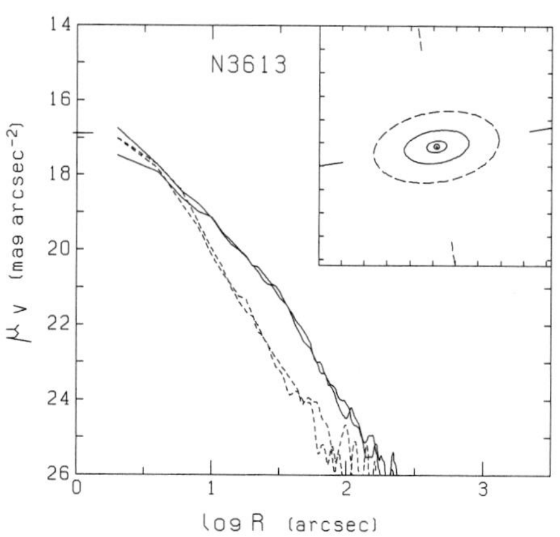

NGC3623

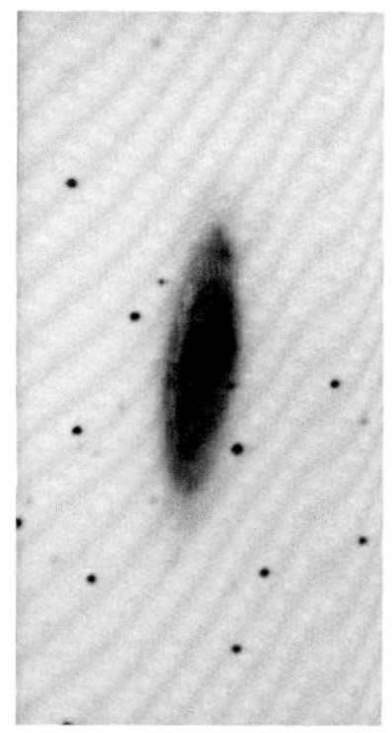

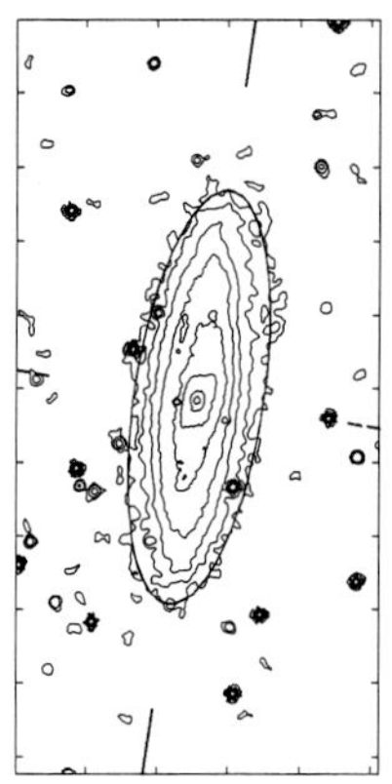

N3623 K3737 2 20.70 256X512

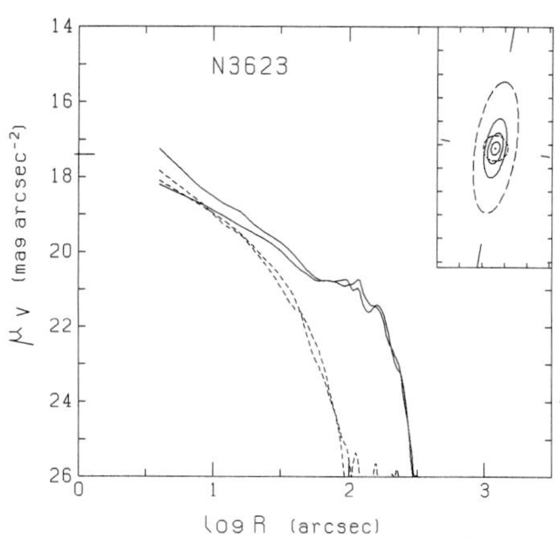

NGC3619

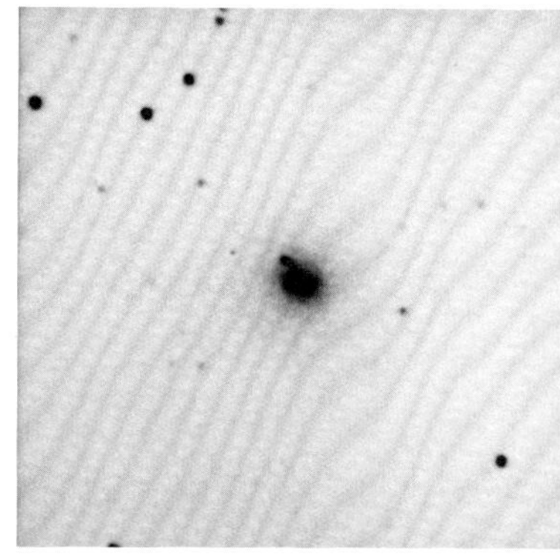

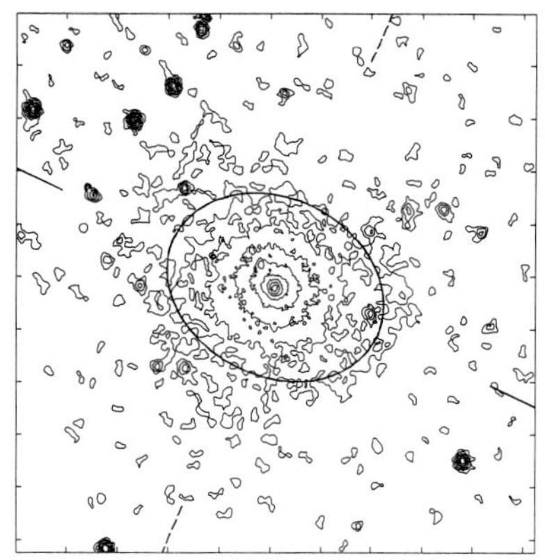

N3619 K4914 1 22.28 512X512

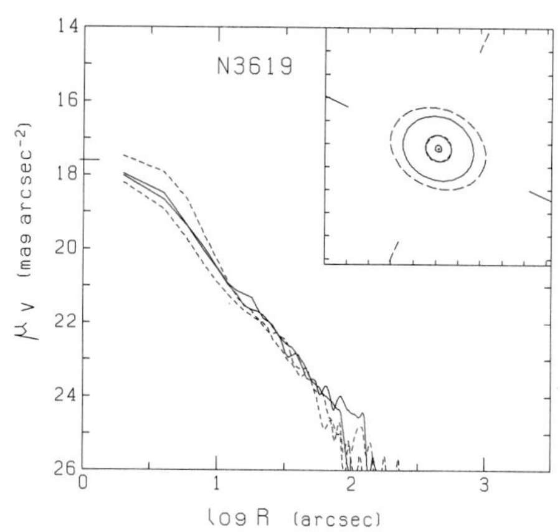

NGC3626

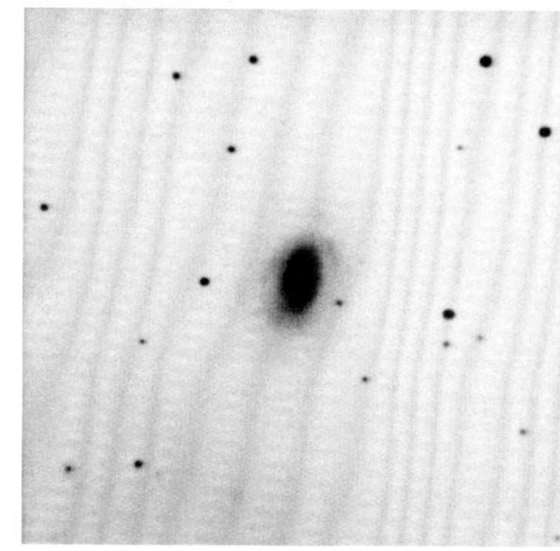

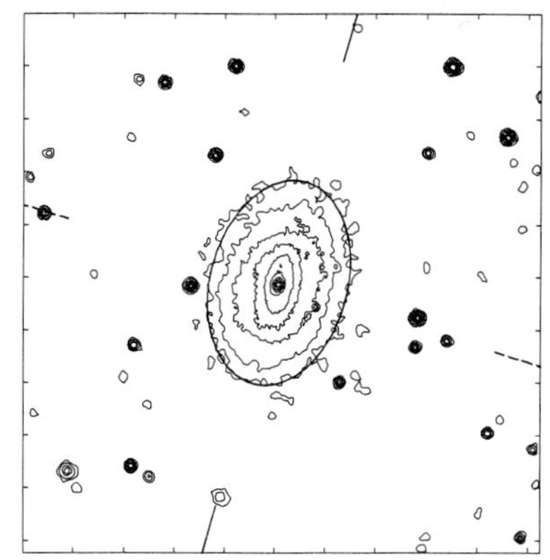

N3626 K3209 1 21.72 512X512

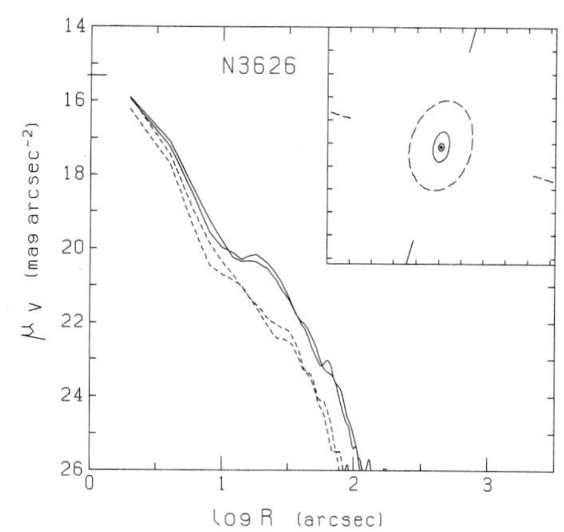

NGC3627

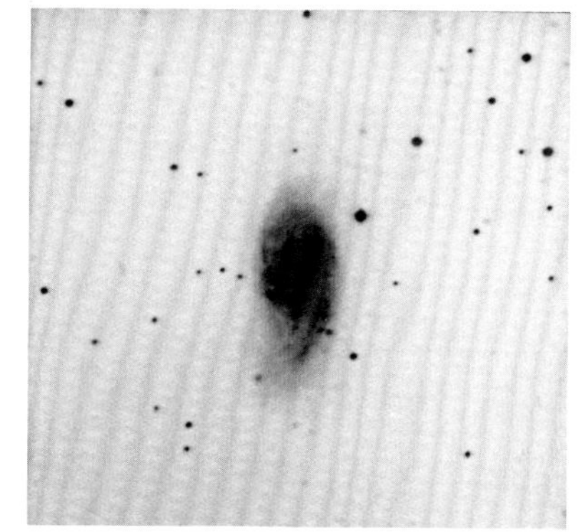

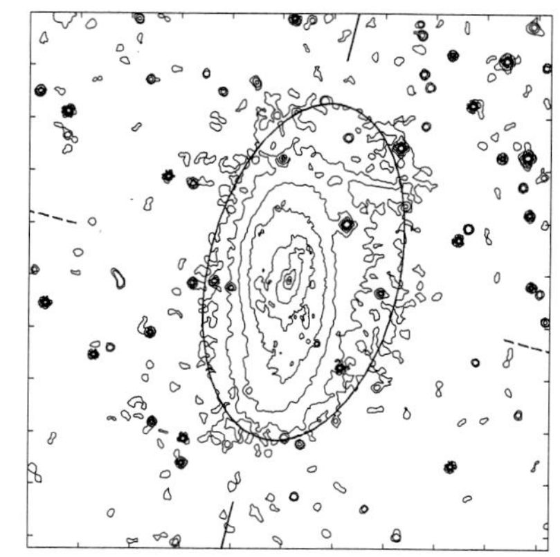

N3627 K3737 2 20.70 512X512

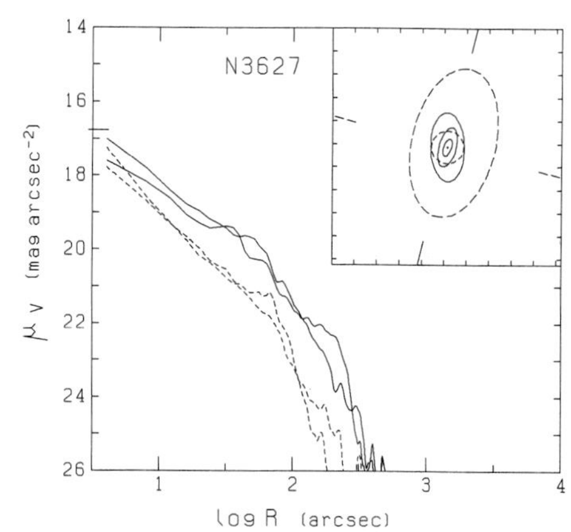

NGC3628

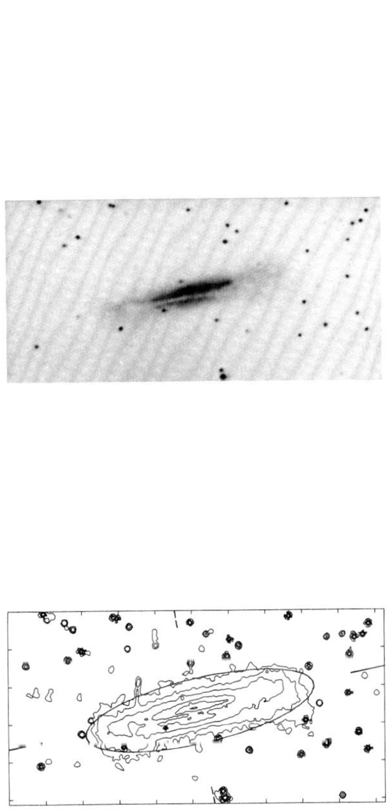

N3628 K3737 3 20.70 512X256

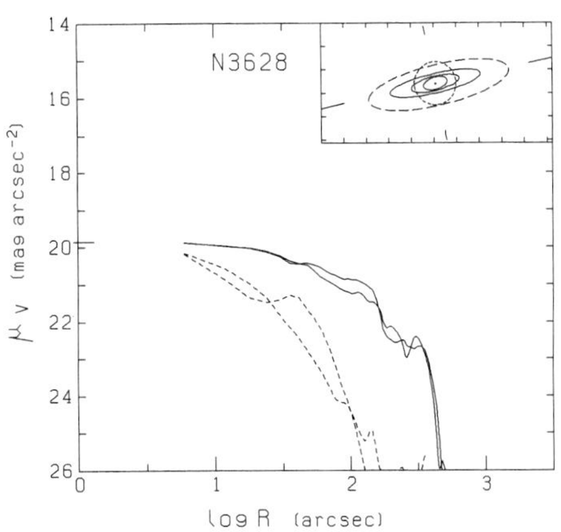

NGC3630

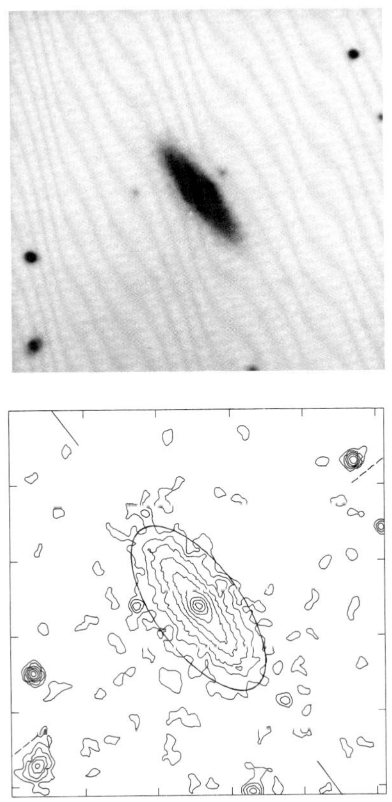

N3630 K5686 1 21.75 256X256

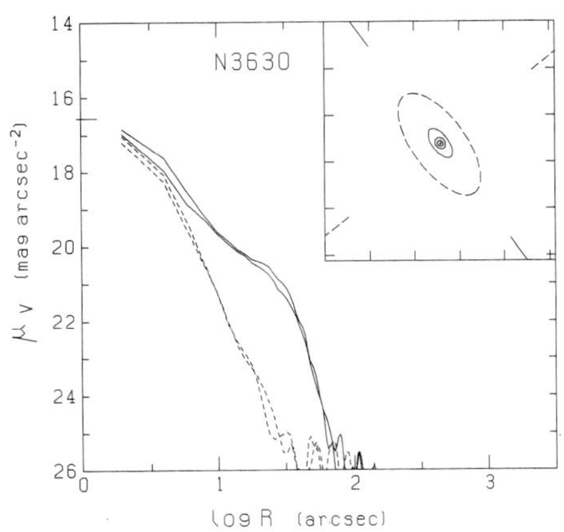

NGC3631

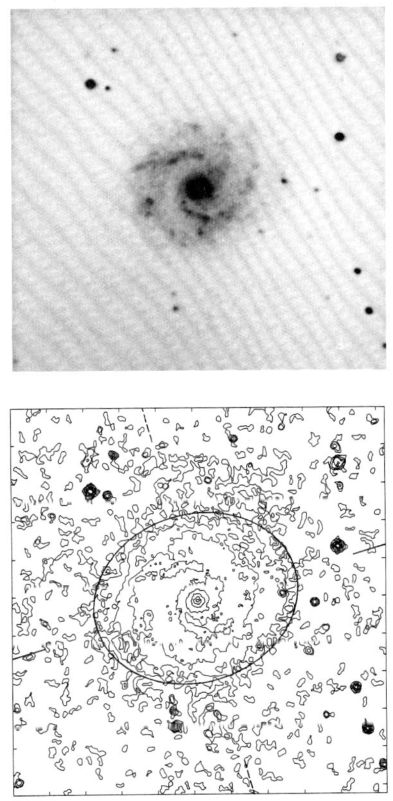

N3631 K4058 1 21.01 512X512

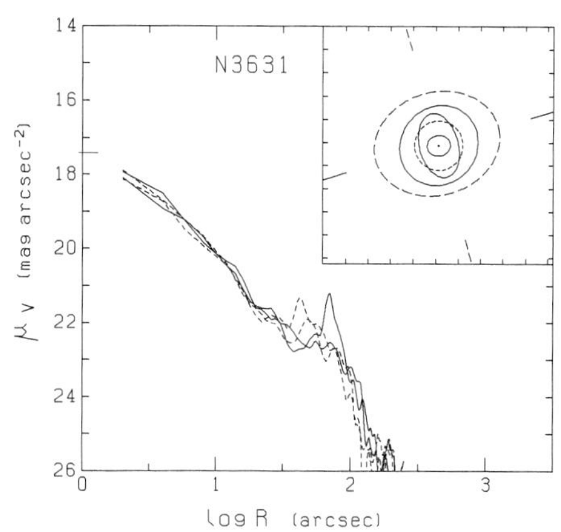

NGC3640

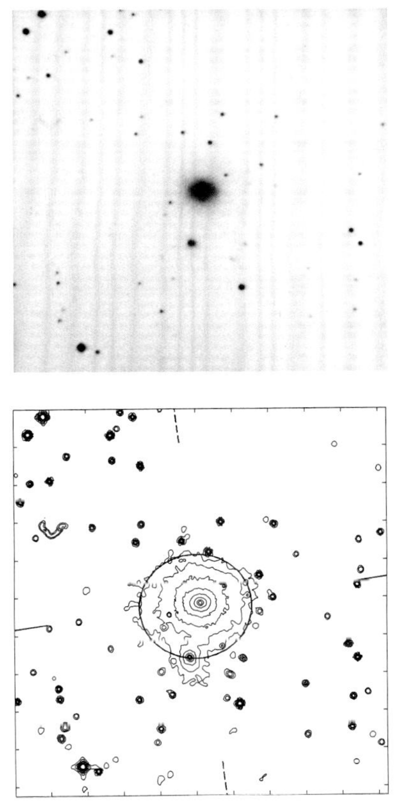

N3640 K5686 2 21.75 512X512

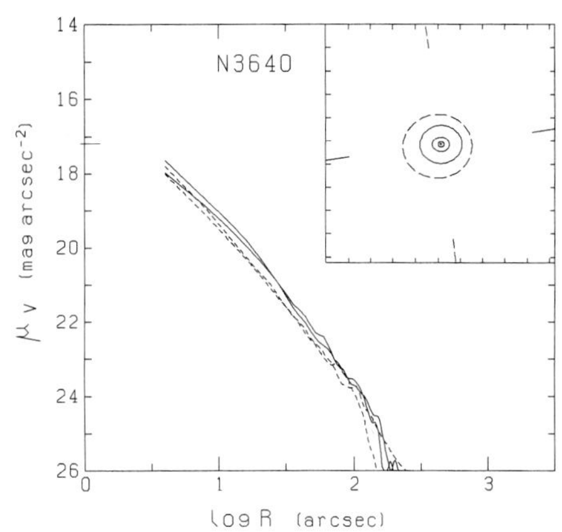

NGC3646

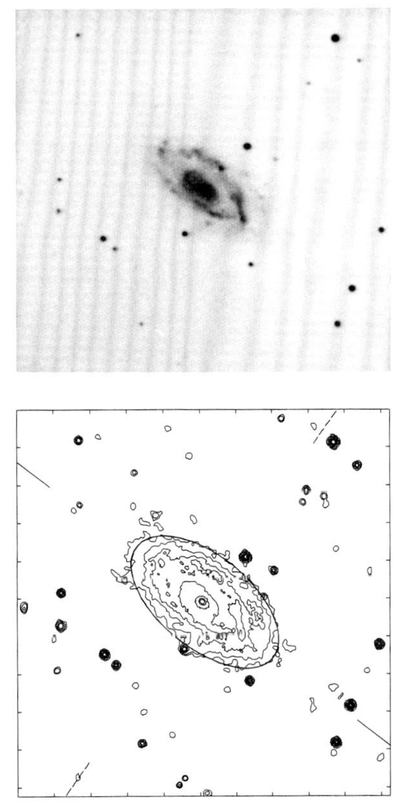

N3646 K3209 1 21.72 512X512

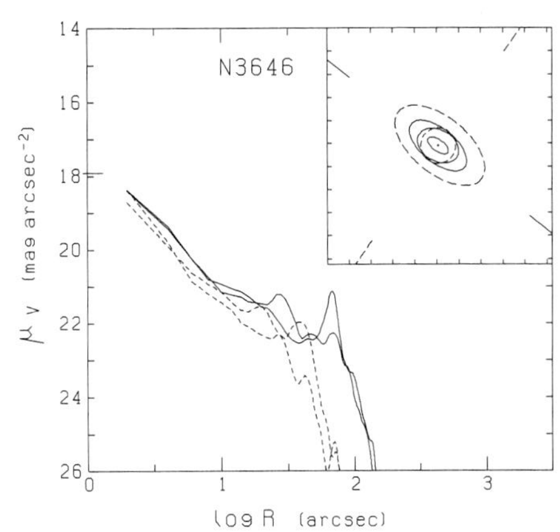

NGC3642

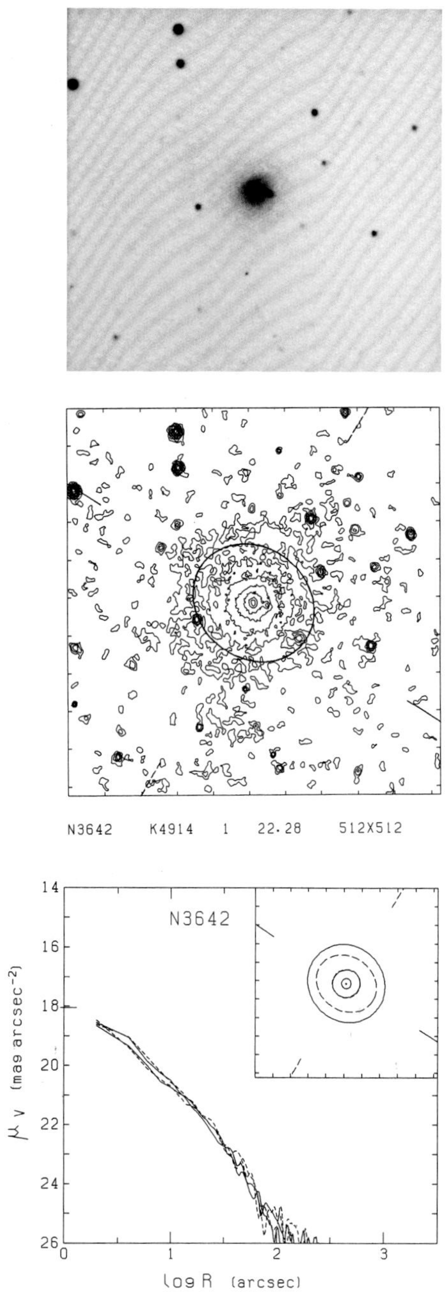

NGC3655

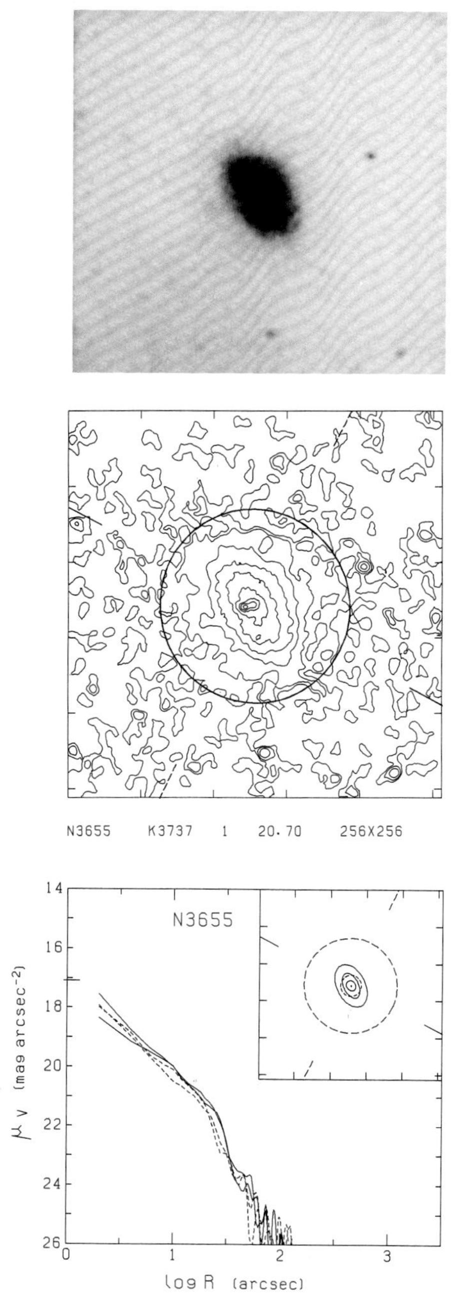

NGC3659

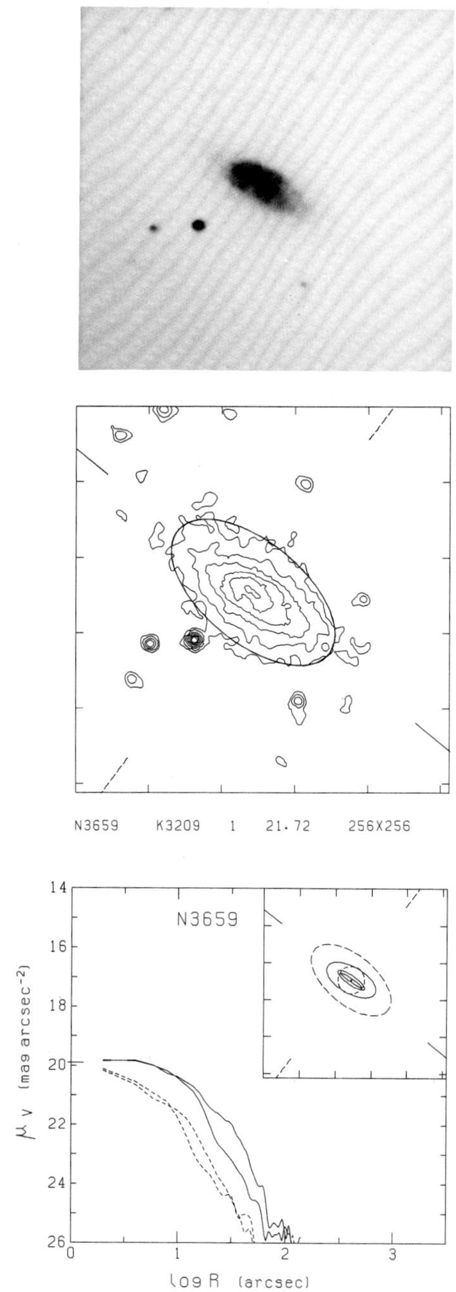

NGC3664

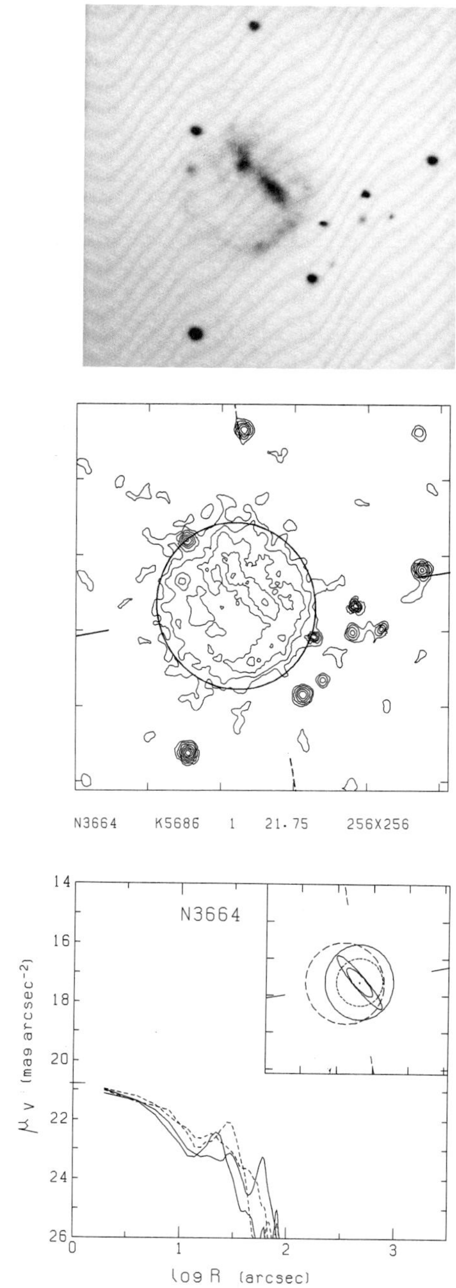

NGC3665

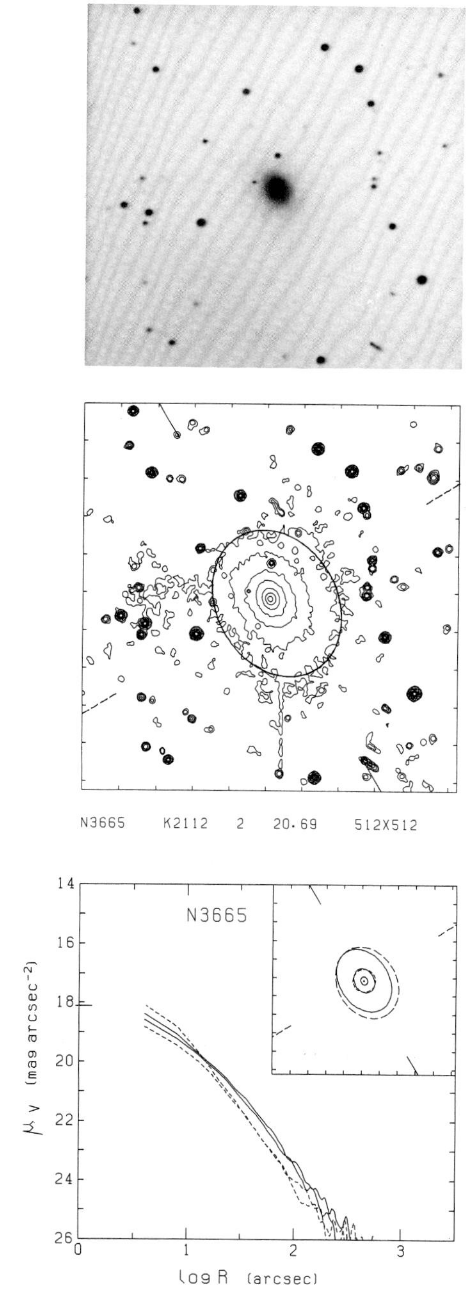

NGC3672

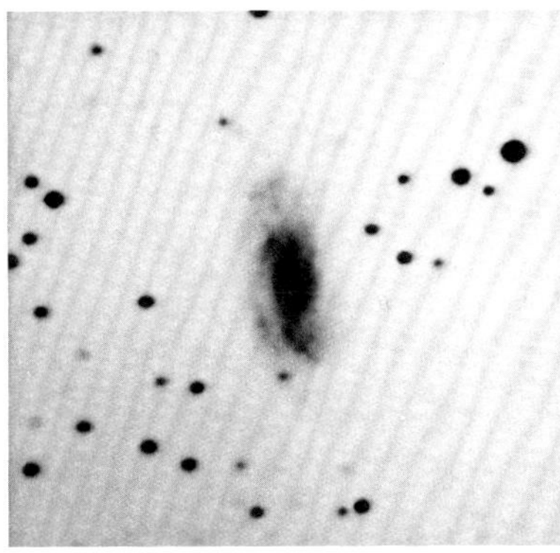

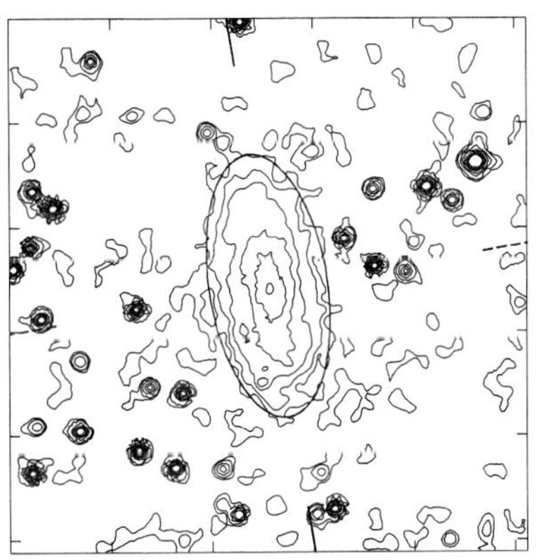

N3672 K5062 2 21.01 256X256

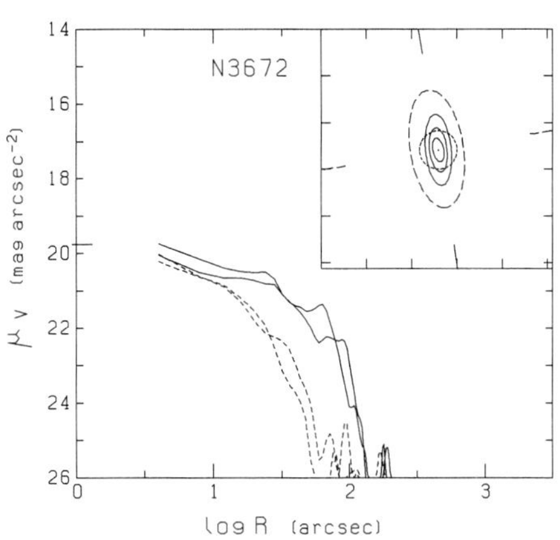

NGC3681

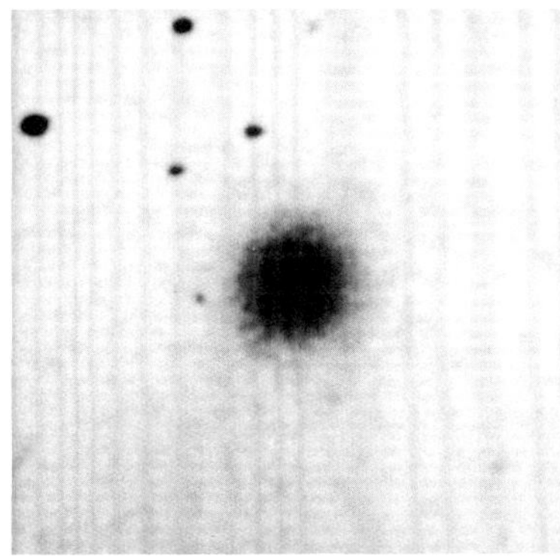

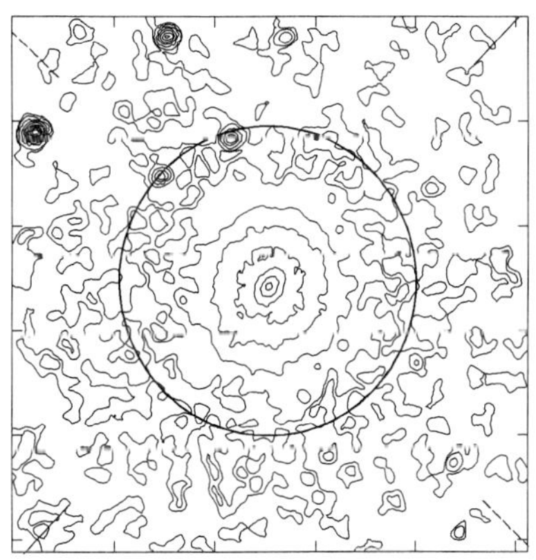

N3681 K3737 1 20.70 256X256

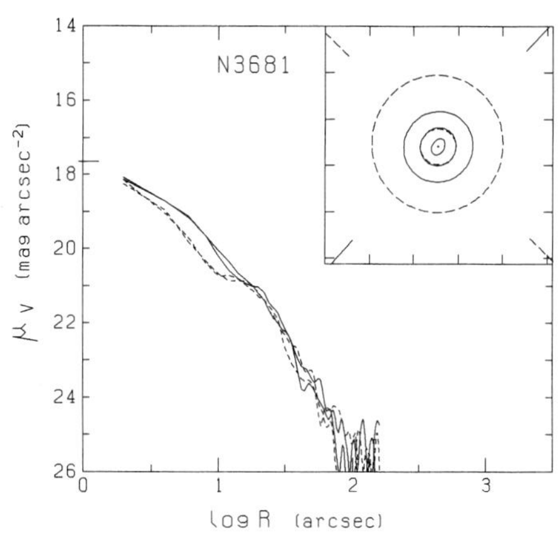

NGC3684

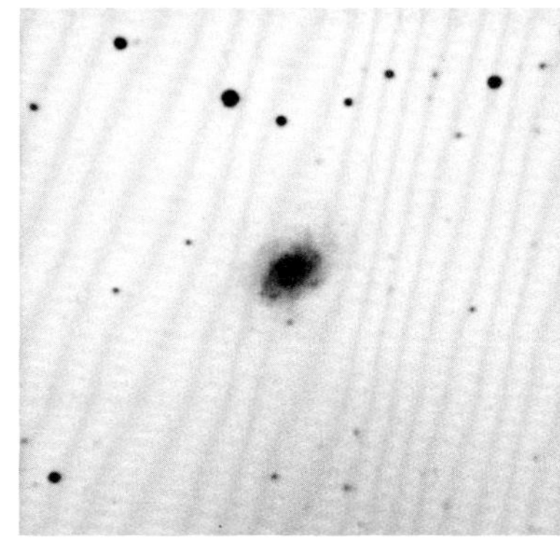

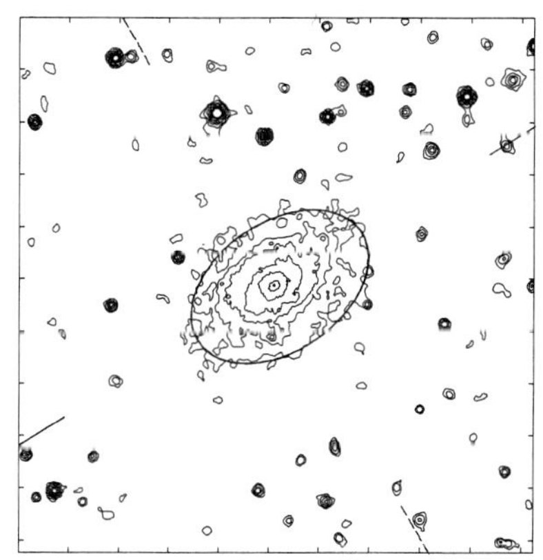

N3684 K3209 1 21.72 512X512

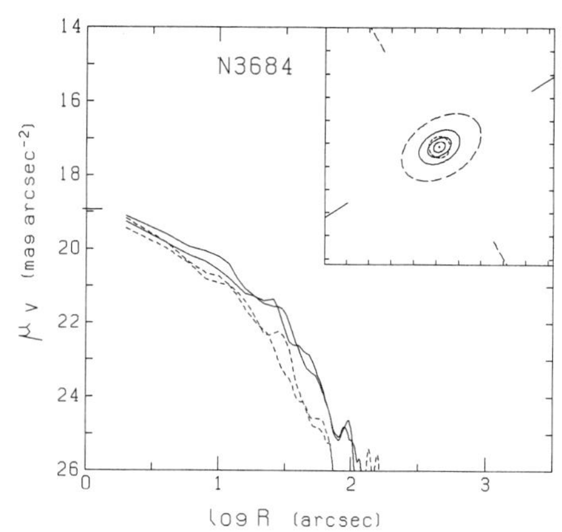

NGC3683

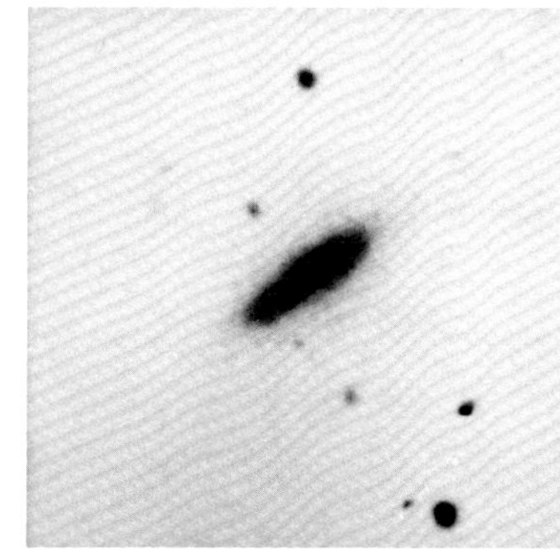

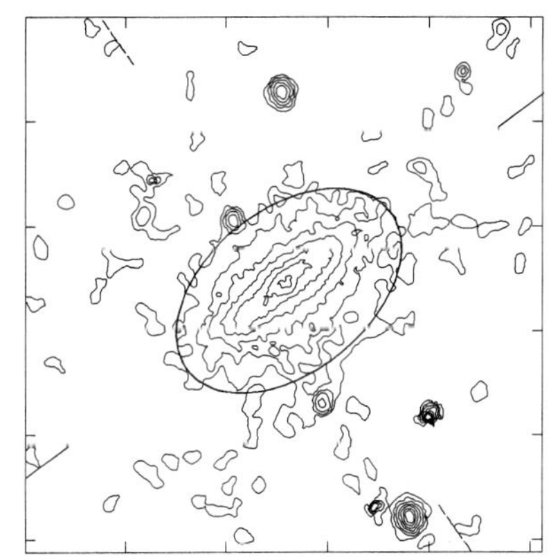

N3683 K4914 1 22.28 256X256

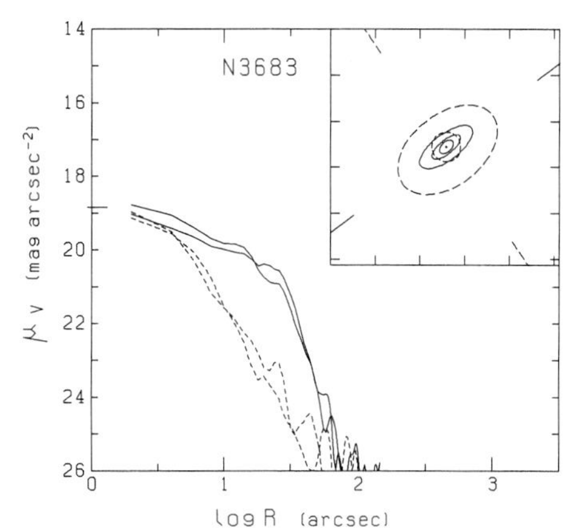

NGC3686

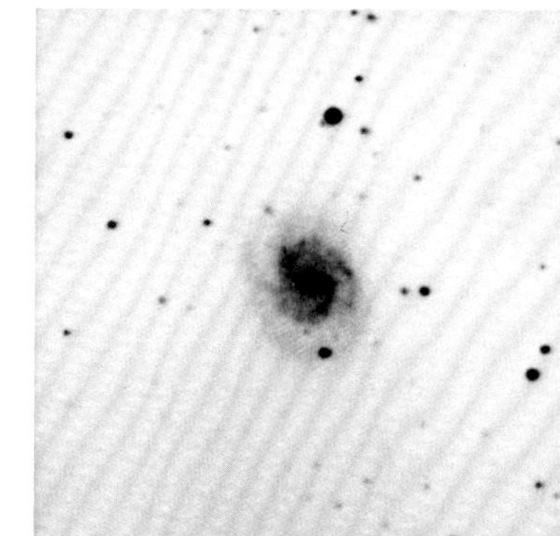

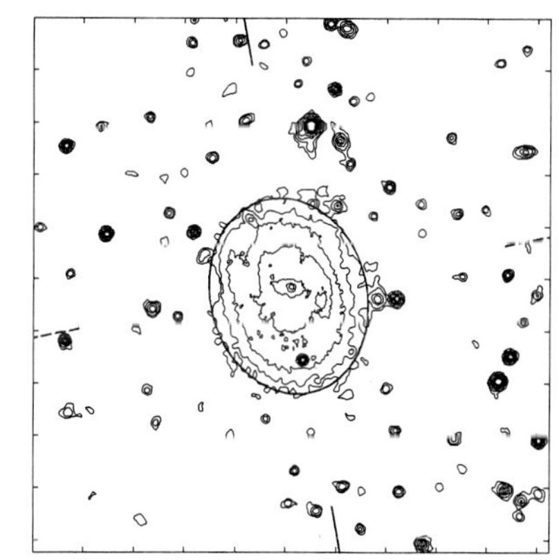

N3686 K3209 1 21.72 512X512

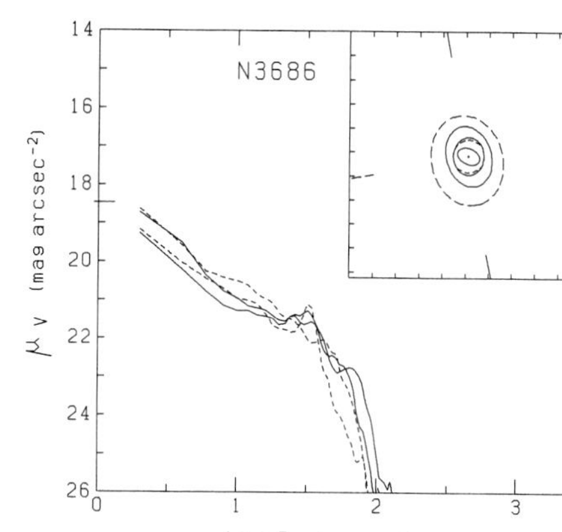

NGC3691

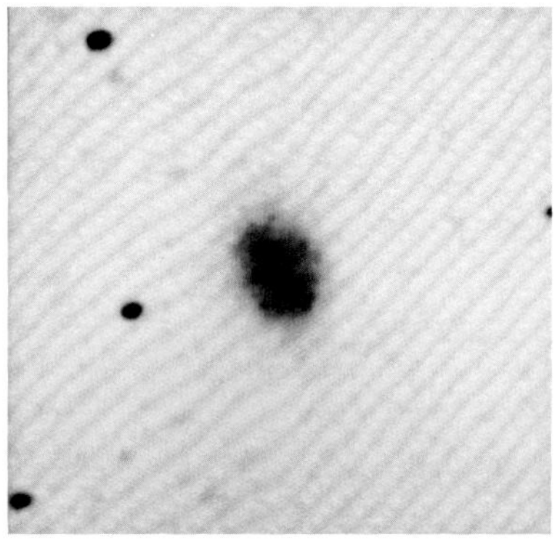

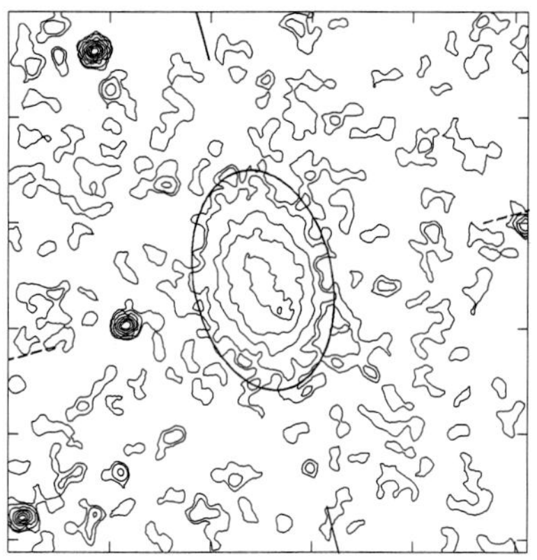

N3691 K3737 1 20.70 256X256

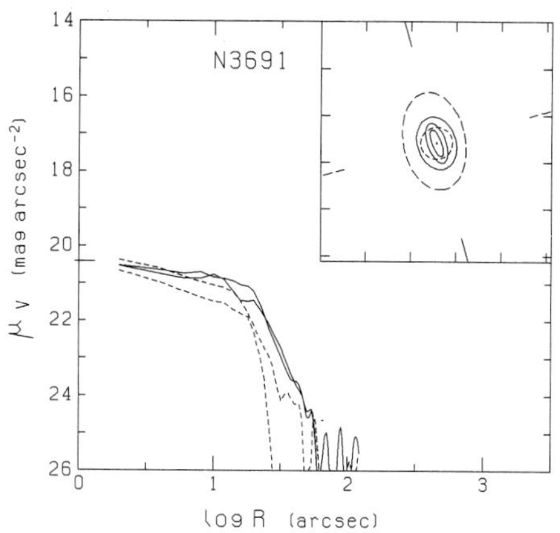

NGC3690

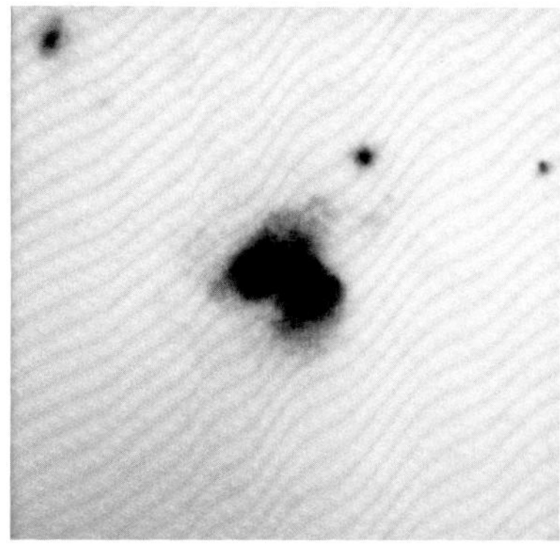

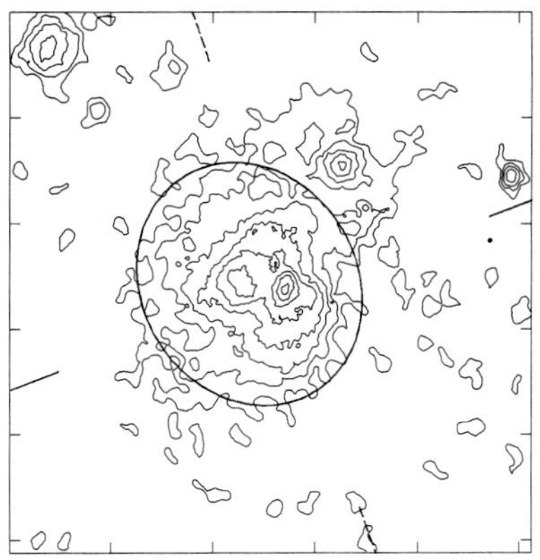

N3690 K4914 1 22.28 256X256

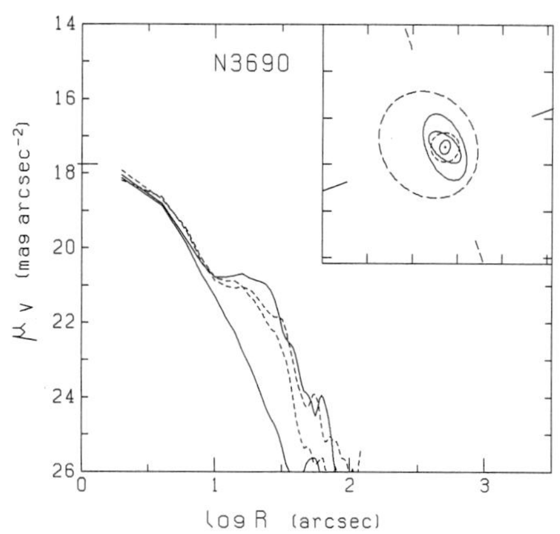

NGC3705

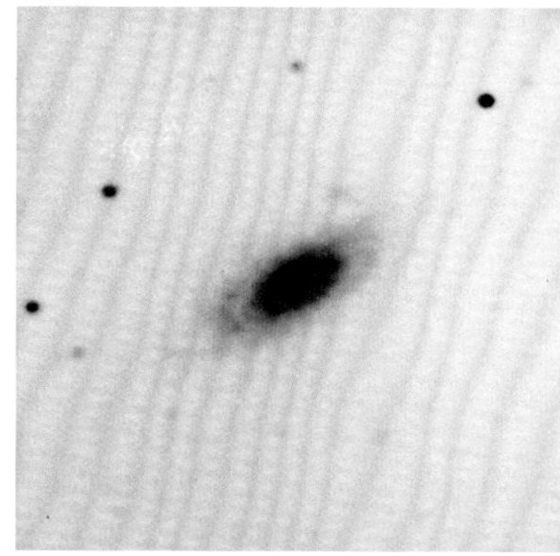

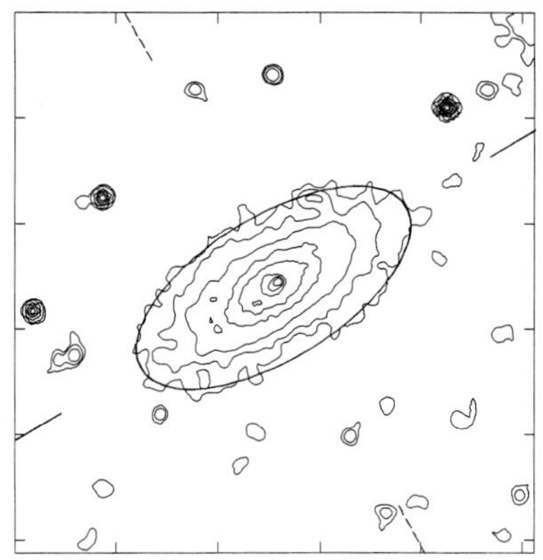

N3705 K3929 2 20.83 256X256

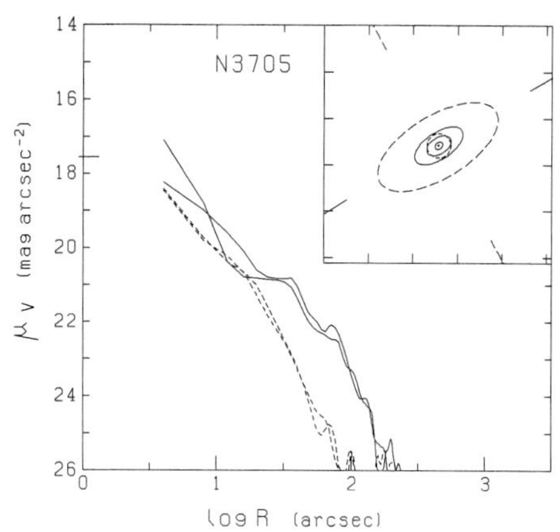

NGC3717

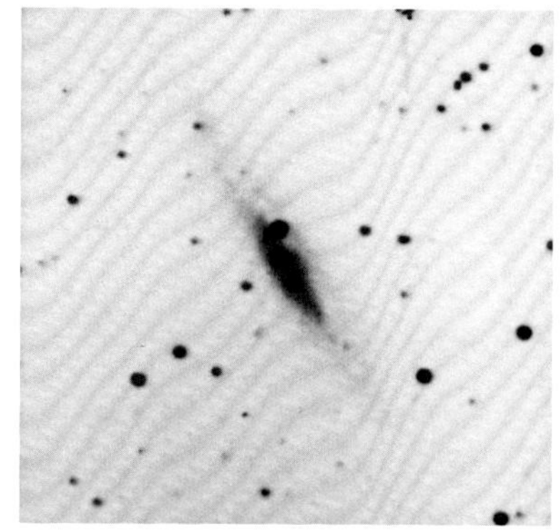

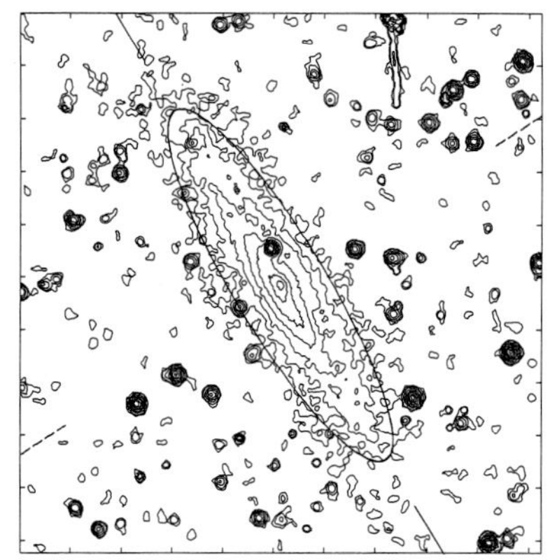

N3717 K5350 1 21.31 512X512

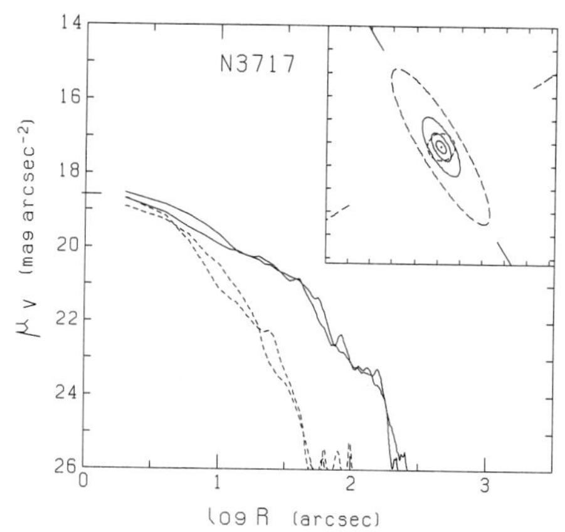

NGC3720

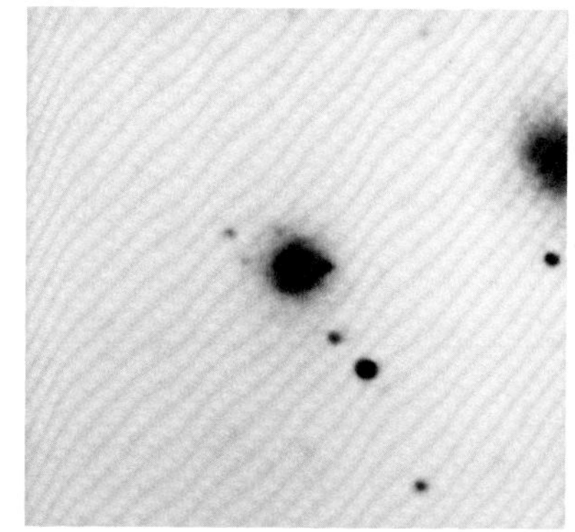

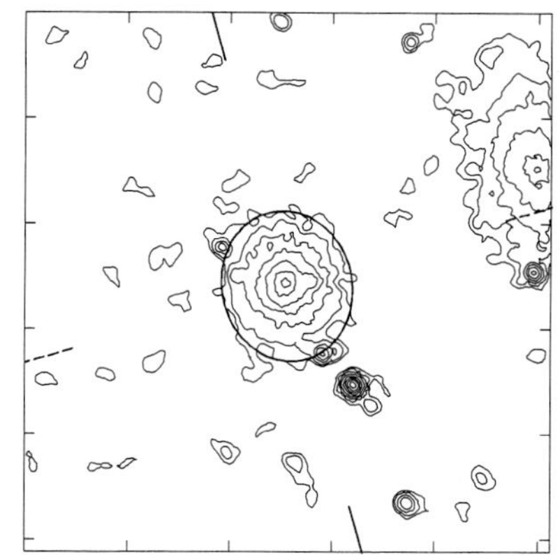

N3720 K5686 1 21.75 256X256

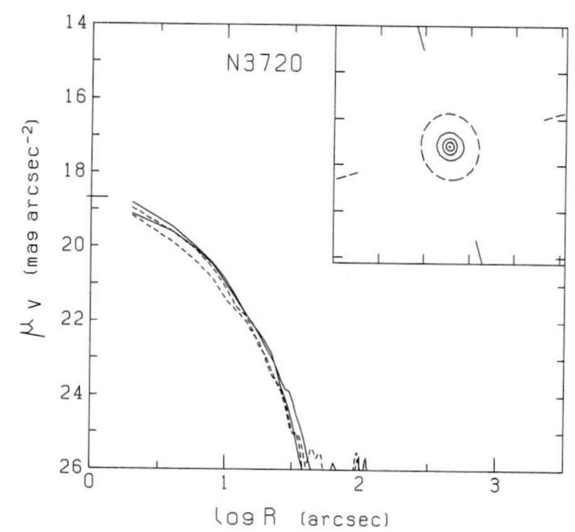

NGC3718

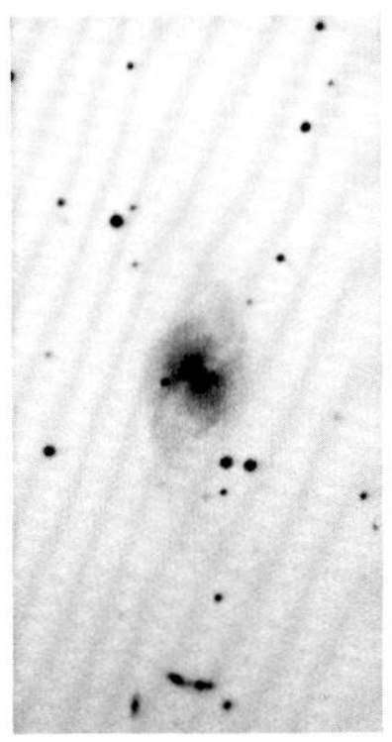

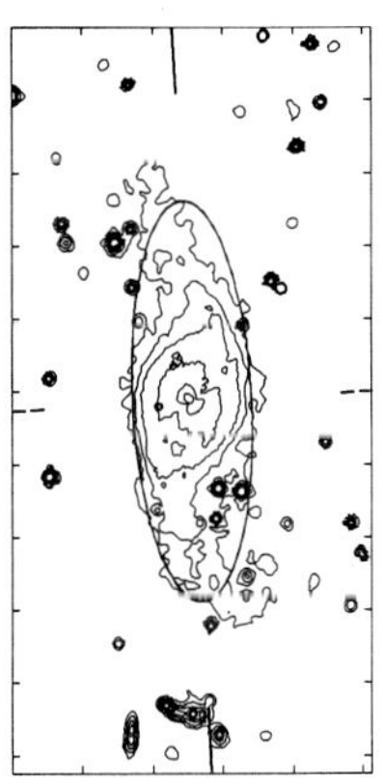

N3718 K2726 2 21.74 256X512

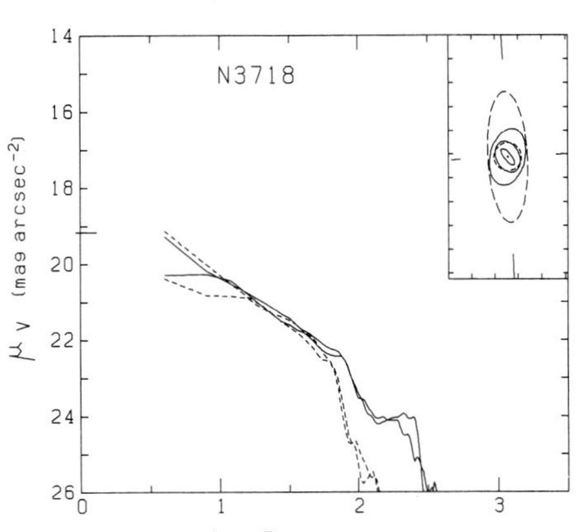

NGC3729

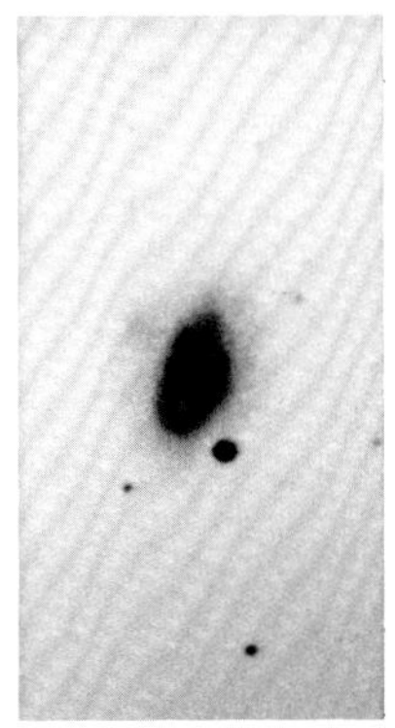

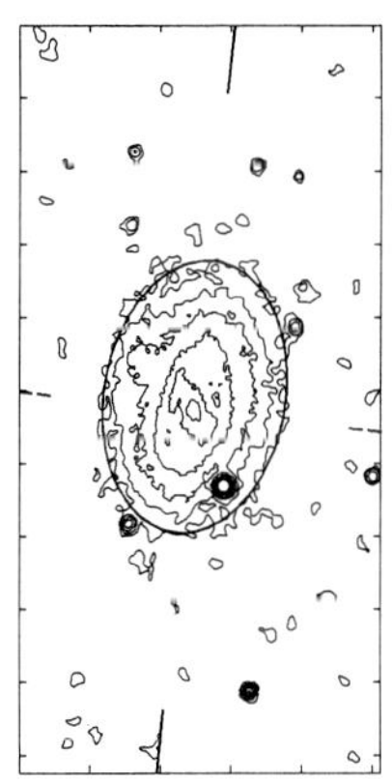

N3729 K2726 1 21.74 256X512

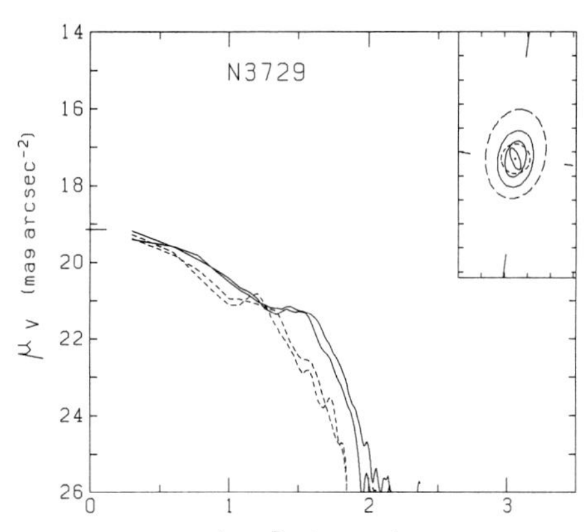

NGC3738

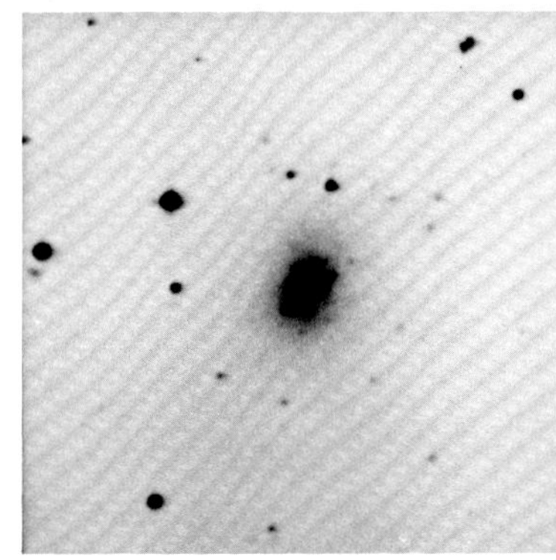

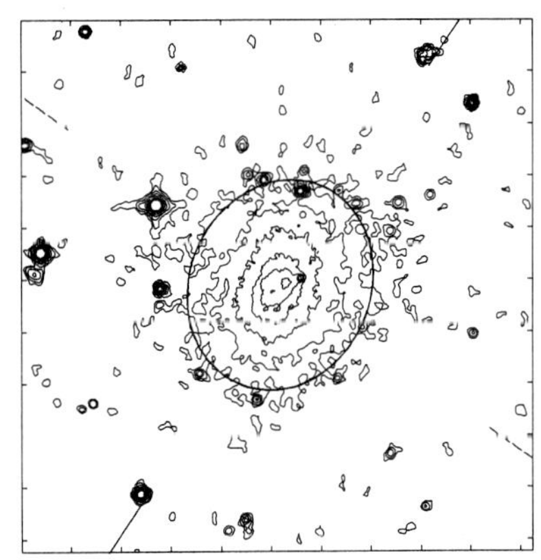

N3738 K2726 1 21.74 512X512

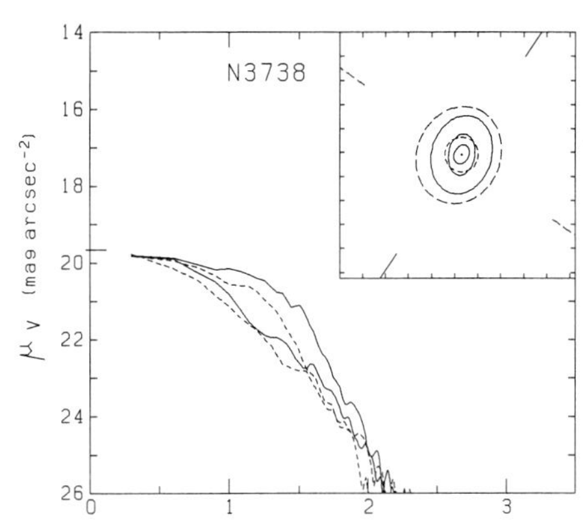

NGC3756

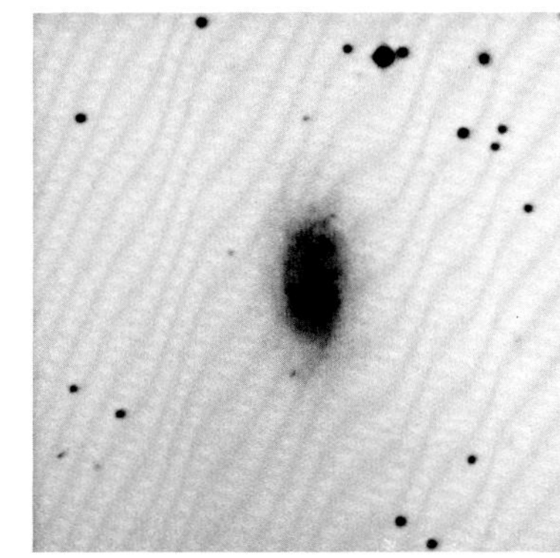

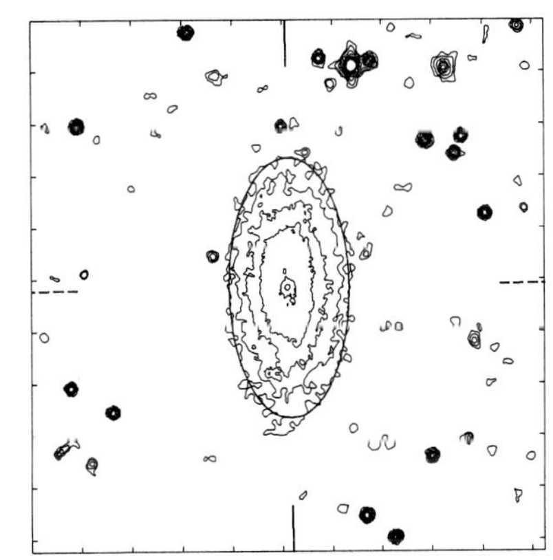

N3756 K2726 1 21.74 512X512

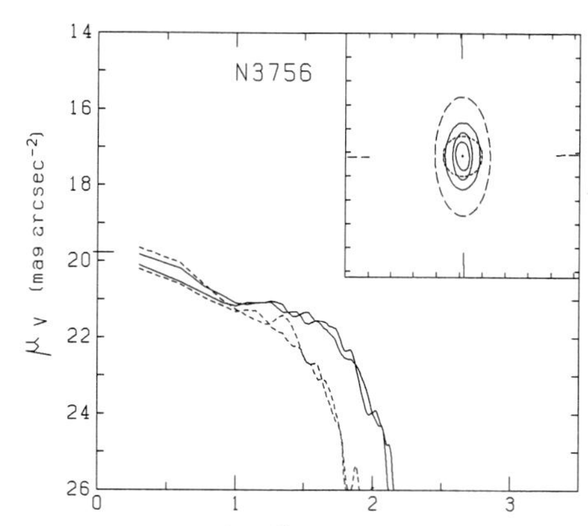

NGC3769

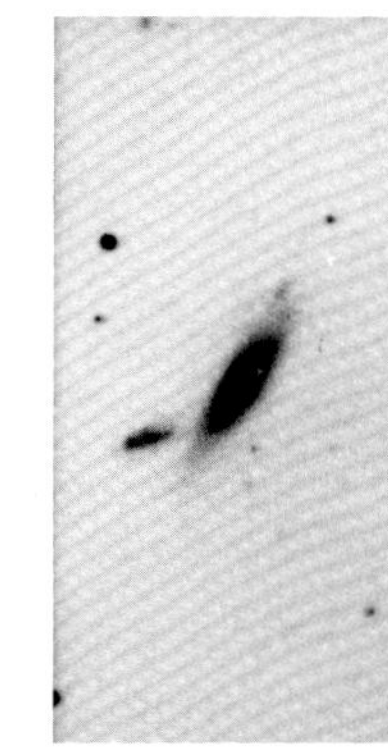

N3769 K4969 1 22.11 256X512

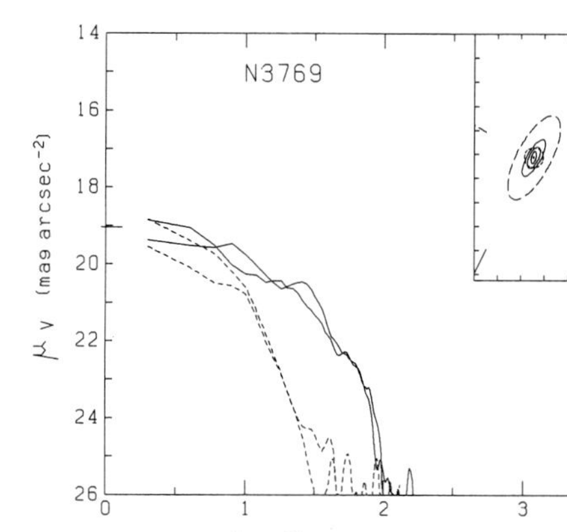

NGC3773

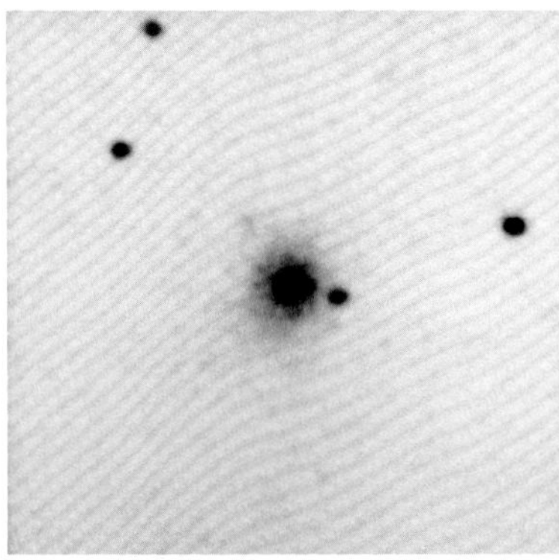

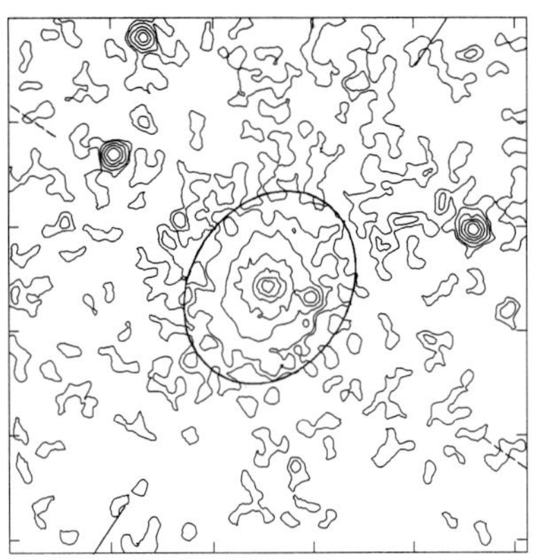

N3773 K3929 1 20.83 256X256

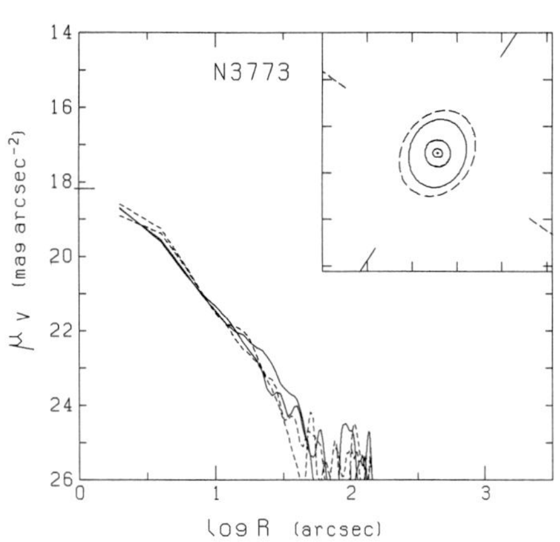

NGC3780

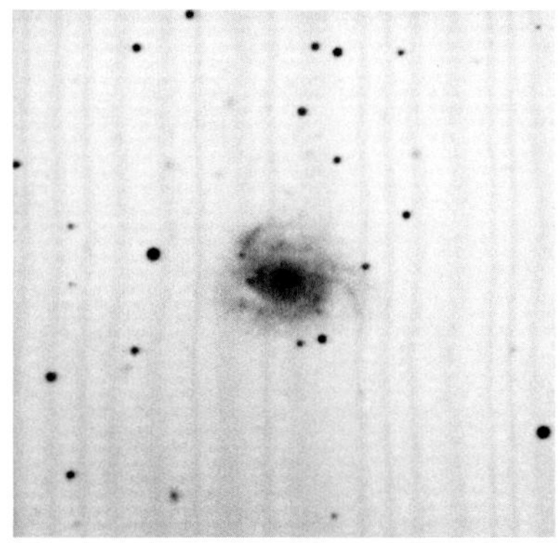

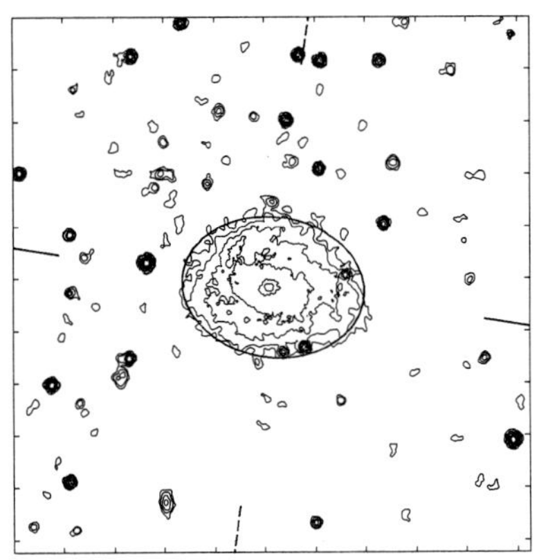

N3780 K2726 1 21.74 512X512

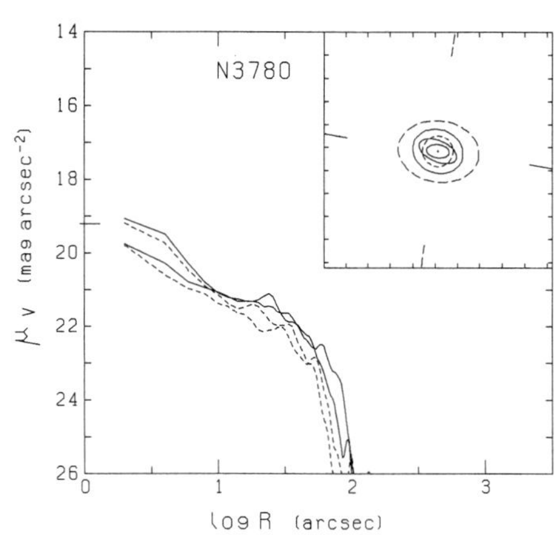

NGC3810

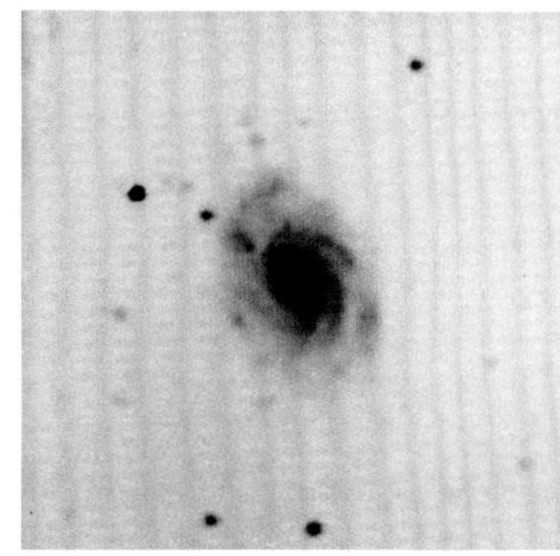

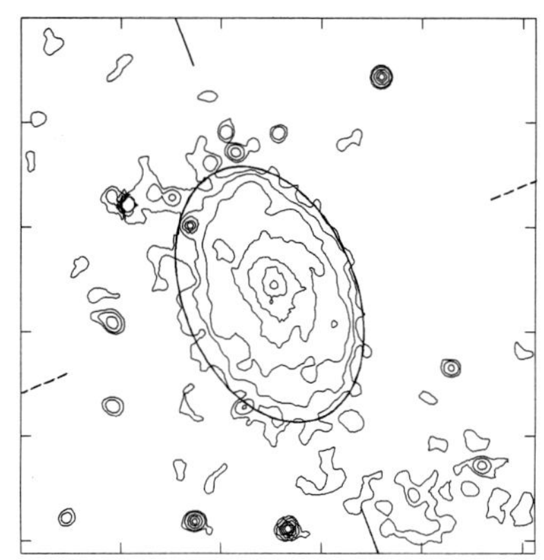

N3810 K3929 2 20.83 256X256

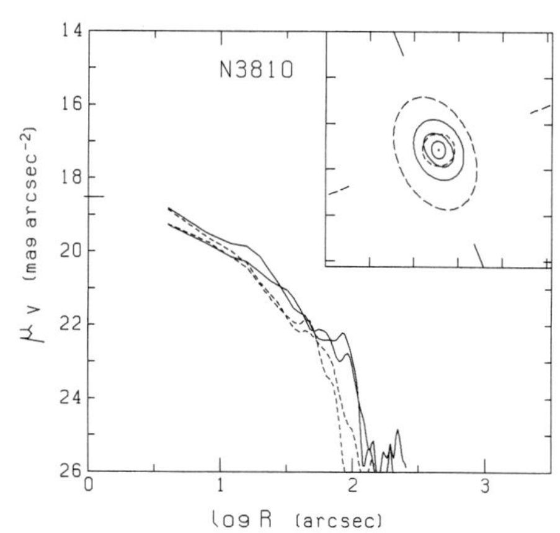

NGC3813

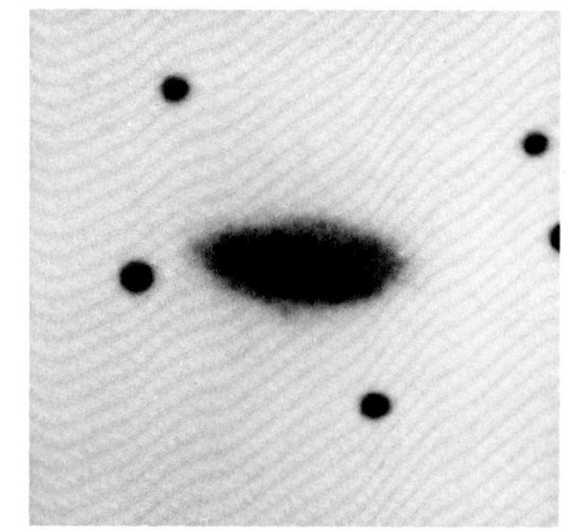

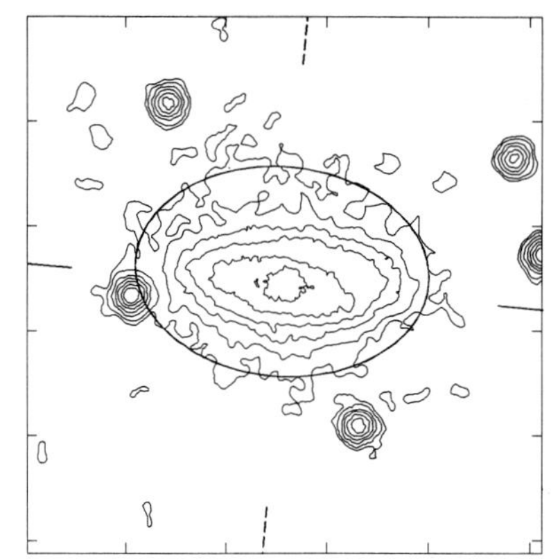

N3813 K2662 1 21.55 256X256

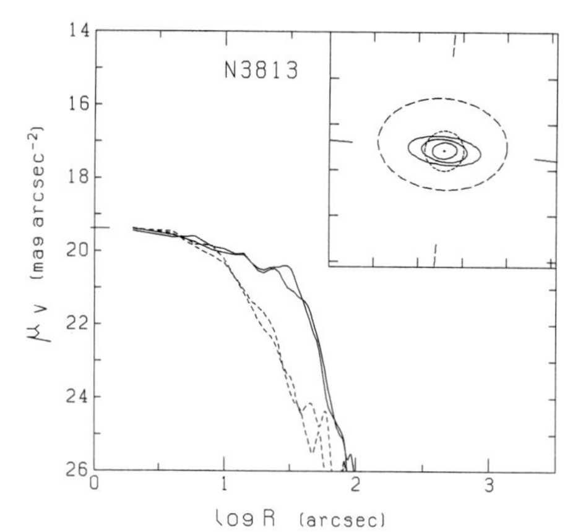

NGC3818

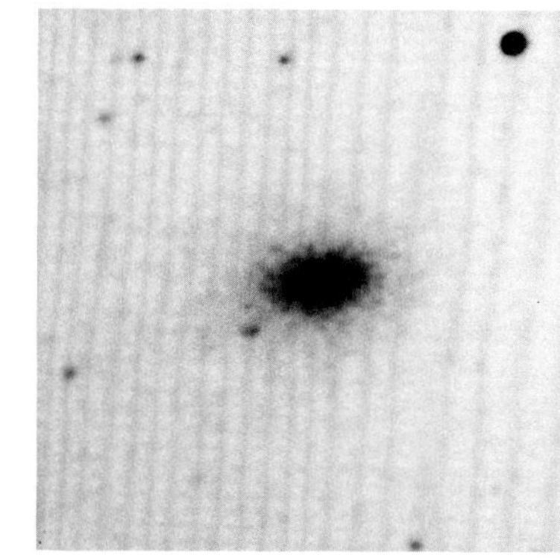

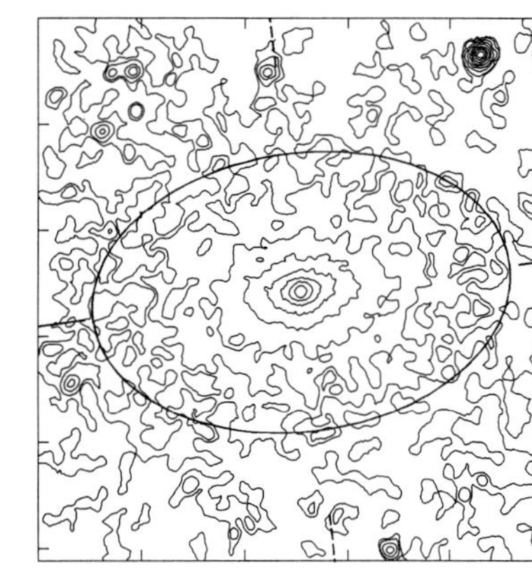

N3818 K4022 1 21.09 256X256

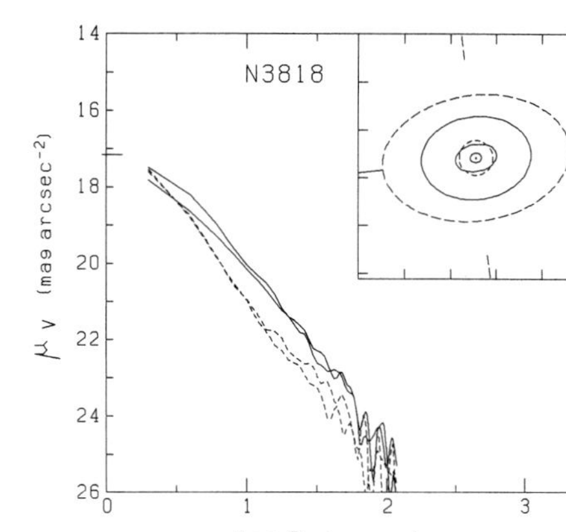

NGC3877

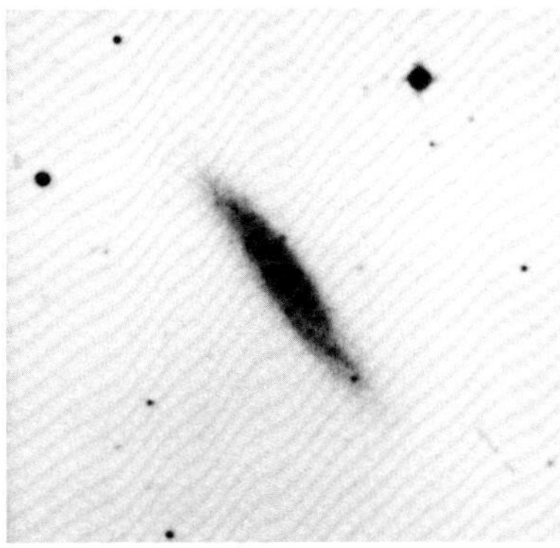

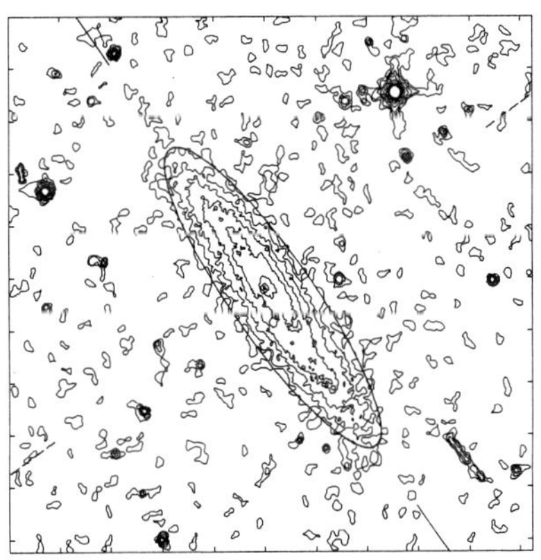

N3877 K4969 1 22.11 512X512

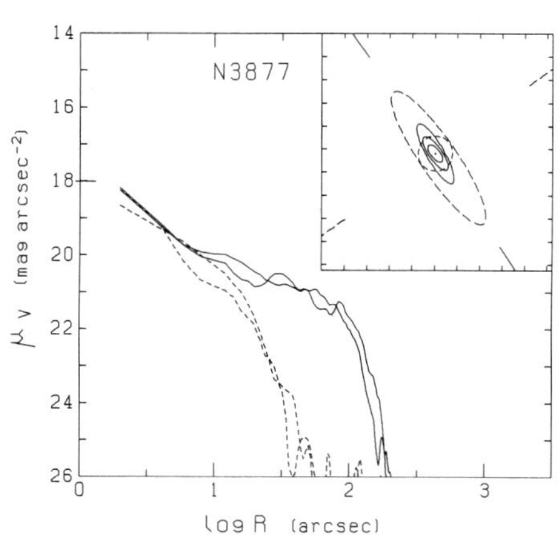

NGC3885

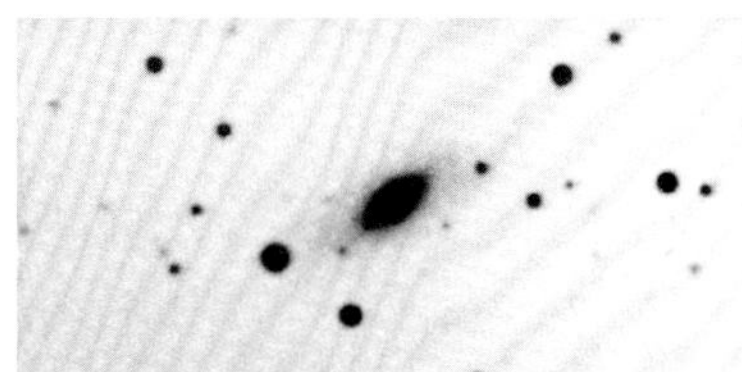

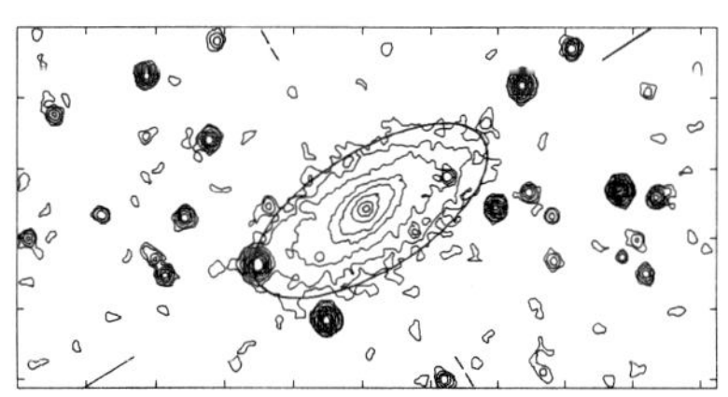

N3885 K5350 1 21.31 512X256

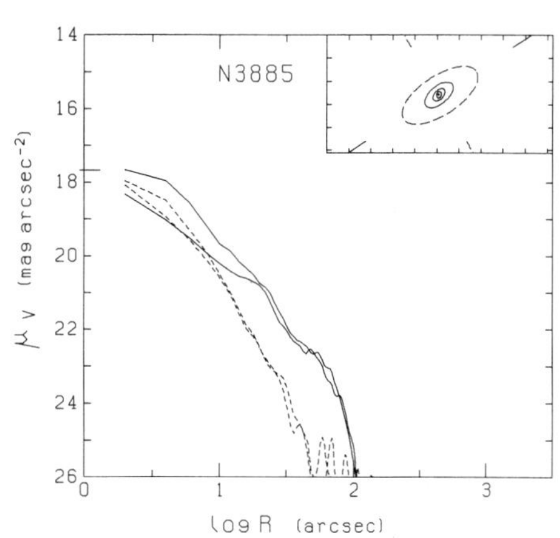

NGC3887

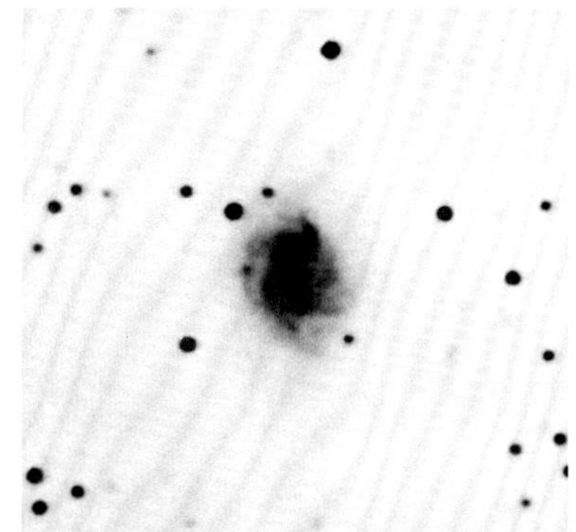

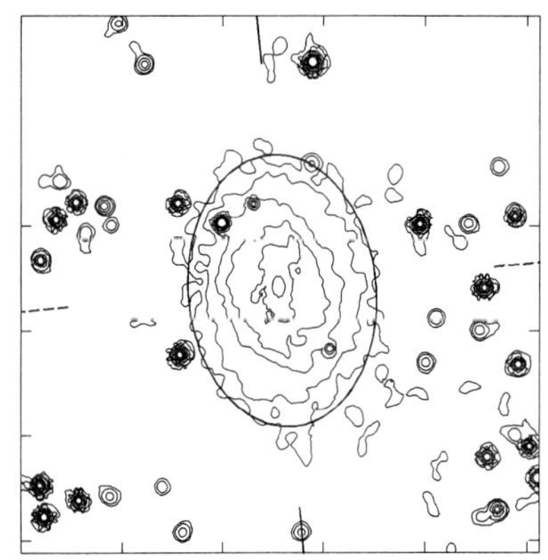

N3887 K4857 2 20.99 256X256

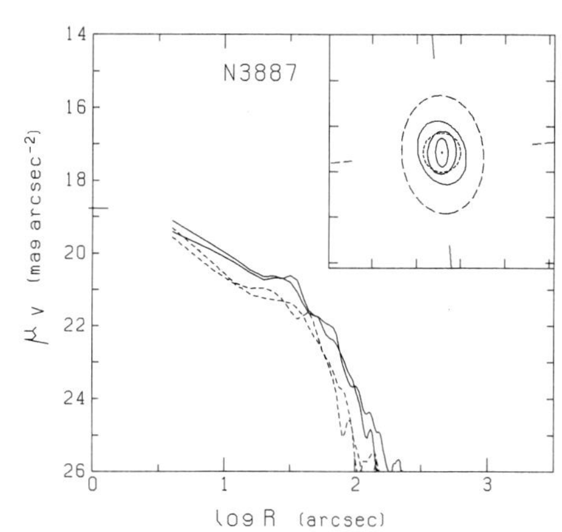

NGC3888

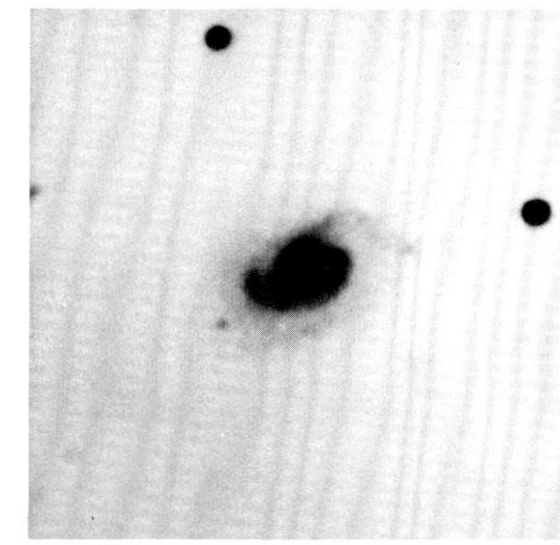

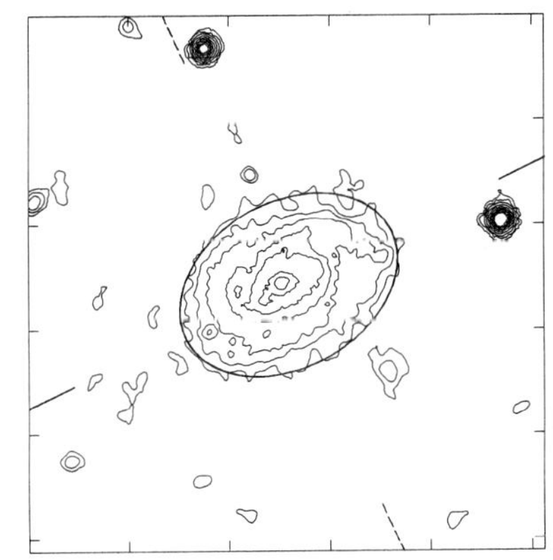

N3888 K2726 1 21.74 256X256

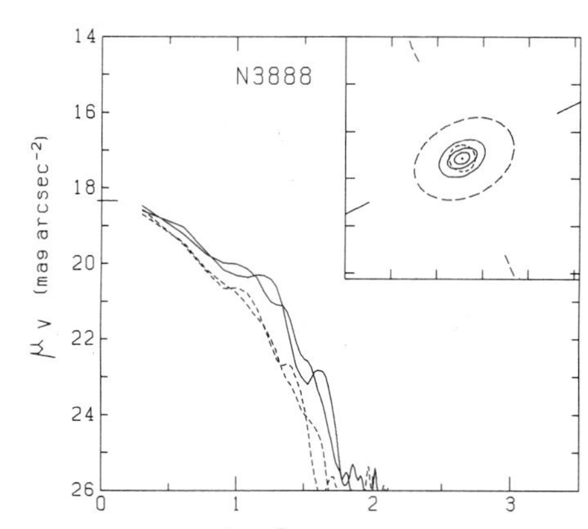

NGC3893

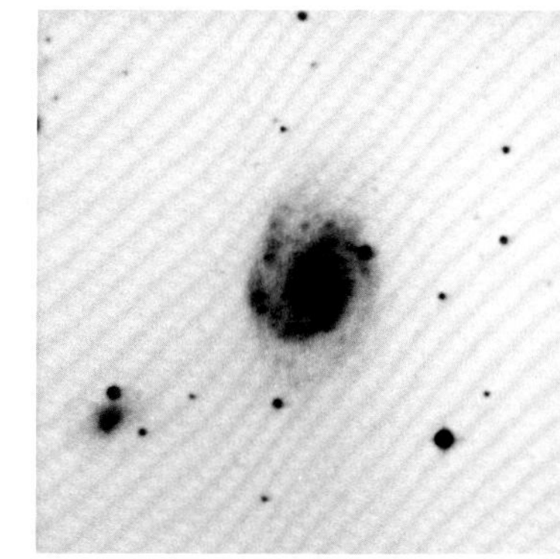

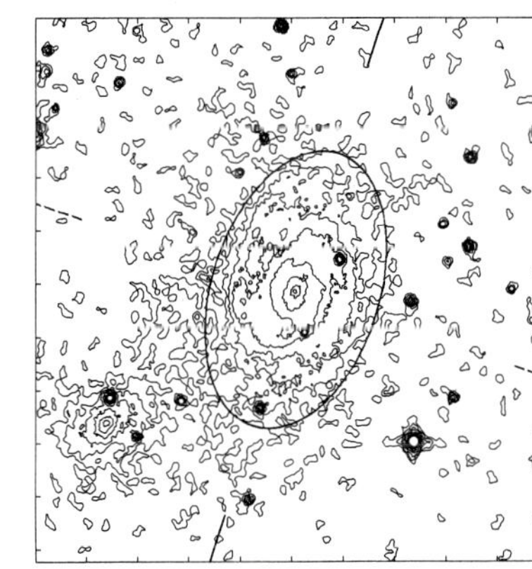

N3893 K4969 1 22.11 512X512

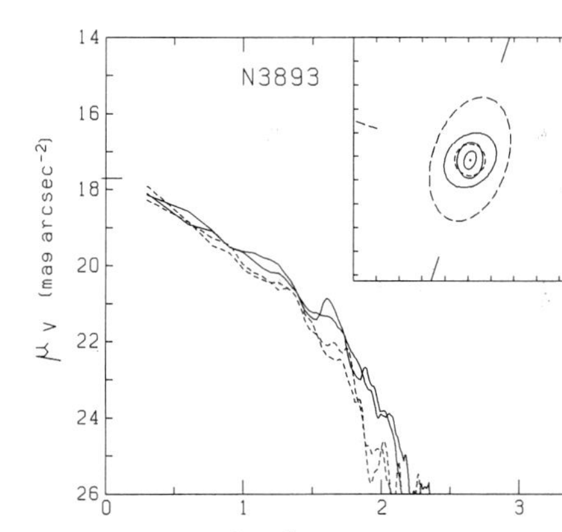

NGC3900

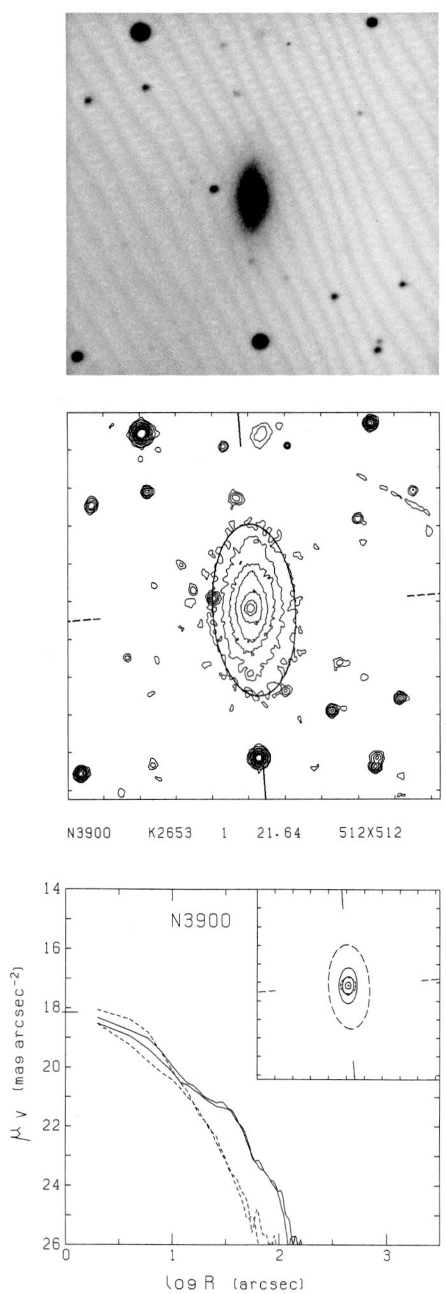

NGC3898

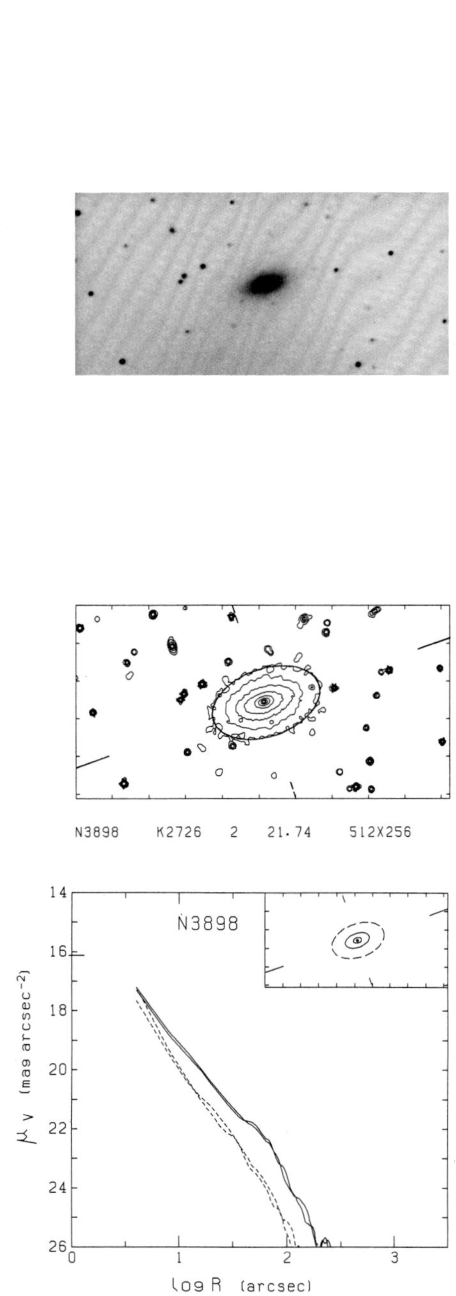

NGC3904

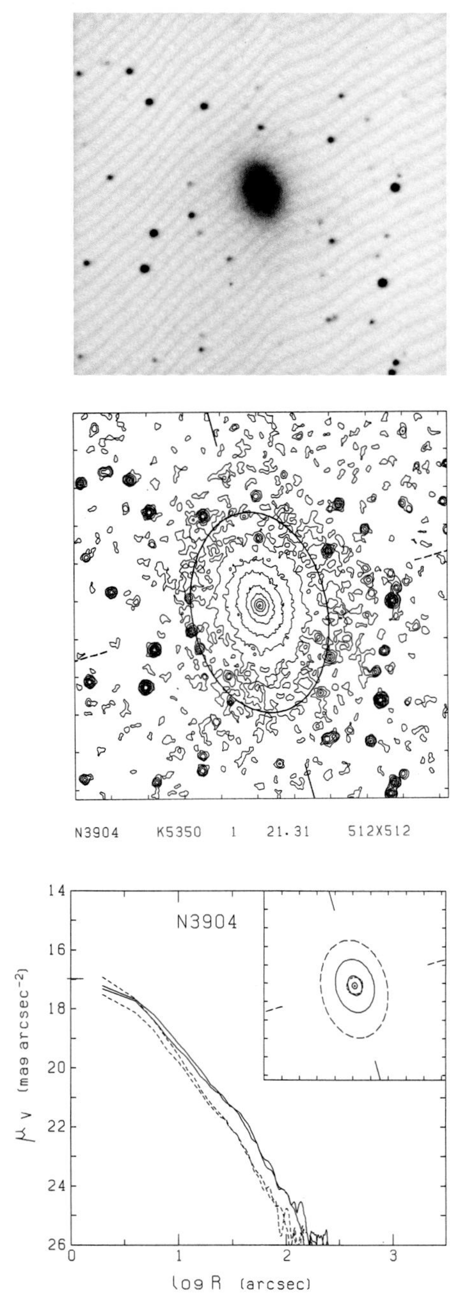

NGC3912

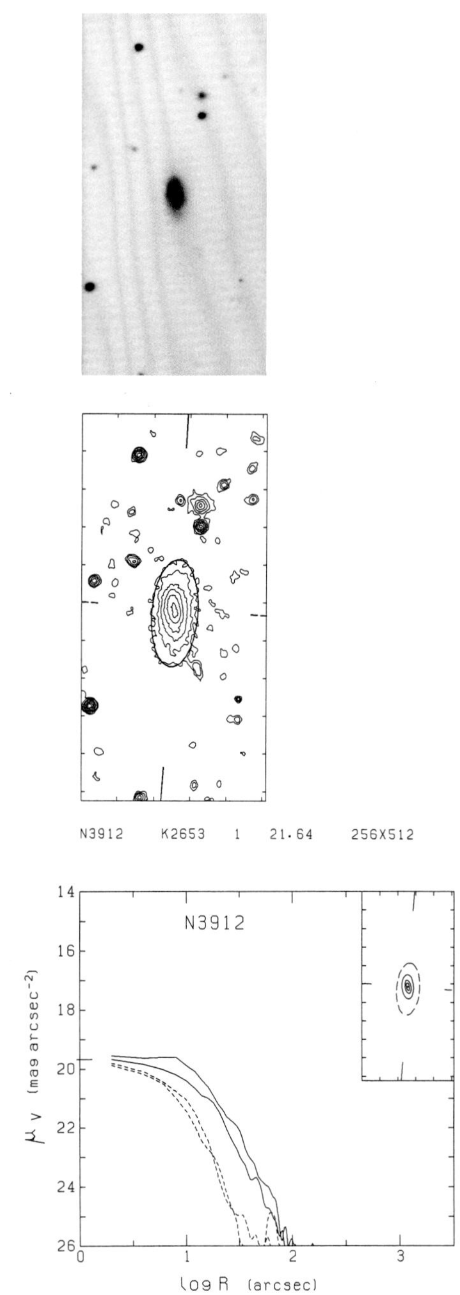

NGC3917

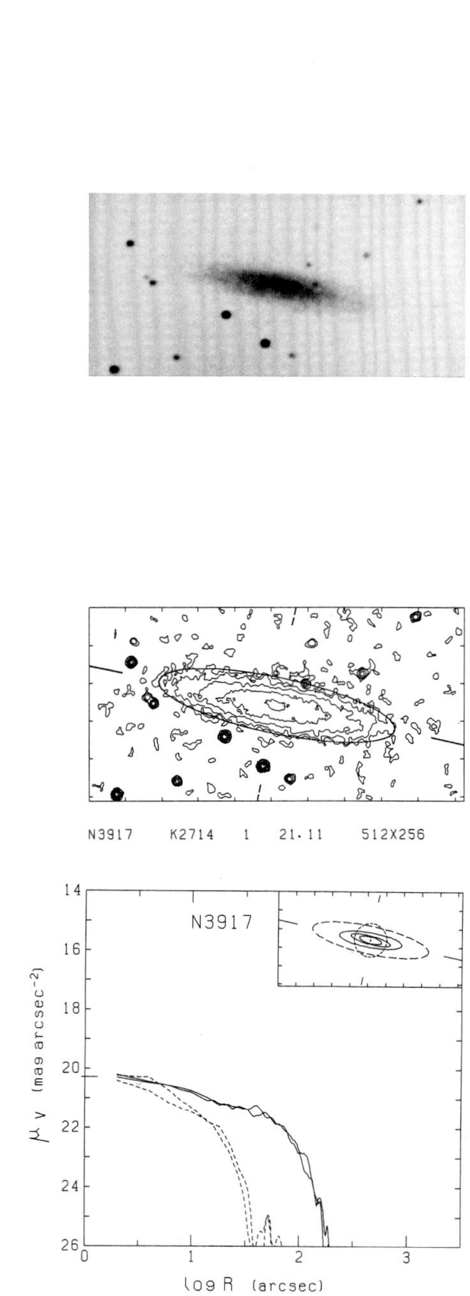

NGC3923

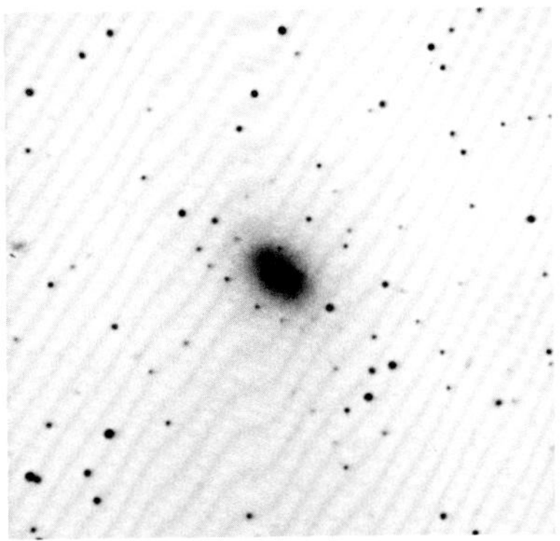

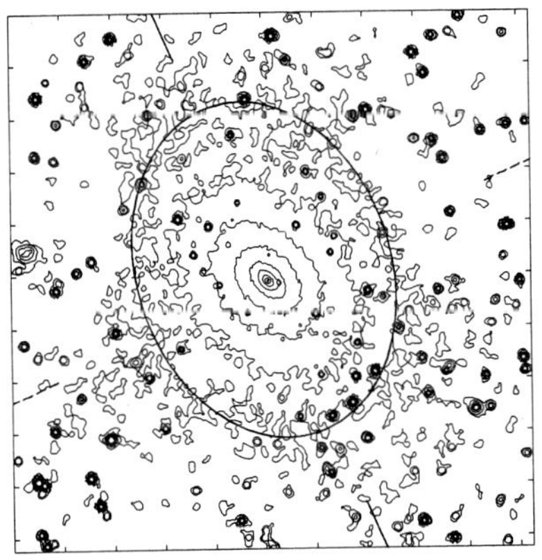

N3923 K4892 2 21.28 512X512

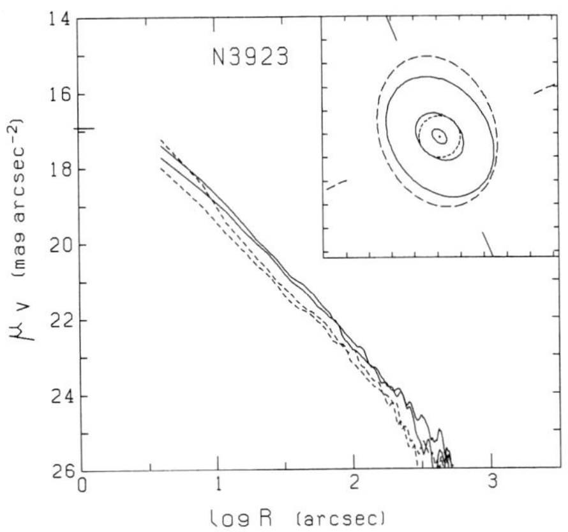

NGC3936

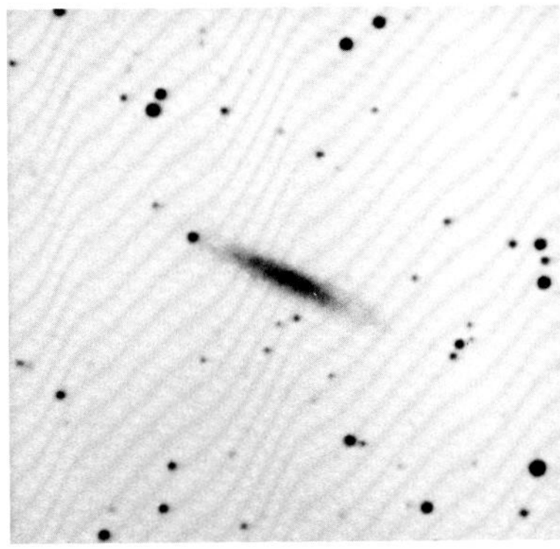

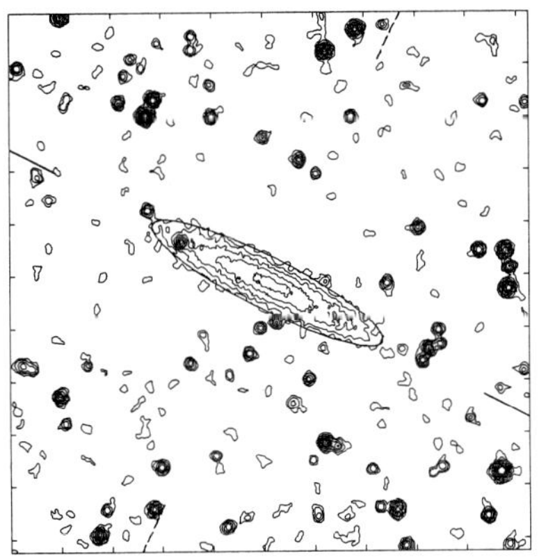

N3936 K5687 1 21.78 512X512

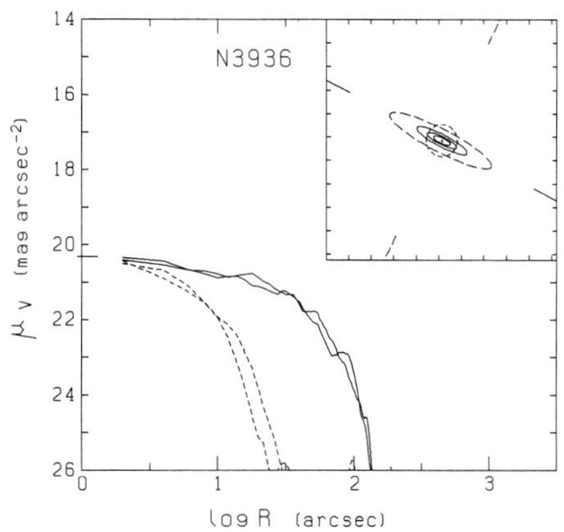

NGC3938

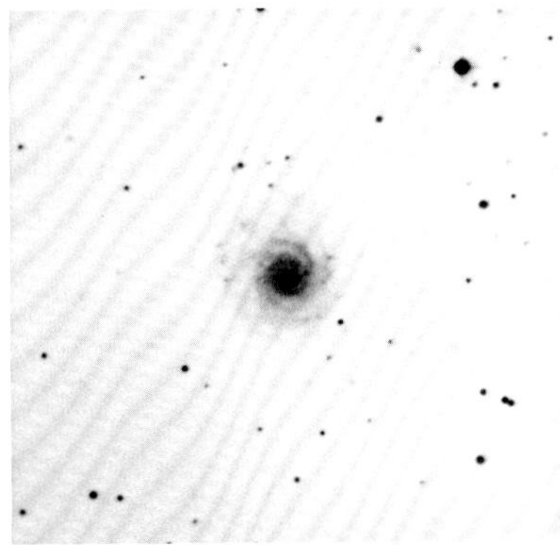

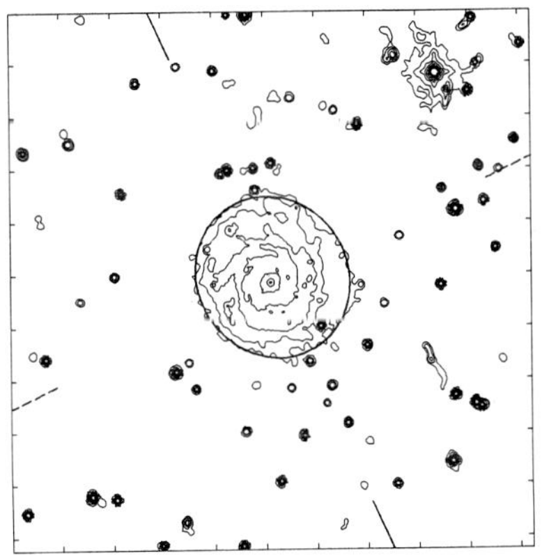

N3938 K3676 2 21.65 512X512

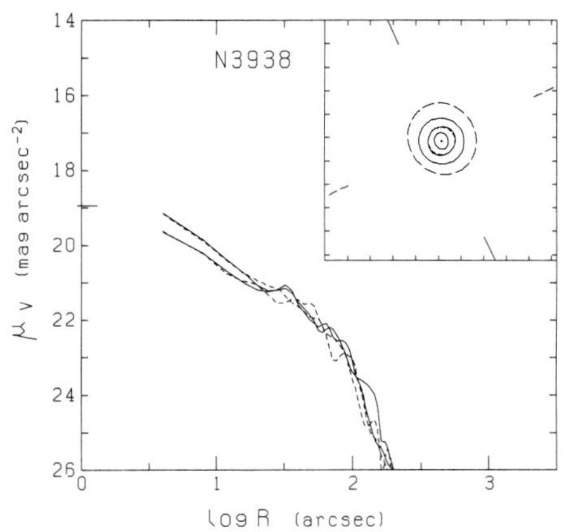

NGC3941

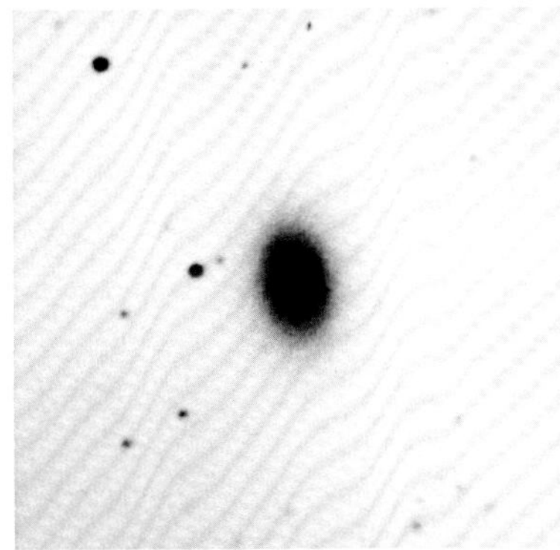

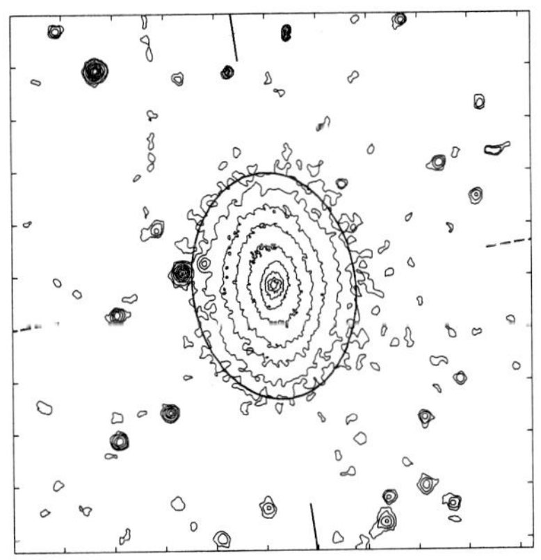

N3941 K2662 1 21.55 512X512

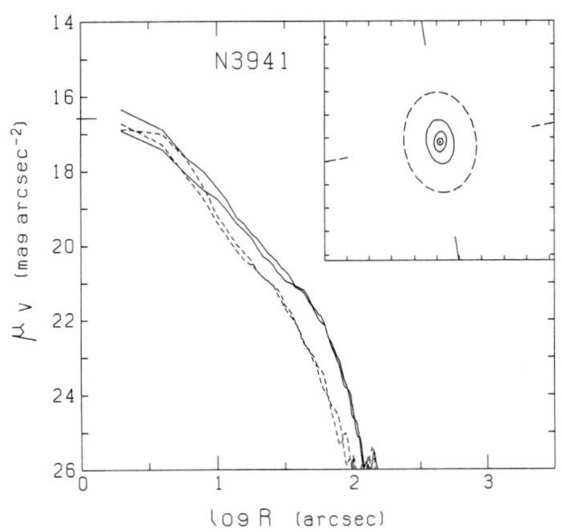

NGC3945

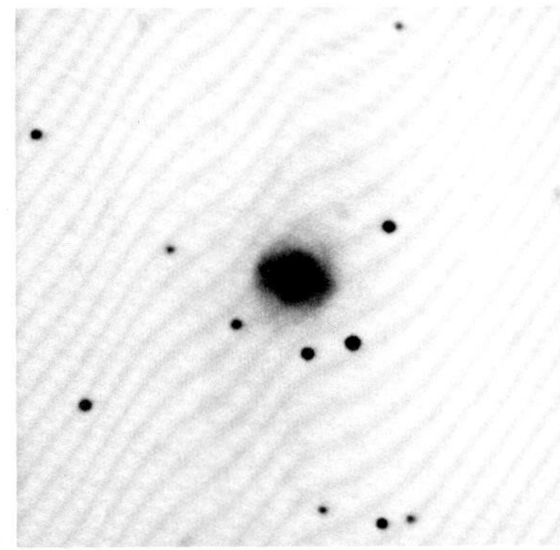

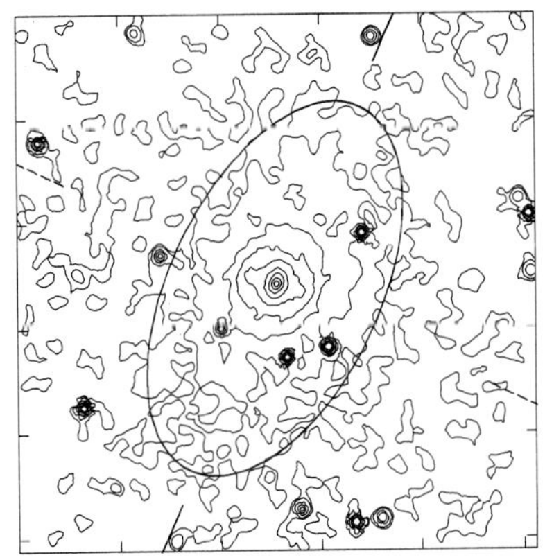

N3945 K5770 2 20.99 256X256

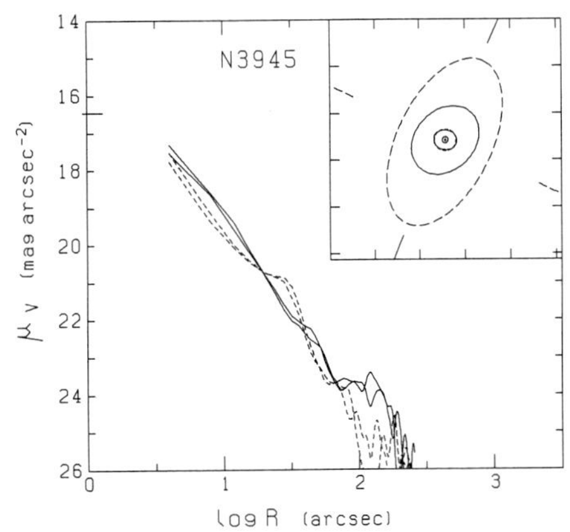

NGC3949

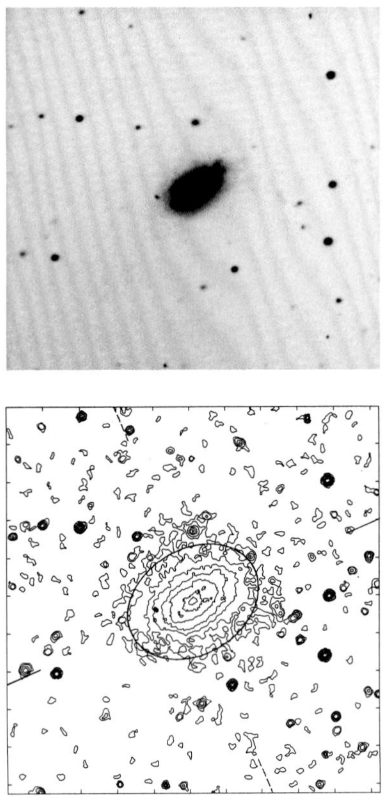

N3949 K2714 1 21.11 512X512

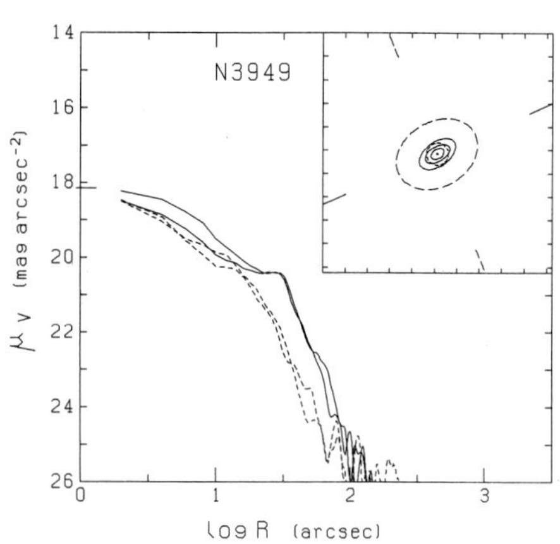

NGC3952

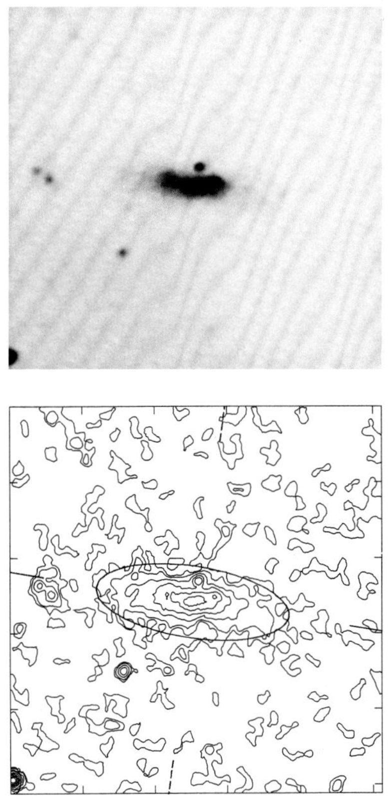

N3952 K4022 1 21.09 256X256

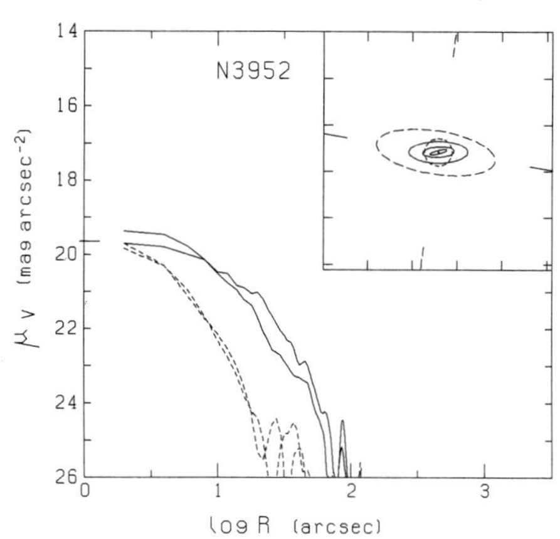

NGC3953

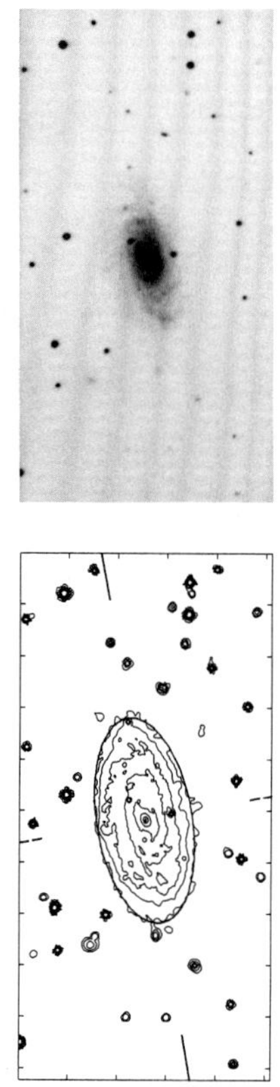

N3953 K2726 2 21.74 256X512

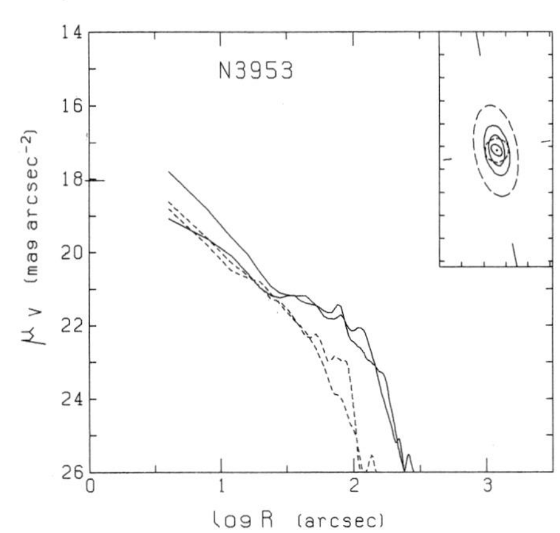

NGC3955

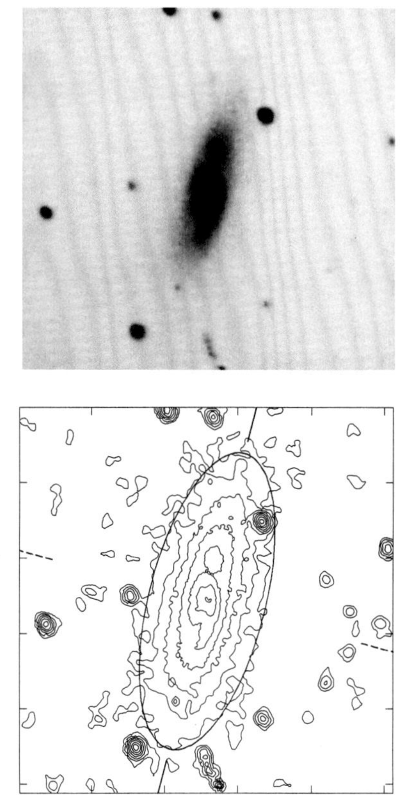

N3955 K5687 1 21.78 256X256

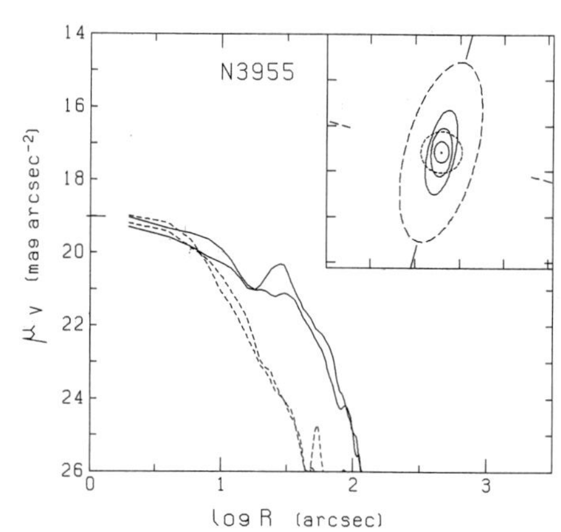

NGC3956

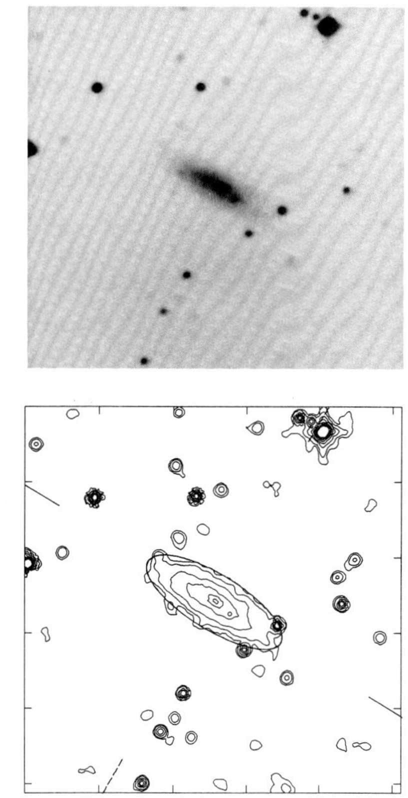

N3956 K4857 2 20.99 256X256

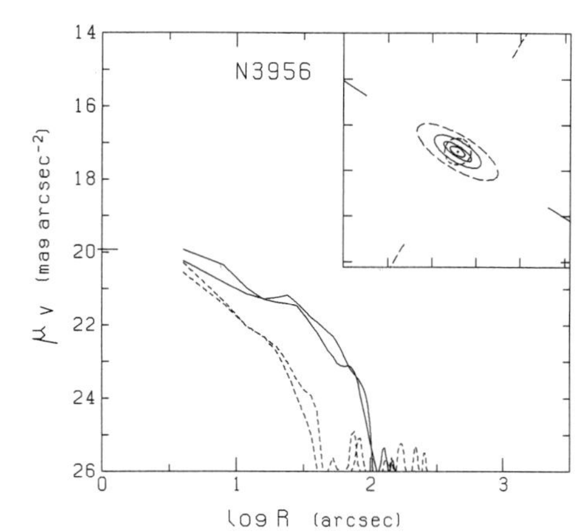

NGC3957

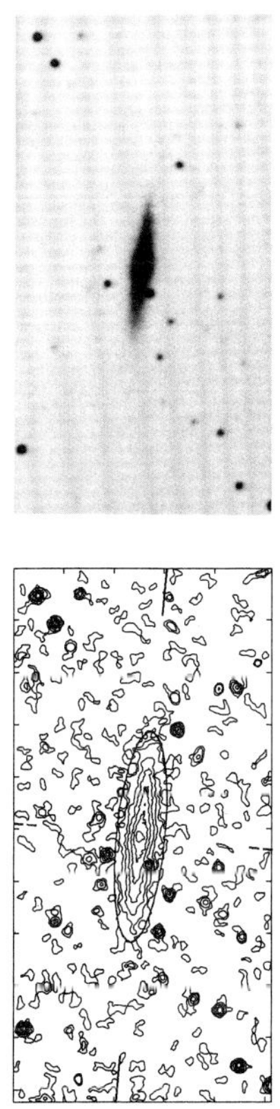

N3957 K4857 1 20.99 256X512

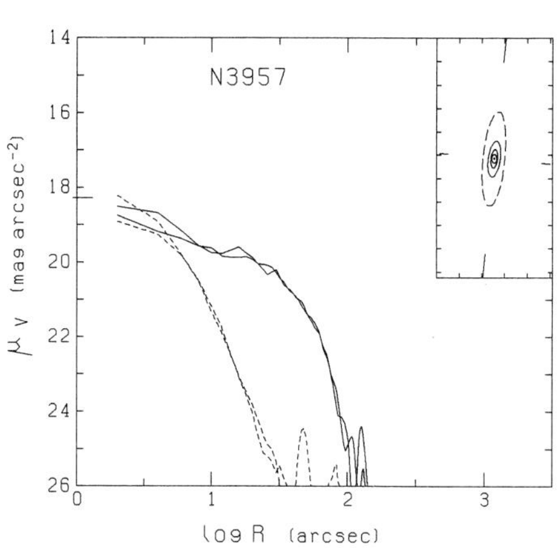

NGC3962

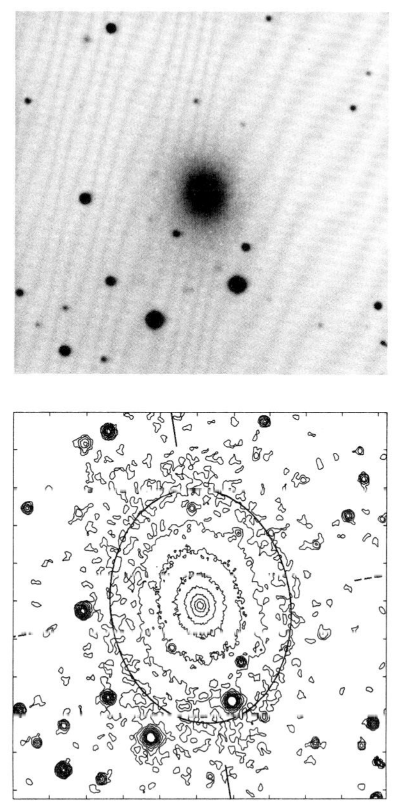

N3962 K5341 1 21.72 512X512

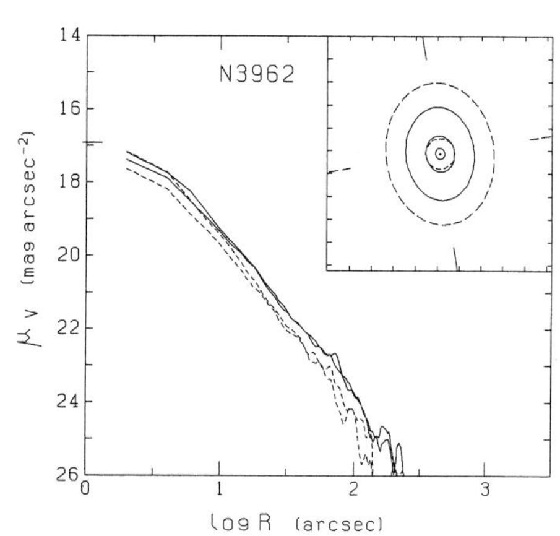

NGC3976

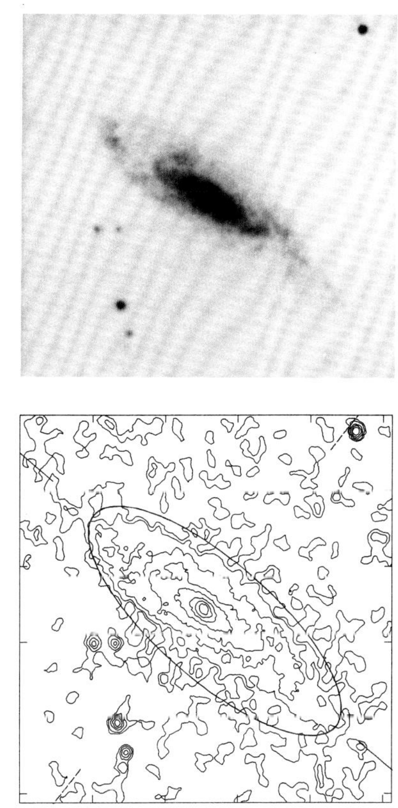

N3976 K4993 1 21.60 256X256

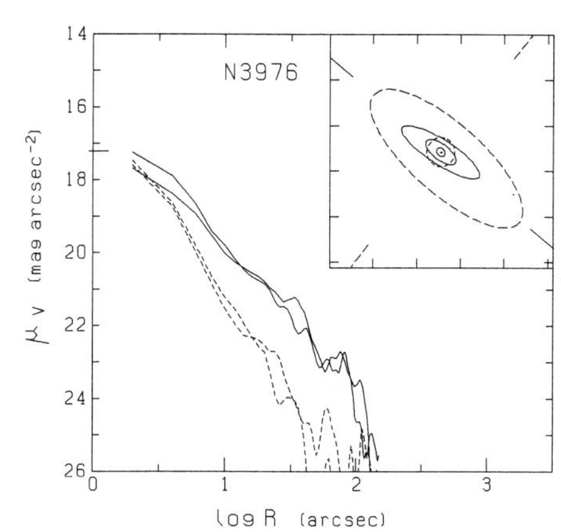

NGC3981

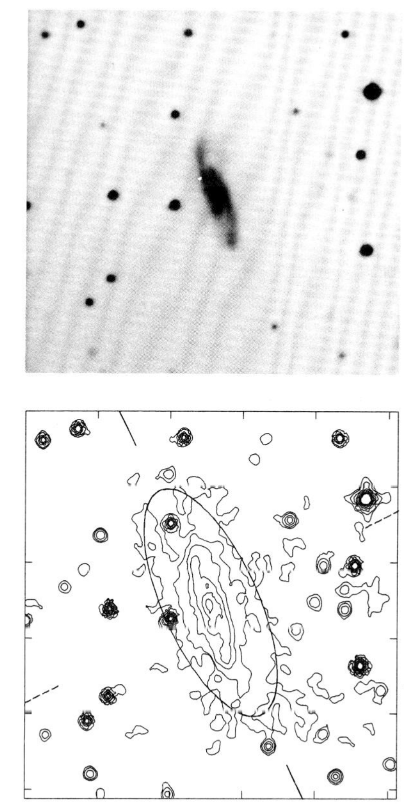

N3981 K4857 2 20.99 256X256

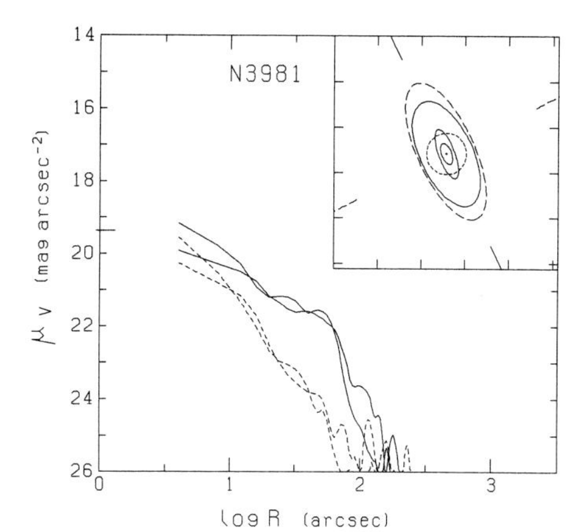

NGC3982

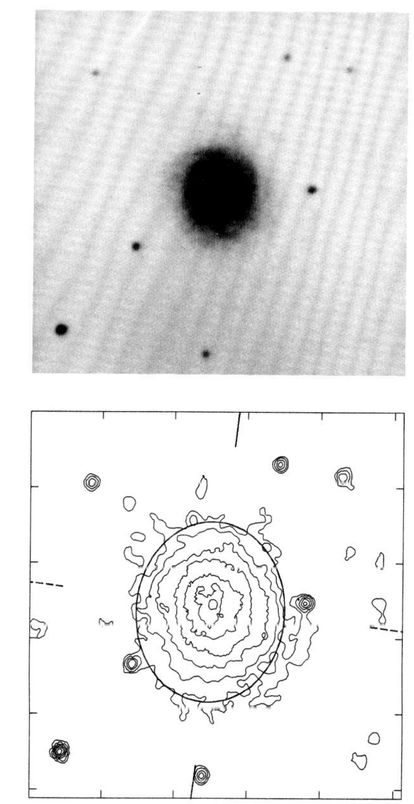

N3982 K2726 1 21.74 256X256

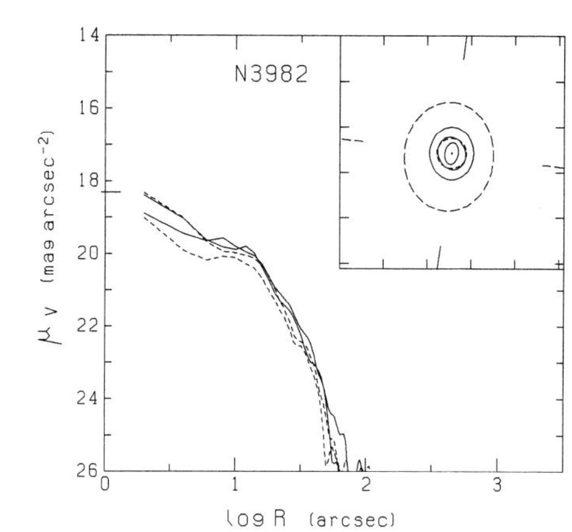

NGC3985

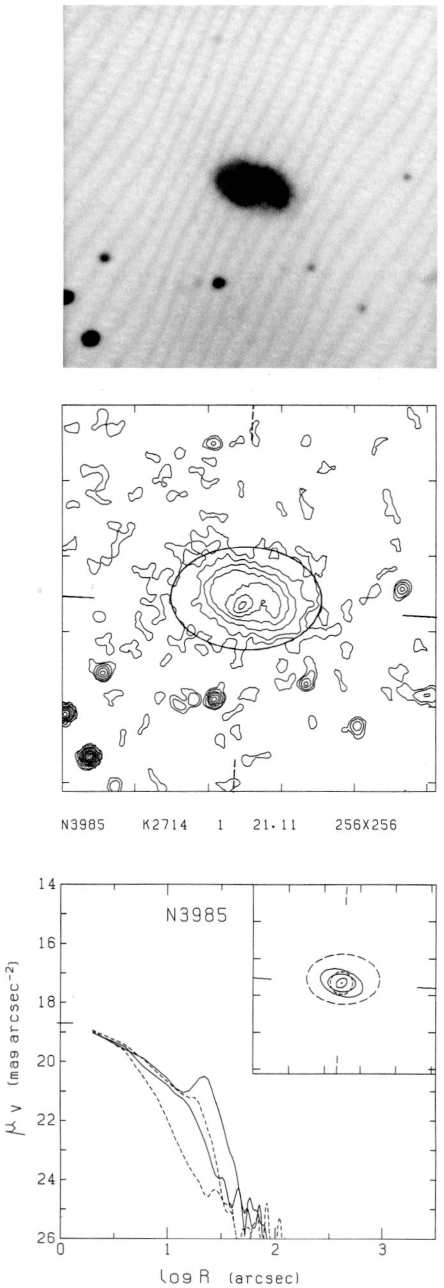

NGC3992

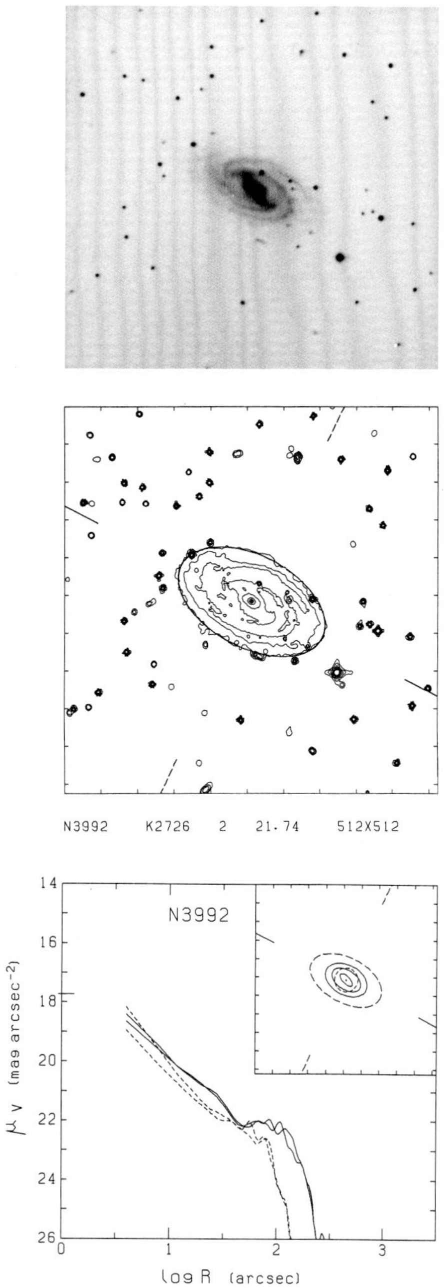

NGC3995

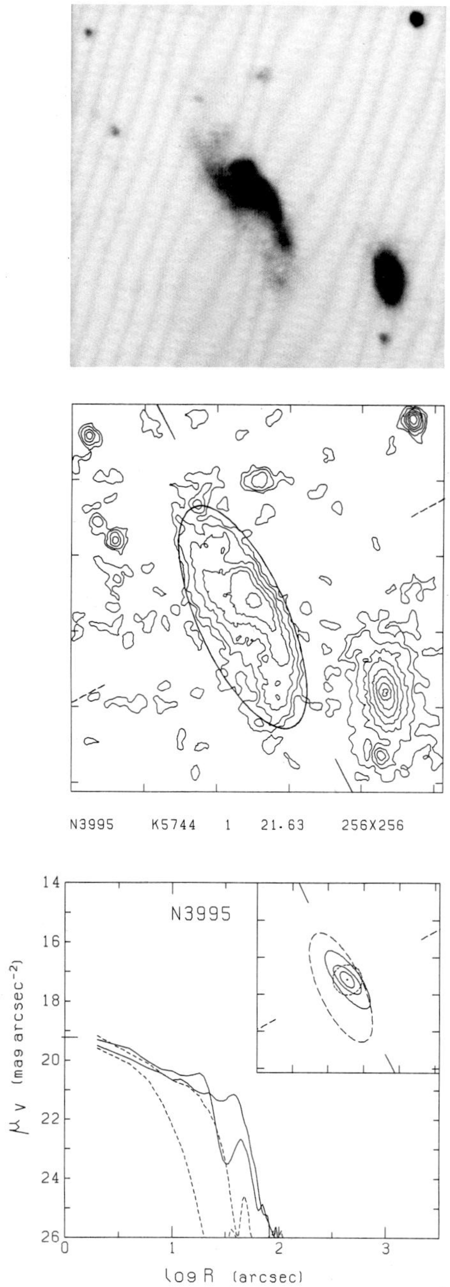

NGC3998

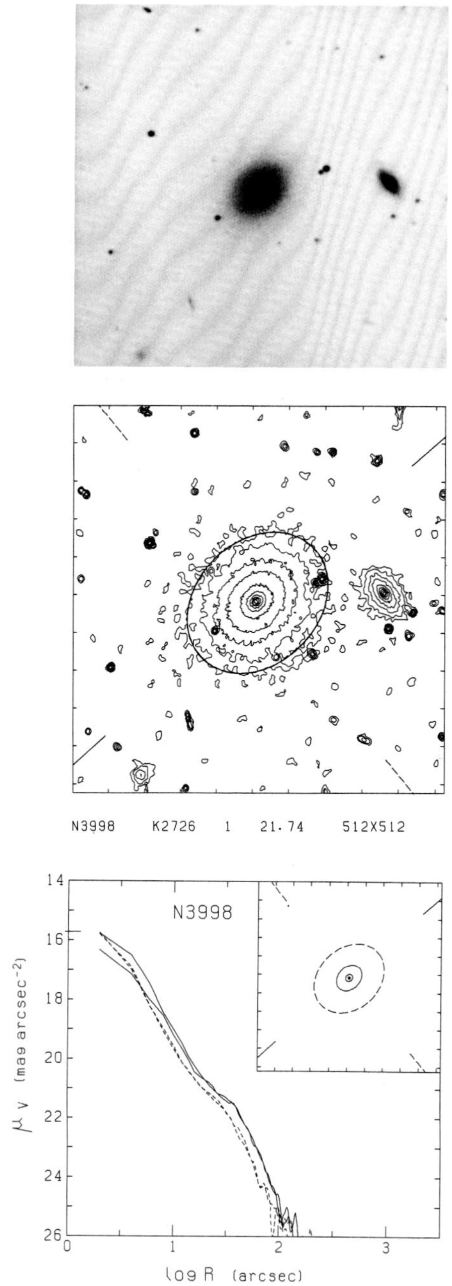

NGC4008

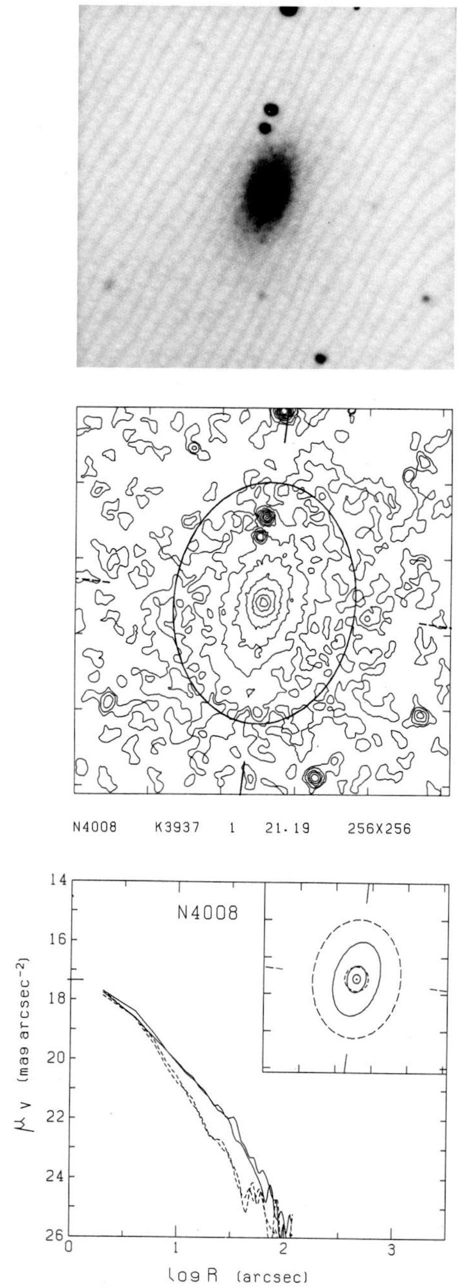

NGC4013

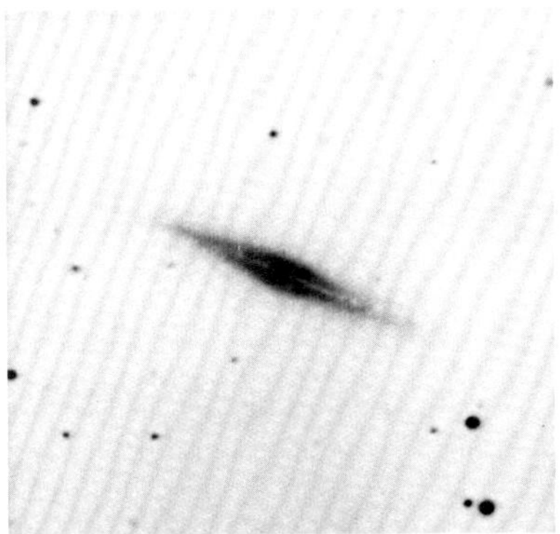

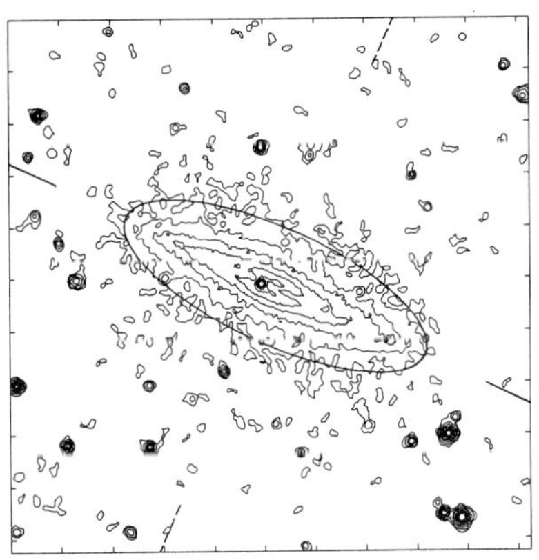

N4013 K3676 1 21.65 512X512

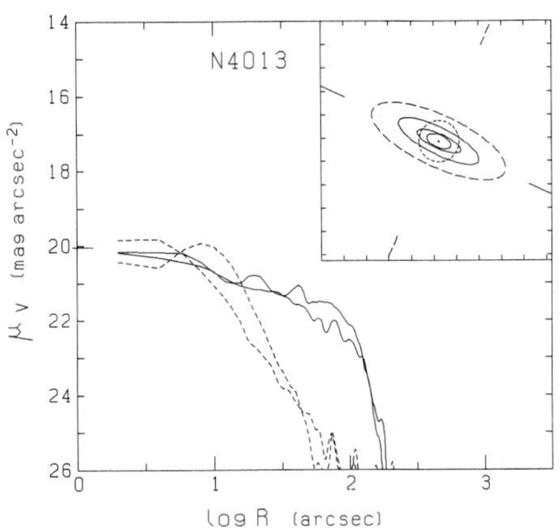

NGC4024

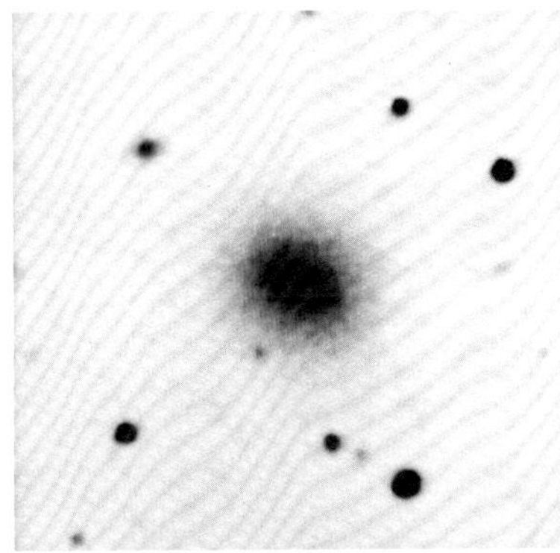

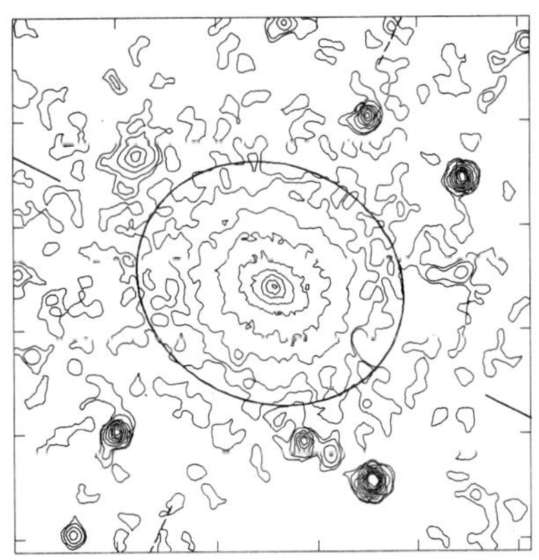

N4024 K4857 1 20.99 256X256

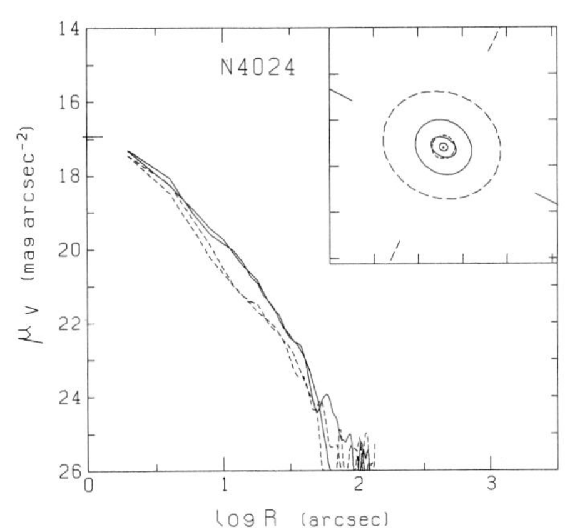

IC 749

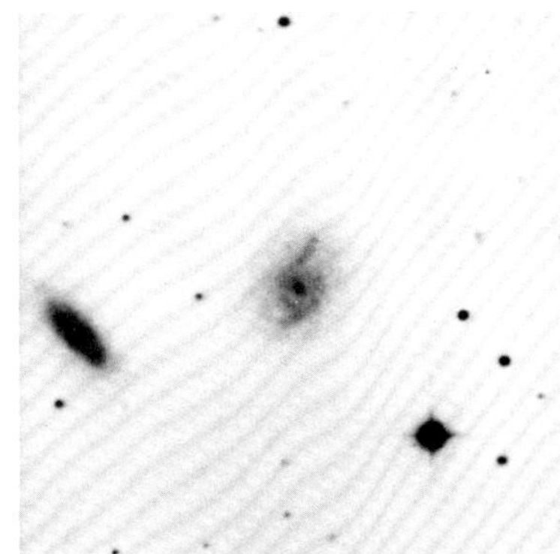

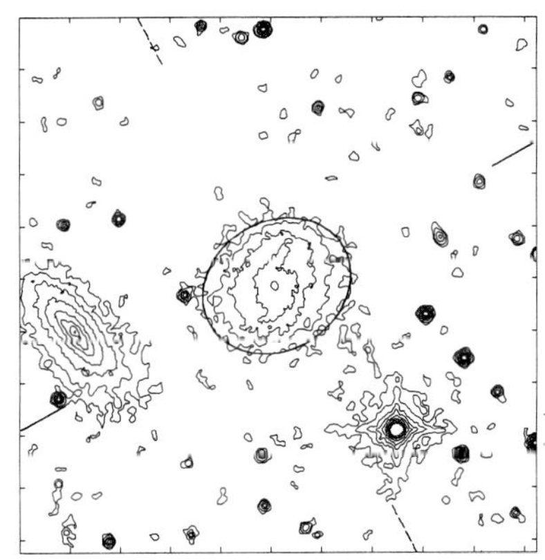

I0749 K3676 1 21.65 512X512

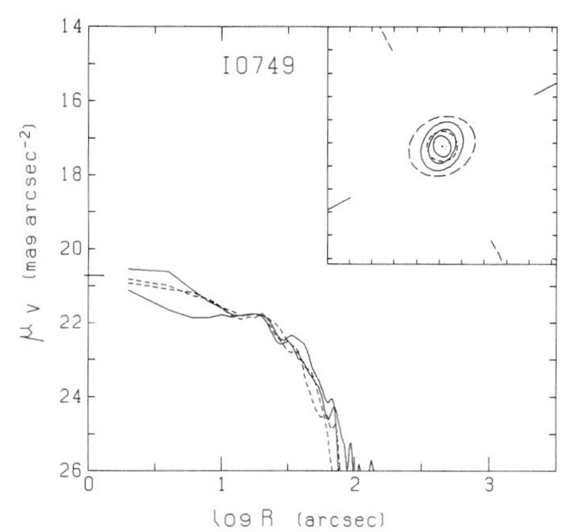

IC 750

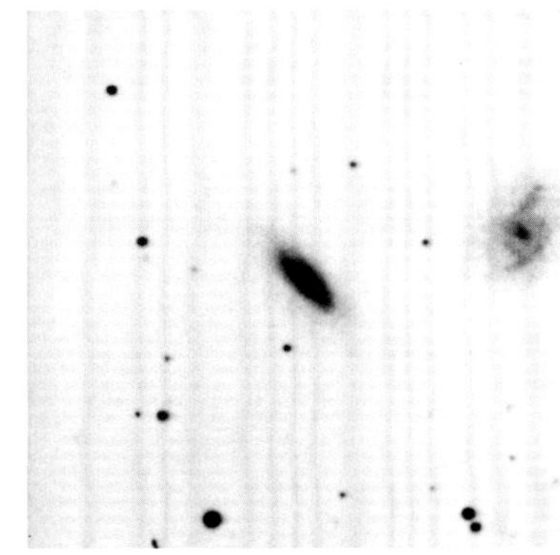

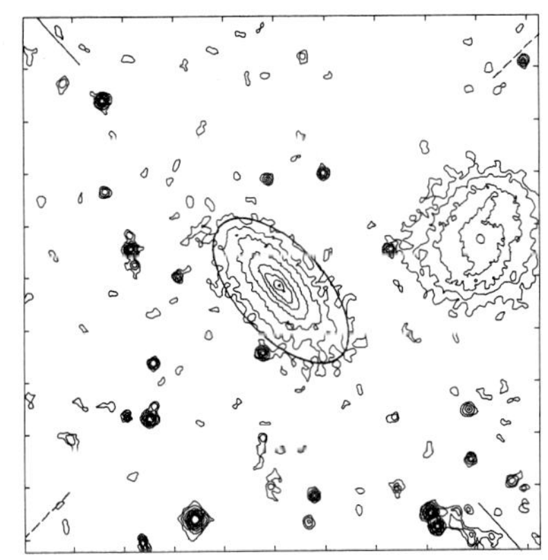

I0750 K3676 1 21.65 512X512

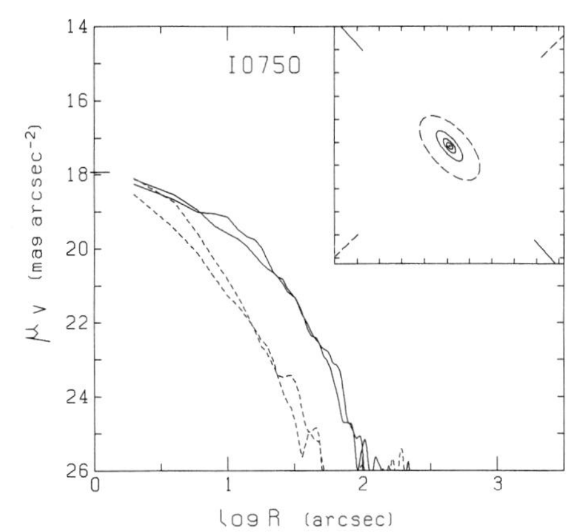

NGC4026

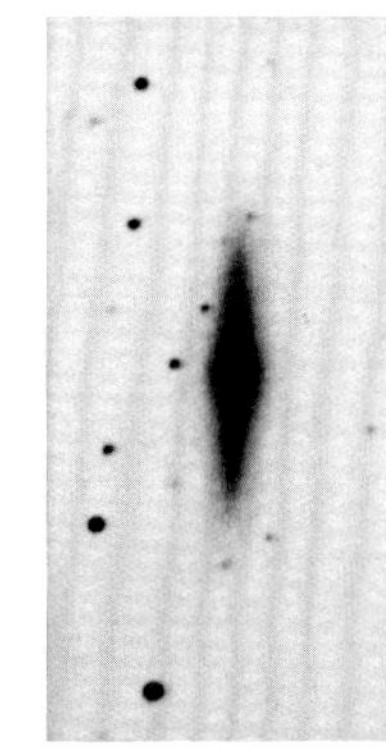

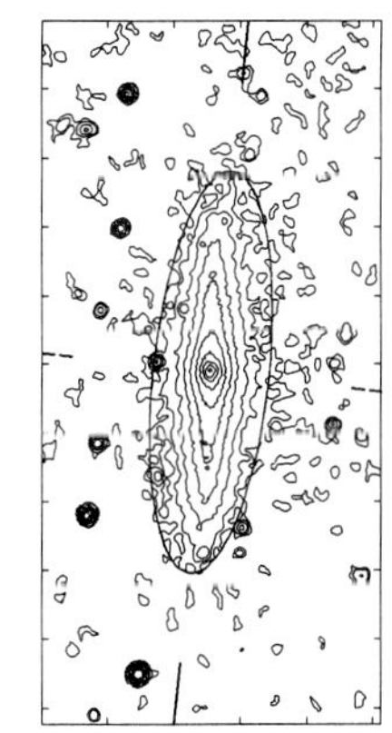

N4026 K2714 1 21.11 256X512

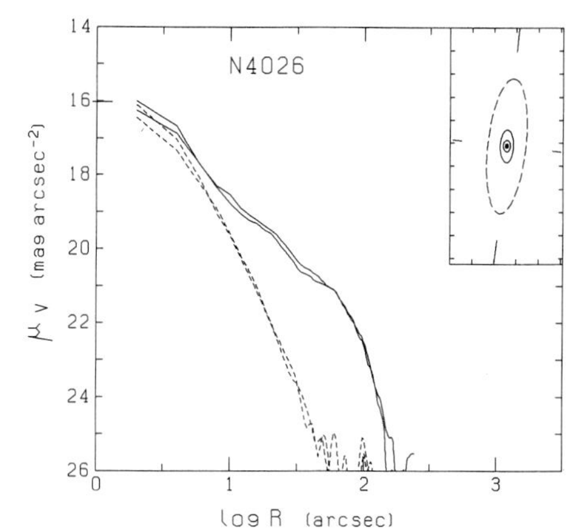

NGC4027

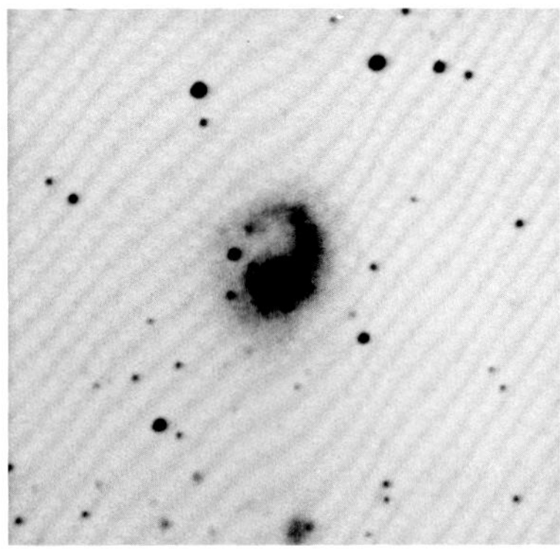

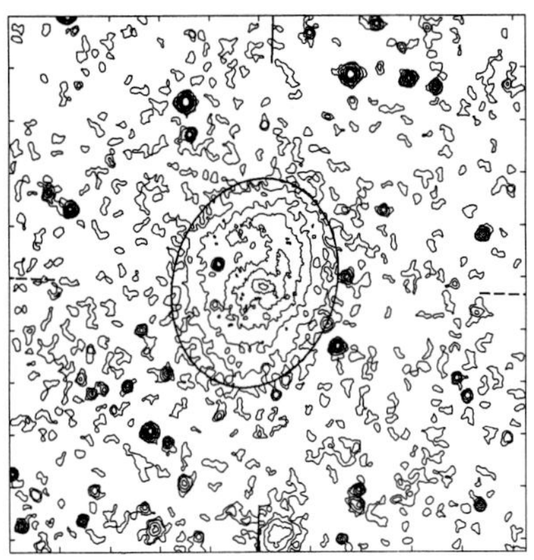

N4027 K4857 1 20.99 512X512

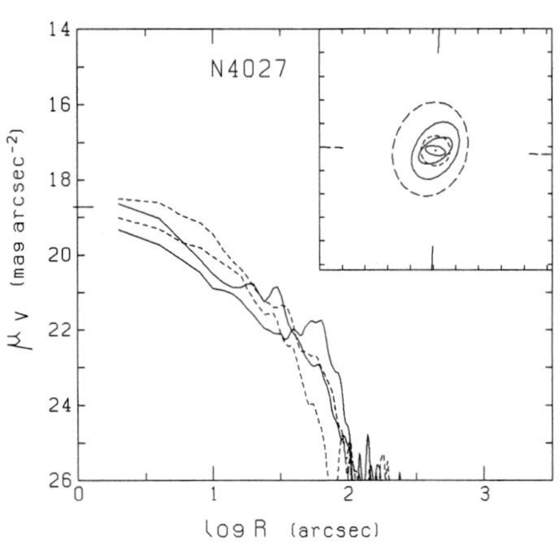

NGC4030

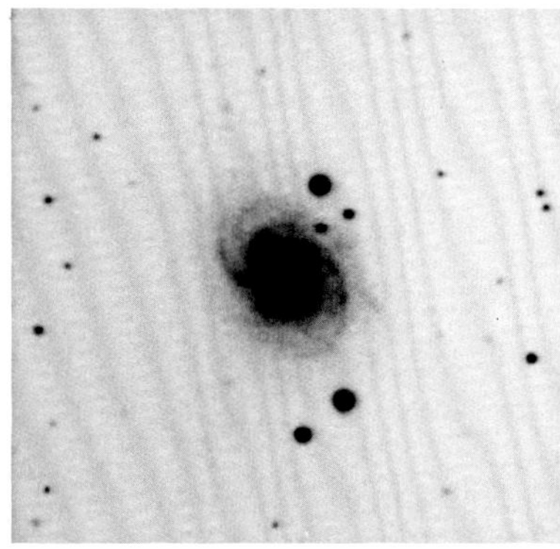

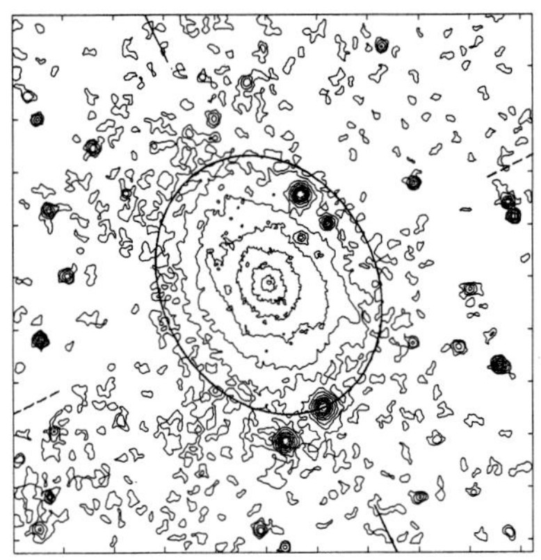

N4030 K3243 1 21.12 512X512

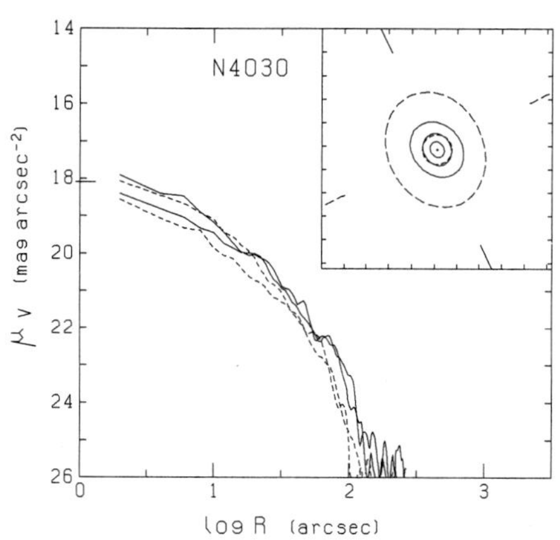

NGC4033

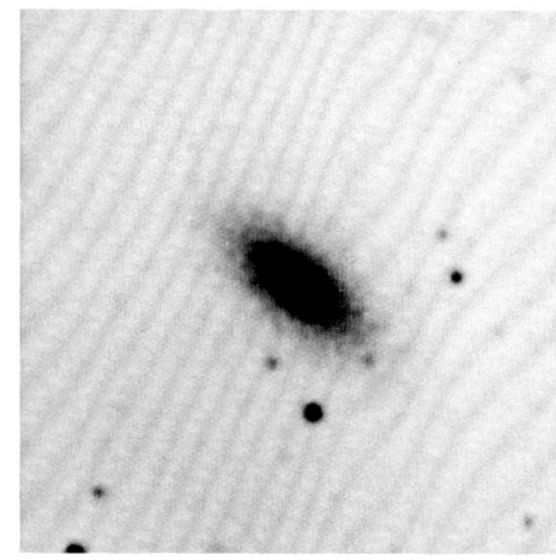

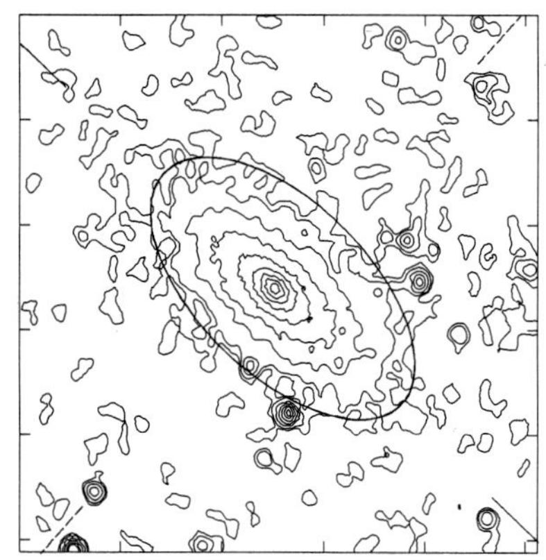

N4033 K4857 1 20.99 256X256

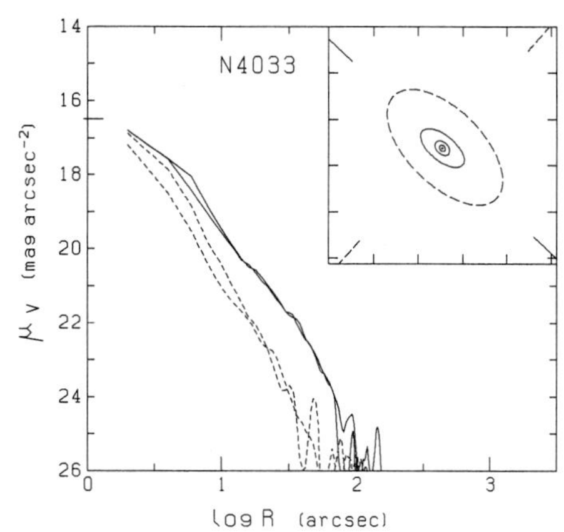

NGC4037

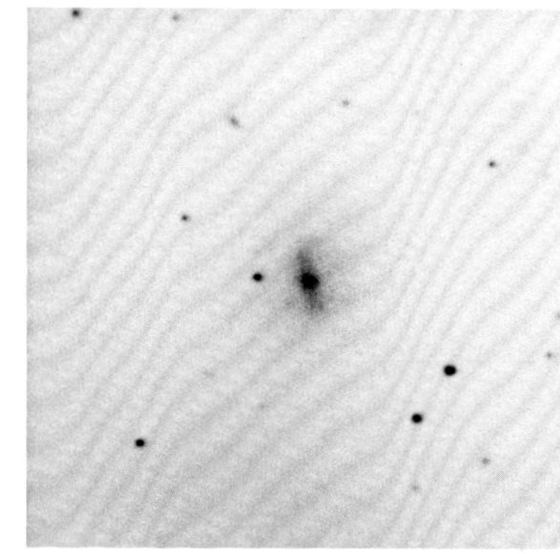

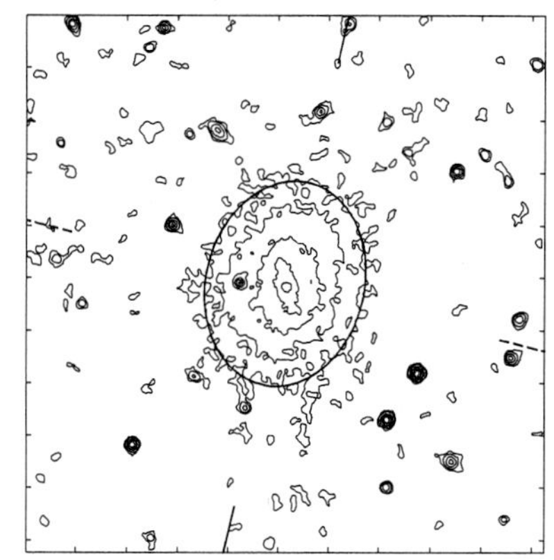

N4037 K3523 1 21.23 512X512

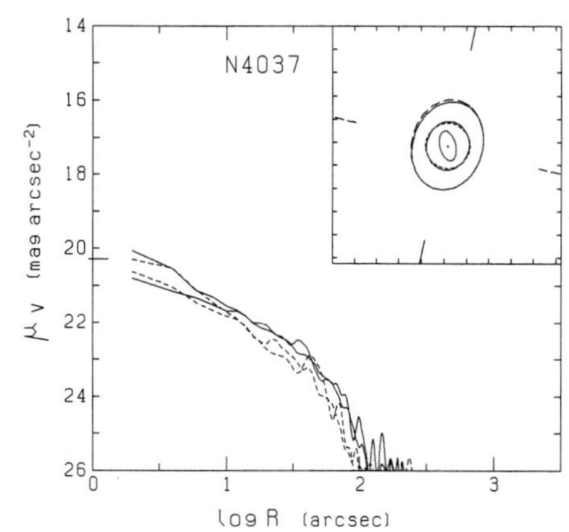

NGC4036

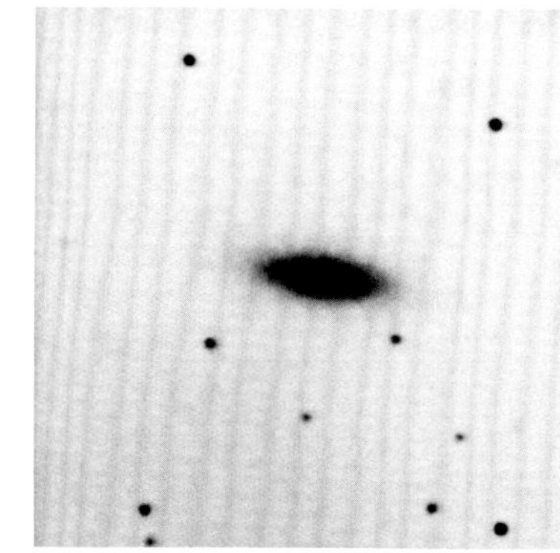

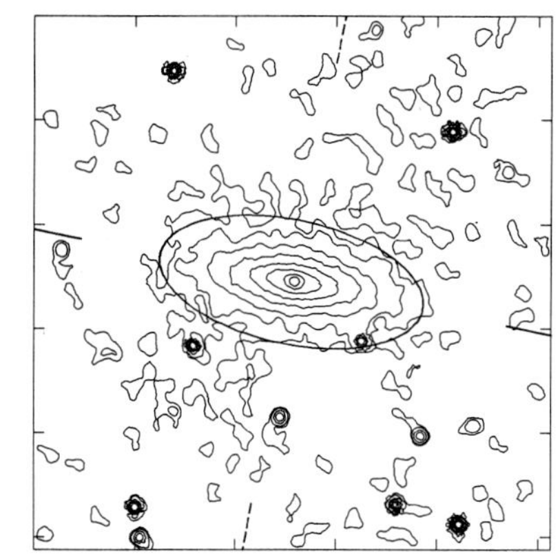

N4036 K5770 2 20.99 256X256

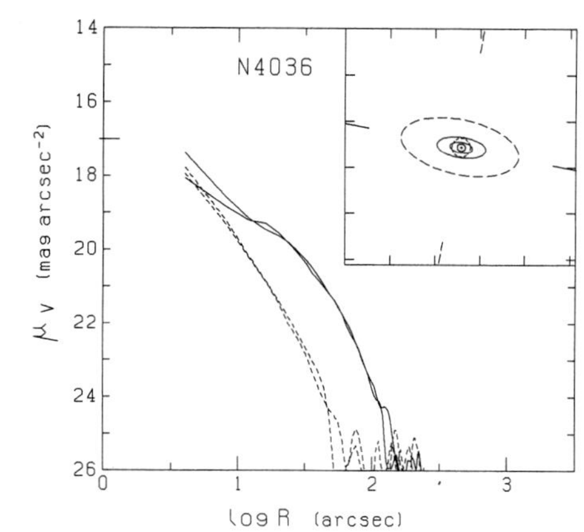

NGC4038

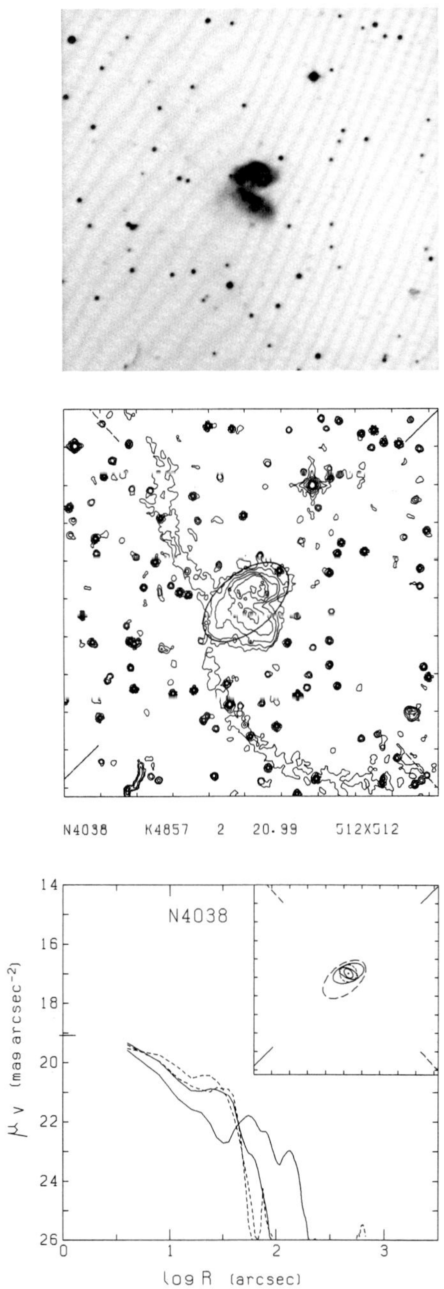

NGC4045

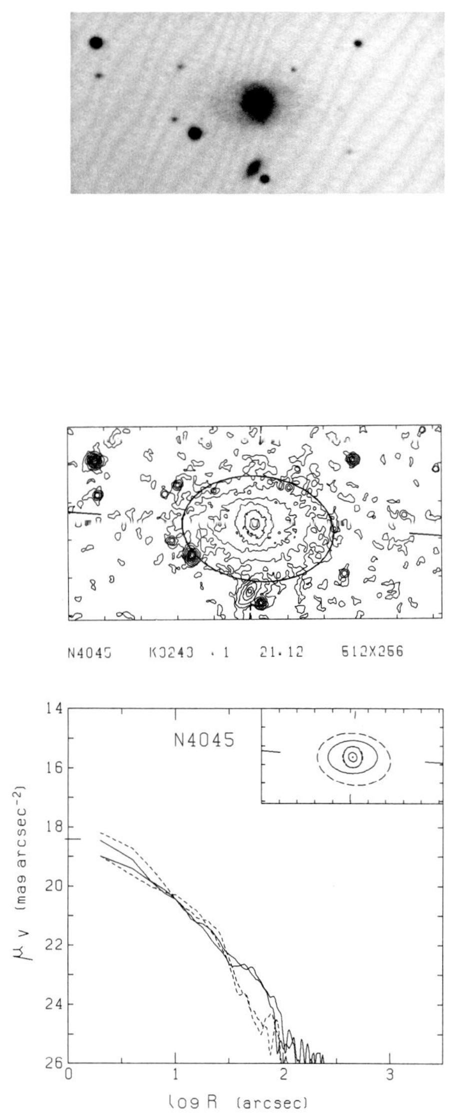

NGC4047

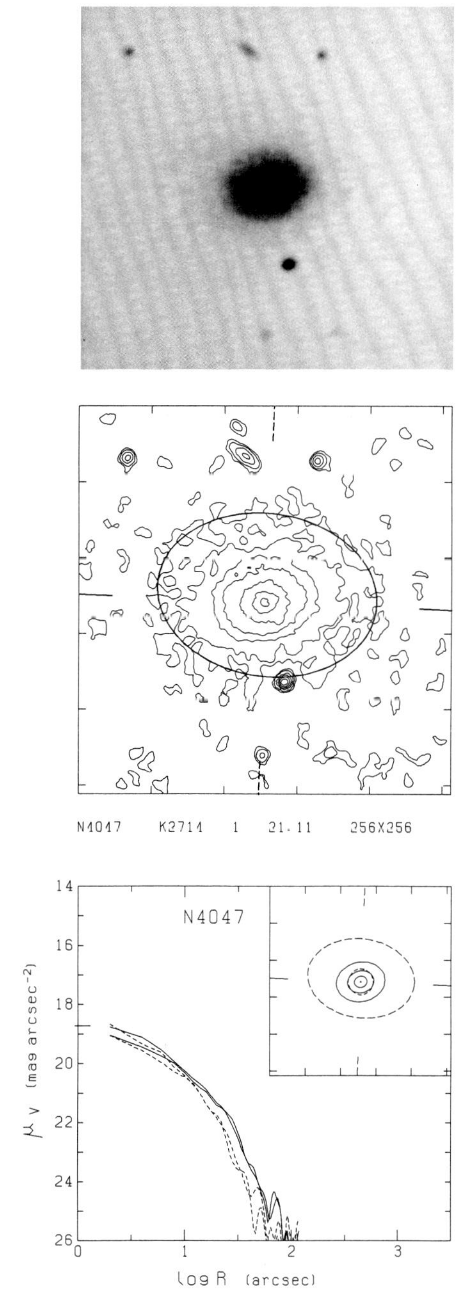

NGC4050

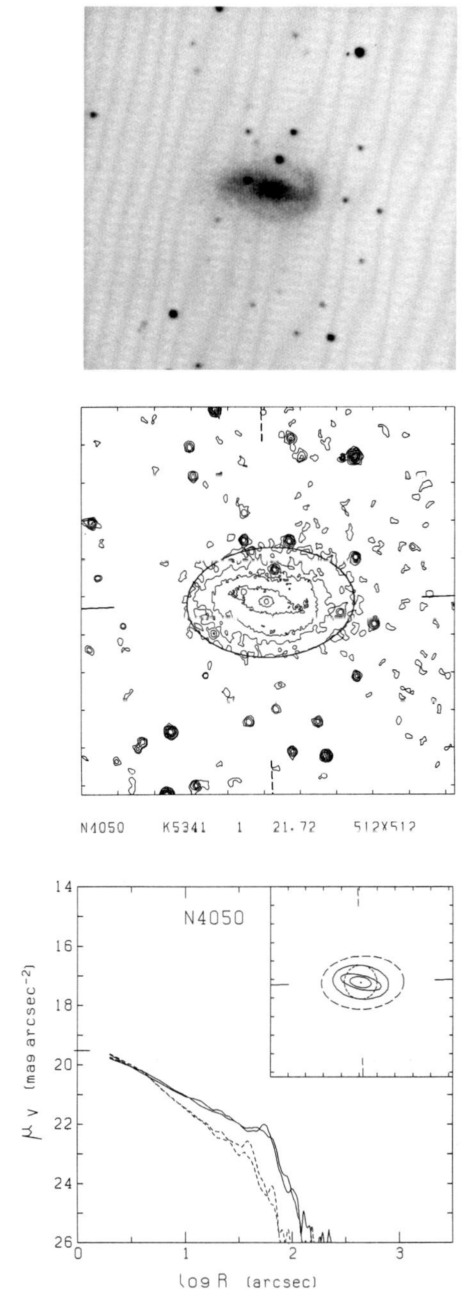

NGC4051

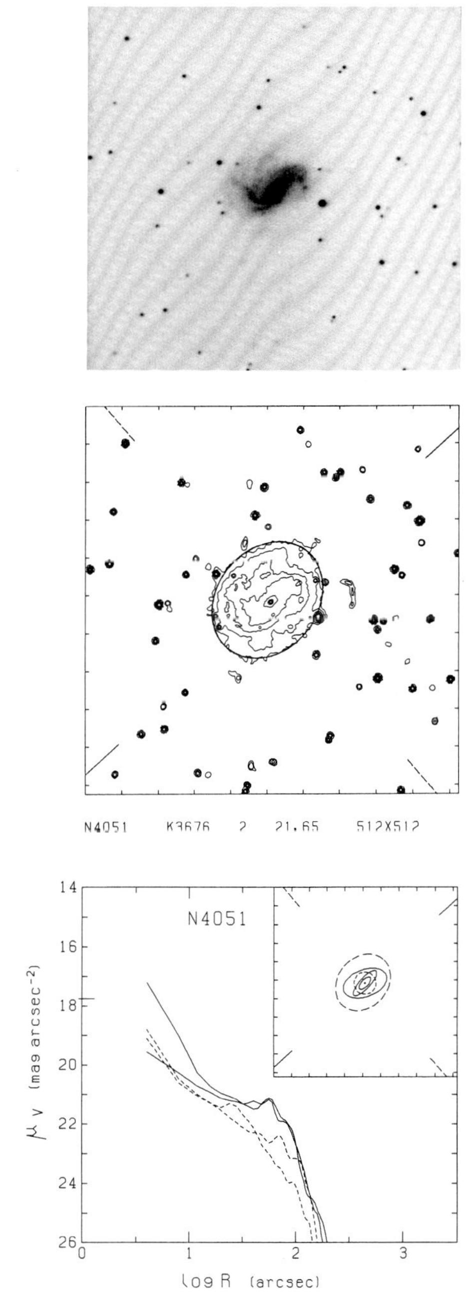

NGC4062

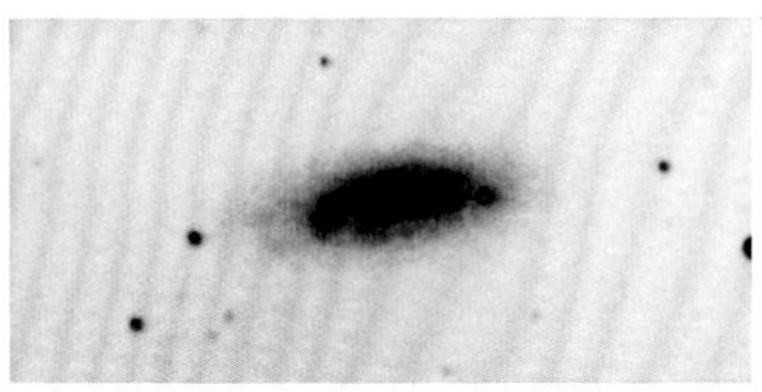

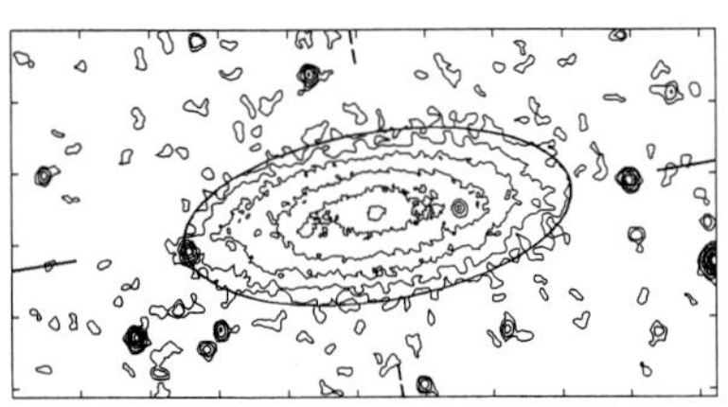

N4062 K5744 1 21.63 512X256

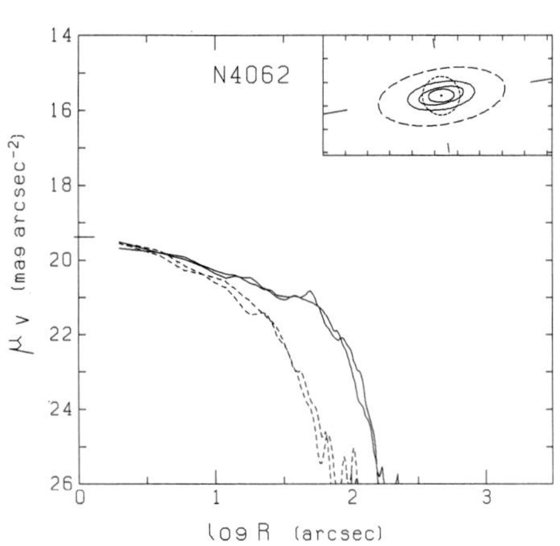

NGC4073

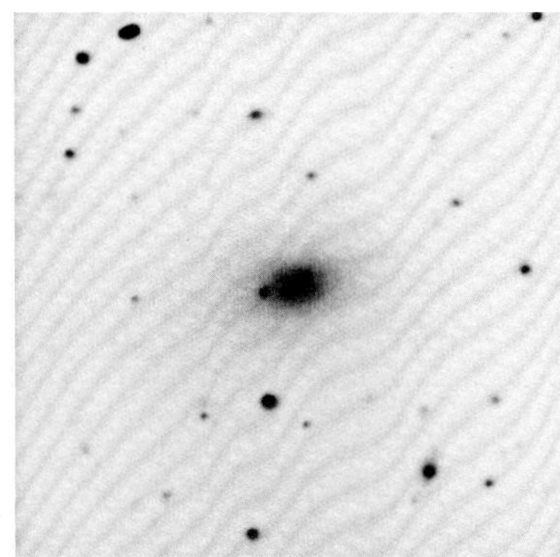

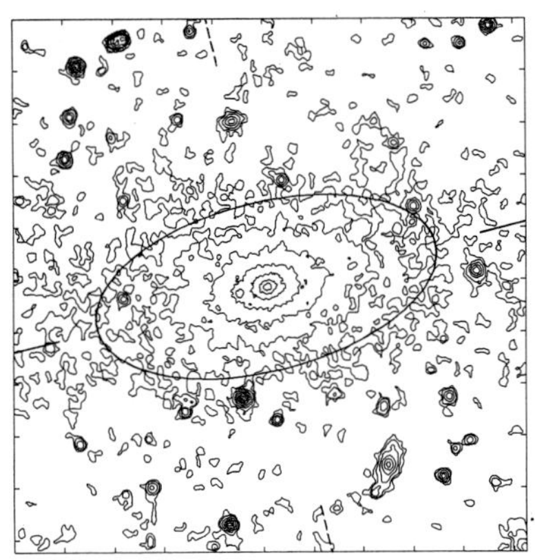

N4073 K3243 1 21.12 512X512

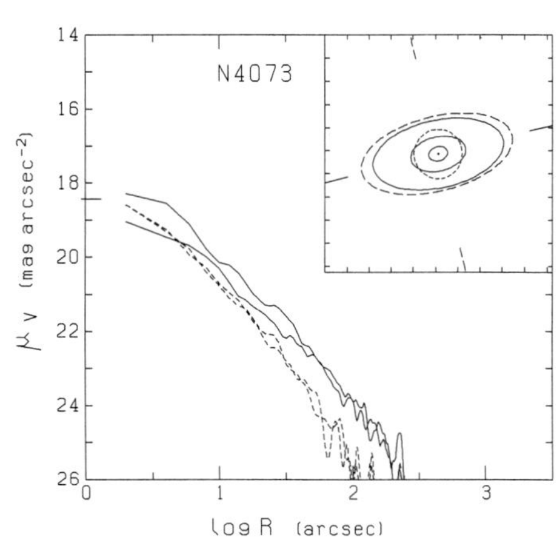

NGC4085

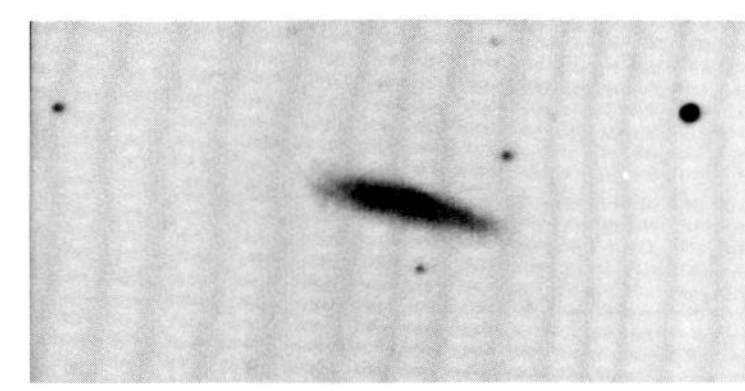

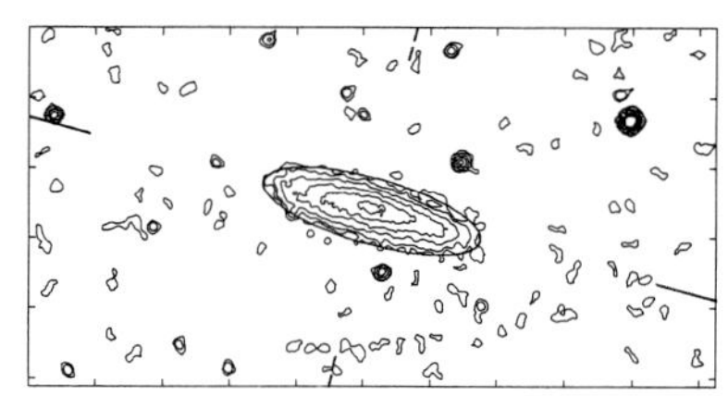

N4085 K2714 1 21.11 512X256

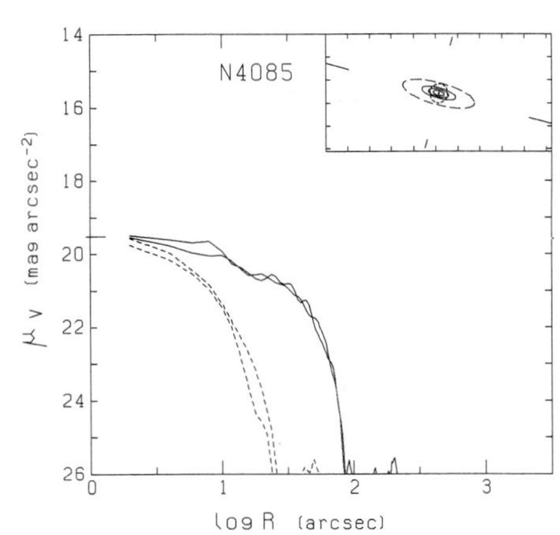

NGC4088

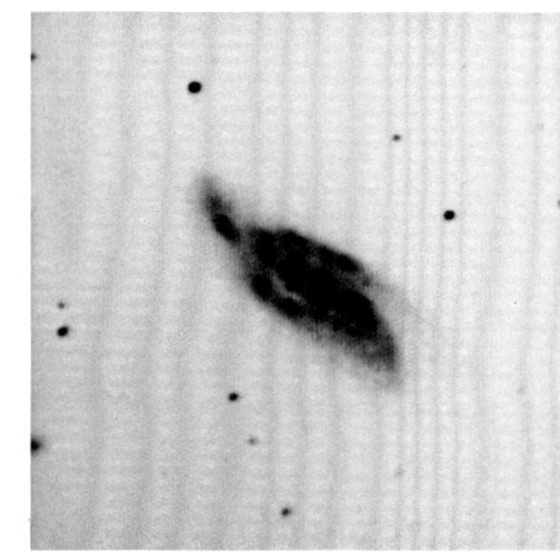

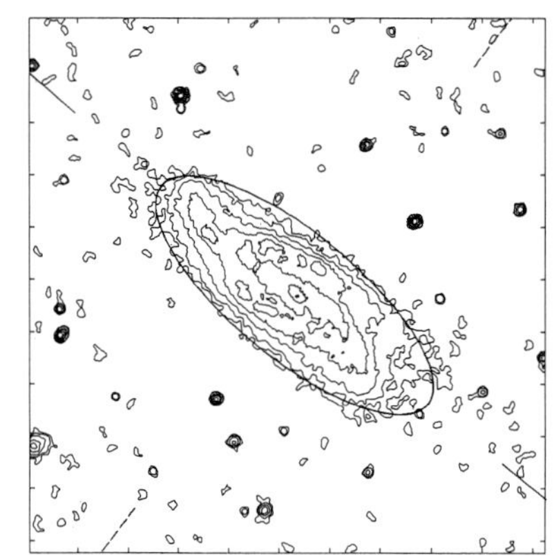

N4088 K2714 1 21.11 512X512

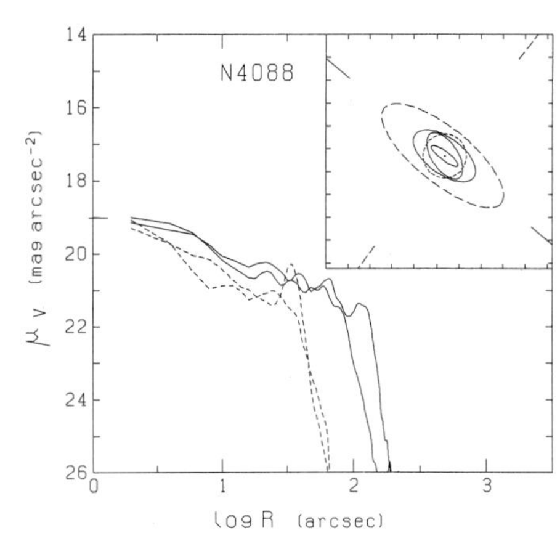

IC2995

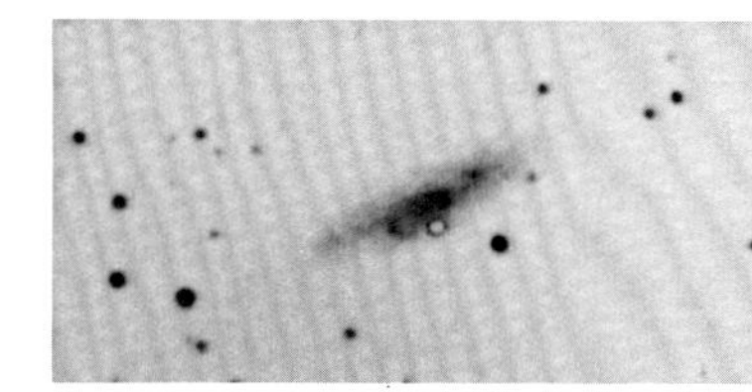

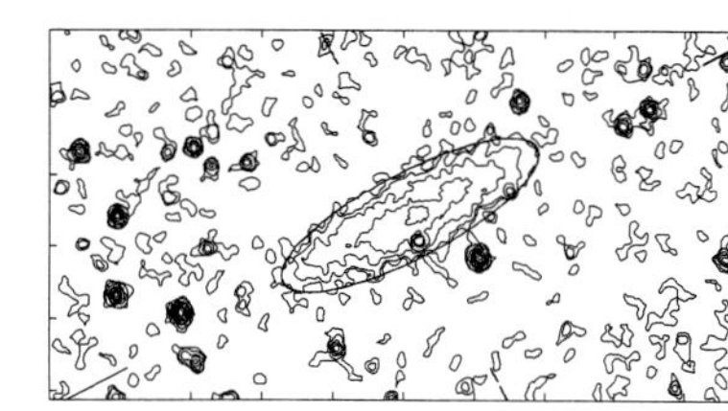

I2995 K4892 1 21.28 512X256

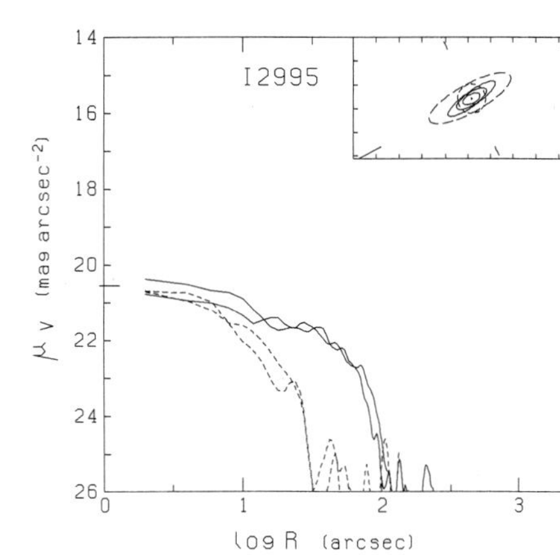

NGC4094

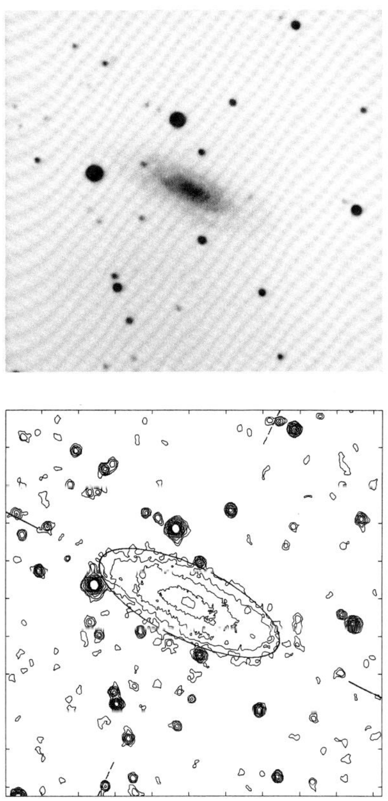

N4094 K5341 1 21.72 512X512

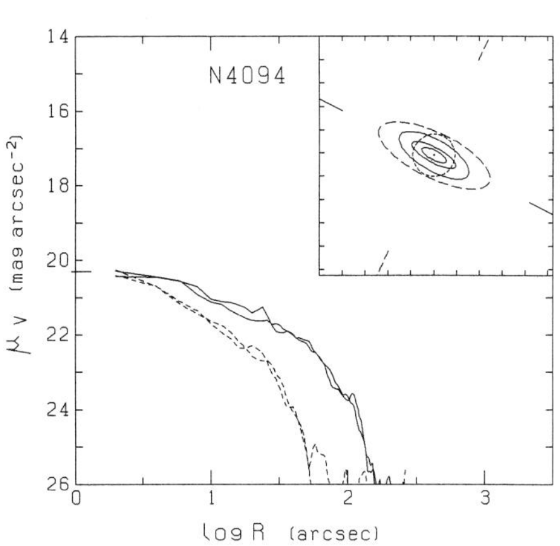

NGC4096

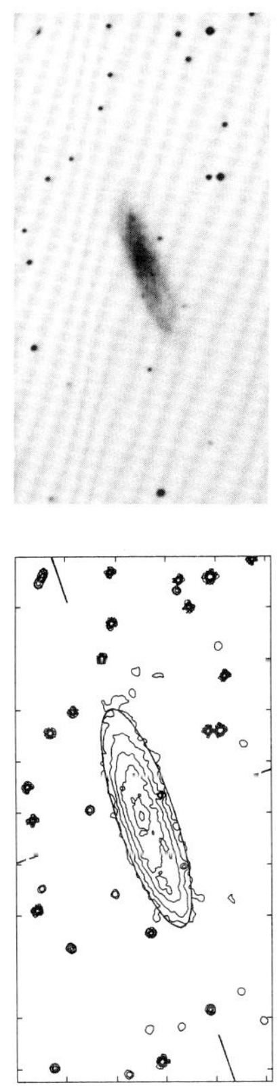

N4096 K2714 2 21.11 256X512

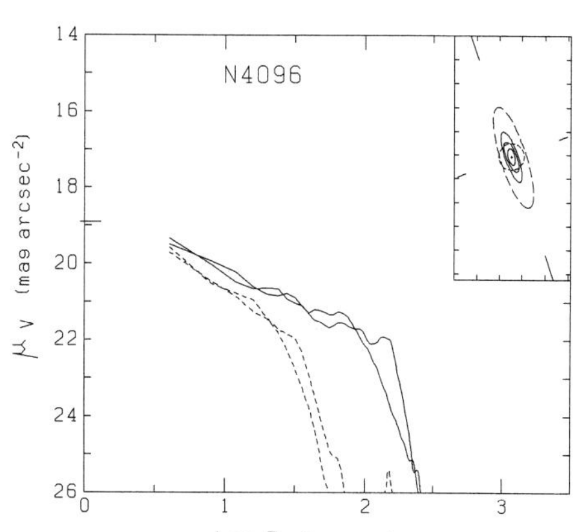

NGC4100

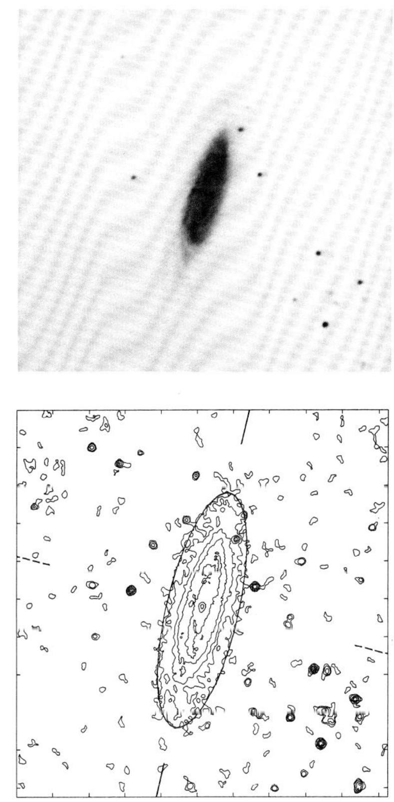

N4100 K2714 1 21.11 512X512

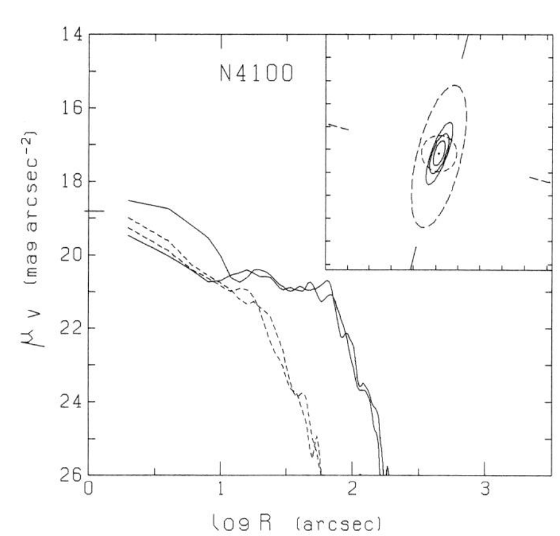

NGC4105

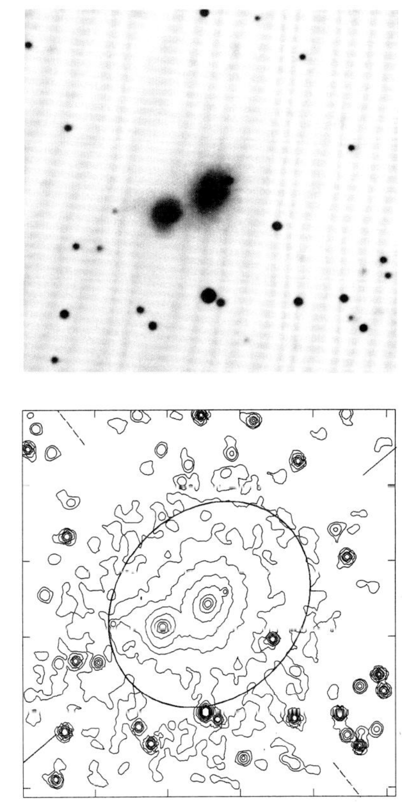

N4105 K4892 2 21.28 256X256

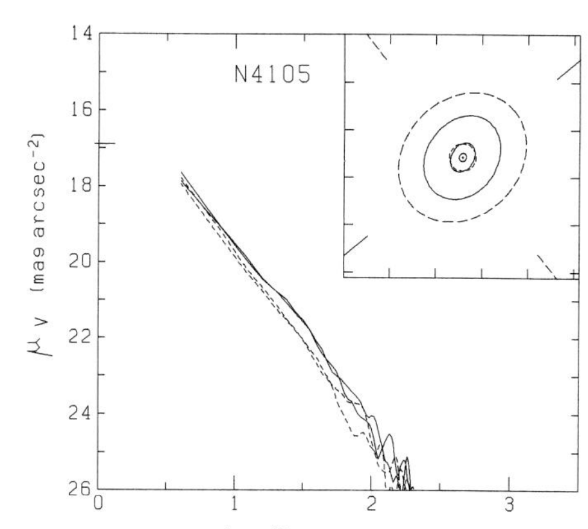

NGC4106

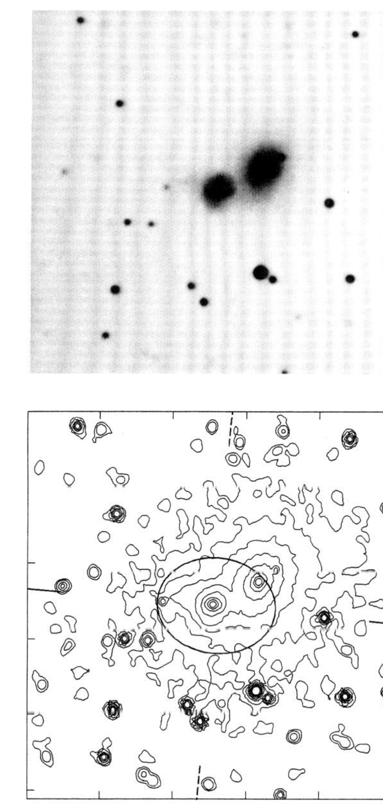

N4106 K4892 2 21.28 256X256

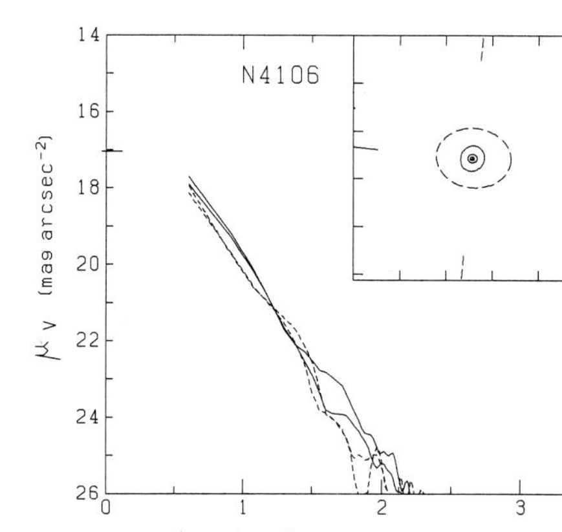

NGC4111

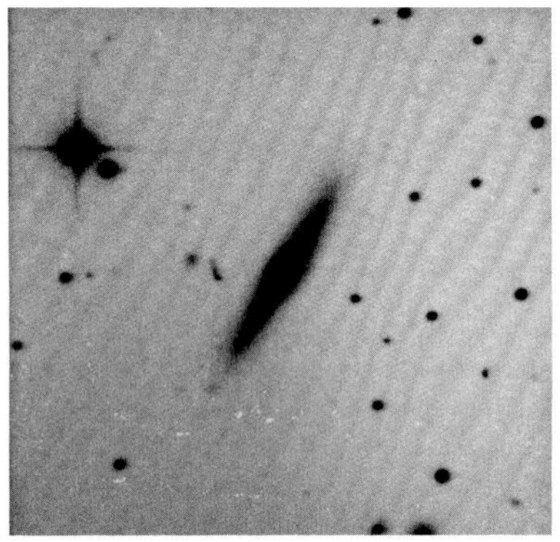

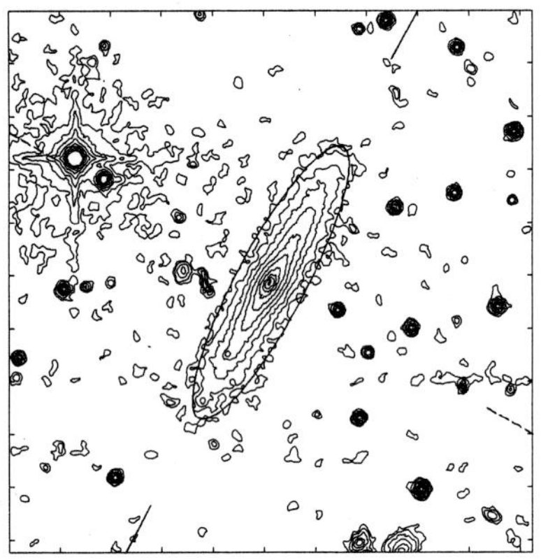

N4111 K3676 1 21.65 512X512

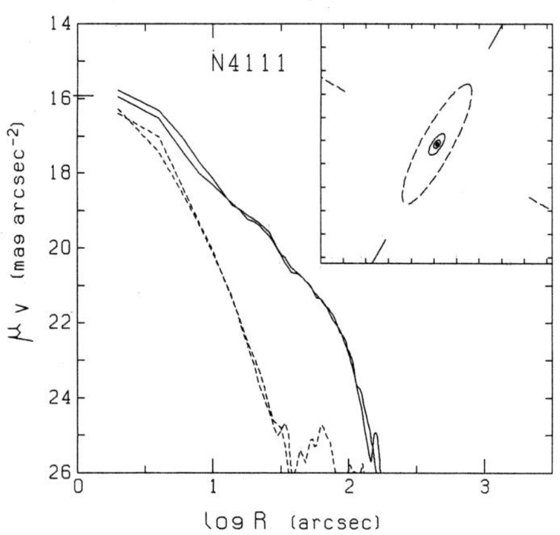

NGC4116

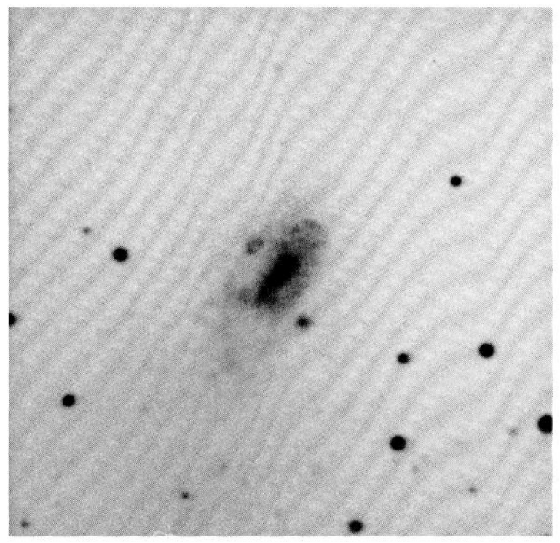

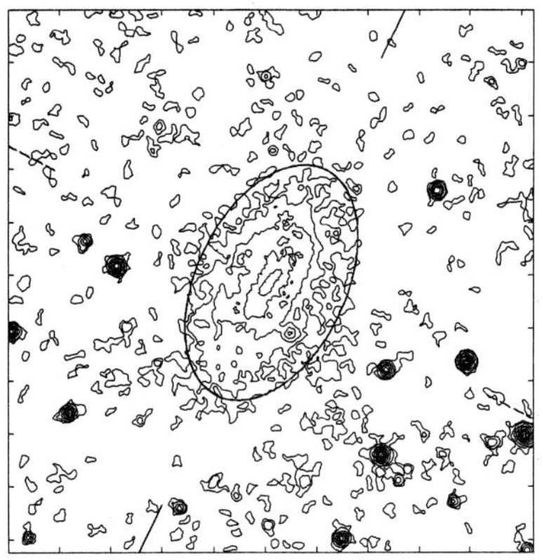

N4116 K3243 1 21.12 512X512

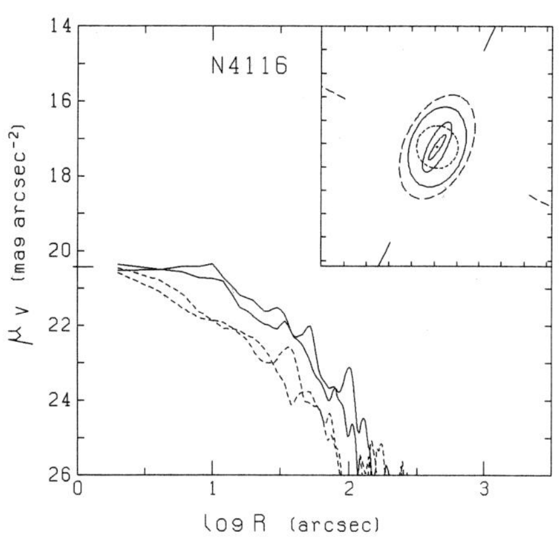

NGC4124

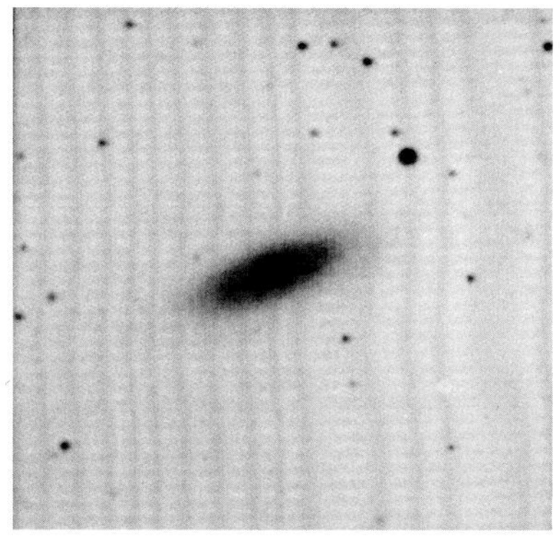

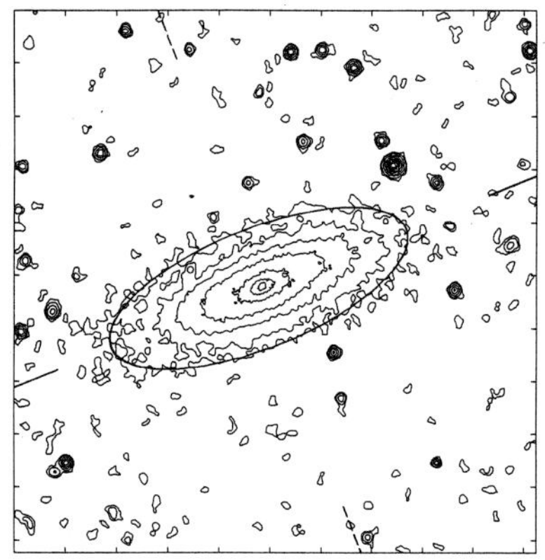

N4124 K3524 1 21.04 512X512

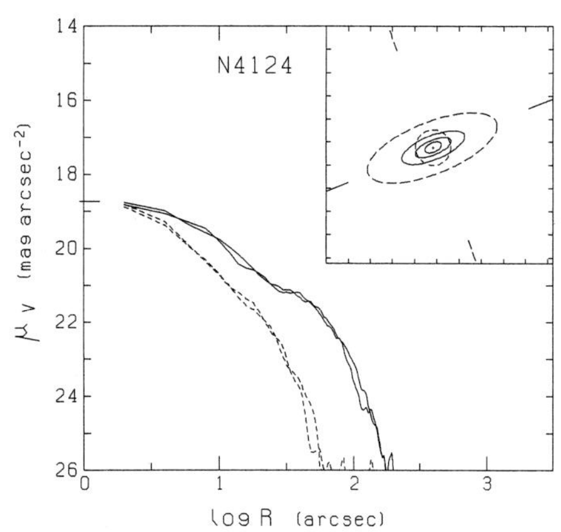

NGC4123

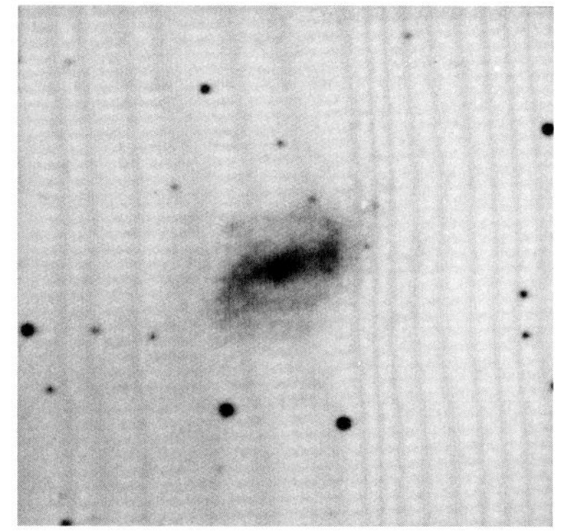

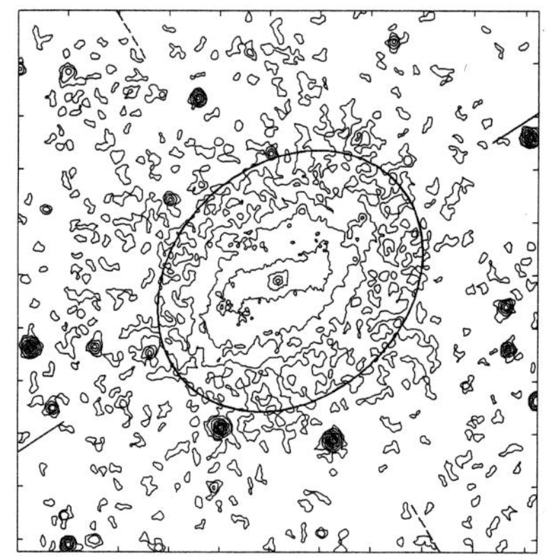

N4123 K3243 1 21.12 512X512

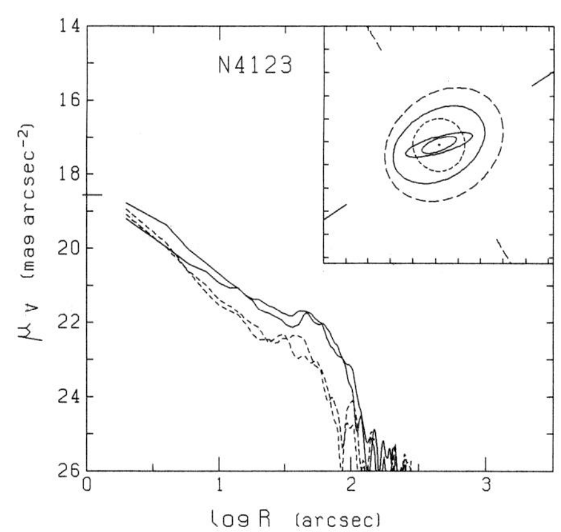

NGC4136

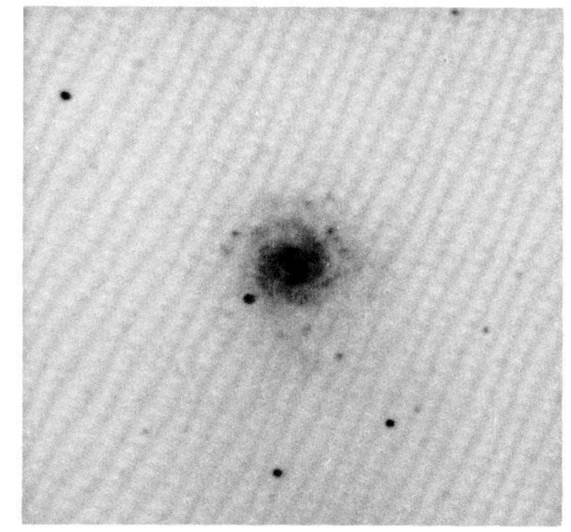

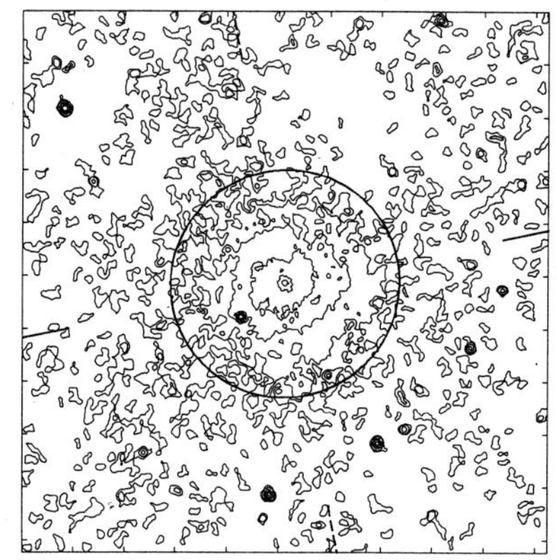

N4136 K3937 1 21.19 512X512

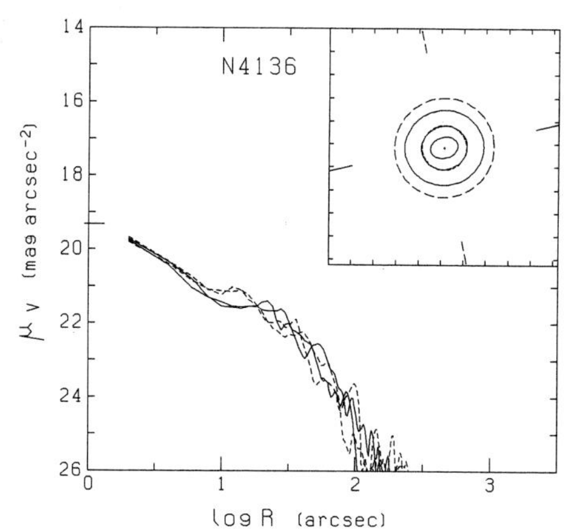

NGC4138

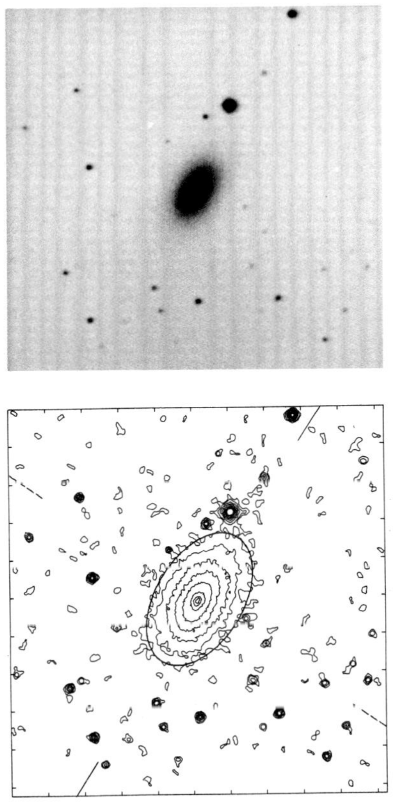

N4138 K3676 1 21.65 512X512

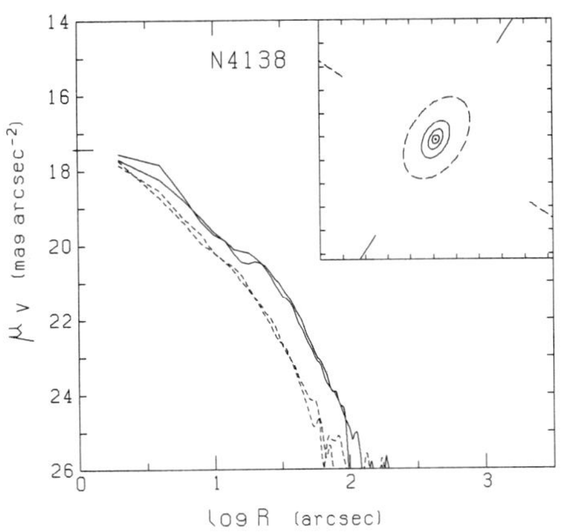

NGC4143

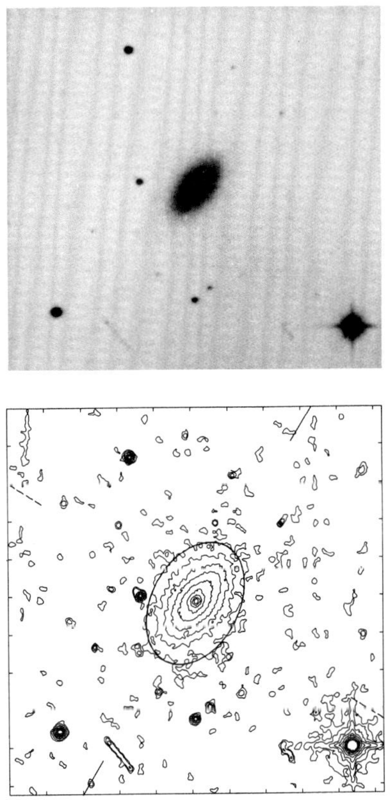

N4143 K3676 1 21.65 512X512

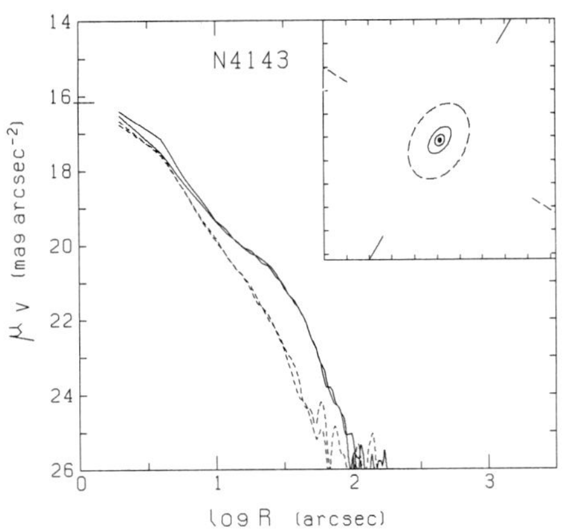

NGC4144

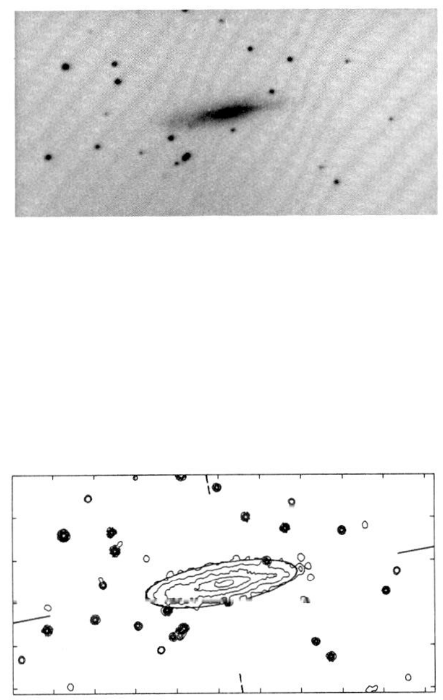

N4144 K3676 2 21.65 512X256

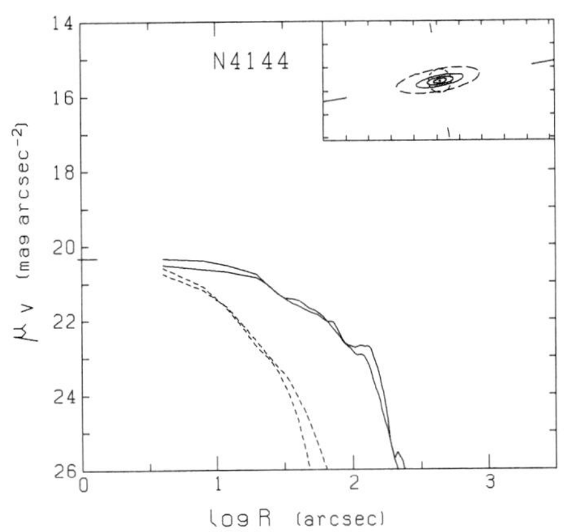

NGC4145

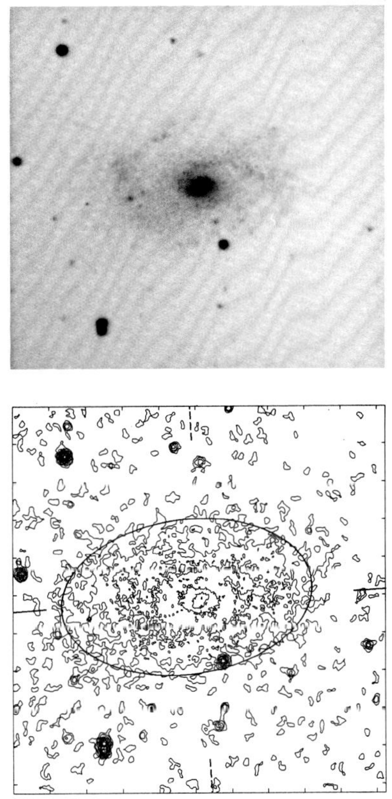

N4145 K4885 1 21.94 512X512

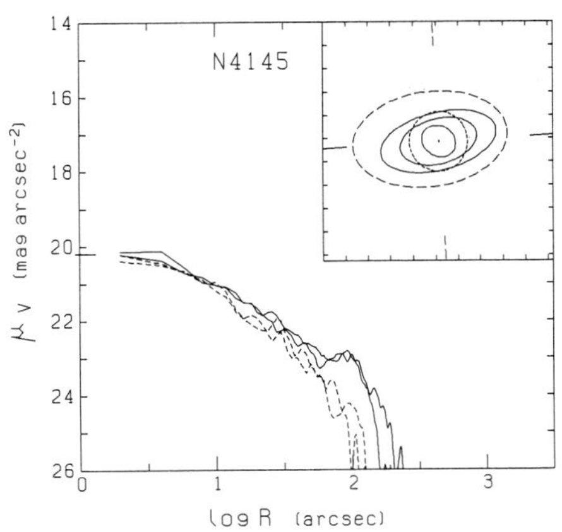

IC 764

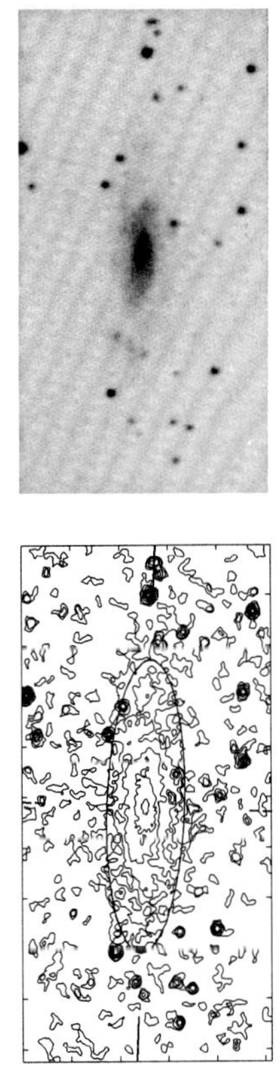

I0764 K4892 1 21.28 256X512

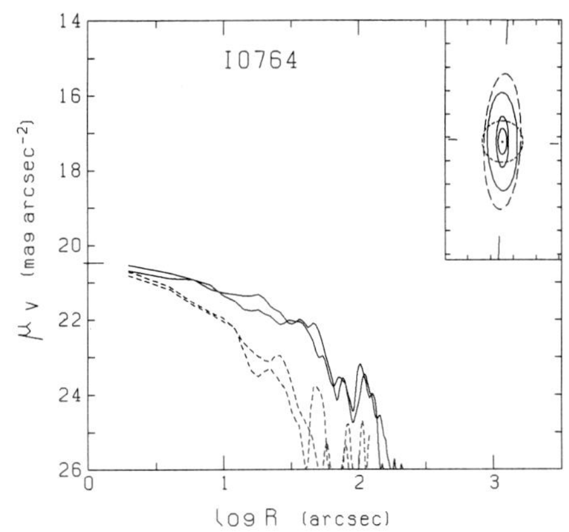

NGC4151

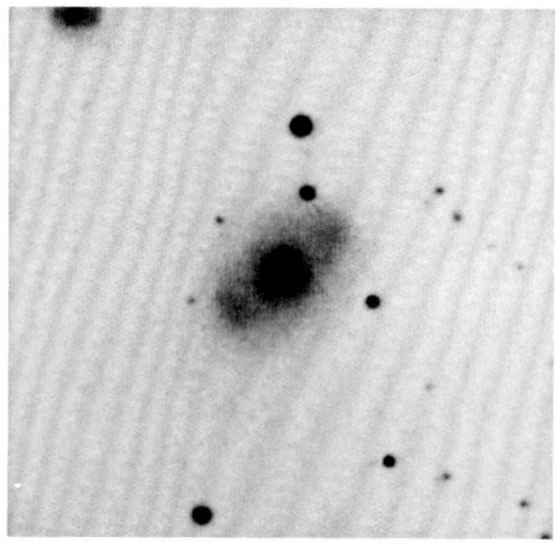

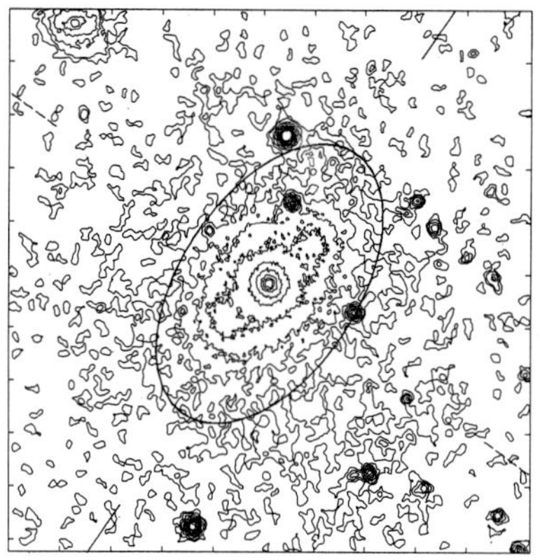

N4151 K4885 1 21.94 512X512

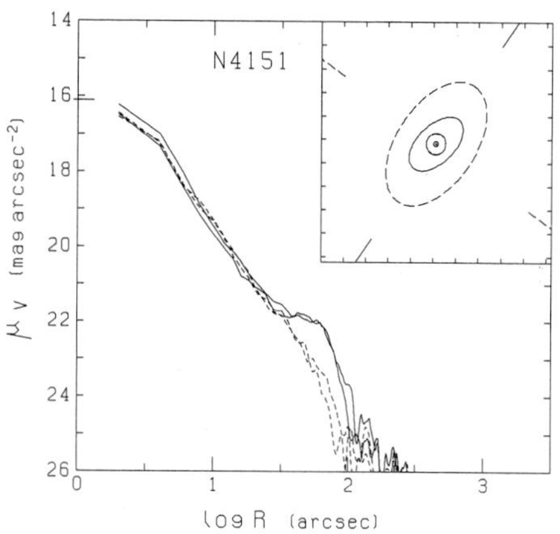

NGC4150

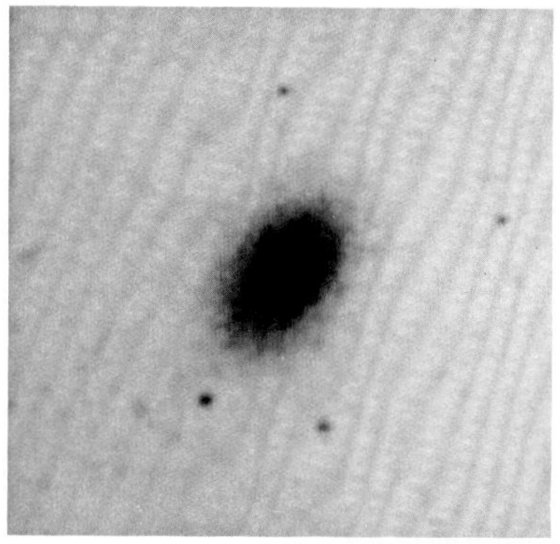

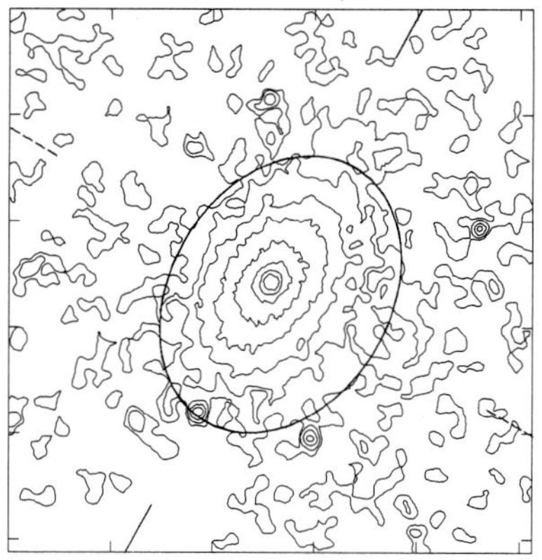

N4150 K3937 1 21.19 256X256

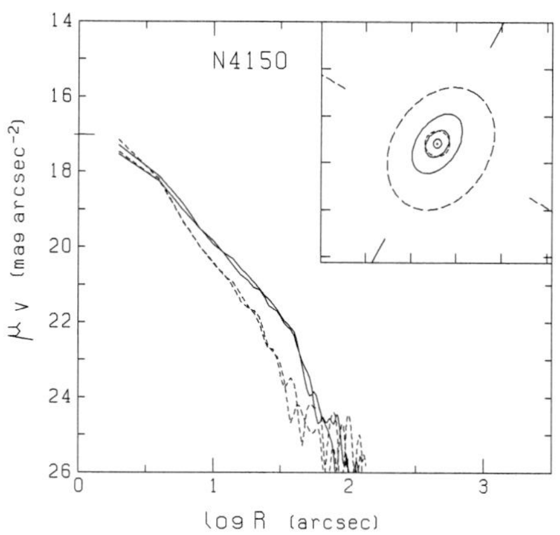

NGC4152

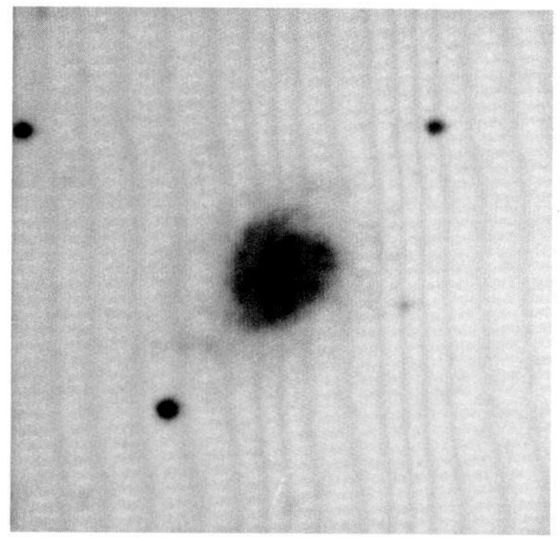

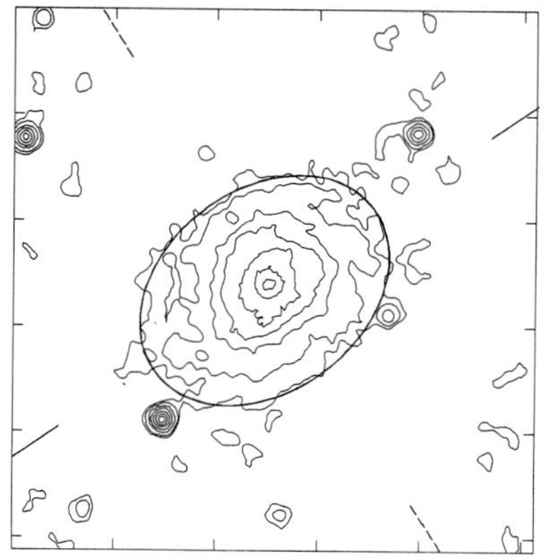

N4152 K3523 1 21.23 256X256

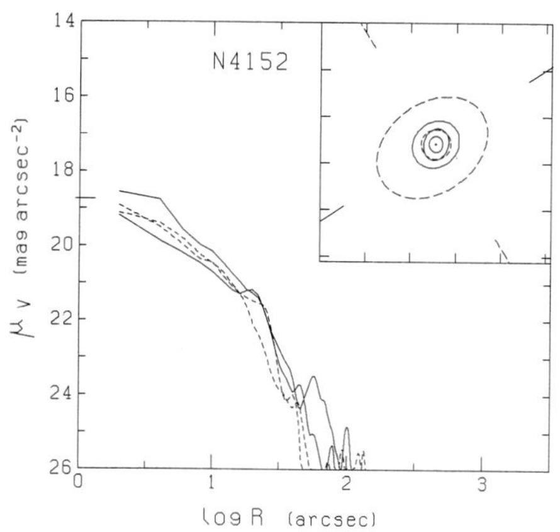

NGC4157

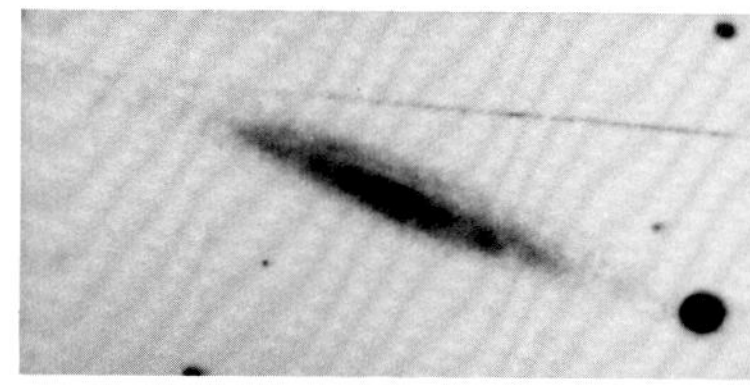

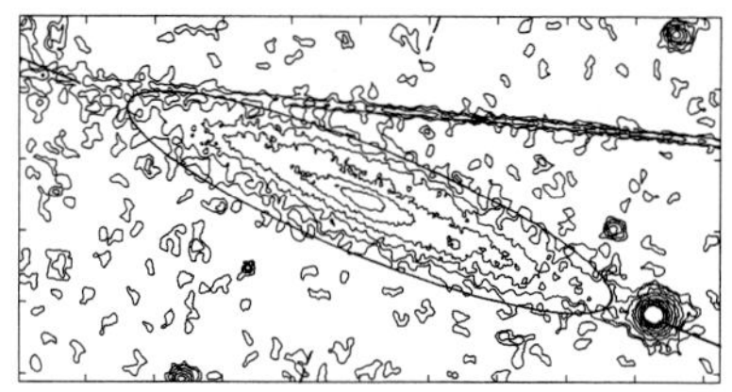

N4157 K4894 1 21.75 512X256

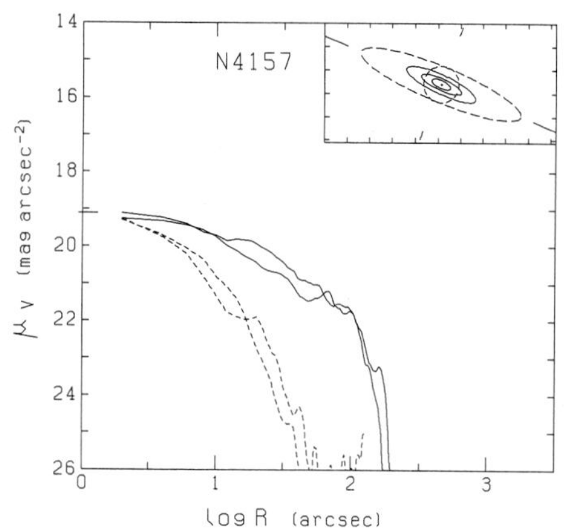

NGC4162

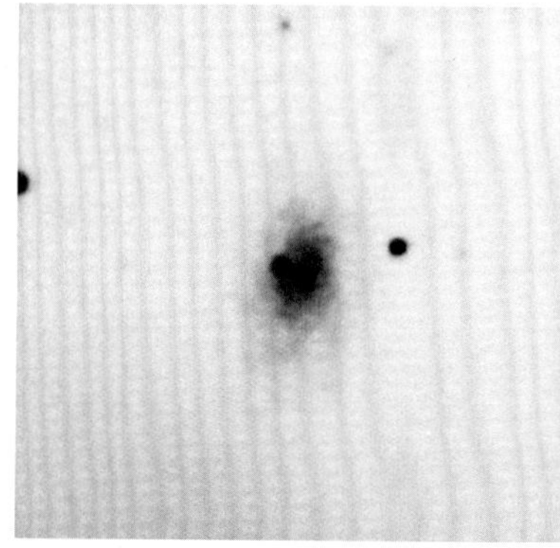

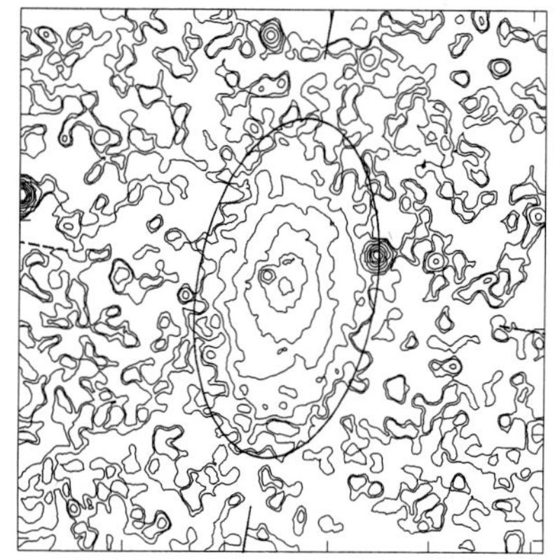

N4162 K3936 1 19.92 256X256

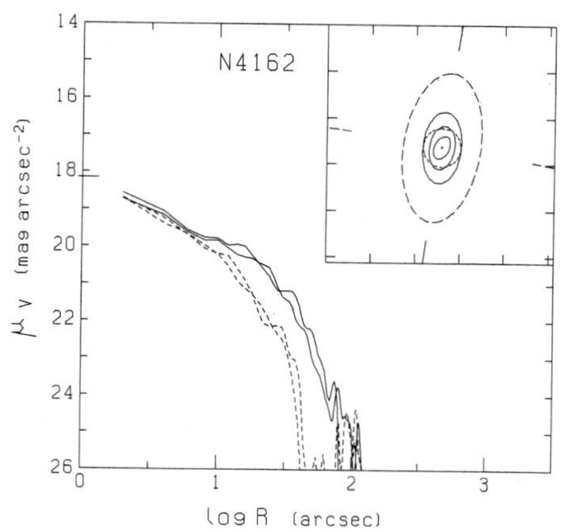

NGC4168

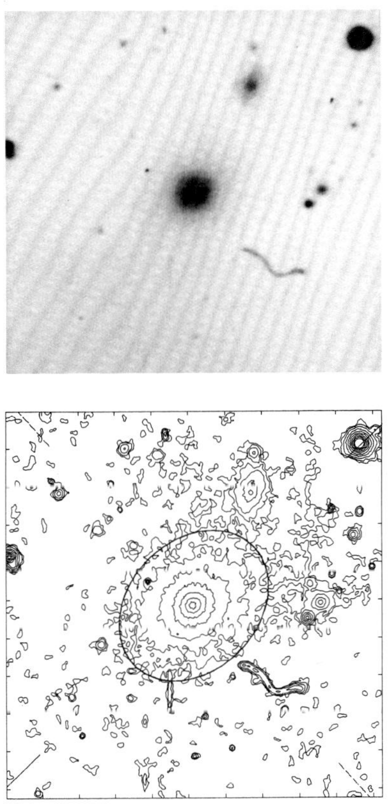

N4168 K2106 1 20.73 512X512

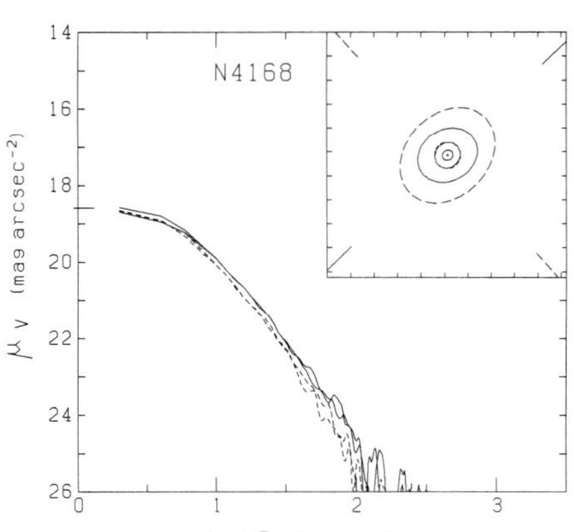

NGC4178

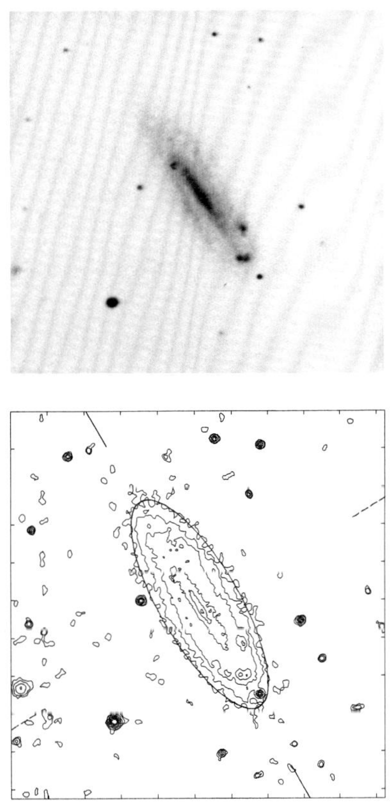

N4178 K2186 1 21.34 512X512

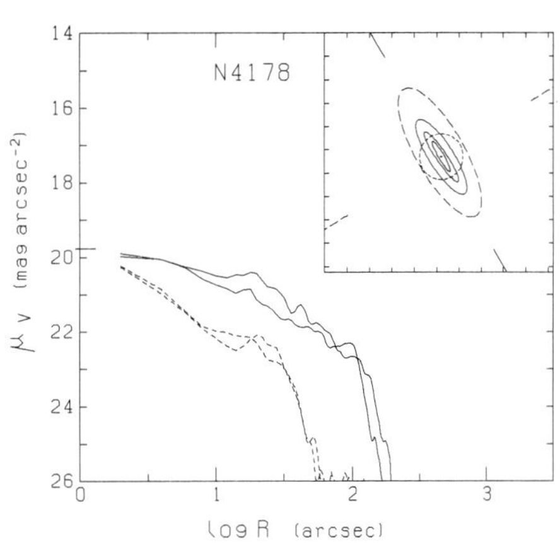

NGC4179

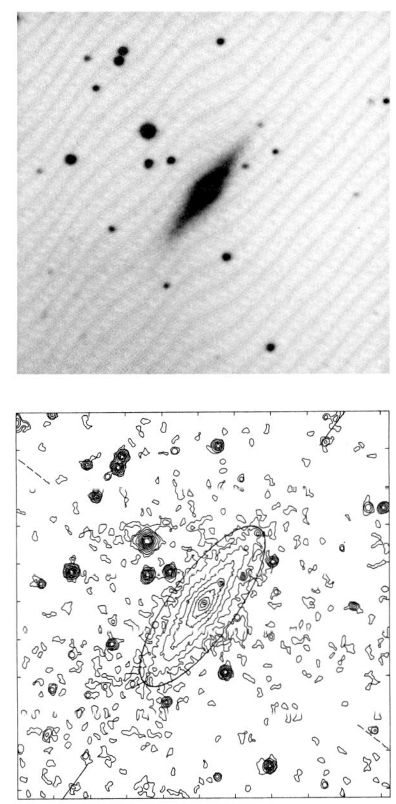

N4179 K3243 1 21.12 512X512

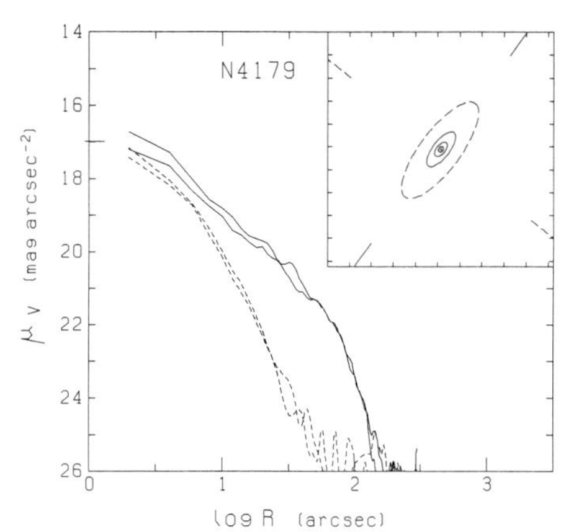

NGC4183

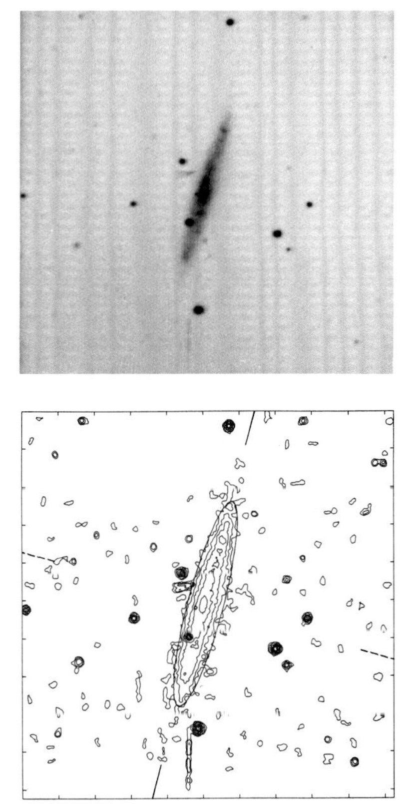

N4183 K3676 1 21.65 512X512

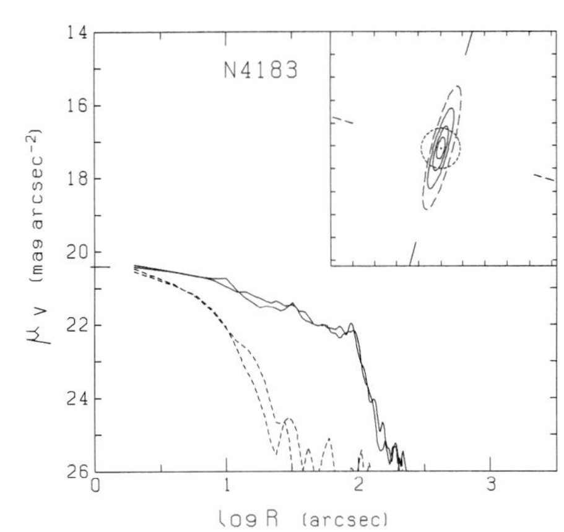

NGC4190

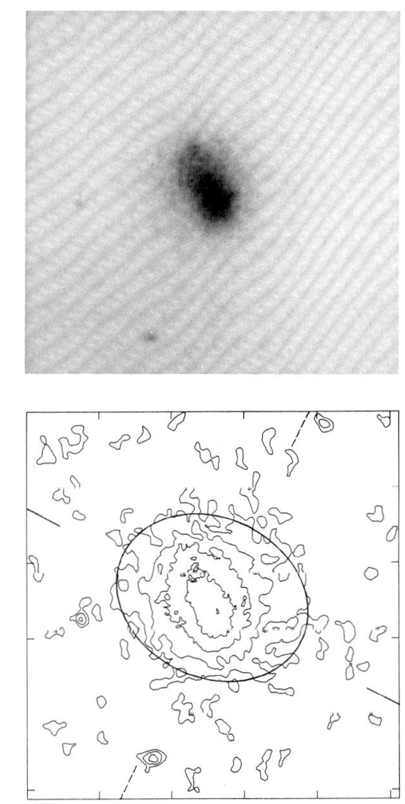

N4190 K5763 1 21.94 256X256

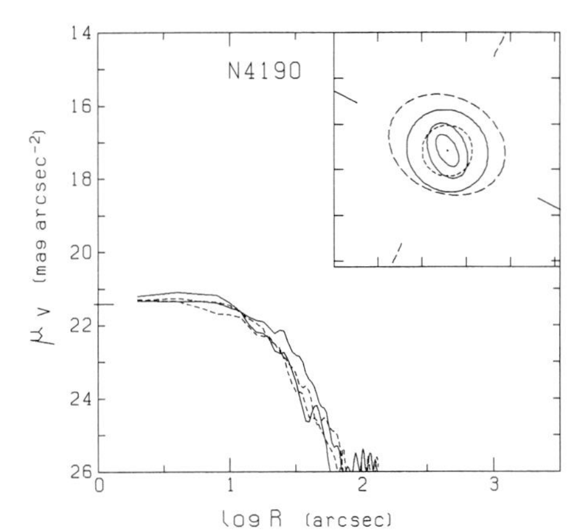

NGC4189

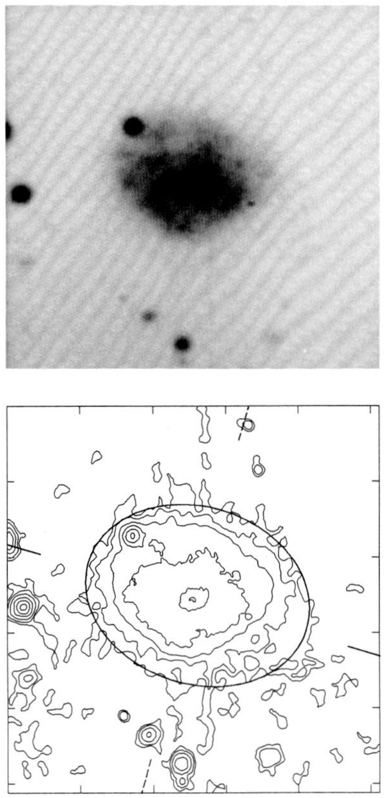

N4189 K2106 1 20.73 256X256

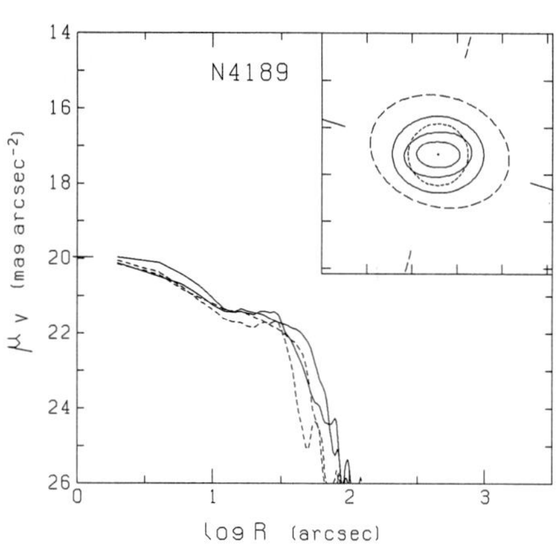

NGC4192

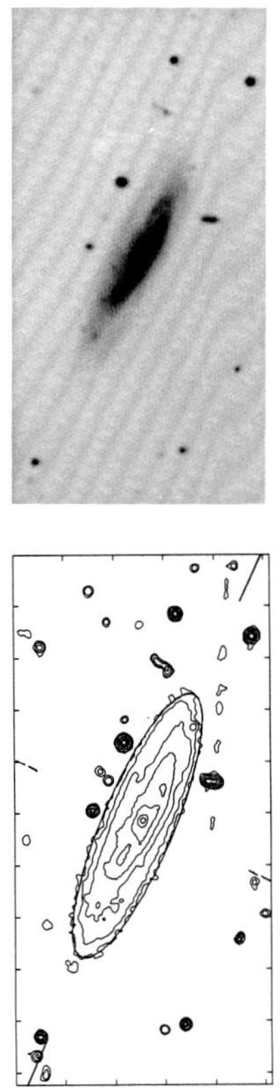

N4192 K2106 2 20.73 256X512

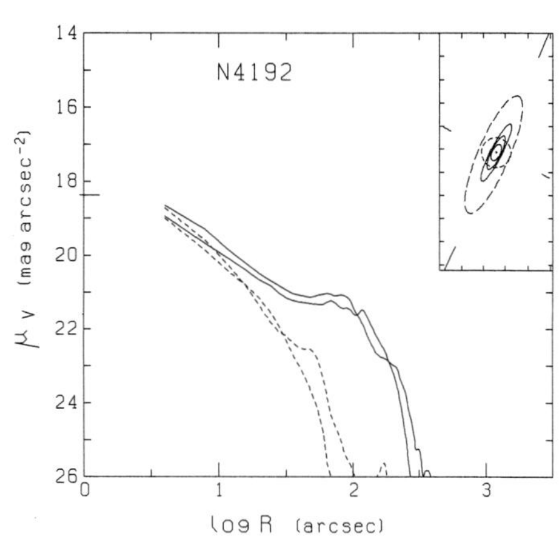

NGC4203

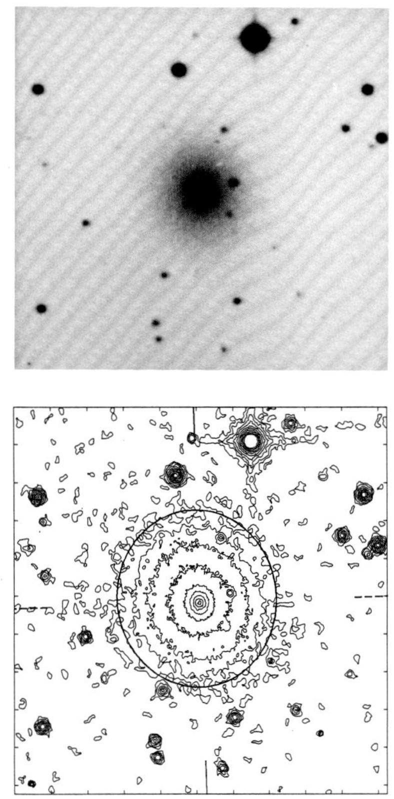

N4203 K5763 1 21.94 512X512

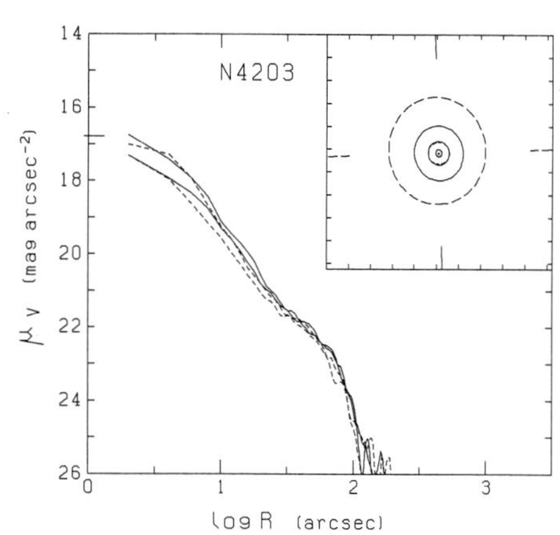

NGC4212

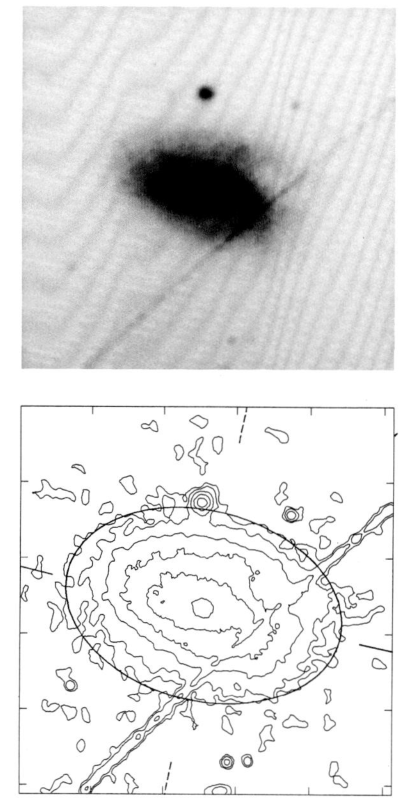

N4212 K2106 1 20.73 256X256

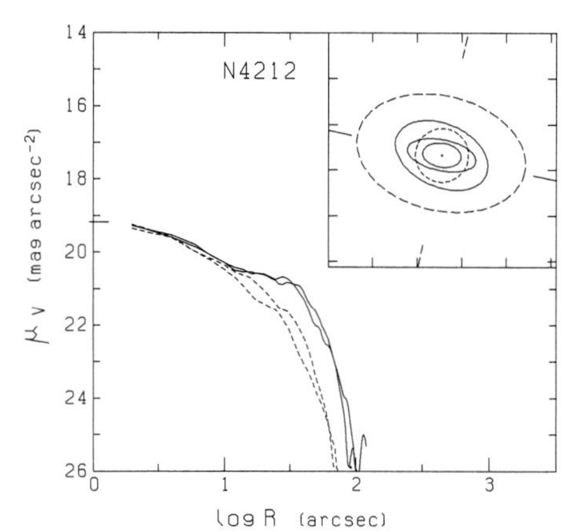

NGC4214

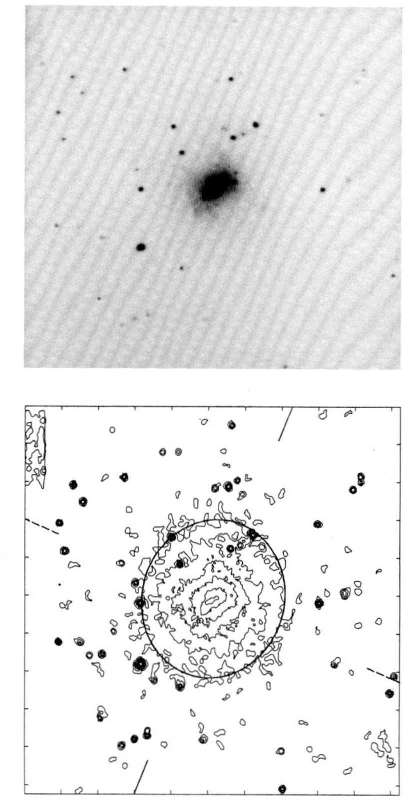

N4214 K4885 2 21.94 512X512

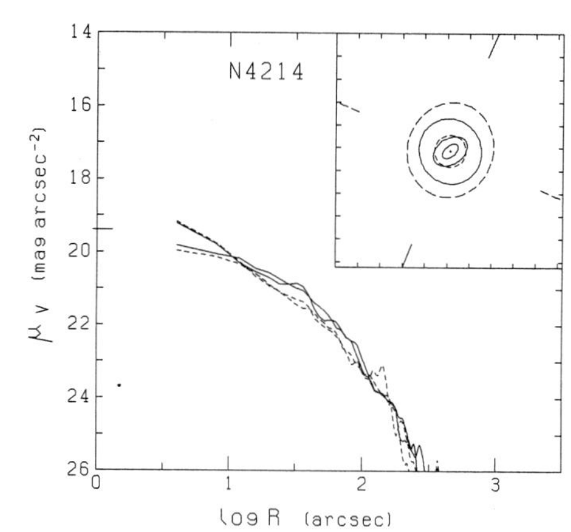

NGC4216

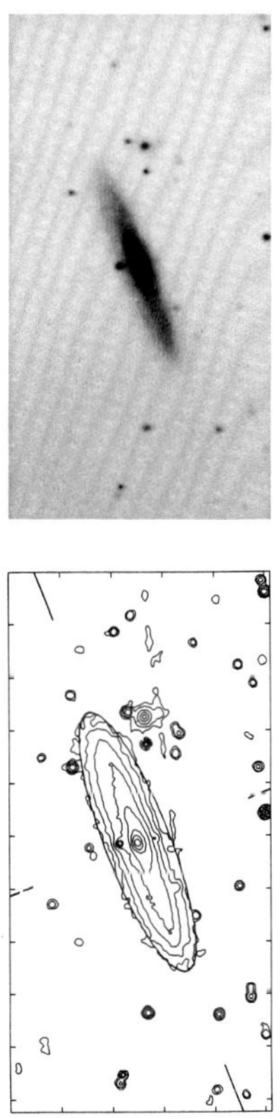

N4216 K2106 2 20.73 256X512

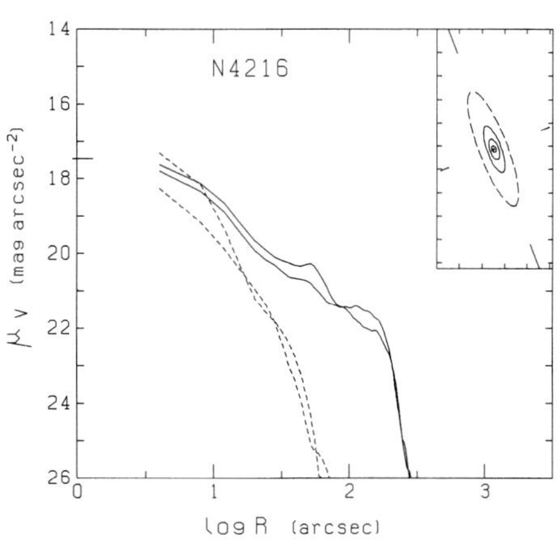

NGC4217

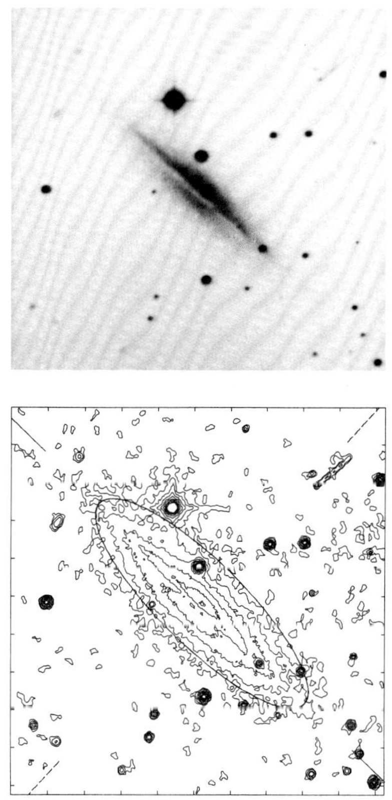

N4217 K3676 1 21.65 512X512

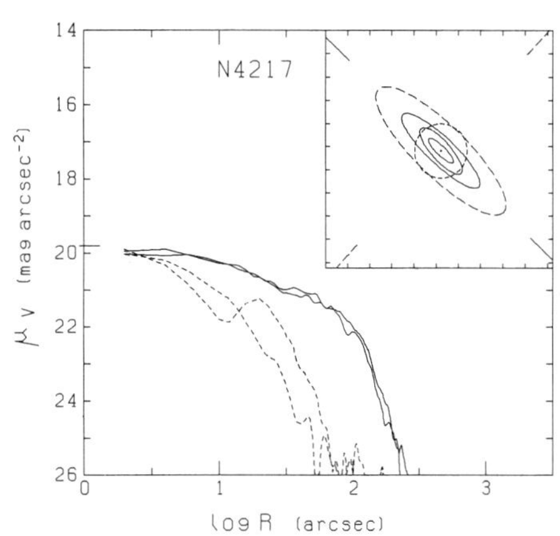

NGC4215

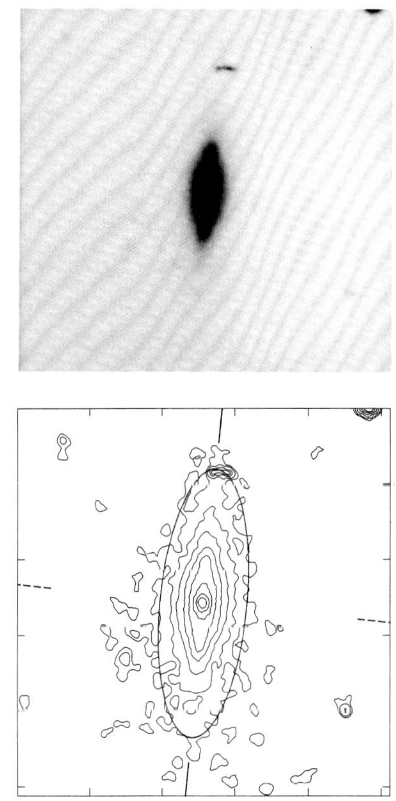

N4215 K3547 1 21.21 256X256

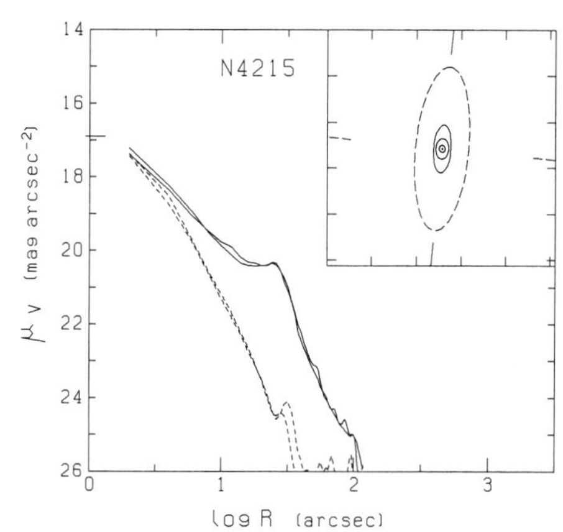

NGC4220

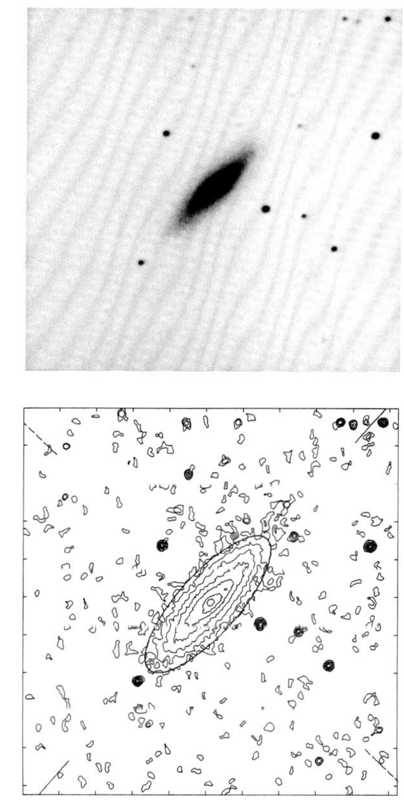

N4220 K2714 1 21.11 512X512

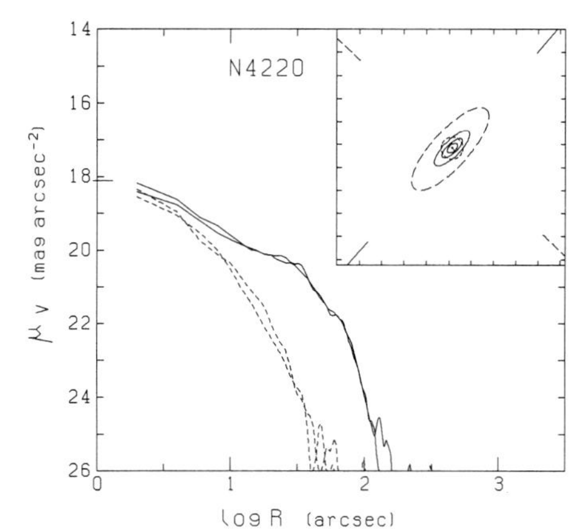

NGC4224

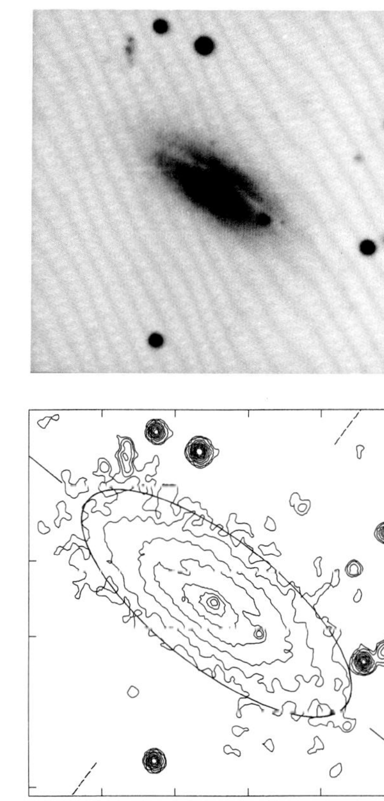

N4224 K3547 1 21.21 256X256

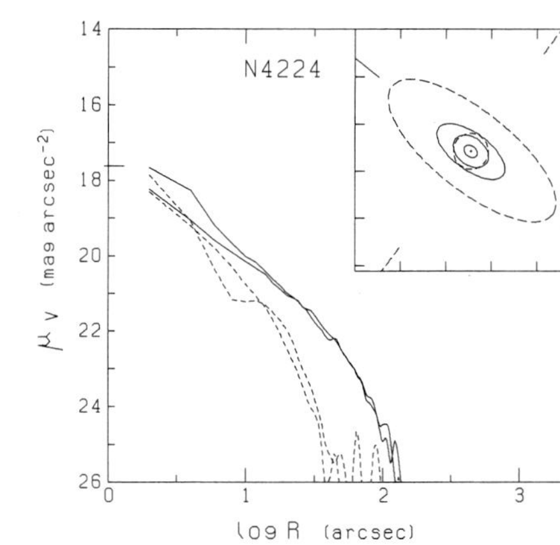

NGC4233

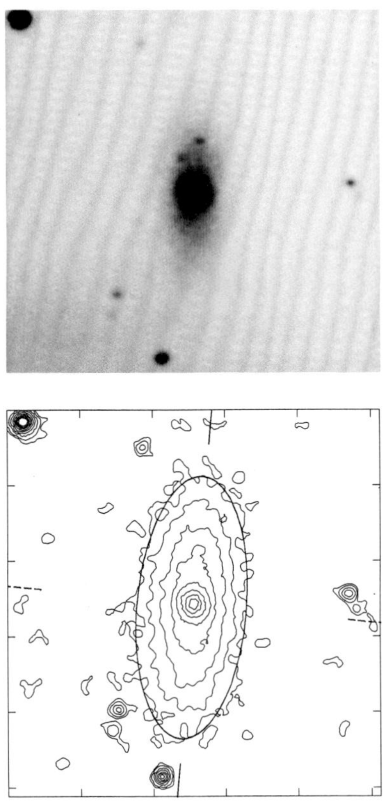

N4233 K2186 1 21.34 256X256

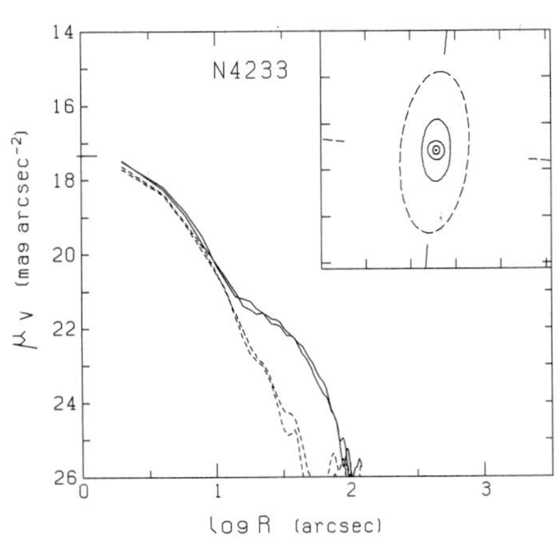

NGC4234

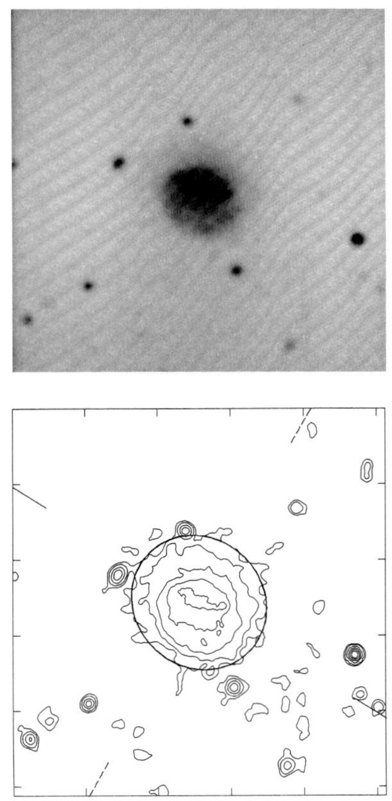

N4234 K3547 1 21.21 256X256

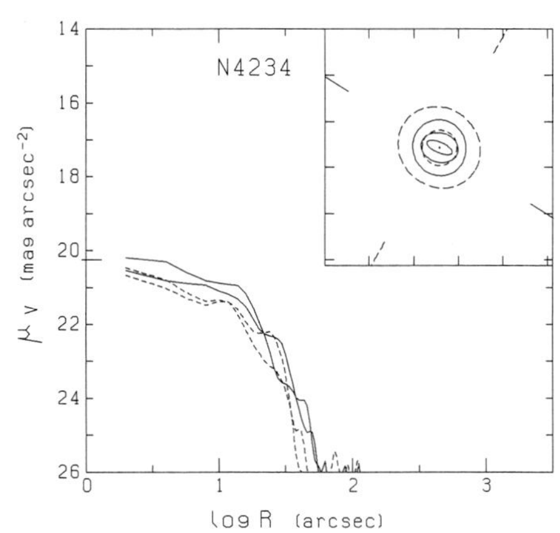

NGC4235

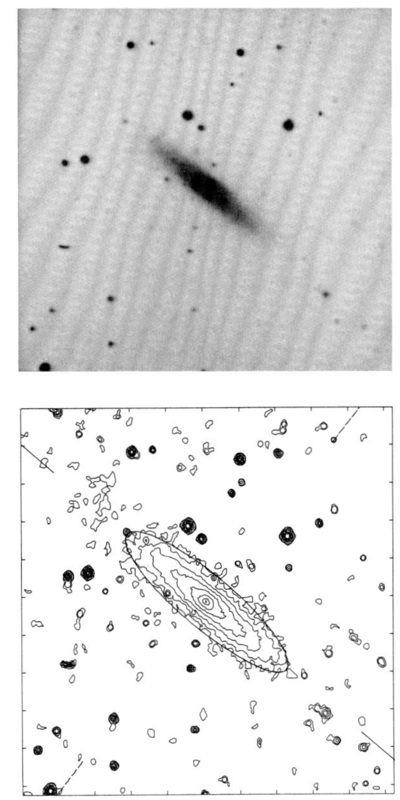

N4235 K3547 1 21.21 512X512

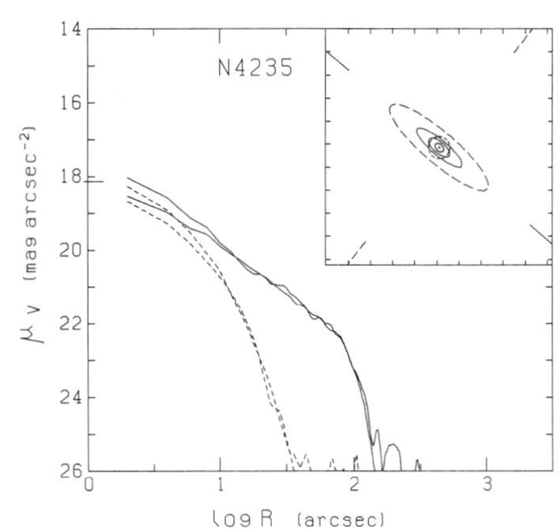

NGC4237

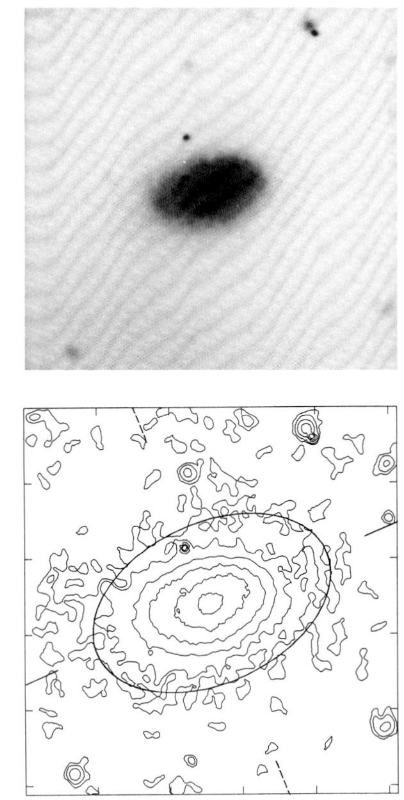

N4237 K2106 1 20.73 256X256

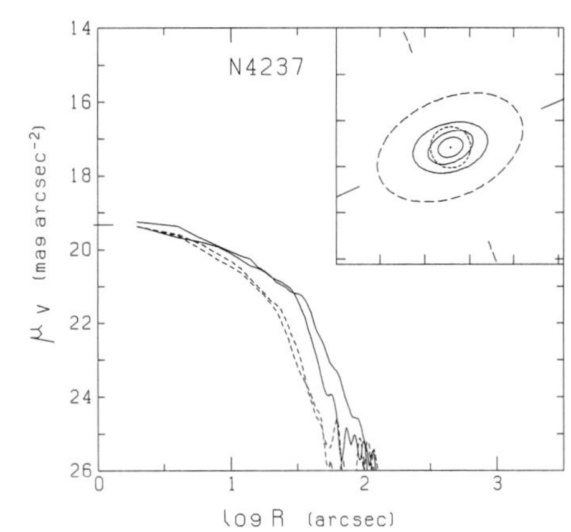

NGC4244

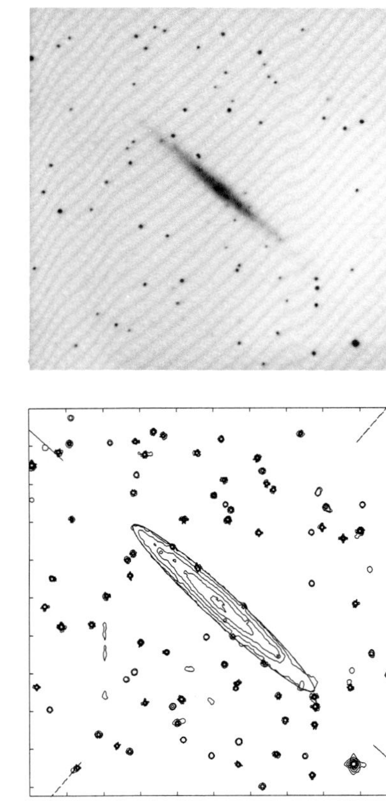

N4244 K3173 3 21.49 512X512

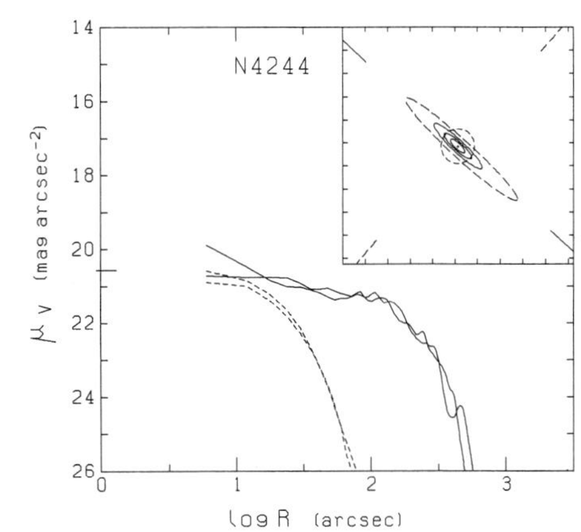

NGC4242

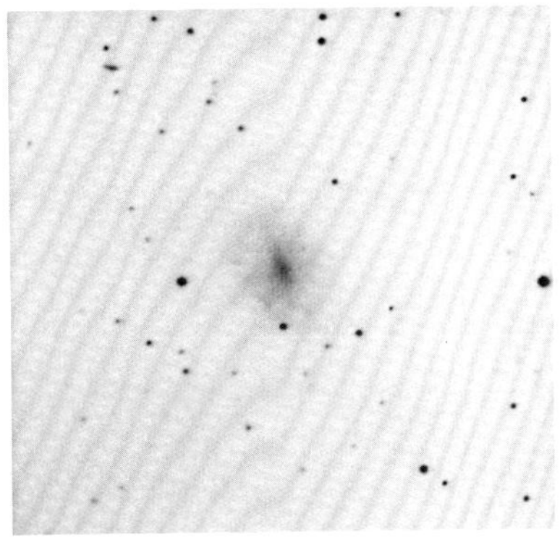

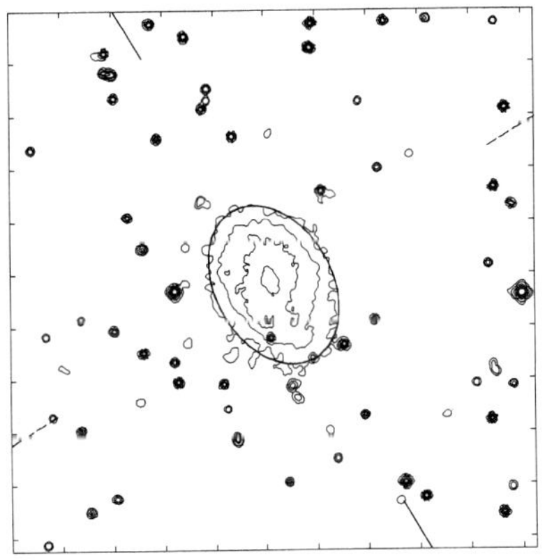

N4242 K3676 2 21.65 512X512

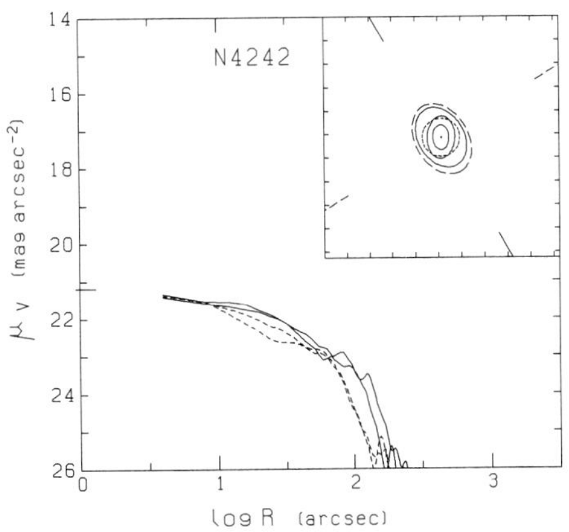

NGC4245

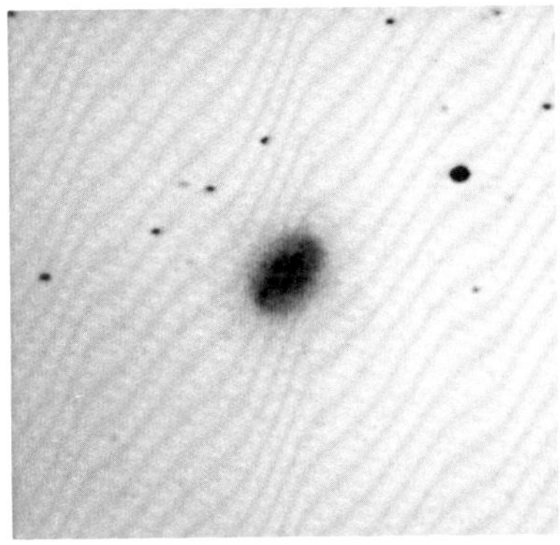

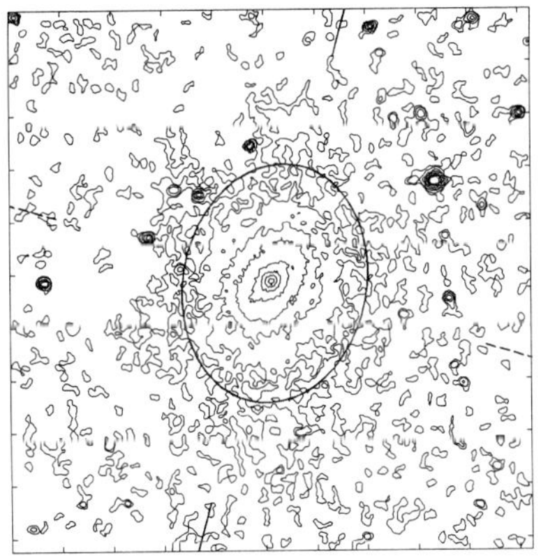

N4245 K4057 1 21.23 512X512

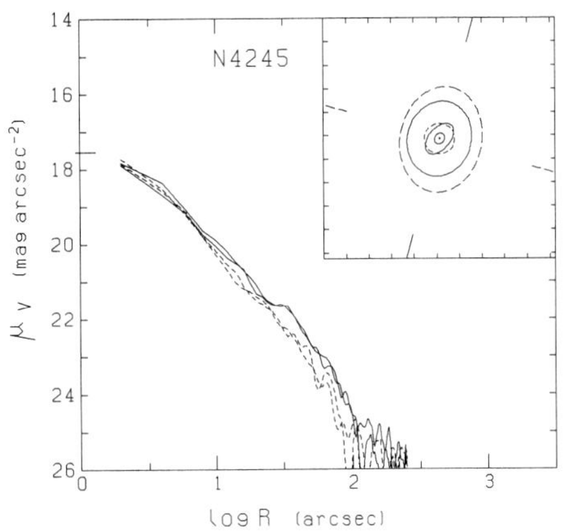

NGC4251

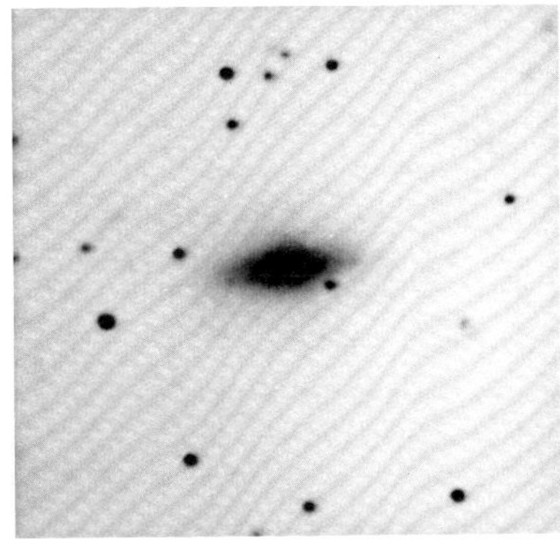

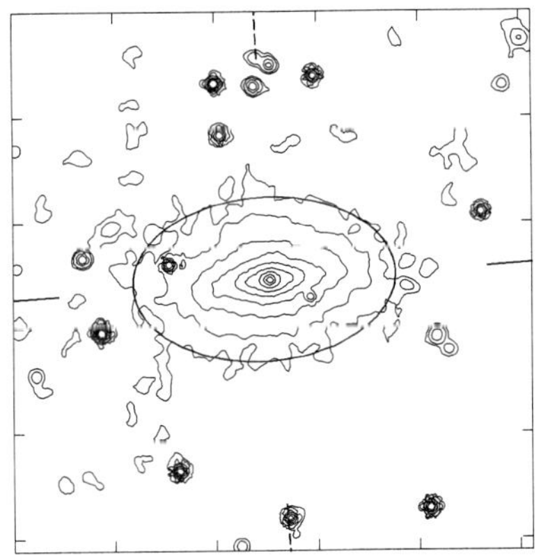

N4251 K3937 2 21.19 256X256

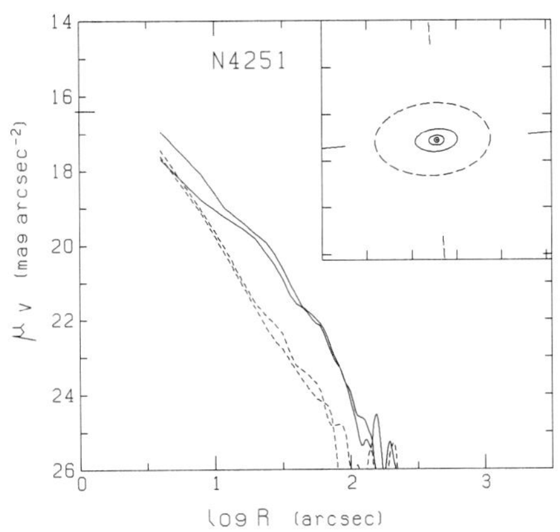

NGC4254

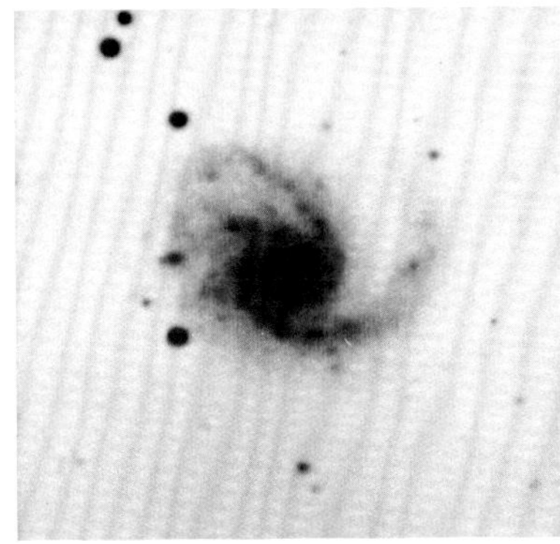

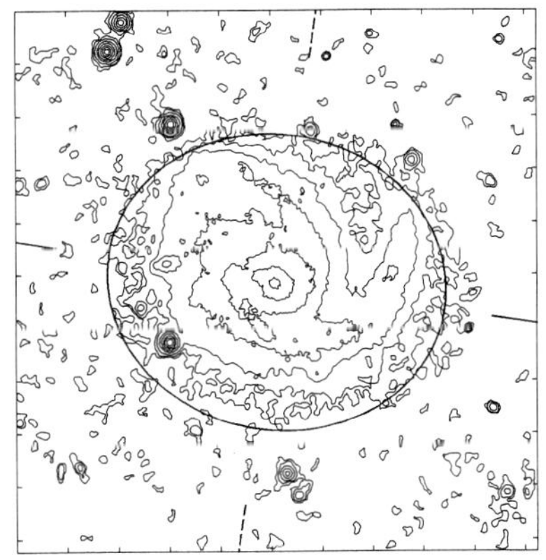

N4254 K2106 1 20.73 512X512

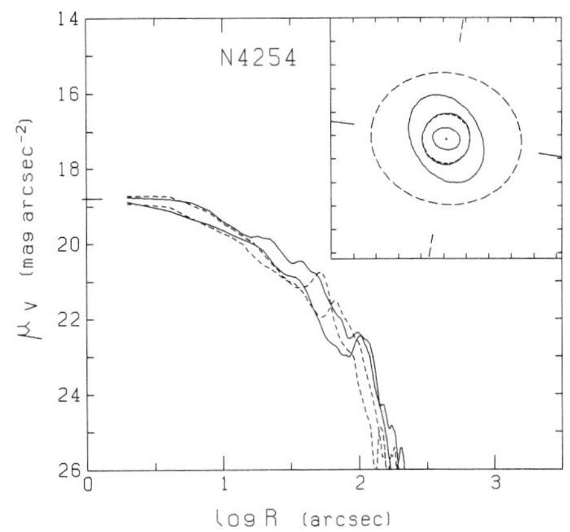

NGC4258

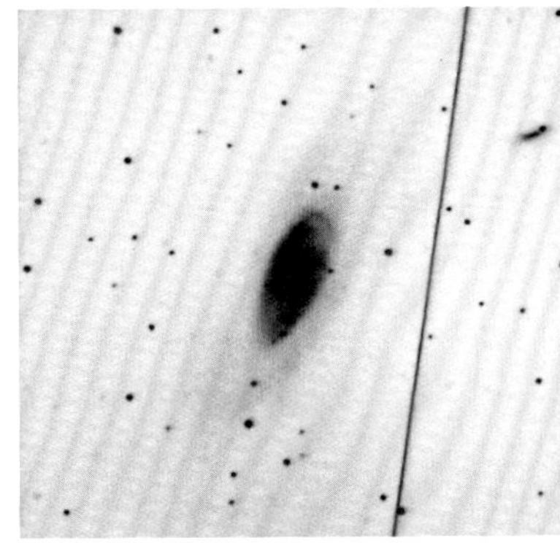

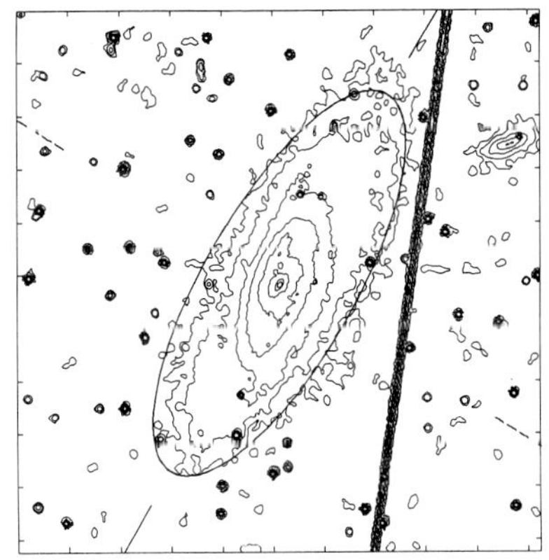

N4258 K4705 3 21.19 512X512

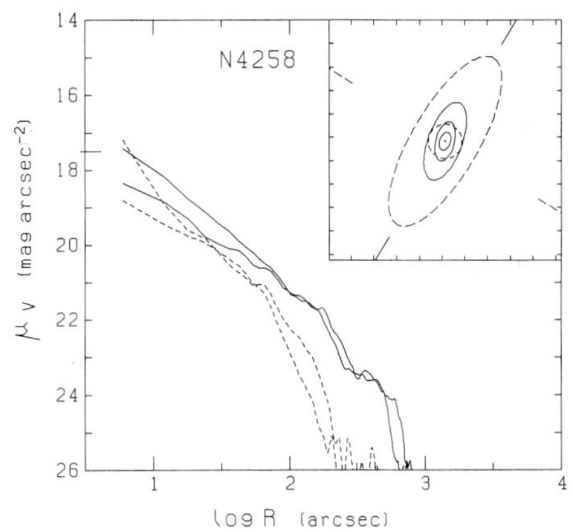

NGC4260

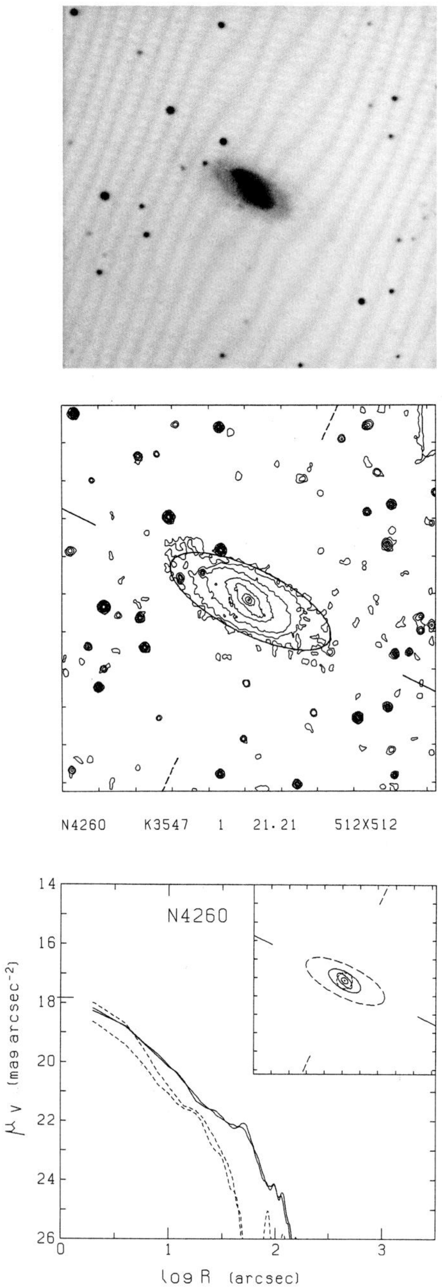

NGC4261

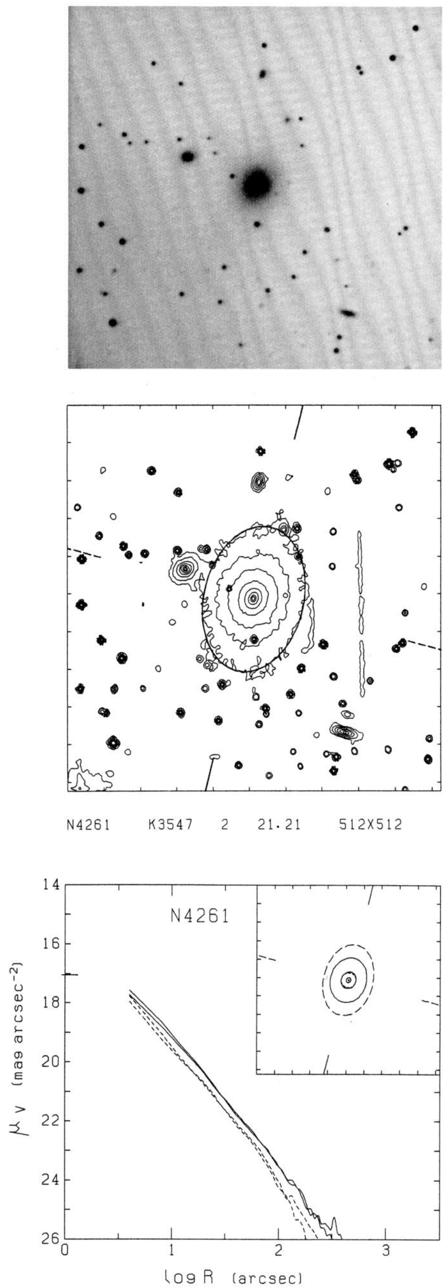

NGC4262

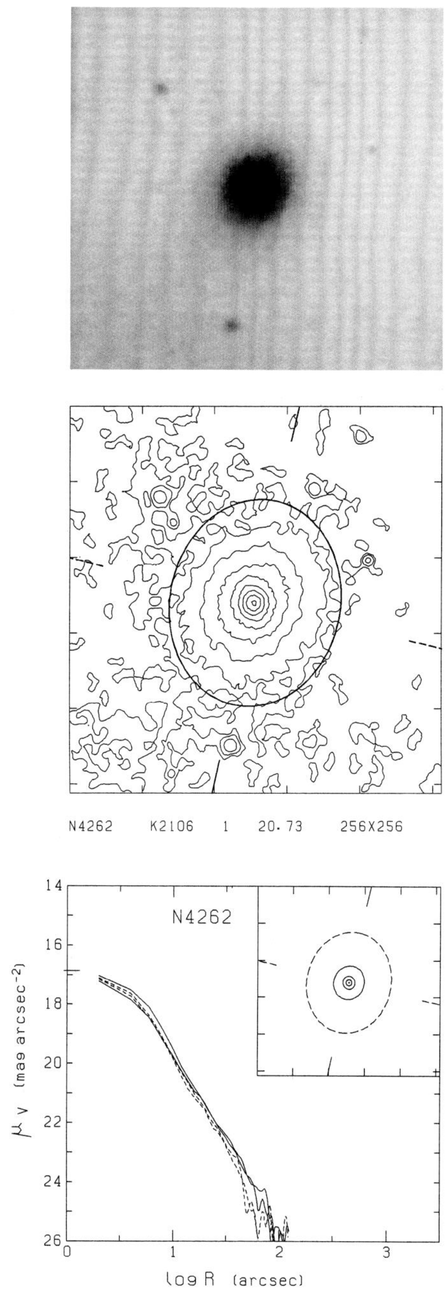

NGC4267

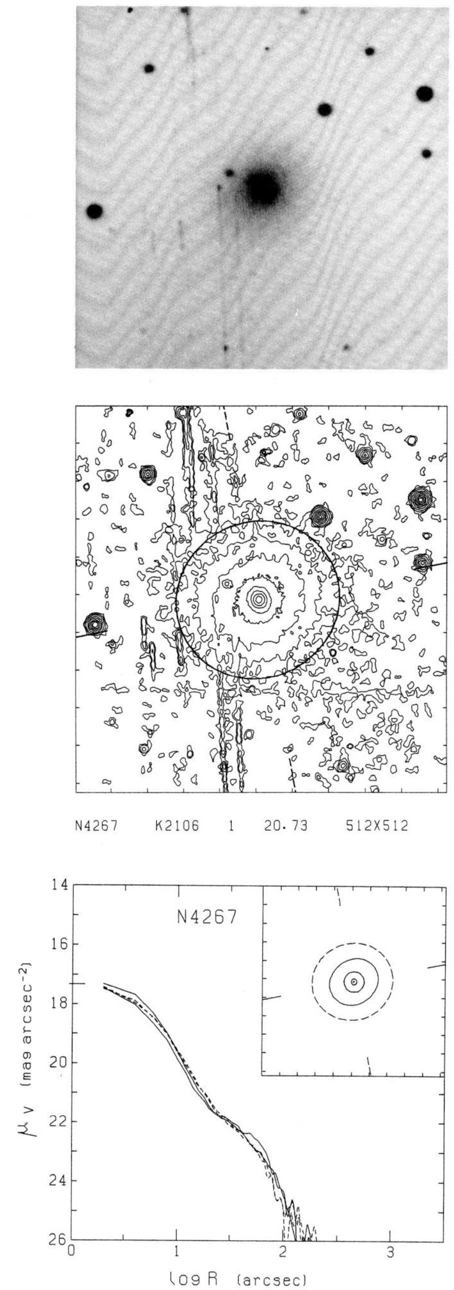

NGC4270

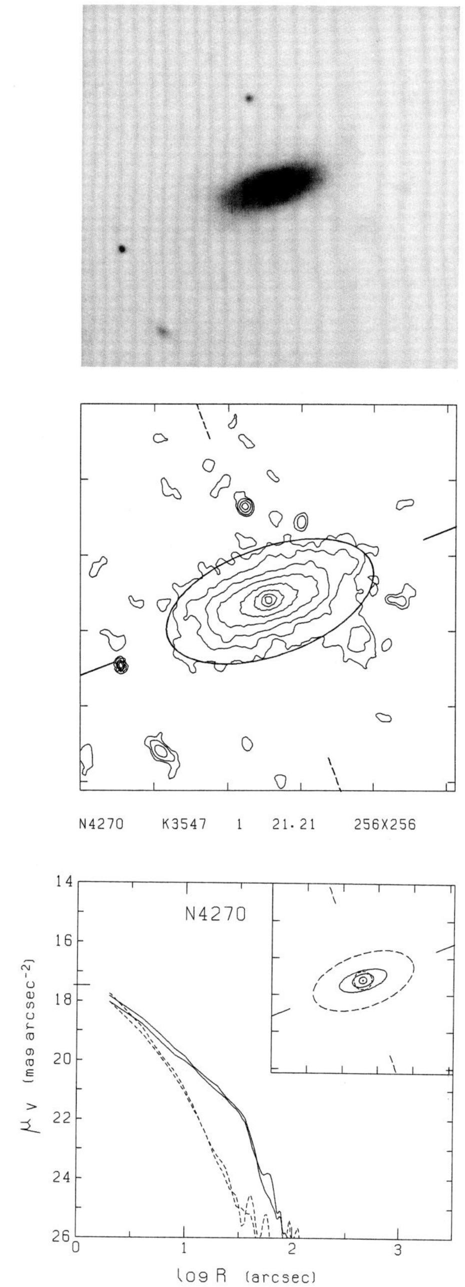

NGC4274

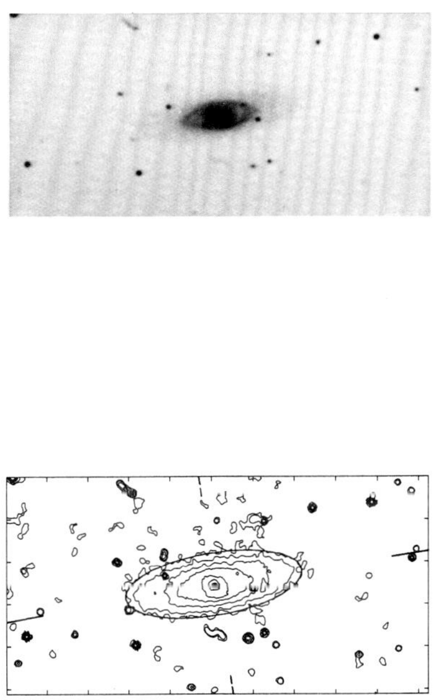

N4274 K3037 2 21.10 512X256

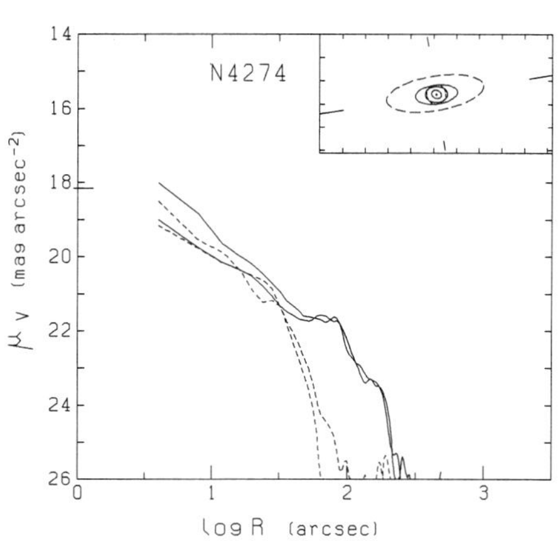

NGC4273

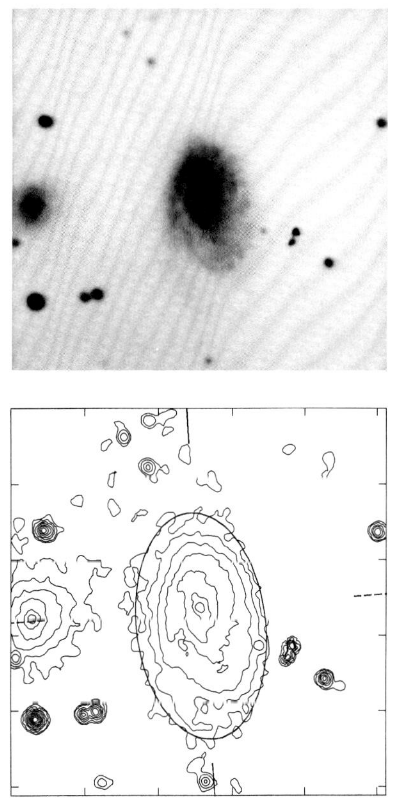

N4273 K3547 1 21.21 256X256

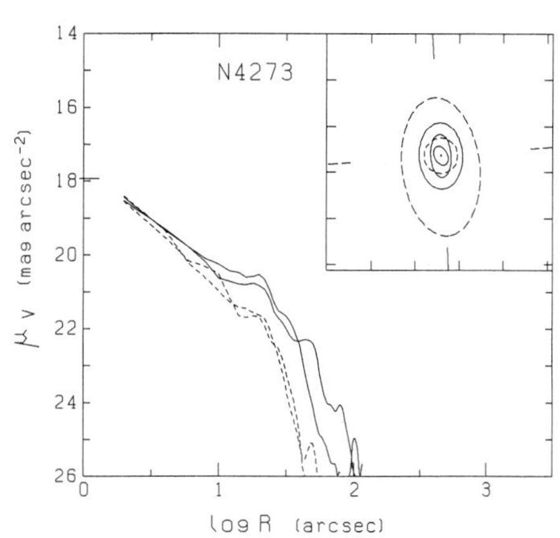

NGC4278

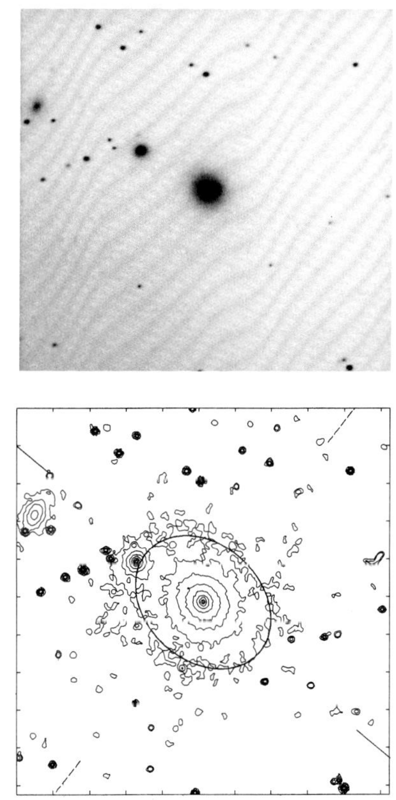

N4278 K4057 2 21.23 512X512

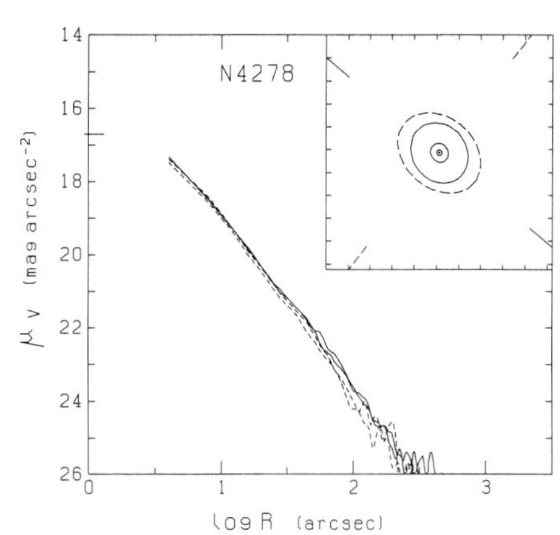

NGC4281

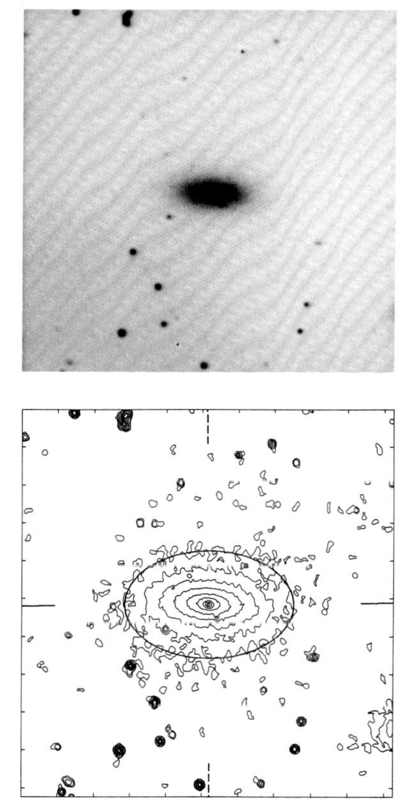

N4281 K3547 1 21.21 512X512

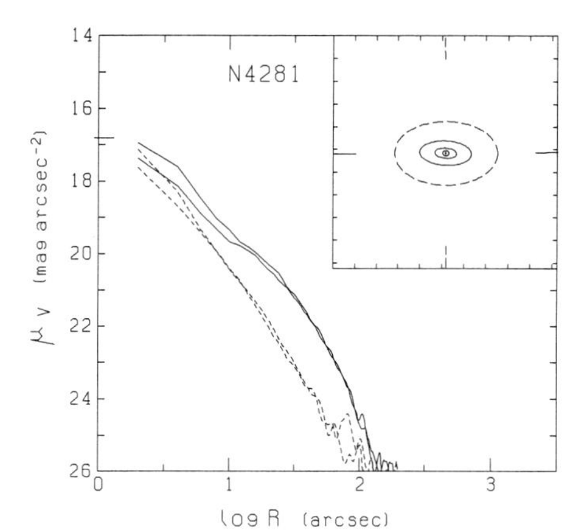

NGC4283

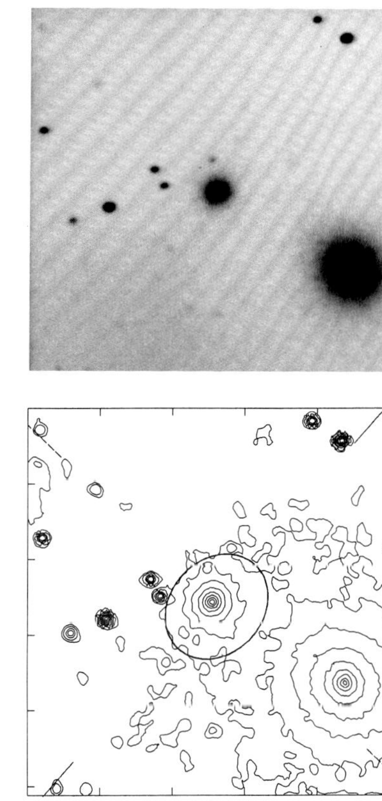

N4283 K4057 2 21.23 256X256

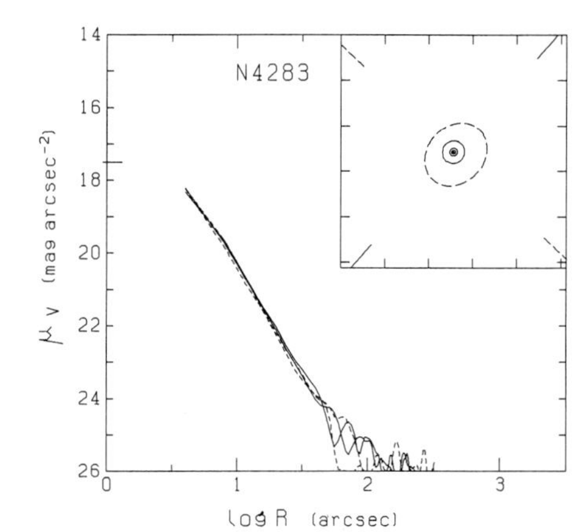

NGC4291

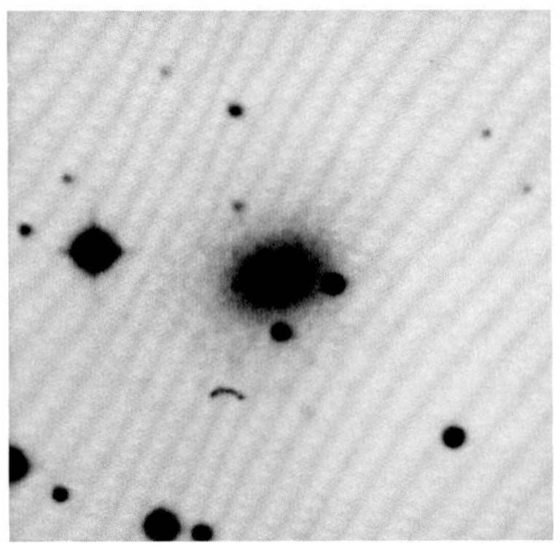

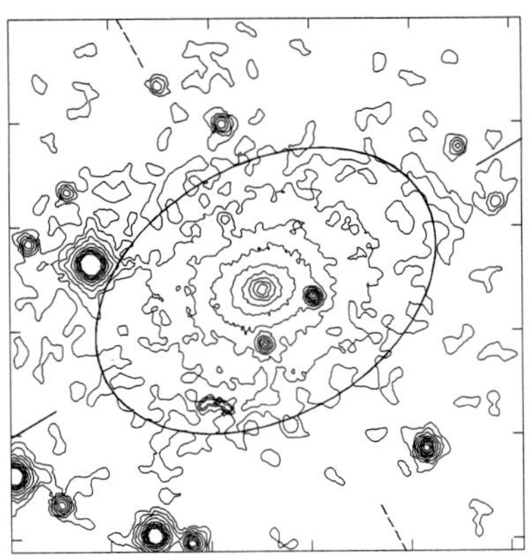

N4291 K5759 1 22.42 256X256

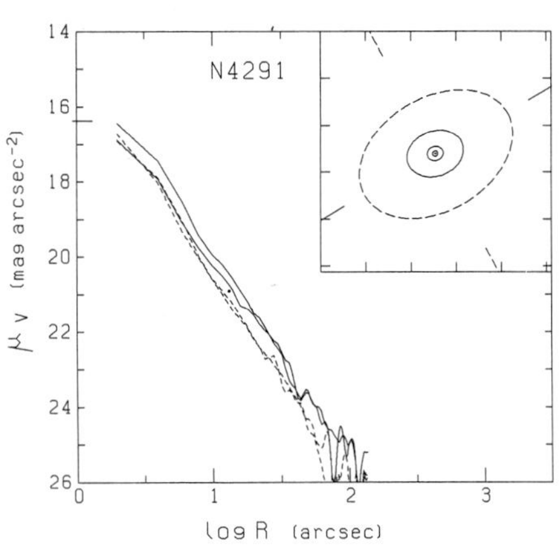

NGC4293

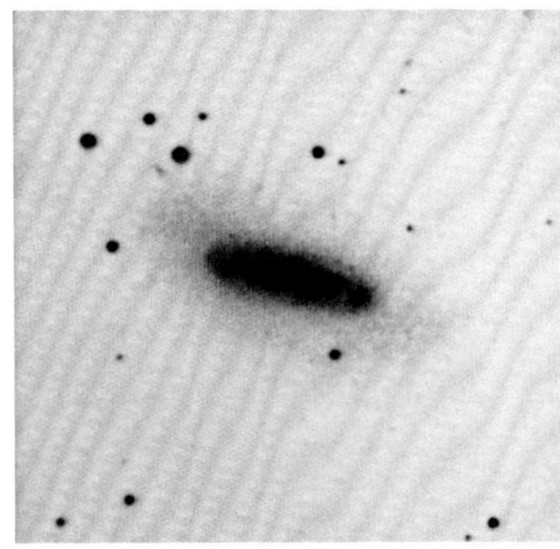

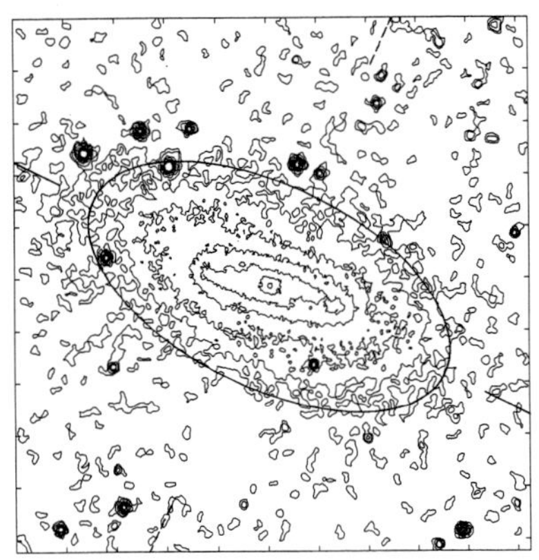

N4293 K4994 1 21.67 512X512

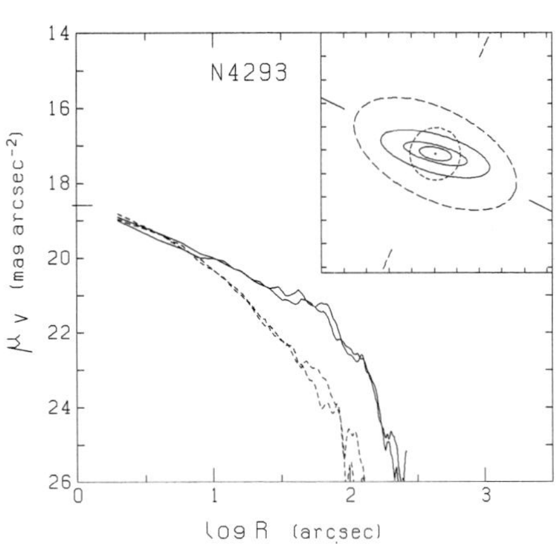

NGC4294

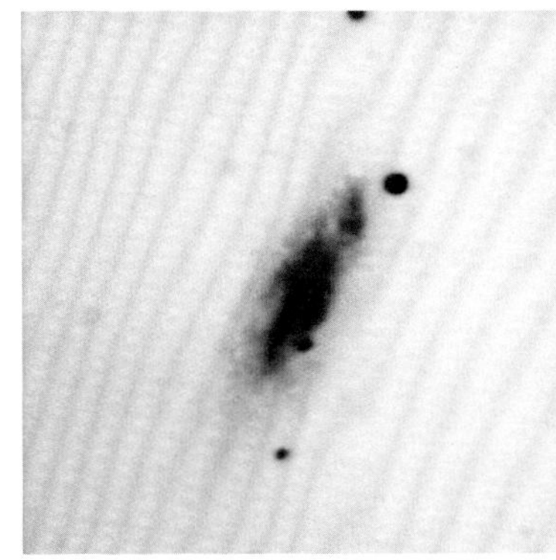

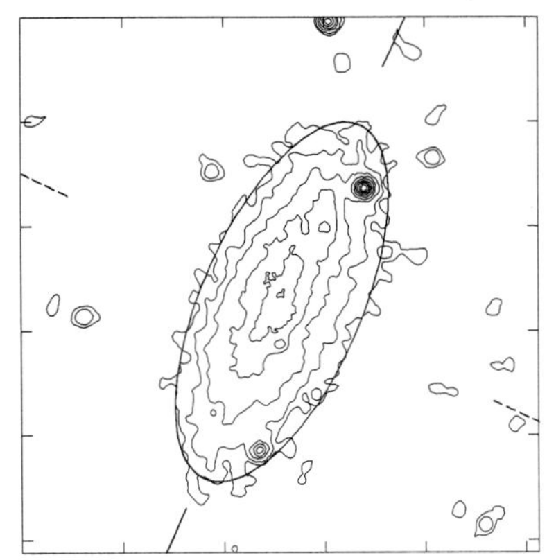

N4294 K2186 1 21.34 256X256

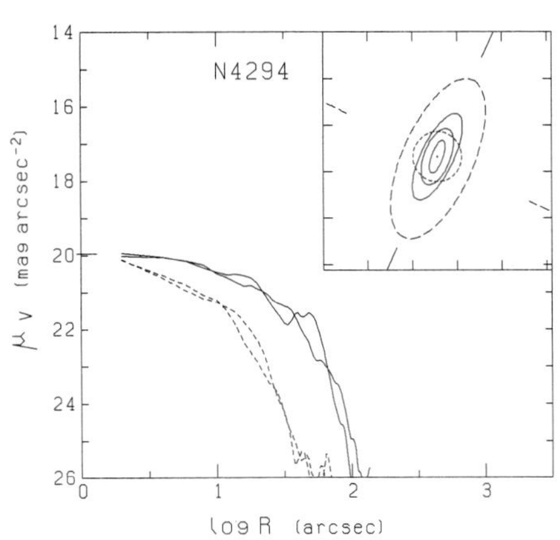

NGC4298

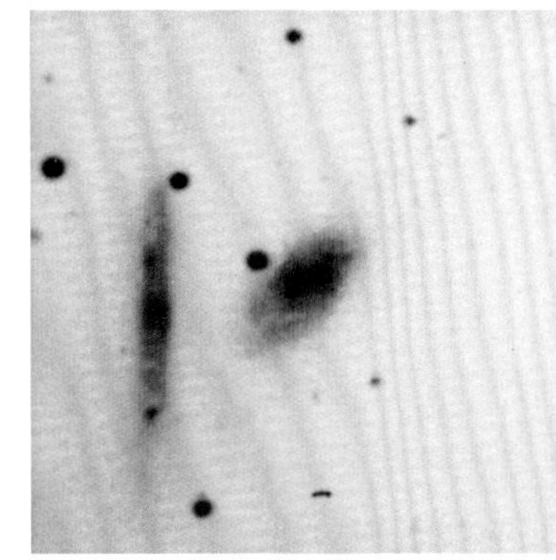

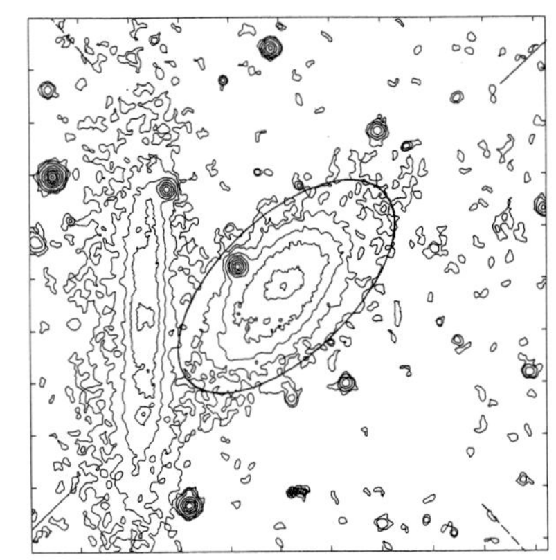

N4298 K2106 1 20.73 512X512

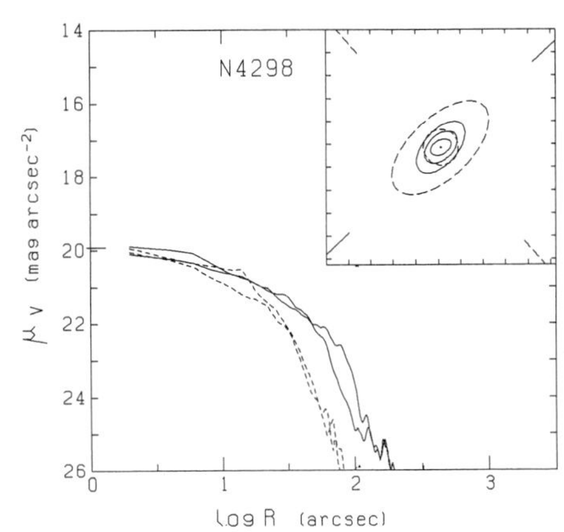

NGC4299

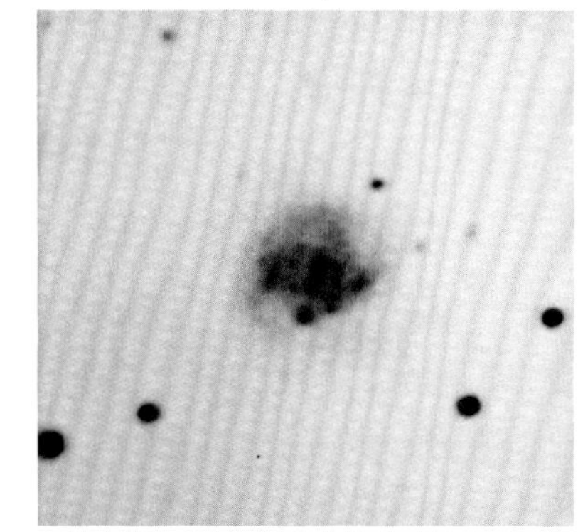

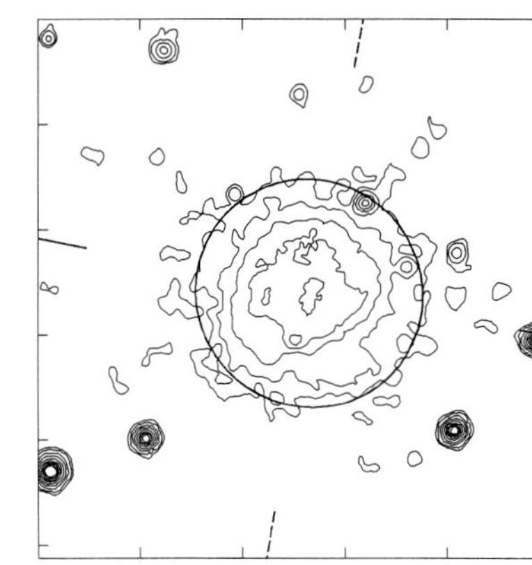

N4299 K2186 1 21.34 256X256

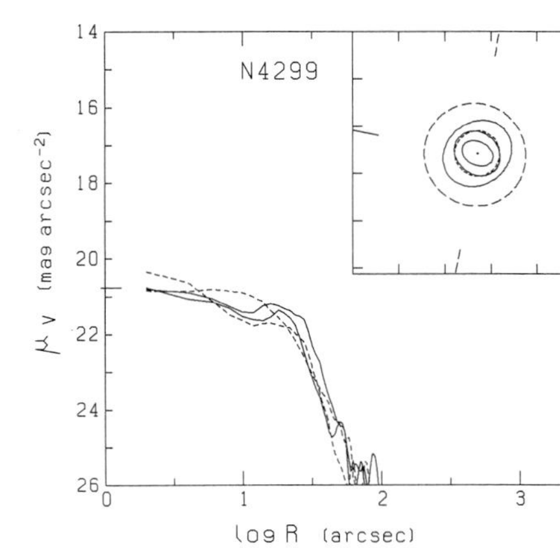

NGC4302

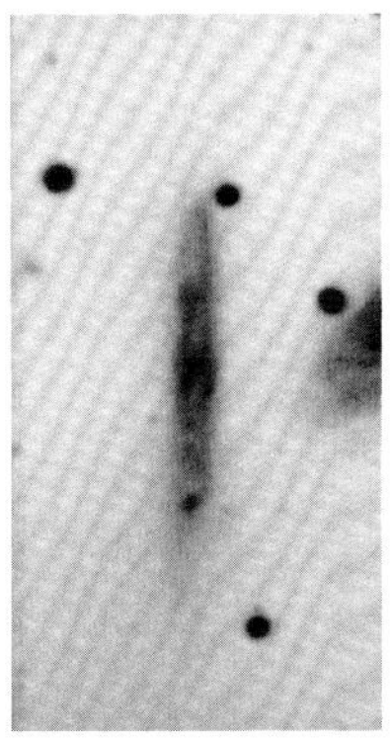

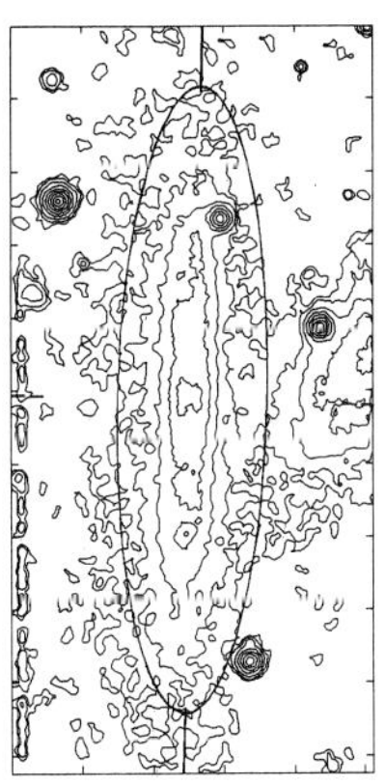

N4302 K2106 1 20.73 256X512

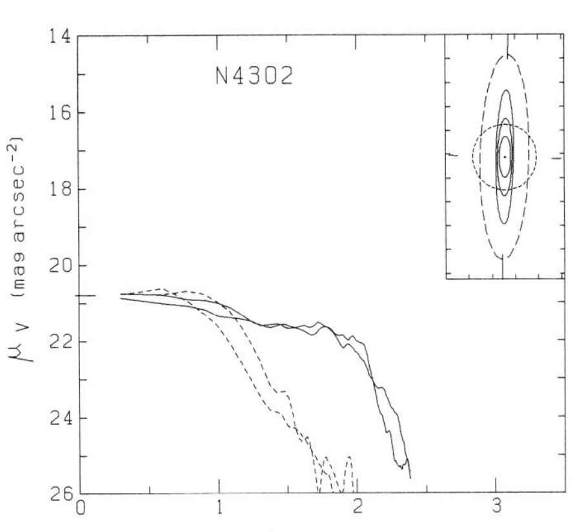

NGC4303

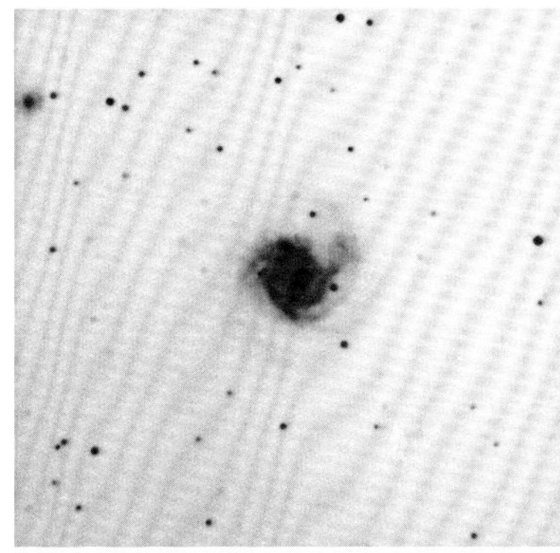

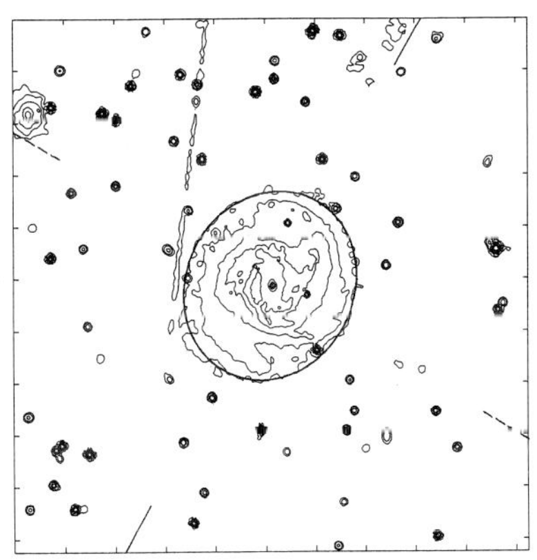

N4303 K3547 2 21.21 512X512

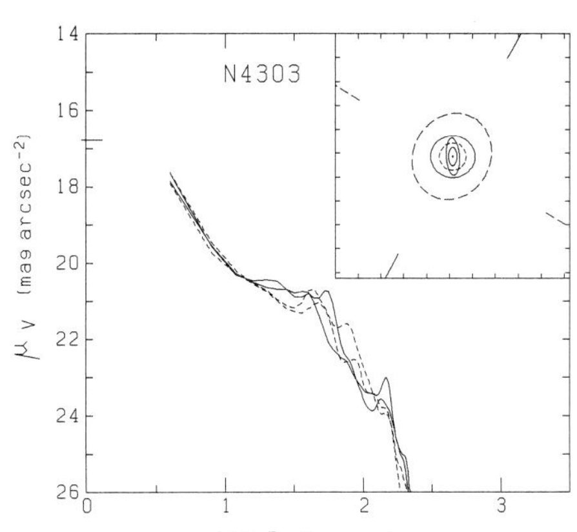

NGC4307

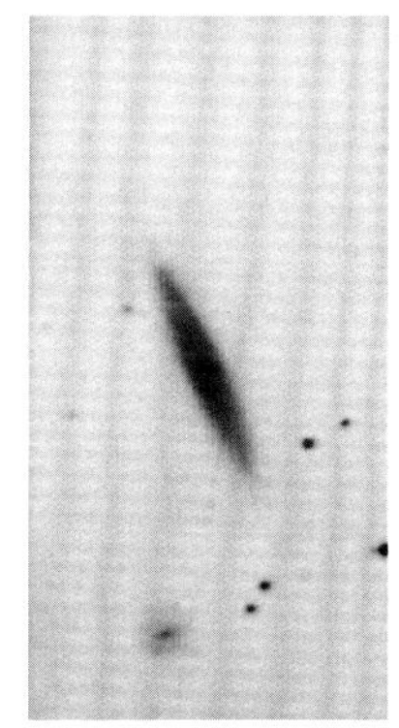

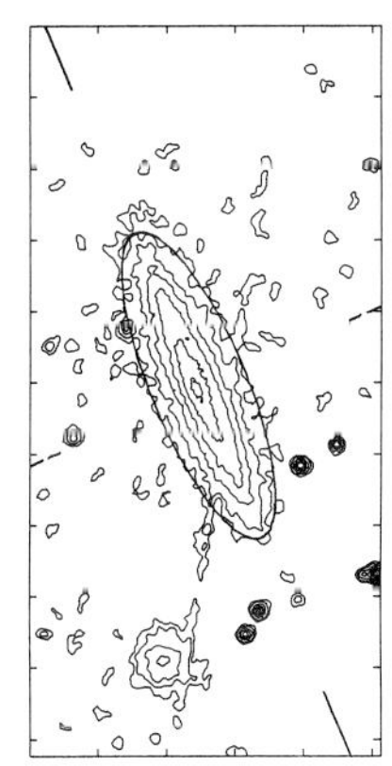

N4307 K2186 1 21.34 256X512

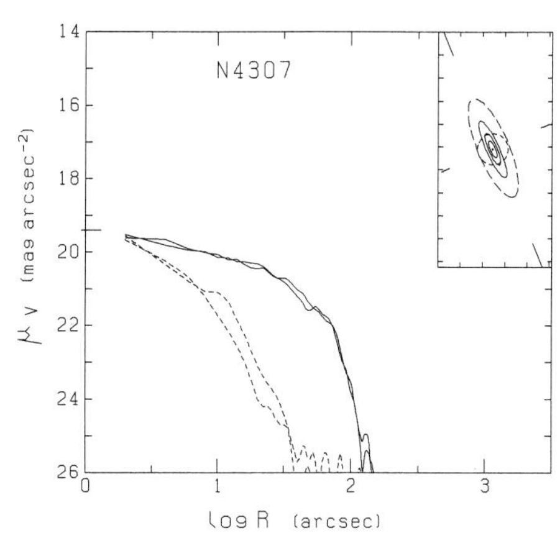

NGC4314

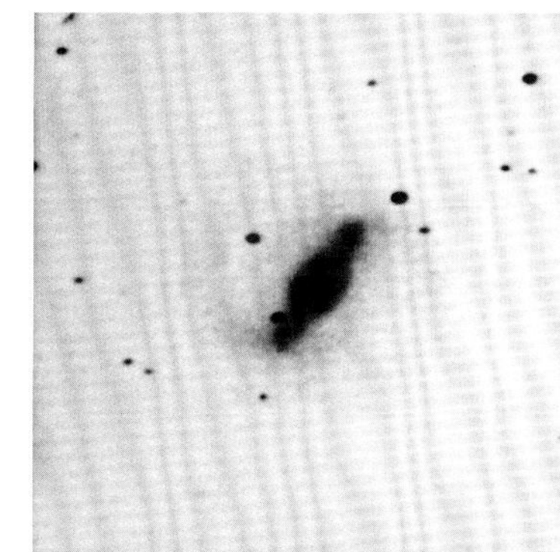

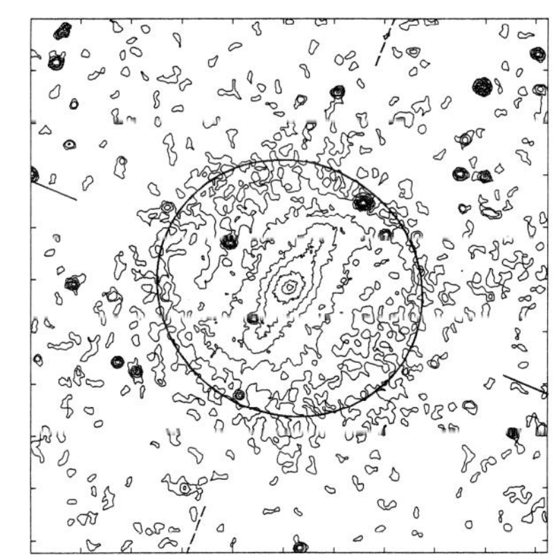

N4314 K4057 1 21.23 512X512

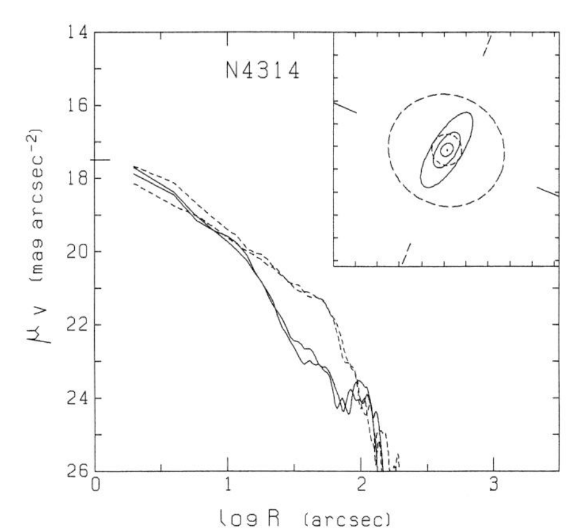

NGC4321

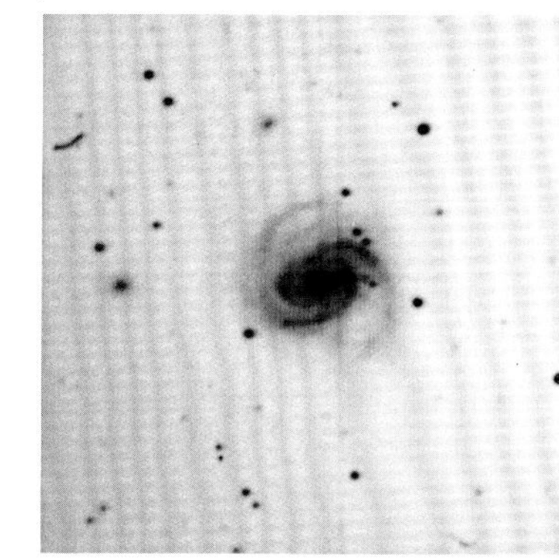

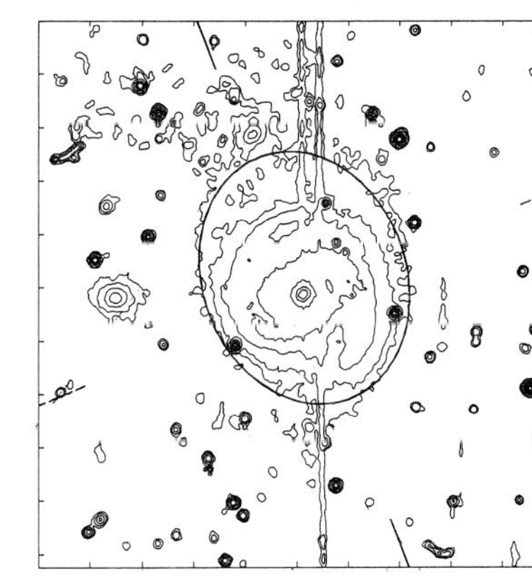

N4321 K2106 2 20.73 512X512

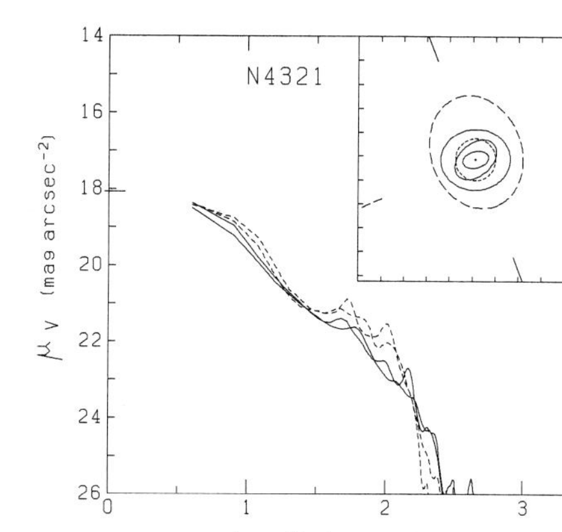

NGC4324

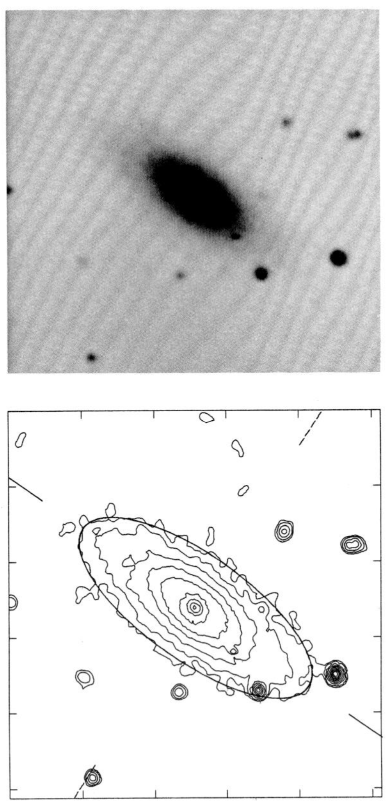

N4324 K3547 1 21.21 256X256

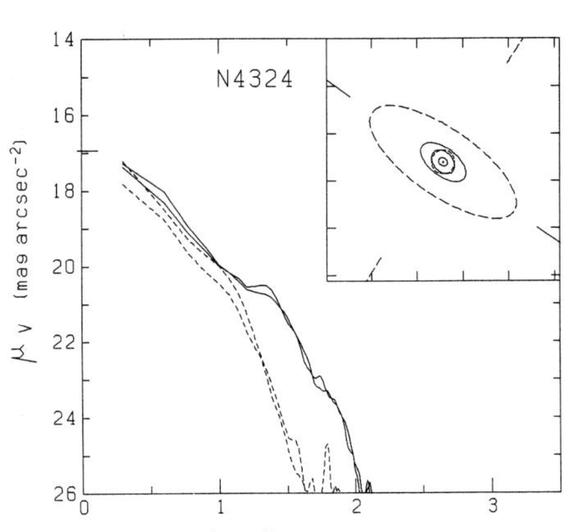

NGC4346

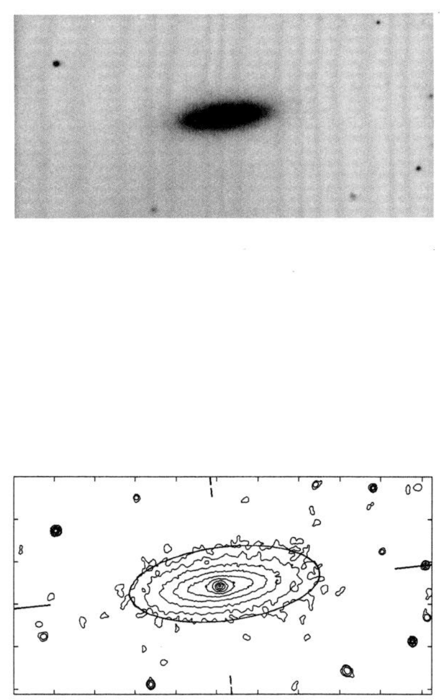

N4346 K3546 1 21.48 512X256

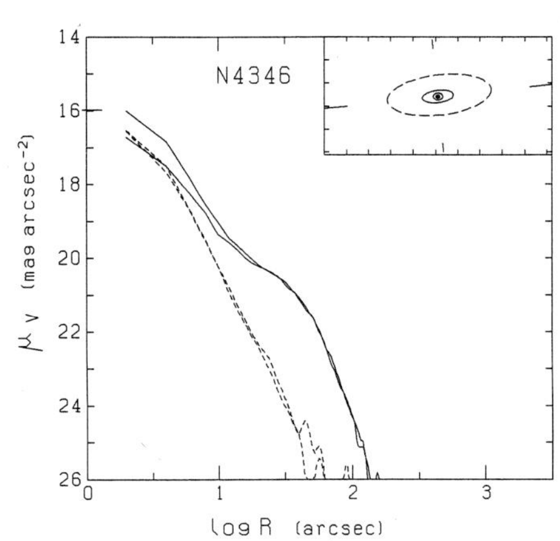

NGC4339

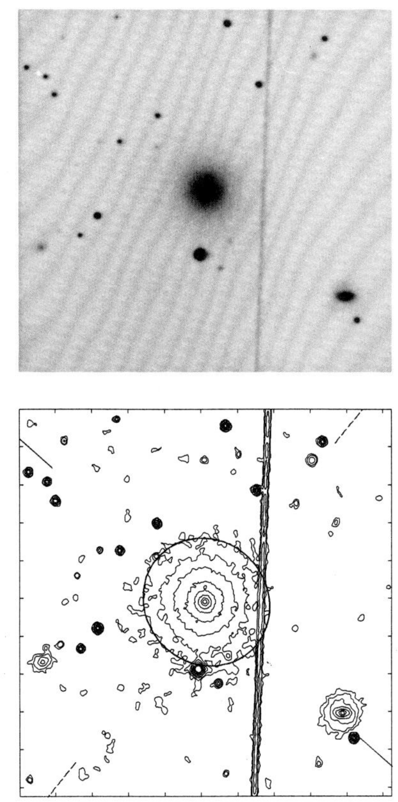

N4339 K3547 1 21.21 512X512

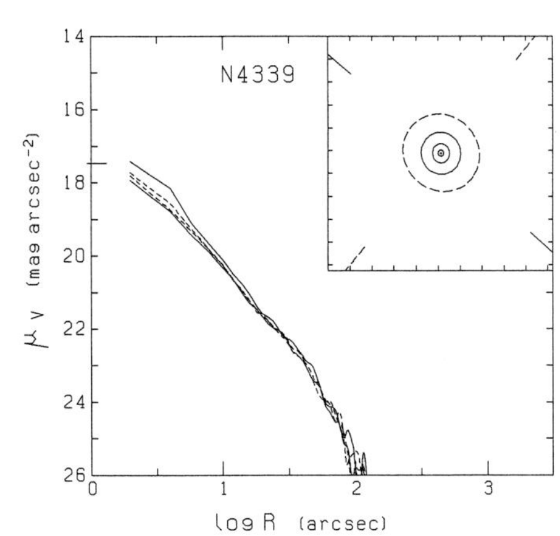

NGC4340

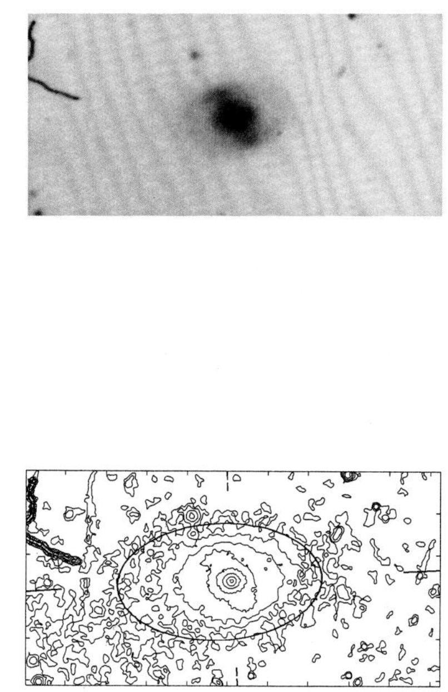

N4340 K2106 1 20.73 512X256

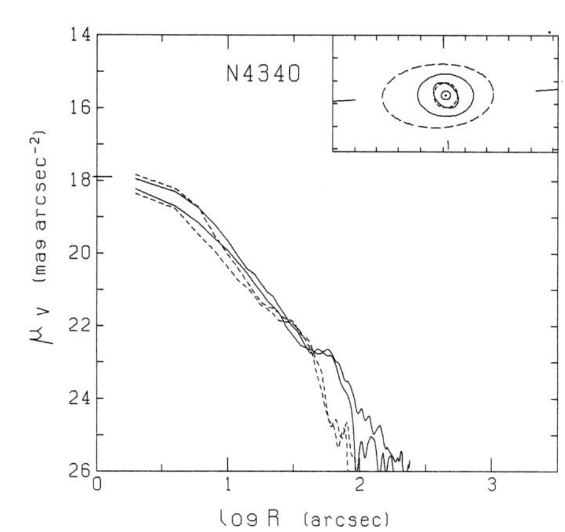

NGC4342

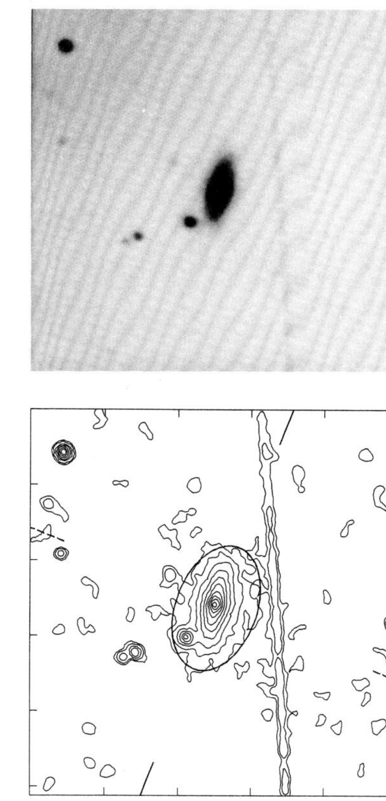

N4342 K3547 1 21.21 256X256

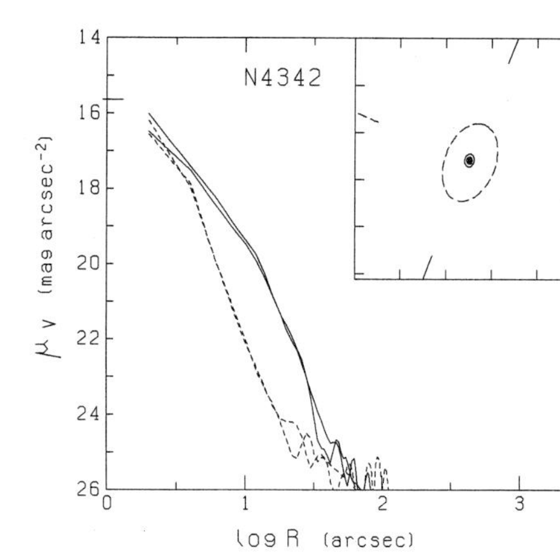

NGC4348

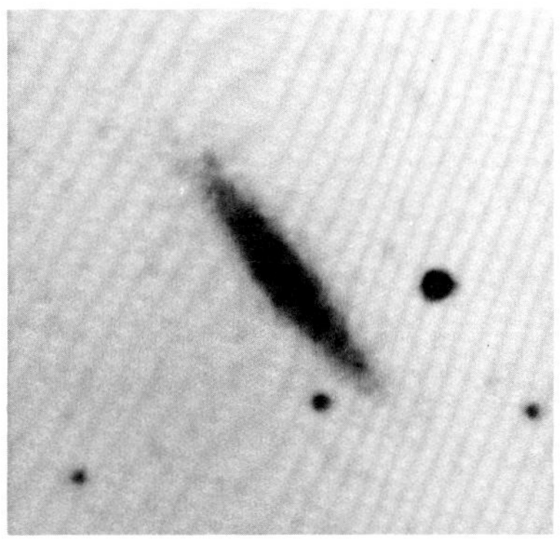

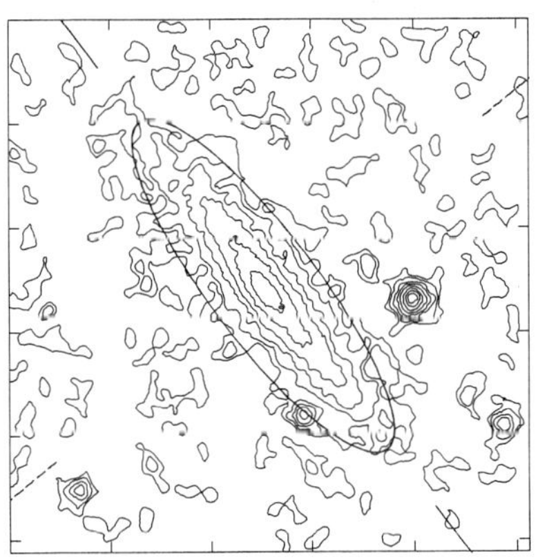

N4348 K5012 1 21.51 256X256

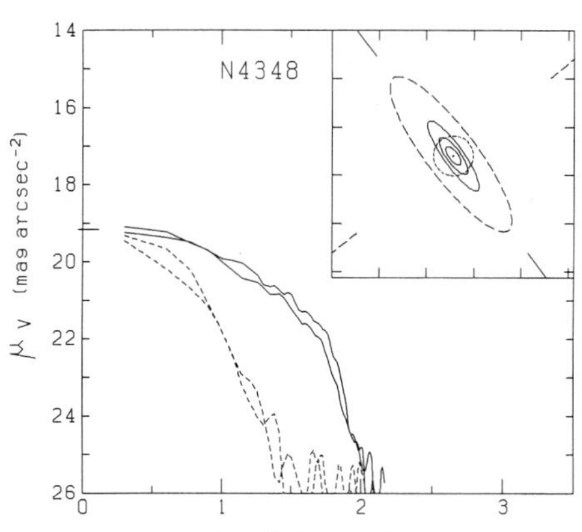

NGC4350

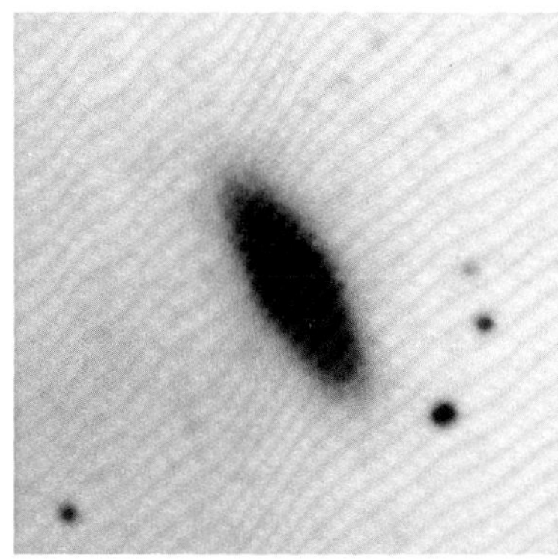

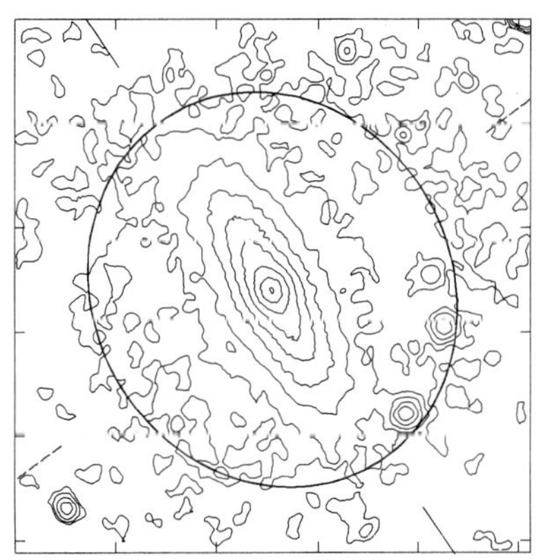

N4350 K2106 1 20.73 256X256

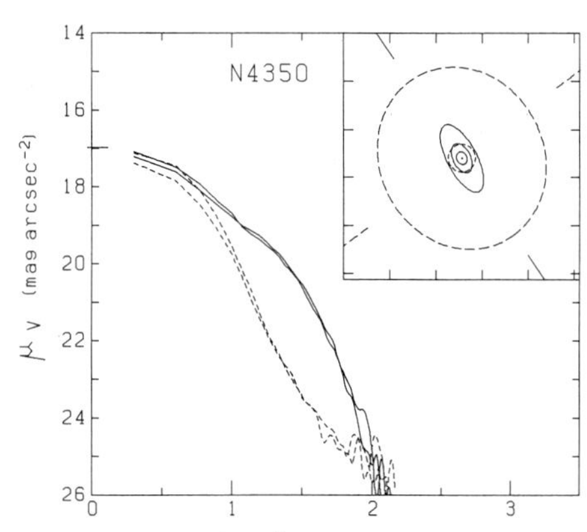

NGC4365

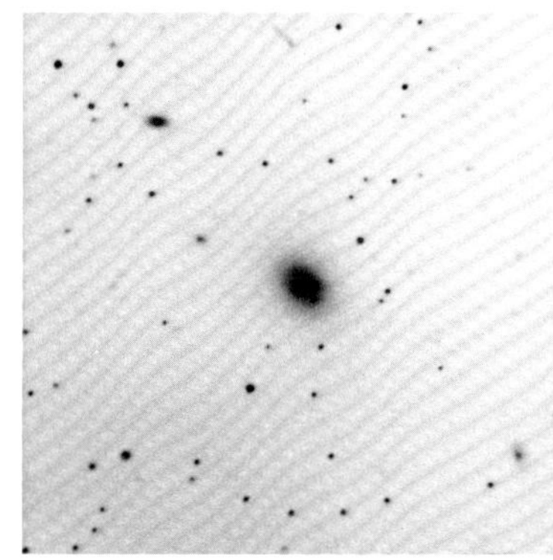

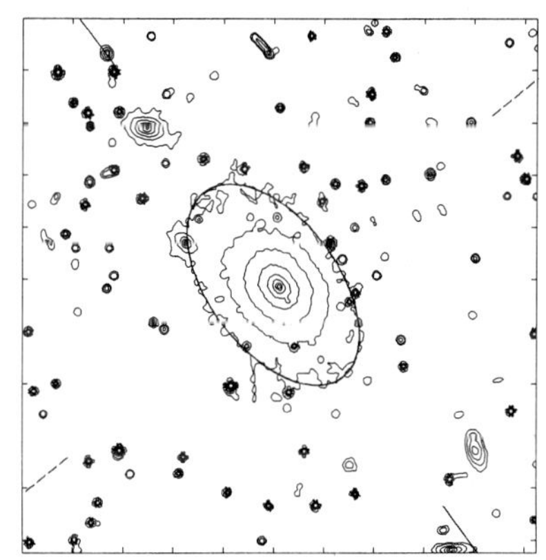

N4365 K2186 3 21.34 512X512

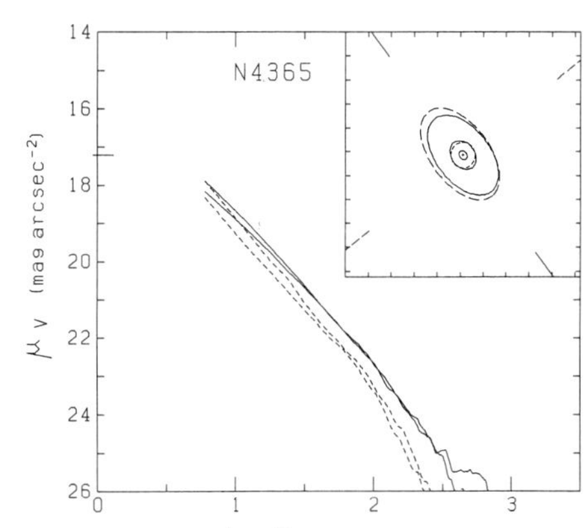

NGC4369

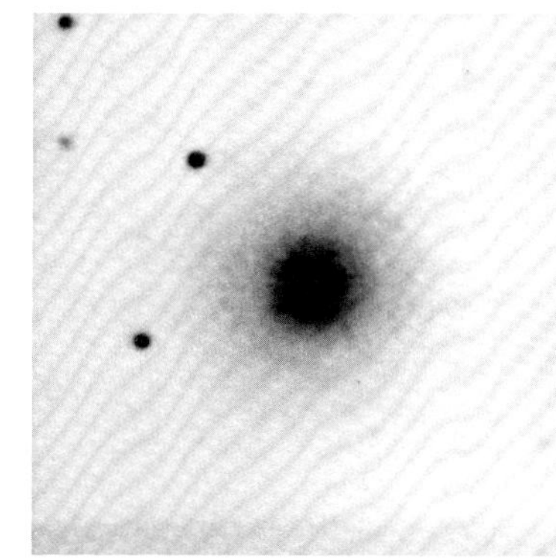

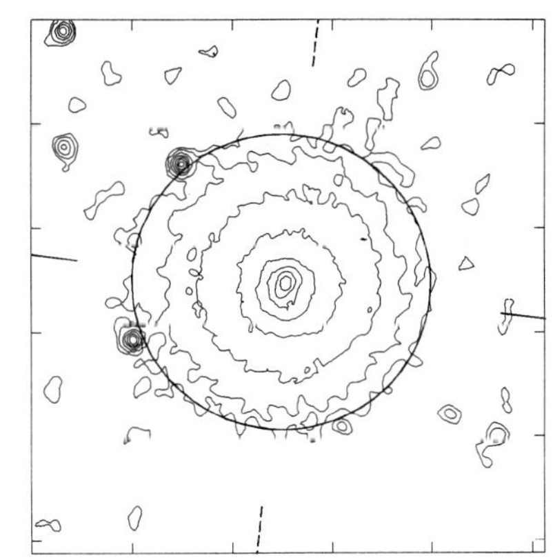

N4369 K3173 1 21.49 256X256

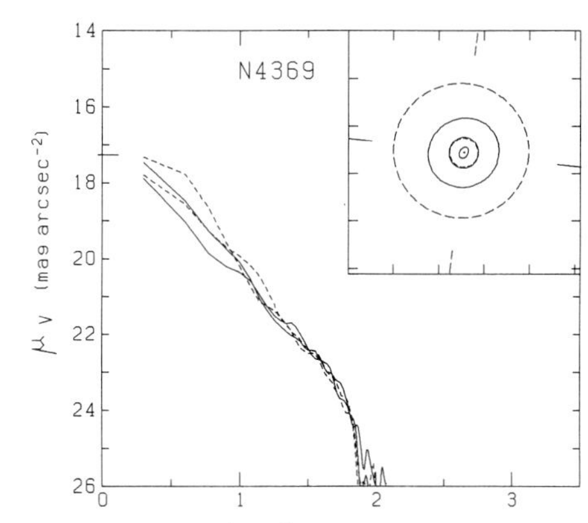

NGC4386

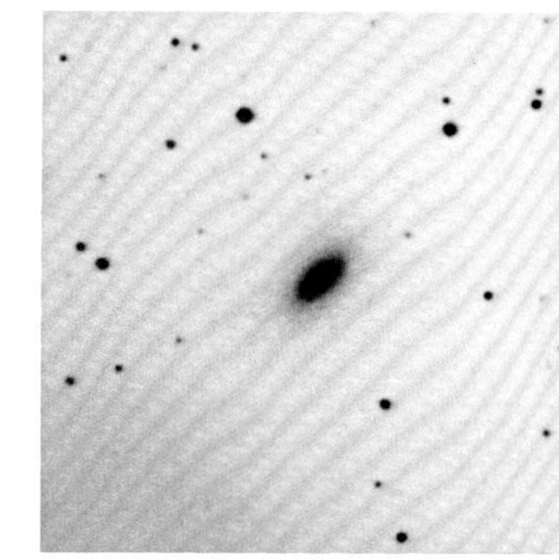

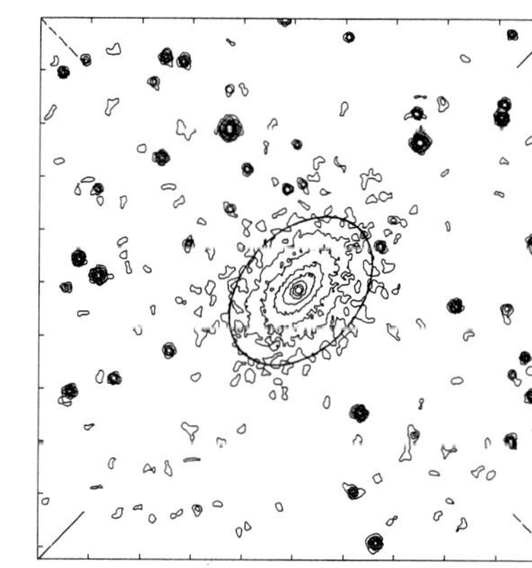

N4386 K5759 1 22.42 512X512

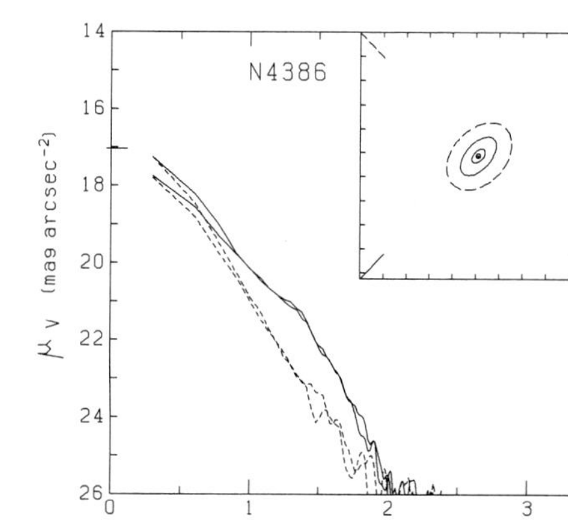

NGC4371

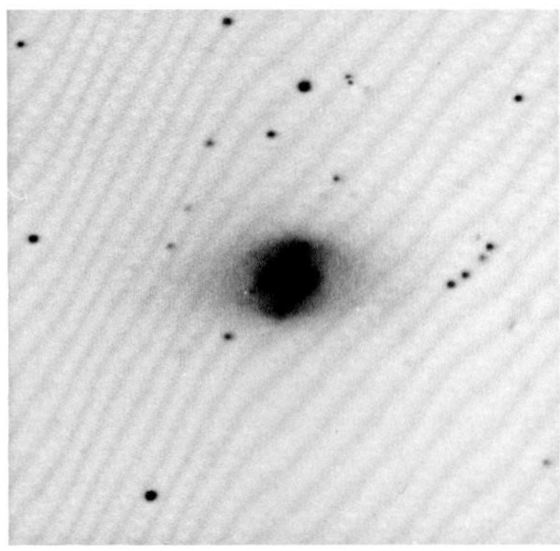

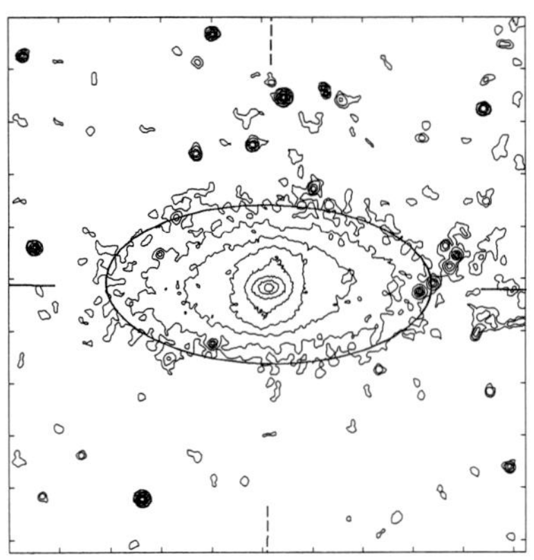

N4371 K2186 1 21.34 512X512

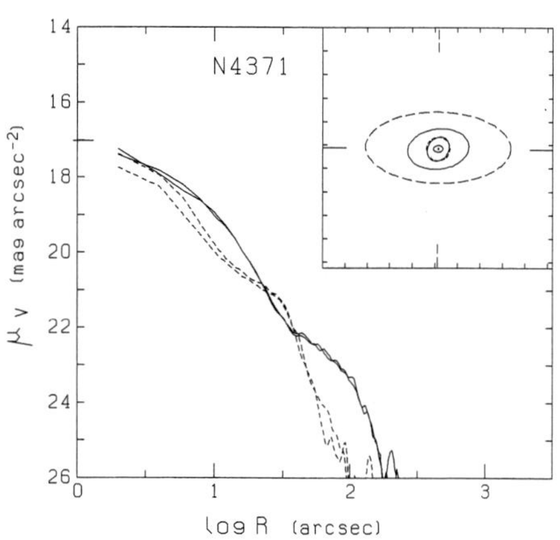

NGC4374

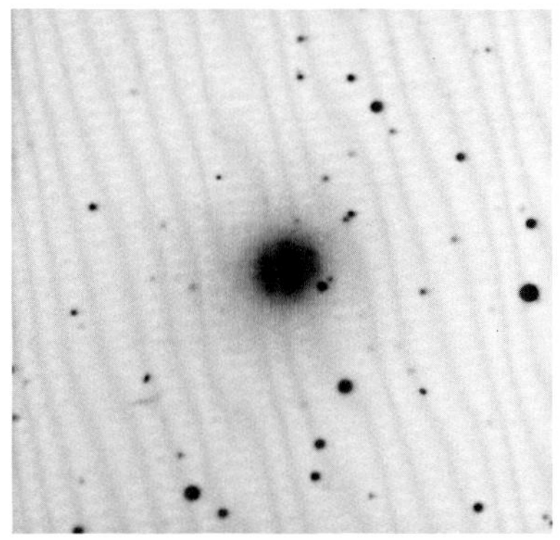

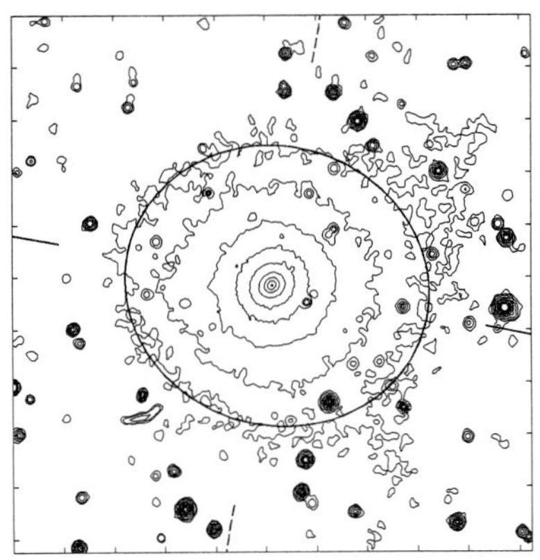

N4374 K2106 2 20.73 512X512

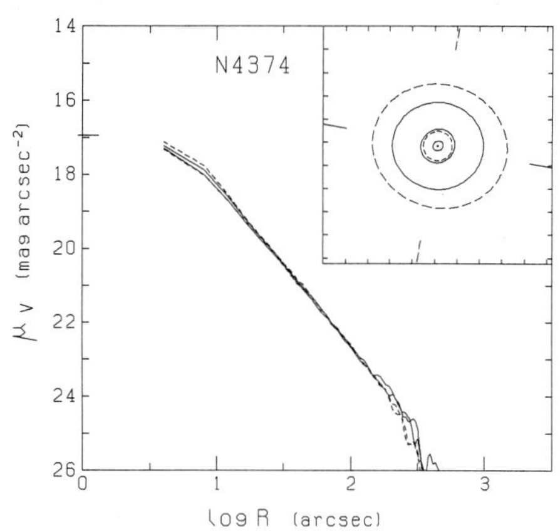

NGC4377

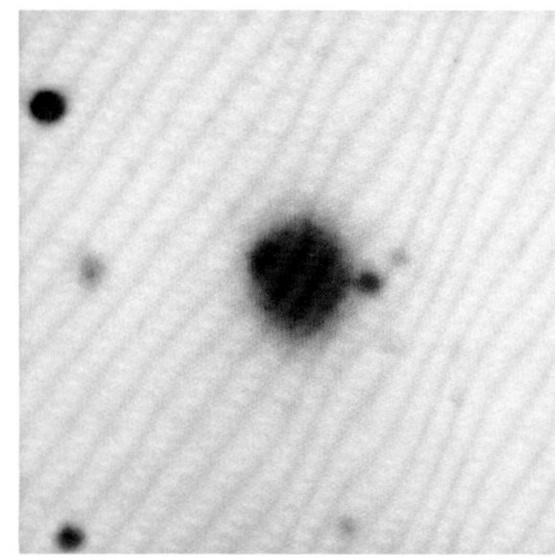

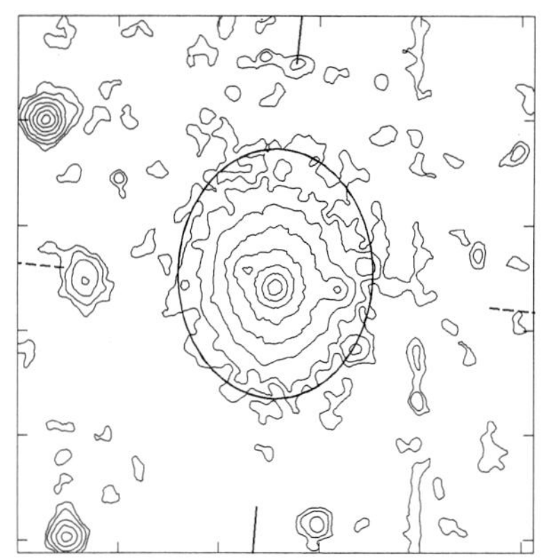

N4377 K2106 1 20.73 256X256

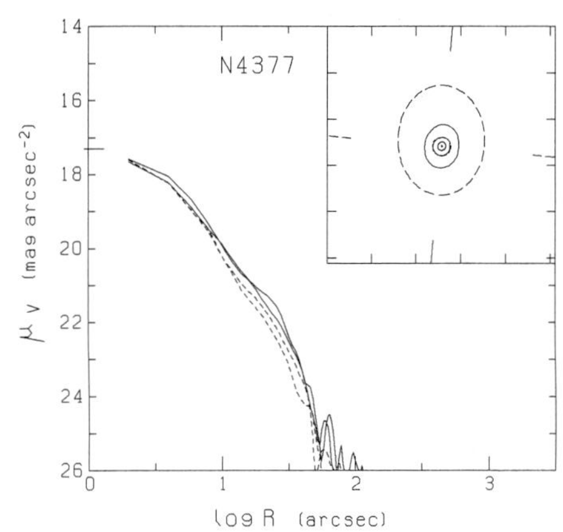

NGC4379

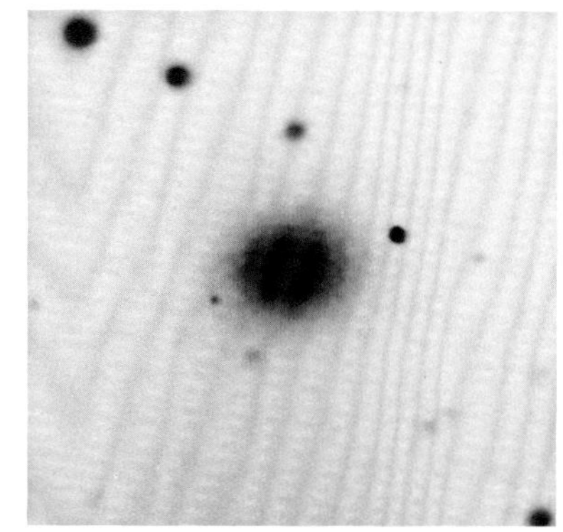

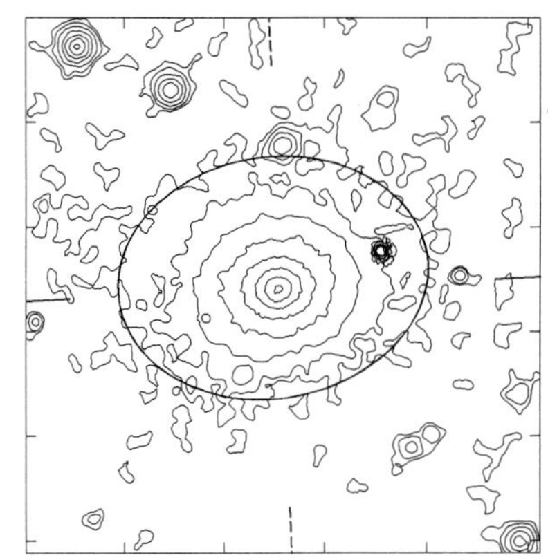

N4379 K2106 1 20.73 256X256

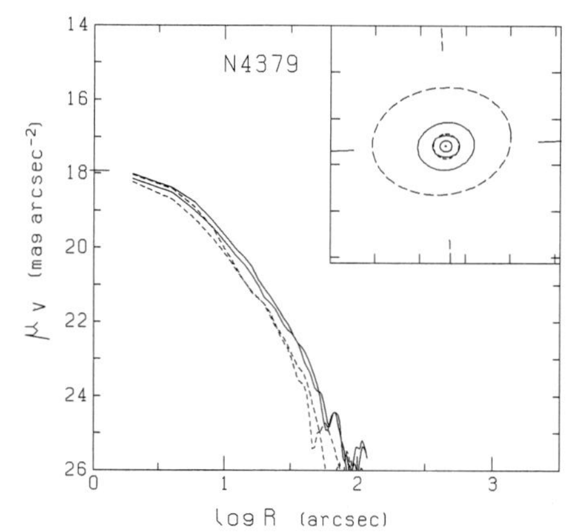

NGC4378

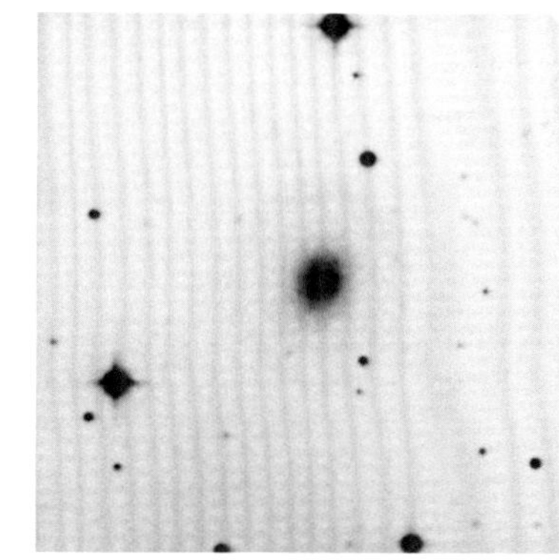

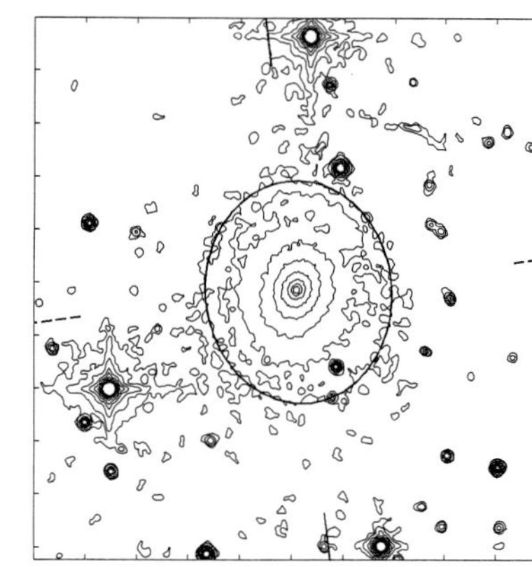

N4378 K3547 1 21.21 512X512

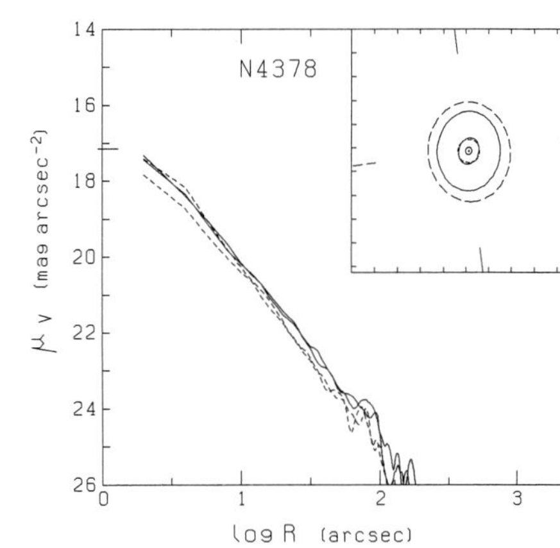

NGC4380

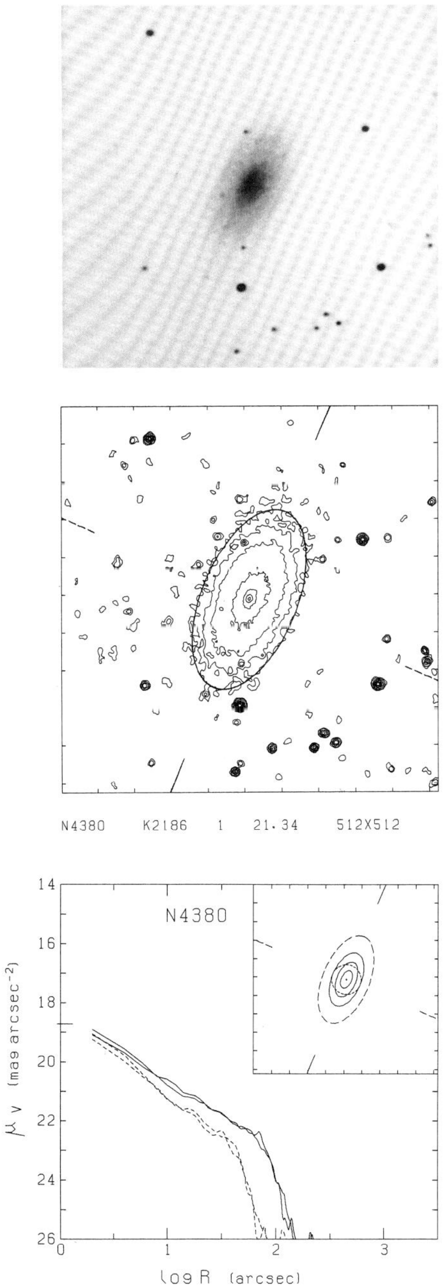

NGC4382

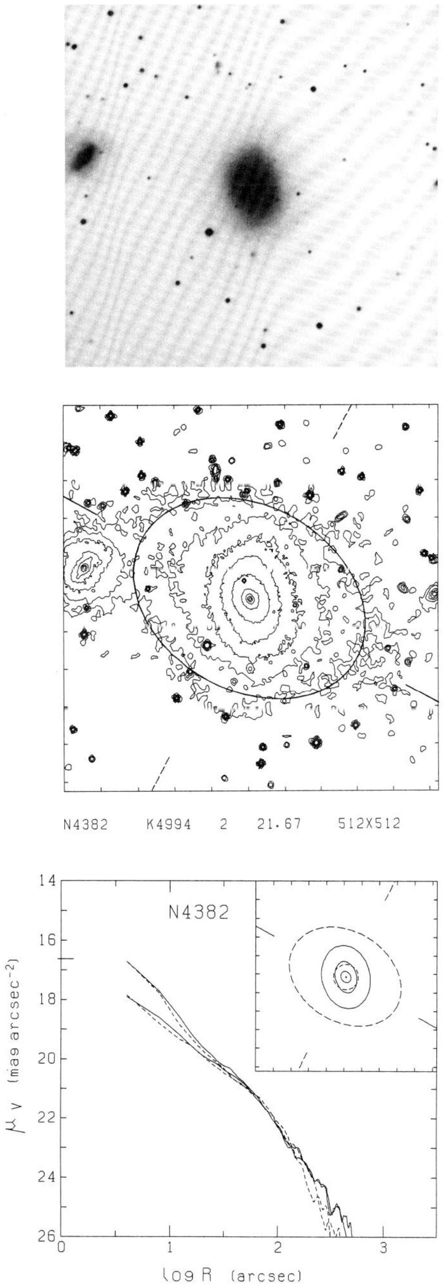

NGC4383

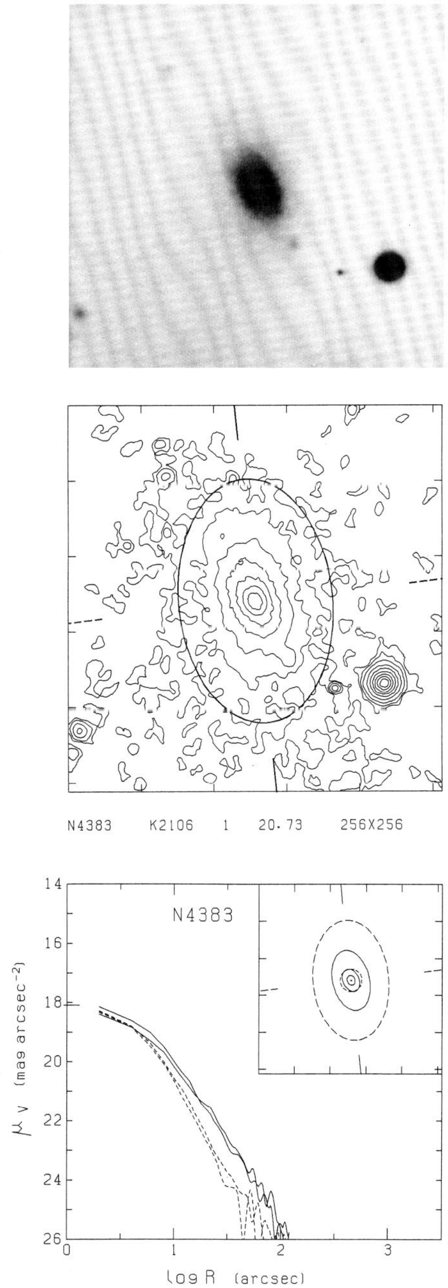

NGC4389

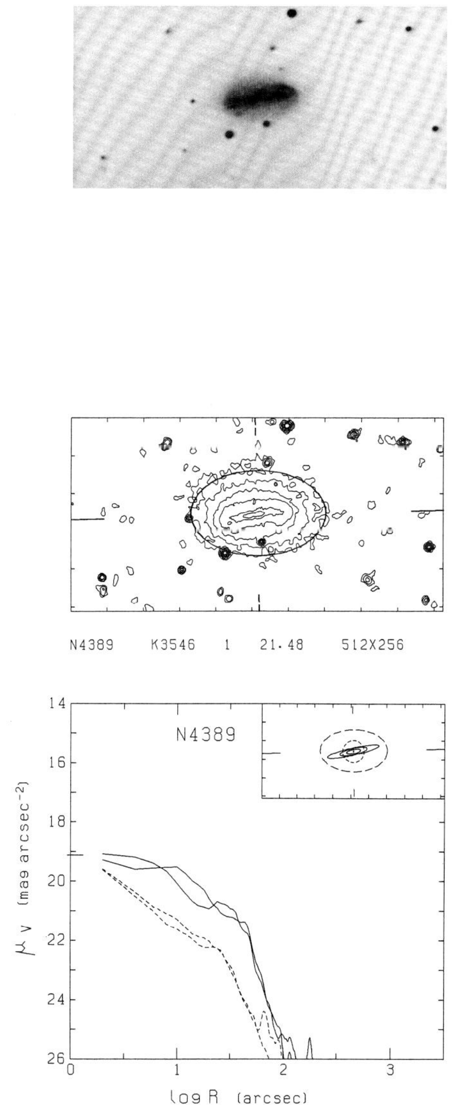

NGC4385

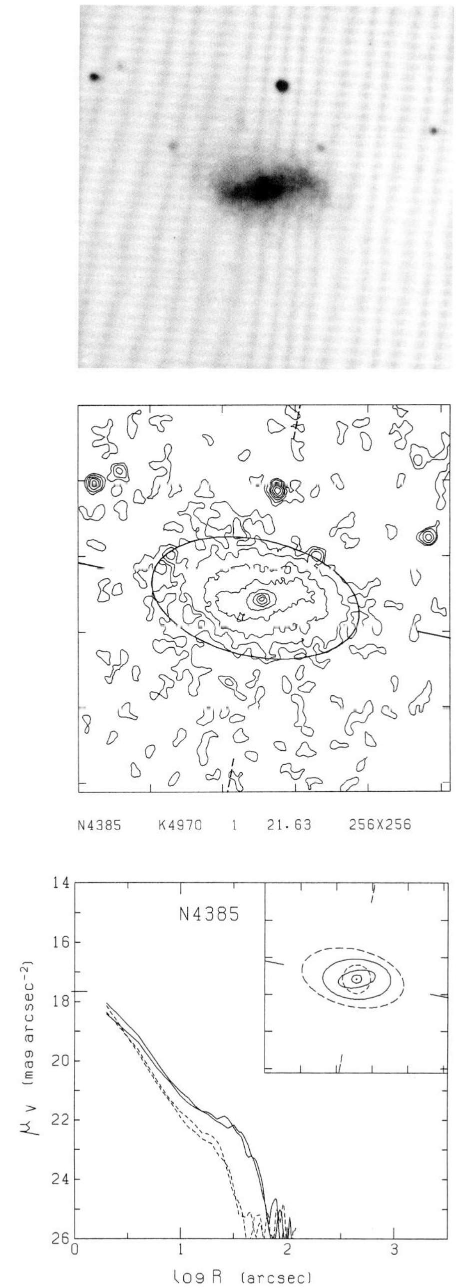

NGC4388

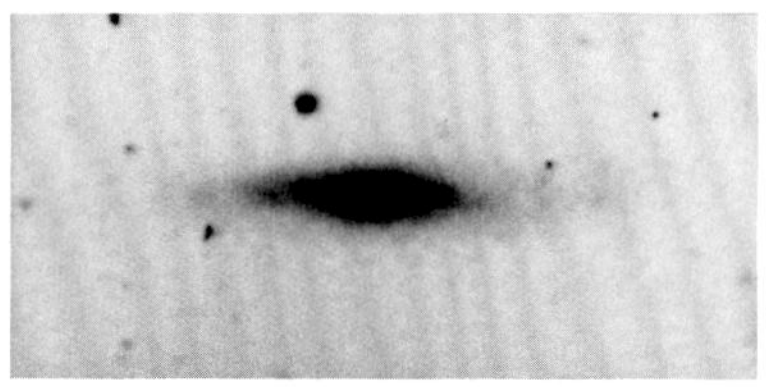

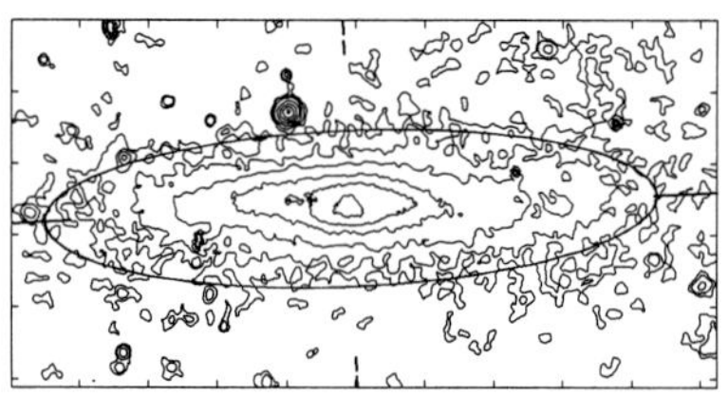

N4388 K2106 1 20.73 512X256

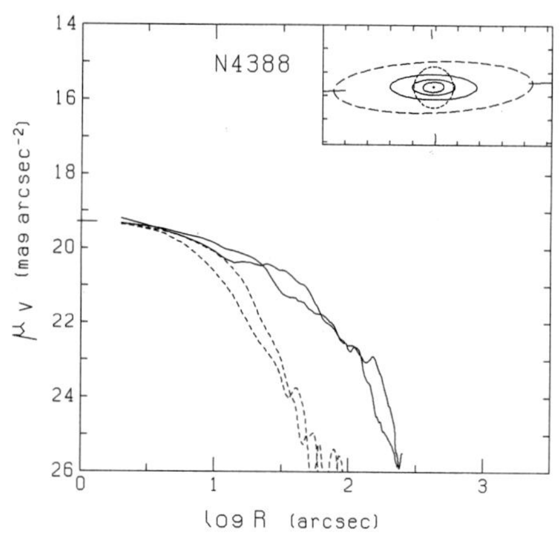

NGC4395

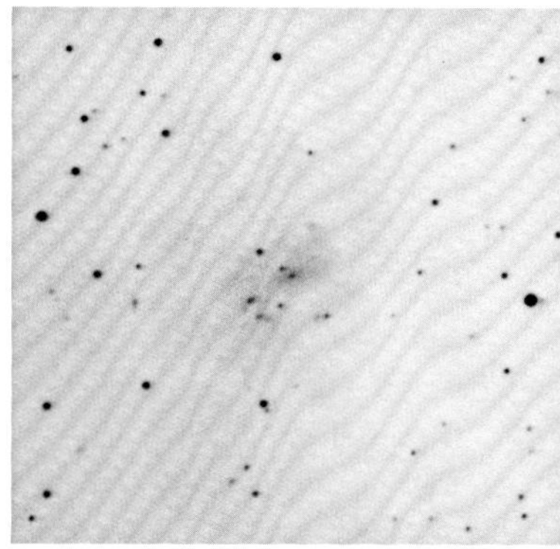

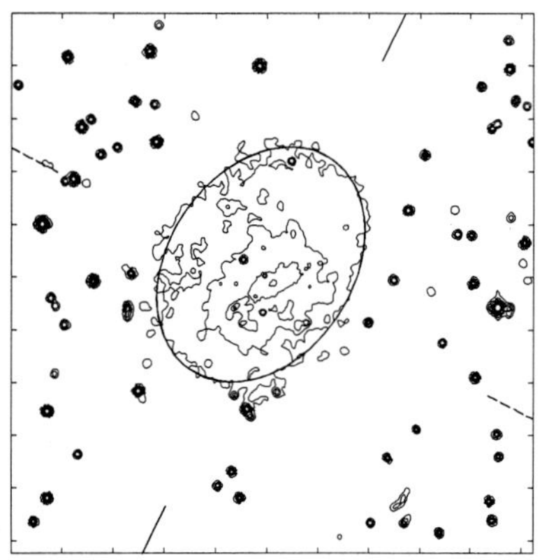

N4395 K5763 3 21.94 512X512

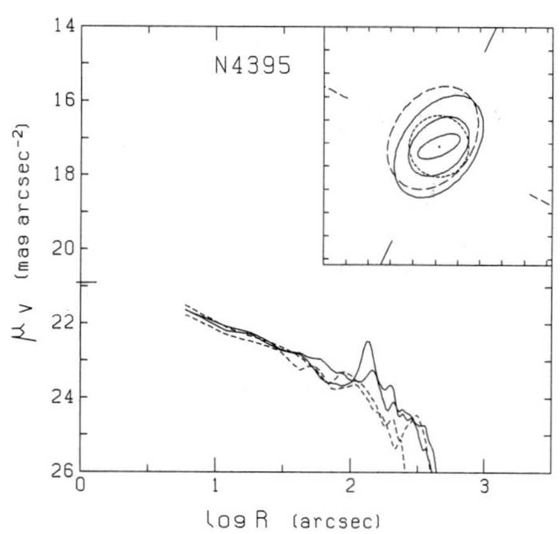

NGC4394

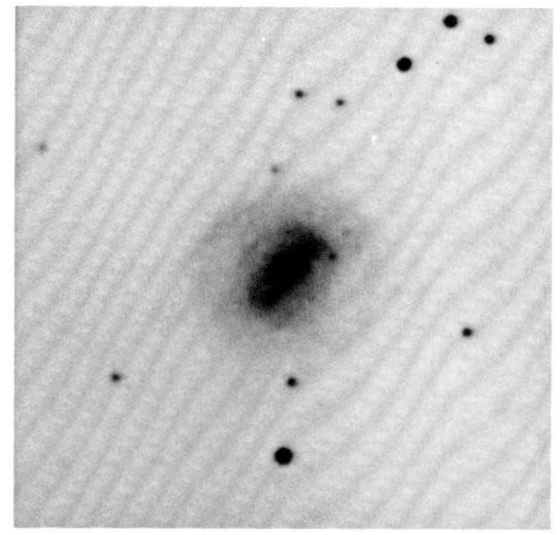

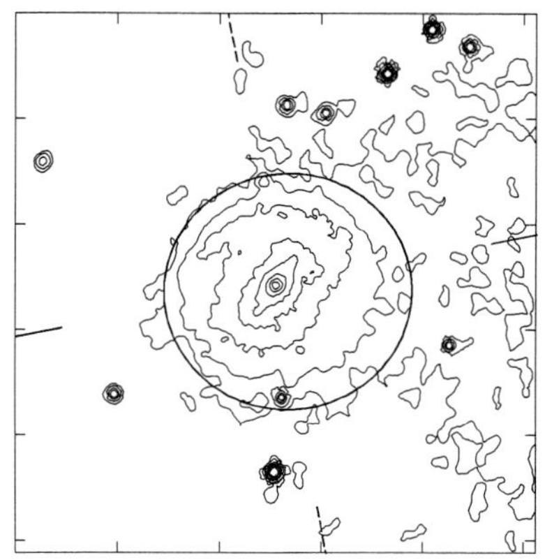

N4394 K4994 2 21.67 256X256

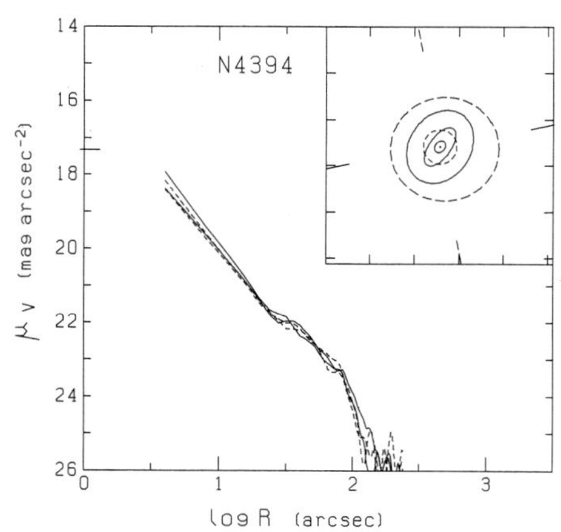

NGC4406

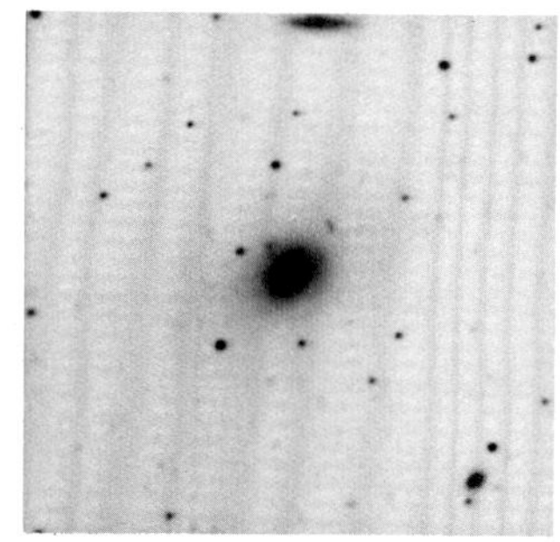

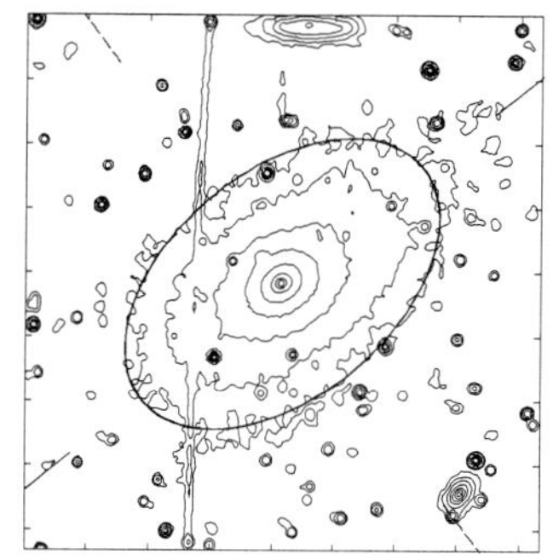

N4406 K2106 3 20.73 416X416

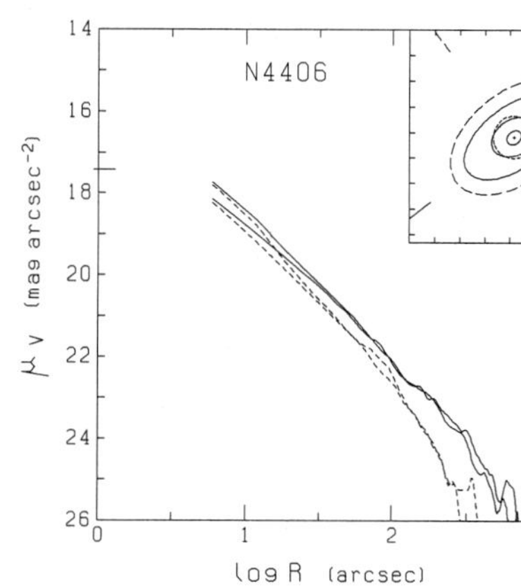

NGC4414

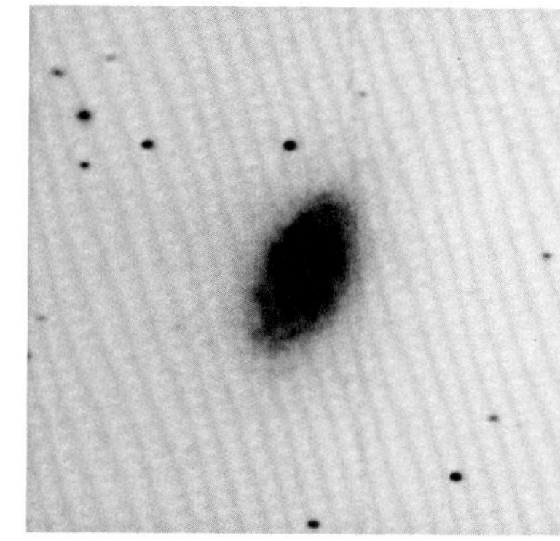

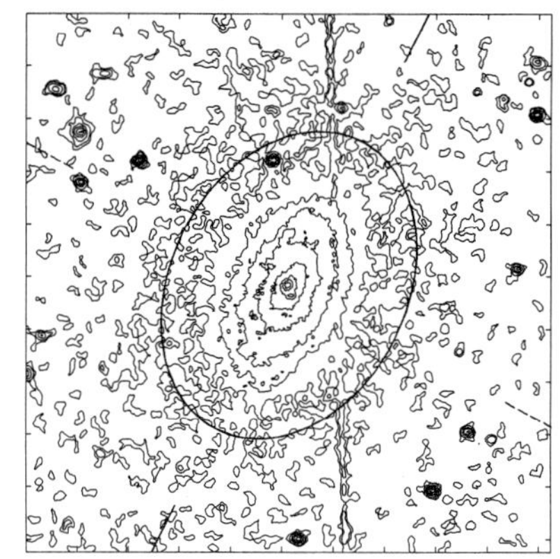

N4414 K4057 1 21.23 512X512

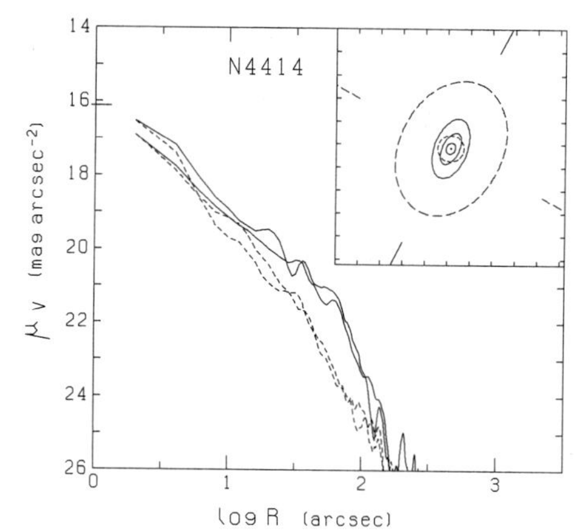

NGC4412

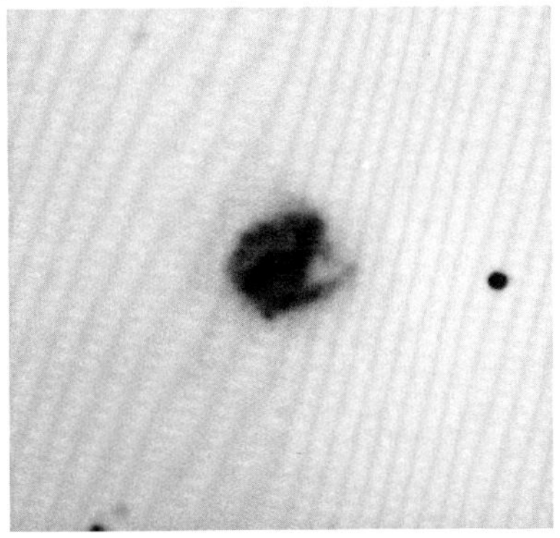

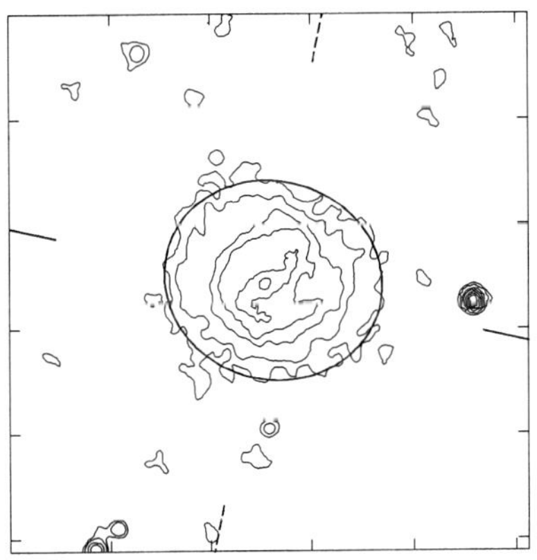

N4412 K3547 1 21.21 256X256

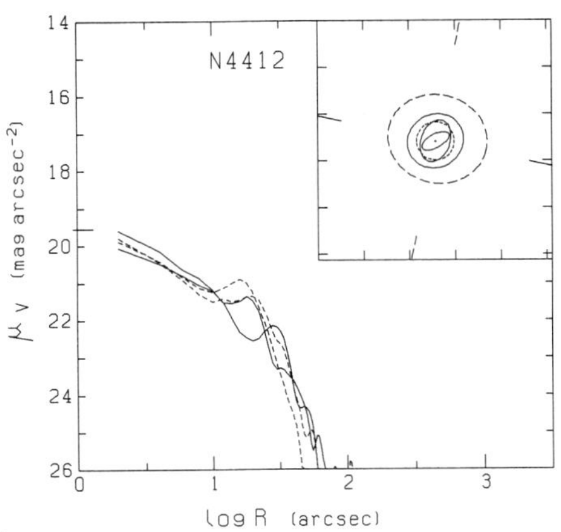

NGC4417

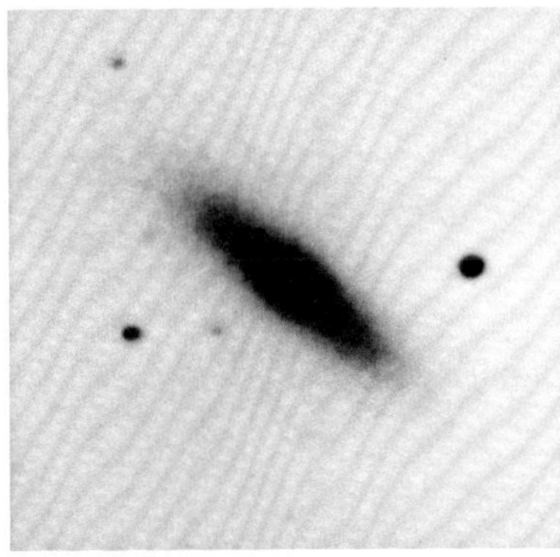

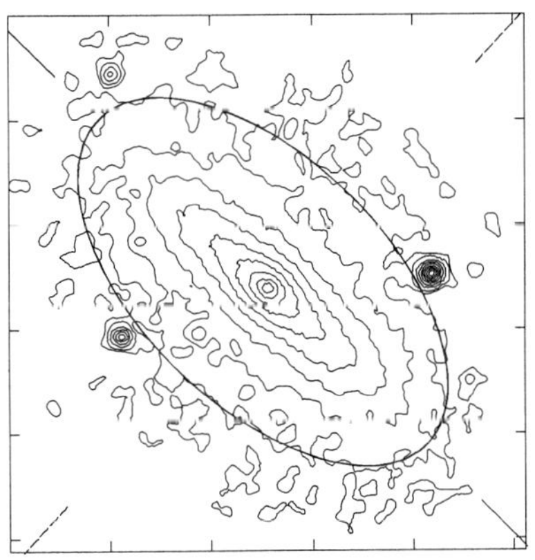

N4417 K2186 1 21.34 256X256

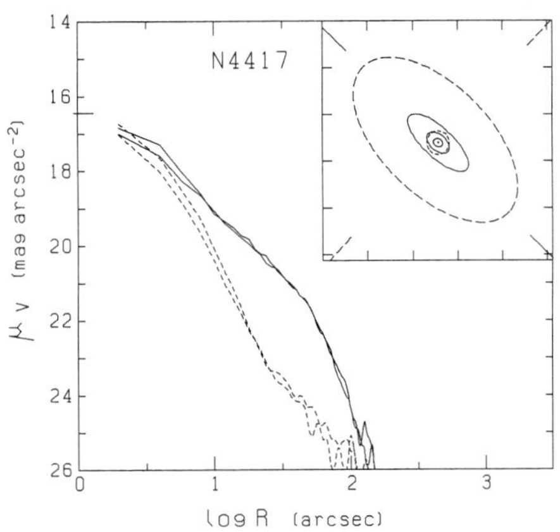

NGC4419

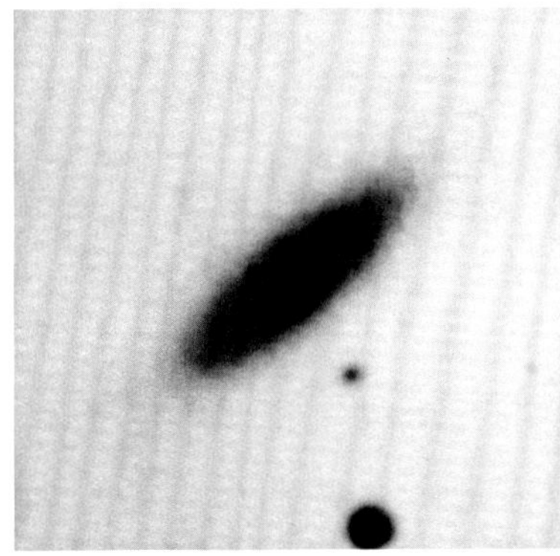

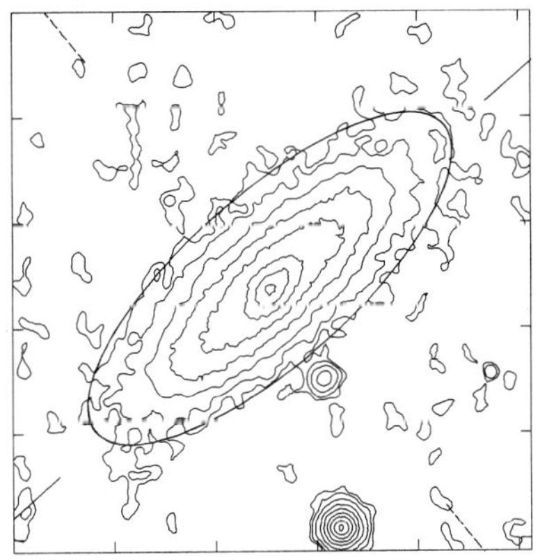

N4419 K2106 1 20.73 256X256

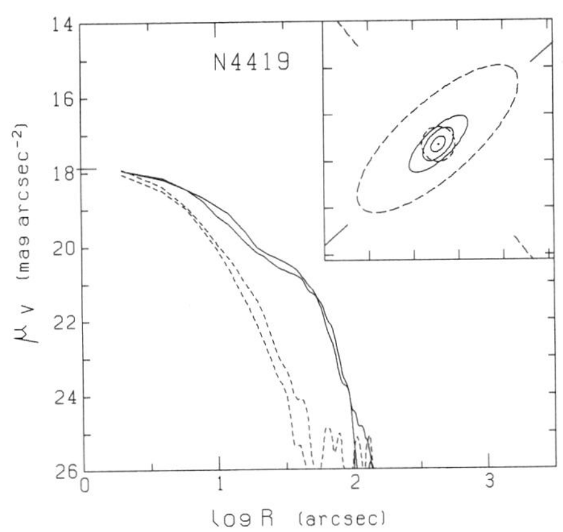

NGC4420

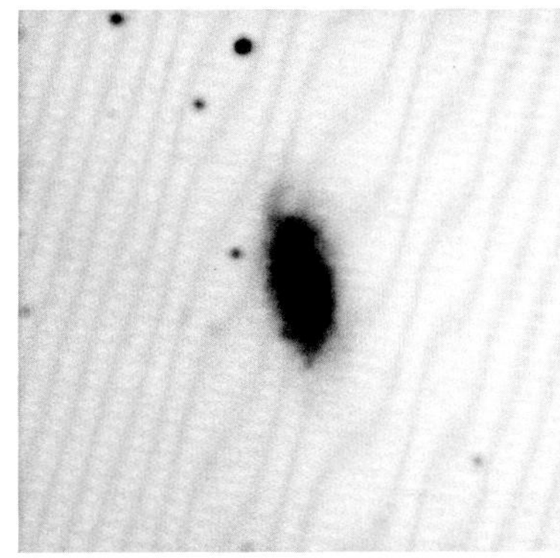

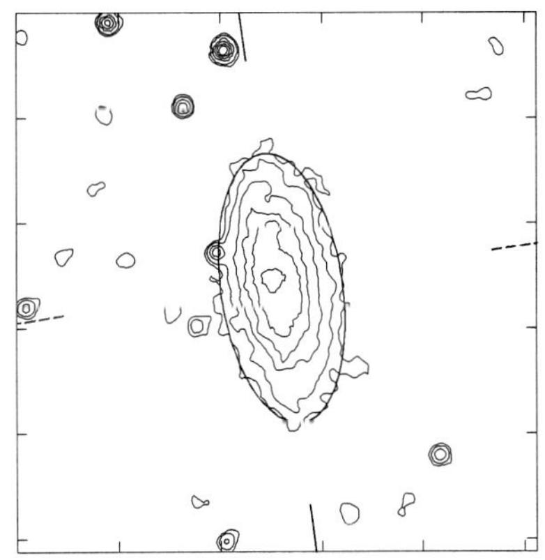

N4420 K3547 1 21.21 256X256

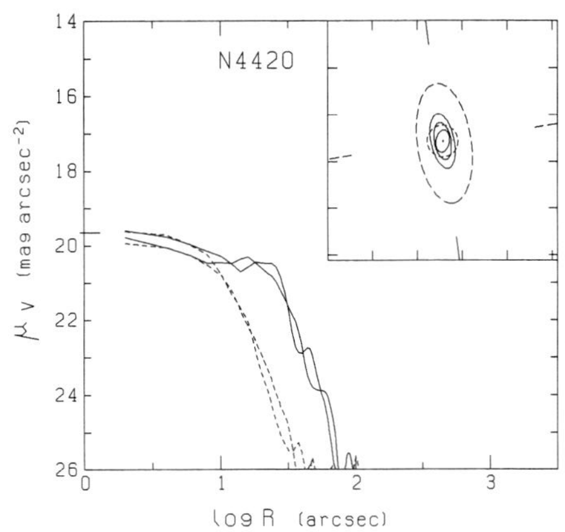

NGC4424

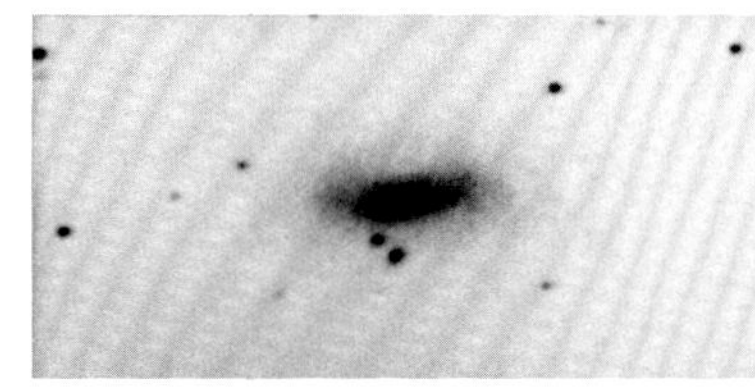

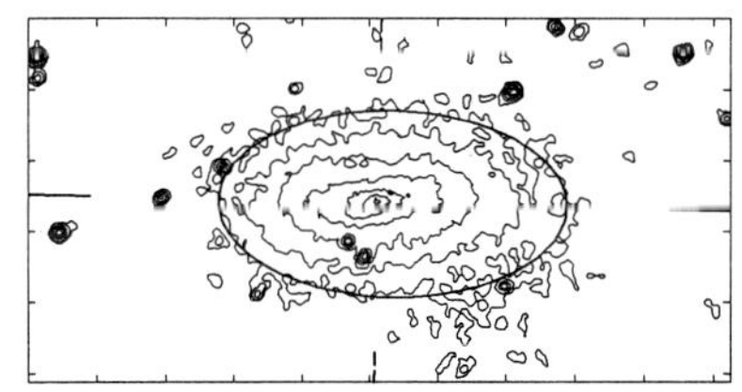

N4424 K2186 1 21.34 512X256

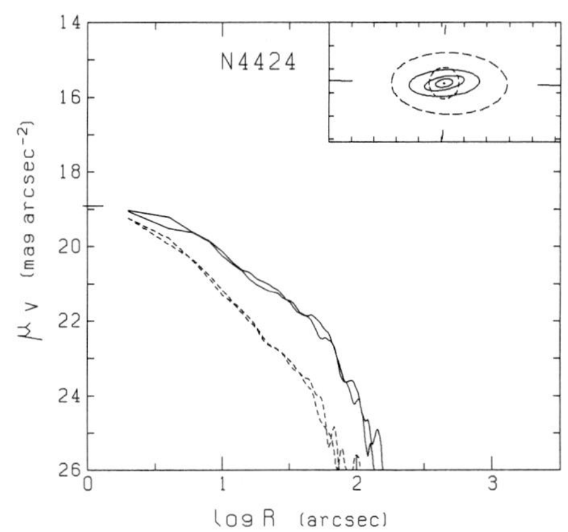

NGC4425

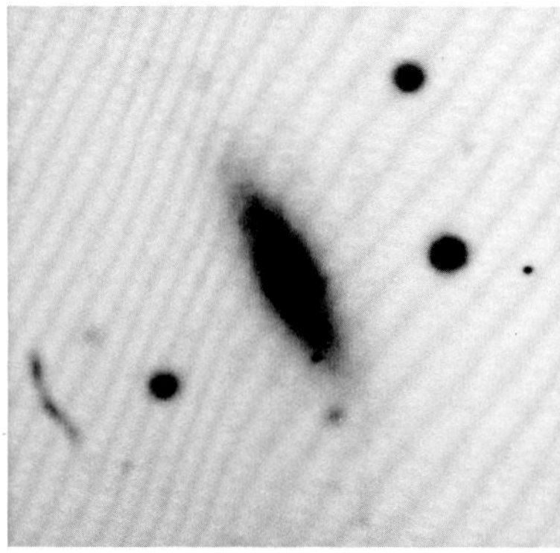

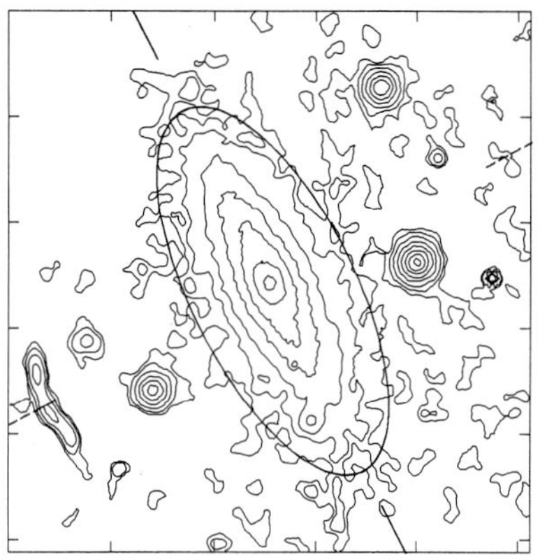

N4425 K2106 1 20.73 256X256

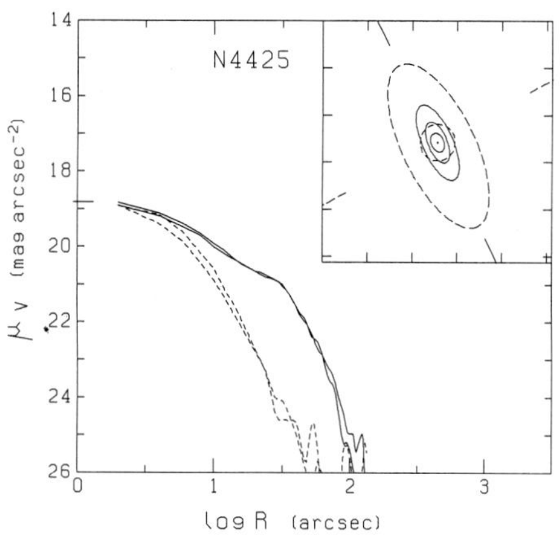

NGC4428

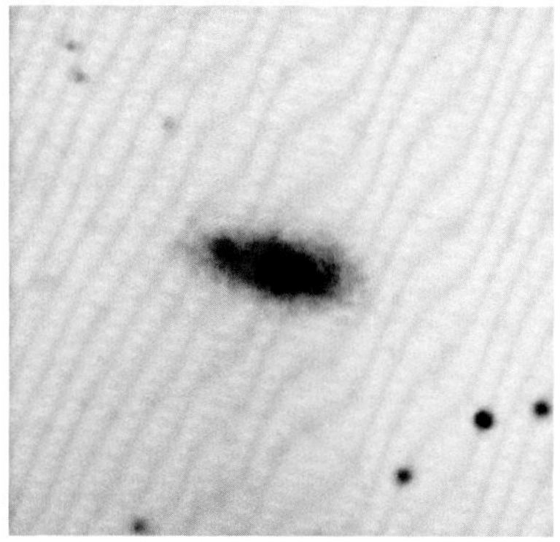

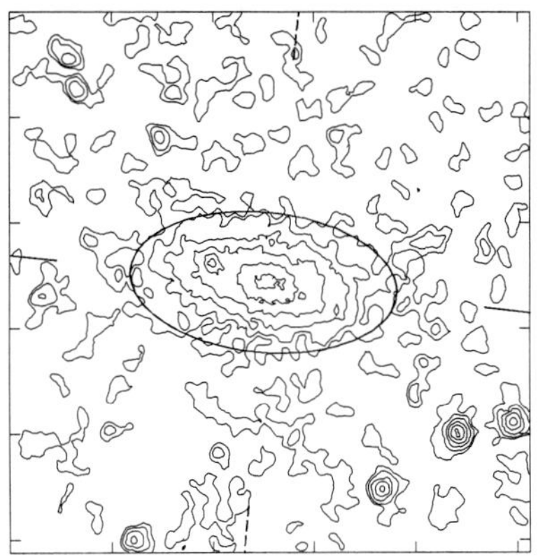

N4428 K5012 1 21.51 256X256

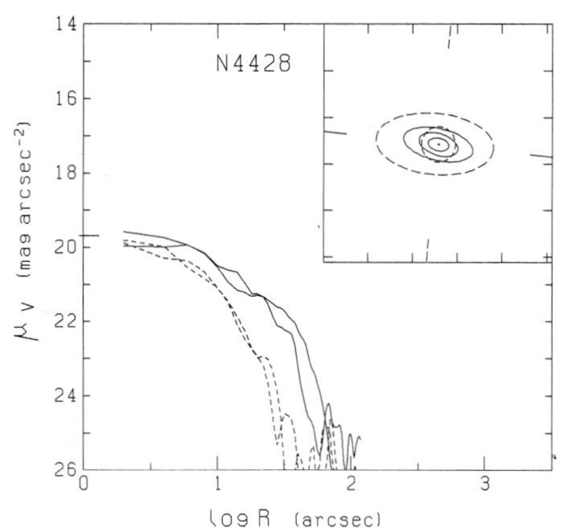

NGC4429

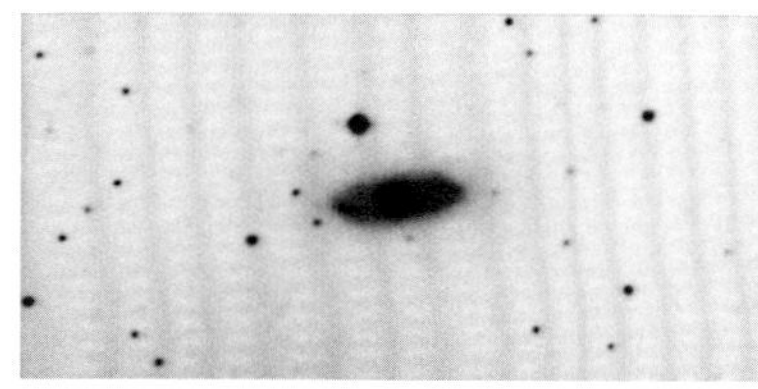

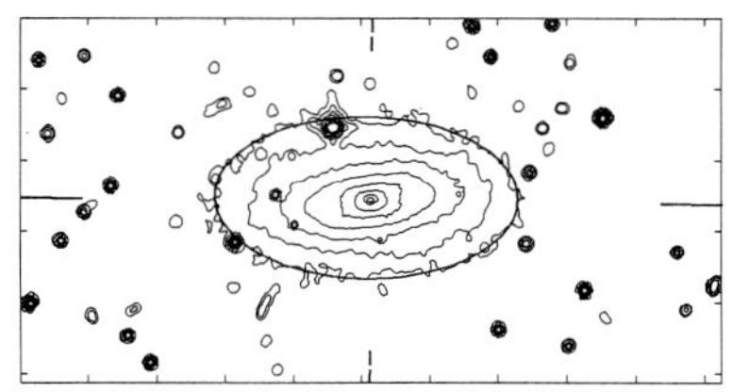

N4429 K2186 2 21.34 512X256

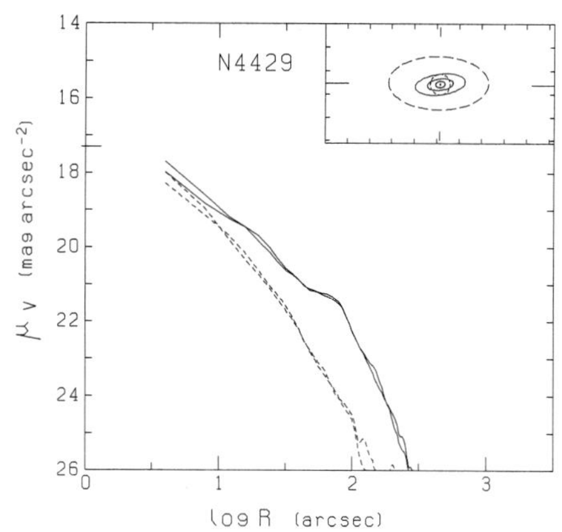

NGC4433

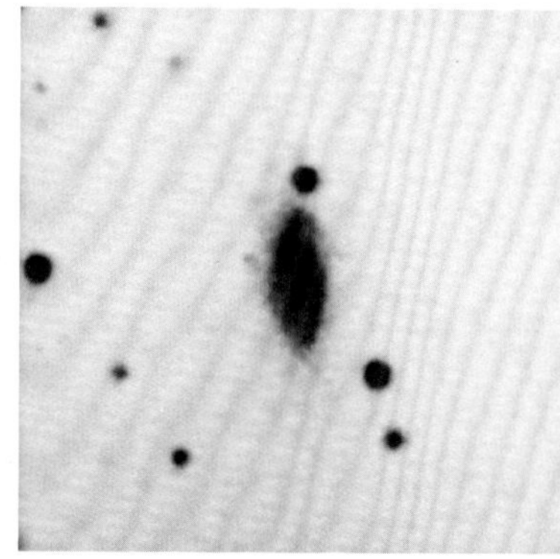

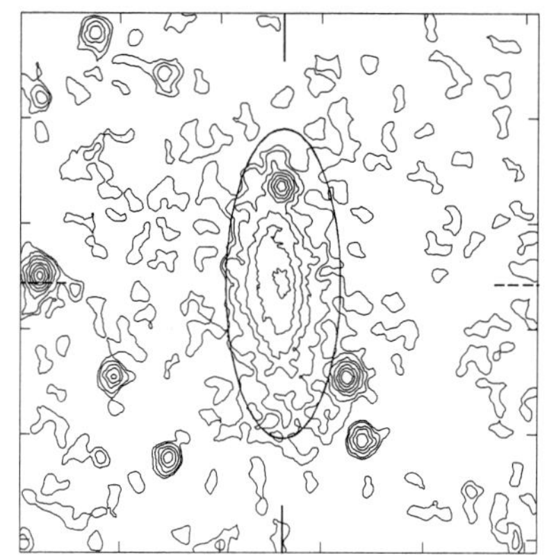

N4433 K5012 1 21.51 256X256

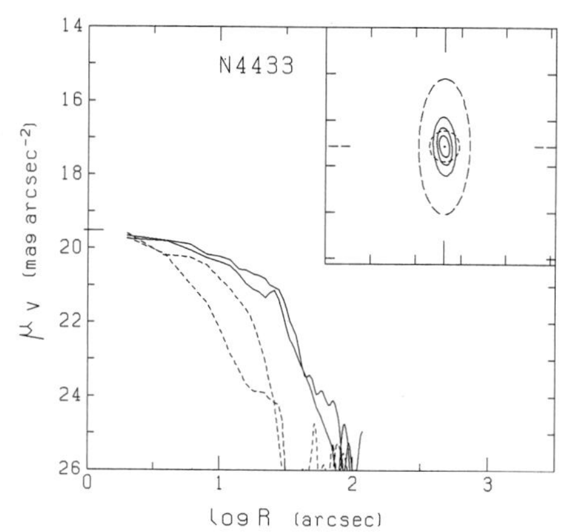

NGC4435

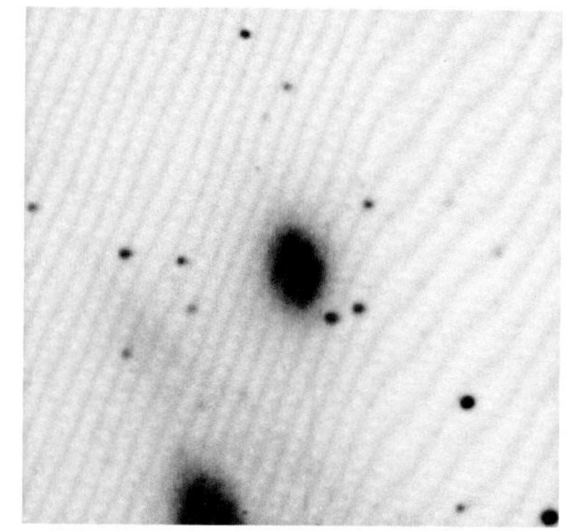

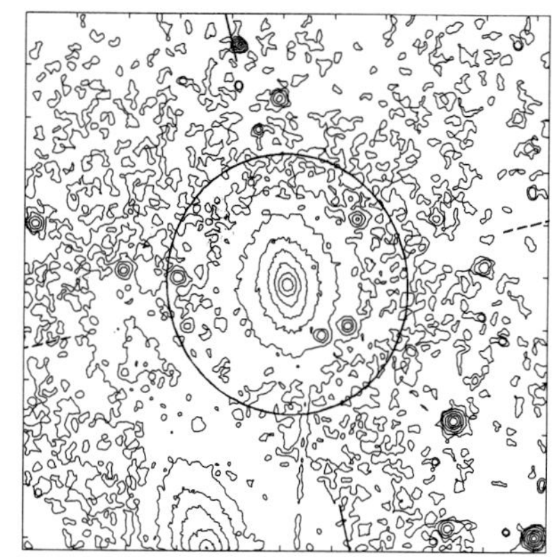

N4435 K2106 1 20.73 512X512

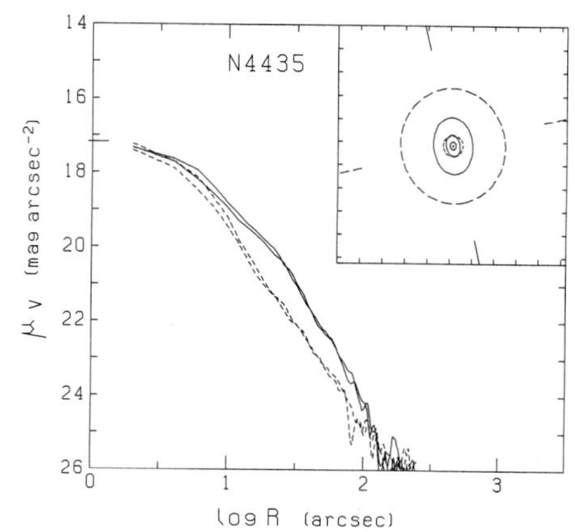

NGC4438

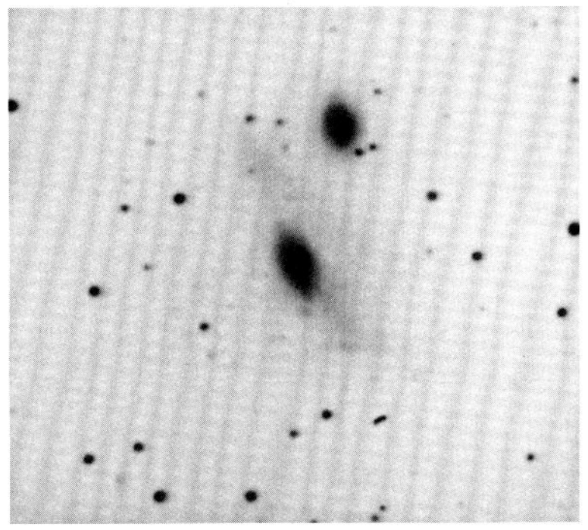

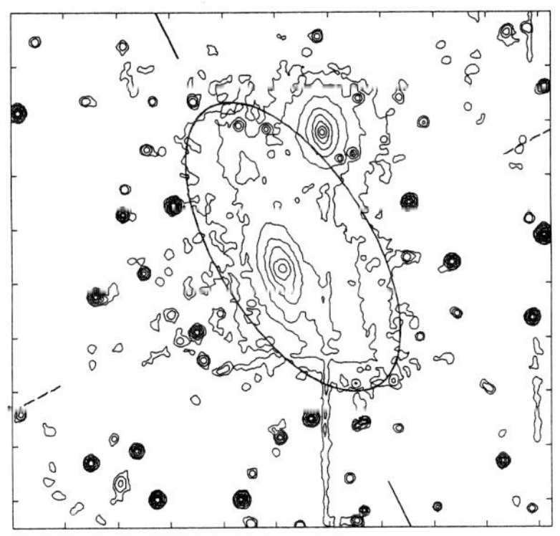

N4438 K2106 2 20.73 512X475

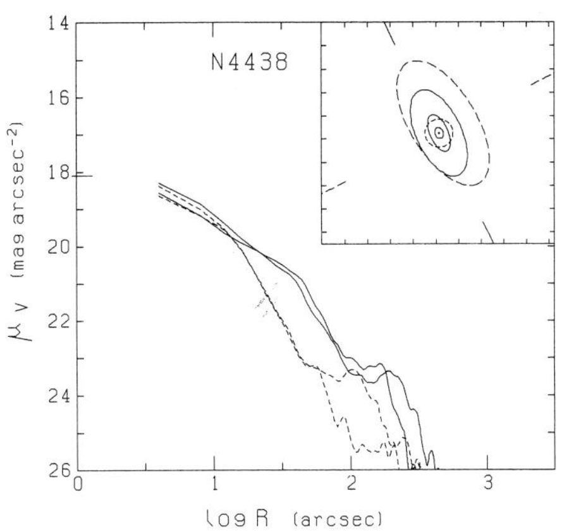

NGC4442

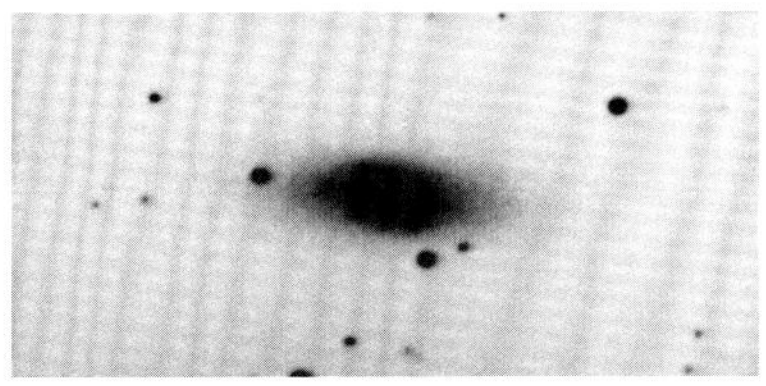

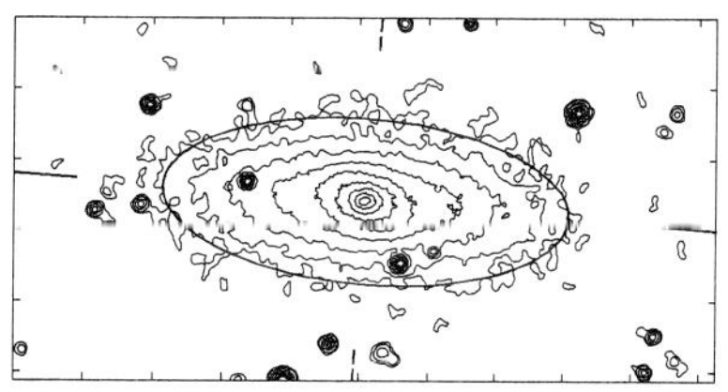

N4442 K2186 1 21.34 512X256

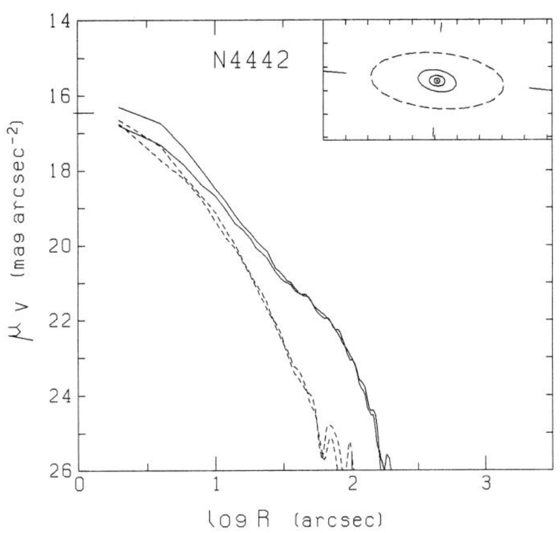

NGC4448

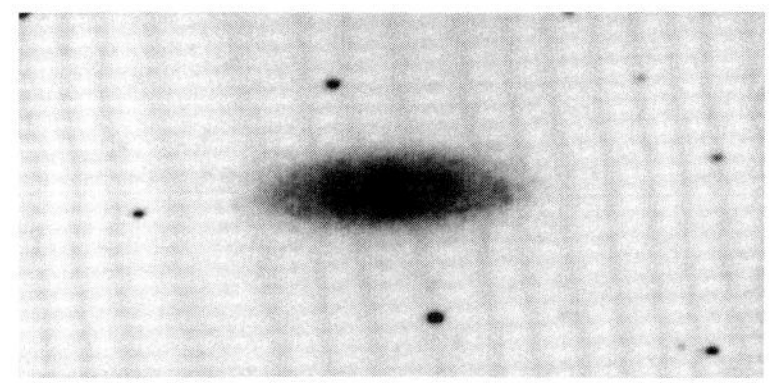

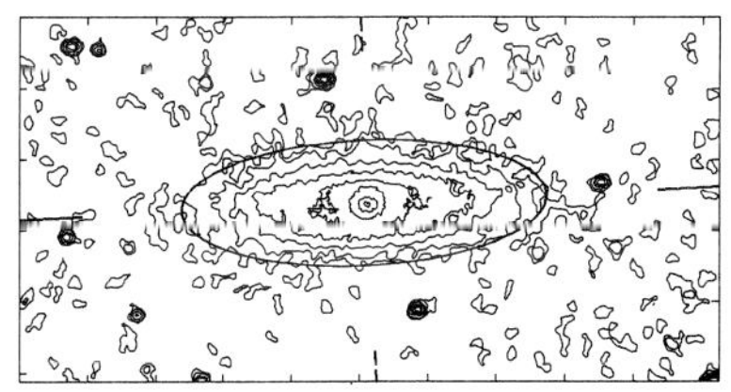

N4448 K4057 1 21.23 512X256

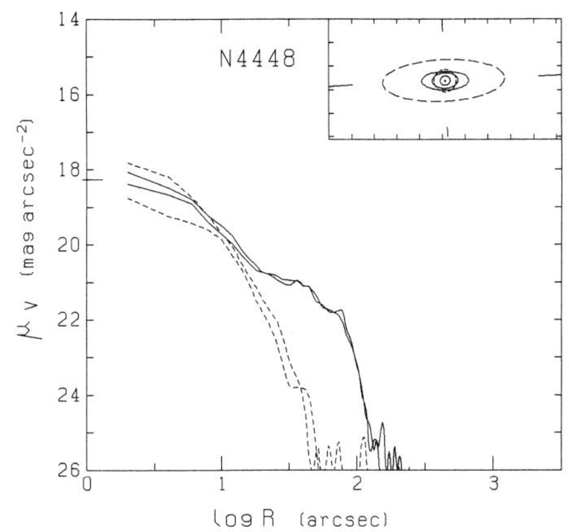

NGC4449

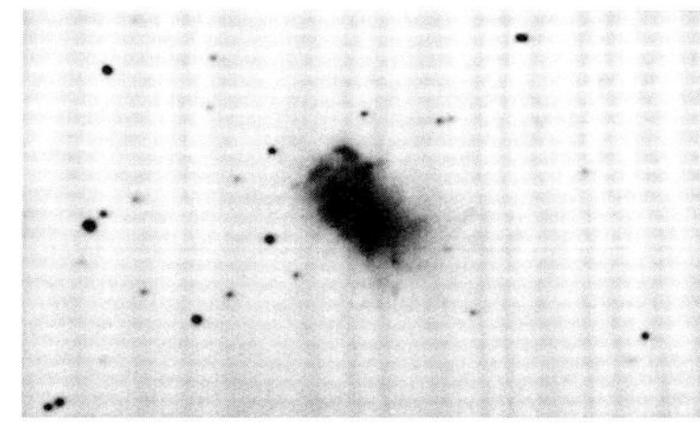

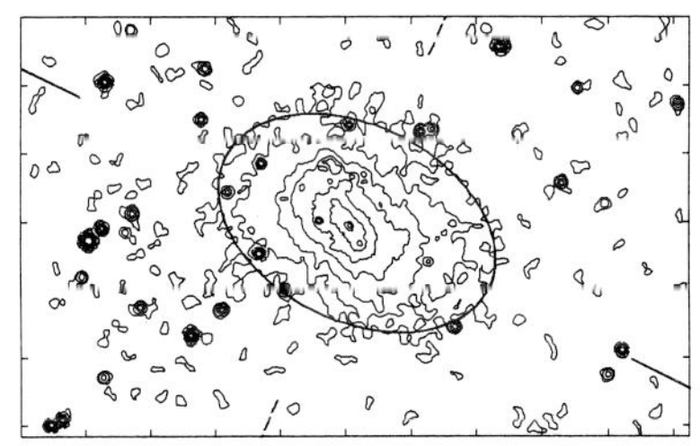

N4449 K4705 2 21.19 512X308

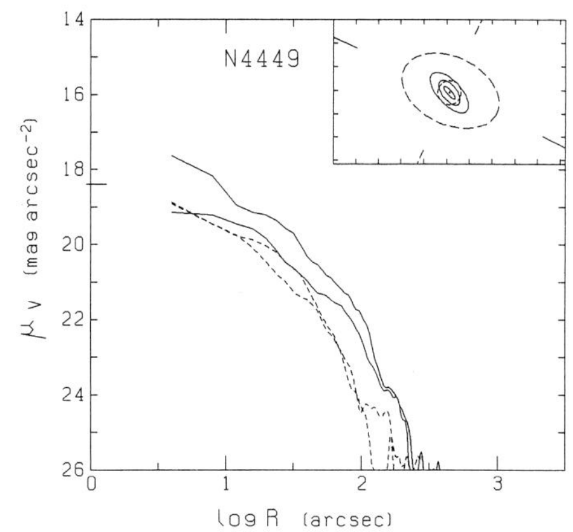

NGC4450

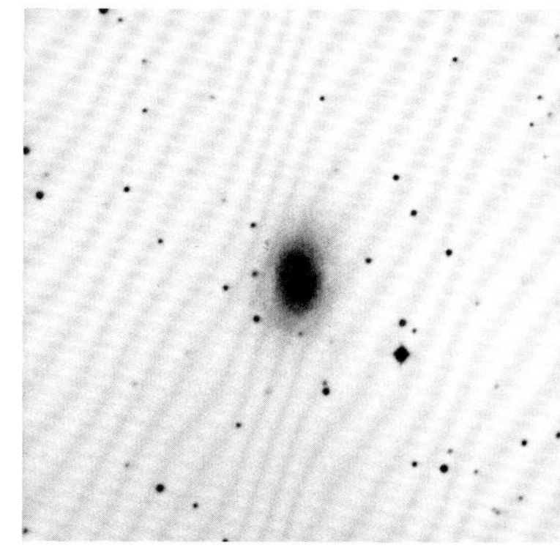

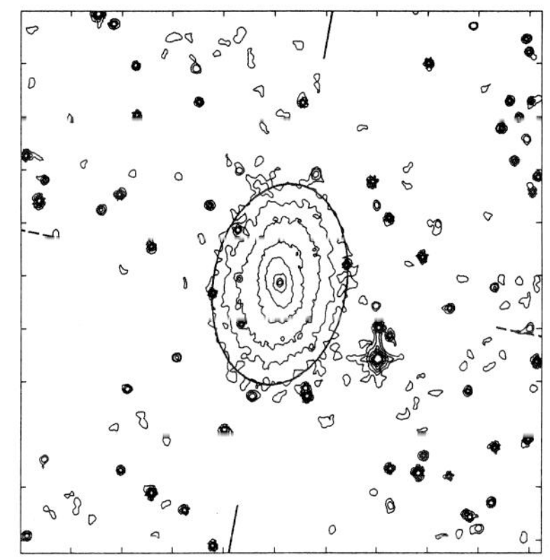

N4450 K4994 2 21.67 512X512

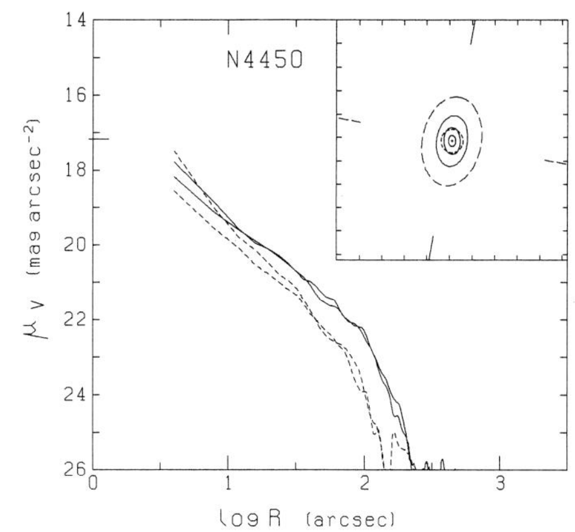

NGC4452

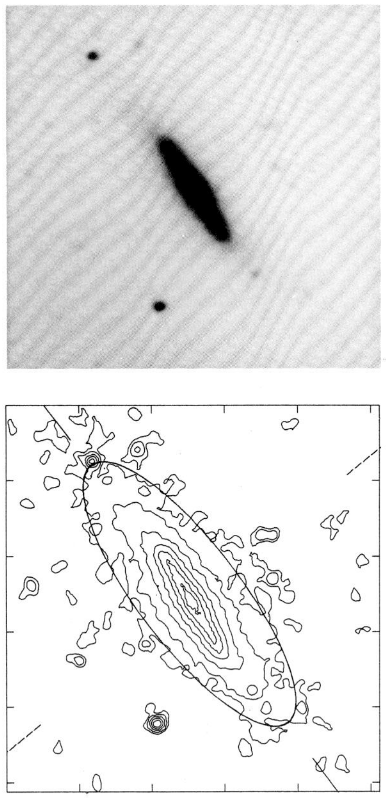

N4452 K2186 1 21.34 256X256

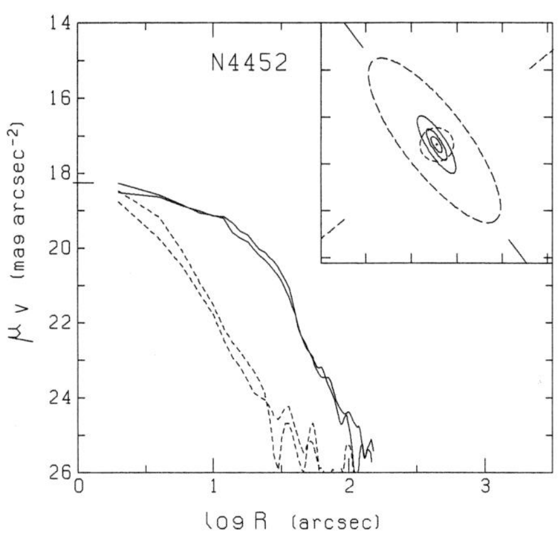

NGC4454

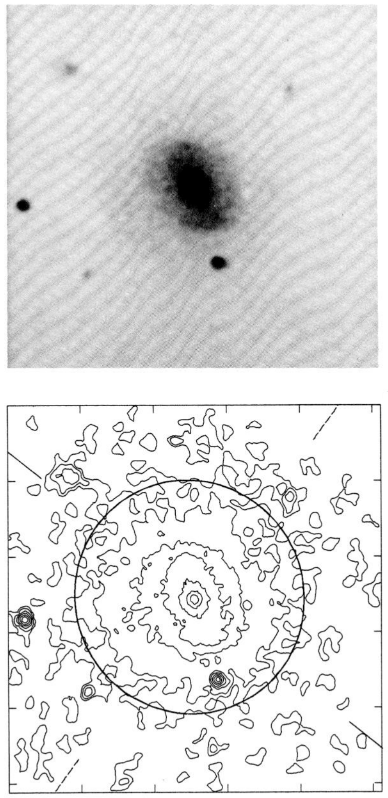

N4454 K4970 1 21.63 256X256

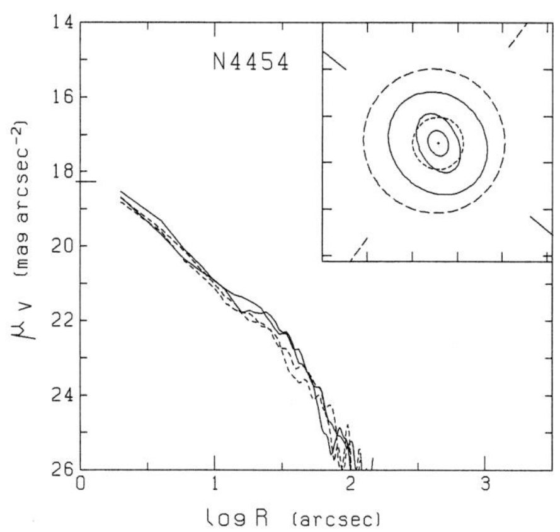

NGC4460

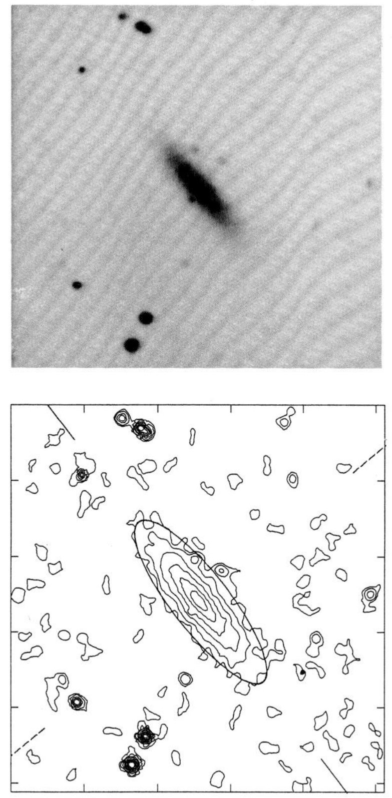

N4460 K4705 2 21.19 256X256

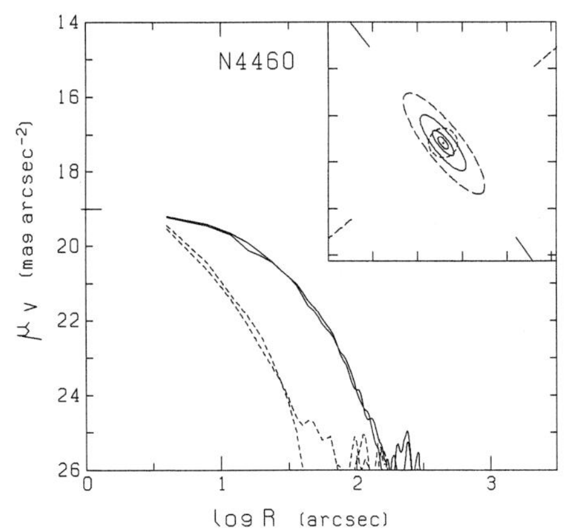

NGC4457

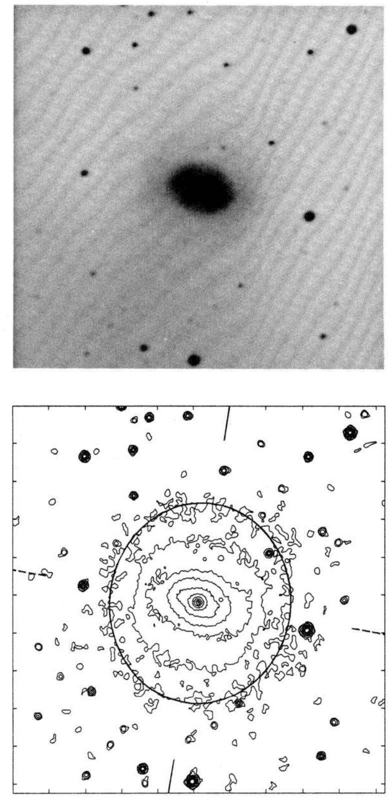

N4457 K3547 1 21.21 512X512

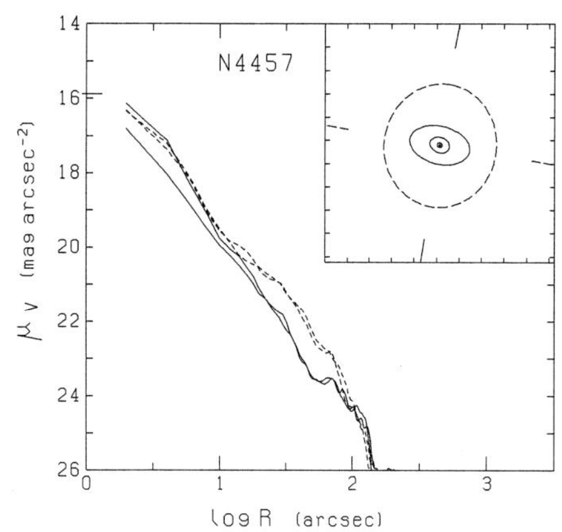

NGC4459

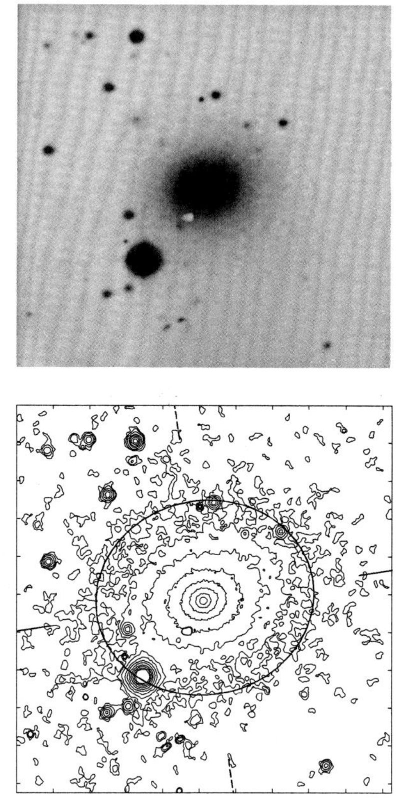

N4459 K2106 1 20.73 512X512

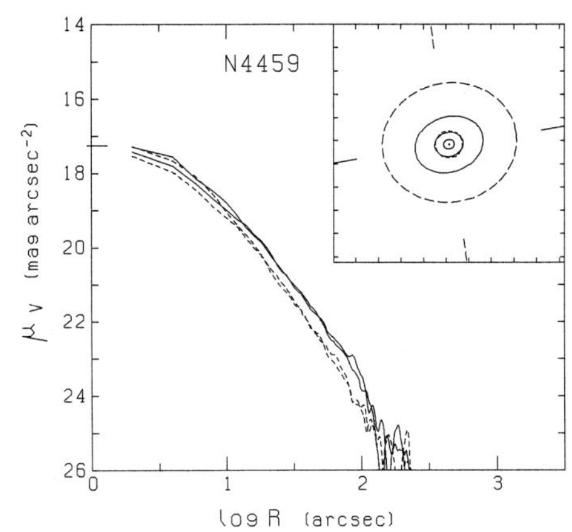

NGC4461

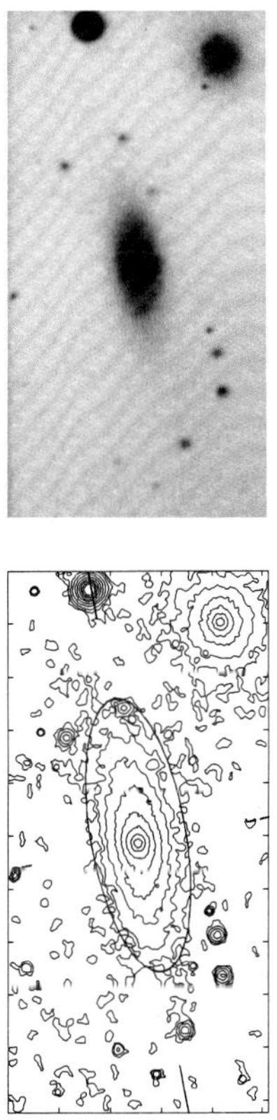

N4461 K2106 1 20.73 256X512

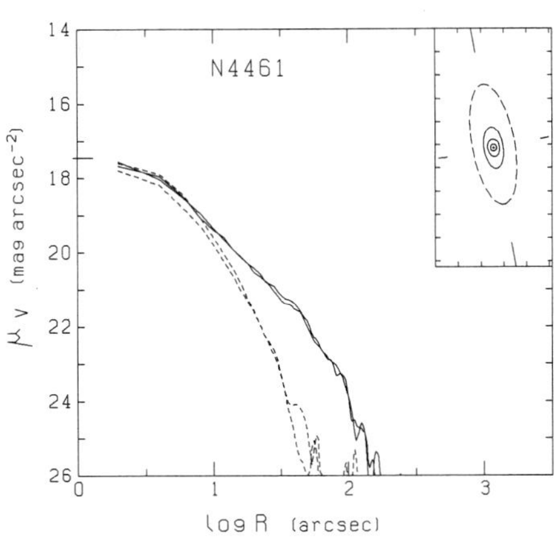

NGC4469

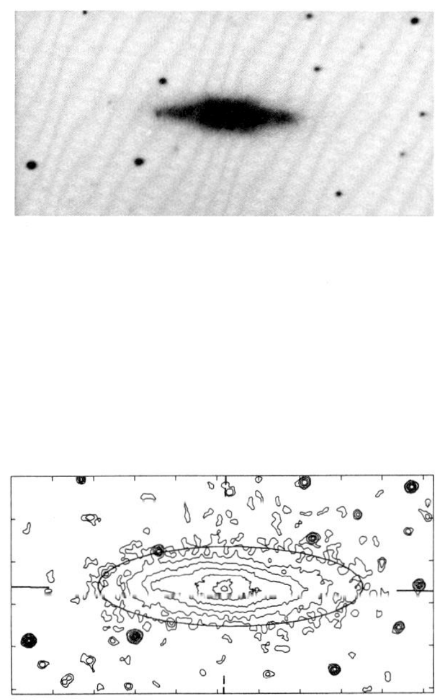

N4469 K2186 1 21.34 512X256

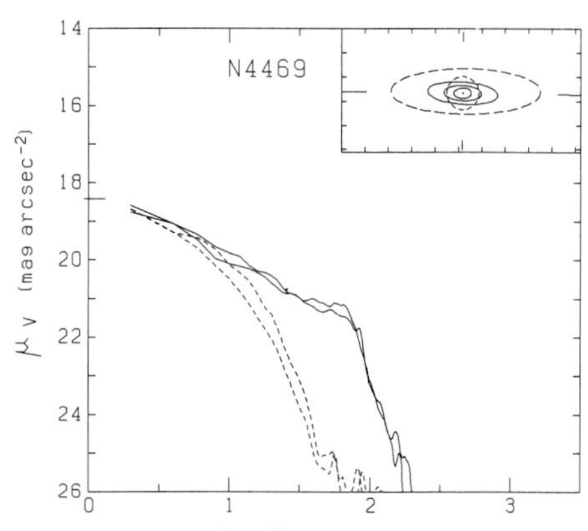

NGC4472

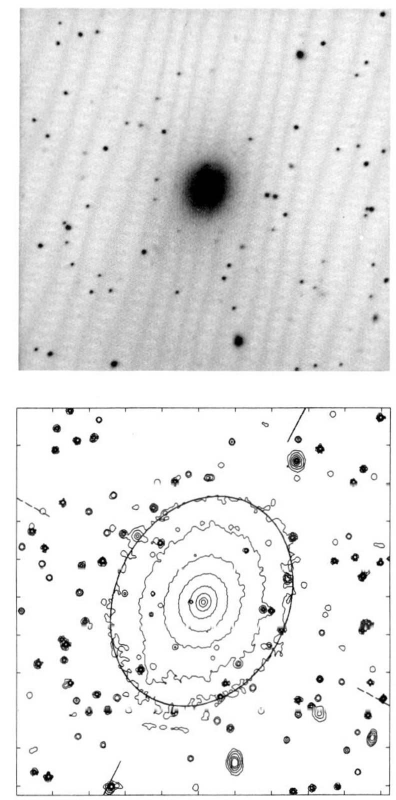

N4472 K2186 3 21.34 512X512

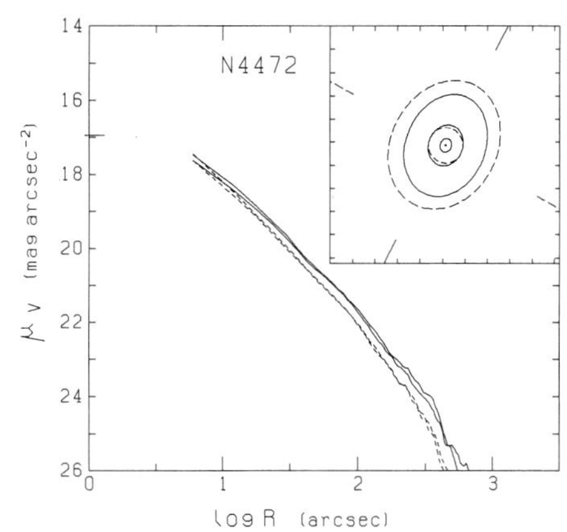

NGC4473

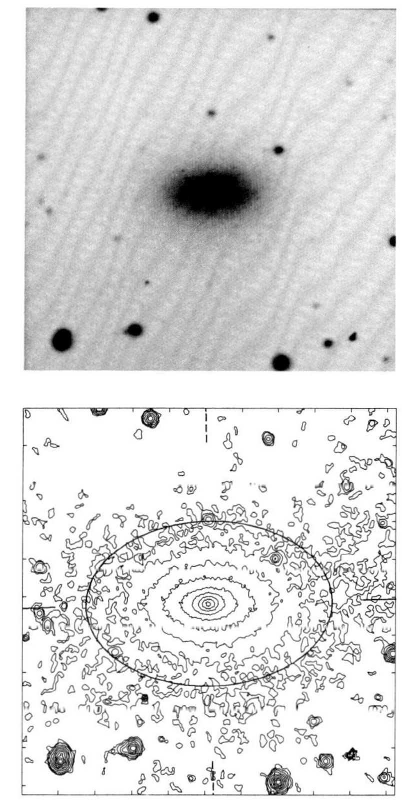

N4473 K2106 1 20.73 512X512

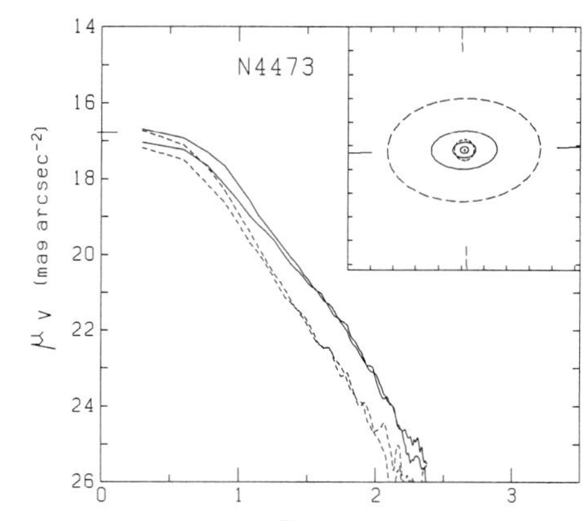

NGC4474

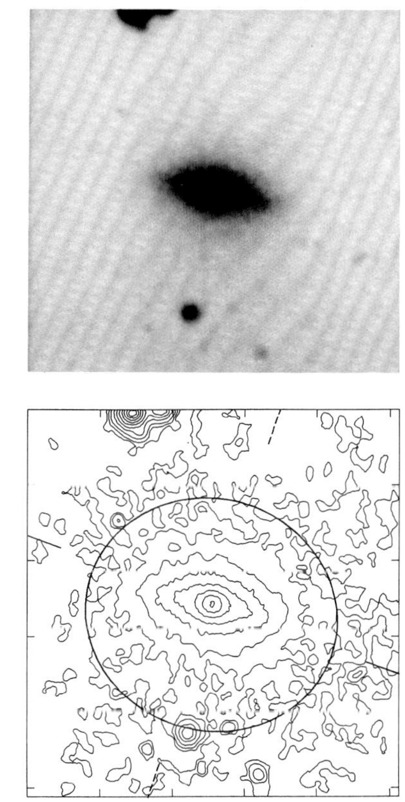

N4474 K2106 1 20.73 256X256

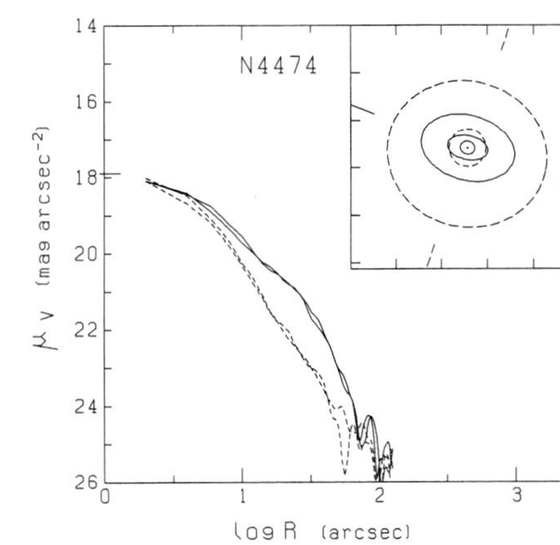

NGC4476

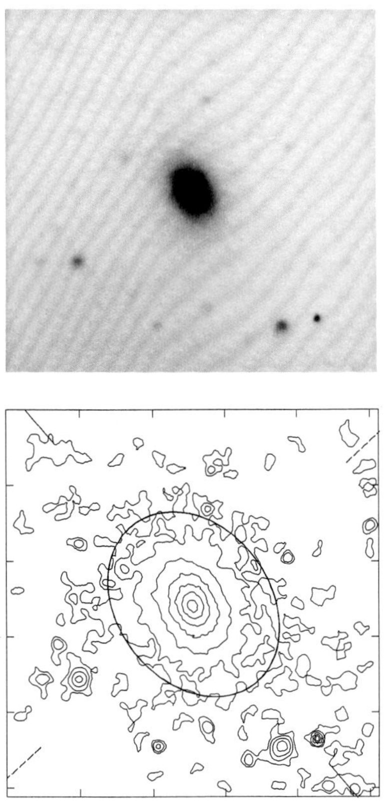

N4476 K2106 1 20.73 256X256

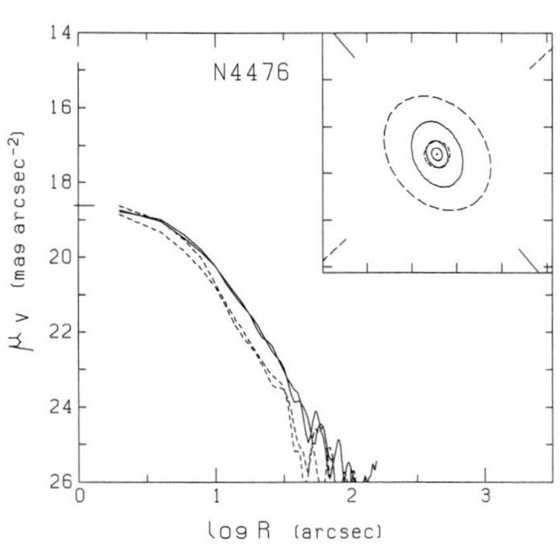

NGC4477

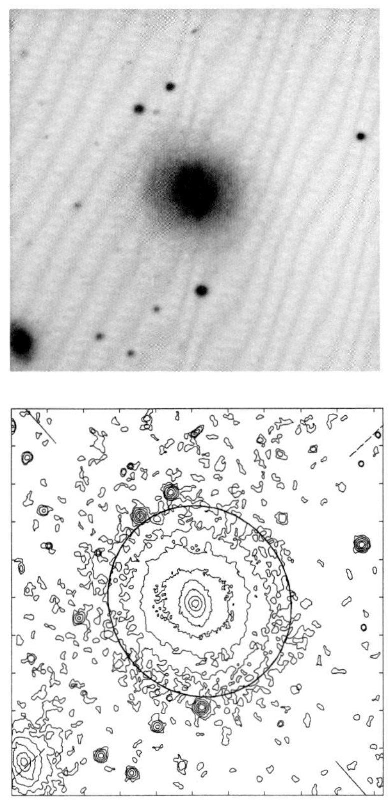

N4477 K2106 1 20.73 512X512

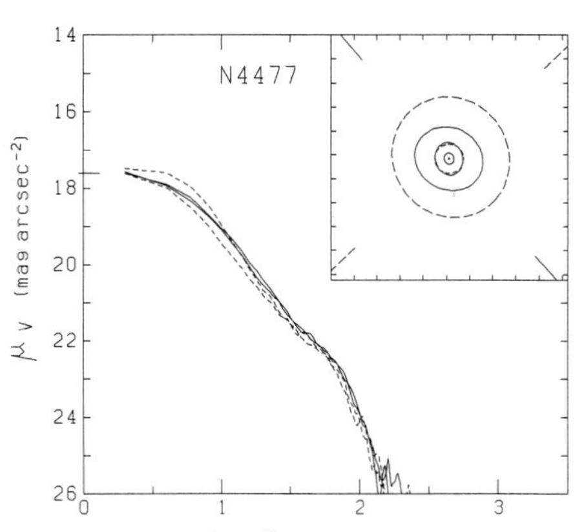

NGC4478

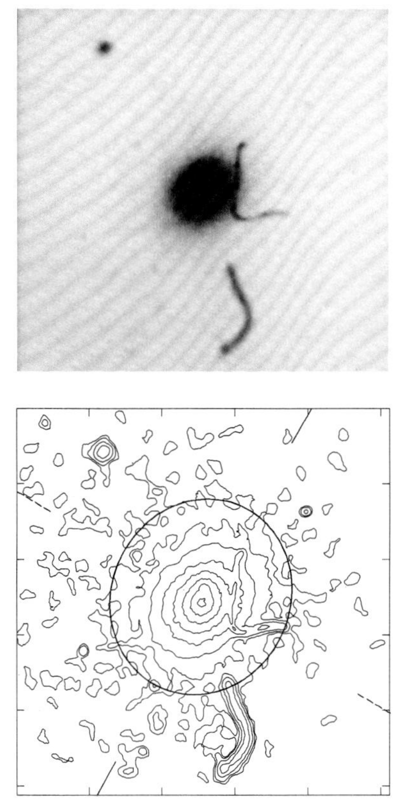

N4478 K2106 1 20.73 256X256

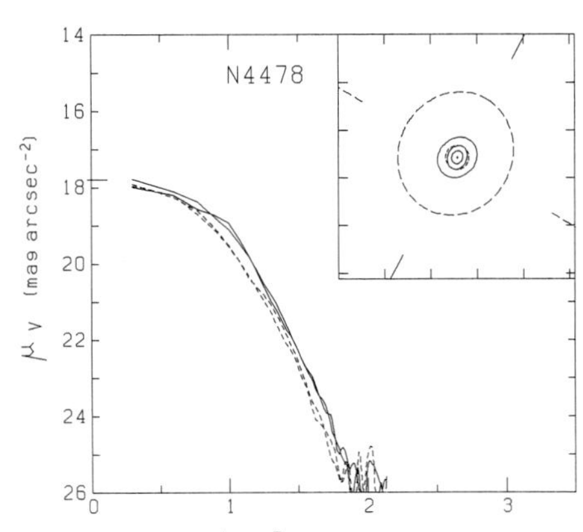

NGC4485

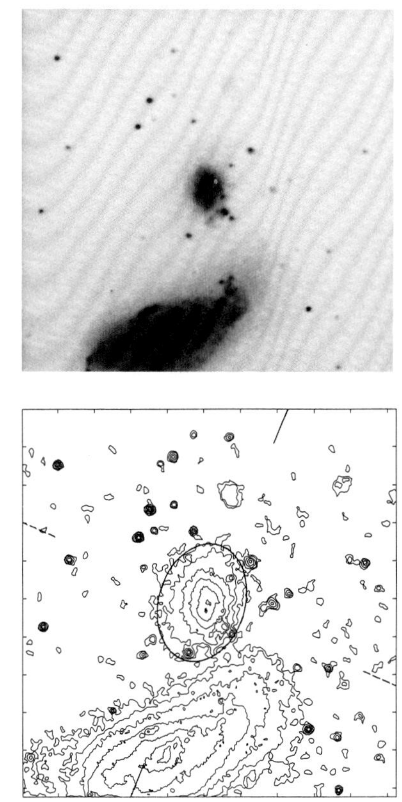

N4485 K3173 1 21.49 512X512

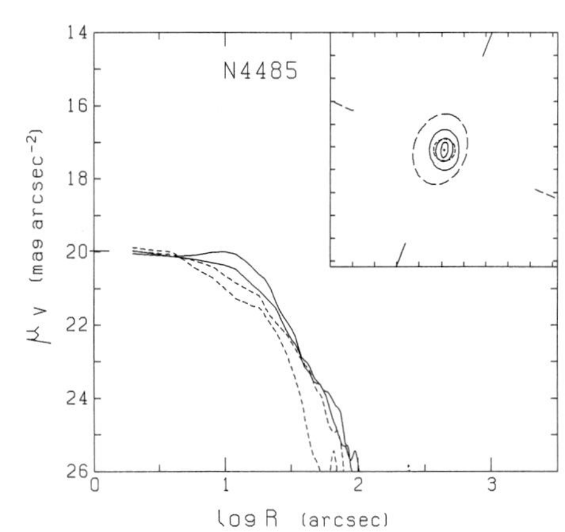

NGC4483

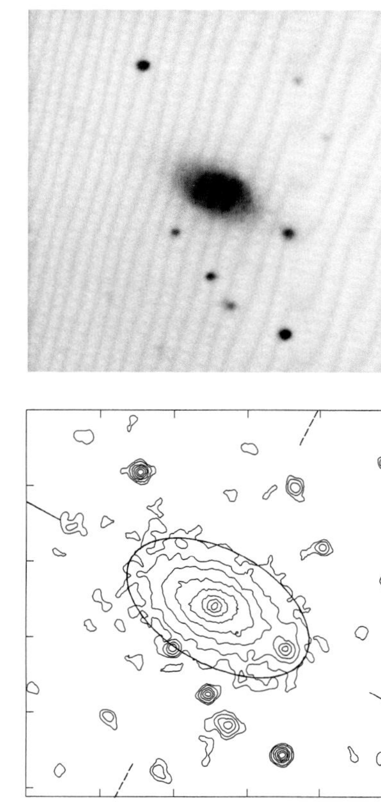

N4483 K2186 1 21.34 256X256

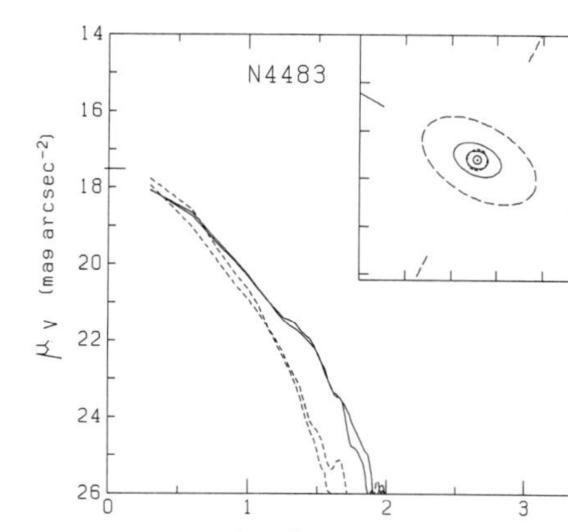

NGC4490

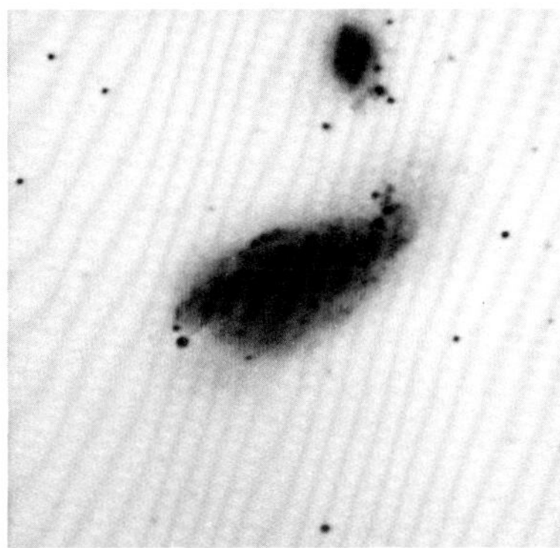

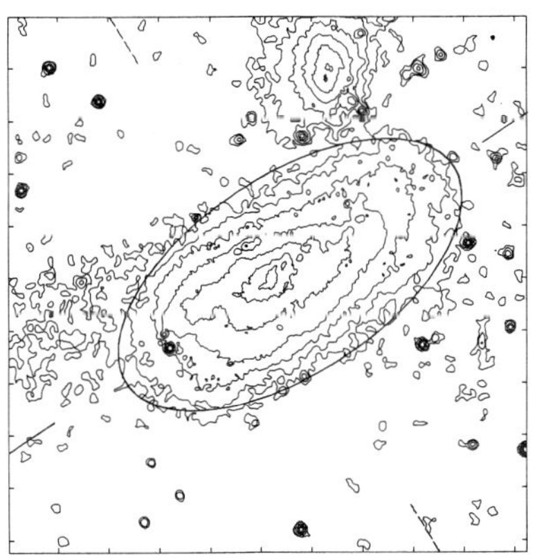

N4490 K3173 1 21.49 512X512

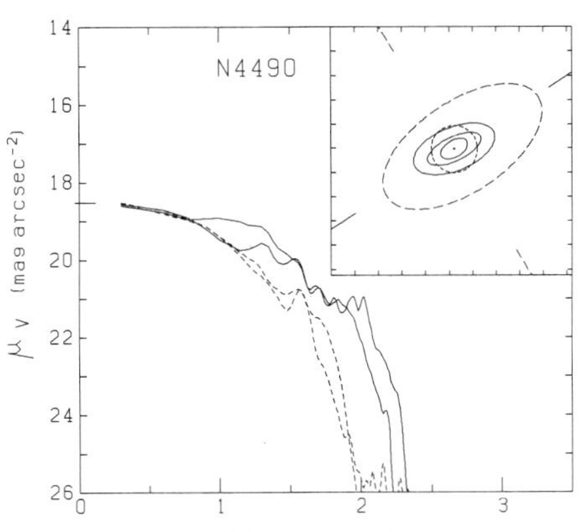

NGC4486

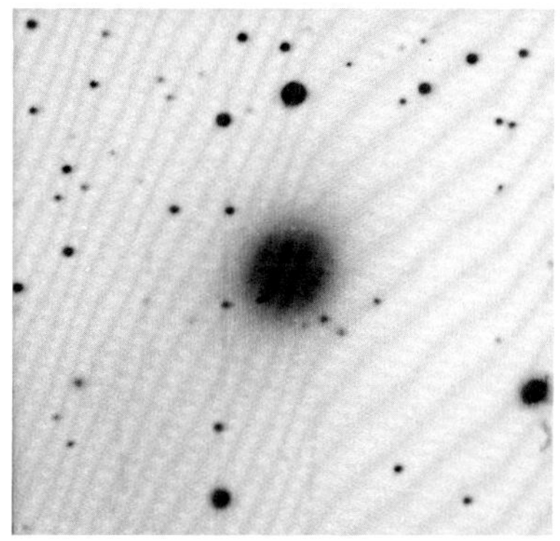

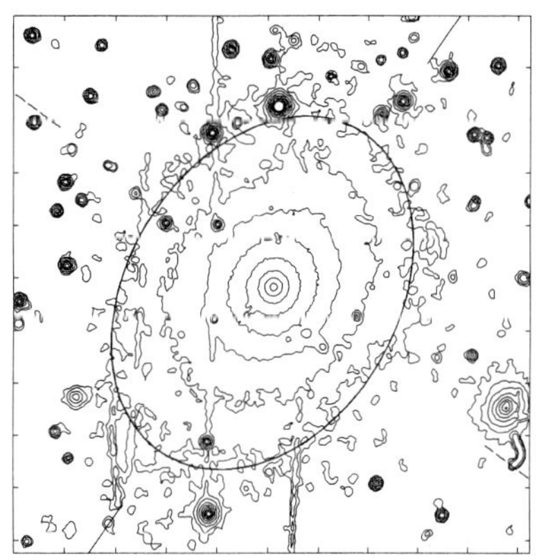

N4486 K2106 2 20.73 512X512

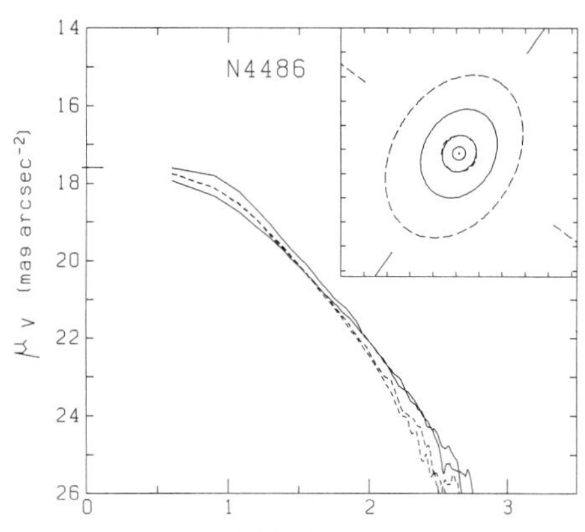

NGC4487

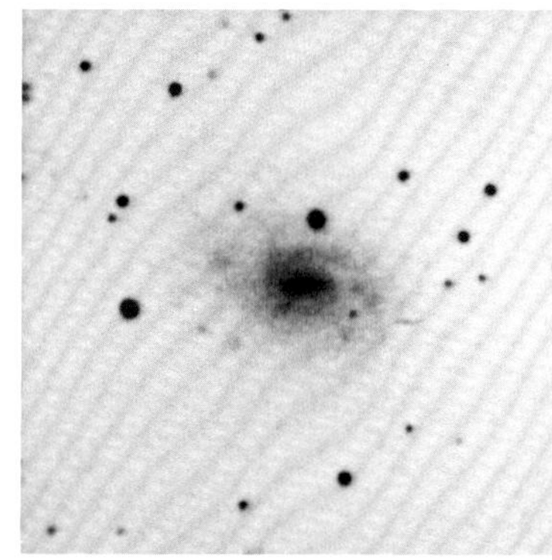

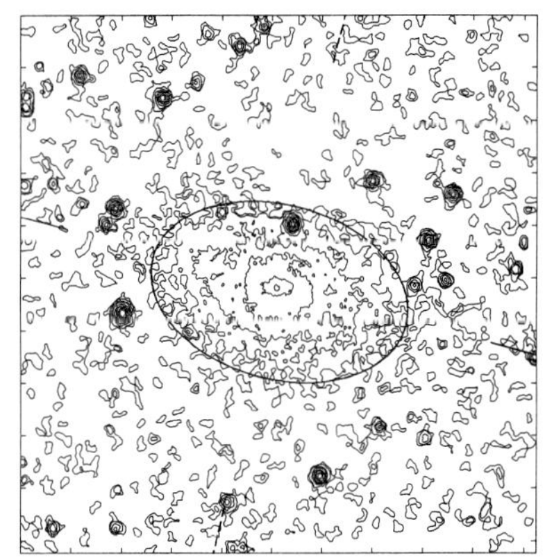

N4487 K5012 1 21.51 512X512

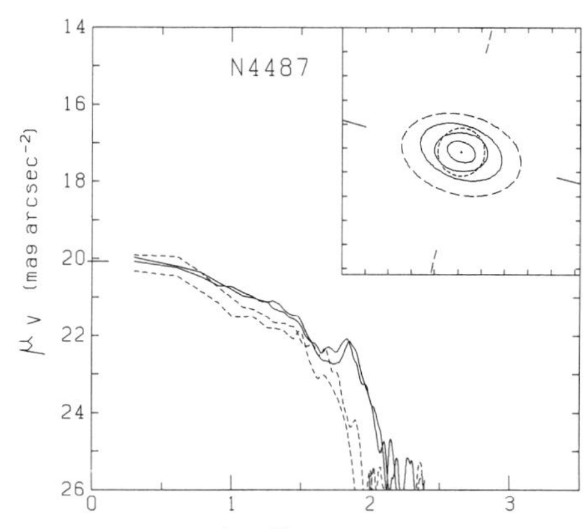

NGC4494

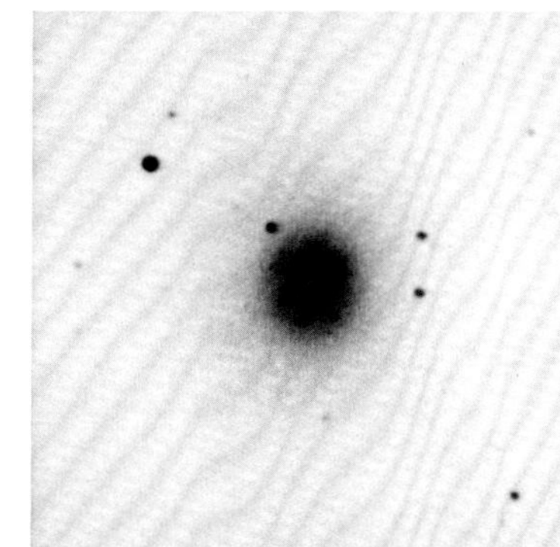

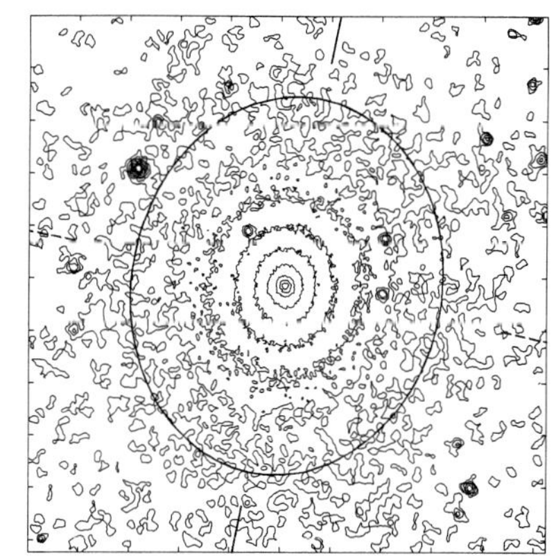

N4494 K4926 1 22.06 512X512

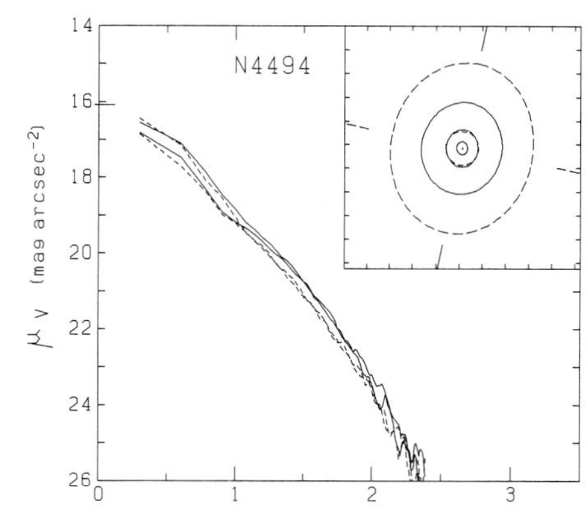

NGC4496

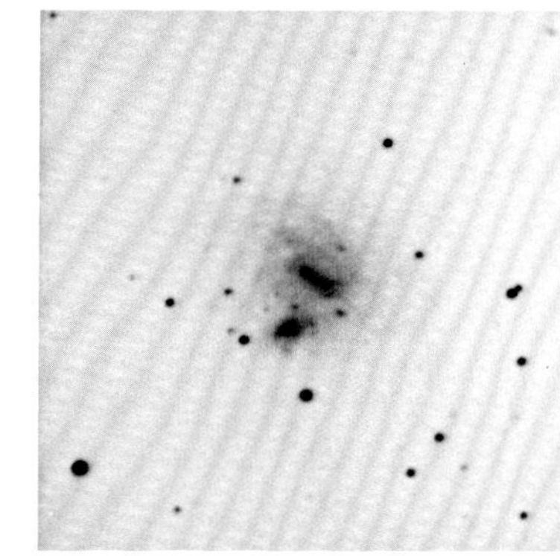

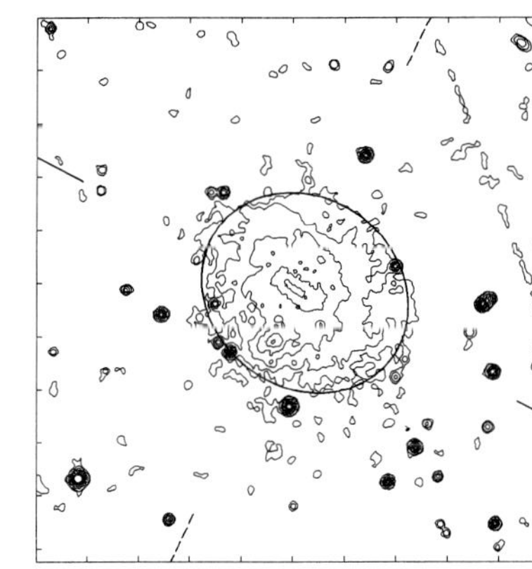

N4496 K3547 1 21.21 512X512

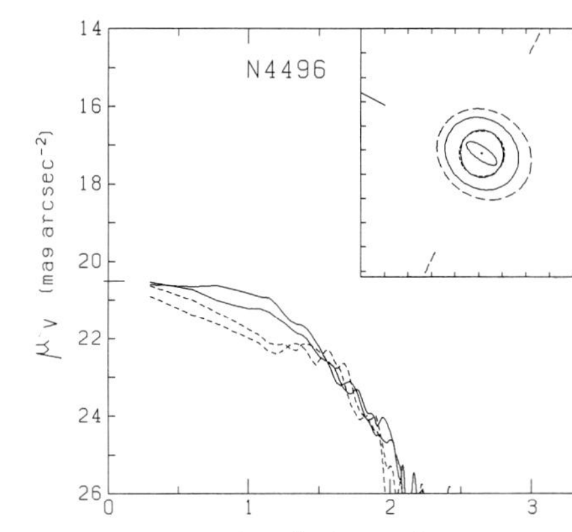

NGC4501

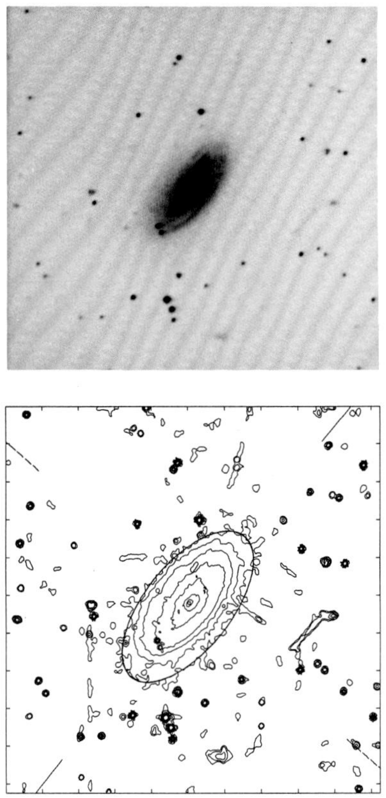

N4501 K2730 2 21.32 512X512

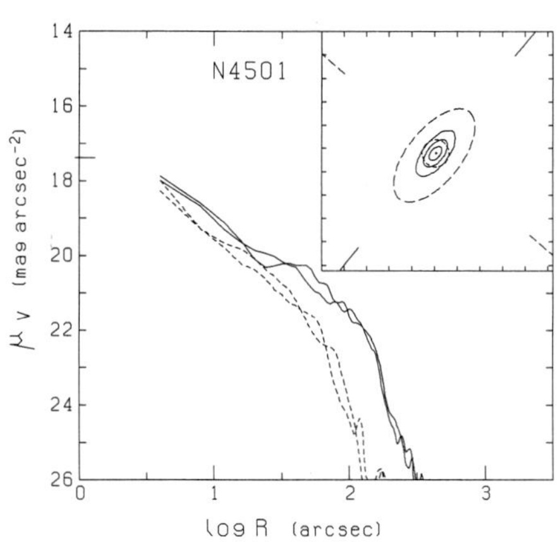

NGC4503

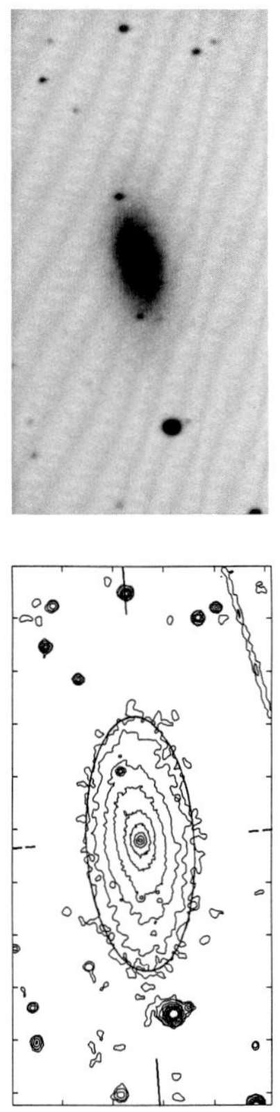

N4503 K2186 1 21.34 256X512

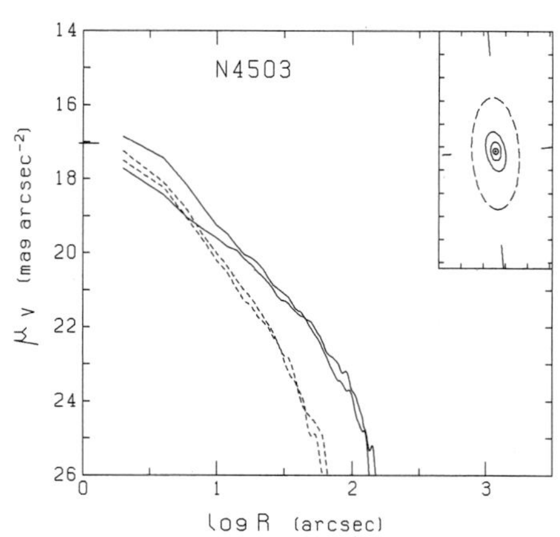

NGC4504

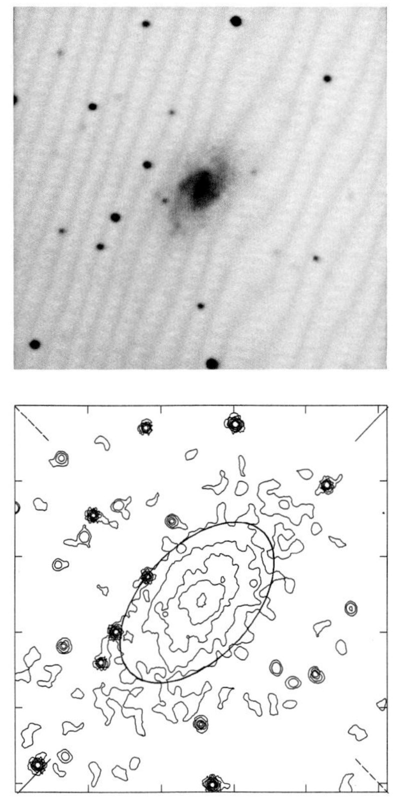

N4504 K4977 2 21.56 256X256

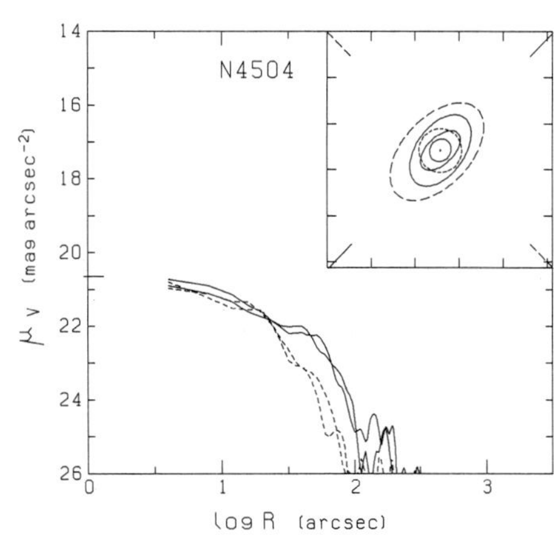

R 80

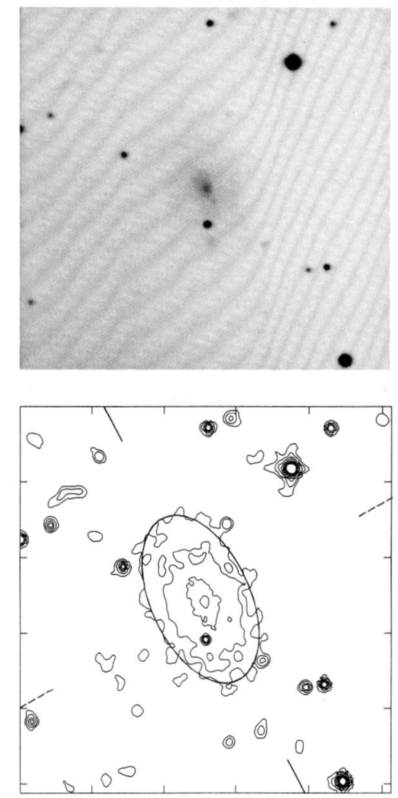

R0080 K4970 2 21.63 256X256

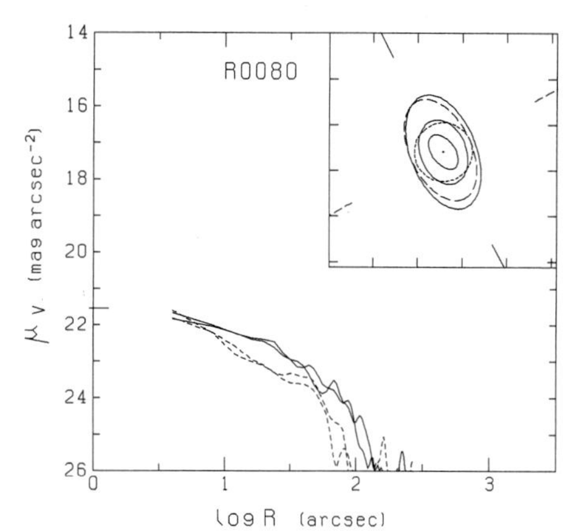

NGC4517

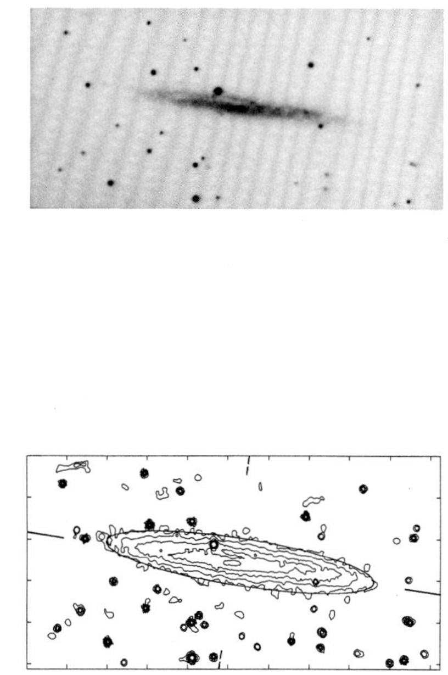

N4517 K4970 2 21.63 512X256

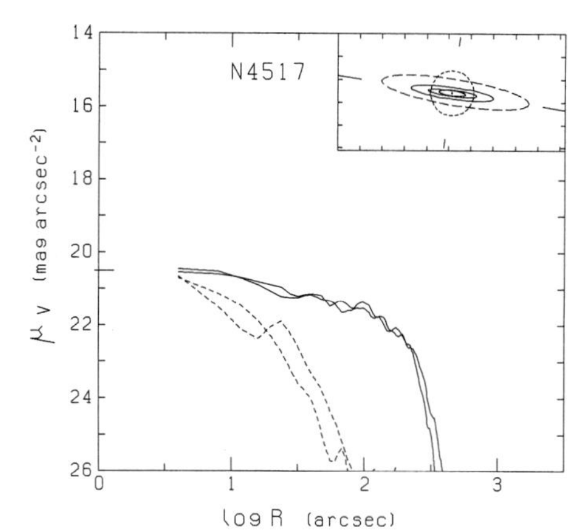

NGC4519

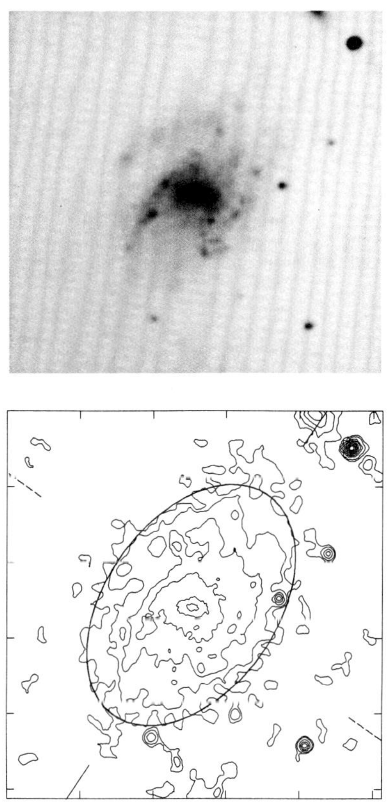

N4519 K2220 1 21.12 256X256

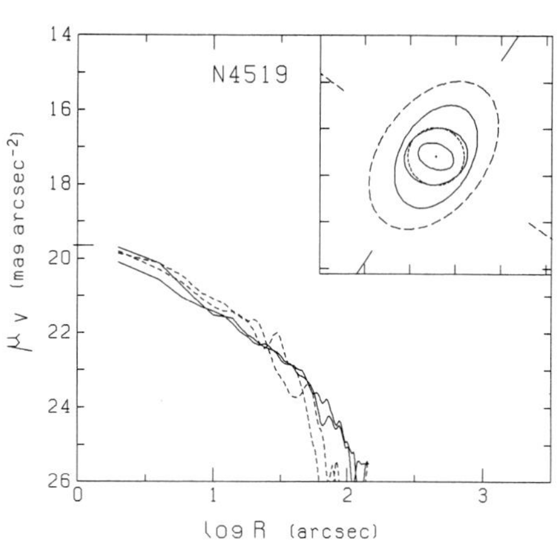

NGC4522

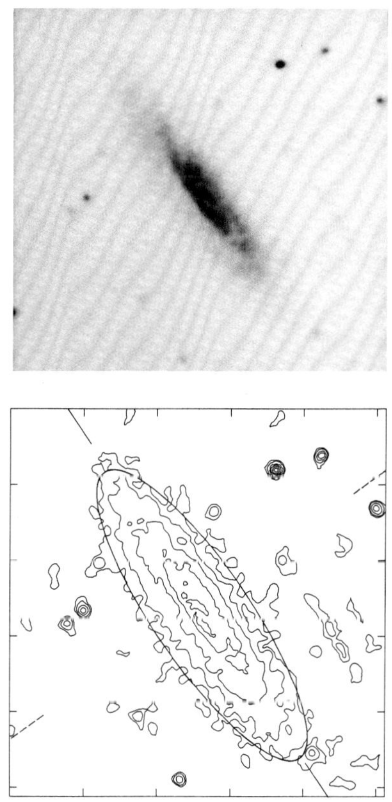

N4522 K2220 1 21 12 256X256

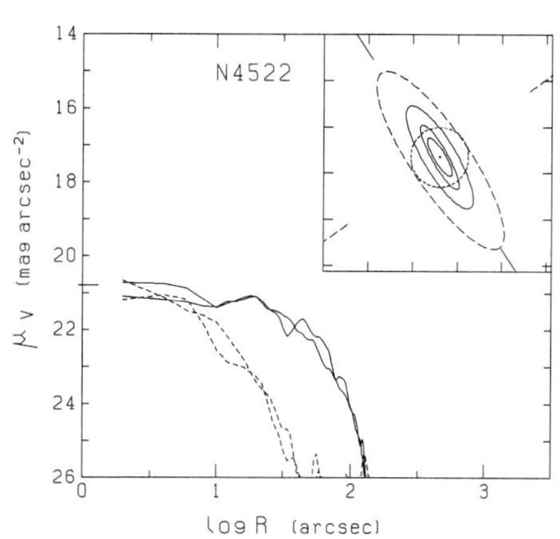

NGC4526

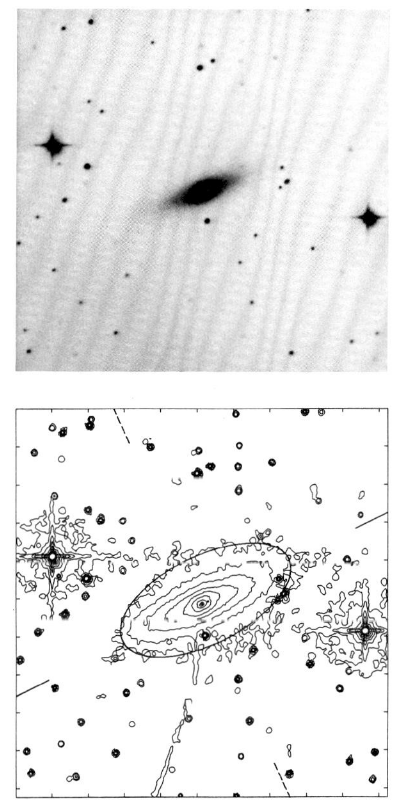

N4526 K2220 2 21 12 512X512

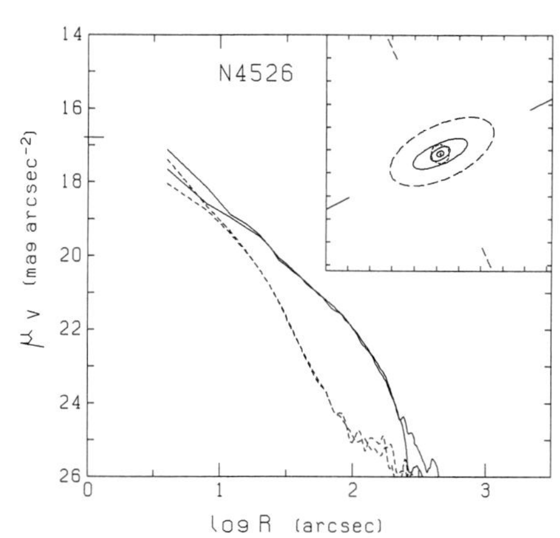

NGC4527

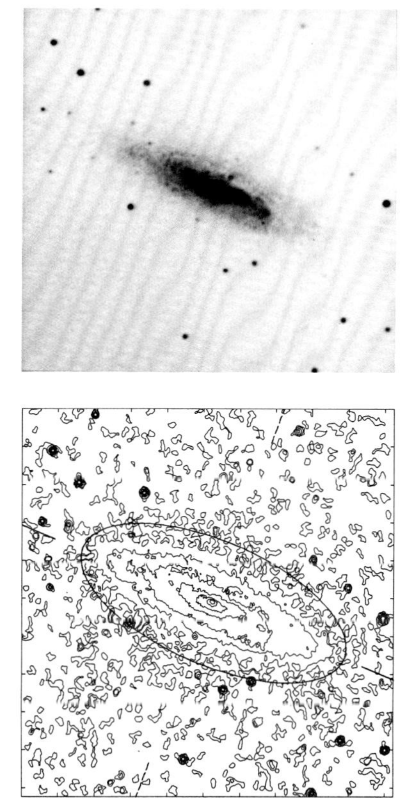

N4527 K3740 1 21.03 512X512

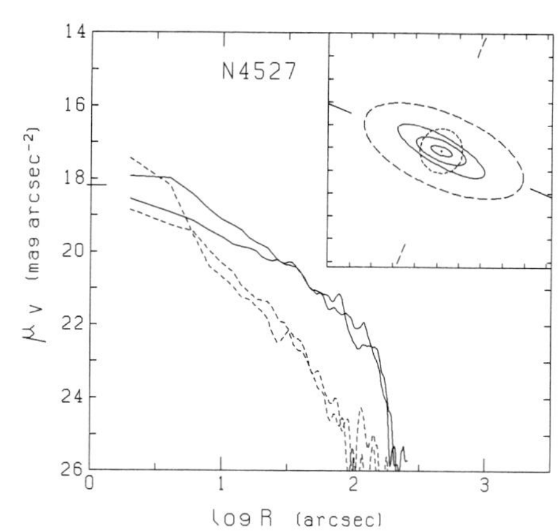

NGC4532

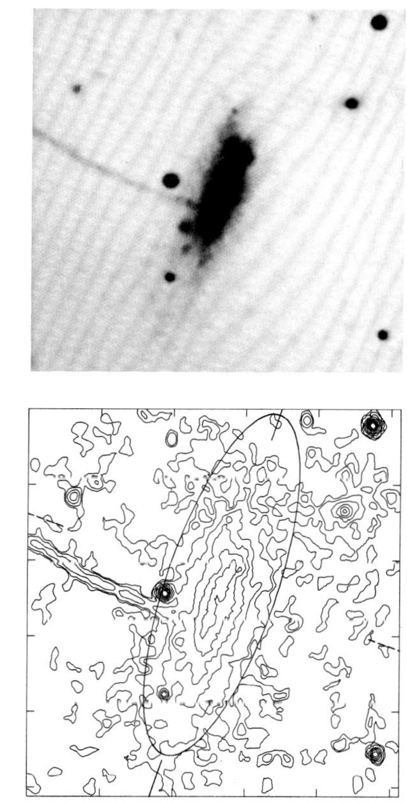

N4532 K3740 1 21.03 256X256

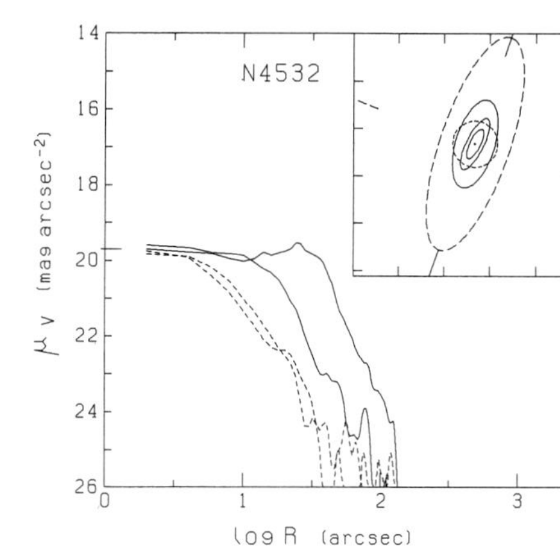

NGC4535

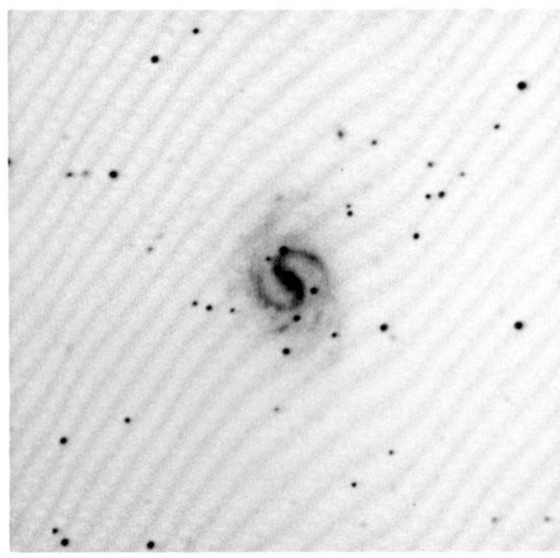

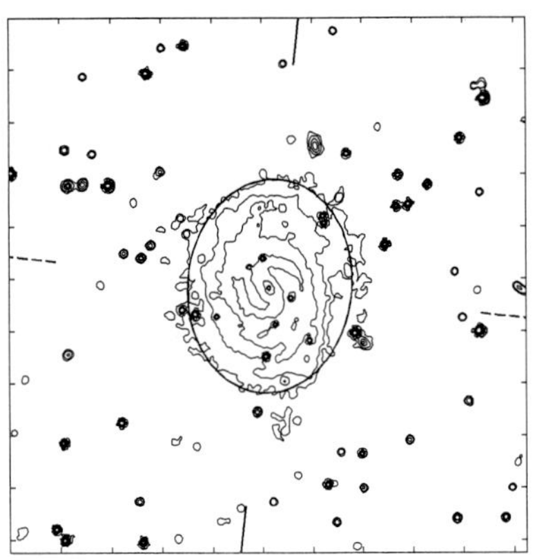

N4535 K2220 2 21.12 512X512

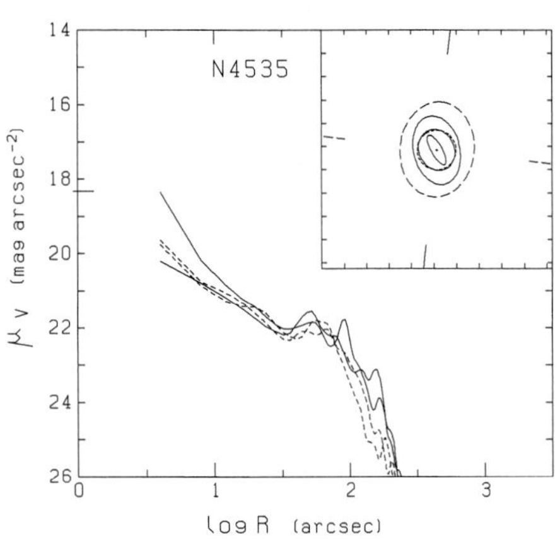

NGC4536

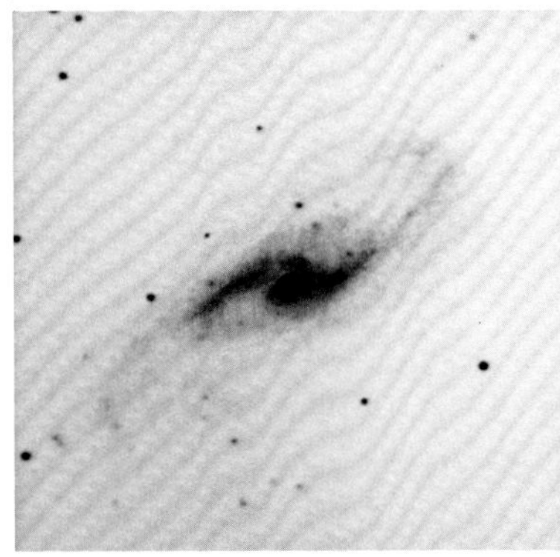

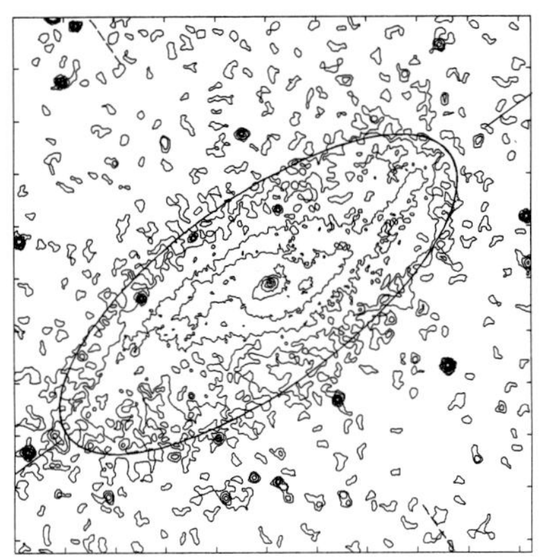

N4536 K4639 1 21.55 512X512

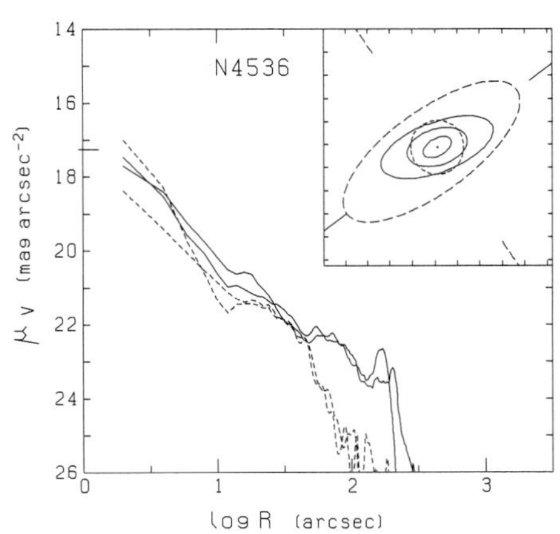

NGC4540

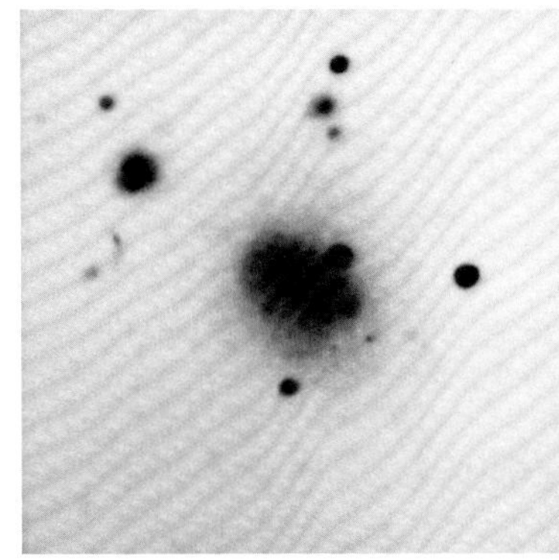

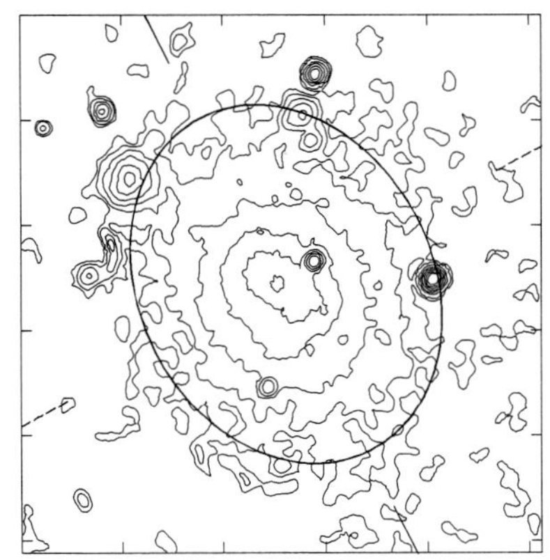

N4540 K2730 1 21.32 256X256

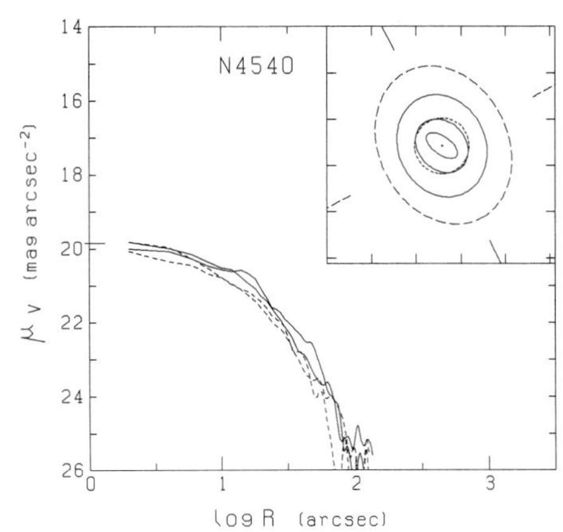

NGC4546

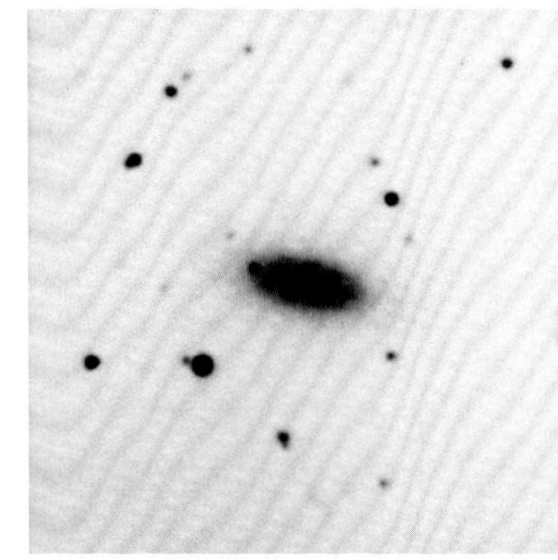

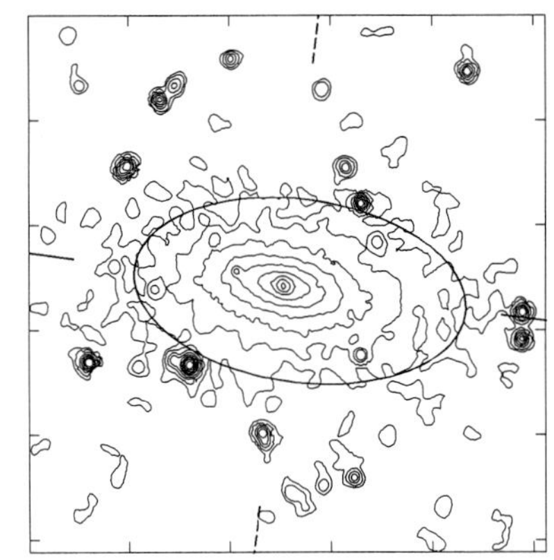

N4546 K5012 2 21.51 256X256

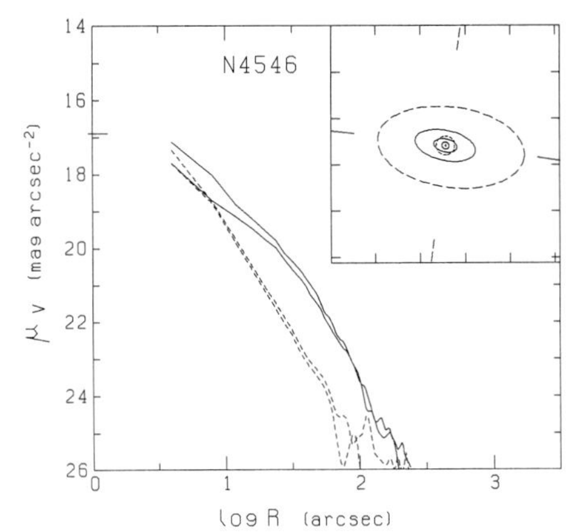

NGC4548

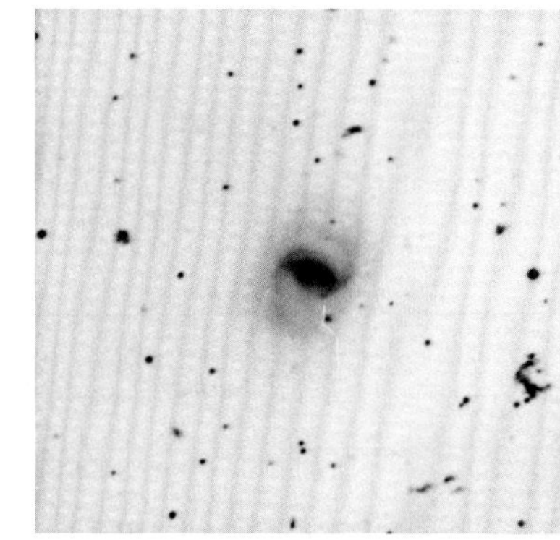

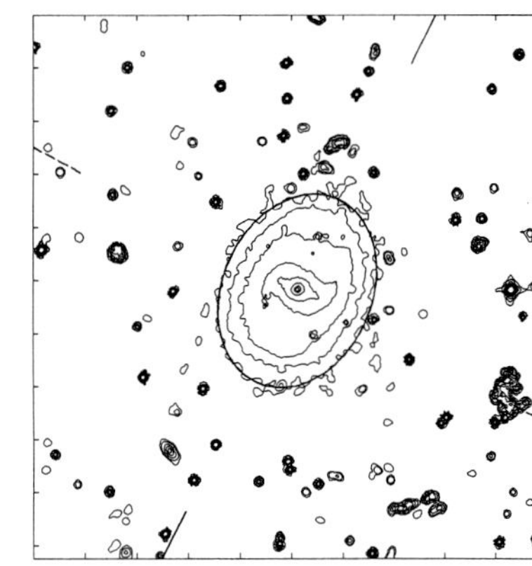

N4548 K2730 2 21.32 512X512

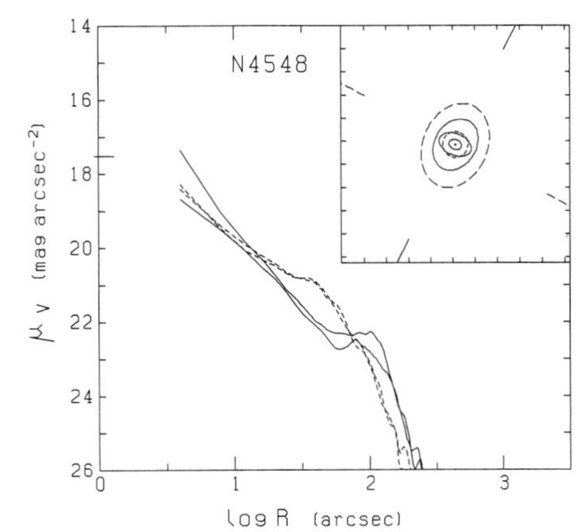

NGC4550

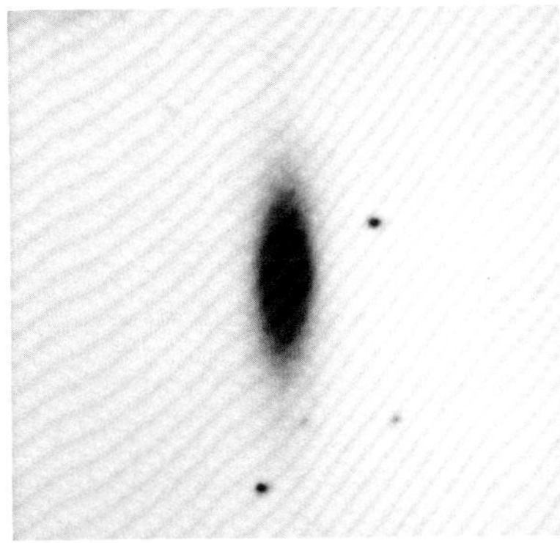

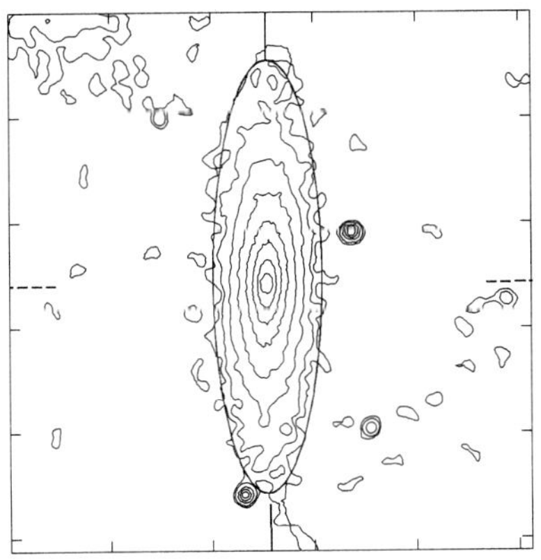

N4550 K2220 1 21.12 256X256

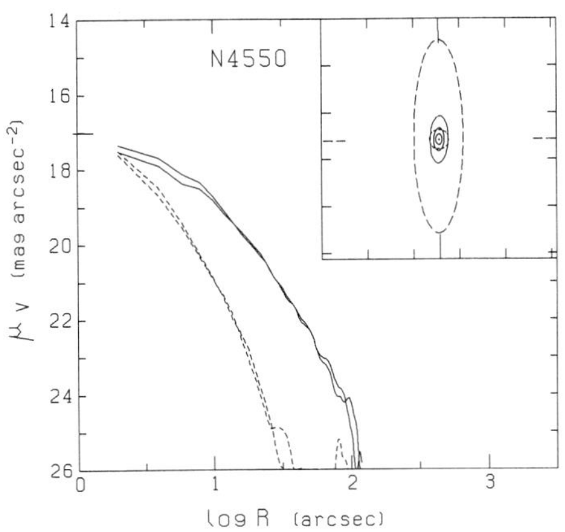

NGC4552

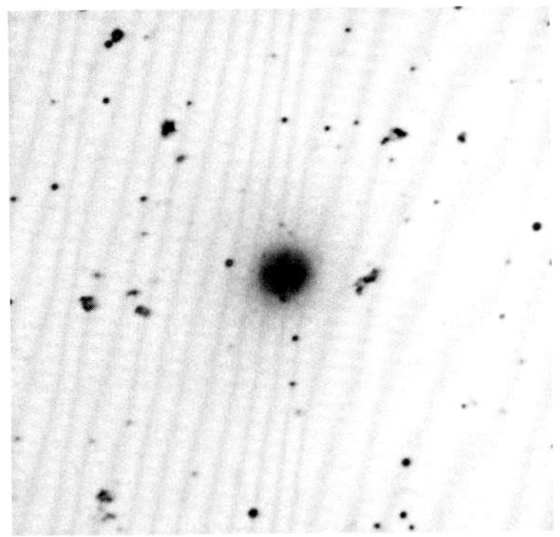

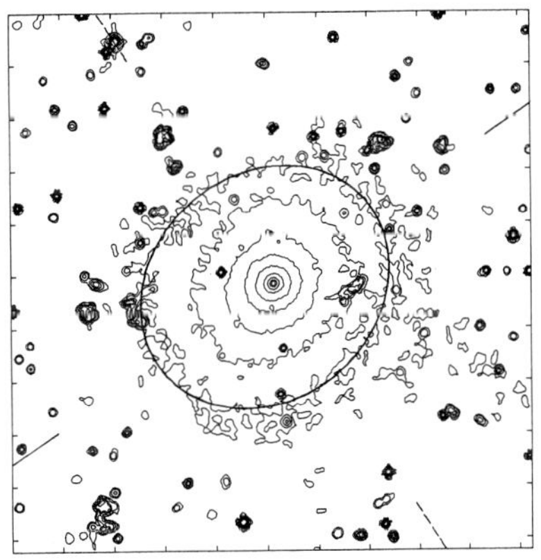

N4552 K2730 2 21.32 512X512

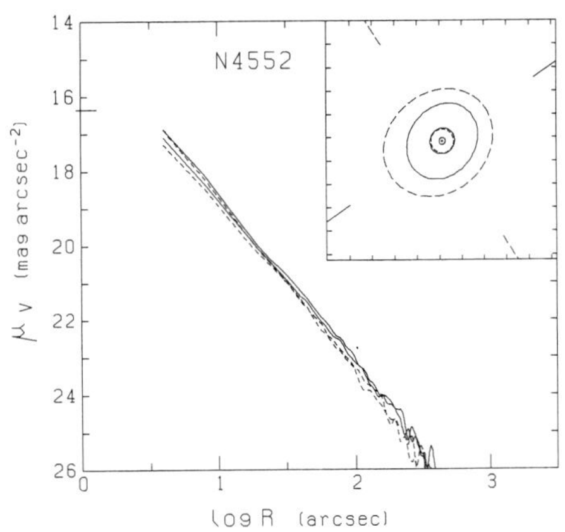

NGC4559

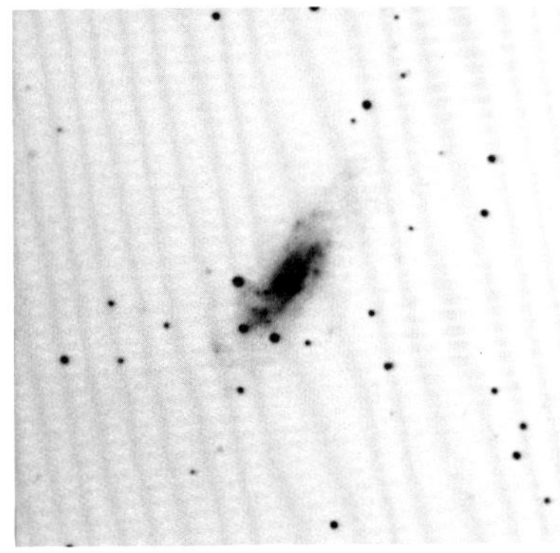

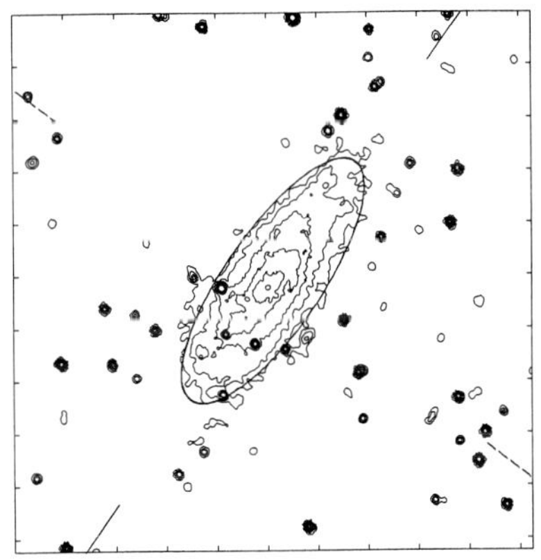

N4559 K4926 2 22.06 512X512

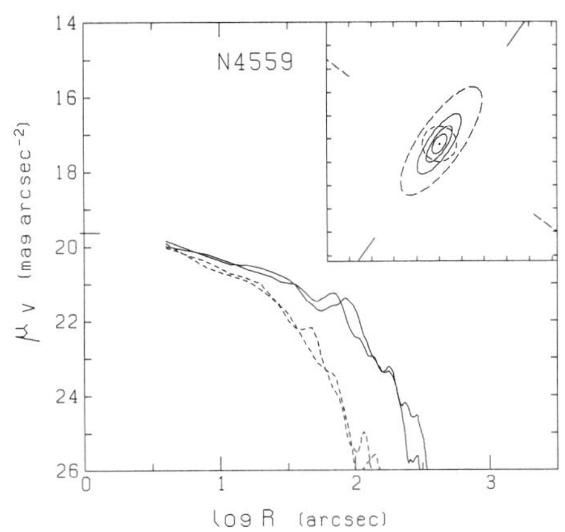

NGC4561

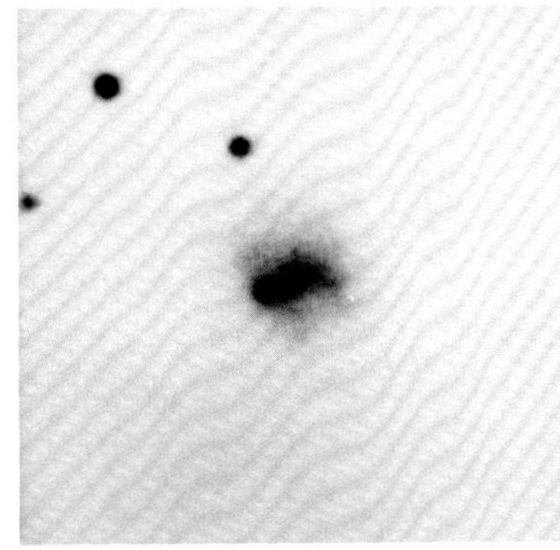

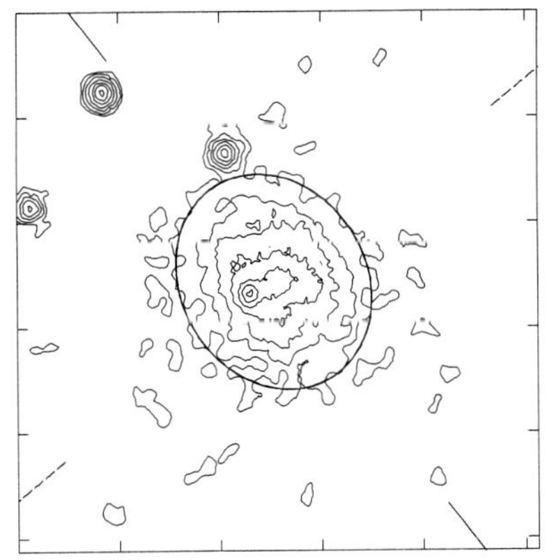

N4561 K5791 1 21.84 256X256

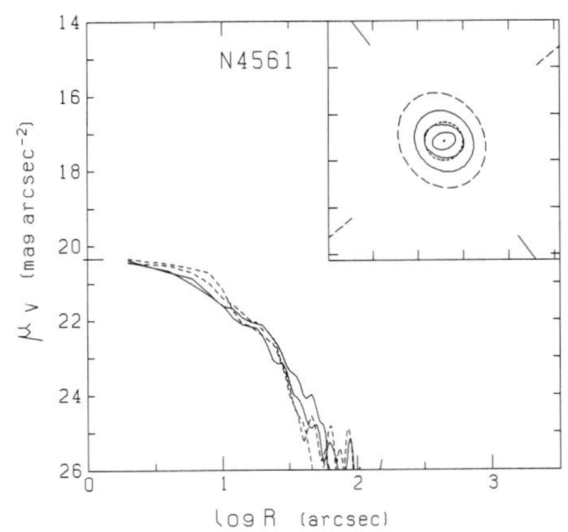

NGC4565

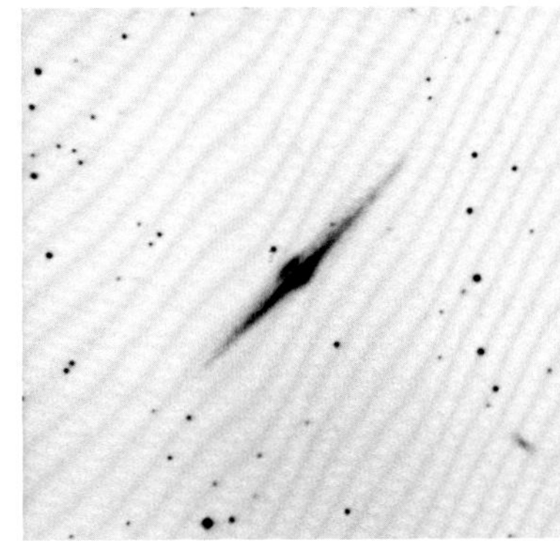

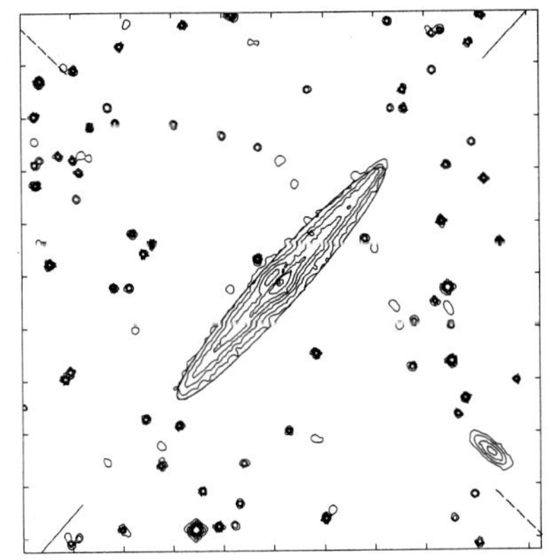

N4565 K4926 3 22.06 512X512

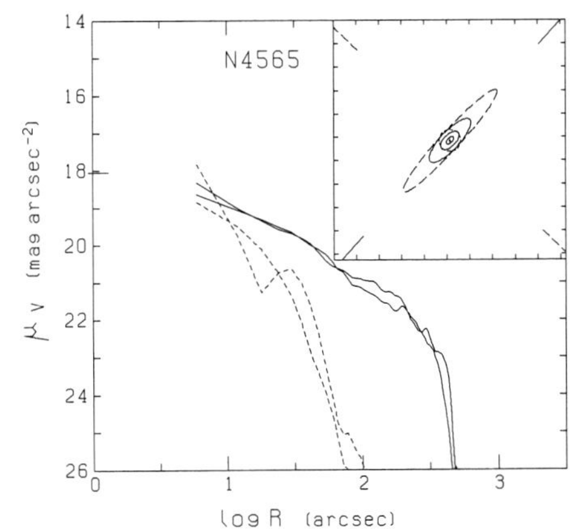

NGC4564

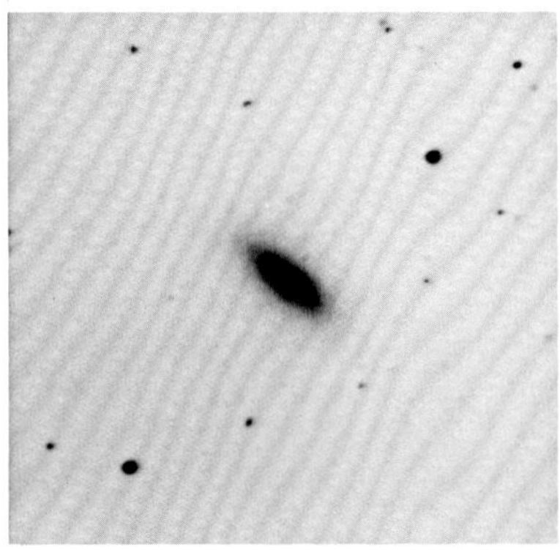

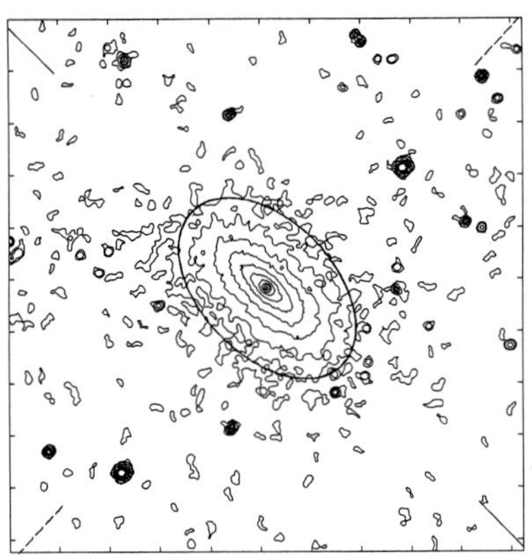

N4564 K2220 1 21.12 512X512

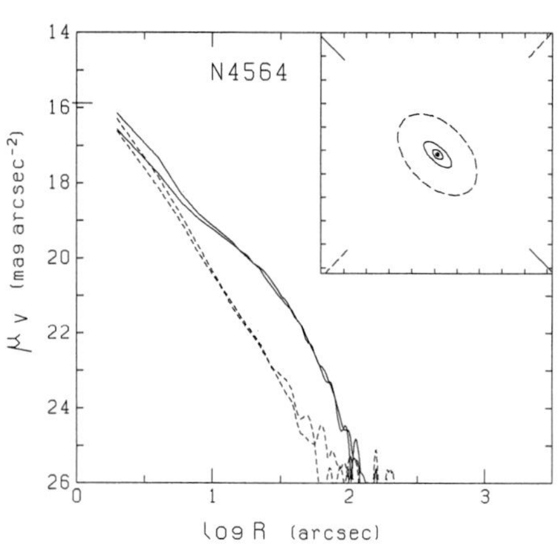

NGC4567

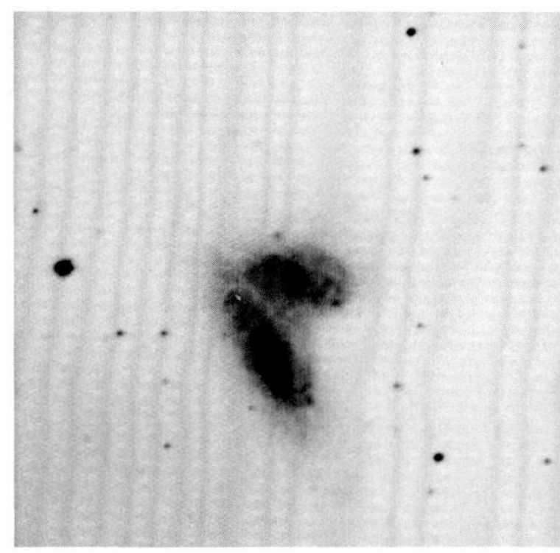

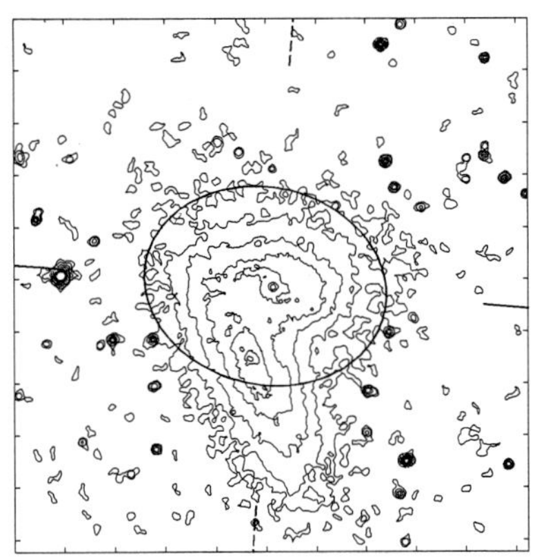

N4567 K2220 1 21.12 512X512

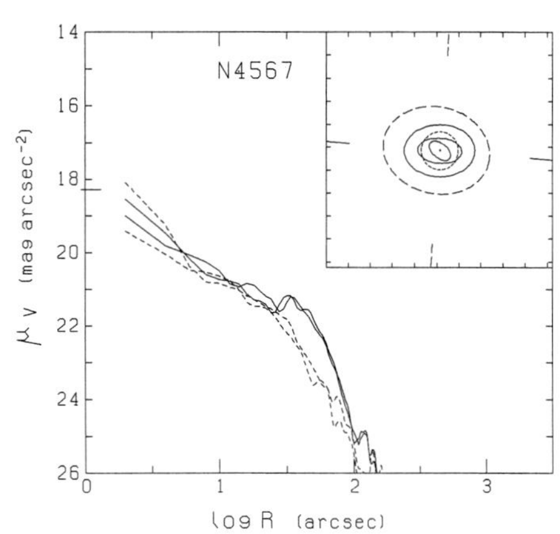

NGC4568

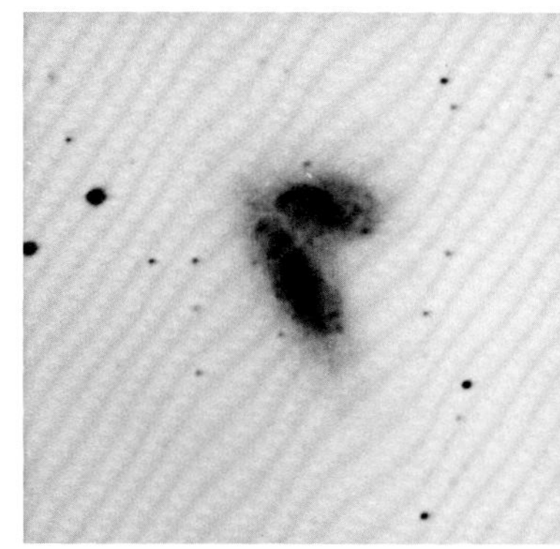

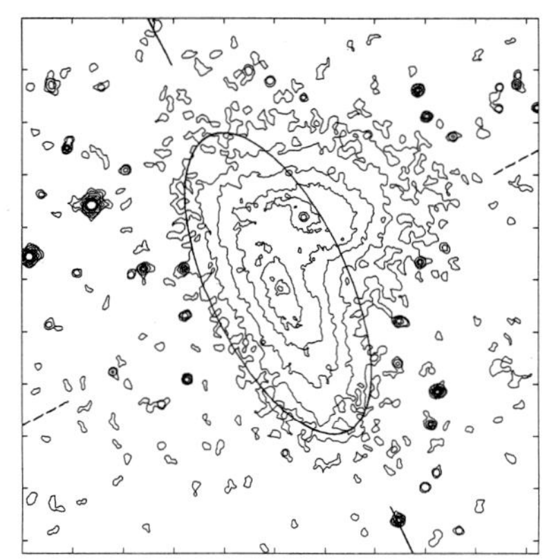

N4568 K2220 1 21.12 512X512

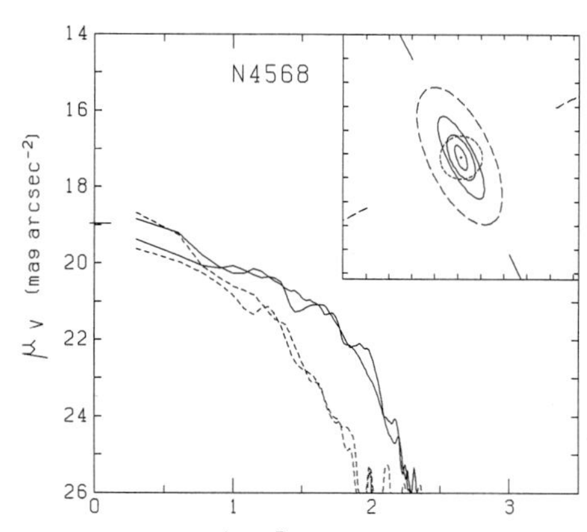

NGC4569

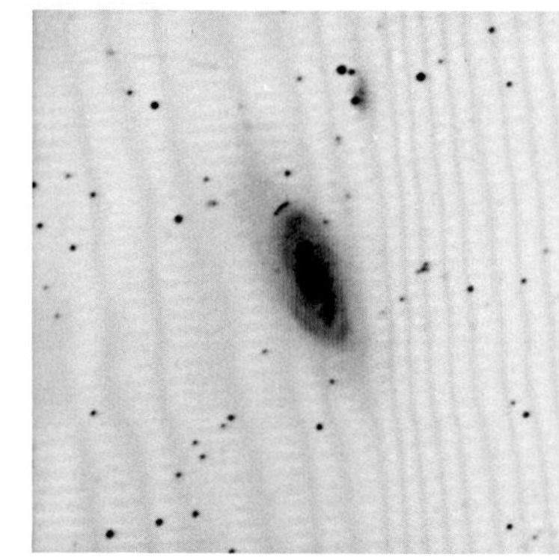

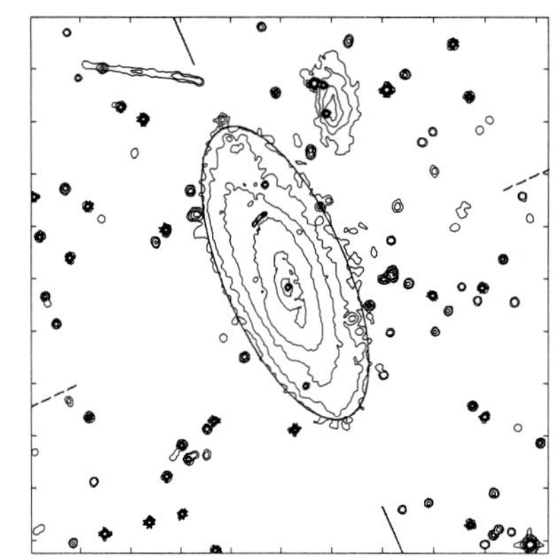

N4569 K2730 2 21.32 512X512

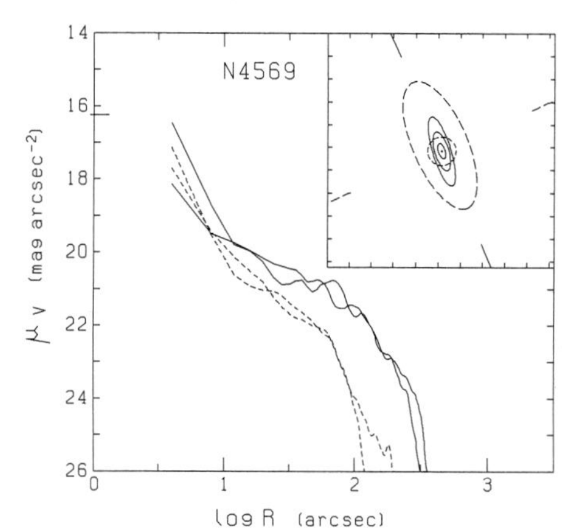

NGC4570

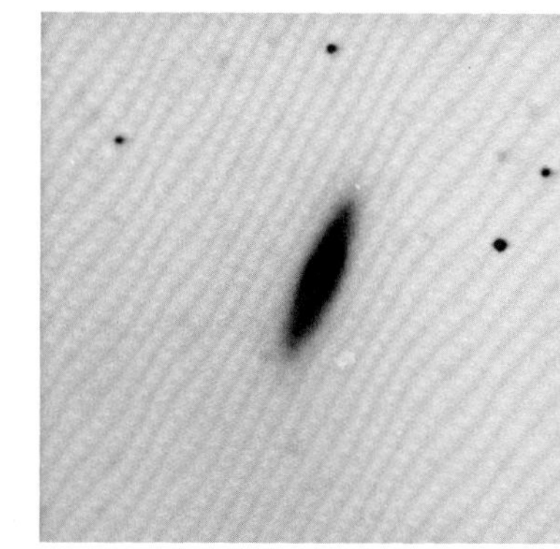

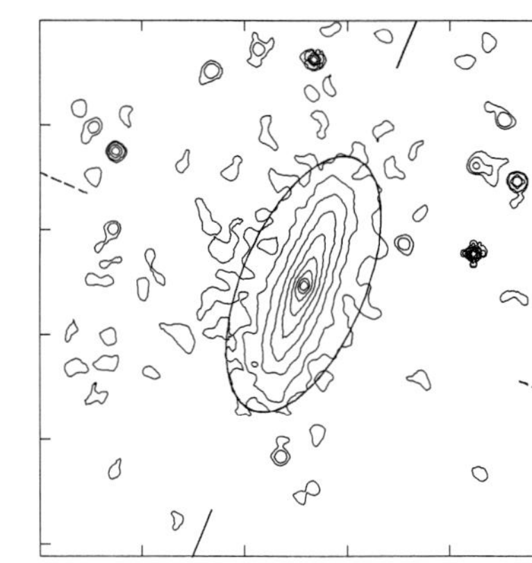

N4570 K3740 2 21.03 256X256

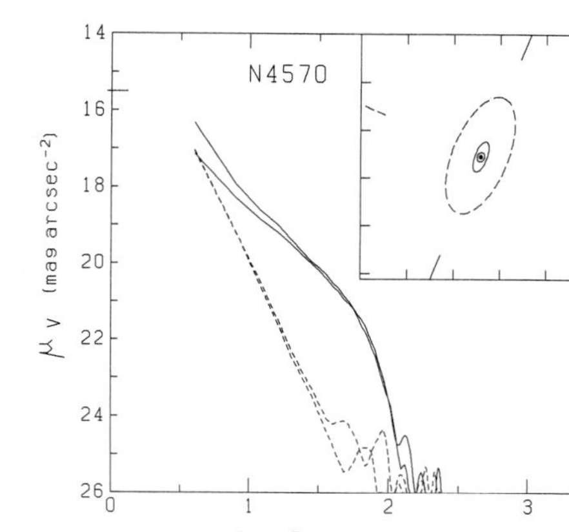

NGC4571

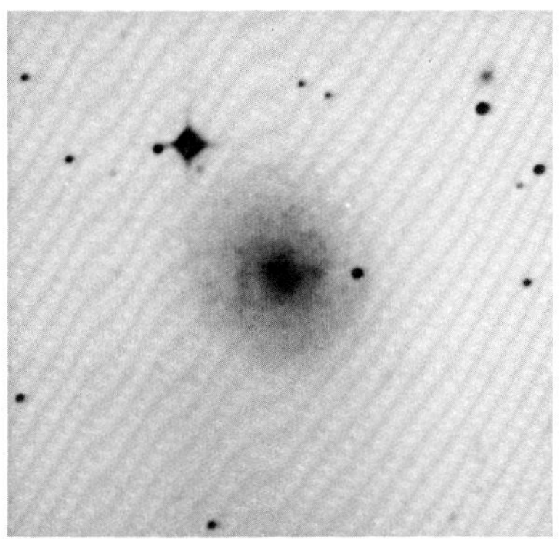

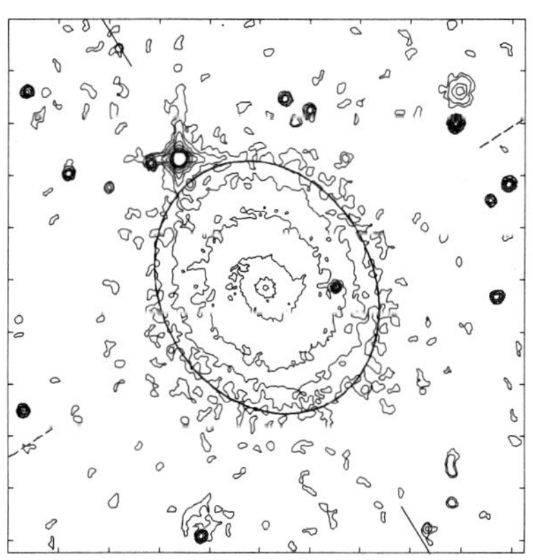

N4571 K2730 1 21.32 512X512

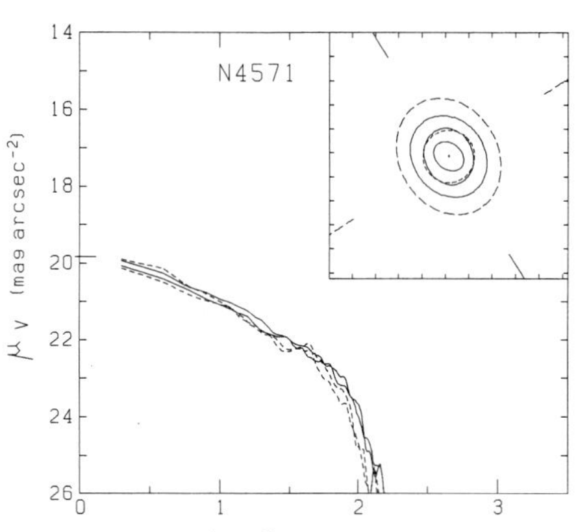

NGC4578

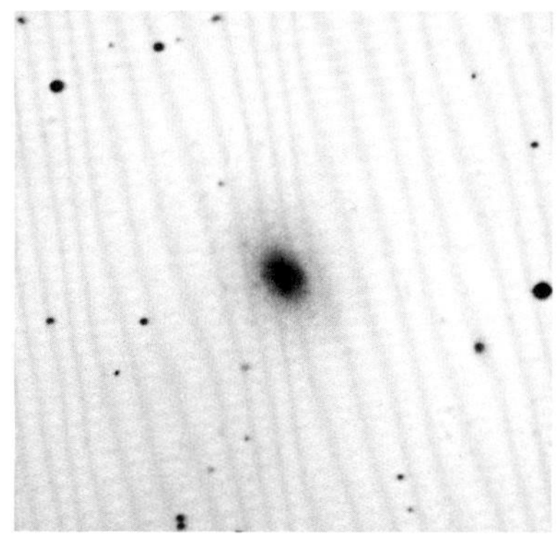

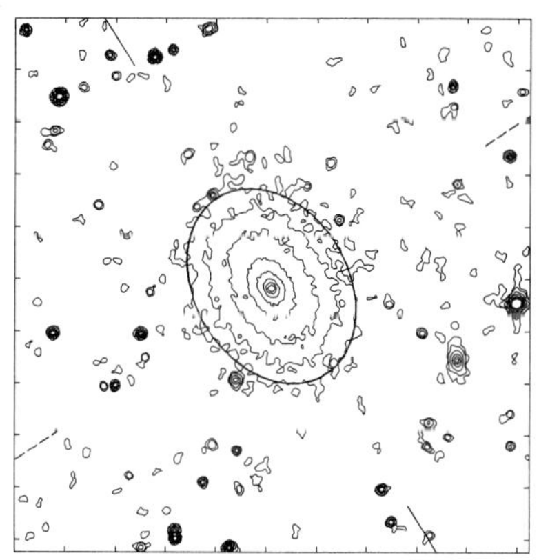

N4578 K2220 1 21.12 512X512

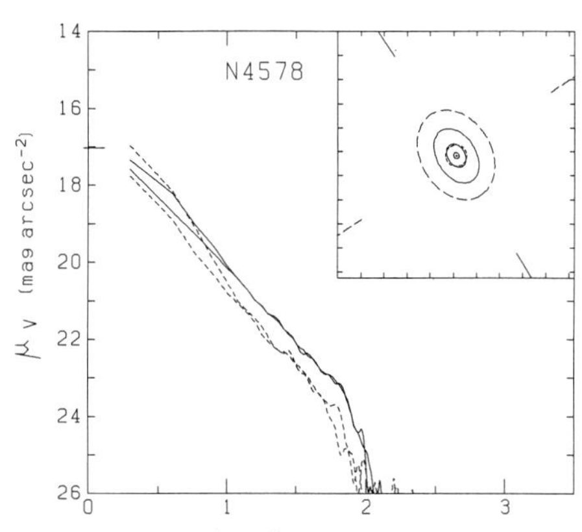

NGC4579

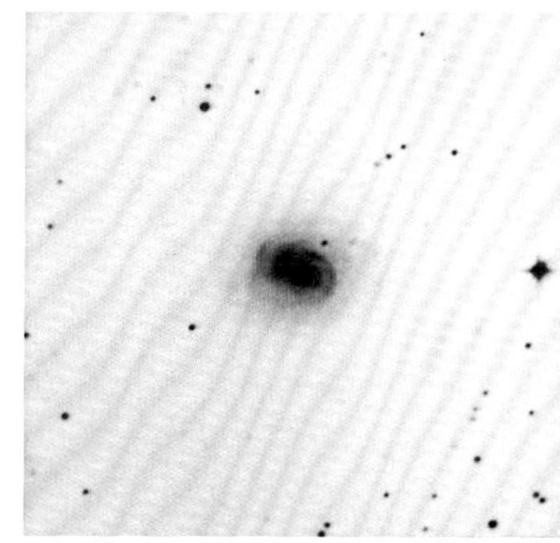

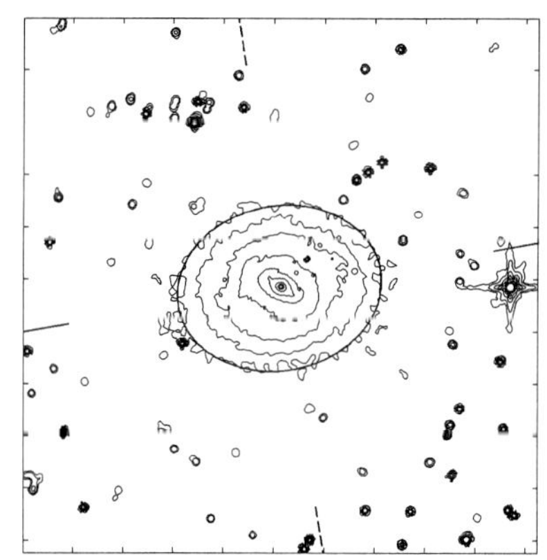

N4579 K2220 2 21.12 512X512

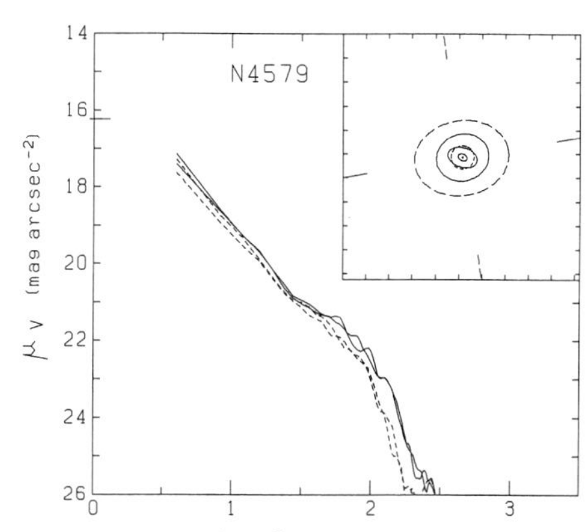

NGC4580

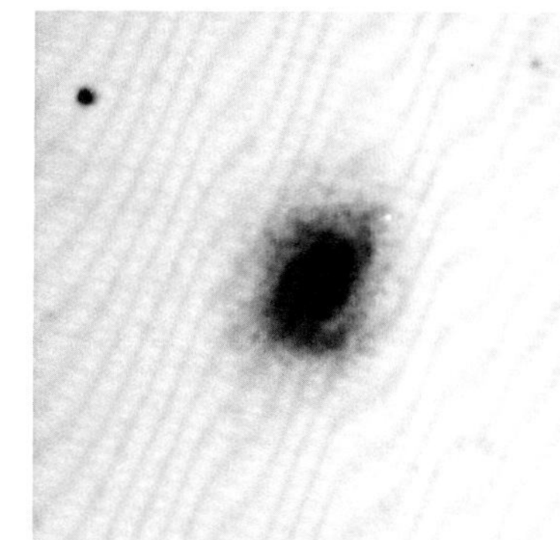

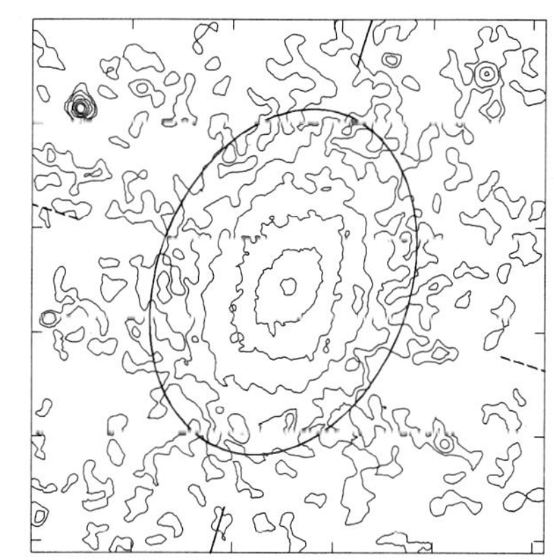

N4580 K3740 1 21.03 256X256

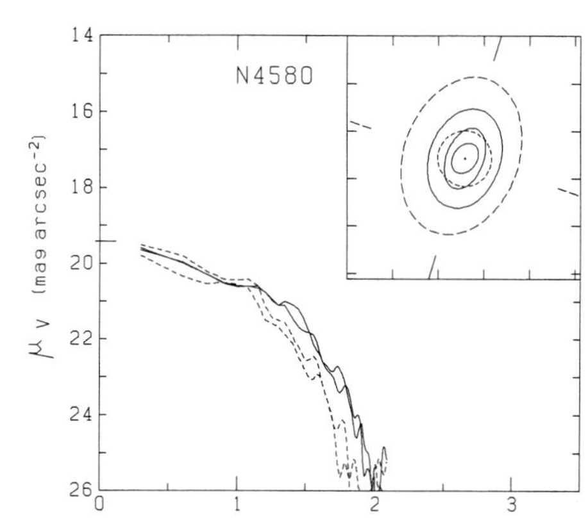

NGC4589

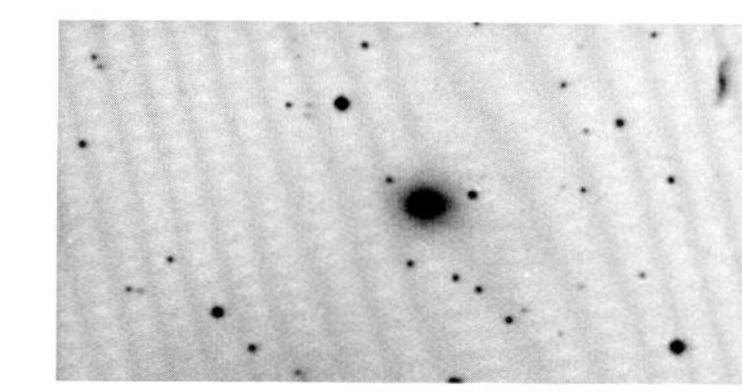

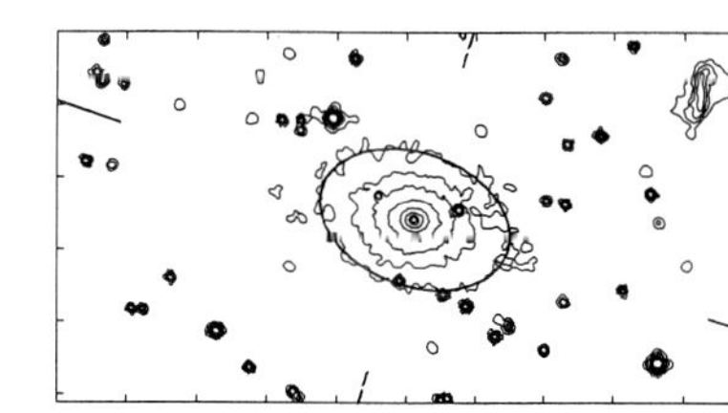

N4589 K5759 2 22.42 512X256

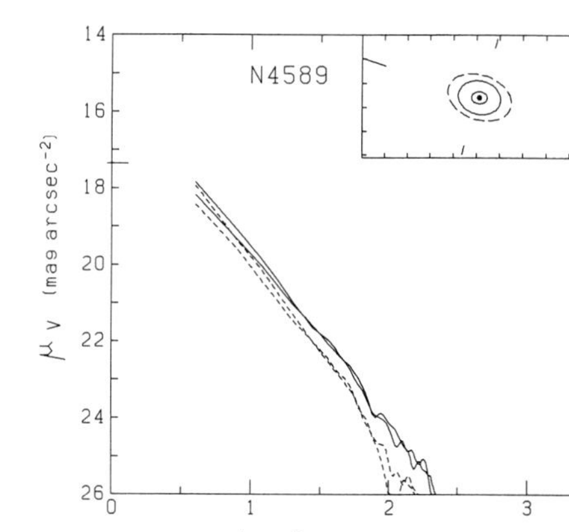

NGC4586

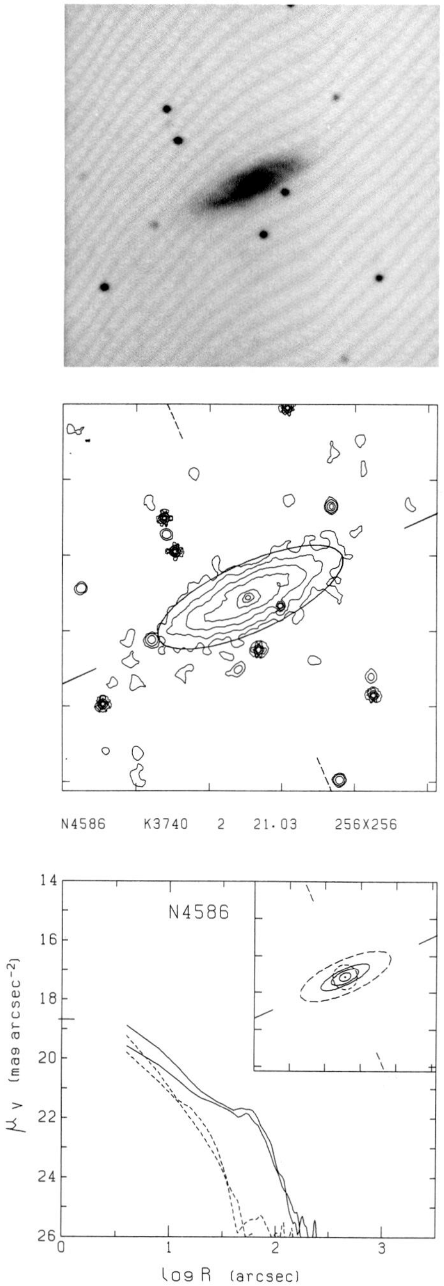

NGC4592

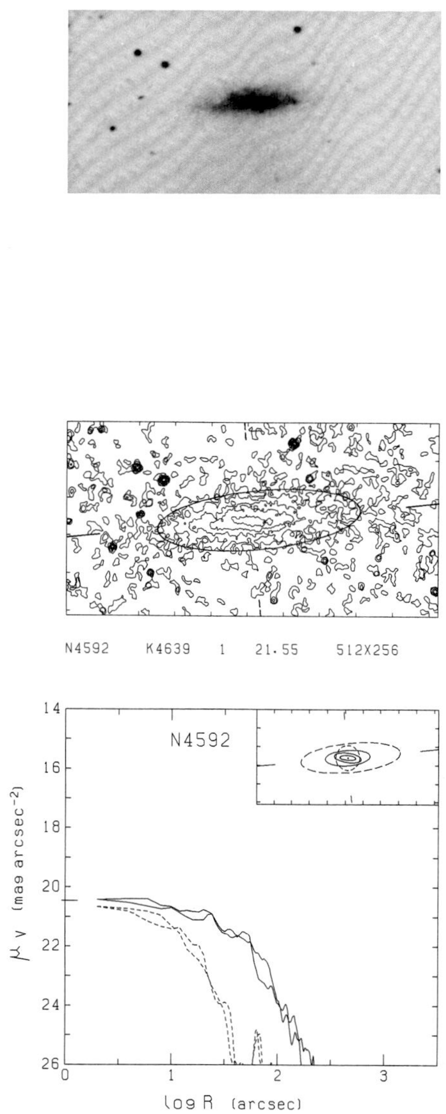

NGC4593

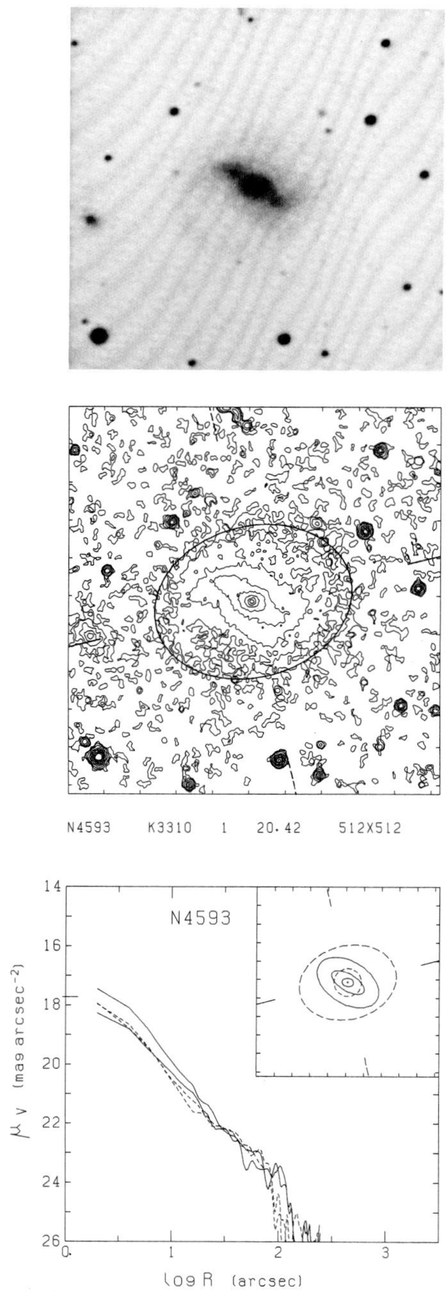

NGC4595

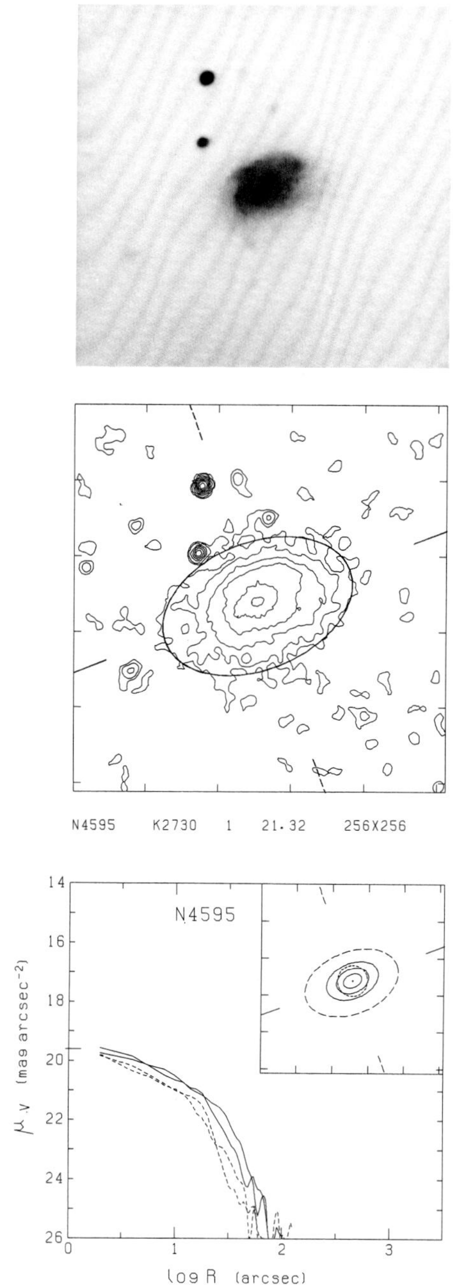

NGC4594

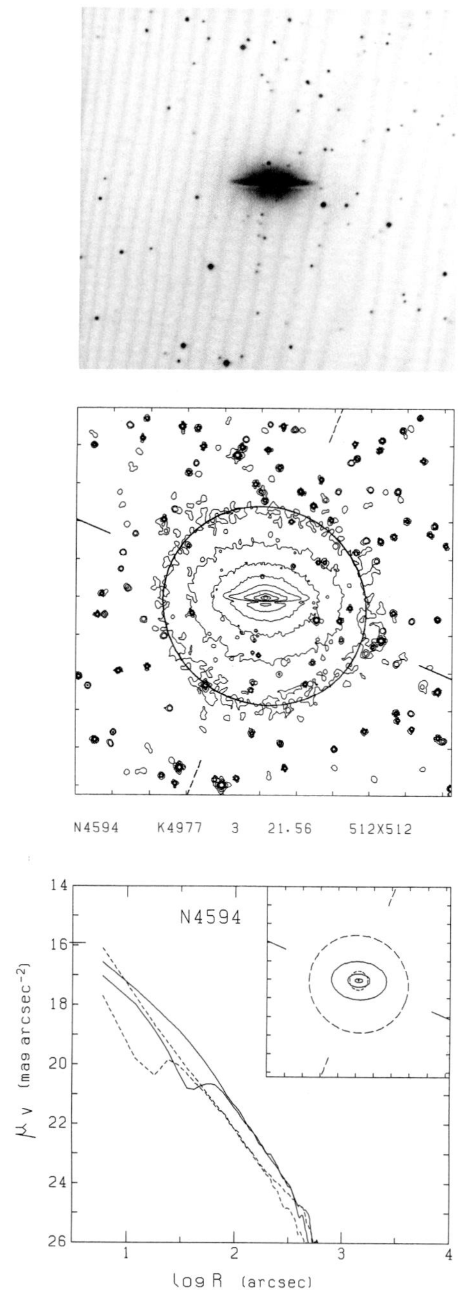

NGC4596

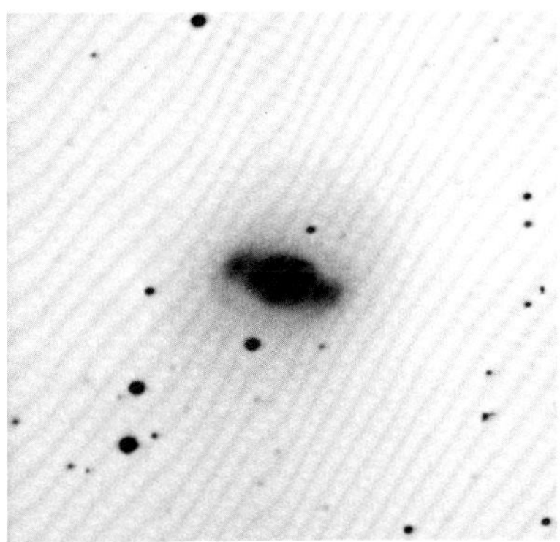

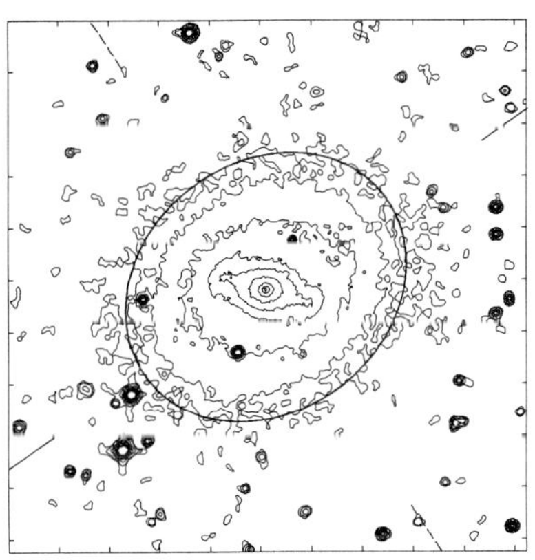

N4596 K2220 1 21.12 512X512

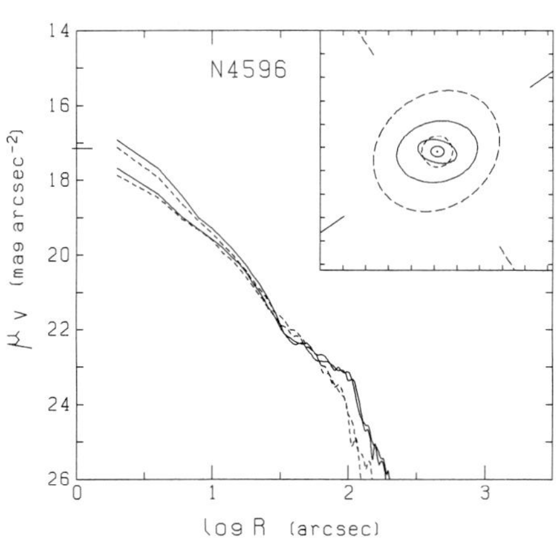

NGC4597

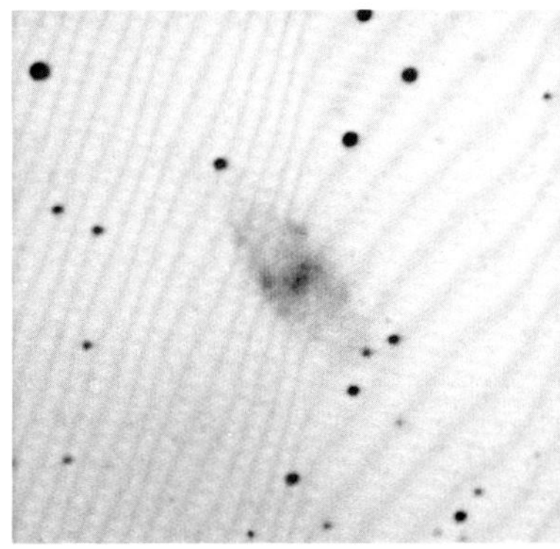

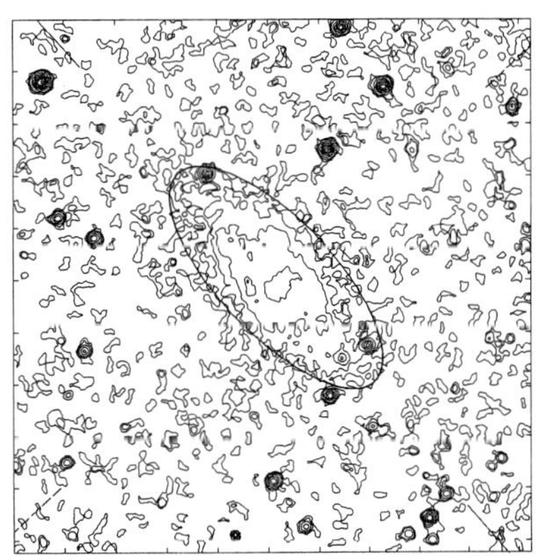

N4597 K3310 1 20.42 512X512

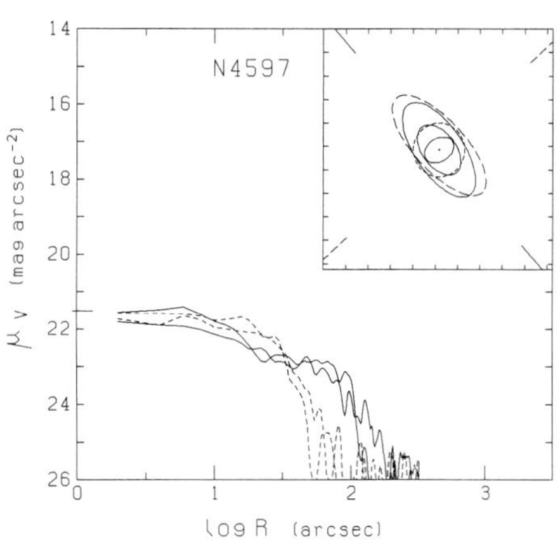

NGC4605

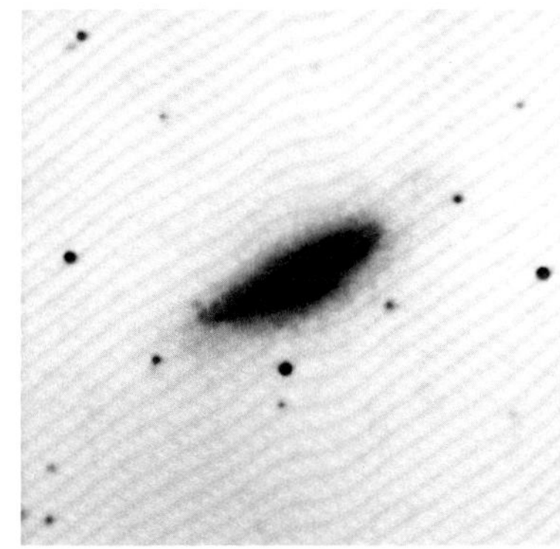

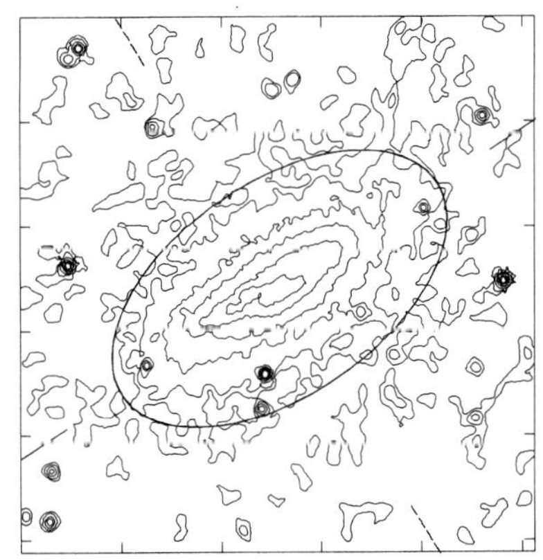

N4605 K5064 2 21.66 256X256

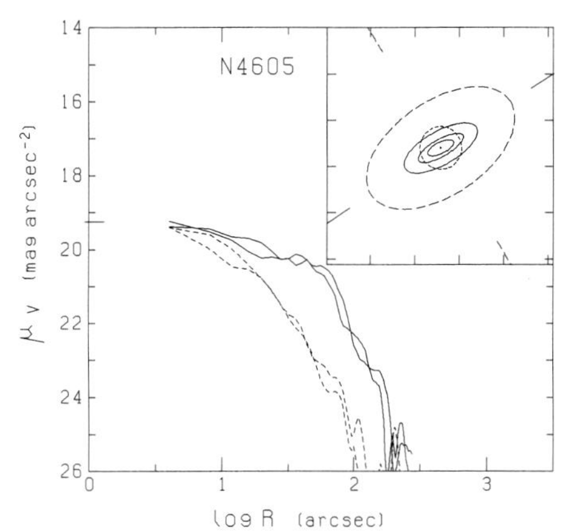

NGC4602

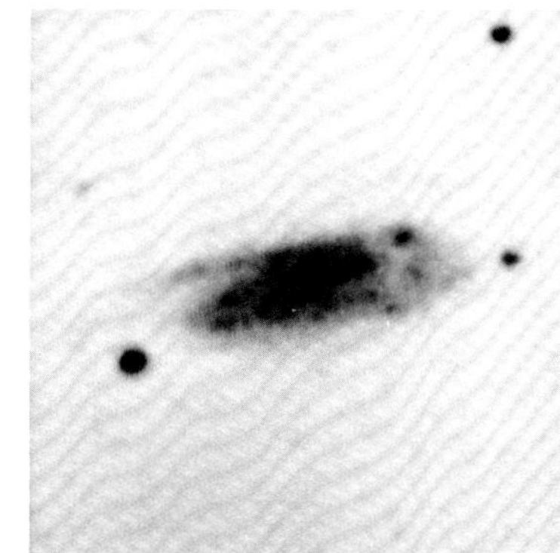

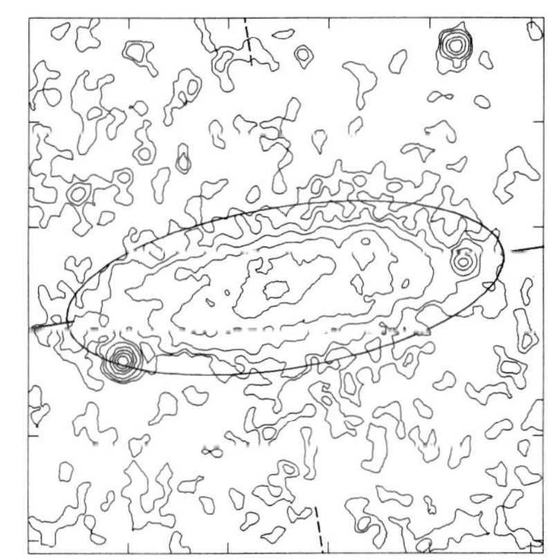

N4602 K3310 1 20.42 256X256

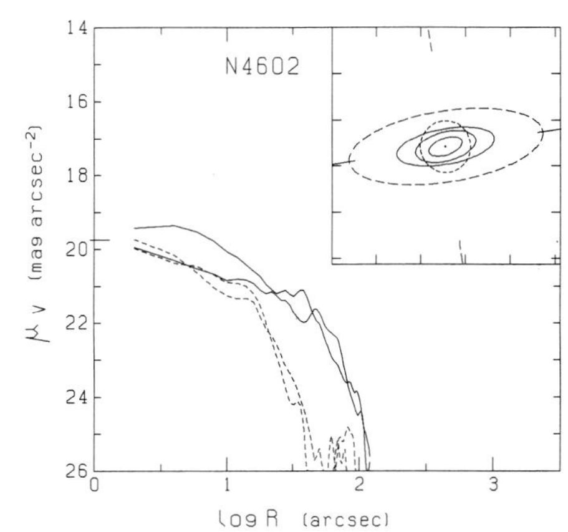

NGC4608

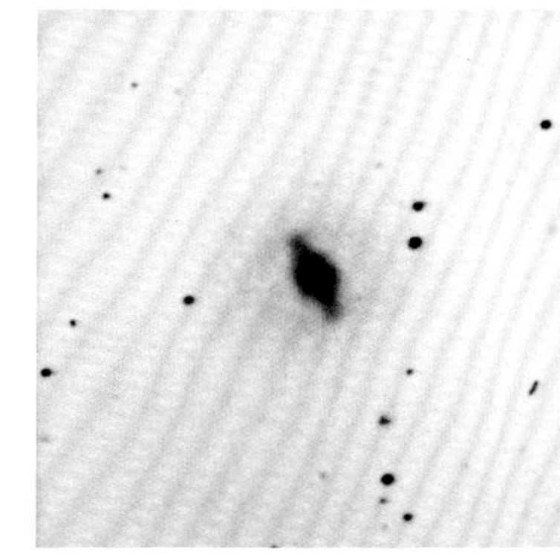

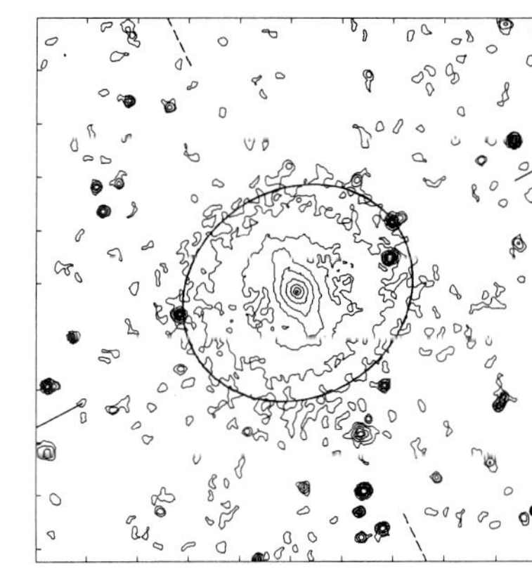

N4608 K2220 1 21.12 512X512

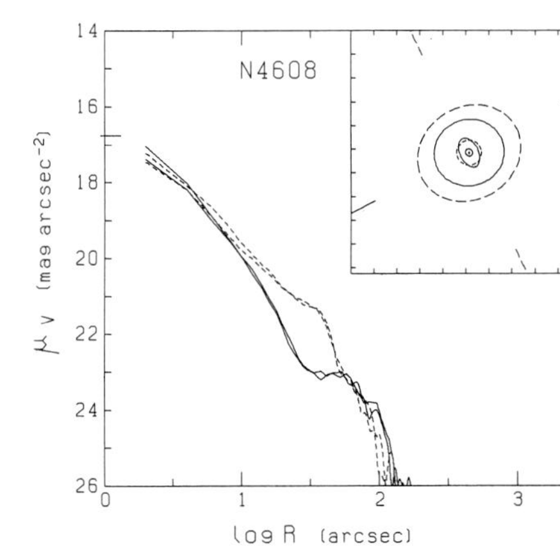

NGC4612

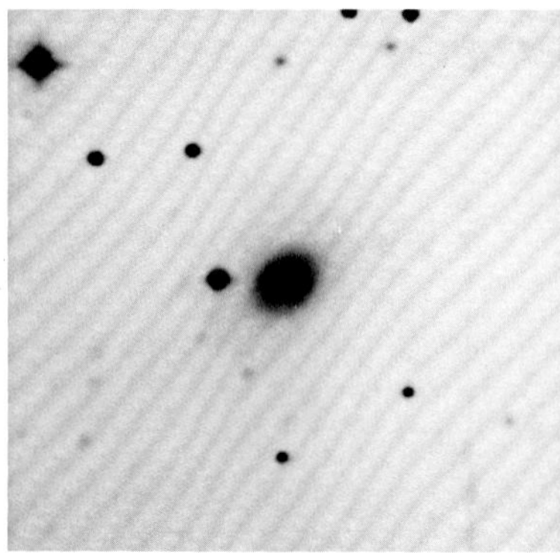

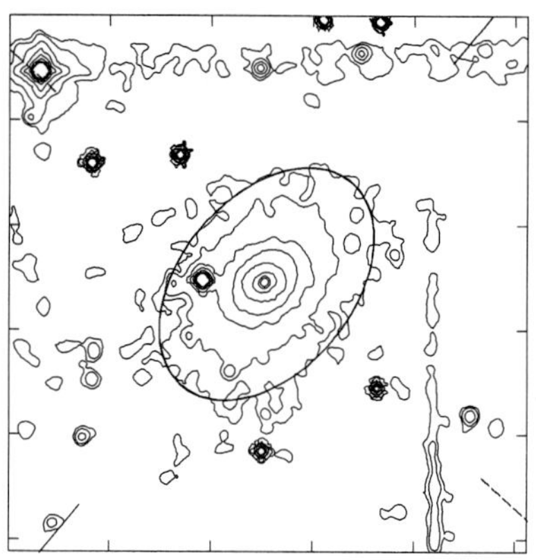

N4612 K3740 2 21.03 256X256

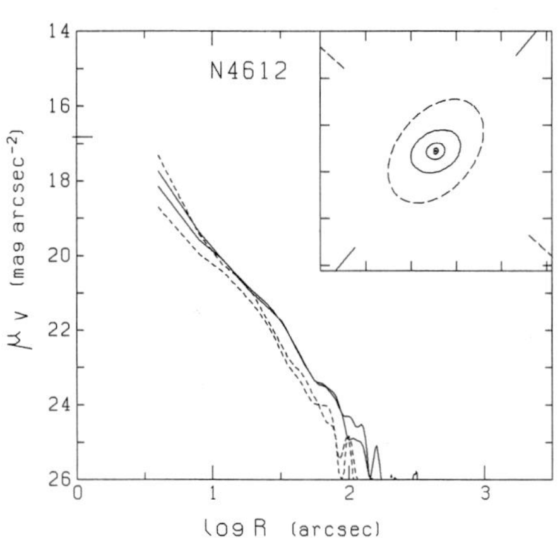

NGC4618

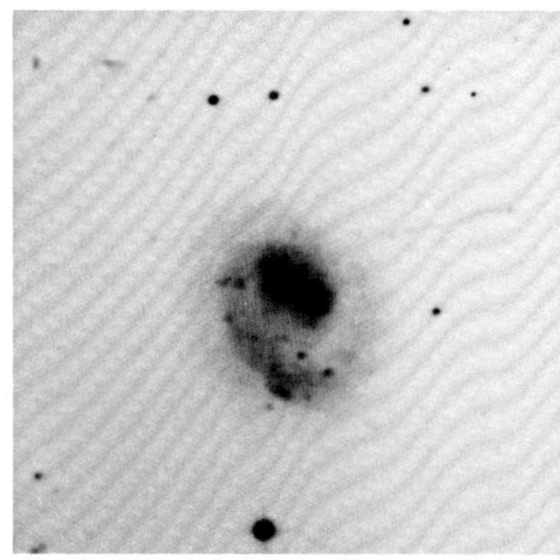

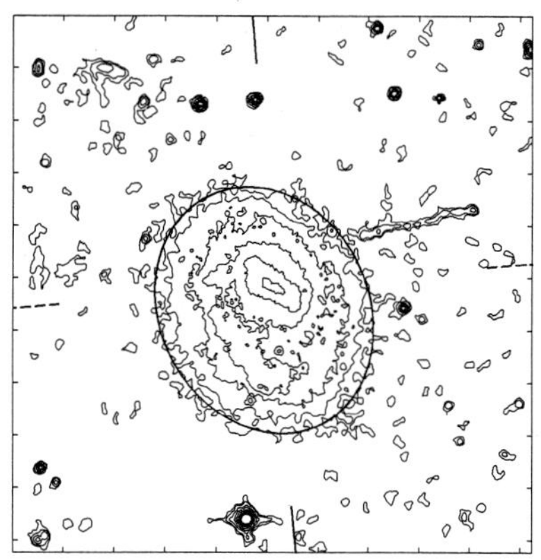

N4618 K3173 1 21.49 512X512

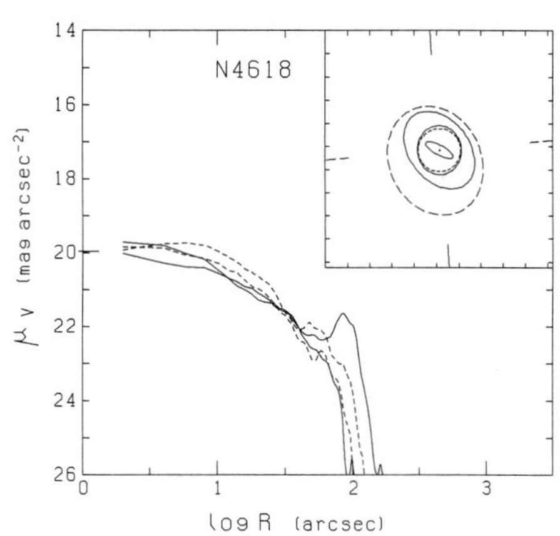

NGC4621

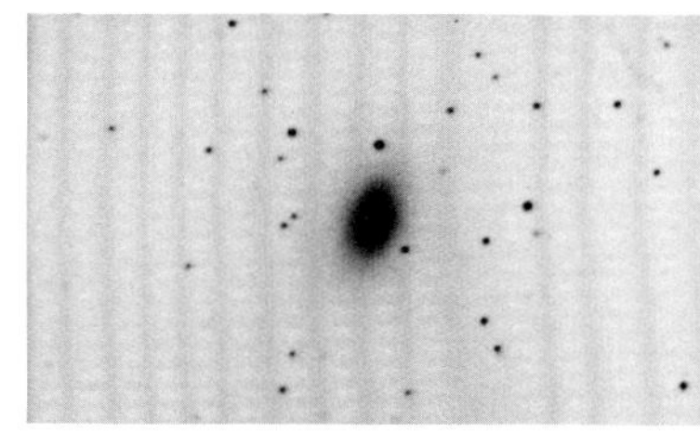

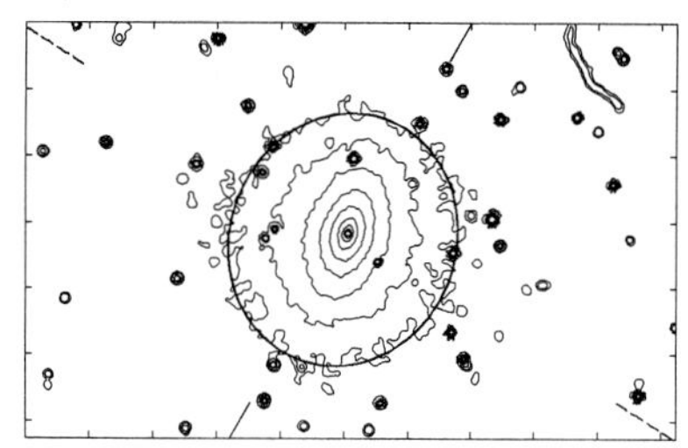

N4621 K2220 2 21.12 512X313

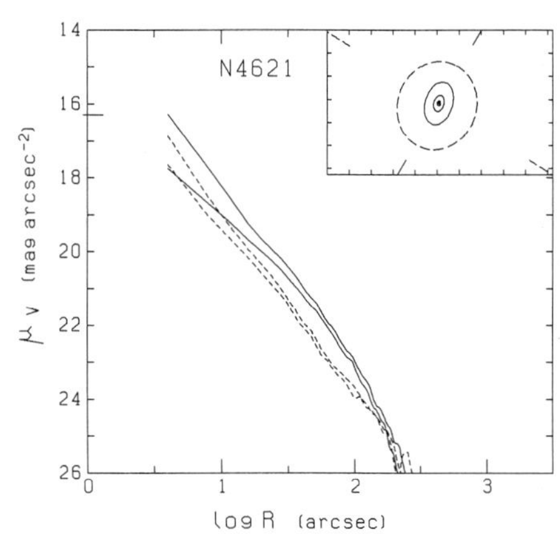

NGC4623

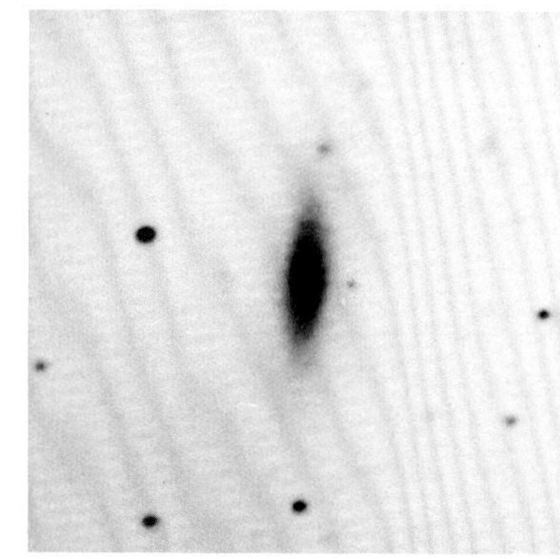

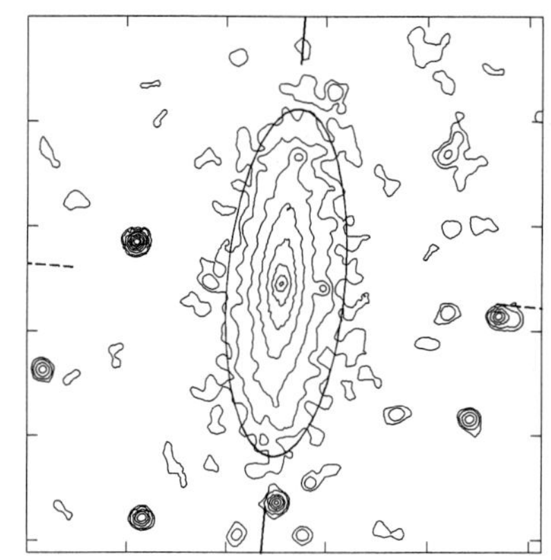

N4623 K2220 1 21.12 256X256

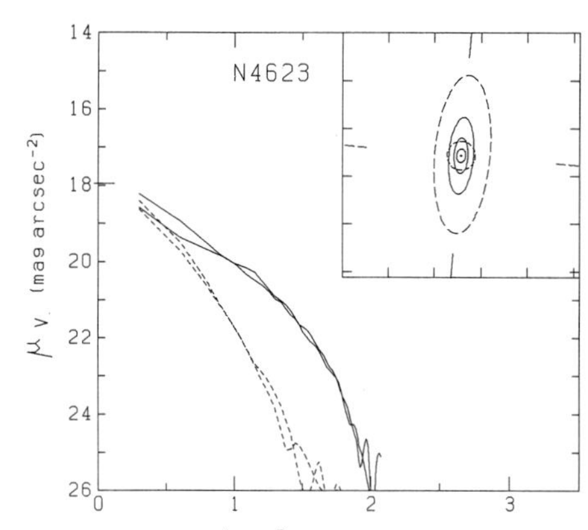

NGC4631

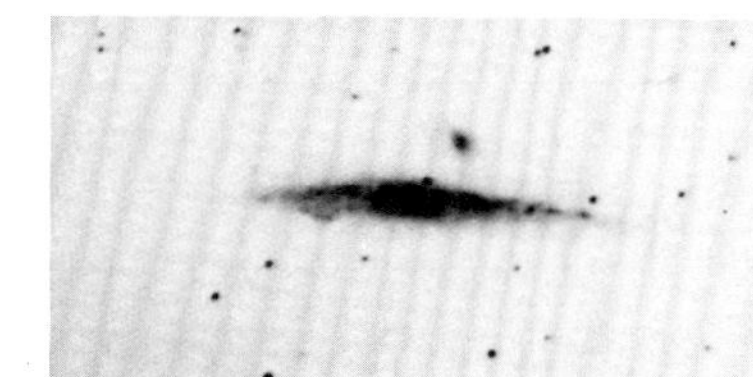

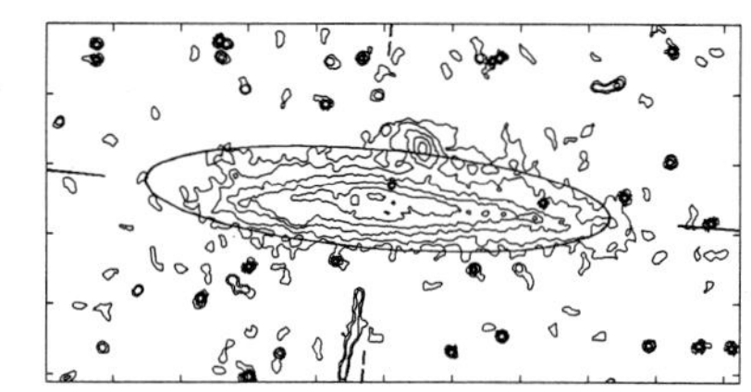

N4631 K5069 3 21.51 512X256

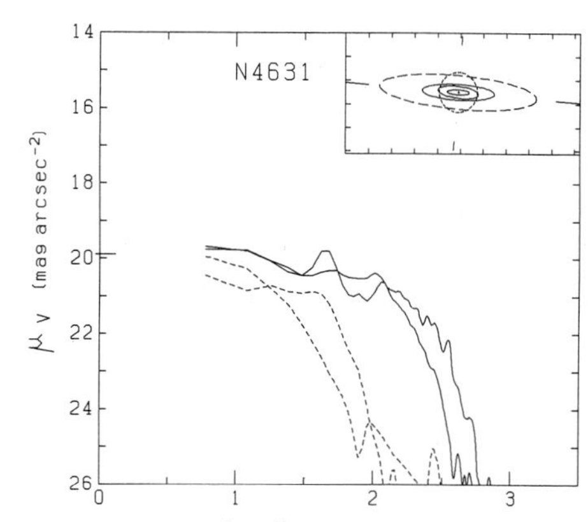

NGC4630

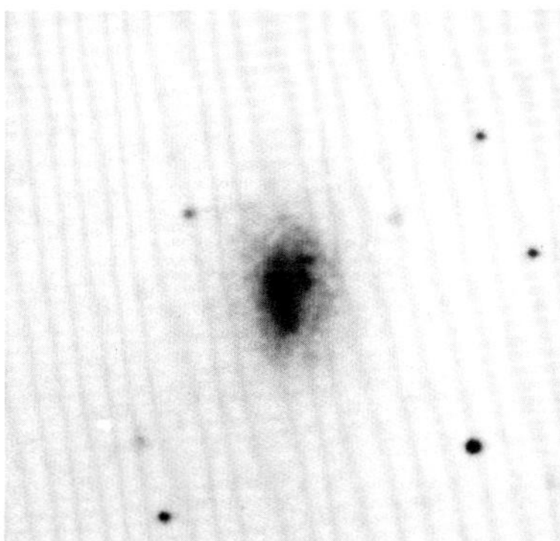

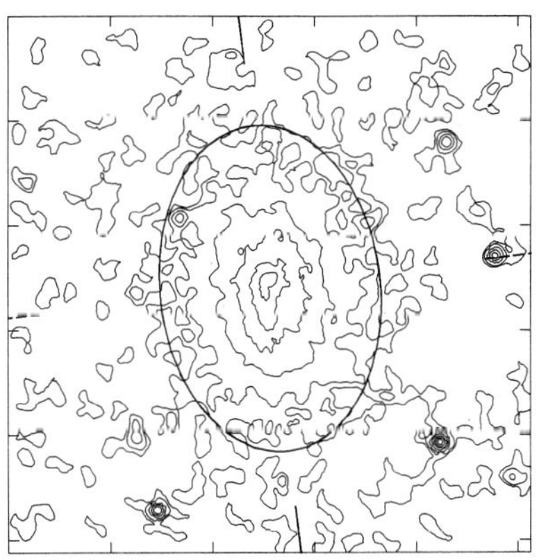

N4630 K3740 1 21.03 256X256

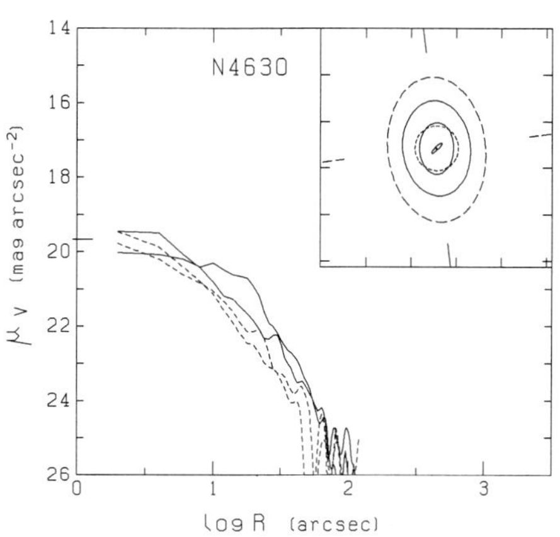

NGC4632

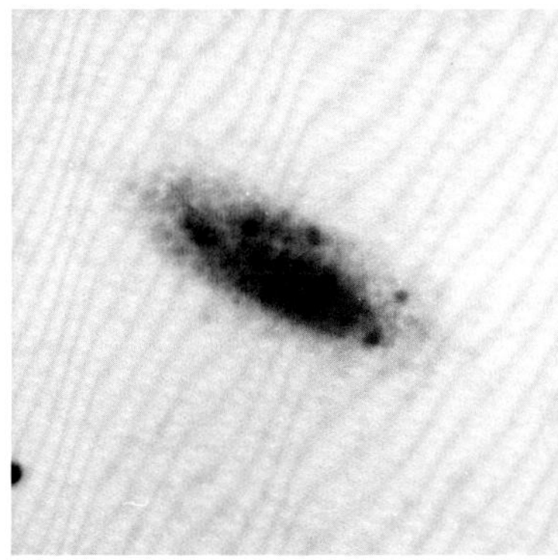

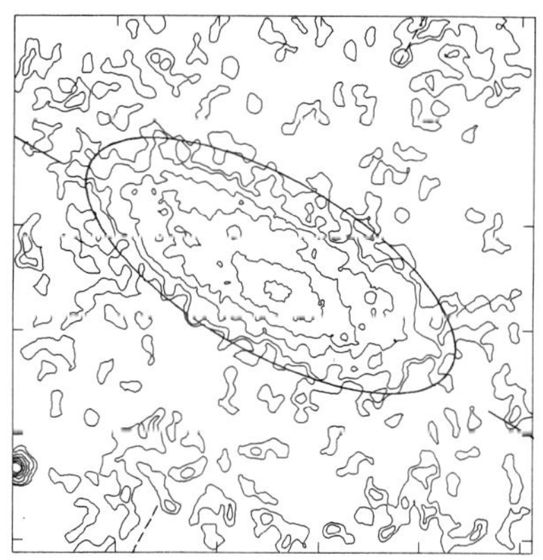

N4632 K4674 1 21.30 256X256

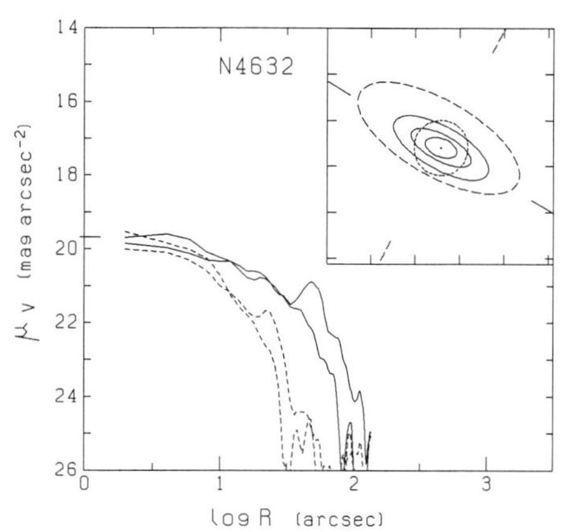

NGC4635

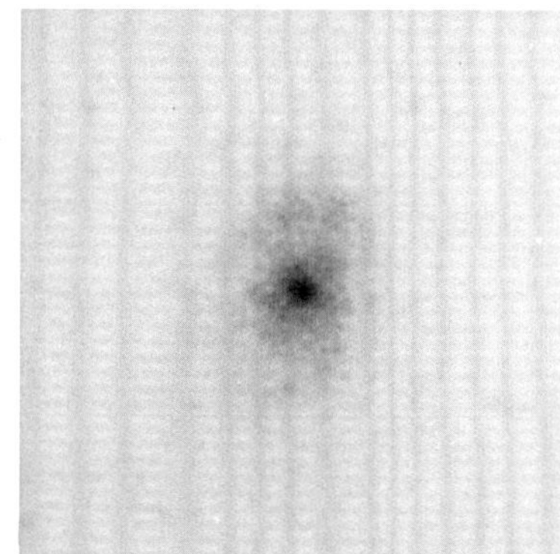

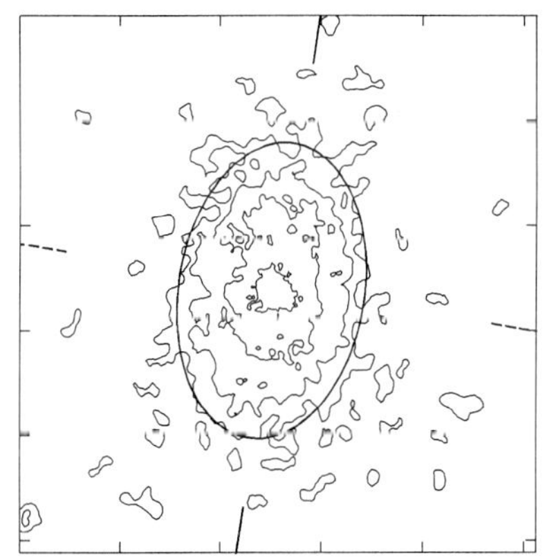

N4635 K5791 1 21.84 256X256

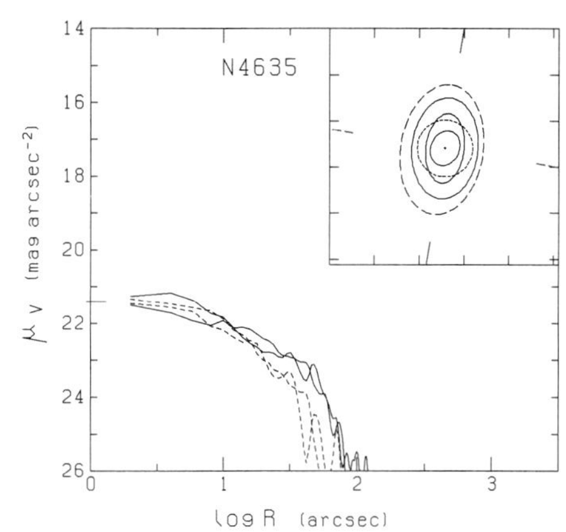

NGC4638

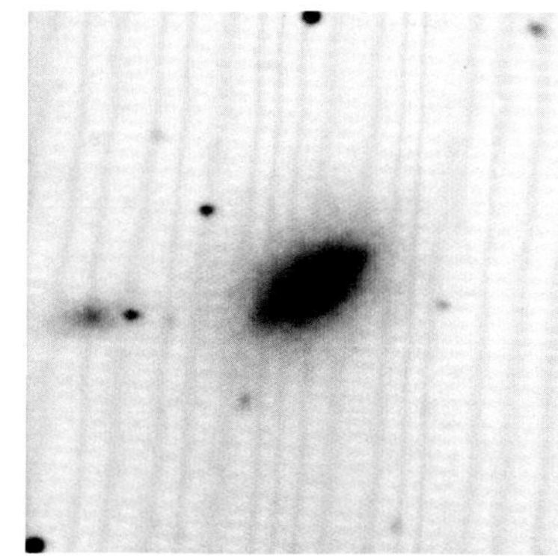

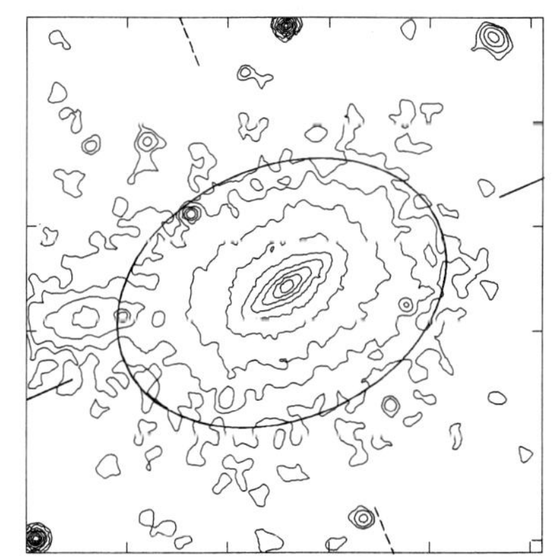

N4638 K2220 1 21.12 256X256

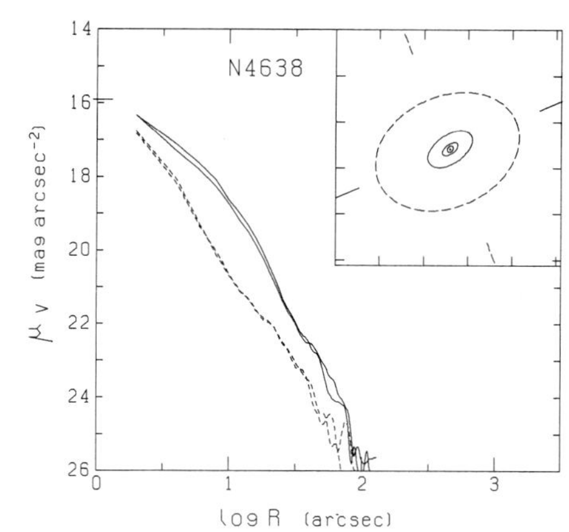

NGC4636

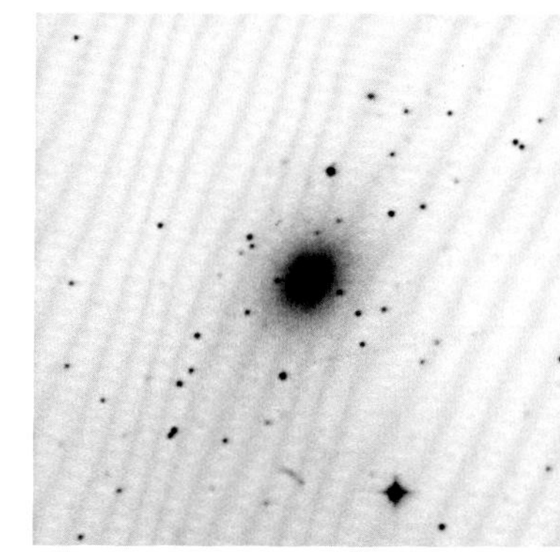

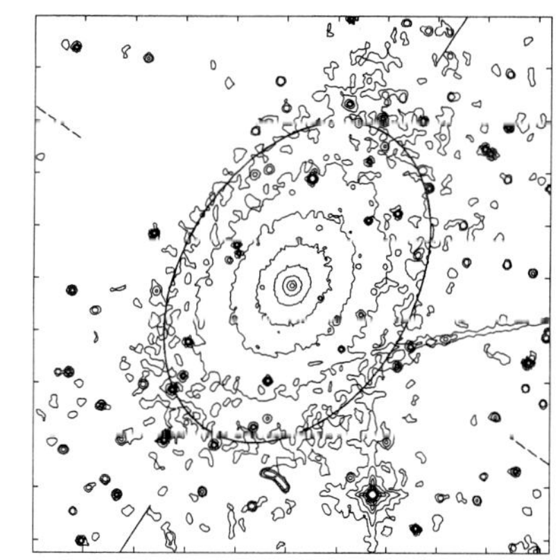

N4636 K3740 2 21.03 512X512

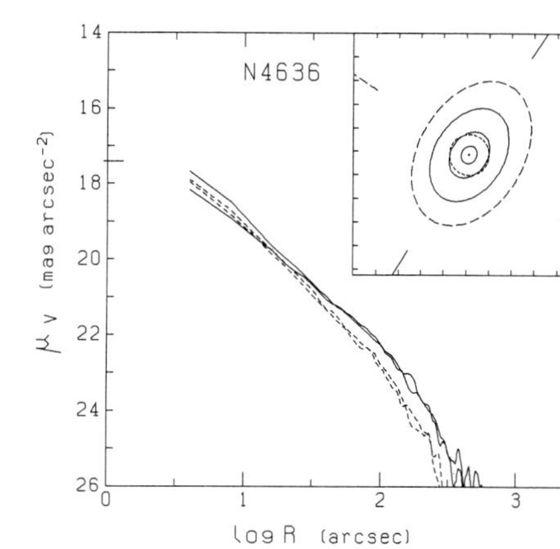

NGC4639

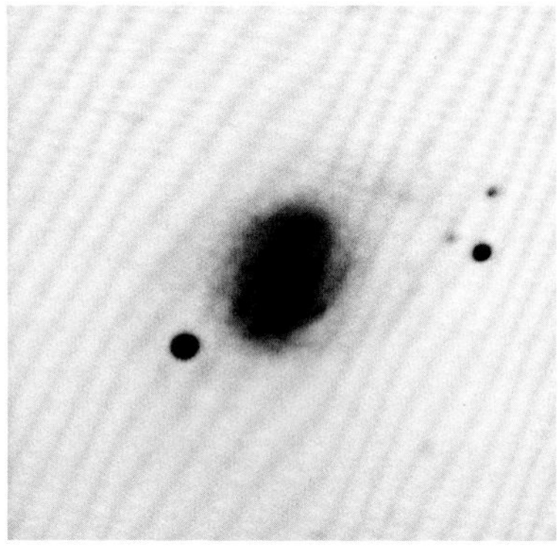

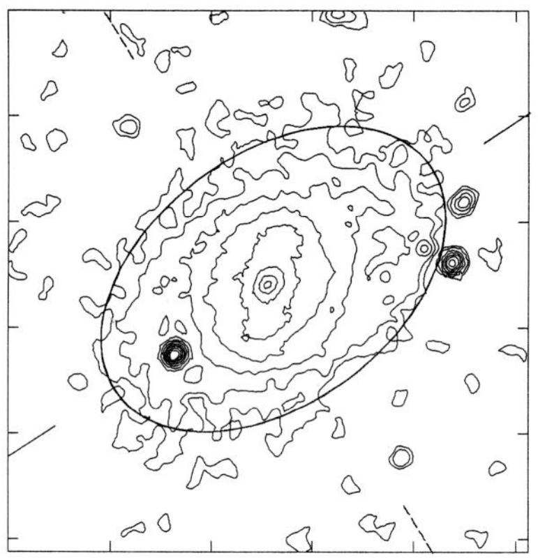

N4639 K2730 1 21.32 256X256

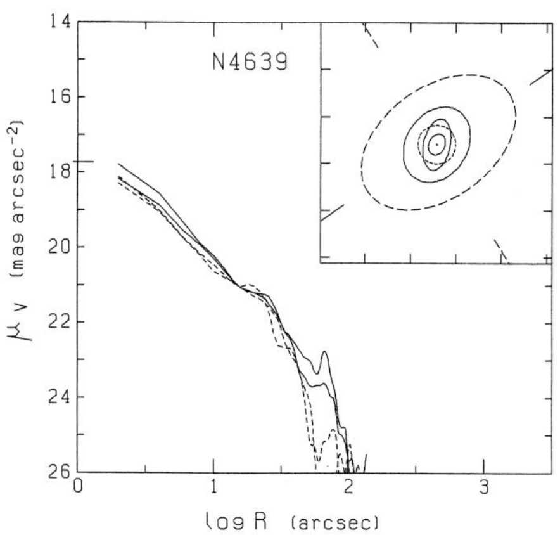

NGC4643

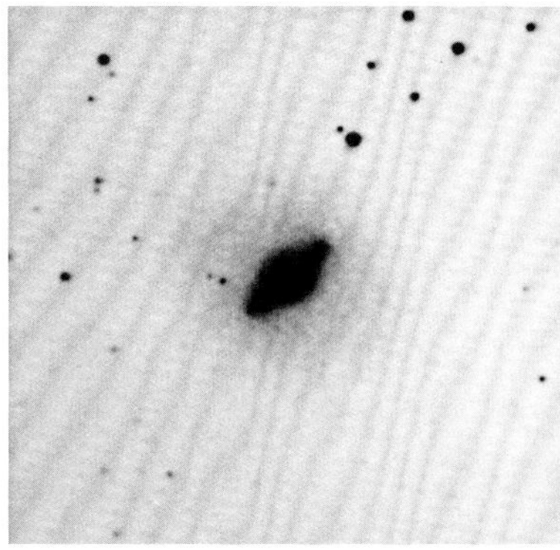

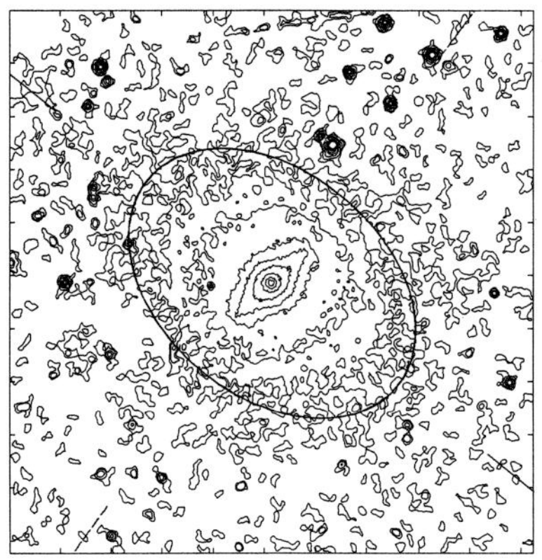

N4643 K3740 1 21.03 512X512

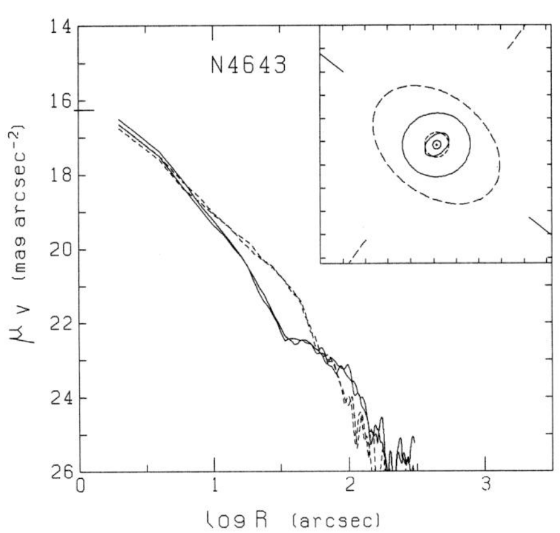

NGC4647

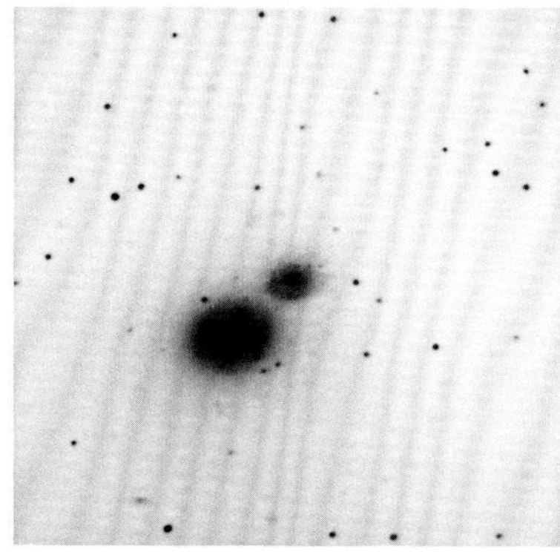

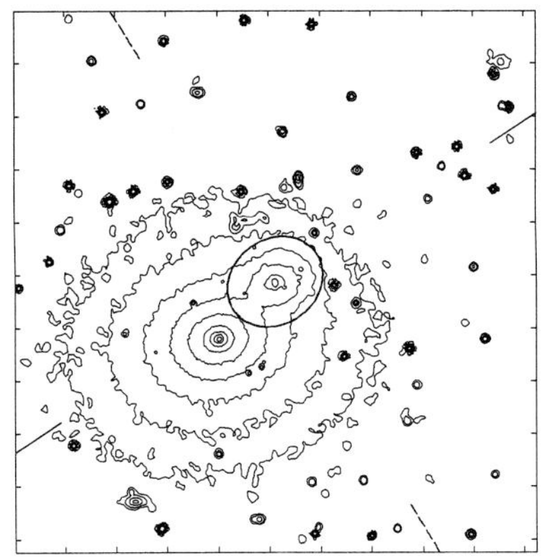

N4647 K2220 2 21.12 512X512

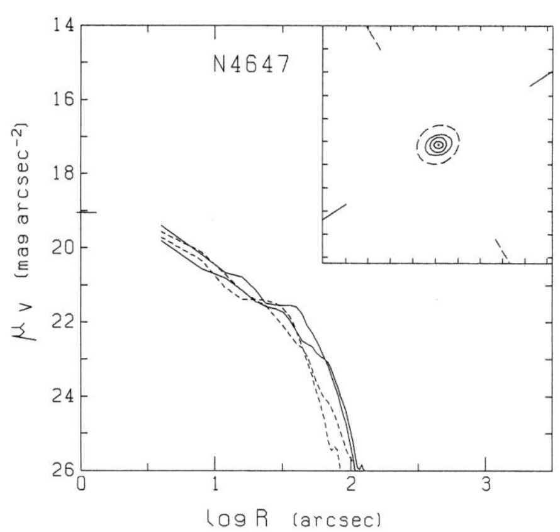

NGC4649

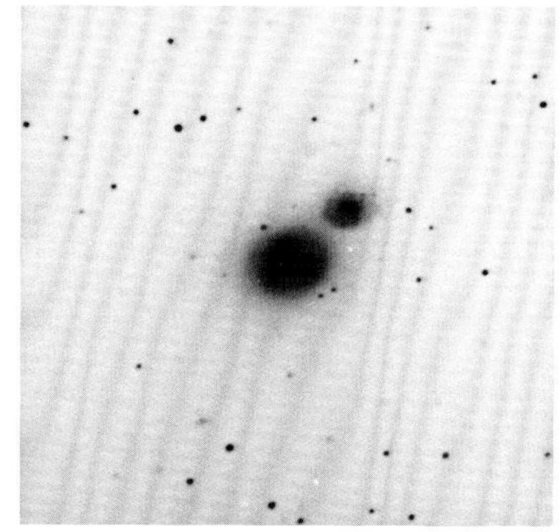

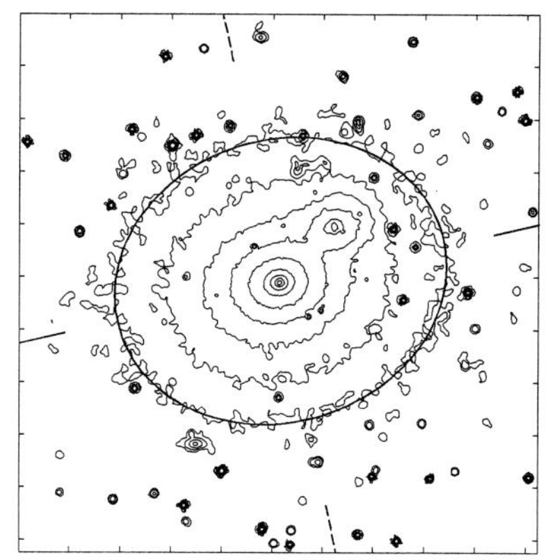

N4649 K2220 2 21.12 512X512

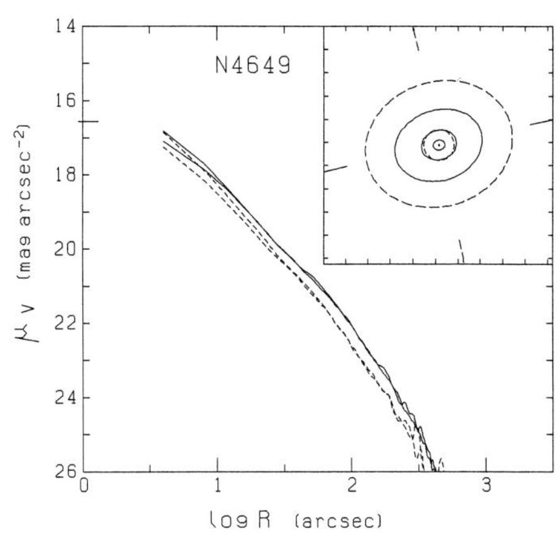

NGC4651

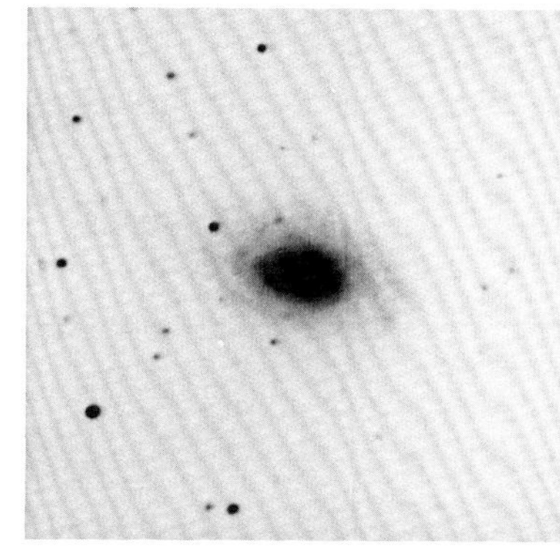

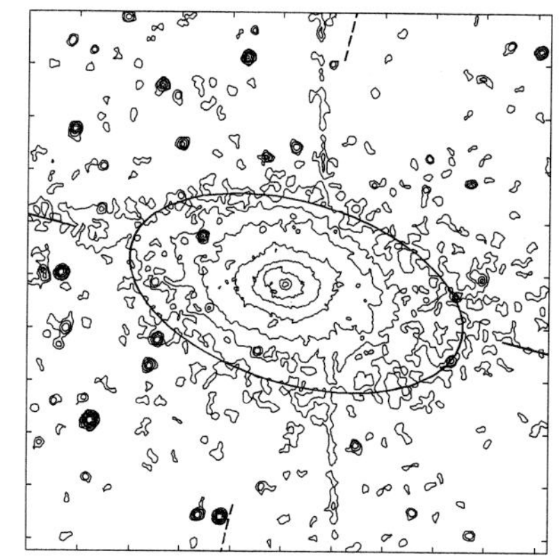

N4651 K2730 1 21.32 512X512

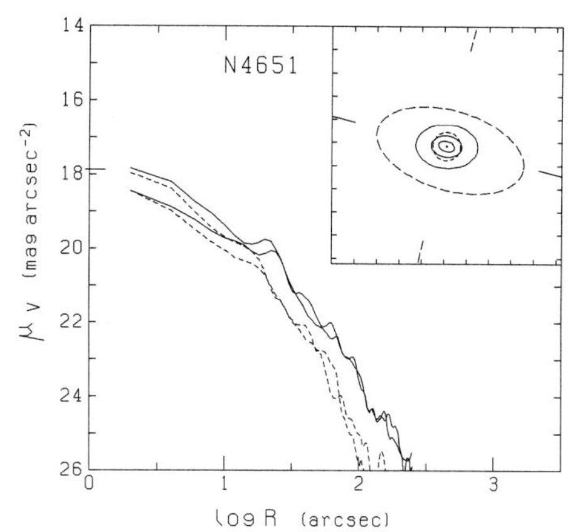

NGC4653

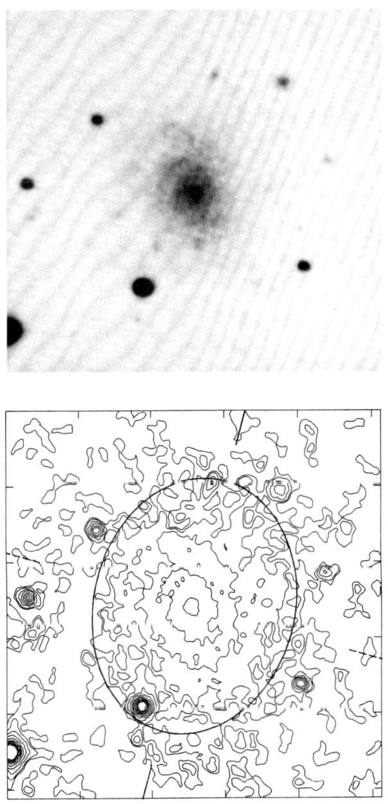

N4653 K4674 1 21.30 256X256

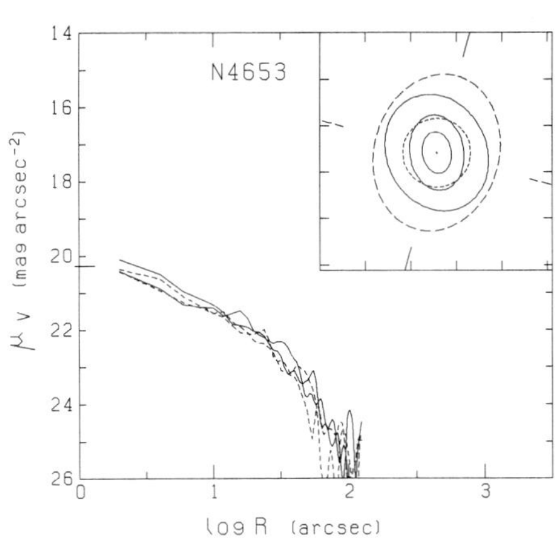

NGC4654

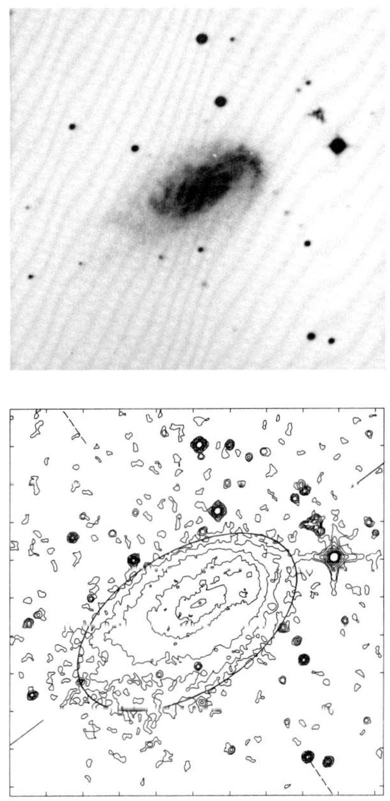

N4654 K2730 1 21.32 512X512

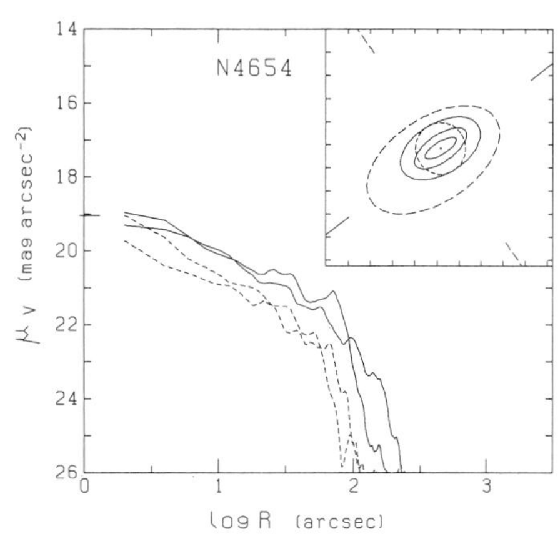

NGC4656

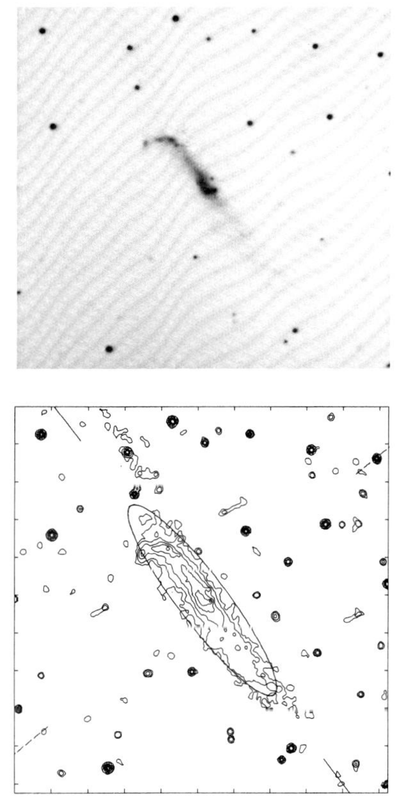

N4656 K3186 2 20.88 512X512

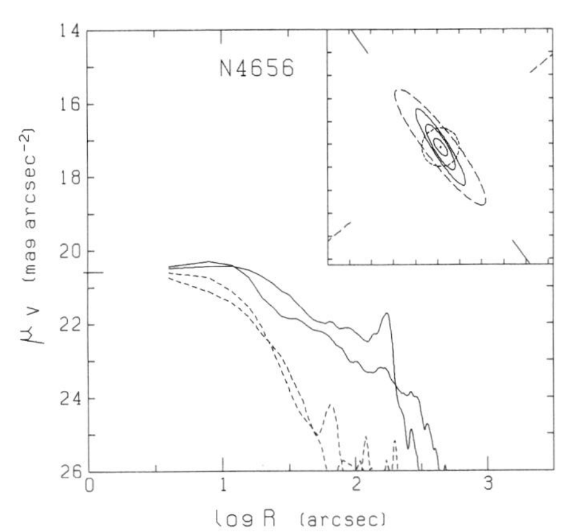

NGC4660

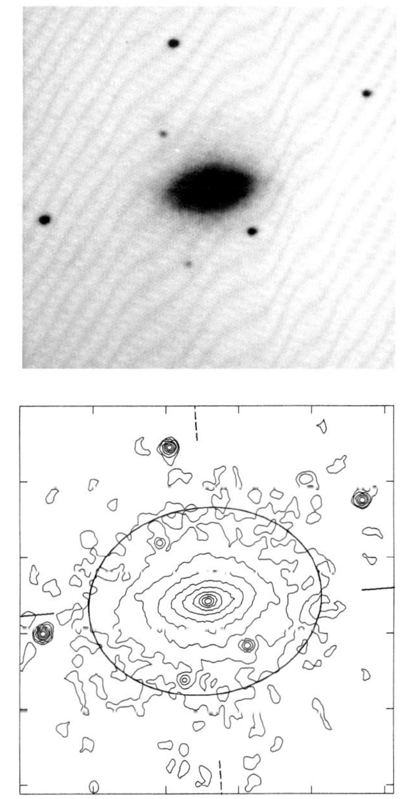

N4660 K2220 1 21.12 256X256

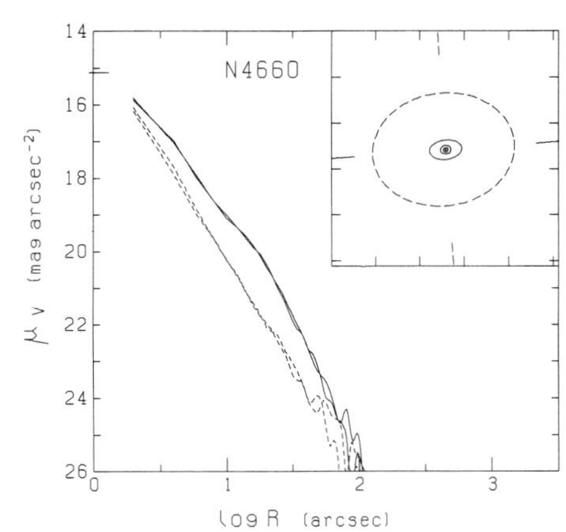

NGC4658

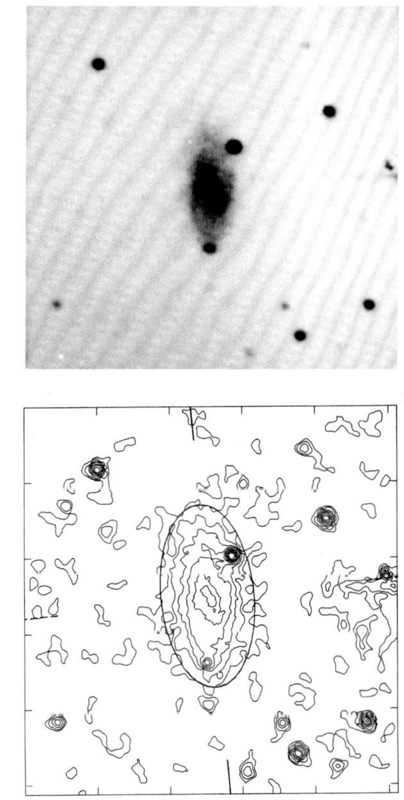

N4658 K4977 1 21.56 256X256

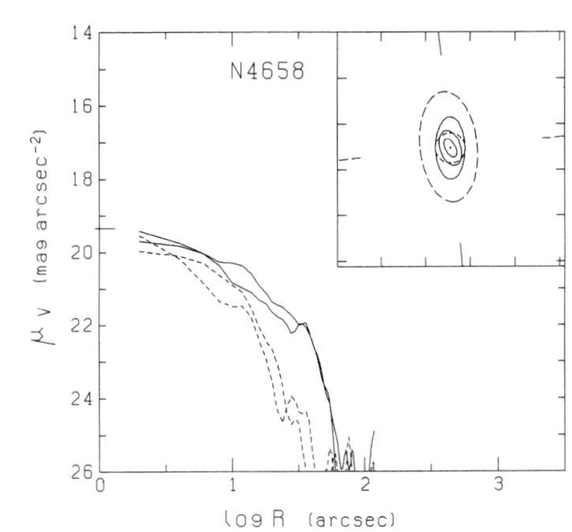

NGC4665

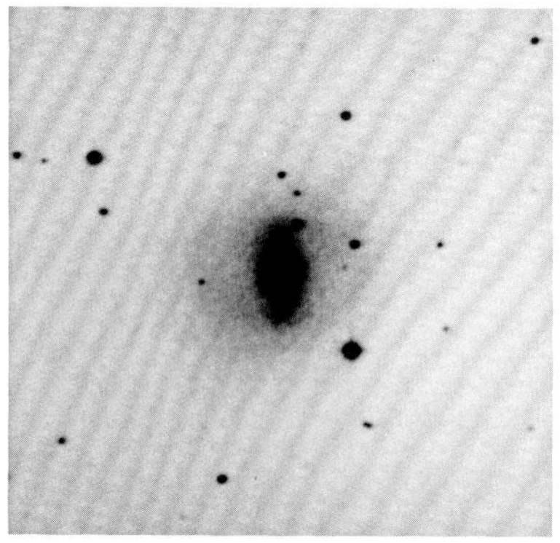

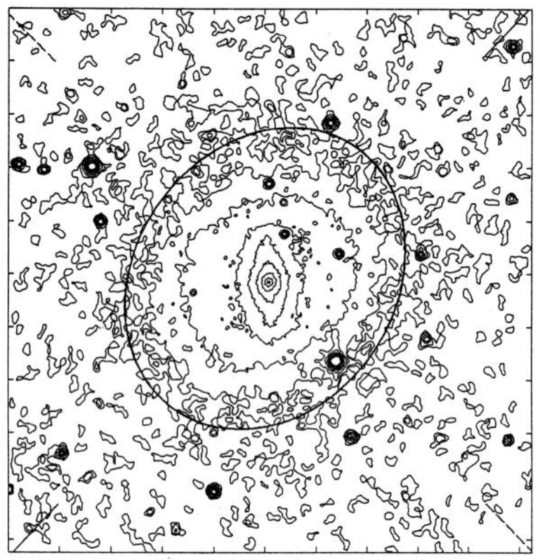

N4665 K3740 1 21.03 512X512

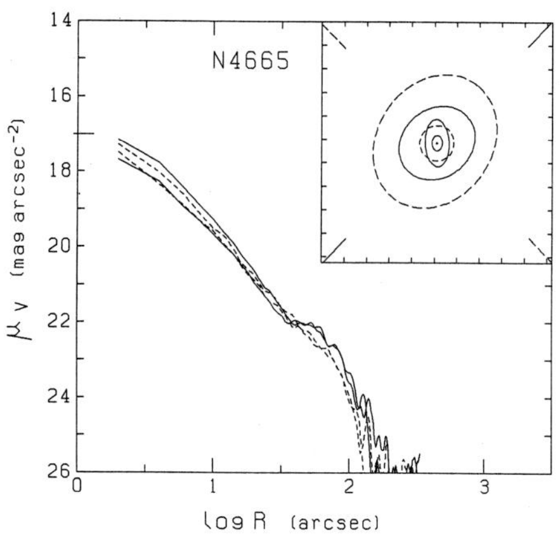

NGC4666

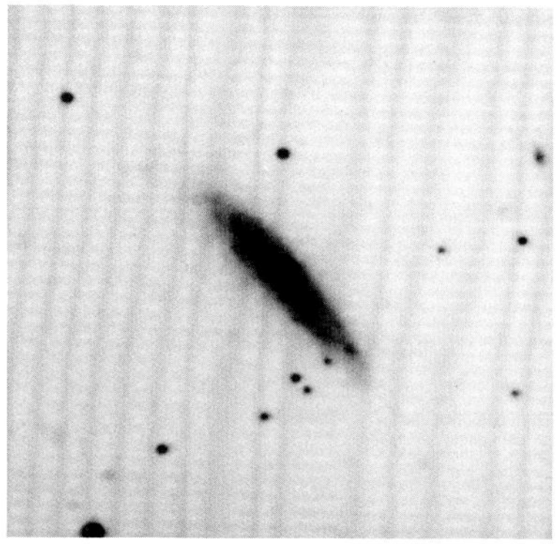

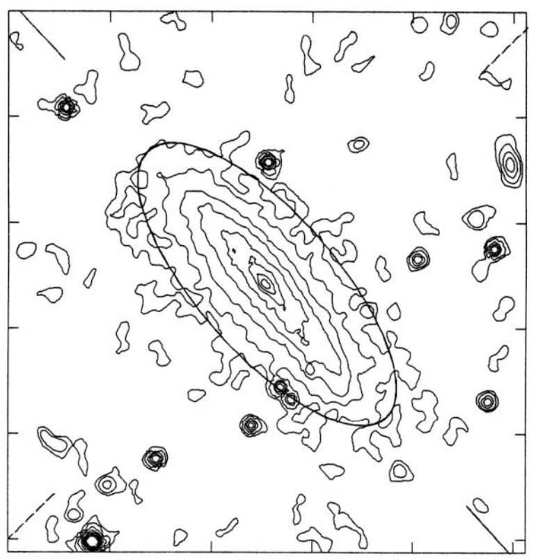

N4666 K4674 2 21.30 256X256

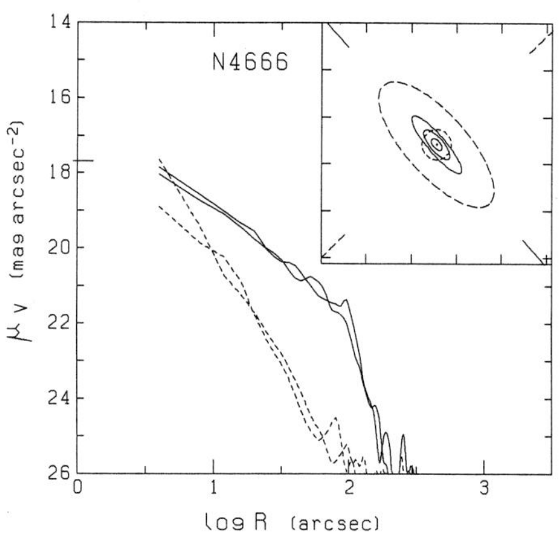

NGC4670

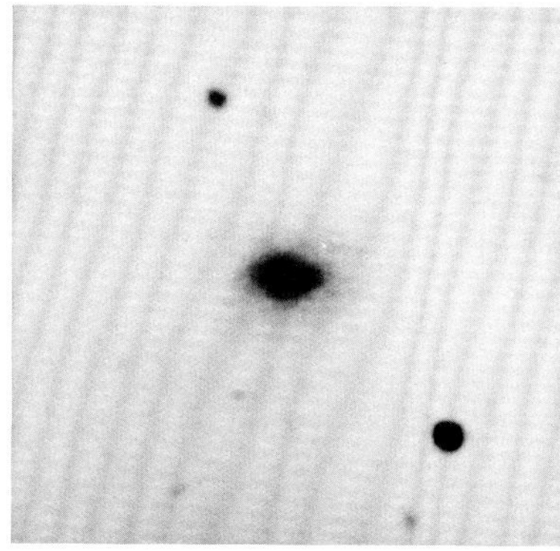

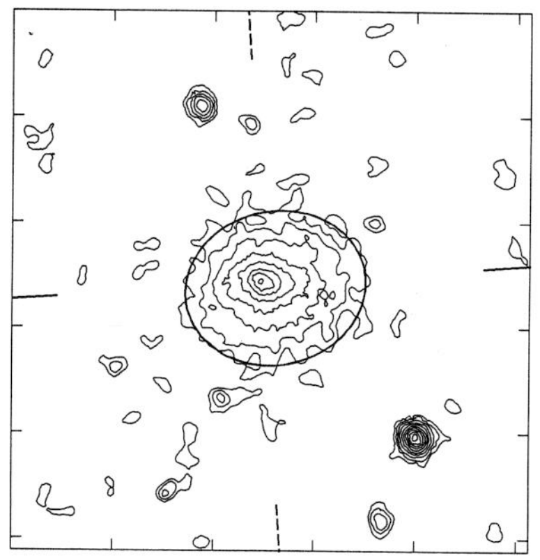

N4670 K4926 1 22.06 256X256

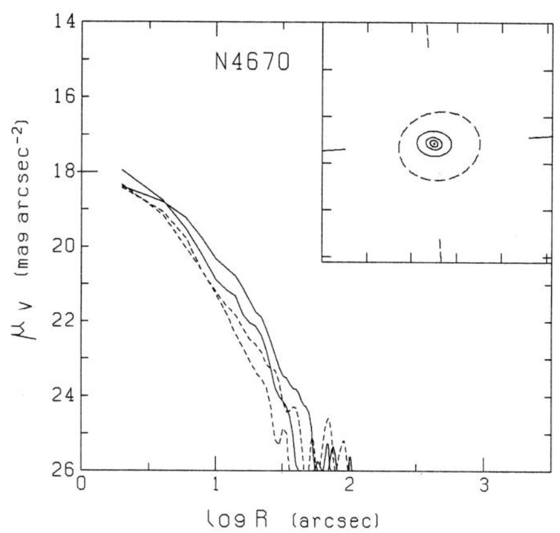

NGC4668

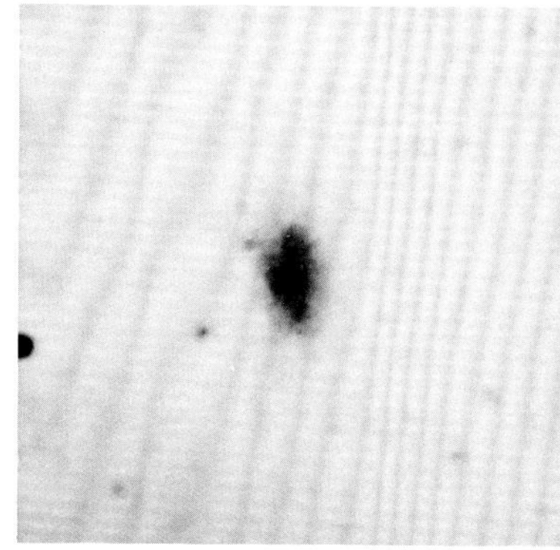

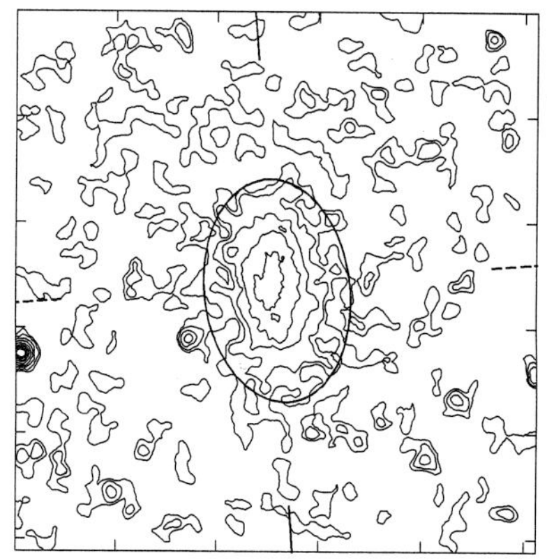

N4668 K4674 1 21.30 256X256

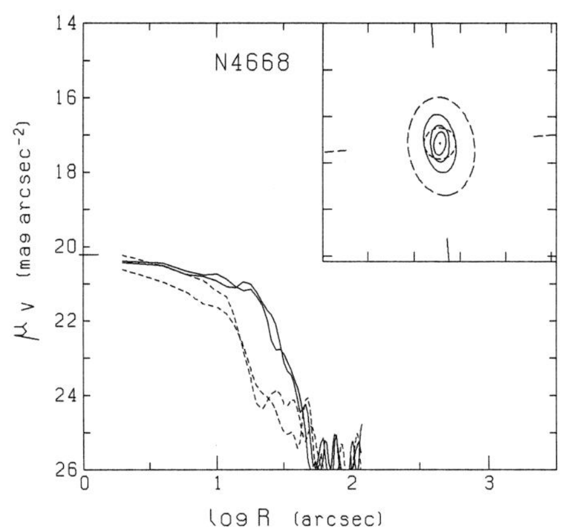

NGC4682

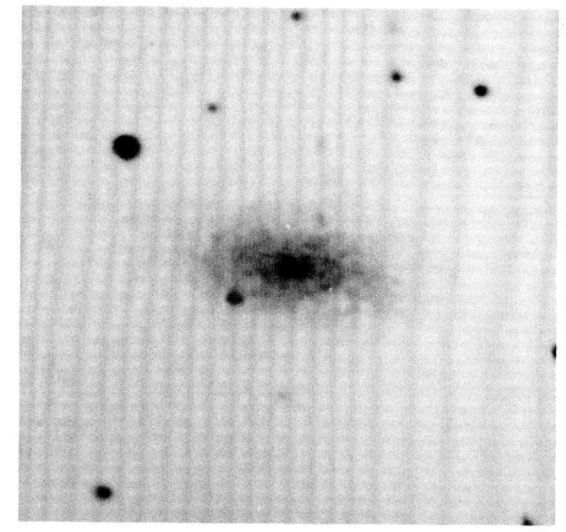

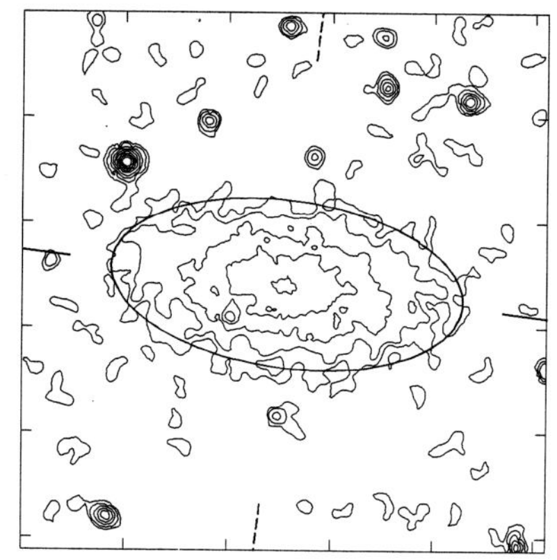

N4682 K4977 1 21.56 256X256

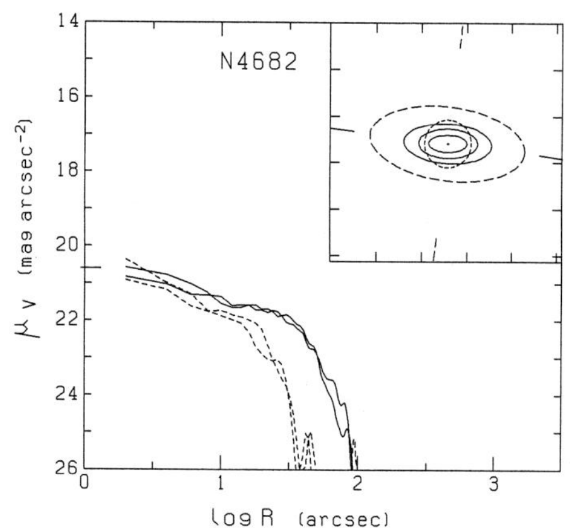

NGC4689

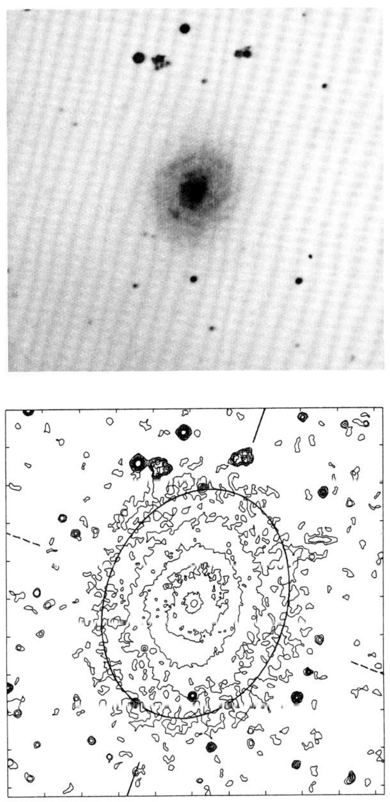

N4689 K2730 1 21.32 512X512

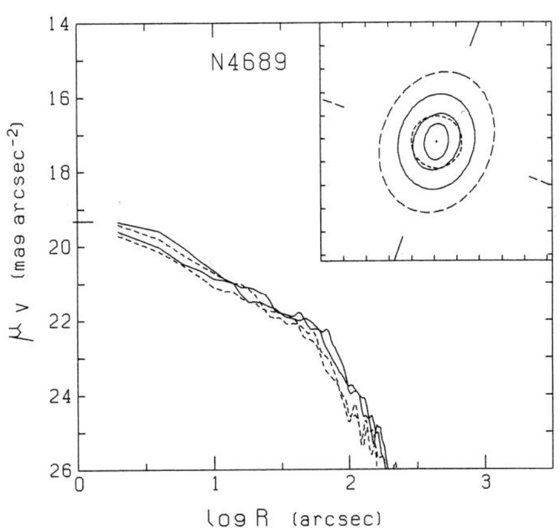

NGC4691

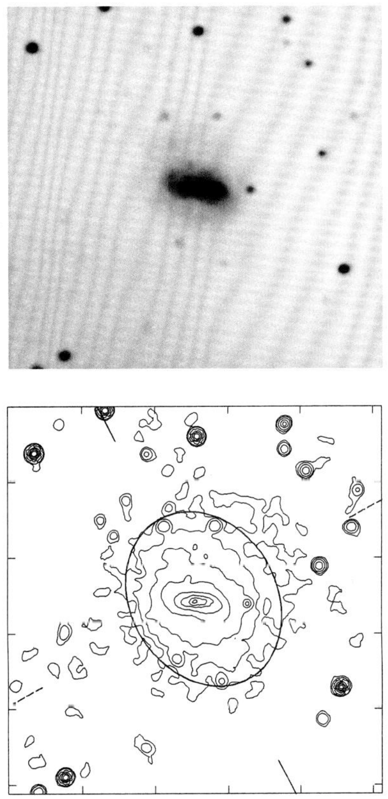

N4691 K3310 2 20.42 256X256

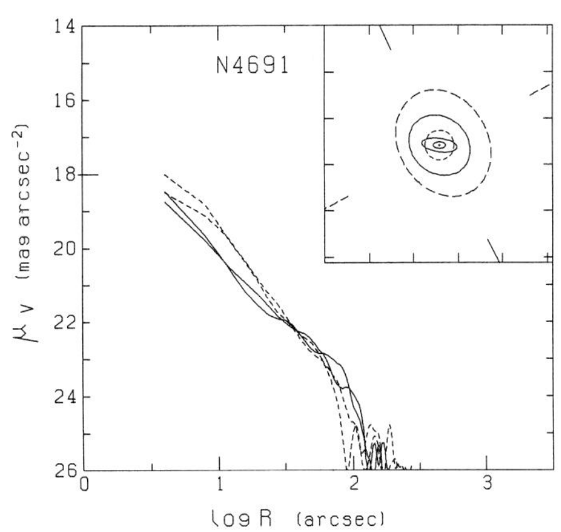

NGC4694

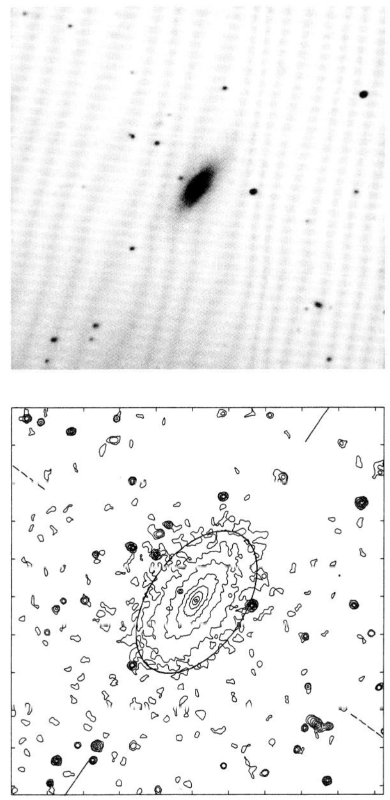

N4694 K2220 1 21.12 512X512

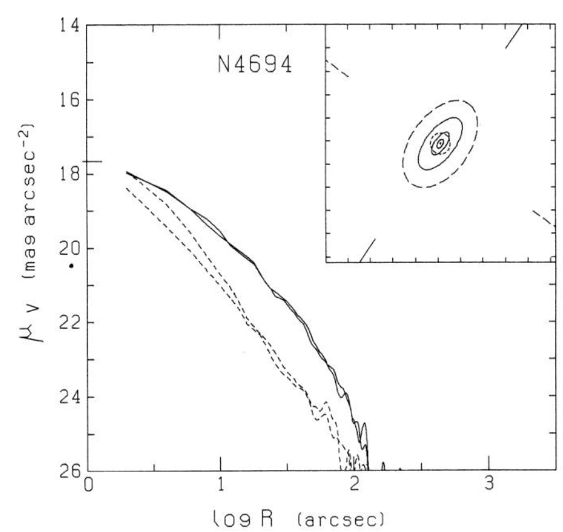

NGC4698

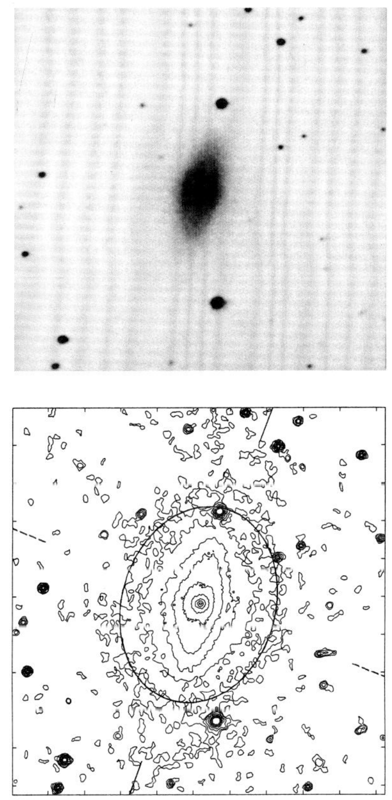

N4698 K2220 1 21.12 512X512

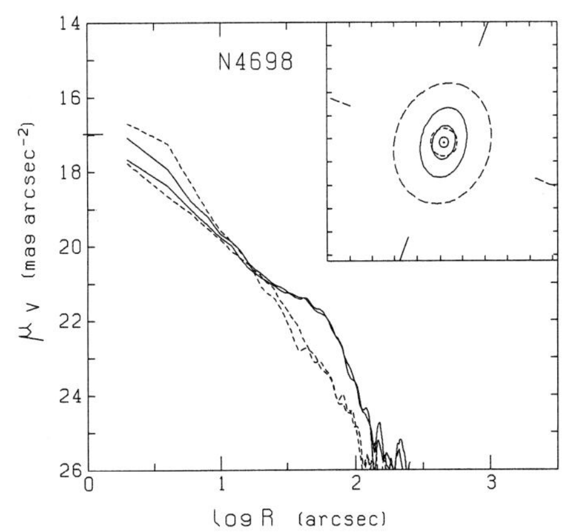

NGC4697

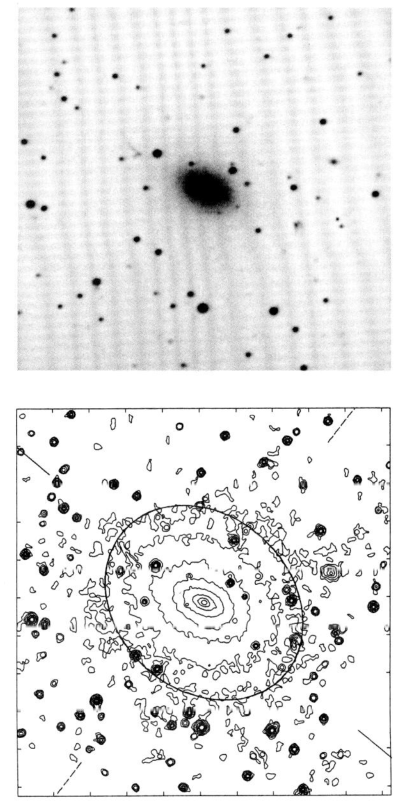

N4697 K3310 2 20.42 512X512

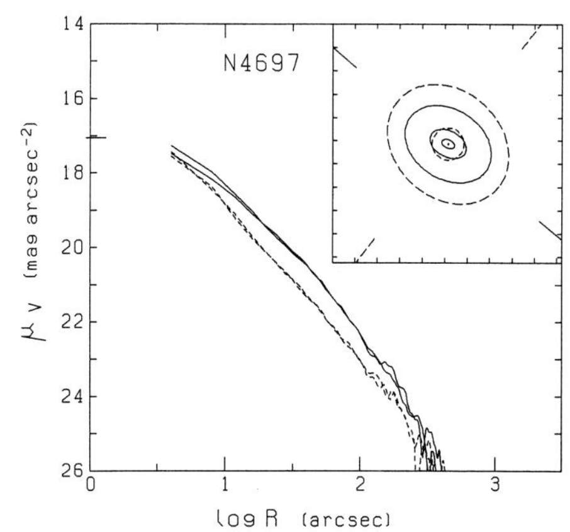

NGC4699

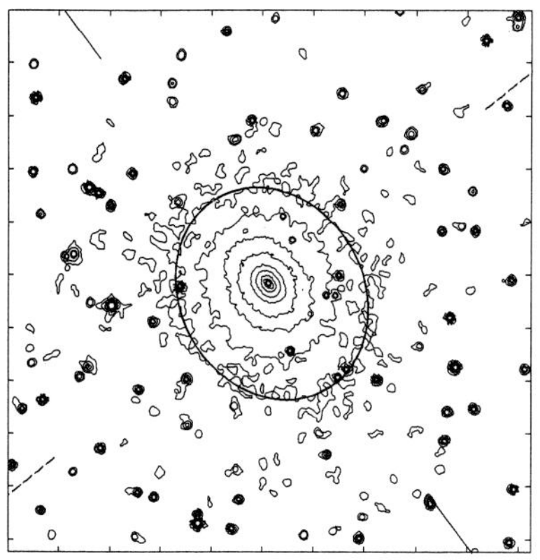

N4699 K4977 2 21.56 512X512

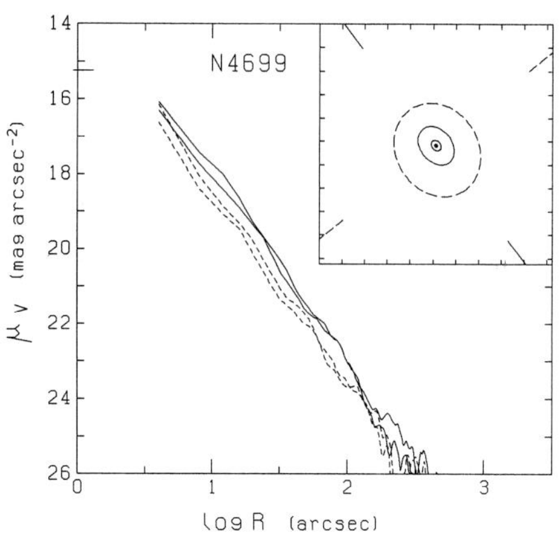

NGC4700

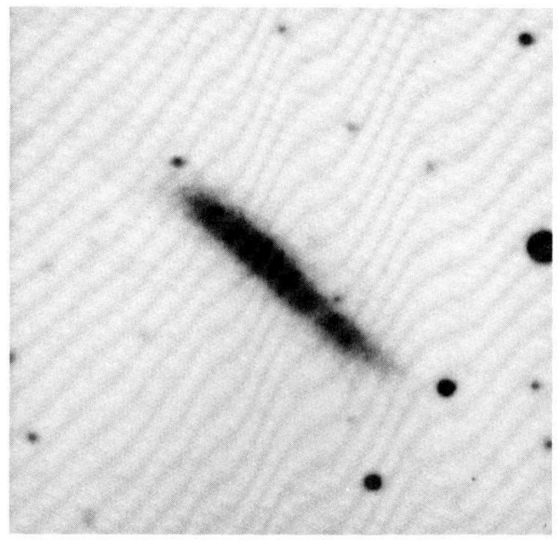

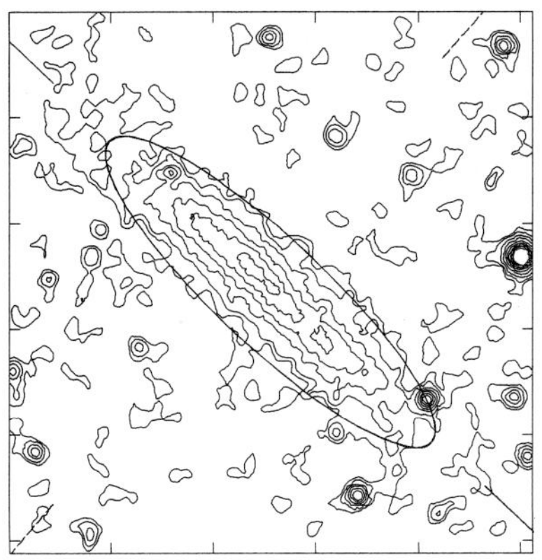

N4700 K4977 1 21.56 256X256

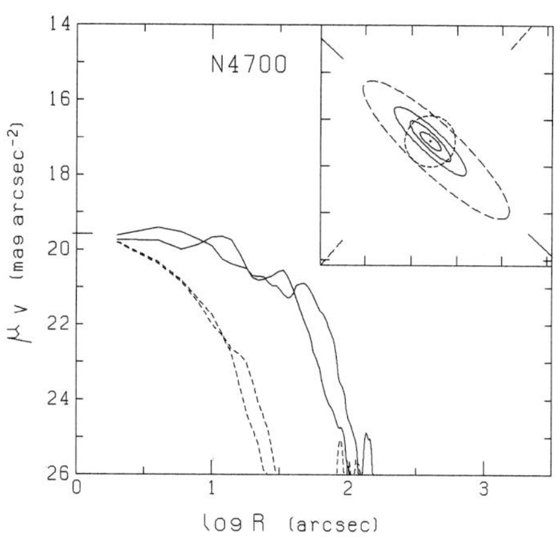

NGC4701

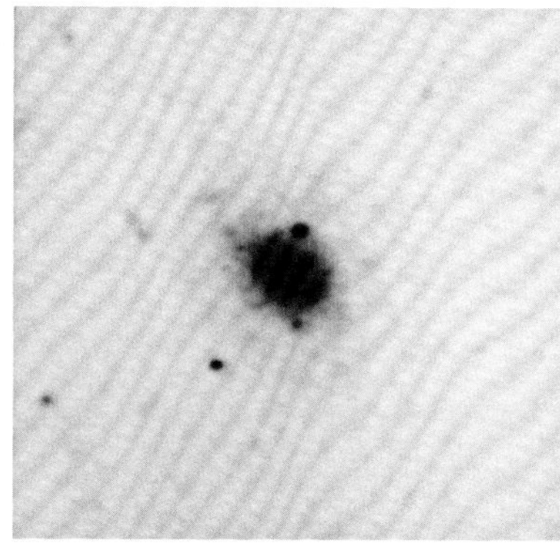

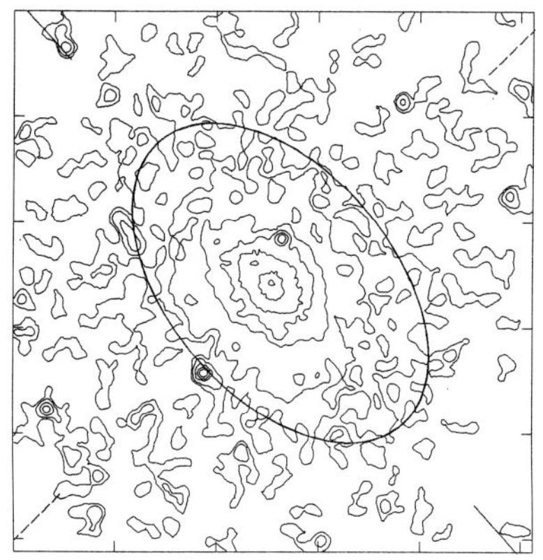

N4701 K3740 1 21.03 256X256

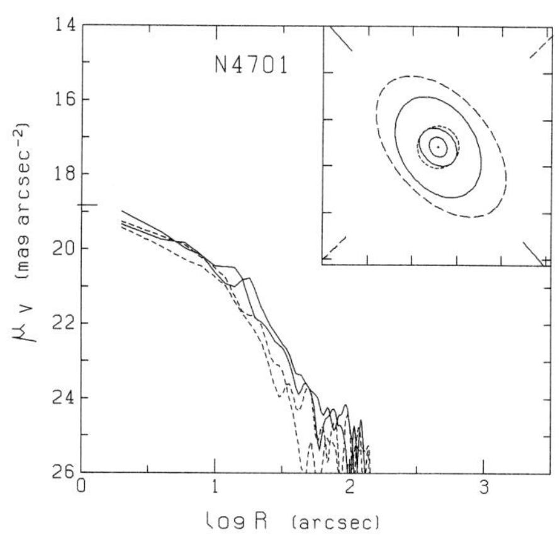

New 3

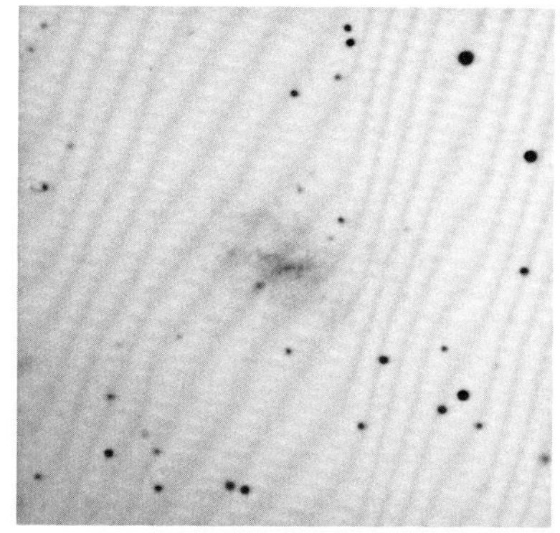

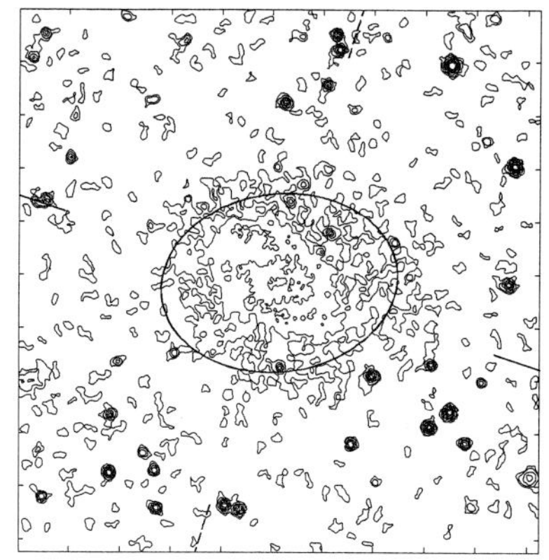

NEW3 K4977 1 21.56 512X512

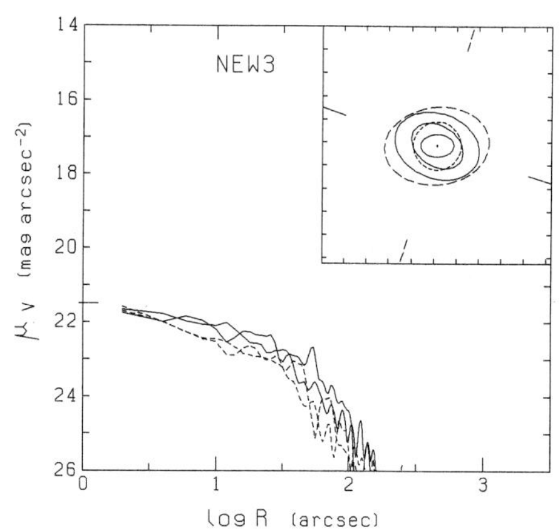

NGC4712

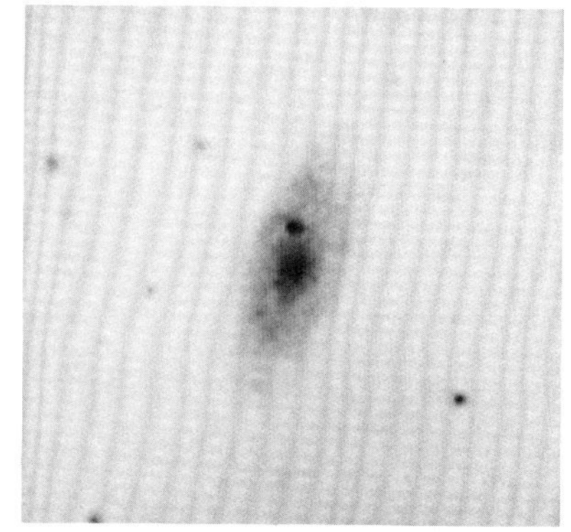

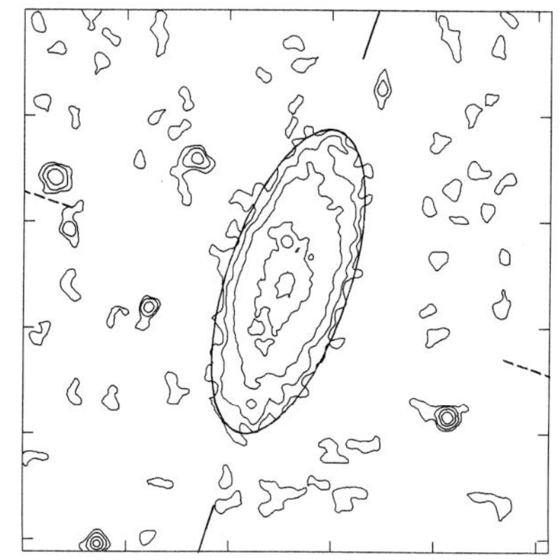

N4712 K3683 1 21.15 256X256

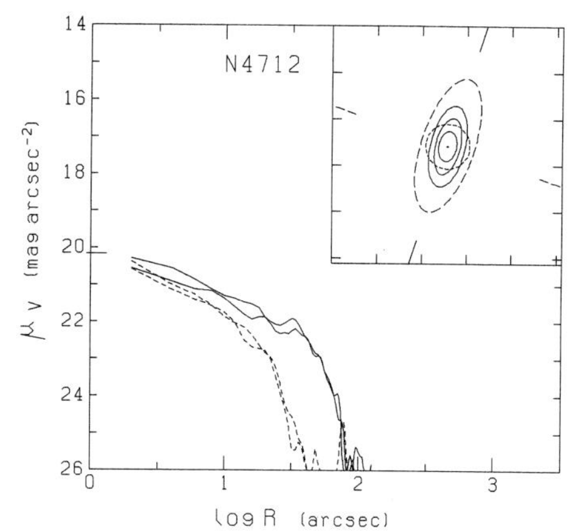

NGC4710

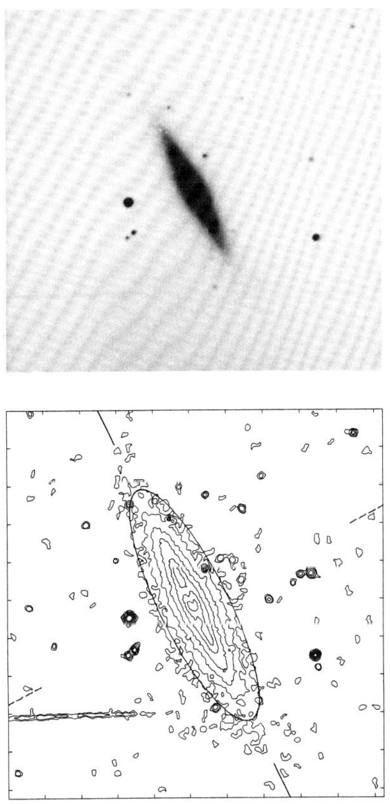

N4710 K2730 1 21.32 512X512

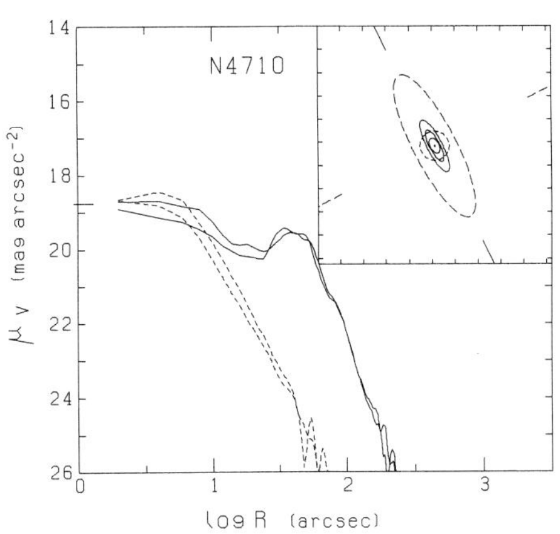

NGC4713

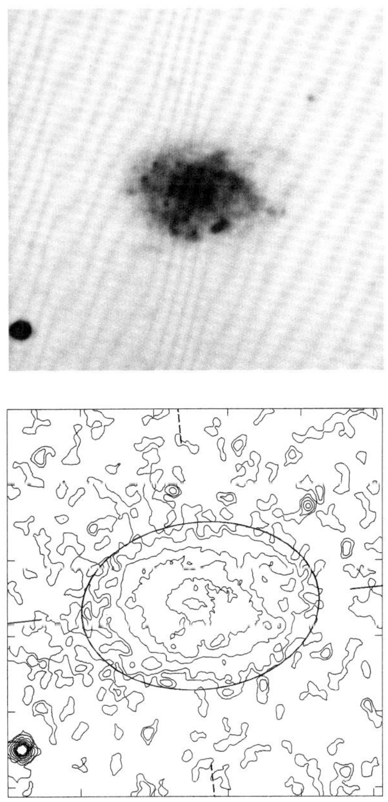

N4713 K3740 1 21.03 256X256

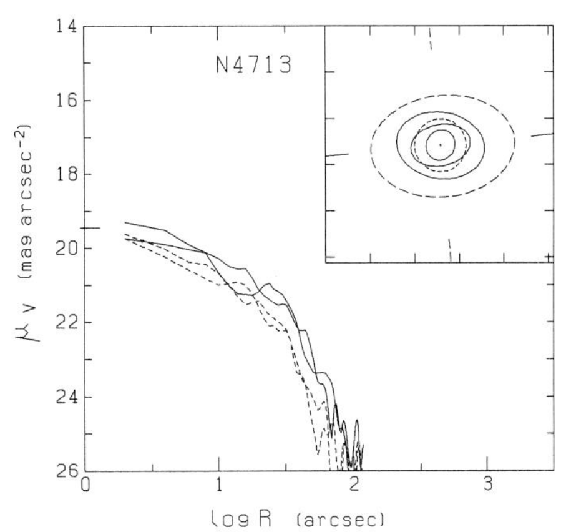

NGC4725

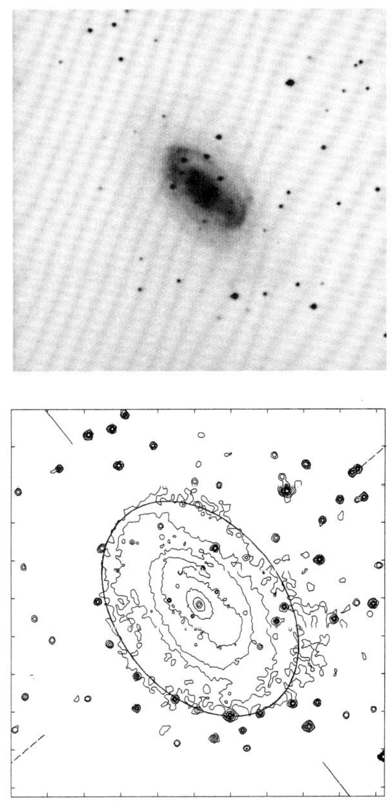

N4725 K3683 2 21.15 512X512

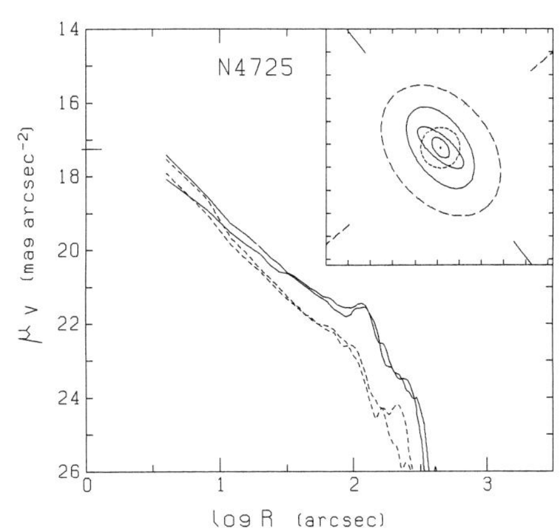

NGC4750

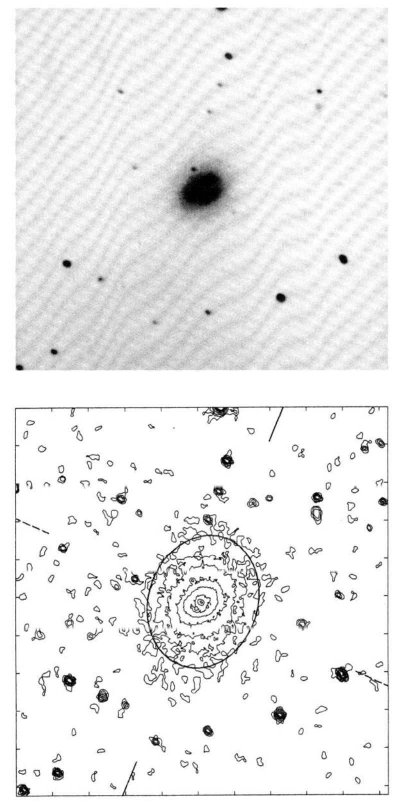

N4750 K5759 1 22.42 512X512

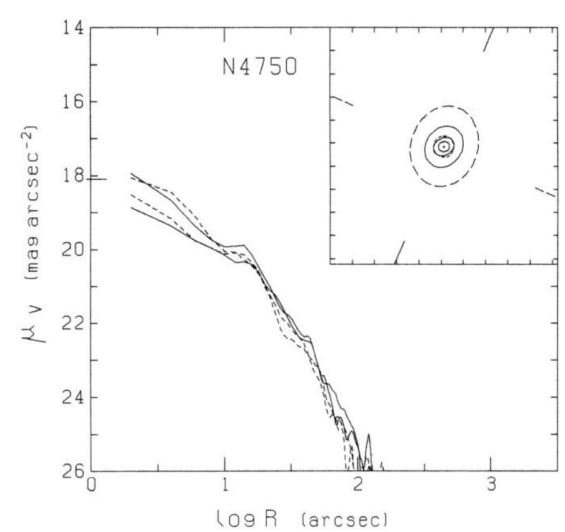

NGC4731

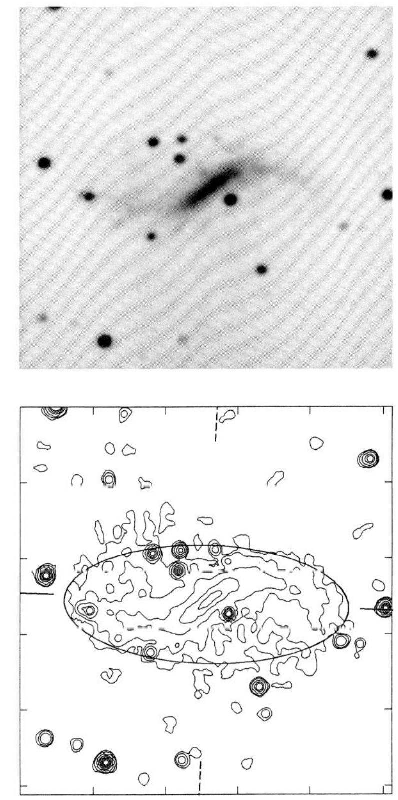

N4731 K3310 2 20.42 256X256

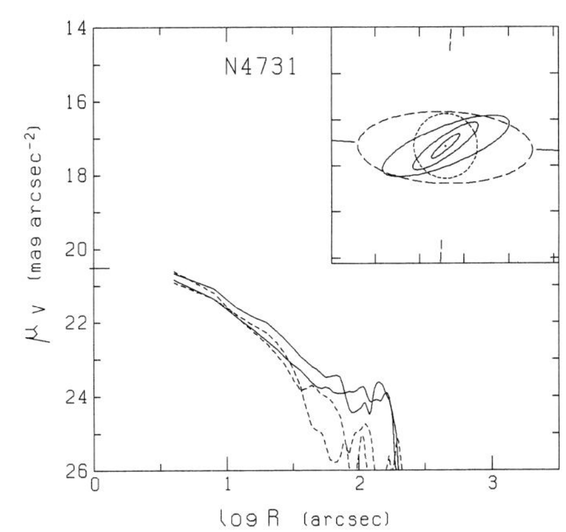

NGC4736

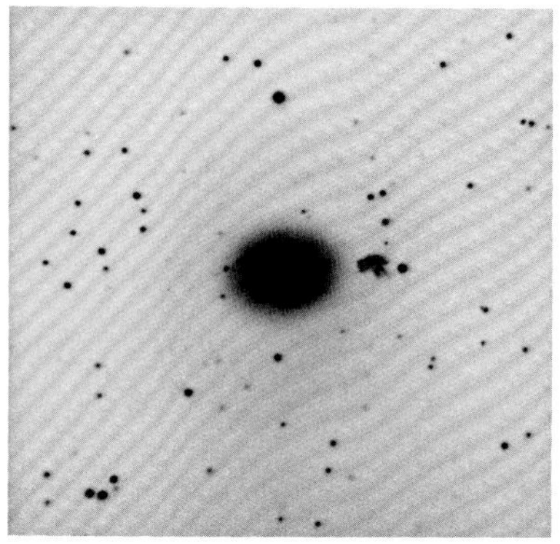

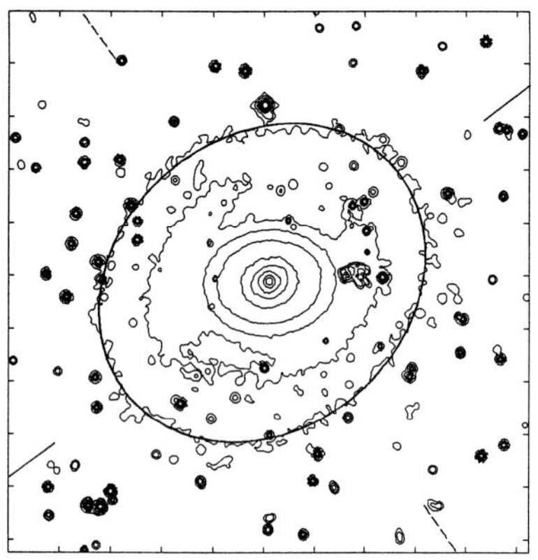

N4736 K2647 3 21.51 512X512

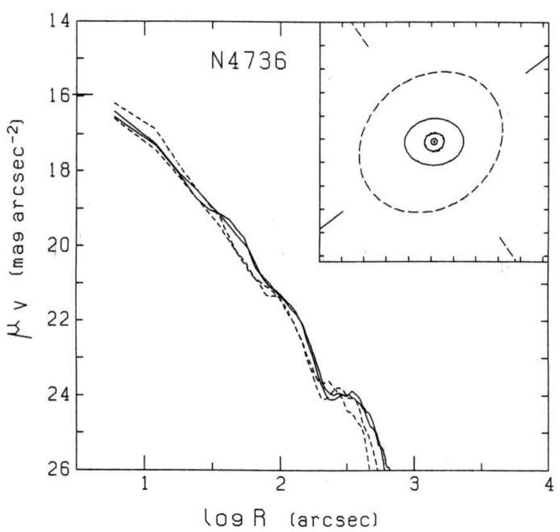

NGC4742

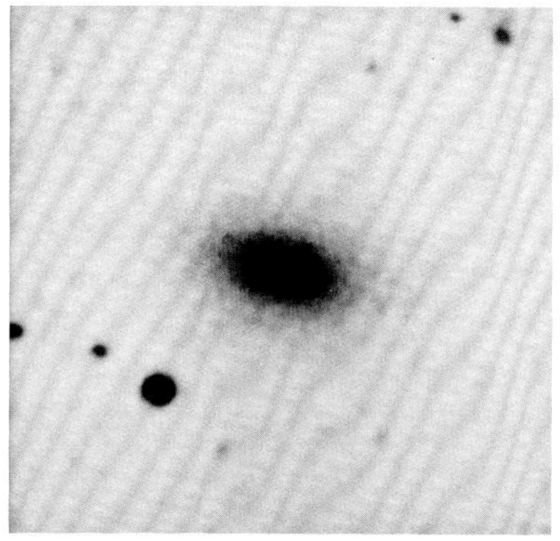

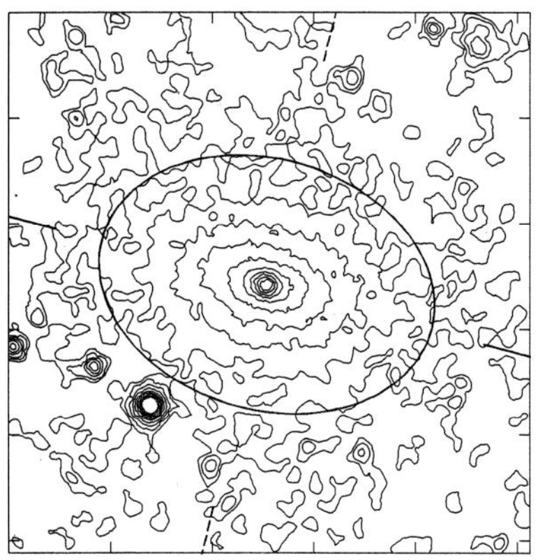

N4742 K4977 1 21.56 256X256

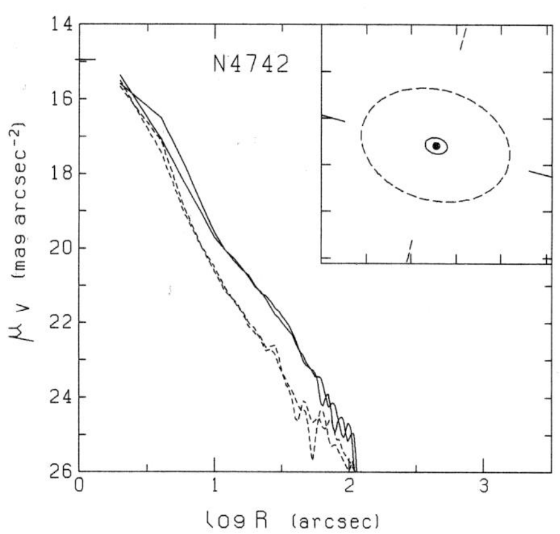

NGC4747

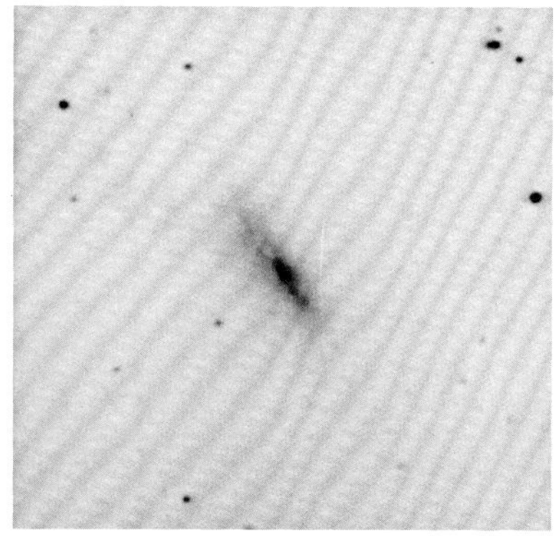

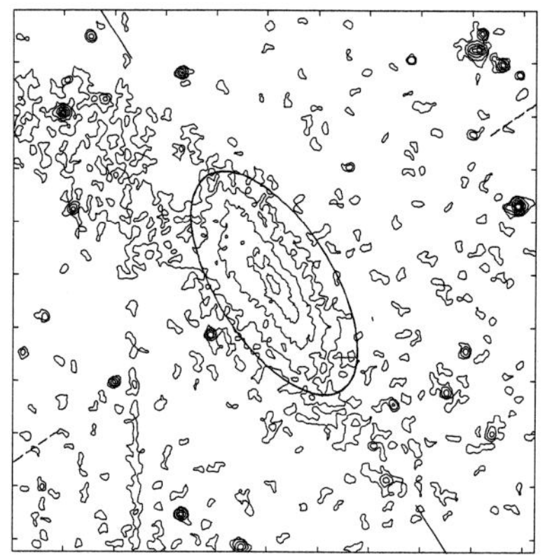

N4747 K3683 1 21.15 512X512

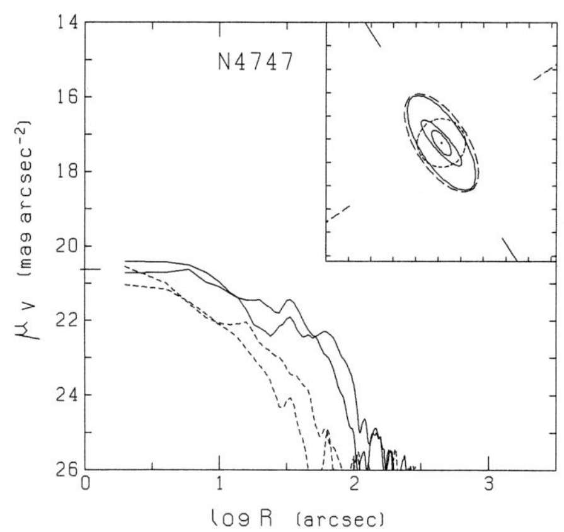

NGC4754

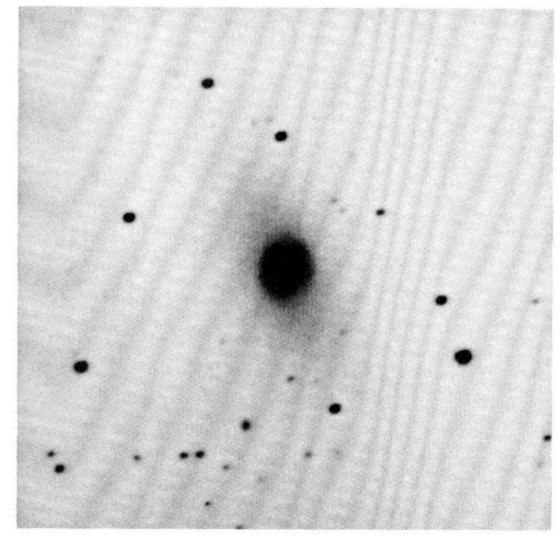

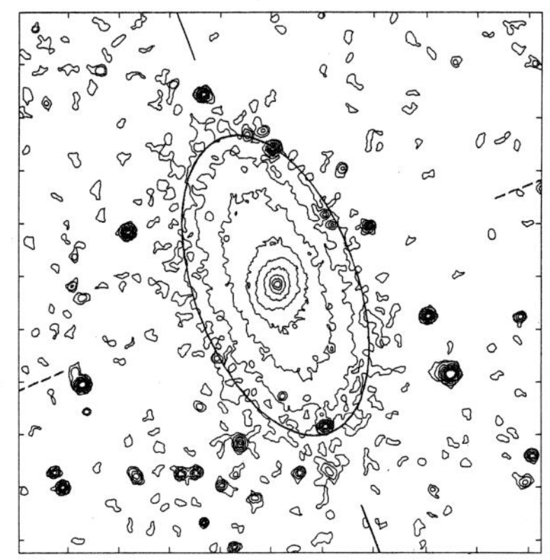

N4754 K2220 1 21.12 512X512

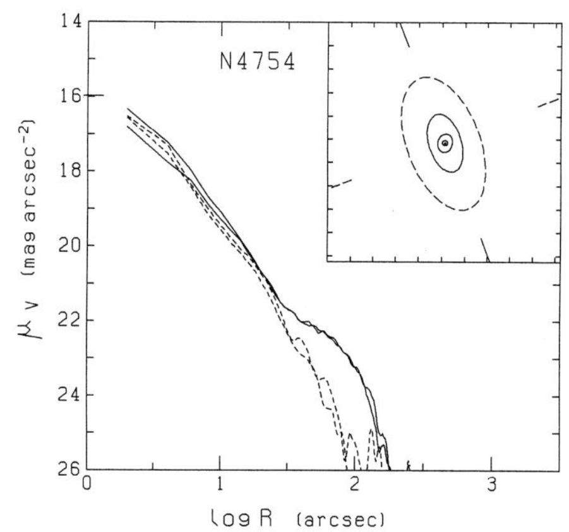

NGC4753

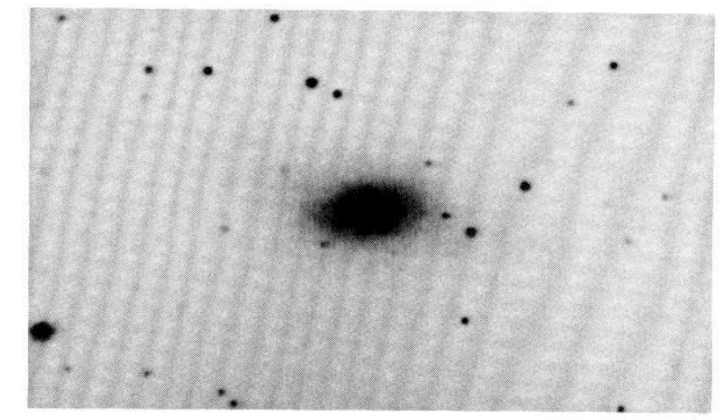

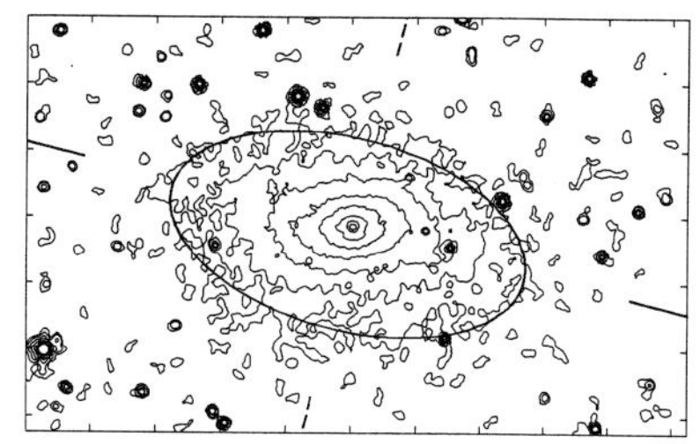

N4753 K4674 2 21.30 512X312

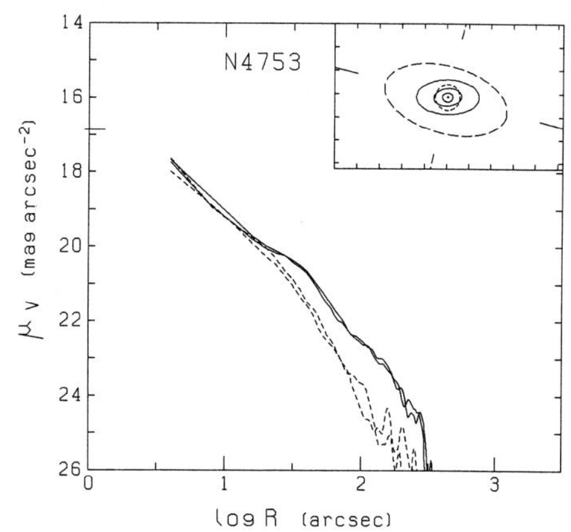

NGC4762

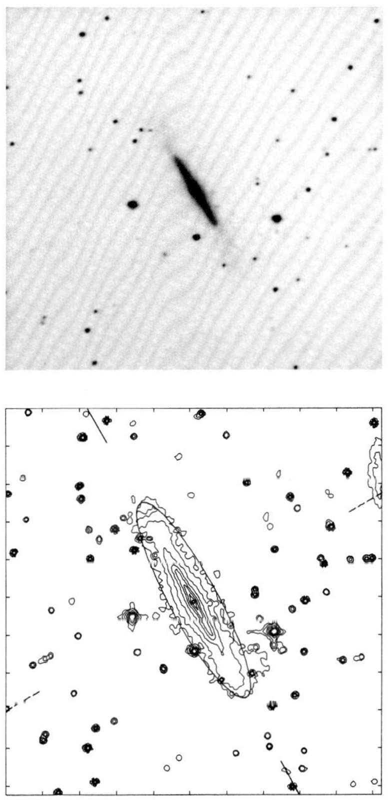

N4762 K2220 2 21.12 512X512

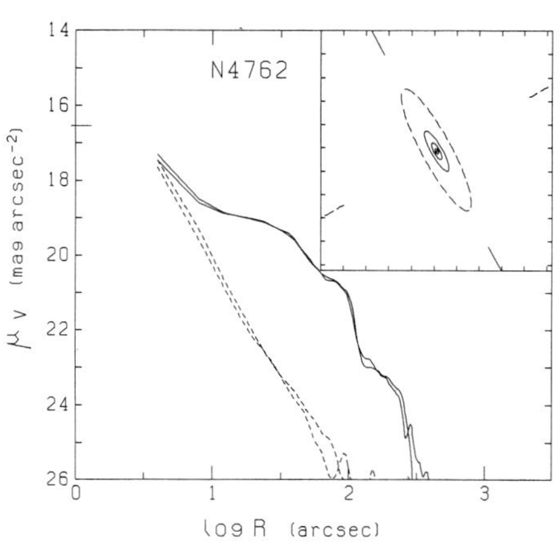

NGC4760

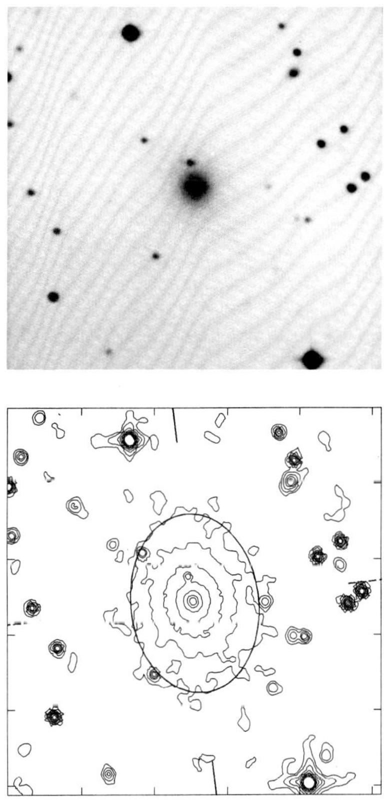

N4760 K4893 2 21.54 256X256

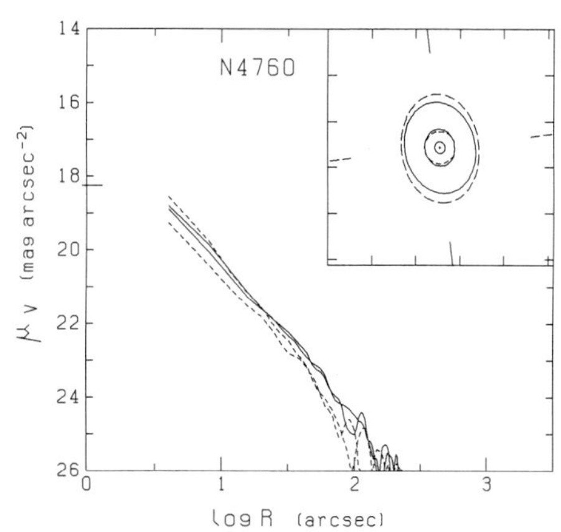

NGC4763

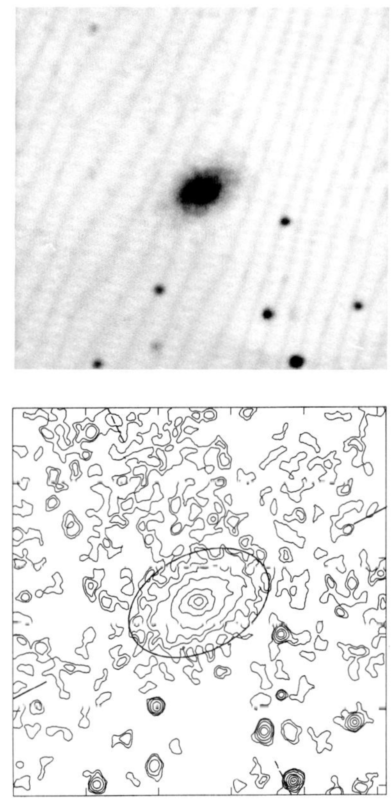

N4763 K3172 1 20.31 256X256

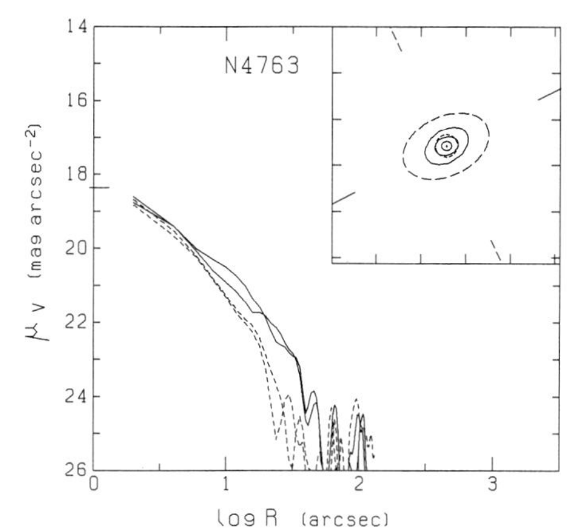

NGC4771

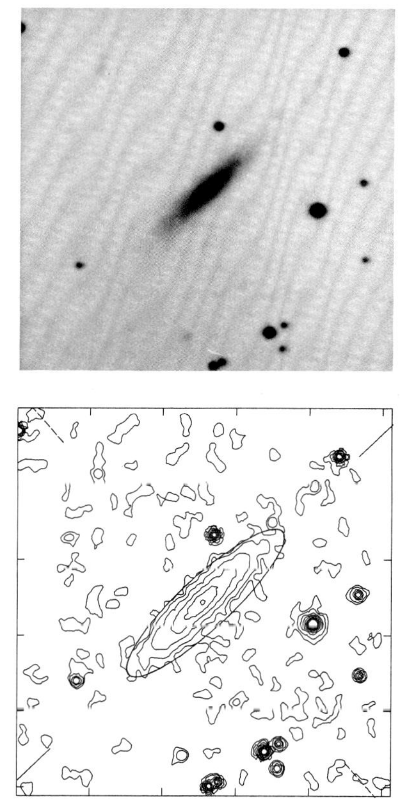

N4771 K4674 2 21.30 256X256

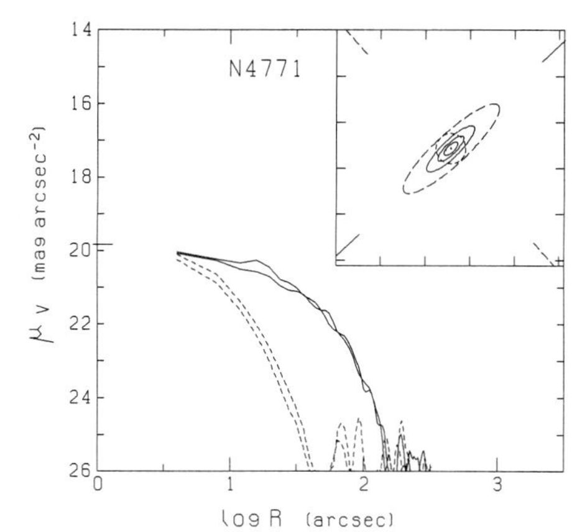

NGC4772

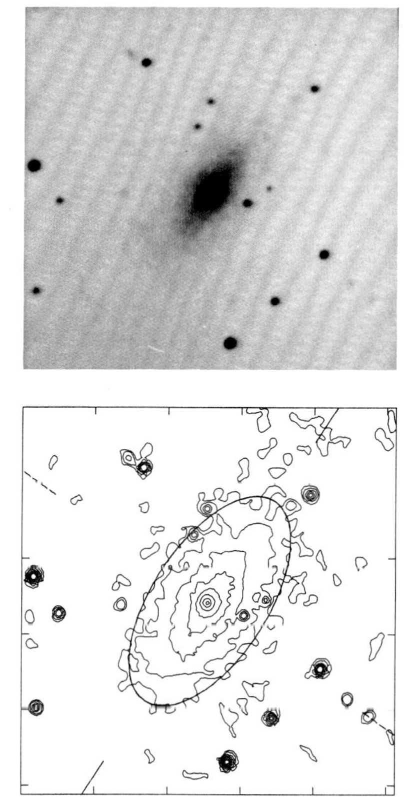

N4772 K5049 2 21.47 256X256

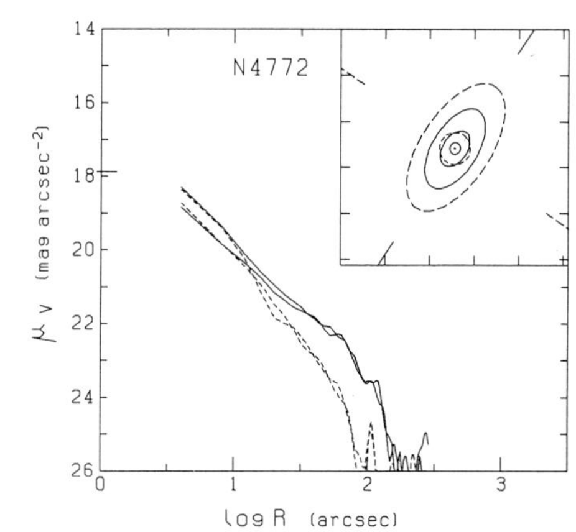

NGC4775

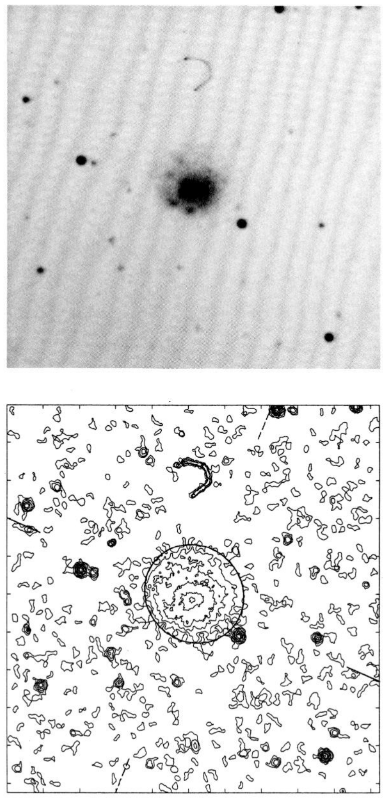

N4775 K4985 1 21.76 512X512

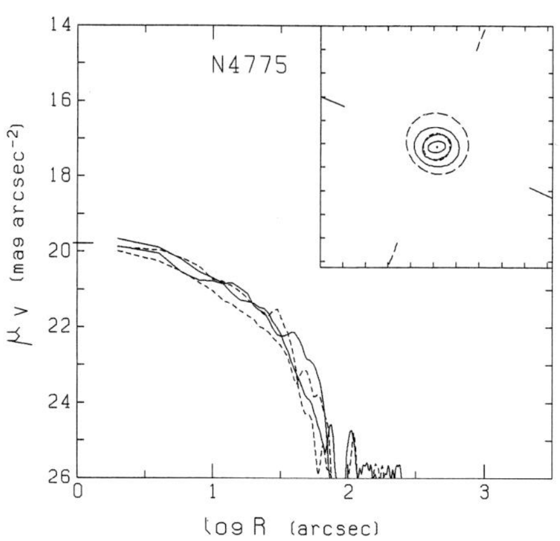

NGC4781

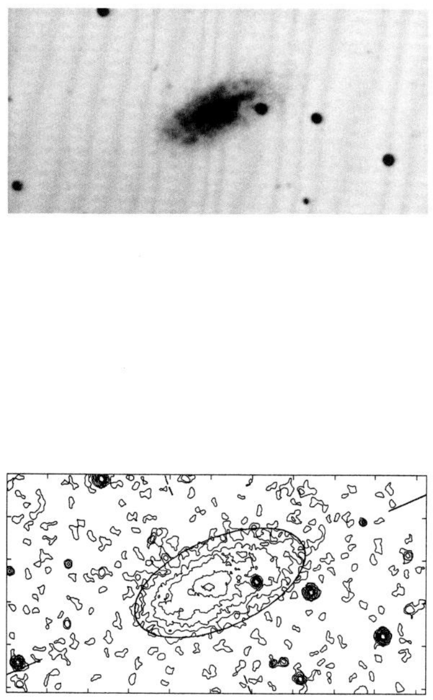

N4781 K4893 1 21.54 512X256

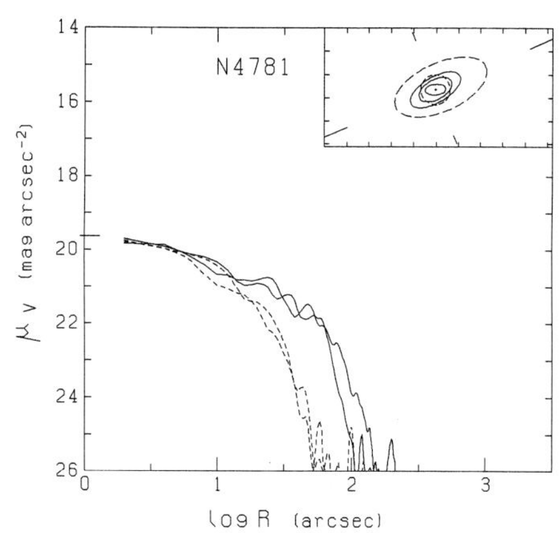

NGC4786

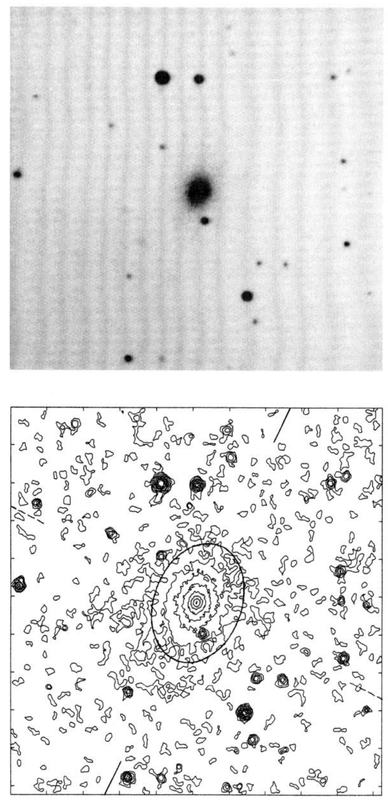

N4786 K4985 1 21.76 512X512

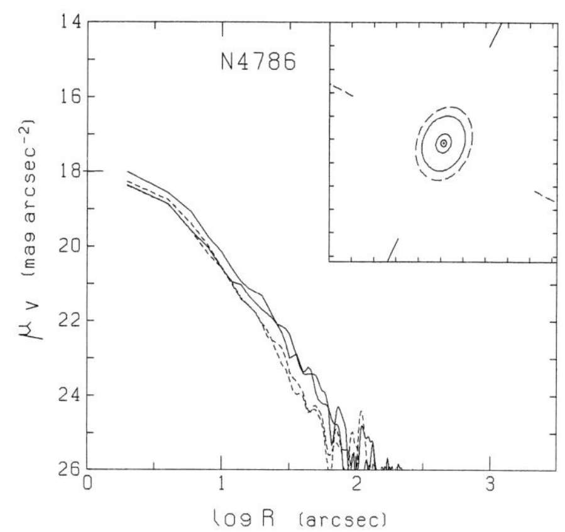

NGC4782

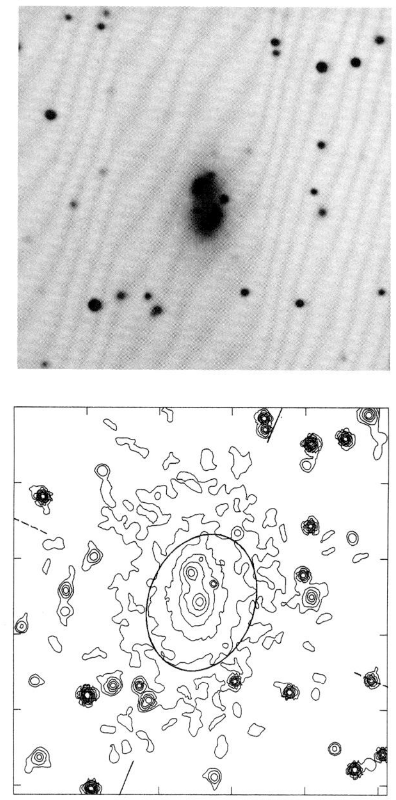

N4782 K4893 2 21.54 256X256

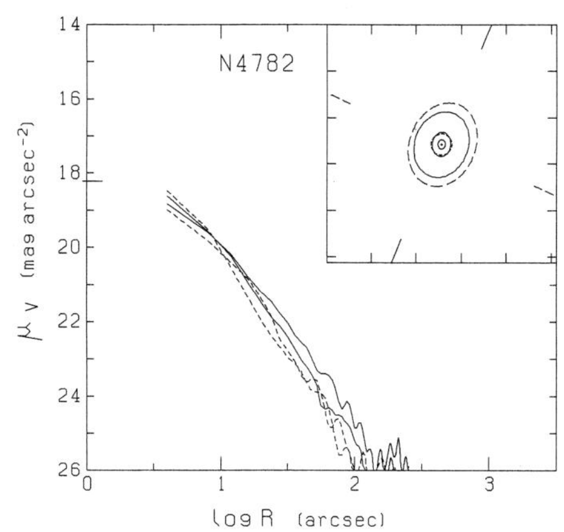

NGC4783

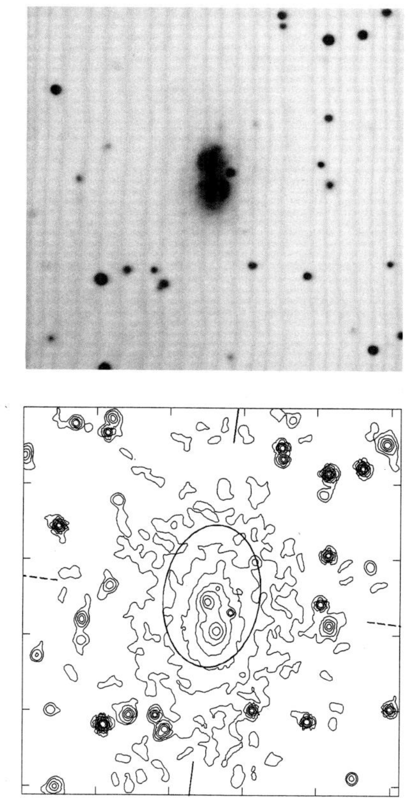

N4783 K4893 2 21.54 256X256

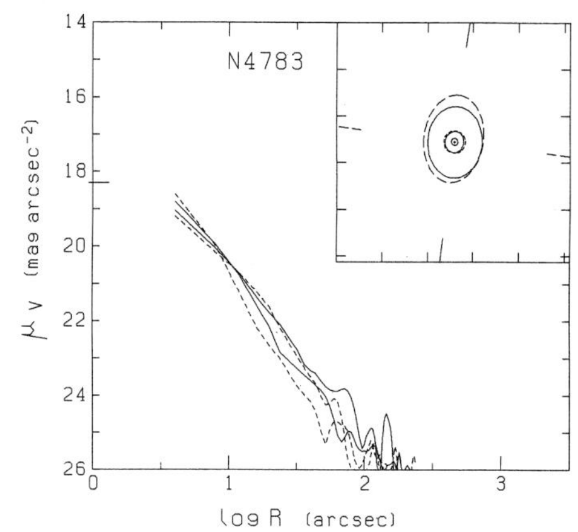

NGC4790

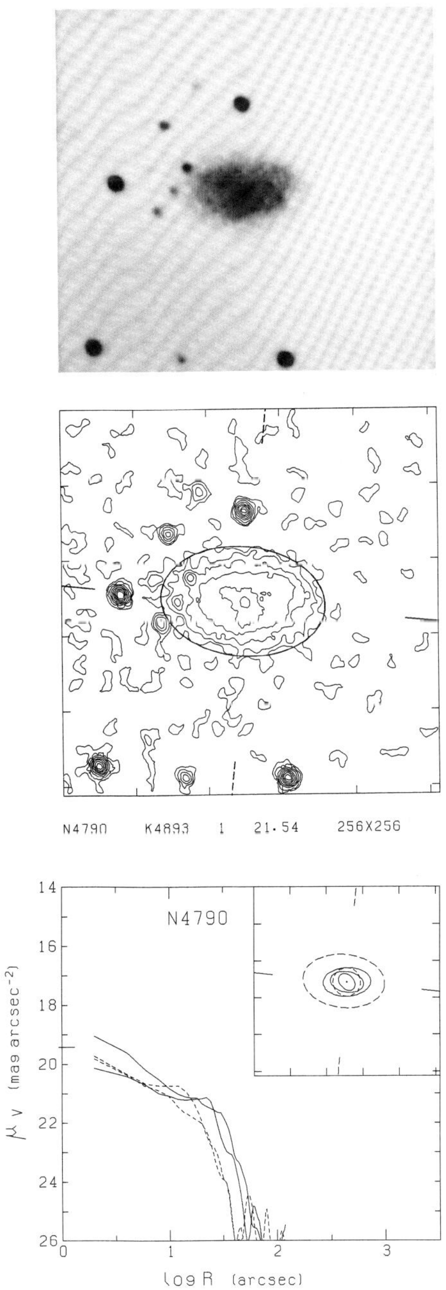

NGC4793

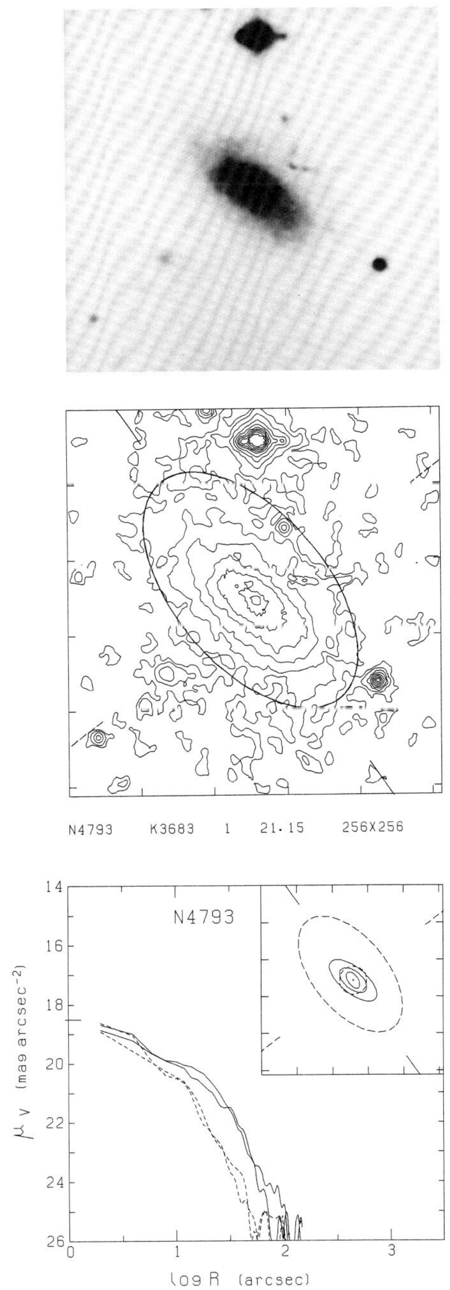

NGC4800

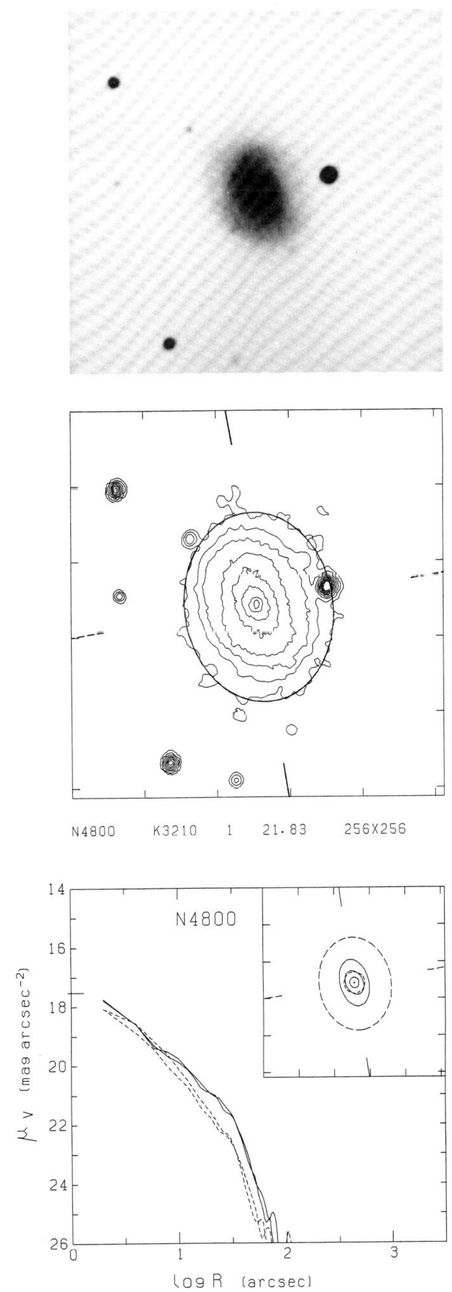

New 4

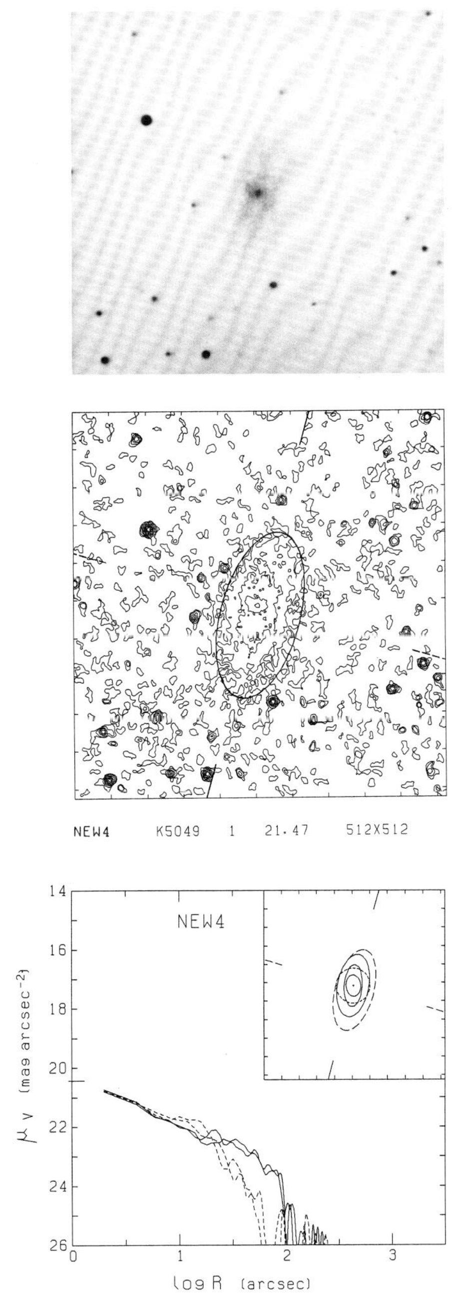

IC 3896

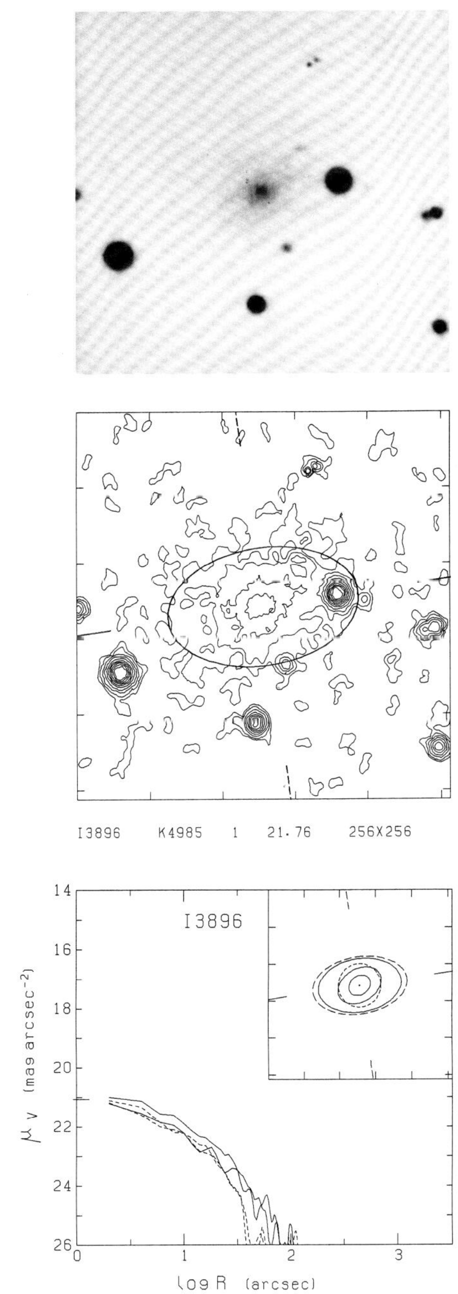

NGC4818

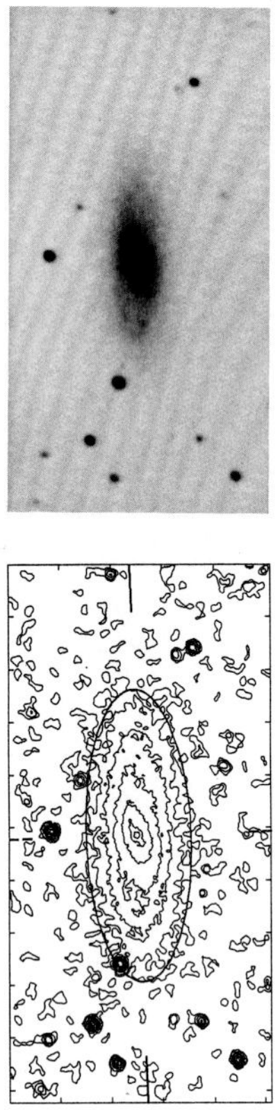

N4818 K4893 1 21.54 256X512

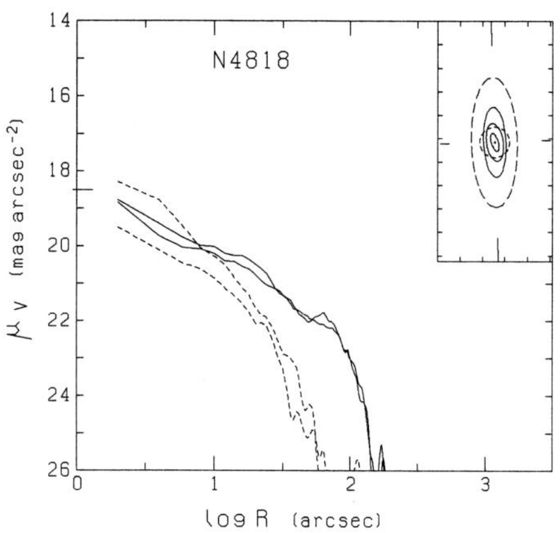

NGC4826

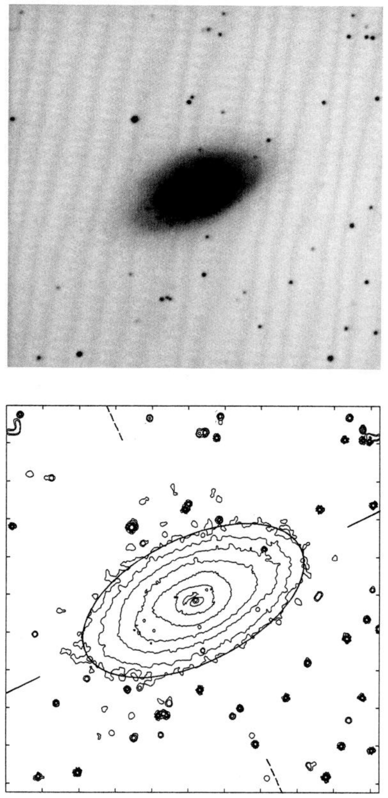

N4826 K2211 2 21.32 512X512

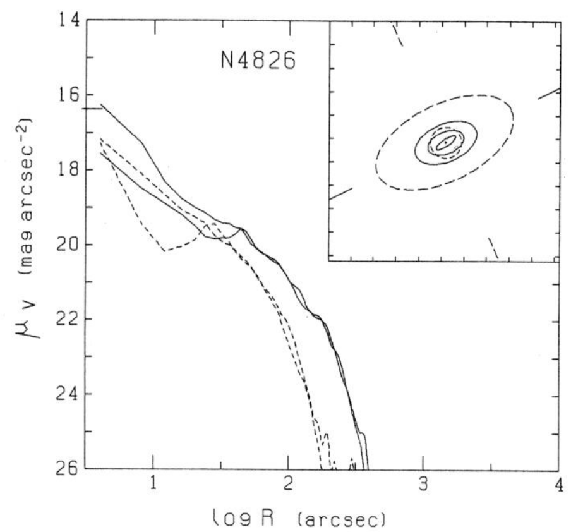

NGC4825

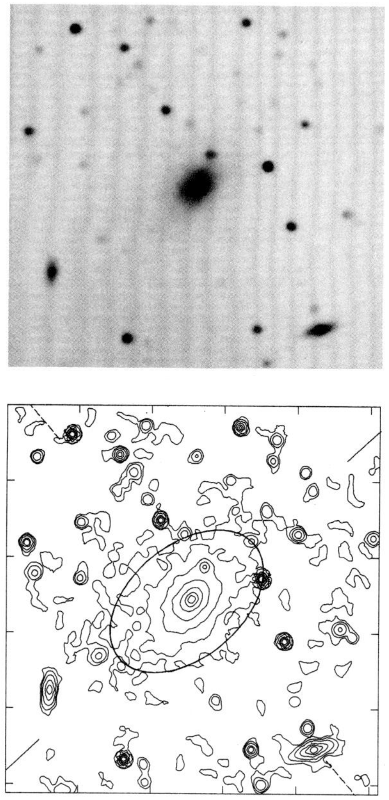

N4825 K3172 2 20.31 256X256

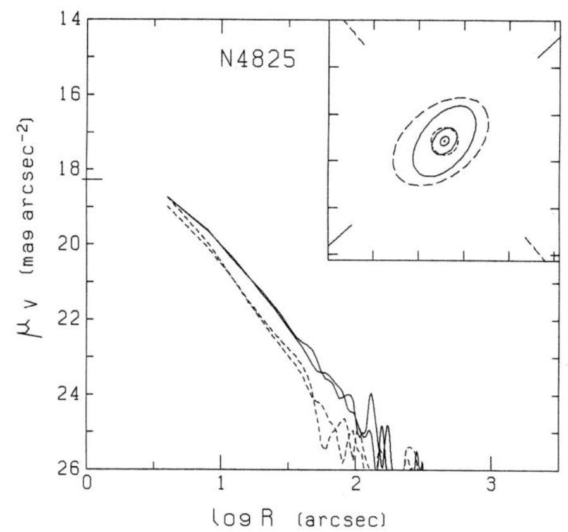

NGC4845

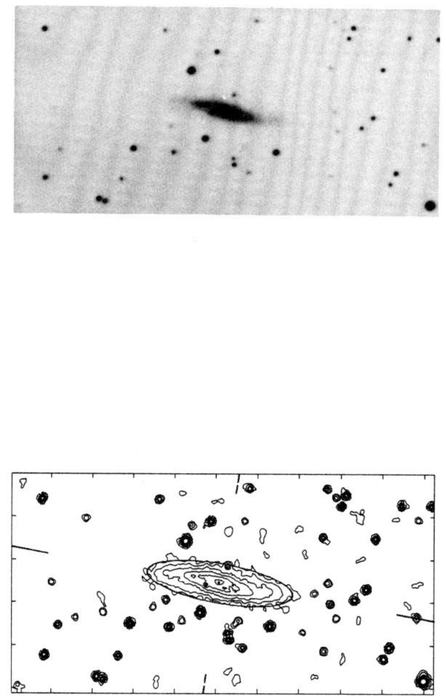

N4845 K5049 2 21.47 512X256

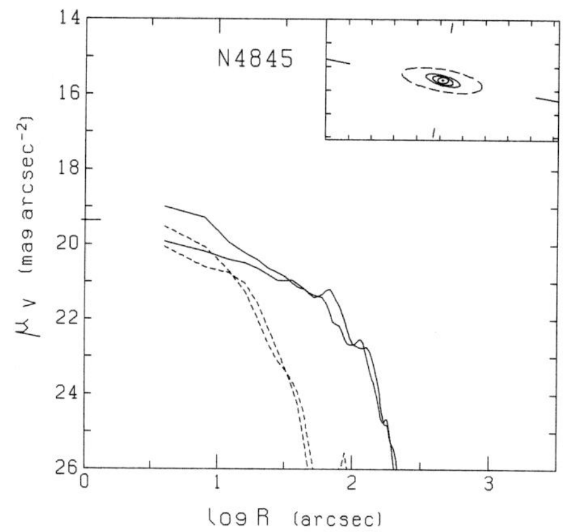

NGC4861

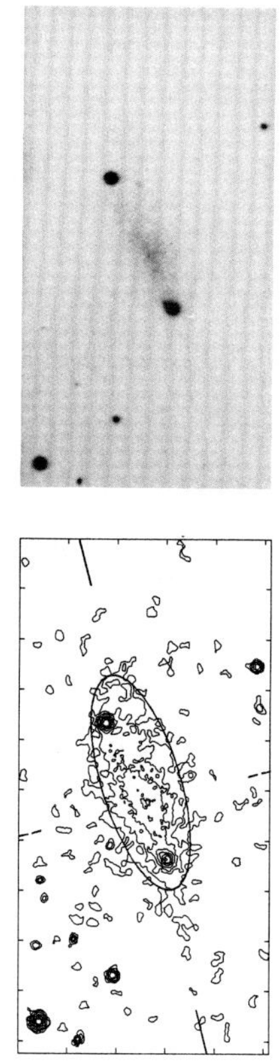

N4861 K4619 1 21.91 256X512

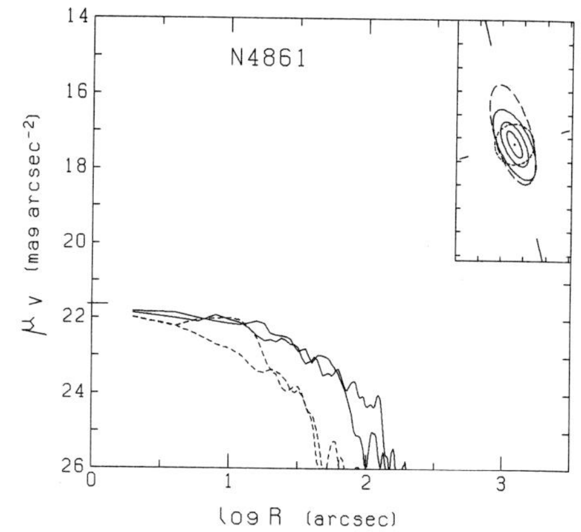

NGC4868

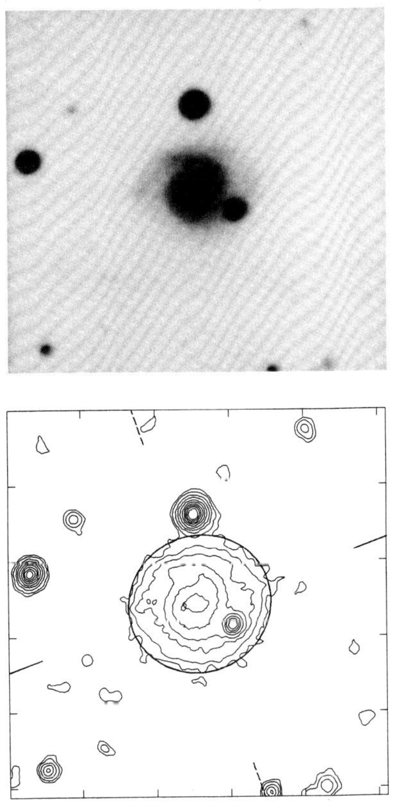

N4868 K2647 1 21.51 256X256

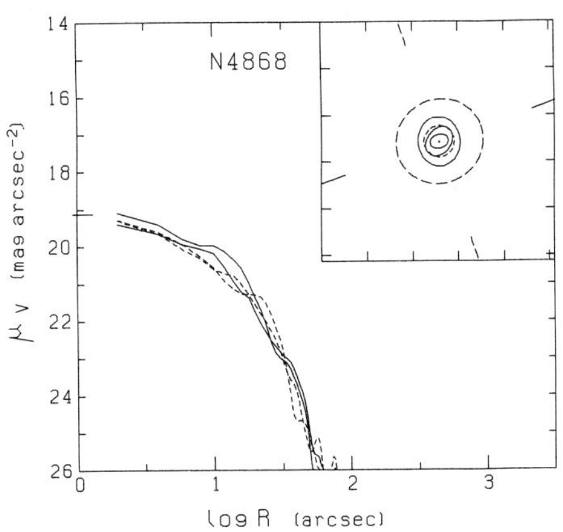

NGC4889

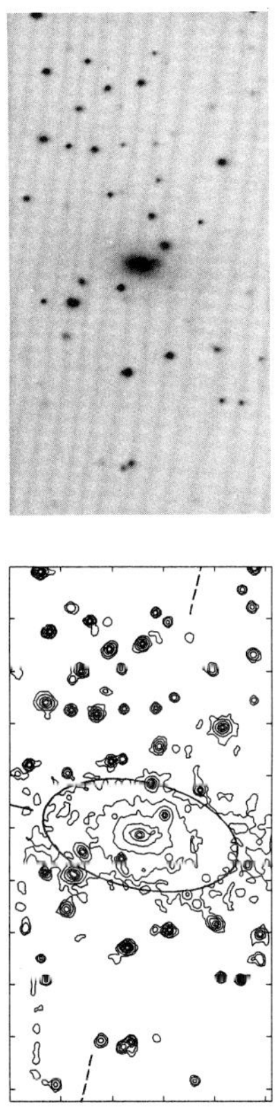

N4889 K3683 2 21.15 256X512

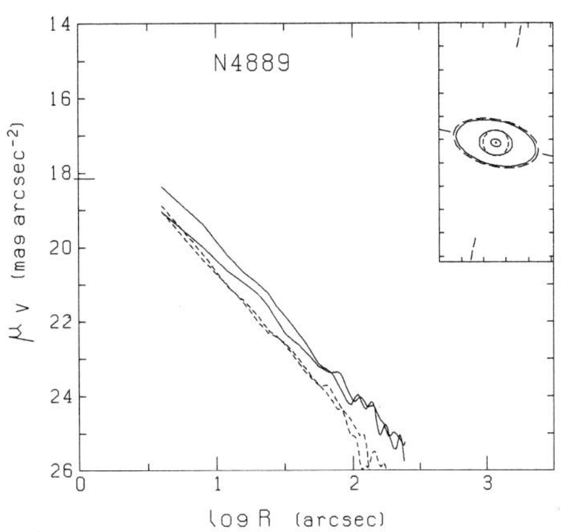

NGC4900

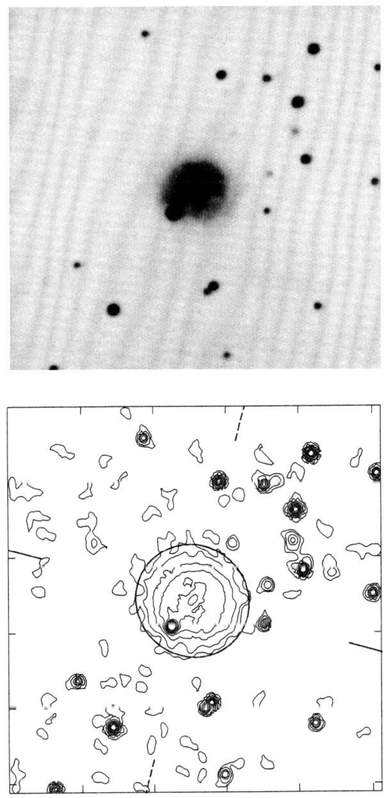

N4900 K5049 2 21.47 256X256

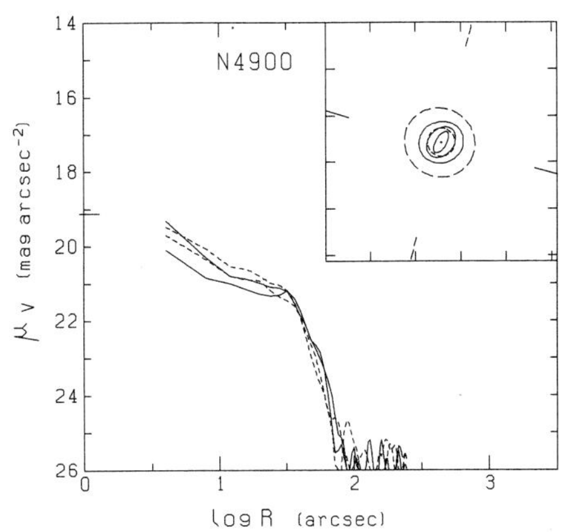

NGC4891

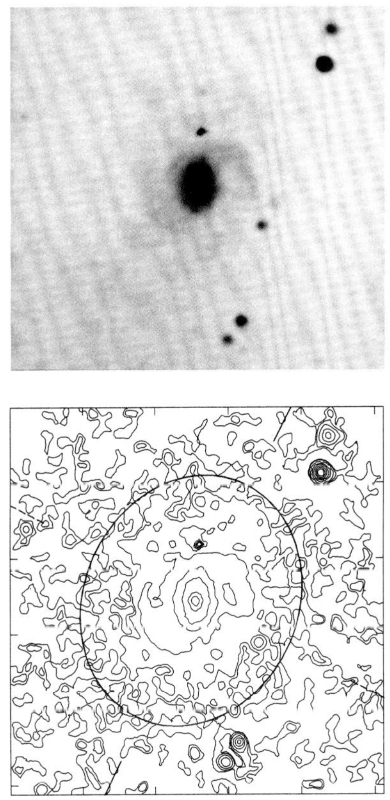

N4891 K3172 1 20.31 256X256

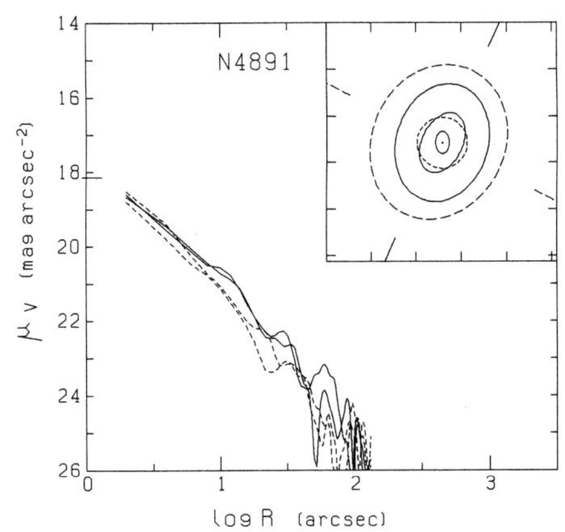

NGC4899

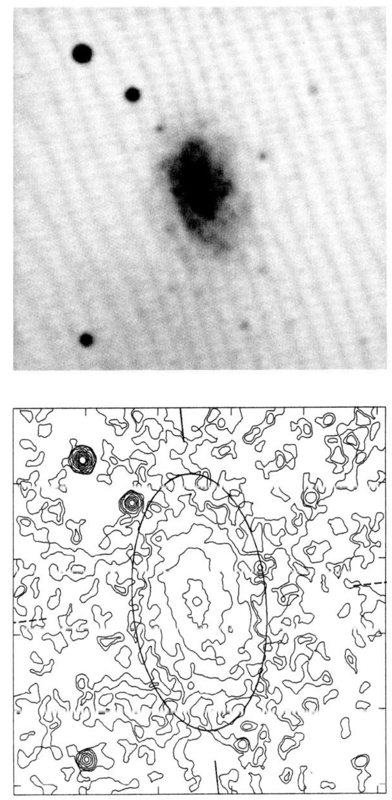

N4899 K3172 1 20.31 256X256

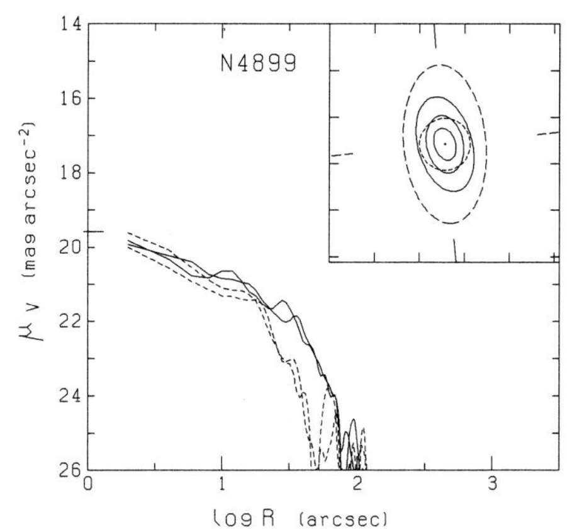

NGC4902

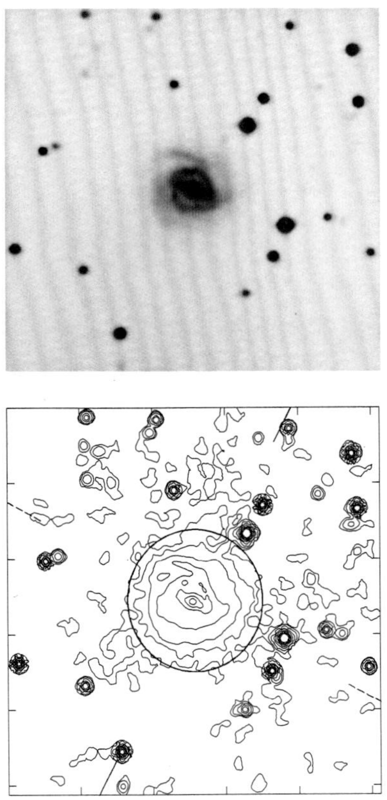

N4902 K3172 2 20.31 256X256

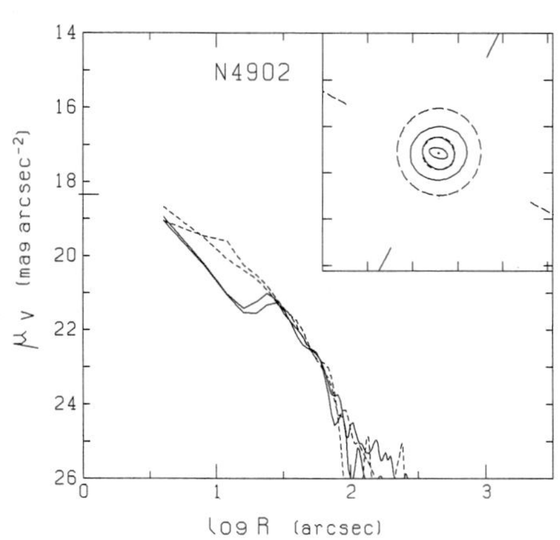

NGC4914

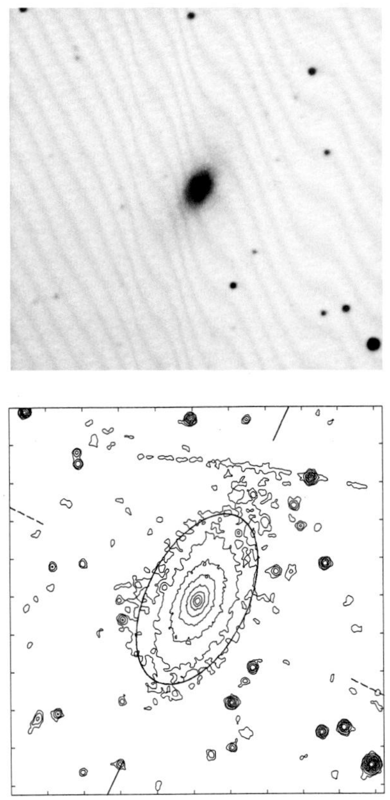

N4914 K2647 1 21.51 512X512

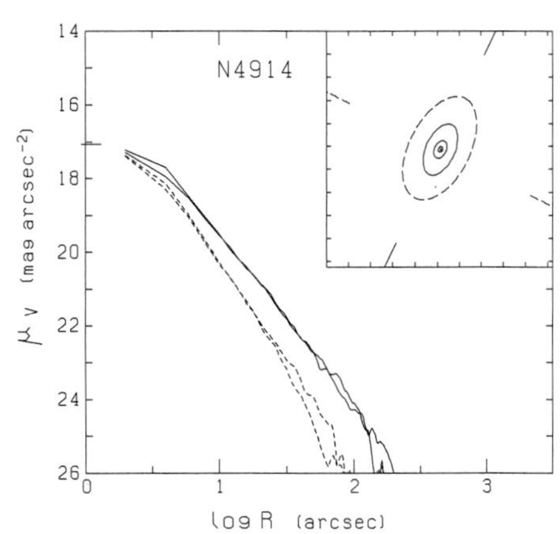

NGC4904

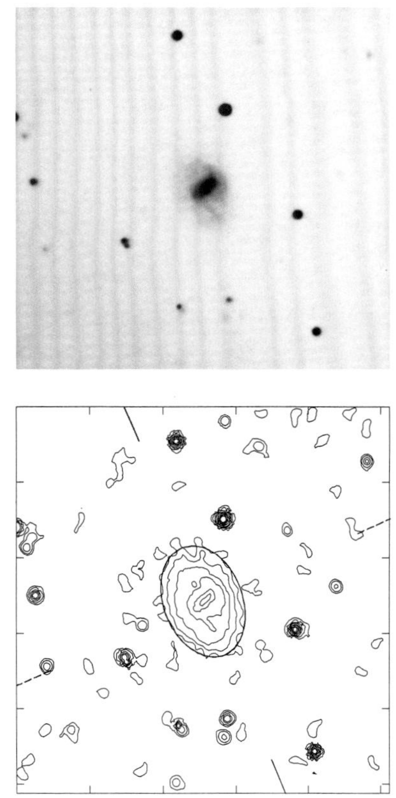

N4904 K5049 2 21.47 256X256

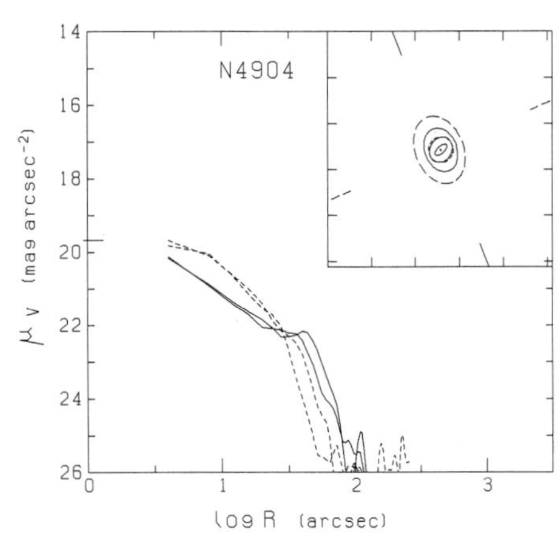

NGC4915

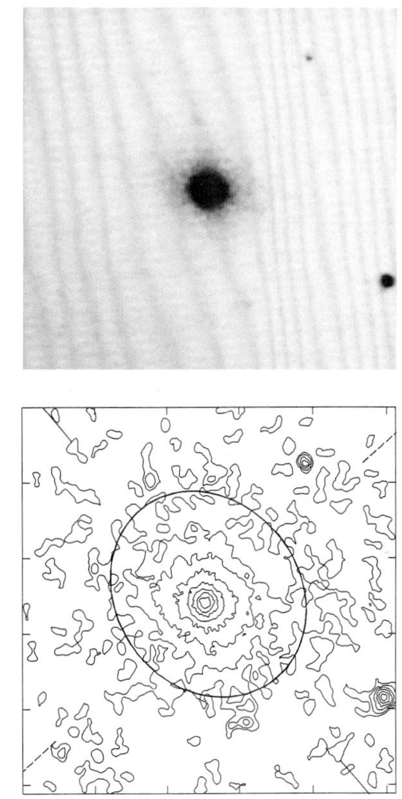

N4915 K4985 1 21.76 256X256

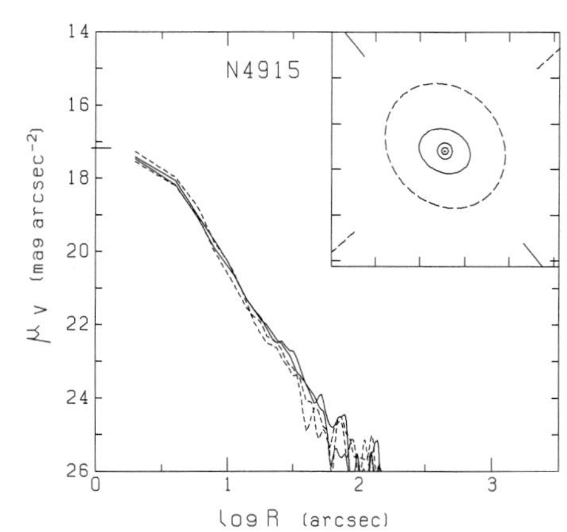

NGC4928

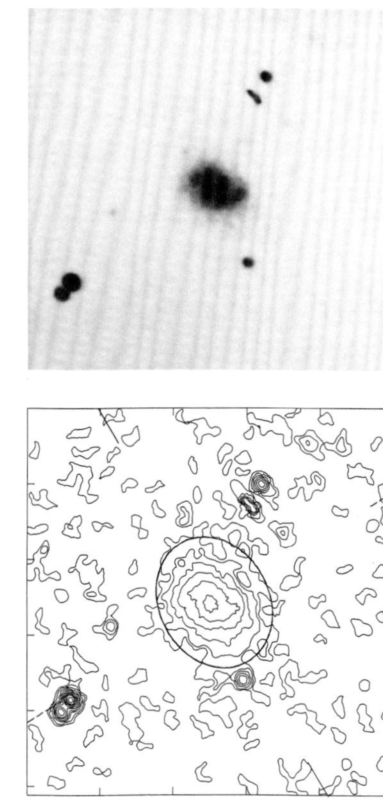

N4928 K4893 1 21.54 256X256

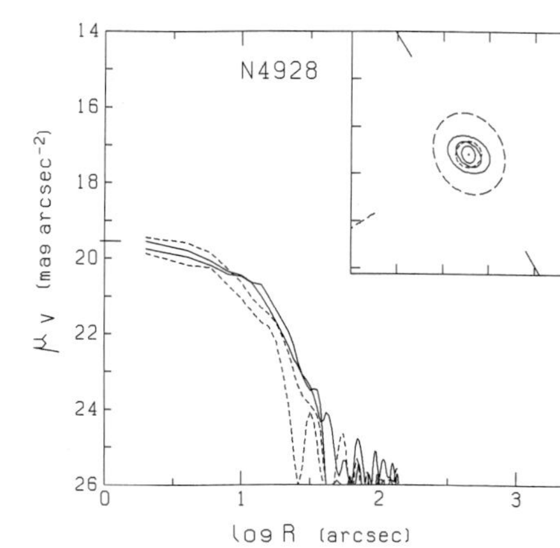

NGC4933

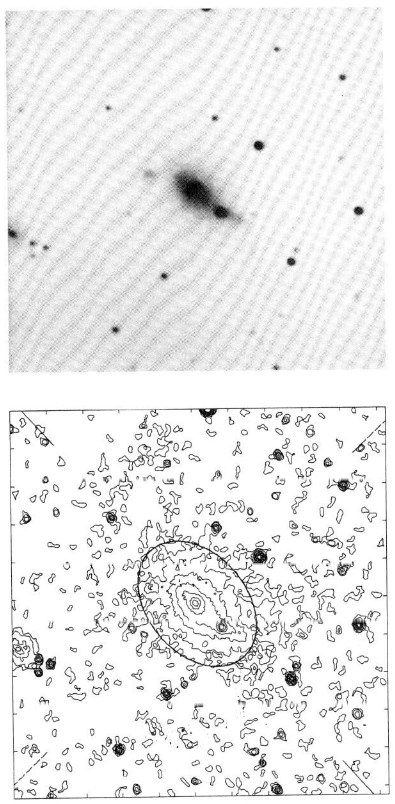

N4933 K4893 1 21.54 512X512

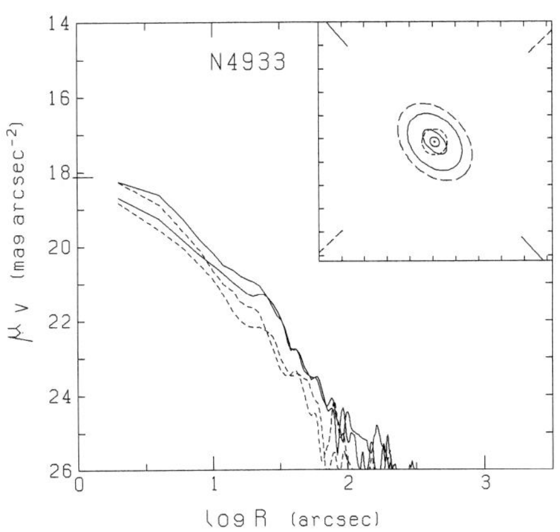

NGC4941

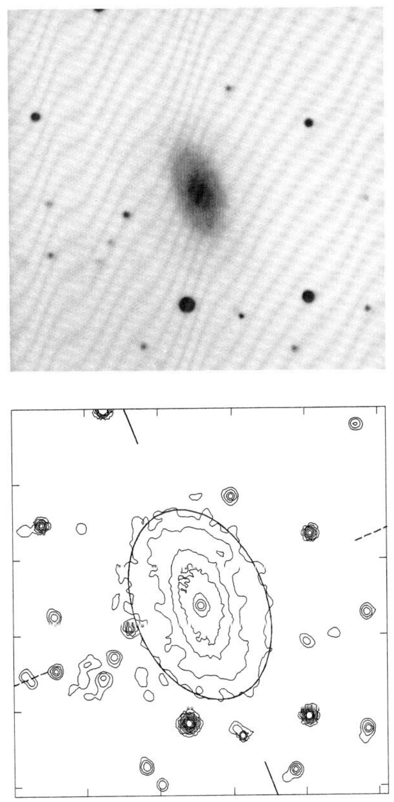

N4941 K4985 2 21.76 256X256

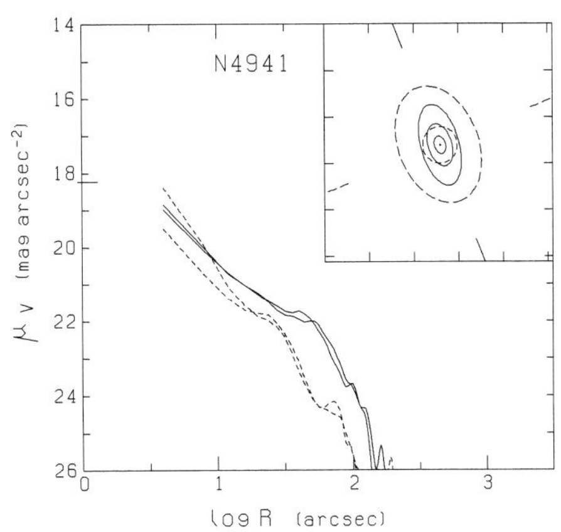

NGC4939

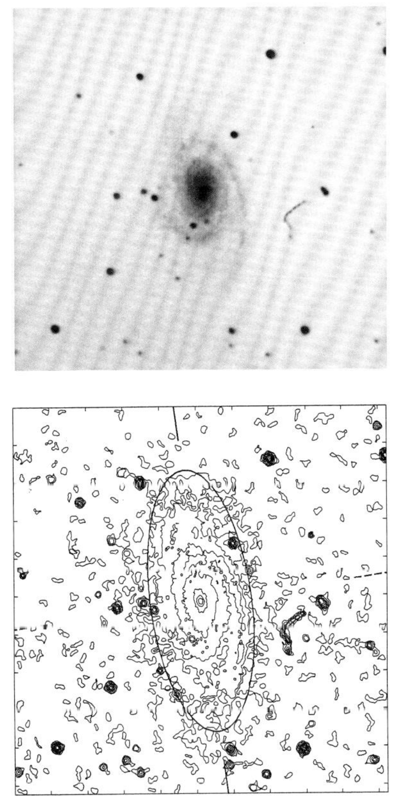

N4939 K4893 1 21.54 512X512

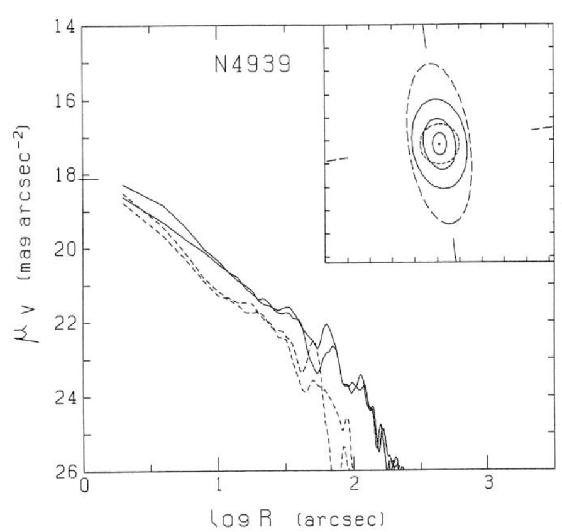

NGC4951

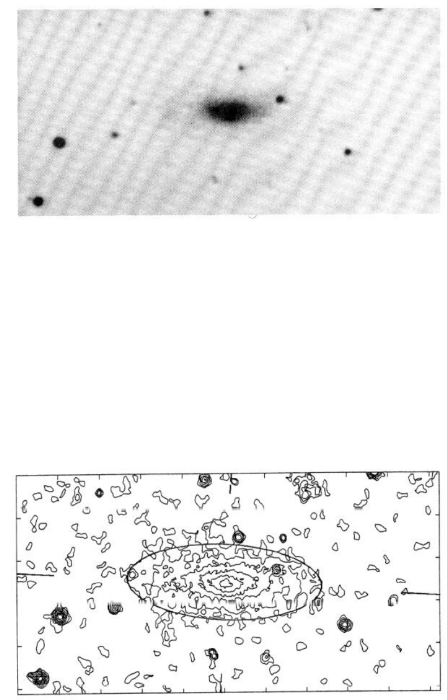

N4951 K4985 1 21.76 512X256

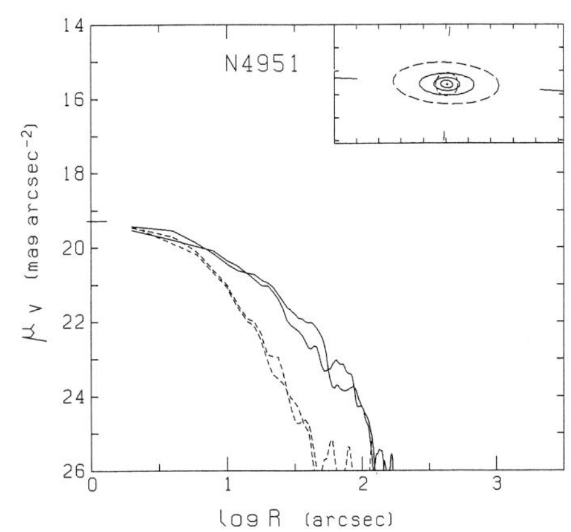

NGC4958

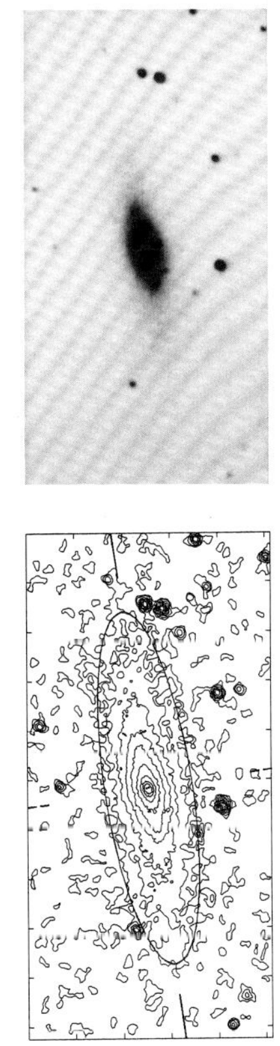

N4958 K4893 1 21.54 256X512

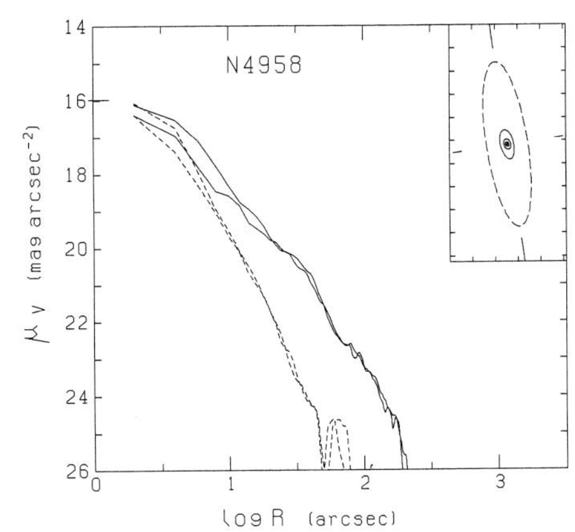

NGC4961

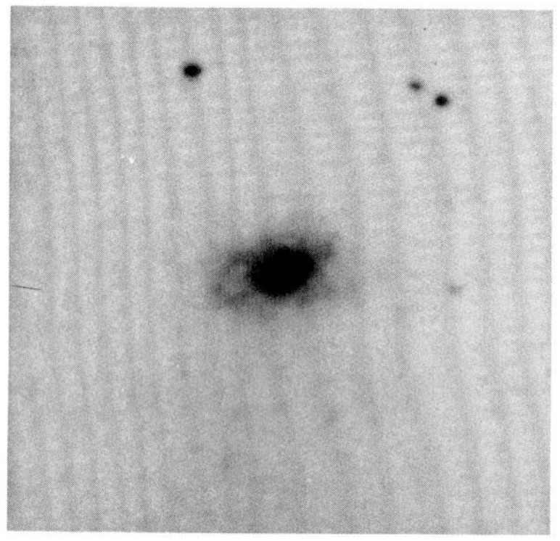

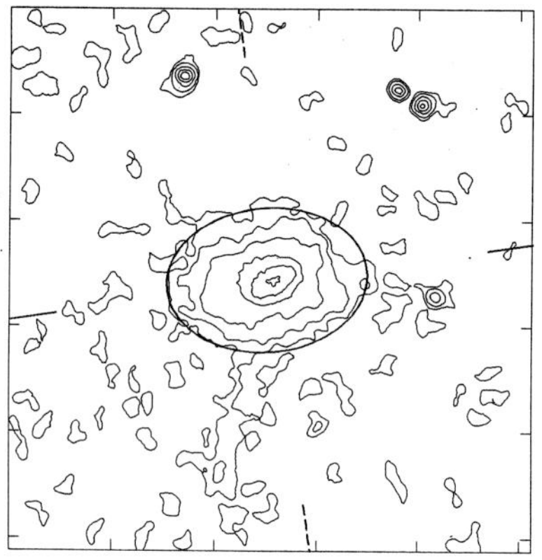

N4961 K3683 1 21.15 256X256

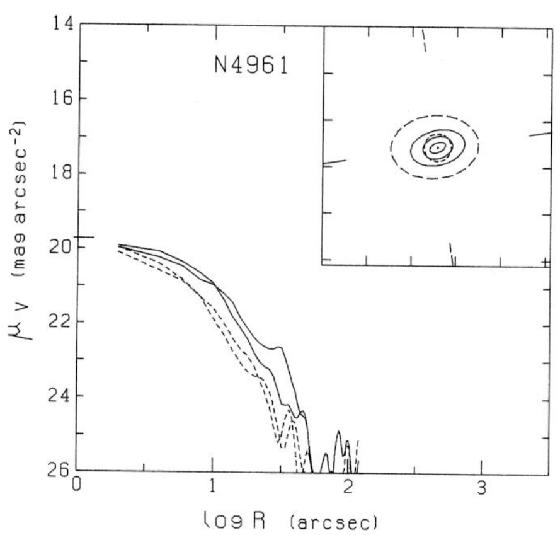

NGC4981

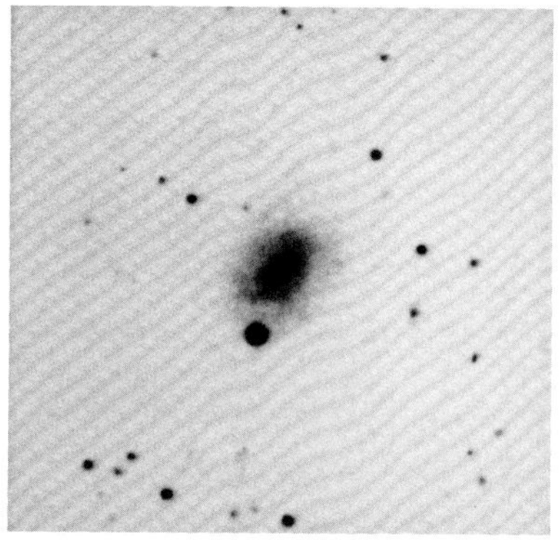

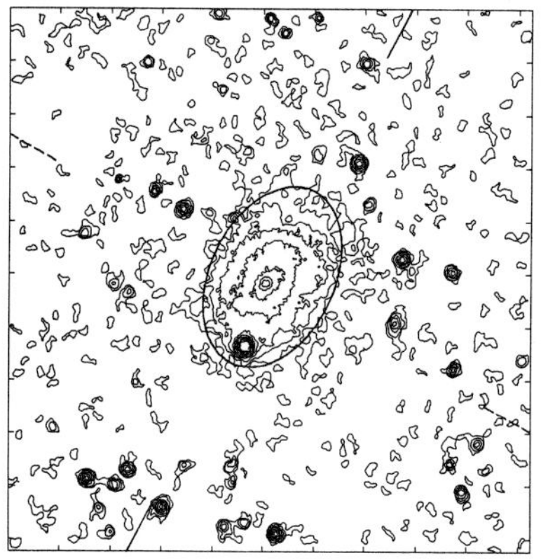

N4981 K4985 1 21.76 512X512

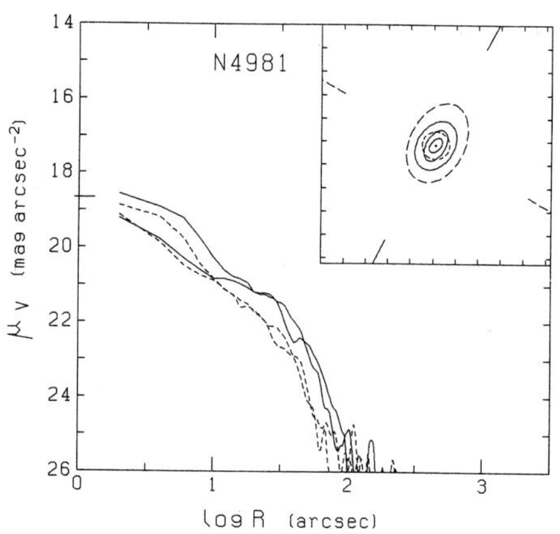

NGC4984

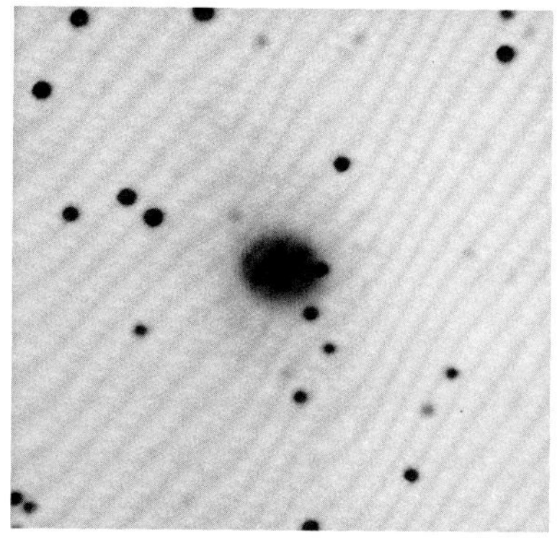

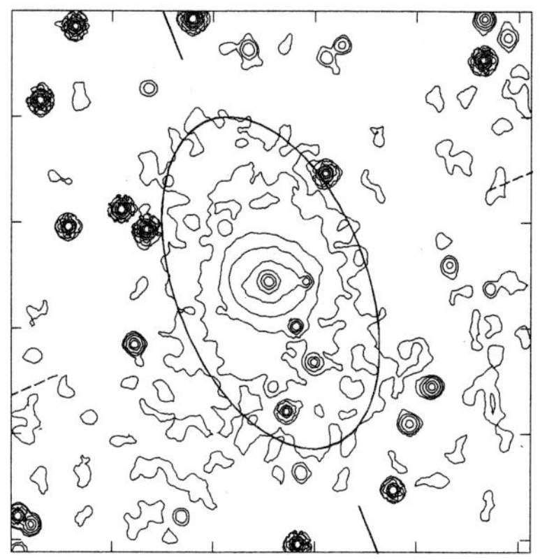

N4984 K3172 2 20.31 256X256

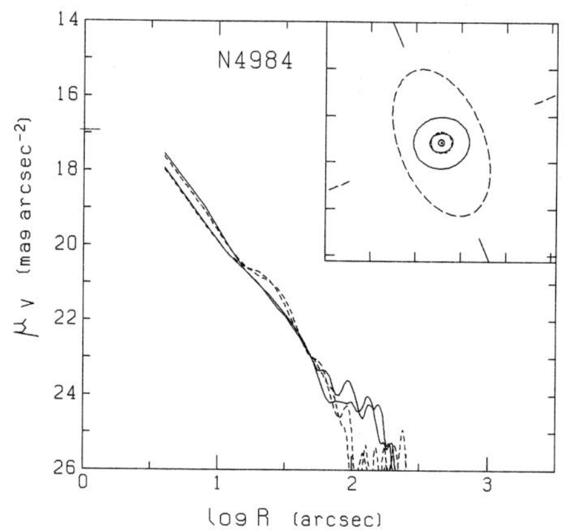

NGC4999

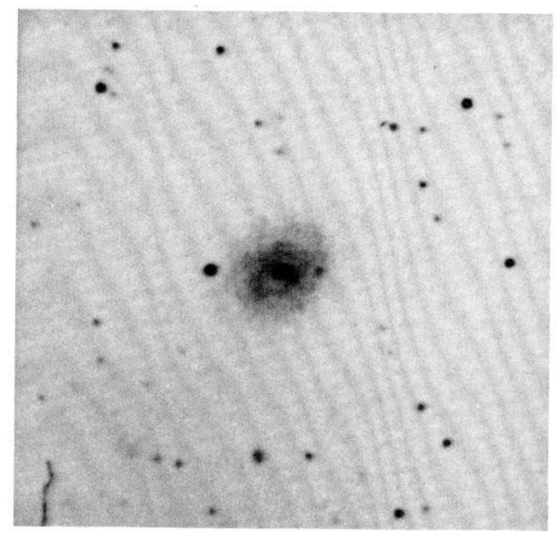

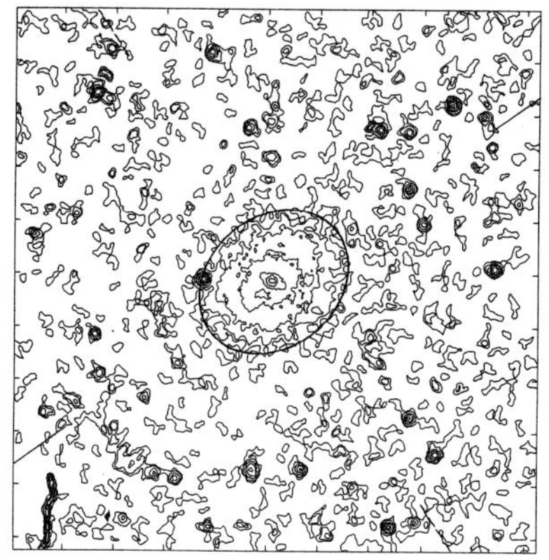

N4999 K5049 1 21.47 512X512

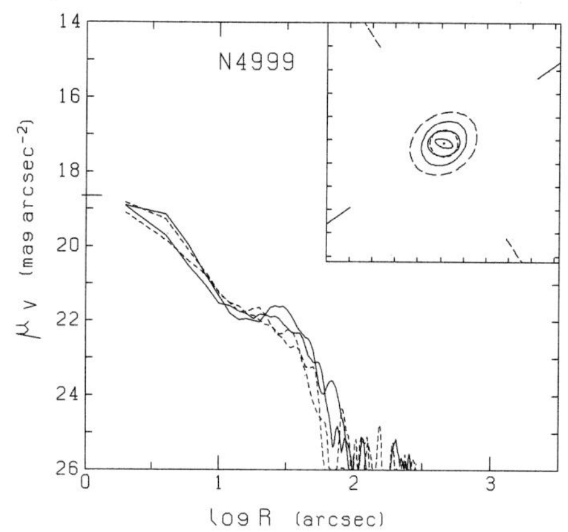

NGC4995

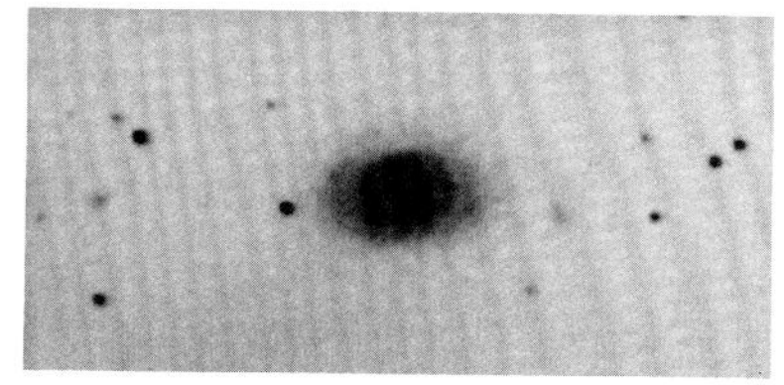

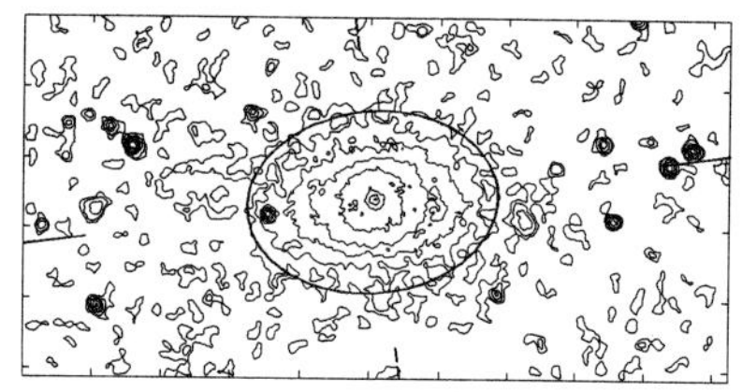

N4995 K4893 1 21.54 512X256

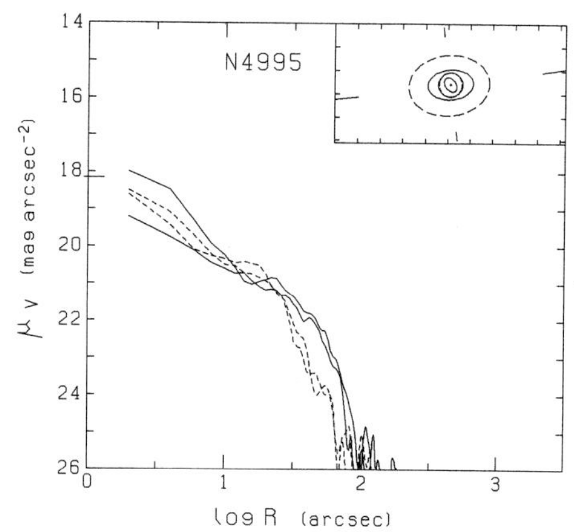

NGC5005

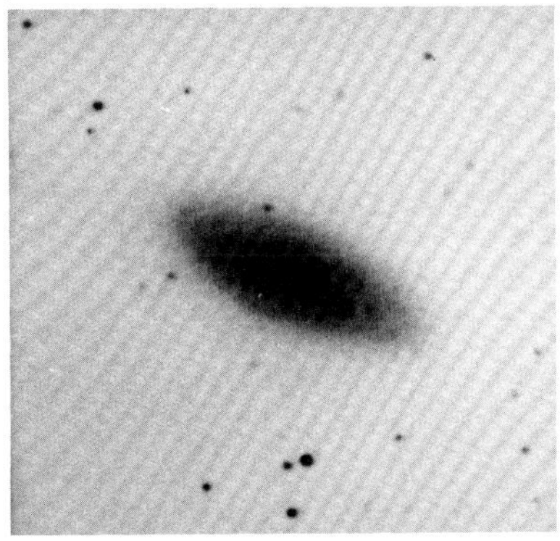

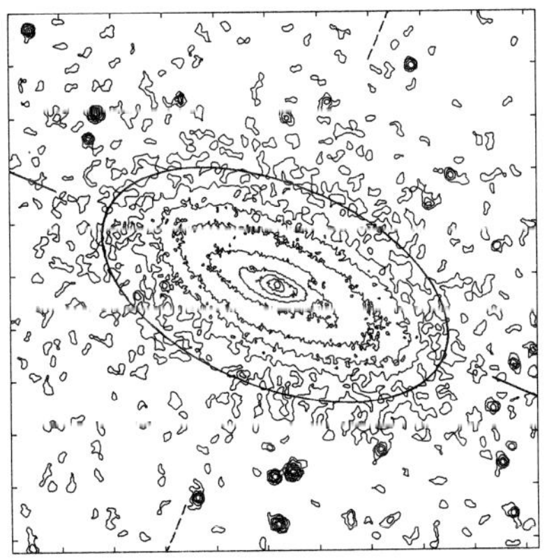

N5005 K4619 1 21.91 512X512

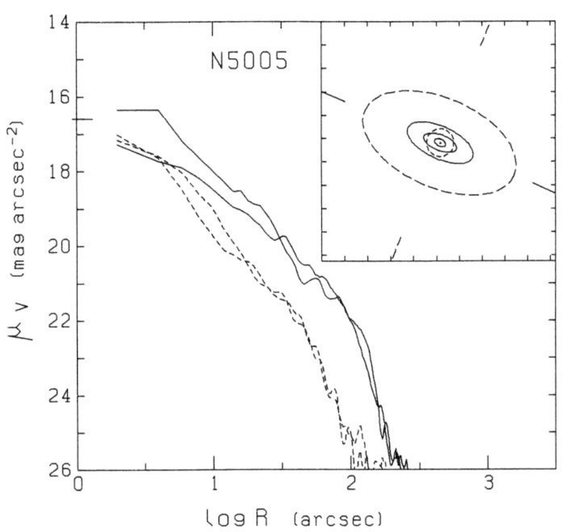

NGC5017

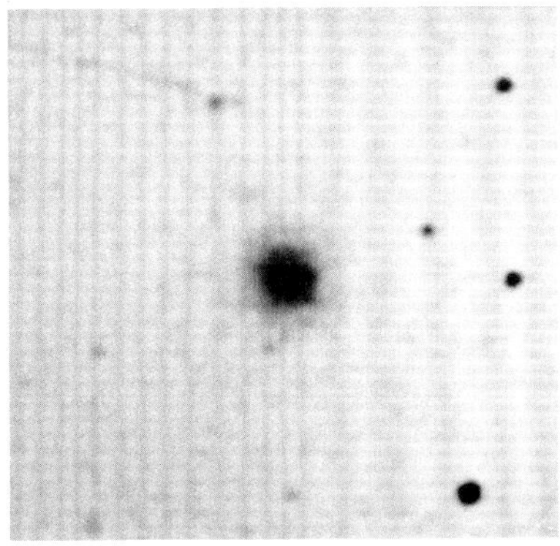

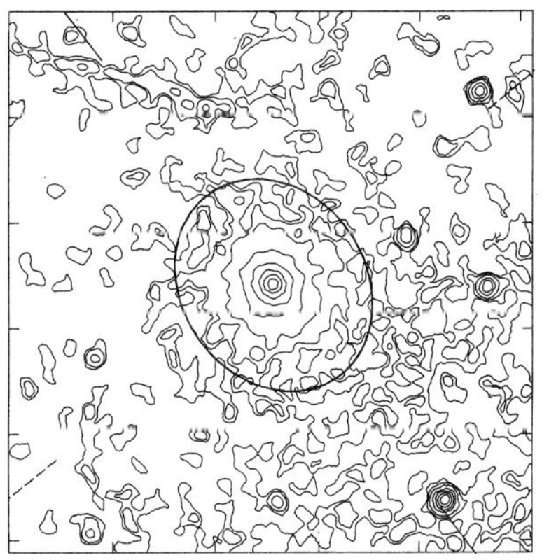

N5017 K3172 1 20.31 256X256

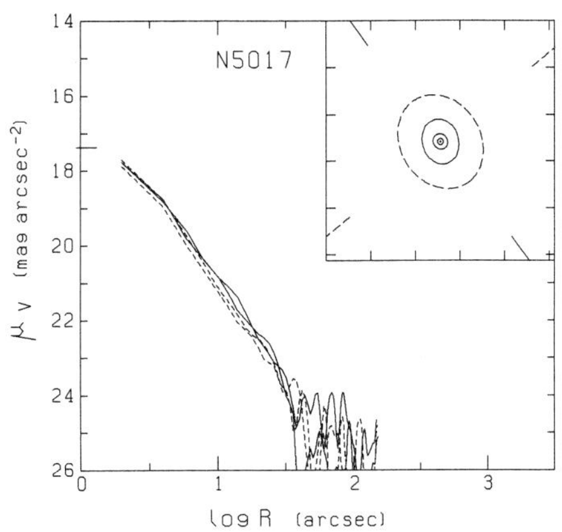

NGC5018

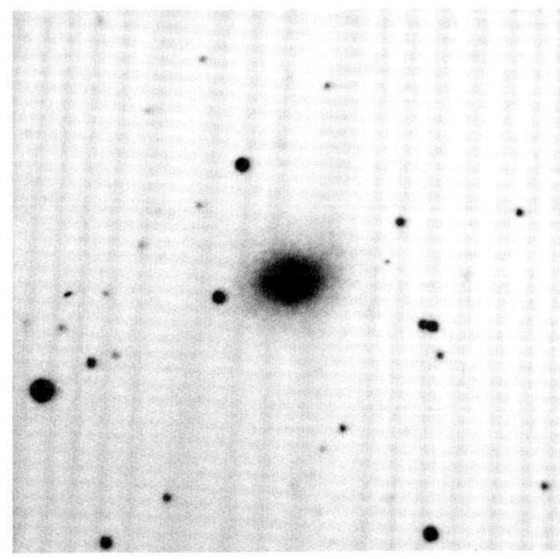

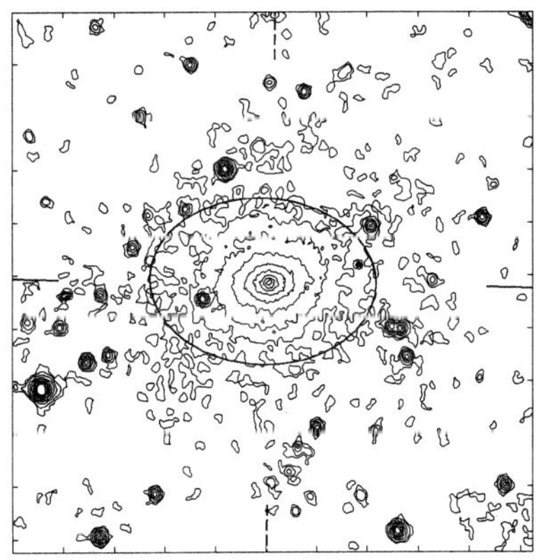

N5018 K5374 1 21.42 512X512

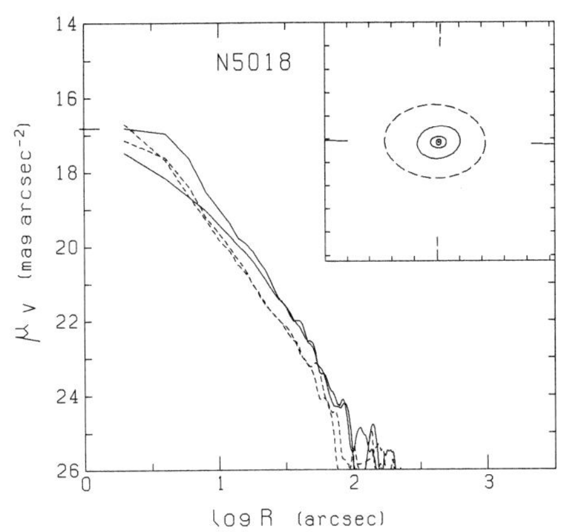

NGC5033

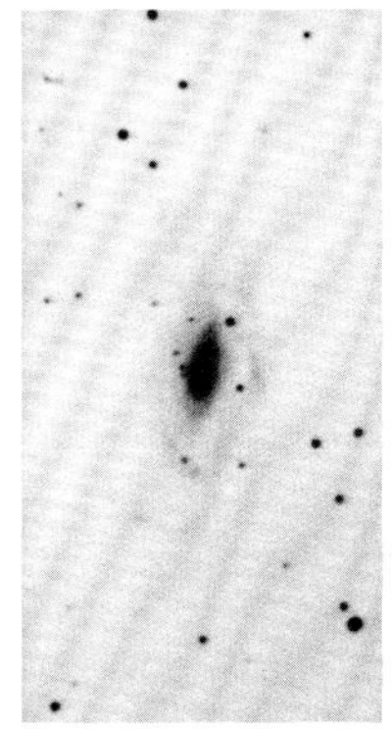

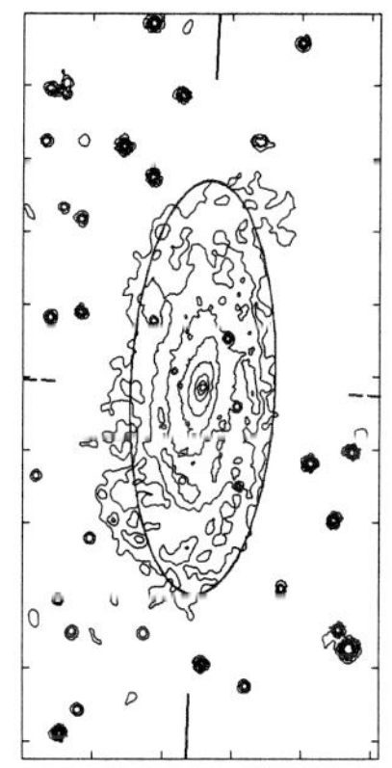

N5033 K4619 2 21.91 256X512

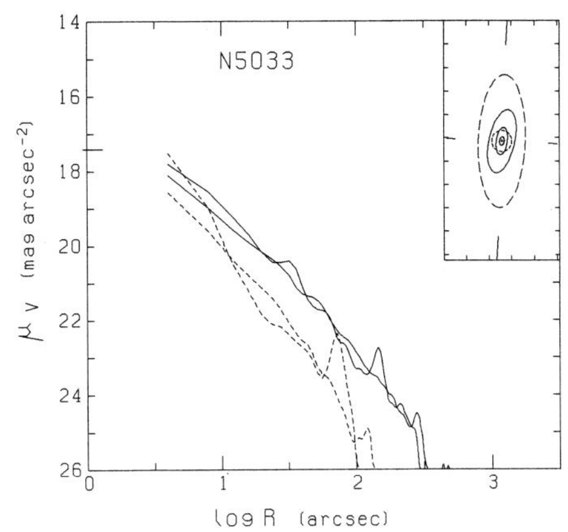

NGC5055

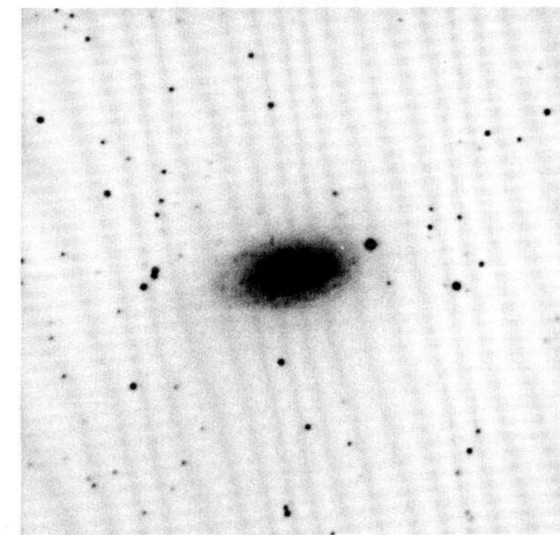

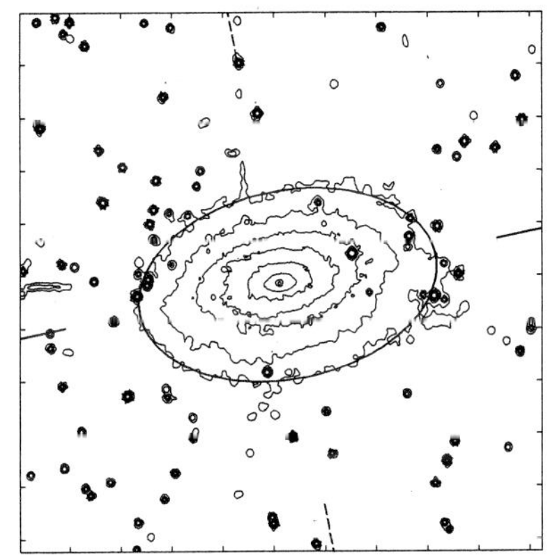

N5055 K5372 3 22.12 512X512

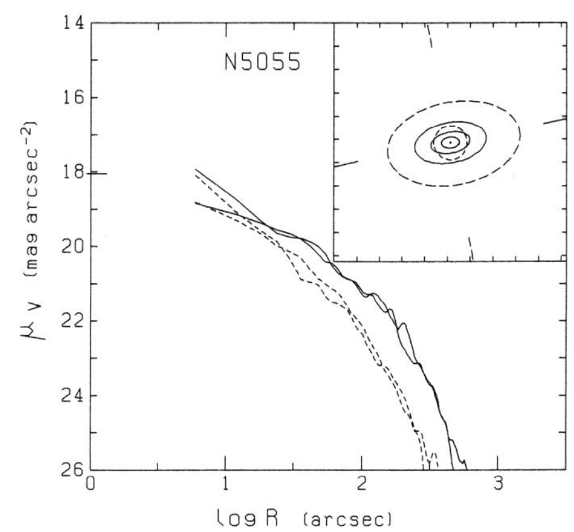

NGC5074

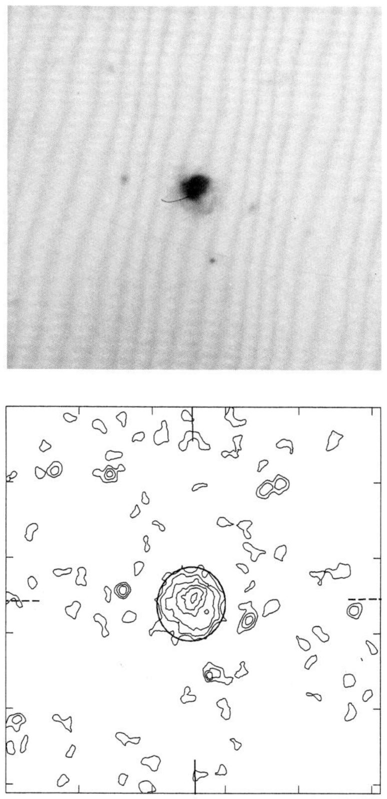

N5074 K5074 1 21.96 256X256

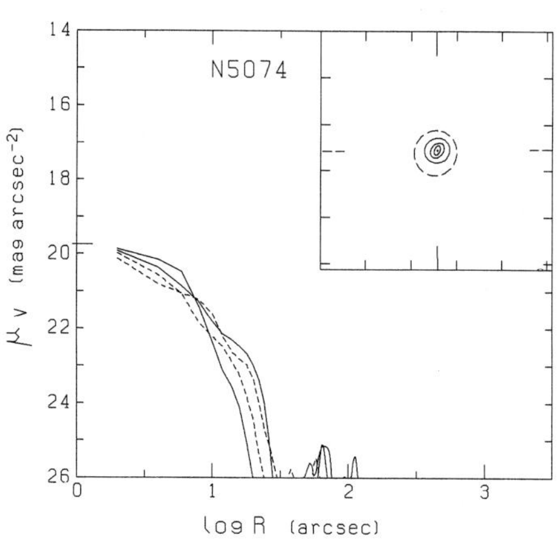

NGC5068

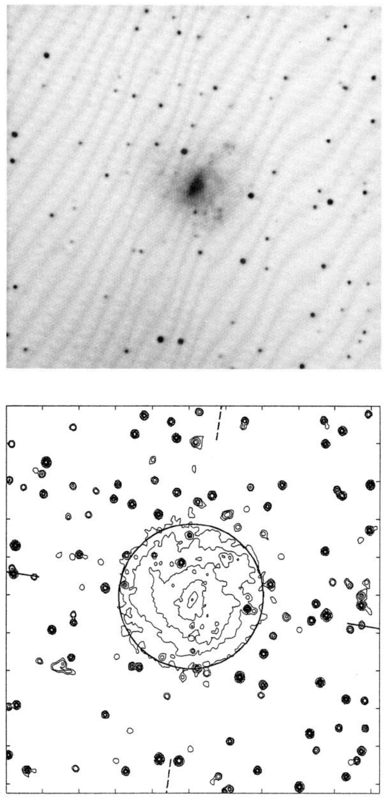

N5068 K5374 2 21.42 512X512

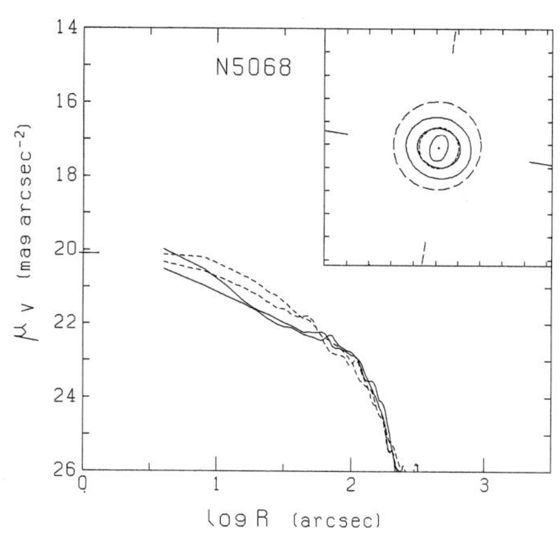

NGC5084

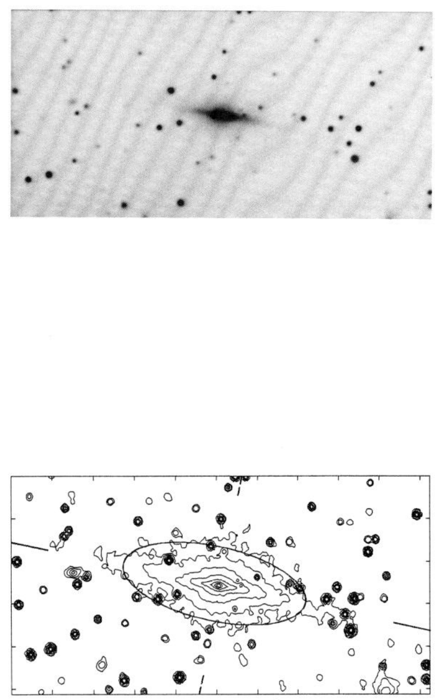

N5084 K5374 2 21.42 512X256

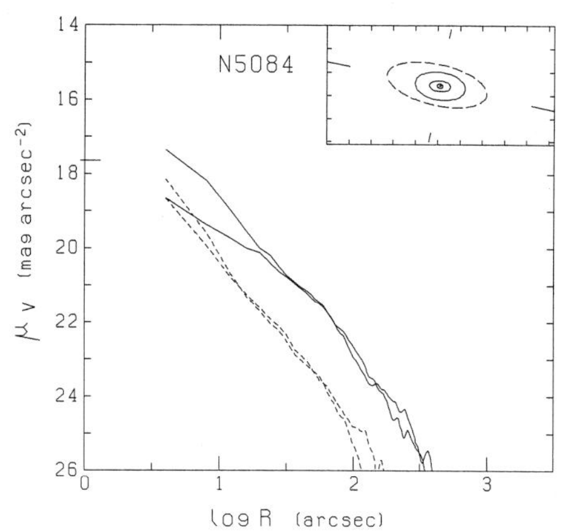

NGC5087

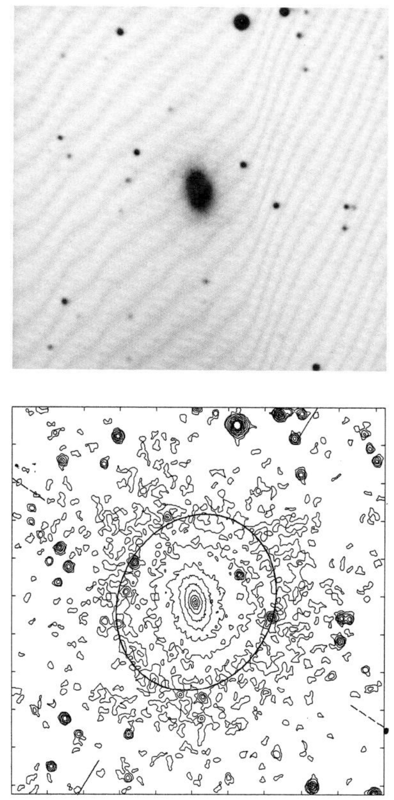

N5087 K5374 1 21.42 512X512

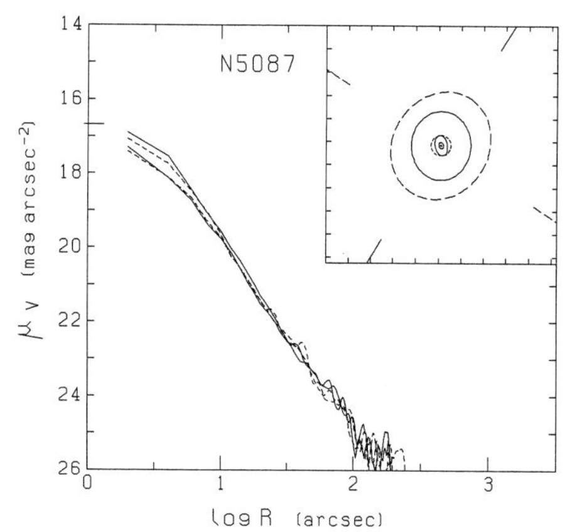

NGC5112

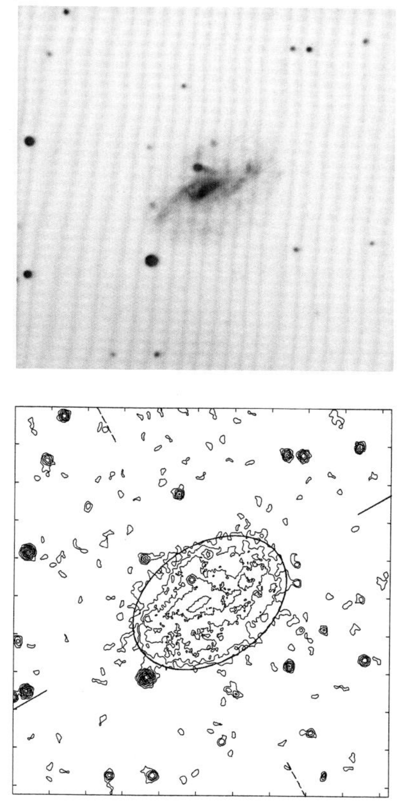

N5112 K5372 1 22.12 512X512

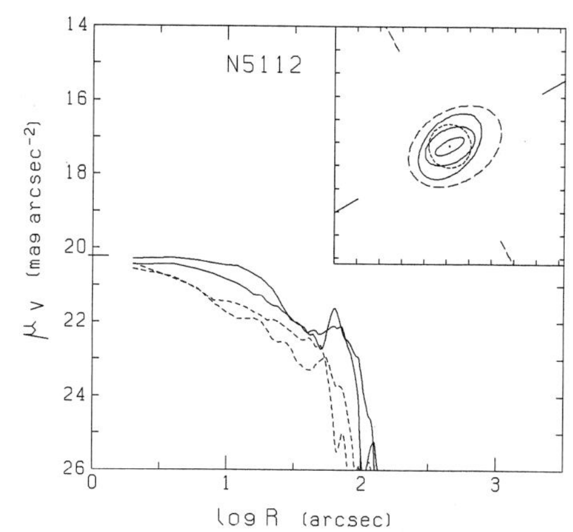

NGC5134

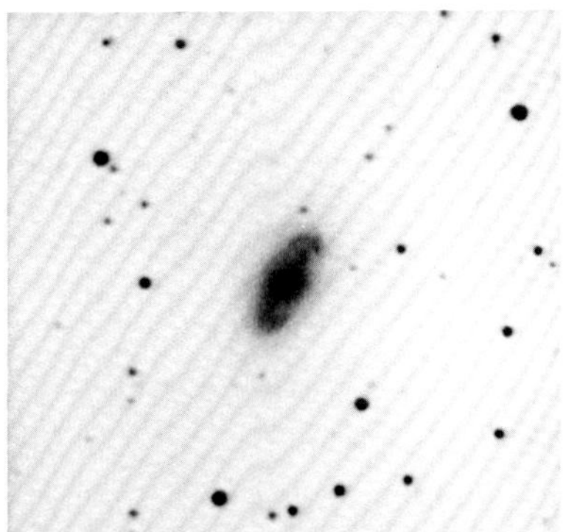

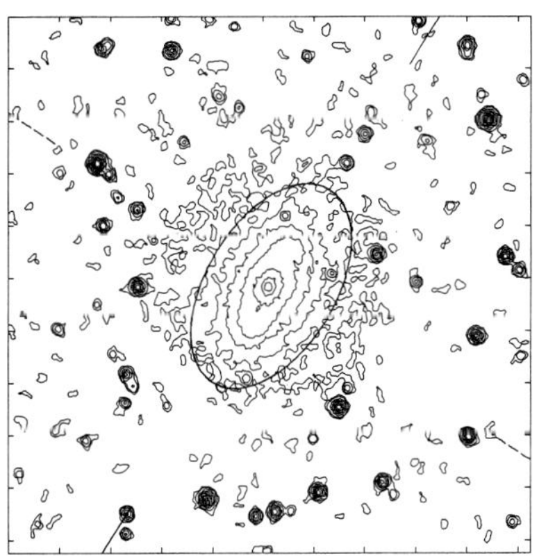

N5134 K5374 1 21.42 512X512

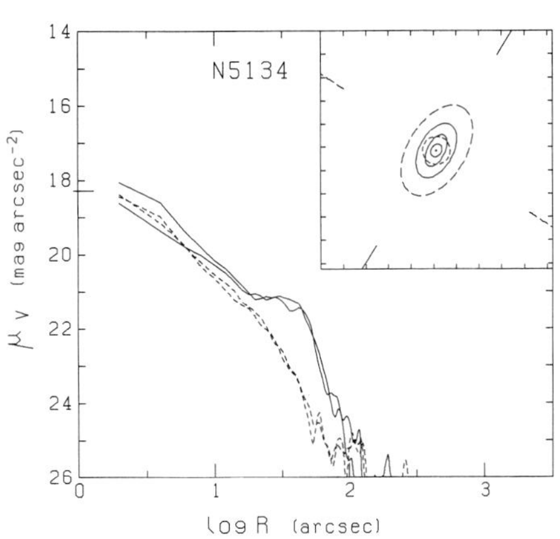

NGC5170

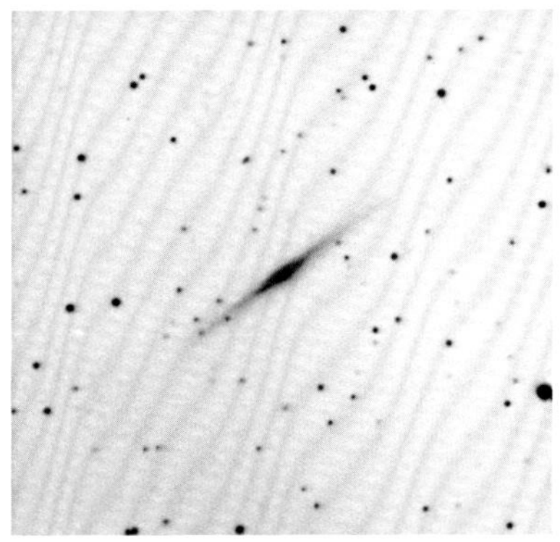

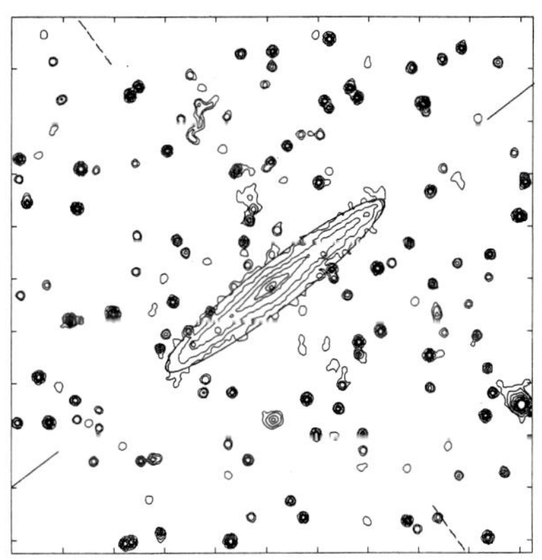

N5170 K5374 2 21.42 512X512

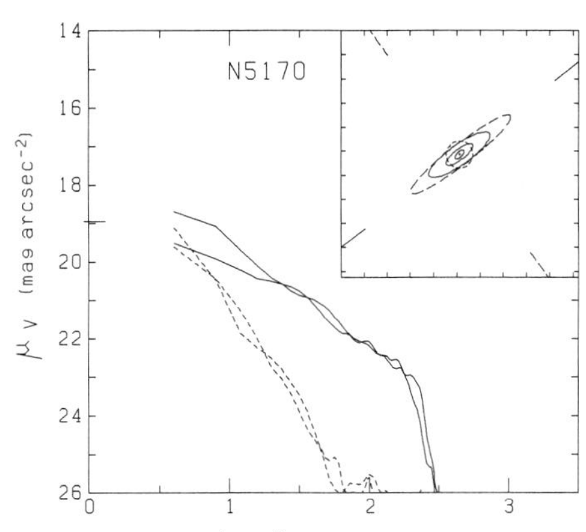

NGC5204

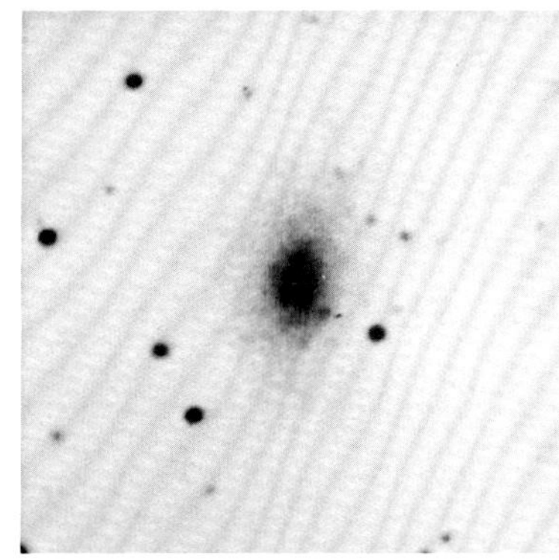

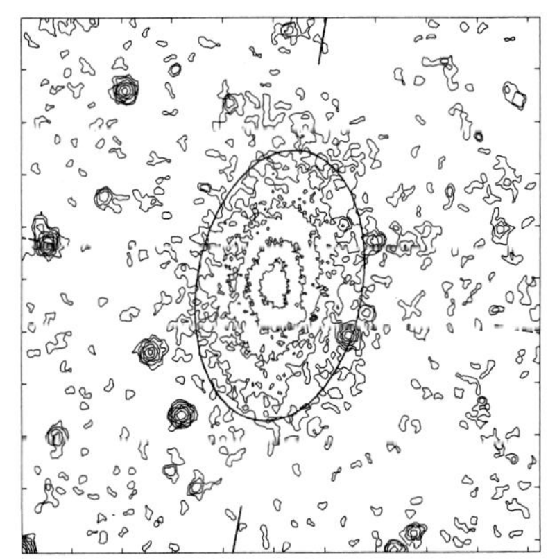

N5204 K5765 1 21.82 512X512

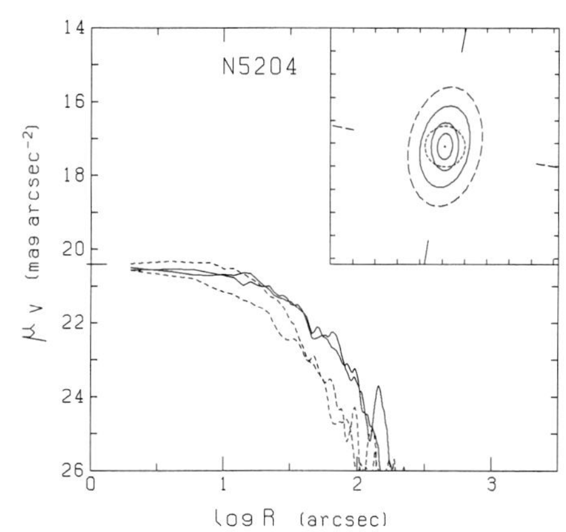

NGC5194

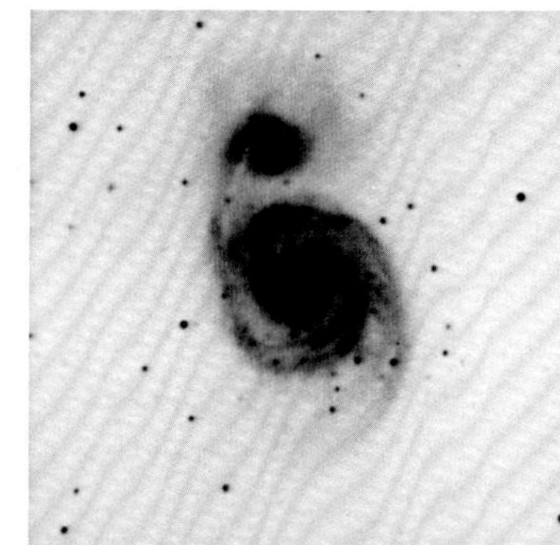

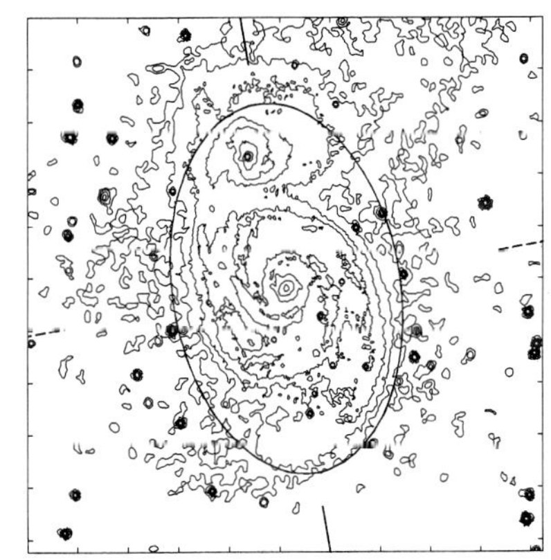

N5194 K4938 2 21.79 512X512

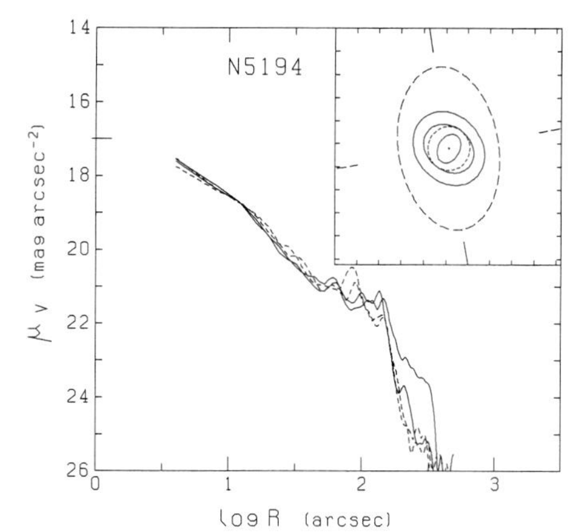

NGC5195

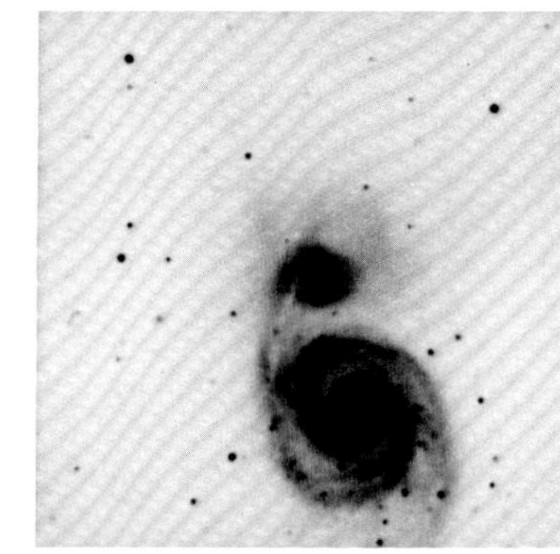

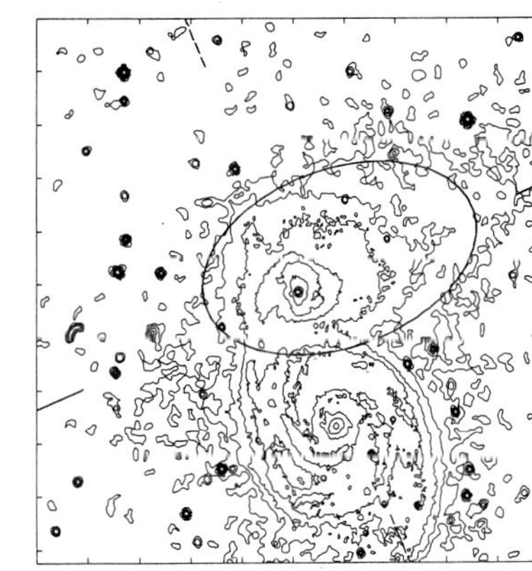

N5195 K4938 2 21.79 512X512

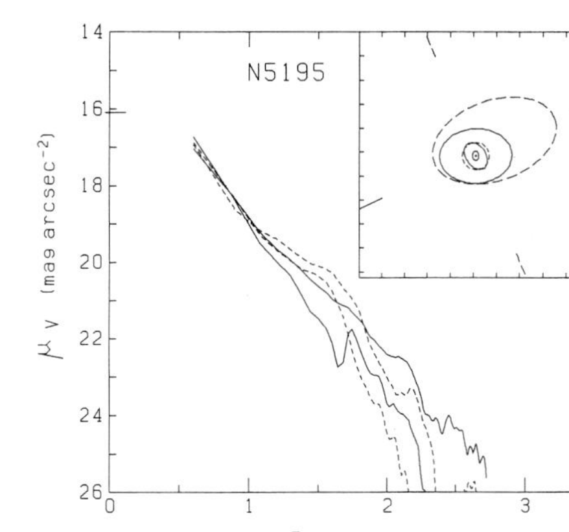

NGC5198

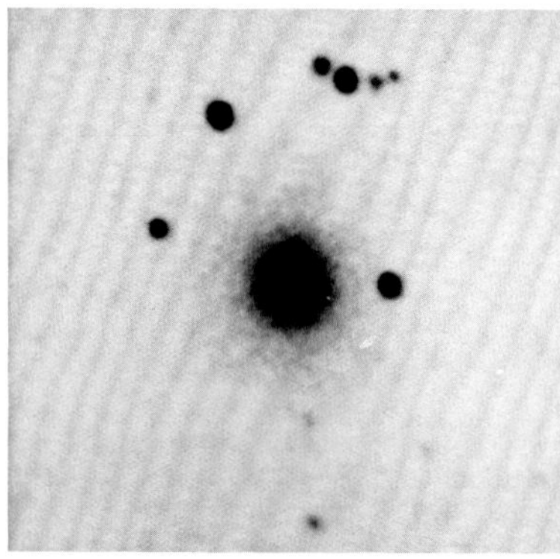

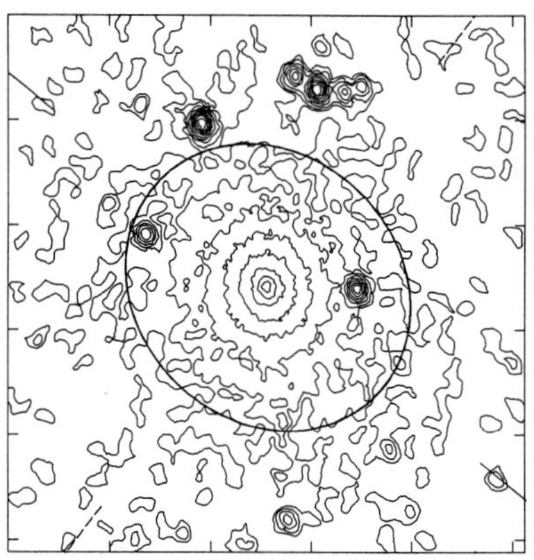

N5198 K4938 1 21.79 256X256

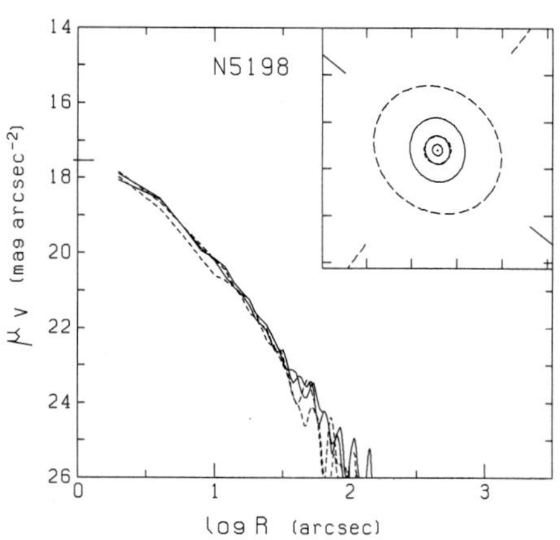

NGC5273

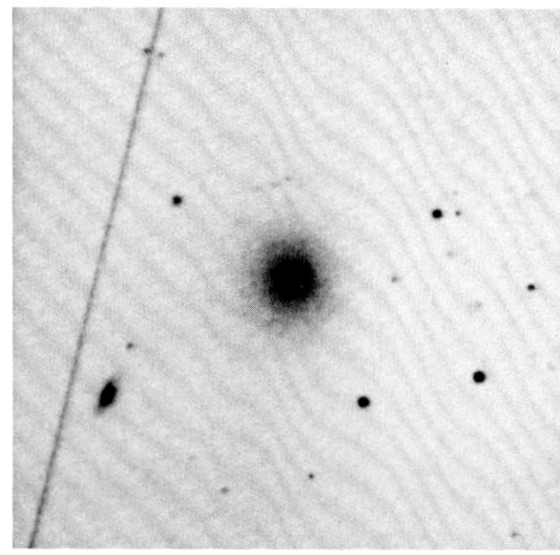

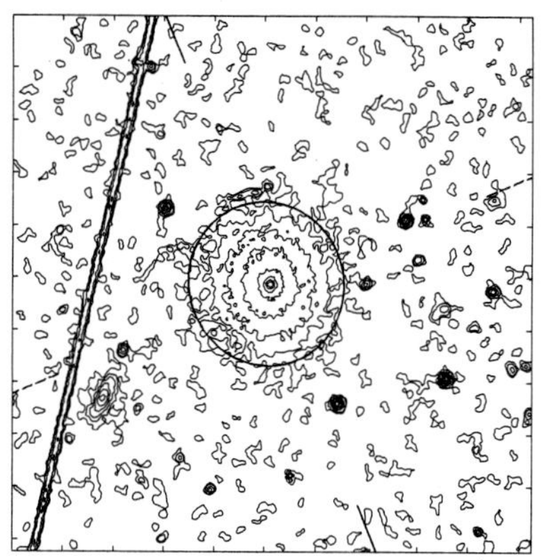

N5273 K5079 1 21.48 512X512

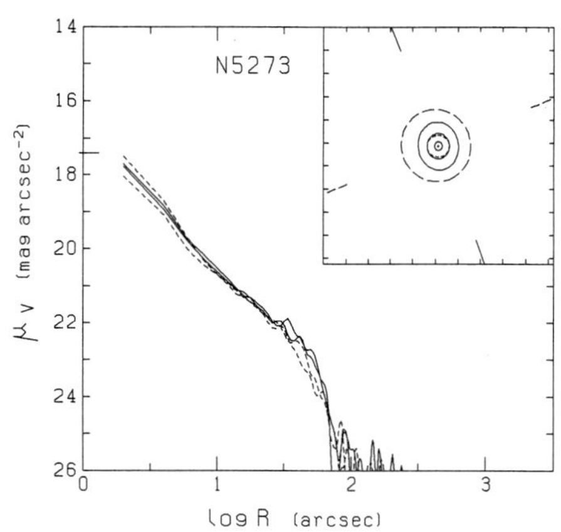

NGC5297

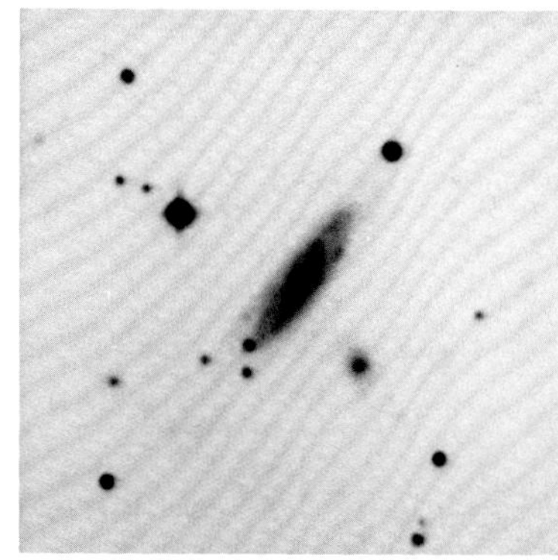

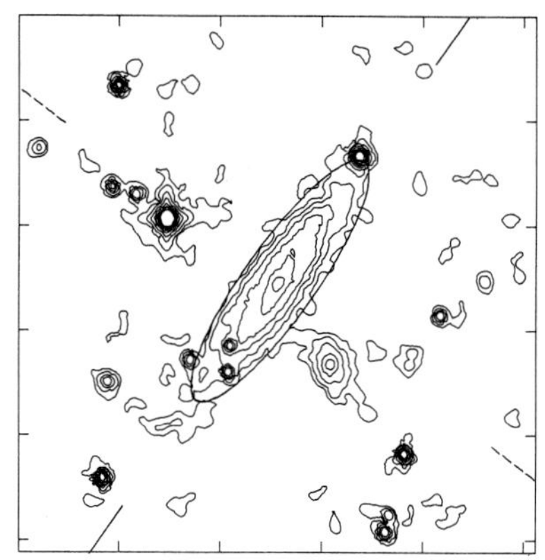

N5297 K4938 2 21.79 256X256

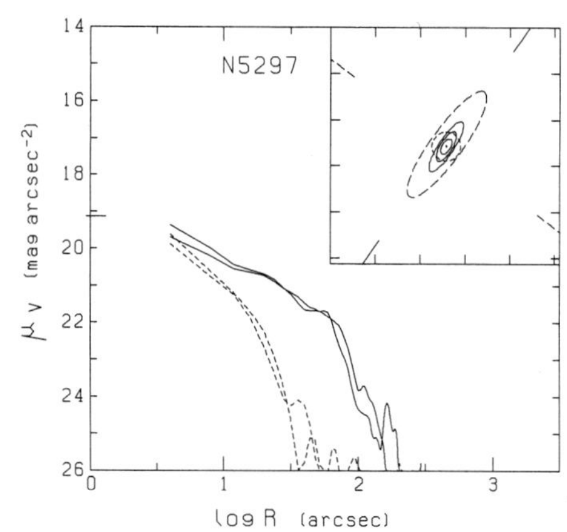

NGC5301

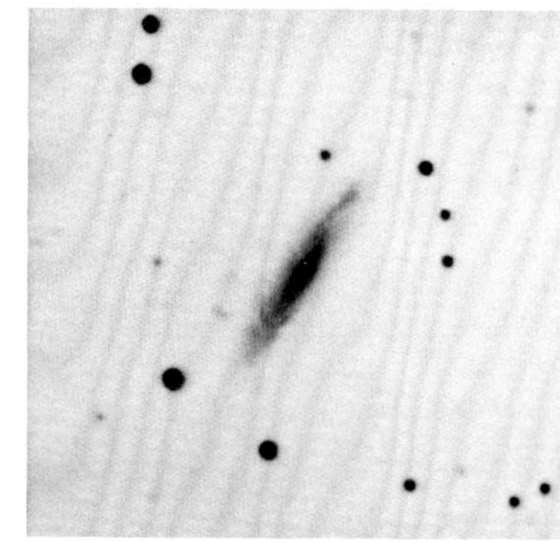

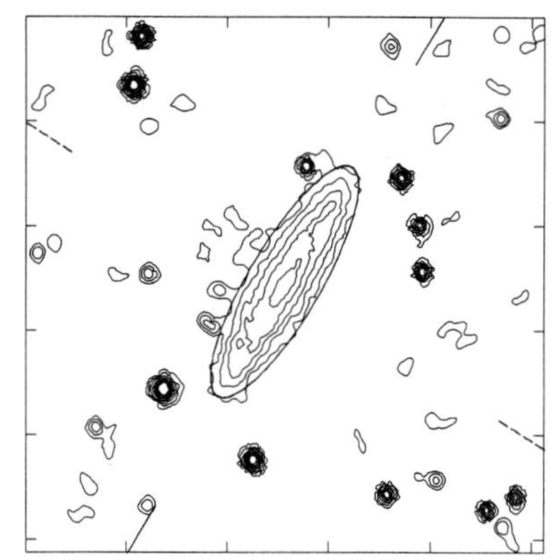

N5301 K4938 2 21.79 256X256

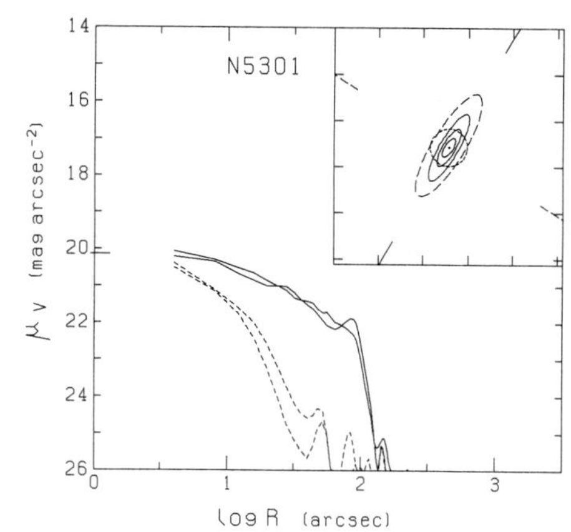

NGC5313

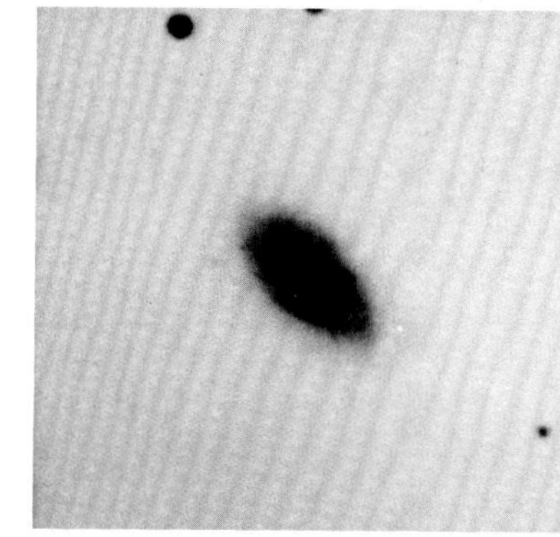

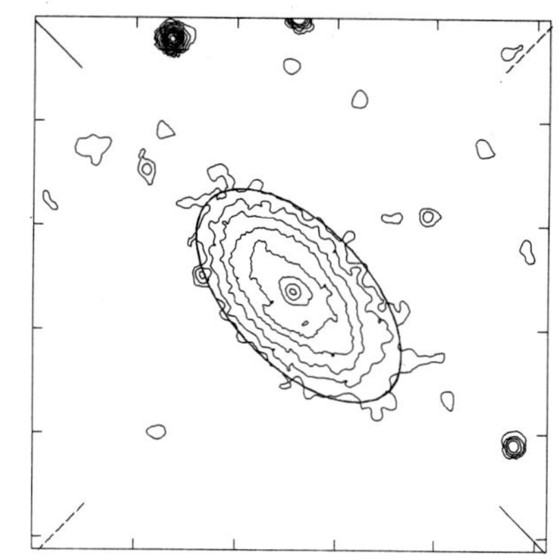

N5313 K5351 1 22.35 256X256

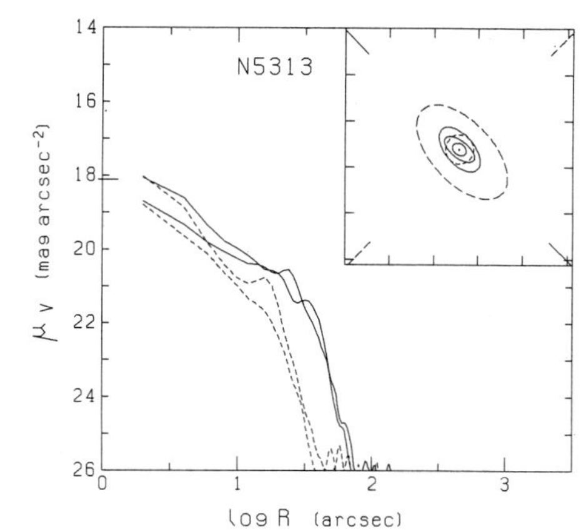

NGC5326

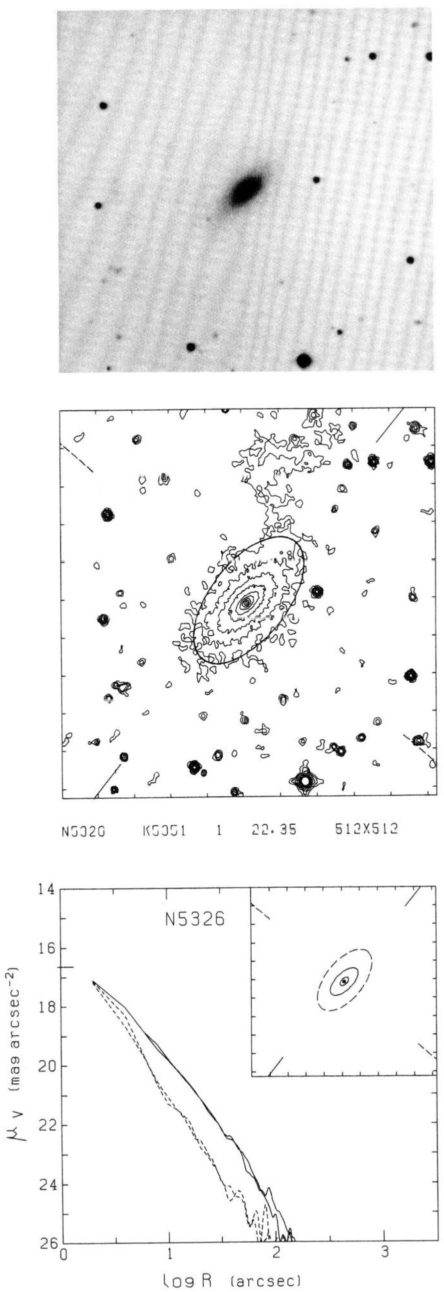

NGC5324

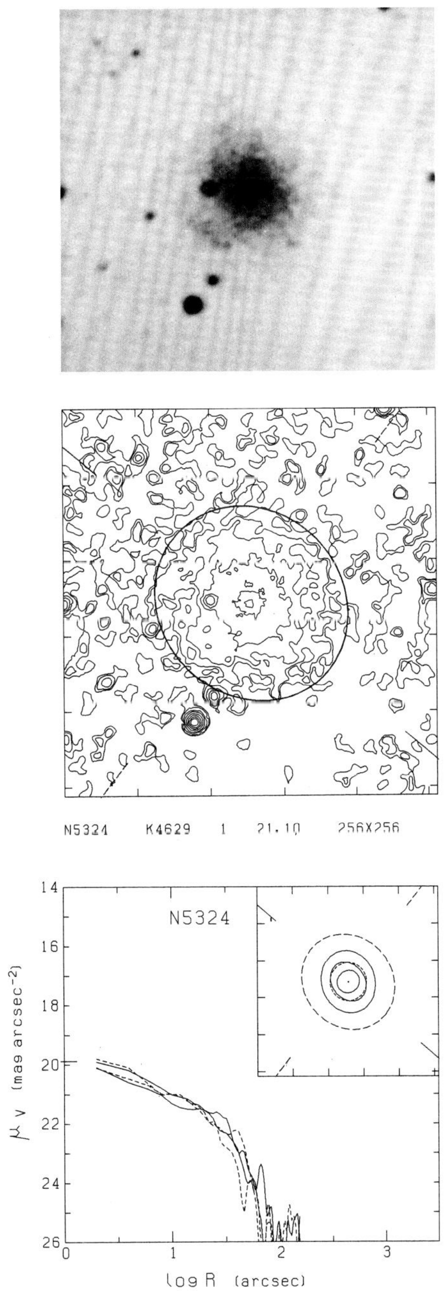

NGC5347

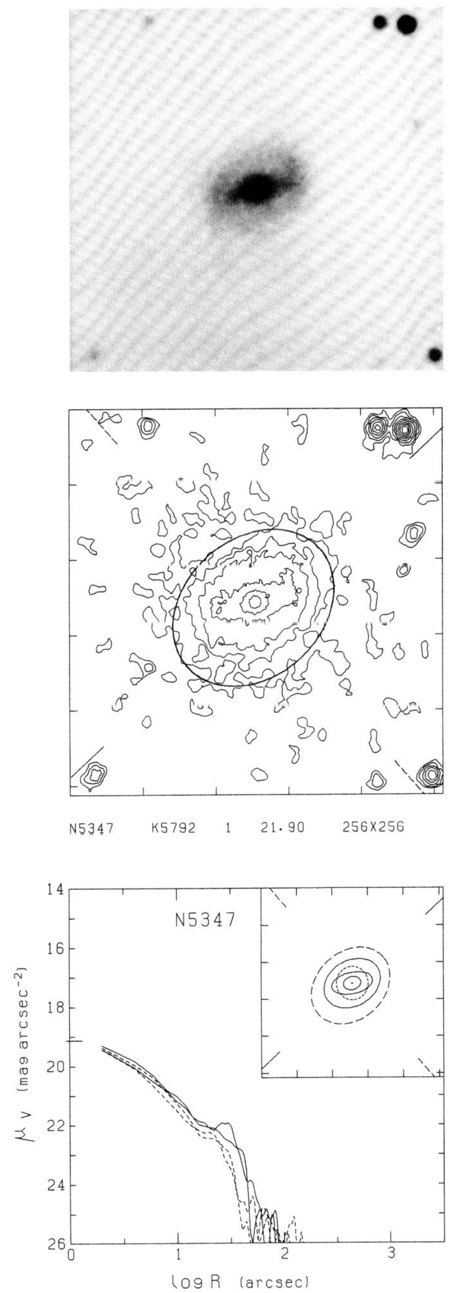

NGC5350

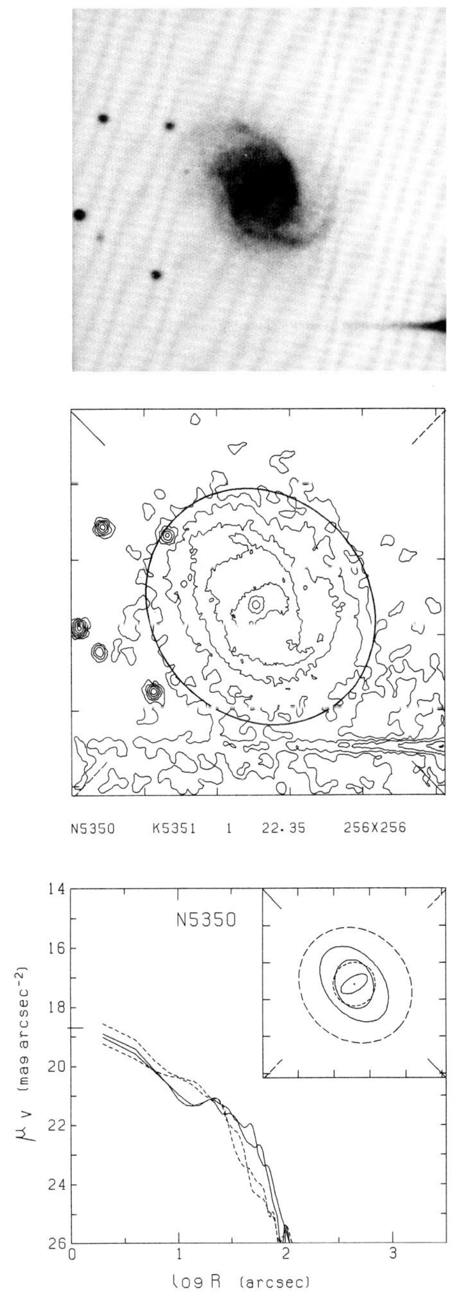

NGC5351

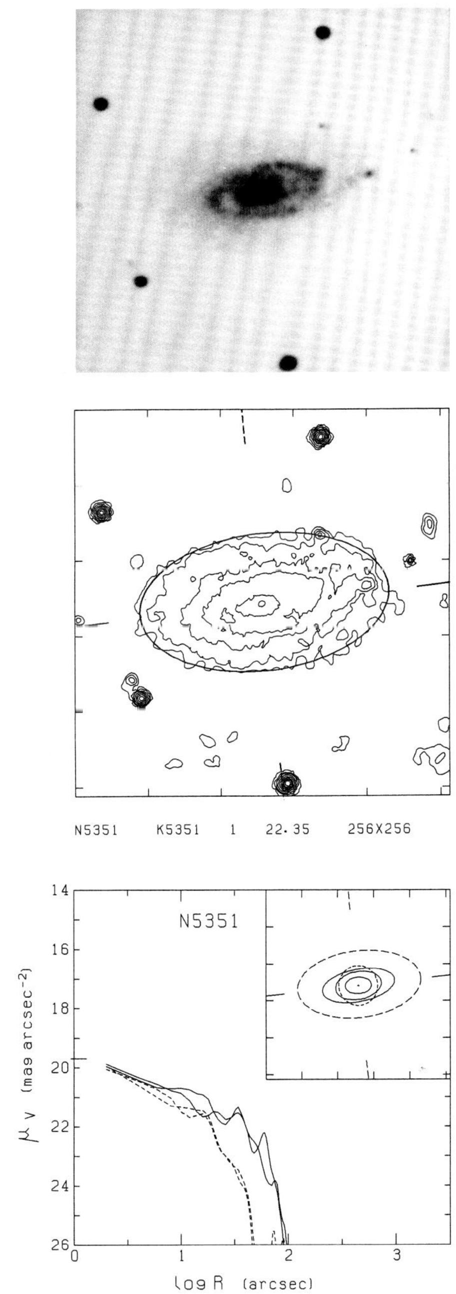

NGC5353

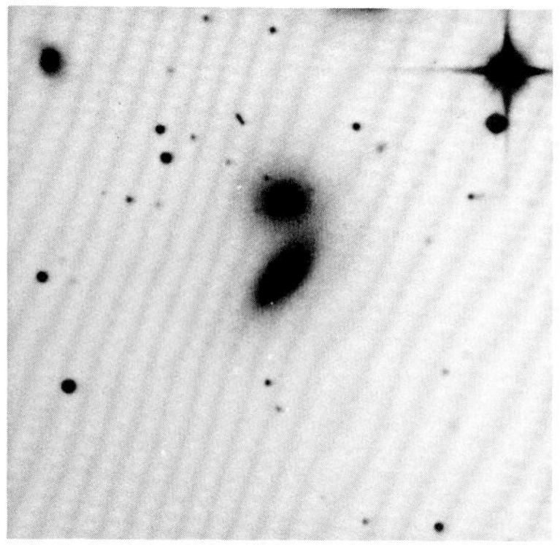

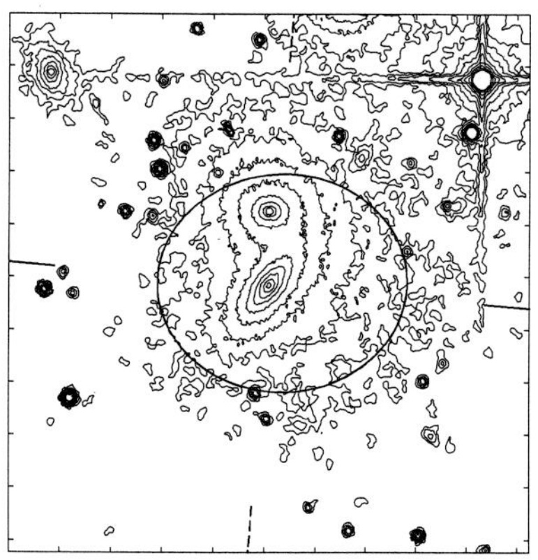

N5353 K5351 1 22.35 512X512

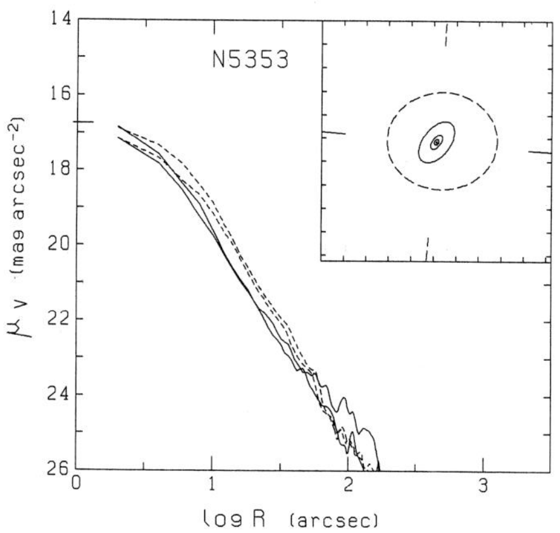

NGC5362

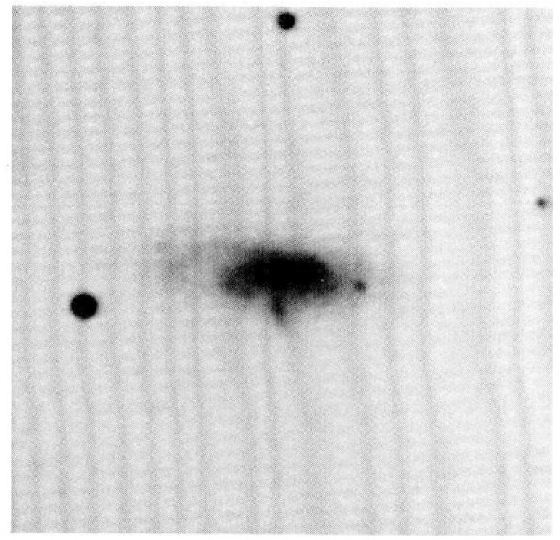

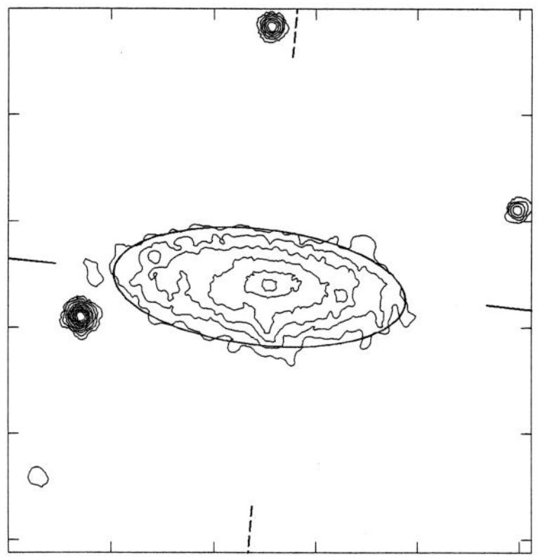

N5362 K5351 1 22.35 256X256

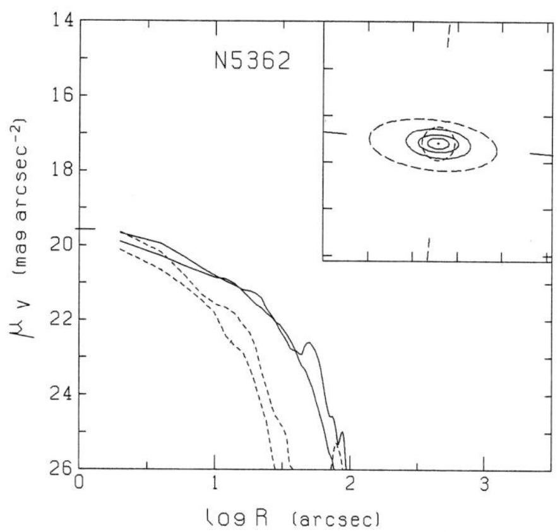

NGC5371

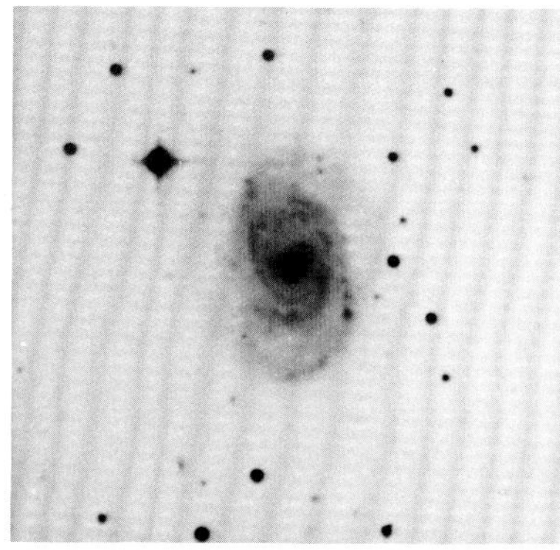

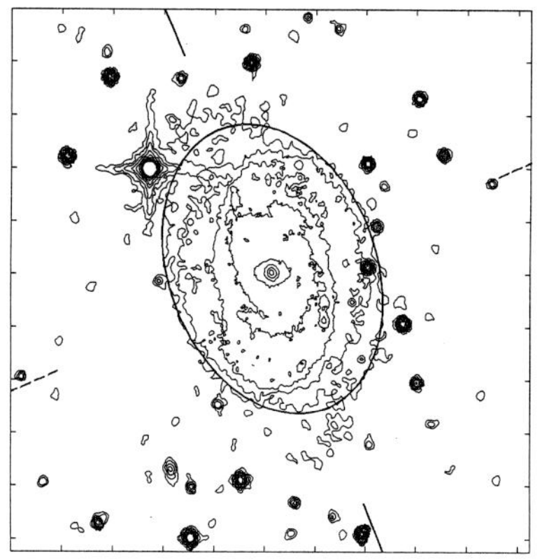

N5371 K5351 1 22.35 512X512

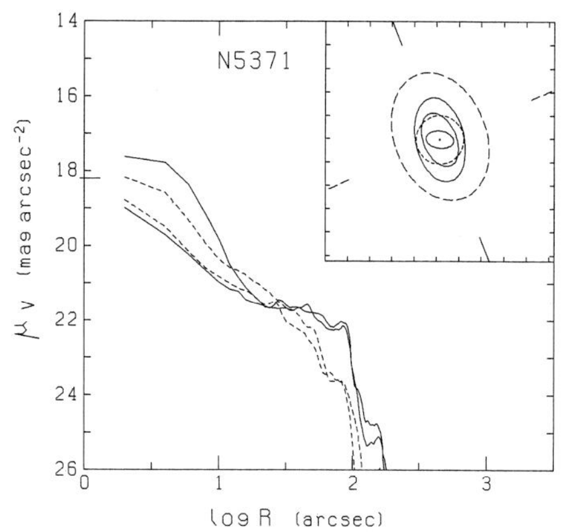

NGC5363

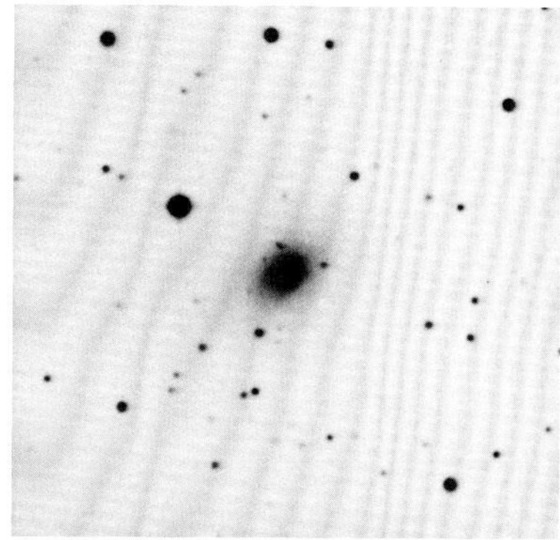

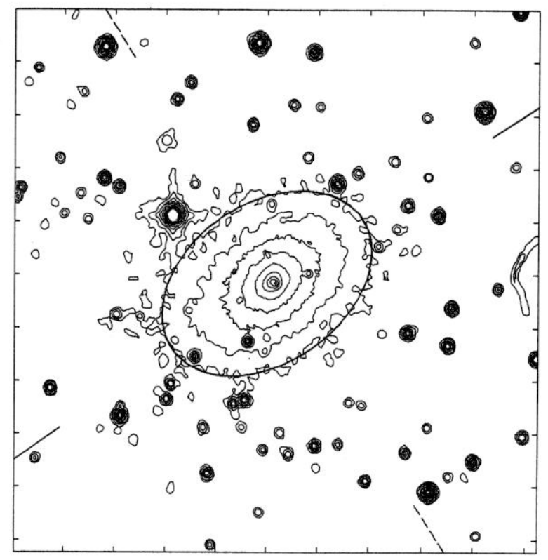

N5363 K5395 2 21.73 512X512

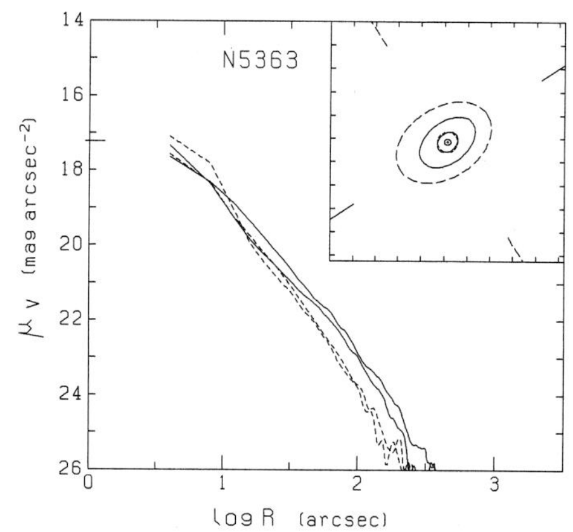

NGC5364

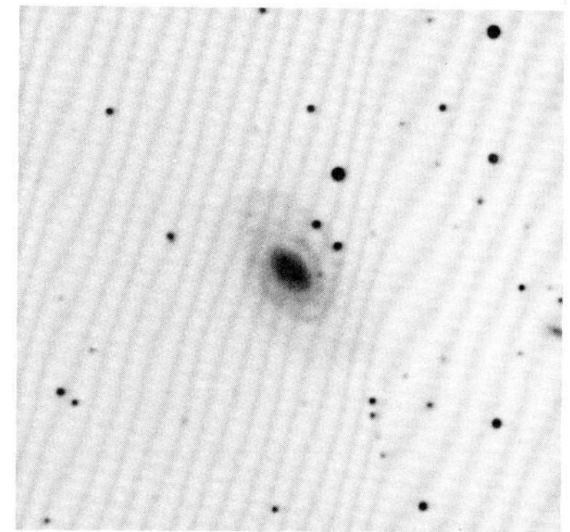

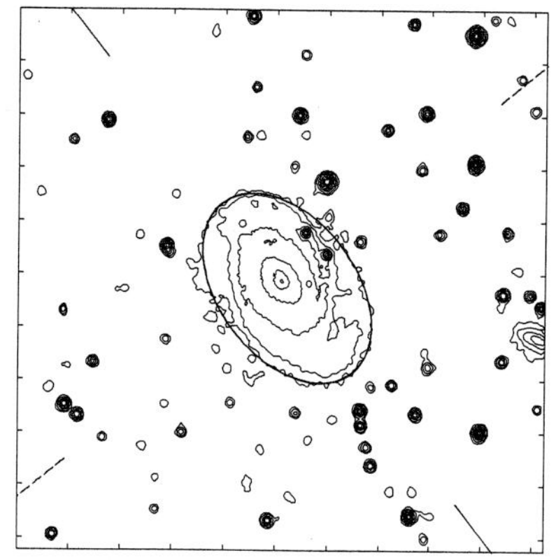

N5364 K5395 2 21.73 512X512

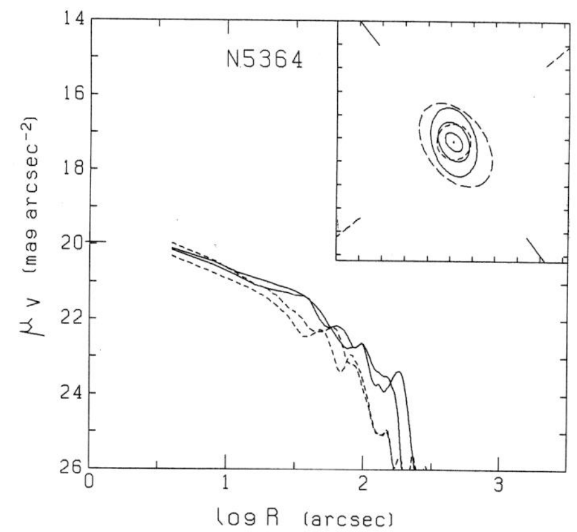

NGC5377

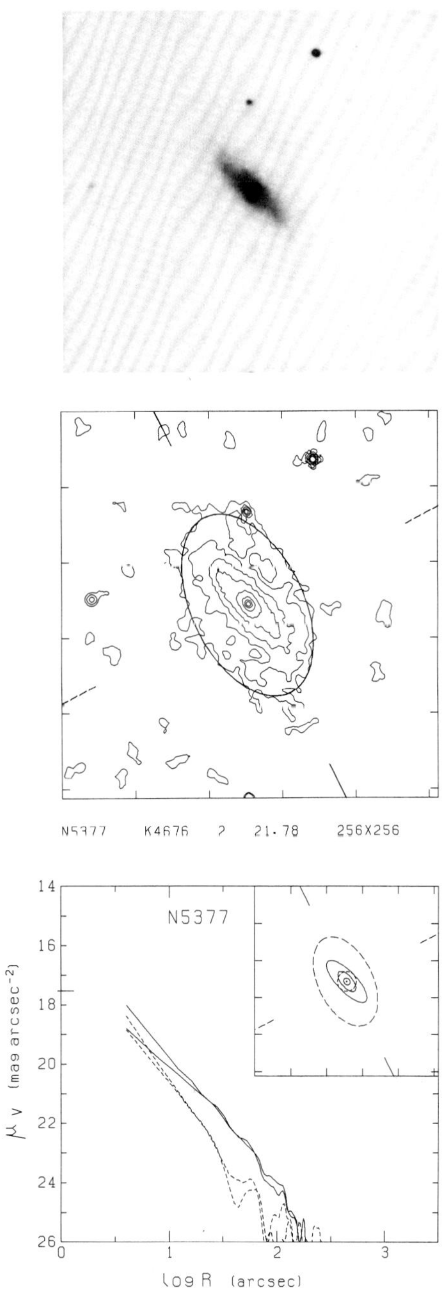

NGC5380

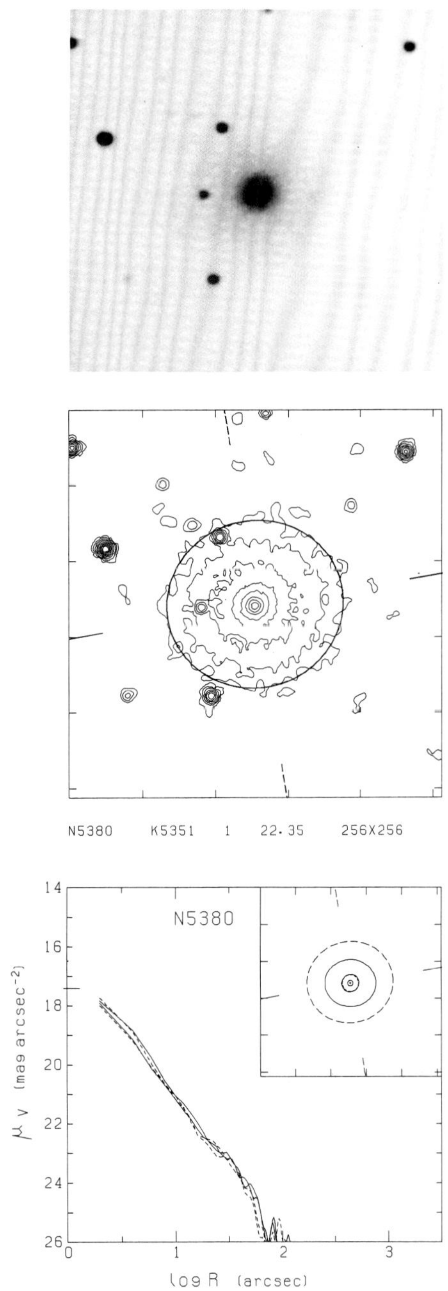

NGC5383

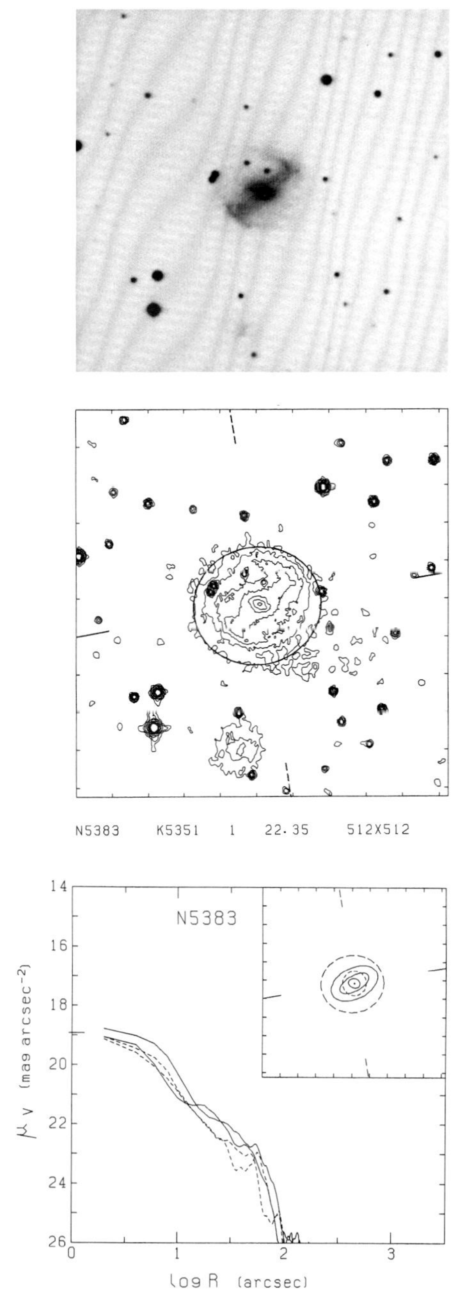

NGC5395

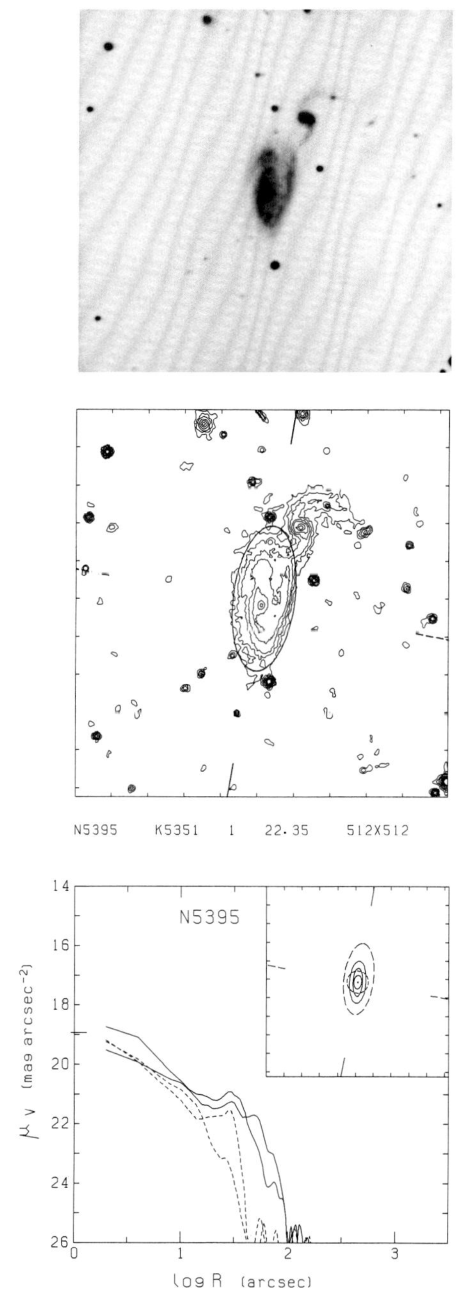

NGC5406

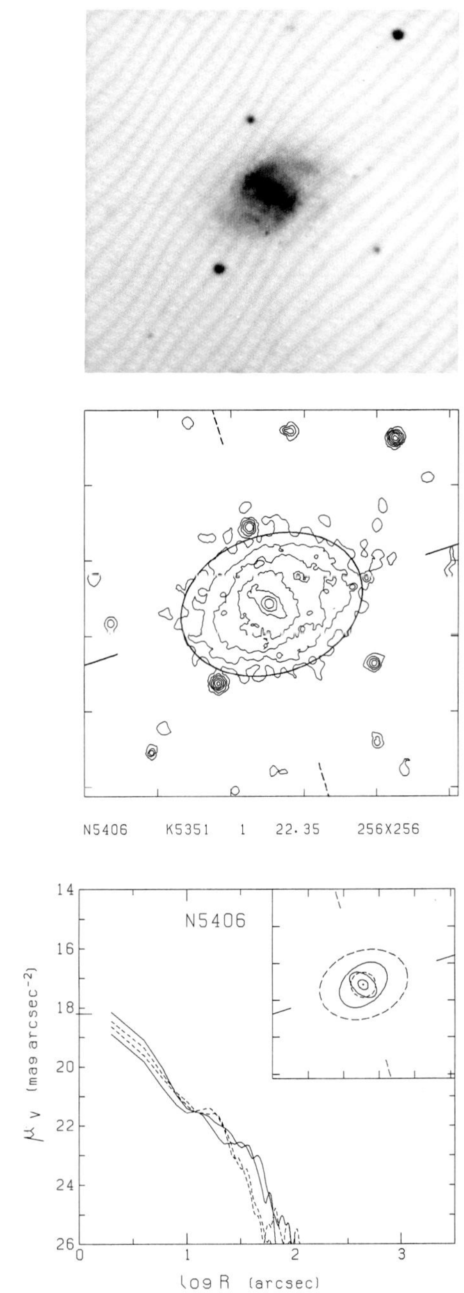

NGC5422

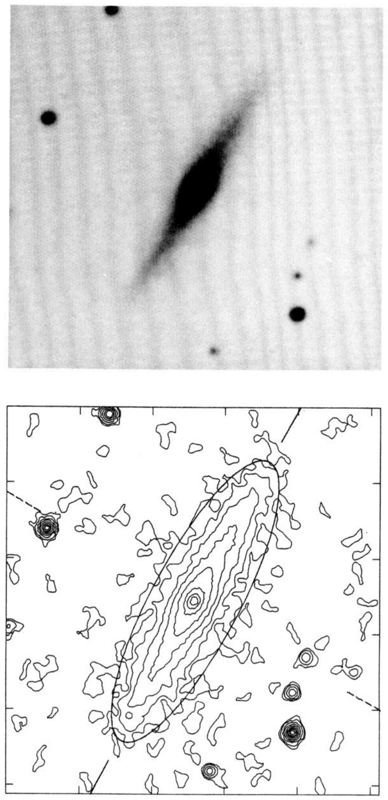

N5422 K3289 1 21.01 256X256

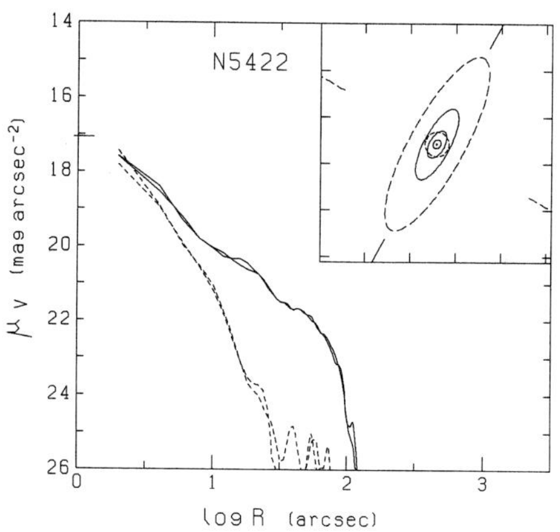

NGC5426

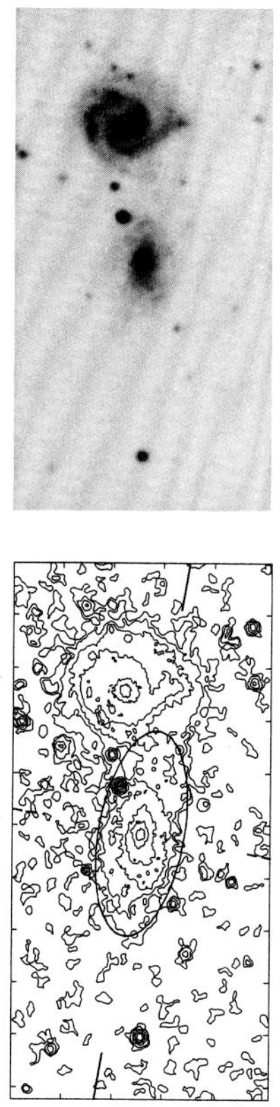

N5426 K4986 1 21.49 256X512

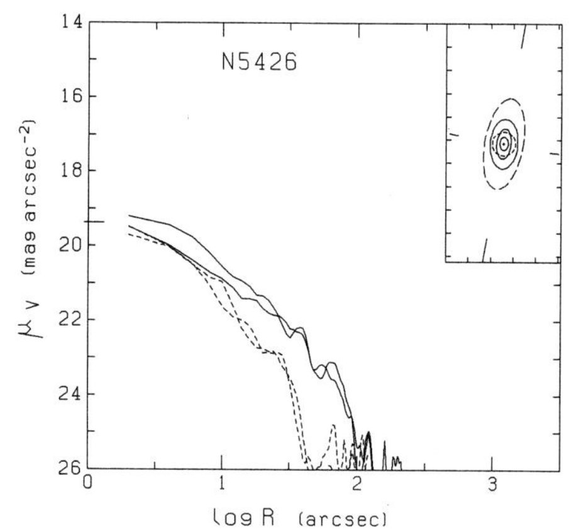

NGC5427

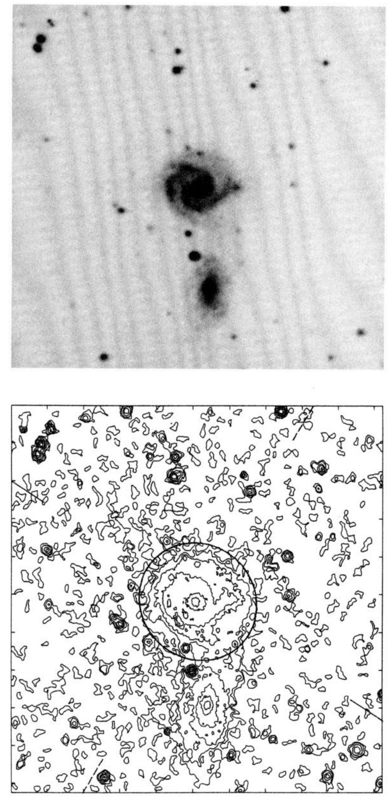

N5427 K4986 1 21.49 512X512

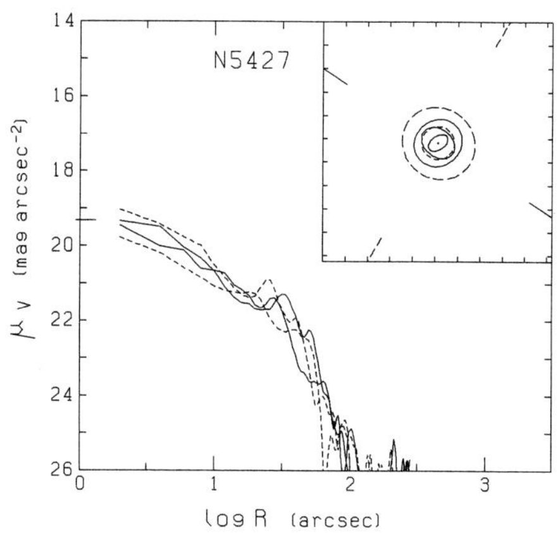

NGC5448

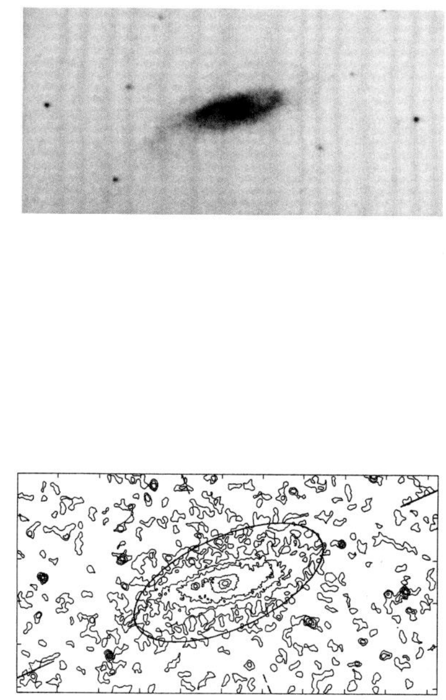

N5448 K4676 1 21.78 512X256

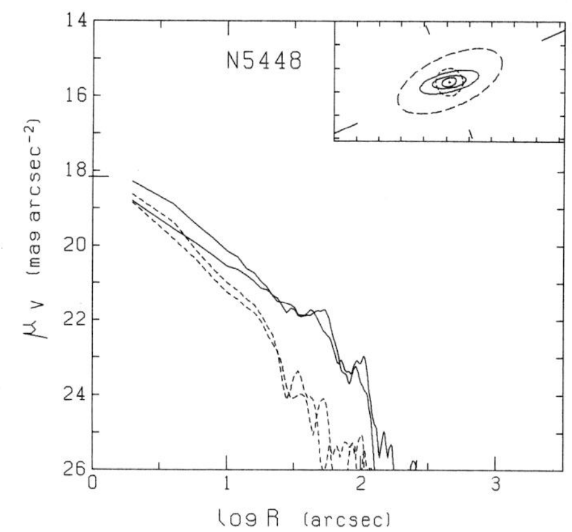

NGC5444

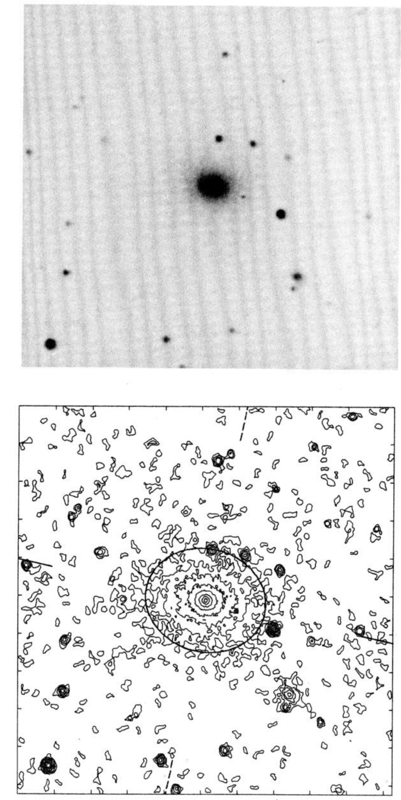

N5444 K5792 1 21.90 512X512

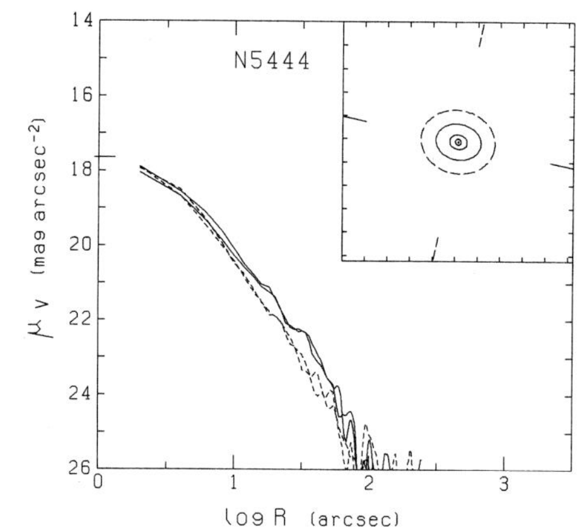

NGC5457

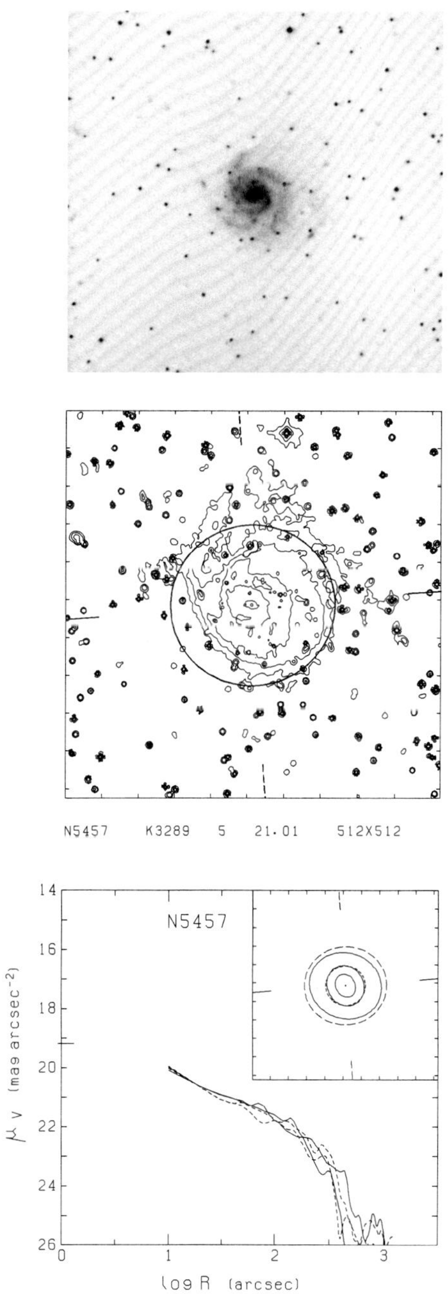

NGC5473

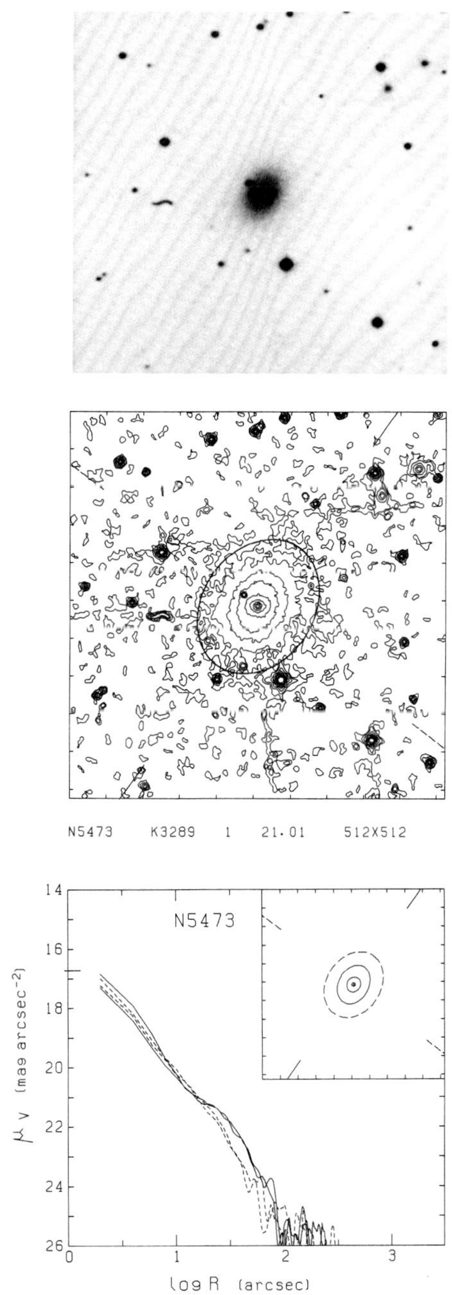

NGC5474

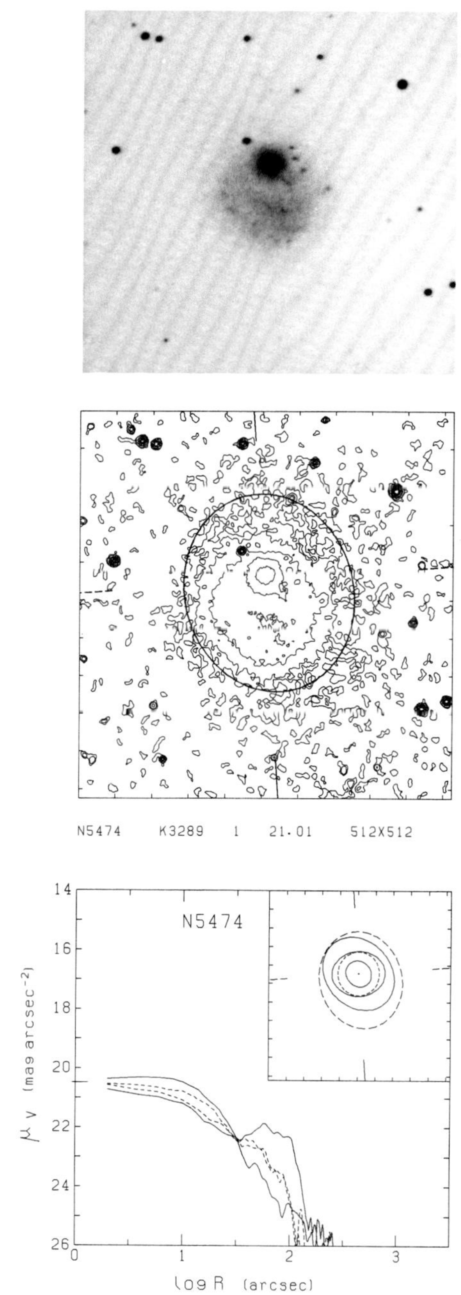

NGC5468

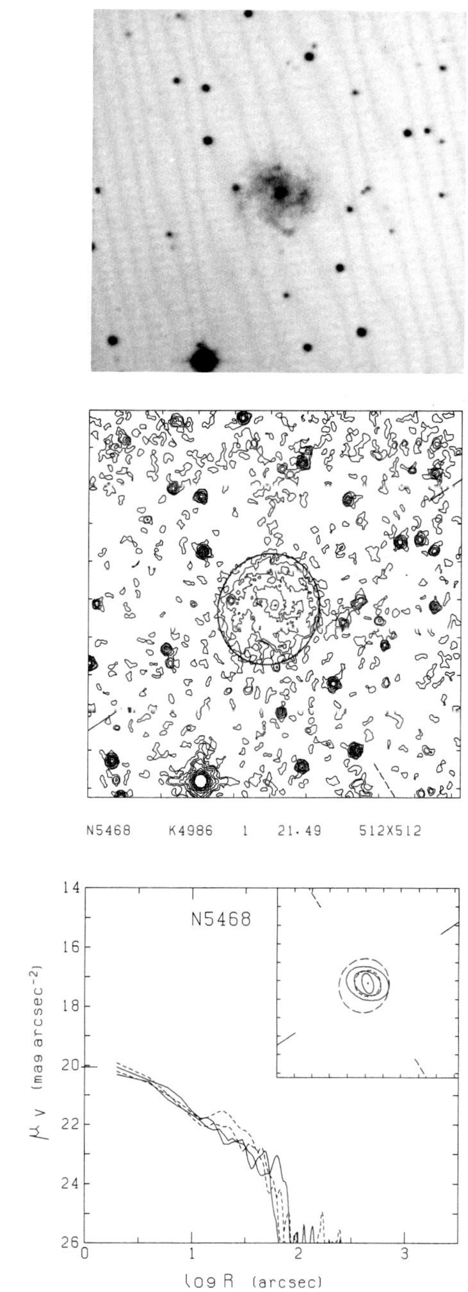

NGC5480

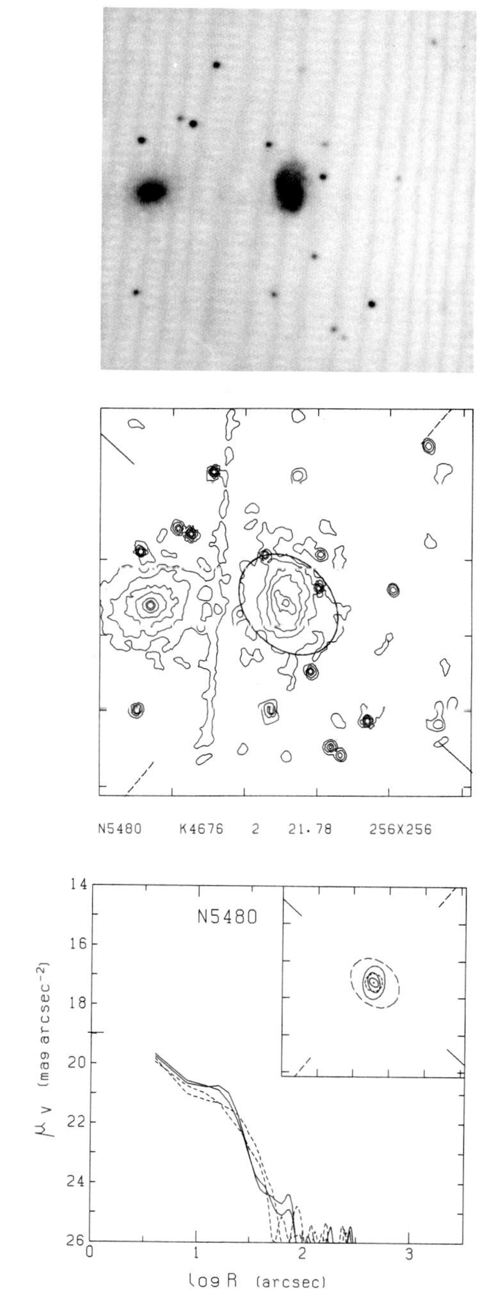

NGC5485

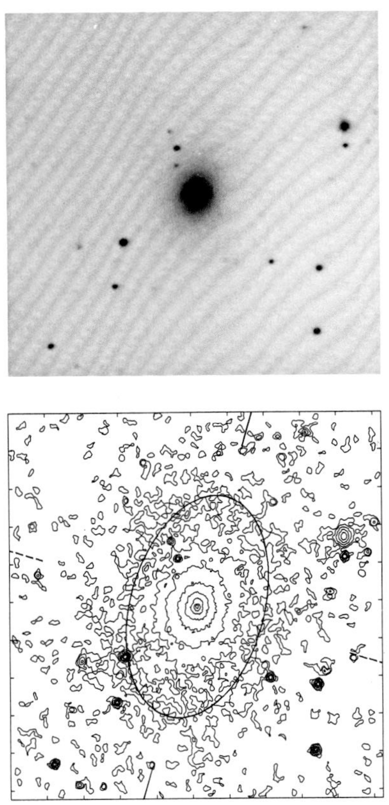

N5485 K3289 1 21.01 512X512

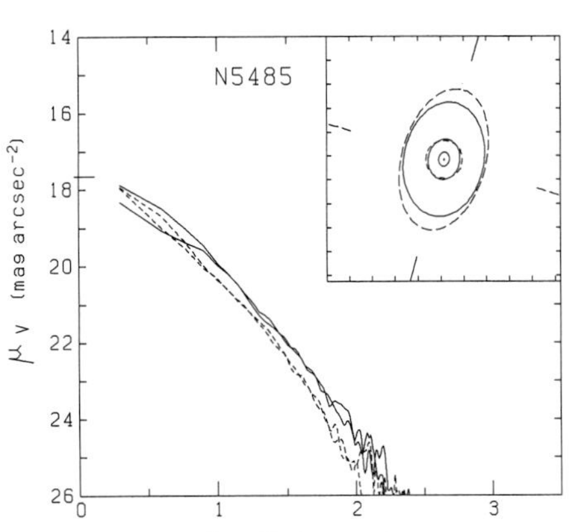

NGC5493

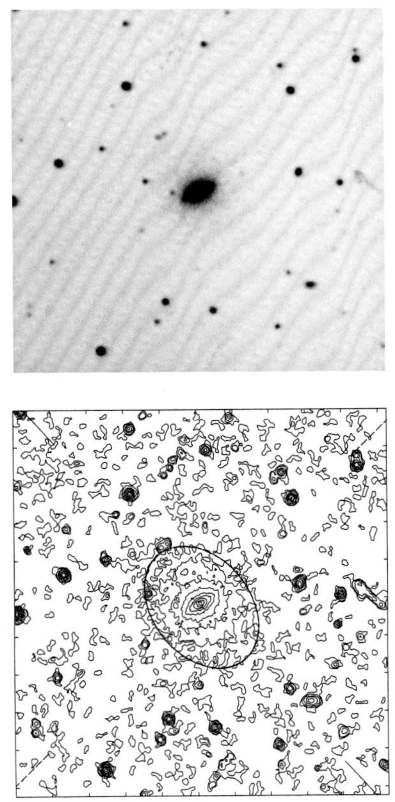

N5493 K4986 1 21.49 512X512

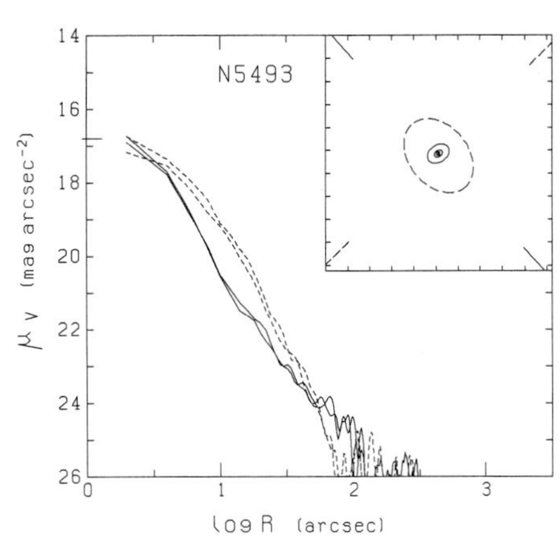

NGC5496

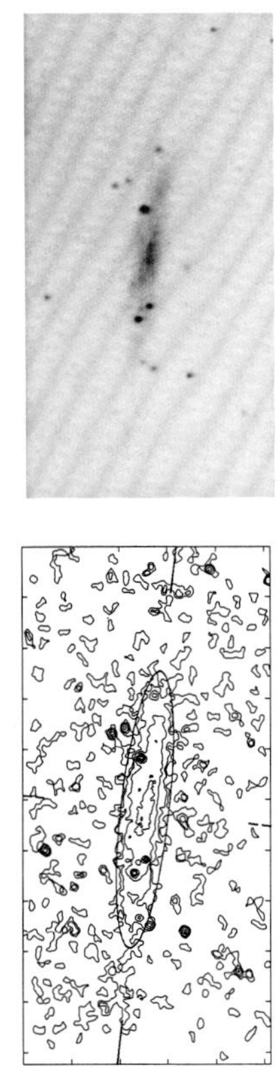

N5496 K3766 1 20.76 256X512

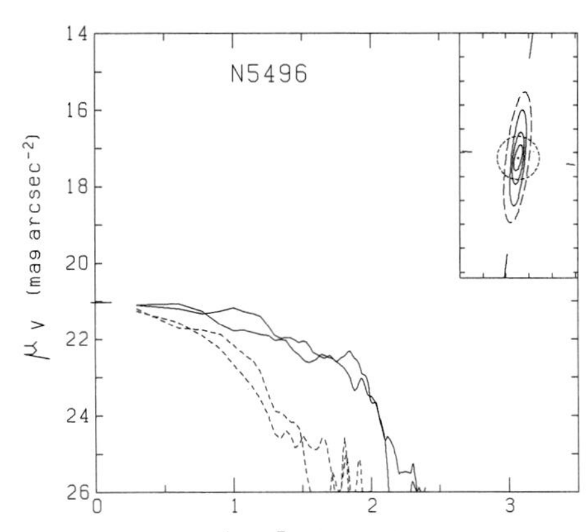

NGC5494

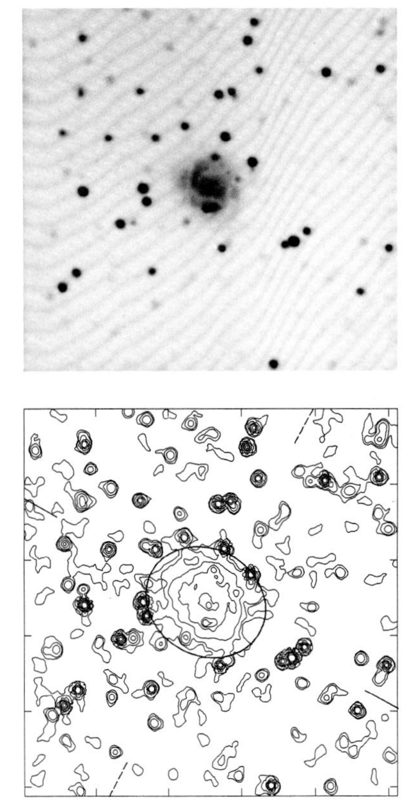

N5494 K4978 2 20.70 256X256

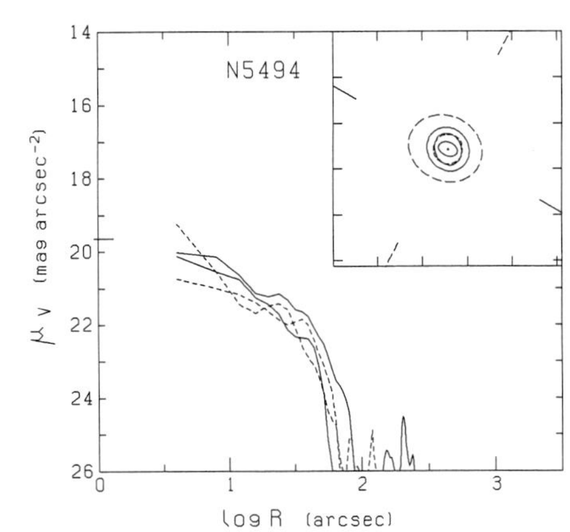

NGC5523

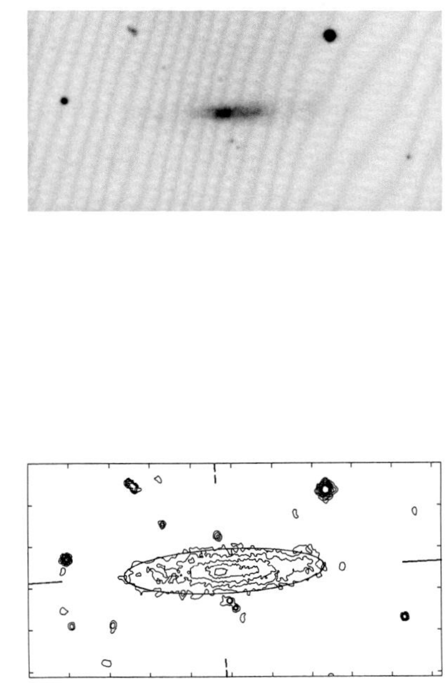

N5523 K4996 1 22.54 512X256

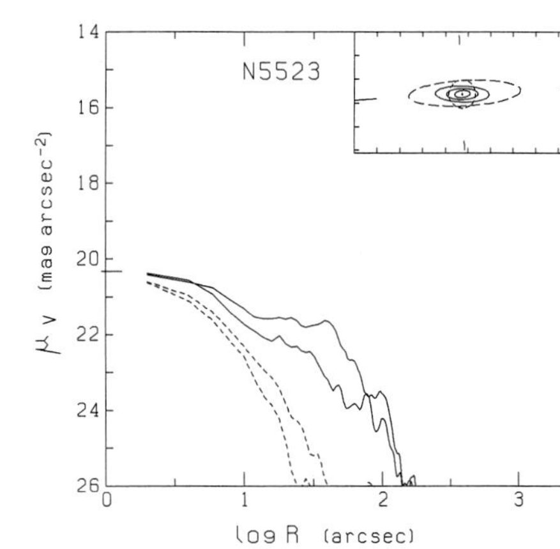

NGC5533

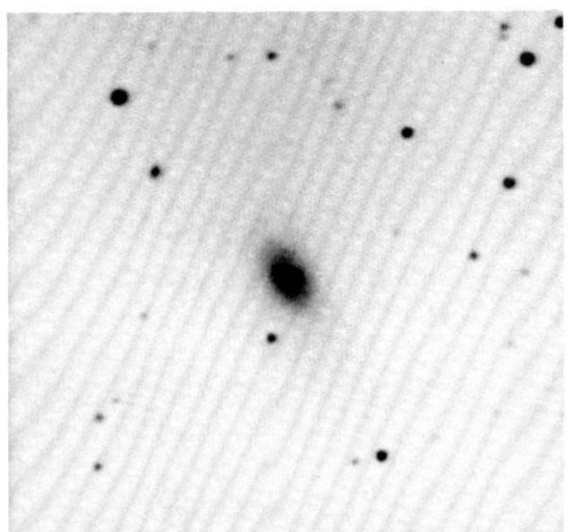
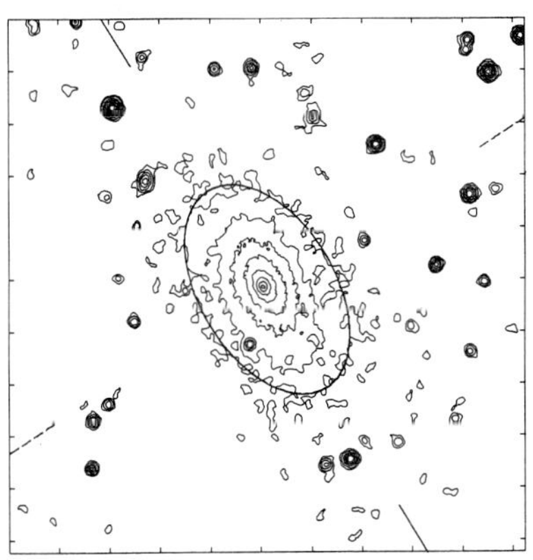

N5533 K2649 1 21.66 512X512

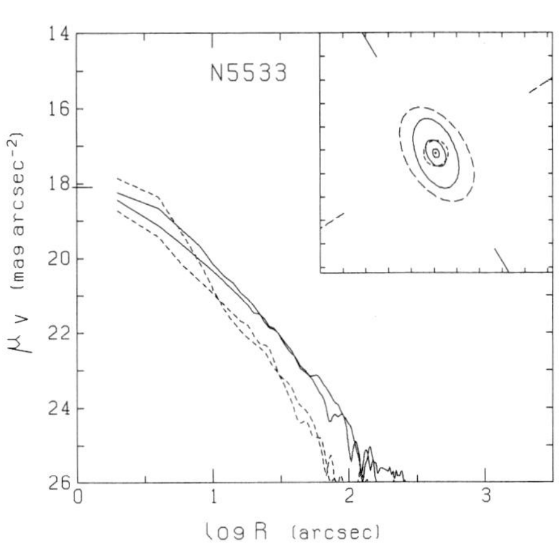

NGC5534

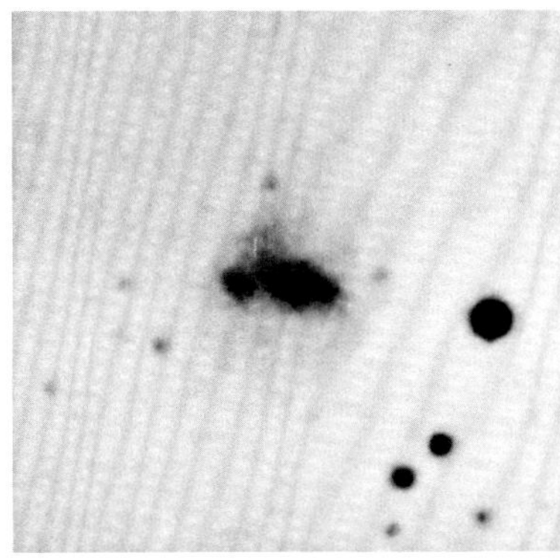
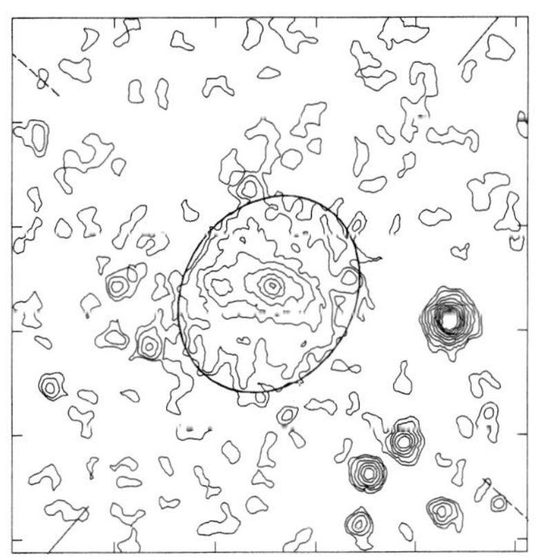

N5534 K4986 1 21.49 256X256

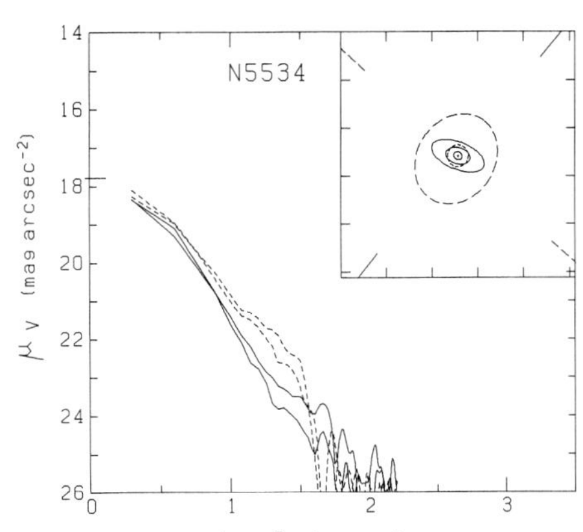

NGC5548

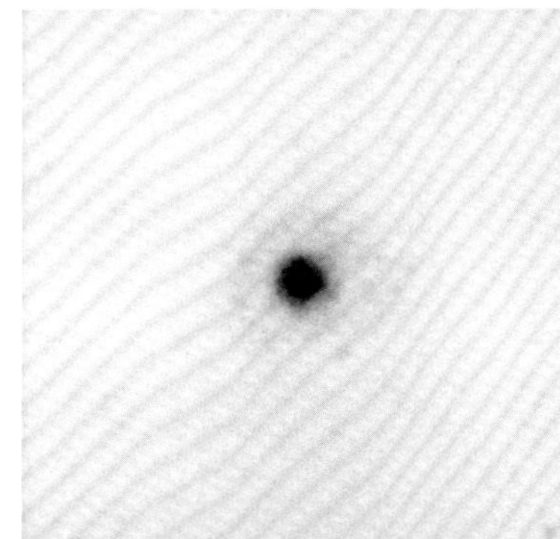
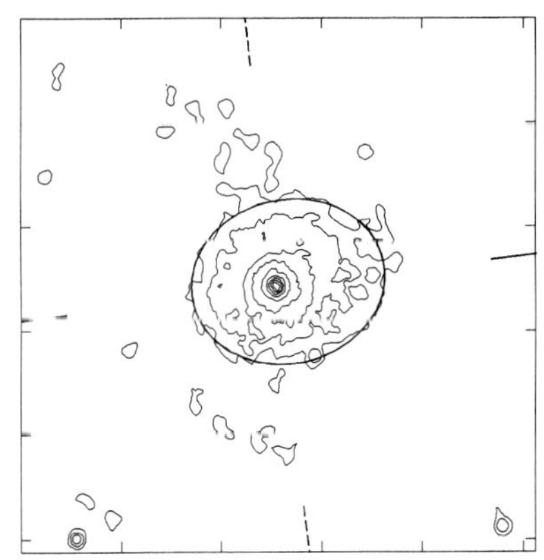

N5548 K4996 1 22.54 256X256

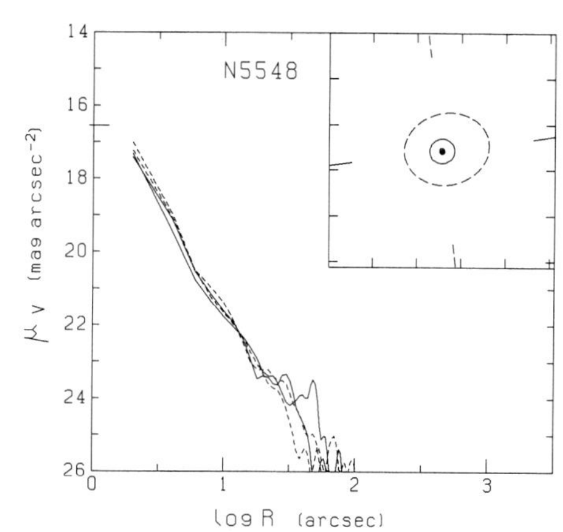

NGC5557

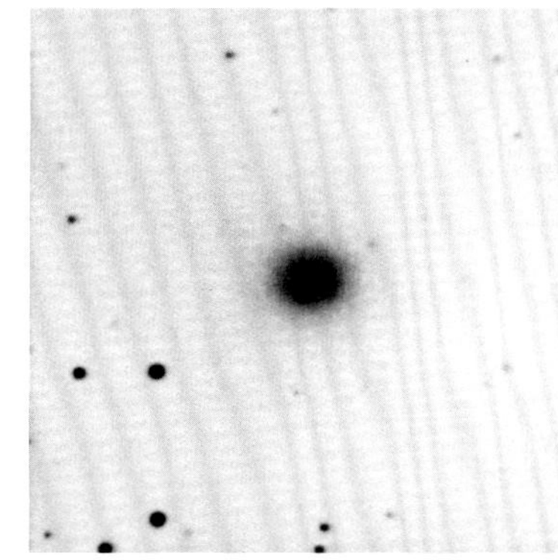
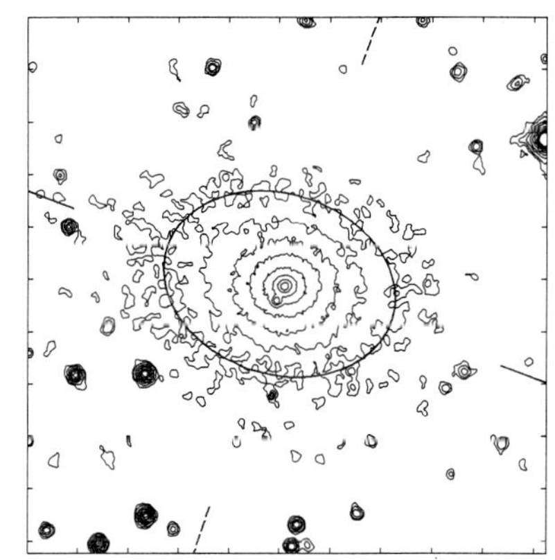

N5557 K2649 1 21.66 512X512

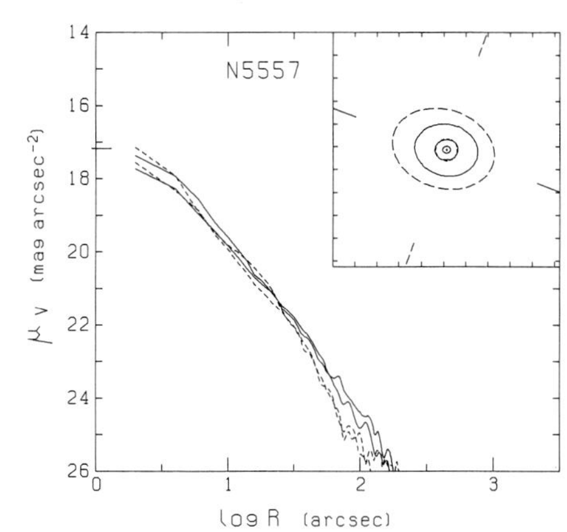

NGC5566

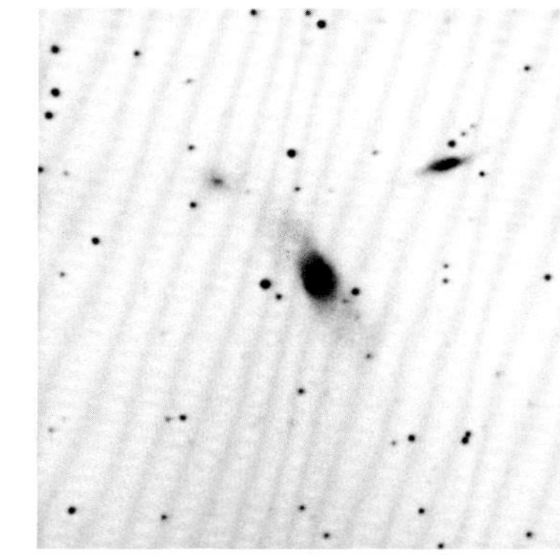
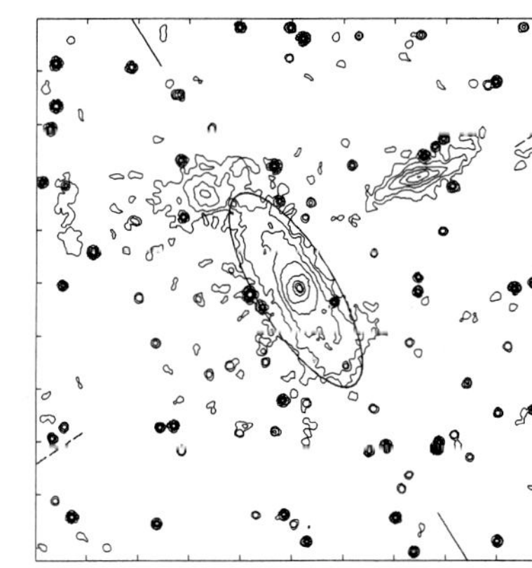

N5566 K4026 2 21.20 512X512

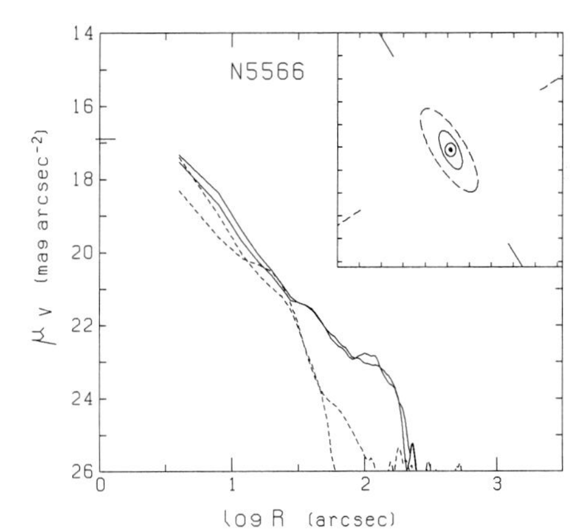

NGC5585

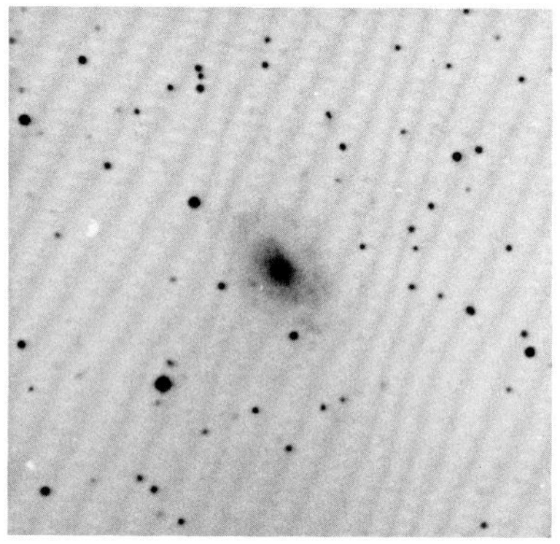

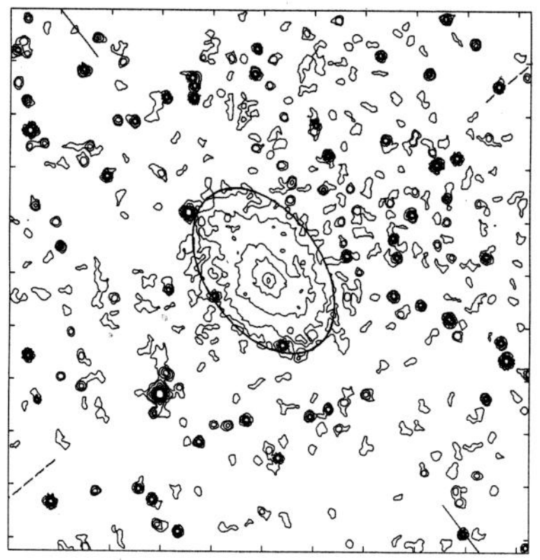

N5585 K4664 2 21.40 512X512

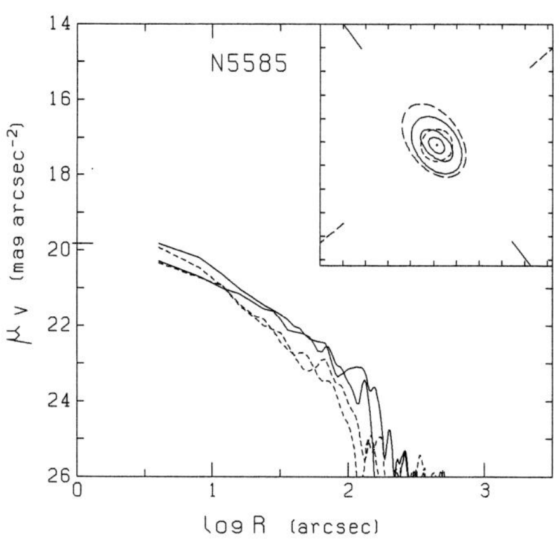

NGC5574

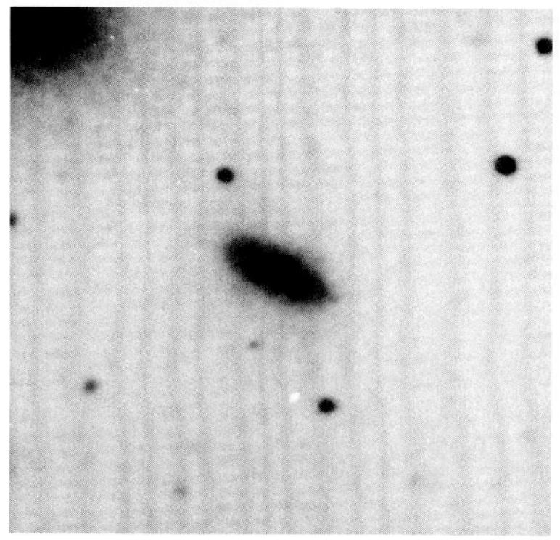

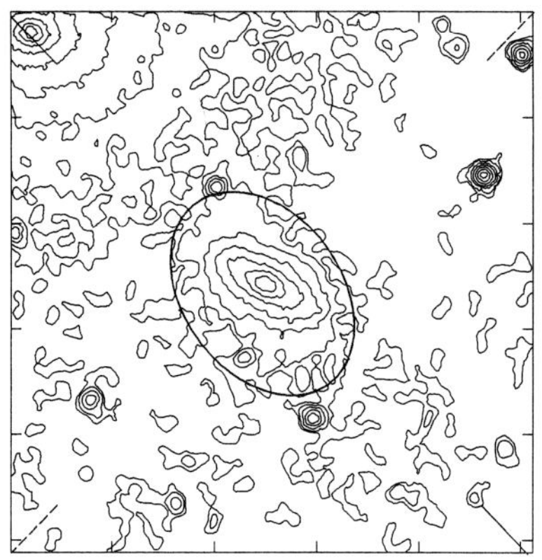

N5574 K4026 1 21.20 256X256

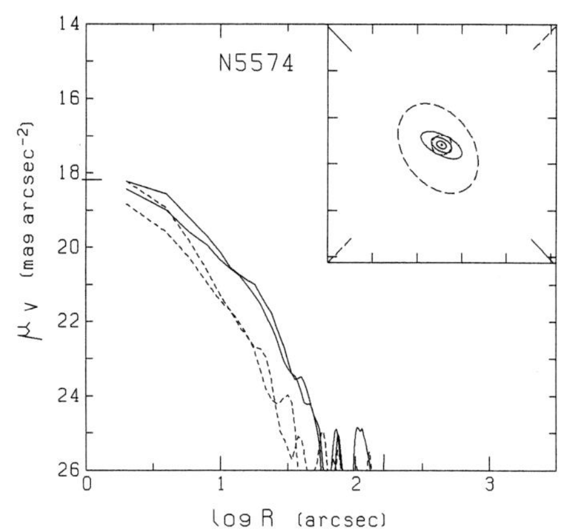

NGC5576

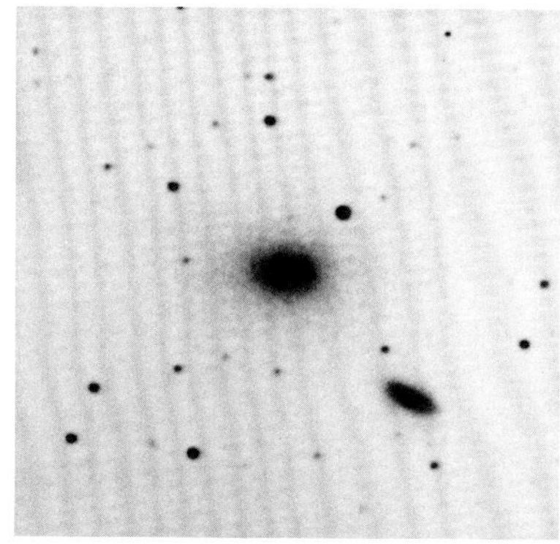

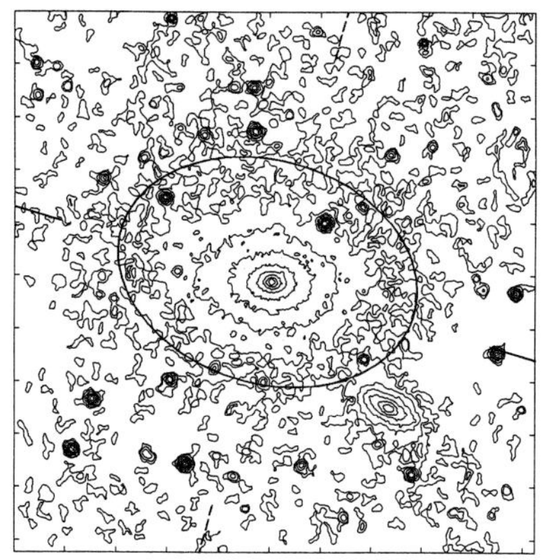

N5576 K4026 1 21.20 512X512

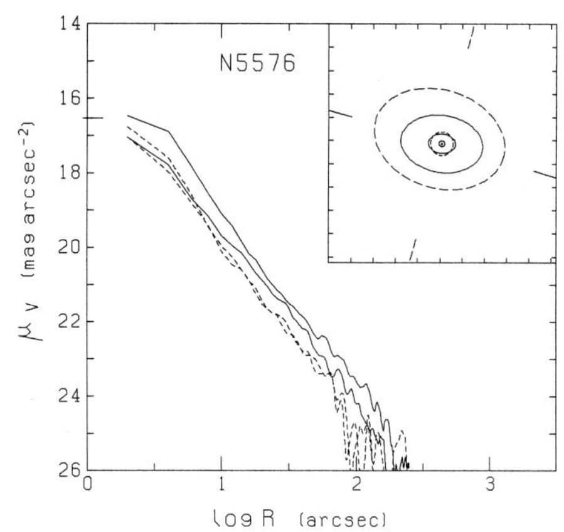

NGC5584

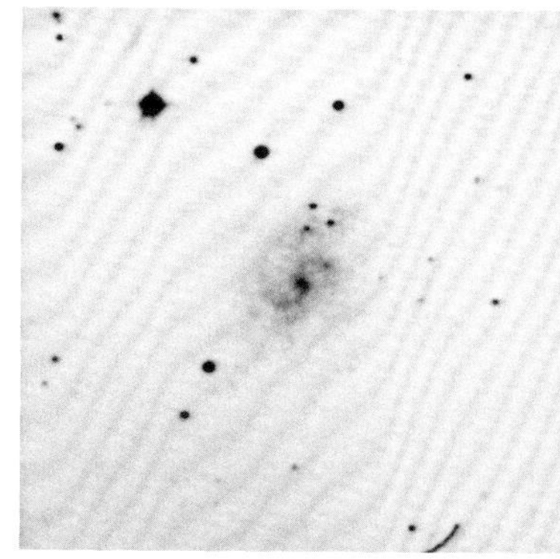

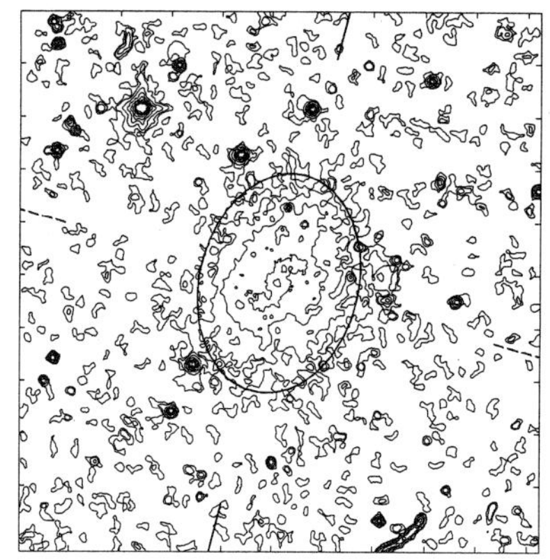

N5584 K3766 1 20.76 512X512

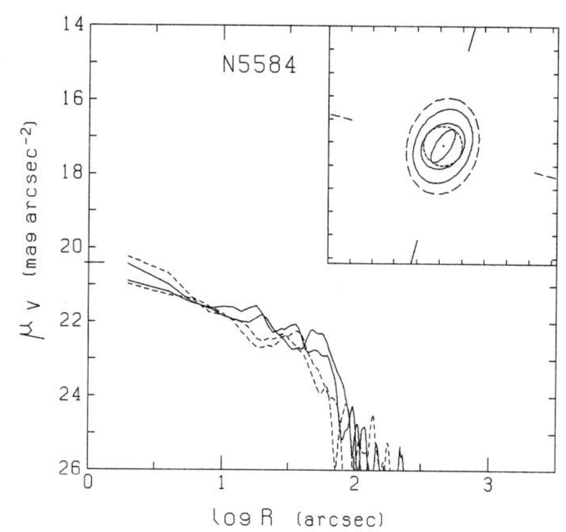

NGC5600

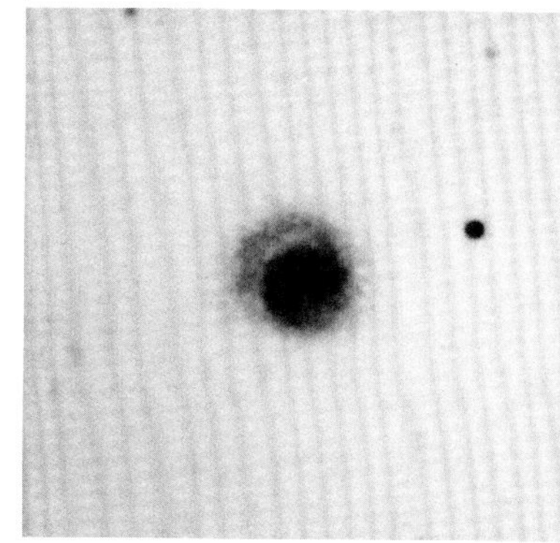

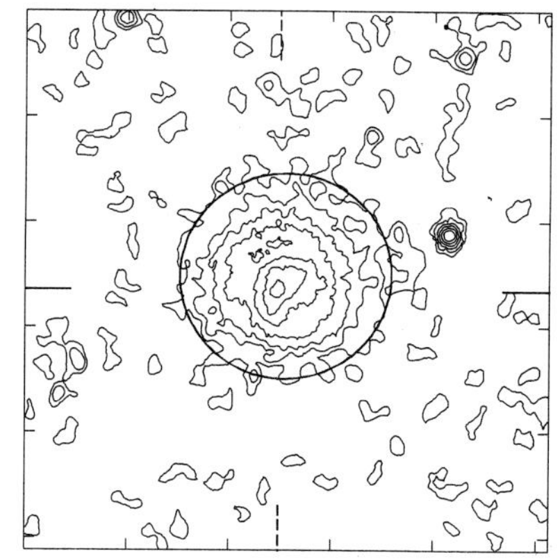

N5600 K5091 1 21.79 256X256

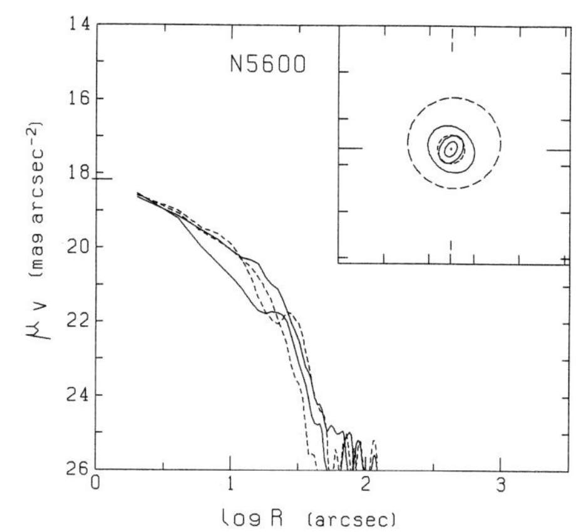

NGC5595

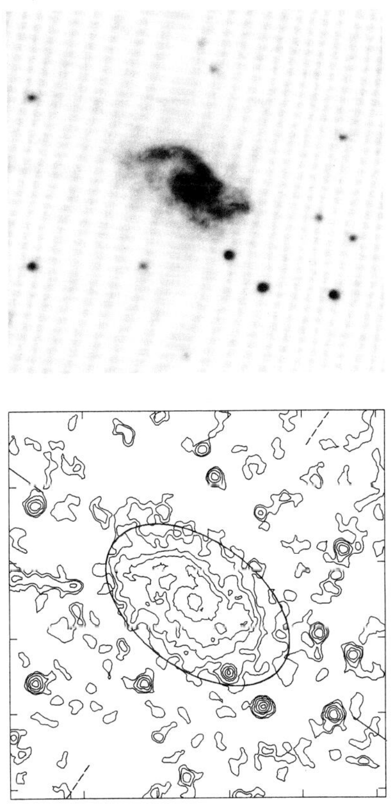

N5595 K5045 1 21.29 256X256

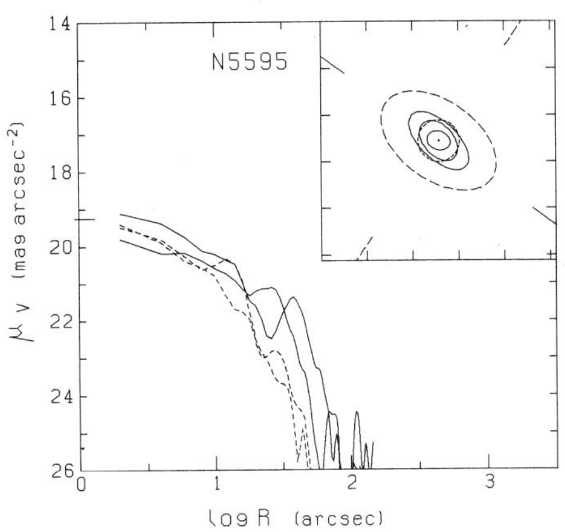

NGC5597

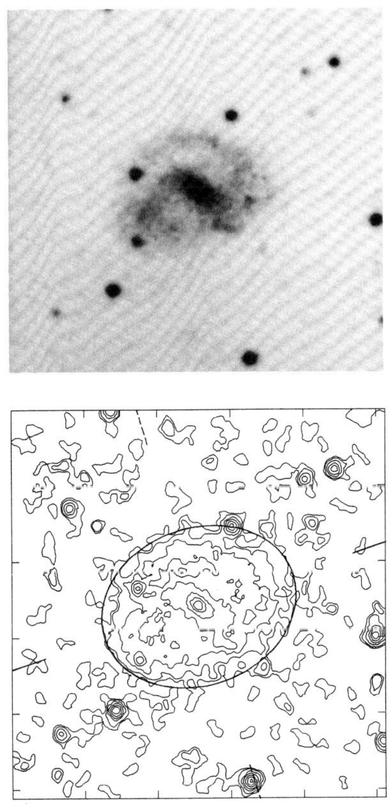

N5597 K5045 1 21.29 256X256

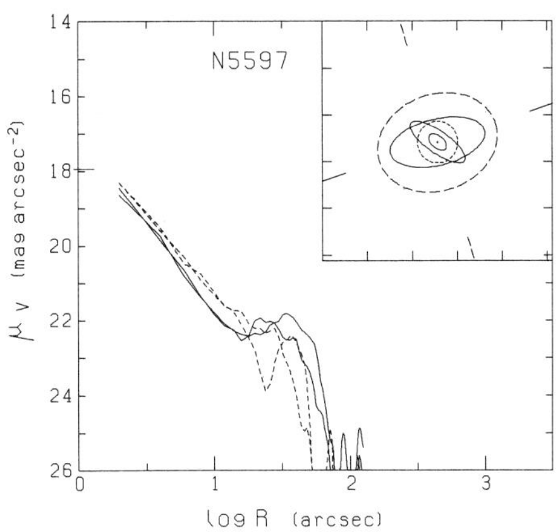

NGC5614

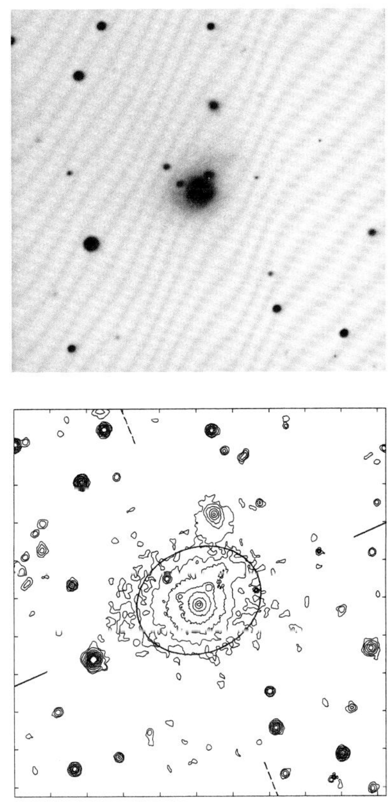

N5614 K2649 1 21.66 512X512

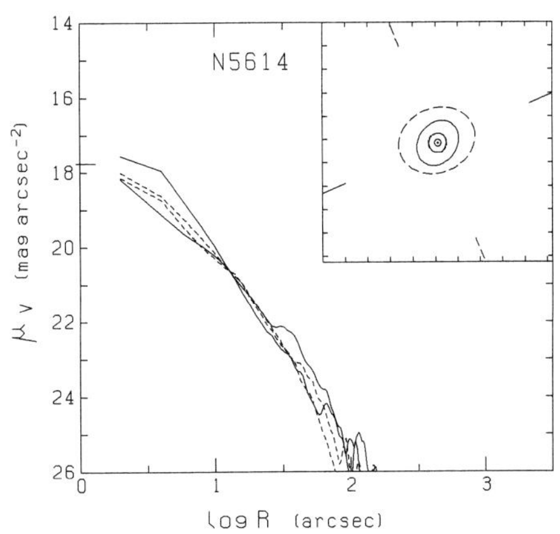

NGC5605

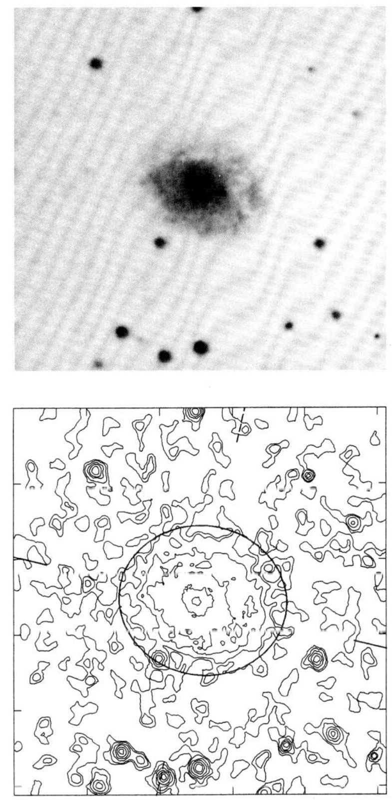

N5605 K5045 1 21.29 256X256

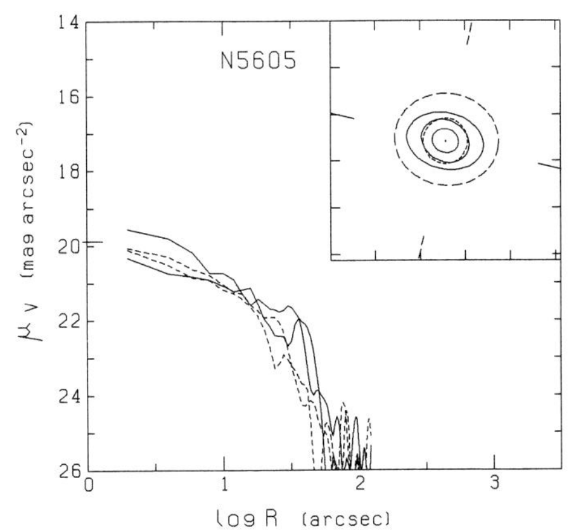

NGC5641

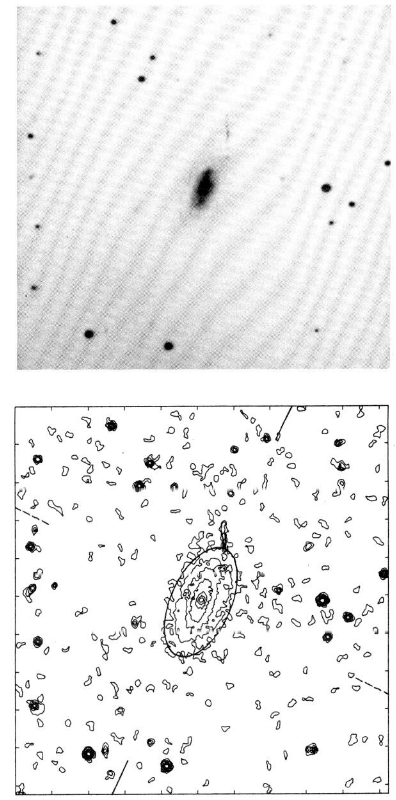

N5641 K4640 1 22.16 512X512

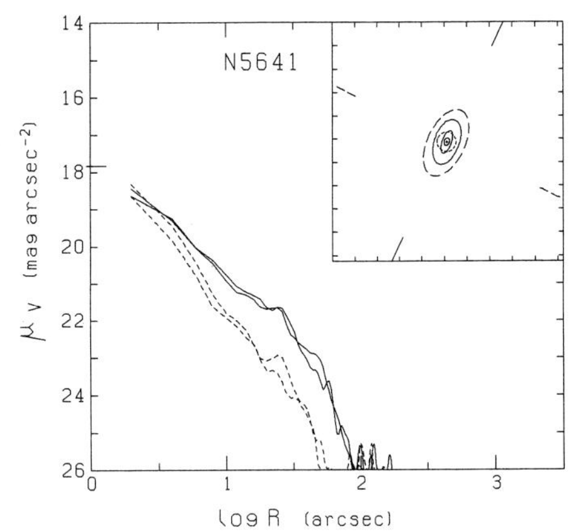

NGC5638

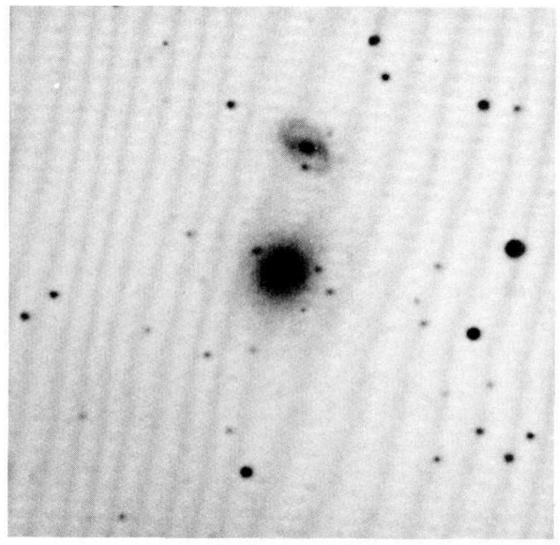

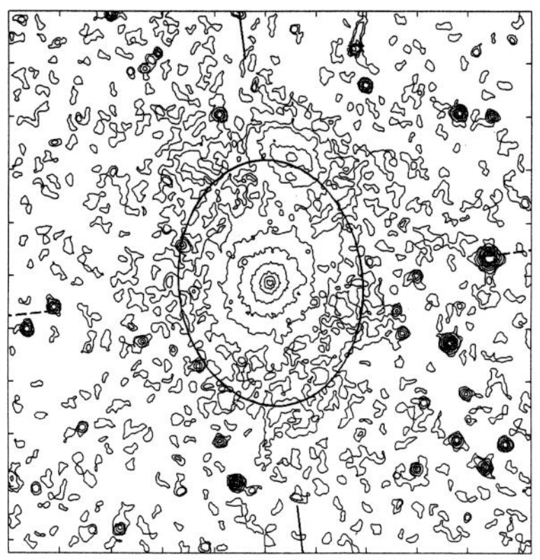

N5638 K4026 1 21.20 512X512

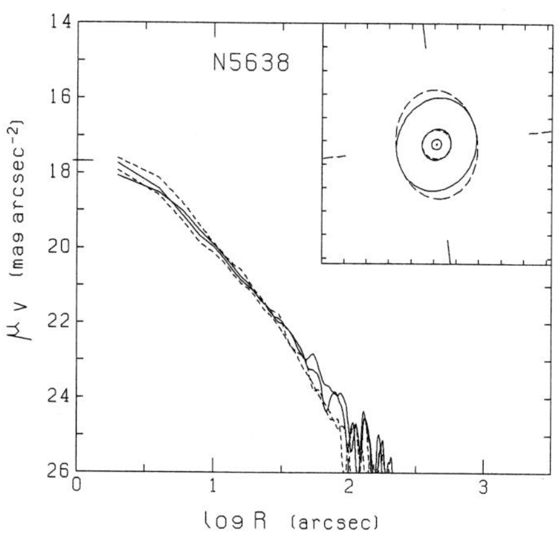

NGC5653

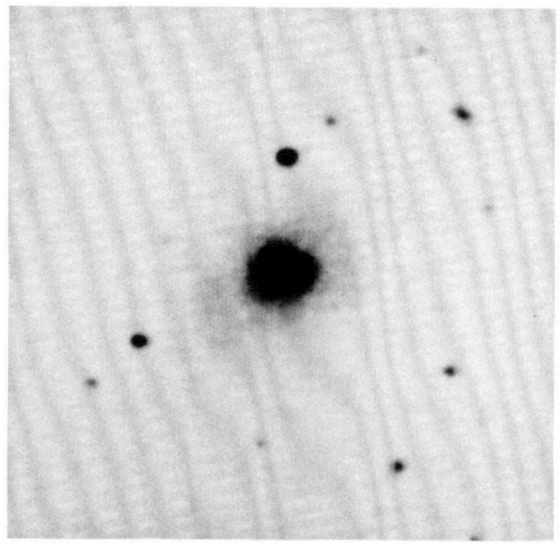

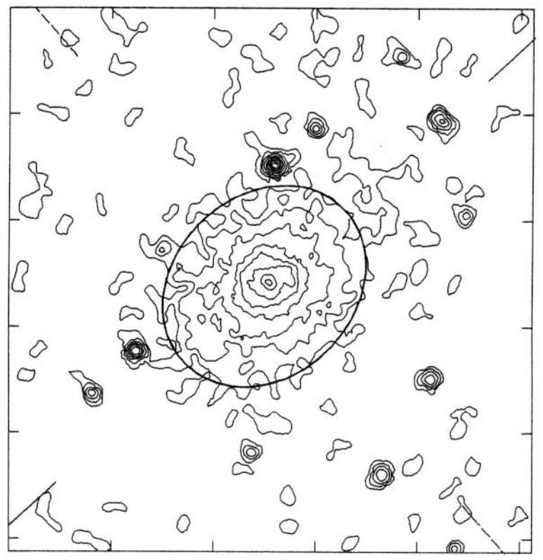

N5653 K4640 1 22.16 256X256

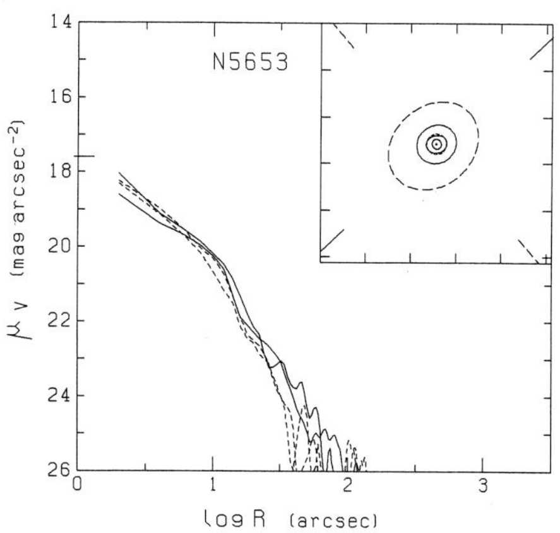

NGC5660

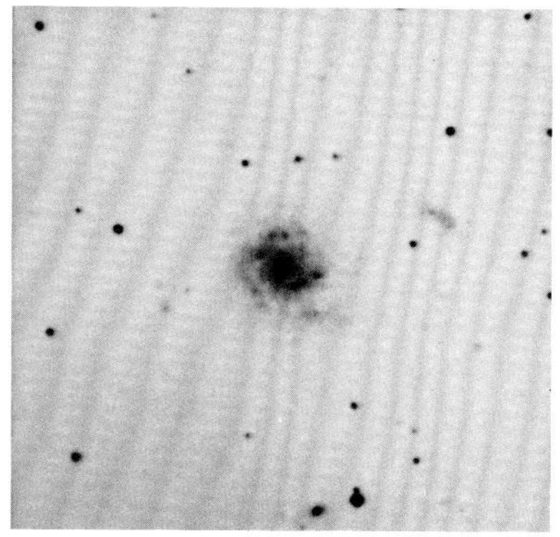

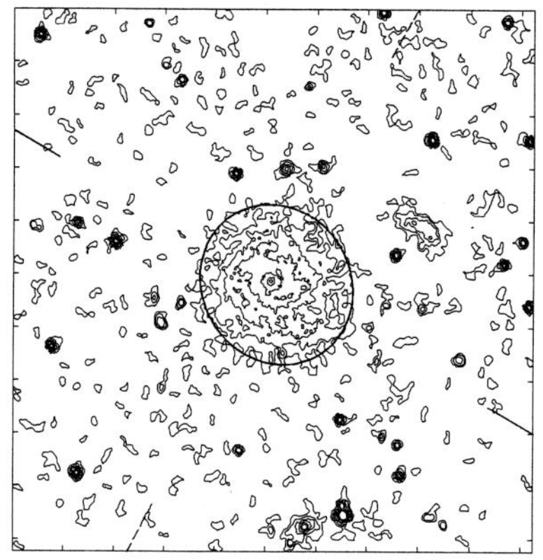

N5660 K4979 1 21.90 512X512

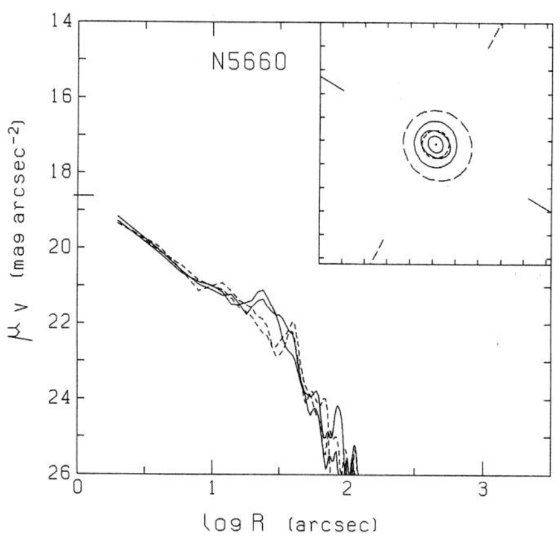

NGC5645

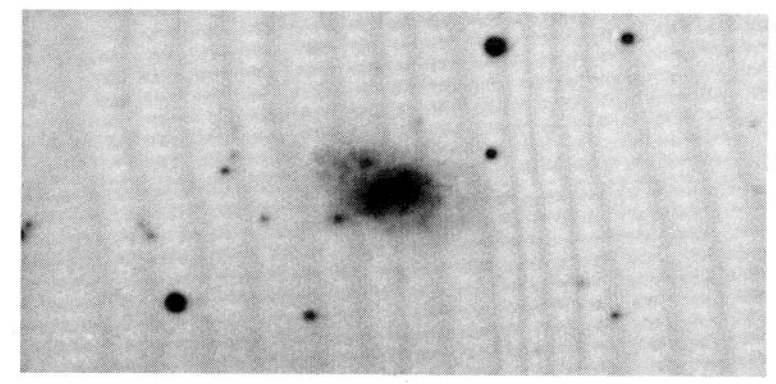

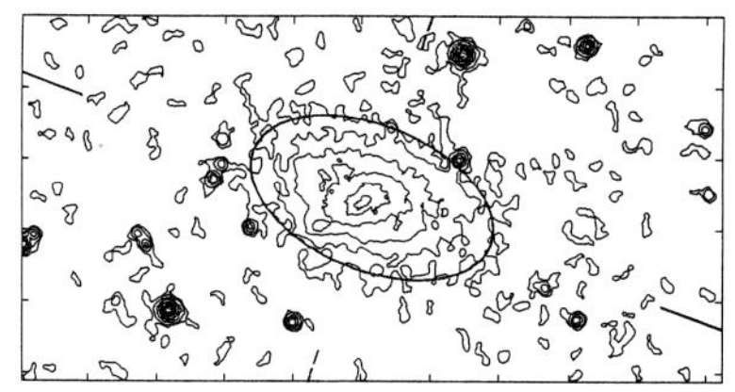

N5645 K3714 1 21.30 512X256

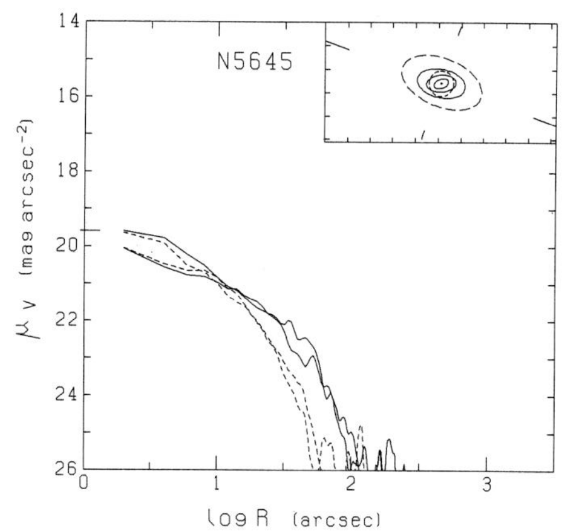

NGC5665

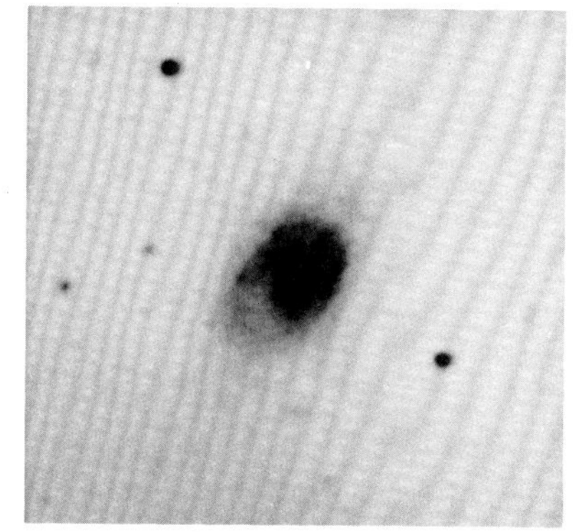

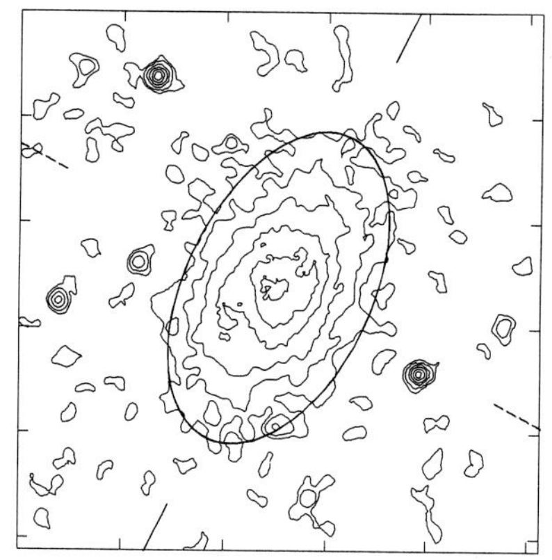

N5665 K3714 1 21.30 256X256

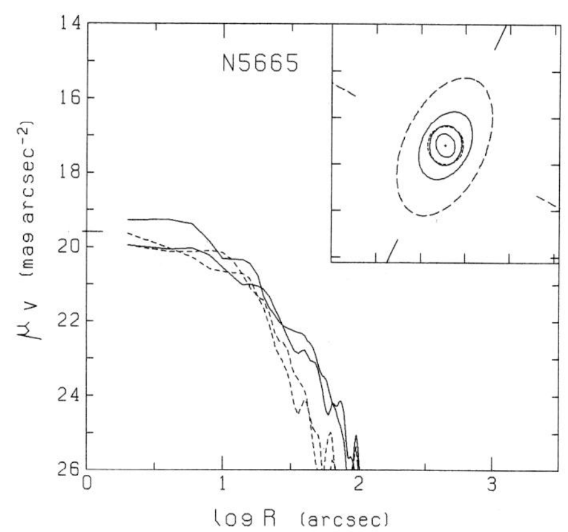

NGC5668

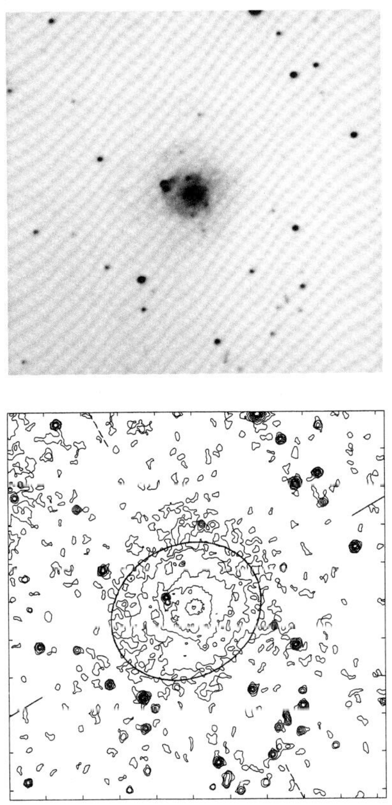

N5668 K3714 1 21.30 512X512

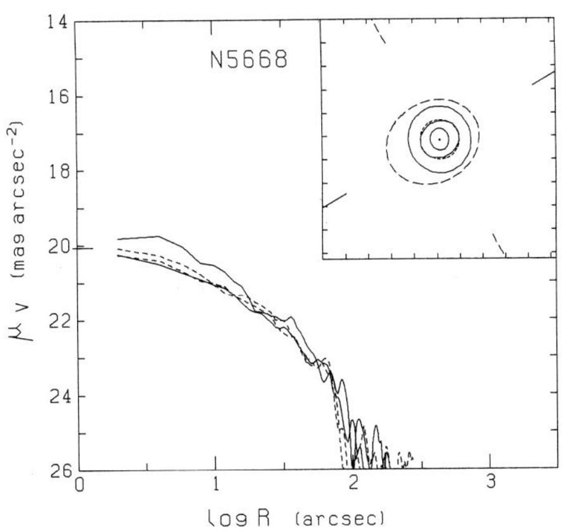

NGC5676

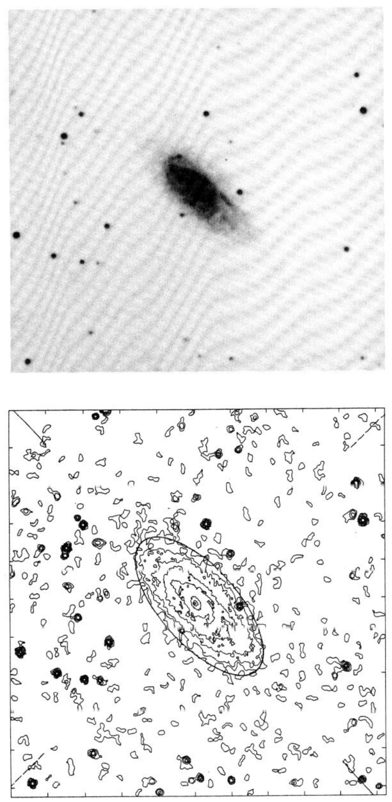

N5676 K4979 1 21.90 512X512

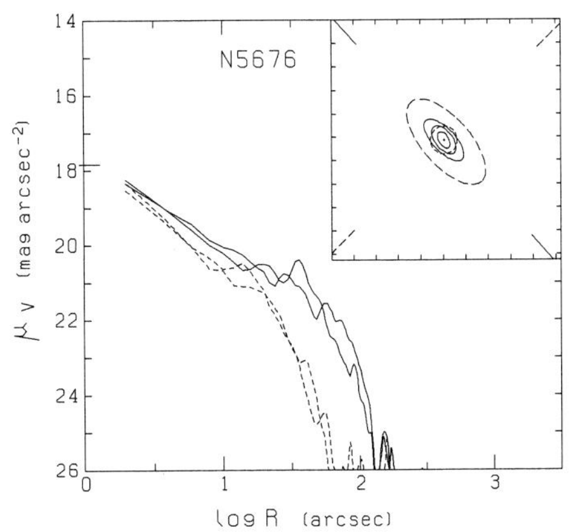

NGC5687

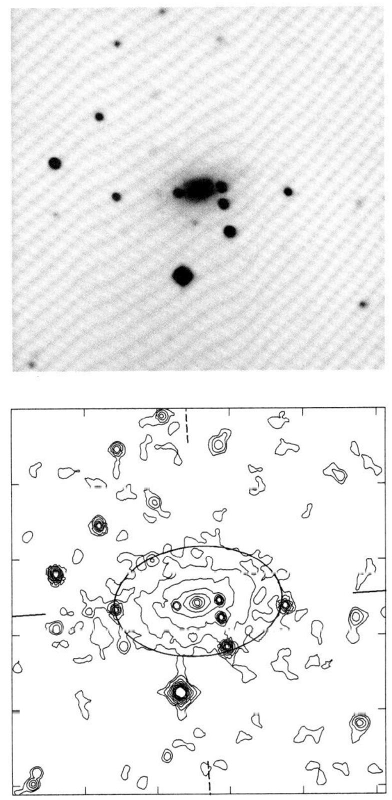

N5687 K4664 2 21.40 256X256

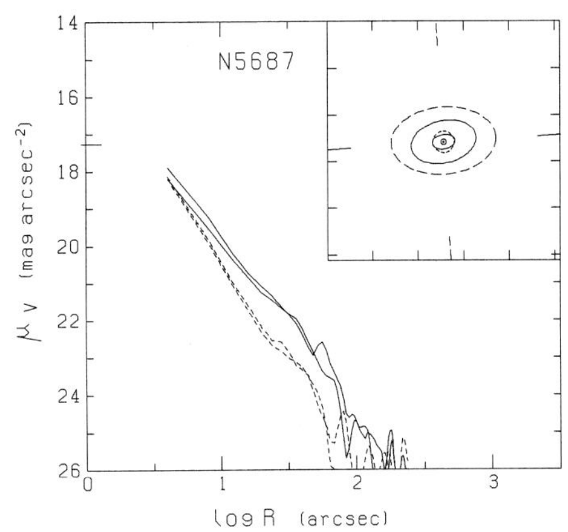

NGC5689

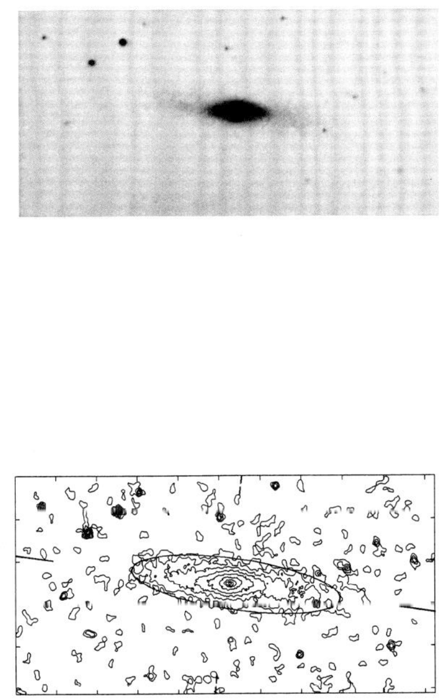

N5689 K4979 1 21.90 512X256

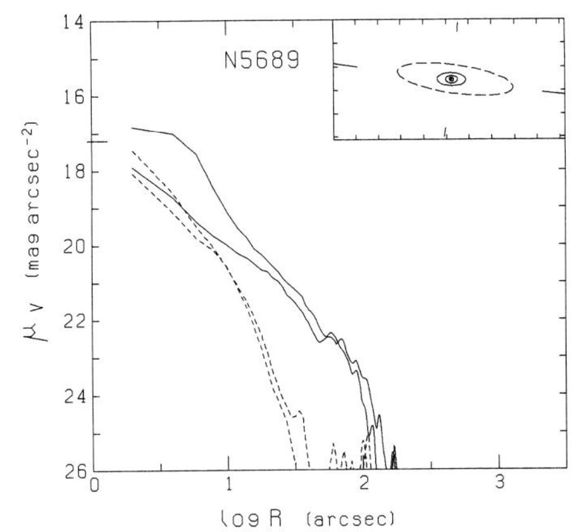

NGC5690

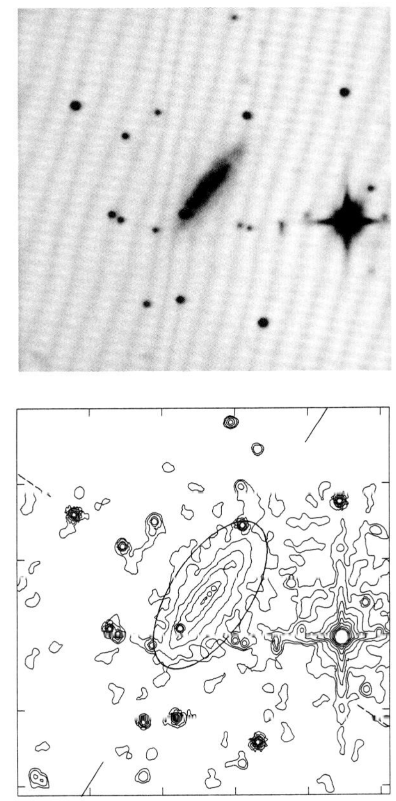

N5690 K4675 2 21.29 256X256

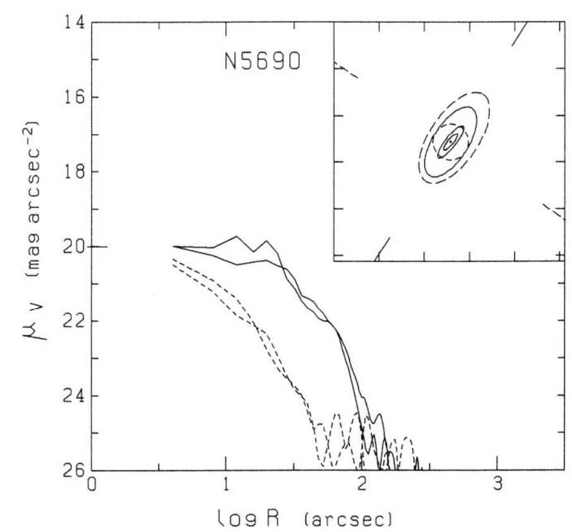

NGC5691

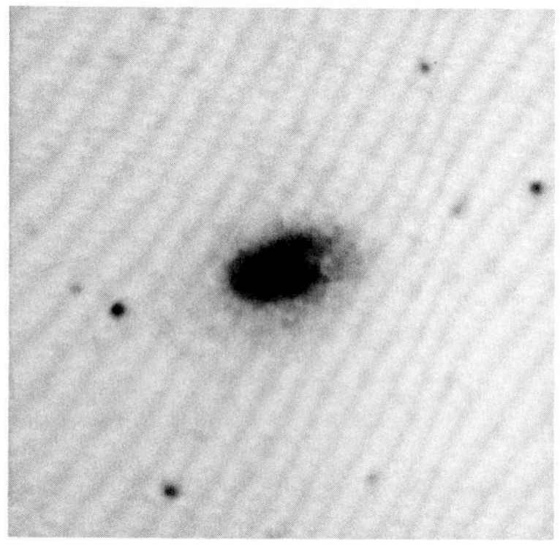

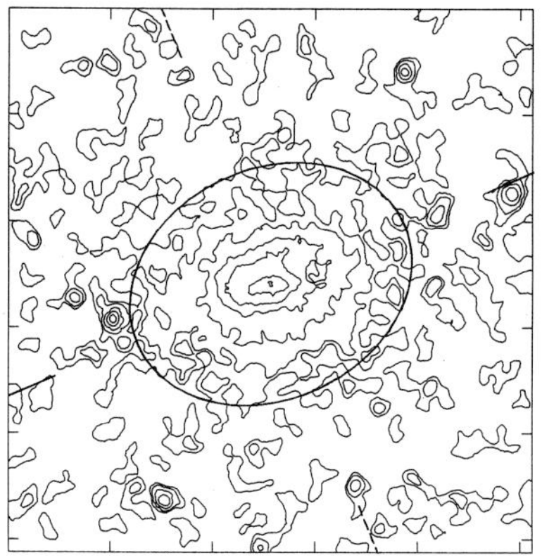

N5691 K4675 1 21.29 256X256

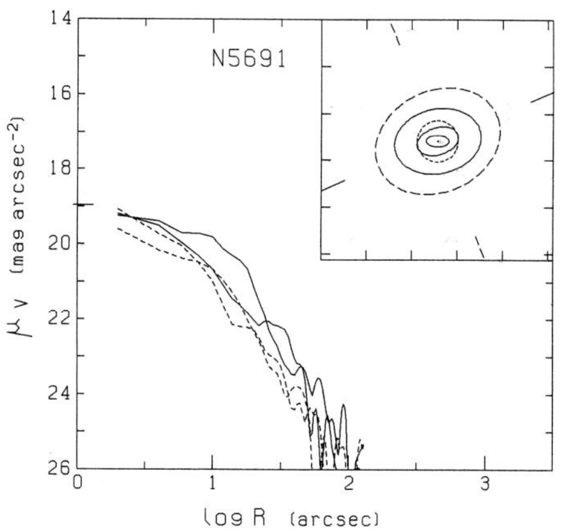

NGC5701

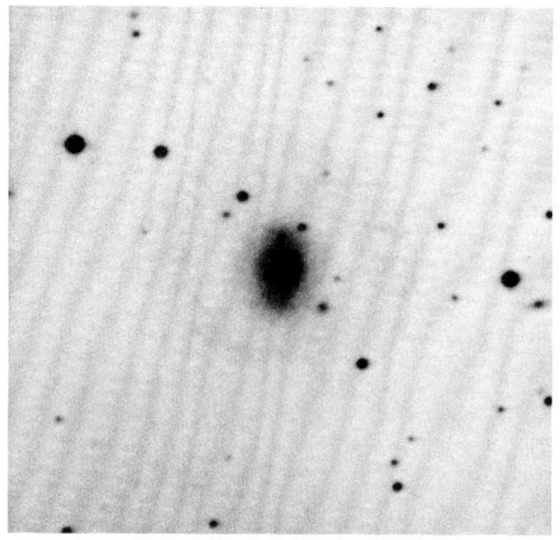

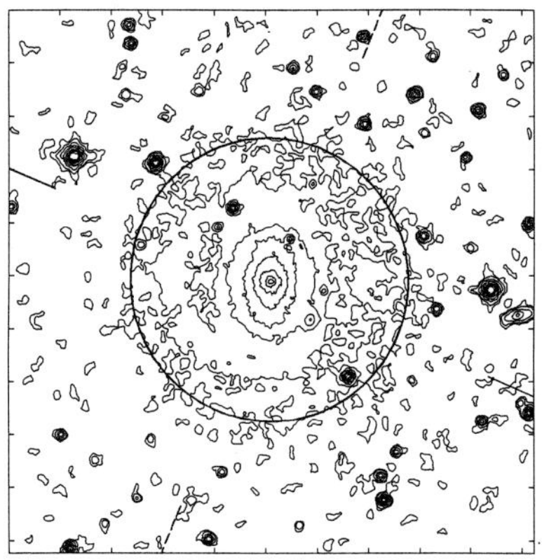

N5701 K3714 1 21.30 512X512

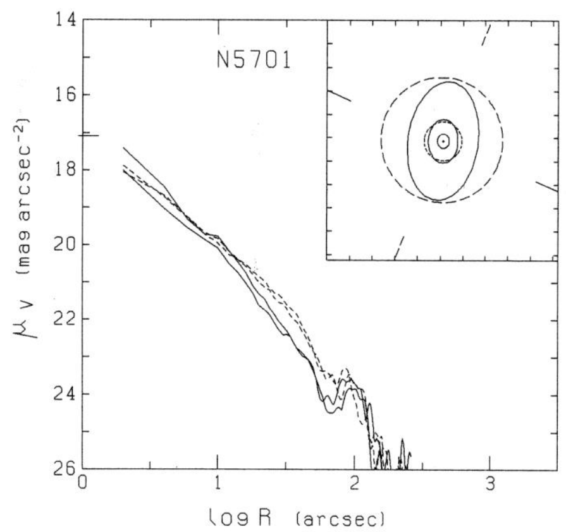

NGC5713

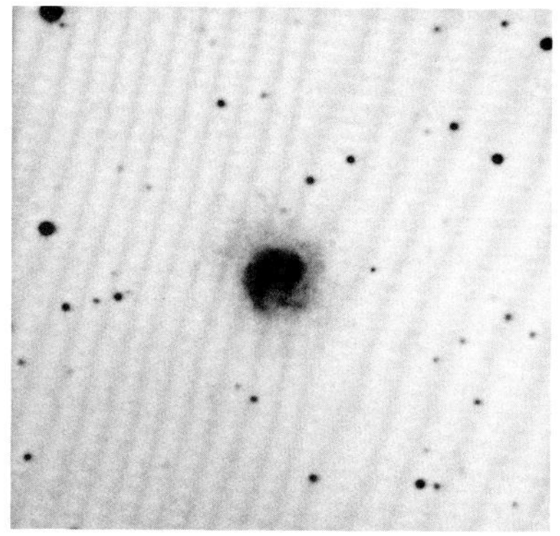

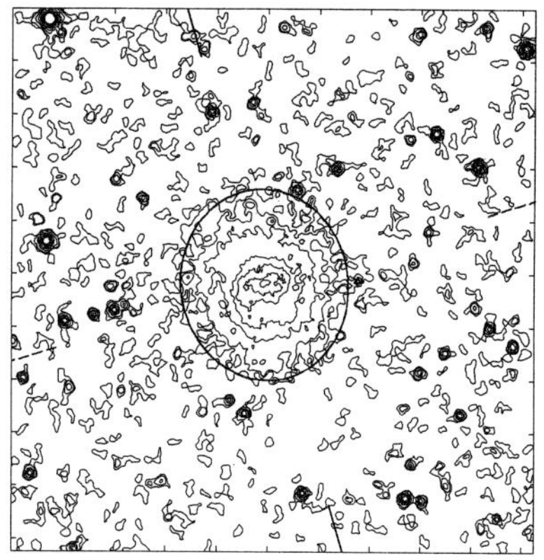

N5713 K4675 1 21.29 512X512

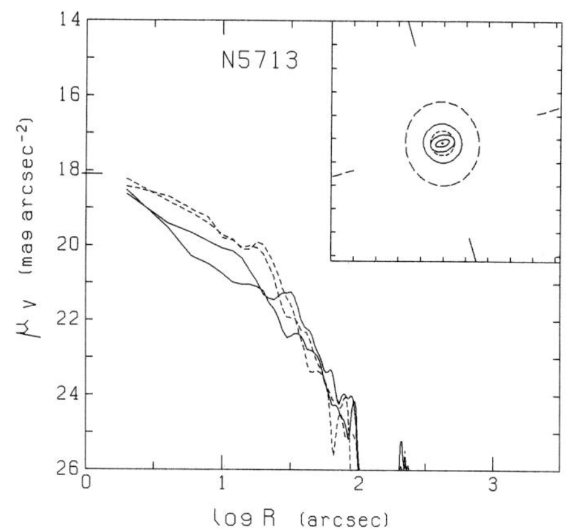

NGC5728

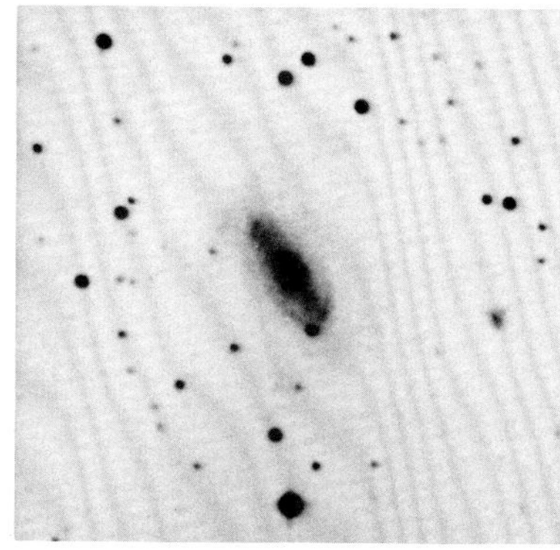

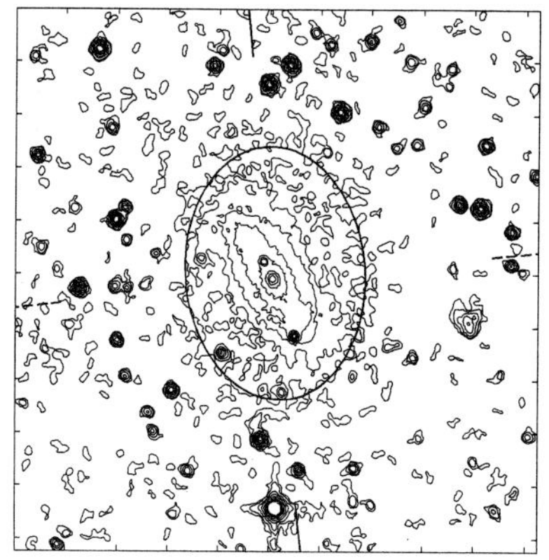

N5728 K5422 1 21.26 512X512

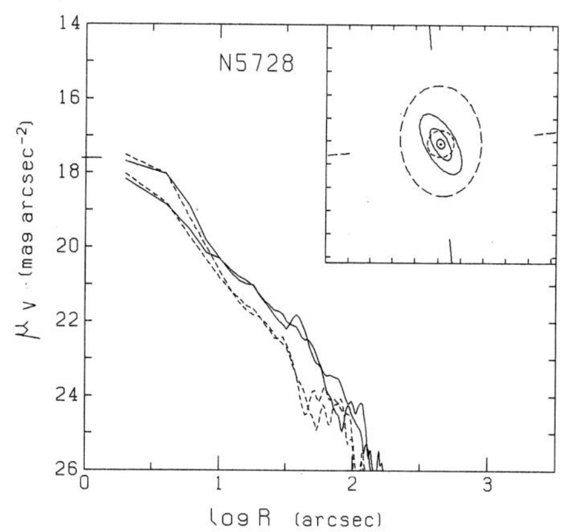

NGC5739

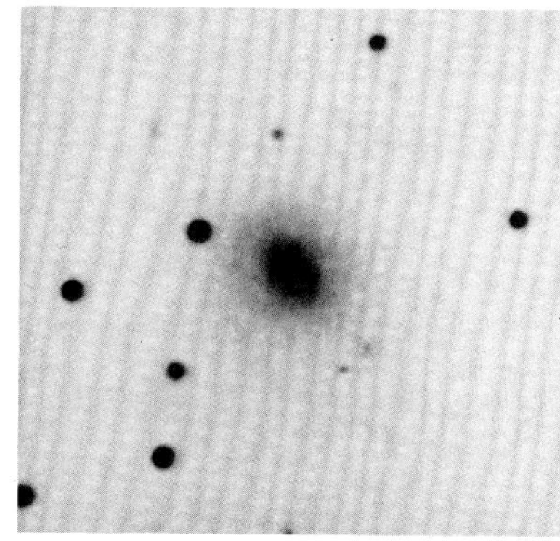

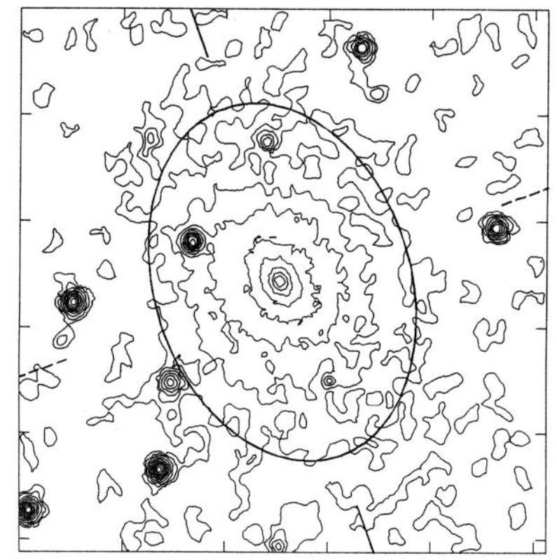

N5739 K5080 1 21.98 256X256

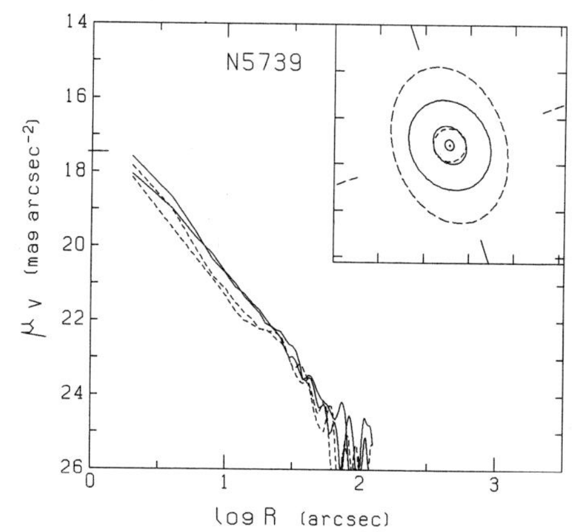

NGC5740

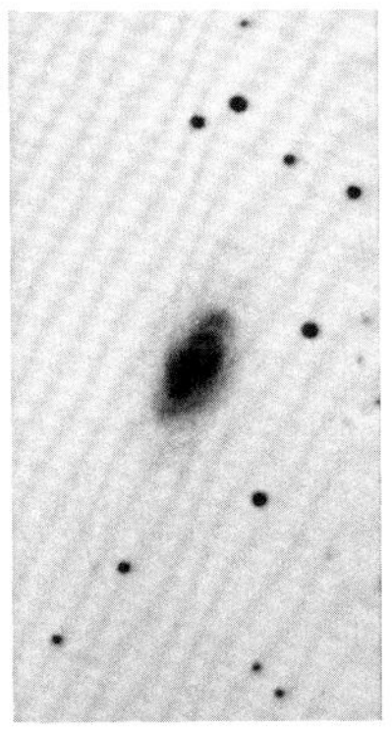

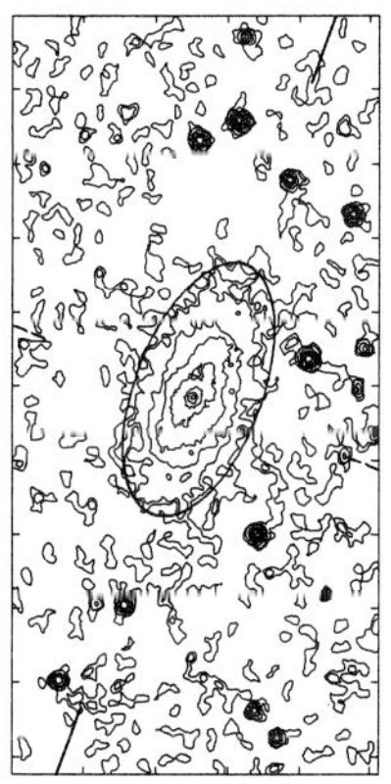

N5740 K4675 1 21.29 256X512

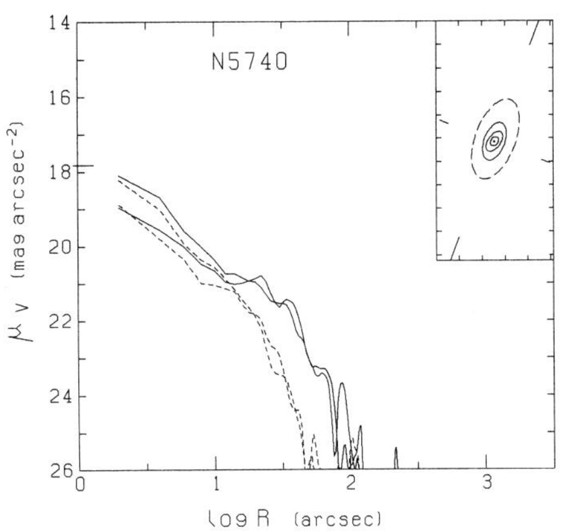

NGC5746

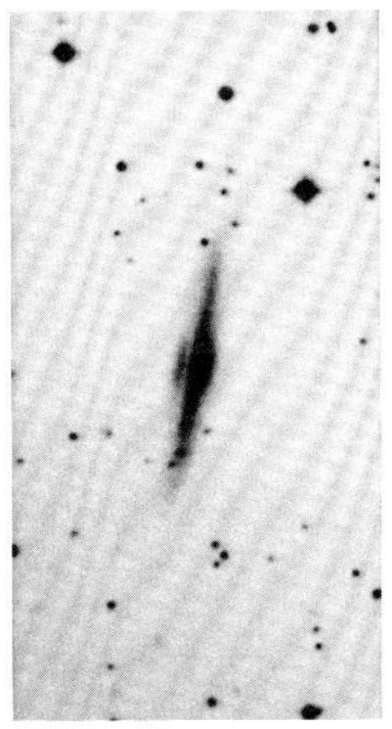

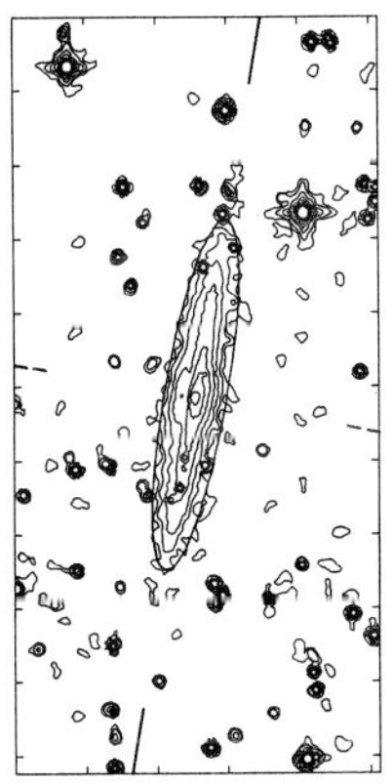

N5746 K4675 2 21.29 256X512

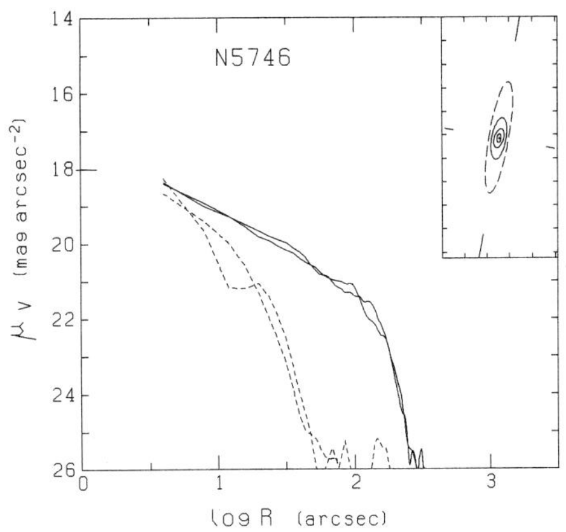

NGC5750

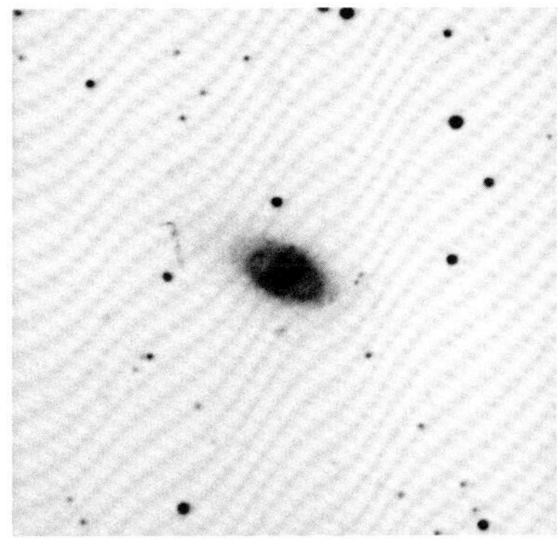

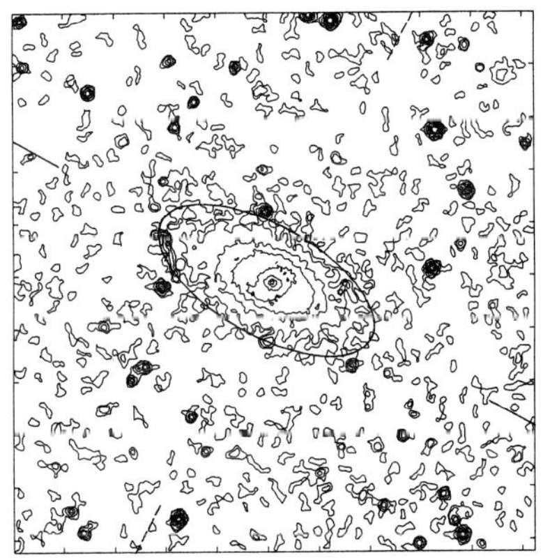

N5750 K4675 1 21.29 512X512

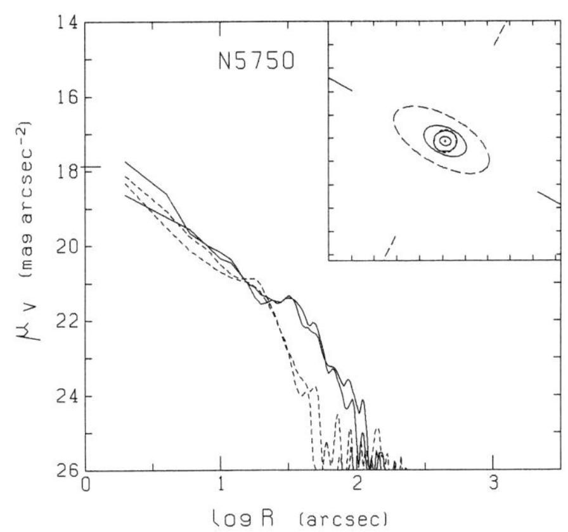

NGC5756

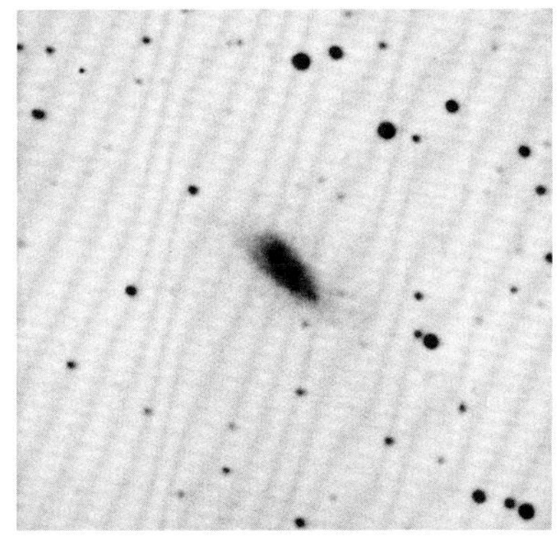

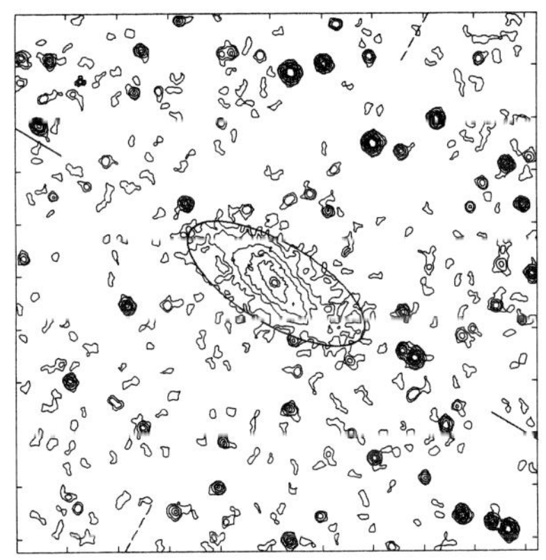

N5756 K5422 1 21.26 512X512

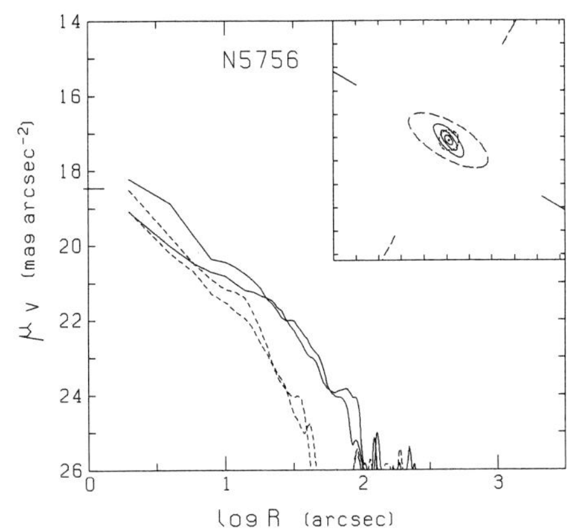

NGC5757

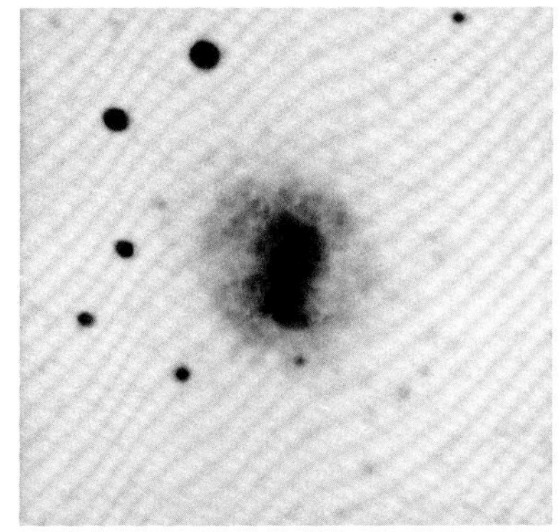

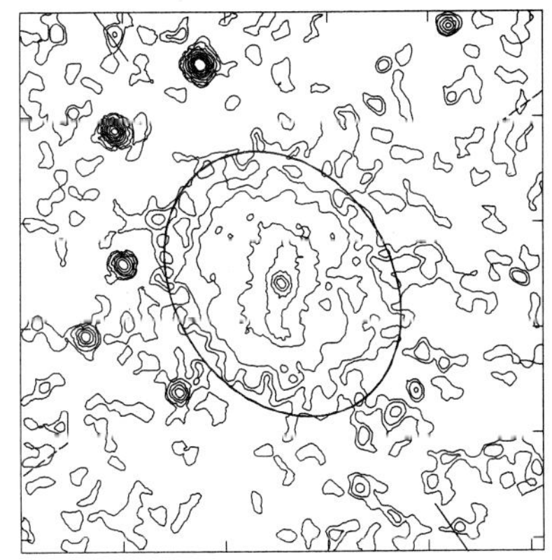

N5757 K5439 1 21.12 256X256

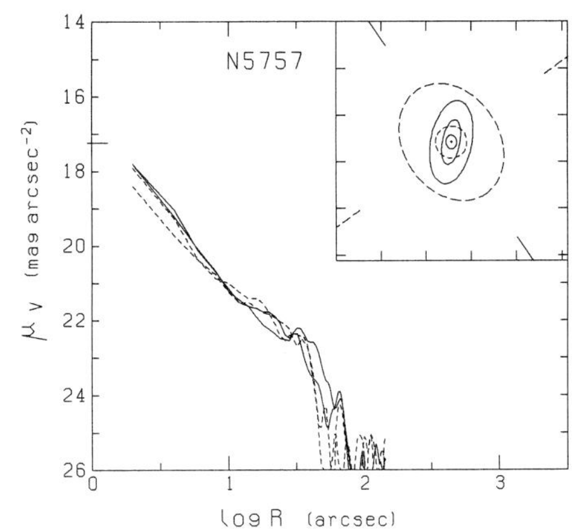

NGC5768

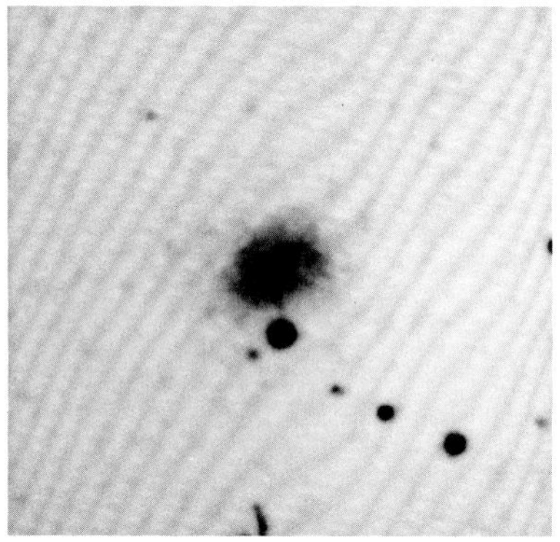

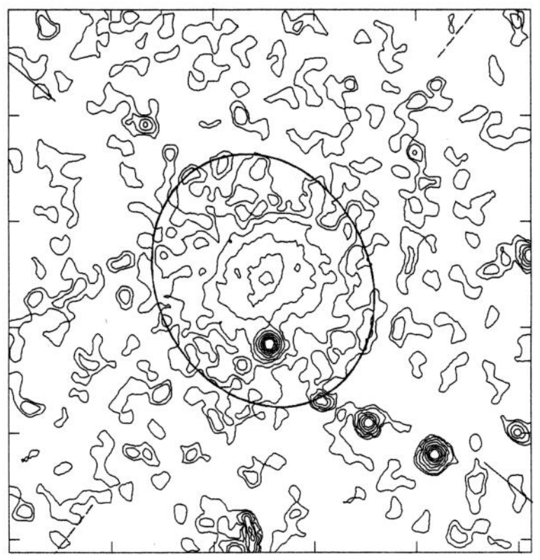

N5768 K4675 1 21.29 256X256

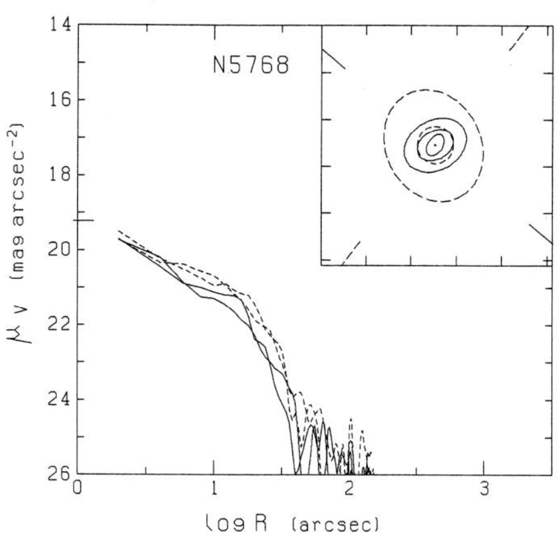

NGC5775

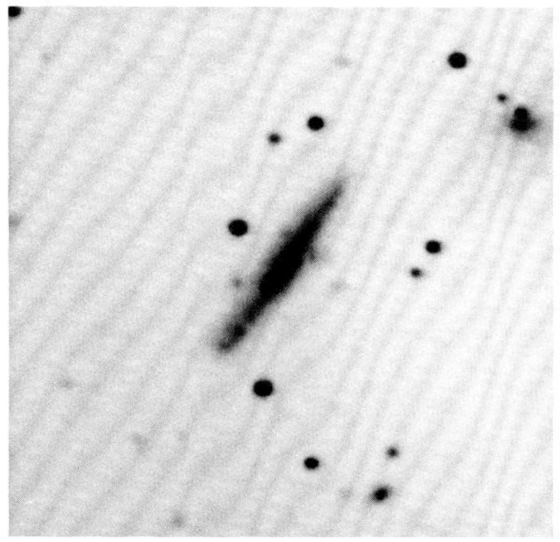

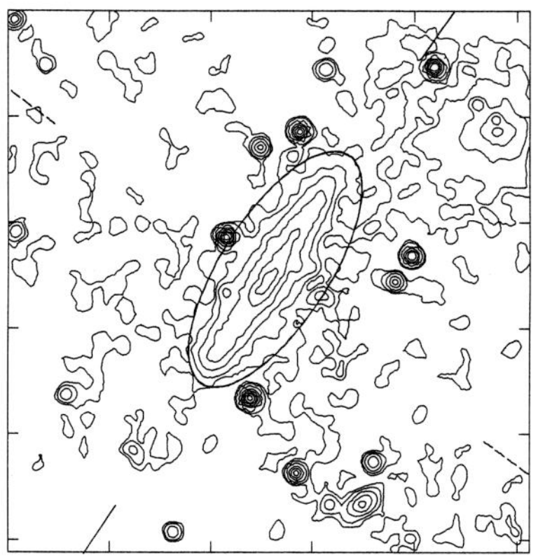

N5775 K4389 2 20.64 256X256

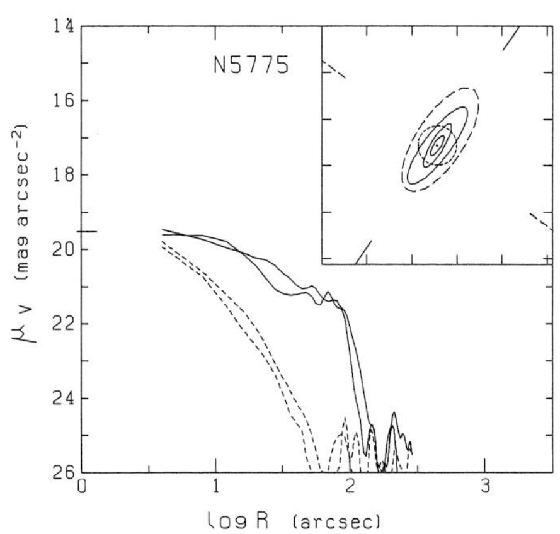

NGC5792

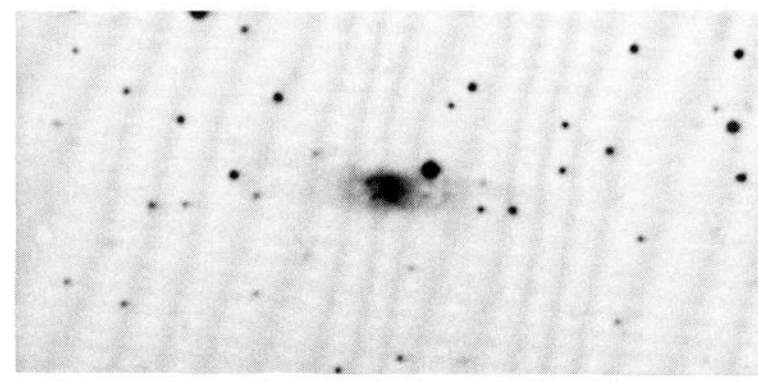

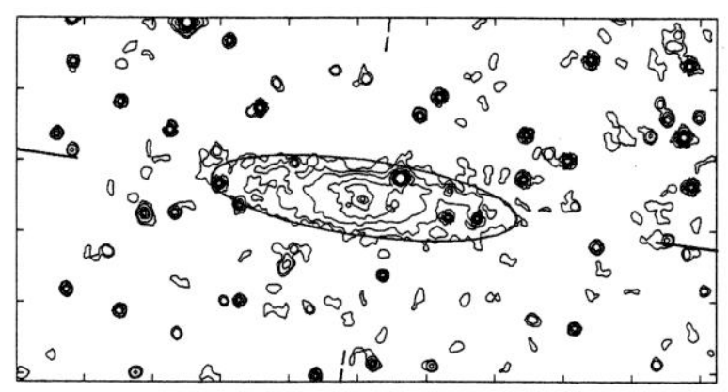

N5792 K4378 2 21.07 512X256

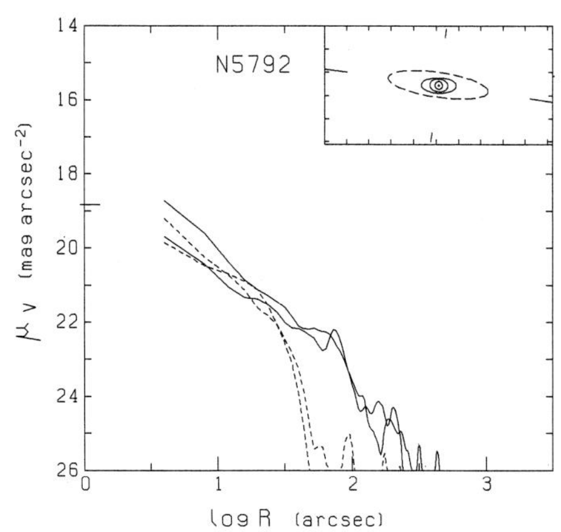

NGC5796

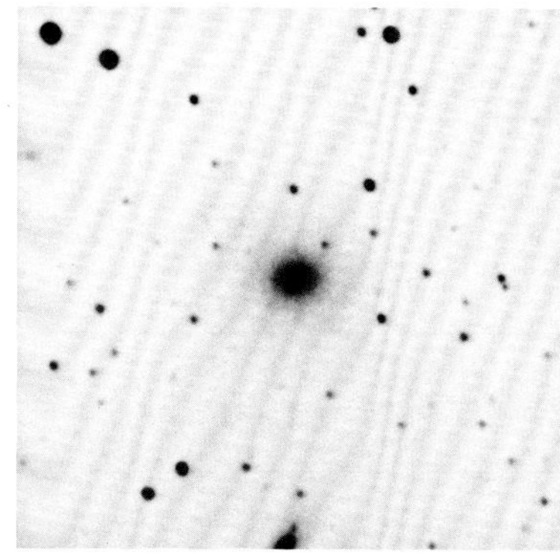

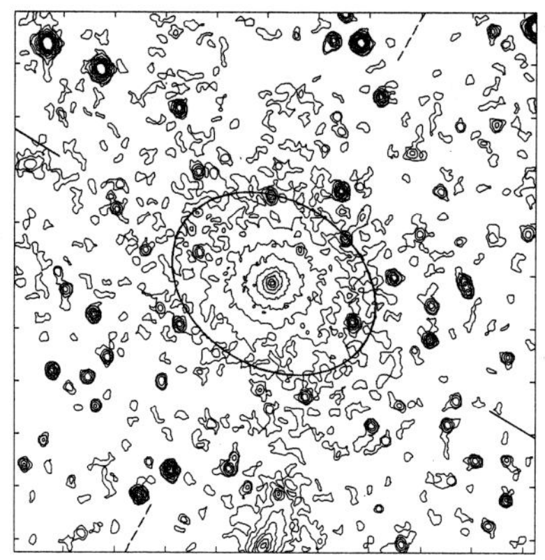

N5796 K5422 1 21.26 512X512

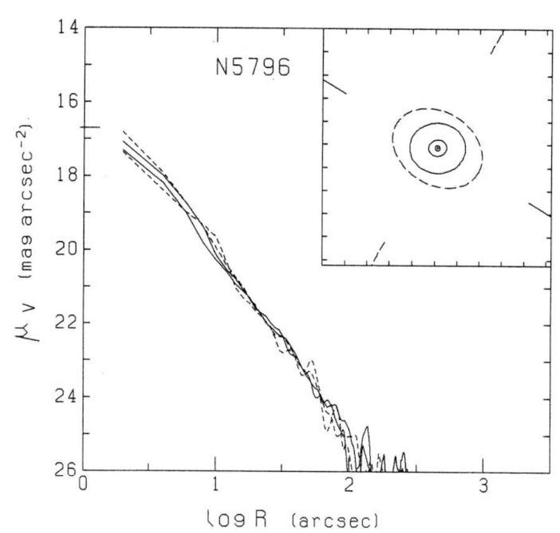

NGC5820

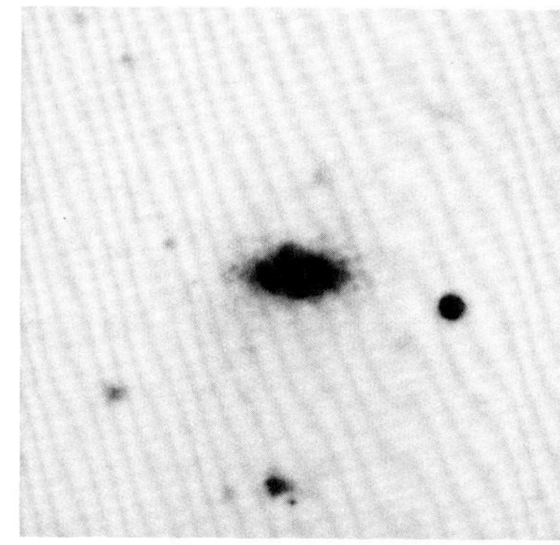

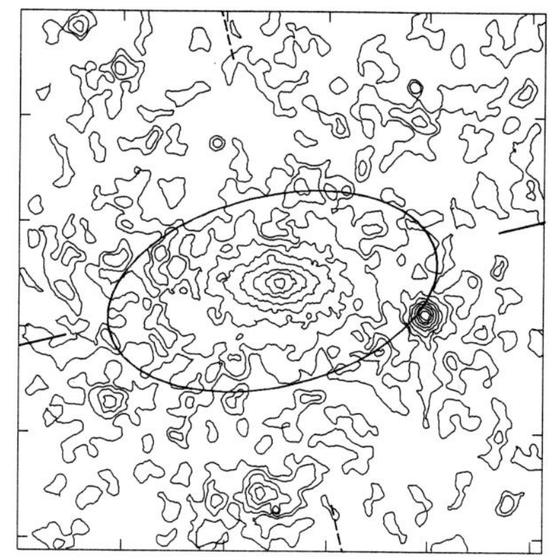

N5820 K4654 1 21.74 256X256

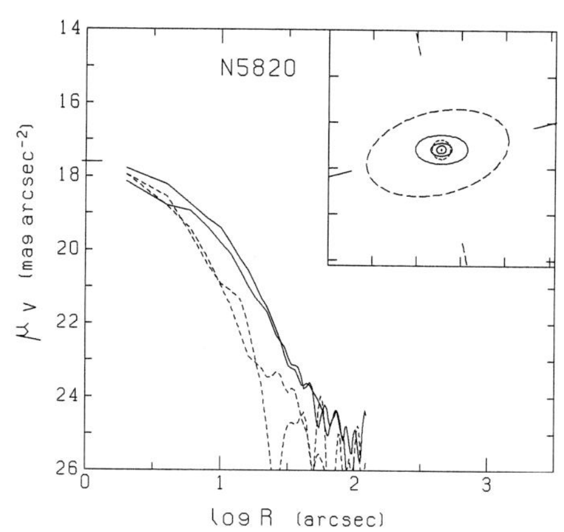

NGC5806

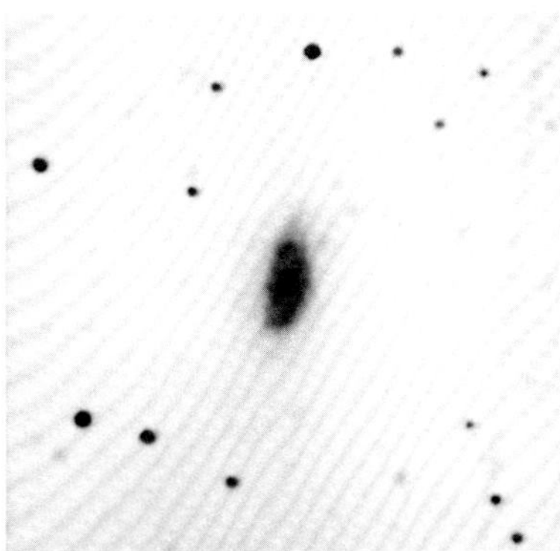

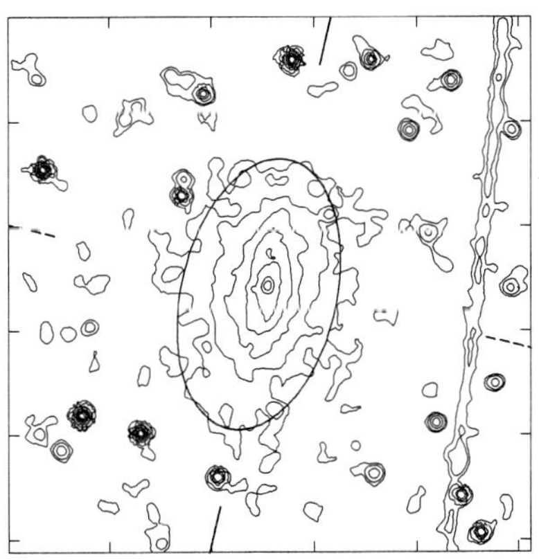

N5806 K4378 2 21.07 256X256

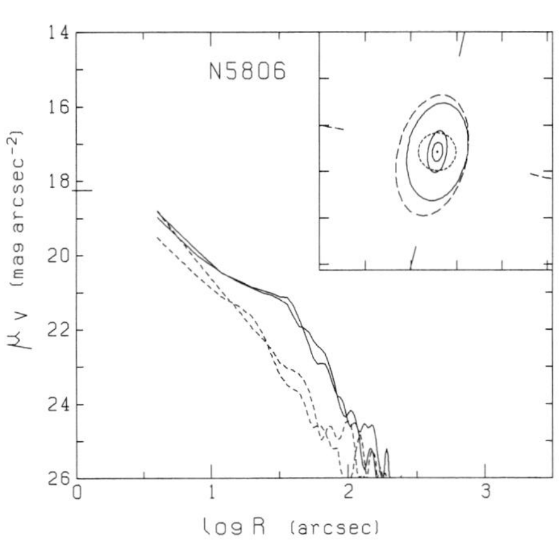

NGC5812

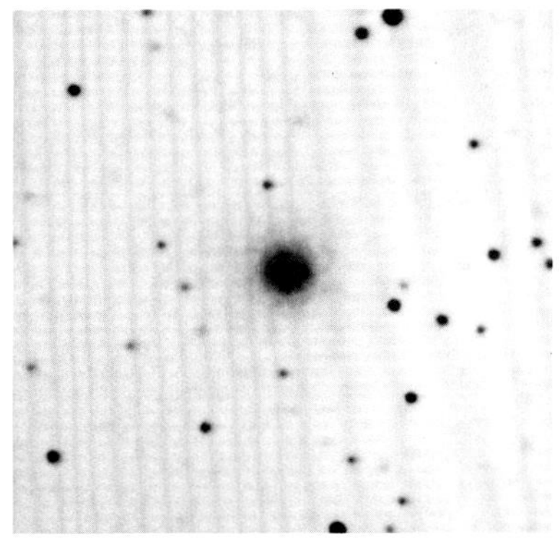

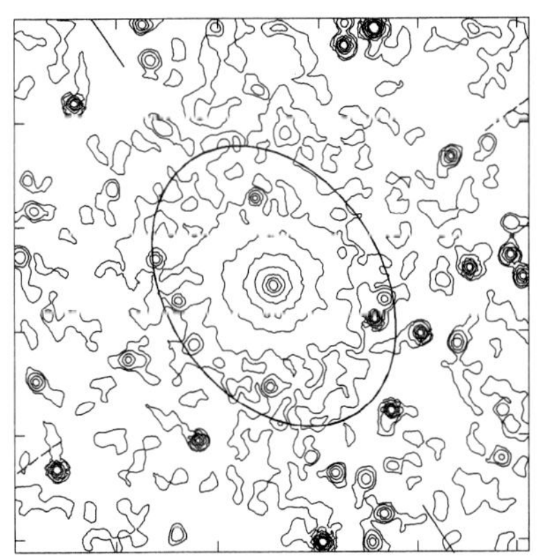

N5812 K5066 2 21.09 256X256

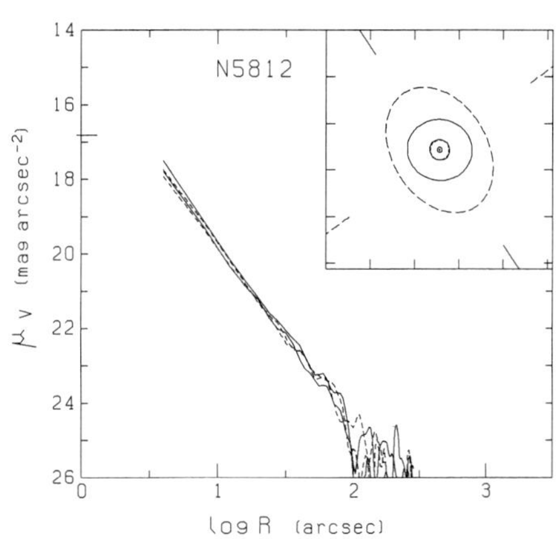

NGC5813

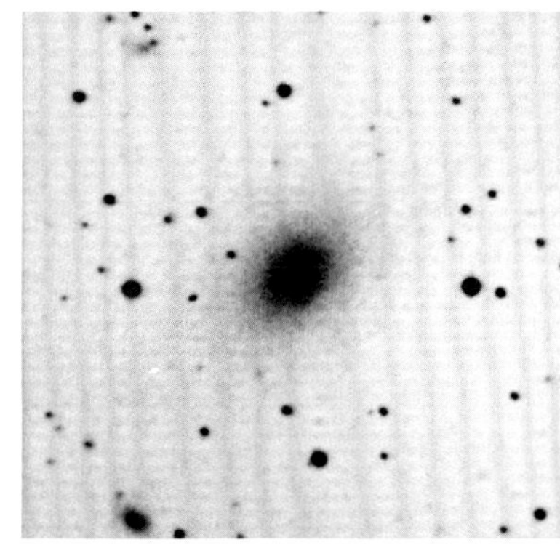

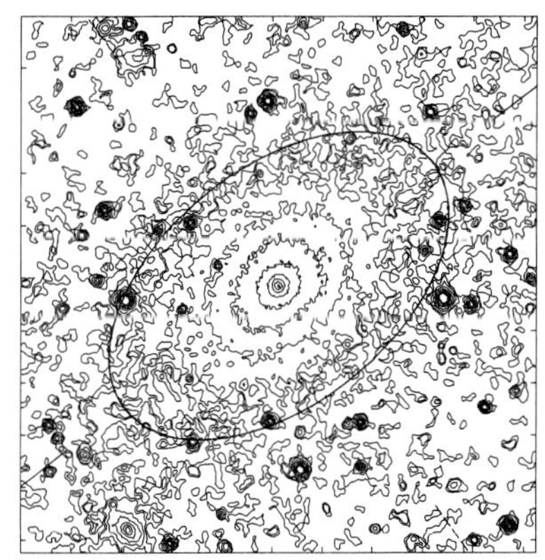

N5813 K4378 1 21.07 512X512

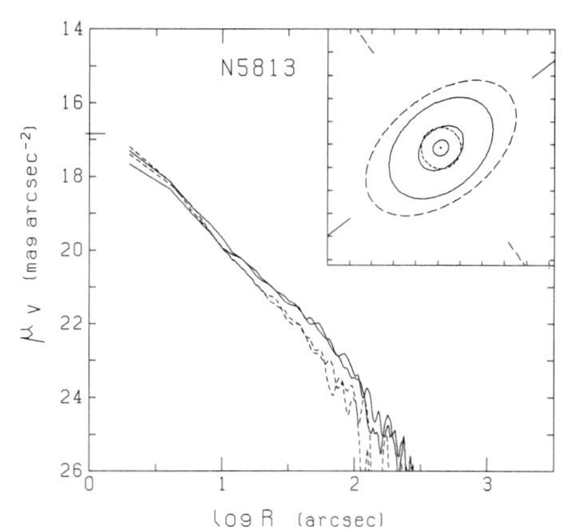

NGC5831

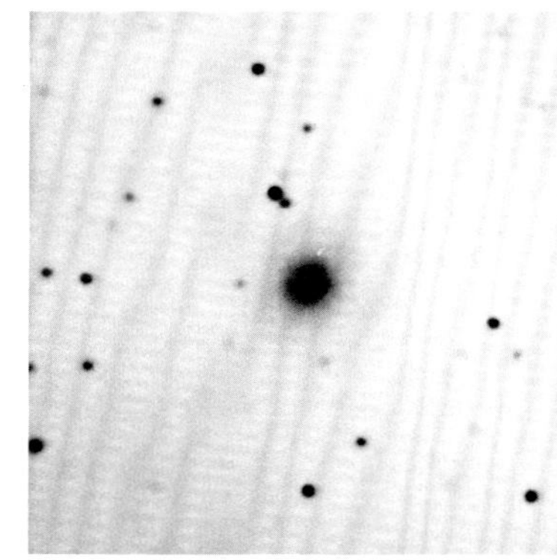

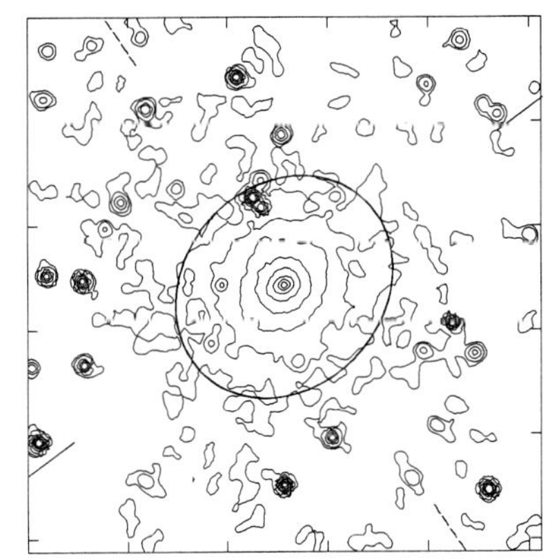

N5831 K4378 2 21.07 256X256

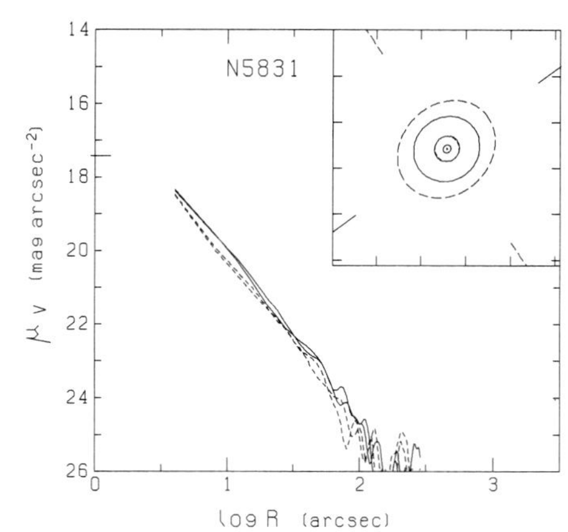

NGC5838

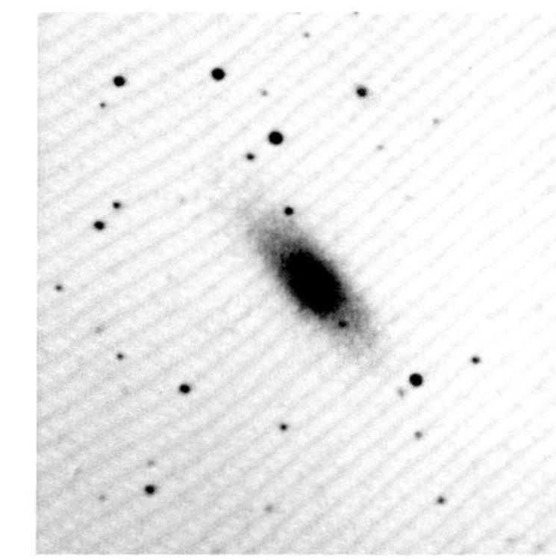

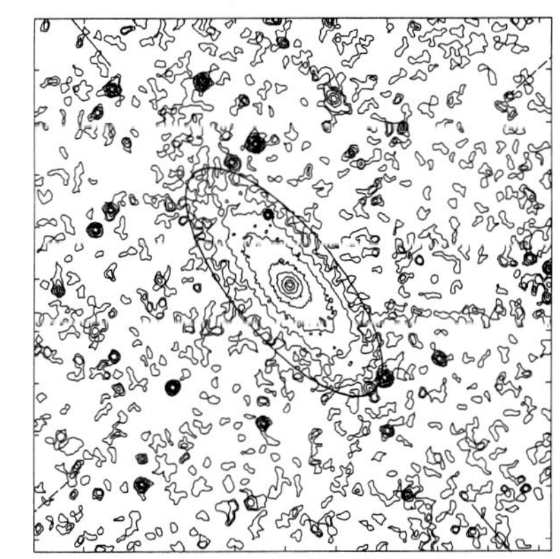

N5838 K4378 1 21.07 512X512

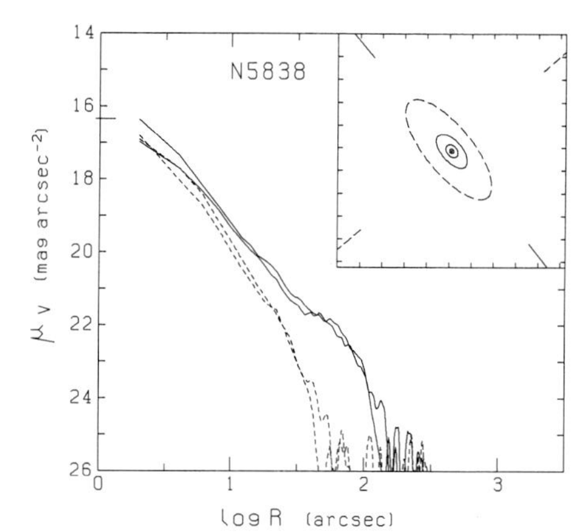

NGC5846

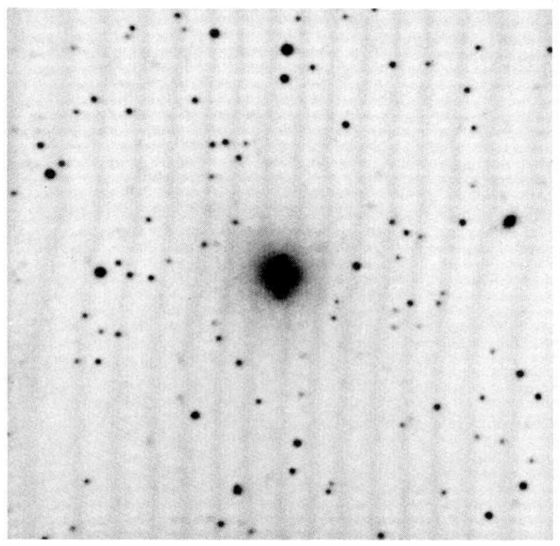

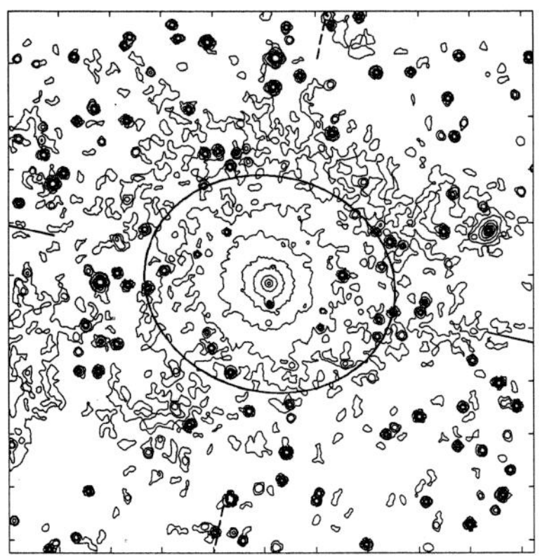

N5846 K4378 2 21.07 512X512

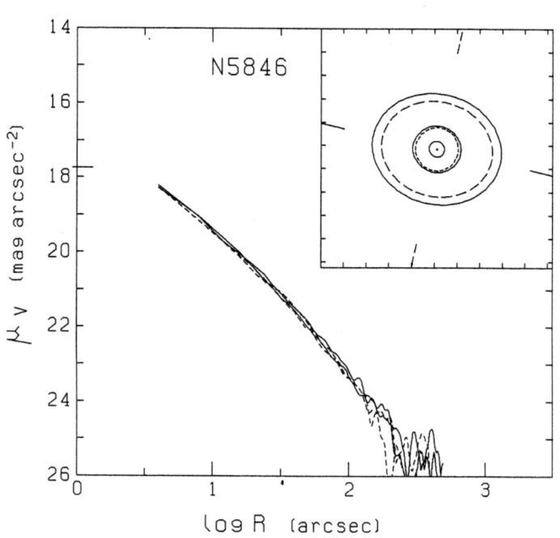

NGC5850

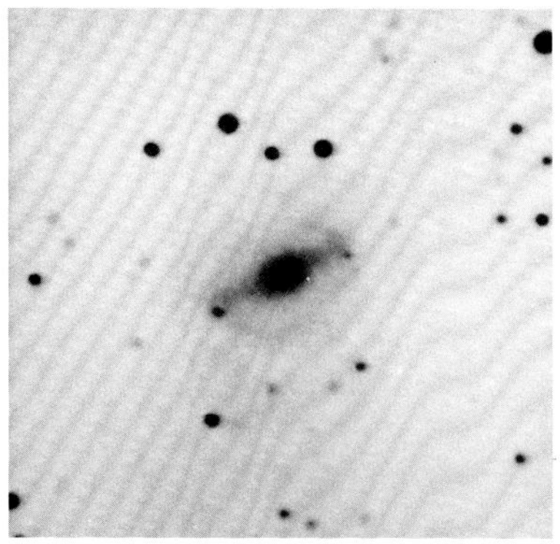

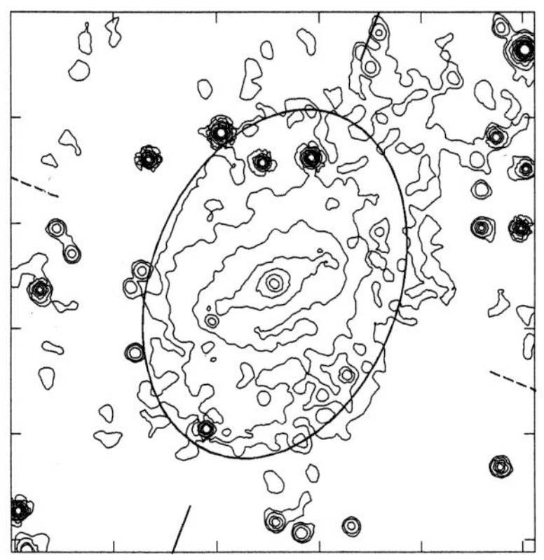

N5850 K4378 2 21.07 256X256

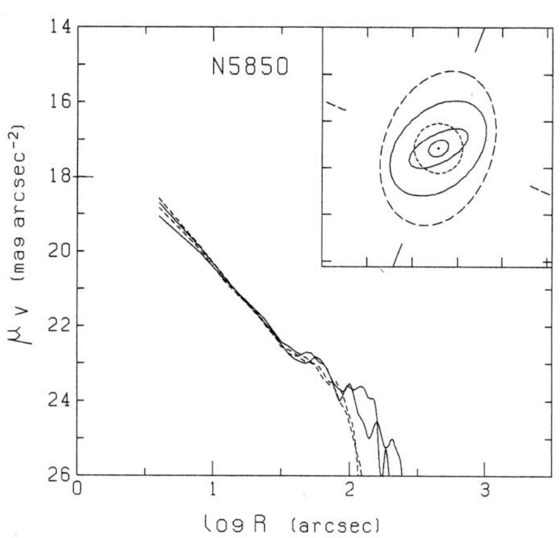

NGC5866

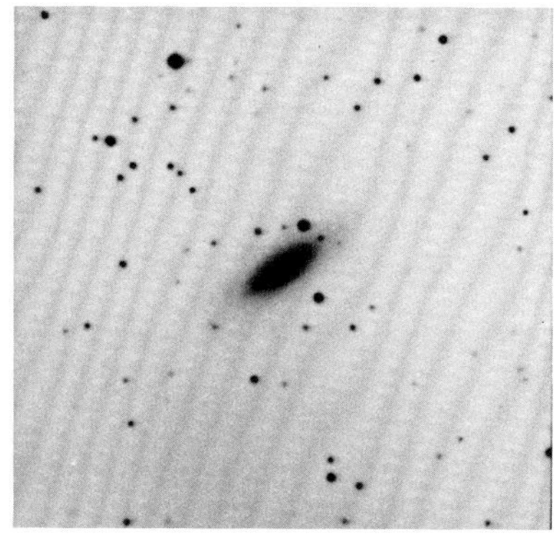

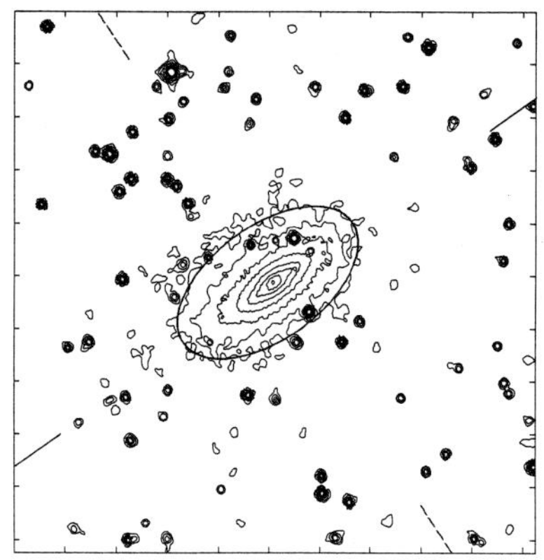

N5866 K4654 2 21.74 512X512

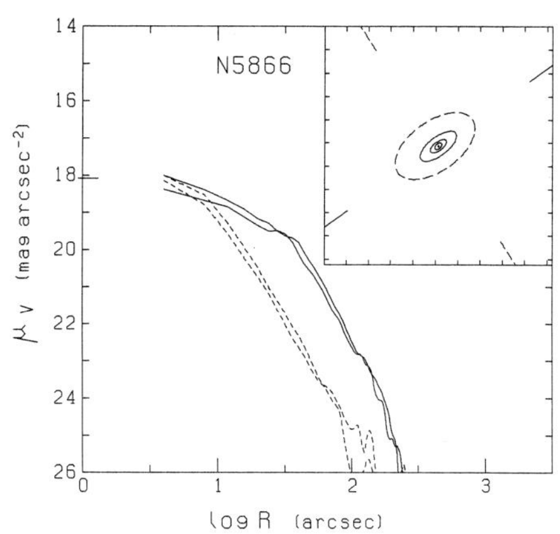

NGC5854

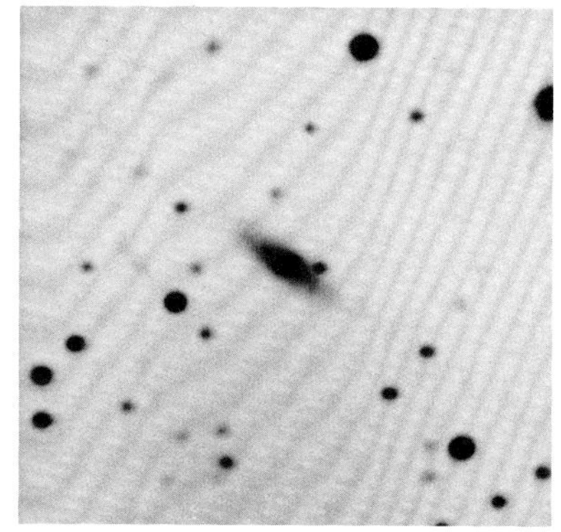

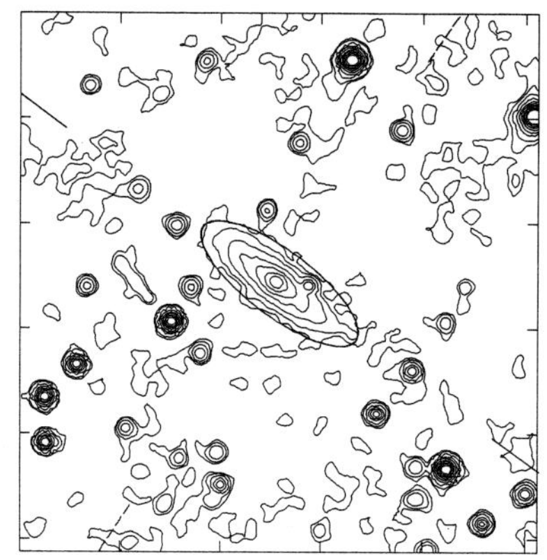

N5854 K4389 2 20.64 256X256

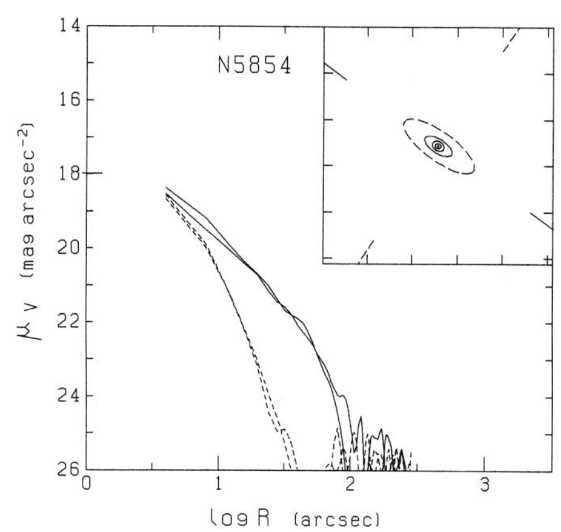

NGC5861

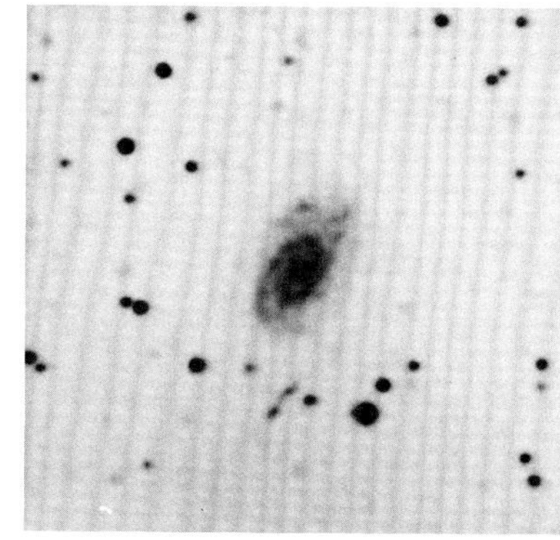

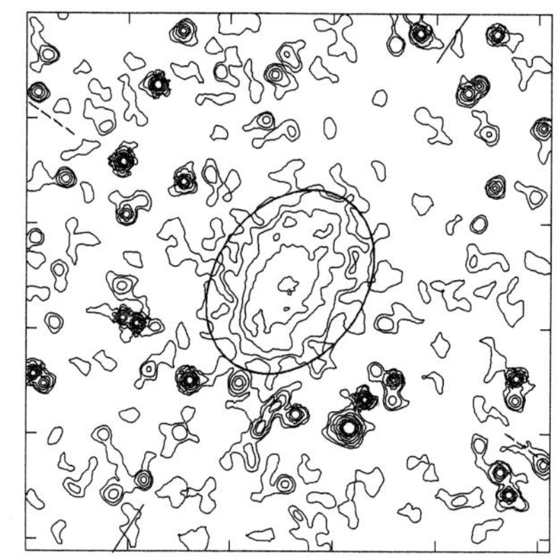

N5861 K5066 2 21.09 256X256

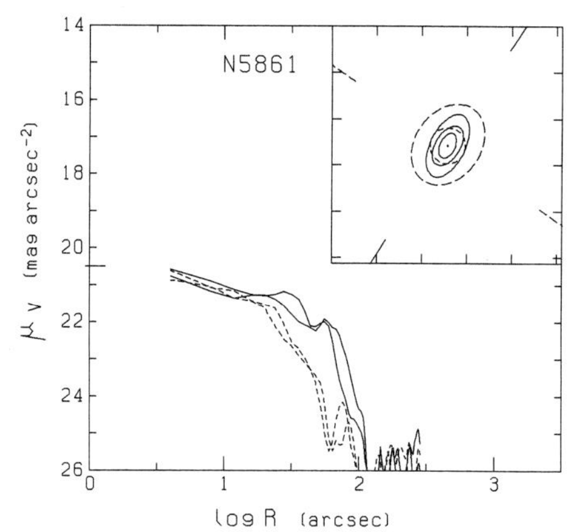

NGC5864

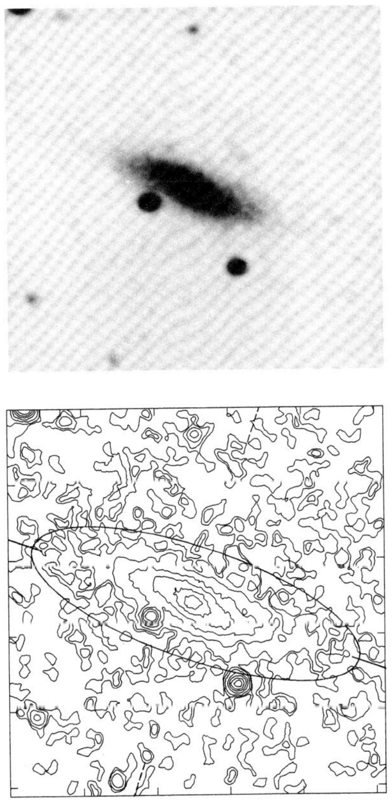

N5864 K4389 1 20.64 256X256

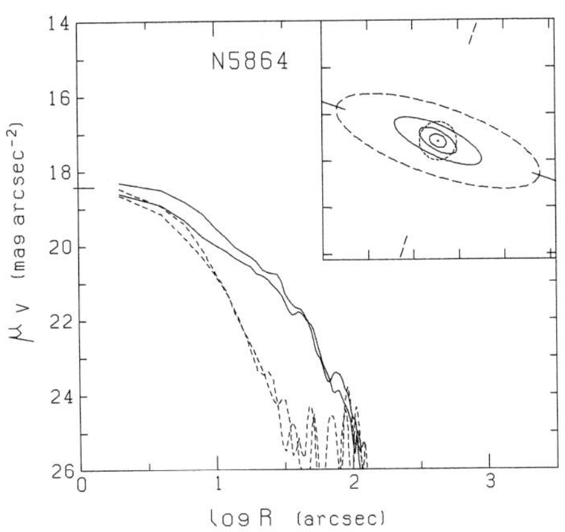

NGC5879

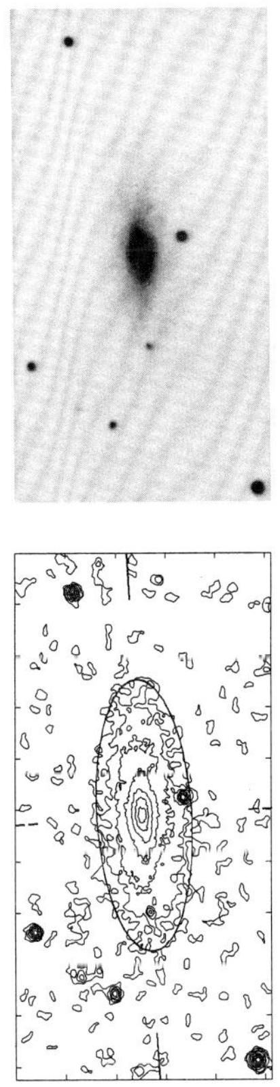

N5879 K4654 1 21.74 256X512

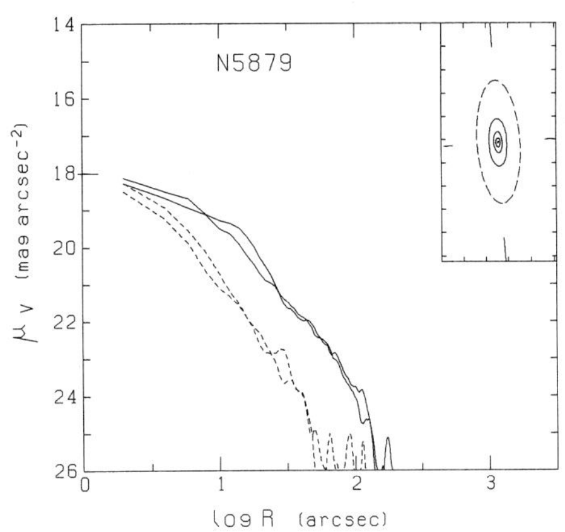

NGC5878

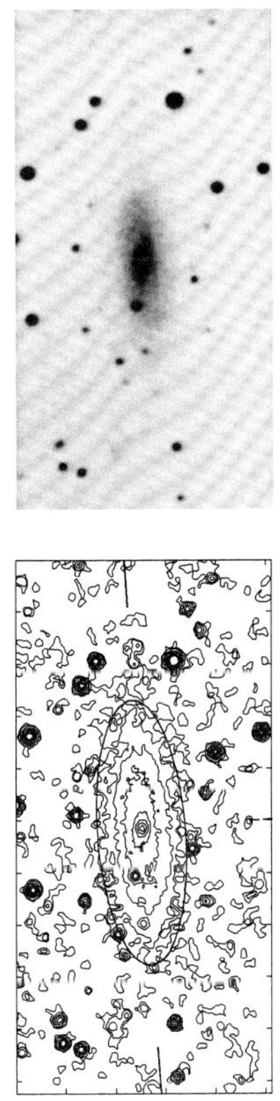

N5878 K4647 1 20.96 256X512

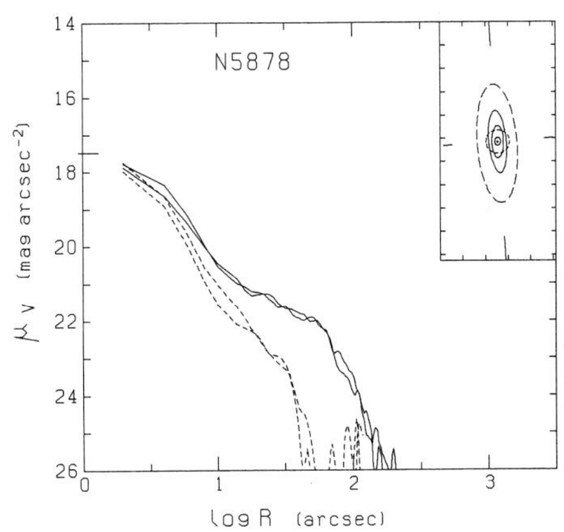

F703

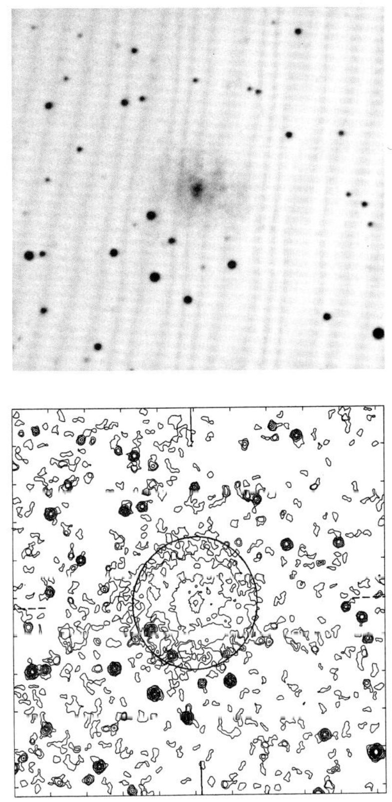

F703 K4647 1 20.96 512X512

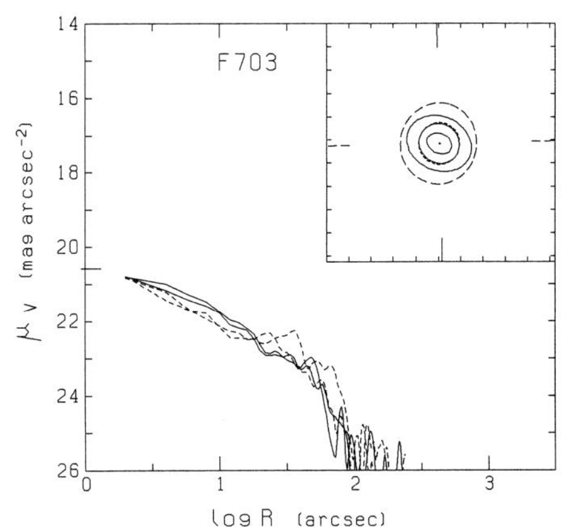

NGC5899

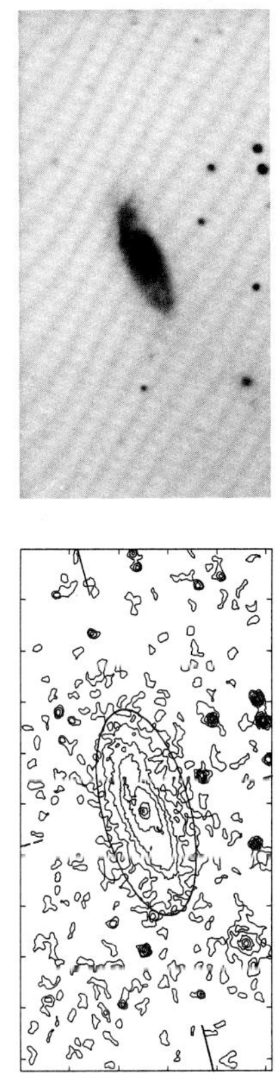

N5899 K5092 1 21.81 256X512

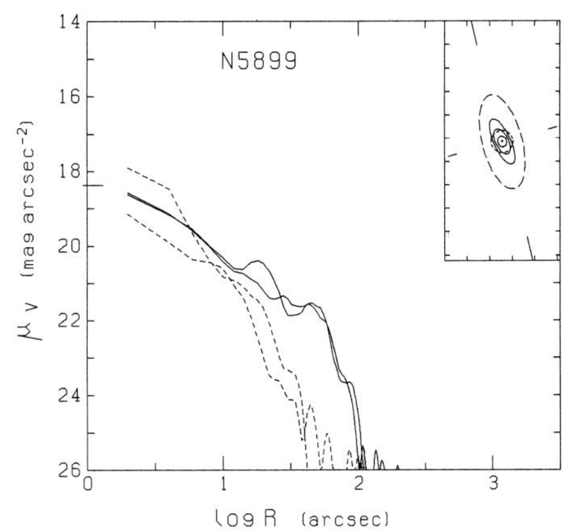

NGC5905

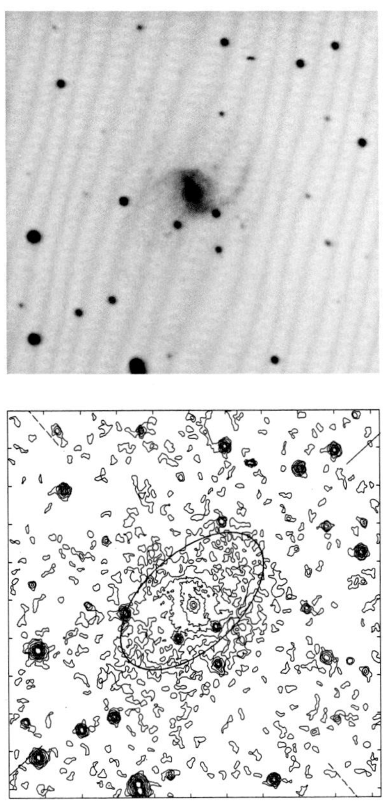

N5905 K4654 1 21.74 512X512

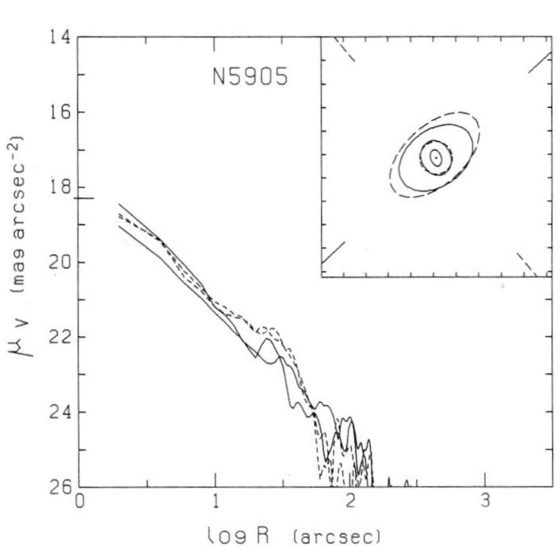

NGC5907

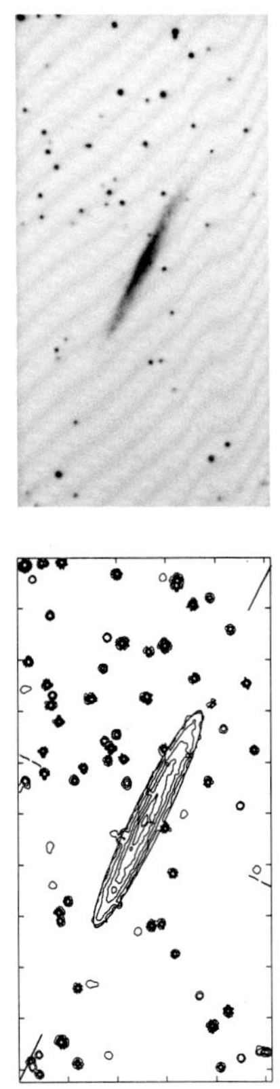

N5907 K4654 3 21.74 256X512

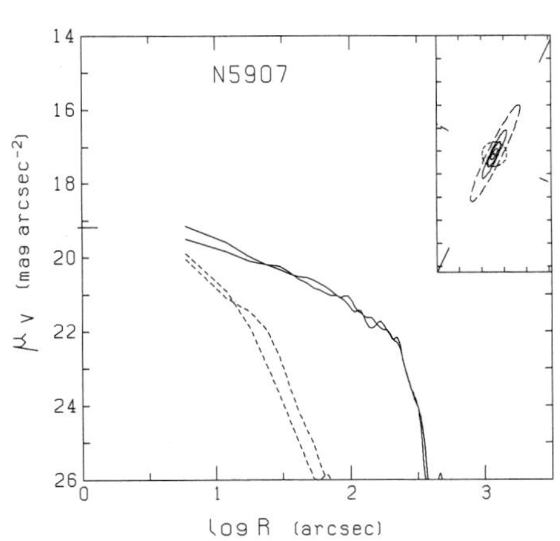

NGC5898

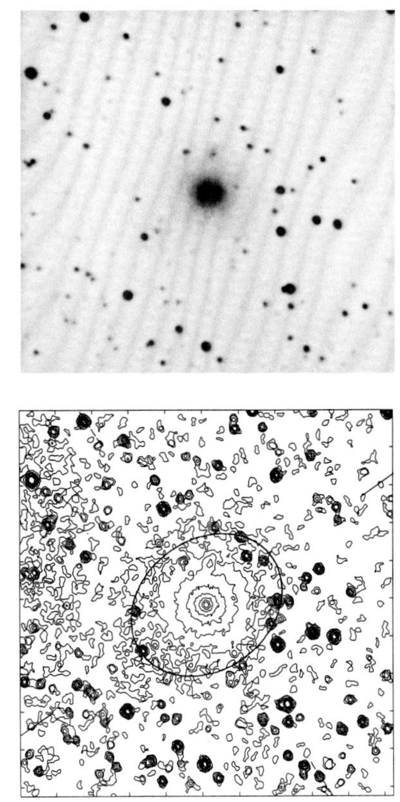

N5898 K5423 1 20.98 512X512

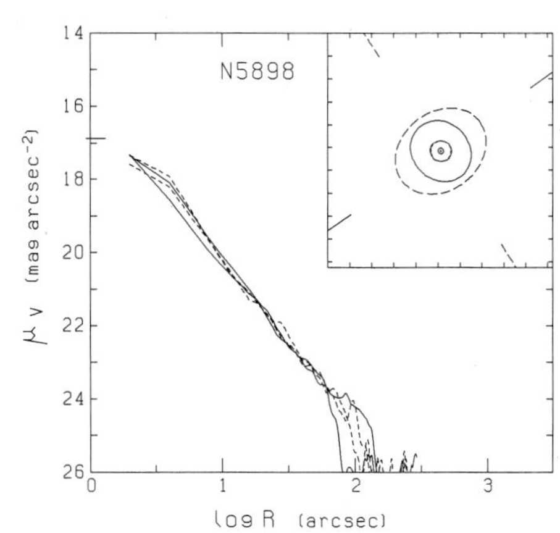

NGC5908

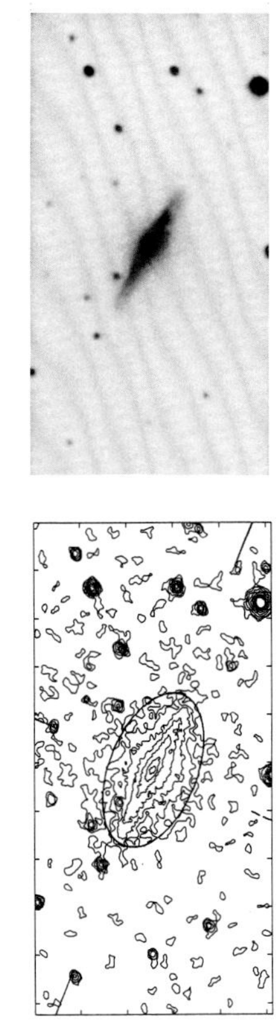

N5908 K4654 1 21.74 256X512

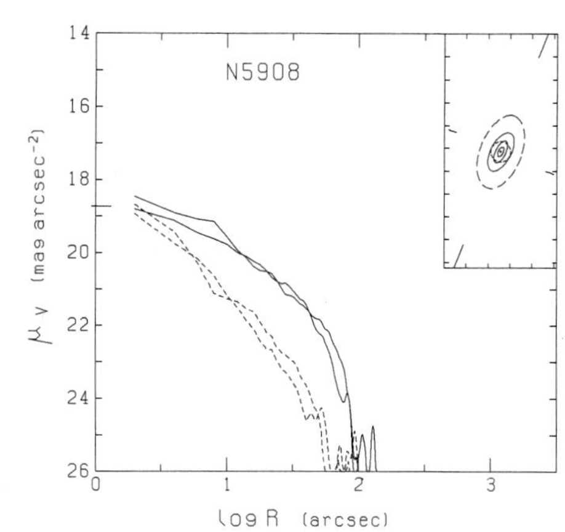

NGC5903

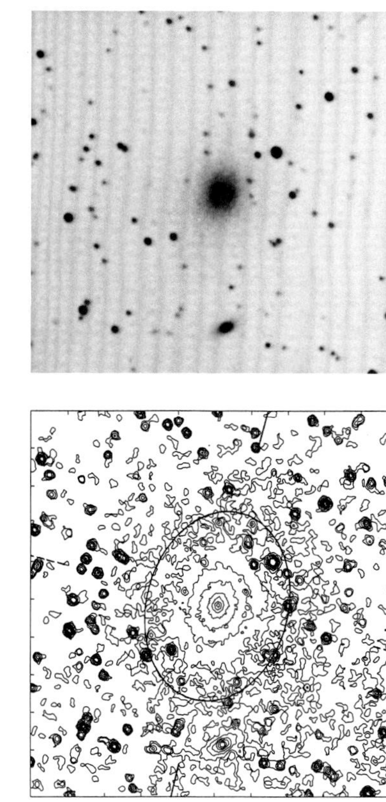

N5903 K5423 1 20.98 512X512

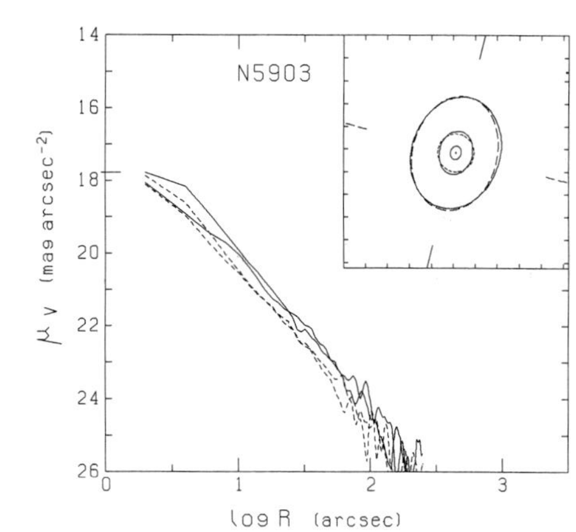

NGC5915

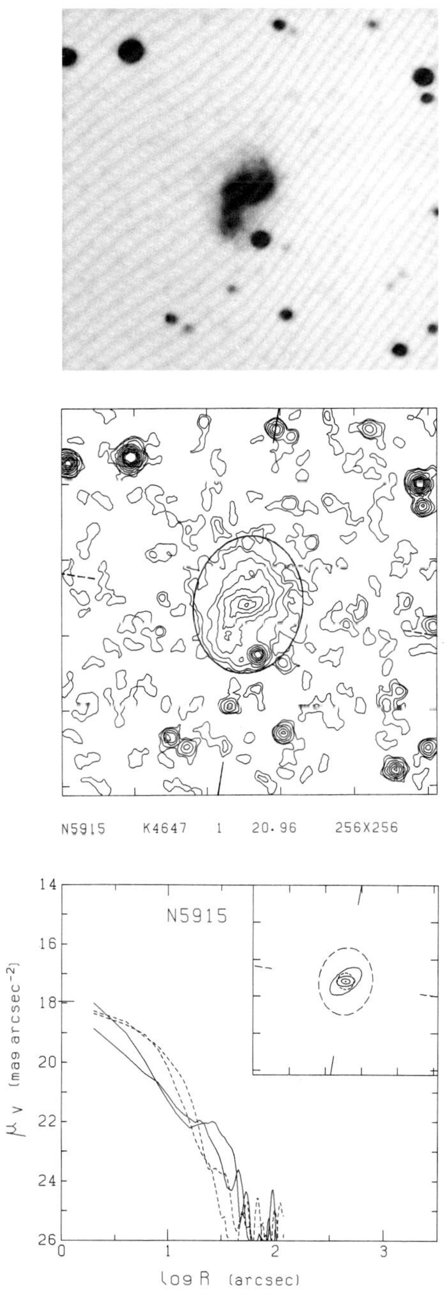

NGC5921

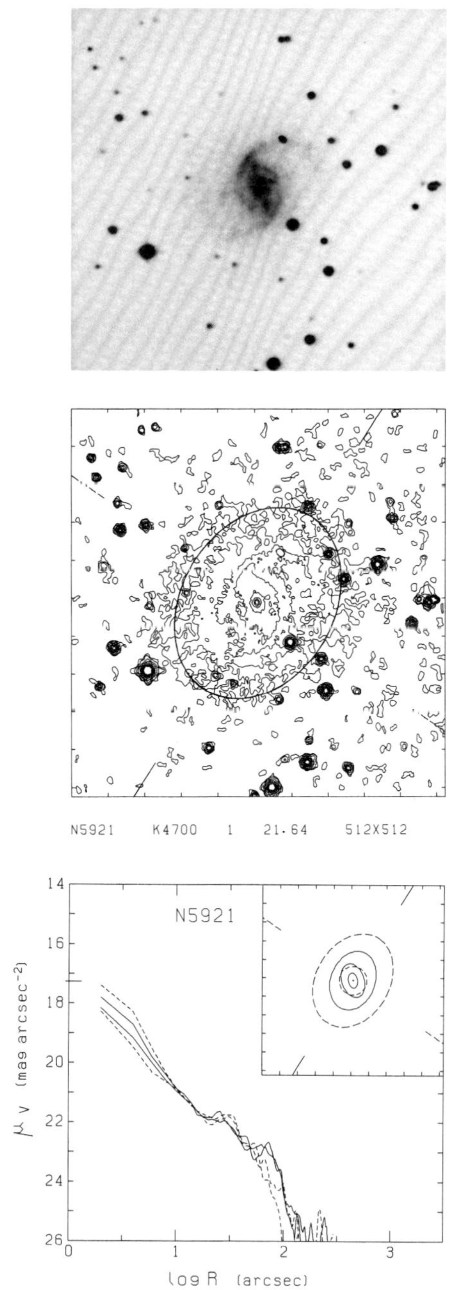

NGC5949

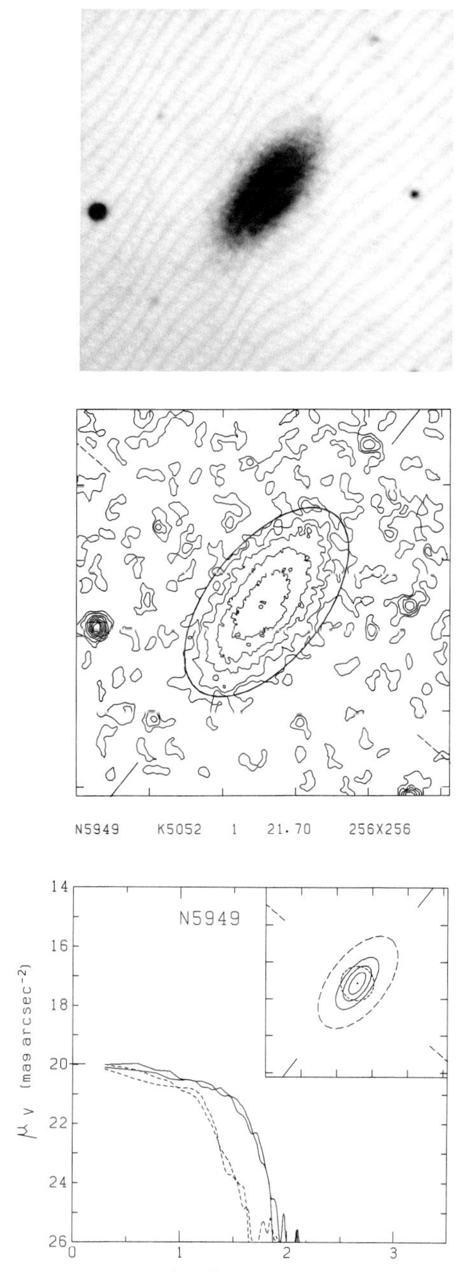

NGC5936

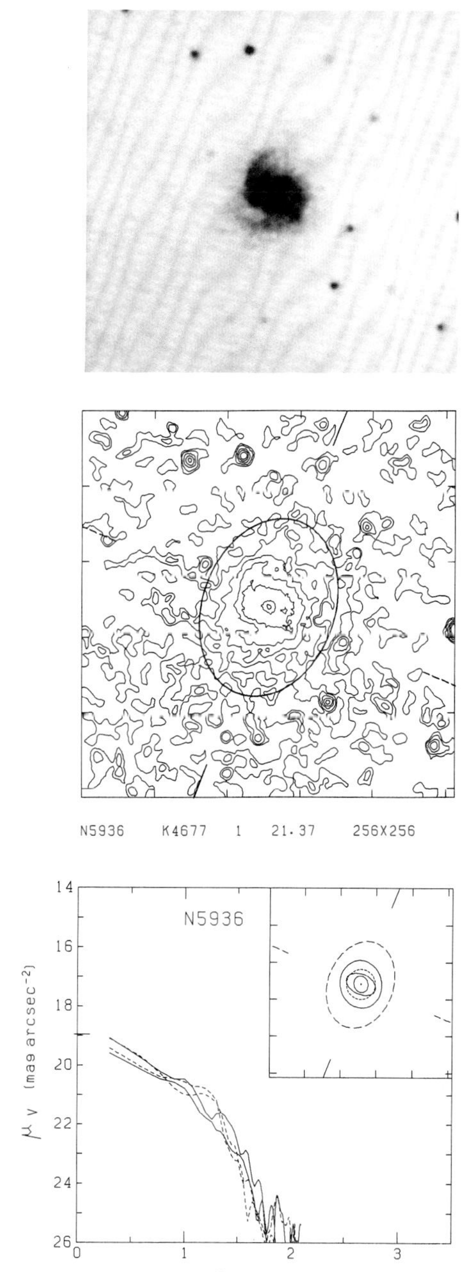

NGC5970

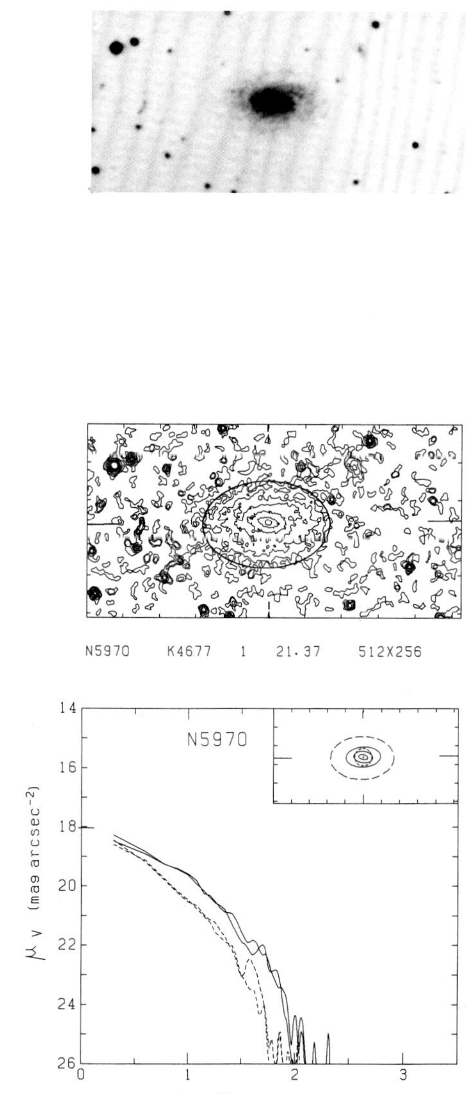

NGC5982

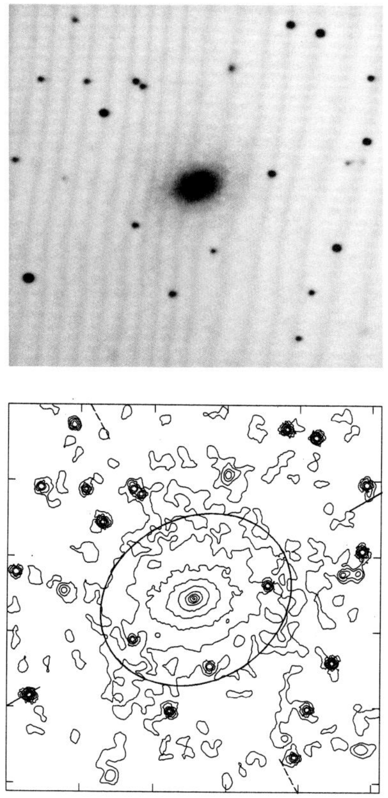

N5982 K4414 2 21.65 256X256

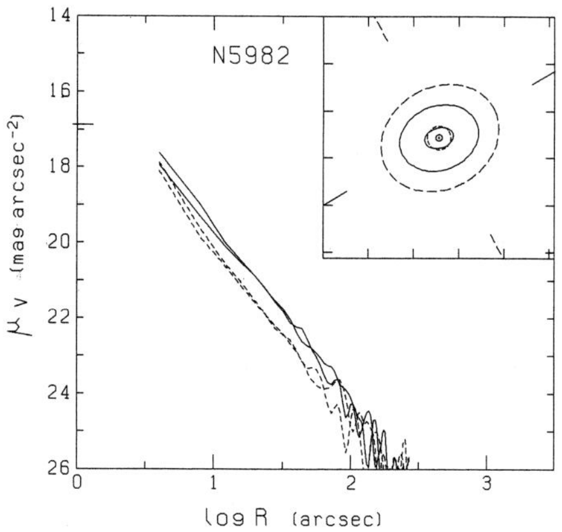

NGC5985

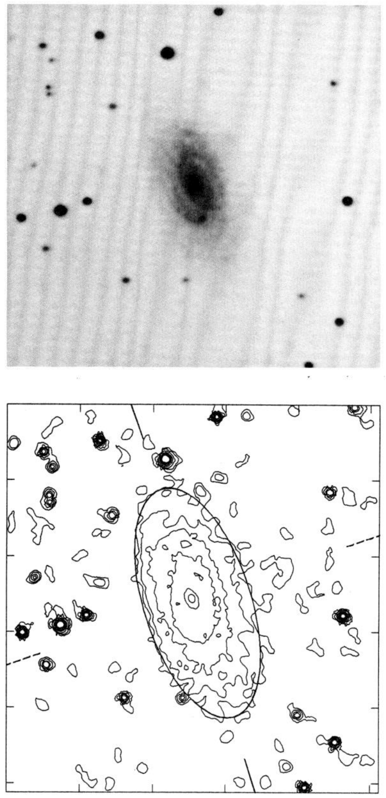

N5985 K4414 2 21.65 256X256

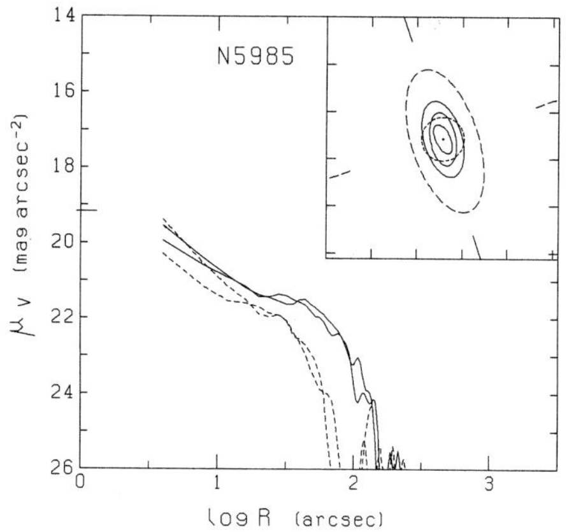

NGC6015

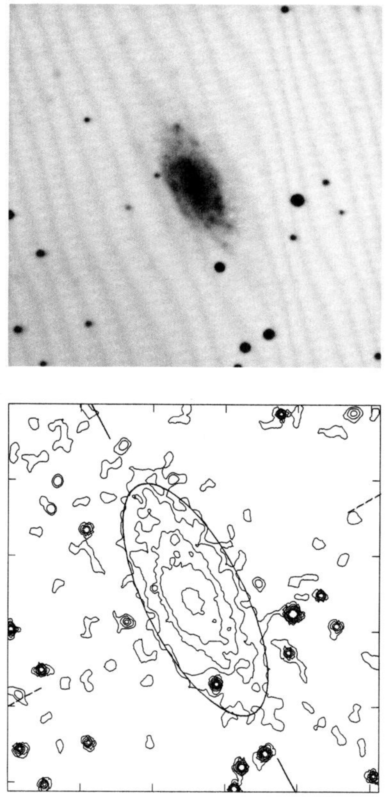

N6015 K4414 2 21.65 256X256

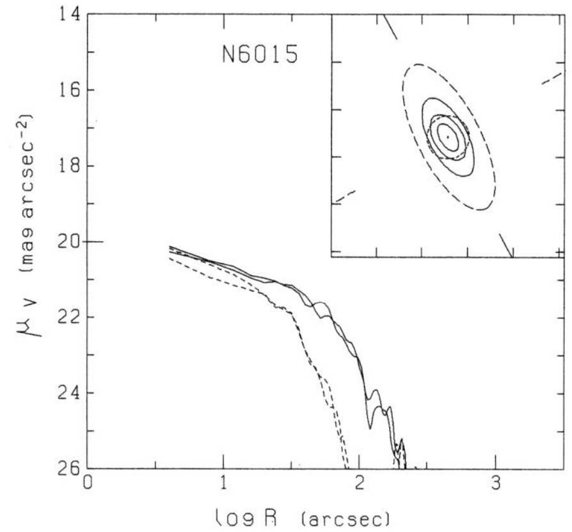

NGC6052

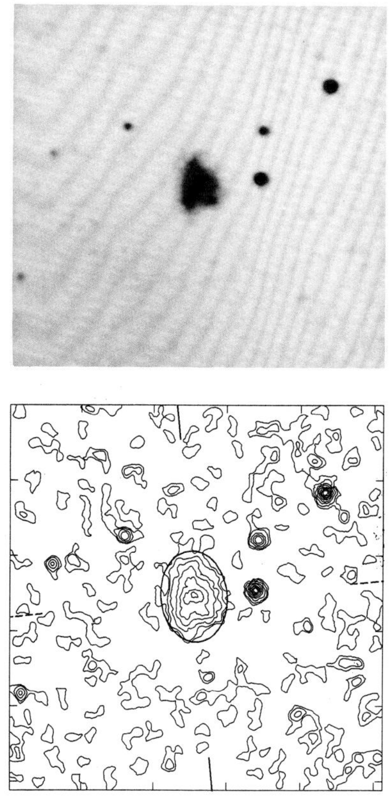

N6052 K4653 1 21.41 256X256

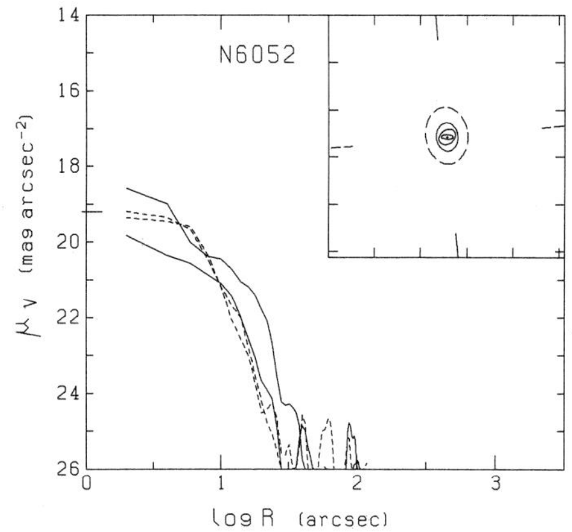

NGC6070

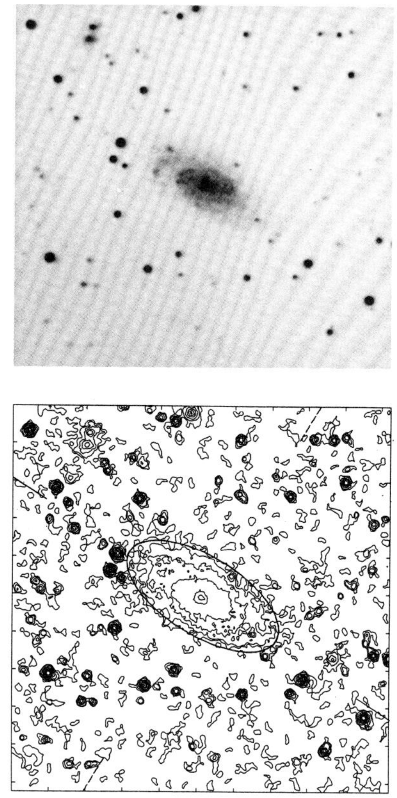

N6070 K5057 1 21.58 512X512

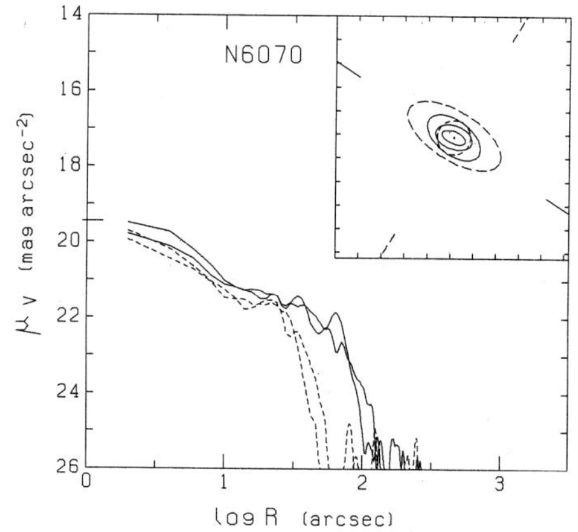

NGC6118

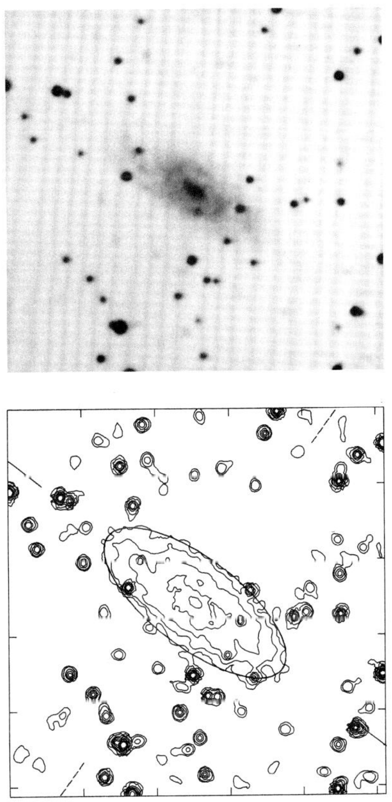

N6118 K6017 2 21.27 256X256

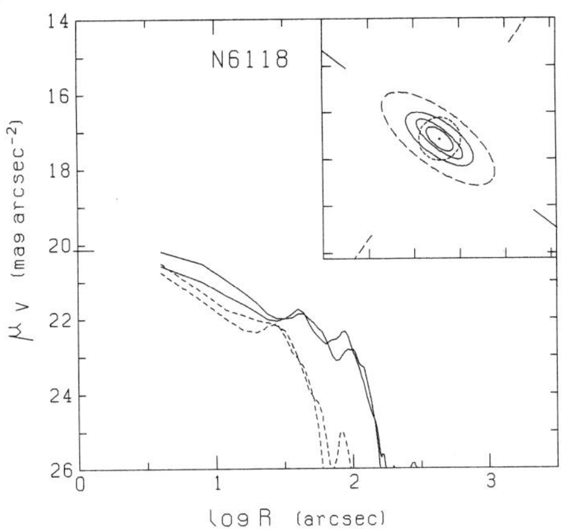

NGC6181

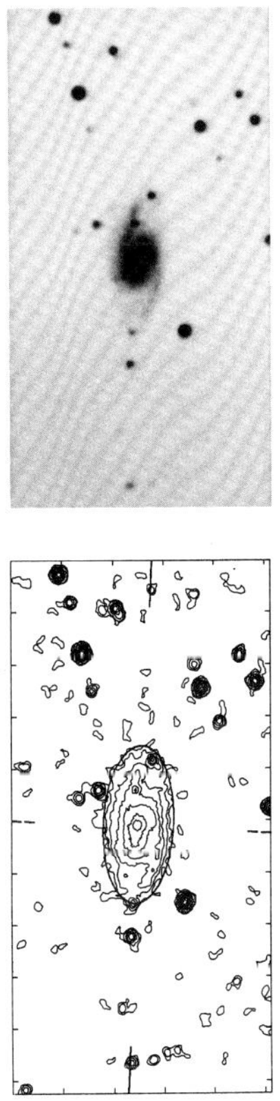

N6181 K5424 1 21.33 256X512

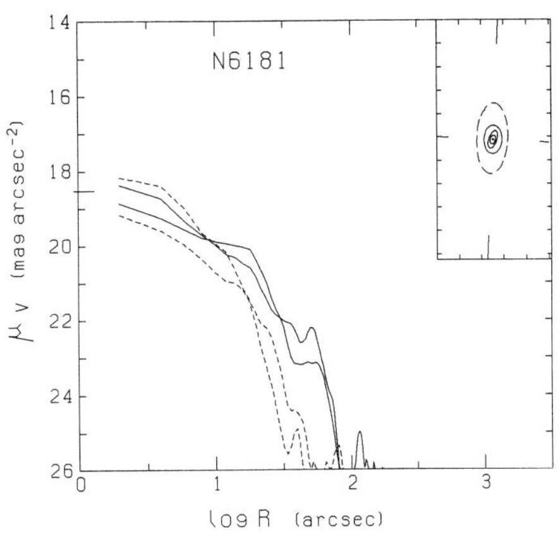

NGC6217

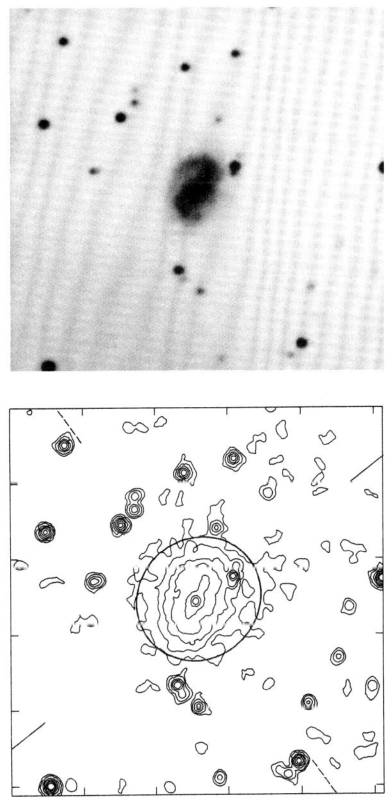

N6217 K4682 2 21.53 256X256

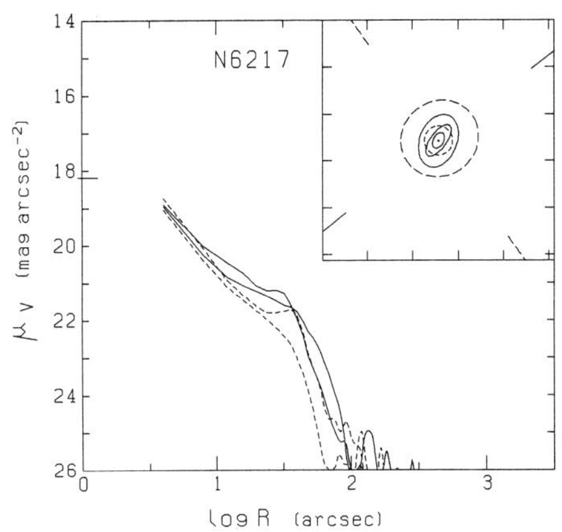

NGC6207

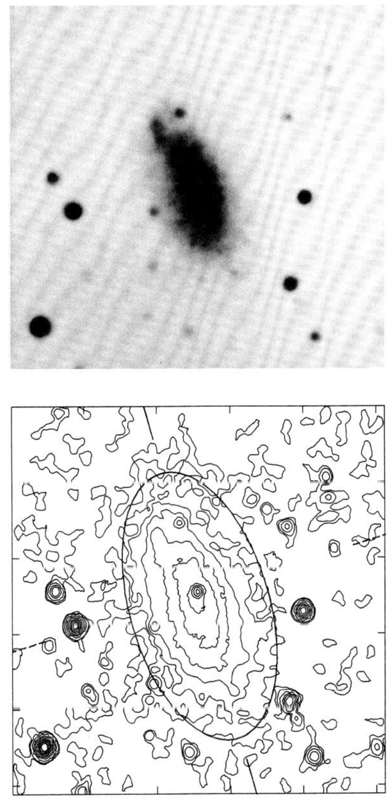

N6207 K2304 1 20.70 256X256

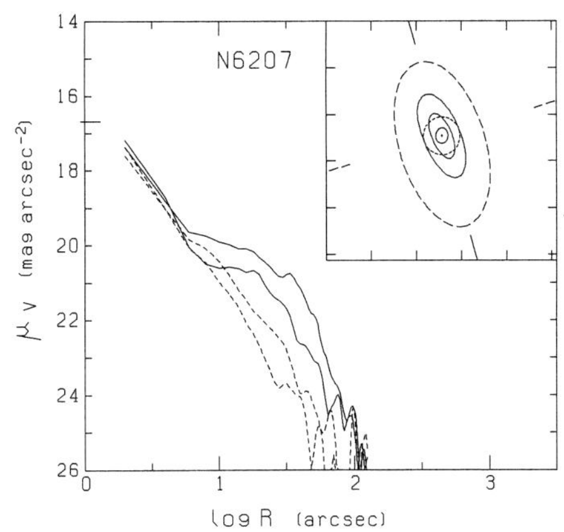

NGC6340

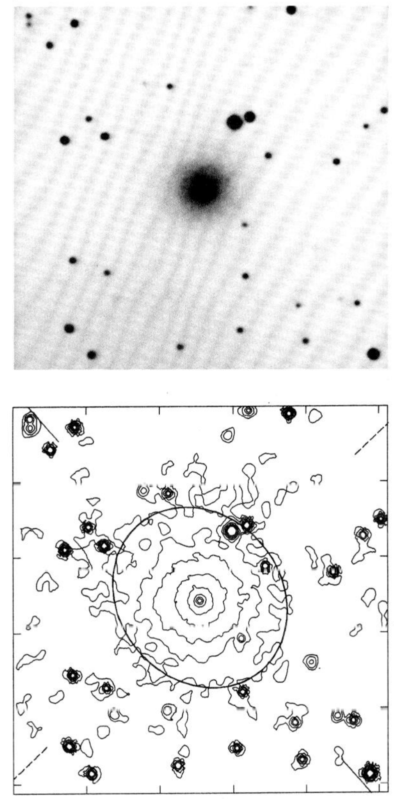

N6340 K4687 2 21.67 256X256

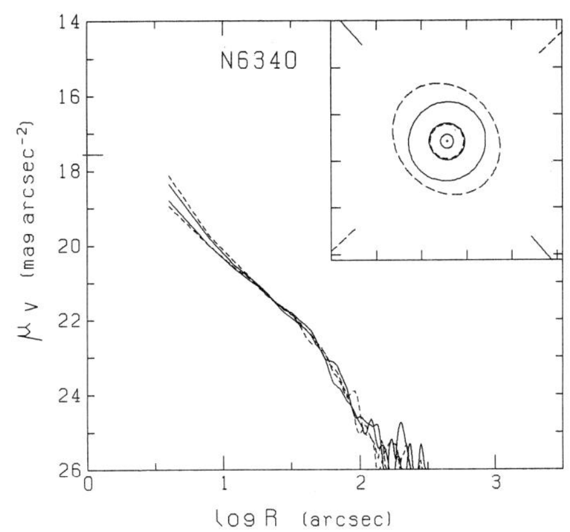

NGC6384

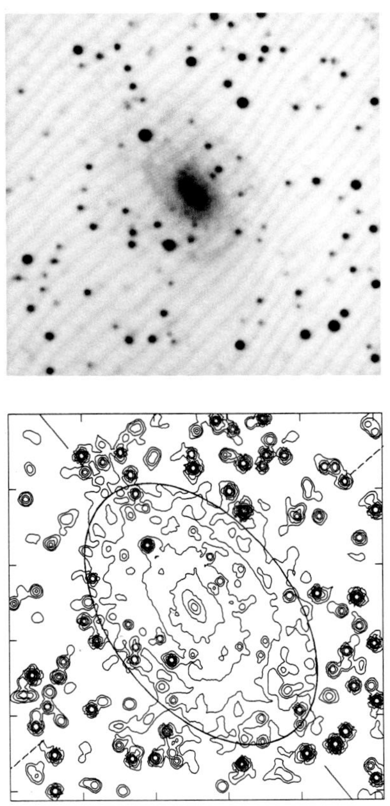

N6384 K5058 2 21.58 256X256

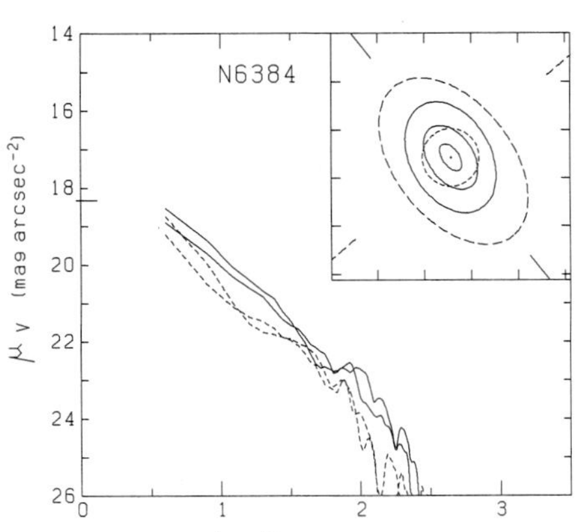

NGC6482

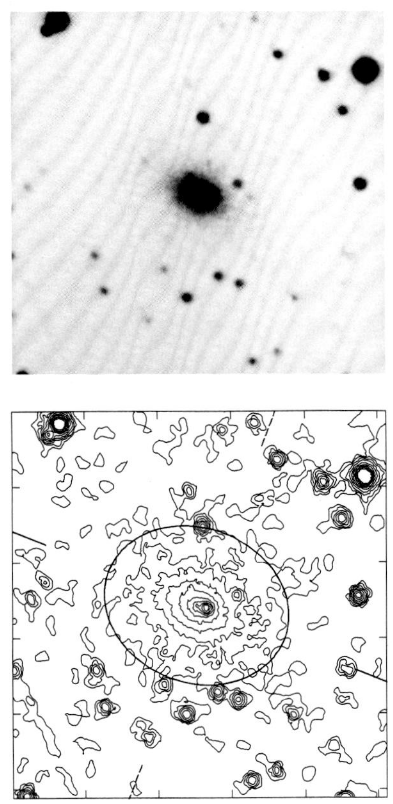

N6482 K5067 1 22.44 256X256

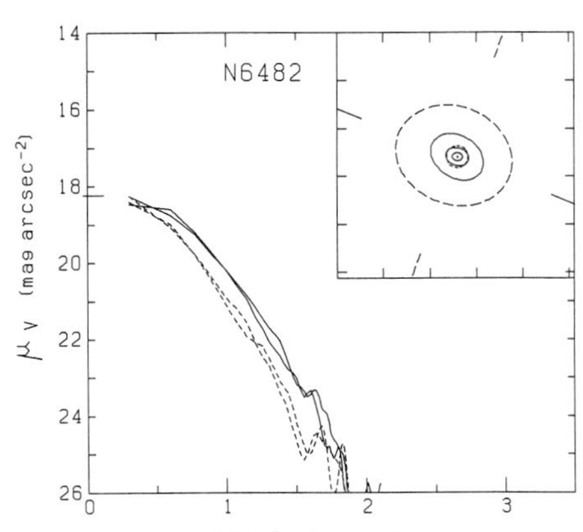

NGC6503

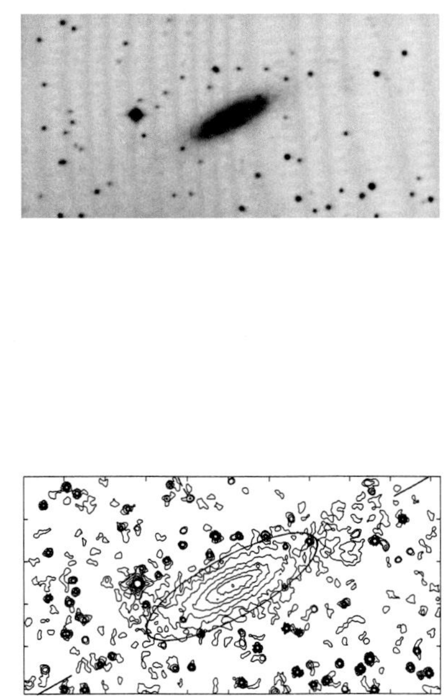

N6503 K4665 2 21.31 512X256

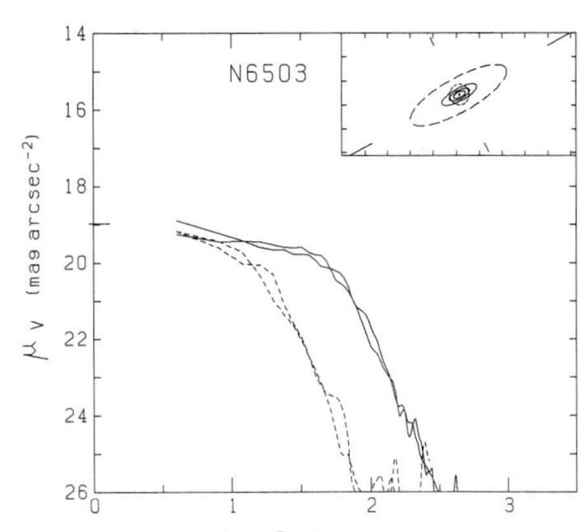

NGC6643

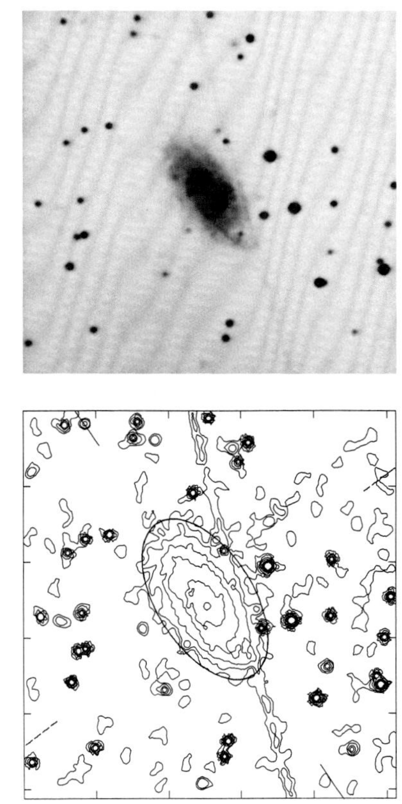

N6643 K4711 2 21.51 256X256

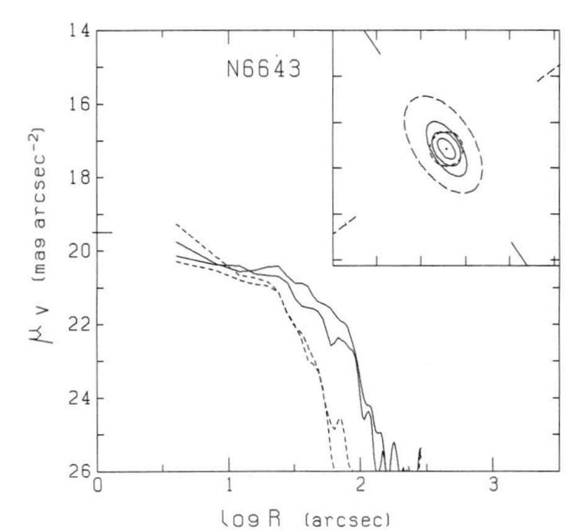

NGC6814

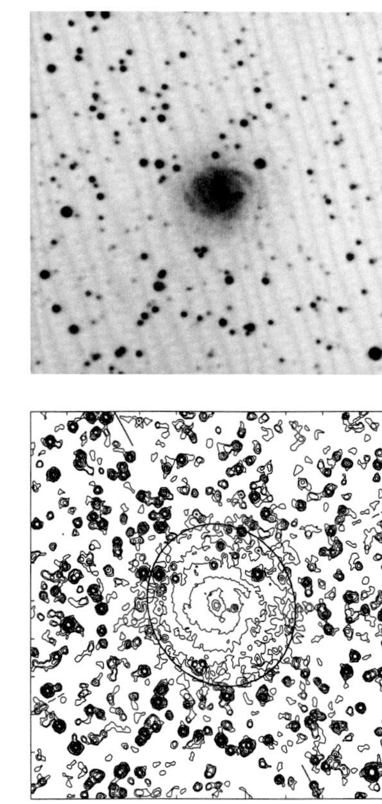

N6814 K5094 1 21.28 512X512

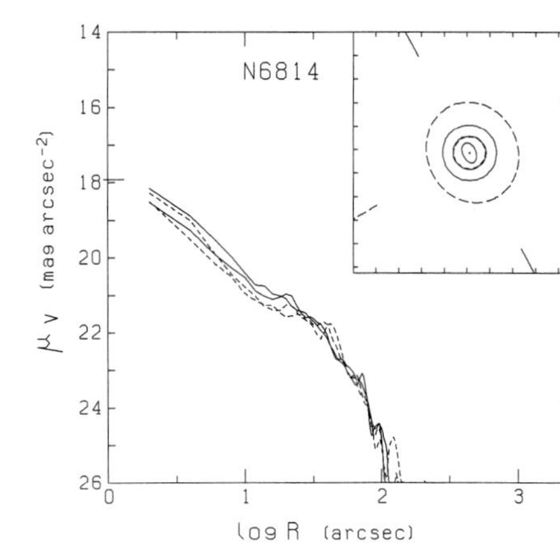

NGC6822

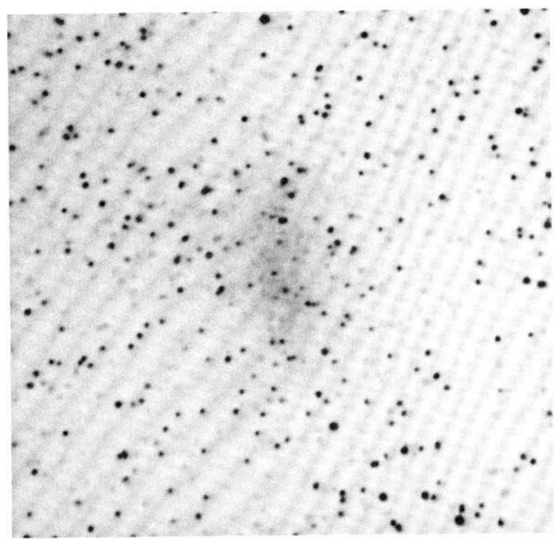

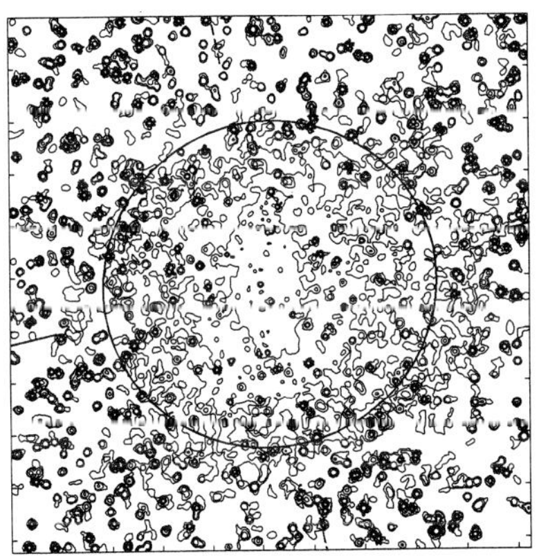

N6822 K5068 3 20.75 512X512

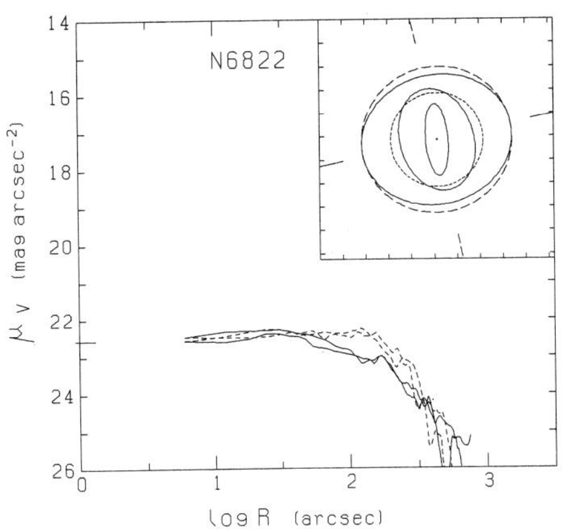

NGC6907

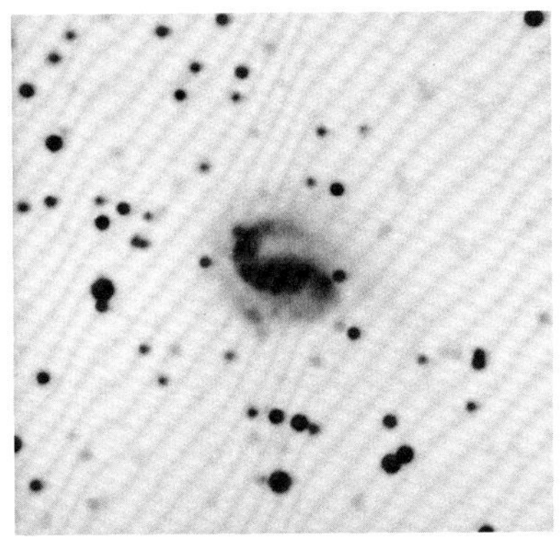

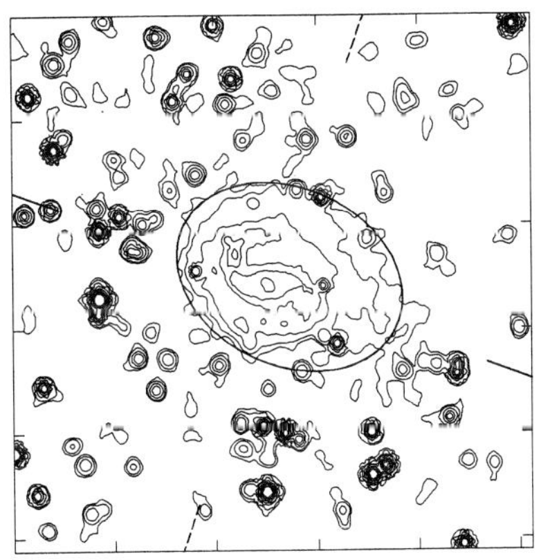

N6907 K5112 2 20.86 256X256

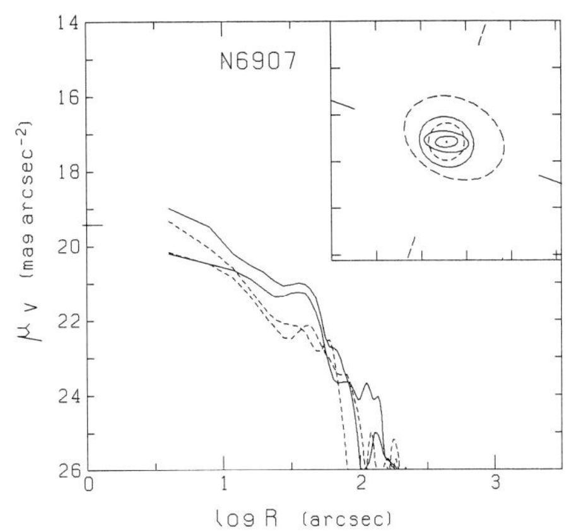

NGC6925

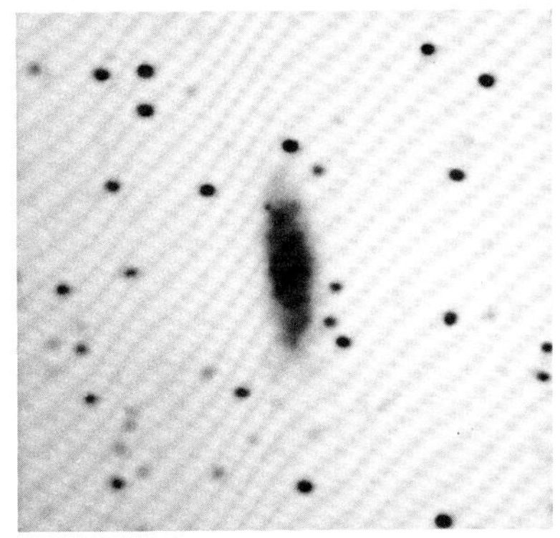

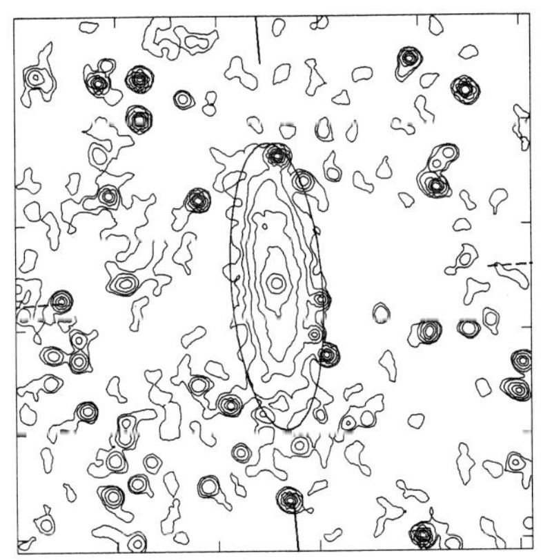

N6925 K5122 2 20.60 256X256

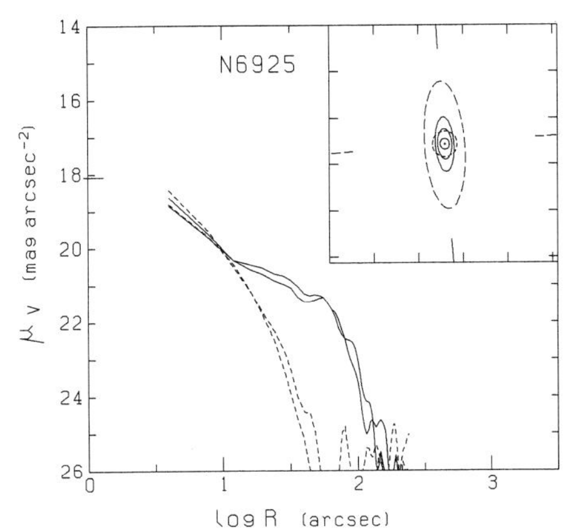

NGC6946

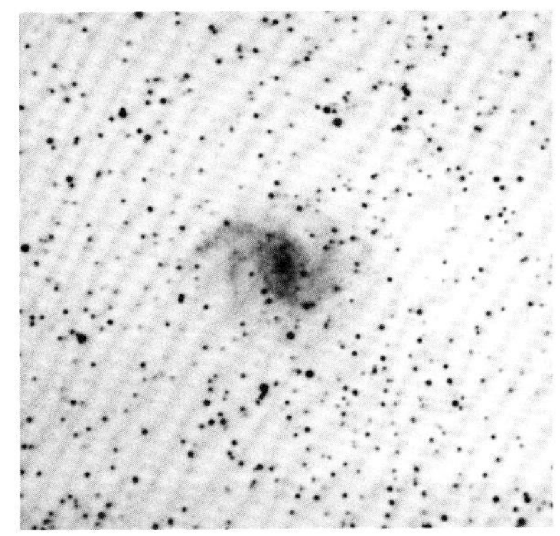

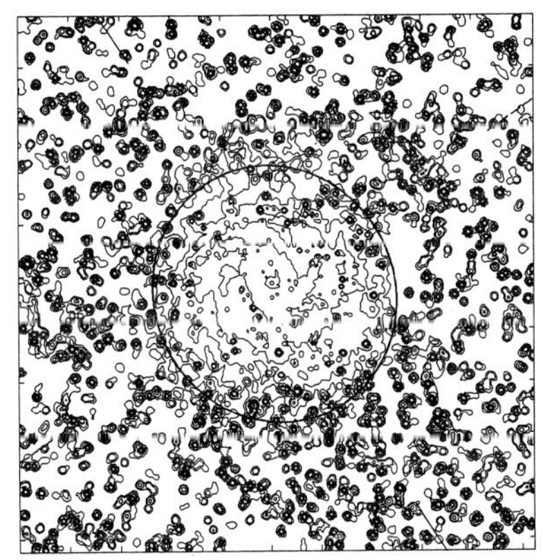

N6946 K5137 3 21.36 512X512

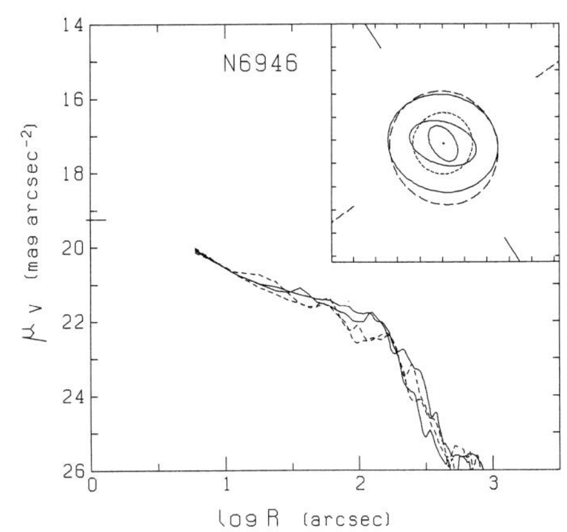

NGC6951

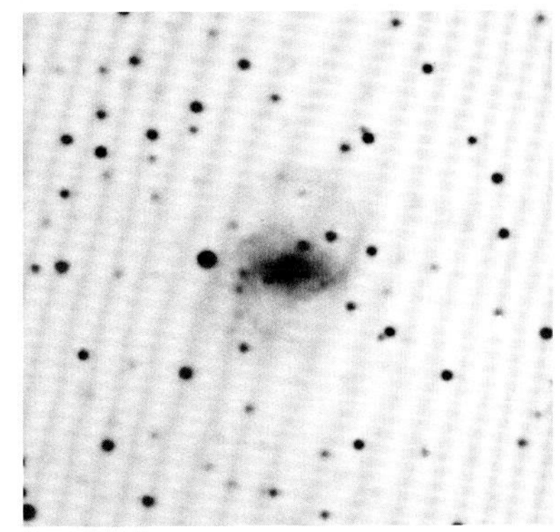

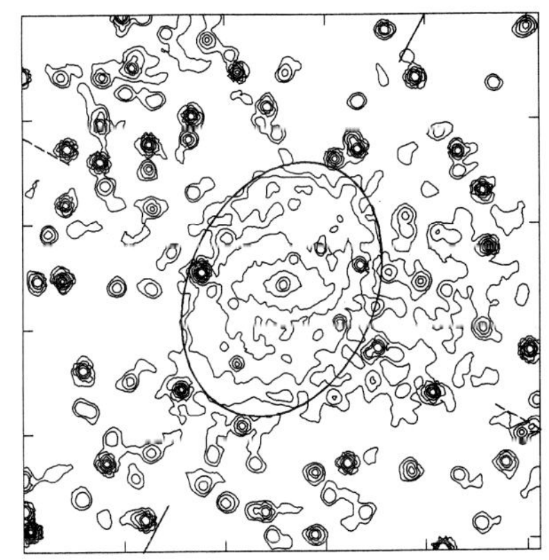

N6951 K5103 2 21.52 256X256

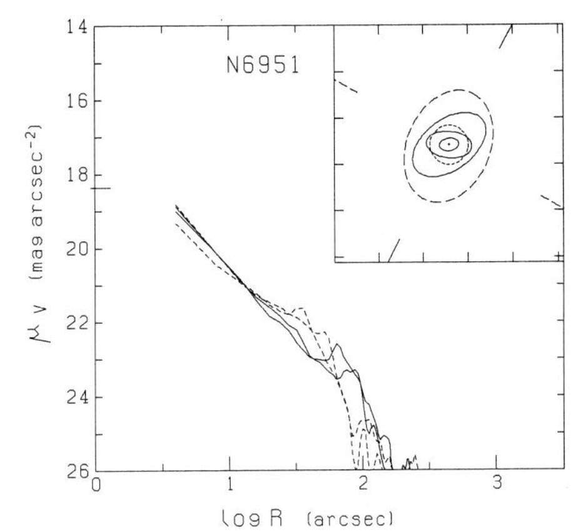

NGC7137

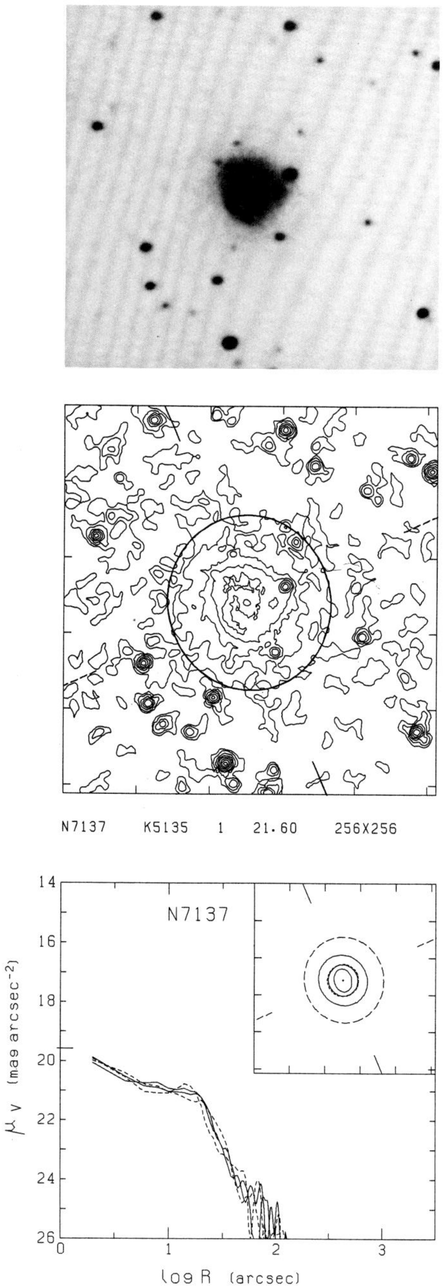

NGC7177

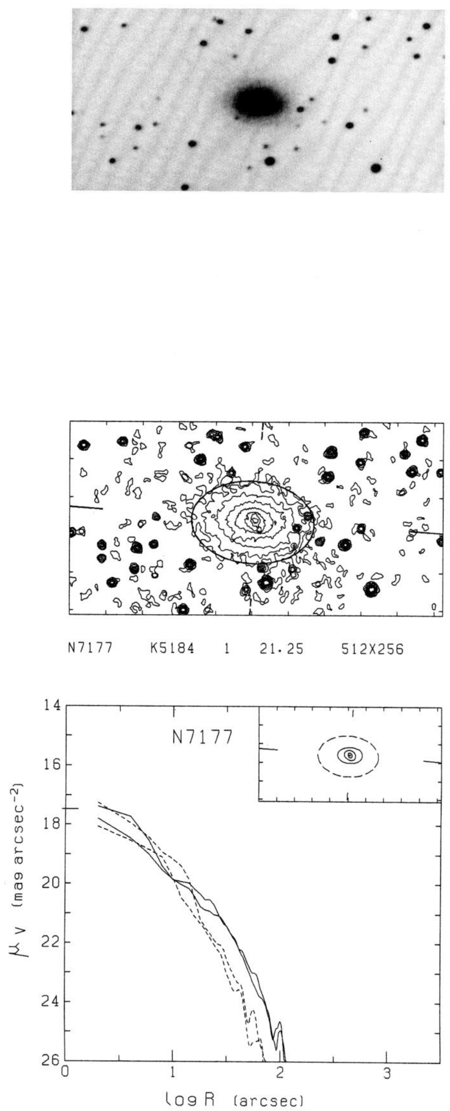

NGC7171

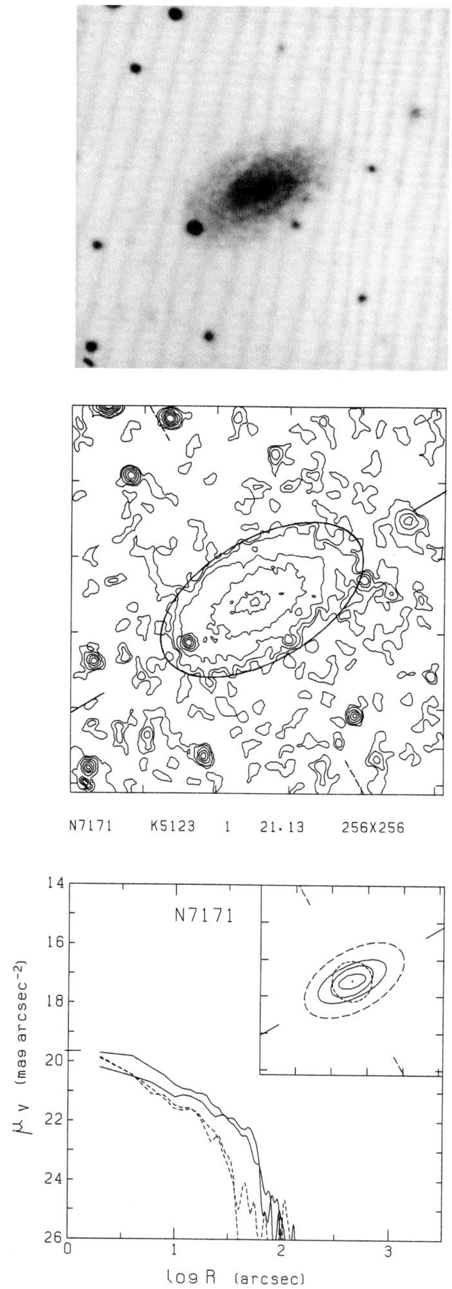

NGC7217

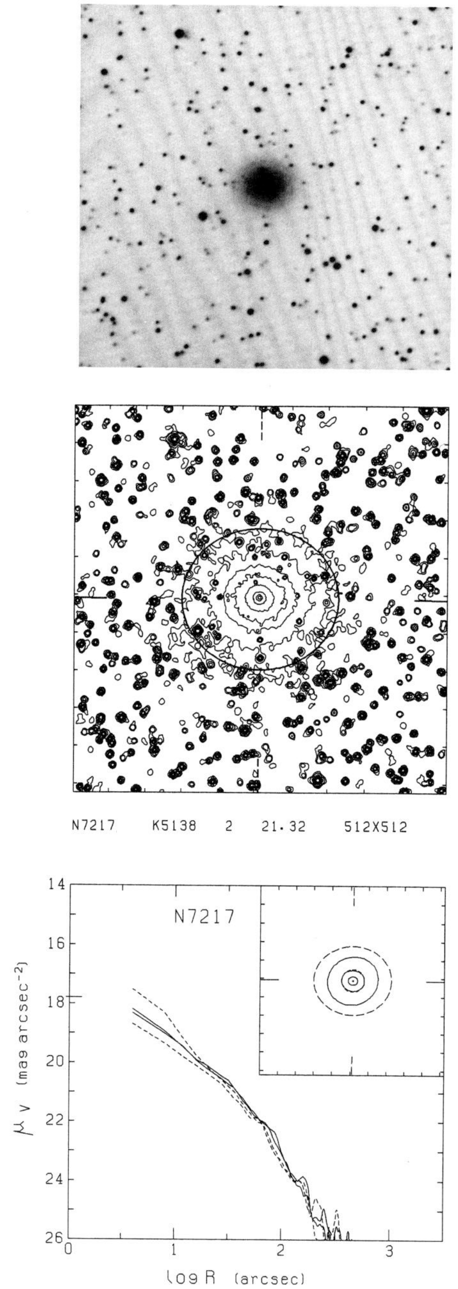

NGC7218

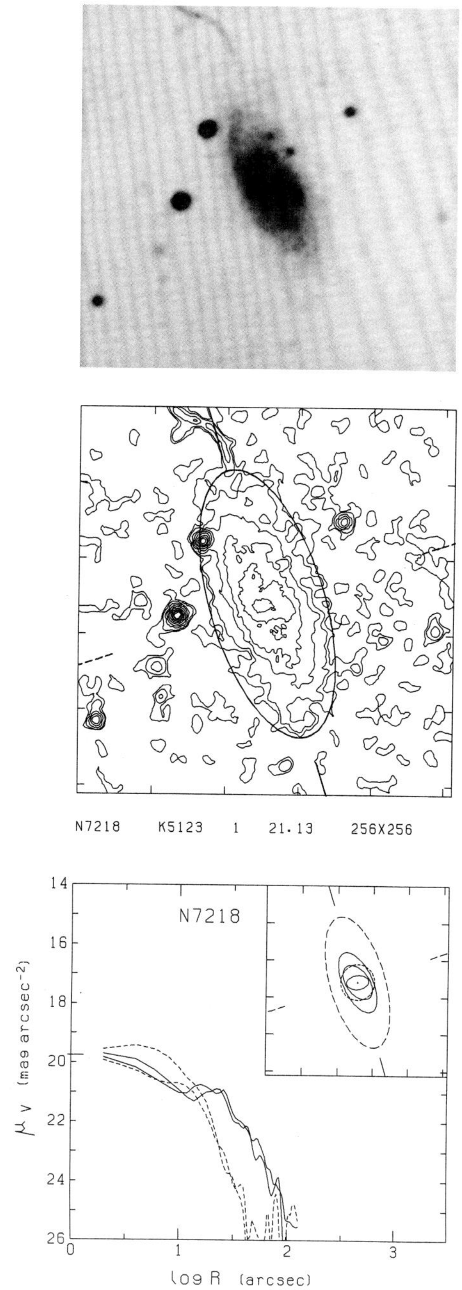

NGC7300

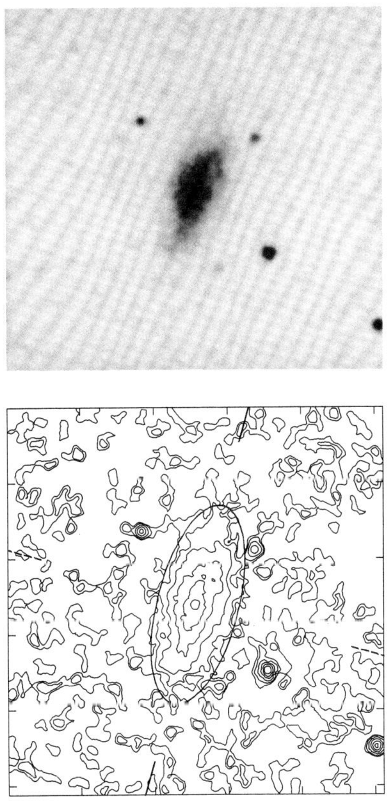

N7300 K4788 1 20.60 256X256

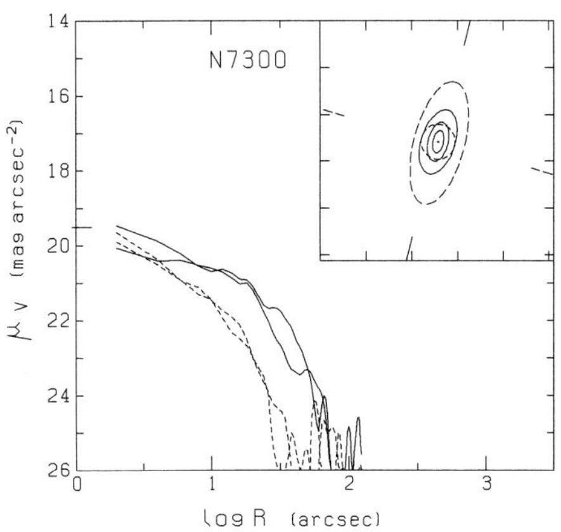

NGC7309

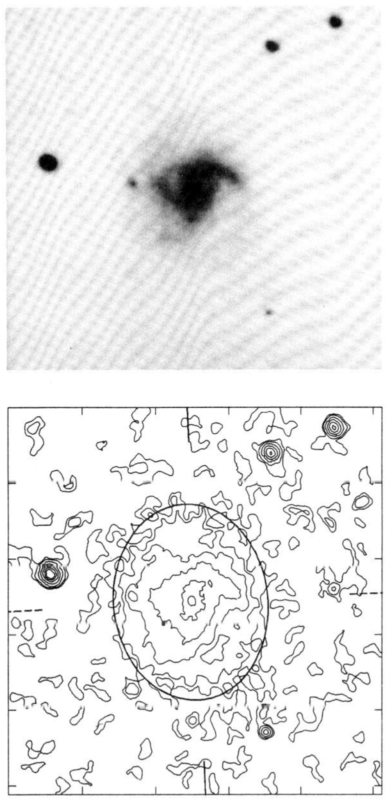

N7309 K5139 1 20.96 256X256

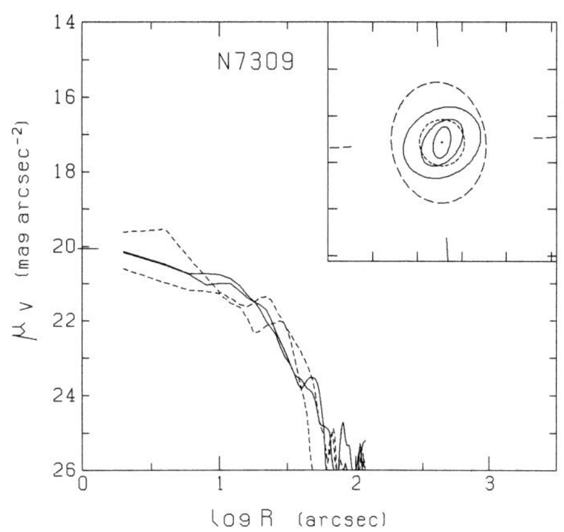

NGC7331

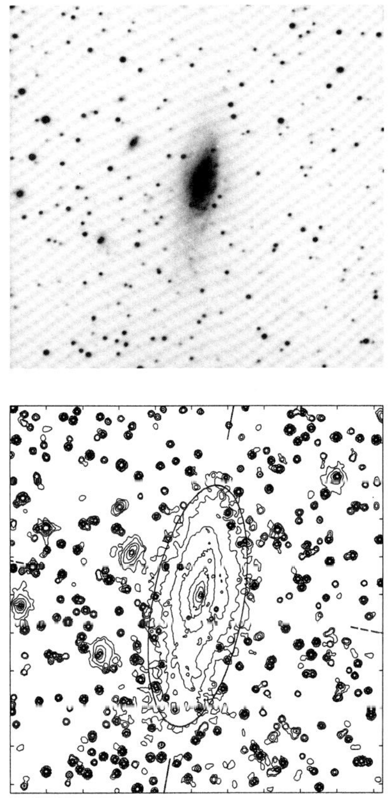

N7331 K3394 2 21.00 512X512

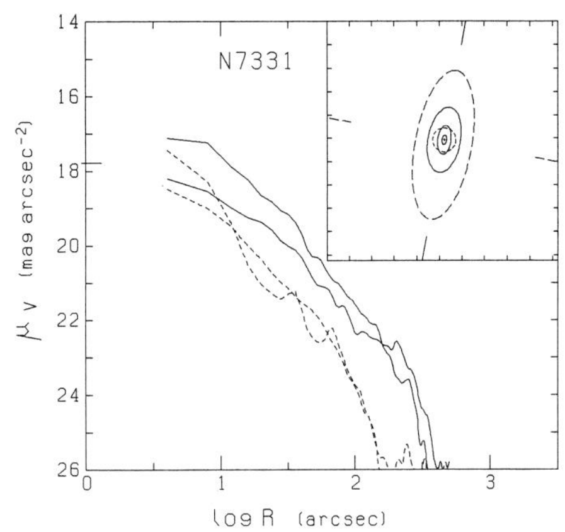

NGC7332

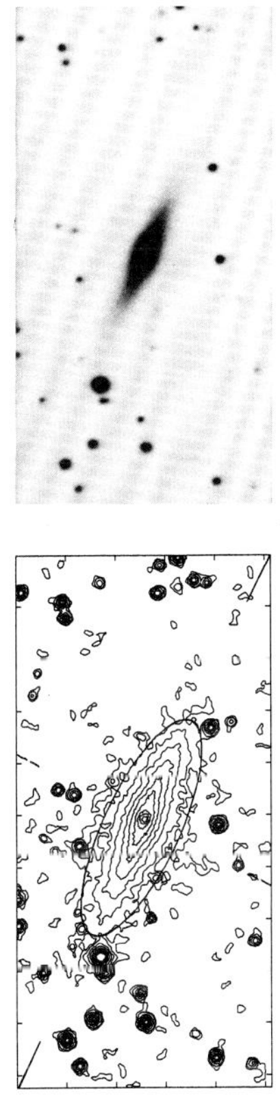

N7332 K3396 1 21.18 256X512

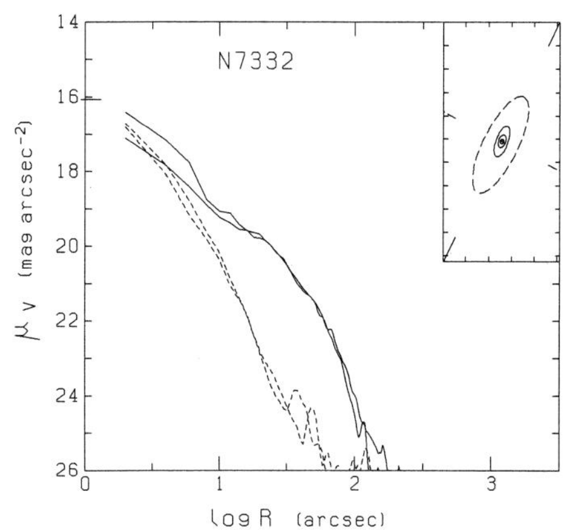

NGC7361

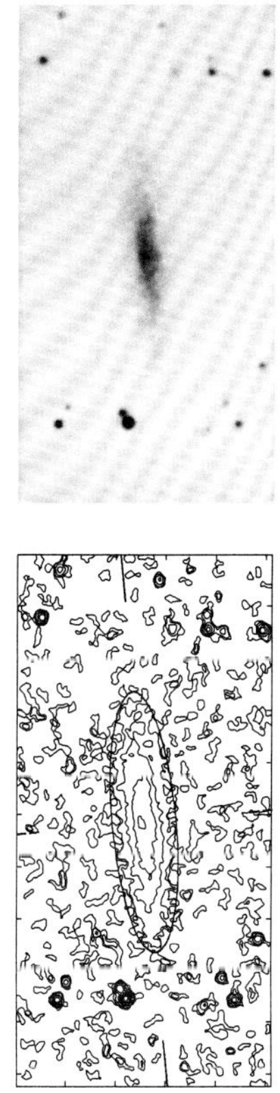

N7361 K5525 1 20.55 256X512

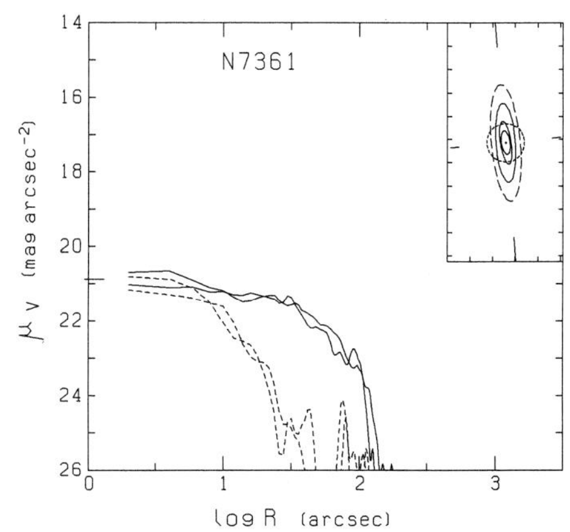

NGC7371

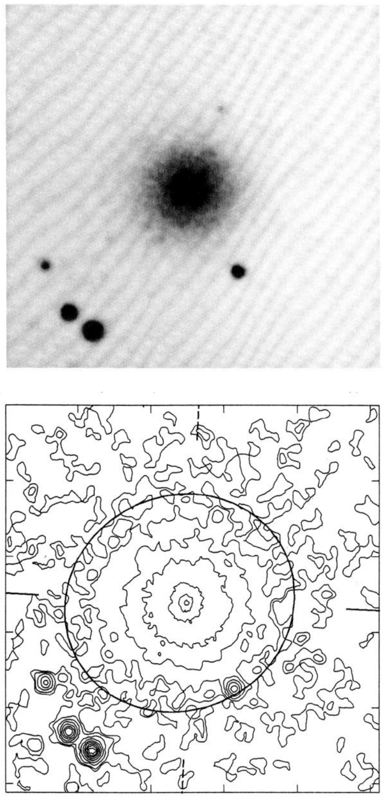

N7371 K5139 1 20.96 256X256

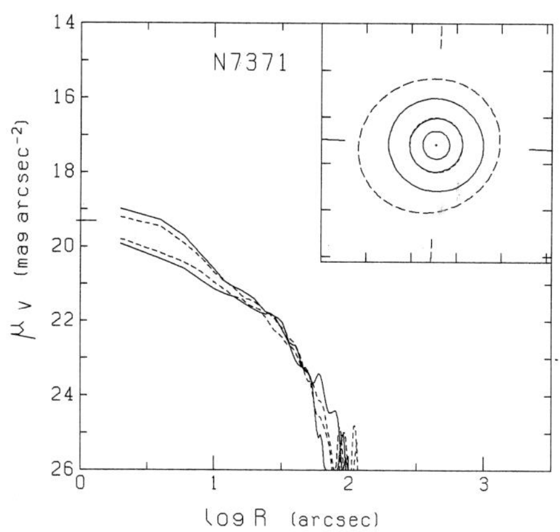

NGC7377

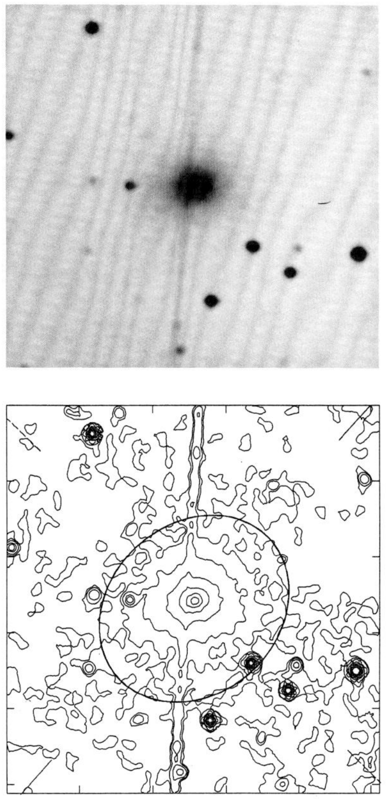

N7377 K5540 2 20.16 256X256

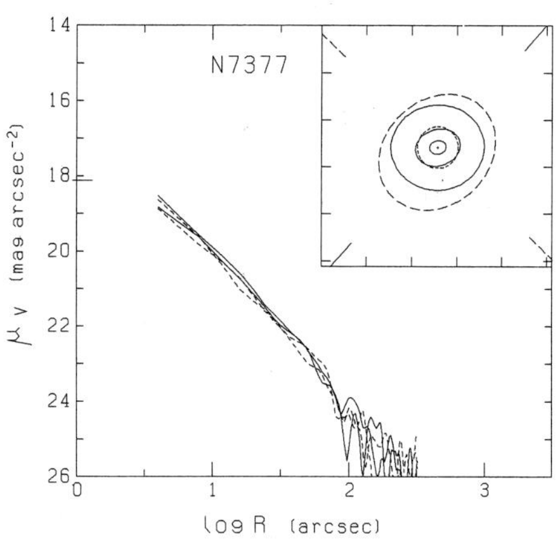

NGC7448

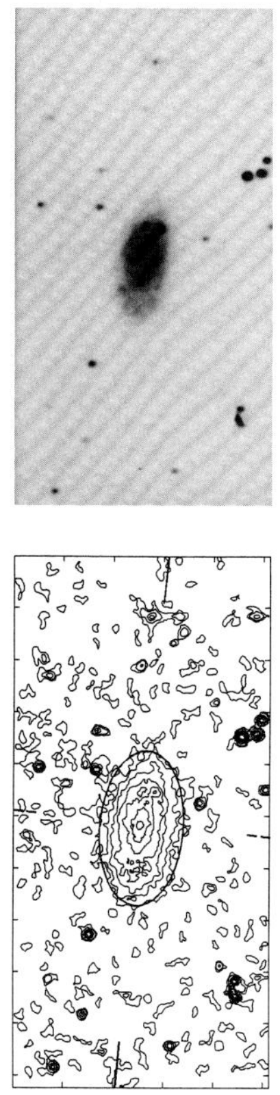

N7448 K5185 1 21.20 256X512

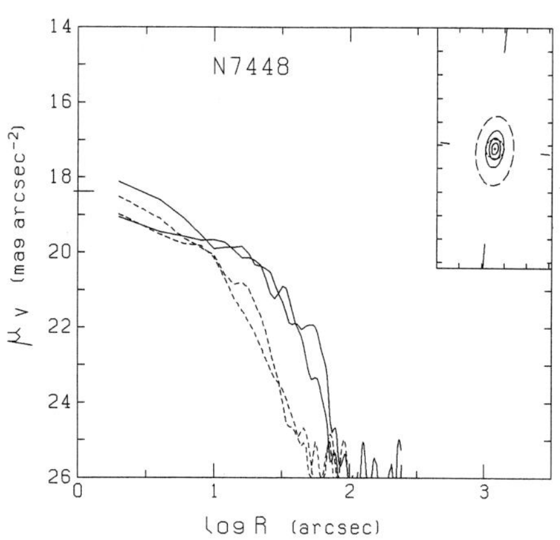

NGC7469

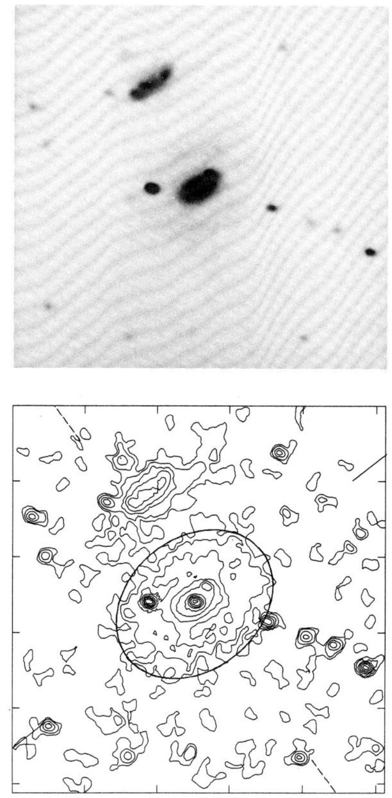

N7469 K4486 1 21.25 256X256

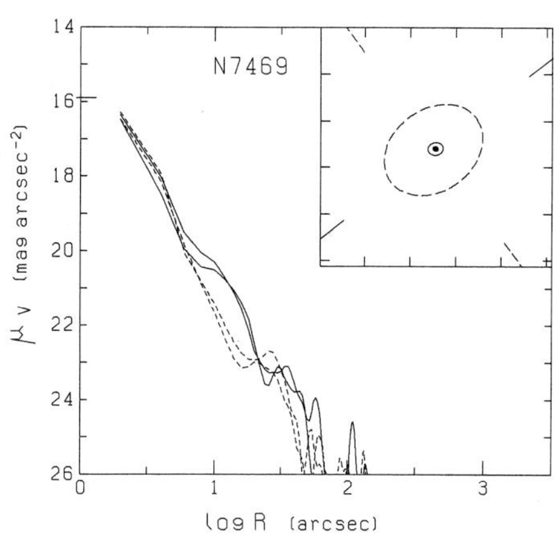

NGC7479

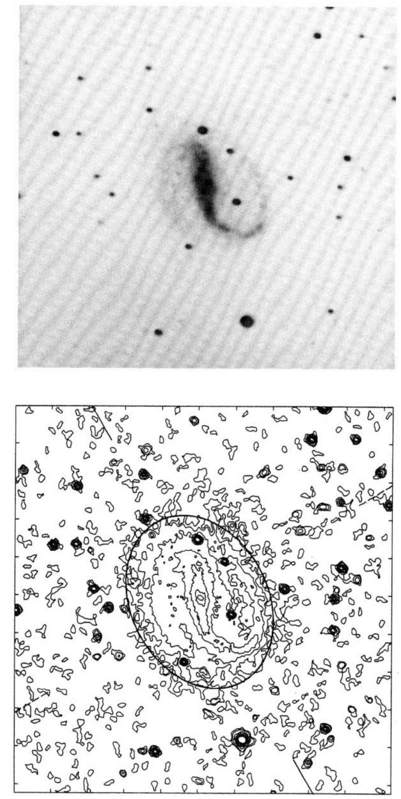

N7479 K4486 1 21.25 512X512

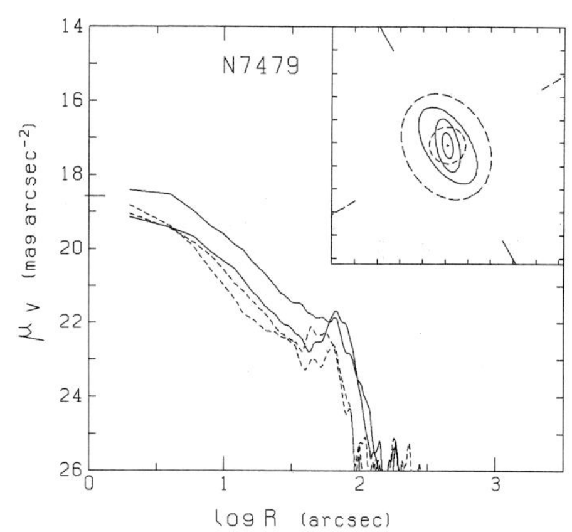

NGC7541

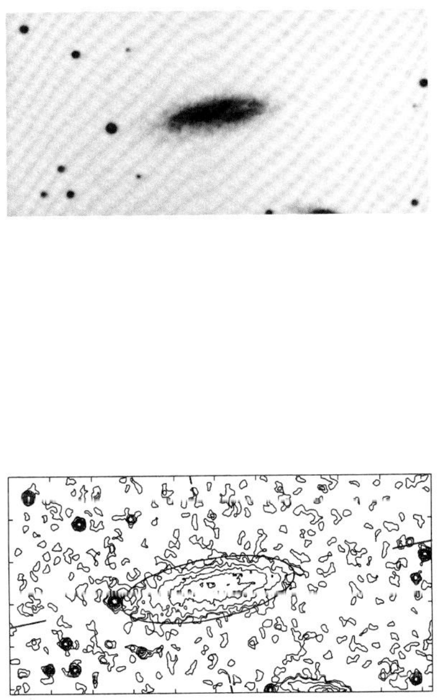

N7541 K4789 1 20.96 512X256

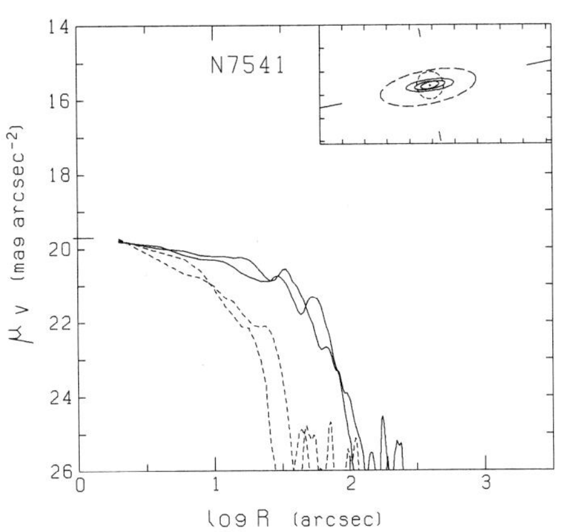

NGC7619

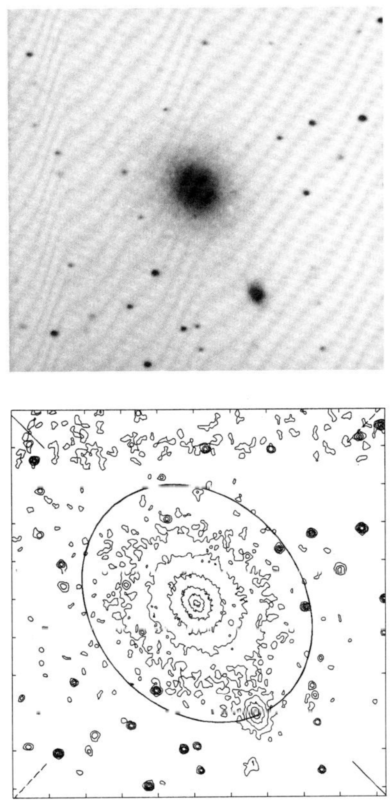

N7619 K2974 1 20.92 512X512

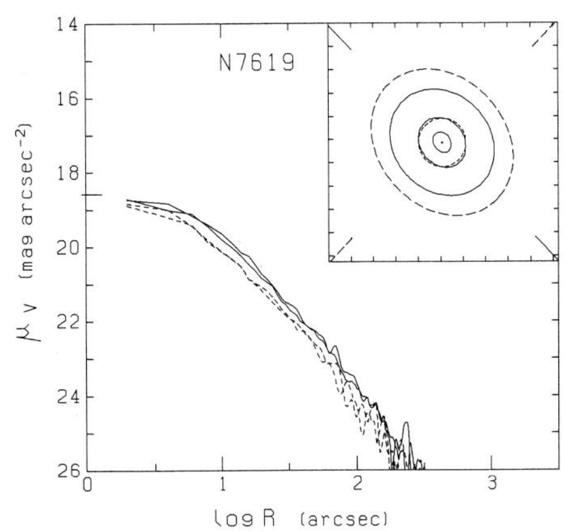

NGC7626

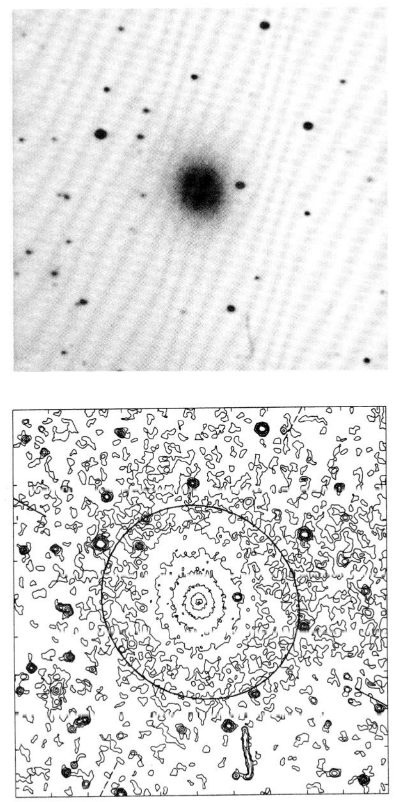

N7626 K2974 1 20.92 512X512

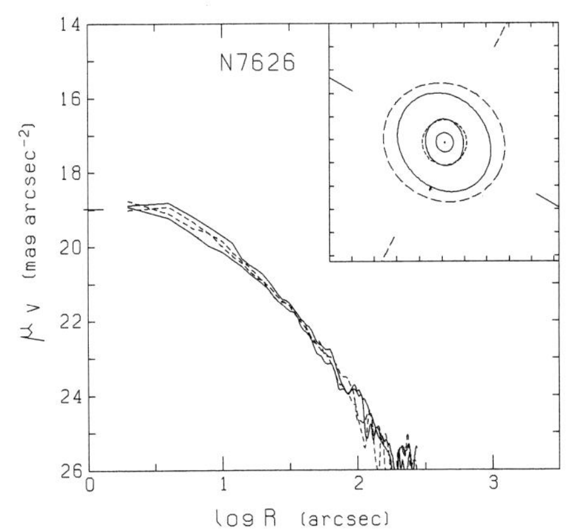

NGC7640

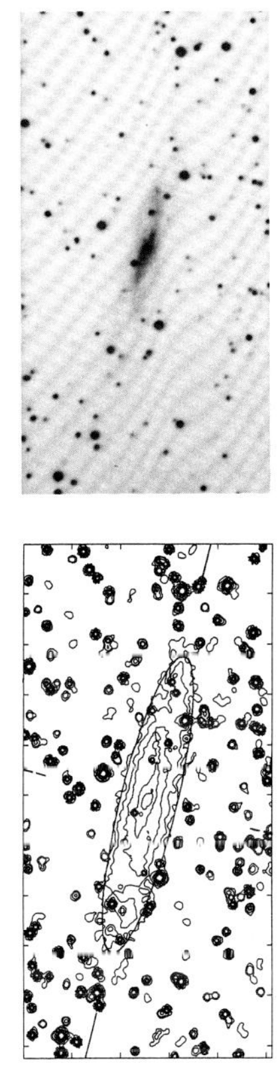

N7640 K5121 2 21.43 256X512

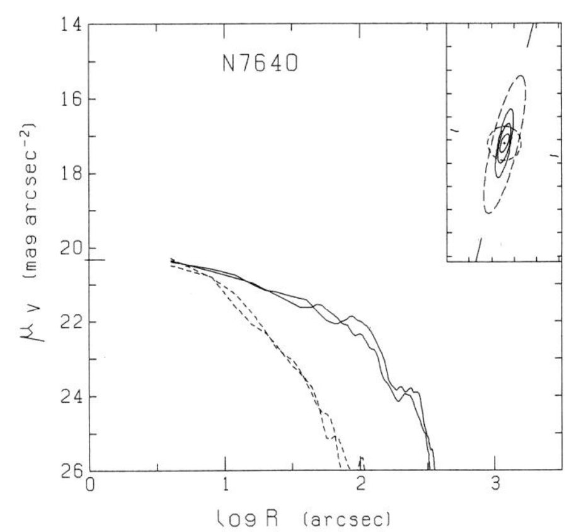

NGC7679

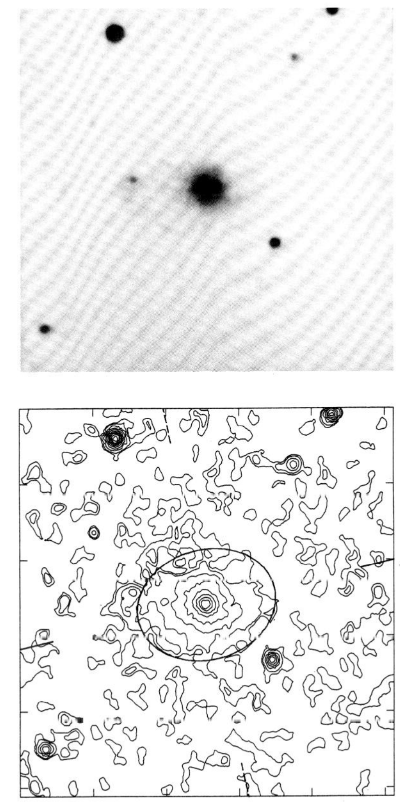

N7679 K4789 1 20.96 256X256

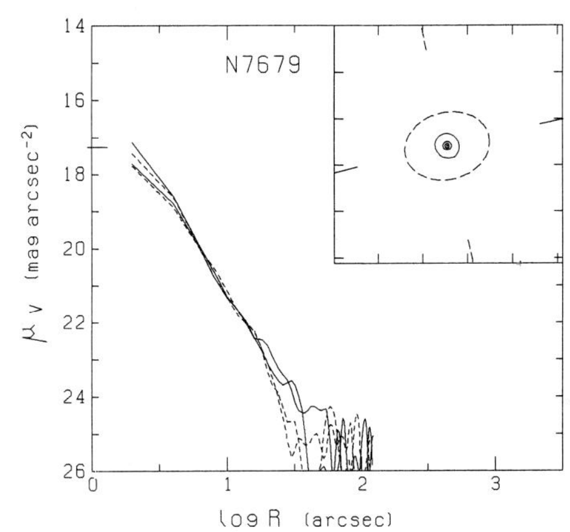

NGC7716

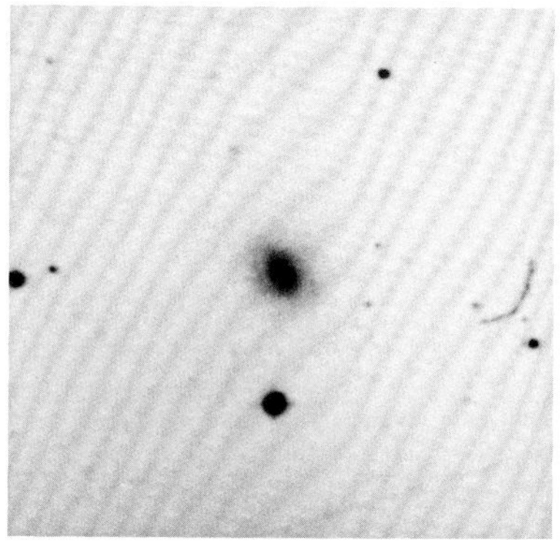

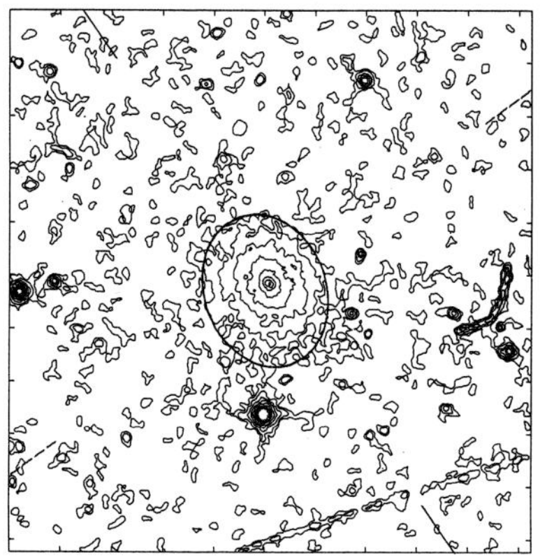

N7716 K5140 1 21.28 512X512

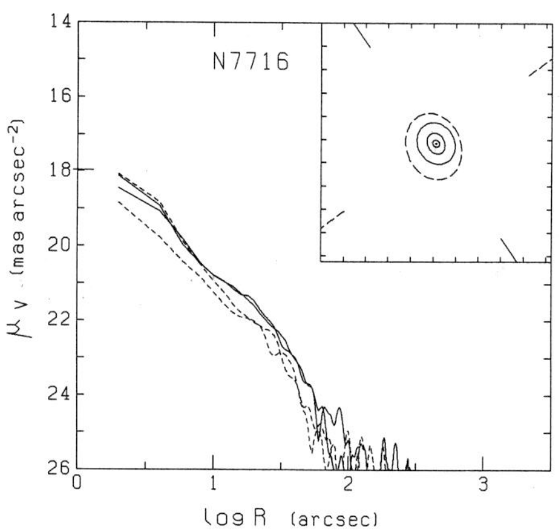

NGC7741

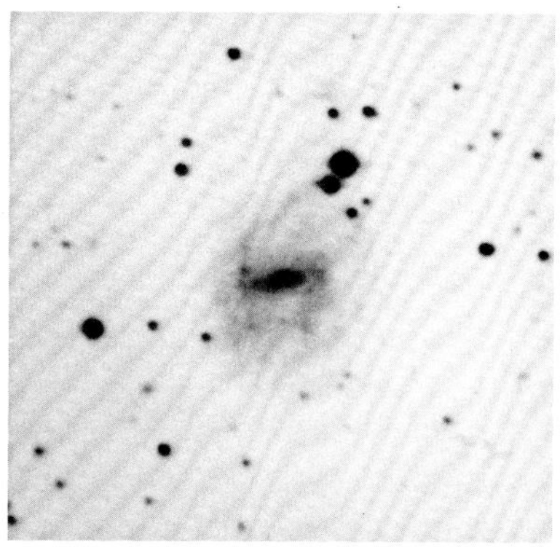

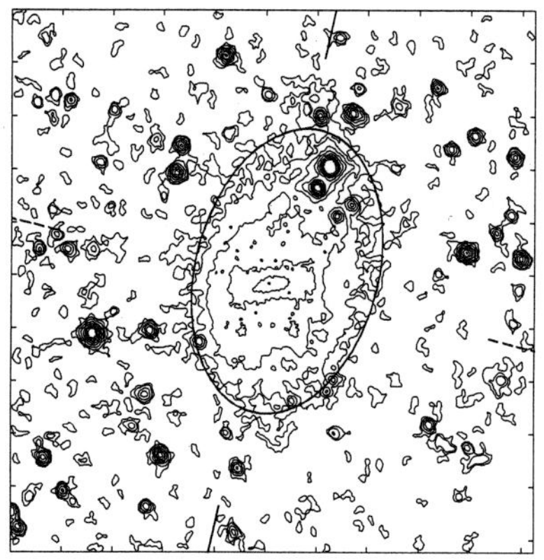

N7741 K5177 1 21.07 512X512

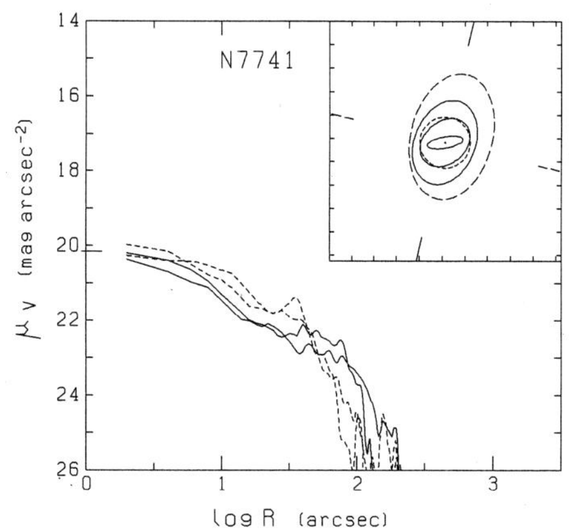

NGC7743

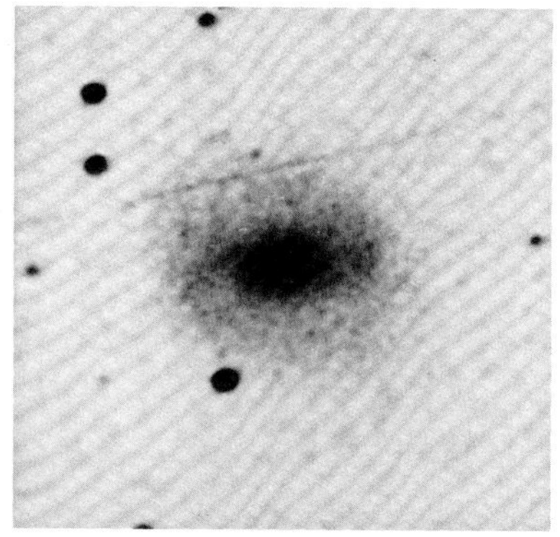

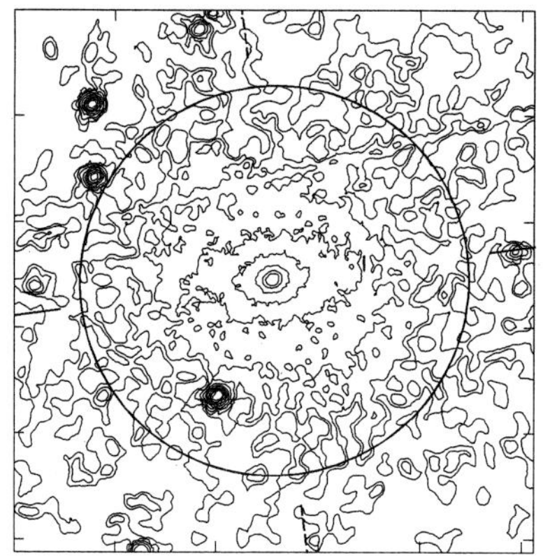

N7743 K4758 1 21.88 256X256

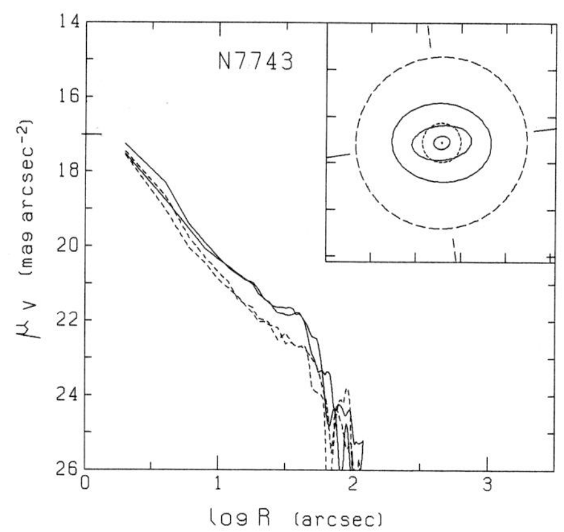

NGC7769

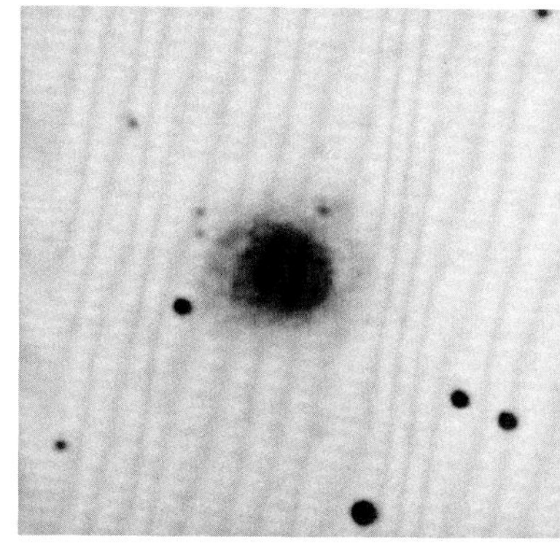

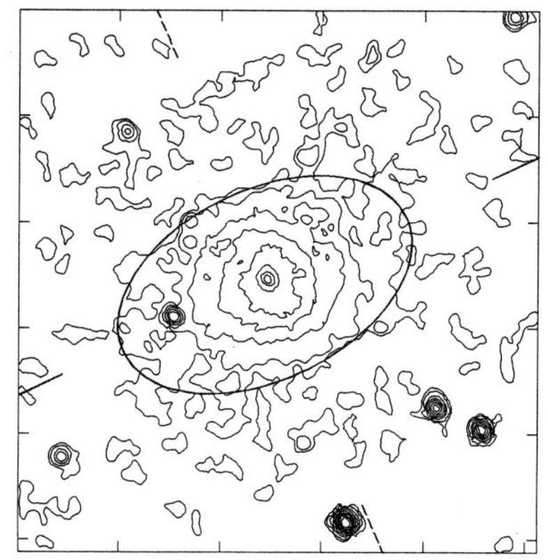

N7769 K5186 1 21.44 256X256

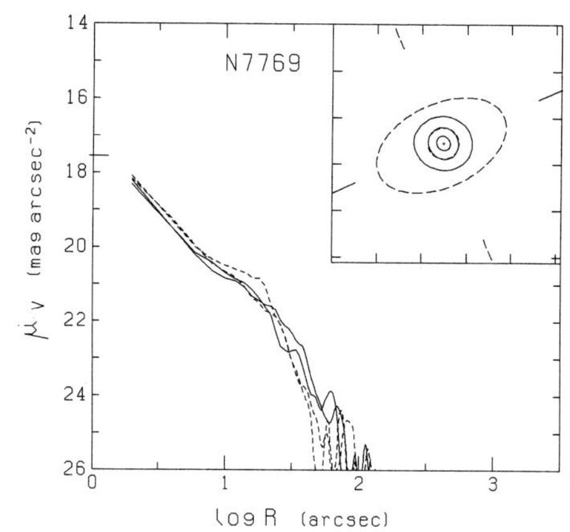

NGC7782

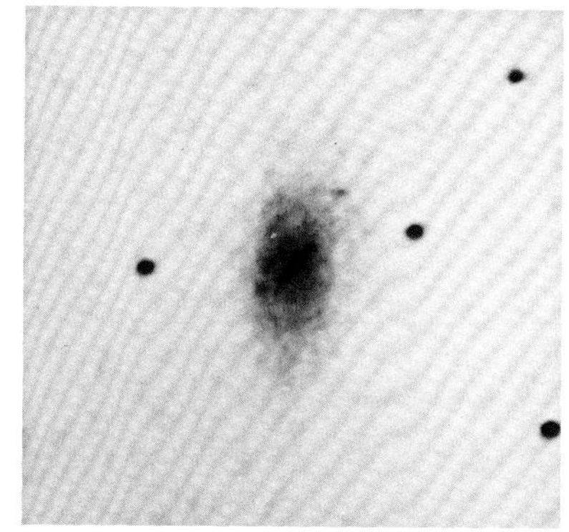

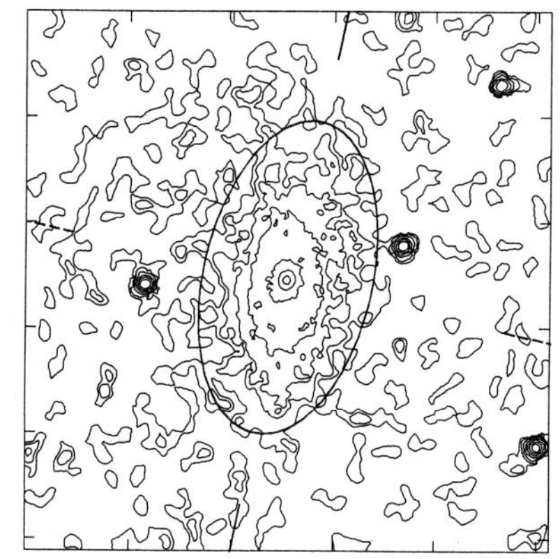

N7782 K4758 1 21.88 256X256

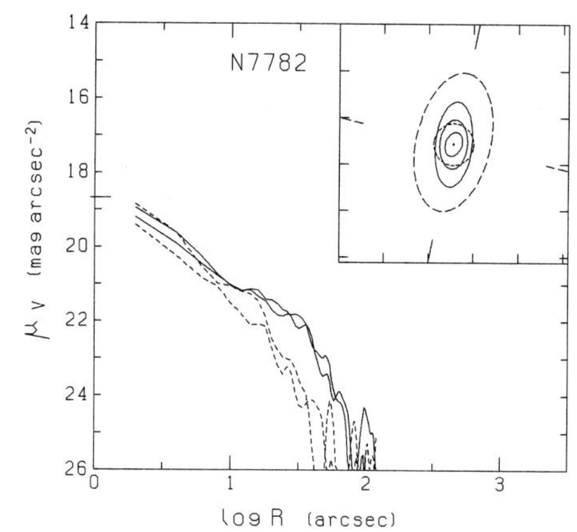

NGC7785

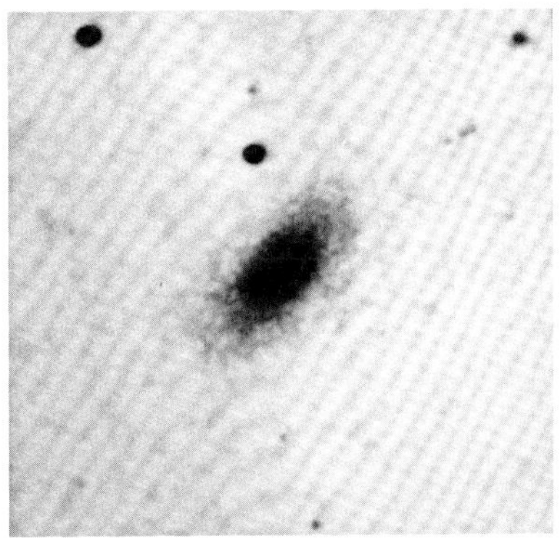

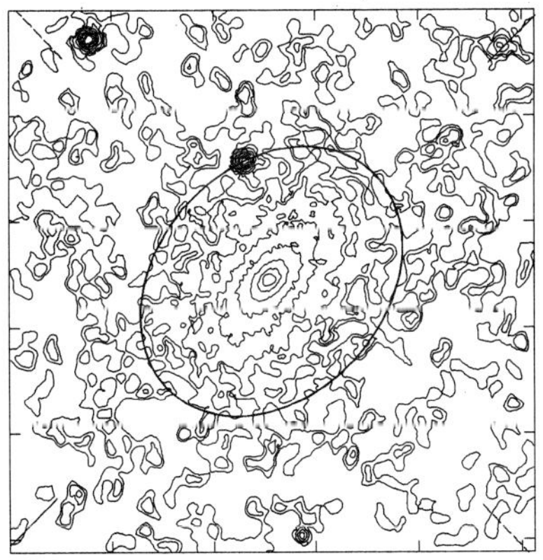

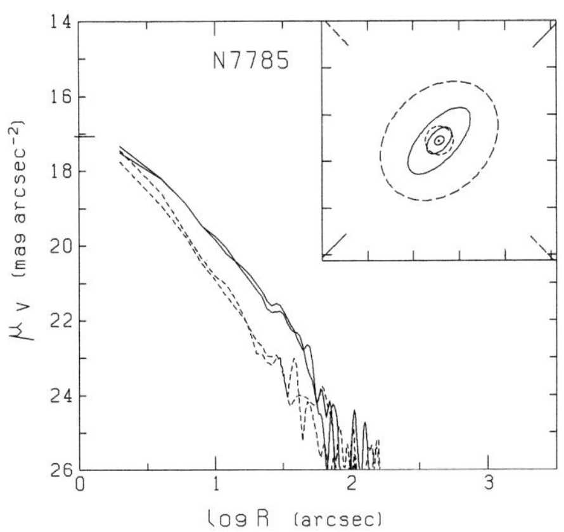

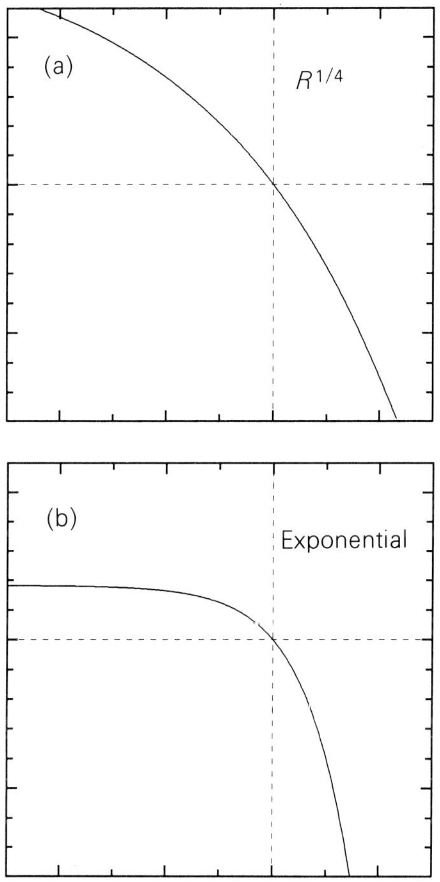

Fig. 3.1. Surface brightness profiles of model galaxies which follow a $R^{1/4}$-law surface brightness distribution (a) and an exponential surface brightness distribution (b).

Part IV Supplementary Data

Table 4.1 Record of the plates and sky brightnesses

Plate (1)	α_c (1950) (2)	δ_c (3)	Date (4)	Seeing (5)	Exp. (6)	Dev. (7)	D_{sky} (8)	N_G (9)	μ_{sky} (10)	σ_p (11)	N_{cal} (12)	N_{pe} (13)
K1962	10 24.0	+40 00	1978/12/30	7	60.0	P	0.37	3	21.85	0.10	2	17
K2007	3 12.0	+45 00	1979/01/23	3	28.7	P	0.32	1	21.71	0.07	1	4
K2034	9 0.0	+05 00	1979/01/30	4	41.1	P	0.27	2	21.66	0.06	2	19
K2071	3 40.0	−15 00	1979/02/20	4	35.0	P	0.39	3	20.82	0.06	3	17
K2106	12 20.0	+15 00	1979/03/21	7	40.0	P	0.37	32	20.73	0.12	30	358
K2112	11 11.8	+40 00	1979/03/22	4	54.4	P	0.35	1	20.69	0.03	1	6
K2186	12 20.0	+10 00	1979/05/02	5	40.0	P	0.40	17	21.34	0.13	16	101
K2211	13 0.0	+20 00	1979/05/20	4	30.0	P	0.19	1	21.32	0.06	1	6
K2220	12 40.0	+10 00	1979/05/23	2	30.0	P	0.24	22	21.12	0.14	17	132
K2304	16 40.0	+35 00	1979/10/11	2	30.0	P	0.25	1	20.70	0.08	1	6
K2416	1 13.0	+01 40	1979/11/19	5	30.0	P	0.22	3	21.08	0.06	3	8
K2488	0 0.0	+15 00	1979/12/14	3	40.0	P	0.42	1	21.12	0.04	2	11
K2536	8 40.0	+20 00	1979/12/24	3	40.0	P	0.28	1	21.86	0.13	2	12
K2605	9 20.0	−10 00	1980/02/07	5	33.8	P	0.38	2	20.58	0.10	2	12
K2647	12 48.0	+40 00	1980/02/17	5	50.0	P	0.30	3	21.51	0.10	2	9
K2649	14 20.0	+35 00	1980/02/17	4	38.6	P	0.26	3	21.66	0.04	3	15
K2653	11 40.0	+25 00	1980/02/19	4	50.0	P	0.29	2	21.64	0.13	1	4
K2662	11 40.0	+33 09	1980/02/20	4	50.0	D	0.66	2	21.55:	0.08	2	4
K2714	12 0.0	+50 00	1980/03/13	3	30.0	P	0.29	10	21.11	0.23	11	32
K2726	11 44.0	+55 00	1980/03/16	2	30.0	P	0.21	11	21.74	0.15	5	15
K2730	12 40.0	+15 00	1980/03/16	2	30.1	P	0.22	12	21.32	0.10	8	53
K2891	3 20.0	−20 00	1980/09/15	3	26.9	P	0.19	6	21.01	0.05	5	50
K2913	0 34.0	+48 18	1980/10/03	3	50.0	D	0.57	3	21.12	0.10	2	19
K2953	8 20.0	+20 00	1980/11/07	3	50.0	D	0.77	1	20.93	0.05	2	14
K2970	3 12.0	+40 00	1980/11/09	3	50.0	D	0.47	2	21.21	0.10	1	44
K2974	23 13.3	+08 20	1980/11/10	2	50.0	D	0.38	2	20.92	0.13	2	27
K2978	2 40.0	+00 00	1980/11/10	3	50.0	D	0.37	5	21.34	0.15	7	89
K3135	10 32.9	+41 39	1981/02/04	4	60.0	P	0.63	1	21.41:	0.00	1	1
K3172	13 0.0	−15 00	1981/02/14	3	40.1	P	0.24	7	20.31	0.11	9	37
K3173	12 24.0	+40 00	1981/02/14	3	40.0	P	0.20	5	21.49	0.11	5	19
K3186	12 40.0	+30 00	1981/03/02	4	50.0	P	0.38	1	20.88	0.17	1	3
K3193	9 36.0	+70 00	1981/03/06	5	50.0	P	0.42	7	21.21	0.13	6	33
K3194	10 47.0	+13 20	1981/03/06	5	50.0	P	0.47	12	21.12	0.11	10	129
K3209	11 20.0	+20 00	1981/03/11	3	28.1	P	0.22	8	21.72	0.13	4	27
K3210	12 48.0	+45 00	1981/03/11	3	45.0	P	0.34	1	21.83:	0.03	1	2
K3226	9 40.0	−30 00	1981/03/28	4	40.0	P	0.47	3	20.11	0.06	3	45
K3243	12 6.7	+01 40	1981/04/07	4	35.0	D	0.63	6	21.12	0.07	5	22
K3254	10 40.0	+05 00	1981/04/21	4	23.0	P	0.06	1	21.05:	0.00	1	1
K3289	13 52.3	+55 00	1981/05/01	2	30.0	D	0.18	5	21.01	0.12	5	24
K3310	12 40.0	−05 00	1981/05/30	5	50.0	P	0.66	6	20.42	0.09	5	31
K3388	0 39.1	+41 39	1981/10/25	4	40.1	D	0.15	3	21.01	0.20	3	56
K3394	22 40.0	+35 00	1981/10/27	2	30.1	D	0.51	1	21.00	0.08	1	11
K3396	22 40.0	+25 00	1981/10/27	3	30.0	D	0.50	1	21.18	0.08	1	20
K3398	1 40.0	+15 00	1981/10/27	2	40.1	D	0.26	1	20.89	0.07	1	5
K3523	12 0.0	+15 00	1981/12/24	4	40.1	P	0.45	2	21.23	0.05	1	3
K3524	12 0.0	+10 00	1981/12/24	4	40.0	P	0.49	1	21.04	0.06	1	15
K3533	1 30.0	+30 00	1981/12/26	2	40.1	D	0.18	1	21.32	0.27	1	2
K3546	12 9.7	+46 39	1981/12/27	3	45.1	P	0.19	2	21.48A	—	0	0
K3547	12 20.0	+05 00	1981/12/27	3	43.1	P	0.29	18	21.21	0.10	16	90
K3567	1 20.0	+10 00	1982/01/11	3	30.0	P	0.27	1	20.88	0.07	1	30
K3615	8 0.0	+75 00	1982/02/12	4	60.0	P	0.46	3	21.20	0.11	3	7
K3628	10 20.0	+30 00	1982/02/15	3	40.0	P	0.37	4	21.27	0.08	3	12
K3676	12 0.0	+45 00	1982/03/21	3	50.0	P	0.22	12	21.65	0.14	10	36
K3680	10 0.0	−05 00	1982/03/22	4	50.0	D	0.41	1	20.55	0.07	1	31
K3681	9 20.0	+20 00	1982/03/22	4	50.0	D	0.40	1	20.84	0.04	1	14
K3683	12 52.4	+28 19	1982/03/22	3	50.0	D	0.28	6	21.15	0.07	4	22
K3714	14 33.3	+06 40	1982/03/29	3	40.1	D	0.30	4	21.30	0.10	3	15
K3737	11 20.0	+15 00	1982/04/23	3	40.0	P	0.30	8	20.70	0.14	8	46
K3740	12 40.0	+05 00	1982/04/23	2	40.1	P	0.28	12	21.03	0.09	8	45
K3766	14 13.3	+01 40	1982/05/23	3	40.0	D	0.33	2	20.76	0.10	4	32
K3859	2 48.0	+40 00	1982/11/18	4	22.6	P	0.33	1	20.59	0.09	2	16
K3929	11 40.0	+10 00	1982/12/19	3	25.0	P	0.37	3	20.83	0.09	3	10
K3936	12 12.7	+23 19	1982/12/20	4	25.0	P	0.34	1	19.92	0.03	1	3
K3937	12 7.6	+28 19	1982/12/20	3	25.1	P	0.31	5	21.19	0.08	6	45
K4022	11 46.7	−03 20	1983/04/08	4	30.0	P	0.27	2	21.09	0.06	2	4
K4026	14 26.7	+03 20	1983/04/08	3	40.0	P	0.18	4	21.20	0.09	4	34
K4057	12 20.4	+30 00	1983/05/09	2	45.0	D	0.30	6	21.23	0.10	8	49
K4058	11 12.3	+55 00	1983/05/09	3	45.0	D	0.35	2	21.01	0.25	4	7
K4221	2 33.3	−08 20	1983/12/04	5	45.0	P	0.33	6	21.29	0.13	4	34
K4378	15 0.0	+00 00	1984/05/07	4	40.0	P	0.17	7	21.07	0.09	6	70
K4389	15 0.0	+05 00	1984/05/21	6	40.0	P	0.20	3	20.64	0.07	2	8
K4414	15 50.0	+61 37	1984/08/15	3	40.0	P	0.08	3	21.65	0.12	3	26
K4486	23 0.0	+10 00	1984/11/14	2	50.0	P	0.16	2	21.25	0.10	1	11
K4487	0 40.0	+00 00	1984/11/14	2	50.1	P	0.16	3	21.48	0.11	2	11
K4508	4 40.0	−20 00	1984/11/24	5	50.0	P	0.18	1	21.05	0.17	1	3
K4513	3 40.0	−25 00	1984/11/27	6	50.0	P	0.25	5	21.19	0.15	5	54
K4515	8 47.7	+50 00	1984/11/27	4	50.2	P	0.14	3	22.05	0.13	4	22
K4517	9 50.0	+33 19	1984/11/27	4	50.0	P	0.13	3	21.91	0.09	4	28
K4518	0 40.0	−10 00	1984/11/28	2	50.0	P	0.27	3	20.96	0.07	3	44
K4519	3 20.0	−30 00	1984/11/28	2	50.0	P	0.29	3	21.11	0.11	3	32

Table 4.1 (continued)

Plate (1)	α_c (1950) (2)	δ_c (3)	Date (4)	Seeing (5)	Exp. (6)	Dev. (7)	D_{sky} (8)	N_G (9)	μ_{sky} (10)	σ_p (11)	N_{cal} (12)	N_{pe} (13)
K4520	3 40.0	−05 00	1984/11/28	2	50.0	P	0.22	4	21.77	0.09	3	15
K4521	5 0.0	−10 00	1984/11/28	2	50.0	P	0.22	1	21.64	0.15	2	13
K4522	6 20.0	−25 00	1984/11/28	1	50.0	P	0.23	2	21.23	0.06	1	11
K4523	9 0.0	+75 00	1984/11/28	2	50.1	P	0.15	3	21.85	0.14	3	5
K4524	8 24.0	+50 00	1984/11/28	1	50.0	P	0.12	2	22.02	0.10	3	9
K4525	3 40.0	−30 00	1984/11/29	3	50.0	P	0.30	3	21.02	0.10	3	30
K4526	5 0.0	−15 00	1984/11/29	3	50.0	P	0.20	2	21.54	0.05	1	8
K4528	8 24.0	+60 00	1984/11/29	3	50.0	P	0.15	2	22.14	0.15	2	12
K4529	9 12.0	+50 00	1984/11/29	3	60.1	P	0.06	1	21.89	0.10	1	11
K4531	4 0.0	−20 00	1984/11/30	2	50.0	P	0.18	2	21.40	0.19	2	13
K4533	5 40.0	−20 00	1984/11/30	2	50.0	P	0.16	1	22.23	0.16	1	6
K4536	6 0.0	+80 00	1984/11/30	3	50.0	P	0.12	1	21.61	0.11	1	6
K4549	4 6.7	−01 40	1984/12/17	5	50.0	P	0.24	1	21.20:	0.00	1	1
K4580	3 0.0	−25 00	1985/01/23	4	50.0	P	0.45	3	20.48	0.05	3	37
K4581	9 0.0	+60 00	1985/01/23	2	40.1	P	0.20	3	21.79	0.11	4	29
K4582	9 20.0	−15 00	1985/01/23	3	50.0	P	0.17	4	21.81	0.09	4	19
K4593	9 2.8	+43 19	1985/01/26	6	50.2	P	0.19	2	21.70	0.12	2	16
K4618	10 40.0	+35 00	1985/02/17	4	50.0	P	0.19	4	21.59	0.13	4	7
K4619	13 0.0	+35 00	1985/02/17	5	50.1	P	0.11	3	21.91	0.11	2	20
K4629	14 0.0	−05 00	1985/03/21	6	50.0	P	0.16	1	21.10:	0.03	1	3
K4639	12 26.7	+01 40	1985/03/23	2	50.0	P	0.19	2	21.55	0.10	1	4
K4640	14 20.0	+30 00	1985/03/23	2	50.0	P	0.11	2	22.16	0.08	1	6
K4647	15 20.0	−15 00	1985/03/27	4	50.0	P	0.21	3	20.96	0.10	2	14
K4653	16 0.0	+20 00	1985/03/28	2	50.1	P	0.14	1	21.41	0.12	1	10
K4654	15 8.1	+56 38	1985/04/01	1	50.0	P	0.15	6	21.74	0.13	5	24
K4664	14 24.0	+55 00	1985/04/16	4	50.0	P	0.10	2	21.40:	0.01	1	2
K4665	17 36.0	+70 00	1985/04/16	4	30.0	P	0.08	1	21.31	0.06	1	11
K4674	12 39.0	+00 00	1985/04/18	3	48.2	P	0.18	6	21.30	0.11	4	14
K4675	14 40.0	+00 00	1985/04/18	3	50.0	P	0.15	7	21.29	0.12	4	27
K4676	14 0.0	+50 00	1985/04/18	3	50.0	P	0.14	3	21.78	0.15	3	9
K4677	15 33.2	+11 40	1985/04/18	3	50.0	P	0.15	2	21.37	0.05	2	5
K4682	16 30.0	+80 00	1985/04/20	5	45.0	P	0.12	1	21.53	0.15	1	5
K4687	17 23.2	+73 15	1985/04/21	3	50.0	P	0.09	1	21.67	0.10	1	11
K4700	15 20.0	+05 00	1985/05/11	2	50.1	P	0.11	1	21.64	0.19	1	8
K4705	12 24.0	+45 00	1985/05/16	3	50.0	P	0.16	3	21.19	0.11	5	13
K4711	18 0.0	+75 00	1985/05/17	3	50.0	P	0.09	1	21.51	0.07	2	9
K4758	23 46.0	+08 10	1985/09/20	2	50.0	P	0.05	3	21.88	0.13	4	28
K4779	0 7.5	+26 39	1985/11/09	1	50.0	P	0.15	2	21.76	0.20	3	15
K4782	2 20.0	+35 00	1985/11/10	2	50.0	P	0.16	3	21.76	0.10	2	12
K4783	3 40.0	−20 00	1985/11/10	4	50.0	P	0.27	7	21.07	0.11	8	51
K4784	4 38.0	−05 00	1985/11/10	4	50.0	P	0.23	5	21.37	0.09	7	43
K4786	7 12.0	+85 00	1985/11/10	3	50.1	P	0.24	3	21.00	0.09	3	26
K4787	7 0.0	+20 00	1985/11/10	3	50.0	P	0.31	1	20.98	0.12	3	9
K4788	22 28.0	−12 30	1985/11/11	4	50.1	P	0.50	1	20.60	0.09	2	8
K4789	23 20.0	+05 00	1985/11/11	3	50.0	P	0.26	2	20.96	0.18	2	16
K4821	0 54.0	−07 50	1985/12/10	3	50.0	P	0.31	7	20.72	0.10	5	39
K4823	3 0.0	−15 00	1985/12/10	5	50.0	P	0.23	3	21.07	0.06	3	34
K4854	2 20.0	+00 00	1986/01/07	3	50.0	P	0.31	3	21.07	0.11	3	36
K4857	11 52.0	−19 00	1986/01/07	2	60.0	P	0.28	8	20.99	0.09	4	12
K4880	10 48.0	+45 00	1986/01/16	5	50.0	P	0.15	3	21.87	0.13	2	5
K4883	10 24.0	+21 30	1986/01/17	4	50.0	P	0.22	8	21.88	0.12	7	76
K4885	12 0.0	+39 00	1986/01/17	5	50.0	P	0.18	3	21.94	0.14	4	48
K4886	10 40.0	+25 00	1986/01/18	5	50.0	P	0.17	2	21.62:	0.00	1	1
K4890	10 0.0	−30 00	1986/01/19	2	50.0	P	0.28	5	20.76	0.09	6	61
K4891	10 40.0	−30 00	1986/01/19	4	50.0	P	0.20	4	21.02	0.12	4	34
K4892	12 0.0	−28 00	1986/01/19	4	50.1	P	0.20	5	21.28	0.13	3	16
K4893	13 0.0	−10 00	1986/01/19	4	50.1	P	0.18	11	21.54	0.13	7	37
K4894	12 24.0	+50 00	1986/01/20	6	50.0	P	0.11	1	21.75	0.10	4	12
K4914	11 24.0	+58 00	1986/02/04	4	50.0	P	0.10	7	22.28	0.16	5	11
K4926	12 39.0	+27 00	1986/02/06	5	50.0	P	0.07	4	22.06	0.07	3	22
K4933	7 48.0	+60 00	1986/02/08	5	50.0	P	0.19	1	21.98	0.05	1	4
K4938	13 36.0	+46 00	1986/02/08	4	36.2	P	0.10	5	21.79	0.12	4	22
K4939	4 10.0	−30 00	1986/02/10	5	50.0	P	0.54	1	20.23	0.05	1	10
K4941	7 0.0	+75 00	1986/02/10	3	50.0	P	0.22	1	21.77	0.07	1	10
K4942	8 6.0	+50 00	1986/02/10	2	50.0	P	0.17	1	21.44	0.01	1	2
K4945	7 12.0	+70 00	1986/02/12	4	50.0	P	0.23	1	21.53	0.04	1	6
K4946	8 0.0	+45 00	1986/02/12	4	50.5	P	0.21	1	21.66	0.07	2	9
K4947	9 40.0	+61 00	1986/02/12	4	50.0	P	0.14	3	21.89	0.09	2	8
K4969	11 36.0	+50 00	1986/03/04	3	50.0	P	0.10	3	22.11	0.07	3	7
K4970	12 20.0	+00 00	1986/03/04	3	50.2	P	0.25	4	21.63	0.17	3	13
K4977	12 40.0	−10 00	1986/03/06	2	50.0	P	0.21	8	21.56	0.14	3	30
K4978	14 0.0	−30 00	1986/03/06	2	42.5	P	0.23	1	20.70	0.07	4	15
K4979	14 24.0	+50 00	1986/03/06	2	50.0	P	0.14	3	21.90	0.14	3	12
K4985	13 0.0	−05 00	1986/03/07	4	50.0	P	0.22	7	21.76	0.17	5	21
K4986	14 10.0	−06 00	1986/03/07	6	50.0	P	0.23	5	21.49	0.14	3	14
K4988	5 36.0	+70 00	1986/03/08	3	50.1	P	0.30	1	21.61	0.19	1	2
K4992	10 40.0	+55 00	1986/03/08	2	50.0	P	0.16	5	22.01	0.15	3	20
K4993	12 0.0	+05 00	1986/03/08	2	50.0	P	0.24	1	21.60	0.07	3	8
K4994	12 20.0	+20 00	1986/03/08	2	50.0	P	0.22	4	21.67	0.06	4	34
K4996	14 20.0	+25 00	1986/03/08	2	50.0	P	0.13	2	22.54	0.15	1	27

Table 4.1 (continued)

Plate (1)	α_c (1950) (2)	δ_c (3)	Date (4)	Seeing (5)	Exp. (6)	Dev. (7)	D_{sky} (8)	N_G (9)	μ_{sky} (10)	σ_p (11)	N_{cal} (12)	N_{pe} (13)
K5011	10 47.0	+18 19	1986/03/31	5	50.0	P	0.12	1	22.26A	—	0	0
K5012	12 24.0	−05 30	1986/03/31	4	50.0	P	0.16	5	21.51	0.04	4	11
K5045	14 20.0	−15 00	1986/04/12	4	50.0	P	0.25	3	21.29	0.02	2	5
K5047	16 20.0	+00 00	1986/04/12	5	50.0	P	0.20	1	21.27	0.14	1	6
K5049	13 0.0	+00 00	1986/04/13	4	50.0	P	0.22	6	21.47	0.06	5	13
K5052	15 20.0	+65 00	1986/04/13	5	50.0	P	0.13	1	21.70:	0.00	1	1
K5057	16 0.0	+00 00	1986/04/16	5	50.0	P	0.13	1	21.58	0.05	1	3
K5058	17 26.7	+06 40	1986/04/16	4	46.1	P	0.11	1	21.58	0.05	1	5
K5062	11 20.0	−10 00	1986/05/07	4	50.0	P	0.33	1	21.01	0.13	2	7
K5064	12 36.0	+60 00	1986/05/07	2	50.1	P	0.09	1	21.66	0.05	1	5
K5066	15 0.0	−09 00	1986/05/07	3	50.0	P	0.20	2	21.09	0.22	2	6
K5067	17 47.3	+23 19	1986/05/07	3	50.0	P	0.08	1	22.44	0.09	1	12
K5068	19 47.0	−13 50	1986/05/07	3	50.1	P	0.25	1	20.75	0.08	2	12
K5069	12 48.0	+33 19	1986/05/08	3	50.0	P	0.14	1	21.51:	0.24	1	2
K5074	13 20.0	+30 00	1986/05/09	4	60.0	P	0.16	1	21.96:	0.00	1	1
K5079	13 40.0	+35 00	1986/05/12	4	50.1	P	0.12	1	21.48	0.06	1	9
K5080	14 48.0	+40 00	1986/05/12	3	50.0	P	0.09	1	21.98:	0.00	1	1
K5091	14 20.0	+15 00	1986/06/08	4	50.1	P	0.10	1	21.79	0.03	1	3
K5092	15 20.9	+41 39	1986/06/08	2	50.0	P	0.12	1	21.81	0.07	1	5
K5094	19 40.0	−10 00	1986/06/08	3	50.0	P	0.20	1	21.28	0.07	1	18
K5103	20 40.0	+65 00	1986/08/28	2	50.0	P	0.12	1	21.52	0.13	1	12
K5112	20 20.0	−25 00	1986/09/03	6	50.0	P	0.41	1	20.86	0.13	1	3
K5121	23 12.0	+40 00	1986/09/05	3	50.1	P	0.17	1	21.43:	0.05	1	2
K5122	20 40.0	−30 00	1986/09/05	4	50.0	P	0.32	1	20.60	0.06	3	25
K5123	22 0.0	−15 00	1986/09/05	3	60.0	P	0.27	2	21.13	0.08	2	9
K5135	21 40.0	+20 00	1986/09/29	3	50.0	D	0.20	1	21.61	0.07	1	6
K5137	20 24.0	+60 00	1986/10/04	3	60.0	D	0.38	1	21.36	0.15	1	11
K5138	22 0.0	+30 00	1986/10/04	4	60.0	D	0.33	1	21.32	0.10	1	13
K5139	22 40.0	−10 00	1986/10/04	4	60.0	D	0.40	2	20.96	0.11	2	12
K5140	23 40.0	+00 00	1986/10/04	4	60.0	D	0.28	1	21.28	0.06	1	6
K5141	1 0.0	+35 00	1986/10/04	2	60.0	D	0.25	1	21.36	0.10	2	9
K5142	2 0.0	+35 00	1986/10/04	3	60.0	D	0.19	2	21.42	0.13	3	18
K5177	23 40.0	+25 00	1986/10/30	5	50.0	D	0.33	1	21.07	0.10	1	10
K5180	4 40.0	+00 00	1986/10/30	4	50.0	D	0.33	1	21.27	0.07	4	23
K5182	8 40.0	+30 00	1986/10/30	3	40.0	D	0.26	1	21.32	0.16	1	14
K5184	21 53.0	+18 19	1986/10/31	3	50.0	P	0.37	1	21.25	0.10	1	11
K5185	23 0.0	+15 00	1986/11/02	4	50.0	P	0.17	1	21.20	0.14	1	10
K5186	23 40.0	+20 00	1986/11/02	4	50.0	P	0.17	1	21.44:	0.14	1	2
K5235	5 0.0	−25 00	1986/12/05	5	50.0	P	0.46	1	20.98	0.15	1	3
K5341	12 0.0	−15 30	1987/01/27	5	60.0	P	0.22	3	21.72	0.11	2	18
K5348	9 12.8	−21 40	1987/01/28	3	50.0	P	0.21	4	21.65	0.08	3	31
K5350	11 40.0	−28 00	1987/01/28	3	50.0	P	0.18	3	21.31	0.07	4	26
K5351	13 50.0	+40 00	1987/01/28	3	50.0	P	0.14	11	22.35	0.18	8	32
K5356	9 0.0	+20 00	1987/01/29	3	50.0	P	0.19	2	21.75	0.17	2	14
K5357	10 6.8	−11 40	1987/01/29	3	45.1	P	0.14	1	21.31	0.05	1	8
K5361	9 4.0	+55 00	1987/01/30	4	60.0	P	0.16	1	22.22:	0.00	1	1
K5369	9 40.0	+05 00	1987/02/05	6	50.1	P	0.20	2	21.93	0.07	2	14
K5370	11 0.0	+15 00	1987/02/05	7	50.1	P	0.27	1	22.01	0.10	2	18
K5372	13 12.0	+40 00	1987/02/07	4	50.0	P	0.13	2	22.12	0.09	1	8
K5374	13 20.0	−20 00	1987/02/08	5	50.1	P	0.27	6	21.42	0.13	6	26
K5389	9 0.0	+80 00	1987/03/02	4	60.0	P	0.31	3	21.38	0.13	2	14
K5395	14 0.0	+05 00	1987/03/06	4	50.0	P	0.19	2	21.73	0.14	2	18
K5422	14 48.0	−16 00	1987/03/29	5	60.0	P	0.27	3	21.26	0.02	1	4
K5423	15 20.0	−25 00	1987/03/29	5	50.0	P	0.32	2	20.98	0.06	2	14
K5424	16 40.0	+20 00	1987/03/29	4	50.0	P	0.20	1	21.33	0.09	1	20
K5439	14 40.0	−20 00	1987/04/20	3	60.0	P	0.16	1	21.12:	0.32	1	3
K5518	1 20.0	+15 00	1987/10/22	5	50.0	P	0.21	2	21.76	0.08	2	8
K5525	22 40.0	−30 00	1987/10/27	5	60.0	P	0.43	1	20.55	0.02	1	2
K5526	0 40.0	−30 00	1987/10/27	5	60.0	P	0.35	4	20.68	0.08	4	46
K5527	2 40.0	+30 00	1987/10/27	5	60.0	P	0.20	1	21.59	0.06	1	7
K5534	2 20.0	−20 00	1987/10/28	2	60.0	P	0.22	1	21.08	0.06	1	9
K5535	3 20.0	−10 00	1987/10/28	2	60.0	P	0.18	1	21.48	0.04	1	9
K5540	22 47.2	−21 40	1987/11/09	3	50.0	P	0.52	1	20.16	0.07	2	19
K5557	0 40.0	−25 00	1987/11/17	5	50.0	P	0.45	1	20.35	0.14	3	22
K5575	1 47.5	+26 39	1987/11/21	4	60.0	P	0.22	2	21.52	0.13	3	11
K5576	2 0.0	+30 00	1987/11/21	3	60.0	P	0.26	1	21.63:	0.07	1	3
K5578	2 40.0	+20 00	1987/11/21	3	50.0	P	0.27	1	21.24	0.11	2	12
K5579	4 40.0	+65 00	1987/11/21	3	50.0	P	0.19	1	21.74	0.05	1	7
K5643	3 0.0	+25 00	1987/12/23	3	53.6	P	0.32	1	20.99:	0.07	1	3
K5648	7 0.0	−30 00	1987/12/24	4	60.0	P	0.48	1	20.60	0.11	1	6
K5649	9 20.0	+35 00	1987/12/24	3	60.0	P	0.24	3	21.81	0.09	2	20
K5652	8 48.0	+33 19	1987/12/26	4	50.0	P	0.29	1	21.19	0.03	1	11
K5653	6 47.5	−26 39	1987/12/26	4	50.0	P	0.33	1	20.90	0.06	2	14
K5657	9 12.0	+40 00	1987/12/27	4	50.0	P	0.20	3	21.67	0.07	1	16
K5667	2 0.0	+05 00	1988/01/14	2	50.0	P	0.28	2	21.47	0.10	2	17
K5668	3 0.0	−10 00	1988/01/14	2	50.0	P	0.25	2	21.49	0.08	1	5
K5679	5 0.0	−06 00	1988/01/20	2	60.0	P	0.25	2	21.82	0.11	3	25
K5681	6 7.2	−21 40	1988/01/20	3	60.0	P	0.24	4	21.50	0.16	4	37
K5683	8 32.7	−23 19	1988/01/20	3	60.0	P	0.27	1	21.42	0.07	1	8
K5686	11 22.0	+03 00	1988/01/20	2	60.0	P	0.21	5	21.75	0.09	5	24

Table 4.1 (continued)

Plate (1)	α_c (1950) (2)	δ_c (3)	Date (4)	Seeing (5)	Exp. (6)	Dev. (7)	D_{sky} (8)	N_G (9)	μ_{sky} (10)	σ_p (11)	N_{cal} (12)	N_{pe} (13)
K5687	11 52.0	−25 00	1988/01/20	3	60.0	P	0.20	2	21.78	0.07	1	5
K5721	7 20.0	+65 00	1988/02/17	4	60.0	P	0.25	2	21.78	0.17	2	11
K5734	7 30.0	+80 00	1988/02/19	3	60.0	P	0.23	1	21.57	0.04	1	5
K5735	8 0.0	−10 00	1988/02/19	4	60.0	P	0.26	1	21.34:	0.09	1	2
K5740	9 44.0	+00 00	1988/02/21	5	60.0	P	0.18	2	21.94	0.03	2	8
K5741	9 53.0	+16 40	1988/02/21	5	60.0	P	0.22	1	21.94:	0.00	1	1
K5742	9 40.0	+30 00	1988/02/21	5	60.0	P	0.23	3	21.72	0.08	2	17
K5744	11 52.0	+33 19	1988/02/21	5	54.4	P	0.20	2	21.63	0.09	2	4
K5759	12 24.0	+75 00	1988/04/10	3	60.0	P	0.18	4	22.42	0.09	3	3
K5761	11 0.0	+30 00	1988/04/15	4	60.0	P	0.20	5	21.60	0.13	4	12
K5763	12 20.0	+35 00	1988/04/15	5	60.0	P	0.16	3	21.94	0.12	3	15
K5765	13 12.0	+60 00	1988/04/15	7	60.0	P	0.14	1	21.82	0.10	1	4
K5768	10 1.5	+03 00	1988/04/16	2	60.0	P	0.25	2	20.53	0.08	5	22
K5770	12 0.0	+60 00	1988/04/16	2	60.0	P	0.19	2	20.99	0.12	4	13
K5791	12 40.0	+20 00	1988/05/08	3	60.0	P	0.12	2	21.84:	0.01	1	2
K5792	14 0.0	+35 00	1988/05/08	3	60.0	P	0.13	2	21.90	0.04	1	2

Note:

Column 1: Plate number
Column 2: Right ascension in hours and minutes at the epoch of 1950.0
Column 3: Declination in degrees and minutes at the epoch of 1950.0
Column 4: Date of observation
Column 5: Seeing size in arc seconds
Column 6: Exposure time in minutes
Column 7: Developer used; P for Pandol and D for D-19
Column 8: Sky density above fog in diffuse density
Column 9: Number of sample galaxies included in the plate
Column 10: Sky brightness in units of mag arcsec^{-2}. The colon indicates the uncertainty larger than σ_P given in column (11). The character 'A' indicates the adjusted value as explained in section 1.8 in Part I.
Column 11: Standard deviation of the sky brightness in units of mag arcsec^{-2}
Column 12: Number of calibrators included in the plate
Column 13: Total number of photoelectric magnitudes used in the determination of the sky brightness

Table 4.2 Frame parameters and sky brightness for individual galaxies

Galaxy (1)	Plate (2)	Frame Size (3)	PS (4)	θ (5)	μ_{sky} (6)	σ_G (7)	N_{pe} (8)
NGC7814	K2488	512 × 512	2	0.0	21.11	0.04	8
NGC 16	K4779	256 × 256	1	−0.1	21.86	0.08	5
NGC 23	K4779	256 × 256	1	0.0	22.00	0.10	3
NGC 147	K2913	512 × 512	3	−0.7	21.30	0.20	6*
NGC 151	K4518	256 × 256	2	0.4	20.99	0.05	17
NGC 148	K5526	256 × 256	1	1.0	20.68	0.06	12
NGC 150	K5526	512 × 256	1	1.0	20.63	0.07	7
NGC 157	K4518	512 × 512	1	0.3	20.94	0.08	25
NGC 185	K2913	512 × 512	3	0.4	21.15	0.10	13
NGC 205	K3388	512 × 512	3	−0.2	21.09	0.29	15
NGC 221	K3388	512 × 512	2	0.1	20.98	0.16	36
NGC 224	K3388	400 × 400	30	0.1	21.01	0.05	5
NGC 227	K4487	256 × 256	1	0.0	21.58	0.04	3
NGC 237	K4487	256 × 256	1	0.0			
NGC 245	K4487	256 × 256	1	0.0			
NGC 254	K5526	256 × 256	1	−0.6	20.74	0.07	14
NGC 253	K5557	512 × 512	6	−0.5	20.43	0.05	6
NGC 255	K4518	256 × 256	1	−0.2	20.96	0.02	2
NGC 268	K4821	256 × 256	1	0.2			
NGC 274	K4821	256 × 256	1	0.2	20.90	0.03	3
NGC 275	K4821	256 × 256	1	0.2	20.64	0.02	2
NGC 278	K2913	512 × 512	1	2.8	21.06	0.07	6
NGC 289	K5526	256 × 256	1	−1.3	20.63	0.08	13
NGC 309	K4821	256 × 256	1	0.0	20.67	0.05	13
NGC 337	K4821	256 × 256	1	−0.1	20.64	0.05	8
NGC 357	K4821	256 × 256	1	−0.2	20.78	0.07	13
New 1	K4821	256 × 256	2	−0.3			
NGC 404	K5141	512 × 512	2	1.0	21.32	0.02	8
NGC 428	K2416	512 × 512	1	0.0	21.01	0.00	2
NGC 473	K5518	256 × 256	1	−0.2	21.78	0.06	4
NGC 514	K5518	512 × 512	1	0.1	21.73	0.08	4
NGC 521	K2416	512 × 512	1	0.1	21.12	0.04	3
NGC 524	K3567	512 × 512	1	0.1	20.88	0.07	30
NGC 533	K2416	512 × 512	1	0.1	21.09	0.06	3
NGC 598	K3533	512 × 512	12	0.1	21.32	0.27	2
NGC 628	K3398	512 × 512	2	−0.4	20.89	0.07	5
NGC 670	K5575	256 × 256	1	−0.3	21.64	0.07	4
NGC 672	K5575	512 × 512	1	−0.3	21.46	0.05	5
NGC 718	K5667	512 × 512	1	−0.2	21.51	0.09	3
NGC 741	K5667	512 × 512	1	−0.1	21.46	0.09	14
NGC 750	K5142	256 × 256	2	−0.8	20.76	0.16	3*
NGC 753	K5142	256 × 256	1	−0.8	21.48	0.12	4
NGC 777	K5576	256 × 256	1	−0.3	21.63	0.07	3
NGC 890	K4782	512 × 512	1	−0.1	21.79	0.09	9
NGC 908	K5534	512 × 256	2	−0.1	21.08	0.06	9
NGC 925	K4782	512 × 512	2	0.6	22.10	0.60	4*
NGC 936	K4854	512 × 512	1	0.0	21.08	0.09	25
NGC 941	K4854	256 × 256	1	0.0	20.92	0.07	6
NGC 949	K4782	256 × 256	1	1.1	21.86	0.31	4*
NGC 955	K4854	256 × 256	1	0.0	21.18	0.03	5
NGC 976	K5578	256 × 256	1	−0.8	21.32	0.07	4
NGC 972	K5527	512 × 512	1	−1.1	21.59	0.06	7
NGC 991	K4221	512 × 512	1	0.0			
NGC1022	K4221	512 × 512	1	−0.1	21.23	0.02	10
NGC1035	K4221	256 × 256	1	−0.1			
NGC1023	K3859	512 × 256	2	−1.7	20.56	0.08	12
NGC1042	K4221	512 × 512	1	−0.2	21.24	0.03	3
NGC1052	K4221	512 × 512	1	0.2	21.39	0.09	14
NGC1055	K2978	512 × 512	2	0.0	21.14	0.06	4
NGC1068	K2978	512 × 512	2	0.0	21.43	0.11	47
NGC1073	K2978	512 × 512	1	0.0	21.25	0.05	11
NGC1084	K4221	512 × 512	1	−0.4	21.18	0.16	7
NGC1087	K2978	512 × 512	1	0.0	21.15	0.15	8
NGC1090	K2978	512 × 256	1	0.0	21.28	0.06	4
NGC1140	K5668	256 × 256	1	0.3	21.49	0.08	5
NGC1156	K5643	512 × 512	1	−0.3	20.99	0.07	3
NGC1172	K4823	256 × 256	2	0.0	21.05	0.04	8
NGC1169	K2007	512 × 512	1	−2.1	21.71	0.07	4
NGC1187	K4580	512 × 512	2	0.0	20.49	0.04	9
NGC1175	K2970	256 × 256	1	−1.7	21.61	0.17	4*
NGC1199	K4823	512 × 512	1	−0.1	21.04	0.05	11
NGC1201	K4580	256 × 256	2	−0.2	20.47	0.06	20
NGC1209	K4823	512 × 256	1	−0.2	21.11	0.05	15
NGC1241	K5668	512 × 512	1	−0.4			
NGC1255	K4580	256 × 256	2	−1.2	20.34	0.04	17*
NGC1292	K4519	256 × 256	2	0.5			
NGC1275	K2970	512 × 512	1	0.7	21.21	0.10	44
NGC1297	K2891	256 × 256	2	0.3	21.06	0.06	4
NGC1300	K2891	512 × 313	2	0.2	21.00	0.03	13
NGC1325	K2891	256 × 256	2	−0.2	22.08	1.26	4*
NGC1332	K2891	256 × 256	2	−0.3	21.01	0.04	29
NGC1337	K5535	512 × 512	1	−0.2	21.47	0.04	9
NGC1339	K4519	256 × 256	1	−0.8	21.10	0.05	4
NGC1344	K4519	512 × 512	2	−0.8	21.01	0.04	14
NGC1353	K2891	256 × 256	2	−0.8	20.91	0.01	2
NGC1357	K2071	512 × 512	1	0.6	20.81	0.06	8
NGC1358	K4520	256 × 256	1	0.2	21.23	0.15	6*
IC 1953	K4783	256 × 256	2	0.7	20.86	0.05	6
NGC1359	K2891	256 × 256	1	−1.0	20.95	0.02	2
NGC1366	K4525	256 × 256	1	1.0	21.42	0.06	14*
NGC1371	K4513	512 × 512	1	0.8	21.13	0.08	10
NGC1376	K4520	256 × 256	1	0.1			
NGC1385	K4513	512 × 512	1	0.5	21.01	0.04	6
NGC1395	K4513	512 × 512	2	0.4	21.39	0.07	8
NGC1398	K4513	512 × 512	2	0.3	21.06	0.04	13
NGC1400	K4783	256 × 256	2	0.2	21.20	0.07	10
NGC1406	K4525	256 × 512	1	0.3	20.89	0.02	2
NGC1407	K4783	512 × 512	2	0.2	21.04	0.02	7
NGC1415	K4513	512 × 512	1	0.1	21.30	0.06	17
NGC1417	K4520	256 × 512	1	0.0	21.66	0.06	2
NGC1421	K2071	256 × 512	1	0.0	20.79	0.03	6
NGC1425	K4525	512 × 512	1	0.0	21.00	0.09	11
NGC1426	K4783	256 × 256	2	−0.1	21.09	0.08	11
NGC1439	K4783	256 × 256	2	−0.2	21.10	0.06	9
NGC1440	K4783	256 × 256	2	−0.2	21.28	—	1*
NGC1452	K4783	256 × 256	2	−0.3	21.08	0.03	4
NGC1453	K4520	512 × 512	1	−0.1	21.75	0.03	10
NGC1461	K2071	512 × 512	1	−0.4	20.88	0.03	3
NGC1507	K4549	256 × 512	1	0.0	21.20	—	1
NGC1518	K4531	512 × 512	1	−0.4	21.21	0.03	6
NGC1521	K4531	512 × 512	1	−0.5	21.57	0.08	7
NGC1537	K4939	256 × 256	2	−0.2	20.23	0.05	10
NGC1569	K5579	512 × 512	1	−3.2	21.74	0.05	7
NGC1600	K4784	512 × 512	2	0.2	21.32	0.06	18
NGC1625	K4784	256 × 256	1	0.1	21.37	0.01	2
NGC1637	K4784	256 × 256	2	0.0	21.36	0.07	10
NGC1638	K5180	512 × 512	1	0.0	21.32	0.06	9
NGC1640	K4508	256 × 256	2	0.0	21.05	0.17	3
NGC1659	K4784	256 × 256	1	−0.1	21.46	0.05	4
NGC1667	K4784	256 × 256	1	−0.2	21.40	0.02	3

Table 4.2 (continued)

Galaxy (1)	Plate (2)	Frame Size (3)	PS (4)	θ (5)	μ_{sky} (6)	σ_G (7)	N_{pe} (8)
NGC1700	K5679	512 × 512	1	0.1	21.90	0.06	13
NGC1726	K5679	512 × 512	1	0.1	21.79	0.05	9
NGC1744	K5235	256 × 512	2	0.2	20.98	0.15	3
NGC1784	K4521	512 × 512	1	−0.1	21.43	0.07	4
HA85−1	K4526	512 × 512	1	−0.6			
NGC1832	K4526	512 × 512	1	−0.6	21.53	0.05	8
NGC1964	K4533	512 × 512	1	0.7	22.23	0.16	6
NGC1961	K4988	512 × 512	1	0.1	21.61	0.19	2
NGC2139	K5681	256 × 256	1	0.8	21.34	0.03	11
NGC2179	K5681	256 × 256	1	0.1	21.69	0.06	5
NGC2196	K5681	512 × 512	1	−0.3	21.47	0.08	15
NGC2146	K4536	512 × 512	1	2.6	21.61	0.11	6
NGC2207	K5681	512 × 512	1	−0.6	21.71	0.12	6
NGC2217	K4522	512 × 512	1	0.0	21.23	0.06	11
NGC2223	K4522	512 × 512	1	−0.3	21.39	0.05	3*
NGC2280	K5653	256 × 512	2	0.5	20.89	0.06	12
NGC2325	K5648	512 × 512	1	−0.1	20.60	0.11	6
NGC2268	K4786	256 × 256	2	−2.8	21.03	0.05	7
NGC2314	K4941	256 × 256	1	0.9	21.77	0.07	10
NGC2339	K4787	256 × 256	1	0.5	21.00	0.14	5
NGC2276	K4786	256 × 256	2	−0.4	20.86	0.13	4
NGC2347	K5721	256 × 256	1	−2.0	21.78	0.03	5
NGC2300	K4786	512 × 256	2	0.9	21.02	0.06	15
NGC2336	K5734	256 × 512	2	−2.8	21.57	0.04	5
NGC2366	K4945	512 × 512	2	2.7	21.53	0.04	6
NGC2403	K5721	512 × 512	4	2.7	21.78	0.23	6
NGC2441	K3615	256 × 256	1	−3.3	20.89	—	1*
NGC2460	K4933	512 × 512	1	1.0	21.97	0.05	4
NGC2500	K4942	256 × 256	2	−1.5	21.44	0.01	2
NGC2525	K5735	256 × 256	2	−0.1	21.34	0.09	2
NGC2523	K3615	512 × 512	1	2.2	21.16	0.01	2
NGC2537	K4946	512 × 512	1	1.7	21.67	0.07	7
NGC2541	K4524	256 × 512	2	−2.5	21.95	0.07	5
NGC2545	K2953	256 × 256	1	−0.7	20.96	0.03	3
NGC2549	K4528	256 × 512	1	−2.0	22.21	0.11	9
NGC2552	K4524	512 × 512	1	−1.6	22.13	0.03	2
NGC2551	K3615	256 × 256	1	4.6	21.06	0.02	2
NGC2613	K5683	512 × 512	1	0.2	21.42	0.07	8
NGC2608	K5182	256 × 256	1	−1.0	21.33	0.16	14
NGC2639	K4515	512 × 512	1	−1.5	21.94	0.04	2
NGC2633	K4523	256 × 256	1	−4.2	21.83	0.10	2
NGC2646	K4523	256 × 256	1	−3.6	21.58	0.29	2*
NGC2654	K4528	512 × 512	1	4.6	21.93	0.04	3
NGC2672	K2536	512 × 512	1	0.6	21.82	0.10	10
NGC2655	K5389	512 × 512	2	−2.7	21.47	0.07	8
NGC2683	K5652	512 × 512	2	0.2	21.19	0.03	11
NGC2681	K4515	512 × 512	1	0.4	22.08	0.11	11
NGC2685	K4581	512 × 512	1	−1.8	21.87	0.09	15
NGC2693	K4515	512 × 512	1	1.1	22.12	0.07	6
NGC2713	K2034	512 × 512	1	−0.1	21.70	0.05	8
NGC2701	K5361	512 × 512	1	−1.7	22.22	—	1
NGC2712	K4593	256 × 256	1	−1.1	21.75	0.11	10
NGC2715	K5389	512 × 512	1	0.5	21.25	0.06	6
NGC2749	K5356	512 × 512	1	0.2	21.87	0.11	7
NGC2742	K4581	512 × 256	1	0.8	21.73	0.08	3
NGC2764	K5356	256 × 256	1	0.5	21.64	0.14	7
NGC2732	K5389	256 × 256	1	1.7	21.77	0.18	3*
NGC2775	K2034	512 × 512	2	0.2	21.63	0.03	11
NGC2768	K4581	512 × 256	2	1.7	21.68	0.06	8
NGC2748	K4523	256 × 256	1	1.9	21.65	—	1
NGC2776	K4593	512 × 512	1	1.1	21.61	0.05	6
NGC2784	K5348	512 × 256	2	0.2	21.69	0.06	17
NGC2782	K5657	512 × 256	1	−0.2	21.67	0.06	16
NGC2793	K5649	256 × 256	1	−0.9			
NGC2811	K4582	256 × 512	1	0.4	21.83	0.03	8
NGC2815	K5348	512 × 512	1	−0.1	21.61	0.03	10
NGC2798	K5657	256 × 256	1	0.3	21.45	0.46	3*
NGC2787	K3193	512 × 512	1	−5.0	21.26	0.05	6
NGC2835	K5348	512 × 512	2	−0.3	21.53	0.09	4
NGC2832	K5649	256 × 512	1	−0.5	21.85	0.07	14
NGC2848	K4582	256 × 256	1	0.1	21.66	0.05	4
NGC2841	K4529	512 × 512	2	1.3	21.89	0.09	11
NGC2844	K5657	256 × 256	1	1.1	21.67	0.14	6*
NGC2855	K2605	256 × 256	2	0.0	20.61	0.16	4
NGC2865	K5348	512 × 512	1	−0.8	21.92	0.10	8*
NGC2859	K5649	512 × 512	1	0.2	21.73	0.08	6
NGC2889	K2605	256 × 256	1	−0.2	20.56	0.04	8
NGC2880	K4947	512 × 512	1	−3.1	21.85	0.11	4
NGC2902	K4582	256 × 256	1	−0.6	21.91	0.03	4
NGC2903	K3681	512 × 512	3	0.8	20.84	0.04	14
NGC2907	K4582	512 × 512	1	−0.6	21.81	0.04	3
NGC2962	K5369	512 × 512	1	0.0	21.93	0.07	13
NGC2950	K4947	512 × 512	1	−0.2	21.92	0.03	4
NGC2967	K5740	512 × 512	1	0.0	21.94	0.03	4
NGC2964	K5742	512 × 256	1	0.0	21.71	0.06	8
NGC2968	K5742	512 × 512	1	0.0	21.73	0.09	9
NGC2976	K3193	512 × 512	1	1.7	21.37	0.21	4
NGC2997	K3226	512 × 512	2	−0.4	20.10	0.05	23
NGC2990	K5369	256 × 256	1	0.1	21.83	—	1
NGC3001	K3226	256 × 256	2	−0.5	20.12	0.10	8
NGC3003	K4517	512 × 256	1	−0.6	21.95	0.06	3
NGC2985	K3193	512 × 512	1	2.3	21.37	—	1
NGC3021	K4517	256 × 256	1	−0.3			
NGC3038	K4890	256 × 256	2	1.4	20.71	0.06	14
NGC3032	K5742	256 × 256	1	1.2	22.10	0.07	6*
NGC3041	K5741	512 × 256	1	−0.2	21.94	—	1
NGC3044	K5740	512 × 512	1	0.0	21.94	0.01	4
NGC3031	K3193	512 × 512	4	3.6	21.06	0.13	4
NGC3034	K3193	512 × 512	2	3.7	21.19	0.04	3
NGC3056	K3226	256 × 256	1	−1.5	20.59	0.06	2*
NGC3043	K4947	256 × 256	1	2.8			
NGC3067	K4517	512 × 256	1	0.7	21.85	0.09	8
NGC3089	K4890	256 × 256	1	0.3	20.84	—	1
NGC3065	K3193	256 × 256	1	5.1	21.62	0.07	4*
NGC3095	K4890	256 × 256	2	0.3	20.75	0.08	29
NGC3077	K3193	512 × 512	2	5.5	21.18	0.06	15
NGC3115	K3680	512 × 512	3	−0.1	20.55	0.07	31
NGC3125	K4890	256 × 256	1	−0.5	20.76	0.07	11
NGC3145	K5357	512 × 512	1	0.0	21.31	0.05	8
NGC3158	K1962	512 × 512	1	−2.1	21.84	0.09	16
NGC3166	K5768	512 × 256	2	0.1	20.55	0.06	8
NGC3169	K5768	512 × 512	2	0.1	20.59	0.05	6
NGC3175	K4890	256 × 256	2	−1.6	20.91	0.06	4
NGC3177	K4883	256 × 256	1	−0.9	21.83	0.05	5
NGC3185	K4883	256 × 256	1	−0.8	21.97	0.04	2
NGC3184	K1962	512 × 512	2	−1.4	21.49	0.14	4*
NGC3190	K4883	512 × 512	1	−0.8	22.04	0.09	9
NGC3193	K4883	512 × 512	1	−0.8	21.92	0.08	12
NGC3226	K4883	512 × 512	1	−0.3	21.81	0.06	4
NGC3227	K4883	512 × 512	1	−0.3	21.87	0.12	34

Table 4.2 (continued)

Galaxy (1)	Plate (2)	Frame Size (3)	PS (4)	θ (5)	μ_{sky} (6)	σ_G (7)	N_{pe} (8)
NGC3245	K3628	512 × 512	1	0.6	21.34	0.16	6*
NGC3254	K3628	512 × 512	1	0.8	21.08	—	1
NGC3274	K3628	256 × 256	1	1.2	21.23	0.01	3
NGC3277	K3628	512 × 512	1	1.3	21.32	0.05	8
NGC3287	K4883	256 × 256	1	0.7			
NGC3294	K1962	512 × 512	1	1.5	21.96	—	1
NGC3301	K4883	512 × 512	1	0.9	21.80	0.11	10
NGC3309	K4891	256 × 256	2	0.7	21.11	0.04	18
NGC3312	K4891	256 × 256	2	0.7	21.14	—	1
NGC3310	K4992	512 × 512	1	−0.9	22.05	0.10	11
NGC3319	K3135	512 × 512	1	0.6	21.41	—	1
NGC3320	K4880	256 × 256	1	−2.0	20.01	—	1*
NGC3338	K3194	512 × 256	2	−0.4	20.95	—	1
NGC3344	K4886	512 × 512	2	0.1	21.62	0.00	1
NGC3346	K3194	512 × 512	1	−0.3			
NGC3351	K3194	512 × 512	2	−0.3	21.10	0.09	11
NGC3353	K4992	256 × 256	1	0.5	22.16	0.03	4
NGC3367	K3194	512 × 512	1	−0.2	20.88	0.16	3
NGC3368	K3194	512 × 512	2	−0.2	21.01	0.07	12
NGC3370	K5011	256 × 256	1	−0.2			
NGC3377	K3194	512 × 512	1	−0.1	21.19	0.09	17
NGC3379	K3194	512 × 512	2	−0.1	21.12	0.06	41
NGC3384	K3194	512 × 512	2	−0.1	21.08	0.05	17
NGC3390	K4891	256 × 512	1	−0.7	20.84	0.05	8
NGC3389	K3194	512 × 512	1	−0.1	21.19	0.08	4
NGC3395	K4618	256 × 256	1	1.0	21.47	0.04	2
NGC3396	K4618	256 × 256	1	1.0	21.61	0.06	2
NGC3412	K3194	512 × 512	1	0.1	21.25	0.09	11
NGC3414	K5761	512 × 512	1	−1.4	21.62	0.09	4
NGC3423	K3254	512 × 512	1	0.2	21.05	—	1
NGC3415	K4880	256 × 256	1	0.2	21.97	0.07	3
NGC3430	K4618	512 × 512	1	1.4	21.55	0.04	2
NGC3432	K4618	512 × 512	1	1.4	23.06	—	1*
NGC3437	K4886	256 × 256	1	1.0			
NGC3449	K4891	256 × 256	2	−1.3	21.00	0.06	7
NGC3445	K4992	256 × 256	1	2.4	21.67	—	1*
NGC3448	K4992	512 × 256	1	2.4	21.81	0.08	5
NGC3455	K5370	256 × 256	1	−0.5	22.69	—	1*
NGC3458	K4992	256 × 256	1	2.7			
NGC3478	K4880	256 × 256	1	1.5	21.73	0.00	2
NGC3485	K3194	512 × 512	1	0.6			
NGC3489	K3194	512 × 512	1	0.6	21.13	0.09	12
NGC3486	K5761	512 × 512	2	−0.3	21.54	0.14	4
NGC3504	K5761	512 × 512	1	0.1	22.14	0.11	6*
NGC3510	K5761	256 × 512	1	0.1	21.71	0.07	3
NGC3512	K5761	256 × 256	1	0.2	21.50	—	1
NGC3549	K4058	256 × 256	1	−0.9	20.82	0.00	2
NGC3556	K4914	512 × 256	2	−3.3	22.37	0.12	3*
NGC3593	K3737	512 × 512	1	−0.5	20.73	0.04	8
NGC3596	K3737	512 × 512	1	−0.5			
NGC3605	K3209	512 × 512	1	−0.5	21.87	0.10	4
NGC3607	K3209	512 × 512	2	−0.5	21.63	0.12	10
NGC3608	K3209	512 × 512	2	−0.5	21.76	0.08	11
NGC3611	K5686	256 × 256	1	−0.1	21.79	0.04	4
NGC3610	K4914	512 × 512	1	−1.8	22.41	0.03	3
NGC3613	K4914	512 × 512	1	−1.8	22.31	0.09	2
NGC3623	K3737	256 × 512	2	−0.2	20.67	0.09	12
NGC3619	K4914	512 × 512	1	−1.6	22.30	0.00	2
NGC3626	K3209	512 × 512	1	−0.2	22.00	0.10	8*
NGC3627	K3737	512 × 512	2	−0.2	20.66	0.07	14
NGC3628	K3737	512 × 256	3	−0.2	20.71	0.38	3
NGC3630	K5686	256 × 256	1	−0.1	22.05	—	1
NGC3631	K4058	512 × 512	1	1.2	21.35	—	1
NGC3640	K5686	512 × 512	2	0.0	21.68	0.02	10
NGC3646	K3209	512 × 512	1	−0.1			
NGC3642	K4914	512 × 512	1	−1.0	21.99	0.08	2
NGC3655	K3737	256 × 256	1	0.0	20.68	0.07	3
NGC3659	K3209	256 × 256	1	0.1			
NGC3664	K5686	256 × 256	1	0.0	21.72	0.05	4
NGC3665	K2112	512 × 512	2	1.6	20.69	0.03	6
NGC3672	K5062	256 × 256	2	−0.1	20.93	0.02	5
NGC3681	K3737	256 × 256	1	0.3	20.67	0.13	2
NGC3684	K3209	512 × 512	1	0.4	21.34	—	1*
NGC3683	K4914	256 × 256	1	0.2			
NGC3686	K3209	512 × 512	1	0.4	21.70	0.05	2
NGC3691	K3737	256 × 256	1	0.4			
NGC3690	K4914	256 × 256	1	0.4	22.34	0.11	2
NGC3705	K3929	256 × 256	2	−0.5	20.87	0.04	4
NGC3717	K5350	512 × 512	1	1.3	21.20	0.03	5
NGC3720	K5686	256 × 256	1	0.1	21.82	0.05	5
NGC3718	K2726	256 × 512	2	−2.9	23.77	—	1*
NGC3729	K2726	256 × 512	1	−2.6	23.74	—	1*
NGC3738	K2726	512 × 512	1	−2.2	21.44	0.00	2
NGC3756	K2726	512 × 512	1	−2.0	21.60	—	1
NGC3769	K4969	256 × 512	1	−0.2			
NGC3773	K3929	256 × 256	1	−0.2	20.99	—	1
NGC3780	K2726	512 × 512	1	−1.5			
NGC3810	K3929	256 × 256	2	−0.1	20.77	0.06	5
NGC3813	K2662	256 × 256	1	−0.2	21.46	—	1
NGC3818	K4022	256 × 256	1	0.1	21.08	0.06	3
NGC3877	K4969	512 × 512	1	1.4	22.12	0.05	4
NGC3885	K5350	512 × 256	1	−0.5	21.34	0.03	3
NGC3887	K4857	256 × 256	2	0.6	21.00	0.02	2
NGC3888	K2726	256 × 256	1	0.2	21.71	0.09	4
NGC3893	K4969	512 × 512	1	1.9	21.37	—	1*
NGC3900	K2653	512 × 512	1	0.7	21.64	0.13	4
NGC3898	K2726	512 × 256	2	0.5	21.82	0.06	4
NGC3904	K5350	512 × 512	1	−0.8	21.35	0.05	12
NGC3912	K2653	256 × 512	1	0.8			
NGC3917	K2714	512 × 256	1	−2.3	21.22	0.07	2
NGC3923	K4892	512 × 512	2	1.4	21.30	0.06	7
NGC3936	K5687	512 × 512	1	0.2			
NGC3938	K3676	512 × 512	2	−1.7	21.57	—	1
NGC3941	K2662	512 × 512	1	1.4	21.58	0.07	3
NGC3945	K5770	256 × 256	2	−2.0	21.05	0.08	4
NGC3949	K2714	512 × 512	1	−1.7	21.01	—	1
NGC3952	K4022	256 × 256	1	−0.1	21.14	—	1
NGC3953	K2726	256 × 512	2	1.5	23.12	—	1*
NGC3955	K5687	256 × 256	1	0.1	21.78	0.07	5
NGC3956	K4857	256 × 256	2	0.0			
NGC3957	K4857	256 × 512	1	0.0	21.11	0.01	2
NGC3962	K5341	512 × 512	1	0.5	21.78	0.07	12
NGC3976	K4993	256 × 256	1	−0.1	21.78	—	1
NGC3981	K4857	256 × 256	2	−0.1	21.11	—	1
NGC3982	K2726	256 × 256	1	2.0			
NGC3985	K2714	256 × 256	1	−1.1			
NGC3992	K2726	512 × 512	2	2.3	22.15	0.05	2*
NGC3995	K5744	256 × 256	1	0.4	21.60	0.08	3
NGC3998	K2726	512 × 512	1	2.3	21.87	0.04	4
NGC4008	K3937	256 × 256	1	−1.4	21.60	—	1*

Table 4.2 (continued)

Galaxy (1)	Plate (2)	Frame Size (3)	PS (4)	θ (5)	μ_{sky} (6)	σ_G (7)	N_{pe} (8)
NGC4013	K3676	512 × 512	1	−0.7	22.38	0.07	2*
NGC4024	K4857	256 × 256	1	−0.3	21.34	—	1*
IC 749	K3676	512 × 512	1	−0.7	21.48	0.05	3
IC 750	K3676	512 × 512	1	−0.7	21.56	0.01	3
NGC4026	K2714	256 × 512	1	−0.6	21.48	0.14	4
NGC4027	K4857	512 × 512	1	−0.4	20.94	0.06	7
NGC4030	K3243	512 × 512	1	−0.1	21.01	0.10	2
NGC4033	K4857	256 × 256	1	−0.5	21.44	—	1*
NGC4037	K3523	512 × 512	1	−0.1			
NGC4036	K5770	256 × 256	2	−0.2	20.97	0.08	5
NGC4038	K4857	512 × 512	2	−0.6			
NGC4045	K3243	512 × 256	1	0.0	21.16	0.02	6
NGC4047	K2714	256 × 256	1	0.1			
NGC4050	K5341	512 × 512	1	0.0	21.60	0.03	6
NGC4051	K3676	512 × 512	2	0.1	21.61	0.07	7
NGC4062	K5744	512 × 256	1	1.3	21.72	—	1
NGC4073	K3243	512 × 512	1	0.0	21.10	0.04	3
NGC4085	K2714	512 × 256	1	0.5	21.01	—	1
NGC4088	K2714	512 × 512	1	0.6	20.80	0.27	5
IC 2995	K4892	512 × 256	1	−0.4			
NGC4094	K5341	512 × 512	1	−0.2			
NGC4096	K2714	256 × 512	2	0.7	21.07	0.10	5
NGC4100	K2714	512 × 512	1	0.7	21.10	0.01	2
NGC4105	K4892	256 × 256	2	−0.5	21.21	0.12	5
NGC4106	K4892	256 × 256	2	−0.5	21.32	0.18	4
NGC4111	K3676	512 × 512	1	0.8	21.80	0.08	12
NGC4116	K3243	512 × 512	1	0.0	21.12	0.01	2
NGC4124	K3524	512 × 512	1	0.2	21.04	0.06	15
NGC4123	K3243	512 × 512	1	0.0	21.49	0.03	2*
NGC4136	K3937	512 × 512	1	−0.1			
NGC4138	K3676	512 × 512	1	1.2	21.57	0.03	3
NGC4143	K3676	512 × 512	1	1.3	21.87	0.32	2*
NGC4144	K3676	512 × 256	2	1.3	21.50	0.02	3
NGC4145	K4885	512 × 512	1	1.2	21.78	0.01	3
IC 764	K4892	256 × 512	1	−0.9			
NGC4151	K4885	512 × 512	1	1.3	21.95	0.13	39
NGC4150	K3937	256 × 256	1	0.0	21.24	0.06	6
NGC4152	K3523	256 × 256	1	0.5	21.23	0.05	3
NGC4157	K4894	512 × 256	1	−3.0	21.70	0.03	3
NGC4162	K3936	256 × 256	1	−0.3	19.92	0.03	3
NGC4168	K2106	512 × 512	1	−0.7	20.68	0.03	7
NGC4178	K2186	512 × 512	1	−0.4	21.16	0.08	2
NGC4179	K3243	512 × 512	1	0.0	21.12	0.08	9
NGC4183	K3676	512 × 512	1	1.9	21.67	0.07	2
NGC4190	K5763	256 × 256	1	−1.3			
NGC4189	K2106	256 × 256	1	−0.6	20.56	—	1
NGC4192	K2106	256 × 512	2	−0.6	20.63	0.04	6
NGC4203	K5763	512 × 512	1	−1.1	22.05	0.05	7
NGC4212	K2106	256 × 256	1	−0.4	20.68	0.09	2
NGC4214	K4885	512 × 512	2	2.1	21.84	0.12	3
NGC4216	K2106	256 × 512	2	−0.4	20.68	0.05	7
NGC4217	K3676	512 × 512	1	2.4	21.66	—	1
NGC4215	K3547	256 × 256	1	−0.1	21.33	0.03	3
NGC4220	K2714	512 × 512	1	2.6	21.08	0.04	4
NGC4224	K3547	256 × 256	1	−0.1	21.24	0.12	5
NGC4233	K2186	256 × 256	1	−0.2	21.47	0.11	4
NGC4234	K3547	256 × 256	1	−0.1	21.17	0.20	2
NGC4235	K3547	512 × 512	1	−0.1	21.27	0.09	10
NGC4237	K2106	256 × 256	1	−0.3	20.52	—	1
NGC4244	K3173	512 × 512	3	−1.4	21.44	0.11	5
NGC4242	K3676	512 × 512	2	2.7	21.45	—	1
NGC4245	K4057	512 × 512	1	−0.7	21.30	0.07	5
NGC4251	K3937	256 × 256	2	0.9	21.23	0.02	4
NGC4254	K2106	512 × 512	1	−0.2	20.58	0.05	10
NGC4258	K4705	512 × 512	3	−1.3	21.29	0.14	4
NGC4260	K3547	512 × 512	1	−0.1	21.12	0.08	2
NGC4261	K3547	512 × 512	2	−0.1	21.15	0.05	17
NGC4262	K2106	256 × 256	1	−0.2	20.91	0.07	19
NGC4267	K2106	512 × 512	1	−0.2	20.74	0.08	15
NGC4270	K3547	256 × 256	1	−0.1	21.27	0.06	3
NGC4274	K3937	512 × 256	2	1.2	21.08	0.07	8
NGC4273	K3547	256 × 256	1	−0.1	21.13	0.07	5
NGC4278	K4057	512 × 512	2	−0.4	21.18	0.04	14
NGC4281	K3547	512 × 512	1	0.0	21.22	0.08	9
NGC4283	K4057	256 × 256	2	−0.3	21.37	0.05	9
NGC4291	K5759	256 × 256	1	−1.4	22.54	—	1
NGC4293	K4994	512 × 512	1	−0.1	21.65	0.07	9
NGC4294	K2186	256 × 256	1	−0.1	21.34	0.11	3
NGC4298	K2106	512 × 512	1	−0.1	21.01	—	1*
NGC4299	K2186	256 × 256	1	0.0	21.26	0.09	4
NGC4302	K2106	256 × 512	1	−0.1			
NGC4303	K3547	512 × 512	2	0.0	21.13	0.09	8
NGC4307	K2186	256 × 512	1	0.0			
NGC4314	K4057	512 × 512	1	0.0	21.10	0.03	4
NGC4321	K2106	512 × 512	2	0.0	20.65	0.10	13
NGC4324	K3547	256 × 256	1	0.0	21.25	0.06	5
NGC4346	K3546	512 × 256	1	2.1			
NGC4339	K3547	512 × 512	1	0.0	21.18	0.05	8
NGC4340	K2106	512 × 256	1	0.1	20.79	0.08	13
NGC4342	K3547	256 × 256	1	0.0			
NGC4348	K5012	256 × 256	1	0.1	21.52	0.02	2
NGC4350	K2106	256 × 256	1	0.1	20.88	0.09	13
NGC4365	K2186	512 × 512	3	0.1	21.28	0.05	13
NGC4369	K3173	256 × 256	1	−0.3	21.65	0.09	3
NGC4386	K5759	512 × 512	1	−0.4	22.39	—	1
NGC4371	K2186	512 × 512	1	0.1	21.41	0.08	9
NGC4374	K2106	512 × 512	2	0.2	20.73	0.11	38
NGC4377	K2106	256 × 256	1	0.2	20.83	0.03	5
NGC4379	K2106	256 × 256	1	0.2	20.79	0.03	6
NGC4378	K3547	512 × 512	1	0.1	21.19	0.03	6
NGC4380	K2186	512 × 512	1	0.1	21.22	0.01	3
NGC4382	K4994	512 × 512	2	0.2	21.66	0.06	13
NGC4383	K2106	256 × 256	1	0.2	20.74	0.03	4
NGC4389	K3546	512 × 256	1	2.4			
NGC4385	K4970	256 × 256	1	0.0	21.78	0.08	6
NGC4388	K2106	512 × 256	1	0.2	20.69	0.05	5
NGC4395	K5763	512 × 512	3	0.5	21.87	0.03	5
NGC4394	K4994	256 × 256	2	0.3	21.68	0.03	4
NGC4406	K2106	416 × 416	3	0.2	20.69	0.06	24
NGC4414	K4057	512 × 512	1	0.4	21.29	0.04	3
NGC4412	K3547	256 × 256	1	0.1	21.08	—	1
NGC4417	K2186	256 × 256	1	0.2	21.49	0.06	10
NGC4419	K2106	256 × 256	1	0.3	20.72	0.09	3
NGC4420	K3547	256 × 256	1	0.1			
NGC4424	K2186	512 × 256	1	0.2	21.23	—	1
NGC4425	K2106	256 × 256	1	0.3	20.71	0.06	8
NGC4428	K5012	256 × 256	1	0.0	21.51	0.02	2
NGC4429	K2186	512 × 256	2	0.2	21.31	0.06	8
NGC4433	K5012	256 × 256	1	0.0	21.49	0.03	3
NGC4435	K2106	512 × 512	1	0.3	20.85	0.05	8

Table 4.2 (continued)

Galaxy (1)	Plate (2)	Frame Size (3)	PS (4)	θ (5)	$\mu_{\rm sky}$ (6)	σ_G (7)	N_{pe} (8)
NGC4438	K2106	512 × 475	2	0.3	20.70	0.08	8
NGC4442	K2186	512 × 256	1	0.2	21.43	0.06	10
NGC4448	K4057	512 × 256	1	0.7	21.15	0.02	3
NGC4449	K4705	512 × 308	2	0.3	21.14	0.06	6
NGC4450	K4994	512 × 512	2	0.5	21.70	0.04	8
NGC4452	K2186	256 × 256	1	0.3	21.41	0.01	2
NGC4454	K4970	256 × 256	1	0.0	21.56	0.10	4
NGC4460	K4705	256 × 256	2	0.4			
NGC4457	K3547	512 × 512	1	0.1	21.33	0.10	5
NGC4459	K2106	512 × 512	1	0.4	20.79	0.04	11
NGC4461	K2106	256 × 512	1	0.4	20.77	0.05	11
NGC4469	K2186	512 × 256	1	0.3	21.29	—	1
NGC4472	K2186	512 × 512	3	0.3	21.20	0.07	22
NGC4473	K2106	512 × 512	1	0.5	20.84	0.05	16
NGC4474	K2106	256 × 256	1	0.5	20.79	0.07	6
NGC4476	K2106	256 × 256	1	0.5	20.69	0.06	14
NGC4477	K2106	512 × 512	1	0.5	20.79	0.05	13
NGC4478	K2106	256 × 256	1	0.5	20.74	0.08	17
NGC4485	K3173	512 × 512	1	0.7	21.48	0.05	3
NGC4483	K2186	256 × 256	1	0.4	21.39	0.05	3
NGC4490	K3173	512 × 512	1	0.7	21.45	0.06	4
NGC4486	K2106	512 × 512	2	0.5	20.60	0.08	54
NGC4487	K5012	512 × 512	1	−0.1			
NGC4494	K4926	512 × 512	1	−1.1	22.02	0.09	8
NGC4496	K3547	512 × 512	1	0.2			
NGC4501	K2730	512 × 512	2	−0.7	21.22	0.02	7
NGC4503	K2186	256 × 512	1	0.4	21.46	0.05	6
NGC4504	K4977	256 × 256	2	0.4			
R 80	K4970	256 × 256	2	0.0			
NGC4517	K4970	512 × 256	2	0.0	21.41	0.03	3
NGC4519	K2220	256 × 256	1	−0.4	20.61	0.12	2*
NGC4522	K2220	256 × 256	1	−0.4			
NGC4526	K2220	512 × 512	2	−0.4	20.94	0.04	10
NGC4527	K3740	512 × 512	1	−0.2	21.06	0.09	7
NGC4532	K3740	256 × 256	1	−0.2	21.71	—	1*
NGC4535	K2220	512 × 512	2	−0.4	20.93	0.05	5
NGC4536	K4639	512 × 512	1	0.0	21.55	0.10	4
NGC4540	K2730	256 × 256	1	−0.5			
NGC4546	K5012	256 × 256	2	−0.2	21.52	0.04	4
NGC4548	K2730	512 × 512	2	−0.5	21.35	0.07	8
NGC4550	K2220	256 × 256	1	−0.3	21.20	0.06	8
NGC4552	K2730	512 × 512	2	−0.4	21.36	0.04	13
NGC4559	K4926	512 × 512	2	−0.6	22.12	0.31	2*
NGC4561	K5791	256 × 256	1	−0.5	21.84	0.01	2
NGC4565	K4926	512 × 512	3	−0.6	22.09	0.06	7
NGC4564	K2220	512 × 512	1	−0.3	21.33	0.11	11
NGC4567	K2220	512 × 512	1	−0.3	21.74	0.00	2*
NGC4568	K2220	512 × 512	1	−0.3	21.34	—	1
NGC4569	K2730	512 × 512	2	−0.4	21.41	0.12	8
NGC4570	K3740	256 × 256	2	−0.1	21.61	0.10	6*
NGC4571	K2730	512 × 512	1	−0.4	24.93	—	1*
NGC4578	K2220	512 × 512	1	−0.2	21.17	0.08	6
NGC4579	K2220	512 × 512	2	−0.2	21.10	0.12	11
NGC4580	K3740	256 × 256	1	−0.1	21.02	0.02	3
NGC4589	K5759	512 × 256	2	2.8	22.32	—	1
NGC4586	K3740	256 × 256	2	−0.1	21.07	0.02	2
NGC4592	K4639	512 × 256	1	0.1			
NGC4593	K3310	512 × 512	1	0.1	20.49	0.03	6
NGC4595	K2730	256 × 256	1	−0.2			
NGC4594	K4977	512 × 512	3	0.1	21.48	0.13	16
NGC4596	K2220	512 × 512	1	−0.1	21.06	0.09	9
NGC4597	K3310	512 × 512	1	0.1			
NGC4605	K5064	256 × 256	2	0.4	21.66	0.04	5
NGC4602	K3310	256 × 256	1	0.0	20.28	—	1
NGC4608	K2220	512 × 512	1	−0.1	22.45	—	1*
NGC4612	K3740	256 × 256	2	0.0			
NGC4618	K3173	512 × 512	1	2.4	21.50	0.10	4
NGC4621	K2220	512 × 313	2	0.0	21.20	0.09	10
NGC4623	K2220	256 × 256	1	0.0	21.09	0.09	4
NGC4631	K5069	512 × 256	3	−1.1	21.51	0.24	2
NGC4630	K3740	256 × 256	1	0.0			
NGC4632	K4674	256 × 256	1	0.0			
NGC4635	K5791	256 × 256	1	0.0			
NGC4638	K2220	256 × 256	1	0.0	21.36	0.09	5
NGC4636	K3740	512 × 512	2	0.0	20.96	0.09	13
NGC4639	K2730	256 × 256	1	0.0	21.85	—	1*
NGC4643	K3740	512 × 512	1	0.0	21.08	0.08	7
NGC4647	K2220	512 × 512	2	0.0	21.04	0.29	5*
NGC4649	K2220	512 × 512	2	0.0	21.01	0.08	18
NGC4651	K2730	512 × 512	1	0.1	21.34	0.06	5
NGC4653	K4674	256 × 256	1	0.0			
NGC4654	K2730	512 × 512	1	0.1	21.19	0.06	7
NGC4656	K3186	512 × 512	2	0.2	20.87	0.17	3
NGC4660	K2220	256 × 256	1	0.1	21.45	0.06	5*
NGC4658	K4977	256 × 256	1	−0.1	21.48	—	1
NGC4665	K3740	512 × 512	1	0.1	21.03	0.05	7
NGC4666	K4674	256 × 256	2	0.0	21.35	0.09	4
NGC4670	K4926	256 × 256	1	0.4	22.07	0.04	7
NGC4668	K4674	256 × 256	1	0.0	21.26	—	1
NGC4682	K4977	256 × 256	1	−0.2			
NGC4689	K2730	512 × 512	1	0.3	21.24	—	1
NGC4691	K3310	256 × 256	2	−0.1	20.44	0.03	7
NGC4694	K2220	512 × 512	1	0.2	21.12	0.08	7
NGC4698	K2220	512 × 512	1	0.3	21.01	0.08	7
NGC4697	K3310	512 × 512	2	−0.1	20.36	0.04	14
NGC4699	K4977	512 × 512	2	−0.3	21.67	0.04	13
NGC4700	K4977	256 × 256	1	−0.3	21.23	—	1*
NGC4701	K3740	256 × 256	1	0.1	21.14	0.04	4
New 3	K4977	512 × 512	1	−0.3			
NGC4712	K3683	256 × 256	1	−0.6	21.16	—	1
NGC4710	K2730	512 × 512	1	0.5	21.28	0.02	4
NGC4713	K3740	256 × 256	1	0.2	20.93	0.04	2
NGC4725	K3683	512 × 512	2	−0.5	21.07	0.04	4
NGC4750	K5759	512 × 512	1	5.9			
NGC4731	K3310	256 × 256	2	−0.2	20.58	0.11	3
NGC4736	K2647	512 × 512	3	0.1	21.46	0.09	5
NGC4742	K4977	256 × 256	1	−0.4	22.06	0.04	3*
NGC4747	K3683	512 × 512	1	−0.4	21.20	—	1
NGC4754	K2220	512 × 512	1	0.4	21.13	0.06	5
NGC4753	K4674	512 × 312	2	0.0	21.24	0.07	7
NGC4762	K2220	512 × 512	2	0.5	21.10	0.04	5
NGC4760	K4893	256 × 256	2	0.4	21.54	0.07	6
NGC4763	K3172	256 × 256	1	0.6	20.16	0.07	4
NGC4771	K4674	256 × 256	2	0.0			
NGC4772	K5049	256 × 256	2	0.0	21.49	0.02	5
NGC4775	K4985	512 × 512	1	0.2	21.43	—	1*
NGC4781	K4893	512 × 256	1	0.4	21.38	—	1
NGC4786	K4985	512 × 512	1	0.2	21.72	—	1
NGC4782	K4893	256 × 256	2	0.3	21.50	0.08	9
NGC4783	K4893	256 × 256	2	0.3	21.54	0.04	4

Table 4.2 (continued)

Galaxy (1)	Plate (2)	Frame Size (3)	PS (4)	θ (5)	μ_{sky} (6)	σ_G (7)	N_{pe} (8)
NGC4790	K4893	256 × 256	1	0.3			
NGC4793	K3683	256 × 256	1	0.0	21.72	—	1*
NGC4800	K3210	256 × 256	1	0.8	21.83	0.03	2
New 4	K5049	512 × 512	1	0.0			
IC 3896	K4985	256 × 256	1	0.1	19.45	0.16	7*
NGC4818	K4893	256 × 512	1	0.3	21.38	0.06	3
NGC4826	K2211	512 × 512	2	−0.5	21.32	0.06	6
NGC4825	K3172	256 × 256	2	0.3	20.41	0.11	2
NGC4845	K5049	512 × 256	2	0.0	21.44	0.04	2
NGC4861	K4619	256 × 512	1	−0.5	19.63	0.30	4*
NGC4868	K2647	256 × 256	1	1.4			
NGC4889	K3683	256 × 512	2	0.6	21.17	0.07	16
NGC4900	K5049	256 × 256	2	0.0	21.47	0.07	4
NGC4891	K3172	256 × 256	1	0.1	20.40	0.08	2
NGC4899	K3172	256 × 256	1	0.1	20.33	0.03	2
NGC4902	K3172	256 × 256	2	0.1	20.23	0.07	8
NGC4914	K2647	512 × 512	1	1.7	21.98	—	1*
NGC4904	K5049	256 × 256	2	0.0	21.38	—	1
NGC4915	K4985	256 × 256	1	0.0	21.93	0.03	4
NGC4928	K4893	256 × 256	1	0.0			
NGC4933	K4893	512 × 512	1	−0.1			
NGC4941	K4985	256 × 256	2	0.0	21.66	0.31	3
NGC4939	K4893	512 × 512	1	−0.1	21.48	0.09	7
NGC4951	K4985	512 × 256	1	−0.1			
NGC4958	K4893	256 × 512	1	−0.1	21.74	0.05	7
NGC4961	K3683	256 × 256	1	1.3	21.48	—	1*
NGC4981	K4985	512 × 512	1	−0.1			
NGC4984	K3172	256 × 256	2	−0.4	20.41	0.06	7
NGC4999	K5049	512 × 512	1	0.0	21.56	—	1
NGC4995	K4893	512 × 256	1	−0.3	21.26	0.22	3*
NGC5005	K4619	512 × 512	1	1.2	21.91	0.12	13
NGC5017	K3172	256 × 256	1	−0.7	20.41	0.03	3
NGC5018	K5374	512 × 512	1	0.8	21.43	0.06	9
NGC5033	K4619	256 × 512	2	1.6	21.90	0.10	7
NGC5055	K5372	512 × 512	3	0.3	22.12	0.09	8
NGC5074	K5074	256 × 256	1	−0.5	21.96	—	1
NGC5068	K5374	512 × 512	2	0.3	21.13	0.07	2
NGC5084	K5374	512 × 256	2	0.2	21.36	0.08	5
NGC5087	K5374	512 × 512	1	0.2	21.55	0.07	7
NGC5112	K5372	512 × 512	1	1.2	22.55	—	1*
NGC5134	K5374	512 × 512	1	−0.2	21.27	—	1
NGC5170	K5374	512 × 512	2	−0.6	21.38	0.05	2
NGC5204	K5765	512 × 512	1	3.4	21.82	0.10	4
NGC5194	K4938	512 × 512	2	−1.5	21.70	0.03	8
NGC5195	K4938	512 × 512	2	−1.5	21.92	0.10	7
NGC5198	K4938	256 × 256	1	−1.4	22.04	—	1*
NGC5273	K5079	512 × 512	1	0.0	21.48	0.06	9
NGC5297	K4938	256 × 256	2	1.5	21.78	0.09	4
NGC5301	K4938	256 × 256	2	1.5	21.76	0.06	3
NGC5313	K5351	256 × 256	1	−0.4			
NGC5326	K5351	512 × 512	1	−0.2	22.53	0.09	2
NGC5324	K4629	256 × 256	1	0.2	21.10	0.03	3
NGC5347	K5792	256 × 256	1	−1.3	21.90	0.04	2
NGC5350	K5351	256 × 256	1	0.2	22.41	0.12	3
NGC5351	K5351	256 × 256	1	0.2	22.35	0.18	2
NGC5353	K5351	512 × 512	1	0.2	22.58	0.08	7
NGC5362	K5351	256 × 256	1	0.4			
NGC5371	K5351	512 × 512	1	0.6	22.31	0.08	8
NGC5363	K5395	512 × 512	2	−0.1	21.83	0.07	10
NGC5364	K5395	512 × 512	2	−0.1	21.59	0.06	8
NGC5377	K4676	256 × 256	2	−1.1	21.73	0.05	5
NGC5380	K5351	256 × 256	1	0.8			
NGC5383	K5351	512 × 512	1	0.8	22.20	0.03	4
NGC5395	K5351	512 × 512	1	1.0	22.12	0.12	4
NGC5406	K5351	256 × 256	1	1.3	22.23	0.04	2
NGC5422	K3289	256 × 256	1	1.4	20.93	—	1
NGC5426	K4986	256 × 512	1	0.2	21.42	0.20	3*
NGC5427	K4986	512 × 512	1	0.2	21.18	0.09	2*
NGC5448	K4676	512 × 256	1	0.2	21.92	0.18	3
NGC5444	K5792	512 × 512	1	0.2			
NGC5457	K3289	512 × 512	5	1.9	20.97	0.12	6
NGC5473	K3289	512 × 512	1	2.2	21.12	0.09	5
NGC5474	K3289	512 × 512	1	2.2	20.83	0.09	2
NGC5468	K4986	512 × 512	1	0.2	21.27	0.02	2
NGC5480	K4676	256 × 256	2	0.9	21.63	—	1
NGC5485	K3289	512 × 512	1	2.7	21.01	0.07	10
NGC5493	K4986	512 × 512	1	0.0	21.67	0.06	3
NGC5496	K3766	256 × 512	1	0.0	20.66	—	1
NGC5494	K4978	256 × 256	2	−1.2	20.75	0.05	2
NGC5523	K4996	512 × 256	1	−0.8	20.88	—	1*
NGC5533	K2649	512 × 512	1	−0.9	21.62	—	1
NGC5534	K4986	256 × 256	1	−0.1	21.48	0.08	9
NGC5548	K4996	256 × 256	1	−0.5	22.54	0.15	27
NGC5557	K2649	512 × 512	1	−0.5	21.66	0.05	10
NGC5566	K4026	512 × 512	2	−0.1	21.24	0.07	11
NGC5585	K4664	512 × 512	2	−1.2	21.40	0.01	2
NGC5574	K4026	256 × 256	1	−0.1	21.15	0.08	8
NGC5576	K4026	512 × 512	1	−0.1	21.21	0.08	12
NGC5584	K3766	512 × 512	1	0.0			
NGC5600	K5091	256 × 256	1	0.1	21.79	0.03	3
NGC5595	K5045	256 × 256	1	−0.1	21.27	—	1
NGC5597	K5045	256 × 256	1	−0.1	21.26	0.35	3*
NGC5614	K2649	512 × 512	1	0.3	21.67	0.01	4
NGC5605	K5045	256 × 256	1	−0.2	21.29	0.02	4
NGC5641	K4640	512 × 512	1	0.9			
NGC5638	K4026	512 × 512	1	0.0	21.13	0.12	3
NGC5653	K4640	256 × 256	1	1.0	22.16	0.08	6
NGC5660	K4979	512 × 512	1	0.8	21.93	0.22	2
NGC5645	K3714	512 × 256	1	−0.1	21.29	—	1
NGC5665	K3714	256 × 256	1	−0.1			
NGC5668	K3714	512 × 512	1	−0.1	21.17	0.20	2
NGC5676	K4979	512 × 512	1	1.3	21.85	0.08	8
NGC5687	K4664	256 × 256	2	1.9	25.21	—	1*
NGC5689	K4979	512 × 256	1	1.9	22.06	0.04	2
NGC5690	K4675	256 × 256	2	0.0			
NGC5691	K4675	256 × 256	1	0.0	21.24	0.07	6
NGC5701	K3714	512 × 512	1	0.1	21.32	0.04	12
NGC5713	K4675	512 × 512	1	0.0	21.15	0.14	4
NGC5728	K5422	512 × 512	1	0.6	21.50	0.14	4*
NGC5739	K5080	256 × 256	1	−1.2	21.98	—	1
NGC5740	K4675	256 × 512	1	0.0	21.51	0.10	3*
NGC5746	K4675	256 × 512	2	0.0	21.34	0.07	7
NGC5750	K4675	512 × 512	1	0.0	21.33	0.11	10
NGC5756	K5422	512 × 512	1	0.2	21.26	0.02	4
NGC5757	K5439	256 × 256	1	−0.4	21.12	0.32	3
NGC5768	K4675	256 × 256	1	0.0			
NGC5775	K4389	256 × 256	2	−0.2	21.20	0.41	2*
NGC5792	K4378	512 × 256	2	0.0	20.89	—	1*
NGC5796	K5422	512 × 512	1	−0.6	21.62	0.12	3*
NGC5820	K4654	256 × 256	1	−2.3	21.91	0.03	6

Table 4.2 (continued)

Galaxy (1)	Plate (2)	Frame Size (3)	PS (4)	θ (5)	μ_{sky} (6)	σ_G (7)	N_{pe} (8)
NGC5806	K4378	256 × 256	2	0.0	21.19	—	1
NGC5812	K5066	256 × 256	2	0.1	21.37	0.11	2
NGC5813	K4378	512 × 512	1	0.0	21.11	0.09	14
NGC5831	K4378	256 × 256	2	0.0	21.11	0.05	14
NGC5838	K4378	512 × 512	1	0.0	21.09	0.09	5
NGC5846	K4378	512 × 512	2	0.0	20.99	0.07	24
NGC5850	K4378	256 × 256	2	0.0	21.08	0.06	12
NGC5866	K4654	512 × 512	2	−0.6	21.68	0.09	10
NGC5854	K4389	256 × 256	2	0.1	20.61	0.09	4
NGC5861	K5066	256 × 256	2	−0.3	20.95	0.09	4
NGC5864	K4389	256 × 256	1	0.2	20.67	0.02	4
NGC5879	K4654	256 × 512	1	0.1	21.62	0.12	2
NGC5878	K4647	256 × 512	1	0.6	21.13	0.06	3
F 703	K4647	512 × 512	1	0.6			
NGC5899	K5092	256 × 512	1	−1.3	21.81	0.07	5
NGC5905	K4654	512 × 512	1	1.2			
NGC5907	K4654	256 × 512	3	1.4	21.68	0.03	5
NGC5898	K5423	512 × 512	1	0.5	20.99	0.05	7
NGC5908	K4654	256 × 512	1	1.5	21.71	—	1
NGC5903	K5423	512 × 512	1	0.5	20.96	0.07	7
NGC5915	K4647	256 × 256	1	0.1	20.92	0.05	11
NGC5921	K4700	512 × 512	1	0.0	21.64	0.19	8
NGC5949	K5052	256 × 256	1	1.7	21.70	—	1
NGC5936	K4677	256 × 256	1	−0.3	21.37	0.03	2
NGC5970	K4677	512 × 256	1	0.1	21.38	0.07	3
NGC5982	K4414	256 × 256	2	−2.7	21.74	0.09	13
NGC5985	K4414	256 × 256	2	−2.5	21.57	0.04	8
NGC6015	K4414	256 × 256	2	0.1	21.54	0.08	5
NGC6052	K4653	256 × 256	1	0.3	21.41	0.12	10
NGC6070	K5057	512 × 512	1	0.0	21.58	0.05	3
NGC6118	K5047	256 × 256	2	0.0	21.27	0.14	6
NGC6181	K5424	256 × 512	1	−0.8	21.33	0.09	20
NGC6217	K4682	256 × 256	2	1.2	21.53	0.15	5
NGC6207	K2304	256 × 256	1	0.2	20.70	0.08	6
NGC6340	K4687	256 × 256	2	−2.9	21.67	0.10	11
NGC6384	K5058	256 × 256	2	0.1	21.58	0.05	5
NGC6482	K5067	256 × 256	1	0.2	22.44	0.09	12
NGC6503	K4665	512 × 256	2	3.3	21.31	0.06	11
NGC6643	K4711	256 × 256	2	5.1	21.47	0.05	6
NGC6814	K5094	512 × 512	1	0.0	21.28	0.07	18
NGC6822	K5068	512 × 512	3	0.3	20.68	0.05	2
NGC6907	K5112	256 × 256	2	−0.2	20.86	0.13	3
NGC6925	K5122	256 × 256	2	1.1	20.54	0.02	7
NGC6946	K5137	512 × 512	3	2.1	21.36	0.14	11
NGC6951	K5103	256 × 256	2	−0.8	21.52	0.13	12
NGC7137	K5135	256 × 256	1	0.5	21.61	0.07	6
NGC7177	K5184	512 × 256	1	0.4	21.25	0.10	11
NGC7171	K5123	256 × 256	1	0.1	21.17	0.05	7
NGC7217	K5138	512 × 512	2	0.7	21.32	0.10	13
NGC7218	K5123	256 × 256	1	−0.5	21.01	0.03	2
NGC7300	K4788	256 × 256	1	0.0	20.60	—	1
NGC7309	K5139	256 × 256	1	0.4	21.04	0.07	3
NGC7331	K3394	512 × 512	2	−0.7	21.00	0.08	11
NGC7332	K3396	256 × 512	1	−0.5	21.18	0.08	20
NGC7361	K5525	256 × 512	1	0.1	20.55	0.02	2
NGC7371	K5139	256 × 256	1	−0.1	20.93	0.10	9
NGC7377	K5540	256 × 256	2	0.2	20.15	0.07	17
NGC7448	K5185	256 × 512	1	−0.2	21.20	0.14	10
NGC7469	K4486	256 × 256	1	0.0	21.81	0.07	24*
NGC7479	K4486	512 × 512	1	0.1	21.25	0.10	11
NGC7541	K4789	512 × 256	1	−0.2	20.83	0.07	10
NGC7619	K2974	512 × 512	1	0.2	20.84	0.08	10
NGC7626	K2974	512 × 512	1	0.2	20.97	0.12	17
NGC7640	K5121	256 × 512	2	1.2	21.43	0.05	2
NGC7679	K4789	256 × 256	1	0.1	21.18	0.03	6
NGC7716	K5140	512 × 512	1	0.0	21.28	0.06	6
NGC7741	K5177	512 × 512	1	0.1	21.07	0.10	10
NGC7743	K4758	256 × 256	1	−0.1	21.96	0.11	10
NGC7769	K5186	256 × 256	1	0.7	21.44	0.14	2
NGC7782	K4758	256 × 256	1	0.2	21.74	0.05	8
NGC7785	K4758	256 × 256	1	0.2	21.90	0.08	6

Note:

Column 1 : Galaxy name
Column 2 : Source plate number
Column 3 : Frame size in pixels
Column 4 : Pixel size in units of arcsec
Column 5 : The angle between north and the upward direction of the frame. When north is inclined to the right (left), the sign of the angle is + (−).
Column 6 : Sky brightness in units of mag arcsec^{-2} for the calibrator determined in the absolute calibration
Column 7 : Standard deviation of the residuals in units of mag arcsec^{-2} for the calibrator computed in the absolute calibration (see section 1.5 in Part I)
Column 8 : Number of photoelectric magnitudes used in the absolute calibration. The asterisk indicates that the sky brightness for the calibrator was not used in the computation of the mean sky brightness for the plate.

Table 4.3 Multiply observed galaxies

Galaxy (1)	V_T (2)	ΔV_T (3)	log R_e (4)	Δlog R_e (5)	N (6)
NGC1344	10.44	0.01	1.48	0.04	2
NGC1353	11.44	0.03	1.48	0.02	2
NGC1359	12.63	0.02	1.48	0.0	2
NGC1366	11.88	0.10	0.82	0.04	2
NGC1667	12.23	0.13	1.18	0.01	2
NGC1726	11.54	0.02	1.43	0.05	2
NGC2139	11.74	0.09	1.44	0.01	2
NGC2551	12.18	0.13	0.90	0.05	2
NGC2639	11.65	0.01	1.13	0.03	2
NGC2654	12.06	0.05	1.15	0.03	2
NGC2798	12.33	0.02	1.06	0.01	2
NGC2907	11.53	0.06	1.41	0.02	2
NGC2964	11.33	0.02	1.38	0.01	2
NGC2968	11.57	0.01	1.45	0.01	2
NGC3044	11.89	0.02	1.61	0.02	2
NGC3056	11.38	0.09	1.14	0.01	2
NGC3294	11.43	0.12	1.55	0.02	2
NGC3319	11.46	0.14	1.90	0.02	2
NGC3412	10.77	0.10	1.22	0.05	2
NGC3455	13.12	0.00	1.27	0.01	2
NGC3485	12.20	0.19	1.48	0.01	2
NGC3489	10.38	0.03	1.16	0.05	2
NGC3556	10.40	0.01	1.90	0.02	2
NGC3683	12.30	0.07	1.23	0.02	2
NGC3686	11.38	0.14	1.56	0.03	2
NGC3780	11.83	0.08	1.55	0.02	2
NGC3893	10.71	0.05	1.53	0.01	2
NGC3917	11.78	0.04	1.65	0.0	2
NGC3923	9.45	0.04	1.92	0.03	2
NGC3936	12.42	0.09	1.52	0.01	2
NGC3938	10.40	0.09	1.79	0.02	2
NGC3941	10.42	0.01	1.23	0.02	2
NGC3949	11.11	0.00	1.39	0.01	2
NGC3953	10.24	0.09	1.72	0.01	2
NGC4045	11.74	0.03	1.45	0.0	2
NGC4214	10.09	0.02	1.86	0.01	2
NGC4217	11.17	0.05	1.75	0.03	4
NGC4220	11.38	0.02	1.40	0.02	2
NGC4251	10.75	0.00	1.18	0.01	2
NGC4258	8.47	0.02	2.03	0.02	2
NGC4274	10.50	0.01	1.61	0.01	2
NGC4346	11.01	0.02	1.02	0.03	2
NGC4389	11.83	0.01	1.48	0.01	2
NGC4618	10.80	0.05	1.69	0.02	2
NGC4736	8.11	0.01	1.69	0.01	2
NGC4868	12.47	0.13	1.21	0.01	2
NGC4914	11.46	0.14	1.24	0.01	2
NGC5566	10.63	0.05	1.38	0.05	2
NGC5574	12.47	0.03	1.06	0.02	2
NGC5576	10.70	0.01	1.41	0.01	2
NGC5690	11.76	0.05	1.57	0.02	2

Note:

Column 1 : Galaxy name
Column 2 : Average value of V_T in units of magnitudes
Column 3 : Difference of V_T in units of magnitudes
Column 4 : Average value of log R_e in units of arcsec
Column 5 : Difference of log R_e in units of arcsec
Column 6 : Number of observations